European Handbook of Plant Diseases

THIS HANDBOOK IS

DEDICATED TO THE MEMORY OF

Juliet Elizabeth Smith

1 OCTOBER 1967–9 APRIL 1986

European Handbook of Plant Diseases

EDITED BY

I.M. SMITH
Director-General, European and Mediterranean
Plant Protection Organization

J. DUNEZ
Director, Plant Pathology Station
INRA-Bordeaux, France

D.H. PHILLIPS
Formerly Principal Pathologist
Forest Commission of Great Britain

R.A. LELLIOTT
Formerly Director
Harpenden Laboratory, UK

S.A. ARCHER
Lecturer in Plant Pathology
Imperial College, London, UK

REGIONAL CONSULTANTS

CHAPTERS 1–4

M.F. CLARK, Institute for Horticultural Research
East Malling, UK

G.P. MARTELLI, Plant Pathology Institute
Bari, Italy

CHAPTERS 5–17

H.B. GJAERUM, Norwegian Plant Protection
Institute, Ås, Norway

E.C. TJAMOS, Benaki Phytopathological
Institute, Athens, Greece

F. VIRÁNYI, Research Institute for
Plant Protection, Budapest, Hungary

BLACKWELL SCIENTIFIC PUBLICATIONS

OXFORD LONDON EDINBURGH

BOSTON PALO ALTO MELBOURNE

This handbook is sponsored by the British Society for Plant Pathology and the European and Mediterranean Plant Protection Organization with the support, for the illustrations, of Du Pont de Nemours (SA)

Editorial offices:
Osney Mead, Oxford, OX2 0EL
8 John Street, London, WC1N 2ES
23 Ainslie Place, Edinburgh, EH3 6AJ
52 Beacon Street, Boston, Massachusetts 02108, USA
667 Lytton Avenue, Palo Alto, California 94301, USA
107 Barry Street, Carlton, Victoria 3053, Australia

First published 1988

Set by Enset (Photosetting)
Midsomer Norton, Bath, Avon
and printed and bound in
Great Britain

DISTRIBUTORS

USA and Canada
Blackwell Scientific Publications Inc
PO Box 50009, Palo Alto
California 94303

Australia
Blackwell Scientific Publications
(Australia) Pty Ltd
107 Barry Street, Carlton, Victoria 3053

British Library
Cataloguing in Publication Data

European handbook of plant diseases.
1. Plant diseases—Europe
I. Smith, I.M.
632'.3'094 SB605.E85

ISBN 0-632-01222-6

Library of Congress
Cataloging-in-Publication Data

European handbook of plant diseases.
Includes bibliographies and index.
1. Plant diseases—Europe—Handbooks, manuals, etc.
2. Micro-organisms, Phytopathogenic—Europe—Handbooks, manuals, etc. 3. Micro-organisms, Phytopathogenic—Control—Europe—Handbooks, manuals, etc. 4. Micro-organisms, Phytopathogenic—Europe—Host plants—Handbooks, manuals, etc.
I. Smith, I.M. (Ian M.)
SB655.E97 1988 632'.3'094 86-26916

ISBN 0-632-01222-6

Contents

Preface

The aim of the *European Handbook of Plant Diseases* is to provide a comprehensive treatment of the diseases of cultivated plants and major forest and amenity trees in Europe. The emphasis of this treatment is on the biology of the pathogen, the epidemiology of the disease, and on economic importance and control. Nevertheless, it was decided to arrange the Handbook following the systematic classification of pathogens, rather than on a host-plant basis. Each approach has its advantages and disadvantages, and the arrangement chosen by the editors follows the example of such eminent forerunners as Brooks (1953) and Viennot-Bourgin (1949).

In compiling the Handbook, pathogens are first assessed and classed as of major, intermediate or minor importance in Europe. This assessment has to allow for the practical importance of the disease and the importance of the crop—thus a very minor disease of grapevine might receive about the same treatment as a major disease of violas. Wild plants (other than important forest or amenity trees), and garden plants not commercially grown on a significant scale, are not included. Our aim is to have each major and intermediate pathogen treated by a European plant pathologist thoroughly familiar in practice with the diseases concerned, and thus to produce a manual based as far as possible on recent local experience, rather than on a world-scale bibliographic view. In fact, the editors have called on a few non-European authors and have mostly compiled themselves the information on the minor organisms, without necessarily direct experience of them. To try to correct for any regional bias, all the material has been reviewed by regional consultants who have tried to ensure a well-balanced approach.

As a rule, only pathogens occurring in Europe are included. However, certain non-European organisms can be considered to present a specific threat to Europe, and are considered as 'quarantine organisms' for the continent. Brief mention is accordingly made of pathogens on the EPPO A1 list of quarantine organisms (EPPO 1982).

Each chapter concludes with a bibliography. In general, an attempt has been made to concentrate on general, easily accessible references. However, the international authorship of the Handbook is reflected in the inclusion of references from many sources and in many languages. In general, at least one reference is given for every organism included. For forest pathogens, this can be taken to be Phillips & Burdekin (1982), a recent comprehensive work co-authored by our co-editor, or Lanier *et al.* (1978).

Both authors and editors have had very frequent occasion to refer to certain serial publications, which have been extremely useful in the preparation of the Handbook. The CMI *Descriptions of Pathogenic Fungi and Bacteria* (CMI 1964–1985), the CMI/AAB and AAB *Descriptions of Plant Viruses* (CMI/AAB 1970–1984; AAB 1985) and the CMI *Distribution Maps of Plant Diseases* (CMI 1945–1985) are cited in the text by number. The reader is urged to refer to them for further details on morphology and geographical distribution in particular. The American Phytopathological Society's Compendia of Plant Diseases are similarly cited by crop, and the EPPO Data Sheets on Quarantine Organisms by number. The editors recognize here their great indebtedness to the multiple authors of these series, whose contribution is not explicitly cited.

In matters of nomenclature, much use has been made of the checklists for scientific names of common parasitic fungi produced by Boerema & Verhoeven (1972–1980). In addition, two works which appeared during editing have much helped to clarify taxonomic position and synonymy, especially among the ascomycetes: the 7th edition of the CMI Dictionary of the Fungi (Hawksworth *et al.* 1983) and the Annotated Checklist of British Ascomycotina (Cannon *et al.* 1985). The last word has certainly not been said on the assignment of genera to the various ascomycete groups, but the editors at least felt that they had a much more coherent framework to work with than before these texts appeared. Two other books which appeared very recently have not been taken into account but provide very valuable information, and in particular numerous figures: Ellis & Ellis (1985) and Brandenburger (1985). One publication which has been much used without this appearing directly in the text at all is the BAYER list of pathogens and their computer codes (Bayer 1979). This six-character coding system has been used as the basic reference field in a microcomputer data base, run with the software dBase-II (Ashton-Tate), to keep track of the 2700 organisms and 4200 names encountered (though not all used) in editing the Handbook. Finally, we mention a number of other texts which have been frequently used, and may also be cited in the references (Moore 1959; Khokhryakov 1966; Messiaen & Lafon 1970; O'Rourke 1976; Kranz *et al.* 1977; Neergaard 1977; Dixon 1982).

The editors have, in their systematic arrangement, taken account of two points which are by no means new, but which have not yet been widely assimilated in the phytopathological literature. Firstly, plant viruses are presented in their groups following Matthews (1982)—current practice in the virology literature but hardly yet in pathology. Secondly, the teleomorph/anamorph terminology has been used for fungi, teleomorph names have been used wherever possible, and anamorphs without a known teleomorph have been grouped with their most probable relatives. No use has been made of the classification of imperfect fungi. It is true that a certain number of fungi remained for which no probable relatives were apparent; these have been grouped at the end of Chapter 13. As a better understanding of 'the whole fungus' evolves, including its capacity as a plant pathogen, it will be possible finally to relate all so-called imperfect fungi to a single mycological classification. In the meantime, we have no doubt made a certain number of errors in trying to work to this principle but feel it is worthwhile in order to uphold it. The editors are responsible for the decisions on nomenclature and systematic arrangement. It may also be noted that, in citing authorities for names, the list proposed by Hawksworth (1980) has been followed for abbreviations. Other authorities have not been abbreviated.

It has been necessary in this Handbook to refer by name to many fungicides for the control of plant diseases. The editors are not in so doing asserting that these compounds are registered for the stated use in any European country. The Royal Society of Chemistry's *European Directory of Agrochemical Products* (RSC 1984) is an excellent publication in which to verify the situation for a given compound in most European countries. In addition, it must be recognized that authors have not been consistent in the degree of detail given on fungicides used against a pathogen. The editors have not sought to correct this, since this Handbook is not a treatment manual. The reader is referred to works such as BCPC's *Pest and Disease Control Handbook* (Scopes & Ledieu 1983) or ACTA's *Index Phytosanitaire* (ACTA 1985).

The initial concept of the *European Handbook of Plant Diseases* owes much to Dr Bryan Wheeler, who, as chairman of the Federation of British Plant Pathologists in 1980, provided the necessary stimulation for work to start on the project which he himself had hoped to realize. The British Society for Plant Pathology, which succeeded the Federation, has maintained its sponsorship. The Handbook is also sponsored by the European and Mediterranean Plant Protection Organization, whose Secretariat has provided indispensable and patient support in its preparation. The inclusion of colour illustrations has been made possible by the generous support of Du Pont Nemours (SA).

THE ARRANGEMENT OF THE TEXT

The diseases are treated by pathogen, arranged in chapters by major systematic groups. Within each chapter, entries are grouped (if appropriate) according to further systematic groups above the genus (order, family). Finally, the genera are arranged broadly alphabetically, as are the species within each genus. However, the reader should use the index to locate a pathogen rather than rely on an alphabetic sequence which is frequently, for various reasons, disturbed. Each entry is in principle independent and self-explanatory, unless it specifically includes cross-references to other pathogens. However, material in the introduction to each chapter is not repeated in individual entries.

Each entry for a major pathogen is divided into sections as follows:

1) Name, synonyms and anamorph names.
2) Basic description of the pathogen. This contains essential noteworthy features chosen by the author, especially distinguishing characters from other pathogens. It does not attempt to be comprehensive or consistent. The CMI descriptions are recommended in particular for detail which cannot be given here.
3) Host plants. Unless otherwise indicated, only natural hosts are given. The information in this section has served as the basis for the host-plant index, which also provides Latin names of the common and familiar cultivated plants the editors have chosen to refer to only by English name.

4) Host specialization, or information on host-specialized formae speciales, varieties, races or pathotypes. The term 'physiologic race' has been rejected as bearing no relation to modern conceptions. If no evidence of such specialization exists, the section is generally omitted.
5) Geographical distribution. Obtaining accurate and up-to-date distributional information can be very difficult, partly through under-recording, and partly through difficulty in assessing old or isolated records. The information given is essentially drawn from the literature and has not been directly verified in any way. Distribution in Europe is generally presented in more detail than for the rest of the world, and it should not be assumed that information on non-European distribution is in any way comprehensive. The CMI maps are recommended for further details.
6) Diseases. The main aim of this section is to describe symptoms, of which the disease name is a short-hand form. Commonly used English disease names are given but, except in the special case of viruses, have not been indexed. No particular name is preferred. Readers may refer to BSPP's *Names of Common British Plant Diseases* (BSPP 1984), which also includes names in several other languages, or else the APS's Multilingual Compendia (Miller & Pollard 1976–1977).
7) Epidemiology. Whereas the earlier sections generally summarize established information, this and the later sections should reflect recent research work and the current agricultural situation.
8) Economic importance.
9) Control.
10) Special research interest. This section, often omitted, reflects the fact that many pathogens or diseases have an importance in phytopathological research and/or concepts which may be quite independent of economic importance.

In some cases, when one pathogen causes very distinct diseases on different hosts, sections 6–10 have been repeated for each. The arrangements for viruses are somewhat different (see also Introduction, Chapter 1). For organisms of lesser importance, the same basic sequence has mostly been respected but the text is no longer divided into separate sections. Literature citations have been placed at an appropriate point in the text, but will often have a much wider relevance to many aspects concerning the pathogen. The bibliography as a whole is placed at the end of the chapter.

The chapters were edited as follows: 1–4 (JD), 5 (RAL & IMS), 6–7 (RAL), 8–13 (IMS), 14 (SA), 15–16 (SA & IMS), 17 (IMS). DHP oversaw all sections concerned with forest pathogens. Each individual entry has its author indicated at the end. If no name is given, the editor of the chapter can be taken to have been the author. All authors' names and addresses are given on pp. xi–xiv.

REFERENCES

AAB (1985) *Descriptions of Plant Viruses* nos 296–309 Association of Applied Biologists, NVRS, Wellesbourne.

ACTA (1985) *Index Phytosanitaire. Produits insecticides, fongicides, herbicides* (ed. G. Dubois) (21st edn). Association de Coordination Technique Agricole, Paris.

APS (1977–1985) *Compendia of: Wheat* (1977), *Alfalfa* (1979), *Corn* (1980), *Cotton* (1981), *Potato* (1981), *Elm* (1981), *Soybean* (1982), *Barley* (1982), *Turfgrass* (1983), *Rose* (1983), *Strawberry* (1984), *Peanut* (1984), *Pea* (1984) *Diseases*. American Phytopathological Society, St. Paul.

Bayer (1979) *Verzeichnis der Wichtigsten Krankheitserreger bzw.-ursachen.* Bayer, Leverkusen.

Boerema G.H. & Verhoeven A.A. (1972–1980) Check-list for scientific names of common parasitic fungi. Series 1a–1b: Fungi on trees and shrubs. Series 2a–2d: Fungi on field crops. *Netherlands Journal of Plant Pathology* **78** (suppl. 1); **79,** 165–179; **82,** 193–214; **83,** 165–204; **85,** 151–185; **86,** 199–228.

Brandenburger W. (1985) *Parasitische Pilze an Gefäss-pflanzen in Europa.* Gustav Fischer, Stuttgart.

Brooks F.T. (1953) *Plant Diseases* (2nd edn). Oxford University Press.

BSPP (1984) *Names of Common British Plant Diseases and their Causes. Phytopathological Paper* no. 28. Commonwealth Agricultural Bureaux, Slough.

Cannon P.F., Hawksworth D.L. & Sherwood-Pike M.A. (1985) The *British Ascomycotina: An Annotated Checklist.* Commonwealth Agricultural Bureaux, Slough.

CMI (1945–1985) *Distribution Maps of Plant Diseases* nos 1–566. Commonwealth Agricultural Bureaux, Slough.

CMI (1964–1986) *Descriptions of Pathogenic Fungi and Bacteria* nos 1–870. Commonwealth Agricultural Bureaux, Slough.

CMI/AAB (1970–1984) *Descriptions of Plant Viruses* nos 1–295. Commonwealth Agricultural Bureaux, Slough.

Dixon G.R. (1982) *Vegetable Crop Diseases* (2nd edn). Macmillan, London.

Ellis M.B. & Ellis J.P. (1985) *Microfungi on Land Plants. An Identification Handbook.* Croom Helm, London.

EPPO (1978–1985) Data sheets on quarantine organisms. *Bulletin OEPP/EPPO Bulletin* **8** (2); **9** (2); **10** (1); **11** (1); **12** (1); **13** (1); **14** (1), 3–72; **16** (1), 13–78.

EPPO (1982) *Recommendations on New Quarantine Measures* (2nd edn). *Bulletin OEPP/EPPO Bulletin* **12** (special issue).

Hawksworth D.L. (1980) Recommended abbreviations for the names of some commonly cited authors of fungi. *Review of Plant Pathology* **59,** 473–480.

Hawksworth D.L., Sutton B.C. & Ainsworth G.C. (1983) *Ainsworth and Bisby's Dictionary of the Fungi* (7th edn). Commonwealth Agricultural Bureaux, Slough.

Khokhryakov M.K. (1966) *Opredelitel' Boleznei Rastenii.* Kolos, Leningrad.

Kranz J., Schmutterer H. & Koch W. (1977) *Diseases, Pests and Weeds in Tropical Crops.* Paul Parey, Berlin.

Lanier L., Joly P., Bondoux P. & Bellemère A. (1978) *Mycologie et Pathologie Forestières* (2 volumes). Masson, Paris.

Matthews R.E.F. (1982) Classification and nomenclature of plant viruses. *Intervirology* **17,** 1–200.

Messiaen C.M. & Lafon R. (1970) *Les Maladies des Plantes Maraîchères.* INRA, Paris.

Miller P.R. & Pollard H.L. (1976–1977) *Multilingual Compendium of Plant Diseases.* Vols. I–II. American Phytopathological Society, St. Paul.

Moore W.C. (1959) *British Parasitic Fungi.* Cambridge University Press.

Neergaard P. (1977) *Seed Pathology* (2 volumes). Macmillan, London.

O'Rourke C.J. (1976) *Diseases of Grasses and Forage Legumes in Ireland.* An Foras Talúntais, Carlow.

Phillips D.H. & Burdekin D.A. (1982) *Diseases of Forest and Ornamental Trees.* Macmillan, London.

RSC (1984) *European Directory of Agrochemical Products.* Vol. 1. *Fungicides.* Royal Society of Chemistry, Nottingham.

Scopes N. & Ledieu M. (1983) *Pest and Disease Control Handbook.* BCPC, Croydon.

Viennot-Bourgin G. (1949) *Les Champignons Parasites des Plantes Cultivées.* Masson, Paris.

Contributors

A.N. ADAMS *East Malling Research Station, Maidstone ME19 6BJ, UK*

B. AGULHON *Institut Technique de la Vigne, Domaine de la Bastide, Route de Générac 30000 Nîmes, France*

J. ALBOUY *Station de Pathologie Végétale, CNRA, Route de St Cyr, 78000 Versailles, France*

B. ALLIOT *ENSA, 9 place Viala, 34060 Montpellier, France*

S.A. ARCHER *Department of Pure and Applied Biology, Imperial College, South Kensington, London SW7 2BB, UK*

K. ÅRSVOLL *Statskonsulenten i Plantevern, Boks 70, 1432 Ås-NLH, Norway*

N. BALDWIN *Biology Department, Liverpool Polytechnic, Liverpool L3 3AF, UK*

M. BAR-JOSEPH *Agricultural Research Organization, The Volcani Center, P.O. Box 6, 50250 Bet Dagan, Israel*

D.J. BARBARA *East Malling Research Station, Maidstone ME19 6BJ, UK*

G. BARTELS *Institut für Pflanzenschutz in Ackerbau und Grünland, Biologische Bundesanstalt, Messeweg 11/12, 3300 Braunschweig, FRG*

L. BECZNER *Research Institute for Plant Protection, P.O. Box 102, 1525 Budapest, Hungary*

B.H.H. BERGMAN *Laboratorium voor Bloembollen-onderzoek, Vennestraat 22, Postbus 85, 2160 Ab Lisse, Netherlands*

R. BERNHARD *Centre de Recherches de Bordeaux, INRA, Domaine de la Grande Ferrade, 33140 Pont-de-la-Maye, France*

A. BERTACCINI *Istituto di Patologia Vegetale, Università di Bologna, via Filippo Re 8, 40126 Bologna, Italy*

B. BESSELAT *Service de la Protection des Végétaux, Chemin d'Artigues, 33150 Cenon, France*

A. BOLAY *Station Fédérale de Recherches Agronomiques de Changins, Route de Duillier, 1260 Nyon, Switzerland*

G. BOMPEIX *Laboratoire de Pathologie Végétale, Université Pierre et Marie Curie, Place Jussieu, 75230 Paris Cedex 05, France*

P. BONDOUX *Station de Pathologie Végétale, INRA, Beaucouzé, 49000 Angers, France*

D. BOUBALS *ENSA, Station de Recherches Viticoles, 9 Place Viala, 34060 Montpellier Cedex, France*

D. BOUHOT *Station de Recherches sur la Flore Pathogène dans le Sol, INRA, 17 rue Sully, BV 1540, 21034 Dijon Cedex, France*

J.M. BOVÉ *Station de Biologie, INRA, BP 131, 33140 Pont-de-la-Maye, France*

N. BRADSHAW *ADAS Block 4 Government Buildings, 66 Ty Glas Road, Llanisten, Cardiff CF4 5ZB, UK*

C.M. BRASIER *Forest Research Station, Alice Holt Lodge, Wrecclesham, Farnham GU10 4LH, UK*

S.T. BUCZACKI *National Vegetable Research Station, Wellesbourne, Warwick CV35 9EF, UK*

Y. BUGARET *Station de Recherches de Bordeaux, INRA, Domaine de la Grande Ferrade, 33140 Pont-de-la-Maye, France*

J. BULIT *Centre de Recherches de Bordeaux, INRA, Domaine de la Grande Ferrade, 33140 Pont-de-la-Maye, France*

D.J. BUTT *East Malling Research Station, Maidstone ME19 6BJ, UK*

R.J.W. BYRDE *Long Ashton Research Station, Bristol BS18 9AF, UK*

M. CAMBRA *IVIA, Centro de Levante, Apartado Oficial, Moncada (Valencia), Spain*

S. CARNEGIE *Department of Agriculture and Fisheries for Scotland, East Craigs, Edinburgh EH12 8NJ, UK*

A.J.H. CARR *Welsh Plant Breeding Station, Plas Gogerddan, Aberystwyth SY23 3EB, UK*

M.V. CARTER *University of Adelaide, Waite Agricultural Research Institute, Glen Osmond SA 5064, Australia*

R. CASSINI *Station de Pathologie Végétale, INRA, Route de St Cyr, 78000 Versailles, France*

E. CASTELLANI *Istituto di Patologia Vegetale, Unversità di Torino, Via P. Giuria 15, 10126 Torino, Italy*

A. CAUDWELL *Station de Physiopathologie Végétale, INRA, Domaine d'Epoisses, BP 1540, 21034 Dijon Cedex, France*

A. CHITZANIDIS *Benaki Phytopathological Institute, Kifissia, Athens, Greece*

J. CHOD *Department of Virology, Research Institute for Plant Production, Ruzyne 507, 16106 Praha, Czechoslovakia*

B.C. CLIFFORD *Welsh Plant Breeding Station, Plas Gogerddan, Aberystwyth SY23 3EB, UK*

S. COHEN *Agricultural Research Organization, Volcani Center, P.O. Box 6, 50250 Bet Dagan, Israel*

J.R. COLEY-SMITH *Department of Botany, The University, Hull HU6 7RX, UK*

M. CONTI *Laboratorio de Fitovirologia Applicata, via O. Vigliani 104, 10135 Torino, Italy*

R.T.A. COOK *Harpenden Laboratory, Hatching Green, Harpenden AL5 2BD, UK*

J.I. COOPER *Unit of Invertebrate Virology, South Parks Road, Oxford OX1 3RB, UK*

M.T. COUSIN *Station de Pathologie Végétale, CNRA, Route de St Cyr, 78000 Versailles, France*

Y. COUTEAUDIER *Station de Recherches sur la Flore, Pathogène dans le Sol, INRA, 17 rue Sully, BV 1540, 21034 Dijon Cedex, France*

I.R. CRUTE *National Vegetable Research Station, Wellesbourne, Warwick CV35 9EF, UK*

A. DALMASSO *Station de Nématologie, INRA, BP 78, 06606 Antibes, France*

D.L. DAVIES *East Malling Research Station, Maidstone ME19 6BJ, UK*

P.J.G.M. DE WIT *Laboratory of Phytopathology, Agricultural University, Binnenhaven 9, 6709 PD Wageningen, Netherlands*

C. DELATOUR *Laboratoire de Pathologie Forestière, CNRF, Champenoux, 54280 Seichamps, France*

R.P. DELBOS *Station de Recherches de Bordeaux, INRA, Domaine de la Grande Ferrade, 33140 Pont-de-la Maye, France*

B. DELÉCOLLE *Station de Pathologie Végétale, INRA, Domaine St Maurice, 84140 Montfavet, France*

R. DELON *Institut Expérimental du Tabac de la SEITA, Domaine de la Tour, BP 168, 24108 Bergerac, France*

C. DENNIS *Food Canners Research Association, Chipping Campden, GL55 6LD, UK*

J.C. DESVIGNES *CTIFL, Domaine de Lanxade, Prigonrieux, 24130 La Force, France*

J.C. DEVERGNE *Station de Pathologie Végétale, INRA, Villa Thuret, BP 78, 06602 Antibes, France*

J. DICKENS *Harpenden Laboratory, Hatching Green, Harpenden AL5 2BD, UK*

G.R. DIXON *University of Strathclyde and West of Scotland Agricultural College, Auchincruive, Ayr KA6 5HW, UK*

J.M. DUNCAN *Scottish Crop Research Institute, Invergowrie, Dundee DD2 5DA, UK*

J. DUNEZ *Station de Recherches de Bordeaux, INRA, Domaine de la Grande Ferrade, 33140 Pont-de-la-Maye, France*

M. EBBEN *Glasshouse Crops Research Institute, Worthing Road, Littlehampton BN17 6LP, UK*

H. EHLE *Biologische Bundesanstalt, Messeweg 11/12, 3300 Braunschweig, FRG*

H. EPTON *Department of Botany, University of Manchester, Manchester M13 9PL, UK*

H. FEHRMANN *Institut für Pflanzenpathologie und Pflanzenschutz, Universität Göttingen, 3400 Göttingen, FRG*

R.W. FULTON *Department of Plant Pathology, University of Wisconsin, Madison WI 53706, USA*

W. GABRIEL *Institut Ziemniaka, Bonin, 75016 Koszalin, Poland*

C.E.M. GARRETT *East Malling Research Station, Maidstone ME19 6BJ, UK*

K. GEBRE-SELASSIÉ *Station de Pathologie Végétale, INRA, Domaine St Maurice, 84140 Montfavet, France*

M. GERLAGH *Instituut voor Plantenziektenkundig Onderzoek, Binnenhaven 12, 6700 GW Wageningen, Netherlands*

D. GINDRAT *Station Fédérale de Recherches Agronomiques de Changins, Route de Duillier, 1260 Nyon, Switzerland*

H.B. GJAERUM *Department of Plant Pathology, Norwegian Plant Protection Institute, Boks 70, 1432 Ås-NLH, Norway*

A. GRANITI *Istituto di Patologia Vegetale, via Giovanni Amendola 165/A, 70126 Bari, Italy*

E. GRIFFITHS *Department of Agricultural Botany, University College of Wales, Penglais, Aberystwyth SY23 3DD, UK*

C. GROSCLAUDE *Station de Pathologie Végétale, INRA, Domaine St Maurice, 84140 Montfavet, France*

M. GRUNTZIG *Martin Luther Universität, Sektion Pflanzenproduktion Virusserologie, Emil-Abderhalden Strasse 25, 4020 Halle, GDR*

J.J. GUILLAUMIN *Station de Pathologie Végétale, INRA, 12 av. du Brézet, 63039 Clermont Ferrand Cedex, France*

M.N. GVRITISHVILI *Georgian Institute for Plant Protection, Chavchavadze 82, 380062 Tbilisi, Georgian SSR, USSR*

G. HAMDORF *Landespflanzenzuchtamt Mainz, Essenheimer Strasse 144, 6500 Mainz-Bretzenheim, FRG*

D. HARIRI *Station de Pathologie Végétale, CNRA, Route de St Cyr, 78000 Versailles, France*

D.C. HARRIS *East Malling Research Station, Maidstone ME19 6BJ, UK*

J.G. HARRISON *Scottish Crop Research Institute, Invergowrie, Dundee DD2 5DA, UK*

G. HIDE *Plant Pathology Department, Rothamsted Experimental Station, Harpenden AL5 2JQ, UK*

F.M. HUMPHERSON-JONES *National Vegetable Research Station, Wellesbourne, Warwick CV35 9EF, UK*

W. HUTH *Institut für Virologie, Biologische Bundesanstalt, Messeweg 11/12, 3300 Braunschweig, FRG*

T.M. JEVES *Department of Biology, Liverpool Polytechnic, Liverpool L3 3AF, UK*

A.T. JONES *Scottish Crop Research Institute, Invergowrie, Dundee DD2 5DA, UK*

V.W.L. JORDAN *Long Ashton Research Station, Bristol BS18 9AF, UK*

B. JOUAN *INRA, Domaine de la Motte au Vicomte, BP 29, 35650 Le Rheu, France*

E.V.R. JULIO *Centro Nacional de Protecçao da Producçao Agricola (CNPPA), Quinto do Marques, 2780 Oeiras, Portugal*

H. KEGLER *Institut für Phytopathologie, Theodor Römer Weg 4, 4320 Aschersleben, GDR*

W. KENNEL *Universität Hohenheim, Schuhmacherhof, 7980 Ravensburg 1, FRG*

H. KLEINHEMPEL *Institut für Phytopathologie, Theodor Römer Weg 4, 4320 Aschersleben, GDR*

C. KNIGHT *National Institute of Agricultural Botany, Huntingdon Rd, Cambridge CB3 0LE, UK*

M. KOJIMA *Faculty of Agriculture, Niigata University, Niigata 950-21, Japan*

A. KOWALSKA-NOORDAM *Department of Genetics, Meochow, 05-832 Rozalin, Poland*

H. KRCZAL *Institut für Pflanzenschutz in Obstbau, Biologische Bundesanstalt, Postfach 73, 6901 Dossenheim/Heidelberg, FRG*

L. KUNZE *Institut für Pflanzenschutz in Obstbau, Biologische Bundesanstalt, Postfach 73, 6901 Dossenheim/Heidelberg, FRG*

R. LAFON *Station de Recherches de Bordeaux, INRA, Domaine de la Grande Ferrade, 33140 Pont-de-la-Maye, France*

E. LANGERFELD *Institut für Botanik, Biologische Bundesanstalt, Messeweg 11/12, 3300 Braunschweig, FRG*

H. LAPIERRE *Station de Pathologie Végétale, CNRA, Route de St Cyr, 78000 Versailles, France*

D. LAPWOOD *Rothamsted Experimental Station, Harpenden AL5 2JQ, UK*

A. LEBRUN *10 Grande Rue, Andelu, 78770 Thoiry, France*

H. LECOQ *Station de Pathologie Végétale, INRA, Domaine St Maurice, 84140 Montfavet, France*

R.A. LELLIOTT *New Wood House, Beacon, Ilminster TA19 9AH, UK*

J.M. LEMAIRE *Station de Pathologie Végétale, INRA, Domaine St Maurice, 84140 Montfavet, France*

B.G. LEWIS *School of Biological Sciences, University of East Anglia, Norwich NR4 7TJ, UK*

H. LOT *Station de Pathologie Végétale, INRA, Domaine St Maurice, 84140 Montfavet, France*

J. LOUVET *Station de Recherches sur la Flore Pathogène dans le Sol, INRA, 17 rue Sully, BV 1540, 21034 Dijon Cedex, France*

J.A. LUCAS *Department of Botany, University of Nottingham, Nottingham NG7 2RD, UK*

J. LUISETTI *Station de Phytobactériologie, INRA, Beaucouzé, 49000 Angers, France*

T. LUNDSGAARD *Department of Plant Pathology, Royal Veterinary and Agricultural University, Thorvaldsensvej 40, 1871 Copenhagen, Denmark*

H.A. MAGNUS *Norwegian Plant Protection Institute, Boks 70, 1432 Ås-NLH, Norway*

P. MANTLE *Department of Biochemistry, Imperial College, South Kensington, London SW7 2BB, UK*

F. MARANI *Istituto di Patologia Vegetale, Università di Bologna, via Filippo Re 8, 40126 Bologna, Italy*

H. MARCELIN *Institut Technique du Vin, Chambre d'Agriculture, 19 avenue de Grande-Bretagne, 66025 Perpignan Cedex, France*

G. MARCHOUX *Station de Pathologie Végétale, INRA, Domaine St Maurice, 84140 Montfavet, France*

C. MAROQUIN *Station de Phytopathologie de l'Etat, Avenue Maréchal Juin 13, 5800 Gembloux, Belgium*

G.P. MARTELLI *Istituto di Patologia Vegetale, Via Giovanni Amendola 165/A, 70126 Bari, Italy*

P. MAS *Station de Pathologie Végétale, INRA, Domaine St Maurice, 84140 Montfavet, France*

A. MATTA *Istituto di Patologia Vegetale, Università di Torino, Via P. Giuria 15, 10126 Torino, Italy*

R.B. MAUDE *National Vegetable Research Station, Wellesbourne, Warwick CV35 9EF, UK*

Y. MAURY *Station de Pathologie Végétale, CNRA, Route de St Cyr, 78000 Versailles, France*

S. MEUNIER *Laboratoire de Virologie, Université de Louvain, Place Croix du Sud 3, 1348 Louvain la Neuve, Belgium*

C.S. MILLAR *Department of Forestry, University of Aberdeen, St Machar Drive, Old Aberdeen AB9 2UU, UK*

P.M. MOLOT *Station de Pathologie Végétale, INRA, Domaine St Maurice, 84140 Montfavet, France*

D. MONCOMBLE *Comité Interprofessionnel du Vin de Champagne, 5 rue Henri-Martin, 51204 Epernay, France*

M. MONSION *Station de Recherches de Bordeaux, INRA, Domaine de la Grande Ferrade, 33140 Pont-de-la-Maye, France*

P. MORENO *IVIA, Centro de Levante, Apartado Oficial, Moncada (Valencia), Spain*

G. MORVAN *Station de Pathologie Végétale, INRA, Domaine St Maurice, 84140 Montfavet, France*

G.S. NAGY *Plant Protection Institute, Herman Otto ut 15, P.O. Box 102, 1525 Budapest, Hungary*

L. NAVARRO *IVIA, Centro de Levante, Apartado Oficial, Moncada (Valencia), Spain*

P. NEERGAARD *Danish Government Institute of Seed Pathology for Developing Countries, 78 Ryvangs Allé, 2900 Hellerup, Denmark*

C.G. PANAGOPOULOS *Athens College of Agricultural Sciences, Votanikos, Athens 301, Greece*

A. PAPPAS *Benaki Phytopathological Institute, Kifissia, Athens, Greece*

G. PERROTTA *Istituto di Patologia Vegetale, Via Valdisavoia 5, Catania, Italy*

D.H. PHILLIPS *Grange Cottage, 35 The Street, Wrecclesham, Farnham GU10 4QS, UK*

J. POLÁK *Institute of Plant Protection, Department of Virology, 6 Ruzyne, 16606 Praha, Czechoslovakia*

A. PORTA-PUGLIA *Istituto Sperimentale per la Patologia Vegetale, Via C.G. Bertero 22, 00156 Roma, Italy*

D. PRICE *Glasshouse Crops Research Institute, Worthing Road, Littlehampton BN17 6LP, UK*

R.H. PRIESTLEY *National Institute of Agricultural Botany, Huntingdon Rd, Cambridge CB3 0LE, UK*

G. PROESELER *Institut für Phytopathologie, Theodor Römer Weg 4, 4320 Aschersleben, GDR*

P. PSALLIDAS *Benaki Phytopathological Institute, Kifissia, Athens, Greece*

C. PUTZ *Station de Pathologie Végétale, INRA, 28 rue de Herrlisheim, BP 507, 68201 Colmar, France*

A. RAGAZZI *Istituto di Patologia e Zoologia Forestale e Agraria, Piazzale delle Cascine 28, 50144 Firenze, Italy*

C.J. RAWLINSON *Rothamsted Experimental Station, Harpenden AL5 2JQ, UK*

P. RICCI *Station de Pathologie Végétale, INRA, Villa Thuret, BP 78, 06602 Antibes, France*

M.J. RICHARDSON *Department of Agriculture and Fisheries for Scotland, East Craigs, Edinburgh EH12 8NJ, UK*

J. RICHTER *Landesanstalt für Pflanzenschutz, Reinsburgstrasse 107, 7000 Stuttgart 1, FRG*

M. RIDÉ *Station de Phytobactériologie, INRA, Beaucouzé, 49000 Angers, France*

A.V. ROBERTS *North East London Polytechnic, Faculty of Science—Department of Biology, Romford Road, London E15 4LZ, UK*

F. ROLL-HANSEN *Norwegian Forest Research Institute, Forest Pathology, P.O. Box 62, 1432 Ås-NLH, Norway*

C. ROUSSEL *Service de la Protection des Végétaux, Chemin d'Artigues, 33150 Cenon, France*

D.J. ROYLE *Long Ashton Research Station, Bristol BS18 9AF, UK*

L.F. SALAZAR *Centro Internacional de la Papa, P.O. Box 5969, Lima, Peru*

R. SAMSON *Station de Phytobactériologie, INRA, Beaucouzé, 49000 Angers, France*

F.R. SANDERSON *Crop Research Division, DSIR, Private Bag, Christchurch, New Zealand*
P. SCHILTZ *Institut Expérimental du Tabac de la SEITA, Domaine de la Tour, BP 168, 24108 Bergerac, France*
H.H. SCHIMANSKI *Institut für Phytopathologie, Theodor Römer Weg 4, 4320 Aschersleben, GDR*
E. SCHWARZBACH *Probstdorfer Saatzucht, 2301 Gross Enzersdorf, Austria*
P.R. SCOTT *Plant Breeding Institute, Maris Lane, Trumpington, Cambridge CB2 2LQ, UK*
E. SEEMÜLLER *Institut für Pflanzenschutz in Obstbau, Biologische Bundesanstalt, Postfach 73, 6901 Dossenheim/Heidelberg, FRG*
R. SHATTOCK *School of Plant Biology, University College of North Wales, Denial Road, Bangor LL57 2UW, UK*
P. SIGNORET *ENSA, 9 place Viala, 34060 Montpellier, France*
V. SMEDEGAARD-PETERSEN *Department of Plant Pathology, Royal Veterinary and Agricultural University, Thorvaldsensvej 40, 1871 Copenhagen, Denmark*
I.M. SMITH *EPPO, 1 rue le Nôtre, 75016 Paris, France*
P.M. SMITH *Parkhill, Watersfield, Pulborough, West Sussex RH20 1NF, UK*
D.M. SPENCER *Plant Pathology and Microbiology Department, Glasshouse Crops Research Institute, Worthing Road, Littlehampton BN16 3PU, UK*
D. ŠUTIĆ *Faculty of Agriculture, Centre for Plant Virus Diseases, Zemun-Beograd, Yugoslavia*
P. TALBOYS *East Malling Research Station, Maidstone ME19 6BJ, UK*
E.C.TJAMOS *Benaki Phytopathological Institute, Kifissia, Athens, Greece*
I.A. TOMLINSON *National Vegetable Research Station, Wellesbourne, Warwick CV35 9EF, UK*
M. TOŠIĆ *Faculty of Agriculture, Centre for Plant Virus Diseases, Zemun-Beograd, Yugoslavia*
R. TRAMIER *Station de Pathologie Végétale, INRA, Villa Thuret, BP 78, 06602 Antibes, France*
T. TURCHETTI *Istituto di Patologia e Zoologia Forestale e Agraria, Piazzale delle Cascine 28, 50144 Firenze, Italy*
J.J. TUSET *IVIA, Centro de Levante, Apartado Oficial, Moncada (Valencia), Spain*
J.K. UYEMOTO *Department of Plant Pathology, University of California, Rt. 1 Box 1240, Davis CA 95616, USA*
J. VACKE *Department of Virology, Research Institute for Plant Production, Ruzyne 507, 16106 Praha, Czechoslovakia*
L. VAJNA *Research Institute for Plant Protection, P.O. Box 102, 1525 Budapest, Hungary*
K. VERHOEFF *Phytopathologisch Laboratorium Willie Commelin Scholten, Javalaan 20, 3742 CP Baarn, Netherlands*
M. VERHOYEN *Laboratoire de Virologie, Université de Louvain, Place Croix du Sud 3, 1348 Louvain la Neuve, Belgium*
F. VIRÁNYI *Research Institute for Plant Protection, P.O. Box 102, 1525 Budapest, Hungary*
J. VÖRÖS *Research Institute for Plant Protection, P.O. Box 102, 1525 Budapest, Hungary*
A. VUITTENEZ *Station de Pathologie Végétale, INRA, BP 507, 68021 Colmar, France*
I. WEBER *Institut für Phytopathologie, Theodor Römer Weg 4, 4320 Aschersleben, GDR*
H.L. WEIDEMANN *Biologische Bundesanstalt, Messeweg 11/12, 3300 Braunschweig, FRG*
B. WILLIAMSON *Scottish Crop Research Institute, Invergowrie, Dundee DD2 5DA, UK*
P. YOT *Station de Pathologie Végétale, INRA, Auzeville BP 12, 31320 Castanet Tolosan, France*
J.C. ZADOKS *Department of Phytopathology, Agricultural University, Binnenhaven 9, 6709 PD Wageningen, Netherlands*

1: Viruses I
General; Major Virus Groups

This and the following three chapters describe over 250 plant diseases known or thought to be due to viruses, present in or significant for Europe. This handbook does not deal predominantly with the characteristics of the causal agent, i.e. the virus itself (for which the reader is referred especially to the CMI/AAB virus descriptions), but rather, for each virus, with the disease caused. Chapters 1 and 2 cover diseases which have been clearly demonstrated to be caused by a virus so well identified and characterized that it can be classified in a group (Matthews 1982). Within each group, the viruses are presented in alphabetical order, the name of the main host plant in which the disease was originally described being quoted first. Synonyms of the virus name, and common names of the diseases caused, are given and indexed. The virus groups themselves are also presented alphabetically. Chapter 3 covers diseases associated with more or less well characterized viruses, but where some doubts remain on grouping or aetiology. Finally, Chapter 4 relates to diseases whose virus aetiology is only probable.

The following few abbreviations have been used: DNA—deoxyribonucleic acid; RNA—ribonucleic acid; M-Wt—molecular weight; d—dalton; ELISA—enzyme-linked immunosorbent assay; IEM—immunoelectron microscopy; ISEM—immunosorbent electron microscopy; SDS—sodium dodecyl sulphate. In addition, acronyms for virus names have been given, following the list of Van Regenmortel (1982)*. Attention is being given at international level to the development of a fully consistent system of acronyms, readily extendible to new virus names. However, this was not yet available at the time of publication.

This first chapter includes 10 major virus groups (with the most important and/or numerous members, at least in Europe): carlaviruses, closteroviruses, cucumoviruses, ilarviruses, luteoviruses, nepoviruses, potexviruses, potyviruses, tobamoviruses, viroids.

*In a few cases, acronyms not featuring in the list have been used, if they could be constructed without ambiguity from elements already used in the list. In other cases, it has not been possible to do this. For example, HMV is attributed to henbane mosaic virus and so cannot be used for hop mosaic virus; the list provides no abbreviation for hop.

MAJOR GROUPS

THE CARLAVIRUS GROUP

Elongated, slightly flexuous particles (600–700× 12–15 nm) with one positive-sense, single-stranded RNA with M-Wt $2.3–3.0\times10^6$ and usually one single coat polypeptide species (M-Wt 33000 d). The particle contains 5% RNA. Aphid-transmitted in the non-persistent (stylet-borne) manner. Type member is carnation latent virus (CMI/AAB 259).

Carnation latent virus (CLV) has been identified in Europe in both carnation and potato but causes few or no symptoms. Aphid-borne in the non-persistent manner (CMI/AAB 61), it is indexed by mechanical inoculation onto *Chenopodium* spp. (chlorotic local lesions and systemic mottle), by ELISA or IEM, and eliminated by heat treatment and meristem-tip culture.

Chrysanthemum mosaic virus and chrysanthemum virus B are assumed to be similar if not identical strains, distributed world-wide. Symptoms are mild leaf mottling and slight loss of flower quality on susceptible cultivars; most cultivars are symptomless when infected (CMI/AAB 110; Horst *et al.* 1977). Aphid-borne (stylet-borne manner), it is indexed by mechanical inoculation to *Petunia* (inoculated leaves show chlorotic spots within 3–4 weeks), by ELISA or IEM. Control is by propagation of healthy plants; virus-free plants are obtained from infected plants by heat treatment or meristem-tip culture.
MONSION

Garlic latent virus is a probable member of the carlavirus group (Cadilhac *et al.* 1976) and is widely distributed. In France all apparently healthy cultivars are infected with this and an unidentified potyvirus. It is possibly related serologically to shallot latent virus.

Mechanically transmitted to onion and leek, it provokes a systemic but symptomless infection (Delécolle & Lot 1981).
DELECOLLE

Hop latent virus is extremely common in Europe and the USA as well as occurring in Australia. Infection remains symptomless in all known hop cultivars but the virus may cause systemic chlorotic flecking, chlorosis and distortion in some seedlings; the effect on yield is unknown but probably slight. Hop latent virus (HLV) is aphid-borne (*Phorodon humuli*) in the non-persistent manner, and mechanically transmitted to a narrow range of hosts. No reliable herbaceous indicator has been described. Serological indexing by ELISA is effective and readily distinguishes hop latent virus from the other carlaviruses found in hop: hop mosaic virus and American hop latent virus (q.v.) (Adams & Barbara 1982; CMI/AAB 261). For control, hop latent virus-free stock of some cultivars is available but not widely distributed; rapid spread by aphids makes control difficult unless well isolated from old plantings.
BARBARA & ADAMS

American hop latent virus probably belongs to the carlavirus group. In Europe it is found only in introductions from N. America where it is common. Infection remains symptomless in all hop cultivars tested; the effect on yield is unknown. It is mechanically transmitted, and can be experimentally transmitted by aphids in the non-persistent manner but there is no evidence of natural spread in Europe. *Datura stramonium* or *Chenopodium quinoa* can be used as indicators; serological indexing by ELISA is effective (Adams & Barbara 1982; CMI/AAB 262). Thorough quarantine testing of introductions (especially those from N. America) should prevent the American hop latent virus from becoming established in Europe.
BARBARA & ADAMS

Hop mosaic virus

Basic description

Hop mosaic virus has straight or slightly flexuous filamentous particles *c.* 14×650 nm with single RNA species of M-Wt *c.* 3.0×10^6, comprising 5–7% of the particle and a single peptide of M-Wt *c.* 34 000. Purified preparations show a single sedimenting component (Adams & Barbara 1980; CMI/AAB 241).

Host plants and diseases

Hop mosaic virus occurs naturally in hop and occasionally in weed species. It is mechanically transmissible to a moderate range of hosts. Symptoms are seen only in hop cultivars of the true Golding type. In these sensitive cultivars it induces chlorotic vein-banding, recurving of the leaf margins, stunting and a tendency for bines to fall away from their support; affected plants yield poorly and usually die prematurely. Sensitive cultivars are no longer widely grown but symptomless infection is common in tolerant cultivars. The effect on yield in these cultivars is not known.

Strains

Isolates causing only mild symptoms have been reported and may be responsible for several poorly characterized diseases.

Diagnosis

No diagnostic herbaceous hosts are known and indexing is best performed serologically (by ELISA). This is reliable and readily distinguishes hop mosaic virus from other filamentous viruses in hop. Leaf dip electron microscopy may be used but it is necessary to use antisera (for 'decoration' or 'trapping') to differentiate hop mosaic virus from other carlaviruses in hop.

Distribution and transmission

Hop mosaic virus is probably common wherever tolerant hop cultivars are grown throughout the world. It is aphid-transmitted in the non-persistent manner; the damson-hop aphid (*Phorodon humuli*) is probably the most important vector in the field. Spread may be rapid. Propagation of infected but symptomless material is also important in disseminating the virus.

Economic importance

Despite the widespread occurrence of hop mosaic virus there is no estimate of its effect on tolerant cultivars. Isolated plantings of sensitive cultivars remain substantially free from infection and although individual plants are severely affected the overall loss is probably small. There is limited evidence that where tolerant cultivars are being introduced to farms or regions which hitherto grew only sensitive cultivars mosaic disease is becoming more prevalent.

Control

Isolated plantings remain uninfected and hop mosaic virus-free stocks of some tolerant cultivars are available for planting near sensitive cultivars. However, the rapid spread and apparent lack of deleterious effects in tolerant material means that the only control routinely attempted is removal of diseased plants from plots of sensitive cultivars.

BARBARA & ADAMS

Lily symptomless virus (LSV) (syn. lily curl stripe virus) occurs world-wide. Symptoms (mottle, stripe, distortion and stunting) can appear but are not usually noticeable. It is aphid-borne in the stylet-borne manner. Affected crops are lilies and tulips (CMI/AAB 96). Dip preparations stained with phosphotungstic acid or chop preparations are useful for quick identification, or ELISA for a routine assay (Beijersbergen & van der Hulst 1980a, 1980b). Virus-free plants are obtained by meristem-tip culture.

ALBOUY

Narcissus latent virus (NaLV) was discovered in Europe but is very probably present wherever nerine, narcissus and bulbous irises are grown. Alone, it induces light and dark green mottling near the leaf tips of intolerant cultivars (CMI/AAB 170). In addition to the two natural hosts in Amaryllidaceae and one in Iridaceae, the virus infects species in Amaranthaceae, Solanaceae and Leguminosae. Detection is by mechanical transmission to *Tetragonia expansa* or *Phaseolus vulgaris* (local lesions) or by serological methods. The virus is aphid-borne by *Acyrthosiphon pisum*, *Aphis gossypii* and *Myzus persicae*.

Pea streak virus (PeaSV) is serologically related to red clover vein mosaic virus (q.v.). Reported in N. America and the FRG it could be established in several other countries. Symptoms on pea are necrotic streaks on stems and petioles, veinal necrosis, wilting and death of the plants (Hagedorn & Walker 1949). Transmitted by *Acyrthosiphon pisum* in the non-persistent manner, the virus is indexed by mechanical inoculation onto *Chenopodium giganteum* (inoculated leaves show necrotic lesions with chlorotic margins) and *Gomphrena globosa* (inoculated leaves show local yellow lesions after 2–3 days). Also detected by serology. Control is by propagation of resistant cultivars (CMI/AAB 112).

CHOD

Poplar mosaic virus (PopMV)

(Syn. poplar latent virus)

Basic description

PopMV has flexuous filamentous particles *c.* 12–15 × 685 nm (CMI/AAB 75). Sedimentation coefficient of the particles is 165S. The associated single-stranded RNA has M-Wt 2.3×10^6. Two proteins have been associated with the purified particles, a major component having 37 000 d and a minor one of *c.* 30 000 d (Boccardo & Milne 1976; Luisoni *et al.* 1976).

Host plants and diseases

PopMV naturally infects poplar species in the taxonomic sections Aigeiros and Tacamahaca but not those in sections Turanga, Leuce or Leucoides. All clones of the commercially important trans-American (*P. deltoides* × *P. trichocarpa*) and Eur-American (*P. nigra* × *P. deltoides*) hybrids are infectible (Cooper 1979). Species in some 20 dicotyledonous families are experimentally infectible.

Chlorotic-yellow spots extending along the fine leaf veins are characteristic but many clones have slight, transient or no foliar symptoms. The specific gravity and strength of wood can be adversely affected.

Strains

Some isolates differ in temperature optima for growth or pathogenicity (Cooper & Edwards 1981; Van der Meer *et al.* 1980a). A carlavirus from *Lonicera* that did not infect poplar experimentally has antigenic and other properties in common with isolates of poplar mosaic virus obtained in the Netherlands (Van der Meer *et al.* 1980b).

Identification and diagnosis

Bioassays (veinal necrosis) in *Nicotiana megalosiphon* after mechanical inoculation with sap from foliage, bark or buds in winter are about as sensitive as the indirect ELISA test. The latex flocculation test is rapid but less sensitive than either of these methods. Electron microscopy and serology have been used to identify the virus particles in leaf extracts. Ultrastructural changes in leaf cells of poplar associated with PopMV are not helpful in diagnosis (Atkinson & Cooper 1976).

Geographical distribution and transmission

This virus is found wherever poplars are grown commercially, with the possible exception of S. Africa and S. America. Transmission is largely attributable to vegetative propagation from infected stock. The virus is not seed-borne and no vector has yet been identified, even though virus-free plants become infected within a few years (Cooper & Edwards 1981).

Economic importance and control

Clones differ in vigour but virus-free poplars from shoot meristems reportedly grow twice as fast as their infected counterparts. Thermotherapy or meristem-tip propagation has produced virus-free plants for experimental assessment but these have not been adopted commercially.
COOPER

Potato virus M (PVM) (syn. potato paracrinkle virus) is often symptomless in potato but can cause severe mosaic, crinkling, rolling of leaves, necrosis and stunting (CMI/AAB 87). It is sap-transmissible, detected by mechanical inoculation to *Lycopersicon chilense* (systemic leaf deformation) and bean cv. Red Kidney (local lesions), and easily detectable by ELISA (Wetter & Milne 1981). It occurs world-wide and in Europe is prevalent in areas of continental climate (due to higher temperatures and efficient transmission in the non-persistent manner mainly by *Aphis nasturtii*). Control by clonal selection of healthy plants and application of insecticides appears effective. Virus-free plants have been obtained by combining heat treatment and meristem-tip culture.
GABRIEL & KOWALSKA-NOORDAM

Potato virus S (PVS)

Basic description

PVS has filamentous particles *c.* 650×12 nm. It is strongly antigenic and is distantly related serologically to other carlaviruses (APS Compendium; CMI/AAB 60).

Host plants and diseases

Host range is restricted, PVS occurring naturally only in potato. Infected plants are symptomless or show slight chlorosis, deepening of veins, rugosity and more open growth habit. Green or brown spots may appear on yellowing leaves (Chrzanowska 1976).

Strains

In Europe only minor variants occur. Properties of strains from Andean regions seem to be different from those occurring elsewhere (APS Compendium).

Identification and diagnosis

PVS is easily detected serologically; ELISA is reliable for detecting the virus in tubers. *Chenopodium quinoa* (chlorotic local lesions) and *Nicotiana debneyi* (systemic interveinal mottle and vein banding) are reliable test plants (Kowalska & Waś 1976).

Geographical distribution and transmission

PVS is probably the most common virus of potato in the world. It is propagated in tubers but not in true seeds (Goth & Webb 1975). It is easily transmitted mechanically. In nature transmission in the non-persistent manner by aphids is the main method of virus propagation (Turska 1980). The most effective vector is *Aphis nasturtii*, although the virus is also transmitted by *Myzus persicae*, *Aphis fabae* and *Rhopalosiphum padi*.

Economic importance and control

Many commercial cultivars are totally infected. Affected plants usually show no obvious symptoms and depression of yield of tubers is generally low (maximum 20%). Control is by clonal selection combined with serological testing. Rate of reinfection may be reduced by using resistant cultivars and the application of insecticides. Totally infected cultivars may be freed from PVS by meristem-tip culture. Hypersensitivity to PVS may be used in the future to improve the level of resistance to virus infection.
GABRIEL & KOWALSKA-NOORDAM

Red clover vein mosaic virus (RCVMV) induces vein chlorosis, mosaic, streaking and occasional stunting in *Trifolium pratense* and various legumes. It is responsible for severe yield losses in pea, causing stunting, and even killing early infected plants. It is widely distributed in Europe, N. America and S. Africa (CMI/AAB 22) and is aphid-borne in the non-persistent manner by several species, *Acyrthosiphon pisum*, *Myzus persicae* and *Cavariella aegopodii*. It is serologically related to several members of the carlavirus group (carnation latent, chrysanthemum mosaic, potato virus S, pea streak, passiflora latent, cactus virus 2, for example). Detection is by mechanical transmission onto *Gomphrena globosa* (local lesion) or by serology (Wetter & Milne 1981).

Shallot latent virus (SLV). Infected shallot plants usually remain symptomless. In the field, mixed infection is frequent: in onion with leek yellow stripe virus (LYSV) and onion yellow dwarf virus (q.v.); in leek with LYSV, causing more severe symptoms (severe chlorotic or white streaking); some cultivars can be killed (Paludan 1980). SLV is mechanically transmissible to onion and *Allium fistulosum*, which remain symptomless (CMI/AAB 250). Identified in Europe, it occurs especially in the Netherlands, Denmark, Belgium and France but is probably present world-wide (Bos 1982). Aphid-borne in the non-persistent manner by *Myzus ascalonicus*, it is detected by transmission onto *Chenopodium quinoa* or *Ch. amaranticolor* (local lesions), or by serology.
DELECOLLE

THE CLOSTEROVIRUS GROUP

Elongated, very flexuous particles (600–2000 × 12 nm) with one positive-sense, single-stranded RNA (M-Wt $2.2–4.7 \times 10^6$ d) and one coat polypeptide (23–27 000 d). Most are aphid-transmitted in the semi-persistent manner. Nevertheless, due to differences in particle length and in natural transmission, 3 subgroups have been described: subgroup A (particles 730 nm; RNA 2.5×10^6 d, natural transmission unknown), subgroup B (particles 1250–1450 nm, RNA 4.2×10^6 d, aphid-transmitted), subgroup C (particles 1650–2000 nm, RNA 6.5×10^6 d, aphid-transmitted). Aphid transmission often depends on a helper factor or helper virus (Bar-Joseph *et al.* 1979; CMI/AAB 260).

Apple chlorotic leaf spot virus (ACLSV)

(Syn. apple latent virus 1, pear ring mosaic virus, pear ring pattern mosaic virus)

Basic description

ACLSV is classified in subgroup A of the closterovirus group according to its particle length (CMI/AAB 250). The virion appears as an elongated particle (720 × 12 nm) comprising one RNA molecule of 2.5×10^6 d. Coat protein is made up of a single polypeptide *c.* 23 500 d. Purified ACLSV suspensions are moderately immunogenic but immunogenicity depends on the strain.

Host plants

ACLSV was originally described from apple (Mink & Shay 1959) after transmission to *Malus platycarpa*; it was subsequently identified in other pome and stone fruits. All rosaceous fruit trees and probably also most ornamental trees of this family can be naturally infected. Outside the Rosaceae, the host range is very restricted. The virus can also be transmitted artificially to a few herbaceous host plant species.

Diseases

Various diseases induced by ACLSV have been reported in fruit trees.

Apple. Most commercial cultivars are symptomlessly infected. Only some cultivars, seedlings (R 12740-7A) or species (*Malus platycarpa*) show symptoms and can be used as indicator plants (ACLSV causes a line pattern symptom in *Malus platycarpa*).

Pear. Leaf mottle symptoms and ring mosaic (or ring pattern mosaic) are associated with ACLSV. Many cultivars remain symptomless when infected but some commercially important ones (Beurré Hardy, Beurré Bosc, Williams) are susceptible; some growth reduction can also be observed.

Peach. Most commercial cultivars are symptomlessly infected but in a few very susceptible cultivars or in some peach seedlings (Elberta, GF 305) ACLSV induces a dark green sunken mottle of leaves. Graft incompatibilities can be observed occasionally when some particular plum cultivars are used as rootstocks.

Plum and prune. Pseudopox disease of plums is caused by ACLSV; grooves and deformations develop on the fruits especially on blue cultivars—this symptom can be mistaken for plum pox infection but, by contrast with PPV, no premature dropping of fruits occurs. Bark split of plum (Fig. 1) causes very severe losses in some prune-growing areas: after flattening of the branches, deep splits develop on the branches and trunks, and very often branches break.

Apricot. Several types of fruit symptoms of graft incompatibility have been described in apricot which are very likely due to ACLSV. Two fruit symptoms, viruela (or apricot pseudopox) and butteratura, have been described in Spain and in Italy respectively. Viruela-infected fruits of susceptible cultivars show deformities, depressions and some discolorations. Both viruela and butteratura symptoms can be increased by cold, wet weather in spring and early

Fig. 1. Apple chlorotic leafspot virus: bark splits on plum.

summer. In the presence of butteratura, fruits develop brown necrotic blotches; the flesh beneath is brown and some symptoms of abnormal lignification are present on the stones which occasionally show holes and cracks. With some cultivars or some rootstocks, graft incompatibility develops 6–12 months after grafting when the cultivar or the rootstock is infected. Incompatibility occurs especially when *Prunus domestica*, *P. persica* or *P. armeniaca* are used as rootstocks. When grafted on *P. cerasifera*, incompatibility does not appear consistently or is delayed.

Quince. ACLSV induces typical stunting (quince stunt).

Strains

ACLSV shows a great diversity of strains with regard to symptomatology, consequently biological indexing is difficult and often unreliable (Marenaud *et al.* 1976). Most strains appear capable of infecting all hosts but some apple strains appear to be very difficult to transmit to stone fruits. Symptomatology is sometimes highly strain-specific: only one ACLSV strain induces bark split on prune, while most strains from *Prunus* do not induce clear symptoms after transmission onto susceptible apple cultivars. Two serotypes have been described (Chairez & Lister 1973); they are very closely related and an antiserum to one strain (even to a strain which could seem very distant from the biological point of view) cross reacts very clearly with all the known ACLSV strains (Detienne *et al.* 1980).

Identification and diagnosis

Biological indexing is by chip budding onto susceptible indicators, but the technique is often unreliable as most indicators only react to certain strains (Marenaud *et al.* 1976). The best indicators are peach seedlings or *P. tomentosa* (which, respectively, develop dark green sunken mottle and line pattern symptom after 1–3 months in the greenhouse) or quince C7/1 (which shows chlorotic leaf spots and deformations in the field 1 year after inoculation). Serological indexing with ELISA (Detienne *et al.* 1980) or immune electron microscopy (Kerlan *et al.* 1981) are recommended as being highly sensitive and much less strain-specific than biological indexing. The virus is uniformly distributed in trees and can be detected in leaves, flowers, fruits and seeds.

Geographical distribution

ACLSV occurs world-wide and is very common in all fruit-tree and woody ornamental species of Rosaceae.

Transmission

Apart from one report of possible nematode transmission (Fritzsche & Kegler 1968), no means of natural transmission has been observed. The virus is spread by vegetative propagation of infected material. Seed transmission may occur in apricot.

Economic importance and losses

Damage is difficult to estimate because of the diversity of virus strains and great differences in susceptibility between cultivars and species. Apple is not usually affected, although ACLSV together with other diseases (stem grooving or stem pitting) reduces growth and yield. Similarly with pear, only a few (but important) commercial cultivars are affected. Significant losses are observed (reduction of growth and yield) on mixed infection (with pear vein yellows, quince sooty ring spot).

In stone fruits, severe symptoms develop and serious losses have been reported in several areas, especially in plum and prune (the French Prune d'Ente P707 is the most susceptible cultivar to the bark split-ACLSV strain) or apricot. Several hundred hectares of prune have reportedly been replaced in France; in Spain, production was reduced by 56% in one susceptible cultivar cultivated on more than 50 000 ha. In many nurseries, productivity is reduced because of effects of the virus on bud take and graft compatibility.

In general, except for the severe diseases of stone fruits, the impact of ACLSV is more due to its wide occurrence and the number of susceptible species than to the severity of symptoms.

Control

Most commercial cultivars and rootstocks were found to be infected, so virus-free material was obtained through heat therapy. Very good results were obtained with this treatment or with shoot-tip grafting (Navarro *et al.* 1982).

Special research interest

Natural spread is so slow and uncertain that further investigations are needed to decide whether the virus has an aphid vector, and to confirm its classification as a closterovirus. Moreover ACLSV is a very good model to start investigating mechanisms of virus-induced graft-incompatibility.

DELBOS & DUNEZ

Beet yellows virus (BYV)

(Syn. beet necrotic yellows virus, beet etch yellowing virus, beat virus 4)

Basic description

The virion is a filamentous flexuous particle *c.* 1250 × 10 nm, comprising a single RNA molecule of 4.15 × 10^6 (CMI/AAB 13). Coat protein has a single polypeptide with M-Wt *c.* 23 000 d. $S_{20,W}$ of the particle is 130S. Virus suspensions are good immunogens. The particle length and aphid transmissibility make it a member of closterovirus subgroup B.

Host plants

BYV infects over 150 species of herbaceous plants. Cultivated hosts include *Beta, Spinacia, Papaver, Tetragonia.* The virus also infects most wild *Beta* species and some weed plants of the Chenopodiaceae, Cruciferae, Polygonaceae, Caryophyllaceae and Compositae. Some species (for example *Bilderdykia convolvulus, Plantago lanceolata, Capsella bursa-pastoris, Senecio vulgaris, Stellaria media* and *Thlaspi arvense*) are important natural sources of infection.

Diseases

Severe symptoms of yellowing are observed in beet (sugar beet, fodder beet, red beet, spinach beet), spinach and poppy. Similar symptoms are caused by beet mild yellowing virus and the related strain, beet western yellows virus. In beet, the first symptom is vein-clearing or vein-yellowing of the young leaves. Later the leaves become thick, leathery and brittle. Leaves approaching maturity may show small translucent 'pinpoint' spots. Small reddish or brown necrotic spots develop on older yellowed leaves. The combination of necrotic spots and yellowing often gives the leaves a bronze cast. Necrotic areas can enlarge from the apex causing premature leaf death. In spinach, young leaves show chlorotic spotting and vein-clearing, leaves became yellow, necrotic and distorted, and plants are dwarfed. Some infected plants die. On poppy, leaves become yellow and necrotic and plants are dwarfed.

Strains

Several strains have been described on the basis of their pathogenicity to sugar beet and indicator plants (Polák 1971a) and of their isoelectric points and serological differences (Polák 1971b). The isoelectric points of five strains vary from pH 4.7 to 5.4. Strains of the virus are closely related and the antigenic variation appears limited. Strains differ in their effect on sugar beet—necrotic strains are more pathogenic than mild ones.

Identification and diagnosis

The virus may be identified by visual inspection of beet plants, but indexing is essential to distinguish from beet mild yellowing virus. The two viruses often occur together. Several sensitive and reliable techniques can be used to identify BYV, including transmission onto *Chenopodium capitatum* and *Chenopodium foliosum* plants in the greenhouse (acute stunting, distortion and vein-clearing of young leaves within 2–3 weeks, infected plants usually die prematurely). Serological techniques can be applied: the microprecipitin drop test (Polák *et al.* 1975), the latex agglutinin technique (Fuchs 1976) and ELISA. All provide an adequate and reliable means of detecting BYV; ELISA also seems to be reliable for detecting the virus in beet roots during winter.

Geographical distribution

BYV occurs world-wide (Europe, Asia, N. and S. America, Africa, Australia). The virus is present in all European countries producing beet. In some years it causes severe damage especially across the central zone of Europe.

Transmission

BYV is transmitted by aphids in a semi-persistent manner (Brĉák & Polák 1976). Out of 35 aphid species reported to be vectors of BYV, the most efficient is *Myzus persicae*. The second, but less efficient vector of the virus is *Aphis fabae*. Seed transmission has not been proved. The virus is transmissible mechanically to beet and other indicator plants (Polák 1971a). Several species of dodder can transmit BYV.

Economic importance

BYV alone or together with beet mild yellowing virus (beet western yellows virus) causes a very important virus disease of sugar beet in Europe. The losses of root yields in some European countries reach 15–30% in some years with a slight (2–3%) reduction in sucrose content. Virus infection also decreases the yield and quality of sugar beet seed. After early experimental infection of beet the yield reduction can be 40–60%.

Control

The effectiveness of control measures depends mainly on: prediction and chemical control of vectors, especially in the case of early infection (Watson *et al.* 1975); planting seed crops away from root crops; elimination of other sources of infection (overwintering weeds); breeding sugar beet for virus resistance (Björling 1969); destruction of volunteers; siting of fodder beet clamps.

Special research interest

Special attention has been paid to the epidemiology of the virus (Jadot 1976). BYV now provides a classic example of forecasting (Watson *et al.* 1975).
POLÁK

Beet pseudoyellows virus was first described in California on glasshouse crops, not causing any economically significant disease and has since been found in N. and S. America, W. and S. Europe, and Australia. Yellows, chlorotic spotting and stunting symptoms are seen on sugar beet, glasshouse cucumbers, lettuce and weeds (Chenopodiaceae, Compositae, Umbelliferae, Solanaceae). The symptoms on beet in particular are interveinal yellowing of the leaves, thickening and brittleness of the leaves and dwarfing of the plants. The disease is transmitted semi-persistently only by the whitefly *Trialeurodes vaporariorum* (Duffus 1965). There is no mechanical transmission. Indexing is by vector transmission to *Capsella bursa-pastoris*, lettuce, flax or *Taraxacum officinale* (inoculated plants show chlorosis, interveinal yellowing, reddening and thickening of leaves). The virus associated with beet pseudo-yellows was not purified or characterized.

Around 1980, a damaging yellows disease transmitted by *T. vaporariorum* was described on lettuce and cucumber in France and the Netherlands (Van Dorst *et al.* 1983). It is apparently present also in Italy, Spain and Germany. The agent seems to be very similar to the beet pseudoyellows agent, but full description is still lacking because of the difficulties of virus purification. Thread-like particles of about 1000 nm have been observed in ultrathin sections and erratically in crude sap. The nature and localization of these particles suggests that the agent is probably a closterovirus. Control is by insecticide treatment against the vectors and elimination of weed hosts in glasshouses.
POLÁK & LOT

Beet yellow stunt virus (BYSV) belongs to closterovirus subgroup B. Described from N. America it very probably occurs in Europe as well. It is common on the weed *Sonchus oleraceus* in California. Symptoms are severe twisting, cupping and epinasty of one or two intermediate-aged leaves of beet. Leaves are mottled and yellow, plants are stunted and often collapse and die. Beet and lettuce are occasionally infected. It is aphid-borne in a semi-persistent manner (Duffus 1972; CMI/AAB 207) and indexed by serological precipitin test or inoculation by aphids onto lettuce (severe stunting, extreme yellowing) or *Senecio vulgaris* (purple spots on the interveinal areas). The aphid vector *Nasonovia lactucae* can be controlled chemically and infection sources such as *S. oleraceus* eliminated.
POLÁK

Carnation necrotic fleck virus (CNFV)

Basic description

This is a flexuous elongated virus (Poupet *et al.* 1975) belonging to closterovirus subgroup B. Isolates of different origins have been described under the names carnation streak virus and carnation yellow fleck virus (Smookler & Loebenstein 1974). However, carnation necrotic fleck virus (Inouye & Mitsuhata 1973; CMI/AAB 136) is the accepted name.

Host plants and diseases

Infected carnations show chlorotic, then necrotic spots, flecks and streaks on the basal leaves and along the flowering stems (Fig. 2). A reddish-purple discoloration is characteristic of the disease. Symptoms which appear in early spring are much more severe in plants also infected by carnation mottle virus. In some cases, leaf yellowing is followed by necrosis and wilt, but most often a partial recovery of the infected plant occurs. There are large differences in the behaviour of cultivars (Devergne & Cardin 1971).

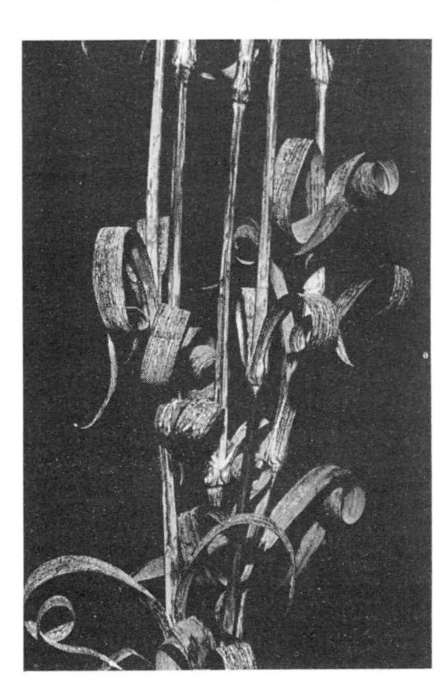

Fig. 2. Carnation necrotic fleck virus: streaks on carnation leaves.

Identification and diagnosis

Except during the short periods of symptom expression, infections can be detected only by indexing. As carnation necrotic fleck is not easily transmitted mechanically, bioassays on test-plants have to use aphid vectors (*Myzus persicae*) and are time-consuming. Nevertheless, *Dianthus barbatus*, *Silene armeria* and *Saponaria vaccaria* have been used for diagnosis. Infected *Saponaria* develop the most typical symptoms (vein clearing, leaf roll, stunt, wilt) (Poupet *et al.* 1975). ELISA now appears to be the most rapid and reliable means for detection even in the symptomless parts of infected plants.

Transmission

CNFV is aphid-transmitted in a semi-persistent manner, but it seems that the disease does not spread significantly in nature.

Geographical distribution and economic importance

CNFV is prevalent in the Mediterranean region. Flowers from infected plants are unaffected, but the virus has a harmful effect on foliage which impairs the quality of cut flowers and decreases yield.

Control measures

Healthy plants can be obtained by meristem-tip culture; regeneration must be followed by several indexings for at least a year. The elimination of carnation mottle virus generally leads to the disappearance of CNFV too.

DEVERGNE

Citrus tristeza virus (CTV)

(Syn. citrus quick decline virus)

Basic description

CTV is a member of closterovirus subgroup C. Virions are highly flexuous, filamentous particles *c.* $2000 \times 10–12$ nm comprising a single positive strand RNA with an estimated M-Wt $6.4–6.9^6$ d. The coat protein has a single polypeptide with M-Wt *c.* 25 000 d. Cytopathological effects are confined to cells of the phloem which contain virus aggregates and group-characteristic vesicles (Bar-Joseph *et al.* 1979).

Host plants

CTV infects most *Citrus* species. Cultivars and hybrids and some citrus relatives such as *Aeglopsis chevalieri*, *Afraegle paniculata* and *Pamburus missiones*. The only hosts outside the Rutaceae are certain *Passiflora* species (Bennet & Costa 1949; CMI/AAB 33).

Diseases

Tristeza is principally a disease of sweet orange and certain other cultivars when grafted on the sour orange rootstock (Fig. 3); however several citrus species such as sour lime (*Citrus aurantifolia*), Alemow (*C. macrophylla*), sweet lime, etrog citron, grapefruits and Pera sweet orange show stem pitting and decline whether they are used as seedlings, scions or rootstocks. These diseases are variously known as lime disease, grapefruit stem-pitting, citrus yellow shoot and Hassaku dwarf. On sweet orange, symptoms depend on the rootstock combination inoculation time and virus strain. Sweet orange (Sw.O.) grafted on sour orange (S.O.) rootstock may suddenly wilt, decline or die (quick decline) or develop a characteristic overgrowth at the bud union and honeycomb-pitting in the S.O. stock, just below the union. Trees grafted on susceptible rootstocks such as *C. macrophylla* or sweet lime show stunting and decline. Fruits often ripen early at the outset of the disease and are reduced in size. Leaves on declined trees often have a bronzed appearance.

Fig. 3. Citrus tristeza virus: dieback of sweet orange grafted on sour orange.

Young leaves on infected seedlings of sour lime are small, develop yellow flecks along their veins (vein clearing) and may become cupped. The intensity of the vein clearing depends on strain severity which can vary from extremely mild (few small flecks) to extremely severe reactions (vein suberization).

Strains prevalent in E. Asia and S. Africa cause severe stem pitting, stunting and the production of small fruits in grapefruit trees (McClean 1977).

Strains

Numerous CTV strains have been described on the basis of their pathogenicity and aphid transmissibility. They differ in virulence on the common Sw.O./S.O. combinations and may have various effects on different seedling or scionic combinations. Seedlings of grapefruit, S.O. and lemon show yellowing and stunting when inoculated with a group of isolates named seedling yellows (SY). Most workers now regard SY as a symptom of severe CTV isolates. Serological tests have not revealed antigenic differences between CTV isolates.

Identification and diagnosis

The virus may be identified by indexing on sour lime seedlings (Wallace 1968). Symptoms of vein clearing appear 3–6 weeks after inoculation. Laboratory methods now available for CTV diagnosis include SDS immunodiffusion (Gonsalves *et al.* 1978), IEM, ELISA, radioimmunoassay (RIA) (Garnsey *et al.* 1981), double-strand RNA profiles (Dodds & Bar-Joseph 1983) and nucleic acid hybridization with cloned cDNA (Rosner *et al.* 1983). These procedures provide rapid diagnosis and can be used singly and/or in combination depending on specific requirements.

Geographical distribution

The virus is widely distributed in most citrus-growing areas of the world except some parts of the Mediterranean basin and a few other isolated areas. CTV has been prevalent in Spain since 1956 and has more recently been found in Israel.

Transmission

CTV is transmitted semi-persistently by aphids. *Toxoptera citricida, Aphis gossypii* and *A. citricola* are the principal vectors of CTV. Recent reports from different locations indicate very high transmission of certain CTV strains by *A. gossypii* (Bar-Joseph & Loebenstein 1973; Roistacher *et al.* 1980).

Economic importance

Tristeza has caused extensive damage to citrus crops in Spain and threatens citrus production of S.O. rootstocks in Israel and other Mediterranean countries. Summarizing the world-wide damage during the last 50 years it has been estimated that about 40 million trees were lost or became unproductive due to CTV. In addition, new stem-pitting strains have been extremely destructive to grapefruit, Sw.O. and lime in several areas. The actual losses vary considerably according to the particular horticultural and economic situation in each epidemic area, e.g. marketing opportunities, economic value of the infected cultivar, tree density, age of infected grove, rate of disease spread and rate and efficiency of replanting.

Control

CTV decline is best controlled by excluding virus-infected material from citrus-growing areas by quarantine measures (Mather 1968) or by eradicating trees which carry the virus. In virus-invaded areas tristeza-tolerent scionic combinations should be used. Pre-immunization with mild strains (cross protection) is now used in commercial orchards to control the decline caused by severe stem-pitting isolates in Pera Sw.O. groves in Brazil (Costa & Muller 1980).

Special research interest

Research interests include factors affecting differential transmissibility by *A. gossypii*, selection of resistant rootstocks, mechanisms of cross protection and developing strain-specific diagnostic methods.
BAR-JOSEPH

Grapevine viruses A and B (see 'Grapevine leafroll', Ch. 3).

Heracleum latent virus (HLV) has particles 600 nm long and belongs to closterovirus subgroup A. It infects the wild umbellifer *Heracleum sphondylium* in England and the Netherlands. Though of no economic significance it is of special interest because it is aphid-borne in the semi-persistent manner in the presence of a helper virus (*Heracleum virus* 6), which is another closterovirus (subgroup C) with particles 1600 nm long (CMI/AAB 228). It is detected by mechanical transmission on *Chenopodium quinoa* or by serology.

Lilac chlorotic leafspot virus (LCLV) is a member of closterovirus subgroup C (particles 1540×12.5 nm) and was first detected in symptomless *Syringa* species. It is associated with chrome yellow line patterns or spots when in the UK and with slight chlorosis if infecting in the Netherlands. No vector is known and it is not seed-borne. Indexing is not routine but is possible by mechanical inoculation to *Chenopodium* species. No specific control exists other than propagation from indexed virus-free plants (CMI/AAB 202).
COOPER

THE CUCUMOVIRUS GROUP

Isometric particles 29 nm in diameter containing 4 molecules of positive-sense, single-stranded RNA. One molecule of RNA_1 (1.2×10^6 d), one molecule of RNA_2 (1.1×10^6 d), one molecule of RNA_3 (0.8×10^6 d) and one molecule of RNA_4 (0.35×10^6 d) are respectively encapsidated in 3 different particles. RNA_1, RNA_2 and RNA_3 are required for infectivity. RNA_4 is a monocistronic RNA containing the gene for the coat protein. One coat polypeptide species (24 000 d). Several virus isolates also encapsidate a short single-stranded RNA of M-Wt 0.1×10^6 d: this small RNA is a satellite RNA which can be associated with special symptom expression. These viruses are aphid transmitted in the non-persistent manner. Type member is cucumber mosaic virus.

Cucumber mosaic virus (CMV)

Basic description

CMV is the type member of the cucumovirus group (CMI/AAB 213). The virus has a tripartite genome: ssRNA molecules (1.0; 0.89; 0.68; 0.33, plus possibly the satellite RNA 0.1×10^6) are packaged in at least three classes of isometric particles with extremely close buoyant densities. The protein subunit has a M-Wt of *c.* 24 500 d (Kaper & Waterworth 1981). Particles are labile and the immunogenicity of the virus is poor, but this can be enhanced by fixation with formaldehyde.

Host plants

CMV has an extremely wide host range—there are thought to be 775 susceptible species in 86 families (Douine *et al.* 1979).

Diseases

The diseases caused by CMV are more serious on plants grown in the open, especially during the summer, than in protected crops. The affected crops in temperate areas are mainly vegetables and flowers

Fig. 4. Cucumber mosaic virus: breaking of gladiolus flowers.

(Fig. 4). Fruit trees and tropical crops are rarely affected.

In cucumber, melon and courgette, CMV causes green or yellow mosaic of leaves, leaf distortion and stunting of the plant. The virus generally induces a severe yield reduction mainly due to decreased fruit set. Fruit symptoms may appear on cucumber, (mosaic), courgette (mosaic, distortion) and occasionally on melon.

In tomato CMV causes two types of symptom: i) mosaic and fern leaf with an occasional shoe-string appearance to the leaves. Many flowers remain sterile; ii) necrotic areas appear on leaves, petioles and stem and may cause total collapse. This typical syndrome is linked to the presence of the specific RNA satellite of the virus (Kaper & Waterworth 1977; Jacquemond & Lot 1981).

In pepper, ringspotting on inoculated leaves is followed by systemic mosaic. CMV stops growth while fruits show ringspots or deformation and fall prematurely (Plate 1). Plants infected early may be totally fruitless. Systemic infection of tobacco causes mild general mottle or severe green mosiac, and some strains may produce leaf distortion.

In spinach, symptom intensity varies with the age of the plant. The plants may show various symptoms including mottling, stunting, yellowing and malformation of the younger leaves (spinach blight).

In celery, leaflets first show interveinal mosaic which later develops into whitish areas; they generally become crinkled and filiform. Pronounced curling of the younger petioles gives the heart a characteristic open and flattened appearance.

Strains

About 70 isolates of CMV are reported as strains. For example, Lovisolo & Conti (1969) did not find any identical isolates based on host range and symptoms, after testing 60 isolates obtained from different localities. A number of strains have been assigned to two groups according to their biological properties of symptomatology, thermosensitivity *in vivo* and serological or electrophoretical properties (Devergne & Cardin 1975; Marrou *et al.* 1975). The finding of two main groups of CMV strains, based on nucleic acid homology, parallels the earlier divisions (Piazzolla *et al.* 1979).

Identification and diagnosis

As with most other viruses, it is extremely difficult to identify CMV from the symptoms in affected crops. Very often, the set of symptoms obtained using a restricted host range of test plants (*Vigna unguiculata*, tobacco, melon) may induce errors. Rapid results may be obtained directly from plant extracts using immunodiffusion, with or without detergent. Ten ng of virus may be detected using double antibody sandwich ELISA. A number of specific antisera have been obtained. Serological differentiation has been established using gel diffusion and ELISA with semi-purified antigen (Devergne & Cardin 1974; Devergne *et al.* 1981). The two major serotypes may also be detected easily by immunodiffusion with detergent on leaf extracts (Lot & Lecoq, unpublished results).

Geographical distribution

World-wide, but especially in temperate regions (e.g. Europe).

Transmission and epidemiology

CMV is transmitted by aphids in a non-persistent manner. Among 60 aphid species reported as vectors, the most efficient are *Myzus persicae*, *Aphis gossypii*, *A. craccivora* and *A. fabae*. The virus can be acquired by all instars within 5–10 s; their ability to transmit declines after about 2 min and is usually lost within 2 h. Transmission through seed occurs to varying degrees in *c.* 20 species, including some weeds;

transmission and persistence in seeds of *Stellaria media* may be important in the epidemiology of the virus (Tomlinson *et al.* 1970). The two main virus populations are not distributed at random and some wild or cultivated species are infected specifically by only one strain type while others may contain both (Quiot *et al.* 1979).

Economic importance

CMV-induced diseases are probably most important economically in cucurbits, tomato, pepper, spinach, celery, carrot, tobacco, and in flowers such as gladioli and lilies, and various ornamentals.

Control measures

Sources of resistance or tolerance have been discovered and used in most cultivated species: for example, melon, cucumber, squash, spinach, pepper, tobacco or related species (*Cucurbita* sp., *Lactuca saligna, Solanum peruvianum*). In most cases, some strains are found capable of infecting the tolerant lines. In the absence of absolute resistance to CMV, partial resistance may contribute to some extent to the protection of the crops: for example, resistance to virus transmission by *A. gossypii* in *Cucumis melo* (Lecoq *et al.* 1979), resistance to virus movement in *Capsicum annuum* (Pochard 1977). The benefits of partial resistance are increased when used with certain cultural practices (Lecoq & Pitrat 1983).
MARCHOUX & LOT

Peanut stunt virus (PSV) (CMI/AAB 92) was described in the USA but has also been reported in Japan and identified in Europe on bean (Douine & Devergne 1978). Compared to cucumber mosaic virus, peanut stunt virus has a host range restricted (with the exception of tobacco) to legumes. It causes mottle in clover and lucerne and pronounced stunting, epinasty, leaf distortions and malformations of fruits in beans and peanuts. Several isolates have been described which are appreciably different from the biological and serological points of view. Relationships have been described with strains of cucumber mosaic virus and tomato aspermy virus (Devergne & Cardin 1975). It can be detected by mechanical inoculation on *Nicotiana tabacum* (chlorotic local rings and systemic mosaic), *Phaseolus aureus* (systemic mosaic) or *Vigna sinensis* (chlorotic local lesions and systemic mottling) or by serology. It has also been isolated from *Robinia*, in which it seems to be associated with *Robinia* mosaic (CMI/AAB 65).

Tomato aspermy virus (TAV) (syn. chrysanthemum aspermy virus) is widespread in chrysanthemums throughout the world, but rarer in tomatoes where it causes leaf distortion, stunting and seedless fruits (CMI/AAB 79). In chrysanthemum it is often responsible for symptomless infections but can also induce dwarfing (especially when associated with chrysanthemum mosaic virus) and flower distortion (Horst *et al.* 1977). It is one of the serotypes of the cucumovirus group (Devergne & Cardin 1975). Detection is by mechanical inoculation on *Petunia hybrida* or *Nicotiana tabacum* systemic leaf mottling and distortions) or by serology (ELISA).
MONSION & DUNEZ

THE ILARVIRUS GROUP

Quasi-isometric particles different in size (26–35 nm diameter) and containing 4 molecules of positive-sense, single-stranded RNA. One molecule of RNA_1 (1.1×10^6 d), one molecule of RNA_2 (0.9×10^6 d), one molecule of RNA_3 (0.7×10^6 d) and one molecule of RNA_4 (0.3×10^6 d) are respectively encapsidated in 3 different particles. RNA_4 is a monocistronic RNA for coat protein (25000 d). RNA_1, RNA_2 and RNA_3 are infective only in the presence of the coat protein or the RNA_4. Some viruses of the group have only isometric particles differing in diameter; others have larger particles which are obviously anisomorphic and bacilliform. Some viruses of this group are pollen-borne. The type member is tobacco streak virus (CMI/AAB 275).

Apple mosaic virus (ApMV)

(Syn. hop virus A, rose mosaic virus)

Basic description

ApMV has isometric (*c.* 25 nm diameter) to bacilliform particles, sedimenting as three components, with four RNA species and probably a single peptide of M-Wt *c.* 25000 d (a minor component of 19000 has been reported). It is serologically related to prunus necrotic ringspot virus (PNRSV) and isolates intermediate between them occur in hop and rose. They might therefore best be regarded as serotypes of one virus. This description refers to ApMV types as originally described and also includes the intermediate 'Hop C' isolates (CMI/AAB 83, 275).

Host plants and diseases

Most known natural hosts are rosaceous. ApMV is mechanically transmissible to a wide host range but often only with difficulty from rosaceous hosts. Petals are frequently a better source of inoculum than leaves. Typically ApMV induces either a chlorotic mosaic or ring- and line-pattern on leaves, but it is often carried symptomlessly (Fulton 1981, CMI/AAB 83).

In apple it is responsible for mosaic, a disease that was at one stage of more frequent occurrence than today. In plum, almond (Fig. 5) and peach it induces a line pattern symptom (European plum line pattern and peach line pattern)—both ApMV and PNRSV occur inthese three plants and are often difficult to distinguish by symptoms alone. The same applies to rose mosaic, which in Europe is more commonly found to be due to PNRSV (though the original isolate, rose mosaic virus, is identical to ApMV). It should be noted that in N. America, plum line pattern is due to another virus (**American plum line pattern virus**), which is also thought to be an ilarvirus (CMI/AAB 280) and considered a quarantine pest in Europe (EPPO 28).

Fig. 5. Apple mosaic virus: mosaic on almond.

Both hop A and C strains are common in hop but are usually symptomless. However, large temperature variations may induce chlorotic/necrotic line-patterns.

ApMV occurs rarely in cherry and horse chestnut (chestnut mosaic), but is commoner in birch (*Betula* spp.). It is apparently common in *Rubus* in the USA and has been recorded in this host in the FRG, but its importance is unclear. As they could only infect European *Rubus* by pollen transmission from introduced infected plants, these strains present a quarantine threat to Europe (EPPO 147).

Strains

Many isolates with differing biological properties are known. Most are serologically similar but isolates intermediate between PNRSV and ApMV occur in hop and rose.

Identification and diagnosis

Grafting to woody indicators is widely used for rosaceous hosts. The following are currently recommended: *Malus pumila* Lord Lambourne or Golden Delicious, *Prunus persica* seedling GF305 or Elberta, *Prunus* hybrid Shiro Plum. Mechanical inoculation (usually to *Cucumis sativus*) and serology (e.g. by ELISA) are also used but in some hosts ApMV may not be detected at all times, especially by inoculation.

Transmission

No means of spread other than natural root grafts and propagation of infected material is known. Nevertheless spread has been recorded in some hosts, e.g. in *Rubus* in the USA (rapid) and in hop (slow).

Geographical distribution, economic importance and control

ApMV occurs world-wide in temperate regions. It may cause significant yield losses (e.g. up to 20% of the saleable active ingredient in hop) but its occurrence has been greatly reduced by the use of virus-free stocks of many of its major hosts. Infected *Aesculus* species have been deliberately propagated as variegated ornamentals.

BARBARA

Asparagus virus 2 (syn. asparagus virus C, asparagus latent virus), described by Uyeda & Mink in 1981, is distributed world-wide. Frequently symptomless, when associated with asparagus virus 1, it causes loss in vigour and productivity, or 'decline' symptoms. It is seed-borne in commercial seed, with nearly 60% infection (Bertaccini *et al.* 1982). Indexing is by mechanical inoculation onto *Nicotiana*

tabacum cv. White Burley and by agar gel double diffusion test. Control is by propagation of healthy plants obtained from virus-free seeds or meristem-tip culture (CMI/AAB 288).
BERTACCINI & MARANI

Black raspberry latent virus (BRLV) (CMI/AAB 106) is probably a strain of tobacco streak virus (q.v.) and is an important constituent of the virus degeneration complex on *Rubus* in N. America. It does not occur in *Rubus* in Europe and, being only seed- and pollen-transmissible, could only infect European *Rubus* from introduced infected plants, against which appropriate quarantine measures should be taken (EPPO 147).
SMITH

Citrus variegation virus (CVV)

Basic description

Citrus variegation was shown to be a transmissible disease in 1939. Citrus crinkly leaf was reported in 1943 as a separate virus disease, but the citrus crinkly leaf virus (CCLV) is in fact a strain of CVV (Yot-Dauthy & Bové 1968). Particles are isometric, differing in diameter (Gonsalves & Garnsey 1975).

Host plants and diseases

Only *Citrus* species are naturally infected. In general, the virus causes a slight stunting of the tree, but some CVV-type isolates cause a severe stunting in satsuma mandarins. In addition to chlorotic spots and crinkling on the leaves, CVV-type isolates produce chlorotic areas and CCLV-type isolates cause warping and pocketing. CVV-type isolates may cause fruit deformations.

Identification and diagnosis

The virus is indexed by graft inoculation to lemon or sour orange indicator seedlings and by ELISA (Garnsey *et al.* 1983).

Transmission

The disease is propagated naturally through budwood. CCLV-type isolates are seed-transmitted in low percentages. Experimentally, the virus can be mechanically transmitted to citrus and to non-citrus hosts such as *Cucumis sativus*, *Vigna sinensis*, *Phaseolus aureus*, etc. (Desjardins 1980).

Geographical distribution, economic importance and control

Although of world-wide distribution the disease is not of great economic importance, except in the case of a few severe isolates. It is controlled by propagating healthy material, which can be recovered by thermotherapy, nucellar embryogeny and *in vitro* shoot-tip grafting.
CAMBRA & NAVARRO

Elm mottle virus (EMotV) (syn. lilac white mosaic virus) naturally infects *Syringa* and *Ulmus* species (CMI/AAB 139). It induces white mosaic and ringspots in *Syringa* leaves and is associated with chlorotic mottling and line-pattern symptoms in *Ulmus* leaves. No vector is known. It is seed-borne in elm. Detected by mechanical inoculation of sap to *Chenopodium quinoa* (faint chlorotic local lesions and weak systemic mosaic). No control measures are known other than propagation of virus-free plants.
JONES

Lilac ring mottle virus (LRMV) is only detected in the Netherlands, and causes chlorotic rings and line patterns in *Syringa vulgaris*. It is transmitted in seeds of *Chenopodium* species (not *S. vulgaris*) but no vector is known. Indexing is not routine but is possible by mechanical inoculation of *Chenopodium quinoa* or by serological testing. No specific control exists other than propagation from indexed virus-free plants (CMI/AAB 201).
COOPER

Prune dwarf virus (PDV)

(Syn. cherry chlorotic ringspot virus, sour cherry yellows virus, peach stunt virus)

Basic description

PDV is a multicomponent virus with five types of particles differing in size and with sedimentation coefficients of 75, 81, 85, 99 and 113S. Most PDV virions have a quasi-isometric shape (diameter 19–26 nm), but some of the faster sedimenting ones are bacilliform; the protein of PDV has a molecular weight of about 24 000 d (Halk & Fulton 1978; CMI/AAB 275). PDV is a labile virus, rapidly inactivated in undiluted plant sap. In buffer solutions it loses half of its infectivity within a few hours at room temperature unless polyphenol oxidase inhibitors are added as stabilizers. The virus is moderately immunogenic (CMI/AAB 19).

Host plants

All species of the genus *Prunus*, especially sweet and sour cherry and *P. mahaleb* (Gilmer *et al.* 1976) are hosts of PDV. The virus can be transmitted mechanically to young cucumbers and a wide range of other herbaceous plants. *Cucurbita maxima* cv. Buttercup is a good source for virus purification.

Diseases

In sweet cherry, PDV causes chlorotic spots, rings and diffuse mottling on leaves in spring. Sometimes necrotic flecks also occur, but the symptoms of PDV appear only on leaves that are almost completely expanded in contrast to those of prunus necrotic ringspot virus (PNRSV). Later in the year no further symptoms develop. Infected trees often remain symptomless in the following years.

In sour cherry the same symptoms are seen as in sweet cherry during spring; leaf yellowing and leaf abscission may occur in June (Fig. 6). This sour cherry yellows is favoured by mean temperatures of about 15°C during the post-bloom period. Additional infection with PNRSV intensifies the yellows symptoms. In diseased trees the number of fruiting spurs is reduced and the trees may develop a willowy growth habit with bare spaces on the twigs. Cv. Montmorency is very susceptible, while cv. Schattenmorelle very seldom shows yellows. However in this cultivar, growth of young trees is markedly reduced by infection.

Fig. 6. Prune dwarf virus: leaf cast of sour cherry (sour cherry yellows disease).

In plum, cv. Italian Prune and a few other cultivars (e.g. Krikon, Emma Leppermann) have shortened shoot internodes in spring and the tree exhibits dwarfed growth. The leaves are narrow like willow leaves and have a rugose surface. At higher temperatures newly developed leaves grow almost normally.

In peach, shoots develop shortened internodes during spring and their leaves are dark green and more erect than those of non-infected trees. From June onwards diseased plants grow normally, but after infection with severe strains of PDV the plants remain smaller than healthy ones and yield less fruit.

Strains

With regard to the symptoms in cherry a number of strains have been distinguished from the type strain (Fulton's virus B). Some of them are: i) chlorotic necrotic ring spot strain (Kegler 1965). In spring necrotic spots appear on a few leaves, later some leaves develop concentric etched rings; ii) necrotic leaf mottle strain (Kunze *et al.* 1984). Recurrent chlorotic mottle with necrotic spots on leaves developed in June, severe reduction of growth in sour cherries; iii) yellow mottle strain (Ramaswamy & Posnette 1972). Rings and lines on ornamental cherries; iv) ring mottle strain (Ramaswamy & Posnette 1971). Recurrent chlorotic rings and lines on sweet cherry.

Serological characterization of these strains has not yet been possible, but differences in electrophoretic mobility were demonstrated between the type strain, the necrotic leaf mottle strain and the yellow mottle strain (Kunze *et al.* 1984). Among different isolates of PDV, great variation occurs with regard to the symptoms in herbaceous plants; however, the appearance and type of symptom are also considerably influenced by environmental conditions.

Diagnosis and identification

Latent infections with PDV can be detected by graft transmission to woody indicators, mechanical inoculation of young cucumber seedlings or by serological tests using ELISA. The indicators react as follows: i) *Prunus serrulata* Shirofugen: inoculated buds from virus-infected trees cause local necrosis within 5 weeks. The reaction is similar to that caused by PNRSV but less severe. The reaction of Shirofugen to the ring mottle and yellow mottle strains of PDV is delayed until the following growing season; ii) peach seedling: stunting as growth starts but this is not always very obvious. Intensity of the reaction depends on the source of the inoculated virus. The ring mottle strain does not cause symptoms; iii) cucumber: chlorotic local lesions on inoculated cotyledons, systemic mottle and rosette of leaves. There is no clear distinction from the reaction to PNRSV.

PDV can be distinguished from PNRSV and identified exactly by serological tests or by transmission to herbaceous hosts which develop characteristic symptoms, i.e. *Cucurbita maxima* cv. Buttercup (yellow mottle and veinbanding on systemically infected leaves), *Sesbania exaltata* (dark local lesions on cotyledons), *Momordica balsaminea* (systemic mosaic), *Tithonia speciosa* (yellow systemic spots and lines, but not by all isolates). PDV does not infect *Chenopodium quinoa* in contrast to PNRSV.

Geographical distribution and epidemiology

PDV occurs world-wide. It is seed- and pollen-transmissible, especially in *Prunus avium, P. cerasus* and *P. mahaleb*. Seed of diseased trees can be infected up to 75% and seedlings up to 50%. Furthermore the virus can be transmitted to healthy trees by virus-infected pollen during bloom. Natural spread of PDV occurs readily in sweet and sour cherry orchards when the trees have grown for more than 4 years. PDV has not yet been observed to spread in plums. No animal vectors of the virus have been found (Gilmer *et al.* 1976).

Economic importance

In Europe PDV is very common in sweet cherry, reducing yield up to 35%. Combination with PNRSV can further increase the damage. In young sour cherries PDV greatly reduces growth, especially the necrotic leaf mottle strain. As infected nursery plants are below the standard for marketing, PDV symptoms are seldom found in young orchards. In older trees infection with the necrotic leaf mottle strain can drastically lower yield. Sour cherry yellows is a very important disease in America where it can cause yield losses exceeding 50% in the susceptible cv. Montmorency. In Europe however, the yellows reaction of the common cv. Schattenmorelle is fairly mild. In plum visible damage occurs only on a small number of cultivars including Italian Prune. In peach, PDV in combination with PNRSV or strawberry latent ringspot virus can become economically important by considerable reduction of growth and yield.

Control

The production of healthy nursery plants using virus-tested propagation material for cultivars and rootstocks is essential for an effective control of PDV. For seedling rootstocks this includes the production of seed in isolated blocks of mother seed trees which are indexed annually for pollen-borne viruses with Shirofugen and ELISA. New cherry orchards should be planted only with virus free trees. Interplanting of young trees between older infected ones is not recommended. Virus-free material can be obtained by heat treatment from plants infected with PDV and by separate propagation of the tips grown during the heat treatment.

KUNZE

Prunus necrotic ringspot virus (PNRSV)

Basic description

PNRSV has isometric (*c.* 23 nm) to bacilliform particles with three main components of sedimentation coefficients 72, 90 and 95S. There is a major polypeptide of 196 amino acids with a molecular weight of *c.* 2.5×10^4 d but several minor components of lower molecular weight have been reported. The virus is moderately immunogenic. It should be noted that PNRSV has been described separately from apple mosaic virus (q.v.) (CMI/AAB 5, 83), although as they are serologically related and isolates intermediate between them occur in hop and rose they can be considered to be serotypes of a single virus. This description concerns only PNRSV (CMI/AAB 5). The 'hop C' isolates are of the intermediate type and are not included here (see apple mosaic virus).

Host plants

PNRSV infects most, if not all, *Prunus* species and many *Rosa* species. The virus is mechanically transmissible to several herbaceous hosts, although often only with difficulty from its natural hosts. Many cultures maintained in herbaceous hosts become milder and are eventually lost.

Strains

Many isolates with different biological characteristics have been described and some of these are sufficiently distinct to be called strains. Where appropriate these are indicated in the section on diseases.

Diseases

The many apparently distinct diseases described are too numerous to detail here and only major strains and syndromes are listed. Fuller descriptions of diseases are given by Németh (1979) and the USDA Handbook on stone fruit diseases (USDA 1976). PNRSV typically induces shock symptoms in the first year after infection and chronic symptoms subsequently although in some instances the shock symptoms may recur annually. There may be apparent recovery and latent infections are common.

On sweet and sour cherry, PNRSV causes cherry line pattern, cherry rugose mosaic, lace leaf, necrotic ringspot, stecklenberger and tatterleaf diseases. Shock symptoms are chlorotic or necrotic lines, rings and spots. The centres of necrotic rings or spots often fall away to give a 'shot-hole' effect. Young trees may be killed. Some isolates induce shock symptoms followed by apparent recovery whereas in other cases the shock symptoms recur annually. The distinct cherry rugose mosaic strain induces chronic symptoms of leaf distortions and enations on the abaxial surface of the leaf adjacent to the midrib.

Sour cherry line mosaic and sour cherry fruit necrosis are diseases induced by synergistic mixtures of PNRSV and other viruses (cherry leaf roll and apple chlorotic leaf spot viruses, respectively).

On plum PNRSV causes plum decline, plum line pattern and plum oak leaf diseases with chlorotic lines and oak-leaf patterns on leaves. There may be significant loss of vigour. Infection is often symptomless. Similar line patterns may be induced by apple mosaic virus or other viruses.

Almond calico disease is caused by a distinct strain. White (or light yellow bleaching to white) chlorotic spots or blotches occur on variable amounts of the foliage. Calico spots may be widely dispersed on the leaf or aggregated in discrete areas. There may also be failure of blossom and leaf buds to grow. PNRSV also causes almond bud failure disease. This has been described separately from almond calico but the aetiology is uncertain. The disease is always associated with the almond calico strain of PNRSV and is probably caused by this virus. Almond line pattern and almond necrotic ringspot diseases have syndromes like those of plum and cherry with similar names.

Peach line pattern may be induced by either PNRSV or apple mosaic virus with fine irregular chlorotic line patterns, often symmetrical and usually light green. There may be more general line patterns, rings and oak-leaf markings. In peach mule's ear and willow leaf diseases, leaves remain erect near the terminals of willowy shoots and are retained after the normal leaves have fallen. Lateral buds fail to grow in the year after they are formed. Symptoms similar to almond calico are often seen in spring and the same strain of PNRSV is probably responsible.

Peach ringspot, necrotic ringspot and necrotic leaf spot diseases cover a range of necrotic and chlorotic leaf spotting and blotching, often associated with bare shoots and fruit symptoms.

Infection of apricot appears relatively uncommon. Apricot necrotic ringspot and line pattern are recorded diseases induced by PNRSV.

Rose mosaic may be induced by PNRSV (Fig. 7) or by apple mosaic virus and is characterized by a variety of chlorotic ring and line patterns on leaves, with reduced vigour and flower quality. In many cases, PNRSV is more common than apple mosaic virus but the original isolate from the USA is of the apple mosaic virus serotype.

Fig. 7. Prunus necrotic ringspot virus: oak-leaf symptoms on rose.

Diagnosis

Grafting to woody indicators is used widely for indexing *Prunus* species, either in the glasshouse or outside. The following are currently recommended by the ISHS Working Group on Fruit Tree Virus Diseases.

Prunus serrulata Shirofugen. Local necrosis and gumming around the inserted bud. This does not distinguish PNRSV from prune dwarf virus; and

similar gumming may also be caused by *Pseudomonas syringae* pv. *morsprunorum*.
Prunus persica seedlings GF305 (or if not available Elberta). Necrotic spots, rings or lines on newly formed leaves. No symptoms in the second year.
Prunus tomentosa IR473/1 or IR474/1. Leaf mottle on newly formed leaves.
Prunus avium F12/1 has been widely used but is no longer recommended. Severe necrosis and tattering of the leaf lamina on the first-formed leaves.

Mechanical transmission to herbaceous hosts may be used for indexing and to differentiate prune dwarf virus from PNRSV. *Cucurbita maxima* cv. Buttercup is recommended. Transmission is often unreliable from leaf tissue and petals are usually a better source of inoculum.

Serological detection by ELISA is rapid and sensitive. PNRSV may often be detected in tissues from which mechanical transmission is not possible but virus may not be detectable throughout the season.

Geographical distribution, transmission and epidemiology

PNRSV occurs world-wide in temperate regions, and probably wherever *Prunus* or *Rosa* are grown. Natural spread between mature plants is often via pollen. In cherry and almond, most, if not all, of the virus is carried on the exterior of the pollen grains. PNRSV is also seed-borne. Transmissions by mites and by nematodes have been reported but the biological significance of these is not clear. Rates of natural spread are often slow (e.g. less than 2% per year in plum in Spain) but rapid spread is possible (e.g. more than 10% per year in cherry in Bulgaria). Propagation of latently infected material has been very important in disseminating PNRSV. In rose, the disease is mainly spread through infected vegetatively propagated rootstocks (*Rosa manetti*, *Rosa indica major*).

Economic importance

PNRSV is common and widespread in many of its hosts and is probably a major cause of crop loss despite often being carried latently. Reported effects on crop yields range from none in Italian Prune to total fruit loss in some years, with 80% of trees dying in peach. Overall crop losses of 15% for a large area have been estimated for PNRSV in sweet cherry.

Control

PNRSV-free planting material of many cultivars has been produced either by indexing and selection or by heat-therapy. The use of these stocks has provided the best means of control. Isolation of new plantings from older infected stocks is often necessary to prevent re-infection although the precise requirement will vary with the crop. In rose, use of seedling rootstocks reduced the incidence of the disease.

BARBARA

Tobacco streak virus (TSV)

Basic description and host range

TSV, the type member of the ilarvirus group, has a wide natural and experimental host range. It has been found as a pathogen in tobacco, dahlia, cotton, red clover, strawberry, *Melilotus alba*, tomato, asparagus, rose, pea, phaseolus beans, potato, soybean, grapevine, *Rubus* spp., pawpaw (CMI/AAB 307), sweet peppers (Gracia & Feldman 1974), cowpea, artichoke (Costa & Tasaka 1971) and numerous ornamentals. Symptoms vary from the characteristic tan necrotic lines and rings in tobacco, with subsequent recovery, to mottling, stunting, and, in soybean, bud blighting.

Identification and diagnosis

TSV exists in many strains varying in the symptoms they cause. A coarse toothing of 'recovered' leaves of Turkish tobacco is usually diagnostic, but some strains do not cause this, so specific antisera must be used for reliable identification. Serological differences exist among TSV strains, but they all share common antigens. Agar gel double diffusion tests are convenient for crude extracts, although the virus in some hosts may need to be concentrated by centrifugation to obtain visible reactions (Fulton 1972).

Transmission

The disease is spread by *Thrips tabaci* or *Frankliniella* species (Costa & Lima Neto 1976; Kaiser *et al.* 1982). It is seed-transmitted in certain hosts (*Datura stramonium*, *Phaseolus vulgaris*, *Chenopodium quinoa*, *Melilotus alba*, and *Glycine max*).

Geographical distribution, economic importance and control measures

TSV infection of one or other host occurs nearly world-wide. In most hosts TSV is an incidental pathogen and control measures have not been formulated. Indexing and care to avoid propagation from infected plants would probably be effective.
FULTON

THE LUTEOVIRUS GROUP

These viruses form isometric particles, 25–30 nm diameter with one positive-sense, single-stranded RNA (2×10^6 d) and one coat polypeptide (24 000 d). They are confined to phloem tissues and are responsible for very severe diseases with frequent yellows symptoms. Aphid-transmitted in a persistent manner, several cases depend on a helper factor or a helper virus. Type member is barley yellow dwarf virus.

Barley yellow dwarf virus (BYDV)

(Syn. cereal yellow dwarf virus)

Basic description

BYDV is the type member of the luteovirus group. Virions sediment at 115–120S; they comprise a single RNA molecule of 1.8×10^6 d for a vector non-specific type (Mehrad *et al.* 1979) and 1.9×10^6 d for the RPV and RMV vector-specific types (Gildow *et al.* 1983). Coat proteins of the MAV and PAV types have M-Wt 23 500 and 24 450 respectively. The virus is strongly immunogenic (CMI/AAB 32).

Host plants

BYDV infects many Gramineae and a few members of the Cyperaceae have also been experimentally infected. Most small-grain cereals are very susceptible (wheat, barley, oats, rice). Maize and rye cultivars are generally more tolerant. Generally, ryegrass infection is symptomless. No transmission of BYDV to dicotyledons has been demonstrated.

Strains

At the present time, four distinct vector and serological types of BYDV have been described in Europe. Two vector non-specific types are prevalent, transmitted by both *Rhopalosiphum padi* and *Sitobion avenae* (= *Macrosiphum avenae*). The two types are distantly related serologically and are more efficiently transmitted by *R. padi* (cereal severe type) and by *S. avenae* (cereal mild type) respectively. The latter is transmitted inefficiently by *Metopolophium dirhodum* and erratically by *R. maidis*. Two more vector-specific types have been characterized; one is mainly transmitted by *S. avenae* (MAV-type) and for most isolates also by *M. dirhodum* (Maroquin unpublished). Another type is transmitted specifically by *R. padi* (RPV-type) (Jones & Catherall 1970). Isolates predominantly transmitted by *S. avenae* are generally mild, but severe ones are also known (Jones & Catherall 1970). Investigation of these different types isolated in Europe with monoclonal antibodies will probably explain links between immunological properties and rates of transmission by vectors as proposed by Gildow & Rochow (1980), and also the differences between European types and those from other continents.

Diagnosis

Plant indexing (by vector transmission) is carried out on susceptible cultivars (oat cv. Coast Black, barley cv. Capri); 15–20 days are needed to obtain clear symptoms in growth rooms at 18–20°C. This test allows the aggressiveness of the isolate to be characterized. The ELISA test has been used by Denéchère *et al.* (1979) and Torrance & Jones (1982) with chromogenic or fluorogenic substrates respectively. Studies with monoclonal antibodies (Hsu *et al.* 1984) are in progress.

Diseases

BYDV generally causes reduced root growth, leaf discoloration and dwarfing in small-grain cereals (Panayoutou 1979). Yield and seed germination rate are affected (Panayoutou 1978). Immunity to BYDV has not yet been found in any cereal species.

Most barley cultivars are susceptible. In Scotland early sown (August) barley is often severely affected and many plants die in November (Holmes 1985). The first symptom observed in the field is yellowing at the leaf tip. Several cultivars, most of them from Ethiopia, have good tolerance to BYDV, determined by the Yd_2 gene and according to plant phenotype (Boulton & Catherall 1980). Two other apparently different genes have been described: Yd_1 (Suneson & Ramage 1957) and recently another still undesignated gene from cv. Post.

In most wheat cultivars leaves turn yellow or red in their distal or near distal parts. This symptom is clearly observed on appearance of the flag leaf. Many tolerant cultivars described in America (Cisar *et al.*

1982) and tested in France are susceptible (Auriau *et al.* unpublished).

Oats infected with BYDV show reddish leaves. Tolerant cultivars have been described in America (Weerapat *et al.* 1974).

As in barley, leaves of rice become yellow and the plants are stunted. Tolerant cultivars have been identified—cv. Venaria is either resistant or highly tolerant.

Geographical distribution, transmission and epidemiology

BYDV is known to be present throughout the world except in tropical areas. It is a circulative aphid-borne virus transmitted in the persistent manner by at least 11 species in Europe (A'Brook & Dewar 1980). This persistent virus-vector relationship is an important feature of BYDV epidemiology because the virus source may be far away from cereal fields. The presence of infected weeds, cultivated grasses, maize and cereal volunteers are another important feature of the epidemiology. The proportion of infective migrant aphids is variable: it is usually low in Britain but higher in France and Belgium in autumn reaching 50%. No seed transmission has been demonstrated even when the virus has been detected in barley seeds.

Economic importance and control

BYDV is the most severe disease of winter barley. Incidence on wheat and oats is often underestimated. Early sown crops are exposed to severe inoculum pressure: they may require 1–3 insecticide sprays. In the future, better protection should be obtained by use of granulated insecticides at planting time. If necessary, late-sown crops may be sprayed. This decision should be made by the local Plant Protection Service, on the basis of data from suction traps (Euraphid), which give estimates of the population of infectious aphids migrating (infectivity index of Plumb & Lennon 1982), and also of data on secondary spread.

LAPIERRE & MAROQUIN

Bean leaf roll virus (syn. pea leaf roll virus) causes leaf rolling and interveinal yellows of old leaves as well as decrease in pod number in broad bean. It causes chlorosis and stunting in pea, French bean, chickpea, cowpea and lentil. Common in Europe and the Middle East, it has also been reported in the USA.

It is aphid-borne in a persistent manner by several species, especially *Acyrthosiphon pisum*, but transmitted mechanically. Detection is by ELISA and control by use of resistant cultivars (CMI/AAB 286).

Beet western yellows virus (BWYV)

Basic description

The particles of BWYV (c. 26 nm in diameter) occur in low numbers in affected plants and are probably confined to phloem tissue. Leaf yellowing symptoms are induced by phloem degeneration, senescence and the collapse of chloroplasts. Beet mild yellowing virus (BMYV), malva vein yellows virus, radish yellows virus and turnip yellows virus appear to be closely related strains of this one virus (CMI/AAB 89).

Host plants and main diseases

Strains of BWYV and BMYV cause mild or more serious diseases of many cultivated plants including beet, endive, flax, lettuce, spinach, sunflower, radish and turnip. The most important diseases are those of sugar beet and lettuce. In sugar beet the outer and middle leaves turn yellow or yellow-orange and become thickened and brittle and are often invaded by weakly parasitic fungi (e.g. *Alternaria alternata*). In field-grown lettuce, symptoms appear only after plants have been infected for 40–50 days when the outermost leaves develop a slight or intense interveinal yellowing (Fig. 8) usually associated with dark brown marginal leaf necrosis.

BWYV and BMYV may be prevalent in the over-wintered common weeds *Capsella bursa-pastoris* and *Senecio vulgaris* as well as in winter rape (Helen G. Smith, pers. comm.). Other weed hosts of BMYV

Fig. 8. Beet western yellows virus: discoloration and yellowing of lettuce.

include *Stellaria media, Sinapis arvensis, Plantago lanceolata, P. major* and *Veronica* spp., *Cardamine hirsuta, Galium aparine*, and *Portulaca oleracea* (Tomlinson, unpublished).

Although sugar beet and lettuce can be experimentally infected with BMYV and BWYV respectively under glasshouse conditions, disease symptoms are usually very mild, or plants remain symptomless.

Strains

Isolates have been distinguished on the basis of host range and symptom intensity (CMI/AAB 89). There appear to be pathogenic differences between American and most European isolates so far studied. Thus, most American BWYV isolates infect lettuce and sugar beet but in Europe BMYV isolates do not infect lettuce and, with one exception, lettuce-infecting BWYV isolates do not infect sugar beet. The exception is the sugar beet-infecting isolate of BWYV found in Czechoslovakia by Polák (1979) to infect lettuce cv. Kamýk. Whether this isolate infects a wide range of lettuce cultivars was not stated. All isolates, however, infect *Capsella bursa-pastoris, Claytonia perfoliata* and *Senecio vulgaris*. Based on serological neutralization infectivity tests, American isolates of BWYV could not be distinguished from English isolates of BMYV (Duffus & Russell 1975). Several workers have now shown serological relationships between other luteoviruses originally thought to be distinct viruses (Rochow & Duffus 1981).

Diagnosis

The virus can be diagnosed by visual inspection of the leaves of infected plants but this is not entirely satisfactory as infected plants may be symptomless or develop symptoms resembling those caused by certain mineral deficiencies. Diagnosis by electron microscopy of sap of infected leaves and visualization of virus particles usually fails. Normally, diagnosis can only be satisfactorily demonstrated by transferring aphids (*Myzus persicae*) from infected plants to *Claytonia perfoliata* which then develop the characteristic symptoms (infected plants develop pink, orange or red coloured leaves 15–25 days after infection). Another useful indicator species is *Crambe abyssinica*, the leaves of which become red and curled after infection. IEM has been demonstrated but only with concentrated, purified BWYV. Since 1981, the incidence of BMYV in English sugarbeet crops has been studied using the ELISA technique.

Geographical distribution

Sugar beet-infecting strains (BMYV) have been reported throughout Europe. Lettuce-infecting strains (BWYV) have been described in Britain, France, the Netherlands and FRG.

Economic importance

The disease of sugar beet known as 'virus yellows', probably one of the most serious crop diseases in northern Europe, is caused by BMYV, or by beet yellows virus, or a mixture of the two viruses. Root yield losses in England from 1970 to 1975 ranged from less than 1% to 18% (Heathcote 1978). Significant losses have been reported from Hungary, the Netherlands, USSR, Sweden, Turkey and FRG. In France, average losses are 2 tons per ha (C. Putz, pers. comm.).

Transmission

The virus is not sap-transmissible but transmitted persistently for up to 50 days by several aphid species, including *Myzus persicae, M. ascalonicus* and *Aulacorthum solani*. There is a minimum acquisition feeding period of 5 min and a latent period of 12–24 h. Seed transmission may occur at a very low level (Fritzsche *et al.* 1983).

Control

Methods for control of virus yellows in sugar beet are based on: i) elimination of overwintering sources and separation of root and seed crops; ii) use of granular aphicides applied with the seed at sowing and foliar applications of insecticide when aphid numbers in the crops increase; iii) growing cultivars least affected by infection if they become commercially available. In some countries, growers liaise with the sugar-marketing firms (e.g. British Sugar plc in England) for specialist advice on when to apply insecticides (MAFF 1981).

In lettuce there is considerable variation in the reaction of different cultivars to infection by BWYV. Crisp and cos cultivars are more tolerant (less reactive) than butterhead cultivars (Tomlinson & Ward 1982b). With transplanted crops of butterhead lettuce, however, control can be achieved by incorporating the fungicide carbendazim into the seeded peat blocks at the rate of 0.05 g per block (4.3 cm cube) (Tomlinson *et al.* 1976; Tomlinson & Ward 1982a). Carbendazim is absorbed and translocated to

the leaves where it prevents chloroplast degeneration and leaf yellowing following BWYV infection (Tomlinson & Webb 1978). Carbendazim is also insecticidal and protects young plants from aphid colonization for 3 weeks.
TOMLINSON

Carrot red leaf virus (CMI/AAB 249) is distributed world-wide and frequently associated with carrot mottle virus which is dependent on it for its transmission by aphids. It causes reddening and yellowing of leaves and some stunting of the plants. With carrot mottle virus, these symptoms are accentuated. The carrot motley dwarf disease caused by this virus complex has been reported in Australia, Japan, New Zealand, Europe and Canada. The virus is transmitted in the persistent manner by *Cavariella aegopodii*, but is not seed-transmitted. Detection is by transmission to *Anthriscus cerefolium* (chervil) where the virus induces a mild to severe stunting and curling of leaves. Similar symptoms occur when carrot mottle is present but symptoms can be more severe with carrot red leaf alone. Celery and parsley are immune to most isolates but can be infected by carrot mottle virus (q.v.).
MARCHOUX

Celery yellow spot virus is a possible member of the luteovirus group. Naturally infected celery plants develop circular white spots. The virus also infects parsnips which develop mild mottling. Reported in the USA and England, the virus is aphid-borne in the persistent manner by *Hyadaphis foeniculi* (Smith 1972).

Potato leaf roll virus (PLRV)

Basic description

PLRV is a member of the luteovirus group (Rowhani & Stace-Smith 1979; Takanami & Kubo 1979; Rochow & Duffus 1981). The virus particles are isometric, *c.* 25 nm in diameter and confined to phloem tissue (Kojima *et al.* 1969; CMI/AAB 291). The virion comprises a positive single-stranded RNA to which is attached a small genome-linked protein (Mayo *et al.* 1982), and one coat protein (M-Wt 26 300 d) (Rowhani & Stace-Smith 1979). Virus suspensions are good immunogens (Murayama & Kojima 1974).

Host plants

Besides potato PLRV infects mainly Solanaceae, but also plants such as *Amaranthus* spp., *Celosia argentea*, *Gomphrena globosa* and *Nolana lanceolata* (Natti *et al.* 1953).

Strains

Several strains are distinguished on the basis of severity of reaction of *Physalis floridana* (Webb *et al.* 1952), but there is little detailed information on European strains.

Diseases

PLRV-infected potato plants generally produce smaller tubers than healthy ones. Symptoms of primary infection appear mainly in young leaves at the top of the plants which are usually pale yellow. The leaflets are often rolled, especially at the base. Symptoms of secondary infection are always more severe; the whole plants look erect and may be smaller than healthy ones. The older leaves are rolled and those higher up are chlorotic. Basal leaves, in particular, are stiff and leathery. Infected tomato leaves are very slightly rolled but they become rigid and somewhat leathery.

Geographical distribution and transmission

PLRV probably occurs wherever potatoes are grown. It is transmitted by aphids in a persistent (circulative) manner. *Myzus persicae* seems to be the most efficient of the 10 or more aphid species known to be vectors. The virus is not sap-transmissible.

Economic importance

By its prevalence and the severity of the disease, PLRV is one of the most important potato viruses, frequently reducing tuber yields by 50–80%. The average loss due to PLRV has been estimated at 10%, which amounts to about 20 million tonnes of potatoes each year over the whole world.

Control

The three major methods of control are: i) elimination of virus source plants (e.g. weeds and volunteers); ii) application of insecticides to reduce aphid populations; and iii) the use of certified seed potatoes. For the second purpose, especially, application of granulated systemic insecticides in the field with fertilizer at planting time may be effective. In the Netherlands, primary infection is avoided by systematic haulm destruction.

Special research interest

The main research interest concerns the aetiology of this disease; PLRV isolates from Canada, Europe and Japan are serologically related (Rowhani & Stace-Smith 1979; Kojima 1981), but beet western yellows virus has been isolated from potatoes with typical potato leaf roll symptoms in the USA (Duffus 1981). Should we consider that both viruses are involved in the leaf roll syndrome in Europe and Japan, too?
KOJIMA & LAPIERRE

Raspberry leaf curl virus is possibly a member of the luteovirus group. It is aphid-borne (circulative manner). Widespread in N. America (Stace-Smith 1984), it is a quarantine pest for Europe (EPPO 31). Leaves of diseased raspberry are curled and slightly yellowed. Fruiting laterals are shortened and shoots may proliferate producing a rosette. Fruits are small and crumbly. Indexing is by graft transmission to *Rubus phoenicolasius* and *R. henryi* and control by propagation of healthy plants or control of aphids. The European disease of raspberry formerly called raspberry leaf curl is due to infection by raspberry ringspot virus (q.v.).
KRCZAL

Strawberry mild yellow edge virus (SMYEV)

Basic description

SMYEV is a possible member of the luteovirus group. Isometric virus particles of 23–28 nm in diameter were isolated from chronically infected strawberries (Martin & Converse 1982).

Host plants and diseases

SMYEV infects only *Fragaria* spp. Commercial cultivars are usually symptomless carriers. When symptoms occur, they consist of a slight marginal chlorosis in the young leaves. SMYEV in combination with mottle (SMV) or/and crinkly virus (SCV) causes the disease called yellow edge or xanthosis. Characteristic symptoms are dwarfing of the plants, small cupped young leaves with marginal chlorosis and premature reddening of the older leaves. Symptom expression varies with the sensitivity of the commercial cultivar and the interacting viruses (Mellor & Frazier 1970).

Identification and diagnosis

Indexing by stolon or excised-leaf grafting to *Fragaria vesca* (clones, UC1, UC4, UC5) or *F. vesca* var. *semperflorens* (alpine seedlings). Symptoms appear within 2–4 weeks after inoculation. Symptoms of SMYEV alone are initial vein chlorosis and chlorotic spotting on the young leaves. The leaflets are slightly cupped and have slightly chlorotic margins. The older leaves die prematurely. Symptoms of mild strains consist of a slight chlorotic spotting on young leaves and some interveinal necrosis of the older ones. In combination with other viruses the symptoms are intensified and the vigour of plants is severely reduced. An antiserum has been prepared against an American SMYEV isolate.

Geographical distribution, transmission and economic importance

SMYEV occurs world-wide in all important strawberry growing areas. It is transmitted by aphids in a persistent (circulative) manner. Vectors are *Chaetosiphon fragaefolii, C. thomasi* and *C. jacobi* (Mellor & Frazier 1970; Krczal 1980). Economic importance is mainly due to the synergistic effect with other viruses. The yield of sensitive cultivars can be reduced to 25% by the yellow edge complex.

Control measures

Production and use of healthy plants and the control of vectors. Healthy plants can be obtained by heat therapy and meristem-tip culture.
KRCZAL

THE NEPOVIRUS GROUP

There are three types of isometric particles of the same size (28 nm), distinguished by their sedimentation constants. One (T) is the empty capsid; the two others, M and B, differ in RNA content. The genome consists of two positive-sense, single-stranded RNA molecules, RNA_1 (2.6×10^6 d) and RNA_2 ($1.3–2.4\times10^6$ d). M particles contain one molecule of RNA_2. In some nepoviruses, B contains one molecule of RNA_1 (tomato black ring virus); in the others, B particles are heterogeneous and indistinguishable, some contain 1 molecule of RNA_1 and the others 2 molecules of RNA_2 (tobacco ringspot virus). Both RNA_1 and RNA_2 are required for infectivity but RNA_1 autoreplicates in protoplasts.

RNA_1 contains the gene for a virus-encoded replicase and RNA_2 contains the coat-protein cistron. One coat polypeptide species (55–60 000 d). Soil-transmitted by two genera of nematodes (*Xiphinema* and *Longidorus*). Infectivity is retained for weeks or months but lost after moulting. Type member is tobacco ringspot virus (CMI/AAB 185).

Arabis mosaic virus (ArMV)

Basic description

ArMV has 3 types of isometric particles which sediment at 53S, 93S and 126S respectively. The hop strain (ArMV-H) may have a low M-Wt (< 100 000) nucleic acid (S-NA) (Davies & Clark 1983), in addition to the two genomic viral RNAs.

Host plants and diseases

ArMV occurs naturally in many species of wild and cultivated plants and is associated with diseases in crops as diverse as cherry, raspberry, strawberry, cucumber, lettuce, celery, rhubarb, *Forsythia* (yellow net disease), grapevine, hop and *Narcissus* (CMI/AAB 16). Symptoms in infected plants vary greatly between host, cultivar, etc. See also 'Ash mosaic' (Ch. 3). Most strains infect all commonly used herbaceous test plants. However, ArMV-H has a relatively narrow host range, only infecting hops and a few standard herbaceous test plants.

All hop cultivars are susceptible. Affected plants have shoots with a sparse, bare appearance ('bare-bine') early in the growing season. Plants later become severely stunted and fail to climb the supporting strings. The leaves roll upwards and may develop vein clearing and/or enations on the underside (nettlehead disease).

Strains

In addition to ArMV-H, several strains from different sources exist e.g. strawberry and ivy. All are closely related serologically and distantly related to grapevine fan leaf virus. Within ArMV-H, nettlehead isolates possess one or more species of S-NA. The number of species of S-NA, and their molecular weight may vary between different isolates. Non-nettlehead isolates only induce the initial bare-bine symptoms and are apparently free of S-NA.

Identification and diagnosis

Several herbaceous test plants develop characteristic symptoms (CMI/AAB 16). However, ArMV frequently occurs with strawberry latent ringspot virus, which has the same vector. Both viruses induce similar symptoms in herbaceous test plants, but can readily be distinguished serologically.

On hops, nettlehead symptoms vary greatly in severity and tend to be inconspicuous at temperatures > 18°C. ArMV can be detected serologically by gel diffusion or with greater sensitivity by ELISA. Isolates of ArMV-H possessing S-NA can be distinguished by polyacrylamide gel electrophoresis of viral RNA (Davies & Clark 1983).

Geographical distribution and transmission

ArMV is widespread in Europe and has also been reported in Canada and New Zealand. ArMV-H infection is prevalent in hop-growing regions of England and Belgium and also occurs elsewhere in Europe and Tasmania. In general, ArMV is transmitted by the dagger nematodes *Xiphinema diversicaudatum* and *X. coxi*. The virus is also seed-borne (Thresh & Pitcher 1978). The hop strain has been widely disseminated by vegetative propagation, and the same is liable to apply to ArMV in other vegetatively propagated crops.

Economic importance and control

Infection with ArMV, sometimes in association with other viruses, can cause serious losses in hop, raspberry and strawberry. Hops affected with nettlehead are virtually useless, whereas hops affected only by bare-bine produce some crop. The planting of virus-free material is a basic control measure. Use of sites free from *X. diversicaudatum* is recommended. In areas where infective populations of the vector occur, spread of the virus may be limited by soil fumigation and/or fallow before planting with healthy material. Fumigation is most effective with crops possessing a shallow root system e.g. strawberry.

DAVIES

Artichoke Italian latent virus (AILV) (CMI/AAB 176) has been reported from Italy, Greece and Bulgaria. Most artichoke cultivars are invaded symptomlessly but some show generalized yellowing and stunting. Besides artichoke, several cultivated (chicory, grapevine, lettuce, gladiolus, pelargonium) and wild plants are infected in nature, some with symptoms. Two serological variants occur in Italy and Greece which are transmitted by *Longidorus apulus* and *L. fasciatus*, respectively. Indexing is by mechanical inoculation to tobacco (chlorotic/necrotic

ringspots and line patterns), *Gomphrena globosa* (necrotic, whitish ring-shaped local lesions), French bean cv. la Victoire (chlorotic/necrotic local lesions, systemic necrosis followed by recovery) and serology (immunodiffusion, IEM).
MARTELLI

Artichoke yellow ringspot virus (CMI/AAB 271) occurs in Italy and Greece, causing bright chrome-yellow blotches, ringspots and line-pattern on artichoke leaves accompanied by mild malformations. In nature the virus also infects tobacco, French and broad beans and a number of weeds and shrubs in which it induces yellow discolorations. It is transmitted at high percentage through seed, by pollen to pollinated plants and through propagating material. The vector, if any, is unknown. The experimental host range is wide. Indexing is by mechanical inoculation to tobacco (local and systemic chlorotic rings, lines and oak leaf pattern) and *Gomphrena globosa* (reddish local lesions, systemic yellow ringspot and line pattern) and by IEM (Rana *et al.* 1980).
MARTELLI

Cherry leaf roll virus (CLRV)

Basic description

CLRV is a probable member of the nepovirus group, whose particles have sedimentation coefficients of 115–118S and 128–130S. The 115S component contains 38% ssRNA of 2.02×10^6 d whereas the 128S component contains 42% ssRNA of 2.2×10^6 d (Murant *et al.* 1981). Neither of these components is infectious alone (Jones & Duncan 1980). Some isolates of CLRV have a non-infectious third component of 52S. Each component contains a single protein of 54 000 d. The virus particles are moderate immunogens (CMI/AAB 306).

Host plants and diseases

CLRV naturally infects species in the following genera: *Betula, Cornus, Juglans, Ligustrum, Olea, Prunus, Ptelea, Rheum, Rubus, Sambucus* and *Ulmus* (Cooper 1979). Experimentally, numerous herbaceous plants are infected, including one monocot. CLRV causes leaf rolling and death in *Prunus*, chlorotic-yellow foliar blotches or yellow net in *Sambucus* (golden elderberry virus), foliar mosaic in *Ulmus* (elm mosaic virus), and yellow oak leaf patterns in *Betula*. It is associated with a blackline at scion/stock union of *Juglans regia* grafted on *J. nigra, J. hindsii* or cv. Paradox. Chrome yellow blotches can also be observed on walnut foliage (walnut yellow mosaic virus, walnut ringspot virus) and also leaf necrosis or chlorosis of *Rubus, Ptelea* and *Ligustrum.*

Strains

Isolates from each genus are usually distinct serologically and also in the nucleotide sequences of their genomic RNA as measured by complementary DNA hybridization.

Identification and diagnosis

Symptoms are neither diagnostic nor consistent and serological tests are more useful than biological criteria in distinguishing CLRV from other viruses. Ouchterlony gel diffusion tests and ELISA are reliable but the virus is sometimes unevenly distributed in the infected plant (Delbos *et al.* 1984).

Geographical distribution and transmission

CLRV occurs in wild/amenity plants in GDR, Finland, UK and Yugoslavia (and in the USA) as well as in commercially propagated tree fruits in GDR, France, Hungary, Italy, Poland, the Netherlands and UK. Longidorid nematodes transmit CLRV experimentally but are probably unimportant in the field (Jones *et al.* 1981). Effective transmission occurs through either pollen or ovules to infect the seed. Circumstantial evidence implies that pollen facilitates virus transmission to plants pollinated.

Economic importance and control

CLRV seems uncommon in commercial top or bush fruit except in the case of walnuts. It is controlled by removal and replacement of moribund cherries or walnuts.
COOPER

Cherry rasp leaf virus (CRLV) is a probable member of the nepovirus group. The nematode vector is *Xiphinema americanum* (Stace-Smith & Hansen 1976). The virus infects apple, inducing flat apple disease, and peach and cherry where it is responsible for rasp leaf and enations, which are often difficult to differentiate from the rasp leaf symptoms induced by raspberry ringspot virus, arabis mosaic virus, tomato black ring virus or myrobalan latent ringspot virus. This virus has so far been reported only in N. America (with doubtful old records in New Zealand and S. Africa). The virus is considered a quarantine organism for Europe (EPPO 159). It is indexed by ELISA.

Chicory yellow mottle virus (ChYMV) causes bright yellow mottle, ringspot and line pattern in the leaves of chicory, and foliar alterations in parsley. These are the only natural host plants of the virus. It is reported in S. Italy. No vector has been described. Detection is by transmission to *Phaseolus vulgaris* (erratic local lesions; systemic chlorotic mottling, yellow spots and blotches) or by serology (CMI/AAB 132).

Grapevine Bulgarian latent virus (GBLV) has two nucleoprotein components containing one molecule of RNA_1 (2.2×10^6 d) and one molecule of RNA_2 (2.1×10^6 d) respectively. It is regarded as a likely member of the nepovirus group. It is latent in some cultivars of *Vitis vinifera* although occasionally isolated from plants of Cabernet Sauvignon, Cardinal, Julski Bisser and Bologar with reduced growth and fanleaf-like symptom. It causes delayed budbreak, pale green foliage, and straggly fruit clusters in *Vitis labrusca* cv. Concord. It is reported from Bulgaria and New York State (USA). Nematode transmission is not yet established (CMI/AAB 186).

Grapevine fanleaf virus (GFLV)

Basic description

GFLV was the first virus to be experimentally transmitted from grapevine to herbaceous test plants; it could then be identified as a member of the nepovirus group (Dias & Harrison 1963; CMI/AAB 28). Particles are isometric *c.* 30 nm in diameter, with an angular outline and a poorly resolved surface structure. Purified virus preparations consist of three centrifugal components (T, M and B) with sedimentation coefficients of 50S (T), 86S (M) and 120S (B). T particles do not contain RNA, M particles contain one molecule of RNA_2 (M-Wt 1.4×10^6 d) and B particles contain one molecule of RNA_1 (M-Wt 2.4×10^6 d) or two molecules of RNA_2. Translation of viral RNAs in rabbit reticulocyte lysates gives two products with M-Wts 220000 (from RNA_1) and 125000 (from RNA_2). The RNA_2 coded polypeptide is cleaved into two proteins one of which, with a M-Wt of 58000, is the coat protein (Morris-Krsinich *et al.* 1983). The virus is a good immunogen. Antisera with titres up to 1/2048 (in gel diffusion) are readily obtained.

Host plants

The grapevine (both *Vitis vinifera* and American *Vitis* species) is the only known natural host. Experimental host range is moderately wide comprising over 30 species in 7 families.

Strains

No serological variants are known. Virus isolates from widely separated geographical areas are antigenically very similar. Biological variants occur which consist of distorting and yellowing strains (as indicated by the type of symptoms induced in grapevine). Minor differences can also be detected in the response of experimentally infected herbaceous hosts.

Diseases

Fanleaf is characterized by a wide range of symptoms depending primarily on the infecting strain and on possible interference with other infectious agents. Syndromes comparable to those induced by GFLV can also be caused in grapevine by other nepoviruses, e.g. arabis mosaic, raspberry ringspot, strawberry latent ringspot and tomato black ring viruses (Martelli 1978).

The name of the fanleaf syndrome is derived from a characteristic malformation of the leaves which exhibit wide open petiolar sinuses and gathered primary veins, giving the appearance of an open fan. Leaves may also be variously distorted, asymmetrical and puckered, with acute denticulations and, sometimes, mottling. Canes show abnormal branching, double nodes, short internodes, fasciations and zigzag growth. Bunches are reduced in number and size, ripen irregularly, have poor setting and shot berries. Symptoms develop in early spring and remain visible throughout the vegetative season, although some masking may occur in summer.

Some variants cause yellow mosaic (Plate 2) where chrome-yellow discolorations develop in early spring. Foliar alteration varies from a few scattered yellow spots, rings or lines to extended mottling or total yellowing. Little malformation of leaves and canes is produced, but bunches are small with shot berries. In hot climates, summer masking of symptoms may be very pronounced.

Apparently, GFLV sometimes causes vein banding. Symptoms consist of chrome-yellow or greenish flecking along the main veins of mature leaves spreading some way into interveinal areas. Discolorations appear in late spring or early summer on a few leaves and persist throughout the vegetative cycle. Fruit setting is extremely poor, bunches are straggly and yield is often nil. Krake & Woodham (1983) suggest that vein banding is a composite disease originating from mixed infections of GFLV and the agent of yellow speckle, a graft-transmissible disease of grapevine latent in many cultivars.

Identification and diagnosis

Although GFLV infections are in many cultivars readily detected by visual inspection, indexing is recommended. This can be done by: i) grafting to *Vitis rupestris* St. George which, under glasshouse conditions, develops chlorotic spots, rings and lines on the leaves within 4–6 weeks (shock symptoms) and, later, malformations of the blade (nettle leaf); ii) mechanical transmission to *Chenopodium amaranticolor* and *C. quinoa* (chlorotic local lesions followed by mottling, vein clearing and deformation) and *Gomphrena globosa* (chlorotic/reddish local lesions and twisting of the uppermost pair of leaves); iii) serological testing especially by ELISA and/or IEM (Bovey *et al.* 1982).

Geographical distribution

Now occurring world-wide, GFLV probably shares its centre of origin with *V. vinifera* (south of Caucasus and Asia Minor) from where it spread to the Mediterranean basin, then across the world, with the natural host and the primary vector *Xiphinema index*.

Transmission

GFLV is transmitted in nature by the longidorid nematode *Xiphinema index* (Hewitt *et al.* 1958) and experimentally by *X. italiae* (Cohn *et al.* 1970). *X. index* is a far more efficient vector. It acquires the virus from roots of infected vines and retains it for up to 8 months in the absence of host plants and 3 months when feeding on GFLV-immune hosts. Cuticular linings of the oesophageal lumen are the sites of virus retention. Adults and juveniles transmit both yellowing and distorting strains equally well and spread at a rate of *c.* 1.5 m/year. No seed transmission occurs in grapevine. Infected propagating material is responsible for long-distance spread.

Economic importance

GFLV may cause: i) progressive decline and death of grapevine; ii) low quality and quantity of yield; iii) shortening of the productive life of the vineyard; iv) reduced rooting ability of propagation material; v) poor graft take; vi) decreased resistance to adverse climatic conditions. Vuittenez (1970) reckons that a 50% reduction in weight is the average crop loss caused by GFLV to susceptible *V. vinifera* cultivars in Europe, especially in restricted areas suitable for premium wines, where grapes are cultivated almost continuously (Dalmasso & Vuittenez 1977).

Control

Several lines of action can be taken. For selection and production of virus-free stocks by heat therapy, living plants are exposed to 38°C for 4–6 weeks, followed by removal and rooting of shoot tips under mist (Goheen *et al.* 1965) or culturing excised tips *in vitro* (Ottenwaelter *et al.* 1973), tip micrografting (Bass & Vuittenez 1977) or meristem-tip culture (Barlass *et al.* 1982). Heat treated GFLV-free stocks perform very well. The plants are homogeneous in morphology and productivity, yield is increased by 40–70%, sugar content is higher and the quality of wine is very satisfactory (Bovey 1982). However morphological changes may result in some *V. vinifera* cultivars following prolonged *in vitro* culture. Elimination of root-'reservoirs' and nematode vectors through prolonged fallow breaks the ecological cycle of the nematode-virus complex. More rapid eradication can be obtained by soil fumigation prior to replanting. Breeding for tolerance or resistance to virus and/or vector is a more recent approach (Bouquet 1980).

Special research interest

The main research interest lies in the development of resistant or tolerant rootstocks by intra-generic (*Vitis*×*Vitis*) or inter-generic (*Vitis*×*Muscadinia*) crosses (Goheen pers. comm.; Bouquet 1980), as well as studies on interference between strains of GFLV and arabis mosaic virus (Vuittenez *et al.* 1976) and selection or production of avirulent strains to be used for premunition.

VUITTENEZ & MARTELLI

Myrobalan latent ringspot virus (MLRSV) is a probable member of the nepovirus group (CMI/AAB 160) and seems to be distantly related to some strains of tomato black ring virus. It has only been reported in S.W. France. Latent in many *Prunus* species and cultivars including myrobalan (*Prunus cerasifera*), it causes short internodes and rosetting in peach and leaf enation in sweet cherry. Serological indexing is

used (ELISA). Nematode transmission is suspected but not yet established.
DELBOS & DUNEZ

Olive latent ringspot virus. See olive latent virus 1 (Ch. 3).

Raspberry ringspot virus (RRV)

Basic description

RRV (CMI/AAB 198) has two RNAs (RNA_1:2.4×10^6 d and RNA_2:1.4×10^6 d) encapsidated in two nucleoprotein components M and B. Component M is homogeneous with particles containing one molecule of RNA_2; component B is heterogeneous comprising two types of particles indistinguishable from the sedimentation point of view. These contain either one RNA_1 molecule or two RNA_2 molecules. Virus particles are isometric, *c.* 28 nm in diameter. Purified virus suspensions are highly immunogenic.

Host plants and diseases

The host range is very wide; the virus mainly infects perennial plants and has been frequently reported in small fruit and fruit trees. In raspberry, RRV generally induces ringspotting and sometimes pronounced downward leaf curling (Scottish leaf curl). Susceptible cultivars have stunted brittle canes and often die within 2 years (Murant 1970). Spoonleaf symptoms have been observed on redcurrant (redcurrant ringspot virus) and mottling on strawberry. Severe outbreaks causing obvious patches in crops have been reported on some raspberry cultivars (Malling Jewel) (Harrison 1958). In America a quite different virus (raspberry leaf curl virus, q.v.) causes similar symptoms. Pfeffinger disease of cherry (rosetting and rasp leaf) is caused by RRV; generally symptoms are more severe when RRV is associated with Prunus necrotic ringspot virus. RRV has also been identified in peach and grapevine.

Strains

Two major strains have been described, the Scottish and English strains: they show differences in antigenic characteristics and vector specificity. Another strain, the Lloyd George yellow blotch strain (Murant *et al.* 1968) is serologically similar to the Scottish strain but infects raspberry cultivars which are known to be immune to it.

Identification and diagnosis

Most commonly used herbaceous test plants can be infected. The most suitable test plants are *Chenopodium quinoa* and *C. amaranticolor*, and *Nicotiana tabacum*. RRV is the only nepovirus which gives local lesions on *C. amaranticolor* without becoming systemic on this plant. ELISA is the most suitable technique for rapid and sensitive detection.

Geographical distribution, transmission and control

Reported in Europe, Turkey and the USSR, RRV is nematode-borne: two species of nematode have been described, *Longidorus elongatus* and *L. macrosoma* transmitting the Scottish and English strains respectively. Seed transmission occurs in raspberry and strawberry; in some hosts more than 50% of the progeny seedlings are infected. Control measures comprise use of virus-free material, control of nematodes by rotation or nematicides and the use of resistant raspberry cultivars.
DUNEZ

Satsuma dwarf virus (SDV) causes dwarf disease of satsuma orange (Yamada & Sawamura 1952). The M component (119S) contains RNA of M-Wt 1.7×10^6 d and the B component (129S) RNA of 1.9×10^6 d. Antiserum with titre of 1/8000 in complement fixation has been prepared. Susceptible host species include citranges, *Citrus excelsa*, *C. junos*, clementine, grapefruit, lemon, mandarins, natsudaidai, *Poncirus trifoliata*, rough lemon, satsuma, sour orange, sweet orange, tangelos and small acid lime. A few non-rutaceous species are also susceptible (e.g. tobacco, cowpea, sesame, French bean). Other Rutaceae (Etrog citron, Hassaku, Tahiti lime, Orlando tangelo) are symptomlessly infected, as are *Chenopodium amaranticolor*, cucumber and *Nicotiana glutinosa*. Small boat or spoon-shaped leaves represent the typical symptoms on citrus. Under natural environmental conditions (in Japan), these symptoms occur only on spring flushes of growth, not on summer or autumn flushes (Tanaka 1980). Of the indicator plants sesame develops necrotic local lesions on inoculated leaves; vein clearing, necrosis, curling and malformation on systemically infected leaves. Cowpea and bean show chlorotic local lesions on inoculated leaves; mottling, vein clearing and necrosis on systemically infected leaves (CMI/AAB 208).

SDV is widely distributed in satsuma orchards in Japan. In the Izmir region of Turkey 2% of satsuma

trees carry dwarf in addition to tristeza (Azeri 1973; pers. comm. 1978). Satsuma dwarf may also be present in Yugoslavia and the USSR, and is liable to spread into Europe. SDV causes marked stunting of trees, reduction in the number and size of leaves and shoots, and small-sized fruits with thick peel, all of which results in severe yield decrease. It is naturally transmitted by budwood, and experimentally by graft inoculation, and mechanical means.
BOVE

Strawberry latent ringspot virus (SLRSV)

Basic description

SLRSV is considered to be a possible member of the nepovirus group (CMI/AAB 126). Its genome consists of two RNAs with M-Wt 2.6×10^6 d (RNA_1) and 1.6×10^6 d (RNA_2). SLRSV shows only one type of indistinguishable nucleoprotein containing either one molecule of RNA_1 or two molecules of RNA_2. Particles are isometric, *c.* 28 nm diameter. Purified virus suspensions are good immunogens.

Host plants and diseases

SLRSV has a wide host range, with a predilection for perennial plants and occurs naturally in many species of wild and cultivated plants, e.g. *Robinia pseudacacia* (mosaic), *Euonymus europaeus* (yellow mottle), *Aesculus carnea* (line pattern), rose (mottling, ringspotting and occasional stunting). In strawberry and raspberry, it is associated with mottling and decline, depending on the cultivar. It has also been found in asparagus, blackberry, blackcurrant, redcurrant, narcissus, rhubarb (rhubarb virus 5), elderberry, grapevine, cherry, plum, peach and olive. Many plants are symptomlessly infected but, when mixed infections occur, a synergistic effect can be observed (Scotto la Massese *et al.* 1973; Ragozzino *et al.* 1982).

Strains

Several isolates have been described, many of which are serologically indistinguishable from the type strain, i.e. the Hampshire strawberry isolate.

Identification and diagnosis

Out of 167 species of dicotyledonous plants inoculated mechanically, the virus infected 127 belonging to 27 families (Schmelzer 1969). Mechanical transmission of the virus is easy and the most suitable hosts are *Chenopodium quinoa*, *C. amaranticolor* and *C. murale*, *Datura stramonium*, tobacco and tomato (Murant 1981). Clear identification of the virus on the basis of the induced symptoms is often difficult. The ELISA test is the best detection technique especially in view of the limited antigenic variation of the virus.

Fig. 9. Association of strawberry latent ringspot virus and prune dwarf virus, causing peach decline.

Geographical distribution, transmission and economic importance

SLRSV occurs throughout Europe and has been reported once in Canada. Severe outbreaks have been described in raspberry, generally coinciding with abundance of the nematode vector (Taylor & Thomas 1968). SLRSV is spread naturally by two nematode species *Xiphinema diversicaudatum* and *X. coxi*. It is also seed-borne in some species (Murant 1981). It is generally of little importance except when associated with other viruses—in mixed infection with prune dwarf virus, SLRSV induces growth reduction, rosetting and dieback of peach trees (Fig. 9) (Scotto la Massese *et al.* 1973).

Control

Elimination of infected plants, planting of virus-free material and use of nematicides (when the vector is present) are the most efficient methods. Distribution of the vector to a depth of 50 cm or more makes treatment difficult, and necessitates careful application of nematicides.
DUNEZ

Tomato black ring virus (TBRV)

Basic description

TBRV is a bipartite genome virus: the two RNAs with M-Wt 1.6 and 2.7×10^6 d are respectively distributed in two separate particles M and B. Besides the two viral RNAs, some strains contain a satellite RNA of 0.5×10^6 d. Particles are isometric, about 28 nm in diameter. The virus is strongly immunogenic (CMI/AAB 38).

Host plants and diseases

Host range is very wide. The virus has been identified in many naturally infected species of monocotyledonous and dicotyledonous plants (CMI/AAB 38; Murant 1981). The main diseases described are: ringspot of bean, sugar beet, raspberry and strawberry, vein yellowing of celery and *Forsythia,* bouquet and pseudo aucuba of potato, mosaic of *Robinia pseudoacacia,* yellow mosaic and stunting of grapevine (infected by the potato bouquet strain or by the Hungarian grapevine chrome mosaic strain), leaf necrosis, defoliation and dieback of cacao (cacao necrosis). Many plants are symptomlessly infected. Most herbaceous host plants commonly used in virus research can be infected artificially.

Strains

A great diversity of strains has been reported based on symptomatology and serology. From the serological point of view, despite clearcut differences, typical TBRV, beet ringspot, potato pseudo aucuba, potato bouquet, lettuce ringspot and celery yellow vein are clearly related. More distant are cacao necrosis and Hungarian grapevine chrome mosaic.

Identification and diagnosis

Mechanical transmission onto susceptible host plants (*Chenopodium quinoa* and *C. amaranticolor, Nicotiana tabacum* and *N. clevelandii, Petunia hybrida*) causes clearcut symptoms which are often similar to those induced by other viruses (especially other nepoviruses) and therefore do not allow clear identification of the virus or the strain. Serological detection is the best testing procedure but the technique should be applied and the results interpreted carefully in view of the great diversity of strains. Especially when ELISA is used, cross reactions with strains which are not closely related can be difficult to observe (Kerlan *et al.* 1982).

Transmission

Nematode transmission has been described for some strains; two vector species have been reported: *Longidorus attenuatus* and *L. elongatus.* Vector specificity is related with the protein capsid. In addition, seed transmission has also been reported in 13 botanical families (the rate of transmission varies from 10% to 100%). Pollen transmission also exists but seems to be of little importance in view of the low competitive ability of infected pollen (Lister & Murant 1967).

Geographical distribution and economic importance

With the exception of particular strains (cacao necrosis in W. Africa; CMI/AAB 173) TBRV seems to be limited to Europe. Only some strains seem to be responsible for significant losses (Hungarian grapevine chrome mosaic in E. Europe); very often infected plants remain symptomless and a reduction in growth or yield is not obvious. The importance of the disease is mainly due to its natural means of propagation and its wide host range on woody crops (grapevine, ornamentals, fruit trees, shrubs, hedges etc.).

Control measures

Elimination of infected plants, planting of virus-free material and nematode treatments when vectors are present are the most efficient control methods. Avoiding soils with previous history of disease is strongly recommended.

Special research interest

TBRV is a good material for investigating virus replication (RNA_1 replicates autonomously in protoplasts) and for locating information on the two RNAs (through the formation of pseudorecombinant strains).

DUNEZ

Tobacco ringspot virus (TRSV) is the type member of the nepovirus group and has a wide host range and many reported strains (CMI/AAB 309). A satellite RNA of M-Wt 0.1×10^6 d with circular molecules has been described (Linthorst & Kaper 1984). Although

occurring in most of the world, TRSV is probably not common in Europe. Its Andean potato calico strain is regarded as a quarantine organism for Europe (EPPO 128). Main symptoms are necrotic ringspots, mild mosaic and bright yellow patterns on leaves (Kahn *et al.* 1962). Indexing is by ELISA and gel diffusion. TRSV is efficiently transmitted by *Xiphinema americanum* and less efficiently by thrips, spider mites, grasshoppers and flea beetles. It is seed-borne in many species. Control is by planting virus-free seeds and avoiding planting in nematode-infested soil. Weed control reduces inoculum sources.
SALAZAR

Tomato ringspot virus, largely confined to N. America, has also been reported in Europe, the USSR and Japan. It is responsible for severe diseases of woody perennial plants, especially in N. America: yellow bud mosaic and stem pitting in almond, peach (Fig. 10), nectarine and plum, rasp leaf in cherry, decline of apricot, union necrosis in apple, yellow vein in grapevine, crumbly fruits in raspberry and chlorosis mosaic in redcurrant (Williams & Holdeman 1970). Tomato ringspot virus also infects many annual plants (Murant 1981). It is only locally established in Europe (EPPO 102), where it has mainly been identified in flower crops (e.g. *Pelargonium*). It is nematode-borne (*Xiphinema americanum*). The ELISA test is suitable for detection (Lister *et al.* 1980). Control is by planting virus-free material, and nematicide treatment if nematodes are present (CMI/AAB 290).
DUNEZ

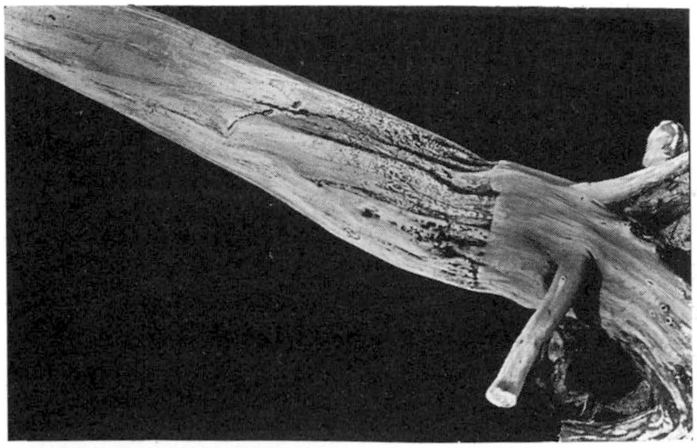

Fig. 10. Tomato ringspot virus (prunus stem pitting virus): abnormalities of the graft union and stem pitting in peach (on American material).

THE POTEXVIRUS GROUP

This group has elongated slightly flexuous particles (470–580 × 13 nm) with one molecule of positive-sense, single-stranded RNA (M-Wt 2.1×10^6 d) and one coat polypeptide species (18–23000 d). They have no known vector but are readily transmitted mechanically. The type member is potato virus X (CMI/AAB 200).

Cactus virus X (CVX) has been reported from Europe and the USA in cultivated cacti. Most infected plants remain symptomless. It is detected by transmission to *Chenopodium quinoa* (necrotic lesions on inoculated leaves followed by systemic mottle) or *C. amaranticolor* (necrotic primary lesions, no systemic infections). A very distant serological relationship has been detected with potato virus X, white clover mosaic, hydrangea ringspot and clover yellow mosaic viruses (CMI/AAB 58).

Clover yellow mosaic virus (ClYMV) (CMI/AAB 111) induces yellow or light green leaf striping in clovers, with occasional stunting and distortion. Yield can be affected. It is also found in lucerne, sweet clover and pea (necrotic streaking). No vector is known. Occurring only in N. America it could be introduced into Europe. Indexing is on *Chenopodium* or by serology. On *Vigna sinensis* it gives indistinct symptoms, in contrast to white clover mosaic virus which induces local lesions.

Cymbidium mosaic virus (CybMV) occurs world-wide and affects all orchids (orchid mosaic, cymbidium black streak). Symptoms are characteristic in some species of *Cymbidium* (mottle and necrosis) but many plants remain symptomless (CMI/AAB 27). Mechanically transmitted it is indexed by IEM, serological precipitation tests and ELISA. Control is by propagation of healthy plants. Virus-free plants are obtained by meristem-tip culture.
ALBOUY

Hydrangea ringspot virus only naturally infects *Hydrangea macrophylla*. It occurs world-wide. Symptoms are chlorotic rings and spots on the leaves, distortion of leaves and decrease in the number of florets per inflorescence on susceptible cultivars (CMI/AAB 114). Many cultivars remain symptomless when infected. Indexing is serological (ELISA or IEM). Control is by propagation of healthy plants.

Virus-free material is produced from virus-infected plants by heat treatment or meristem-tip culture. No vector is known. Since the virus is transmissible mechanically tools should be disinfected during pruning.

Narcissus mosaic virus (NaMV) causes a mild mosaic in trumpet daffodils (Brunt 1966) and must be distinguished from the narcissus yellow stripe virus (q.v.) with which it was previously considered synonymous. The virus, which is not aphid transmitted, is widespread in many cultivars, some of them being symptomless. Detection is by mechanical inoculation or by serology (CMI/AAB 45).

DEVERGNE

Potato aucuba mosaic virus (PAMV) is very probably a member of the potexvirus group, with filamentous particles 580×11 nm (CMI/AAB 98). It causes bright yellow spotting and necrosis on potato leaves. During storage, tubers of many cultivars develop necrosis. The virus occurs world-wide but is uncommon. Mechanically transmissible, it is also aphid-borne in the non-persistent manner by *Myzus persicae* but only in the presence of a helper virus such as potato virus Y or potato virus A. Diagnosis is by transmission to *Capsicum annuum* (necrotic local lesions and epinasty), or *Nicotiana glutinosa* (light green mottle with dark green vein banding), or by serology (Purcifull & Edwardson 1981). The virus is not serologically related to other viruses of the potexvirus group; potato pseudo aucuba disease is due to a strain of tomato black ring virus.

Potato virus X (PVX)

(Syn. potato mild mosaic virus, potato mottle virus)

Basic description

PVX is the type member of the potexvirus group. The virions are flexuous rods *c.* 515×13 nm, with a sedimentation coefficient 117S (CMI/AAB 4). Particles contain *c.* 6% single-stranded RNA with M-Wt 2.1×10^6 d. The RNA is enclosed in about 1400 helically arranged protein subunits, each with M-Wt 27 000 d. PVX is strongly immunogenic. In leaf cells virus particles are dispersed or grouped in aggregates. Amorphous inclusion bodies made of cytoplasmic material, laminate components and virus particles are also formed (Shalla & Shepard 1972). In potato plants PVX is distributed unevenly—regions with high virus concentrations alternate with others containing less or no virus (Weidemann 1981).

Host plants

In natural conditions PVX is known to occur mainly in the Solanaceae (potato, tobacco, tomato, sweet pepper, aubergine, *Hyoscyamus niger*, *Datura stramonium*), though natural infection of plants of other families (*Chrysanthemum sibiricum*, *Taraxacum officinale*, *Galeopsis speciosa*, *Trifolium pratense* and *Melilotus alba*) has been recorded. The virus has a very wide experimental host range. Tobacco (cvs. Samsun, White Burley) is useful for virus propagation and is a susceptible diagnostic species, as is *Datura stramonium*. *Gomphrena globosa* is a good local lesion host commonly used for quantitative infectivity assay.

Diseases

In potato, PVX usually produces no symptoms or very mild interveinal mosaic or mottle. Sometimes deformation or rugosity of leaves, necrotic spots on leaves and stunting of the plant may appear. Some cultivars infected with certain strains develop top necrosis which may result in the death of the whole plant. Symptoms are enhanced by cloudy weather and low (16–20°C) temperature. Symptoms caused by PVX on tomato consist mainly of mottle and, exceptionally, more severe symptoms (necrotic spots and rings, stunting) may appear. On tobacco, the virus produces mottle or necrotic ringspot.

Synergistic effects of double infection of plants with PVX and the other viruses often appear e.g. severe disease may be produced by PVX and potato virus Y on potato and tobacco. The synergistic effect of PVX and tobacco mosaic virus results in the production of so-called streak disease of tomato.

Strains

Many strains and minor variants of PVX have been described. PVX isolates may differ in host range, severity and type of symptoms produced, serological properties, tryptophan and tyrosine contents in the capsid protein, particle stability, isoelectric point and pH stability (Leiser & Richter 1980), as well as thermal inactivation point and tolerance to heat treatment of infected plants. In potato and other solanaceous hosts PVX isolates may vary from those which induce virtually no symptoms to those which cause severe symptoms. Four groups of PVX strains have been distinguished by means of potato cultivars carrying hypersensitivity-inducing genes.

Moreira *et al.* (1980) recently isolated strain X_{HB} in the Andean highlands. This strain differs markedly from previously described PVX strains by infecting potatoes extremely resistant to all other strains of PVX, and by producing symptomless local infection in *Gomphrena globosa*. Such strains are quarantine organisms for Europe (EPPO 128).

Identification and diagnosis

Because PVX often does not produce distinct symptoms, and because symptoms produced by this virus may be masked by simultaneous infection with a second virus, indexing is usually needed for detection and identification of PVX. In leaves the virus is readily detected by means of different serological tests. ELISA seems to be a reliable method for PVX detection in tubers after their dormancy has been broken (de Bokx *et al.* 1980). Diagnosis may also be performed with the indicator plants, *Gomphrena globosa, Datura stramonium*. Treatment of infected plants with some pesticides may reduce the concentration of the virus (Schuster 1977), so testing of potatoes for PVX presence should not be performed shortly after spraying with such chemicals.

Geographical distribution and transmission

PVX occurs wherever potatoes are grown. It is transmitted in potato tubers but not in true seeds. It is very readily transmissible by sap inoculation and by stem and tuber grafting. In the field the virus is spread mainly by mechanical means. PVX is so contagious that it can be transmitted by machinery, workers and animals. Adhering to wood and some clothes, the virus remains infective for up to 6 h (Wright 1974). Tractors are able to carry PVX to healthy plants over a distance of at least 150 m (Winther-Nielsen 1972). Infection may be caused by contact between leaves of healthy and diseased plants. Contact between roots may also allow infection but is of little importance. PVX is not transmitted by aphids. Occasional transmission by some species of chewing insects has been reported. Transmission by fungi is doubtful: an earlier report that zoospores of *Synchytrium endobioticum* are able to transmit the virus has not been confirmed in later experiments (Lange 1978). Some evidence for possible transmission of PVX by *Spongospora subterranea* has been provided (Richardson & French 1978). Seed potatoes may become infected by contact between healthy and affected sprouts. Spread of the virus by the tuber-cutting knife may occur when it passes through an eye.

Economic importance

PVX causes mild diseases of potato. Decrease of yield is dependent on the virus strain and the potato cultivar. Occasionally yield losses of potato caused by PVX infection may exceed 50%, but usually the loss is less than 20%. Mixed infection of PVX with other viruses is much more damaging. The degree of infection is dependent mainly on the manner of basic seed production. Non-selected cultivars may be completely infected.

In many European countries PVX has been completely eliminated from potato seed stocks and is currently of limited economic consideration in this crop. PVX also has very little economic importance in tobacco crops. In tomato the virus is damaging only when accompanied by tobacco mosaic virus.

Control

To control PVX on susceptible potato cultivars healthy starting material must be produced by clonal selection of virus-free plants (based on the results of serological testing). To prevent spread of the virus in the field, mechanical cultivation should be avoided. Selective removal of diseased plants is not an effective way to control PVX as symptomless plants are easily overlooked. The use of resistant cultivars is an alternative approach to the control of PVX. Hypersensitive (field resistant) and extremely resistant cultivars are known. Research into extreme resistance to PVX, determined by a single dominant gene, is promising.

To prevent damage by severe strains of PVX, immunization of the plants with mild virus strains has been tried (Reifman & Romanova 1977). Totally infected potato cultivars may be freed from PVX by meristem-tip culture. For protection of tobacco and tomato against infection with PVX, they should be planted at a distance from potatoes.

GABRIEL & KOWALSKA-NOORDAM

Viola mottle virus (VMV) has been reported in N. Italy. It occurs in *Viola odorata* and causes reduced growth, leaf mottling and white stripes in the flowers. Most host plants are symptomlessly infected. It is distantly related to cactus X virus and hydrangea ringspot virus, and is easily distinguished serologically from cucumber mosaic virus, which is also frequently found in *V. odorata* (CMI/AAB 247).

White clover mosaic virus (WCLMV) causes mosaic and mottle in several clover species and vein clearing, mottle and spots in pea. Most legumes are susceptible (CMI/AAB 41). Reported in N. America, Europe and New Zealand, it is seed-transmitted in clover, and detected by mechanical transmission to *Vigna sinensis* (local lesions), *Phaseolus vulgaris* (vein discoloration).

THE POTYVIRUS GROUP

The group* has elongated flexuous particles (680–900×11 nm) with one molecule of positive-sense, single-stranded RNA (3.0–3.5×10^6 d) and one coat polypeptide species (32–36000 d). Virus infection is often associated with intracellular cytoplasmic and nuclear inclusions, pinwheels, bundles and laminated aggregates. This is a very important and diverse group of plant viruses responsible for many severe and common diseases. They are aphid-transmitted in the non-persistent manner. For some viruses or strains, aphid transmission depends on the presence of a helper factor or helper virus. Type member is potato virus Y (CMI/AAB 245).

Artichoke latent virus occurs in most Mediterranean countries, California (USA) and Brazil. Most artichoke cultivars are symptomlessly infected both in nature and following artificial inoculation with the purified virus, but others show mild mottling and growth reduction. In Tunisia and Spain, the virus is associated with severe degeneration of the plant. It is transmitted non-persistently by aphids and through propagating material, but not through seed (Rana *et al.* 1982). Indexing is by mechanical inoculation to *Gomphrena globosa* (reddish local lesions with necrotic centre within 5–6 days), ELISA or IEM. Virus-free plants can be obtained by meristem-tip culture (Harbaoui *et al.* 1982). Healthy plants give 10–12% more yield.
MARTELLI

Asparagus virus 1 (syn. asparagus virus B) occurs world-wide. It is mostly symptomless in infected asparagus plants, but is associated with asparagus virus 2 (q.v.) in decline symptoms. Aphid-borne (*Myzus persicae*) in a non-persistent manner, it is indexed by mechanical inoculation onto *Chenopodium amaranticolor* (inoculated leaves show necrotic ringspot lesions within 10 days) and by serology, especially using IEM (Bertaccini & Marani 1983; Fujisawa *et al.* 1983). Control is by the propagation of healthy plants obtained from virus-free seeds or meristem-tip culture.
BERTACCINI & MARANI

*The description corresponds to subgroup 1 (aphid-transmitted), now distinguished from subgroups 2 (fungus-transmitted) and 3 (mite-transmitted), in which some previously unclassified viruses are now grouped (cf. Ch. 3).

Bean common mosaic virus (BCMV)

(Syn. bean mosaic virus, bean virus 1)

Basic description

BCMV has filamentous particles *c.* 750×15 nm (CMI/AAB 73). Protein subunits of M-Wt 32800 d accompanied by a minor one of M-Wt 28000 d have been recorded (Moghal & Francki 1976). The estimated amino acid residue number is 290. Cytoplasmic inclusions occur in epidermal strips appearing as pinwheels and scrolls under the electron microscope.

Host plants

In nature BCMV is restricted to *Phaseolus* species, mainly *P. vulgaris* (French bean). Some isolates can be artificially transmitted to other hosts (Bos 1970).

Diseases

BCMV causes common mosaic, vein necrosis and black-root, but the nature and severity of symptoms depends greatly on the cultivar involved, the time of infection and the environmental conditions. Bean cultivars may be tolerant, sensitive or hypersensitive but the kind of reaction depends largely on the virus strain. Tolerant cultivars may show only slight narrowing of systemically infected leaves. Sensitive cultivars may show a reddish rugose mosaic of the lower leaves, a rolling or curling mosaic with dark green areas along the main veins of the upper leaves and mottled or malformed pods. Severe stunting and black-root symptoms occur with typical strains when temperatures are above 30°C, and with virulent strains at 20°C and above. Hypersensitive cultivars react with necrotic local vein necrosis at normal temperature and with black-root symptoms at high temperature. This type of reaction gives field resistance and is genetically dominant.

Strains

Various strains have been described in connection with research on breeding for resistance in *Phaseolus*;

they differ in the symptoms they induce and in the range of cultivars they infect (Drijfhout 1978).

Identification and diagnosis

The virus may be identified by visual inspections during the growing season, or by germination of seed lots in the greenhouse. These methods are laborious and furthermore fail to detect masked infections. ELISA may be more convenient and reliable for detecting the virus, which is moderately antigenic (Jafarpour *et al.* 1979).

Geographical distribution

BCMV is of very wide distribution and can be found everywhere that French beans are grown.

Transmission

The most important source of crop infection is infected seed. Seed transmission occurs at rates as high as 83%, but depends on the cultivar, the temperature before flowering and the precocity of infection; plants infected after flowering do not produce infected seed. BCMV is an aphid-transmitted virus. Several aphid species have been reported to transmit the virus in the non-persistent manner (Kennedy *et al.* 1962), including *Acyrthosiphon pisum*, *Aphis fabae* and *Myzus persicae* (Zettler 1969).

Economic importance

BCMV has been regarded as the most ubiquitous and destructive virus of *Phaseolus vulgaris*. Moderate and severe bean common mosaic in cv. Red Mexican U.I. 34 caused 50% and 64% reductions in the number of pods per plant respectively, and 53% and 68% reductions in seed yield. BCMV-infected plants also tend to produce fewer seeds per pod (Hampton 1975). The tolerant and hypersensitive cultivars greatly reduce BCMV incidence in the field in temperate conditions. In tropical conditions with high temperatures, it always causes hypersensitive necrotic symptoms.

Control

Virus-free seed is especially important for disease control because BCMV is transmitted primarily through seed. Breeding and use of cultivars with II genes also reduces crop loss due to the virus. Disease incidence in an II crop is mostly lower than in one with I^+I^+ because: i) the virus is not seedborne in II plants; ii) aphids can only introduce the virus from a nearby I^+I^+ crop; iii) virus transmission by aphids from II plants with systemic necrosis has never been observed, so further spread within an II crop does not occur. Exclusive cultivation of II cultivars would completely prevent BCMV infection since *Phaseolus vulgaris* is practically the only natural host of the virus (Drijfhout 1978; Van Rheenen & Muigai 1984).

Special research interest

To compare strains, a standard procedure for identification and a set of differential bean cultivars has been established. The differential cultivars are representatives of 11 resistance groups, determined by testing about 450 bean cultivars against 8–10 strains. The virus strains and isolates are classified into 10 pathogenicity groups and subgroups, so that 10 strains are distinguished and the others considered as isolates of those strains (Drijfhout 1978).

VERHOYEN & MEUNIER

Bean yellow mosaic virus (BYMV)

(Syn. bean virus 2, gladiolus mosaic virus, pea mosaic virus)

Basic description

BYMV has filamentous particles *c.* 750×15 nm helically constructed with a pitch of 3.4 nm. It has a sedimentation coefficient of 140–144S and a buoyant density of 1.318–1.325 g cm^{-3} (Huttinga 1975). Protein subunits of M-Wt 33 000 d, accompanied by a minor one of M-Wt 28 000 d have been recorded (Moghal & Francki 1976). The estimated amino acid residue number is 292. Many granular and/or crystalline cytoplasmic inclusions and often intranuclear crystalline inclusions or nucleolar enlargements occur in infected tissues (Bos 1969). The cytoplasmic inclusions appear mainly as pinwheels and laminated aggregates (CMI/AAB 40).

Host plants

BYMV causes diseases in many legumes and also in Iridaceae such as gladiolus, freesia, tritonia and crocosmia (Derks & van den Abeele 1980). It also infects globe artichoke (Russo & Rana 1978). *Nicotiana clevelandii*, *Chenopodium quinoa*, *C. ambrosioides* and *C. capitatum* are susceptible by inoculation in the glasshouse (Kovachevsky 1968).

Diseases

French bean. The surface of the leaflet is slightly irregular and small light yellow spots soon develop in the dark green background. The yellowing gradually spreads over the entire surface, causing the leaflets to become more or less chlorotic. In these early stages the young growth has a tendency to become brittle. The first trifoliate leaflets do not remain curled downwards, but, as they enlarge, become slightly concave on their upper surfaces and take on a glossy appearance. On the third and fourth trifoliate leaves there is a very distinct mottling of yellow-green and dark green areas. Plants become stunted and bushy because of a reduction in the length of the internodes and a proliferation of branches. Maturity is delayed and the production of pods reduced.
Garden pea. The first signs of the disease (originally known as pea mosaic) consist of a faint mottling of the leaves which later becomes more intense owing to the presence of numerous dark green areas, irregular in outline and occurring between the larger veins. Later a pronounced vein-clearing appears and immediately adjacent to the larger veins the dark green tissue often persists. The region between the veins remains green, but is of a lighter shade than in the normal plant. Leaves sometimes show yellow mottling accompanied by malformation of the lamina. Infected plants are slightly distorted and stunted. The pods may be somewhat malformed and distorted, and in some cases reduced in size.
Melilotus alba. Symptoms first appear as small light yellow spots on the leaves. These spots may enlarge and coalesce with others, producing small light green blotches interspersed with dark green areas. Frequently there is a clearing of the veins, with the dark islands situated between them. Severe infection may cause slight dwarfing and ruffling of the leaves.
Broad bean. A mild chlorosis with irregular dark green islands develops with a slight malformation of the leaves. The plant is stunted.
Lentil. BYMV induces stunting, mild mottling of the leaves, smaller and deformed pods.
Lupin. The foliage of diseased blue lupin plants (*Lupinus angustifolius*) at an early stage of infection develops a characteristic bronze colour, with a slight wilting. Necrosis of the stem is usually unilateral, and the apex of the plant tends to bend to the necrotic side, forming a shepherd's crook. Diseased yellow lupin (*Lupinus luteus*) develops vein clearing and a light mottle at an early stage of infection. Later, the mottle becomes more pronounced and the leaflets of the new growth are usually narrow and malformed and tend to stand erect. Proliferation of the axillary buds often occurs, giving the new growth a bunchy appearance. White lupins (*Lupinus albus*) infected with BYMV develop a severe mottle, with various leaf deformations.
Gladiolus. BYMV induces colour-break symptoms in flowers and a slight mottle or yellowish stippling on the leaves (Fig. 11).

Fig. 11. Bean yellow mosaic virus: yellow mosaic of gladiolus leaves.

Freesia. BYMV causes severe yellowing of the leaves, necrosis of the leaf veins and necrotic spots in the corms.
Artichoke. Infected plants may show yellow flecking and mottling of the leaves.

Strains

Bos *et al.* (1974) divided BYMV isolates into 3 distinct groups on the basis of host range, symptoms and reactions on cultivars of bean and pea. The distinction of groups is especially important in breeding for resistance. Bean top necrosis, pea mosaic, red clover necrosis, and cowpea strains are the most important.

Identification and diagnosis

The virus is moderately antigenic. ELISA is the simplest and fastest means of detecting and identifying BYMV and indexing large numbers of legumes (McLaughlin *et al.* 1984). Identification of BYMV by

ELISA is possible in leaves and petals but not in the corms of gladioli. ELISA may be of potential value for producing virus-tested propagation material (Stein *et al.* 1979). Herbaceous hosts useful as differential indicators have been identified (Hampton *et al.* 1978).

Geographical distribution

The virus is reported from most countries of the world where legumes are grown.

Transmission

More than 20 aphid species can transmit the virus in a non-persistent manner, but especially *Acyrthosiphon pisum, Macrosiphum euphorbiae, Myzus persicae* and *Aphis fabae* (Kennedy *et al.* 1962). Transmission through seed is not common but approximately 6% can be expected from *Lupinus luteus* (Corbett 1958) and 3–5% from *Melilotus alba* (Phatak 1974).

Economic importance

BYMV virus causes a 33% reduction in the number of pods per plant and a 41% reduction in seed yield in naturally infected field bean cv. Red Mexican U.I.34 (Hampton 1975). In faba bean, yield losses as high as 96.3% have occurred as a result of severe early infections (Frowd & Bernier 1977). In a survey, 92% of all gladioli tested from the Netherlands were infected with BYMV (Nagel *et al.* 1983).

Control

The most important sources of infection seem to be clover and gladioli, therefore beans should not be planted near these crops. It is advisable to rogue out the first infected plants as soon as they are observed, but if the number of infections increases rapidly this becomes impracticable. Field tests with insecticides and mineral oil for the protection of French beans from BYMV indicate double yields compared with untreated controls (Singh *et al.* 1981). BYMV control with oil has been also applied to iris (Deutsch & Loebenstein 1967). Meristem-tip culture and thermotherapy may be applied to eliminate the virus from vegetatively propagated plant species. BYMV has been successfully eliminated from freesia by Brants & Vermeulen (1965).

VERHOYEN & MEUNIER

Bearded iris mosaic virus (BIMV) is prevalent in many species of rhizomatous iris. Chlorotic elongated areas, spots or rings appear on the leaves, associated with colour breaking of flowers in susceptible cultivars. Many cultivars remain symptomless when infected. Symptoms are enhanced by low temperatures and in the presence of cucumber mosaic virus. Probably occurring world-wide, the virus is aphid-borne in the non-persistent manner (CMI/AAB 147). It is serologically distinct from iris mild mosaic virus and iris severe mosaic virus (Lisa 1980). Beardless iris mosaic could be a severe strain of bearded iris mosaic. Serological indexing using ELISA or IEM is applicable.

Carnation vein mottle virus (CVMV) sometimes causes chlorotic spots and dark green areas along the veins of carnation leaves (Poupet & Marais 1973), but is also responsible for a severe malformation of petals with flower breaking, particularly on some susceptible Mediterranean cultivars (Poupet 1971). The virus (particles *c.* 790 nm long) is sap and aphid-transmitted. It is easily detected in carnation by mechanical indexing on *Chenopodium quinoa* or *Silene armenia* (CMI/AAB 78). Infected carnation cultivars must be regenerated by meristem-tip culture.

DEVERGNE

Carrot mosaic virus occurs in Europe and the UK. On carrot a mosaic of non-uniformly distributed spots without sharp margins appears 12–16 days after infection, and the leaves are slightly deformed. Indexing is by mechanical inoculation onto *Chenopodium quinoa* (causing yellowish local lesions) or *C. giganteum* (light yellow spots and systemic infection) (Chod 1965). The virus is aphid-borne in the non-persistent manner by *Myzus persicae* and *Cavariella aegopodii*, and can be distinguished serologically by the precipitation test.

CHOD

Celery mosaic virus (CeMV) causes curly leaf symptoms on celery, and may, in addition, cause yellow spotting and vein clearing. It occurs in C., S. and W. Europe. Indexing is by mechanical inoculation to celery, coriander, *Ammi majus* or *Chenopodium* (chlorotic and necrotic lesions on *C. giganteum* and *C. quinoa* within 7–10 days). The virus is aphid-borne in the non-persistent manner by 26 species (CMI/AAB 50; Wolf & Schmelzer 1972) and can be diagnosed serologically by the precipitation test. Successful control can be achieved by the enforcement of a celery-free period in the field. Seed-

lings can be usefully protected on the seed bed with oil sprays prior to transplanting.
CHOD

Clover yellow vein virus (ClYVV) causes mild vein yellowing and mottling in several *Trifolium* species, especially white clover. It also infects coriander (vein clearing and mosaic). Found in N. America and England, it is aphid-borne in the non-persistent manner by *Myzus persicae*, *Acyrthosiphon pisum*, *Aulacorthum solani* and *Macrosiphum euphorbiae*. It is serologically distantly related to some other potyviruses: bean common mosaic, bean yellow mosaic, and soybean mosaic viruses for example. Detection is by mechanical transmission onto *Chenopodium quinoa* (local lesions) and by serology (CMI/AAB 131).

Cocksfoot streak virus (CSV) (syn. cocksfoot mosaic virus) occurs in C., W. and N. Europe. It causes mild or severe chlorotic streaking and decreased tillering in *Dactylis glomerata* and other grasses; some species are symptomless when infected. The virus is transmissible by several species of aphids in the non-persistent manner, and is indexed by mechanical inoculation onto *Setaria macrostachia* or *Paspalum membranaceum* (very conspicuous yellow streaks within 2–3 weeks) or serologically by double diffusion test (CMI/AAB 59).
VACKE

Cowpea aphid-borne mosaic virus (CABMV) (CMI/AAB 134) causes a severe mosaic of cowpea, and many Leguminosae. Diseased plants show dark green vein banding or interveinal chlorosis, leaf distortions and stunting. The virus occurs in Africa, Europe and Asia. Related viruses occur in the USA. Aphid-borne in the non-persistent manner by several species, especially *Aphis craccivora*, *A. fabae*, *A. gossypii*, and also seed-borne, it is detected by mechanical transmission to *Chenopodium amaranticolor* (local lesions) or by serology.

Dasheen mosaic virus (DasMV) infects species of 13 genera in the family Araceae, causing leaf mosaic symptoms with occasional distortions of the leaves. It is aphid-borne in the non-persistent manner (CMI/AAB 191). Occurring world-wide its main economic host in Europe is the glasshouse-grown arum lily (*Zantedeschia*). Mechanically transmissible, it is indexed by inoculation to *Chenopodium quinoa* (chlorotic local lesions and systemic mottle) and by ELISA (Rana *et al.* 1983).

Freesia mosaic virus is the commonest of several viruses which often occur in complex infections of freesia. It causes only faint symptoms when alone (some cultivars are symptomless). Bean yellow mosaic virus is associated with freesia mosaic virus in plants showing 'closed' flowers (when the pistil protrudes above the corolla) and fringed petals. Occasionally, cucumber mosiac virus is also found in plants developing leaf mosaic. All these viruses are aphid transmitted. Freesia mosaic virus is a probable member of the potyvirus group. Necrotic spots and strips observed on the leaves are caused by a mixed infection with freesia mosiac and freesia streak viruses.

Iris mild mosaic virus (IMMV). Iris species are the only naturally infected hosts of IMMV. The virus causes a mild mosaic on bulbous irises. It occurs world-wide and is probably present in most commercially important cultivars; these are totally infected (CMI/AAB 324). Aphid-borne in the non-persistent manner, two efficient vectors are *Myzus persicae* and *Aphis gossypii*. It is not seed-borne in iris. Both iris mild mosaic and iris severe mosaic are aphid-borne and have elongated particles but they are serologically distinct and can be separated in differential hosts. Detection is by bioassay on herbaceous hosts and by ELISA. Virus-free plants can be obtained from the infected ones by meristem-tip culture.

Iris severe mosaic virus is probably a member of the potyvirus group. It causes severe leaf chlorosis, discoloration and stunting on bulbous irises (CMI/AAB 116), and strong colour breaking in the flowers. Aphid-borne, it is serologically distinct from iris mild mosaic virus and can be separated in differential hosts. Detection is by ELISA or by bioassay on herbaceous hosts. Virus-free plants can be obtained by meristem-tip culture.

Leek yellow stripe virus (LYSV)

Basic description, host plants and diseases

LYSV has flexuous filamentous particles *c.* 820 nm (CMI/AAB 240). The host range is very narrow: of 32 *Allium* species tested, 9 were found to be susceptible but most are symptomless and/or very difficult to infect. Most are rarely naturally infected (e.g. onion, shallot) (Graichen 1978). On leek, symptoms of yellow and chlorotic striping are more or less severe depending on cultivar and time of infection. Frost strongly affects the quality of infected plants. Mixed infection by LYSV and shallot latent virus (SLV) aggravates the symptoms due to LYSV (Paludan 1980).

Identification and diagnosis

Chlorotic local lesions develop after three and two weeks respectively on *Chenopodium amaranticolor* and *C. quinoa*, becoming evident when leaves senesce. Infection on these hosts is not systemic. Small necrotic local lesions which appear more rapidly on *Chenopodium* species may be due to shallot latent virus which often infects leeks. No host plant can be considered a practical diagnostic species. Light microscopy of stained epidermal strips of infected leek leaves readily shows the presence of characteristic inclusion bodies in the cytoplasm. The reliability and sensitivity of ELISA and biological assay on *C. amaranticolor* are currently being compared. IEM is also useful for rapid diagnosis and evidence of double infection with SLV.

Transmission, geographical distribution and economic importance

LYSV is transmitted by some 50 aphid species in the non-persistent manner. Identified in New Zealand, it is widely distributed in Europe. It is considered as one of the major diseases of leek in several European countries.

Control

The only sources of infection by LYSV seem to be leek itself, but a leek-free period is impossible to establish, at least in Europe. Protection of seed beds and selective removal of early infected plants are advised. Of numerous cultivars tested, all were susceptible. A source of immunity has been discovered in *Allium ampeloprasum* (cv. Kurrat from Egypt) which has been recently crossed with *A. porrum* (Bos *et al.* 1978; Van der Meer 1980).
LOT

Lettuce mosaic virus (LMV)

Basic description

LMV (CMI/AAB 9) has filamentous particles measuring 13×750 nm sedimenting as a single component of 143S and containing a single-stranded RNA molecule with M-Wt *c.* 3.0×10^6 d. Coat protein is made up of a single polypeptide with M-Wt 33 000 d.

Host plants

LMV is known to naturally infect 21 species belonging to 9 families (mainly Compositae) and can be mechanically inoculated to 121 species (Horvath 1980). Among the widespread weeds naturally infected are *Capsella bursa-pastoris, Picris echioides, Senecio vulgaris, Sonchus asper* and *Stellaria media.*

Diseases

The most important diseases occur on lettuce and endive in the field, and on spinach. Variable symptoms are observed on all types of lettuce—butterhead, crisphead (European Batavia and Iceberg) and cos lettuce (Romaine). Vein clearing, mosaic and yellow mottling are commonly observed on most cultivars. Severe dwarfing and failure to form a head are frequent, especially on plants infected at a young stage. Necrotic spots or vein necrosis may occur on some cultivars with common strains but are most often due to aggressive pathotypes (LMV-1) or to an incubation period with great variation in temperature. Seed yields are considerably reduced.

On curled leaf endive or broad leaf endive (escarole), chlorotic or yellow spots associated with growth reduction are often due to LMV but symptoms may be confused with those due to turnip mosaic virus. Certain pathotypes severely affect endive causing yellowing, distortion and stunting. With common strains, symptoms (yellowing, flaccidity) on LMV-infected plants appear after first frosts.

Symptoms on spinach consist of bright yellow circular spots which tend to coalesce into a diffuse chlorotic mottle. Plants are stunted and old infected leaves wither prematurely (Provvidenti & Schroeder 1972).

Strains

In California, variants of LMV were discriminated on the basis of symptoms on lettuce, pea and weeds (McLean & Kinsey 1962). A strain isolated from *Picris echioides* (LMV-L) induces a lethal reaction on specific genotypes and provokes severe symptoms on cv. Gallega (Zink *et al.* 1973). Pathotypes inducing severe mosaic and growth reduction on cultivars carrying the 'mo' gene (LMV-1) have also been isolated in France from tolerant cultivars, endive and weeds. These aggressive isolates do not seem to be seed-transmitted in lettuce. Differentiation of serotypes is not recorded. Recent experiments to distinguish between various pathotypes using ELISA and IEM have shown very slight differences.

Identification and diagnosis

The virus cannot be identified unequivocally on the basis of symptoms on cultivated hosts. On *Chenopodium quinoa* and *C. amaranticolor*, local lesions appear 8–10 days after inoculation followed by a systemic reaction on young leaves. Immunodiffusion in the presence of SDS is a rapid method for detection of LMV directly from plant extracts. ELISA can be used from numerous samples and is suitable for indexing seeds (Jafarpour *et al.* 1979; Falk & Purcifull 1983).

Geographical distribution and transmission

LMV is distributed world-wide. It is transmitted by aphids in the non-persistent manner. Several species of aphids are probably responsible for spread of LMV in the fields; at least 8 species have been demonstrated to be vectors in laboratory tests. Among them, *Myzus persicae* and *Macrosiphum euphorbiae* seem to play a decisive role in natural dissemination. True seed transmission occurs in lettuce (3–10%) and in *Senecio vulgaris* (Phatak 1974). Genotype and temperature influence the level of transmission. 'Mo' gene-carrying cultivars do not transmit the virus but a few resistant lines were demonstrated as able to transmit LMV at extremely low rates (Ryder 1973).

Economic importance

LMV is still a widespread damaging virus on lettuce in countries where tolerant cultivars are not grown and/or where production of 'virus-free' seed has not been organized under official inspection. The emergence of new pathotypes could be a major threat, especially if such variants are transmitted through seed.

Control

Use of insecticide compounds is not effective for preventing the spread of LMV, as in most cases with stylet-borne viruses (Gonzalez & Rawlins 1969). In the USA, the use of MTO ('Mosaic Tolerance Zero') seed has considerably reduced LMV infection in most areas. Resistant cultivars are not yet grown but their use is considered as the ultimate solution. In Europe, the acreage of tolerant cultivars increased rapidly with the progressive introduction of the 'mo' gene in most lettuce types. In most European countries the use of certified virus-free (< 0.1%) seeds is not compulsory for growers, so the virus is not controlled in many areas. Official certification of commercial lettuce seed should concern both susceptible and tolerant cultivars.

LOT

Maize dwarf mosaic virus (MDMV)

(Syn. sugarcane mosaic virus, sorghum red stripe virus)

Basic description

MDMV has filamentous particles *c.* 750×13 nm comprising a single RNA molecule of M-Wt 2.7–2.9×10^6 d (Pring & Langenberg 1972; Hill & Benner 1976) and *c.* 2000 protein subunits of M-Wt 36 500 or 28 500 d, estimated by SDS gel electrophoresis or amino acid analysis respectively (Hill *et al.* 1973). The sedimentation coefficient of the particles is 170 ± 5S. (Tošić & Ford 1974).

Host range

The host range comprises more than 250 plant species from more than 60 genera. Besides maize, the most important hosts are species of *Saccharum*, *Sorghum*, *Setaria* and *Panicum*. Weeds like *Sorghum halepense* serve as natural and perennial sources of inoculum under field conditions (CMI/AAB 88).

Diseases

The main symptom of MDMV infection on host plants is mosaic. Some host plants, for example certain species of *Bromus*, *Lolium*, *Eremopyrum*, *Oryza*, *Phalaris*, *Rottboellia*, *Sporobolus* and *Stipa* are symptomless. Others, like some *Sorghum* cultivars, react with local necrosis on leaves inoculated with MDMV strain B. Maize plants are most frequently infected during the second and third months of growth, but new infections can arise until growth ceases. Mosaic symptom appears first on the youngest leaf. Chlorotic spots, dots and streaks starting at the base of the youngest leaf later spread over the whole blade, as well as appearing on new leaves. Chlorotic stripes and lines can also be seen. Some inbred lines are symptomless. Infected maize plants also show a decrease in height (up to 18%), due to the shortening of upper internodes, a delay in growth (1–2 wks); more frequent sterility (up to 25%), and cobs are smaller and partially infertile. Time of infection and the susceptibility of the maize cultivar influence symptom intensity and virus concentration.

On sorghum, different MDMV strains cause local necrosis on inoculated leaves, mosaic, red stripes, and systemic necrosis on new leaves. Some sorghum cultivars can be used for differentiation and identification of MDMV strains, but others are symptomless, and yet others have been shown to be resistant. In many regions, *S. halepense* is almost 100% infected. Mosaic symptoms, usually chlorotic streaks and stripes, are visible early in spring soon after emergence of the plant. Later in the season, mosaic is followed by red striping.

Strains

Two strains have been described so far. Strain A infects *S. halepense* and is widespread in Europe. It is identical to MDMV-A and SCMV-Jg (sugar cane mosaic virus-Jg) described in the USA and elsewhere. Strain B does not infect *S. halepense* and has been found in Europe so far only in France (Signoret 1974). It is identical with MDMV-B and SCMV-E described in the USA and elsewhere (Tošić & Ford 1982). Partial cross protection has been found between strains A and B, as well as between these and other strains of SCMV (Tošić 1981).

Identification and diagnosis

Identification should be based on symptoms, mechanical and aphid transmission, host range and the reaction of differential sorghum cultivars, virus particle morphology, physical and biophysical properties and serology. ELISA has been found to be a very sensitive method by which 1 ng of MDMV/ml can be detected. MDMV-A infects *S. halepense* and causes mosaic and, rarely, necrosis on sorghum cvs Atlas, Rio and SA8735, and more often necrosis on NM31. MDMV-B does not infect *S. halepense* and causes local necrosis on inoculated leaves of cvs Atlas, NM31 and SA8735. This strain causes faint mosaic on sorghum cv. Martin and chlorotic streaks and sparse necrotic spots on sorghum cv. Rio (Tošić & Ford 1982). MDMV infection can be surveyed by aerial photography. The Ektachrome aero-infrared type 8443 has been shown to be suitable.

Geographical distribution

Widespread in S. and S.E. Europe, MDMV also occurs in some Middle European countries. It is also present in N. and S. America, Africa, Asia and Australia.

Transmission

MDMV is mechanically, aphid- and seed-transmitted. More than 10 aphid species have been reported to be vectors of MDMV in a non-persistent manner. Aphids remain viruliferous for up to 6 h. Among the most efficient vectors are *Schizaphis graminum*, *Rhopalosiphon padi* and *Myzus persicae*. Seed transmission of MDMV with maize seed is very low (0.01–0.02%) but may reach 0.2% in some inbred lines.

Economic importance

Most maize hybrids are sensitive to MDMV and infection rates may be 25–50%, and up to 80%. Sweet corn and some inbred lines can be 100% infected. One quarter of maize dwarf mosaic affected plants are sterile. With fertile affected plants, yield is decreased by 42%. This means that if 100% of maize plants are infected with MDMV the total yield of that crop could be reduced by up to 60% (Tošić & Mišović 1967). Changes in quality should also be taken into consideration, for example the oil content of the kernels of affected plants is reduced by 30%. MDMV was involved with other viruses in epidemic lethal necrosis of maize in 1978 in Kansas (USA). MDMV infection increases susceptibility of maize plants to fungi such as *Gibberella zeae*, *Ustilago maydis*, *Puccinia sorghi*, *Setosphaeria turcica* and *Cochliobolus heterostrophus*. A synergistic effect has been shown between MDMV and nematodes as well with *Ostrinia nubilalis*. Sporulation of some fungi is better on MDMV-infected maize plants, and conidia are bigger and of better viability.

When grain sorghum cultivars react to MDMV with mosaic, yield can be reduced by up to 54%. Plants of some sorghum cultivars showing mosaic and red stripe symptoms die prematurely.

Control

Maize mosaic can be controlled by breeding resistant hybrids or by eradication of *S. halepense* and other sources of virus. It has been shown that different maize inbred lines and hybrids differ in susceptibility to MDMV and some of them are very resistant or immune. Resistance appears to be partially dominant. Early sowing may correlate with lower incidence of MDMV infection, but results are contradictory. This may depend on seasonal vector activity.

TOŠIĆ

Malva vein-clearing virus is a possible potyvirus; it is aphid-borne and causes yellow mosaic along the veins of various ornamental Malvaceae (*Lavatera, Althaea, Anada* and *Hibiscus*) (Martelli *et al.* 1969). It occurs in Europe and the USA.

Narcissus yellow stripe virus (NYSV) causes the longest known and most important *Narcissus* virus disease, consisting of a severe streaking and curling of the leaves and breaking of flowers (Brunt 1971). The virus, probably a potyvirus, is sap-transmissible, but also slowly spread by aphids (non-persistent manner). Virus particles (755 nm long) are readily detected in negatively stained sap from infected daffodils. Infected plants showing conspicuous symptoms are easily recognized in *Narcissus* fields and are removed by rogueing (CMI/AAB 76).

Onion yellow dwarf virus (OYDV)

Basic description

OYDV has flexuous filamentous particles *c.* 770 nm long and 16 nm in diameter (CMI/AAB 158). Typical inclusions (pinwheels) of the potyvirus group can be detected in infected cells.

Host plants and diseases

OYDV causes yellow striping, stunting, leaf curling and flaccidity in most onion and shallot cultivars. Some Spanish onion cultivars and grey shallot types are resistant. In garlic, OYDV only causes yellow striping and stunting, more or less severe depending on the cultivar. Of other *Allium* species, *A. roseum* is very susceptible, *A. ampeloprasum* and *A. polyanthum* are tolerant, *A. fistulosum* is susceptible or tolerant, and there is no information on *A. schoenoprasum*. Finally, both *Narcissus tazetta* and *N. jonquilla* develop symptoms (Bos 1982).

Strains

Some shallot isolates are more aggressive and can infect resistant onion cultivars. Garlic isolates (sometimes called garlic mosaic virus) infect onion with difficulty and symptoms are less severe. In France, garlic isolates seem to be serologically closely related to the Netherlands OYDV onion strain (Delécolle & Lot 1981).

Identification and diagnosis

The virus may be identified by visual symptoms in onion, shallot and garlic crops. Indexing is difficult because few *Allium* species are susceptible and, except seeded onion, are vegetatively propagated so that mixed infections are common. *Chenopodium quinoa* is immune to OYDV. IEM is an easy, rapid and reliable method for detecting OYDV and differentiating it from other viruses such as leek yellow stripe virus, or latent viruses such as shallot latent virus.

Transmission

The disease is propagated by aphids (50–60 species) (Kennedy *et al.* 1962) or vegetatively in onion, garlic and shallot bulbs. Whether seed-transmission in onion is viable remains controversial (Hardtl 1972). The virus is easily mechanically transmissible.

Geographical distribution, economic importance and control

OYDV occurs world-wide, causing seed loss, reduced seed germination, impaired storage in onion, and reduced bulb yields (20–40%) in onion, garlic and shallot. Control measures include breeding for resistance (onion); propagation of healthy clones (shallot and garlic) obtained by mass and clonal selection or by meristem-tip culture; rogueing in onion, garlic and shallot crops and multiplication under conditions that preclude re-infection. It should be noted that in New Zealand, a garlic potyvirus named garlic yellow streak virus seems serologically distantly related to OYDV and LYSV (Mohamed & Young 1981).
DELÉCOLLE

Papaya ringspot virus (PRSV) (CMI/AAB 292) has a strain, watermelon mosaic virus 1, occurring world-wide on cucurbits but mainly in tropical and subtropical countries, and in Europe in the Mediterranean basin. It is aphid-borne in the non-persistent manner. Symptoms on cucumber, melon and courgette are generally severe (mosaic, leaf deformation, fruit distortion). Indexing is by mechanical inoculation of resistant melon B 633 (necrotic local lesions within 1–2 weeks), observation of cytoplasmic inclusion bodies in epidermal strips or serological methods (SDS immunodiffusion, ELISA, IEM) (Purcifull & Hiebert 1979). A high level of resistance is available in melon (Webb 1979).
LECOQ

Parsnip mosaic virus (PanMV) occurs commonly in England. It causes a mild mosaic of parsnip (CMI/

AAB 91) and is of minor economic importance. It often occurs in mixed infections with parsnip yellow fleck and parsnip mottle viruses. This virus is transmitted by several aphids *Cavariella aegopodii*, *C. theobaldi* and *Myzus persicae*, and is detected by mechanical transmission onto *Chenopodium amaranticolor* or *C. quinoa* (local lesions).
MARCHOUX

Pea seed-borne mosaic virus (PSbMV)

Basic description

PSbMV (CMI/AAB 146) has elongated virus particles 770×12 nm with a sedimentation coefficient of 154S. The coat is made up of one protein subunit with a M-Wt of 34 000 d (Huttinga 1975). The virus is a moderately good immunogen.

Host plants and diseases

Cultivars of *Pisum* are the main natural hosts for PSbMV where it causes various degrees of stunting, leaf narrowing, downward rolling of leaflets (pea leaf roll virus) and a transient clearing and swelling of leaf veins. Sometimes flowers are distorted giving rise to distorted pods. Ovule development is uneven and pods often contain one or two seeds. Field and broad beans are more severely affected, displaying rolling and yellowing of the upper leaves. Plants are stunted and necrosis of many growing points occurs. Lucerne can be a winter host for the virus and also its main vector. Most non-leguminous hosts are symptomlessly infected.

Identification and diagnosis

Mechanical inoculation onto *Chenopodium amaranticolor* results in the development of small local lesions on inoculated leaves. In addition serological detection using ELISA is applicable.

Geographical distribution, transmission and control measures

PSbMV is very probably distributed world-wide because of the international exchange of seeds of many cultivars. The virus is seed-borne. Commercial seed lots containing up to 90% infected seeds have been found (Knesek & Mink 1970). The virus can also be transmitted through a low percentage of seeds of several *Vigna* species. In addition, the virus is transmitted in a non-persistent manner by several aphids, especially *Acyrthosiphon pisum*, *Macrosiphum euphorbiae* etc. (Aapola & Mink 1973). Control is by use of virus-free seed.
DUNEZ

Plum pox virus (PPV)

(Syn. sharka virus)

Basic description

The virion of PPV (CMI/AAB 70; Kerlan & Dunez 1976) appears as a filamentous particle *c.* 750×15 nm containing a single RNA molecule of 3.5×10^6 d. Virus suspensions are good immunogens.

Host plants

PPV infects all fruit tree species of the genus *Prunus* except cherry (sour and sweet). Almond trees may be infected occasionally. The virus also infects most wild *Prunus* species, especially *Prunus spinosa* which is an important natural source of infection. The virus can be artificially transmitted onto *P. mahaleb* (but infection remains localized), *Sorbus domestica* and several herbaceous plants.

Diseases

Severe symptoms are observed on apricot, plum and peach trees; the virus causes pox disease on fruits and frequently also clearcut leaf symptoms (Lansac *et al.* 1979).

Apricot. Chlorotic pale green lines, rings or spots develop on the leaves during spring and remain until the middle of summer. Fruits show chlorotic, yellow rings with frequently severe deformations, irregular grooves and more rarely internal gummosis. Clearcut rings and spots can be observed on the stones (yellow at first, these turn brown when the stones dry) (Fig. 12).

Plum. Diffuse, pale green rings or areas develop on the leaves in spring: in some cultivars, these disappear in summer. The fruits of susceptible cultivars show severe deformation with irregular grooves. Reddish discolorations of the flesh and some brown spots on the stones develop as do internal gummosis and necrosis. Decrease of the sugar content and increase of acidity has been noticed in severely affected cultivars. Fruit drops prematurely, some weeks before ripening, especially in the case of late cultivars.

Peach. Some chlorotic lines and small areas can develop on leaves along the secondary and tertiary

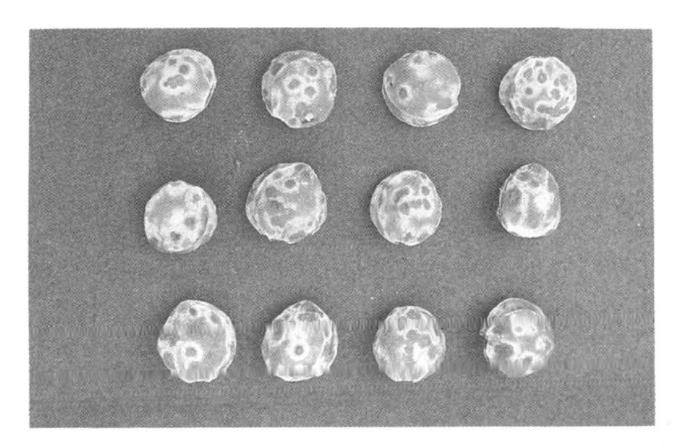

Fig. 12. Plum pox virus: spots on apricot stones.

veins but usually this symptom is difficult to observe in the orchard. Fruit symptoms, which appear clearly 4–6 weeks before maturity, consist of white or green rings (or spots) on white and yellow flesh cultivars respectively (Plate 3). Slight deformations can be observed on highly susceptible cultivars. Similar symptoms develop in nectarines.

Almost all known apricot, plum and peach cultivars are susceptible to PPV but some remain symptomless when infected (Lansac *et al.* 1979). The resistance of two plum cultivars and one apricot cultivar to some virus strains has been described (Šutić & Ranković 1981, 1983). All the usual rootstocks (*P. cerasifera, P. mariana, P. domestica, P. persica,* etc.) are susceptible.

Strains

Several virus strains have been described on the basis of their pathogenicity and aphid transmissibility (Šutić *et al.* 1971; Krczal & Kunze 1976; Massonie & Maison 1976). Two serotypes have been reported; these are closely related and the antigenic variation of the virus appears limited (Kerlan & Dunez 1979).

Identification and diagnosis

The virus may be identified by visual inspection of symptoms during the growing season; indexing is essential however, in view of the occurrence of non specific symptoms, the disappearance of leaf symptoms during the course of the growing season and the existence of many cultivars which remain symptomless when infected. Two sensitive and reliable techniques can be used: i) graft transmission (chip budding) onto peach seedlings (GF 305) in the greenhouse results in the development of typical leaf symptoms (clearing and spotting along the veins; corkscrew twisting of the leaves), within 1–2 months; ii) ELISA provides a very efficient, sensitive and reliable tool for PPV detection. IEM is also a very rapid and sensitive technique (Kerlan *et al.* 1981). The main difficulty results from the uneven distribution of the virus in the infected tree; this means it is necessary to check several samples (at least 5) per infected tree. Samples should consist of well expanded leaves and should preferably be taken from the outer part of the tree.

Geographical distribution

The virus is probably now present in almost all European countries but its spread (from Bulgaria) was mainly observed over the last 20–30 years (EPPO 96). It is prevalent and causes severe damage especially in E. and S.E. Europe. So far, other than in Turkey, Cyprus and Syria, PPV does not seem to occur outside of Europe.

Transmission

PPV is transmitted by aphids in a non-persistent, stylet-borne manner. Out of ten aphid species reported to be vectors of PPV, the most efficient are *Myzus persicae, Brachycaudus helychrisi, B. cardui* and *Phorodon humuli*. Some reports indicate some possible seed transmission.

Economic importance

From the prevalence and severity of the disease, PPV can be said to cause one of the most important virus diseases in Europe. In Poland and Greece 70% (100% in some areas) of the trees are infected. In Yugoslavia, 15 million plum trees are infected; in Bulgaria, 80–100% of the fruit falls prematurely. In many countries, infected fruits can no longer be exported and even in countries where the disease has been found but is not established, the exportation of budwood and rootstocks has become difficult.

Control

The possibility of controlling the disease depends mainly on the level of infection. In highly infected areas where removal of infected trees is no longer possible, fruit growers replace susceptible cultivars by ones less susceptible or by tolerant ones which yield profitable crops. In countries where the disease has been observed only in restricted areas (France, the Netherlands, Switzerland, UK) a severe policy of eradication has been applied. New plantations of virus-free material have been established in non-infected areas. Special attention has to be paid to potential virus reservoirs in wild *Prunus* species, especially *P. spinosa*. Preliminary attempts at cross protection have been made. Chemical control of vectors may help to slow down the rate of spread of the virus.

Special research interest

Despite their easy graft transmissibility, some PPV strains are not aphid transmitted. Mechanisms involving a helper factor, well known in the potyvirus group, can probably explain the lack of transmissibility of certain strains. Interference between PPV and widespread apple chlorotic leaf spot virus may influence the susceptibility of plum to PPV. Some genes for resistance to PPV seem to exist (Šutić & Ranković 1981, 1983) and could be used for breeding commercial hybrids resistant to prevalent PPV strains; this research could also include breeding for resistance to aphids (Maison *et al.* 1983).
DUNEZ & ŠUTIĆ

Potato virus A (PVA)

Basic description

PVA has flexuous filamentous particles *c.* 730–750 nm long, with a sedimentation coefficient of 150 S. The coat protein subunit consists of a single polypeptide species with M-Wt *c.* 33000 d. PVA is moderately immunogenic (CMI/AAB 54).

Host plants and diseases

The host range of PVA includes species of the following genera of Solanaceae: *Nicandra, Lycium, Physalis, Solanum, Hyoscyamus, Datura, Nicotiana, Petunia*. Hosts also include 2 genera of Leguminosae: *Trigonella* and *Melilotus* (Edwardson 1974; Spaar & Hamann 1977). The disease on potato is variously known as mild mosaic, veinal mosaic or common mosaic. Depending on the virus strain and the potato cultivar, symptoms can range from mild to severe. Veinal mosaic is most common. The veins are more or less deepened, the leaf surface becomes rough, and the leaf margins are crinkled. In some cultivars infection remains latent.

Strains

PVA strains differ in the severity of symptoms in potato cultivars (Beemster & Rozendahl 1972) and in host range (Kowalska *et al.* 1981).

Geographical distribution and transmission

PVA occurs world-wide in potato growing areas. It is naturally transmitted by aphids in a non-persistent manner. The most important vector is *Myzus persicae*, but other aphid species are also successful vectors, e.g. *Aphis nasturtii, Aphis frangulae, Aulacorthum circumflexus* and *Macrosiphum euphorbiae*. Potato plants in the field serve as virus sources.

Economic importance and control

PVA occurs less frequently in potato crops than other potato viruses. Infections cause severe diseases, especially in combination with potato viruses X and Y. The yield reduction of infected plants can be up to 40% (Dědić 1975) but virus spread can be reduced by oil spray (Sharma & Varma 1982). Resistant and hypersensitive cultivars are available. Replacement of infected stocks by healthy ones is recommended.
WEIDEMANN

Potato virus Y (PVY)

Basic description

PVY is the type member of the potyvirus group (CMI/AAB 242). The virus particles are helically constructed, flexible and filamentous in shape, measuring *c.* 730×11 nm. They sediment as a single component with a coefficient of 150S. The protein subunit consists of a single polypeptide of M-Wt *c.* 33000 d. PVY is strongly immunogenic.

Host plants

Natural host range is probably limited to members of the Solanaceae, but the virus has been transmitted mechanically to members of other families (e.g. Chenopodiaceae, Leguminosae, Compositae). The host range comprises more than 100 species.

Main diseases

PVY causes economically important diseases in potato, tobacco, pepper and tomato. Multiple infections with other viruses are very common. In potato, as in other hosts, symptoms vary according to the strain (see below). For PVY^N strains, primary (first year) symptoms are usually slight (potato mottle), often late in the growing season; secondary symptoms are sometimes more obvious, ranging from mild to severe mottle; necrosis is rare. For PVY^{O} strains, primary symptoms are mottling and necrosis of leaflets, followed by the death of leaves which either drop from the plants or remain hanging from the stem (potato leaf-drop streak). Secondarily infected plants are dwarfed with crinkled and puckered leaves (potato rugose mosaic); necrotic symptoms are usually less severe than after primary infection. For PVY^{C} strains, primary and secondary symptoms are mottle, crinkle and necrotic streaks on leaves, petioles and stems (potato stipple-streak). Necrosis may also occur in tubers.

In tobacco, PVY^{O} strains cause vein-clearing and epinasty, followed by mottle and vein banding (tobacco vein banding). PVY^{N} strains cause similar primary symptoms, with a later appearance of necrotic lesions scattered on the leaves. Primary veins of fully developed leaves become brown, the leaves collapse prematurely against the stem and only a small bunch of leaves remains at the top of the plants (tobacco veinal necrosis). Stem necrosis may appear especially near the base (PVY^{N} strains). In pepper, PVY causes mosaic with mottling and crinkling in the apical leaves and dark vein banding in fully expanded leaves; there are no typical fruit symptoms.

Strains

The following three strain groups can be distinguished according to symptoms on tobacco, potato and *Physalis floridana*. PVY^{O} are the common strains; PVY^{N} the necrotic or tobacco veinal necrosis strains, and PVY^{C} the stipple streak strains (including potato virus C). However, some strains do not belong to any of these groups. Serological differences between strains exist (but they are not consistent between the members of the strain groups mentioned above). Commonly, serological differences between strains cannot be demonstrated by cross reactions using precipitin tests and conventional antisera (Gugerli & Fries 1983), but can by the cross-absorption technique (Leiser & Richter 1979) or by ELISA tests with monoclonal antibodies (Gugerli & Fries 1983).

Identification and diagnosis

Rapid identification is best achieved using serological immunodiffusion tests. With carefully selected antisera, good results can be obtained after degradation of leaf extracts with pyrrolidine using the radial immunodiffusion test (Shepard 1972; Richter *et al.* 1979) or addition of sodium dodecyl sulphate to the agar gel (Purcifull & Batchelor 1977) for the double diffusion test. In potato certification schemes, the microprecipitin drop test, the radial immunodiffusion test or, more commonly, plant tests with detached leaves of assay species (*Solanum demissum*×*S. tuberosum* 'A6', *S. demissum* 'Y' or *S. chacoense* 'TEL') are used. In the latter case necrotic local lesions appear after inoculation with almost all strains. Recently, direct double-antibody sandwich ELISA proved a more reliable tool for mass diagnosis (Gugerli & Gehriger 1980; Vetten *et al.* 1983).

Geographical distribution and transmission

Occurring world-wide, PVY is transmitted naturally in a non-persistent manner by several aphid species. *Myzus persicae* is the most efficient vector. Transmission of PVY and other potyviruses by aphids may depend on the presence of a helper component which is present in extracts of infected plants. The helper component is a virus-coded protein (Thornbury Pirone 1983).

Economic importance

PVY decreases yield of infected potato plants by 10–80%, depending on virus strain, potato cultivar and time of infection. Although less severe in yield reduction, PVY^{N} strains ruined some approved potato cultivars in Europe during an epidemic in the fifties. In tobacco, PVY^{N} strains are very destructive in sensitive cultivars and may cause the complete loss of a crop. In pepper and tomato, yield losses occur especially in the case of mixed infections with other viruses (cucumber mosaic virus, tobacco mosaic virus).

Control

PVY is largely a problem of outdoor cultivation. The production of virus-free planting material has formed the basis for effective control measures. The most important are: i) spraying with light mineral oils or use of reflective surfaces and sticky yellow sheets to reduce spread of PVY in the field; ii) avoiding growing susceptible Solanaceae near other susceptible crops; iii) early lifting of potatoes after destroying haulms of seed-potato crops to restrict virus spread into the tubers; iv) breeding for resistance. Successful breeding programmes for resistant cultivars have now started with potato, tobacco and pepper.

Special research interest

Lines of interest include the mechanism of aphid transmission and involvement of the helper factor, studies on various virus-encoded inclusion proteins, genome mapping (Hiebert & Dougherty 1981).
RICHTER

Ryegrass mosaic virus (RyMV) is possibly a potyvirus. Recorded from Europe and N. America (Paliwal & Tremaine 1976), it causes systemic mosaic symptoms on many Gramineae; some strains of *Lolium multiflorum* are severely damaged. In the field, symptomless infections are possible. Transmission by *Abacarus hystrix* (Eriophyidae) has been reported. Indexing is by mechanical inoculation to *Lolium multiflorum, L. perenne, Dactylis glomerata* or *Avena sativa*, by ELISA or IEM. Inclusions (pinwheels and bundles) are seen in leaf parenchyma cells (CMI/AAB 86). Control is by spraying pesticides (aldicarb, endosulfan, fenpropathrin, menazon) against the mites, and by meristem-tip culture and the selection of resistant plants.
KLEINHEMPEL

Soybean mosaic virus (SoyMV) occurs world-wide, causing mild to severe mosaic on susceptible soybean cultivars (CMI/AAB 93). Virulent strains cause necrosis on cultivars with resistance to less virulent strains. Most strains are indexed by mechanical inoculation onto *Phaseolus vulgaris* cv. Top Crop (detached leaves develop necrotic local lesions 48 h after inoculation at 30°C with continuous lighting), by ELISA or solid phase radio immunoassay. The virus is seed and aphid-borne (non-persistent manner). Control is by sowing healthy seed; methods are available for seed certification. Several resistance genes have been identified (Irwin & Goodman 1981).
MAURY

Sugar beet mosaic virus (SuBMV)

Basic description

SuBMV has filamentous infectious particles *c.* 730×13 nm (Jafri 1972). The virus is moderately immunogenic; it is serologically related to bean yellow mosaic virus and potato virus Y (Bercks 1960; Grüntzig & Fuchs 1979a).

Host plants

The host range comprises about 100 plant species in 26 dicotyledonous families, mainly belonging to the Chenopodiaceae, Solanaceae and Leguminosae, which are infected naturally and/or experimentally.

Diseases

On sugar beet, the youngest leaves show a vein-clearing 8–12 days after inoculation. Later this symptom changes to a mosaic of yellowish, white, and dark green leaf spots. The leaves are often puckered, curled and crinkled. Especially in the case of beet seedling plants the leaf stems are shortened. Infected plants can be stunted. Inclusion bodies are found in the cytoplasm and chloroplasts. Both nucleus and plastids exhibit changes of shape and size after infection (Hoefert 1969; Jafri 1972). On spinach the initial symptoms consist of a distinct upward curving of the youngest leaves with small golden-yellow spots. The latter often coalesce to give large chlorotic areas forming a mosaic. Older leaves become progressively chlorotic and necrotic, beginning at the tip. Infected plants are stunted.

Strains

Strains of significantly different virulence occur in the USA (Shepherd *et al.* 1966) but are unknown in Europe (Grüntzig & Fuchs 1979b).

Identification and diagnosis

A reliable diagnostic species is *Chenopodium quinoa* (Grüntzig 1983). Many discoloured to brownish local lesions appear within 7–8 days. Serological detection of SuBMV in leaves is possible with either the latex test or ELISA (Grüntzig 1983).

Geographical distribution and transmission

SuBMV occurs world-wide in major beet growing areas, especially in temperate regions (CMI/AAB 53). The virus is transmitted in a non-persistent manner by about 30 aphid species. The most efficient are *Myzus persicae* and *Aphis fabae*. The virus is sap-transmissible but not seed-transmissible.

Economic importance and control

In Europe SuBMV is usually of little economic importance to root crops (Hartleb 1975). However, in some seed-production areas, it causes severe losses, up to 40–50% as the seeds of infected plants germinate poorly. Control is possible by the separ-

ation of root and seed crops by the chemical control of vectors and by the removal of weed hosts.
GRÜNTZIG

Tobacco etch virus (TEV) is common in N., C. and S. America, and has the potential to be very damaging in Europe. It causes diseases on several solanaceous crops. In tobacco it causes narrowing, mottling and necrotic etching of leaves; mottling, stunting and fruit distortions on pepper, and stunting of tomato plants with yield reductions of over 30%. Diagnosis is by bioassay after mechanical inoculation onto *Nicotiana tabacum* (inoculated leaves develop chlorotic spots or necrotic rings; systemic symptoms are vein clearing, necrotic etching and chlorotic mottle) or *Datura stramonium* (systemic mottle, leaf distortion and vein banding). TEV is strongly immunogenic and serological detection is applicable (CMI/AAB 258).

Fig. 13. Tulip breaking virus: flower breaking in tulip.

Tulip breaking virus (TBV)

Basic description

TBV (CMI/AAB 71) has flexuous particles *c.* 740 nm long with a single RNA molecule of M-Wt 3.5×10^6 d. The virus is a good immunogen.

Host plants

TBV naturally infects species of *Tulipa*, *Lilium* and *Fritillaria* (Liliaceae). It can be mechanically sap-transmitted with some difficulty and only between these genera. Transmission can also occur by bulb-grafting. TBV has been reported not to infect dicotyledonous plants. One isolate from *Lilium longiflorum* has been transmitted to *Chenopodium amaranticolor* (Alper *et al.* 1982).

Strains

Severe strains (STBV) and mild strains (MTBV) have been distinguished according to the severity of flower break symptoms in tulips (CMI/AAB 71). Other strains of TBV probably exist in lilies.

Diseases

On tulip (Fig. 13), TBV causes different types of colour-breaks (self, full and average break) depending on the cultivar and on the strain of the virus. Flower breaking is very marked in red and purple cultivars but absent in white, yellow, brown and dark red cultivars. 'Average break' is the most common; the petals show discoloration (white or yellow is produced by full break) in certain areas and in other parts the colour is intensified (self-break). At the time of the first introduction of tulips to Europe in the 16th century, colour-breaking was used for ornamental purposes, but it is nowadays considered as a disease. Infected plants without flower symptoms may show leaf mottling (tulip mosaic), the intensity of discoloration varying with time. Sometimes flower stalks show light green spots. Infected plants usually produce smaller flowers and bulbs than do healthy plants.

On lily (*L. longiflorum*, *L. speciosum*), TBV causes mottle and streak mottle. Brown ring patterns are observed on the bulb scales in some cultivars when TBV is associated with lily symptomless virus (Asjes *et al.* 1973). More or less severe leaf mottling, and streak mottle are also found in relation to this complex. The quality of forced plants and flowers is also affected (malformed flowers, vase life reduced).

Identification and diagnosis

The presence of flower symptoms is a reliable and easy criterion for the detection of diseased tulips. However, visual selection on the basis of leaf symptoms alone is very difficult (e.g. in cultivars not showing flower symptoms or plants too young to flower). Serological detection by IEM or ELISA gives reliable results but some problems have been encountered. Special precautions are necessary for the preparation of TBV antiserum (Derks *et al.* 1982) and pre-incubation of sample extracts with cellulase or hemicellulase is necessary to detect TBV reliably. During the growing season TBV is detected in tulip leaves and bulbs, and during storage it can be detected in primarily infected tulip and lily bulbs (Beijersbergen & van der Hulst 1980).

Geographical distribution

The virus has been reported in all temperate regions where aphid vectors are abundant early in the growing season.

Transmission

TBV is transmitted by several species of aphids in the non-persistent manner. The most efficient is *Myzus persicae*; others are *Aphis gossypii, A. fabae* and *Macrosiphum euphorbiae*.

Economic importance

Tulip breaking virus is one of the most prevalent viruses in tulip bulb production. It causes loss of varietal trueness and is a major component of the general quality of tulip bulbs. Stocks of infected bulbs are less valuable commercially.

Control

Visual field inspections in spring are essential for control of TBV in tulip. It is possible to maintain clean stocks if the level of infection is not too high and if thorough field inspection and eradication are carried out in cultivars which show the flower-break symptom. (Primarily infected tulips do not show current season symptoms.) Totally infected cultivars of tulip that show no flower symptoms (white, yellow) are potential virus reservoirs and must be isolated. Virus-free plants are obtained by tissue or meristem culture. Mineral-oil spray reduces the spread of the virus. Serological detection (ELISA) has allowed commercial stocks to be checked for the presence of TBV during storage, and mass screening programmes for detection in bulbs have been carried out.
ALBOUY

Turnip mosaic virus (TurMV)

Basic description

TurMV has particles *c.* 750×12 nm. As is the case with potyviruses, infection induces intracellular pinwheels, bundles, scrolls and laminated aggregates. TurMV is a good immunogen (CMI/AAB 8).

Host plants

TurMV infects most cultivated brassicas and other cruciferous plants including cabbage (cabbage black ringspot virus, cabbage virus A), cauliflower, Brussels sprouts, Chinese cabbage, oilseed rape, rutabaga, turnip, horseradish (horseradish mosaic virus), and watercress (watercress mosaic virus). Ornamental hosts include anemone, petunia, stock (*Matthiola incana*) and wallflower all of which develop variegated or 'broken' flowers (a diagnostic symptom). In Europe the virus is found commonly in vegetatively propagated rhubarb (Tomlinson & Walkey 1967) and watercress. For experimental purposes, TurMV multiplies well in rutabaga, turnip and Tendergreen mustard. Local lesion hosts include tobacco cv. White Burley, *Chenopodium amaranticolor* and *C. quinoa*.

Strains

Variation in symptoms may be caused by different strains, as in stock, Iceland poppy, cauliflower and rutabaga. Some strains (as found in stock and anemone) fail to infect *Brassica* species. Susceptibility of some crisp-head lettuce cultivars is genetically related to their resistance to *Bremia lactucae*. In rutabaga, symptom severity is determined by strain virulence; symptom type (e.g. systemic necrosis or mosaic reactions) is determined by plant genotype (Tomlinson & Ward 1978; 1982b).

Transmission

The virus is sap-transmissible and is also transmitted by many aphid species, notably *Myzus persicae* and *Brevicoryne brassicae*, both in the non-persistent manner. There are no reports of seed transmission.

Diseases, identification and diagnosis

Diagnosis of infected field-grown brassicas may be difficult as a result of dual infection with TurMV and cauliflower mosaic virus (CaMV). This causes a generalized mottle mosaic which is more severe than that caused by either virus alone. TurMV produces conspicuous chlorotic or necrotic spots and rings, or a mosaic, depending on the species and cultivar. CaMV causes vein clearing of young leaves followed by dark green vein banding. With dual infection, electron microscopy of negatively stained leaf extracts usually reveals the filamentous and spherical particles of TurMV and CaMV respectively (Tomlinson & Ward 1981).

Geographical distribution, economic importance and control

TurMV occurs world-wide and is present in most European countries. Since disease incidence is usually greatest when seed beds are near old infected brassica crops, they should be placed away from overwintered brassicas. If aphids are numerous, spraying seed beds with systemic insecticides may be useful to prevent TurMV spread within the seed bed. Use of resistant or tolerant cultivars is recommended as with white cabbage (*Brassica oleracea* var. *capitata alba*) (Walkey & Neeley 1980). Watercress crops should be grown annually from seed, and new rhubarb plantations should be established from virus-free stocks.

TOMLINSON

Watermelon mosaic virus 2 (WMV2)

Basic description

WMV2 has flexuous filamentous particles *c.* 780 nm long. It shows major differences in serological and host range properties from watermelon mosaic virus 1, a quite separate entity now considered as a strain of papaya ringspot virus (q.v.) (Purcifull & Hiebert 1979). WMV2 is serologically distantly related to zucchini yellow mosaic virus (CMI/AAB 293).

Host plants and diseases

The host range is relatively wide; naturally infected species are mainly Cucurbitaceae, but the virus has been found occasionally in Leguminosae. In cucumber, melon and courgette foliar symptoms are vein banding, mosaic and deformation, and possibly mosaic on fruit. Mosaic is also observed in bean (*Phaseolus vulgaris*) and pea (*Pisum sativum*). Large differences in symptom intensity may occur according to the virus isolate and host genotype.

Identification and diagnosis

The reaction of a limited number of indicator plants may be useful for the identification of the virus (cowpea, tobacco: no reaction; *Lavatera trimestris*, *Chenopodium amaranticolor*: necrotic local lesions; differential cultivars of pea and French bean: susceptible or resistant to the virus). Because of frequent mixed infections in natural conditions a reliable diagnosis from field samples is better obtained using standard serological tests (SDS immunodiffusion, ELISA, IEM).

Geographical distribution, transmission and control

WMV2 is distributed world-wide causing a major disease of cucurbits (Luis Arteaga *et al.* 1976; Russo *et al.* 1979; Lisa & Delavalle 1981). It is non-persistently transmitted by several aphid species (e.g. *Myzus persicae*, *Aphis gossypii*). No seed transmission is reported. No specific control method has been developed. As for other aphid-borne viruses use of plastic mulches or oil sprays may provide temporary protection. High levels of resistance have been reported in cucumber, bean and pea (Schroeder & Provvidenti 1971; Provvidenti 1974).

LECOQ

Wisteria vein mosaic virus is a possible potyvirus causing bright yellow discoloration, vein clearing, and diffuse mottling or spotting on the widely grown ornamental climber, *Wisteria floribunda*. Discoloured leaves may also be malformed (Bos 1970). The virus is reported only in Europe.

Zucchini yellow fleck virus has been reported from Italy and Greece, and causes pinpoint yellow spots on the leaves of courgette which often coalesce to form larger blotches, followed by generalized yellowing and necrosis. Plants are slightly stunted, fruits are malformed and yield may be considerably reduced. Natural host range includes cucumber, melon and watermelon. The experimental host range is limited to the Cucurbitaceae. Transmitted non-persistently by aphids, it is indexed by mechanical inoculation to vegetable marrow (necrotic local lesions followed by necrosis and death of seedlings or, if the plant survives, systemic yellow flecking) and *Lagenaria siceraria* (systemic vein clearing) or by IEM (Vovlas *et al.* 1981).

MARTELLI

Zucchini yellow mosaic virus (ZYMV)

ZYMV (Lisa *et al.* 1981) has flexuous filamentous particles *c.* 750 nm long (CMI/AAB 282). It is identical serologically with muskmelon yellow stunt virus (Lecoq *et al.* 1983).

Host plants and diseases

Naturally infected species belong to the Cucurbitaceae, for example, courgette, melon and cucumber. Symptoms in these hosts are generally very severe but may vary according to the virus isolate or the plant cultivar (Lecoq *et al.* 1981; Lesemann *et al.* 1983). Typical symptoms are: on leaves—vein clearing, yellowing, mosaic, enations and shoestringing; on fruits—deformation, alteration of the pulp (Plate 4), and on seeds—deformation. Plants are usually stunted.

Identification and diagnosis

Similarity of the symptoms induced by some ZYMV isolates and by other viruses infecting cucurbits prevents a reliable diagnosis based only on symptomatology. Mechanical inoculation to a limited host range including *Vigna sinensis, Lavatera trimestris* (no reactions), *Chenopodium amaranticolor, C. quinoa, Gomphrena globosa* (local lesions) allows differentiation of ZYMV from other major cucurbit viruses provided that it is not in complex with other viruses. This latter case being very common in field samples, classical serological techniques (SDS immunodiffusion, IEM, ELISA) provide the most appropriate methods for detection and identification.

Geographical distribution, transmission and economic importance

Initially described in the Mediterranean basin, ZYMV appears now to have a world-wide distribution. It is transmitted by several aphid species (*Myzus persicae, Aphis gossypii, Macrosiphum euphorbiae*) in the non-persistent manner. No seed transmission is reported. Diseases produced in courgette and melon are particularly severe and fruit produced by infected plants is generally unmarketable.

Control

No specific control method has been developed. Some cultural practices intended to delay aphid-borne virus spread (use of plastic mulches, oil sprays) may provide temporary protection. Resistance to the virus is being sought in various cucurbits. ZYMV seems to be highly variable however (several pathotypes have already been described); this may rapidly lead to the development of resistance-breaking variants.

LECOQ

THE TOBAMOVIRUS GROUP

These viruses have elongated rigid particles with one molecule of positive-sense, single-stranded RNA (M-Wt 2×10^6 d) and one coat polypeptide species (17–18 000 d). Subgenomic RNAs have been described; one of them is messenger for the coat protein. The viruses are readily mechanically transmissible. The type member is tobacco mosaic virus (CMI/AAB 184).

Cucumber green mottle mosaic virus (CGMMV) (syn. cucumber virus 2) is widely distributed in cucumbers; it has also been observed in watermelons and melons. Symptoms are mottling and distortion of the leaves, and occasionally of the fruits. It is seed-borne and transmitted by foliage contact, handling of plants and soil contamination. There is no satisfactory differential host, so the virus is better identified by serology or electron microscopy. Control is by sanitary measures to prevent mechanical transmission (disinfection of tools, sticks and hands with Teepol or Na_3PO_4 solutions). The virus is mainly carried externally by the seeds, but dry heat treatment of the seeds (70°C for 3 days) is very efficient in preventing its transmission (CMI/AAB 154).

LECOQ

Odontoglossum ringspot virus (ORSV) (CMI/AAB 155) occurs world-wide in cultivated orchids. Many plants are symptomless when infected; others show brown necrotic ringspot and leaf mottling, or breaking in Cattleya flower. Indexing is by mechanical inoculation, IEM, serological precipitation tests and ELISA. Control is by propagation of healthy plants and severe prophylaxy.

ALBOUY

Potato mop-top virus (PMTV)* probably occurs wherever potatoes are grown, notably in the Andean region of S. and N. America and Europe. It causes extreme stunting (mop-top), yellow, brown or chlorotic chevrons in parts of the foliage of diverse potato genotypes (inconsistently), and brown rings/arcs in the tubers of a few cultivars (CMI/AAB 138). It is transmitted by the powdery scab fungus *Spongospora subterranea*. Indexing is by mechanical inocul-

*Now considered to be a furovirus (cf. Ch. 3).

ation of *Chenopodium amaranticolor* or *Nicotiana debneyi*, or by IEM. The vector can be controlled with Zn salts or by diminishing soil pH with sulphur (Cooper *et al.* 1976).
COOPER

Tobacco mosaic virus (TMV)

Basic description

TMV (type strain) is the type member of the tobamovirus group. The virion appears as a rigid tubular particle *c.* 18×300 nm with a sedimentation coefficient of *c.* 149S. TMV is a good immunogen (CMI/AAB 151; CMI/AAB 156).

Host plants

TMV can be transmitted to numerous species in 30 angiosperm families. It is only important economically in the Solanaceae; it can naturally infect tobacco, tomato, pepper, rarely aubergine and wild Solanaceae.

Diseases

Severe symptoms are observed on tobacco, tomato, pepper and sometimes on petunia (Marchoux *et al.* 1983). In tobacco, the first signs of infection are vein clearing, followed by vein banding and mosaic. The infected leaves may also show raised green blisters and sunken yellow areas. In the case of severe early infection, plants are usually stunted and the lamina of the leaf is considerably reduced, giving a filiform appearance.

In tomato, symptoms vary in intensity according to the strain, cultivar, age of plant and climatic conditions. The disease is more serious on protected crops than on plants grown in the open. Mottled areas of light and dark green on the leaves are the most common reaction in summer. In winter, with low light intensity, short days and temperatures below 20°C, plants are often stunted and leaves distorted to 'fern-leaf' with slight mottling. Some strains may cause a striking yellow mottling (aucuba mosaic). Ordinarily, fruits do not show marked symptoms though they may be reduced in number and size. Internal necrosis or 'internal browning' or 'bronzing' in fruits results from infection with certain tomato mosaic strains when the fruits are well-developed but unripe.

Pepper infected by TMV shows greenish-yellow mottling on the young leaves which are slightly curled and irregular in shape. Certain tomato mosaic strains cause a brown streak on the stem and branches, followed by severe leaf necrosis and abscission. Severely affected plants are stunted and fruits are often small, mottled and wrinkled.

Strains

Natural isolates of TMV fall into two main groups: tobacco mosaic virus (type strain) (CMI/AAB 151) and tomato mosaic virus strains (CMI/AAB 156). Tobacco strains principally infect tobacco, and rarely, or with difficulty, tomato. The U1-strain is the type strain used throughout the world. The U2-strain differs from U1-strain in biological, serological and chemical properties. Recently 'pepper unusual strains' belonging to the tobacco strains have been reported from several countries in Europe (Pop 1979; Rast 1979; Gébré-Sélassié *et al.* 1981). A classification based on pepper genotypes is proposed by Boukema (1980). Tomato strains have been classified by their ability to induce symptoms in plants of certain isogenic lines of tomato possessing resistance genes Tm-1, Tm-2 or Tm-2^2. The European system of identification is the most widely used at present (Pelham 1972).

Identification and diagnosis

The virus cannot be identified unequivocally on the basis of visual symptoms on crops. *Datura stramonium, Nicotiana glutinosa* and *N. tabacum* with the hypersensitivity N gene are good diagnostic hosts. Several *Nicotiana* species have been used to distinguish tomato strains from tobacco strains. *N. sylvestris* gives local lesions with tomato strain and systemic disease with tobacco strains. In *N. tabacum* cv. White Burley or Paraguay and in *N. rustica*, most tomato strains induce necrotic local lesions, and not the systemic invasion due to tobacco strains. Electron microscopy and serological techniques are the most rapid, specific and sensitive methods of diagnosis. Serotypes may easily be revealed using immunodiffusion and double antibody sandwich ELISA. Indirect ELISA is better for detection work (Van Regenmortel 1981).

Geographical distribution

TMV occurs world-wide and is common in countries where tomato, pepper or tobacco are grown in the glasshouse and the field.

Transmission

TMV is one of the most infectious and persistent disease agents; it is transmitted and readily spread between plants by contact, and by man and contaminated implements during cultural operations. It can survive for years in plant debris which can serve as sources of infection via the roots. Tomato and pepper strains are transmissible by seed (up to 94%). The virus is carried in the external mucilagenous seed surface, testa, and sometimes in the endosperm, but has not been proved to be present within the embryo. Seed samples with endosperm infection remained infected for at least 9 years.

Economic importance

The incidence of TMV in Solanaceous crops may fluctuate from season to season but severe attacks may cause heavy losses, partly in reducing yield and partly in lowering quality.

Control

Different methods of control have been adequately reviewed by Broadbent (1976). The following practices are efficient enough in most cases: i) use of fresh soil that has not grown tomatoes or peppers, or of steam-disinfected soil for seed beds; ii) washing hands thoroughly with soft soap and running water before and after handling seedlings, and iii) reduction of seed contamination by soaking pepper seeds in 10% (w/v) Na_3PO_4 solutions for 20 minutes, and by heat treatment for 2–4 days at 70°C of tomato seeds, without impairing the rate of germination. Fermentation and acetic acid treatment techniques have also been used successfully.

A nitrous acid mutant (M 11–16) of low pathogenicity has been widely used for protective inoculation of tomato seedlings in European glasshouses (Marrou & Migliori 1971; Rast 1972), but the demand for premunition of tomato plants has decreased considerably due to the development and cultivation of new resistant cultivars. Much breeding work has been done in different countries now that resistance genes have been discovered. Promising resistant cultivars are now available for several different crops. However, some resistant genes, notably in peppers, have recently been overcome by new strains.

MARCHOUX & GÉBRÉ-SÉLASSIÉ

THE VIROID GROUP

Viroids consist of a naked molecule of single-stranded infectious RNA of *c.* 0.12×10^6 d. This RNA appears as linear covalently closed circular molecules. About 15 viroids have been described, up to now only in plants. Several have been sequenced and sequence homologies exist between them. Due to a high percentage of complementarity, the molecule has a highly base-paired structure. Viroids very probably depend completely on cell mechanisms for their replication; their transcription seems to involve the nuclear DNA-dependent RNA-polymerase II which is able to use viroid RNA as a template because of its secondary structure. Transmission occurs in nature mechanically by infected tools and by seed. The type member, described in 1971 as the first viroid, is potato spindle tuber viroid. It is also the type member of one of three viroid groups known so far. The other two correspond respectively to the cadang cadang viroid and the avocado sun blotch viroid (q.v.), not yet known to occur in Europe.

Apple dapple apple disease is probably due to a viroid (Koganezawa *et al.* 1982). Discoloured greenish areas develop after mid-July; dappling becomes intense and especially visible at maturity in red cultivars. Scar skin disease of apple is probably due to the same agent. Scar tissue develops in the skin at the time of pigmentation; at maturity 50% of the skin may present this corky texture and affected fruits remain small and starchy. The disease was described in N. America and is occasionally reported in Europe. Crops of very susceptible cultivars (Red Delicious, McIntosh, Starking, for example) may be worthless when infected. No vector is known. Indexing is by graft transmission to Red Delicious or Starkrimson.

Avacado sun blotch disease was first described in California, and more recently in Australia and the Near East (Thomas & Mohamed 1979). Symptoms of sun blotch involve the fruit (modification of the colour, longitudinal grooves or depressed streaks), young stems, branches and trunk (bark streak, spotty markings) (CMI/AAB 254). Infection is often symptomless. Very probably due to a viroid, the disease is readily transmitted by mechanical means (cutting tools) and grafting; it is also seed and pollen-borne. It can be detected in symptomless trees by graft transmission to avocado seedlings, polyacryl-

amide gel electrophoresis and molecular hybridization using a specific complementary DNA (Palukaitis *et al.* 1981). The cultivation of avocado is beginning to develop in S. Europe and the possible presence of this damaging viroid needs to be carefully checked.
MONSION & DUNEZ

Chrysanthemum chlorotic mottle disease has a restricted area, having been reported in the USA and Denmark. It is probably caused by a viroid, whose presence is associated with rather mild symptoms (leaf mottle, slight reduction of the leaf and flower size and occasionally a mild stunting of the whole plant). Most cultivars are symptomless when infected. The viroid can be transmitted readily by grafting. Presumably most spread of the disease occurs through infected propagating material. Usual detection is by indexing on susceptible cultivars; the most frequently used are Deep Ridge and Yellow Delaware, exhibiting leaf mottle and yellowing which can extend to the whole plant (Niblett *et al.* 1980). Symptoms are particularly intense when the plants are grown at a temperature of 28–30°C and 15 000 lux illumination. Indexing techniques used for other viroids (electrophoresis and molecular hybridization) are not yet applicable.
MONSION & DUNEZ

Chrysanthemum stunt viroid

Basic description

Chrysanthemum stunt was described in 1947 and demonstrated to be caused by a viroid by Diener and Lawson in 1973. It belongs to the same viroid group as potato spindle tuber viroid, sharing 60% sequence homology.

Host plants and diseases

The only naturally infected host is *Chrysanthemum morifolium*; more than half the commercial cultivars show more or less severe symptoms such as growth reduction (as much as 85%), premature blossoming (from a few days to 4 weeks), colour breaking and reduction in size of the flowers (Fig. 14); stems become brittle and root initiation from infected cuttings is consistently reduced.

Identification and diagnosis

In view of non-specific or weak symptoms and the existence of symptomless infected cultivars, indexing is essential. Three techniques can be used: i) grafting (chip-budding) onto a susceptible cultivar (Mistletoe or Bonnie Jean). Systemically infected leaves develop chlorotic spots within 4–8 weeks and symptoms are especially clear if indexing is carried out at 28–30°C; ii) electrophoretic assay is a sensitive technique allowing viroid detection within 2 days from less than 10 mg infected leaves (Monsion *et al.* 1981), and iii) molecular hybridization using a cDNA labelled probe, a technique which has been shown to be a highly sensitive tool for routine detection of several viroids, especially potato spindle tuber viroid. It is also applicable to chrysanthemum stunt viroid (Palukaitis & Symons 1979).

Fig. 14. Chrysanthemum stunt viroid: flower breaking and deformation in chrysanthemum (healthy flower on left).

Transmission

The disease is propagated by infected cuttings. In addition, it is spread mechanically on infected tools during cultural operations, and is seed transmitted (5% or less) (Monsion *et al.* 1973).

Geographical distribution and economic importance

Occurring world-wide, the disease is prevalent in many countries especially in N. America and in Europe. Susceptible cultivars are unmarketable when infected. Because it is so likely to move with internationally traded propagating material, it is regarded as a quarantine organism within Europe (EPPO 92).

Control

Elimination of infected plants, build-up of virus-free stocks and propagation of healthy clones have greatly

reduced incidence of the disease in countries where it became well established. In addition, suitable precautions should be taken during cultural operations. Infected cultivars can be cured by meristem-tip culture, but the percentage of viroid-free plants is low.
MONSION & DUNEZ

Citrus exocortis viroid

Basic description

Citrus exocortis viroid is a naked, single-stranded, covalently closed circular RNA of 371 nucleotide residues with a known sequence (Gross *et al.* 1982). Maximal base-pairing results in an extended rod-like secondary structure. There is 60–73% sequence homology between citrus exocortis viroid, potato spindle tuber viroid and chrysanthemum stunt viroid.

Host plants

According to Weathers (1981), the following plants are susceptible to the disease—*Poncirus trifoliata*, Etrog citron, sweet limes, Rangpur lime and other mandarin limes, citranges, *Gynura aurantiaca*, *Petunia hybrida* and tomato. All *Citrus* species seem to be susceptible to stunting. As to tolerant plants (symptomless carriers), many *Citrus* species tolerate citrus exocortis viroid in that they do not develop bark scaling. Such tolerant (non-scaling) plants are sweet orange, sour orange, rough lemon and grapefruit.

Diseases

With *P. trifoliata*, some citranges and certain mandarin limes, whether used as seedlings or rootstocks citrus exocortis viroid causes stunting (often considerable), bark cracking and scaling (scaly butt), yellow blotching of twigs. On cuban shaddock, some lemons and sweet limes cracking of bark and stunting occurs; on leaves of Harvey lemon, certain clones of Etrog citron (see indexing), gynura, petunia and other herbaceous hosts dwarfing, midrib underside cracking, stem and leaf epinasty, yellow blotching of stems, corky lesions, and vertical cracks on stems all occur (CMI/AAB 226).

Identification and diagnosis

Field diagnosis is based on bark cracking and scaling in *P. trifoliata*, citranges and mandarin limes and on bark cracking of sweet limes and some lemons. Psorosis scaly bark is similar in some respects to exocortis bark scaling but occurs on sweet orange; citrus exocortis viroid does not cause bark scaling on sweet orange.

As many commercial citrus rootstock-scion combinations such as sweet orange, mandarin, clementine, tangelo, or grapefruit grafted on sweet orange, sour orange, rough lemon or tangelo are tolerant to citrus exocortis bark scaling and cracking, indexing is required with all such combinations. Indicator species are Etrog citron clones such as 60–13, 861 or 861 S. The Etrog citron bud is grafted on a fast-growing seedling such as *Citrus volkameriana* or rough lemon. One bark patch of the tree to be indexed is grafted above the Etrog bud, the other below. Only the Etrog bud is forced and allowed to grow. If the inoculum from the candidate tree contained the viroid the Etrog shoot should show characteristic symptoms within 1–3 months, e.g. leaf epinasty and cracking of the lower side of the leaf midrib. Additional symptoms are stem epinasty, yellow blotching of the stem, small corky lesions and vertical cracking of the stem.

Geographical distribution and economic importance

Citrus exocortis viroid occurs world-wide; it is a destructive disease when trees are grafted onto susceptible, bark-scaling rootstocks such as *P. trifoliata* and citranges (Plate 5). With tolerant rootstocks, various degrees of stunting can be observed.

Transmission and control

Natural transmission is by budwood or grafts. The viroid is mechanically transmissible by cutting tools in the field and, in the laboratory, by slashing the bark of healthy plants with knives previously dipped in infective extracts or slashed in the bark of infected Etrog citron plants. For control, use exocortis-free budwood and disinfect tools between cuts into different plants by dipping in 10–15% NaOCl solution.
BOVÉ

Hop stunt viroid is of importance in Europe only in causing a disease of cucumber known as pale fruit, reported to occur naturally in the Netherlands only in glasshouses (Van Dorst & Peters 1974). The most distinctive symptoms are found on the fruits which are pale green in colour, retarded in growth and mostly slightly pear-shaped. Developing leaves may be smaller, blue-green and rugose. On ageing, chlorosis appears and the plant may be somewhat stunted. Spread of the disease in glasshouses occurs mechanically through pruning and harvesting. No vector is known. *Benincasa hispida* is a test plant, its incubation period is short and the symptoms are the most pronounced. On hops in Japan, infection causes reduced growth and cone size. The content of α-acid is reduced to 30–50% of that of viroid-free cones (Momma & Takahashi 1984).

MONSION & DUNEZ

Potato spindle tuber viroid

Basic description

Potato spindle tuber viroid is a 359 nucleotide long, covalently closed circular RNA molecule of M-Wt 120 000 d. (Sänger *et al.* 1976; Sänger 1982). Known as a serious potato disease in the USA since 1922 its original nature was first discovered by Diener & Raymer (1967). It was the first viroid described. On the basis of nucleotide sequencing it belongs to the same group as citrus exocortis viroid and chrysanthemum stunt viroid (CMI/AAB 66).

Host plants and diseases

Though potato spindle tuber viroid can infect (usually symptomlessly) many plant species in the Solanaceae and other families it is naturally found only in potato. Above-ground symptoms on potato are usually subtle and consist of leaflets of smaller size with fluted margins. Angles between stem and petioles are more acute than normal. Tubers are elongated and in some cultivars have pointed ends and a round cross-section (Plate 6). Tuber eyes are more prominent and appear to be more numerous. Several strains are known; they induce more or less severe symptoms and differ by 3 or 4 nucleotides.

Identification and diagnosis

Potato spindle tuber viroid is symptomless in many modern cultivars, so that a diagnostic test has to be used. Inoculation of infective sap onto tomatoes, preferably cv. Rutgers, causes new leaflets to show rugosity, epinasty, down curling and bunchy top, depending on viroid strain (Fernow 1967). Electrophoresis is sensitive for detecting the viroid (Morris & Wright 1975) if infected plants are grown at temperatures above 25°C and nucleic acid bands in polyacrylamide gels are stained with ethidium bromide or silver nitrate. The nucleic acid hybridization spot test using radioactive cDNA is highly sensitive, allowing detection of minute amounts of the viroid in foliage, tuber and seed extracts (Salazar *et al.* 1983).

Transmission

Potato spindle tuber viroid is mainly transmitted mechanically and to a lesser extent by chewing insects. It may also be transmitted by pollen and true seed of potato and other hosts.

Geographical distribution and economic importance

Potato spindle tuber viroid is reliably reported from N. America (USA, Canada) and E. Europe (Poland, USSR, Turkey). It may occur more widely, but is absent from the potato-growing areas of W. Europe where it is considered an important quarantine organism (EPPO 97). It causes marked reductions in yield on some susceptible cultivars giving fewer, smaller tubers.

Control

Use of viroid-free tubers and avoiding mechanical transmission by planting whole (uncut) seed tubers reduces the possibility of dissemination. Eradication of the viroid by low temperature treatment and meristem culture may be used to obtain viroid-free plants from universally infected cultivars (Lizárraga *et al.* 1980). In W. Europe, post-entry quarantine measures are imposed against this viroid.

SALAZAR

REFERENCES

Aapola A.A. & Mink G.I. (1973) Potential aphid vectors of pea seed-mosaic virus in Washington. *Plant Disease Reporter* **57**, 552.

A'Brook J. & Dewar A.M. (1980) BYDV infectivity of alate aphid vectors in West Wales. *Annals of Applied Biology* **96**, 51–58.

Adams A.N. & Barbara D.J. (1980) Host range, purification and some properties of hop mosaic virus. *Annals of Applied Biology* **96**, 201–208.

Adams A.N. & Barbara D.J. (1982) Host range, purification and some properties of two carlaviruses from hop (*Humulus lupulus*): hop latent and American hop latent. *Annals of Applied Biology* **101**, 483–494.

Alper M., Koenig R., Lesemann D.E. & Loebenstein G. (1982) Mechanical transmission of a strain of tulip breaking virus from *Lilium longiflorum* to *Chenopodium* spp. *Phytoparasitica* **10**, 193–199.

Asjes C.J., De Vos Neeltje P. & Van Slogteren D.H.M. (1973) Brown ring formation and streak mottle, two distinct syndromes in lilies associated with complex infections of lily symptomless virus and tulip breaking virus. *Netherlands Journal of Plant Pathology* **79**, 23–35.

Atkinson M.A. & Cooper J.I. (1976) Ultrastructural changes in leaf cells of poplar naturally infected with poplar mosaic virus. *Annals of Applied Biology* **83**, 395–398.

Azeri T. (1973) First report of satsuma dwarf virus disease on satsuma mandarins in Turkey. *Plant Disease Reporter* **57**, 149.

Bar-Joseph M. & Loebenstein G. (1973) Effects of strain, source plant, and temperature on the transmissibility of citrus tristeza virus by the melon aphid. *Phytopathology* **63**, 716–720.

Bar-Joseph M., Garnsey S.M. & Gonsalves D. (1979) The clostero-viruses: a distinct group of elongated plant viruses. *Advances in Virus Research* **25**, 93–168.

Barlass M., Skene K.G.M., Woodham R.C. & Krake L.R. (1982) Regeneration of virus-free grapevines using *in vitro* apical culture. *Annals of Applied Biology* **101**, 291–295.

Bass P. & Vuittenez A. (1977) Amélioration de la thermothérapie des vignes virosées au moyen de la culture d'apex sur milieux nutritifs ou par greffage sur vignes de semis cultivées aseptiquement *in vitro*. *Annales de Phytopathologie* **9**, 539–540.

Beemster A.B.R. & Rozendahl A. (1972) Potato viruses: properties and symptoms. In *Viruses of Potatoes and Seed-Potato Production* (Ed. J.A. de Bokx), pp. 121–124. PUDOC, Wageningen.

Beijersbergen J.C.M. & van der Hulst C.T.C. (1980a) Detection of lily symptomless virus (LSV) in bulb tissue of lily by means of ELISA. *Acta Horticulturae* **109**, 487–493.

Beijersbergen J.C.M. & van der Hulst C.T.C. (1980b) Application of enzymes during bulb tissue extraction for detection of lily symptomless virus by ELISA in *Lilium* spp. *Netherlands Journal of Plant Pathology* **86**, 277–283.

Bennett C.W. & Costa A.S. (1949) Tristeza disease of citrus. *Journal of Agricultural Research* **78**, 207–237.

Bercks R. (1960) Serological relationships between beet mosaic virus, potato virus Y, and bean yellow mosaic virus. *Virology* **12**, 311–313.

Bertaccini A. & Marani F. (1983) Virosi dell'asparago. *L'Italia Agricola* **1**, 24–26.

Bertaccini A., Marani F. & Vischi M. (1982) Trasmissione per seme dell'asparago virus 2. *La Difesa delle Piante* no. 5/6, 415–420.

Björling K. (1969) Virus strains in relation to breeding for resistance to virus diseases of beet. *I.I.R.B.* **4**, 63–72.

Boccardo G. & Milne R.G. (1976) Poplar mosaic virus: electron microscopy and polyacrylamide gel analysis. *Phytopathologische Zeitschrift* **87**, 120–131.

de Bokx J.A., Piron P.G.M. & Maat D.Z. (1980) Detection of potato virus X in tubers with the enzyme linked immunosorbent assay (ELISA). *Potato Research* **23**, 129–131.

Bos L. (1969) Inclusion bodies of bean yellow mosaic virus, some less known closely related viruses and beet mosaic virus. *Netherlands Journal of Plant Pathology* **75**, 137–143.

Bos L. (1970) Identification of three new viruses isolated from *Wisteria* and *Pisum* in the Netherlands and the problems of variation in the potato virus Y group. *Netherlands Journal of Plant Pathology* **76**, 8–46.

Bos L. (1982) Viruses and virus diseases of *Allium* species. *Acta Horticulturae* **127**, 11–25.

Bos L., Kowalska C.Z. & Maat D.Z. (1974) The identification of bean mosaic, pea yellow mosaic and pea necrosis strains of bean yellow mosaic virus. *Netherlands Journal of Plant Pathology* **80**, 173–191.

Bos L., Huijberts N., Huttinga H. & Maat D.Z. (1978) Leek yellow stripe virus and its relation to onion yellow dwarf virus: characterization, ecology and possible control. *Netherlands Journal of Plant Pathology* **84**, 185–204.

Boukema I.W. (1980) Allelism of genes controlling resistance to TMV in *Capsicum*. *Euphytica* **29**, 433–439.

Boulton R.E. & Catherall P.L. (1980) Inheritance and effectiveness of genes in barley that condition tolerance to BYDV. *Annals of Applied Biology* **65**, 153–161.

Bouquet A. (1980) *Vitis*×*Muscadinia* hybridization: a new way in grape breeding for disease resistance in France. In *Proceedings of the 2nd International Symposium on Grapevine Breeding*, pp. 46–61. University of California, Davis.

Bové J.M. & Vogel R. (1980) *Description and Illustration of Virus and Virus-like Diseases of Citrus*. SETCO-IRFA, Paris.

Bovey R. (1982) Control of virus and virus-like diseases of grapevine: sanitary selection and certification, heat therapy, soil fumigation and performance of virus-tested material. In *Proceedings of the 7th Meeting of the International Council for the Study of Viruses and Virus-like Diseases of the Grapevine*, pp. 299–309. Niagara Falls, Ontario.

Bovey R., Brugger J.J. & Gugerli P. (1982) Detection of fanleaf virus in grapevine tissue extracts by enzyme-linked immunosorbent assay (ELISA) and immune electron microscopy (IEM). In *Proceedings of the 7th Meeting of the International Council for the Study of Viruses and Virus-like Diseases of the Grapevine*, pp. 259–275. Niagara Falls, Ontario.

Brants D. & Vermeulen H. (1965) Production of virus-free Freesias by means of meristem culture. *Netherlands Journal of Plant Pathology* **71**, 25–27.

Brčák J. & Polák J. (1976) Aphid transmission of beet yellows virus affected by homologous antiserum. *Biologia Plantarum* **18**, 88–92.

Broadbent L. (1976) Epidemiology and control of tomato mosaic virus. *Annual Review of Phytopathology* **14**, 75–96.

Brunt A.A. (1966) Narcissus mosaic virus. *Annals of Applied Biology* **58**, 13–23.

Brunt A.A. (1971) Occurrence and importance of viruses infecting narcissus in Britain. *Acta Horticulturae* **23**, 292–299.

Cadilhac B., Quiot J.B., Marrou J. & Leroux J.P. (1976)

Mise en évidence au microscope électronique de deux virus différents infectant l'ail et l'échalotte. *Annales de Phytopathologie* **8,** 65–72.

Cardin L. & Devergne J.C. (1975) Occurrence on rose of a virus belonging to prunus necrotic ringspot group (NRSV). Indexing methods on indicators. *Phytopathologia Mediterranea* **14,** 106–112.

Chairez R. & Lister R.M. (1973) Soluble antigens associated with infections with apple chlorotic leaf spot. *Virology* **54,** 506–514.

Chevallier D., Engel A., Wurtz M. & Putz C. (1983) The structure and characterization of a closterovirus, beet yellows virus and a luteovirus, beet mild yellowing virus, by scanning transmission electron microscopy, optical diffraction of electron images and acrylamide gel electrophoresis. *Journal of General Virology* **64,** 2289–2293.

Chod J. (1965) Studies of some ways which carrot mosaic virus can be transmitted. *Biologia Plantarum* **6,** 463–468.

Chrzanowska M. (1976) Symptoms caused by PVS on the potato plants grown in the greenhouse. *Biuletyn Instytutu Ziemniaka* **18,** 65–69.

Cisar G., Brown C.M. & Jedunski H. (1982) Effect of fall or spring infection and sources of tolerance of BYDV of winter wheat. *Crop Science* **22,** 474–477.

Cohn E., Tanne E. & Nitzany F. (1970) *Xiphinema italiae*, a new vector of grapevine fanleaf virus. *Phytopathology* **60,** 181–182.

Cooper J.I. (1979) *Virus Diseases of Trees and Shrubs.* Institute of Terrestrial Ecology, Cambridge.

Cooper J.I. & Edwards M.L. (1981) The distribution of poplar mosaic virus in hybrid poplars and virus detection by ELISA. *Annals of Applied Biology* **99,** 53–61.

Cooper J.I., Jones R.A.C. & Harrison B.D. (1976) Field and glasshouse experiments on the control of potato mop-top virus. *Annals of Applied Biology* **83,** 215–230.

Corbett M.K. (1958) A virus disease of lupins caused by bean yellow mosaic virus. *Phytopathology* **48,** 86–91.

Costa A.S. & Lima Neto V. da C. (1976) Transmissão do virus da necrose branca do fumo por *Frankliniella* sp. *Fitopatologia* **11,** 1–35.

Costa A.S. & Muller G.W. (1980) Tristeza control by cross protection. *Plant Disease Reporter* **64,** 538–541.

Costa A.S. & Tasaka H. (1971) Enfezamento e malformação foliar da alcachofra induzidos pelo virus da necrose branca do fumo. *Biologico* **37,** 176–179.

Dalmasso A. & Vuittenez A. (1977) Problèmes de replantation de la vigne et désinfection du sol dans les pays tempérés. *Bulletin de l'O.I.V.* no. 567, 337–351.

Davies D.L. & Clark M.F. (1983) A satellite-like nucleic acid of arabis mosaic virus associated with hop nettlehead disease. *Annals of Applied Biology* **103,** 439–448.

Dědič P. (1975) Effect of virus A on yield in some varieties of potato. *Ochrona Rostlin* **11,** 127–133.

Delbos R., Bonnet A. & Dunez J. (1984) Le virus de l'enroulement des feuilles du cerisier, largement répandu en France sur noyer, est-il à l'origine de l'incompatibilité de greffage du noyer *Juglans regia* sur *J. nigra? Agronomie* **4,** 333–339.

Delécolle B. & Lot H. (1981) Viroses de l'ail. I. Mise en évidence et essais de caractérisation par immuno-électromicroscopie d'un complexe de trois virus chez différentes populations d'ail atteintes de mosaïque. *Agronomie* **1,** 763–770.

Denéchère M., Cante F. & Lapierre H. (1979) Détection immunoenzymatique du virus de la jaunisse nanisante de l'orge dans son vecteur *Rhopalosiphum padi. Annales de Phytopathologie* **11,** 507–514.

Derks A.F.L.M. & van den Abeele J.L. (1980) Bean yellow mosaic virus in some iridaceous plants. *Acta Horticulturae* **110,** 31–38.

Derks A.F.L.M., van den Abeele J.L. & van Schadewijk A.R. (1982) Purification of tulip breaking virus and production of antisera for use in ELISA. *Netherlands Journal of Plant Pathology* **88,** 87–98.

Desjardins P.R. (1980) Infectious variegation-crinkly leaf. In *Description and Illustration of Virus and Virus-like Diseases of Citrus* (Eds J. Bové & R. Vogel). SETCO-IRFA, Paris.

Detienne G., Delbos R. & Dunez J. (1980) Use and versatility of the immunoenzymatic ELISA procedure in the detection of different strains of apple chlorotic leaf spot virus. *Acta Phytopathologica Academiae Scientiarum Hungaricae* **15,** 39–45.

Deutsch M. & Loebenstein G. (1967) Field experiments with oil sprays to prevent yellow mosaic virus in irises. *Plant Disease Reporter* **51,** 318–319.

Devergne J.C. & Cardin L. (1971) Evolution des symptômes de streak dans les plantations d'oeillet. *Annales de Phytopathologie* **3,** 539.

Devergne J.C. & Cardin L. (1974) Contribution à l'étude du virus de la Mosaïque du Concombre. IV. Essai de classification de plusieurs isolats sur la base de leur structure antigénique. *Annales de Phytopathologie* **5,** 409–430.

Devergne J.C. & Cardin L. (1975) Relation sérologique entre cucumovirus (CMV, TAV, PSV). *Annales de Phytopathologie* **7,** 255–276.

Devergne J.C., Cardin L., Burckard J. & Van Regenmortel M.H.V. (1981) Comparison of direct and indirect ELISA for detecting antigenically related cucumoviruses. *Journal of Virological Methods* **3,** 193–200.

Dias H.F. & Harrison B.D. (1963) The relationship between grapevine fanleaf, grapevine yellow mosaic and arabis mosaic viruses. *Annals of Applied Biology* **51,** 97–105.

Diener T.O. & Lawson R.H. (1973) Chrysanthemum stunt: a viroid disease. *Virology* **51,** 94–101.

Diener T.O. & Raymer W.B. (1967) Potato spindle tuber virus. A virus with properties of a free nucleic acid. *Science* **158,** 378–381.

Dodds J.A. & Bar-Joseph M. (1983) Double stranded RNA from plants infected with closteroviruses. *Phytopathology* **73,** 419–423.

Douine L. & Devergne J.C. (1978) Isolement en France du virus du rabougrissement de l'arachide (peanut stunt virus). *Annales de Phytopathologie* **10,** 79–92.

Douine L., Quiot J.B., Marchoux G. & Archange P. (1979) Recensement des espèces végétales sensibles au virus de la mosaïque du concombre (CMV). Etudes bibliographiques. *Annales de Phytopathologie* **11,** 439–475.

Drijfhout E. (1978) *Genetic Interaction between* Phaseolus

vulgaris *and Bean Common Mosaic Virus with Implications for Strain Identification and Breeding for Resistance*. PUDOC, Wageningen.

Duffus J.E. (1965) Beet pseudo-yellows virus, transmitted by the greenhouse whitefly (*Trialeurodes vaporariorum*). *Phytopathology* **55**, 450–453.

Duffus J.E. (1972) Beet yellow stunt, a potentially destructive virus disease of sugar beet and lettuce. *Phytopathology* **62**, 161–165.

Duffus J.E. (1981) Beet western yellows virus—a major component of some potato leaf roll-affected plants. *Phytopathology* **71**, 193–196.

Duffus J.E. & Russell G.E. (1975) Serological relationships between beet western yellows virus and beet mild yellowing virus. *Phytopathology* **65**, 811–815.

Edwardson J.R. (1974) Host ranges of viruses in the PVY-group. In *Monograph Series Florida Agricultural Experiment Stations* no. 5, 94.

Falk B.W. & Purcifull D.E. (1983) Development and application of an enzyme-linked immunosorbent assay (ELISA) to index lettuce seeds for lettuce mosaic virus in Florida. *Plant Disease* **67**, 413–416.

Fernow K.H. (1967) Tomato as a test plant for detecting mild strains of potato spindle tuber virus. *Phytopathology* **57**, 1347–1352.

Frazier N.W. (1970) *Virus Diseases of Small Fruits and Grapevines*. University of California, Berkeley.

Fritzsche R. & Kegler H. (1968) Nematoden als Vektoren von Viruskrankheiten der Obstgewächse. *Tagungsberichte* **97**, 289–296.

Fritzsche R., Proeseler G., Karl E. & Kleinhempel H. (1983) Nachweis der Samenübertragbarkeit des milden Rübenvergilbungs-Virus. *Phytopathologische Zeitschrift* **106**, 360–364.

Frowd J.A. & Bernier C.C. (1977) Virus diseases of faba beans in Manitoba and their effects on plant growth and yield. *Canadian Journal of Plant Science* **57**, 845–852.

Fuchs E. (1976) Der Nachweis des nekrotischen Rübenvergilbungs-Virus (NRVV) mit dem Latextest. *Archiv für Phytopathologie und Pflanzenschutz* **12**, 273–274.

Fujisawa I., Goto T., Tsuchizaki T. & Iizuka N. (1983) Host range and some properties of asparagus virus 1 isolated from *Asparagus officinalis* in Japan. *Annals of the Phytopathological Society of Japan* **49**, 299–307.

Fulton R.W. (1952) Mechanical transmission and properties of rose mosaic virus. *Phytopathology* **42**, 413–416.

Fulton R.W. (1972) Inheritance and recombination of strain-specific characters in tobacco streak virus. *Virology* **50**, 810–820.

Fulton R.W. (1981) Ilarviruses. In *Handbook of Plant Virus Infections and Comparative Diagnosis* (Ed. E. Kurstak), pp. 377–414. Elsevier, Amsterdam.

Garnsey S.M., Baksh N. & Davino M. (1983) A mild isolate of citrus variegation virus found in Florida citrus. In *Proceedings of the 9th Conference of the International Organization of Citrus Virologists*.

Garnsey S.M., Bar-Joseph M. & Lee R.F. (1981) Applications of serological indexing to develop control strategies for citrus tristeza virus. In *Proceedings of the International Society for Citriculture*.

Gébré-Sélassié K., Dumas de Vaulx R., Marchoux G. & Pochard E. (1981) Le virus de la mosaïque du tabac chez le piment. I. Apparition en France du pathotype P 1–2. *Agronomie* **10**, 853–858.

Gildow F.E. & Rochow W.F. (1980) Role of accessory salivary glands in aphid transmission of barley yellow dwarf virus. *Virology* **104**, 97–108.

Gildow F.E., Ballinger M.E. & Rochow W.F. (1983) Identification of double-stranded RNAs associated with barley yellow dwarf virus infection of oats. *Phytopathology* **73**, 1570–1572.

Gilmer R.M., Nyland G. & Moore J.D. (1976) Prune dwarf. In *Virus Diseases and Non-infectious Disorders of Stone Fruits in North America*. Agriculture Handbook no. 437, pp. 179–190. USDA, Washington.

Goheen A.C., Luhn C.F. & Hewitt W.B. (1965) Inactivation of grapevine viruses *in vivo*. In *Proceedings of the International Conference on Virus Vectors on Vitis*, pp. 255–265. University of California, Davis.

Gonsalves D. & Garnsey S.M. (1975) Nucleic acid components of citrus variegation virus and their activation by coat protein. *Virology* **64**, 23–31.

Gonsalves D., Purcifull D.E. & Garnsey S.M. (1978) Purification and serology of citrus tristeza virus. *Phytopathology* **68**, 553–559.

Gonzalez D. & Rawlins W.A. (1969) Relation of aphid populations to field spread to lettuce mosaic virus in New York. *Journal of Economic Entomology* **62**, 1109–1114.

Goth R.W. & Webb R.E. (1975) Lack of potato virus S transmission via true seed in *Solanum tuberosum*. *Phytopathology* **65**, 1347–1351.

Gracia O. & Feldman J.M. (1974) Tobacco streak virus in pepper. *Phytopathologische Zeitschrift* **80**, 313–323.

Graichen K. (1978) Untersuchungen zum Wirtspflanzenkreis des Porreegelbstreifen virus (LYSV). *Archiv für Phytopathologie und Pflanzenschutz* **14**, 1–6.

Gross H.J., Krupp G., Domdey H., Raba M., Jank P., Lossow C., Alberty H., Ramm K. & Sanger L. (1982) Nucleotide sequence and secondary structure of citrus exocortis and chrysanthemum stunt viroid. *European Journal of Biochemistry* **121**, 249–257.

Grüntzig M. (1983) Der serologische Nachweis des beet mosaic virus. *Archiv für Phytopathologie und Pflanzenschutz* **19**, 61–63.

Grüntzig M. & Fuchs E. (1979a) Untersuchungen zur serologischen Verwandtschaft von beet mosaic virus, potato virus Y, bean yellow mosaic virus und plum pox virus. *Archiv für Phytopathologie und Pflanzenschutz* **15**, 153–159.

Grüntzig M. & Fuchs E. (1979b) Vergleich von Isolaten des beet mosaic virus. *Archiv für Phytopathologie und Pflanzenschutz* **15**, 225–232.

Gugerli P. & Fries P. (1983) Characterization of monoclonal antibodies to potato virus Y and their use for virus detection. *Journal of General Virology* **64**, 2471–2477.

Gugerli P. & Gehriger W. (1980) Enzyme-linked immunosorbent assay (ELISA) for the detection of potato leafroll virus and potato virus Y in potato tubers after artificial break of dormancy. *Potato Research* **23**, 353–359.

Hagedorn D.J. & Walker J.C. (1949) Wisconsin pea streak. *Phytopathology* **39**, 837–847.

Halk E.L. & Fulton R.W. (1978) Stabilization and particle

morphology of prune dwarf virus. *Virology* **91,** 434–443.

Hampton R.O. (1975) The nature of bean yield reduction by bean yellow and bean common mosaic viruses. *Phytopathology* **65,** 1342–1346.

Hampton R., Beczner L., Hagedorn D., Bos L., Inouye T. Barnett O., Musil M. & Meiners J. (1978) Host reactions of mechanically transmissible legume viruses of the Northern temperate zone. *Phytopathology* **68,** 989–997.

Harbaoui Y., Samiyn G., Welvaert W. & Debergh P. (1982) Assainissement viral de l'artichaut par la culture *in vitro* d'apex méristématiques. *Phytopathologia Mediterranea* **21,** 15–19

Hardtl H. (1972) Die Übertragung der Zwiebelgelbstreifigkeit durch den Samen. *Zeitschrift für Pflanzenkrankheiten und Pflanzenschutz* **79,** 694–701.

Harrison B.D. (1958) Further studies on raspberry ring spot and tomato black ring, soil-borne viruses that affect raspberry. *Annals of Applied Biology* **46,** 571–584.

Hartleb H. (1975) Der Befall von Beta-Rüben durch Viruskrankheiten in der Deutschen Demokratischen Republik in den Jahren 1972–1974. *Nachrichtenblatt für den Pflanzenschutz in der DDR* **29,** 45–50.

Heathcote G.D. (1978) Review of losses caused by virus yellows in English sugar-beet crops and the cost of partial control with insecticides. *Plant Pathology* **27,** 12–17.

Hewitt W.B., Raski D.J. & Goheen A.C. (1958) Nematode vector of soil-borne fanleaf virus of grapevines. *Phytopathology* **48,** 586–595.

Hiebert E. & Dougherty W.G. (1981) A genetic map of the potyviral genome. *Phytopathology* **75,** 225.

Hill J.H. & Benner I.H. (1976) Properties of potyvirus RNAs: turnip mosaic, tobacco etch, and maize dwarf mosaic viruses. *Virology* **75,** 419–432.

Hill J.H., Ford R.E. & Benner I.H. (1973) Purification and partial characterization of maize dwarf mosaic virus strain B (Sugarcane mosaic virus). *Journal of General Virology* **20,** 327–339.

Hoefert L.L. (1969) Proteinaceous and virus-like inclusions in cells infected with beet mosaic virus. *Virology* **37,** 498–501.

Holmes S.J.I. (1985) Barley yellow dwarf virus in ryegrass and winter barley in the West of Scotland. *Mitteilungen aus der Biologischen Bundesanstalt für Land- und Forstwirtschaft Berlin-Dahlem* **228,** 38–41.

Horst R.K., Langhans R.W. & Smith S.H. (1977) Effects of chrysanthemum stunt, chlorotic mottle, aspermy and mosaic on flowering and rooting of chrysanthemums. *Phytopathology* **67,** 9–14.

Horvath J. (1980) Viruses of lettuce. II. Host ranges of lettuce mosaic virus and cucumber mosaic virus. *Acta Agronomica Academiae Scientiarum Hungaricae* **29,** 333–352.

Hsu H.T., Aebig J. & Rochow W.F. (1984) Differences among monoclonal antibodies to barley yellow dwarf viruses. *Phytopathology* **74,** 600–605.

Huttinga H. (1975) Properties of viruses of the potyvirus group. 3. A comparison of buoyant density, S value, particle morphology and molecular weight of the coat protein subunit of 10 viruses and virus isolates. *Netherlands Journal of Plant Pathology* **81,** 58–63.

Inouye T. & Mitsuhata K. (1973) Carnation necrotic fleck virus. *Berichte des Ôhara Instituts für Landwirtschaftliche Biologie Okayama Universität* **15,** 195–205.

Irwin M.E. & Goodman R.M. (1981) Ecology and control of soybean mosaic virus. In *Plant Diseases and Vectors: Ecology and Epidemiology* (Eds K. Maramorosch & K.E. Harris), pp. 181–220. Academic Press, New York.

Jacquemond M. & Lot H. (1981) L'ARN satellite du virus de la mosaïque du concombre. I. Comparaison de l'aptitude à induire la nécrose de la tomate d'ARN satellites isolés de plusieurs souches du virus. *Agronomie* **1,** 927–932

Jadot R. (1976) Aspects des épidemies de jaunisse et de mosaïque de la betterave. *Revue de l'Agriculture* **29,** 555–616.

Jafarpour B., Shepherd R.J. & Grogan R.G. (1979) Serologic detection of bean common mosaic and lettuce mosaic viruses in seed. *Phytopathology* **69,** 1125–1129.

Jafri A.M. (1972) Localization of virus and ultrastructural changes in sugar beet leaf cells associated with infection by a necrotic strain of beet mosaic virus. *Phytopathology* **62,** 1103.

Jones A.T. & Catherall P.L. (1970) The effect of different virus isolates on the expression of tolerance to barley yellow dwarf virus in barley. *Annals of Applied Biology* **65,** 147–152.

Jones A.T. & Duncan G.H. (1980) The distribution of some genetic determinants in the two nucleoprotein particles of cherry leaf roll virus. *Journal of General Virology* **50,** 269–273.

Jones A.T., McElroy F.D. & Brown D.J.F. (1981) Tests for transmission of cherry leaf roll virus using *Longidorus, Paralongidorus* and *Xiphinema* nematodes. *Annals of Applied Biology* **99,** 143–150.

Kahn R.P., Scott H.A. & Monroe R.L. (1962) Eucharis mottle strain of tobacco ringspot virus. *Phytopathology* **52,** 1211–1216.

Kaiser W.J., Wyatt S.D. & Pesho G.R. (1982) Natural hosts and vectors of tobacco streak virus in eastern Washington. *Phytopathology* **72,** 1508–1512.

Kaper J.M. & Waterworth H.E. (1977) Cucumber mosaic virus-associated RNA-5 causal agent for tomato necrosis. *Science* **196,** 429–431.

Kaper J.M. & Waterworth H.E. (1981) Cucumoviruses. In *Handbook of Plant Virus Infections and Comparative Diagnosis* (Ed. E. Kurstak), pp. 257–332. Elsevier, Amsterdam.

Kegler H. (1965) Untersuchungen über Ringfleckenkrankheiten der Kirsche. II. Wirtspflanzen und physikalische Eigenschaften von Ringfleckenviren. *Phytopathologische Zeitschrift* **54,** 305–327.

Kennedy J.S., Day M.F. & Eastop V.F. (1962) *A Conspectus of Aphids as Vectors of Plant Viruses.* Commonwealth Institute of Entomology, London.

Kerlan C. & Dunez J. (1976) Some properties of plum pox virus and its nucleic acid and protein components. *Acta Horticulturae* **67,** 185–191.

Kerlan C. & Dunez J. (1979) Différenciation biologique et sérologique de souches de virus de la sharka. *Annales de Phytopathologie* **11,** 241–250.

Kerlan C., Mille B. & Dunez J. (1981) Use of immunosorbent electron microscopy for detecting apple chlorotic leaf spot and plum pox viruses. *Phytopathology* **71,** 400–404.

Kerlan C., Mille B., Detienne G. & Dunez J. (1982) Comparison of immunoelectronmicroscopy, immunoenzymology and gel diffusion for investigating virus strain relationships. *Annales de Virologie (Institut Pasteur)* **133 E**, 3–14.

Knesek J. & Mink G.I. (1970) Incidence of a seed-borne virus in peas grown in Washington and Idaho. *Plant Disease Reporter* **54**, 497.

Koganezawa H., Yanase H. & Sakuma T. (1982) Viroid-like RNA associated with apple scar skin (or dapple apple). *Acta Horticulturae* **130**, 193–197.

Kojima M. (1981) Note on the serological relationship between Japanese and foreign isolates of potato leaf roll virus. *Bulletin of the Faculty of Agriculture of Niigata University* **33**, 73–77.

Kojima M., Shikata E., Sugawara M. & Murayama D. (1969) Purification and electron microscopy of potato leaf roll virus. *Virology* **39**, 162–171.

Kovachevsky I.C. (1968) Das Bohnengelbmosaikvirus in Bulgarien. *Phytopathologische Zeitschrift* **61**, 41–48.

Kowalska A. & Waś M. (1976) Detection of potato virus M and potato virus S on test plants. *Potato Research* **19**, 131–139.

Kowalska A., Skrzeczkowska S., Syller J. & Rudzinska-Langwald A. (1981) New strain of potato virus A isolated from hybrids of *Solanum megistacrolobum*. *Zeszyty Problemowe Postępów Nauk Rolniczych* **244**, 152–165.

Krake L.R. & Woodham R.C. (1983) Grapevine yellow speckle agent implicated in the etiology of vein banding. *Vitis* **22**, 40–50.

Krczal H. (1980) Transmission of the strawberry mild yellow edge and strawberry crinkle virus by the strawberry aphid *Chaetosiphon fragaefolii*. *Acta Horticulturae* **95**, 23–30.

Krczal H. & Kunze L. (1976) Experiments on the transmissibility of sharka virus by aphids. *Acta Horticulturae* **67**, 165–170.

Kunze L., Clark M.F. & Flegg C.L. (1984) Comparison of methods for characterising and distinguishing isolates of prune dwarf virus. *Phytopathologische Zeitschrift* **110**, 251–260.

Kurstak E. (Ed.) (1981) *Handbook of Plant Virus Infections and Comparative Diagnosis*. Elsevier, Amsterdam.

Lange L. (1978) *Synchytrium endobioticum* and potato virus X. *Phytopathologische Zeitschrift* **92**, 132–142.

Lansac M., Bernhard R., Massonie G., Maison P., Kerlan C., Dunez J. & Morvan G. (1979) La Sharka: connaissances actuelles. In *3e Journées sur les Maladies des Plantes*, pp. 452–468. ACTA, Paris.

Lecoq H. & Pitrat M. (1983) Field experiments on the integrated control of aphid-borne viruses in muskmelon. In *Plant Virus Epidemiology* (Eds R.T. Plumb & J.M. Thresh), pp. 170–176. Blackwell Scientific Publications, Oxford.

Lecoq H., Lisa V. & Dellavalle G. (1983) Serological identity of muskmelon yellow stunt and zucchini yellow mosaic viruses. *Plant Disease* **67**, 824–825.

Lecoq H., Pitrat M. & Clément M. (1981) Identification et caractérisation d'un potyvirus provoquant la maladie du rabougrissement jaune du melon. *Agronomie* **1**, 827–834.

Lecoq H., Cohen S., Pitrat M. & Labonne G. (1979) Resistance to cucumber mosaic virus transmission by aphids in *Cucumis melo*. *Phytopathology* **69**, 1223–1225.

Leiser R.-M. & Richter J. (1979) Ein Beitrag zur Frage der Differenzierung von Stammen des Kartoffel-Y-Virus. *Archiv für Phytopathologie und Pflanzenschutz* **15**, 289–298.

Leiser R.M. & Richter J. (1980) Differenzierung von Stämmen des Kartoffel-X-Virus. *Tagungsbericht der Akademie der Landwirtschaftswissenschaften der DDR* **184**, 107–113.

Lesemann D.E., Makkouk K.M., Koenig R. & Natafji Samman E. (1983) Natural infection of cucumbers by zucchini yellow mosaic virus in Lebanon. *Phytopathologische Zeitschrift* **108**, 304–313.

Linthorst H.J.M. & Kaper J.M. (1984) Circular molecule of tobacco ringspot infected tissue. *Virology* **137**, 206–210.

Lisa V. (1980) Two viruses from rhizomatous Iris. *Acta Horticulturae* **110**, 39–53.

Lisa V. & Dellavalle G. (1981) Characterization of two potyviruses in *Cucurbita pepo*. *Phytopathologische Zeitschrift* **100**, 279–286.

Lisa V., Boccardo G., D'Agostino G., Dellavalle G. & D'Aquilio M. (1981) Characterization of a potyvirus that causes zucchini yellow mosaic. *Phytopathology* **71**, 668–672.

Lister R.M. & Murant A.F. (1967) Seed transmission of nematode-borne viruses. *Annals of Applied Biology* **59**, 49–62.

Lister R.M., Allen W.R., Gonsalves D., Gotlieb A.R., Powell C.A. & Stouffer R.F. (1980) Detection of tomato ringspot virus in apple and peach by ELISA. *Acta Phytopathologica Academiae Scientiarum Hungaricae* **15**, 47–55.

Lizárraga R.E., Salazar L.F., Roca W.M. & Schilde-Rentschler L. (1980) Elimination of potato spindle tuber viroid by low temperature and meristem culture. *Phytopathology* **70**, 754–755.

Lovisolo O. & Conti M. (1969) Biological characterization of some isolates of cucumber mosaic virus. *Annales de Phytopathologie* **1**, 367–373.

Luis Arteaga M.P., Quiot J.B. & Leroux J.P. (1976) Mise en évidence d'une souche de watermelon mosaic virus (WMV2) dans le Sud-Est de la France. *Annales de Phytopathologie* **8**, 347–353.

Luisoni E., Boccardo G. & Milne R.G. (1976) Purification and some properties of an Italian isolate of poplar mosaic virus. *Phytopathologische Zeitschrift* **85**, 64–73.

MAFF (1981) Virus yellows of sugar beet. *Ministry of Agriculture Leaflet* no. 323, HMSO, London.

Maison P., Kerlan C. & Massonié G. (1983) Sélection de semis de *Prunus persica* résistants à la transmission du virus de la Sharka par les virginopares aptères de *Myzus persicae*. *Comptes Rendus des Séances de l'Académie d'Agriculture de France* **69**, 337–345.

Marchoux G., Besson L, Cornu A. & Gianinazzi S. (1983) Réaction différentielle de cinq espèces de pétunia à différentes souches du virus de la mosaïque de tabac. *Agronomie* **3**, 279–282.

Marenaud C., Dunez J. & Bernhard R. (1976) Identification and comparison of different strains of apple chlorotic leaf spot virus and possibility of cross protection. *Acta Horticulturae* **67**, 219–229.

Marrou J. & Migliori A. (1971) Essai de protection des cultures de tomate contre le virus de la mosaïque du tabac: mise en évidence d'une spécificité étroite de la prémunition entre souches de ce virus. *Annales de Phytopathologie* **3**, 447–459.

Marrou J., Quiot J.B. & Marchoux G. (1975) Caractérisation par la symptomatologie de quatorze souches du virus de la mosaïque du concombre et de deux autres cucumovirus: tentative de classification. *Mededelingen van de Faculteit Landbouwwetenschapen Rijksuniversiteit Gent* **40**, 107–121.

Martelli G.P. (1978) Nematode-borne viruses of grapevine, their epidemiology and control. *Nematologia Mediterranea* **6**, 1–127.

Martelli G.P., Russo M. & Castellano M.A. (1969) Ultrastructural features of *Malva parviflora* with vein clearing and of plants infected with beet mosaic. *Phytopathologia Mediterranea* **8**, 175–186.

Martin R.R. & Converse R.H. (1982) Purification, some properties, and serology of strawberry mild yellow-edge virus. *Acta Horticulturae* **129**, 75.

Massonie C. & Maison R. (1979) Contribution to the study of sharka transmission by *Myzus persicae*. *Revue de Zoologie Agraire et de Pathologie Végétale* **75**, 31–35.

Mather S.M. (1968) Regulations affecting the control of citrus virus diseases in California. In *Proceedings of the 4th Conference of the International Organization of Citrus Virologists*, pp. 369–373. University of Florida, Gainsville.

Matthews R.E.F. (1982) Classification and nomenclature of viruses. *Intervirology* **17**, 1–199.

Mayo M.A., Barker H., Robinson D., Tamada T. & Harrison B.D. (1982) Evidence that potato leaf roll virus RNA is positive-stranded, is linked to a small protein and does not contain polyadenylate. *Journal of General Virology* **59**, 163–167.

McClean A.P.D. (1977) Tristeza-virus-complex: influence of host species on the complex. *Citrus and Subtropical Fruit Journal* **522**, 4–16.

McLaughlin M.R., Barnett O.W., Gibson P.B. & Burrows P.M. (1984) Enzyme-linked immunosorbent assay of viruses infecting forage legumes. *Phytopathology* **74**, 965–969.

McLean D.L. & Kinsey M.G. (1962) Three variants of lettuce mosaic virus and methods utilized for differentiation. *Phytopathology* **52**, 403–406.

Mehrad M., Lapierre H. & Yot P. (1979) RNA in potato leaf roll virus. *FEBS Letters* **101**, 169–173.

Mellor F.C. & Frazier N.W. (1970) Strawberry mild yellow edge. In *Virus Diseases of Small Fruits and Grapevines* (Ed. N.W. Frazier), pp. 14–16. University of California, Berkeley.

Mink G.I. & Shay J.R. (1959) Preliminary evaluation of some Russian apple varieties as indicators for apple viruses. *Supplement to the Plant Disease Reporter* no. 254, 13–17.

Moghal S.M. & Francki R.I.B. (1976) Towards a system for the identification and classification of potyviruses I. Serology and amino acid composition of six distinct viruses. *Virology* **73**, 350–352.

Mohamed N.A. & Young B.R. (1981) Garlic yellow streak virus—a potyvirus infecting garlic in New Zealand. *Annals of Applied Biology* **97**, 65–74.

Momma T. & Takahashi T. (1984) Developmental morphology of hop stunt viroid-infected hop plants and analysis of their cone yields. *Phytopathologische Zeitschrift* **110**, 1–14.

Monsion M., Bachelier J.C. & Dunez J. (1973) Quelques propriétés d'un viroïde: le rabougrissement du chrysanthème. *Annales de Phytopathologie* **5**, 467–469.

Monsion M., Macquaire G., Bachelier J.C. & Dunez J. (1981) Recent progress in chrysanthemum viroid detection. *Acta Horticulturae* **125**, 227–232.

Moreira A., Jones R.A.C. & Fribourg C.E. (1980) Properties of a resistance-breaking strain of potato virus X. *Annals of Applied Biology* **95**, 93–103.

Morris T.J. & Wright N.S. (1975) Detection on polyacrylamide gels of a diagnostic nucleic acid from tissue infected with potato spindle tuber viroid. *American Potato Journal* **52**, 57–63.

Morris-Krsinich B.A.M., Forster R.L.S. & Mossop D.W. (1983) The synthesis and processing of the nepovirus grapevine fan leaf proteins in rabbit reticulocyte lysate. *Virology* **130**, 523–526.

Murant A.F. (1970) Soil-borne viruses in Rubus. In *Virus Diseases of Small Fruits and Grapevines* (Ed. N.W. Frazier), pp. 132–136. University of California, Berkeley.

Murant A.F. (1981) Nepoviruses. In *Handbook of Plant Virus Infections and Comparative Diagnosis* (Ed. E. Kurstak), pp. 198–238. Elsevier, Amsterdam.

Murant A.F., Taylor C.E. & Chambers J. (1968) Properties, relationships and transmission of a strain of raspberry ringspot virus infecting raspberry cultivars immune to the common Scottish strain. *Annals of Applied Biology* **61**, 175–186.

Murant A.F., Taylor J., Duncan G.H. & Raschke J.H. (1981) Improved estimates of molecular weight of plant virus RNA by agarose gel electrophoresis and electron microscopy after denaturation with glyoxal. *Journal of General Virology* **53**, 321–332.

Murayama D. & Kojima M. (1974) Antigenicity of potato leaf roll virus. *Proceedings of the Japanese Academy* **50**, 322–327.

Nagel J., Zettler F.W. & Hiebert E. (1983) Strains of bean yellow mosaic virus compared to clover yellow vein virus in relation to gladiolus production in Florida. *Phytopathology* **73**, 449–454.

Natti J.J., Kirkpatrick H.C. & Ross F. (1953) Host range of potato leaf roll virus. *American Potato Journal* **30**, 55–64.

Navarro L., Llacer G., Cambra M., Arregui J.M. & Juarez J. (1982) Shoot tip grafting *in vitro* for elimination of viruses in peach plants. *Acta Horticulturae* **130**, 185–192.

Németh M. (1979) *Manual of Viruses, Mycoplasms and Rickettsias of Fruit-trees*. Mezögazdasági Könyvkiadó, Budapest.

Niblett C.L., Dickson B., Horst R.K. & Romaine C.P. (1980) Additional hosts and an efficient purification procedure for four viroids. *Phytopathology* **70**, 610–615.

Ottenwaelter M.M., Hevin M. & Doazan J.P. (1973) Amélioration du rendement de la thermothérapie sur plants en pots par l'utilisation de la culture sur milieu gélosé stérile. *Vitis* **2**, 46–48.

Paliwal Y.C. & Tremaine J.H. (1976) Multiplication, puri-

fication and properties of ryegrass mosaic virus. *Phytopathology* **66,** 406–414.
Paludan N. (1980) Virus attack on leek: survey, diagnosis, tolerance of varieties and winter hardiness. *Tidsskrift Planteavl* **84,** 371–385.
Palukaitis P. & Symons R. (1979) Hybridization analysis of chrysanthemum stunt viroid with complementary DNA and the quantitation of viroid RNA sequences in extracts of infected plants. *Virology* **98,** 238–245.
Palukaitis P., Rakowski A.G., Alexander D.M.E. & Symons R.H. (1981) Rapid indexing of the sun blotch disease of avocado using a complementary DNA probe to avocado sun blotch viroid. *Annals of Applied Biology* **98,** 439–449.
Panayoutou P.C. (1978) Effect of barley yellow dwarf virus infection on the germinability of seeds and establishment of seedlings. *Plant Disease Reporter* **62,** 243–246.
Panayoutou P.C. (1979) Effect of barley yellow dwarf on the vegetative growth of cereals. *Plant Disease Reporter* **63,** 315–319.
Pelham J. (1972) Strain-genotype interaction of tobacco mosaic virus in tomato. *Annals of Applied Biology* **71,** 219–228.
Phatak H.C. (1974) Seed-borne plant viruses—identification and diagnosis in seed health testing. *Seed Science and Technology* **3,** 155.
Piazzolla P., Diaz-Ruiz J.R. & Kaper J.M. (1979) Nucleic acid homologies of eighteen cucumber mosaic virus isolates determined by competition hybridization. *Journal of General Virology* **45,** 361–369.
Plumb R.T. & Lennon E. (1982) Aphid infectivity and infectivity index. In *Report of Rothamsted Experimental Station 1981,* p. 181.
Pochard E. (1977) Méthodes pour l'étude de la résistance partielle au virus de la mosaïque du concombre chez le piment capsicum 77. In *Comptes Rendus 3eme Congrès Eucarpia Avignon,* pp. 93–104.
Polák J. (1971a) Differentiation of strains of sugar beet yellows virus on *Tetragonia expansa* and other indicator plants. *Biologia Plantarum* **13,** 145–154.
Polák J. (1971b) Physical properties and serological relationships of beet yellows virus strains. *Phytopathologische Zeitschrift* **72,** 235–244.
Polák J. (1979) Occurrence of beet western yellows virus in sugar beet in Czechoslovakia. *Biologia Plantarum* **21,** 275–279.
Polák J., Hartleb H. & Opel H. (1975) The diagnosis of beet yellows virus in beet plants within autumn period. *Ochrona Rostlin* **11,** 243–251.
Pop I.V. (1979) Cercetări privind atacul tulpinolor virusului mozaicului tutunului la ardeiul gras cultivat în sere. *Analele Institutului de Cercetări pentru Protecţia Plantelor* **15,** 17–20.
Poupet A. (1971) Une panachure des fleurs d'oeillets méditerranéens cultivés dans le Sud-Est de la France. *Annales de Phytopathologie* **3,** 509–517.
Poupet A. & Marais A. (1973) Isolement et purification du virus de la marbrure des nervures de l'oeillet (carnation vein mottle virus). *Annales de Phytopathologie* **5,** 265–271.
Poupet A., Cardin L., Marais A. & Cadilhac B. (1975) La bigarrure de l'oeillet: isolement et propriétés d'un virus flexueux. *Annales de Phytopathologie* **7,** 277–286.
Pring D.R. & Langenberg W.G. (1972) Preparation and properties of maize dwarf mosaic virus nucleic acid. *Phytopathology* **62,** 253–255.
Provvidenti R. (1974) Inheritance of resistance to Watermelon Mosaic Virus 2 in *Phaseolus vulgaris. Phytopathology* **64,** 1448–1450.
Provvidenti R. & Schroeder W.T. (1972) Natural infection of *Spinacia oleracea* by lettuce mosaic virus. *Plant Disease Reporter* **56,** 281–284.
Purcifull D.E. & Batchelor D.L. (1977) Immunodiffusion tests with sodium dodecyl sulfate (SDS)-treated plant viruses and plant viral inclusions. *Florida Agricultural Experiment Station Bulletin* no. 788.
Purcifull D.E. & Edwardson J.R. (1981) Potexviruses. In *Handbook of Plant Virus Infections and Comparative Diagnosis* (Ed. E. Kurstak), pp. 628–693. Elsevier, Amsterdam.
Purcifull D.E. & Hiebert E. (1979) Serological distinction of watermelon mosaic virus isolates. *Phytopathology* **69,** 112–116.
Quiot J.B., Marchoux G., Douine L. & Vigouroux A. (1979) Ecologie et épidémiologie du virus de la mosaïque du concombre dans le Sud-Est de la France. V Rôle des espèces spontanées dans la conservation du virus. *Annales de Phytopathologie* **11,** 325–349.
Ragozzino A., D'Errico F.P. & de Vincentiis C. (1982) Rasp leaf and decline of sweet cherry in Avelino province. *Acta Horticulturae* **130,** 301–306.
Ramaswamy S. & Posnette A.F. (1971) Properties of cherry ring mottle, a distinctive strain of prune dwarf virus. *Annals of Applied Biology* **68,** 55–65.
Ramaswamy S. & Posnette A.F. (1972) Yellow mottle disease of ornamental cherries caused by a strain of prune dwarf virus. *Journal of Horticultural Science* **47,** 107–112.
Rana G.L., Vovlas C. & Zettler W. (1983) Manual transmission of dasheen mosaic virus from *Richardia* to non-araceous hosts. *Plant Disease* **67,** 1121–1122.
Rana G.L., Russo M., Gallitelli D. & Martelli G.P. (1982) Artichoke latent virus: characterization, ultrastructure and geographical distribution. *Annals of Applied Biology* **101,** 279–289.
Rana G.L., Gallitelli D., Kyriakopoulou P.E., Russo M. & Martelli G.P. (1980) Host range and properties of artichoke yellow ringspot virus. *Annals of Applied Biology* **96,** 177–185.
Rast A.T.B. (1972) M 11-16, an artificial symptomless mutant of tobacco mosaic virus for seedling inoculation of tomato crops. *Netherlands Journal of Plant Pathology* **78,** 110–112.
Rast A.T.B. (1979) Pepper strains of TMV in Netherlands. *Mededelingen van de Faculteit Landbouwwetenschapen Rijksuniversiteit Gent* **44,** 617–622.
Reifman V.G. & Romanova S.A. (1977) Use of the mild strain of potato virus X in protective vaccination experiments. *Trudy Biologicheskogo-Pochvennogo Instituta Dal'nevostochny: Nauchnyi Tsentr Akademii Nauk SSSR* **46,** 146–154.
Richardson D.E. & French W.M. (1978) Evidence for the possible transmission of PVX by *Spongospora subter-*

ranea. In *7th Triennial Conference of the European Association for Potato Research*, pp. 254–255. Warsaw.

Richter J., Haack I. & Eisenbrandt K. (1979) Routinemässige Serodiagnose der Kartoffelviren X, S, M und Y mit dem Radialimmundiffusionstest. *Archiv für Phytopathologie und Pflanzenschutz* **15**, 81–88.

Rochow W.F. & Duffus J.E. (1981) Luteoviruses and yellows diseases. In *Handbook of Plant Virus Infection and Comparative Diagnosis* (Ed. E. Kurstak), pp. 147–170. Elsevier, Amsterdam.

Roistacher C.N.E., Nauer E.M., Kishaba A. & Calavan E.C. (1980) Transmission of citrus tristeza virus by *Aphis gossypii* reflecting changes in virus transmissibility in California. In *Proceedings of the 8th Conference of the International Organization of Citrus Virologists* (Eds Calavan E.C. *et al.*), pp. 76–82. IOCV, Riverside.

Rosner A., Ginzburg I. & Bar-Joseph M. (1983) Molecular cloning of complementary DNA sequences of citrus tristeza virus RNA. *Journal of General Virology* **64**, 1757–1763.

Rowhani A. & Stace-Smith R. (1979) Purification and characterization of potato leaf roll virus. *Virology* **98**, 45–54.

Russo M. & Rana G.L. (1978) Occurrence of two legume viruses in artichoke. *Phytopathologia Mediterranea* **17**, 212–216.

Russo M., Martelli G.P., Vovlas C. & Ragozzino A. (1979) Comparative studies on Mediterranean isolates of watermelon mosaic virus. *Phytopathologia Mediterranea* **18**, 94–101.

Ryder E.J. (1973) Seed transmission of lettuce mosaic virus in mosaic-resistant lettuce. *Journal of the American Society for Horticultural Science* **98**, 610–614.

Salazar L.F., Owens R.A., Smith D.R. & Diener T.O. (1983) Detection of potato spindle tuber viroid by nucleid acid spot hybridization: evaluation with tuber sprouts and true potato seed. *American Potato Journal* **60**, 587–597.

Sänger H.L. (1982) Biology, structure, functions and possible origin of viroids. In *Nucleic Acids and Proteins in Plants* (Eds B. Parthier & D. Boulter), Vol. 2, pp. 368–454. Springer Verlag, Berlin.

Sänger H.L., Klotz G., Riesner D., Gross H.J. & Kleinschmidt A.K. (1976) Viroids are single-stranded covalently closed circular RNA molecules existing as highly base-paired rod-like structures. *Proceedings of the US National Academy of Sciences* **73**, 3852–3856.

Schmelzer K. (1969) Das latente Erdbeerringflecken-virus aus *Euonymus*, *Robinia* und *Aesculus*. *Phytopathologische Zeitschrift* **66**, 1–24.

Schroeder W.T. & Provvidenti R. (1971) A common gene for resistance to bean yellow mosaic virus and watermelon mosaic virus 2 in *Pisum sativum*. *Phytopathology* **61**, 846–848.

Schuster G. (1977) Der Einfluss einiger Herbizide und Fungizide auf die Konzentration der Kartoffel-X-Virus in Blättern von *Nicotiana tabacum* Samsun und Modelversuche zur Bestimmung der Viruskonzentration. *Archiv für Phytopathologie und Pflanzenschutz* **4**, 229–239.

Scotto la Massese C., Marenaud C. & Dunez J. (1973) Analyse d'un phénomène de dégénérescence du Pêcher dans la vallée de l'Eyrieux. *Comptes Rendus des Séances de l'Académie d'Agriculture de France* **59**, 327–339.

Shalla T.A. & Shepard J.F. (1972) The structure and antigenic analysis of amorphous inclusion bodies induced by potato virus X. *Virology* **49**, 654–667.

Sharma S.R. & Varma A. (1982) Control of virus diseases by oil sprays. *Zentralblatt für Bakteriologie 2*. **137**, 329–347.

Shepard J.F. (1972) Gel-diffusion methods for the serological detection of potato viruses X, S and M. *Montana Agricultural Experiment Station Bulletin* no. 662.

Shepherd R.J., Till B.B. & Schaad N. (1966) A severe necrotic strain of the beet mosaic virus. *Journal of the American Society of Sugar Beet Technologists* **14**, 97–105.

Signoret P.A. (1974) Les maladies à virus des Graminées dans le Midi de la France. *Acta Biologica Jugoslavica B* **11**, 115–120.

Singh S.J., Sastry K.S. & Sastry K.S.M. (1981) Field tests with insecticides and mineral oil for the protection of French beans from yellow mosaic virus disease. *Gartenbauwissenschaft* **46**, 88–91.

Smith K.M. (1972) Celery yellow spot virus. In *Textbook of Plant Virus Diseases*, pp. 159–160. Longman, London.

Smookler M. & Loebenstein G. (1974) Carnation yellow fleck virus. *Phytopathology* **64**, 979–984.

Spaar D. & Hamann U. (1977) Kartoffel. In *Pflanzliche Virologie* (Ed. M. Klinkowski), pp. 74–77. Akademie-Verlag, Berlin.

Stace-Smith R. (1984) Red raspberry virus diseases. *Plant Disease* **68**, 274–279.

Stace-Smith R. & Hansen A.J. (1976) Some properties of cherry raspleaf virus. *Acta Horticulturae* no. 67, 193–197.

Stein A., Loebenstein G. & Koenig R. (1979) Detection of cucumber mosaic virus and bean yellow mosaic virus in gladiolus by enzyme-linked immunosorbent assay (ELISA). *Plant Disease Reporter* **63**, 185–188.

Suneson C.A. & Ramage R.T. (1957) Studies on the importance of the yellow dwarf virus. *Agronomy Journal* **49**, 365–367.

Sutić D. & Ranković M. (1981) Resistance of some plum cultivars and individual trees to plum pox (Sharka) virus. *Agronomie* **1**, 617–622.

Šutić D. & Ranković M. (1983) Sensitivity of some stone fruit species to sharka (plum pox) virus. *Zaštita Bilja* **34**, 241–248.

Šutić D., Jordovic M., Rankovic M. & Festic H. (1971) Comparative studies of some sharka (plum pox) isolates. *Annales de Phytopathologie no. hors série*, 184–192.

Takanami Y. & Kubo S. (1979) Nucleic acids of two phloem-limited viruses: tobacco necrotic dwarf and potato leaf roll. *Journal of General Virology* **44**, 853–856.

Tanaka H. (1980) Satsuma dwarf. In *Description and Illustration of Virus and Virus-like Diseases of Citrus*, (Eds J. Bové & R. Vogel). SETCO-IRFA, Paris.

Taylor C.E. & Thomas P.R. (1968) The association of *Xiphinema diversicaudatum* (Micoletski) with strawberry latent ring spot and arabis mosaic viruses in a raspberry plantation. *Annals of Applied Biology* **62**, 147–157.

Thomas B.J. (1980) The detection by serological methods of viruses infecting the rose. *Annals of Applied Biology* **94**, 91–101.

Thomas B.J. (1981) Studies on rose mosaic disease in field-grown roses produced in the United Kingdom. *Annals of Applied Biology* **98,** 419–429.

Thomas W. & Mohamed N.A. (1979) Avocado sun blotch, a viroid disease? *Australasian Plant Pathology* **8,** 1–2.

Thornbury D.W. & Pirone T.P. (1983) Helper components of two potyviruses are serologically distinct. *Virology* **125,** 487–490.

Thresh J.M. & Pitcher R.S. (1978) The spread of nettlehead and related virus diseases of hop. In *Plant Disease Epidemiology* (Eds P.R. Scott & A. Bainbridge). Blackwell Scientific Publications, Oxford.

Tomlinson J.A. & Walkey D.G.A. (1967) The isolation and identification of rhubarb viruses in Britain. *Annals of Applied Biology* **59,** 415–427.

Tomlinson J.A. & Ward C.M. (1978) The reactions of swede (*Brassica napus*) to infection by turnip mosaic virus. *Annals of Applied Biology* **89,** 61–69.

Tomlinson J.A. & Ward C.M. (1981) The reactions of some Brussels sprout F_1 hybrids and inbreds to cauliflower mosaic and turnip mosaic viruses. *Annals of Applied Biology* **97,** 205–212.

Tomlinson J.A. & Ward C.M. (1982a) Chemotherapy of beet western yellows disease. In *Report of the National Vegetable Research Station Wellesbourne for 1981*, pp. 83–84.

Tomlinson J.A. & Ward C.M. (1982b) Selection for immunity in swede (*Brassica napus*) to infection by turnip mosaic virus. *Annals of Applied Biology* **101,** 43–50.

Tomlinson J.A. & Webb M.J.W. (1978) Ultrastructural changes in chloroplasts of lettuce infected with beet western yellows virus. *Physiological Plant Pathology* **12,** 13–18.

Tomlinson J.A., Faithfull E.M. & Ward C.M. (1976) Chemical suppression of the symptoms of two virus diseases. *Annals of Applied Biology* **84,** 31–41.

Tomlinson J.A., Carter A., Dale W.T. & Simpson C.J. (1970) Weed plants as sources of cucumber mosaic virus. *Annals of Applied Biology* **66,** 11–16.

Torrance L. & Jones R.A.C. (1982) Increased sensitivity of detection of plant viruses obtained by using a fluorogenic substrate in ELISA. *Annals of Applied Biology* **101,** 501–509.

Tošić M. (1981) Cross protection among some strains of sugarcane mosaic virus and maize dwarf mosaic virus. *Agronomie* **1,** 83–88.

Tošić M. & Ford R.E. (1974) Physical and serological properties of maize dwarf mosaic and sugarcane mosaic viruses. *Phytopathology* **64,** 312–317.

Tošić M. & Ford R.E. (1982) Sorghum cultivars differentiating sugarcane mosaic and maize dwarf mosaic virus strains. In *2nd International Maize Virus Disease Colloquium and Workshop.* Wooster.

Tošić M. & Mišović M. (1967) Effect of virus mosaic on the growth and development of several varieties and hybrids of maize. *Zaštita Bilja* **18,** 173–181.

Tung J.S. & Knight C.A. (1972) The coat protein subunits of potato virus X and white clover mosaic virus, a comparison of methods for determining their molecular weights and some *in situ* degradation products of potato virus X protein. *Virology* **49,** 214–223.

Turska E. (1980) Ausbreitung des S-Virus durch Blattläuse. *Kartoffelbau* **11,** 384–385.

USDA (1976) *Virus Diseases and Non-Infectious Disorders of Stone Fruits in North America. Agriculture Handbook* no. 437. USDA, Washington.

Uyeda I. & Mink G.I. (1981) Properties of asparagus virus II: a new member of the Ilarvirus group. *Phytopathology* **71,** 1264–1269.

Van der Meer Q.M. (1980) Resistentie tegen preigeelstreep virus in kurrat *(Allium ampeloprasum). Zaadbelangen* **34,** 173–174.

Van der Meer F.A., Maat D.Z. & Vink J. (1980a) Poplar mosaic virus: purification, antiserum preparation, and detection in poplars with the enzyme-linked immunosorbent assay (ELISA) and with infectivity tests on *Nicotiana megalosiphon. Netherlands Journal of Plant Pathology* **86,** 99–110.

Van der Meer F.A., Maat D.Z. & Vink J. (1980b) *Lonicera* latent virus, a new carlavirus serologically related to poplar mosaic virus: some properties and inactivation *in vivo* by heat treatment. *Netherlands Journal of Plant Pathology* **86,** 69–78.

Van Dorst H.J.M. & Peters D. (1974) Some biological observation on pale fruit, a viroid incited disease of cucumber. *Netherlands Journal of Plant Pathology* **80,** 85–96.

Van Dorst H.J.M., Huijberts N. & Bos L. (1983) Yellows of glasshouse vegetables, transmitted by *Trialeurodes vaporariorum. Netherlands Journal of Plant Pathology* **89,** 171–184.

Van Koot Y., Van Slogteren D.U.M., Van Cremer M.C. & Camfferman J. (1954) Virusversdijuselen in Freesias. *Tijdschrift voor Planteziekten* **60,** 157–192.

Van Regenmortel M.H.V. (1981) Tobamoviruses. In *Handbook of Plant Virus Infections and Comparative Diagnosis* (Ed. E. Kurstak), pp. 541–564. Elsevier, Amsterdam.

Van Regenmortel M.H.V. (1982) *Serology and Immunochemistry of Plant Viruses.* Academic Press, New York.

Van Rheenen H.A. & Muigai S.G.S. (1984) Control of bean common mosaic by deployment of the dominant gene I. *Netherlands Journal of Plant Pathology* **90,** 85–94.

Vetten H.J., Ehlers U. & Paul H.L. (1983) Detection of potato viruses Y and A in tubers by enzyme-linked immunosorbent assay after natural and artificial break of dormancy. *Phytopathologische Zeitschrift* **108,** 41–53.

Vovlas C., Hiebert E. & Russo M. (1981) Zucchini yellow fleck virus, a new potyvirus of zucchini squash. *Phytopathologia Mediterranea* **20,** 123–128.

Vuittenez A. (1970) Fanleaf of grapevine. In *Virus Diseases of Small Fruits and Grapevines* (Ed. N.W. Frazier), pp. 217–228. University of California, Berkeley.

Vuittenez A., Kuszala J., Legin R., Stocky G., Pejcinooski P. & Heyd J.C. (1976) Phénomènes d'interaction entre souches de népovirus affectant la vigne. In *Proceedings of the 6th Meeting of the International Council for the Study of Viruses and Virus-like Diseases of the Grapevine*, pp. 59–68. Madrid.

Walkey D.G.A. & Neeley H.A. (1980) Resistance in white cabbage to necrosis caused by turnip and cauliflower mosaic viruses and pepper spot. *Annals of Applied Biology* **95,** 703–713.

Wallace J.M. (1968) Tristeza and seedling yellows. In *Indexing Procedures for 15 Virus Diseases of Citrus Trees. USDA Agricultural Handbook* no. 333, pp. 20–27. USDA, Washington.

Watson M.A., Heathcote G.D., Lauckner F.B. & Sowray P.A. (1975) The use of weather data and counts of aphids in the field to predict the incidence of yellowing viruses of sugar-beet crops in England in relation to the use of insecticides. *Annals of Applied Biology* **81**, 181–198.

Weathers L.G. (1980) Exocortis. In *Description and Illustration of Virus and Virus-like Diseases of Citrus.* (Eds J. Bové & R. Vogel). SETCO-IRFA, Paris.

Webb R.E. (1979) Inheritance of resistance to watermelon mosaic virus 1 in *Cucumis melo. HortScience* **14**, 265–266.

Webb R.E., Larson R.H. & Walker J.C. (1952) Relationships of potato leaf roll virus strains. *University of Wisconsin Madison Research Bulletin* no. 178.

Weerapat P., Sechler D.T. & Poehleman J.M. (1974) BYDV resistance in crosses with Petkus oats. *Crop Science* **14**, 218–220.

Weidemann H.L. (1981) Detection of potato virus X (PVX) in potato plants and tubers by immunofluorescence. *Phytopathologische Zeitschrift* **102**, 93–99.

Wetter C. & Milne R.G. (1981) Carlaviruses. In *Handbook of Plant Virus Diseases and Comparative Diagnosis* (Ed. E. Kurstak), pp. 696–730. Elsevier, Amsterdam.

Williams H.E. & Holdeman G.L. (1970) American currant mosaic. In *Virus Diseases of Small Fruits and Grapevines* (Ed. N.W. Frazier), pp. 95–97. University of California, Berkeley.

Winther-Nielsen P. (1972) Traktorredskaber som spredere af virus X i kartofler. *Tidskrift for Planteavl* **76**, 297–307.

Wolf P. & Schmelzer K. (1972) Untersuchungen an Umbelliferen. *Zentralblatt für Bakteriologie 2* **127**, 665–672.

Wright N.S. (1974) Retention of infectious potato virus X on common surfaces. *American Potato Journal* **51**, 252–253.

Yamada S. & Sawamura K. (1952) Studies on the dwarf disease of satsuma orange, *Citrus unshiu* (preliminary report). *Bulletin of the Horticultural Division of Tokai-Kinki Agricultural Experiment Station* **1**, 61–71.

Yot-Dauthy D. & Bové J.M. (1968) Purification and characterization of citrus crinkly leaf virus. In *Proceedings of the 4th Conference of the International Organization of Citrus Virologists*, pp. 255–263.

Zettler C.W. (1969) The heterogeneity of bean leaves as sources of bean common mosaic virus for aphids. *Phytopathology* **59**, 1109–1110.

Zink F.W., Duffus J.E. & Kimble K.A. (1973) Relationship of a non-lethal reaction to a virulent isolate of lettuce mosaic virus and turnip mosaic susceptibility in lettuce. *Journal of the American Society for Horticultural Science* **98**, 41–45.

2: Viruses II
Minor Virus Groups

This chapter covers 17 very diverse virus groups, which are either small or poorly represented in Europe. With one or two exceptions, the viruses concerned are of intermediate or minor economic importance. The groups concerned are: alfalfa mosaic virus, bromoviruses, carnation mottle virus group, caulimoviruses, comoviruses, dianthoviruses, geminiviruses, hordeiviruses, penamoviruses, reoviruses, rhabdoviruses, sobemoviruses, tobacco necrosis virus, tobraviruses, tomato spotted wilt virus, tombusviruses, tymoviruses.

THE ALFALFA MOSAIC VIRUS GROUP

The group has the characteristics of its only member.

Alfalfa mosaic virus

(Syn. lucerne mosaic virus)

Basic description

AMV has bacilliform particles of different lengths: B (58×18 nm), M (48×18 nm) and Tb (36×18 nm). Three molecules of positive-sense single-stranded RNA of 1.1×10^6 d (RNA_1), 0.8×10^6 d (RNA_2) and 0.7×10^6 d (RNA_3) are encapsidated in B, M and Tb respectively. A fourth RNA (RNA_4: 0.3×10^6 d) which is monocistronic for the coat protein (24 000 d) is also encapsidated, two RNA_4 molecules being packaged in an isometric particle (Ta: 18 nm in diameter). All components have the same coat protein (*c.* 25 000 d). By contrast to cucumoviruses or bromoviruses, which are also tripartite genome viruses where the 3 genomic RNAs are infective, the 3 genomic RNAs of AMV are not infective. Infectivity requires, either the coat protein or the fourth RNA in addition to the 3 genomic RNAs (CMI/AAB 229).

Host plants and diseases

More than 400 species in 50 families of plants can be infected. The virus occurs naturally in many herbaceous and in some woody plants; infection is often symptomless. In lucerne, leaves show mosaic, mottle and malformations and the plants are dwarfed; plants often become symptomless as they mature (Paliwal 1982). Nodulation and winter survival seem to be reduced (Tu & Holmes 1980). On potato, AMV causes calico and tuber necrosis. Infected soybeans also develop calico. AMV is responsible for various symptoms in other crops, for example vein clearing and mottling on tobacco, mosaic in red and white clover, celery and lettuce; yellow mosaic on bean, cowpea and chilli pepper; necrosis and stunting in pea; severe necrosis in tomato; ringspot on petunia; mosaic and stunting on carrot. AMV also infects woody plants: it causes yellow spotting and ring pattern in grapevine (Beczner & Lehoczky 1980, 1981). Significant reductions in yield (up to 50% in lucerne and cowpea) have been observed.

Strains

Several strains have been described; their differentiation is mainly based on the reaction of a selected host range. Some of these strains have been thoroughly investigated: AMV-S from lucerne, AMV-425 from clover, alfalfa yellow spot strain.

Identification and diagnosis

The virus is detected by mechanical inoculation on a selected host range: *Chenopodium amaranticolor* and *C. quinoa* (chlorotic or necrotic local lesions followed by systemic chlorotic and necrotic flecking), *Nicotiana tabacum* (necrotic or chlorotic local lesions, then systemic mild mottle or chlorotic vein banding and ringspots), *Phaseolus vulgaris* (depending on the strain, chlorotic or necrotic local lesions, or no local reaction, but all strains give systemic mild mottle, vein necrosis and leaf distortion). *Ocimum basilicum, Pisum sativum, Vigna unguiculata* are additional diagnostic species (Van Regenmortel & Pinck 1981). The virus is moderately immunogenic and serological detection is applicable. Electron microscopy can also be used for detection.

Transmission

AMV is aphid-borne in the non-persistent manner by

more than 15 aphid species including *Myzus persicae*. It is seed-transmitted in different species; seed transmission is common in lucerne (up to 50% in seeds from infected plants and 10% in commercial seeds). Transmission rate depends on the cultivar and the virus strain.

Geographical distribution, economic importance and control

AMV occurs world-wide and is very common in many countries. In some countries yield reduction and losses are very high, but AMV does not seem, in spite of its ubiquity, to be responsible for major diseases and losses. Control is by the production of virus-free seed and development of resistant cultivars.
DUNEZ

THE BROMOVIRUS GROUP

This group has isometric particles *c.* 26 nm in diameter containing 4 molecules of positive-sense, single-stranded RNA molecules. One molecule of RNA_1 (1.1×10^6 d), one molecule of RNA_2 (1.0×10^6 d), one molecule of RNA_3 (0.7×10^6 d) and one molecule of RNA_4 (0.3×10^6 d) are respectively encapsidated in 3 different particles. RNA_4 is monocistronic for the coat protein. RNA_1, RNA_2 and RNA_3 are required for infection. The viruses are readily mechanically transmitted. The type member is brome mosaic virus (CMI/AAB 215).

Broad bean mottle virus (BBMV) in nature infects broad bean; artificially it can infect most legumes, and induces chlorotic mosaic and vein clearing in its hosts, especially broad bean, French bean and pea (CMI/AAB 101; Lane 1981). It has been reported in Europe and Africa. Detection is by ELISA and mechanical inoculation to *Vicia faba* (vein clearing and blotchy mosaic on non-inoculated tip leaves). No control measures are known except the production of virus-free plants.

Brome mosaic virus (BMV) has been recorded in Africa, N. America and Europe infecting Gramineae (usually symptomlessly) (Edwards *et al.* 1983). Dicotyledonous species are experimentally infectible. Nematodes, beetles, mites and a fungus (*Puccinia graminis* f.sp. *tritici*) have been claimed to be vectors but mechanical transmission during mowing may be as important. The virus has been exhaustively characterized physico-chemically (CMI/AAB 180) but epidemiological knowledge is sparse. Indexing is by mechanical inoculation into *Zea mays* or *Chenopodium* species or by serology (ELISA, IEM or gel diffusion). Control is by propagation of virus-free plants; sources of resistance have been identified in wheat and barley.
COOPER

THE CARNATION MOTTLE VIRUS GROUP

The group has the characteristics of its main member.

Carnation mottle virus (CarMV)

Basic description

CarMV has isometric particles *c.* 28 nm in diameter, containing a single RNA species (17% of the 1.4×10^6 d particle). The protein shell has 240 structural units (M-Wt 26300 d) probably arranged in icosahedral symmetry (T = 4). The virus is very immunogenic.

Host plants and diseases

Natural infections are only found in cultivated carnations. CarMV can also be transmitted experimentally to many plant species, but mechanical inoculation of plants other than Caryophyllaceae or Chenopodiaceae is often difficult, because of inhibitors in carnation sap (Hollings & Stone 1964; CMI/AAB 7). Infected carnations develop a very faint and diffuse mottle on the leaves that is difficult to observe in carnation crops. Furthermore, many infections, especially of the 'attenuated form', are symptomless. In red cultivars, an attenuation of petal coloration, occurring at low temperatures has been attributed to the virus. All present commercial cultivars are susceptible. However, some carnations of the Mediterranean type seem difficult to infect (Poupet *et al.* 1970).

Strains

A PSR strain, isolated from carnation cv. Pink Shibuiya can be distinguished from the type by the necrotic response of several indicator plants (Hollings & Stone 1964). A distant serotype (PR4) has been described by Kemp & Fazekas (1966).

Identification and diagnosis

Natural infections in carnation crops cannot be detected by visual inspections and it is necessary to use indexing methods. Mechanical transmission onto *Chenopodium quinoa* is by far the most sensitive technique (less than 0.01 ng virus/ml): chlorotic local lesions appear within 7 days, sometimes followed by a systemic infection (15 days). Biological assay on *Chenopodium* is the only means of revealing the so-called 'attenuated form' of infection. When virus concentration is very low in some infected carnations, the test plant develops a few large local lesions, 3–4 weeks after inoculation. After serial passage from *Chenopodium* to *Chenopodium*, the typical reaction occurs (Poupet *et al.* 1972). Serological detection of the virus by agar gel double immunodiffusion has been proposed (Devergne & Cardin 1967). The test is rapid but of low sensitivity (5 μg/ml). The ELISA technique appears to be the best method for large scale routine assays because of its feasibility and good sensitivity (about 1 ng/ml) but it does not detect the attenuated form (Devergne *et al.* 1982).

Geographical distribution and transmission

CarMV is widespread in all European countries wherever carnations are grown. Some years ago most commercial cultivars were wholly infected. CarMV is not aphid-transmitted; its spread in carnation crops is mainly by handling and through foliage contact. The virus is maintained by vegetative propagation (Zandwort 1973).

Economic importance and control

The damage caused by CarMV is insidious and was long overlooked. Despite lack of conspicuous symptoms, the virus is responsible for reduced hardiness and poor plant productivity (it increases losses during rooting). CarMV enhances the severity of symptoms caused by other carnation viruses (ringspot, etched ring, necrotic fleck). CarMV occurs in all parts of infected carnations including (at a low concentration) the meristem cells. Consequently, it is the most difficult of the carnation viruses to remove by heat treatment or meristem-tip culture (Maia *et al.* 1969; Hollings *et al.* 1972). Regenerated carnation stocks must be indexed several times during the successive propagation steps before the sale of virus-free cuttings (to avoid reappearance of the attenuated form or re-infection). The sanitation programme involves host indexing on *Chenopodium* in conjunction with ELISA tests (Devergne *et al.* 1982).
DEVERGNE

Cucumber leaf spot virus is distinct from, but in the same group as, carnation mottle virus. It was first described in the GDR. Young leaves of glasshouse cucumbers show light green to yellowish, indistinct spot-like clearings, with brown necrotic centres. Growth of infected plants is strongly inhibited. Cucumber is the only natural host but the virus can be artificially transmitted onto several herbaceous plants. Mechanical transmission is very easy and the virus is seed-transmissible and soil-borne (CMI/AAB 319). No vector is known. Indexing is by mechanical inoculation onto *Chenopodium quinoa* (inoculated leaves show necrotic spots within 7–9 days), or by serology (double diffusion test or IEM), and control is mediated by soil disinfection (Weber *et al.* 1982).
WEBER

THE CAULIMOVIRUS GROUP

This group has isometric particles *c.* 50 nm in diameter with one molecule of double-stranded DNA of about 8000 base pairs. This open circular DNA molecule shows single-strand discontinuities at specific sites. The coat consists of one protein. DNA replication involves a reverse transcription of the major RNA transcript. Caulimoviruses are transmitted by aphids both in the semi-persistent and the non-persistent manner. They have a restricted host range with most members able to infect plants in only 1–3 families. Type member is cauliflower mosaic virus (CMI/AAB 295).

Carnation etched ring virus (CERV)

Basic description

The virus has particles 45 nm in diameter (Hollings & Stone 1961; CMI/AAB 182). In diseased carnations, unidentified 25 and 28 nm particles are often associated with this virus.

Host plants and diseases

Found naturally only in cultivated carnation, this virus causes necrotic rings and flecks, blotches or line patterns which appear mainly on the leaves in spring. Symptom expression depends on temperature and is enhanced in mixed infections with other viruses, particularly carnation mottle virus (CMI/AAB 182).

Identification and diagnosis

The virus can easily be identified by graft transmission to the susceptible cv. Joker which develops very typical necrotic blotches (Hollings & Stone 1961). It is transmissible by mechanical inoculation or by aphid (*Myzus persicae*) to *Silene armeria* (systemic infection) (Hakkaart 1968) and *Saponaria vaccaria* cv. Pink Beauty (red local lesions) (Hakkaart 1974). Symptoms following sap inoculation are more severe when carnation mottle virus is also present. Identification of the virus can also be attempted by electron microscopy, in dip preparations or in thin sections (characteristic inclusions). For routine assay, serological methods are useful: agar gel double diffusion, ELISA, IEM.

Geographical distribution, transmission, economic importance and control

World-wide, the virus causes severe damage in some susceptible cultivars especially in N. Europe. The disease can be spread by aphids. Carnation etched ring cannot be eliminated by thermotherapy alone; healthy clones are obtained by meristem-tip culture (sometimes combined with heat treatment).
DEVERGNE

Cauliflower mosaic virus (CaMV)

Basic description

CaMV (CMI/AAB 243) was described in 1937 and later shown to contain double-stranded DNA (Shepherd *et al.* 1970). CaMV is serologically related to dahlia mosaic and carnation etched ring viruses. The encapsidated genome is circular and of similar size (8 kb), bearing 2–4 single-stranded interruptions depending on the virus or its isolate. The first nucleotide sequence of CaMV DNA was established by Franck *et al.* (1980).

Host plants and diseases

Many species of the Cruciferae are susceptible to CaMV. Among the Solanaceae, *Nicotiana clevelandii* and *Datura stramonium* can be infected by some strains of CaMV. Infection is systemic and symptoms are of mosaic-mottle type and not sufficiently distinguishable from those of several other viruses.

Identification and diagnosis

Characteristic inclusion bodies are induced in the cytoplasm of CaMV-infected cells. They can be detected in epidermal strips by staining with 0.5% phloxine (Fujisawa *et al.* 1967). A more reliable test is electrophoretic analysis of the virion DNA, which can be performed within one day starting from a few grams of infected leaves (Gardner & Shepherd 1980). Viral DNA can also be identified directly in infected leaves by molecular hybridization (Melcher *et al.* 1981). Identification can also be done by mechanical inoculation to *Brassica campestris,* which develops chlorotic local lesions and systemic vein clearing and mottling.

Transmission

In nature the virus is spread by aphids. It is not seed-transmitted. In the laboratory it can be mechanically propagated using an abrasive, either as the complete virion or as its DNA alone.

Geographical distribution

CaMV is common and widely distributed throughout the temperate regions of the world.

Economic importance and control

Insufficient protection of the inflorescence due to reduced leaf area, and reduction in size of the inflorescence, can make the crop unmarketable. Losses are generally limited, but 20–50% reductions in yield have been noted in some cultivars. The disease is controlled by using virus-free seed and by isolating seed beds. Virus-free plants can be obtained by *in vitro* culture from inflorescence tissue.

Special research interest

A model for replication of the CaMV genome has been proposed, involving reverse transcription of its major transcript (Pfeiffer & Hohn 1983; Volovitch *et al.* 1984). Possible use as a gene vector has been investigated (Hohn *et al.* 1982; Howell 1982).
YOT

Dahlia mosaic virus (DMV) induces yellowing and pale green banding along midribs associated in some cultivars with stunting and shortening of internodes. It occurs world-wide. Aphid-transmitted, the virus can be retained by aphids for more than 3 h. Indexing is by mechanical transmission to *Verbesina encelioides* or *Zinnia elegans* and by serology (CMI/AAB 51). The virus is serologically related to cauliflower mosaic virus and carnation etched ring virus (Shepherd & Lawson 1981), and eliminated by meristem-tip culture.

Strawberry vein banding virus (SVBV) (CMI/AAB 219) occurs in N. America, Brazil, Australia and probably in Turkey, Hungary, Ireland and USSR (EPPO 101). Symptoms are vein banding in strawberry, with crinkle, yellows and leaf curl in mixed infections. Aphid-borne in a non-persistent manner it is indexed by grafting or aphid transmission onto the indicator *Fragaria virginiana* UC-12 or *F. vesca* UC6 (chlorotic vein banding), or by ELISA or IEM using cauliflower mosaic virus antiserum. Control is by the

propagation of healthy plants: virus-free mother plants obtained by heat treatment (10 days at 42°C) or meristem-tip culture. The disease referred to as strawberry leaf curl and attributed to strawberry virus 5 is probably a complex of SVBV and the latent A strain of strawberry crinkle virus (q.v.).
POLÁK

THE COMOVIRUS GROUP

This group has two types of isometric particles (M and B) *c.* 28 nm in diameter respectively containing two molecules of positive-sense, single-stranded RNA of 1.4×10^6 d (M) and 2.4×10^6 d (B); a third type of particle consists of an empty capsid. Two coat polypeptide species 23000 and 37000 d. Both RNA species are necessary for infection but RNA_1 alone is able to replicate in protoplasts. M-RNA contains the gene for the coat polypeptides and B-RNA the gene for a virus-encoded replicase. Members of the group are readily mechanically transmitted. Some viruses of the group are transmitted by beetles. Type member is cowpea mosaic virus (CMI/AAB 199).

Andean potato mottle virus (APMV) is found in thc Andean region and Brazil, where it causes strong mottle on sensitive potato cultivars and is transmitted by the beetle *Diabrotica viridula.* This is one of the group of S. American potato viruses considered to necessitate strict quarantine precautions in Europe (CMI/AAB 203; EPPO 128).
SMITH

Bean pod mottle virus (BPMV) infects bean and soybean, causing severe systemic mottle and extensive chlorosis. Pods and seed coats are mottled. It is presently restricted to the USA, but is liable to be introduced into Europe. It is transmitted by chrysomelid beetles but, apparently, not through seeds. Indexing is by mechanical inoculation to soybean and French bean (severe mosaic, mottling of pods and seed coats) and by serology. No control measures breeding programmes (Stace-Smith 1981; CMI/AAB 108).

Broad bean stain virus (BBSV) (CMI/AAB 29) has a wide geographical distribution (Europe, Mediterranean basin, Australia). Broad bean is the only known natural host. Field symptoms consist of severe leaf mosaic, reduced growth, poor fruit setting, malformation of pods and brown discoloration of seeds. Natural transmission is through seeds (up to 40% infection) and curculionid vectors *Apion* and *Sitona.* Indexing is by mechanical inoculation of *Gomphrena globosa* (reddish local lesions) and various Leguminosae (French bean, broad been, pea, mostly reacting with systemic mottling) and serology (immunodiffusion, ISEM). The virus is readily detected in seed embryos by ISEM.
MARTELLI

Broad bean true mosaic virus (BBTMV) is widespread in Europe and N. Africa. It naturally infects broad bean, inducing systemic leaf mottling indistinguishable from that induced by broad bean stain virus, to which the virus is distantly related serologically (CMI/AAB 20; Jones & Barker 1976). It is seed-borne, and incidence in *Vicia* can be high; infected seedlings are severely stunted and produce few seeds. It is transmitted by mechanical inoculation of sap and by *Apion* and *Sitona* weevil species, *A. vorax* being the most efficient vector (Cockbain *et al.* 1975). The virus can also be detected in *Vicia* sap by gel double-diffusion serology. Virus particles are isometric with properties typical of comoviruses. Control is best exerted by using certified virus-free seeds.
JONES

Cowpea mosaic virus (CPMV) (CMI/AAB 197) is the type member of the comovirus group and is found in Africa, Cuba, USA and S. America where it causes mosaic on *Vigna* species. It is transmitted by beetles and is seed-borne. It is not present in Europe, nor likely to infect any European crop, but has been widely used in many European laboratories for investigating replication and expression of the positive-sense RNAs of a bipartite genome virus (Franssen *et al.* 1984). Indexing is by mechanical inoculation to *Chenopodium quinoa* (local yellow lesions followed by systemic mosaic and distortions) and cowpea, or by serology.

Cowpea severe mosaic virus (CPSMV) (CMI/AAB 209) exists as several strains, all differing biologically and biochemically from the related 'common strain', cowpea mosaic virus. It occurs naturally in N. and S. America on leguminous crops and weeds, inducing severe mosaic and reducing crop yields by up to 50%. It is mechanically transmissible, beetle-transmitted and seed-borne (percentage of infected seeds is 3–5%). Detection is by mechanical transmission onto *Chenopodium amaranticolor* (local reaction without systemic invasion), *Vigna unguiculata* or by serology.

Radish mosaic virus (RaMV) (CMI/AAB 121) was

initially isolated from radish but also occurs on most crucifers: turnip, cabbage, cauliflower and mustard. It causes systemic mottle on radish, and mottle, ringspot, line pattern and necrosis of the leaves and stunting on turnip. It is beetle-transmitted. Common in Europe but also recorded in the USA and Japan. Indexing is by mechanical inoculation to *Brassica campestris* (chlorotic/necrotic local lesions followed by systemic mottling) and by serology (Stace-Smith 1981).

Squash mosaic virus (SMV) occurs in various Cucurbitaceae. *Cucurbita* species, melons etc. (CMI/AAB 43). On watermelon, symptoms consist of stunting, leaf distortions with some local chlorosis or necrosis. On melon, leaves are distorted, rugose and mottled with dark green areas. It is widely distributed in the Western hemisphere and also occurs in countries throughout the world (Japan, N. Africa) (Lockhart *et al.* 1982). It is liable to be introduced into Europe where it could presumably damage courgettes, melons and watermelons.

Two serologically distinct strains have been described, referred to as group 1, which infects a broad range of cucurbits, and group 2, which experimentally infects cantaloup melon (Nelson & Knuhtsen 1973a). It is serologically related to radish mosaic virus and bean pod mottle virus. It is transmitted by chrysomelid beetles and, like all comoviruses, by seed: group 1 by the seeds of pumpkin, squash, melon and watermelon (Alvarez & Campbell 1978); group 2 in pumpkin and squash only (Nelson & Knuhtsen 1973b). Transmission may occur by handling and pruning, mainly in glasshouse-grown cucurbits. There is no evidence of pollen transmission. Indexing is by mechanical inoculation of resistant *Cucumis metuliferus* lines or by standard serological methods (ELISA). Control is achieved by testing seed lots to prevent seed transmission (Nolan & Campbell 1984) and in the field by chemical control of vector populations.

LECOQ

THE DIANTHOVIRUS GROUP

This group exhibits one type of isometric virus particle *c.* 32 nm in diameter with two molecules of positive-sense, single-stranded RNA (M-Wt 1.5 and 0.5×10^6 d) and one coat polypeptide species (40000 d). The coat protein gene is on the larger RNA. Readily mechanically transmissible and soil-borne. Type member is carnation ringspot virus.

Carnation ringspot virus (CRSV) causes an extremely damaging disease of carnation. Infected plants are severely distorted and stunted; leaves show numerous chlorotic and necrotic spots (Kassanis 1955). Flowers are distorted and of poor quality. The virus, with isometric particles *c.* 30 nm in diameter is the type member of the dianthovirus group; it is transmissible by foliage contact and sap inoculation (Hollings & Stone 1965a: CMI/AAB 308). Because of the severity of these symptoms, rogueing infected plants is easy. Therefore, the disease has now practically disappeared from well-maintained carnation stocks. The virus is eliminated by heat treatment or meristem-tip culture. CRSV has also been isolated from pear with stony pit or vein yellows, from apple with spy decline and from declining sour cherries; the role of the virus in these diseases has not yet been demonstrated and the susceptibility of these species to CRSV needs confirmation.

DEVERGNE

Red clover necrotic mosaic virus (RCNMV) is a virus with isometric particles *c.* 27 nm in diameter occurring in red clover plants showing severe leaf mottle, distortions and necrotic areas,with moderate to severe stunting. These symptoms are quite obvious in winter but are often masked in summer. Mosaic also develops in infected *Melilotus officinalis*. Reported in Europe, detection is by transmission to *Chenopodium quinoa* (local lesions without systemic infections). The virus is a good immunogen and can be detected serologically. Several isolates have been described. No vector is known and no seed transmission has been reported (CMI/AAB 181).

THE GEMINIVIRUS GROUP

Geminate particles (18×20 nm) consist of two incomplete icosahedra. Some members of this group have two molecules of single-stranded DNA ($0.7–0.8 \times 10^6$ d): both DNA molecules or both particles are necessary for infection. Other members have only one single-stranded DNA species. One coat protein with M-Wt 28–34000 d. Transmitted by leafhopper or whiteflies. On the basis of the number of DNA species and vector species, the geminivirus group could be divided in future into two different subgroups. So far, the type member is maize streak mosaic virus (which does not occur in Europe).

Beet curly top virus (BCTV) has been described from N. and S. America, Iran and India (CMI/AAB 210). It probably occurs in many Mediterranean countries

(EPPO 89). Natural host plants are beet, spinach, tomato, potato (potato green dwarf), cucumber, tobacco, bean, flax, clover, parsley, and some ornamental and weed plants. Symptoms on beet are leaf twisting and general dwarfing and yellowing. Yield and sugar content are greatly decreased (Magyarosy & Duffus 1977). Transmissible by leafhoppers (*Circulifer tenellus* in the USA, *Circulifer opacipennis* in Iran, *Agallia albidulla* in India) in a persistent manner, it is indexed by leafhopper inoculation to *Atriplex serenana* or *Lepidium nitidum* (death of plants within 2 months), or by serology/latex and ELISA techniques. Control is by the use of virus-resistant cultivars of beet, early sowing and insecticide treatment.
POLÁK

Tobacco leaf curl virus causes stunting in tobacco, with small curled, twisted and puckered leaves; leaves sometimes also show green thickenings and enations along the veins. It also induces curling, yellowing and puckering of tomato and pepper leaves. It differs from tomato yellow leaf curl virus which has a much longer period of latency in its whitefly vector. This virus is transmitted by *Bemisia tabaci* with a minimum latent period of 4–9 h (CMI/AAB 232): it occurs mainly in tropical and subtropical areas but has also been reported in temperate regions, Europe and the USA. Detected by vector transmission onto *Datura stramonium*, tomato or tobacco, it is moderately immunogenic and serology can provide an additional detection method.

Tomato yellow leaf curl is likely to be caused by a geminivirus, as indicated by the ultrastructural modifications observed in cells of affected tomato plants from Israel, Tunisia and Senegal. Geminate virus-like particles were also present in partially purified plant extracts which proved infective when fed to whiteflies through membranes (S. Cohen, unpublished information). The disease occurs in the Middle East, large areas of N., W. and E. Africa, India and the Americas (Mexico and Venezuela) and potentially Europe. Affected plants are severely stunted, shoots are erect, leaflets are small, brittle, mis-shapen, curled upwards and strongly chlorotic. Flower abscission may be severe and fruit set very poor or nil. Crop losses may reach 80% or more, being particularly heavy in late summer plantings when vector populations are very high (Nitzany 1975). In nature the virus also infects *Malva nicaeensis* and *Datura stramonium*. Experimental host range is narrow, mainly restricted to Solanaceae. *Phaseolus vugaris* cv. Bulgarit and *Lens esculenta* can be infected artificially but show no symptoms. Transmission occurs neither by contact nor seed, and only with difficulty by sap inoculation in *Nicotiana glutinosa*. Natural transmission in *Bemisia tabaci* is persistent and circulative, but cyclic due to an antiviral mechanism triggered by the virus. Vectors require multiple access feedings on a virus source. Instars acquire and transmit the virus to and through the adult stage. Females are more efficient than males. Indexing is by grafting or vector-mediated inoculation to susceptible tomato cultivars. Control is through the use of pesticide sprays against the vector, cultural practices (mulching, mixed crops, physical barriers) and the use of resistant cultivars (Cohen 1982). Resistance genes have been identified in *Lycopersicon pimpinellifolium* line LA 121 and *L. peruvianum*.
MARTELLI & COHEN

Wheat dwarf virus occurs in Bulgaria, Czechoslovakia, Romania, Sweden and USSR. It naturally infects cereals, mainly winter wheat. Symptoms are severe dwarfing, chlorotic blotches and necrosis on leaves, poorly developed ears and finally yellowing of the whole plant (Vacke 1961). It is transmitted in the persistent manner by leafhoppers *Psammotettix alienus*. Identification is by vector transmission onto spring cultivars of *Triticum aestivum* and by ELISA. Control is by eradication of vectors and sources of infection, and by growing tolerant cultivars.
VACKE

THE HORDEIVIRUS GROUP

This group has the characteristics of its type member.

Barley stripe mosaic virus (BSMV)

Basic description

BSMV is a multi-component virus with rigid elongated virions about 20×100–150 nm. For review, see CMI/AAB 68 and Jackson & Lane (1981).

Host plants and diseases

Natural infections are mostly found in barley and to a lesser extent in wheat. In barley the infection may not be apparent or it may show different grades of chlorotic or necrotic spots and stripes on leaves, depending on the cultivar, BSMV strain and environmental conditions. In severely affected plants, sterile flowers occur.

Identification and diagnosis

Symptom expression is unreliable for diagnosis. A reliable diagnosis can be made only by serology. The following methods, in order of increasing sensitivity, have been developed for seed health testing but may also be used for leaf testing: i) double immunodiffusion using detergent-treated plant tissue extract (Carroll *et al.* 1979); ii) single immunodiffusion in gel incorporating detergent (SDS) and antiserum (Slack & Shepherd 1975); iii) latex agglutination (Lundsgaard 1976); iv) IEM (Brlansky & Derrick 1979); v) ELISA (Lister *et al.* 1981).

Transmission

The primary source in the field is infected seed. Secondary spread is by sap transmission to neighbouring plants.

Geographical distribution and economic importance

World-wide. Due to high seed transmission rates and inapparent infections, BSMV is probably present in most barley gene banks unless special precautions are taken. BSMV is not common in commercial seed lots.

Control measures

Infected material should be eradicated from barley gene banks. Virus-tested seed should be used (EPPO 88).

LUNDSGAARD

THE PENAMOVIRUS GROUP

The group has the characteristics of its only member.

Pea enation mosaic virus (PEMV)

Basic description

The particles are isometric, polyhedral, 28 nm in diameter, contain *c.* 28% RNA and sediment as two centrifugal components at 99S (M) and 112S (B). The genome is bipartite and constituted of two molecules of single-stranded RNA: RNA_1 contained within B particles (M-Wt 1.7×10^6 d), and RNA_2 contained in M particles (M-Wt 1.3×10^6 d). Both RNA species are needed for infectivity. RNA_1 codes for coat protein which has a major polypeptide with M-Wt 22 000 d. Aphid-transmitted isolates have an additional minor polypeptide with M-Wt 28 000 d (Hull & Lane 1973; Hull 1981).

Host plants and diseases

In nature the virus infects legumes but it can be transmitted artificially to several non-leguminous hosts. Pea and broad bean are the major natural hosts. Infection begins with a light vein clearing followed by chlorotic flecks, then translucent spots along the veins and enations on the underside of the leaf blade. Severe infections result in heavy mottling and rolling of the leaves, stunting and distortion of plants, malformation of pods and occasional necrosis of leaves and stems (CMI/AAB 257; Hull 1981).

Identification and diagnosis

Field diagnosis is based on symptoms shown by infected plants. Translucent spots and enations, when shown, are highly characteristic of the disease. Similar symptoms can be reproduced by mechanical inoculation in several greenhouse-grown legumes. Serology (immunodiffusion, IEM, ISEM) is applicable.

Transmission

The virus is transmitted in a persistent circulative manner by several aphid species, *Acyrthosiphon pisum* being the main vector. Acquisition time is short (s or min) but transmission occurs after a latent period and may continue for 4–5 weeks. The virus is not transmitted transovarially nor does it apparently multiply within the vector. Repeated transfers by mechanical transmission induce loss of aphid transmissibility; this is associated with the loss of the 28 000 d protein.

Geographical distribution, economic importance and control

The virus is widespread in north temperate regions. Crop losses to pea and broad bean as high as 50% of the yield have been reported (Hull 1981). Control is by the use of resistant cultivars when available (e.g. pea).

MARTELLI

THE REOVIRUS GROUP

A group of viruses with particles 60–80 nm in diameter, with a two-layered capsid and double-stranded RNA. The virion contains an RNA-dependent RNA polymerase. The viral genome is divided into 10–12 RNA pieces with a total M-Wt of about $15-20\times10^6$ d. Based on insect vector (leafhopper or planthopper), number of RNA pieces and structure of capsid, 3 subgroups have been described with type members; wound tumor virus (subgroup 1), sugar cane Fiji disease virus (subgroup 2, the fijiviruses) and rice ragged stunt virus (subgroup 3). None of these type members occur in Europe. Transmission is in the persistent manner with cases of transovarial transmission (CMI/AAB 294).

Cereal tillering disease virus (CTDV) belongs to subgroup 2. It multiplies only in Gramineae and in its planthopper vector, *Laodelphax striatellus*. No transovarial transmission has been reported. The virus is only known in N. Europe where epidemics on barley and oats are infrequent (Milne & Luisoni 1977). Naturally infected barley shows excess tillering, severe dwarfing, and malformed leaves with serrated margins. Enations observed on experimental infection of maize are not produced on barley or oats (Lindsten 1985). The virus is serologically related to maize rough dwarf virus. No control of the disease seems necessary.
LAPIERRE

Lolium enation virus, in subgroup 2, infects *Lolium perenne* and *L. multiflorum*. Plants show characteristic enations. Reported in the FRG with unknown vectors, it is serologically related to oat sterile dwarf virus (Lesemann & Huth 1975).

Maize rough dwarf virus (MRDV) belongs to subgroup 2 (CMI/AAB 72). It occurs in Europe (Italy, Switzerland, Spain, France, Czechoslovakia) and in Israel where the planthopper vector *Laodelphax striatellus* is present. Leaves of infected plants become darker green with small enations on the lower surface. Few adventitious roots are formed. Early infected maize plants are severely dwarfed and give poor yields (Harpaz 1972). *Digitaria sanguinalis* and *Echinochloa crusgalli* are natural hosts. The virus is serologically related to cereal tillering virus. Control could be obtained by growing resistant cultivars.
SIGNORET

Oat sterile dwarf virus (OSDV) belongs to subgroup 2 (CMI/AAB 217). It occurs in some countries of C., N. and W. Europe. Symptoms are severe stunting, increased tillering, vein swelling, enations on veins and sterility in oats, barley and other Gramineae. It is persistently transmitted by planthoppers *Javesella pellucida* and a few other delphacid species. Diagnosis is by vector inoculation onto *Avena Sativa* and by double diffusion serological assay. The virus is serologically related to lolium enation virus. Control is by cultural practices (deep ploughing) or chemical eradication of vector nymphs.
VACKE

THE RHABDOVIRUS GROUP

The particles are enveloped bacilliform or bullet-shaped ($135-380\times45-95$ nm) with a negative-sense, single-stranded RNA (M-Wt $3.5-5.0\times10^6$ d). RNA-dependent RNA polymerase occurs in the virion, and the nucleocapsid is formed by a helically wound nucleoprotein. More than 30 members have now been described. At least two groups have been described on the basis of protein composition and site assembly: subgroup A (lettuce necrotic yellows) and subgroup B (potato yellow dwarf). Viruses devoid of envelopes have been observed; they could form a third subgroup. Most of them are leafhopper- or aphid-transmitted. Multiplication occurs in the vectors (CMI/AAB 244).

Barley yellow striate mosaic virus, in subgroup A, is identical to wheat chlorotic streak virus, closely related to maize sterile stunt virus, distantly related to northern cereal mosaic virus and unrelated to maize mosaic virus (Milne *et al.* 1986). It occurs in Italy and possibly in France. The host range is restricted to Gramineae; symptoms are striate mosaic or chlorotic stripes, stunting, and leaf narrowing. It is transmitted propagatively by planthoppers and through the eggs of its natural vector, *Laodelphax striatellus* (Conti 1980); it is not sap-transmissible or seed-borne. The virus multiplies in the cytoplasm of plant cells forming characteristic viroplasms (Conti & Appiano 1973). Indexing is by IEM and hopper-transmission to barley and wheat (bright striate mosaic). Control is by insecticide treatments on overwintering sites of planthoppers (CMI/AAB 312).
CONTI

Beet leaf curl virus (CMI/AAB 268; EPPO 90), in subgroup B, is restricted to C. Europe. Symptoms in beet are vein clearing of the youngest leaves, which later curve inwards, and a decrease in root growth.

Transmitted by *Piesma quadratum* (Heteroptera) in the persistent (propagative) manner, it is indexed by vector transmission to beet. *Atriplex, Chenopodium, Spinacia* or *Tetragonia tetragonoides*, but no mechanical transmission occurs. The vector can be controlled by parathion.

PROESELER

Broccoli necrotic yellows virus (BNYV) belongs to subgroup A, and has been described in Europe and N. America. It is found occasionally in cauliflower-headed broccoli and in brussels sprouts. It is transmitted by the aphid *Brevicoryne brassicae*. Diagnosis is by mechanical inoculation (difficult) onto *Datura stramonium* (chlorotic to necrotic local lesions after 10 days; systemic mosaic and vein necrosis after 3 weeks) (CMI/AAB 85; Francki *et al.* 1981). Serology is very suitable for detection.

Cucumber toad skin virus is a tentative member of the rhabdovirus group and has been isolated from cucumbers grown under cover in S.E. France. Leaves are crinkled and veins show a typical sinuous course. Plants are severely affected in their growth and fruit production. No vector has been identified so far. Indexing is by mechanical inoculation of aubergine, *Lavatera trimestris* or *Nicotiana glutinosa* which develop systemic infections. Transmission to cucumber is erratic by mechanical inoculation but readily achieved through grafting.

LECOQ

Eggplant mottled dwarf virus (CMI/AAB 115) belongs to subgroup B and occurs throughout the Mediterranean area. Symptoms are severe stunting of aubergine plants, vein clearing, mottling and crinkling of the leaves, and general sterility. It is found in nature only in aubergine, several cultivars of which are susceptible (Martelli & Cirulli 1969). Alternative hosts are tomato, *Lonicera* and *Pittosporum*. Spread is as though the virus were insect-borne but the vector is unknown. The experimental host range is narrow, virtually limited to Solanaceae. Indexing is by mechanical inoculation to *Nicotiana glutinosa* and/or tobacco (large chlorotic local lesions in 10–15 days followed by systemic vein yellowing).

MARTELLI

Lucerne enation virus is a possible member of the rhabdovirus group, and has bacilliform particles (85×250 nm) found in the nuclei and perinuclear spaces of phloem companion and parenchymal cells (Alliot & Signoret 1972; Alliot *et al.* 1972). It causes a disease characterized by the formation of enations on the underside of the midveins of leaflets which become crinkled. Plants may attain normal height but are bushy. The virus, not mechanically transmissible, is persistently transmitted by *Aphis craccivora* (Leclant *et al.* 1973) in the field. It is widespread in France, Mediterranean countries and the Near East (Cook & Wilton 1984). Control methods have not been established.

ALLIOT

Orchid fleck virus is probably a member of rhabdovirus subgroup B, causing chlorotic and necrotic flecks in *Cymbidium, Dendrobium, Oncidium* and many other orchids. It has been described in Europe, S. America and Japan (CMI/AAB 183). Detection by sap inoculation is difficult, but is carried out on *Dendrobium nobile, Cymbidium alexanderi* or other orchids (chlorotic or necrotic spots first on inoculated leaves and later on upper leaves after 3–4 weeks), on *Nicotiana glutinosa* or *N. tabacum* cultivars (bright yellow), Xanthi nc (chlorotic or necrotic local lesions after 2–3 weeks).

Potato yellow dwarf virus (PYDV) belongs to subgroup B. It causes leaf chlorosis, pit necrosis of stems and stunting of the whole potato plant. A few tubers can develop internal necrosis. Restricted to N. America, it is a quarantine pest for Europe (EPPO 128). It is transmitted by several leafhopper species in a persistent manner, the most important being *Aceratagallia sanguinolenta*. It also infects clover symptomlessly but the plant does act as a reservoir for both virus and vector (CMI/AAB 35). Diagnosis is by transmission to *Nicotiana rustica* (chlorotic local lesions, mottle and leaf necrosis) or *N. glutinosa* (vein clearing and mosaic). It can be detected serologically (Francki *et al.* 1981).

Raspberry vein chlorosis virus (RVCV)

Basic description

RVCV, also known as raspberry chlorotic net virus, is aphid-borne, but is not transmissible by mechanical inoculation of sap, nor has it been purified. Electron microscopy of ultrathin sections of infected raspberry leaves and viruliferous aphids show large, enveloped bacilliform particles *c.* 450×75 nm (CMI/AAB 174). It is classified in subgroup A.

Host plants and symptoms

The disease occurs naturally only in wild and cultivated raspberry. Most cultivars show symptoms on infection but a few N. American cultivars appear to be immune. Symptoms are most obvious on leaves of vegetative canes, appearing as a chlorosis of the minor leaf veins either in patches or evenly throughout the leaf. If especially severe, leaves may be distorted and show epinasty (Cadman 1952).

Diagnosis

Leaf symptoms in raspberry are usually diagnostic but identification may be made difficult if other viruses induce additional symptoms simultaneously. In such instances graft-inoculation to indicator cultivars that are not sensitive to these additional viruses is necessary. Alternatively, aphid transmissions to indicator cultivars will separate RVCV from contaminating viruses, but culture of the aphid vector is difficult. Symptoms in graft-inoculated raspberry indicator cultivars such as Lloyd George and Norfolk Giant may take 3–12 months to appear.

Natural transmission and importance

RVCV is transmitted specifically and in the persistent manner by the small raspberry aphid *Aphis idaei*, in which the virus probably also multiplies. The aphid is common in S. England and continental Europe but the high incidence of the disease in Europe is due largely to the planting of infected stocks. Almost all stocks of some European raspberry cultivars are totally infected with RVCV. When occurring alone in raspberry, RVCV is not lethal, but it does decrease vigour and fruit size in some cultivars; these effects are accentuated by infections with other viruses.

Control

Planting of healthy stock is essential. This may be a difficulty for some cultivars where almost all stocks are infected because, unlike most other aphid-borne viruses in raspberry, RVCV is not eliminated from plants by heat treatment. Recently, however, heat treatment followed by meristem-tip culture was successful in eliminating RVCV from plants. In Scotland, where vectors are usually scarce, rogueing infected plants has helped to restrict spread.

JONES

Rubus yellow net virus is a heat-stable component of the raspberry mosaic complex. The size of the bacilliform particle is 25–31×80–150 nm. It occurs in the UK, some European countries and N. America (Converse 1977) and is aphid-borne (CMI/AAB 188). Symptoms in raspberry are a pale net-like vein chlorosis and slightly downward-cupped leaflets (Converse *et al.* 1970). Indexing is by graft or aphid transmission to *Rubus occidentalis*, and control is by the propagation of healthy plants and aphid control.

KRCZAL

Sowthistle yellow vein virus (SYVV) belongs to subgroup B. Besides causing vein yellowing in the weed *Sonchus oleraceus* (sowthistle), it attacks lettuce. It has been reported in Europe and N. America, and is transmitted by the aphids *Hyperomyzus lactucae* and *Macrosiphum euphorbiae*. Bioassay is on *Chenopodium quinoa* (local lesions), *Nicotiana glutinosa* (chlorotic lesions, vein clearing and leaf cupping) or the *Nicotiana* hybrid, *N. clevelandii*×*N. glutinosa* (vein clearing and leaf cupping) (CMI/AAB 62; Francki *et al.* 1981). Serology is suitable for detection.

Strawberry crinkle virus

Basic description

Strawberry crinkle virus is probably a member of subgroup A. The particles are bacilliform, 190–380 nm×69±6 nm. Several strains can be distinguished according to the severity of the symptoms (CMI/AAB 163), including those formerly known as strawberry latent virus strains A and B. A distinct disease in N. America, attributed to **strawberry latent C virus** for which no particles have been described, should be excluded from Europe (EPPO 163).

Host plants and diseases

The occurrence of the virus is limited to the genus *Fragaria*. None of the species of this genus is known to be immune. Symptoms on commercial cultivars vary in relation to the virus strain and the sensitivity of the strawberry cultivar. Whereas mild strains are symptomless in all cultivars, virulent strains cause distinct symptoms in sensitive ones. Characteristic symptoms are distorted, crinkled leaves with leaflets unequal in size and small irregularly shaped chlorotic spots, often associated with the veins. In complex with strawberry mottle, mild yellow edge, and vein banding viruses, the symptoms are increased in severity.

Identification and diagnosis

Indexing is by stolon or excised-leaf grafting to *Fragaria vesca* (clones UC4, UC5, UC6) or *F. vesca*, var. *semperflorens* (alpine seedlings). First symptoms are small chlorotic spots, which later become reddish or necrotic. Angular epinasty of leaflets, petiole lesions and petal streak are diagnostic of crinkle. In complex with strawberry mottle or strawberry vein banding viruses symptoms are intensified and hence indicators infected with one of these viruses are very useful for detecting mild strains of crinkle.

Transmission

Strawberry crinkle virus is transmitted by the aphids *Chaetosiphon fragaefolii* and *C. jacobi* in a persistent circulative manner. It is acquired and transmitted by *C. fragaefolii* within feeding periods of 24 h. The latent period in the aphid varies from 10 to 19 days. The vectors remain infective for at least 2 weeks (Krczal 1980). In experiments with *C. jacobi* it was demonstrated that the virus multiplies in the aphid.

Geographical distribution and economic importance

The virus occurs world-wide, in all strawberry-growing areas where *C. fragaefolii* is present. Virulent strains severely reduce vigour and productivity of infected plants. Due to the synergistic effect with other viruses (see above), strawberry crinkle virus can be one of the limiting factors for growing strawberries.

Control

Propagation of healthy plants and control of vectors are the essential measures. Virus-free plants can be obtained by heat therapy and meristem-tip culture.
KRCZAL

Wheat striate mosaic virus (WStMV) is a name which has been given to two viruses, one in N. America and one in Europe. The former is a rhabdovirus (subgroup A) associated with a disease described in the USA and Canada (Smith 1972; Jons *et al.* 1981) and transmitted by the leafhopper *Endria inimica* (Bell *et al.* 1978). The virus infects 20 species in the Gramineae. First symptoms appear as faint chlorotic dashes or streaks along the veins; usually necrosis follows chlorosis and leaf destruction can be severe. Plants become stunted and yield partially sterile heads and low quality grain. All durum wheat cultivars are highly susceptible; several winter wheat cultivars are susceptible and some resistant. Most cultivars of barley and oats are moderately susceptible. The virus can be detected serologically. It could potentially spread to Europe, where a similar disease (European wheat striate mosaic) has been described. Again, bacilliform particles have been isolated. The disease is transmitted by two planthoppers (Ammar 1975), *Javesella pellucida* and *J. dubia* (transovarial transmission occurs for both viruses). Symptoms are chlorotic spots and thread-like, broken streaks on the leaves; finally the leaves turn completely yellow and plants become stunted. When affected at the seedling stage, plants usually die within 1–2 months of infection. The virus infects oat (chlorotic streaks, reddening) and barley (serrations).

THE SOBEMOVIRUS GROUP

Isometric virus particles *c.* 30 nm in diameter with one molecule of positive-sense, single-stranded RNA (M-Wt 1.4×10^6 d) and one coat polypeptide (30000 d). Coat protein could be translated from a subgenomic RNA of $0.3–0.4 \times 10^6$ d. Each virus has a rather narrow host range. Readily transmitted mechanically, viruses of this group are also transmitted by beetles and may be soil-borne. The type member is southern bean mosaic virus.

Cocksfoot mottle virus (CoMV) only attacks cocksfoot (*Dactylis glomerata*) (CMI/AAB 23). Diseased plants show yellow streaking and mottling of leaves which often become white or necrotic (Serjeant 1967). Susceptible cultivars infected with CoMV frequently die (Catherall 1985). Three viruses are often associated in cocksfoot: CoMV, cocksfoot mild mosaic virus (with isometric particles) and cocksfoot streak virus (a potyvirus). The indexing on wheat proposed by Serjeant (1967) could be confusing and

the ELISA test is more convenient. CoMV is transmitted by both the adults and the larvae of the beetles *Oulema melanopus* and *O. lichenis*. It persisted up to 15 days in *O. melanopus* after an acquisition feeding period of several days (Serjeant 1967). It is not seed-borne. CoMV is known in Europe (Germany, France, UK), Asia (Japan) and Oceania (New Zealand). In the UK, CoMV may be a serious disease on pluriannual cocksfoot crops. Resistent cultivars have been described in the UK and Japan.
LAPIERRE & HARIRI

Southern bean mosaic virus (SBMV) has isometric particles of *c.* 30 nm in diameter, sedimenting as a single component, the sedimentation coefficient is 115S. Coat protein is made up of a single polypeptide of M-Wt 28 000 d. The virus is highly immunogenic (CMI/AAB 274). Only Leguminosae are susceptible. SBMV causes mosaic and mottle diseases in French bean (Yerkes & Patino 1960), cowpea (Shepherd & Fulton 1962), and soybean. French beans, in addition to mosaic symptoms, can develop leaf puckering, vein necrosis, defoliation and marked symptoms on pods (dark green, water soaked, blotched areas). Several strains have been described, the B (bean), C (cowpea), G (ghana) and M (severe or mexican) strains. The strains can be distinguished by the symptoms and vectors. SBMV can be detected by mechanical inoculation onto local lesion cultivars of bean (e.g. Bountiful, Pinto) or cowpea (*Vigna radiata* or *V. unguiculata*). ISEM and ELISA are also suitable for the detection of the virus in plant extracts. SBMV is not serologically related to the other members of the sobemovirus group. Probably occurring world-wide, the virus has been reported in the warm temperate and tropical regions of the Americas, in Africa, Brazil, India and also in southern Europe (France), where it is potentially damaging. SBMV is transmissible by leaf beetles probably in a circulative manner (Walters 1969). In N. America, strains B and C are transmitted by *Cerotoma trifurcata* and *Epilachna varivestis*. Strain C is transmitted in Nigeria by *Ootheca mutabilis*. The virus is seed-borne in beans (1.5%) and cowpea (5–40%) (Lamptey & Hamilton 1974), so control is by the production of virus-free seeds. Complete immunity against SBMV-B infection in French bean is not known but cultivars reacting with local lesion production are considered resistant for commercial purposes.

Sowbane mosaic virus (SowMV) (CMI/AAB 64) (syn. chenopodium mosaic virus) is widespread, causing chlorotic mottling of *Chenopodium* species, many of which are used as experimental assay hosts for other plant viruses. Several *Chenopodium* species used by virologists (*C. amaranticolor, C. quinoa, C. murale*) may carry the virus symptomlessly. Though transmitted by some leafhoppers, fleahoppers or leaf miners, the virus is spread mainly by seed transmission: it occurs in up to 83% of the seedlings of an infected plant (Dias & Waterworth 1967). Usually plants infected from seed show few or no symptoms. The virus can be detected by mechanical inoculation of *C. amaranticolor* which develop local lesions or by serology. It is of no economic importance in itself but is liable to interfere in virus testing on *Chenopodium*.

Turnip rosette mosaic virus has been described in Scotland and Switzerland and primarily infects members of the Cruciferae, but also some Solanaceae, Compositae and Resedaceae. In turnip it induces pronounced vein clearing and necrosis, leaf malformations and rosetting (CMI/AAB 125). Transmitted by chrysomelid beetles, especially *Phyllotreta nemorium*, it can be detected by mechanical inoculation of *Brassica pekinensis* which gives satisfactory local lesions, or by serology.

THE TOBACCO NECROSIS VIRUS GROUP

The group has the characteristics of its only member.

Tobacco necrosis virus (TNV)

Basic description

The virus particles are isometric *c.* 28 nm in diameter and have a sedimentation coefficient in the range of 112–133S (CMI/AAB 14). The particle contains 18–21% RNA (Uyemoto 1981) in one molecule of positive-sense, single-stranded RNA of M-Wt $1.3-1.6\times10^6$ d. The coat is made up of 180 protein subunits with M-Wt *c.* 33 000 d. TNV is often accompanied by a satellite virus (CMI/AAB 15) which has isometric particles of *c.* 17 nm in diameter and a sedimentation coefficient of 50S. Satellite particles have an RNA molecule of M-Wt *c.* 0.4×10^6 d and 60 protein subunits with M-Wt *c.* 22 000 d. TMV coat protein and satellite coat protein are completely different; satellite RNA needs the help of TNV RNA to replicate and to be encapsidated in its own protein. Several strains of TNV and of satellite have been described (Kassanis & Philips 1970). TNV suspensions are good immunogens.

Host plants and diseases

TNV naturally infects many cultivated and wild plants; besides tobacco, various vegetables and ornamentals it has been described in hop and fruit trees (almond, pear). Plants usually show local or necrotic symptoms. Disease outbreaks have been reported on bean (bean stipple streak disease), tulip (Augusta disease, tulip necrosis) and cucumber (cucumber necrosis).

Identification and diagnosis

TNV is usually obtained only from the roots of many naturally infected species. Systemic infection only occurs in a few cases. Indexing is essential and different techniques can be used: mechanical transmission onto indicator plants, e.g. *Nicotiana tabacum, N. glutinosa, Phaseolus vulgaris, Chenopodium* species (all show necrotic local lesions within 3–10 days); serological tests (precipitin, gel diffusion, ELISA, IEM). Serology is essential for strain identification.

Transmission

TNV is readily sap-transmitted and zoospores of the fungus *Olpidium brassicae* (Teakle 1962) are the natural vector. It is spread in soil by water movement; it has also been isolated from river water (Tomlinson *et al.* 1983). TNV is not seed-borne.

Geographical distribution, economic importance and control

TNV is widely distributed, and occurs in many countries especially in Europe. It is common in irrigated fields and in unsterilized glasshouse soil. It is of economic importance in some ornamental plants and vegetables. Control is by propagation of virus-free plants that can be obtained by meristem-tip culture or from seed. Additional measures are avoiding excessive watering (to impede movement of *Olpidium brassicae*) and sterilizing soil in glasshouses.

Special research interest

The association of the satellite virus and TNV has been studied in detail. TNV is infectious alone; the satellite virus is not. The capacity of different TNV strains to activate several satellite virus strains is of special interest.

UYEMOTO

THE TOBRAVIRUS GROUP

These have two types of elongated rigid particles and two molecules of positive-sense, single-stranded RNA. The long particle (180–215×22 nm) contains one molecule of RNA_1 of M-Wt 2.4×10^6 d; the short particle (46–114×22 nm) contains one molecule of RNA_2 whose size ($0.6–1.4\times10^6$ d) depends on the particle length. RNA_1 is infective but only produces RNA_1 molecules in infected plants. RNA_2 contains the coat protein cistron; both particles (or both RNA_1 and RNA_2) are required for the production of progeny long and short particles. There is one coat polypeptide species (22 000 d). The virus is soil-borne by nematodes (*Paratrichodorus* and *Trichodorus* spp.). The type member is tobacco rattle virus.

Pea early browning virus (PEBV) has been described in Europe and N. Africa. It causes scattered necrotic patches in foliage or pods, severe stunting and early death of pea, and mild mottle in lucerne and broad bean. In the latter, it is synergistic with bean leafroll virus giving stem necrosis/death (CMI/AAB 120; Cockbain *et al.* 1983). It is transmitted by trichodorid nematodes and is also seed-borne. Indexing is by mechanical inoculation to *Chenopodium amaranticolor* or *Phaseolus vulgaris* and electron microscopy. Resistent cultivars are available but resistance-breaking isolates are known. Vector eradication with chemicals is inefficient.

COOPER

Tobacco rattle virus (TRV)

Basic description

TRV is the type member of the tobravirus group. Particles are rigid rods, 22 nm in diameter, with two predominant lengths; they are referred to as long (180–200 nm) and short (50–110 nm) particles. Particles contain 5% RNA, and two RNA species RNA_1 (2.4×10^6 d) and RNA_2 ($0.6–1.4\times10^6$ d) (CMI/AAB 12; Harrison & Robinson 1978).

Host plants and diseases

More than 100 species can be naturally infected. Tobacco shows strong systemic necrosis on leaves and stems. Potato can be severely affected—the tuber flesh shows brown corky tissues (spraing); passage of the virus from infected tubers to their progeny depends on the virus strain. Then, in a few plants growing from infected tubers, symptoms can develop on the shoots; these are generally restricted to stunting and some mottling or yellow ringspotting of the leaves (Dalmasso *et al.* 1971; Harrison & Robinson 1981). Symptoms on potato are very similar to those induced by potato mop top virus. TRV is frequent on bulbous ornamentals (tulip, hyacinth, narcissus, gladiolus, lily). Leaves of most of these plants become mottled and bulbs may develop necrotic spots (hyacinth); tulip petals may have streaks of darker colour.

Strains

Two main classes have been described. One is called M, containing both types of particle; the other is called NM, also common in nature, containing only the RNA_1 unprotected by coat protein. These two isolates may produce similar symptoms; they differ in their stability and ease of sap inoculation. M isolates show a high antigenic variability. Three serotypes have been described but two of them should be combined into a I-II serotype which is more or less heterogeneous. Serotype III contains Brazilian isolates and is distant from I-II isolates.

Identification and diagnosis

Mechanical transmission onto susceptible host plants—*Chenopodium amaranticolor* (necrotic lesions on inoculated leaves), *Cucumis sativus* (chlorotic necrotic lesions on inoculated cotyledons), *Nicotiana tabacum* cv. Samsun NN (necrotic spots and rings on inoculated leaves, then necrotic spots, lines or rings on leaves systemically infected), *Phaseolus vulgaris* (small necrotic lesions on inoculated primary leaves) (Harrison & Robinson 1981). Tobraviruses are in general moderately immunogenic, nevertheless, ELISA provides a good detection tool.

Transmission

The virus is naturally transmitted by several nematode species belonging to the genera *Trichodorus* (Van Hoof 1964) and *Paratrichodorus*. Vectors can retain infectivity for months or years. Seed transmission was detected in 5 out of 15 naturally infected weed species and probably plays a role in the natural spread of the virus (Cooper & Harrison 1973).

Geographical distribution, economic importance and control

TRV has been found in many European countries, N. America, Brazil, Japan and New Zealand. The most important diseases are observed in potato and in ornamentals. Control is by using nematicides; e.g. dichloropropane/dichloropropene (150 l/ha) or dazomet (200 kg/ha) give satisfactory control for 2–3 years. Cultivars resistant to TRV (in potato Belle de Locronan, Viola, Eba, Radosa, Spartan) are now available.

Special research interest

Autonomous replication of the RNA_1 and investigations of genome properties through building of recombinant strains with RNA isolated from two different strains.

DALMASSO & DUNEZ

THE TOMATO SPOTTED WILT VIRUS GROUP

The group has the characteristics of its only member.

Tomato spotted wilt virus (TSWV) has special particles about 85 nm in diameter with a lipoprotein envelope. It has a very wide host range (about 160 dicotyledons and 10 monocotyledons) especially in the Solanaceae, Compositae and Leguminosae (CMI/AAB 39). First described in Australia, and later identified in many countries, it probably occurs world-wide. It is important in tomato (sudden browning of young leaves followed by cessation of growth), and tobacco (leaf chlorosis and distortion, cessation of growth). It also infects potato, aubergine, pepper, pea, French beans, cowpea etc. In lettuce, infection usually starts in leaves on one side of the plant, which became chlorotic, then discoloration spreads to the heart leaves and growth stops on one side. The most affected monocotyledonous crop is pineapple. TSWV is transmitted by four species of thrips (only larvae can acquire the virus and often vectors retain the virus for life). Despite the fact that TSWV is extremely unstable in plant extracts and therefore sap-transmission is difficult, inoculation to selected hosts can provide evidence of its identity: brown local lesions on *Petunia hybrida,* small chlorotic lesions on *Cucumis sativus,* local necrotic lesions, then systemic mosaic and deformation on *Nicotiana tabacum* cv. Samsun NN. Additional detection methods are based on electron microscopy and serology (Francki & Hatta 1981).

THE TOMBUSVIRUS GROUP

These have isometric virus particles *c.* 30 nm with one molecule of positive-sense, single-stranded RNA (M-Wt 1.5×10^6 d) and one coat polypeptide (41000 d). Readily transmitted by mechanical inoculation, acquisition through soil has been shown for some members. The type member is tomato bushy stunt virus.

Artichoke mottled crinkle virus has been recorded from several Mediterranean countries (Morocco, Tunisia, Italy, Malta and Greece), and causes symptoms on artichoke which vary with changing environmental conditions. Infections may be virtually symptomless or characterized by severe crinkling, distortion and chlorotic blotching of the leaves, stunting of the plants, malformation and reduction in the size and number of flower heads (Martelli 1965, 1981). The virus can be recovered from the soil but the vector, if any, is unknown; it is transmitted through propagating material. The experimental host range is wide but infections are mostly localized. Indexing is by mechanical inoculation to basil (dark-brown lesions with lighter centre) and *Datura stramonium* (chlorotic/necrotic local lesions, no systemic infection) and serology (immunodiffusion) and IEM.
MARTELLI

Cymbidium ringspot virus (CybRSV) is not a common virus and has only been reported from *Cymbidium* orchids. Symptoms are chlorotic ring-mottle. Indexing is by mechanical inoculation onto *Nicotiana clevelandii* (CMI/AAB 178; Hollings *et al.* 1977).
ALBOUY

Narcissus tip necrosis virus (NTNV) causes necrotic symptoms (leaf streak and apical necrosis) appearing after flowering in some *Narcissus* cultivars in the Netherlands and UK (CMI/AAB 166). This virus, with isometric particles 30 nm in diameter, has many properties in common with the tombusviruses, but appears not to be serologically related to any serotype of this group. It is only mechanically transmitted to *Narcissus,* the sole host known until now. The virus can be detected by serological tests (immunodiffusion) in symptomless cultivars or when dry, hot conditions prevent symptom expression.
DEVERGNE

Pelargonium flower break virus (PFBV) has only been isolated from *Pelargonium zonale* and is of little importance in commercial crops. Many cultivars remain symptomless when infected and only a few develop flower break. The virus can be identified by IEM (CMI/AAB 130; Stone 1980).

Pelargonium leaf curl virus (PLCV)

(Syn. pelargonium leaf crinkle virus).

Basic description

PLCV (Hollings 1962) is serologically related to tomato bushy stunt virus. Particles are isometric *c.* 30 nm in diameter. Three strains have been distinguished. The virus is a very good immunogen (Hollings & Stone 1974).

Host plants and diseases

PLCV naturally infects pelargonium cultivars causing leaf curl disease or 'crinkle'. During winter and spring infected plants show severe symptoms. Yellow stellate spots develop on the younger leaves, these spots then become necrotic and leaves become distorted and terminal buds can abort. In summer and autumn, plants show no obvious symptoms, but there is a consistent reduction in flower size and in the number and quality of cuttings. The virus can be transmitted experimentally to numerous species of different families (Hollings & Stone 1965b).

Identification and diagnosis

In spring, visual selection of material for propagation is possible but symptomless plants do occur under glasshouse conditions. Indexing is by mechanical inoculation onto *Chenopodium quinoa* or *C. amaranticolor*. Infected leaves develop tiny, whitish local lesions in 4–5 days. Use of buffer containing polyethylene glycol is necessary to prevent inhibition by the pelargonium extract. It is impossible to detect PLCV after April–May. ELSIA is also useful for routine detection assay using a suitable extraction buffer (Albouy & Poutier 1980; Albouy *et al.* 1980).

Transmission

PLCV is propagated by cuttings from infected mother plants. It does not seem to be spread through infected tools but there is some evidence for soil transmission.

Geographical distribution and economic importance

PLCV occurs world-wide and is well established and distributed in all pelargonium-growing countries. However, it is no longer of major importance in commercial nurseries.

Control

Elimination of infected mother plants and the use of virus-free material for propagation (also obtained by heat treatment or meristem-tip culture) have reduced the prevalence of the disease; clean stocks of all principal cultivars exist, and re-infection seems very slow.

ALBOUY

Petunia asteroid mosaic virus (PeAMV) was first recorded in petunia in Italy (Lovisolo 1957) but this is an unimportant, even accidental host in practice. Since 1965 the virus has been found in privet (Italy and England), dogwood (England), grapevine (Germany, Italy and Czechoslovakia), hop (Czechoslovakia) and in the roots of various herbaceous plants. Symptoms are growth reduction, colour breaking of flowers, stellate yellowish spots on the leaves accompanied by puckering and malformation of the blades. It is soil-borne but the vector is unknown (Campbell *et al.* 1975). The artificial host range is wide. Indexing is by mechanical inoculation to petunia (most isolates are systemic) and basil (large dark-brown lesions with lighter centre), by serology (immunodiffusion) or IEM. Experimentally it was found that soil treatment with fungicides reduced virus transmission.

MARTELLI & LOVISOLO

Tomato bushy stunt virus (TBSV)

Basic description, host plants and diseases

TBSV is the type member of the tombusvirus group (Weber *et al.* 1970). It has 16 known natural host plants, including tomato, pelargonium, pepper, spinach, sweet cherry, apple, grapevine and the weed *Stellaria media* (CMI/AAB 69). Another 126 species are experimental host plants (Schmelzer *et al.* 1977). The most frequently affected crops are tomato and sweet cherry. Tomato shows shoot tip necrosis, with development of secondary necrosis, giving plants a bushy appearance. Older leaves become yellowish or purple; young leaves show yellowish or purple rings and line pattern. Plants infected when very young rapidly die. Sweet cherry shows vein necrosis and leaf distortion. Severe shoot stunting leads to the development of leaf rosettes. Sensitive cultivars react with a detrimental canker (Albrechtová *et al.* 1975).

Strains

Several serologically distinct strains are known; type strain, carnation Italian ringspot, aubergine mottled crinkle. Petunia asteroid mosaic virus and pelargonium leaf curl virus (q.v.) have been considered as strains of TBSV.

Identification and diagnosis

TBSV can be diagnosed biologically with *Chenopodium amaranticolor* (whitish necrotic dots) or *Gomphrena globosa* (pale necrotic local lesions and rings). Serological detection and identification in herbaceous and woody host plants is done by gel diffusion test, latex test or ELISA (Fuchs *et al.* 1979).

Geographical distribution and transmission

TBSV is present in several European countries, N. and S. America. No natural vectors are known. Vectorless transmission has been demonstrated (Kegler & Kegler 1981). TBSV is transmissible by mechanical inoculation, grafting and *Cuscuta campestris* (type strain). It is seed-transmissible in apple and cherry.

Economic importance and control

Severe losses occur in vegetable and ornamental crops. Control is by selection and propagation of virus-free plants. Soil disinfection is usually inefficient.

KEGLER

Turnip crinkle virus (TCV) is a possible tombusvirus, but differs greatly from the typical tombusviruses by its physico-chemical properties and its replicative strategy (Martelli 1981). Found in Europe, TCV induces symptoms ranging from mild yellowing and puckering of foliage to stunting and rosetting of affected plants (CMI/AAB 109). It infects about 20 dicotyledonous families. Turnip crinkle is transmitted by flea beetles and can be detected by mechanical transmission onto *Chenopodium amaranticolor* (local lesions) and by serology.

THE TYMOVIRUS GROUP

These have isometric particles 30 nm in diameter with positive-sense, single-stranded RNA (M-Wt 2×10^6 d) and one coat polypeptide (20 000 d). A subgenomic RNA is the active messenger for the coat protein. This group is possibly restricted to dicotyledonous hosts and is beetle transmitted. The type member is turnip yellow mosaic virus (CMI/AAB 214).

Eggplant mosaic virus (EMV) (CMI/AAB 124) is an essentially tropical virus causing mosaic of aubergine and tomato. It is closely related to **Andean potato latent virus (APLV)**, which may simply be a strain of it (Fribourg *et al.* 1977), and causes severe deformation and necrosis in sensitive potato cultivars. Limited to the Andean region, it is a quarantine pest for Europe (EPPO 128).

SMITH

Erysimum latent virus (EryLV) (CMI/AAB 222) was described in the GDR and is confined to Cruciferae, occurring naturally in wild *Erysimum* species. It is sap-transmissible and transmitted in nature by flea beetles. Diagnosis is by mechanical inoculation on *Brassica sinensis* or *B. pekinensis* (chlorotic or necrotic local lesions—only plants showing chlorotic lesions develop systemic symptoms; vein clearing, rings and mottling). The virus is strongly immunogenic and serological detection is applicable. It has all the characteristics of the viruses of the tymovirus group (double membrane-bounded vesicles in chloroplasts etc.). While of no economic significance in itself, it is related to okra mosaic virus and Andean potato latent virus (see above) and this group of viruses requires further study in Europe.

Turnip yellow mosaic virus (TYMV) is the type member of the tymovirus group. With a host range confined almost to the Cruciferae, it is responsible for several very minor mosaic diseases in *Brassica* species. Readily transmissible by mechanical inoculation, it is transmitted in the field by flea beetles. It is reported from several European countries. Diagnosis is by transmission to *Brassica pekinensis* (bright yellow or yellow-green mosaic) or by serology. Special attention has been given to the replication of this virus; a virus-encoded protein present in the replication complex is involved in the viral RNA replication (CMI/AAB 230; Matthews 1981).

REFERENCES

Adams G., Sander E. & Shepherd R.J. (1979) Structural differences between pea enation virus strains affecting transmissibility by *Acyrthosiphon pisum*. *Virology* **92,** 1–14.

Albouy J. & Poutier J.C. (1980) Adaptation de la méthode immunoenzymatique à la détection des viroses du Pelargonium. *Annales de Phytopathologie* **12,** 71–75.

Albouy J., Morand J.C. & Poutier J.C. (1980) Les méthodes de mise en évidence des virus du Pelargonium dans le cadre d'une production de plantes saines. *Mededelingen van de Faculteit Landbouwwetenschapen Rijksuniversiteit Gent* **45,** 359–368.

Albrechtová L., Chod J. & Zimandl B. (1975) Nachweis des Tomatenzwergbusch-Virus (tomato bushy stunt virus) in Süsskirschen, die mit virösem Zweigkrebs befallen waren. *Phytopathologische Zeitschrift* **82,** 25–34.

Alliot B. & Signoret P.A. (1972) La maladie à énations de la luzerne, une maladie nouvelle pour la France. *Phytopathologische Zeitschrift* **74,** 69–73.

Alliot B., Giannotti J. & Signoret P.A. (1972) Mise en évidence de particules bacilliformes de virus associées à la maladie à énations de la luzerne. *Comptes Rendus Hebdomadaires de l'Académie des Sciences de Paris, Série D* **274,** 1974–1976.

Alvarez M. & Campbell R.N. (1978) Transmission and distribution of squash mosaic virus in seeds of cantaloupe. *Phytopathology* **68,** 257–263.

Ammar E.D. (1975) Effect of European wheat striate mosaic, acquired by feeding on diseased plants, on the biology of its planthopper vector *Javesella pellucida*. *Annals of Applied Biology* **79,** 195–202.

Beczner L. & Lehoczky J. (1980) Infection of grapevine by alfalfa mosaic virus. II. Transmission of the virus to herbaceous plants, symptomatology, serology and electron microscopy. *Kertgazdaság* **12,** 33–34.

Beczner L. & Lehoczky J. (1981) Grapevine disease in Hungary caused by alfalfa mosaic virus infection. *Acta Phytopathologica Academiae Scientiarum Hungaricae* **16,** 119–128.

Bell C.D., Omar S.A. & Lee P.E. (1978) Electron microscopic localization of wheat striate virus in its leafhopper vector, *Endria inimica*. *Virology* **86,** 1–9.

Brlansky R.H. & Derrick K.S. (1979) Detection of seedborne plant viruses using serologically specific electron microscopy. *Phytopathology* **69,** 96–100.

Cadman C.H. (1952) Studies in *Rubus* virus diseases. II. Three types of vein chlorosis of raspberries. *Annals of Applied Biology* **39,** 61–68.

Campbell R.N., Lovisolo O. & Lisa V. (1975) Soil transmission of petunia asteroid mosaic strain of tomato bushy stunt virus. *Phytopathologia Mediterranea* **14,** 82–86.

Carroll T.W., Gossel P.L. & Batchelor D.L. (1979) Use of sodium dodecyl sulfate in serodiagnosis of barley stripe mosaic virus in embryos and leaves. *Phytopathology* **69,** 12–14.

Catherall P.L. (1985) Resistances of grasses to two sobemoviruses, cocksfoot mottle and cynosurus mottle. *Mitteilungen aus der Biologischen Bundesanstalt für Land- und Forstwirtschaft Berlin-Dahlem* **228,** 92–93.

Cockbain A.J., Cook S.M. & Bowen R. (1975) Transmission of broad bean stain virus and Echtes Ackerbohnenmosaik-Virus to field beans (*Vicia faba*) by weevils. *Annals of Applied Biology* **81**, 331–339.

Cockbain A.J., Woods R.D. & Calilung V.C.J. (1983) Necrosis in field beans (*Vicia faba*) induced by interactions between bean leaf roll, pea early browning and pea enation mosaic viruses. *Annals of Applied Biology* **102**, 495–499.

Cohen S. (1982) Control of white fly vectors of viruses by color mulches. In *Pathogens, Vectors and Plant Diseases, Approaches to Control* (Eds K.F. Harris & K. Maramorosch) pp. 45–56. Academic Press, New York.

Conti M. (1980) Vector relationships and other characteristics of barley yellow striate mosaic virus (BYSMV). *Annals of Applied Biology* **95**, 83–92.

Conti M. & Appiano A. (1973) Barley yellow striate mosaic virus and associated viroplasms in barley cells. *Journal of General Virology* **21**, 315–322.

Converse R.H. (1977) Rubus virus diseases important in the United States. *HortScience* **12**, 471–476.

Converse R.H., Stace-Smith R. & Cadman C.H. (1970) Raspberry mosaic. In *Virus Diseases of Small Fruits and Grapevines*, pp. 111–119. University of California, Berkeley.

Cook A.A. & Wilton A.C. (1984) Alfalfa enation virus in the Kingdom of Saudi Arabia. *FAO Plant Protection Bulletin* **32**, 139–140.

Cooper J.I. & Harrison B.D. (1973) The role of weed hosts and the distribution and activity of vector nematodes in the ecology of tobacco rattle virus. *Annals of Applied Biology* **73**, 53–66.

Dalmasso A., Joubert J. & Munck-Cardin M.C. (1971) Quelques données sur une maladie de la pomme de terre. *Comptes Rendus des Séances de l'Académie d'Agriculture de France* **57**, 1427–1432.

Devergne J.C. & Cardin L. (1967) Utilisation de la réaction sérologique en immunodiffusion comme test de diagnostic du virus du mottle de l'oeillet. *Annales des Épiphyties* **18**, 85–103.

Devergne J.C., Cardin L. & Bontemps J. (1982) Indexages biologiques et immunoenzymatique (ELISA) pour la production d'oeillets indemnes de virus de la marbrure (CarMV). *Agronomie* **2**, 655–666.

Dias H.F. & Waterworth H.E. (1967) The identity of a seed-borne mosaic virus of *Chenopodium amaranticolor* and *C. quinoa*. *Canadian Journal of Botany* **45**, 1285.

Edwards M.L., Cooper J.I. & Massalski P.R. (1983) Some natural hosts of brome mosaic virus in the United Kingdom. *Plant Pathology* **32**, 91–94.

Franck A., Guilley H., Jonard G., Richards K. & Hirth L. (1980) Nucleotide sequence of cauliflower mosaic virus DNA. *Cell* **21**, 285–294.

Francki R.I.B. & Hatta T. (1981) Tomato spotted wilt. In *Handbook of Plant Virus Infections and Comparative Diagnosis* (Ed. E. Kurstak), pp. 492–512. Elsevier, Amsterdam.

Francki R.I.B., Kitajima E.W. & Peters D. (1981) Rhabdoviruses. In *Handbook of Plant Virus Infections and Comparative Diagnosis* (Ed. E. Kurstak), pp. 456–489. Elsevier, Amsterdam.

Franssen H., Goldbach R. & Van Kammen A. (1984) Translation of bottom component RNA of cowpea mosaic virus in reticulocyte lysate: faithful proteolytic processing of the primary translation product. *Virus Research* **1**, 39–49.

Fribourg C.E., Jones R.A.C. & Koenig R. (1977) Host plant reactions, physical properties and serology of three isolates of Andean potato latent virus from Peru. *Annals of Applied Biology* **86**, 373–380.

Fuchs E., Merker D. & Kegler G. (1979) Der Nachweis des Chlorotischen Blattfleckungs-Virus des Apfels (apple chlorotic leaf spot virus), des Stammfurchungs-virus des Apfels (apple stem grooving virus) und des Tomatenzwergbusch-Virus (tomato bushy stunt virus) mit dem ELISA. *Archiv für Phytopathologie und Pflanzenschutz* **15**, 421–424.

Fujisawa I., Rubio-Huertos M., Matsui C. & Yamaguchi A. (1967) Intracellular appearance of cauliflower mosaic virus particles. *Phytopathology* **57**, 1130–1132.

Gardner R.C. & Shepherd R.J. (1980) Procedure for rapid isolation and analysis of cauliflower mosaic virus DNA. *Virology* **106**, 159–161.

Hakkaart F.A. (1968) *Silene armeria*, a test plant for carnation etched ring virus. *Netherlands Journal of Plant Pathology* **74**, 150–158.

Hakkaart F.A. (1974) Detection of carnation viruses with the test plant *Saponaria vaccaria*. *Acta Horticulturae* **36**, 35–46.

Harpaz I. (1972) *Maize Rough Dwarf*. Israel Universities Press, Jerusalem.

Harrison B.D. & Robinson D.J. (1978) The tobraviruses. *Advances in Virus Research* **23**, 25–77.

Harrison B.D. & Robinson D.J. (1981) Tobraviruses. In *Handbook of Plant Virus Infections and Comparative Diagnosis* (Ed. E. Kurstak), pp. 516–540. Elsevier, Amsterdam.

Hohn T., Richards K. & Lebeurier G. (1982) Cauliflower mosaic virus on its way to becoming a useful plant vector. *Current Topics in Microbiology and Immunology* **96**, 193–236.

Hollings M. (1962) Studies of pelargonium leaf curl virus. I. Host range, transmission and properties in vitro. *Annals of Applied Biology* **50**, 189–202.

Hollings M. & Stone O.M. (1961) Carnation etched ring: a preliminary report on an undescribed disease. In *Report of the Glasshouse Crops Research Institute 1960*, pp. 94–95.

Hollings M. & Stone O.M. (1962) The attenuation of carnation mottle virus in plants. In *Report of the Glasshouse Crops Research Institute 1961*, pp. 100–102.

Hollings M. & Stone O.M. (1964) Investigation of carnation viruses. I. Carnation mottle. *Annals of Applied Biology* **53**, 103–118.

Hollings M. & Stone O.M. (1965a) Investigation of carnation viruses. II. Carnation ring spot. *Annals of Applied Biology* **56**, 73–86.

Hollings M. & Stone O.M. (1965b) Studies of pelargonium leaf curl virus. II. Relationships to tomato bushy stunt and other viruses. *Annals of Applied Biology* **56**, 87–98.

Hollings M. & Stone O.M. (1974) Serological and immunoelectrophoretic relationships among viruses in the tombusvirus group. *Annals of Applied Biology* **80,** 37–48.

Hollings M., Stone O.M. & Barton R.J. (1977) Pathology, soil transmission and characterization of cymbidium ringspot, a virus from cymbidium orchids and white clover (*Trifolium repens*). *Annals of Applied Biology* **85,** 233–248.

Hollings M., Stone O.M. & Smith D.R. (1972) Productivity of virus-tested carnation clones and the rate of reinfection with virus. *Journal of Horticultural Science* **47,** 141–149.

Howell S.H. (1982) Plant molecular vehicles: potential vectors for introducing foreign DNA into plants. *Annual Review of Plant Physiology* **33,** 609–650.

Hull R. (1981) Pea enation mosaic virus. In *Handbook of Plant Virus Infections and Comparative Diagnosis* (Ed. E. Kurstak), pp. 239–256. Elsevier, Amsterdam.

Hull R. & Lane L.C. (1973) The unusual nature of the components of pea enation mosaic virus. *Virology* **55,** 1–13.

Jackson A.O. & Lane L.C. (1981) Hordeiviruses. In *Handbook of Plant Virus Infections and Comparative Diagnosis* (Ed. E. Kurstak), pp. 565–625. Elsevier, Amsterdam.

Jones A.T. & Barker H. (1976) Properties and relationships of broad bean stain virus and Echtes Ackerbohnen-mosaik-Virus. *Annals of Applied Biology* **83,** 231–238.

Jones A.T., Murant A.F. & Stace-Smith R. (1985) Raspberry vein chlorosis virus. In *Virus Diseases of Small Fruits* (Ed. R.H. Converse), USDA Handbook (in press).

Jons V.L., Timian R.G., Gardner W.S., Stromberg E.L. & Berger P. (1981) Wheat striate mosaic virus in the Dakotas and Minnesota. *Plant Disease* **65,** 447–448.

Kassanis B. (1955) Some properties of four viruses isolated from carnation plants. *Annals of Applied Biology* **43,** 3–113.

Kassanis B. & Philips M.P. (1970) Serological relationships of strains of tobacco necrosis virus and their ability to activate strains of satellite virus. *Journal of General Virology* **9,** 119–126.

Kegler G. & Kegler H. (1981) Beiträge zur Kenntnis der vektorlosen Übertragung pflanzenpathogener Viren. *Archiv für Phytopathologie und Pflanzenschutz* **17,** 307–323.

Kemp W.G. & Fazekas L.J. (1966) Differentiation of strains of carnation mottle virus in crude plant extracts by immunodiffusion in agar plates. *Canadian Journal of Botany* **44,** 1261–1265.

Krczal H. (1980) Transmission of the strawberry mild yellow edge and strawberry crinkle virus by the strawberry aphid *Chaetosiphon fragaefolii*. *Acta Horticulturae* **95,** 23–30.

Kurstak E. (Ed.) (1981) *Handbook of Plant Virus Infections and Comparative Diagnosis*. Elsevier, Amsterdam.

Lamptey P.N.L. & Hamilton R.I. (1974) A new cowpea strain of Southern bean mosaic virus from Ghana. *Phytopathology* **64,** 1110.

Lane L.C. (1981) Bromoviruses. In *Handbook of Plant Virus Infections and Comparative Diagnosis* (Ed. E. Kurstak), pp. 334–376. Elsevier, Amsterdam.

Leclant F., Alliot B. & Signoret P.A. (1973) Transmission et épidémiologie de la maladie à énations de la luzerne. Premiers résultats. *Annales de Phytopathologie* **5,** 441–445.

Lesemann D. & Huth W. (1975) Nachweis von maize rough dwarf virus-ähnlichen Partikeln in Enationen von *Lolium*-Pflanzen aus Deutschland. *Phytopathologische Zeitschrift* **82,** 246–253.

Lindsten K. (1985) Occurrence and development of hopper-borne cereal diseases in Sweden. *Mitteilungen aus der Biologischen Bundesanstalt für Land- und Forstwirtschaft Berlin-Dahlem* **228,** 81–85.

Lister R.M., Carroll T.W. & Zaske S.K. (1981) Sensitive serologic detection of barley stripe mosaic virus in barley seed. *Plant Disease* **65,** 809–814.

Lockhart B.E.L., Ferji Z. & Hafidi B. (1982) Squash mosaic virus in Morocco. *Plant Disease* **66,** 1191–1193.

Lovisolo O. (1957) Petunia: nuovo ospite naturale del virus del rachitismo cespuglioso del pomodoro. *Bollettino della Stazione di Patologia Vegetale Roma* **14,** 103–119.

Lundsgaard T. (1976) Routine seed health testing for barley stripe mosaic virus in barley seeds using the latex-test. *Zeitschrift für Pflanzenkrankheiten und Pflanzenschutz* **83,** 278–283.

Magyarosy A.C. & Duffus J.E. (1977) The occurrence of highly virulent strains of the beet curly top virus in California. *Plant Disease Reporter* **61,** 248–251.

Maia E., Beck D. & Gagelli D. (1969) Obtention de clones d'oeillet méditerranéens indemnes de virus. *Annales de Phytopathologie* **1,** 311–319.

Martelli G.P. (1965) L'arricciamento maculato del carciofo. *Phytopathologia Mediterranea* **4,** 58–60.

Martelli G.P. (1981) Tombusviruses. In *Handbook of Plant Virus Infections and Comparative Diagnosis* (Ed. E. Kurstak), pp. 61–90. Elsevier, Amsterdam.

Martelli G.P. & Cirulli M. (1969) Mottled dwarf of eggplant (*Solanum melongena*), a virus disease. *Annales de Phytopathologie* No. hors série, 393–397.

Matthews R.E.F. (1981) Portraits of a virus: turnip yellow mosaic virus. *Intervirology* **15,** 121–144.

Melcher U., Gardner C.O. & Essenberg R.C. (1981) Clones of cauliflower mosaic virus identified by molecular hybridization in turnip leaves. *Plant Molecular Biology* **1,** 63–73.

Milne R.G. & Luisoni E. (1977) A serological investigation of some maize-rough-dwarf-like viruses. *Annales de Phytopathologie* **9,** 337–341.

Milne R.G., Masenga V. & Conti M. (1986) Serological relationships between the nucleocapsids of some planthopper-borne rhabdoviruses in cereals. *Intervirology* **25,** 83–97.

Nelson M.R. & Knuhtsen H.K. (1973a) Squash mosaic virus variability: review and serological comparisons of six biotypes. *Phytopathology* **63,** 920–926.

Nelson M.R. & Knuhtsen H.K. (1973b) Squash mosaic virus variability: epidemiology consequences of differences in seed transmission frequency between strains. *Phytopathology* **63,** 918–920.

Nitzany F. (1975) Tomato yellow leaf curl virus. *Phytopathologia Mediterranea* **14,** 127–129.

Nolan A. & Campbell R.N. (1984) Squash mosaic virus detection in individual seeds and seed lots of cucurbits by enzyme-linked immunosorbent assay. *Plant Disease* **68,** 971–975.

Paliwal Y.C. (1982) Virus diseases of alfalfa and biology of alfalfa mosaic virus in Ontario and Western Quebec. *Canadian Journal of Plant Pathology* **4,** 175–179.

Pfeiffer P. & Hohn T. (1983) Involvement of reverse transcription in the replication of cauliflower mosaic virus: a detailed model and test of some aspects. *Cell* **33,** 781–789.

Poupet A., Beck D. & Maia E. (1970) Observations préliminaires sur le comportement du virus de la marbrure de l'oeillet, chez les oeillets méditerranéens. *Annales de Phytopathologie* **2,** 663–668.

Poupet A., Beck D. & Maia E. (1972) Relations entre le virus de la marbrure de l'oeillet (carnation mottle virus) et différentes plantes-hôtes. I. Mise en évidence et propriétés biologiques et sérologiques d'une "souche attenuée" du virus de la marbrure. *Annales de Phytopathologie* **4,** 325–335.

Russo M., Savino V. & Vovlas C. (1982) Virus diseases of vegetable crops in Apulia. XXVIII. Broad bean stain. *Phytopathologische Zeitschrift* **104,** 115–123.

Schmelzer K., Wolf P. & Gippert R. (1977) Gemüsepflanzen. In *Pflanzliche Virologie, Band* 3 (Ed. M. Klinkowski), pp. 1–137. Akademie Verlag, Berlin.

Serjeant E.P. (1967) Some properties of cocksfoot mottle virus. *Annals of Applied Biology* **59,** 31–38.

Shepherd R.J. & Fulton R.W. (1962) Identity of a seed virus from cowpea. *Phytopathology* **52,** 489.

Shepherd R.J. & Lawson R. (1981) Caulimoviruses. In *Handbook of Plant Virus Infections and Comparative Diagnosis* (Ed. E. Kurstak), pp. 848–878. Elsevier, Amsterdam.

Shepherd R.J., Brüning G.E. & Wakeman R.J. (1970) Double-stranded DNA from cauliflower mosaic virus. *Virology* **41,** 339–347.

Slack S.A. & Shepherd R.J. (1975) Serological detection of seed-borne barley stripe mosaic virus by a simplified radial-diffusion technique. *Phytopathology* **65,** 948–955.

Smith K.M. (1972) American wheat striate mosaic. In *Textbook of Plant Viruses,* pp. 574–576. Longman, London.

Stace-Smith R. (1981) Comoviruses. In *Handbook of Plant Virus Infections and Comparative Diagnosis* (Ed. E. Kurstak), pp. 171–195. Elsevier, Amsterdam.

Stone O.M. (1980) Nine viruses isolated from pelargonium in the United Kingdom. *Acta Horticulturae* **110,** 177–181.

Teakle D.S. (1962) Transmission of tobacco necrosis by a fungus *Olpidium brassicae. Virology* **18,** 224–231.

Tomlinson J.A., Faithfull E.M., Webb M.J.W., Fraser R.S.S. & Seclay N.D. (1983) *Chenopodium* necrosis: a distinctive strain of tobacco necrosis isolated from water. *Annals of Applied Biology* **12,** 135–147.

Tu J.C. & Holmes T.M. (1980) Effect of alfalfa mosaic virus infection on nodulation, forage yield, forage protein and overwintering of alfalfa. *Phytopathologische Zeitschrift* **97,** 1–9.

Uyemoto J.K. (1981) Tobacco necrosis and satellite viruses. In *Handbook of Plant Virus Infections and Comparative Diagnosis* (Ed. E. Kurstak), pp. 123–146. Elsevier, Amsterdam.

Vacke J. (1961) Wheat dwarf virus disease. *Biologia Plantarum* **3,** 228–233.

Van Hoof H.A. (1964) Serial transmission of rattle virus by a single male of *Trichodorus pachydermus. Nematologica* **10,** 141–144.

Van Regenmortel M.H.V. & Pinck L. (1981) Alfalfa mosaic virus. In *Handbook of Plant Virus Infections and Comparative Diagnosis* (Ed. E. Kurstak), pp. 415–422. Elsevier, Amsterdam.

Volovitch M., Modjtahedi N., Yot P. & Brun G. (1984) RNA-dependent DNA polymerase activity in cauliflower mosaic virus-infected plant leaves. *EMBO Journal* **3,** 309–314.

Walters H.J. (1969) Beetle transmission of plant viruses. *Advances in Virus Research* **15,** 339–363.

Weber I., Proll E., Ostermann W.D., Leiser R.M., Stanarius A. & Kegler H. (1982) Charakterisierung des Gurkenblattflecken-Virus (cucumber leaf spot virus), eines bisher nicht bekannten Virus an Gewächshausgurken (*Cucumis sativus*). *Archiv für Phytopathologie und Pflanzenschutz* **18,** 137–154.

Weber K., Rosenbusch J. & Harrison S. (1970) Structure of tomato bushy stunt virus. *Virology* **41,** 763–765.

Yerkes W.D. & Patino G. (1960) The severe bean mosaic virus, a new bean virus from Mexico. *Phytopathology* **50,** 334.

Zandwort R. (1973) The spread of carnation mottle virus in carnations in glasshouses. *Netherlands Journal of Plant Pathology* **79,** 81–84.

3: Viruses III
Ungrouped Viruses and Virus Complexes

Some of the viruses covered by this chapter are very well characterized, but their affinities remain uncertain and they have not yet been classified in groups*. Others are only partly characterized; in particular, some have been clearly detected in diseased plants but their causal association with the symptoms remains to be conclusively proved by the fulfilment of Koch's postulates. Finally, some clearly recognizable diseases are associated with complexes of well- or poorly characterized viruses—the full aetiology remains to be determined. Although some of the viruses concerned may belong to defined groups, it has been found more convenient to include such cases in this chapter, under the disease name.

Agropyron mosaic virus (AgMV)† has filamentous particles *c.* 717×15 nm. It causes light green to yellow mosaic in wheat as well as in the weed *Elymus (Agropyron) repens.* It is found in N. America, Finland and Germany. Only Gramineae are naturally infected, but the virus has been reported to produce local lesions on *Chenopodium quinoa.* Diagnosis is by transmission to some cultivars of barley or *Lolium multiflorum.* Pinwheel inclusions can be seen in parenchyma cells (CMI/AAB 118).

Anemone brown ringspot virus causes chlorotic spots with brown rings 2–4 mm in diameter on the young leaves of anemone. Later, these rings darken, enlarge and coalesce to form dark brown areas with a few pale spots. Total necrosis is seen on leaves. Described in the UK in 1956, a virus has been isolated from these plants, and investigated by mechanical transmission to several herbaceous species (Hollings 1958).

Apple stem grooving virus (ASGV) has virus particles *c.* 620×12 nm (CMI/AAB 31) and is distributed world-wide. Symptoms are stem grooving, brown line and graft union abnormalities in apple cv. Virginia Crab (Uyemoto & Gilmer 1971). It is symptomless in many commercial apple cultivars. Indexing is by grafting onto cv. Virginia Crab, by mechanical inoculation onto *Chenopodium quinoa* (some local lesions followed by systemic mild mottling). It can also be detected serologically. Control is by propagation of virus-free plants. First included in the closterovirus group, it is now considered as a separate virus, synonymous with apple E36 virus, apple latent virus type 2, apple dark green epinasty virus, Virginia Crab stem grooving virus and apple brown line virus.
POLÁK

Arracacha virus B (AVB)‡ (CMI/AAB 270) is an Andean virus with one strain (the oca strain) attacking the S. American tuberous crop plant oca (*Oxalis tuberosa*), and also potato. For this reason, and because it is readily transmitted through true seed and pollen, it is regarded as a quarantine pest for Europe (EPPO 128). ELISA is suitable for routine detection.
SMITH

Ash mosaic is a syndrome widespread throughout Europe but its causes are imperfectly known and probably diverse (Cooper 1979). Arabis mosaic virus has been identified in European ash, causing chlorotic vein banding, oak leaf patterns, chevrons, mild mottle and reduced stem growth. In N. America, a similar syndrome in *Fraxinus pennsylvanica* and *F. americana* is attributable to tobacco ringspot virus or a tobamovirus. Indexing is not routine but is possible by mechanical inoculation into *Nicotiana* or *Chenopodium* species, confirmed by serology. Control is by propagation of healthy plants in soil freed from vectors.
COOPER

Barley yellow mosaic virus (BarYMV)

Basic description

BarYMV§ has filamentous particles to two different

*Some recent group classifications are indicated by footnotes.
†Now considered to be in subgroup 3 of the potyvirus group (cf. Ch. 1).
‡Possibly a nepovirus (cf. Ch. 1).
§Now considered to be in subgroup 2 of the potyvirus group (cf. Ch. 1).

lengths: 250–300 nm and 550–600 nm and width about 15 nm (CMI/AAB 143). They contain two species of RNA with M-Wt 2.8×10^6 and 1.4×10^6 d. SDS-PAGE yields two protein bands: the most conspicuous contains protein of M-Wt 35 000 d, the other 29 000 d. The virus forms cytoplasmic inclusions (pinwheels) similar to these found with potyviruses. BarYMV differs, however, from potyviruses in particle morphology and mode of transmission (Usugi & Saito 1976; Hibino *et al.* 1981; Maroquin *et al.* 1982; Huth *et al.* 1984).

Host plants

BarYMV infects only *Hordeum* species (*H. sativum, H. spontaneum*). In Europe only autumn-sown winter barley is infected naturally. The virus is also sap-transmissible to spring barley. Natural infection of autumn-sown spring barley has been reported from Japan.

Diseases

Symptoms are chlorotic or pale green spots and streaks along the leaf veins (Huth & Lesemann 1978). First symptoms can be observed in December three months after sowing, or after snow break in February or March. Generally the third leaves show the first symptoms. The appearance of symptoms on the rolled young leaves can be used for differentiation from non-virus induced disorders. In some cultivars the chlorotic streaks become necrotic (Hill & Evans 1980). Affected leaves have an erect habit as a consequence of their rolling. Severity of symptoms depends on environmental conditions as well as on the barley cultivars. At temperatures below 5°C the older leaves turn yellow and die rapidly. Yellow discoloured patches of various sizes are often the first sign of virus infection in early spring. These patches can be observed over a period of few days to some weeks depending on climatic conditions. When temperatures rise, the new leaves remain green but show distinct pale-green streaks. Leaves which developed at temperatures above 15–18°C remain symptomless. Therefore at later stages plants often have both old leaves with symptoms and symptomless young ones. BarYMV significantly reduces root growth as well as height and number of fertile tillers: sometimes infected plants reach only half the height of virus-free plants. In warmer W. European countries some highly susceptible cultivars may be found free of any symptoms on soils infested with BarYMV.

Strains

In purification experiments on leaf material which was infected by sap inoculation only one band was obtained following equilibrium centrifugation in CsCl, but two virus bands differing in buoyant densities were formed when material from naturally infected plants was used for purification. Particles in the two bands were morphologically identical but serologically unrelated. In IEM decoration tests most particles of the upper band and only a few of the lower band were strongly decorated by an antiserum to the mechanically transmitted virus. Using an antiserum to a Japanese isolate of BarYMV the results were reversed and it would appear that two types of BarYMV are usually present in naturally infected barley plants, in Germany at least. Both are soil-transmitted but only one is also readily sap-transmissible. In naturally infected plants the proportion of particles of the mechanically transmissible type increased with plant age, and this effect may be influenced by climatic conditions.

Identification and diagnosis

In general BarYMV may be identified by visual inspection. For indexing, sensitive techniques such as ELISA and IEM can be used but the virus can be identified easily only from leaves with symptoms. In symptomless leaves and in roots, virus particles have been detected only with difficulty by IEM.

Geographical distribution

First described in Japan, BarYMV has now been found in Europe (Belgium, UK, France and FRG), but probably occurs more widely than this. It has also been reported from China.

Transmission

BarYMV is a soil-borne virus, probably transmitted by *Polymyxa graminis* (Toyama & Kusaba 1970). In the FRG, plants infected naturally by fungi seem always to contain two virus types (see above), only one of which is readily sap-transmissible. Seed transmission of BarYMV has not been reported.

Economic importance

In Europe BarYMV was first reported in 1978 (Huth & Lesemann 1978). It was found in areas of N. FRG and E. England where barley and other cereals are predominant. Now areas of more than 2000 ha are uniformly infested with BarYMV. In the FRG now almost one-third of the arable land is potentially endangered by the virus (Huth 1984). Rapid spread within fields is facilitated by machinery; virus-contaminated patches are consequently initially enlarged along the direction of cultivation. Virus-contaminated soil can also be carried from field to field as well as over long distances by wind erosion.

In some areas in C. Europe, BarYMV causes one of the most important diseases of barley. The yield losses caused by BarYMV are in the range 10–90% depending on the cultivar, climatic conditions, soil type and virus content of the soil (Huth 1984). Yield is severely reduced, particularly in areas with low temperatures during winter and spring. In general high yield reductions can be anticipated when susceptible barley is sown on heavy clay soils.

Control

Chemical control of the vector reduces the number of affected plants in the year of soil treatment, but disease incidence increases again in the following years. In areas with normally low temperatures during winter and spring, yield losses can be avoided only by growing resistant cultivars. At present nine European barley cultivars and a large number of barley lines are known to be resistant (Takahashi *et al.* 1973; Huth 1982; Huth 1984). In future, breeding of further resistant cultivars would seem to be the only way of avoiding yield losses. The resistance of all German breeding material is most probably governed by a single recessive gene.

In areas with mild climatic conditions, such as in England, some susceptible barley cultivars are tolerant of BarYMV in certain years and often grow symptomlessly, but sowing these cultivars constitutes a high risk in years when winter temperatures are low.
HUTH

Beet necrotic yellow vein virus (BNYVV)

Basic description

BYNNV* is a rod-shaped multi-component plant virus with three predominant length classes for the Japanese isolate (65–105, 270 and 390 nm) (Tamada & Baba 1973; CMI/AAB 144). Four such classes (85, 100, 265, and 390 nm) were distinguished by Putz (1977) for the French strain. The virus is morphologically similar to tobacco mosaic virus and has tentatively been considered a member of the tobamovirus group (CMI/AAB 184). It has a single coat protein (M-Wt *c.* 21 000 d) and four species of single-stranded RNA (apparent M-Wt respectively 2.3×10^6; 1.8×10^6; 0.7×10^6 and 0.6×10^6 d on electrophoresis in polyacrylamide/agarose gel) (Putz 1977). The virus seems moderately immunogenic.

*Now considered to be a furovirus.

Host plants

BNYVV infects sugar and fodder beet, spinach and several other Chenopodiaceae, all hosts of the plasmodiophoromycete fungus *Polymyxa betae* (Keskin *et al.* 1962). BNYVV can also be artificially inoculated to 17 other herbaceous species (Kuszala & Putz 1977).

Strains

Isolates have been placed in four groups on the basis of the lesion type produced in leaves of *Tetragonia expansa* (Tamada *et al.* 1975). French and German strains may be similarly distinguished.

Disease

BNYVV causes the 'rhizomania' disease of sugar beet. Infected plants grow poorly and their leaves become slightly yellowed. Leaf symptoms (vein yellowing followed by vein necrosis) occur rarely, the virus usually being restricted to the roots. The disease is mainly characterized by the abnormal and heavy proliferation of rootlets to produce root bearding (Plate 7).

Geographical distribution

First found in Japan (1970), the virus was described in Italy (1974), France (1974), FRG (1976), Yugoslavia (1978) and Greece (1979). It has recently been detected in most of the European sugar-beet growing countries but not yet in the UK or Scandinavia. The disease probably occurs in the USA (California), the USSR (Kirghiz SSR) and China.

Transmission and epidemiology

BNYVV is soil-borne and transmitted by the fungus *Polymyxa betae* (q.v.; Keskin 1964; Canova 1966), whose resting spores retain infectivity in air-dried soil for several years. Neither seed nor aphid transmission have been reported. Dissemination occurs by soil transport from infested fields and by sugar factory effluents.

Economic importance

Rhizomania is now considered one of the most important diseases of sugar beet; intensive breeding programmes for rhizomania resistance have been initiated in many European countries. The disease only occurs in a limited number of fields in infected countries but each such field is lost for beet growing until such time as satisfactory control measures can be developed.

Control

Steam sterilization or methyl bromide fumigation of virus-infested soil effectively controls the disease but these measures are not practical in large fields. All control measures so far available are prophylactic (longer rotation, improvements in soil structure and drainage, avoiding transport of infested soil). Considerable progress has been made recently in breeding for tolerance to rhizomania.

Special research interest

The strategy for expression of genetic information in this multi-component virus is of considerable interest. Preliminary experiments indicate that RNA_2 carries information for coat protein.
PUTZ

Black raspberry necrosis virus (BRNV)

Basic description

BRNV was the name given in Canada (Stace-Smith 1955) to an aphid-borne agent, symptomless in most raspberry cultivars, but which caused severe apical tip necrosis in black raspberry (*Rubus occidentalis*). In Scotland, a virus mechanically transmitted to *Chenopodium quinoa* plants from raspberry and code-named 52V, was later equated with BRNV. As BRNV is readily eradicated from plants by heat treatment, virus isolates termed heat-labile mosaic component in the USA may also be isolates of BRNV. BRNV is difficult to study but is known to have isometric particles *c.* 30 nm in diameter.

Host plants and diseases

BRNV naturally infects wild and cultivated raspberry in which it produces few if any leaf symptoms but may, in association with other viruses, cause: i) symptomless degeneration of raspberry stocks; ii) bushy dwarf disease of Lloyd George raspberry in mixture with raspberry bushy dwarf virus; and iii) raspberry vein banding mosaic disease in combination with rubus yellow net virus. In Europe, vein banding mosaic disease is characterized by chlorotic 'tramline' areas that run parallel to the main leaf veins; in severely affected plants leaves may be puckered and distorted due to uneven growth of the lamina (Jones & Roberts 1977; Stace-Smith & Jones 1985). When alone in some blackberry cultivars, BRNV may induce faint chlorotic leaf mottling or blotching.

Diagnosis

Some infected raspberry cultivars show a faint leaf mottling but these symptoms are not diagnostic. In Europe, identification is based on either: i) mechanical inoculation to *C. quinoa* test plants (systemic chlorotic flecking and distortion of leaves); or ii) graft inoculation to *R. henryi* or *R. occidentalis* (severe apical tip necrosis). However, neither test is entirely satisfactory as BRNV is difficult to transmit mechanically, especially in summer, and the two *Rubus* indicators react similarly to infection with BRNV and at least two other aphid-borne viruses in raspberry.

Geographical distribution, transmission and economic importance

BRNV is transmitted in a semi-persistent manner by the large raspberry aphid *Amphorophora idaei* which is common throughout Europe. Other aphids less common on raspberry may also transmit BRNV. BRNV spreads rapidly and is widespread in *Rubus* crops in Europe. It is not seed transmitted in raspberry. It is probably the commonest virus infecting raspberry crops.

Control

Chemical control of aphid vectors does not give satisfactory control of BRNV. Although planting healthy stock in isolation from infected plants may delay infection, the best control has been achieved by growing the more modern raspberry cultivars that are resistant to the main aphid vector.
JONES

Broad bean mild mosaic virus (Devergne & Cousin

1966) has isometric particles *c.* 25 nm in diameter and causes symptoms on broad bean that vary from mild mosaic to partial necrosis. It also infects pea. Symptoms are shortened internodes, stunting of the plant, small curled leaves, and necrosis can occur on stems. Discolorations and deformities develop on broad bean and pea seeds. The virus is seed-borne. It can be distinguished from broad bean mottle virus and broad bean true mosaic virus by its biological, physical and antigenic properties. Detection is by mechanical transmission to pea or to *Chenopodium quinoa* (local lesions and mosaic). Recent information on the status of this virus is lacking.

Broad bean wilt virus (BBWV) (CMI/AAB 81) occurs in several European countries, Australia, America, Japan and Africa. In broad bean, it induces retarded growth, leaf deformations, mottle, occasionally ring-like patterns and, synergistically with fungal disease agents, wilting. It also causes nasturtium ringspot. Lamium mild mosaic is distantly related. Isometric particles (with *c.* 25 nm diameter) have been identified in infected plants. Purified preparations contain three types of particles: empty protein shell and two sorts of nucleoprotein containing two RNA species.

The disease affects more than 150 hosts; it is transmitted by sap inoculation onto test plants, and non-persistently by *Myzus persicae* and 21 other aphids. Indexing is by mechanical inoculation to broad bean (systemic vein clearing followed by necrosis of terminal leaves) and *Chenopodium* species (chlorotic lesions followed by systemic mottling and top necrosis); additional detection is by observation of cytoplasmic inclusion bodies in epidermal strips and serology (gel diffusion and ELISA). The disease is controlled by insecticide treatments against the vectors. Inbred lines are used as resistant parents in breeding programmes.

SCHMIDT

Carrot mottle virus (CMotV) (CMI/AAB 137) has been described on the basis of enveloped isometric particles *c.* 52 nm in diameter seen in infected plants. It occurs in nature in mixed infections with carrot red leaf virus: the complex causes yellowing and reddening of carrots with some stunting, a disease known as motley dwarf which can induce serious reductions in yield. Parsley is symptomlessly infected. Found in Australia, New Zealand, Japan and Europe the virus is transmitted by the aphid *Cavariella aegopodii* in a persistent manner. Carrot red leaf virus acts as a helper for the transmission of carrot mottle virus. There is no seed transmission. Detection is by mechanical inoculation onto *Chenopodium quinoa* (pinpoint chlorotic spots on inoculated leaves).

MARCHOUX

Cocksfoot mild mosaic virus (CMMV) is widespread in some European countries, causing mild leaf mottling on many Gramineae. It is transmissible mechanically, and also by aphids and beetles. Isometric particles *c.* 28 nm in diameter have a sedimentation coefficient of 105S; the coat is made up of one protein sub-unit M-Wt 72 000 d and there are two species of RNA with M-Wt 1.5×10^6 d and 0.5×10^6 d (Paul & Huth 1970). Strains have been named as brome stem leaf mottle virus, holcus transitory mottle virus, phleum mottle virus, festuca mottle virus, cocksfoot mosaic and necrosis virus (Torrance & Harrison 1981). Indexing is by mechanical transmission to *Dactylis glomerata* (mottling and chlorotic streaking) and serology. No control measures are known (CMI/AAB 107).

HUTH

Celery latent virus has been described in W. Europe only, and is of uncertain economic importance (Verhoyen *et al.* 1976; Bos *et al.* 1978). It has been isolated from plants of celeriac and is sap-transmissible to *Chenopodium quinoa* and *C. amaranticolor* (local chlorotic lesions and irregular chlorotic flecks and rings). Aphid species tested were unable to transmit the virus. Seed transmission is reported up to 34% in celeriac. Purification has proved difficult but flexuous virus particles (880–940 nm) have been found in preparations. Testing for virus freedom in seed multiplication is advised.

MARCHOUX

Grapevine leaf roll disease

Basic description

Though as yet not fully understood, the aetiology of grapevine leaf roll is very probably viral. This disease may be caused by different viral agents singly or in mixture or, perhaps, in association with prokaryotes (Caudwell *et al.* 1983). Several viruses or virus-like agents have been detected in grapevines with leaf roll symptoms. However, even though leaf roll is very often associated with other disorders like stem-pitting (legno riccio) and corky bark and fleck in nature, no consistent and straightforward association has been found between any of these and leaf roll proper. The viruses detected include: i) a potyvirus isolated in Israel, with filamentous particles 790×13 nm and sedimentation coefficient *c.* 150S. It contains one molecule of single-stranded RNA with M-Wt 3.5×10^6 d, accounting for 5% of the particle weight. The protein coat is made up of a single polypeptide with M-Wt 31000 d; ii) closterovirus-like particles observed in thin sections or leaf dip-preparations from grapevines with leaf roll symptoms in Japan, Italy, Germany, Switzerland, Israel, New Zealand, USA, South Africa and Australia. One of these viruses, named **grapevine virus A (GVA)** (Milne *et al.* 1984) has particles 800×11–12 nm, containing single-stranded RNA with M-Wt 2.55×10^6 d and a coat protein with a single polypeptide with M-Wt of 22000 d. GVA is a good immunogen. A second closterovirus, named **grapevine virus B (GVB)**, serologically unrelated to GVA, has been detected in material from different countries but the particle length has not been determined (Milne *et al.* 1984). In Europe, the USA and South Africa other closteroviruses occur which are also serologically unrelated to GVA. The particles are 1–2 μm long, like those of closteroviruses from Japan and Australia; iii) a possible isometric virus with particles of 26–28 nm has been observed in Japan in grapevines affected by 'Ajinashika disease', a severe form of leaf roll and fleck (Namba *et al.* 1979a). In Italy another virus, whose particles in thin section measure 22–24 nm, was observed in leaf roll-infected vines from several European countries (Castellano *et al.* 1983).

Host plants

Grapevines (both European and American species) are the only known natural hosts. The experimental host range of the Israeli potyvirus and GVA, the only two viruses that are mechanically transmissible though with great difficulty, is narrow and virtually limited to the Solanaceae.

Strains

Mild and severe strains have been reported based on severity of symptoms and incubation time for symptom expression in grape indicators such as cv. Baco Blanc in California. Leaf roll is normally latent in American *Vitis* species and rootstock hybrids (Vuittenez 1958) but a form of the disease is known in France which also induces foliar symptoms in *Vitis riparia* (Bass & Legin 1981).

Diseases

Affected grapevines may show a slight reduction of growth. Major outward manifestations are downward rolling of the leaves accompanied by variously intense reddish or yellow discolorations of the blade. Discoloured areas appear in the interveinal areas of lower leaves in early summer and become progressively stronger and extended so as to cover the whole leaf surface with the exception of the main veins which may retain the green colour. Bunches are smaller than normal and berries mature late and irregularly.

Identification and diagnosis

Indexing on grapevine indicators is the only reliable method for diagnosis. It is done by grafting onto susceptible cultivars such as Pinot Noir, Gamay, Cabernet Sauvignon, Cabernet Franc, Barbera, Mission or the hybrid LN 33, which reproduce disease symptoms 1–3 years from inoculation. The hybrid Baco 22 A, which reacts strongly with chlorosis and stunting in California, has proved insensitive for detection of leaf roll in several European countries. Light or electron microscope observations of sectioned tissue reveal callose accumulation and tubular inclusions in phloem elements of diseased grapevines (Vuittenez & Stocky 1982; Castellano *et al.* 1983), but the diagnostic value of these abnormalities is uncertain. Serological detection of GVA and other closteroviruses in grapes is possible by IEM (Milne *et al.* 1984) and ELISA.

Geographical distribution and transmission

The disease occurs world-wide, without exception, wherever grapevines are grown. The rate of natural spread is virtually nil, although in certain areas leaf roll-like symptoms do spread very slowly (Dimitrijevic 1973).

In France, a recently observed disease of cv. Chardonnay named 'nervures jaunes'—possibly an association of leaf roll and 'bois noir'—is spreading in the new plantings of some areas, affecting up to 5% of young vines within 3–4 years (Caudwell *et al.* 1983). Grape to grape transmission with dodder (*Cuscuta campestris*) was experimentally obtained in Australia (Woodham & Krake 1983) but this does not have epidemiological significance. GVA was experimentally transmitted to herbaceous hosts and grapevine by the mealybugs *Pseudococcus longispinus* (Rosciglione *et al.* 1983), *P. ficus* and *P. citri*. Infected propagating material is responsible for long distance spread.

Economic importance

Leaf roll reduces the quantity of the yield by 10–70%, lowers sugar content by up to 13° Oechsle, and reduces graft take and rooting ability of the canes. German must production from 1924 to 1930, estimated at 10 million hl, would have been higher than 17 million hl if the vines were not so heavily affected by leaf roll.

Control

Visual selection in many *V. vinifera* cultivars helps to reduce disease incidence. However, disease-free stocks are best obtained, although not always readily, by prolonged heat treatment at 38°C (60–120 days), removal and rooting of shoot tips under mist, heat treatment *in vitro* according to Galzy's method (Valat & Mur 1976), micrografting (Bass & Legin 1981) or meristem-tip culture (Barlass *et al.* 1982).

Special research interest

The main research interest lies in the aetiological definition of the disease and the role, if any, played by the viruses associated with it, especially those which are phloem-limited. As some of these viruses (e.g. closteroviruses), by analogy with established diseases of other crops, may induce alterations of the woody cylinder, the mutual relationship of leaf roll, corky bark and stem pitting requires thorough investigation.
MARTELLI & VUITTENEZ

Lettuce big-vein virus could be the first virus with elongated particles (350×18 nm) containing double-stranded RNA (Dodds & Mirkov 1983). It occurs in N.W. and C. Europe and in N. America. The chief symptom of big-vine on lettuce is a pale yellowish or blanched vein banding of the leaves which is usually most obvious near the base of outer leaves. This is often accompanied by a distortion and blistering of affected leaves until the host has a stark appearance with whitened veins which stand out against the greener interveinal tissues. The sharp transition from a blanched area to normal green tissue is a good diagnostic feature. There is often no, or only slight, reduction in size and hearting of affected plants and frequently symptoms do not appear until a plant is mature. There may be some confusion between the symptoms of lettuce big-vein disease and those caused by lettuce mosaic virus (vein clearing and some diffuse mottling of leaves). Symptoms are most prominent on cos and butterhead types and less so on crisphead cultivars. The disease is transmitted by zoospores of the fungus *Olpidium brassicae*, and the virus can be indexed by transmission of infectious zoospores onto susceptible cultivars of lettuce, endive or *Sonchus oleraceus*. Some cultivars are tolerant (Thompson, Sea Green). Control is mainly aimed at the vector (q.v.; White 1983).
POLÁK

Maize white line mosaic virus (CMI/AAB 283) has isometric particles of 30–35 nm resembling morphologically, but not serologically, those of tombusviruses. It is widespread in N.E. USA, has been found once in Italy and is serologically related to the 'mosaique d'anneaux foliaires' reported in France. Though particles are highly concentrated in maize, the virus has not been transmitted by sap-inoculation or insects. It causes stunting and bright mosaic, interspersed with characteristic white lines (but is frequently symptomless). Yield losses can be 5–45%. It also naturally infects wheat and some grasses. It is soil-borne, probably through a fungus (*Olpidium* or *Polymyxa*). It can be indexed by ELISA or IEM, and controlled by applying benomyl or carbofuran to the soil or by rotation with non-grass crops.
CONTI

Melon necrotic spot virus is distributed world-wide in melons and cucumbers grown in glass houses (Bos *et al.* 1984). It induces chlorotic lesions, turning into necrotic spots, occasionally leading to leaf desiccation. Streaks appear on petioles or stems. It is soil-borne (transmitted by *Olpidium* species) and can be mechanically transmitted by tools used for pruning; a Californian isolate has also been reported to be seed-borne (Gonzalez-Garza *et al.* 1979) and transmitted by cucumber beetles (*Diabrotica* species). The virus is quite different from tobacco necrosis virus and its cucumber necrosis strain. Serologically and physico-chemically, the virus is similar to if not identical with the melon necrotic spot virus described in Japan. Indexing is by mechanical inoculation to melon, cucumber or watermelon (inoculated leaves or cotyledons show necrotic lesions within 1 week). The virus has not been reported to infect plants other than cucurbits. Control is by soil disinfection, prevention of mechanical transmission,

grafting on resistant rootstocks or breeding for resistance (melons) (CMI/AAB 302).
LECOQ

Oat blue dwarf virus (OBDV) (CMI/AAB 123) is an unclassified virus* with isometric particles 28–30 nm in diameter, reported in some countries of C. and N. Europe and N. America. Symptoms are stunting, diameter, reported in some countries of central and N. Europe and N. America. Symptoms are stunting, excess tillering, bluish-green foliage, enations on veins and sterility of florets in oats and other grasses; stunting and crinkle in flax. Some susceptible grasses and dicotyledons are symptomless hosts. OBDV is transmitted by leafhoppers *Macrosteles laevis* and *M. fascifrons* in the persistent manner. Identification is by vector inoculation onto *Linum usitatissimum* and by ELISA.
VACKE

Oat mosaic virus† (CMI/AAB 145) has rod-shaped particles 600–750×12–14 nm, resembling those of potyviruses, from which it differs, like barley yellow mosaic virus, in being soil-borne and transmitted by *Polymyxa graminis*. It only naturally infects species in the genus *Avena*. The virus causes a disease of autumn-sown oats, with mottling symptoms on the first flush of growth in the spring, tending to disappear with further growth. Symptoms vary with the oat cultivar, the virus strain and the environment. Yields of susceptible cultivars can be reduced by up to 50%. The virus occurs in Europe and N. America, and is not transmitted through seeds. Detection is by mechanical inoculation of *Avena sativa*; all cultivars develop mosaic 2–6 weeks after inoculation.

Oat soil-borne stripe virus (OSBSV) is similar or identical to oat golden stripe virus (Plumb *et al.* 1977). It has rigid particles 140 and 300 nm long, and is serologically related to wheat soil-borne mosaic virus by indirect ELISA, but clearly distinct by sandwich ELSIA (Hariri & Lapierre 1985). The virus is probably transmitted by *Polymyxa graminis* which is present in the roots of infected plants. The disease is observed on winter oats only. The first symptoms of discoloration appear in April. In June and July the flag leaf and the leaf below show a brilliant yellow striping leading to necrotic spots. On sap inoculation *Nicotiana clevelandii* and *Chenopodium amaranticolor* give necrotic local lesions, and *N. debneyi* chlorotic lesions if the plants are grown at 15°C. Oats can be infected by watering seedlings of a susceptible cultivar, in whose roots the virus can be detected by ELISA within 7 days. OSBSV has been detected only in the UK and France. As a control measure, treatment of infested soils would be expensive. Of five cultivars tested only one (cv. Crin Noir) showed good resistance.
HARIRI & LAPIERRE

*Now considered to be in the maize rayado fino virus group.
†Now considered to be in subgroup 2 of the potyvirus group (cf. Ch. 1).

Olive latent virus 1 and olive latent virus 2. Olive latent virus 1 has isometric particles and olive latent virus 2 has predominantly bacilliform particles; both have been isolated by sap-inoculation from olive trees in Italy, along with **olive latent ringspot virus** (a nepovirus; Savino *et al.* 1983). Various other viruses (strawberry latent ringspot virus, arabis mosaic virus, cherry leaf roll virus, cucumber mosaic virus) have also been obtained. These mechanically transmissible viruses were recovered from apparently symptomless trees (Martelli 1981), with the exception of olive latent virus 1, which causes mild fasciations and apical bifurcation of twigs and leaves of cv. Paesana, and strawberry latent ringspot virus, which is associated with deformations of leaves and fruits of cv. Ascolana.
MARTELLI

Parsnip yellow fleck virus (PYFV) (CMI/AAB 129)is an RNA-containing virus with particles *c.* 30 nm in diameter. It infects Umbelliferae, causing vein yellowing, yellow flecks and mosaic symptoms on parsnip. Reported in the UK and Germany, it is aphid-transmitted by *Cavariella aegopodii* in the semi-persistent manner only when the aphids are also carrying the helper virus, **anthriscus yellows virus** (Elnagar & Murant 1976). This second virus is mainly responsible for yellows on chervil: it has virus particles 22 nm in diameter. Detection of parsnip yellow fleck virus is by mechanical inoculation to *Chenopodium quinoa* (local lesions) and serology.

Pelargonium line pattern virus causes veinal flecks and line patterns in pelargonium leaves. After infection, plants remain symptomless for a long period of time. Serologically different from pelargonium ring pattern virus (Stone 1980), the virus is transmissible to *Nicotiana clevelandii*, after which ELISA or IEM can be applied for identification.

Pelargonium ring pattern virus has been isolated from *Pelargonium zonale* and causes ring patterns in spring on susceptible cultivars. It is the most common virus disease in all pelargonium cultivars (Stone 1980). No vector is known and there is a long incubation period in infected plants. Transmissible to *N. clevelandii* and *Chenopodium quinoa*, the virus can be identified by ELISA or IEM. Virus-free plants are obtained by meristem-tip culture.

Pelargonium ringspot virus causes clear ringspot symptoms in *Pelargonium peltatum* and line pattern on *P. zonale* in spring. Symptoms are usually very slow to develop in infected plants. Inoculated seedlings develop symptoms only 15 months after inoculation (Stone 1980). Widespread in old cultivars, it is transmissible to *Nicotiana clevelandii* after which ELISA or IEM can be applied for identification.

Pelargonium zonate spot virus has quasi-isometric particles 25–35 nm in diameter resembling those of the ilarviruses, with 2 RNA species of M-Wt 1.25 and 0.95×10^6 d respectively, encapsidated in two different particles. The capsid is very probably made up of a single polypeptide of M-Wt *c.* 23 000 d. The virus causes concentric chrome yellow bands in *Pelargonium zonale* leaves, and can also infect tomato (yellowish ring and line patterns on leaves, stunting of the plant, sterility). It is found in S. Italy. No vector is known. On *Nicotiana glutinosa*, it is seed- and pollen-borne. It can be detected by mechanical inoculation of cucumber (local lesions) and is weakly immunogenic (CMI/AAB 272).

Potato virus T (PVT) (CMI/AAB 187) is serologically related to apple stem grooving virus. Occurring in the Andean region of S. America, it causes relatively mild leaf mottling in potato. Readily transmitted through true seed and pollen, it is regarded as a quarantine pest for Europe (EPPO 128).
SMITH

Raspberry bushy dwarf virus (RBDV)

(Syn. loganberry degeneration virus)

Basic description

RBDV has quasi-isometric particles of *c.* 33 nm diameter with three RNA species of M-Wt *c.* 2.0, 0.9 and 0.4×10^6 d (glyoxylated). There is a single peptide of M-Wt *c.* 29 000 comprising 76% of the particle. For electron microscopy particles disrupt in phosphotungstate stains but are stable in uranyl salts (CMI/AAB 165).

Host plants and diseases

The only known natural hosts are *Rubus* species but RBDV is mechanically transmissible to a moderate range of herbaceous hosts. Its association with various raspberry diseases is complex and some uncertainties remain. In association with the aphid-borne black raspberry necrosis virus, RBDV induces 'bushy dwarf' or 'symptomless decline' of raspberry cv. Lloyd George (producing few, stunted canes and a tendency to autumn fruiting). In other cultivars RBDV is associated with 'crumbly fruit' (aborted drupelets) and loss of vigour and yield but it is often carried symptomlessly. Some *Rubus* species react to graft inoculation with transient chlorotic mottles, water-marks or line patterns. RBDV is probably the cause of loganberry degeneration and has recently been substantiated as the causal agent of raspberry yellows disease. Yellows is a disorder affecting some raspberry cultivars in many areas of the world. Initial symptoms are a striking bright yellow vein banding on parts of some or all leaves of infected plants. This often recurs annually on the newly emerging canes but may not be apparent on later-formed leaves on the same stems. A similar pronounced vein banding has been recorded in the early growth of seedlings arising in *Rubus* breeding programmes. Seedlings frequently recover and subsequently grow apparently normally. In chronic infections the major symptom is much more general chlorosis, especially of leaves on the fruiting canes. The seasonal occurrence of symptoms is often very variable. Once recognized the vein banding is quite distinctive, but the more general chlorosis may be confused with physiological disorders. The association of RBDV with this disease is essentially illustrated by the following points: i) most cultivars recorded as yellows-susceptible are also RBDV-susceptible (the two exceptions have not been adequately assessed for resistance to RBDV); ii) there is a close association of yellows with RBDV infection in individual plants; iii) mechanical inoculation of some cultivars with purified preparations of RBDV induced yellows; iv) the modes of transmission of yellows and RBDV are similar. The converses of i) and ii) are not true in that many cultivars recorded as RBDV-susceptible are not known to show yellows and many RBDV-infected plants are symptomless.

Strains

Most isolates are serologically indistinguishable except some from black raspberry in the USA.

Recently isolates which differ from the type isolate in their *Rubus* host range have been discovered (see *Control* below).

Identification and diagnosis

Chenopodium quinoa is a good herbaceous indicator (transient chlorotic local lesions and systemic chlorotic rings and line patterns); transmission is most reliable in the spring and autumn. Serological detection (by ELISA) is rapid and reliable. For raspberry yellows, raspberry cv. Norfolk Giant has been the recommended indicator but symptom expression is erratic and the Scottish Crop Research Institute selection 6813/74 appears to be more reliable.

Geographical distribution and transmission

RBDV probably occurs world-wide wherever susceptible *Rubus* cultivars are grown. RBDV is seed transmitted (up to 77% in raspberry) but in established plants spread occurs via pollen, by which means both maternal parent and seed may become infected. No vector is known. Propagation of infected material has been important in disseminating RBDV.

Economic importance and control

The spread of RBDV has been controlled by the use of resistant cultivars. High resistance to infection by type isolates via graft-inoculation is conferred by a single dominant gene, designated *Bu*, and cultivars with gene *Bu* have remained RBDV-free in the field for many years. Other resistance mechanisms also appear to be involved as some cultivars, e.g. Malling Leo, do not readily become infected despite lacking *Bu*. Where resistant cultivars are planted RBDV is largely confined to older low-resistance cultivars and its importance has diminished. Elsewhere RBDV continues to cause significant loss of yield and reduction in fruit quality. Strains of RBDV capable of infecting cultivars with gene *Bu* are now known to occur in the UK, USSR and possibly FRG. Some cultivars previously thought immune to RBDV are apparently quickly killed by these isolates and they may become more important in the future. Raspberry yellows is widespread and common in susceptible cultivars (e.g. Norfolk Giant in the UK, Marcy in New Zealand) and although infected plants continue to grow and crop there must be significant overall yield loss.

BARBARA

Sunflower mosaic disease may be due to a number of viruses, including cucumber mosaic virus. One virus with elongated particles has been described specifically from sunflower, although there is no particular information on it in Europe. It causes a mild mosaic followed by a severe leaf necrosis resulting in the death of the upper part of the plant. Synergism has been observed with some fungi (Gupta *et al.* 1977). The rapid spread of the virus suggests a vector. Cytoplasmic inclusions (long rods, pinwheels and circular lamellate inclusions) suggest a potyvirus.

Tobacco stunt virus has isometric particles about 25 nm in diameter. It causes vein clearing and necrosis of tobacco leaf veinlets. Specific necrotic banding is seen in the stem at ground level. The whole plant becomes stunted and rosetted. Transmitted by *Olpidium brassicae*, the virus is unrelated to tobacco necrosis virus (Hiroki 1975) (CMI/AAB 313).

Watercress yellow spot virus has been described in France. It has isometric particles *c.* 27 nm in diameter, and causes yellow spots and sometimes ringspots on watercress (Spire 1962). No vector is known.

Wheat soil-borne mosaic virus (SBWMV)* occurs in America, Asia and locally in Europe. Systemic mosaic symptoms are most prominent on early spring growth of wheat, rye and barley (APS Compendium). It is transmitted by the fungus *Polymyxa graminis* and is indexed by serology with ELISA or electron microscopy. The virus is sap-transmitted with difficulty onto wheat, some grasses and *Chenopodium*; the optimal temperature for symptoms is 16°C. Crystalline inclusions are frequently found. Particle lengths differ with strains; most frequent lengths are 110–160 and 280–300 nm; the diameter is 20 nm. The best means of control is by the use of resistant cultivars (CMI/AAB 77) (Plate 8).

PROESELER

Wheat spindle streak mosaic virus (WSSMV)

Basic description

WSSMV (CMI/AAB 167) has filamentous particles ranging in leaf-dip preparations from 190–1975 nm or more, and with a diameter of 12–15 nm. After puri-

*Now considered to be a furovirus.

fication some shorter particles are also obtained (Slykhuis & Polák 1971). According to Usugi & Saito (1979), wheat yellow mosaic virus described in Japan in 1968 seems to be almost identical to WSSMV. They are serologically related and cross protection has been observed. WSSMV is very close to barley yellow mosaic virus in being potyvirus-like* but fungus-transmitted.

Host plants, diseases and transmission

The virus affects mainly winter-sown wheats (*Triticum aestivum* and *T. durum*). Many species of Gramineae have been tested but they failed to develop symptoms. Infections occur mainly in the autumn and symptoms develop in the spring. The virus requires 30–90 days incubation at 5–15°C, so the disease is less severe on late-planted wheat. Symptoms are typically produced on lower, older leaves of wheat during cool periods in the spring. The mosaic is characterized by chlorotic to necrotic spindle-shaped dashes or streaks. At first, the disease appears in patches through the field then in later years it may spread uniformly over the entire field. At temperatures above 15°C, new growth appears healthy. If day temperatures infrequently exceed 20°C, symptoms continue to develop on new growth, appearing even on the flag leaf. Under these conditions, the chlorotic spindles may coalesce into large indistinct mosaic patterns. Very intensely chlorotic areas turn brown, often giving the appearance of septoria leaf blotch. Diseased plants remain slightly stunted. Cylindrical inclusions (pinwheels) occur in the leaf cells. WSSMV survives in soil in association with the soil fungus *Polymyxa graminis* (Slykhuis & Barr 1978) which invades the roots of wheat plants and transmits the virus. Mechanical transmission is possible. WSSMV is not seed-borne.

Geographical distribution, economic importance and control

WSSMV occurs in Canada (Ontario), some northern states of the USA, India and France. It is liable to spread further in Europe (Signoret *et al.* 1977). Losses are influenced by environmental conditions and have been evaluated in the range of 2–18%. Cultural practices, except delaying autumn plantings, appear to have little influence on disease incidence, which increases if wheat is grown repeatedly on the same ground; soil may remain infective for a long time. Differences in varietal reaction to the disease have been reported in Canada, France and USA. Soil treatment with heat or some chemicals is effective but not economically feasible.

SIGNORET

Wheat streak mosaic virus (WSMV)† occurs in Africa (Egypt), America, Asia and Europe but is most important in the central great plains of the USA. Natural host plants are winter wheat, certain millets, maize, barley and some other cereals and grasses. Symptoms vary with wheat cultivar, strain of virus and other factors; plants are usually stunted, with discontinuous parallel green-yellow leaf streaks (CMI/AAB 48). It is transmitted by *Eriophyes tulipae*, *E. tosichella* and mechanical inoculation. Indexing is by sap-inoculation, ELISA or IEM, or in the mites, with fluorescent antibodies (Slykhuis 1980). It has filamentous particles *c.* 700×15 nm and frequently forms cytoplasmic inclusions (membranous pinwheels and crystals). Control is by cultural practices, carbofuran against the mites, and breeding of resistant wheat cultivars by chromosome substitution from *Agropyron* species.

PROESELER

REFERENCES

Barlass M., Skene K.G.M., Woodham R.C. & Krake L.R. (1982) Regeneration of virus-free grapevines using *in vitro* apical culture. *Annals of Applied Biology* **101**, 291–295.

Bass P. & Legin R. (1981) Thermothérapie et multiplication *in vitro* d'apex de Vigne. Application à la séparation ou à l'élimination de diverses maladies de type viral et à l'évaluation des dégâts. *Comptes Rendus des Séances de l'Académie d'Agriculture de France* **67**, 922–933.

Bos L., Diaz-Ruiz J.R. & Maat D.Z. (1978) Further characterization of celery latent virus. *Netherlands Journal of Plant Pathology* **84**, 61–79.

Bos L., Van Dorst H.J.M., Huttinga H. & Maat D.Z. (1984) Further characterization of melon necrotic spot virus causing severe diseases in glasshouse cucumbers in the Netherlands and its control. *Netherlands Journal of Plant Pathology* **90**, 55–69.

Canova A. (1966) Ricerche virologische nella rizomania della bietola. *Annali dell'Academia Nazionale di Agricoltura (Bologna)* **78**, 37–46.

Castellano M.A., Martelli G.P. & Savino V. (1983) Virus-like particles and ultrastructural modifications in the phloem of leafroll-affected grapevines. *Vitis* **22**, 23–39.

*Now considered to be in subgroup 2 of the potyvirus group (cf. Ch. 1).

†Now considered to be in subgroup 3 of the potyvirus group (cf. Ch. 1).

Caudwell A., Larrue J., Badour C., Polge C., Bernard R. & Leguay M. (1983) Développement épidémique d'un enroulement à nervures jaunes, transmissible par la greffe, dans le vignoble de Champagne. *Agronomie* **3**, 1027–1036.

Cooper J.I. (1979) *Virus Diseases of Trees and Shrubs.* Institute of Terrestrial Ecology, Cambridge.

Devergne J.C. & Cousin R. (1966) Le virus de la mosaïque de la fève et les symptômes d'ornementation sur les grains. *Annales des Epiphyties* **17** (no hors-série, Etudes de Virologie), 147–161.

Dimitrijevic B. (1973) Some observations on natural spread of grapevine leafroll disease in Yugoslavia. *Rivista di Patologia Vegetale* **9** (suppl.), 114–119.

Dodds J.A. & Mirkov T.E. (1983) Association of double-stranded RNA with lettuce big vein disease. In *4th International Congress of Plant Pathology*, Abstract 113. Melbourne.

Elnagar S. & Murant A.F. (1976) Relations of the semi-persistent viruses parsnip yellow fleck and anthriscus yellows with their vector, *Cavariella aegopodii. Annals of Applied Biology* **84**, 169–181.

Gonzalez-Garza R., Gumpf D.J., Kishaba A.N. & Bohn G.W. (1979) Identification, seed transmission and host range pathogenicity of a California isolate of melon necrotic spot virus. *Phytopathology* **69**, 340–345.

Gupta K.C., Roy A.N. & Gupta M.N. (1977) Synergism between *Alternaria alternata* and sunflower mosaic virus. *Proceedings of the Indian National Science Academy B* **43**, 130–132.

Hariri D. & Lapierre H. (1985) Purification and serologial properties of soil-borne oat stripe virus. *Mitteilungen aus der Biologischen Bundesanstalt für Land- und Forstwirtschaft Berlin-Dahlem* **228**, 72.

Hibino H., Usugi T. & Saito J. (1981) Comparative electron microscopy of inclusions associated with five soil-borne filamentous viruses of cereals. *Annals of the Phytopathological Society of Japan* **47**, 810–819.

Hill S.A. & Evans E.J. (1980) Barley yellow mosaic virus. *Plant Pathology* **29**, 197–199.

Hiroki C. (1975) Host range and properties of tobacco stunt virus. *Canadian Journal of Botany* **53**, 2425–2434.

Hollings M. (1958) Anemone brown ring, a virus disease. *Plant Pathology* **7**, 95–98.

Huth W. (1982) Evaluation of sources of resistance to barley yellow mosaic virus in winter barley. *Zeitschrift für Pflanzenzüchtung* **89**, 158–164.

Huth W. (1984) Die Gelbmosaikvirose der Gerste in der Bundesrepublik Deutschland—Beobachtungen seit 1978. *Nachrichtenblatt des Deutschen Pflanzenschutzdienstes* **36**, 49–55.

Huth W. & Lesemann D.E. (1978) Eine für die Bundesrepublik Deutschland neue Virose an Wintergeste. *Nachrichtenblatt des Deutschen Pflanzenschutzdienstes* **30**, 184–185.

Huth W., Lesemann D.E. & Paul H.L. (1984) Barley yellow mosaic virus: electron microscopy, serology, and other properties of two types of the virus. *Phytopathologische Zeitschrift* **111**, 37–54.

Jones A.T. & Roberts I.M. (1977) *Rubus* host range, symptomatology and ultrastructural effects of a British isolate of black raspberry necrosis virus. *Annals of Applied Biology* **86**, 381–387.

Keskin B. (1964) *Polymyxa betae* n.sp., ein Parasit in den Wurzeln von *Beta vulgaris. Archiv für Microbiologie* **46**, 348–374.

Keskin B., Gaertner A. & Fuchs W.H. (1962) Über eine die Wurzel von *Beta vulgaris* befallende Plasmodiophoraceae. *Berichte der Deutsche Botanischen Gesellschaft* **75**, 275–279.

Kuszala M. & Putz C. (1977) La rhizomanie de la betterave sucrière en Alsace. Gamme d'hôtes et propriétés biologiques du beet necrotic yellow vein virus. *Annales de Phytopathologie* **9**, 435–446.

Maroquin C., Cevalier M. & Rassel A. (1982) Premières observations sur le virus de la mosaïque jaune de l'orge en Belgique. *Bulletin des Recherches Agronomiques de Gembloux* **17**, 157–172.

Martelli G.P. (1981) Le virosi dell'olivo: esistono? *Informatore Fitopatologico* **31** (1–2), 97–100.

Milne R.G., Conti M., Lesemann D.E., Stellmach G., Tanne E. & Cohen J. (1984) Closterovirus-like particles of two types associated with diseased grapevines. *Phytopathologische Zeitschrift* **110**, 360–368.

Namba S., Yamashita S., Doi Y. & Yora K. (1979a) A small spherical virus associated with the Ajinashika disease of Koshu grapevine. *Annals of the Phytopathological Society of Japan* **45**, 70–73.

Namba S., Yamashita S., Doi Y., Yora H., Terai J. & Yano R. (1979b) Grapevine leafroll virus, a possible member of the closterovirus. *Annals of the Phytopathological Society of Japan* **45**, 497–502.

Paul H.L. & Huth W. (1970) Untersuchungen über das cocksfoot mild mosaic virus. *Phytopathologische Zeitschrift* **69**, 1–8.

Plumb R.T., Catherall P.L., Chamberlain J.A. & Macfarlane I. (1977) A new virus of oats in England and Wales. *Annales de Phytopathologie* **9**, 365–370.

Putz C. (1977) Composition and structure of beet necrotic yellow vein virus. *Journal of General Virology* **35**, 397–401.

Putz C., Collot D. & Peter R. (1977) Caractérisation biochimique de la sous-unité protéique du virus de la rhizomanie, beet necrotic yellow vein virus. *Comptes Rendus Hebdomadaires des Séances de l'Académie des Sciences D* **284**, 1951–1954.

Rosciglione B., Castellano M.A., Martelli G.P., Savino V. & Cannizzaro F. (1983) Mealybug transmission of grapevine virus A. *Vitis* **22**, 331–347.

Savino V., Gallitelli D. & Barba M. (1983) Olive latent ringspot virus, a newly recognised virus infecting olive in Italy. *Annals of Applied Biology* **103**, 243–249.

Signoret P.A., Alliot B. & Poinso B. (1977) Présence en France du wheat spindle streak mosaic virus. *Annales de Phytopathologie* **9**, 377–379.

Slykhuis J.T. (1980) Mites. In *Vectors of Plant Pathogens* (Eds K.F. Harris & K. Maramorosch), pp. 325–356. Academic Press, New York.

Slykhuis J.T. & Barr D.J.S. (1978) Confirmation of *Polymyxa graminis* as a vector of wheat spindle streak mosaic virus. *Phytopathology* **68**, 639–643.

Slykhuis J.T. & Polák Z. (1971) Factors affecting manual

transmission, purification and particle lengths of wheat spindle streak mosaic virus *Phytopathology* **61**, 569–574.

Spire D. (1962) Etude de la maladie des taches jaunes du cresson (*Nasturtium officinale* R.Br.). *Annales des Epiphyties* **13**, 39–45.

Stace-Smith R. (1955) Studies on Rubus virus diseases in British Columbia. II. Black raspberry necrosis. *Canadian Journal of Botany* **33**, 314–322.

Stace-Smith R. & Jones A.T. (1985) Black raspberry necrosis virus. In *Virus Diseases of Small Fruits* (Ed. R.H. Converse), USDA Handbook (in press).

Stone O.M. (1980) Nine viruses isolated from Pelargonium in the United Kingdom. *Acta Horticulturae* **110**, 177–181.

Takahashi R., Hayashi J., Inouye T., Moriya J. & Hirao C. (1973) Studies on resistance to yellow mosaic disease in barley. I. Tests for varietal reactions and genetic analysis of resistance to the disease. *Berichte des Ôhara Instituts für Landwirtschaftliche Biologie Okayama Universität* **16**, 1–17.

Tamada T. & Baba T. (1973) Beet necrotic yellow vein virus from rhizomania-affected sugar beet in Japan. *Annals of the Phytopathological Society of Japan* **39**, 325–331.

Tamada T., Abe H. & Baba T. (1975) Beet necrotic yellow vein virus and its relation to the fungus, *Polymyxa betae*. In *Proceedings of the 1st Intersectional Congress of IAMS, Tokyo 1974* **3**, 313–320.

Torrance L. & Harrison B.D. (1981) Properties of Scottish isolates of cocksfoot mild mosaic virus and their comparison with others. *Annals of Applied Biology* **97**, 285–295.

Toyama A. & Kusaba T. (1970) Transmission of soil-borne barley yellow mosaic virus. II. *Polymyxa graminis* as vector. *Annals of the Phytopathological Society of Japan* **36**, 223–229.

Usugi T. & Saito Y. (1976) Purification and serological properties of barley yellow mosaic virus and wheat yellow mosaic virus. *Annals of the Phytopathological Society of Japan* **42**, 12–20.

Usugi T. & Saito Y. (1979) Relationship between wheat yellow mosaic virus and wheat spindle streak mosaic virus. *Annals of the Phytopathological Society of Japan* **45**, 397–400.

Uyemoto J.K. & Gilmer R.M. (1971) Apple stem-grooving virus: propagating hosts and purification. *Annals of Applied Biology* **69**, 17–21.

Valat C. & Mur G. (1976) Thermothérapie du cardinal rouge. *Progrès Agricole et Viticole* **93**, 202.

Verhoyen M., Esparza-Duque J., Matthieu J.L. & Horvat F. (1976) Identification du virus latent du céleri en Belgique. *Parasitica* **32**, 158–166.

Vuittenez A. (1958) Transmission par greffage d'une virose du type 'enroulement foliaire' commune dans les vignobles de l'Est et du Centre-Est de la France. *Comptes Rendus des Séances de l'Académie d'Agriculture de France* **44**, 313–316.

Vuittenez A. & Stocky G. (1982) Ultrastructure de vignes infectées par deux maladies de type viral: la 'mosaïque des nervures' et la 'feuille rouge'. In *Proceedings of the 7th Meeting of the International Council for the Study of Viruses and Virus-like Diseases of the Grapevine*, pp. 191–204. Niagara Falls, Ontario.

White J.G. (1983) The use of methyl bromide and carbendazim for the control of lettuce big-vein disease. *Plant Pathology* **32**, 151–157.

Woodham R.C. & Krake L.R. (1983) Investigation on transmission of grapevine leafroll, yellow speckle and fleck diseases by dodder. *Phytopathologische Zeitschrift* **106**, 193–198.

4: Viruses IV
Diseases Assumed to be Due to Viruses

Many plant diseases show all the characteristics (symptoms, transmissibility) of virus diseases but the virus particles have never been isolated. Indeed, the viral diseases of known aetiology were once characterized only at this level. This chapter covers these cases, some of which could still be due to complexes of known viruses or even to mycoplasma-like or rickettsia-like organisms (cf. Chapter 5). Their viral aetiology is probable, but only assumed. Many are of considerable economic importance, especially the great variety of graft-transmissible diseases of woody plants (soft fruits, stone fruits, pome fruits, citrus) for which no viruses have yet been purified or characterized. They are arranged alphabetically by disease name, with alternative disease names ('synonyms') also indexed.

Apple green crinkle disease, probably due to a virus, involves the appearance of depressed areas on apple fruits, some weeks after flowering. Fruits of susceptible cultivars (Golden Delicious, Granny Smith) become severely malformed as the season advances (Kristensen 1963) but infected trees of many cultivars remain symptomless. Vascular tissues below swellings or depressions are green coloured and distorted. Symptoms are often restricted to a few branches of the infected tree. The disease is graft-transmissible (no vector is known) and is indexed by graft inoculation onto Golden Delicious (Dunez *et al.* 1982).

Apple leaf pucker disease causes deep yellow or light green flecking on the leaves, mainly associated with veins and veinlets, as well as some puckering and leaf distortion. Fruit symptoms develop only with low temperatures in the early summer. These include russet rings on some cultivars (e.g. Golden Delicious)—apple russet ring disease—and blotch on others (e.g. Stayman). No vector is known. Indexing is by graft transmission to cv. Golden Delicious (Dunez *et al.* 1982). Some reports have proposed that the disease could be caused by apple chlorotic leaf spot virus (Kegler *et al.* 1979).

Apple ringspot disease has symptoms restricted to fruits. When growth has finished, light brown areas develop; as fruits ripen these areas become more distinct and a brown halo or series of concentric rings develop around the lesion. The skin around the initial lesion or between concentric rings is of a light green or yellow colour (Lemoine 1979a). Important commercial cultivars are susceptible, e.g. Granny Smith, and Golden Delicious, which is used as an indicator (Dunez *et al.* 1982). No vector is known.

Apple rough skin disease and **apple star crack disease** may be caused by strains of the same pathogen, or by different pathogens; the situation remains unclear (Schmid 1963). Symptoms are rough, brown, corky patches or star-shaped cracks on the skin and an irregular fruit shape (Campbell & Hughes 1975). Susceptible cultivars (e.g. Cox's Orange Pippin) show dieback of shoots during winter (Hamdorf 1965). Symptoms depend on cultivar, strain and weather conditions. Indexing is by double budding using Belle de Boskoop (rough skin) and Cox's Orange Pippin or Golden Delicious (star crack) as indicators (Dunez *et al.* 1982) with M9 as rootstock (symptoms after 2–3 crops). Control is by using disease-free stock and scion material for propagation. Virus-free trees are obtained by heat treatment.

HAMDORF

Apple spy epinasty and decline disease. See pear stony pit disease.

Apricot ring pox disease probably has the same causal agent as **cherry twisted leaf disease**. It causes twisted leaf-like symptoms after transmission to Bing sweet cherry and conversely cherry twisted leaf causes ring pox after transmission to Tilton apricot (Hansen *et al.* 1976). Chlorotic spots, streaks and rings appear on the leaves, becoming necrotic. The resulting necrotic areas fall, producing holes. Discoloured and necrotic areas develop on the fruit surface or inside the flesh. These areas form reddish or black spots, rings and pits. Discoloration and necrosis can extend deep into the flesh. Some major commercial cultivars are affected. No vector is known. Indexing is by grafting onto Tilton apricot or Bing cherry.

Artichoke mosaic is a poorly characterized disease recorded from Sicily some 30 years ago (Gigante 1951). Symptoms consist of a rather intense mosaic mottle with bright yellow spots and blotches randomly scattered on the leaf surface. A mechanically transmissible virus was present in symptomatic plants but its properties were not investigated. Similar, though milder, symptoms are frequently observed in artichoke fields of S. Italy. Several mechanically transmissible viruses have been isolated from diseased plants, e.g. bean yellow mosaic, broad bean wilt, artichoke latent viruses and *Cynara* rhabdovirus (Martelli *et al.* 1981). None of these viruses was able to reproduce the field syndrome in artificially inoculated artichoke seedlings. Hence, a proper aetiological definition of artichoke mosaic has not yet been achieved.
MARTELLI

Beet yellow net disease has been described in W. Europe and N. America. Symptoms are chlorotic spots, vein clearing and yellow net symptoms on younger sugar beet leaves, thickening and brittleness of leaves. It is aphid-borne by *Myzus persicae*, *Aphis fabae* and other species, in a persistent manner (Watson 1962). Indexing is by aphid inoculation onto beet. There are no recent experimental data. According to Polák (1970) yellow net could be the first symptoms caused on beet leaves by severe necrotic strains of beet yellows virus (q.v.).
POLÁK

Blackcurrant infectious variegation disease occurs in England and some European countries. It is graft-transmissible but no virus has been isolated. In sensitive cultivars (Daniels September, Laxtons Nigger), symptoms consist of a bright chrome or pale yellow mosaic on the early leaves. In summer the leaves show a vein pattern formed by a yellow banding of the main veins (Thresh 1970). Indexing is by graft transmission to blackcurrant seedlings (Baldwin) and control is by propagation of healthy plants.
KRCZAL

Blackcurrant yellows disease occurs in England but the pathogen is unknown. First symptoms develop in spring following the year of infection: these are slight chlorotic flecks on the leaves; later in June and July, large sectors of the leaves show olive-green mosaic. Diseased bushes are stunted and their crop is reduced. Because spread is very slow, the disease is of practically no economic importance (Thresh 1970). Indexing is by graft transmission to cv. Amos Black and control is by propagation of healthy plants.
KRCZAL

Cherry rusty mottle disease is probably caused by a virus, but it has not yet been purified or identified (Posnette & Cropley 1961). Some commercial cultivars develop slight symptoms, but many remain symptomless. The rusty vein symptom has been recorded on cvs Frogmore, Gaucher, Roundel, Van and Bing. Mazzard F 12-1 is the most sensitive cultivar and strains can readily be distinguished by the severity of symptoms produced in the year after inoculation. Field experiments have shown that rusty mottle reduces the growth of Mazzard F 12-1 plants by 23% (height and girth). The growth of scion cultivars is less affected but the yield of fruit of infected Napoleon and Early Rivers trees is only three-quarters that of healthy controls. Fruit is normal in size and time of ripening. The disease has been reported in England, France, FRG, Netherlands, Czechoslovakia and other European countries. It should be distinguished from an American rusty mottle, restricted to W. USA, and from **cherry necrotic rusty mottle disease** (EPPO 91), described in N. America but also found in a few European countries (especially the UK). The disease is transmitted by the use of infected propagating material; it is infrequent and does not cause severe damage. Indexing on Mazzard F12-1 remains the most reliable technique for detection. The vein-clearing symptom appears in July the year following grafting. In August–September rust spots appear around the small yellow veins on the yellowing leaves.
DESVIGNES

Cherry green ring mottle disease

Basic description

The causal agent of this disease is probably a virus but it has not yet been purified or identified. The disease was described by Rasmussen *et al.* (1951). Because it has not been transmitted mechanically very little is known of its properties.

Host plants and diseases

Green mottle is seen naturally in Montmorency, English Morello and some cultivars of sour cherry. It may readily be identified in fruiting Montmorency cherries by the presence of chlorotic constrictions along the leaf veins and by the occurrence of yellow leaves prominently marked with deep green blotches or rings in July. Leaf symptoms are often associated with discoloration and necrotic pitting of Montmorency fruits. Infected trees of oriental flowering cherry (*P. serrulata*) show epinasty of the foliage, bark splitting and sometimes dieback. Sweet cherry, peach, nectarine and apricot are symptomless carriers of the disease. Peach cvs Halehaven, Sunhaven, Richaven, Glohaven and Sunding are also frequently infected (Parker *et al.* 1976).

Identification and diagnosis

The disease is easily detected on *Prunus serrulata* cvs Shirofugen or Kwanzan. If grafting is done at the beginning of the growing period, characteristic symptoms often appear 1–2 months later.

Geographical distribution, transmission and economic importance

The disease is widespread and has been reported in America, Australia, Europe, New Zealand and Japan. Natural spread in sour cherry orchards has been repeatedly observed. The rate of spread is relatively slow and some of the local spread may result from root grafts between adjacent trees. Though frequently latent in some *Prunus* species, the disease causes significant damage on Montmorency fruits and ornamental flowering cherry.
DESVIGNES

Citrus cachexia-xyloporosis disease

Basic description

Cachexia (Childs 1950) and xyloporosis (Reichert & Perlberger 1934) are one and the same disease. The transmissible nature of the disease was demonstrated first for cachexia (Childs 1952). Hence the name 'cachexia' has priority. The agent has not yet been identified but is easily graft-transmissible by bark patches. It is also transmitted mechanically by knife cuts from citron to citron; transmission can be prevented by sodium hypochlorite treatment of the infective knife (Roistacher *et al.* 1980). These properties are reminiscent of viroids, but the viroid nature of cachexia has not been proved.

Host plants

Susceptible citrus species or cultivars are: *Citrus macrophylla,* clementine, kumquat, mandarin, Rangpur lime, satsuma, sweet lime, tangelos. The following commercial citrus species or cultivars are symptomless carriers: sweet orange, grapefruit, lemons, acid limes. The following rootstocks are also tolerant: *Poncirus trifoliata,* citranges, citrumelos, rough lemon, sour orange.

Disease

Symptoms are restricted to the trunk, limbs and possibly shoots. They consist of stem pitting and bark pegging. Pegs on the cambial side of bark tissue project into corresponding elongate, narrow pits in the cambial face of the wood. Considerable gum usually accumulates in the crushed phloem region of the bark and in xylem near pits. The amount of pitting and gumming depends on cultivar, strain and age of the infected tree.

Identification and diagnosis

Cachexia is readily diagnosed in the field by removing a patch of bark tissue immediately above the bud union line in the case of a susceptible scion, or close to soil level for susceptible rootstocks. Stem pitting and bark pegging and gumming are diagnostic for cachexia. The disease can be confused with cristacortis on plants such as tangelos (cvs Orlando, Wekiwa) which are susceptible to both diseases. However, cristacortis pits are 5–10 times larger than those of cachexia (see cristacortis, p. 105). The symptoms of cachexia and gummy bark (Nour-Eldin 1980) are very similar but the host range is completely different. Gummy bark but not cachexia is seen on sweet orange and rough lemon; cachexia, but not gummy bark is observed on mandarin, tangelo and sweet lime (Childs 1980). A good indicator species is Parsons special mandarin, which is very susceptible to the cachexia agent. The mandarin bud is grafted on a fast-growing rootstock such as rough lemon. One bark patch (inoculum) of the tree to be indexed (candidate tree) is grafted above, the other below the mandarin bud. Only the latter is forced and allowed to grow. It is essential that the plants be grown in a warm glasshouse (average maximum and minimum temperatures are 33°C and 23°C). The developing mandarin shoot is cut back, when required, to about 20 cm above the bud union. One year after inoculation, conspicuous gum deposits are present at bud union (Roistacher *et al.* 1973).

Geographical distribution and transmission

Cachexia occurs wherever citrus is grown. It is

naturally transmitted by infected budwood and can be transmitted mechanically on pruning knives (Roistacher *et al.* 1980).

Economic importance and control

Cachexia is economically important on susceptible cultivars grown either as top or rootstock. It is controlled by using healthy budwood.
BOVÉ

Citrus concave gum disease/Citrus blind pocket disease

Basic description

Described by Fawcett & Bitancourt (1943), the disease is characterized by concavities on trunk and/or limbs; the name 'concave gum' refers to rather broad concavities, 'blind pocket' to narrow ones. These two symptoms are generally believed to be different expressions of the same disease (strains?). The agent is graft-transmissible and is assumed to be virus-like.

Host plants and diseases

Sweet orange, mandarin, clementine, tangelos and grapefruit are susceptible, while sour orange, sweet lime, *Poncirus trifoliata*, citranges, citrons, lemons, rough lemon and acid limes are symptomless carriers. Conspicuous broad concavities are seen on the trunk and, with certain cultivars such as Washington Navel orange and Orlando tangelo, gum can be seen oozing through cracks in bark at the concavities (concave gum). Cross-sections show gum-stained wood layers alternating with unstained layers. The broad concave depressions are accompanied by abruptly depressed concavities resembling deep vertical invaginations, with the two long straight sides touching each other (blind pocket). The two types of depressions can occur on the same tree. Young leaves often show the so-called 'psorosis young leaf symptoms'. These are of two types—vein flecking and oak leaf pattern. Both types occur on the same tree, and sometimes on the same leaves. These symptoms are typical, but not specific, since they may also be associated with scaly bark psorosis, cristacortis, impietratura, and crinkly leaf-infectious variegation (Klotz 1980).

Identification and diagnosis

Indexing is done by bark inoculation to mandarin or sweet orange seedlings. Leaf symptoms (vein flecking, oak leaf pattern) will appear on young leaves during the first flushes of growth in the glasshouse, provided it is kept cool. For trunk and limb symptoms, the inoculated seedlings must be grown in the field, sometimes for several years, before concavities appear.

Transmission

Transmission is through budwood and by bark inoculation. Pollen collected from a concave-gum-affected Parson Brown sweet orange tree, and inserted under the bark of Orlando tangelo seedlings, was able to transmit psorosis young leaf symptoms. However, no trunk symptoms have appeared within the 13 years of being inoculated (Vogel & Bové 1980).

Geographical distribution, economic importance and control

The disease occurs wherever citrus is grown: many severe cases are observed in Mediterranean countries. Injury develops slowly but trees may be severely damaged. Control is by the use of healthy budwood.
BOVÉ

Citrus cristacortis disease

Basic description

The disease was first described as sour orange or Tarocco orange stem pitting (Vogel & Bové 1964). It was later named cristacortis (Vogel & Bové 1968). The agent is easily graft-transmissible and assumed to be virus-like. Certain strains produce symptoms on clementine, others do not.

Host plants and diseases

Citrus pectinifera, clementine, grapefruit, mandarin, rough lemon, satsuma, siamelos, sour orange, sweet lime, sweet orange, tangelos and tangors are susceptible, while bergamot, citranges, citrons, *Citrus hystrix*, *C. volkameriana*, lemons, *Poncirus trifoliata* and acid limes are symptomless. On the trunk, limbs and shoots, vertical depressions or pockets are seen in the wood (stem pitting) with corresponding pegs on the cambial side of the bark. With certain species such as tangelos, gum-like material stains the tissues at the top of the peg and the bottom of the pit. Young pits in the wood can occur without externally visible depressions. Also, depressions, even severe ones, can disappear with time as a consequence of radial growth. Traces of previous depressions and pits are left in the wood and can be seen in cross-sections as radially oriented clear lines. New pits and pegs develop anywhere on the tree, from the trunk to pencil-thin shoots. Vein-flecking symptoms and oak leaf pattern on young leaves are indistinguishable from those occurring with scaly bark psorosis, concave gum/blind pocket, impietratura and crinkly leaf-infectious variegation (Vogel 1980).

Identification and diagnosis

Indicator plants are Orlando, Webber or Williams tangelos grafted on sour orange. A tangelo bud and two inoculum bark patches from the tree to be indexed (candidate tree) are grafted at the same time on a sour orange seedling, one bark patch below, the other above the tangelo bud. With an infective inoculum, vertical depressions or pockets will appear on the stem and shoots of the tangelo scion as well as on the sour orange stock. Generally these stem-pitting symptoms will begin to develop one year after inoculation, in the meantime, the young leaf symptoms will have appeared. With some species, such as clementine, stem pitting may appear only after several years. Test plants without stem-pitting symptoms are discarded after five years. Plants should be grown in the field since cristacortis stem-pitting symptoms are poor in the glasshouse. Cristacortis and cachexia can be indexed with the same Orlando tangelo on sour orange test plants since tangelo reacts differently to the two diseases and sour orange is only susceptible to cristacortis.

Transmission

Transmission is through budwood propagation and graft inoculation. Pollen from a cristacortis-affected Orlando tangelo tree inserted under the bark of healthy Orlando tangelo seedlings transmits both young leaf symptoms and the typical cristacortis stem pitting. However, natural spread of cristacortis by pollen has never been observed (Vogel & Bové 1980).

Geographical distribution, economic importance and control

Cristacortis is essentially found in Mediterranean countries and islands, in the Near and Middle East, but also in North Yemen where budwood from Italy was used many years ago. The spectacular stem-pitting symptoms do not really affect yields. Control is by the use of healthy budwood.

BOVÉ

Citrus gummy bark disease was first described as phloem discoloration of sweet orange (Nour Eldin 1956, 1968). The agent is readily graft-transmissible and is thought to be virus-like. Only sweet orange and rough lemon are susceptible; other species appear to be symptomless. The most conspicuous symptoms are observed on sweet orange. The bark is impregnated with gum particularly near the bud union, but the discoloration can extend 60 cm or more above the union line. Stem pitting accompanies gum impregnation of bark. Gummy bark on sweet orange very closely resembles cachexia on mandarin or tangelo. However the two diseases have a completely different host range with no overlap, and are therefore thought to be quite distinct. In the field, gummy bark can easily be recognized by scraping the bark of sweet orange trees above bud union to reveal circumferential streaks of reddish-brown gum-impregnated tissues (Nour Eldin 1980). Indexing is by side-grafting sweet orange on sour orange or rough lemon rootstock in the field (reddish brown impregnated tissue above bud union after 3–5 years).

The disease is transmitted by bud propagation and graft inoculation. It occurs throughout most of the Near and Middle East, but in Europe it only occurs in Greece. It is not a very destructive disease, but infected trees are usually stunted to some extent. Control is by the use of healthy budwood.

BOVÉ

Citrus impietratura disease

Basic description, host plants and disease

The agent is readily graft-transmissible and is assumed to be virus-like. Bergamot, *Citrus volkameriana*, clementine, grapefruit, lemon, rough lemon, sour orange, sweet orange and tangelo are susceptible, while Chinotto, citron and kumquat are symptomless. Fruits on infected trees have gum pockets in the albedo, visible externally as protuberances or sunken depressions, or else only on cutting or peeling. The axis may also be gum-impregnated. The affected fruit often has a hard, stone-like consistency (impietratura, from the Italian for 'stone'), is reduced in size and tends to drop. Fruit symptoms are very variable in intensity from year to year.

Impietratura is one of the diseases which have in common the symptoms they cause on young leaves—

vein flecking and oak leaf pattern. The other diseases are concave gum/blind pocket, cristacortis, scaly bark psorosis and crinkly leaf-infectious variegation. The symptoms induced by the impietratura agent on fruit are almost identical with those due to boron deficiency, but it has been shown that infected plants contain adequate boron levels and application of this element on impietratura-affected trees has no effect on the disease (Catara & Scaramuzzi 1980).

Identification and diagnosis

Indexing is by bark graft near flowers on 3-year *C. volkameriana* lemon seedlings (albedo gumming and fruit hardening after 6–18 months) or on 3–4 year field-grown grapefruit (bark graft near a set fruit).

Transmission, geographical distribution, economic importance and control

Transmission is by bud propagation and graft inoculation. The disease is found essentially in the citrus-growing countries of the Mediterranean area, the Near East (notably Cyprus) and the Middle East, as well as in S. Africa, Texas (USA) and Venezuela. It is not destructive and trees remain apparently healthy but losses may result from off-season fruit drop and unmarketable fruit. Control is by the use of healthy budwood.

BOVÉ

Citrus psorosis disease

Basic description

First described in Florida under the name psorosis, the disease is often called scaly bark or scaly bark psorosis. Fawcett (1932) gave the name psorosis 'A' to the commonly occurring form of the disease in order to distinguish it from a less common but more severe type which he described as psorosis 'B'. The agent is readily graft-transmissible and is assumed to be virus-like. Strains are suspected.

Host plants

Grapefruit, mandarin, sweet orange and tangelo show the scaly bark symptom. Psorosis young leaf symptoms are seen on the above species as well as on citron, lemon, acid and sweet limes, pummelo, rough lemon and sour orange.

Disease

The characteristic symptom of psorosis is bark scaling on trunks and limbs. Bark scaling on field trees seldom appears before the trees are 6 years old, the average time being 12–15 years. There are, however, strains of the agent such as California strain 339, which induce severe scaling very quickly in the glasshouse on young sweet orange seedlings even though the inoculum used for the graft inoculation is normal bark, not lesion bark. This severe reaction, of the psorosis 'B' type, is also obtained with regular psorosis 'A' strains when the bark used for inoculation is taken within a psorosis 'A' scaling lesion.

Trees with psorosis 'A' bark scaling also show the so-called psorosis young leaf symptoms first described by Fawcett (1933). These symptoms known as vein flecking, appear during the growth flushes and vary greatly in extent and degree on individual trees at a given time. They occur in the region of the veinlets and represent small, elongated cleared patches of lighter colour than the rest of the leaf. This flecking may be general over the entire leaf, or may affect only parts of the leaf. At times most young leaves show symptoms, at others only few leaves are affected (Wallace 1959). Some of the small flecks may coalesce to form conspicuous blotches which may display a zonate or oak leaf pattern centred on the midrib. Vein flecking is sometimes considered to be specific to psorosis 'A', while oak leaf pattern would be restricted to concave gum. This view is now challenged. It is sufficient to say that psorosis 'A', concave gum/blind pocket, cristacortis, impietratura and crinkly leaf-infectious variegation show similar 'psorosis' young leaf symptoms and cannot be distinguished on the basis of these symptoms (Roistacher 1980). In the case of the more severe psorosis 'B' strains, or when a form of psorosis similar to psorosis 'B' is induced experimentally by inoculation of lesion bark, mature leaf symptoms can be observed (see below).

Identification and diagnosis

Indexing is by grafting two bark patches on 2–4 pencil-thickness seedlings of sweet orange, sweet tangor or mandarin, cut back 25 cm from soil to force new growth. Glasshouse temperatures should not be too high. Symptoms are: i) vein flecking and/or oak leaf pattern on young immature leaves; ii) shock effect on newly developing axillary shoots which curve downwards and show leaf necrosis and drop. New growth arises from below and may then show typical young leaf symptoms (Wallace 1978).

Young, healthy sweet orange seedlings which are inoculated by means of normal (non-lesion) bark from a psorosis 'A' (scaly bark)-infected tree develop young leaf symptoms, but no bark lesions appear until after 6 years. In contrast, if infection results from a piece of lesion bark, lesions sometimes begin to develop within two months and spread quite rapidly throughout the seedling (psorosis 'B' reaction). In addition, mature leaves will show irregular, blotchy patterns with brownish eruptions on the undersurface (mature leaf symptoms) (Wallace 1978).

When lesion bark is grafted on sweet orange seedlings that have been pre-inoculated with non-lesion bark from a psorosis 'A' tree, these seedlings will not develop the early, severe psorosis 'B' reaction as they are cross protected (Wallace 1957). When questionable scaly bark symptoms are encountered, they can be shown to be due to the psorosis 'A' agent if inoculation of bark from the scaling lesion is able to induce psorosis 'B', and if non-lesion bark offers cross protection against psorosis 'B'.

Transmission

Transmission is by bud propagation and bark inoculation. Natural spread of a psorosis-like disease in the Entre Rios province of Argentina has been reported (Pujol & Beñatena 1965). The relationship of this disease to classical psorosis 'A' is uncertain.

Geographical distribution, economic importance and control

Citrus psorosis occurs wherever citrus is grown. It is a major cause of tree decline and low yield in many orchards grown from non-certified, psorosis 'A' infected budlines. Control is by the use of healthy budwood. Procedures have been described for scraping and disinfection of bark lesions for prolonging the life of affected trees (Wallace 1959, 1978).
BOVÉ

Citrus ringspot disease

Basic description, host plants and disease

Citrus ringspot was first described by Wallace & Drake (1968). It is assumed to be caused by a virus, and affects a wide range of citrus species, hybrids and relatives. Field symptoms consist of yellow blotches, chlorotic vein banding or chlorotic rings in mature leaves; sometimes young twigs, thorns and fruits may also show yellow blotches and chlorotic rings. In some countries bark scaling (normally attributed to psorosis) is also a consistent symptom. Infected trees are occasionally symptomless in the field.

Identification and diagnosis

Chlorotic rings in leaves, twigs or fruits are specific symptoms in the field. Indexing is necessary for the diagnosis of symptomless infections, either by grafting buds, twigs, bark or leaf patches to grapefruit, sweet orange or mandarin seedlings (spotting and flecking on new leaves after 4–6 weeks, chlorotic rings 1–2 weeks later; some isolates give a shock necrotic reaction) or else by mechanical transmission to herbaceous indicators (Garnsey & Timmer 1980). *Chenopodium quinoa* gives chlorotic to necrotic local lesions in 3–5 days.

Transmission, distribution, importance and control

The disease is mainly propagated by infected budwood. There is limited evidence of natural spread by unknown means (Timmer & Garnsey 1980). The disease occurs in most citrus-producing countries of America and the Mediterranean basin; it is economically important when associated with severe bark scaling. Control is by the use of virus-free budwood for new plantings and quarantine measures to prevent the introduction of naturally spreading isolates.
MORENO & NAVARRO

Citrus vein enation disease (Wallace & Drake 1953) and **citrus woody gall disease** (Fraser 1958) are caused by the same agent (Wallace & Drake 1960), thought to be a virus. Vein enation symptoms are globular irregular tumours on lateral veins on the lower leaf surface, with corresponding depressions on the upper leaf surface. Only Mexican lime and sour orange consistently and diagnostically develop these symptoms under field conditions. Woody gall symptoms are tumours on the trunk, branches and roots; initially as slight swellings of small areas that enlarge slowly into irregularly shaped galls that eventually may reach a 'cauliflower' appearance. These symptoms have only been observed in rough lemon, *Citrus volkameriana*, Rangpur lime and Mexican lime. Other citrus species are symptomless carriers. Indexing is by graft inoculation to Mexican lime seedlings in the glasshouses at 20–25°C (vein enation on a new growth after 5–8 weeks). The disease is

transmitted by the aphids *Myzus persicae*, *Toxoptera citricidus* or *Aphis gossypii*. Experimentally it is transmitted by grafting and by dodder. The disease occurs widely, but in Europe only in Spain (Ballester *et al.* 1979). It is only important in countries where rough lemon is extensively used as a rootstock (S. Africa, Peru, etc.). Control is by use of virus-free propagative budwood and tolerant rootstocks. Healthy plants can be obtained by thermotherapy and *in vitro* shoot-tip grafting.
NAVARRO & PINO

Dandelion yellow mosaic disease involves yellow mottling of the weed *Taraxacum officinale*, which otherwise grows normally. The disease is transmissible to lettuce; after veinal and interveinal necrotic etching, the leaves become thick, stunted and deformed (Smith 1972). It is aphid-transmissible, readily so to lettuce but with great difficulty to *T. officinale*. Described in England, a similar disease has been described in E. Europe and in Scandinavia (Brcak 1979).

Euonymus infectious variegation disease is a yellow blotching, vein clearing syndrome affecting the ornamental shrub *Euonymus japonicus* (and also wild *E. europaeus*) in Europe and N. America. It is graft-transmissible, but its viral aetiology has not been proved. Strawberry latent ringspot virus, cucumber mosaic virus and 2 or 3 less well characterized viruses have been identified in plants with variegation symptoms. There is no specific means of control.
COOPER

Fig mosaic disease is recognized in all continents except S. America; it takes the form of green-yellow blotches, premature leaf fall and fruit spotting affecting at least 16 species of *Ficus*. Transmission is by grafting but the suspected vectors are eriophyid mites (*Aceria ficus*) (Flock & Wallace 1955; Proeseler 1972). The causal agent is unknown. Some fig trees are claimed to be resistant and the fact that seedlings are healthy suggests absence of seed transmission. However, disease avoidance does not seem practicable.
COOPER

Gooseberry vein banding disease, widespread in Europe (Posnette 1970), is graft and aphid-transmissible (in the stylet-borne manner), but the characteristics of the pathogen are unknown. Host range seems to be limited to *Ribes*. Yield and growth of infected plants is reduced. In gooseberry the main veins of the first leaves show translucent pale-yellow vein banding. On extension growth not all veins may be affected. Leaves are often distorted. In blackcurrant, first leaves show pale-yellow vein banding, forming a fern pattern. In redcurrant, first leaves show yellow vein banding; later the vein reticulum may be cleared or banded by translucent tissue. (Adams 1979). Indexing is by aphid or graft-transmission to sensitive cultivars, such as Leveller. Control is by the propagation of healthy plants and the control of aphids.
KRCZAL

Grapevine corky bark disease

Basic description

The aetiology is still unknown, although it is likely to be viral. A virus with filamentous closterovirus-like particles more than 1500 nm long is associated with the corky bark-stem pitting complex in California (USA). Reports from California and Mexico indicate that corky bark and stem pitting (legno riccio) are the same disease. In these countries, the probability that vines with stem pitting index positive for corky bark is higher than 70% (Teliz *et al.* 1982). In Italy, this probability seems lower than 5%.

Host plants and disease

European and American *Vitis* species are the only known natural hosts. Most European cultivars show only a reduction of vigour when infected. Symptoms, when visible, consist of delayed bud opening, irregular maturation of the wood, and soft and rubbery canes with longitudinal cracks at the base. Leaves are smaller than normal and are shed later than in healthy vines. In red-berried cultivars, leaves turn reddish and roll downward like those of vines affected by leaf roll. Many cultivars show pits and grooves of the xylem (Beukman & Goheen 1970).

Identification and diagnosis

The disease is readily transmissible by grafting. The hybrid LN 33 (Couderc 1613×Thompson seedless) is a specific indicator. It reacts, 1–2 years after inoculation, with severe reduction of growth, reddening of the leaves and abnormal swellings of the basal internodes of the canes, which become corky and develop longitudinal cracks. The xylem is pitted and grooved (Beukman & Goheen 1970).

Transmission

Infected budwood is the primary means of spread. Natural spread of the corky bark-stem pitting complex has been observed in the State of Aguacalientes, Mexico, but the vector and acquisition sources are unknown (Teliz *et al.* 1982).

Geographical distribution, economic importance and control

The disease has been recorded in many countries. It probably occurs world-wide (Bovey *et al.* 1980) and very heavy crop losses are known to occur on susceptible cultivars. The yield of cv. Cardinal vines in Mexico was reduced up to 76% and the longevity of the vineyard was also reduced (Teliz *et al.* 1982). Disease-free stocks can be obtained by the prolonged treatment (more than 90 days) of buds at 38°C grafted to healthy rootstocks, or by the removal of shoot tips and rooting under mist (Goheen 1977).
CONTI & MARTELLI

Grapevine enation disease is reported in Europe, USA, S. Africa and Australia. Affected vines are slow to break dormancy, and have short internodes and small leaves. Enations appear on the underside of basal leaves; the upper leaf surface can also be affected (Prota & Garau 1976). Field spread also seems to occur (Graniti & Martelli 1970).

Grapevine stem pitting disease

Basic description

The aetiology is still unknown, although it is likely to be viral. Different forms of stem pitting may exist, with different aetiological agents. One of these, first described in Italy as legno riccio (Graniti & Martelli 1970) may be related to, or the same as, the corky bark agent. Grapevine virus A (GVA) (see grapevine leaf roll disease) was originally isolated in Italy from a vine with stem pitting (Conti *et al.* 1980). Inconsistent association of GVA and other closteroviruses serologically unrelated to it with stem pitting symptoms has been reported from different countries (Milne *et al.* 1984).

Host plants and disease

European and American *Vitis* species are the only known natural hosts. Most *Vitis vinifera* cultivars and rootstock hybrids are susceptible. In a Bulgarian varietal collection, Abracheva (1981) found that 86% of 646 different cultivars were visibly infected. Diseased plants are undersized, less vigorous than normal and may show delayed bud opening in spring. Some decline and die within a few years of planting. Grafted vines often show a swelling above the bud union and a marked difference between the relative diameter of scion and rootstock. The woody cylinder is typically indented with pits and grooves which correspond to peg and ridge-like protrusions on the cambial face of the bark. These alterations may occur on the scion, rootstock or both (Anonymous 1979).

Identification and diagnosis

The disease is readily transmissible by grafting. Several American *Vitis* species and their hybrids are satisfactory indicators, i.e. LN 33, *V. rupestris, V. riparia* × *V. berlandieri* Kober 5BB or 157–11, *V. rupestris* × *berlandieri* 1103 P. Pitting on the woody cylinder appears 1–3 years after inoculation.

Transmission

Primary means of spread is through infected propagating material. No natural spread has been observed in Europe or elsewhere, except in Mexico with reference to the corky bark-stem pitting complex (see grapevine corky bark disease).

Geographical distribution, economic importance and control

The disease probably occurs world-wide, records existing from many countries (Bovey *et al.* 1980). Crop losses can be as high as 30–50%. Disease-free stocks can be obtained by prolonged (more than 150 days) heat treatment at 38°C, removal of shoot tips and grafting or culturing *in vitro* (Legin *et al.* 1979).
MARTELLI & CONTI

Malus platycarpa dwarf disease is graft-transmissible but the causal agent is not yet known. It affects *Malus* and *Chaenomeles,* usually together with other latent apple viruses. It is probably distributed world-wide, and can cause 30% or more growth reduction in the indicator species *M. platycarpa.* It is lethal to *M. floribunda* and *M. sargentii.* There are no symptoms in commercial apple cultivars, but growth may be influenced in combination with other latent apple viruses (Luckwill & Campbell 1963). Indexing is by inoculation of *M. platycarpa.* Healthy propagation

material from infected plants can be obtained by heat treatment followed by tip propagation.
KUNZE

Malus platycarpa scaly bark disease is probably caused by a virus. It occurs in Europe, N. America, S. Africa, Japan and New Zealand. The natural host range is limited to *Malus*. No symptoms are seen in commercial apple cultivars and rootstocks (Luckwill & Campbell 1963). It is graft-transmissible; no other means of transmission is known. Indexing is by graft-inoculation onto the woody indicator species *Malus platycarpa*. Symptoms consisting of a roughened scaly appearance of the bark become evident on the new growth towards the middle of the first summer after inoculation. Control is by heat treatment at 37°C for 2–3 weeks, followed by tip propagation.
SCHIMANSKI

Olive yellow mosaic disease is graft-transmissible only to olive in Italy. It is possibly caused by a virus. Another graft-transmissible disease of olive, **olive sickle leaf disease**, has been recorded only from California (USA) and Israel.
MARTELLI

Peach latent mosaic disease

Basic description

The disease, the causal agent of which has not yet been identified, was described by Desvignes (1980) on peach cultivars introduced into France from N. America. The symptoms produced are similar to those reported for peach mosaic, peach calico and peach blotch diseases in the USA (Pine 1976) and peach yellow mosaic disease in Japan (Kishi *et al.* 1973). **American peach mosaic disease** is regarded to be of quarantine significance for Europe (EPPO 27). Further research will clarify the aetiology of these similar diseases.

Host plants and diseases

The only natural host is peach. Other *Prunus* species are immune. Commercial peach cultivars are generally infected with latent strains. Symptoms are: delay in leaf emergence, flowering and maturity, leaf distortion, discoloration and irregular deformation of fruits with cracked sutures. Infected clones are more liable to frost damage, deficiencies, malnutrition and canker diseases. Sometimes mosaic, blotch, vein banding or calico appear on leaves; occasional stem pitting is observed.

Identification and diagnosis

Cross protection in peach seedlings, using a standard severe strain as a challenge, is used to detect the virus quite rapidly (3 months) in the glasshouse.

Transmission, distribution, importance and control

The disease has been artificially transmitted by the aphid *Myzus persicae* (whereas American mosaic is said to be transmitted by the mite *Eriophyes insidiosus*). It has quite frequently been detected (30%) in peach cultivars introduced from N. America, and may occur more or less widely in Europe, Japan and China. Its importance is hard to evaluate until the situation for this whole group of similar diseases has been clarified. Control is by the removal of infected trees.
DESVIGNES

Pear blister canker disease affects certain susceptible pear cultivars (Williams), causing phloem necrosis. It is latent in many cultivars and no vector is known. Detection is by graft transmission to susceptible cultivars (Dunez *et al.* 1982).

Pear stony pit disease/Pear vein yellows disease

Basic description

These diseases are very probably due to the same agent which is also probably the cause of **apple spy epinasty and decline disease, apple stem pitting disease** and **quince sooty ringspot disease.** The agent is possibly a virus but no particles have been described (Kegler *et al.* 1979). Carnation ringspot virus could not be confirmed as the agent of pear stony pit (Kegler *et al.* 1976).

Host plants and diseases

Apple, pear and quince are the natural hosts. Commercial apple cultivars and rootstocks remain symptomless. Other *Pyrus* species have been experimentally infected (*P. betulaefolia, P. calleryana, P. ussuriensis*). On pear leaves, mild strains cause pale green and later yellowish banding on short sections of leaf veinlets (vein yellows). More severe strains cause a conspicuous red mottling and flecks on the veins (pear red mottle). Symptoms appear no earlier than July. Highly sensitive cultivars are Beurré Bosc, Doyenné du Comice, Nouveau Poiteau; Köstliche von Charneu and Winter Nelis are highly tolerant (Fridlund 1976). Typical fruit symptoms are observed on pear and quince trees. Owing to the development of sclerenchymatous tissues and later necrotic cells at the base of the pits on the fruit skin, these areas grow less than the surrounding normal ones, thus leading to fruit deformation (Fig. 15). Dry matter, ash, potassium, calcium, ascorbic acid and malic acid contents are higher in diseased fruits than in comparable healthy ones (Kegler *et al.* 1961). Rough bark symptoms have also been observed.

Fig. 15. Pear stony pit disease: deformations of pear fruit.

Identification, diagnosis and strains

Pear stony pit has been indexed on pear cv. Beurré Bosc and pear vein yellows on cv. Nouveau Poiteau. Final diagnosis may take three seasons. The apple indicator clone Spy 227 reacts with stem pitting symptoms. Severe strains cause red mottle in pear, stem pitting and decline in Spy 227 and green crinkle in Kola Crab apples; milder ones cause vein yellows in pear. *Pyronia veitchii* reacts with epinasty to all strains (Kegler *et al.* 1979), and with tip necrosis and stem pitting to severe strains. Distinct strains cause stony pit in Beurré Durondeau pear and *P. veitchii* (Lemoine 1979b).

Transmission

The disease is graft-transmissible and propagated with infected budwood and rootstocks. No natural vector has been described.

Geographical distribution, economic importance and control

The disease probably occurs wherever apple and pear are grown, though the pear stony pit symptom has only been recorded from Europe and N. America. Yield losses in pear may be up to 15% (cv. Beurré Hardy) or 30–42% (cv. Conference); growth reductions up to 55% have been noted (Cropley & Posnette 1973). Stony pit is only seen sporadically, but infected trees may show 18–94% damaged fruits. Control is by heat treatment at 37–38°C for 4–5 weeks, followed by tip propagation.

KEGLER & SCHIMANSKI

Raspberry curly dwarf disease, probably caused by a virus, occurs in Scotland and is only graft-transmissible. Symptoms in diseased plants of cv. Lloyd George are numerous weak canes with curled leaves (Prentice & Harris 1950). Indexing is by graft transmission to 'Baumforth's seedling B', and control is by propagation of healthy plants.

KRCZAL

Raspberry leaf mottle (RLM) disease/ Raspberry leaf spot (RLS) disease

Basic description

The agents, probably viruses, responsible for raspberry leaf mottle and leaf spot are distinct, but show similarities in vector relations and response to thermotherapy; they also induce similar symptoms in *Rubus* indicator plants. Because of these similarities, in many published reports it is not clear which of the two diseases is being discussed. It is convenient therefore to consider them together. The diseases induced in some raspberry cultivars have received various names including raspberry mosaic 2, raspberry chlorotic spot and leaf spot mosaic. Neither agent has been purified, nor have virus-like particles been seen in thin sections of infected plants. They are not mechanically transmissible to plants or seed-borne in raspberry. They can be eradicated from plants by heat treatment (Jones 1982).

Host plants and diseases

The two agents infect wild and cultivated *Rubus* species, often occurring together in the same plant, usually in association with other viruses. Most wild and cultivated raspberry plants are infected symptomlessly but some cultivars show distinct angular chlorotic/yellow spots randomly distributed over the leaf. Symptoms are usually most severe on fruiting canes where spots may merge to form large chlorotic areas and leaves may be distorted. Diseased plants become stunted and may die within 3 years of infection.

Identification and diagnosis

In the absence of symptoms caused by other viruses, the symptoms induced in sensitive raspberry cultivars are diagnostic. However, identification of the causal agent in affected plants or in symptomless plants is dependent on symptoms in RLM and RLS-sensitive indicator cultivars following graft inoculation. The RLM agent induces symptoms in raspberry cvs Malling Delight and Malling Landmark and infects, but induces no symptoms in cvs Glen Cova and Norfolk Giant, whereas the reverse is true for the RLS agent. Symptoms in indicators may take 2–12 months to appear in graft-inoculated plants (Jones & Murant 1975).

Transmission and economic importance

RLM and RLS are transmitted, probably in a semi-persistent manner by the large raspberry aphid *Amphorophora idaei*. Both agents and vector are common in Britain and throughout Europe. However, high summer temperatures in some areas of Europe inhibit aphid reproduction and virus spread. Although many raspberry cultivars are infected symptomlessly, such infections can decrease vigour and berry size, especially in combination with other viruses. Most cultivars sensitive to infection have some resistance to the aphid vector so that incidence in such crops is usually small. However, plants that do become infected usually die within 3 years.

Control

Chemical control of aphid vectors does not give satisfactory control of infection. Planting healthy stock in isolation from likely disease sources may delay infection, and in areas of high aphid populations with close proximity to disease sources, planting tolerant cultivars may be useful. The best control has been achieved by growing cultivars with strong resistance to the vector (Jones 1979).

JONES

Rose degeneration diseases, in which hypothetical viruses have sometimes been implicated, have been observed on various rose cultivars in the past 20 years. These disorders are characterized by a more or less severe stunting or dwarfing of plants, accompanied by a partial wilt or dieback of some shoots. Growth abnormalities, mainly a proliferation of axillary buds, are also typical of these diseases. Leaves are small, chlorotic, brittle and plants are more severely affected in spring (particularly after grafting) and often recover in summer.

In Europe, the names 'rose streak' (Schmelzer 1967), 'rose stunt' (UK), 'rose bud proliferation' (Netherlands; Bos & Perquin 1975), 'frisure' (France; Devergne & Goujon 1975) have been used to describe several diseases which have many of these features in common. In some aspects, they also resemble the initial 'rose wilt' described in Italy and Australia more than 50 years ago, and much more recently other degeneration diseases occurring in the USA ('spring dwarf', 'leaf curl'), New Zealand or South Africa ('little leaf', 'wilt'). In spite of symptom similarities, it is unlikely that all these disorders have the same origin or are caused by one pathogen. In some cases, the symptoms are associated with a particular rootstock (*Rosa multiflora, Rosa indica major*) and, consequently, might result from an incompatibility between some rose cultivars and the rootstock. However, especially when it is propagated vegetatively, the rootstock could be carrying a latent pathogen such as a virus. Double-budding experiments and the apparent cure of heat-treated plants seem to confirm this hypothesis. However, until now, no virus has been identified from plants showing symptoms of degeneration, or transmitted by inoculation to other test plants.

DEVERGNE

Strawberry mottle disease, probably due to a virus, occurs world-wide and is aphid-borne (in the stylet-borne manner). Infected strawberry cultivars remain symptomless, but depending on the virus strain, yield is reduced by 20–30%. In combination with other viruses, the productivity of plants is still further reduced. Indexing is by aphid or graft transmission to *Fragaria vesca*. Symptoms appear within 1–3 weeks (mottling of leaflets, chlorosis of veins, distortion of leaves, crown proliferation). Control is by propa-

gation of healthy plants and the control of aphids. Virus-free plants are obtained from infected ones by heat treatment or meristem-tip culture (Mellor & Frazier 1970).
KRCZAL

Strawberry necrosis disease is caused by an agent, probably a virus, which has been partly characterized. It has been isolated by mechanical transmission from symptomless strawberry cv. Herzbergs Triumph and from *Fragaria vesca* inoculated by grafting. The dilution endpoint is 10^{-3} to 10^{-4}, the thermal inactivation point 48–50°C, and longevity *in vitro* 1–2 days. At present no further information is available on this virus which is of minor economic importance. Indexing is by mechanical transmission to *Phaseolus vulgaris* (necrotic discoloration of the veins or the adjacent tissue, all veins are partially or completely damaged within 2–3 days) or *Vigna sinensis* (leaf surface brown and necrotic) (Maassen 1961). Control is by the propagation of healthy plants.
KRCZAL

REFERENCES

Abracheva P. (1981) La sensibilité de certaines varietés de vigne à la maladie du bois strié (legno riccio). *Phytopathologia Mediterranea* **20,** 203–205.

Adams A.M. (1979) The effect of gooseberry vein banding virus on the growth and yield of gooseberry and redcurrant. *Journal of Horticultural Science* **54,** 23–25.

Anonymous (1979) Il legno riccio della vite in Italia. *Informatore Fitopatologico* **29,** 3–18.

Ballester J.F., Pina J.A. & Navarro L. (1979) Estudios sobre el 'vein enation-woody gall' de los agrios en España. *Anales del Instituto Nacional de Investigaciones Agrarias, Protección Vegetal* **12,** 127–138.

Beukman E.F. & Goheen A.C. (1970) Grape corky bark. In *Virus Diseases of Small Fruits and Grapevines* (Ed. N.W. Frazier), pp. 207–209. University of California, Berkeley.

Bos L. & Perquin F.W. (1975) Rose bud proliferation, a disorder of still unknown etiology. *Netherlands Journal of Plant Pathology* **81,** 187–198.

Bové J. & Vogel R. (1980) *Description and Illustration of Virus and Virus-like Diseases of Citrus.* SETCO-IRFA. Paris.

Bovey R., Gärtel W., Hewitt W.B., Martelli G.P. & Vuittenez A. (1980) *Virus and Virus-like Diseases of Grapevines.* Payot, Lausanne.

Brcak J. (1979) Czech and Scandinavian isolates resembling dandelion yellow mosaic virus. *Biologia Plantarum* **21,** 298–301.

Campbell A.I. & Hughes L.F. (1975) Symptoms of star crack virus on the fruit and shoot growth of apple cultivars. *Acta Horticulturae* **44,** 245–250.

Catara A. & Scaramuzzi G. (1980) Impietratura. In *Description and Illustration of Virus and Virus-like Diseases of Citrus* (Eds J. Bové & R. Vogel). SETCO-IRFA, Paris.

Childs J.F.L. (1950) The cachexia disease of Orlando tangelo. *Plant Disease Reporter* **34,** 295–298.

Childs J.F.L. (1952) Cachexia, a bud-transmitted disease and the manifestation of phloem symptoms in certain varieties of citrus, citrus relatives and hybrids. *Proceedings of the Florida State Horticultural Society* **64,** 47–51.

Childs J.F.L. (1980) Cachexia-Xyloporosis. In *Description and Illustration of Virus and Virus-like Diseases of Citrus* (Eds J. Bové & R. Vogel). SETCO-IRFA, Paris.

Conti M., Milne R.G., Luisoni E. & Boccardo G. (1980) A closterovirus from a stem pitting-diseased grapevine. *Phytopathology* **70,** 394–399.

Cropley R. & Posnette A.F. (1973) The effect of virus on growth and cropping of pear trees. *Annals of Applied Biology* **73,** 39–43.

Desvignes J.C. (1980) Different symptoms of the peach latent mosaic. *Acta Phytopathologica Academiae Scientiarum Hungaricae* **15,** 183–190.

Devergne J.C. & Goujon C. (1975) Etude d'anomalies de croissance chez le rosier de serre. *Annales de Phytopathologie* **7,** 71–79.

Dunez J., Stouffer R., Posnette A.F., Fulton R., Fridlund P., Hansen J., Kristensen H.R., Kunze L., Meijneke C.A.R., Németh M. & Šutić D. (1982) Detection of virus and virus-like diseases of fruit trees. *Acta Horticulturae* **130,** 319–330.

Fawcett H.S. (1932) New angles on treatment of bark diseases of citrus. *California Citrograph* **17,** 406–408.

Fawcett H.S. (1933) New symptoms of psorosis, indicating a virus disease of citrus. *Phytopathology* **23,** 930.

Fawcett H.S. & Bitancourt A.A. (1943) Comparative symptomatology of psorosis varieties on citrus in California. *Phytopathology* **33,** 837–864.

Flock R.A. & Wallace J.M. (1955) Transmission of fig mosaic by the eriophyid mite *Aceria ficus. Phytopathology* **45,** 52–54.

Fraser L. (1958) Virus diseases in Australia. *Proceedings of the Linnean Society of New South Wales* **83,** 9–19.

Frazier N.W. (1970) *Virus Diseases of Small Fruits and Grapevines.* University of California, Berkeley.

Fridlund P.R. (1976) Pear vein yellows virus symptoms in greenhouse-grown pear cultivars. *Plant Disease Reporter* **60,** 891–894.

Garnsey S.M. & Timmer L.W. (1980) Mechanical transmissibility of citrus ringspot virus isolates from Florida, Texas and California. In *Proceedings of the 8th Conference of the International Organization of Citrus Virologists,* pp. 174–179. University of California, Riverside.

Gigante R. (1951) Il mosaico del carciofo. *Bollettino della Stazione di Patologia Vegetale di Roma* **7,** 177–181.

Goheen A.C. (1970) Grape leafroll. In *Virus Diseases of Small Fruits and Grapevines* (Ed. N.W. Frazier), pp. 209–212. University of California, Berkeley.

Goheen A.C. (1977) Virus and virus-like diseases of grapes. *HortScience* **12,** 465–569.

Graniti A. & Martelli G.P. (1970) Enations. In *Virus Diseases of Small Fruits and Grapevines* (Ed. N.W. Frazier), pp. 243–245. University of California, Berkeley.

Hamdorf G. (1965) Star crack and rough skin of apple. Transmission studies and varietal host range. *Zaštita Bilja* **16,** 293–298.

Hansen A.J., Parish C.L. & Pine T.S. (1976) Apricot ring pox. In *Virus Diseases and Non-Infectious Disorders of Stone Fruits in North America. Agriculture Handbook no. 437*, pp. 45–49. USDA, Washington.

Jones A.T. (1979) Further studies on the effect of resistance to *Amphorophora idaei* in raspberry (*Rubus idaeus*) on the spread of aphid-borne viruses. *Annals of Applied Biology* **92,** 119–123.

Jones A.T. (1982) Distinctions between three aphid-borne latent viruses of raspberry. *Acta Horticulturae* **129,** 41–48.

Jones A.T. & Murant A.F. (1975) Etiology of a mosaic disease of 'Glen Clova' red raspberry. *Horticultural Research* **14,** 89–95.

Kegler H., Kleinhempel H. & Verderevskaja T.D. (1976) Investigations on pear stony pit virus. *Acta Horticulturae* **67,** 209–218.

Kegler H., Opel H. & Herzmann H. (1961) Untersuchungen über Virosen des Kernobstes. III. Zur Histologie und Physiologie steinfrüchtiger Birnen. *Phytopathologische Zeitschrift* **41,** 42–54.

Kegler H., Verderevskaja T.D. & Fuchs E. (1979) Untersuchungen über Wechselbeziehungen verschiedener Kern- und Steinobstvirosen. *Archiv für Gartenbau* **27,** 325–336.

Kishi K., Takanahi K. & Abiko K. (1973) Studies on virus diseases of stone fruits. IV. On yellow mosaic, oil blotch and star mosaic of peach. *Bulletin of the Horticultural Research Station A* no. 12. 197–208.

Klotz L.J. (1980) Concave gum blind pocket. In *Description and Illustration of Virus and Virus-like Diseases of Citrus* (Eds J. Bové & R. Vogel). SETCO-IRFA, Paris.

Kristensen H.R. (1963) Apple green crinkle. In *Virus Diseases of Apples and Pears, Technical Communication no. 30,* pp. 31–34. Bureau of Horticulture, East Malling.

Legin R., Bass P. & Vuittenez A. (1979) Premiers résultats de guérison par thermothérapie et culture *in vitro* d'une maladie de type cannelure (legno riccio) produite par le greffage du cultivar Servant de *Vitis vinifera* sur le porte greffe *Vitis riparia*×*V. berlandieri* Kober 5BB. Comparaison avec diverses viroses de la vigne. *Phytopathologia Mediterranea* **18,** 207–210.

Lemoine J. (1979a) La maladie des taches annulaires: son apparition en France sur la pomme Granny Smith (ring spot). *Phytoma* no. 312, 15.

Lemoine J. (1979b) Manifestation de symptômes de stony pit sur les fruits d'indicateurs d'espèces différentes à partir de certaines sources de vein yellows du poirier. *Annales de Phytopathologie* **11,** 519–523.

Louie R., Gordon D.T., Knoke J.K., Gingery R.E., Bradfute O.E. & Lipps P.E. (1982) Maize white line mosaic virus in Ohio. *Plant Disease* **66,** 167–170.

Luckwill L.C. & Campbell A.I. (1963) Platycarpa dwarf; platycarpa scaly bark. In *Virus Diseases of Apples and Pears, Technical Communication no. 30* pp. 59–60, 61–62. Bureau of Horticulture, East Malling.

Maassen H. (1961) Untersuchungen über ein von der Erdbeere isoliertes mechanisch übertragbares Nekrosevirus. *Phytopathologische Zeitschrift* **41,** 271–282.

Martelli G.P., Russo M. & Rana G.L. (1981) Virological problems of Cynara species. In *3-o Congresso Internazionale Studi Carciofo*, pp. 895–927.

Mellor F.C. & Frazier N.W. (1970) Strawberry mottle. In *Virus Diseases of Small Fruits and Grapevines* (Ed. N.W. Frazier), pp. 4–8. University of California, Berkeley.

Milne R.G., Conti M., Lesemann D.E., Stellmach G., Tanne E. & Cohen J. (1984) Closterovirus-like particles associated with diseased grapevines. *Phytopathologische Zeitschrift* **110,** 360–368.

Nour-Eldin F. (1956) Phloem discoloration of sweet orange. *Phytopathology* **46,** 238–239.

Nour-Eldin F. (1968) Gummy bark. In *Indexing Procedures for 15 Virus Diseases of Citrus Trees. Agriculture Handbook no. 333.* USDA, Washington.

Nour-Eldin F. (1980) Gummy bark. In *Description and Illustration of Virus and Virus-like Diseases of Citrus* (Eds J. Bové & R. Vogel). SETCO-IRFA, Paris.

Parker K.G., Fridlund P. & Gilmer R.M. (1976) Green ring mottle. In *Virus Diseases and Noninfectious Disorders of Stone Fruits in North America. Agriculture Handbook no. 437,* pp. 193–199. USDA, Washington.

Pine T.S. (1976) Peach blotch, peach calico, peach mosaic. In *Virus Diseases and Noninfectious Disorders of Stone Fruits in North America. Agriculture Handbook no. 437,* pp. 61–70. USDA, Washington.

Polák J. (1970) On the identification of the strains of the beet yellows virus in Czechoslovakia. *Ochrona Rostlin* **6,** 167–174.

Posnette A.F. (1970) Gooseberry vein banding. In *Virus Diseases of Small Fruits and Grapevines* (Ed. N.W. Frazier), pp. 79–91. University of California, Berkeley.

Posnette A.F. & Cropley R. (1961) European rusty mottle disease of sweet cherry. In *East Malling Research Station Report for 1960,* 85–86.

Posnette A.F. & Cropley R. (1963) Spy 227 epinasty and decline. In *Virus Diseases of Apples and Pears, Technical Communication no. 30,* pp. 75–76. Bureau of Horticulture, East Malling.

Prentice J.M. & Harris R.V. (1950) Mosaic disease of the raspberry in Great Britain, III. Further experiments in symptom analysis. *Journal of Horticultural Science* **25,** 122–127.

Proeseler G. (1972) Beziehungen zwischen Virus, Vektor und Wirtpflanze am Beispiel des Feigen-Mosaik Virus und *Aceria ficus* (Eriophyidae). *Acta Phytopathologica Academiae Scientiarum Hungaricae* **7,** 297–300.

Prota U. & Garau R. (1976) About an unusual symptomatology of enation disease of grapevine in Sardinia. In *Proceedings of the 6th Meeting of the International Council for the Study of Viruses and Virus-like Diseases of the Grapevine,* p. 30. Madrid.

Pujol A.R & Beñatena H.N. (1965) Study of psorosis in Concordia, Argentina. In *Proceedings of the 3rd Con-*

ference of the International Organization of Citrus Virologists, pp. 170–174. University of Florida, Gainesville.

Rasmussen E.J., Berkeley G.H., Cation D., Hildebrand E.M., Keitt G.W. & Duain Moore J. (1951) In *Virus Diseases and other Disorders with Virus-like Symptoms of Stone Fruits in North America. Agriculture Handbook no. 10*, pp. 159–161. USDA, Washington.

Reichert I. & Perlberger P. (1934) Xyloporosis, the new citrus disease. *Bulletin of the Jewish Agency for Palestine Agricultural Experiment Station (Rehovoth)* **12,** 1–50.

Roistacher C.N. (1980) Psorosis A. In *Description and Illustration of Virus and Virus-like disease of Citrus* (Eds J. Bové & R. Vogel). SETCO-IRFA, Paris.

Roistacher C.N., Blue R.L. & Calavan E.C. (1973) A new test for citrus cachexia. *Citrograph* **58,** 261–262.

Roistacher C.N., Nauer E.M. & Wagner R.L. (1980) Transmissibility of cachexia, sweet mottle, psorosis, tatterleaf and infectious variegation viruses on knife blades and its prevention. In *Proceedings of the 8th Conference of the International Organization of Citrus Virologists*, pp. 225–229. University of California, Riverside.

Schmelzer K. (1967) Die Strichelkrankheit der Rose (rose streak) in Europa. *Phytopathologische Zeitschrift* **58,** 92–95.

Schmid G. (1963) Transmission experiments on the virus disease causing star crack and rough skin of apples. *Phytopathologia Mediterranea* **2,** 124–126.

Smith K.M. (1972) Dandelion yellow mosaic. In *A Textbook of Plant Virus Diseases*, pp. 261–263. Longman, London.

Teliz D., Valle P., Goheen A.C. & Luevano S. (1982) Grape corky bark and stem pitting in Mexico. I. Occurrence, natural spread, distribution, effects on yield and evaluation of symptoms in 128 grape cultivars. In *Proceedings of the 7th Meeting of the International Council for the Study of Viruses and Virus-like Diseases of the Grapevine*, pp. 51–67. Niagara Falls, Ontario.

Thomas B.J. (1980) Some degeneration and dieback diseases of the rose. In *Glasshouse Crops Research Institute Report for 1979*, pp. 178–190.

Thresh J.M. (1970) Blackcurrant yellows, infectious variegation of blackcurrant. In *Virus Diseases of Small Fruits and Grapevines* (Ed. N.W. Frazier), pp. 88–89. University of California, Berkeley.

Timmer L.W. & Garnsey S.M. (1980) Natural spread of citrus ringspot virus in Texas and its association with psorosis-like diseases in Florida and Texas. In *Proceedings of the 8th Conference of the International Organization of Citrus Virologists*, pp. 167–173. University of California, Riverside.

Vogel R. (1980) Cristacortis. In *Description and Illustration of Virus and Virus-like Diseases of Citrus* (Eds J. Bové & R. Vogel). SETCO-IRFA, Paris.

Vogel R. & Bové J.M. (1964) Stem pitting sur bigaradier et sur oranger Tarocco en Corse: une maladie à virus. *Fruits* **19,** 264–274.

Vogel R. & Bové J.M. (1968) Cristacortis, a virus disease inducing stem pitting on sour orange and other citrus species. In *Proceedings of the 4th Conference of the International Organization of Citrus Virologists*, pp. 221–228. University of Florida, Gainesville.

Vogel R. & Bové J.M. (1980) Pollen transmission to citrus of the agent inducing cristacortis stem pitting and psorosis young leaf symptoms. In *Proceedings of the 8th Conference of the International Organization of Citrus Virologists*, pp. 188–190. University of California, Riverside.

Wallace J.M. (1957) Virus-strain interference in relation to symptoms of psorosis disease of citrus. *Hilgardia* **27,** 223–246.

Wallace J.M. (1959) A half century of research on psorosis. In *Citrus Virus Diseases* (Ed. J.M. Wallace), pp. 5–21. University of California, Berkeley.

Wallace J.M. (1978) Virus and virus-like diseases. In *The Citrus Industry* (Eds W. Reuther, E.C. Calavan & G.E. Carman), pp. 67–184. University of California.

Wallace J.M. & Drake R.J. (1953) A virus-induced vein enation in citrus. *Citrus Leaves* **33,** 22–24.

Wallace J.M. & Drake R.J. (1960) Woody galls on citrus associated with vein enation virus infection. *Plant Disease Reporter* **44,** 580–584.

Wallace J.M. & Drake R.J. (1968) Citrange stunt and ringspot, two previously undescribed virus diseases of citrus. In *Proceedings of the 4th Conference of the International Organization of Citrus Virologists*, pp. 177–183. University of Florida, Gainesville.

Watson M. (1962) Yellow-net virus of sugar beet. I. Transmission and some properties. *Annals of Applied Biology* **50,** 451–460.

5: Mollicutes and Rickettsia-like Organisms

MOLLICUTES

The Mollicutes are very small prokaryotes totally devoid of all walls and bounded only by a trilaminar unit plasma membrane. They may be parasites, commensals or saprophytes and many are pathogens of man, animals (including insects) and plants. The genome size is about 5×10^8 or 1×10^9 d, among the smallest recorded in the prokaryotes and the molecular percentage G+C content of their DNA is low (~ 23 to ~ 46%). Two genera, *Spiroplasma* and *Acholeplasma*, are certainly associated with plants as part of the normal plant flora or in the case of the former also as pathogens. Both are cultivable. A third group, the mycoplasma-like organisms (MLOs), morphologically resemble members of the Mollicutes but, since none has yet been cultured (despite a few doubtful reports), there has been no confirmation of their identity within the Mollicutes, let alone with individual species or genera.

Cells of the genus *Spiroplasma* are pleomorphic, varying from helical and branched uni-helical filaments to spherical or ovoid. The helical forms mostly occur during exponential growth. They are motile and have intracellular fibrils, though they are without flagella or other prokaryotic organs of locomotion. Cells divide by binary fission. Cholesterol is required for growth. They have been found in mice, in the haemolymph and gut of insects and ticks, and in the phloem of plants and on plant surfaces. *S. citri* is a plant pathogen and several spiroplasmas occur in a biological cycle involving plant phloem and mainly homopterous insects. Corn stunt disease, present only in N. America, is also known to be due to a spiroplasma (APS Compendium), recently named *S. kunkelii*.

Cells of the genus *Acholeplasma* are spherical and filamentous and do not require cholesterol for growth. They are apparently parasites of a wide range of animals and occur in sewage, compost, soil and plant materials. Although no clear role for them in plant disease has yet been established, some multiply and persist in leafhoppers, including those known to be vectors of plant mycoplasma diseases.

MLOs are often filamentous and branched and are regularly observed by light or electron microscopy within the sieve tubes of plants affected by yellows diseases and in the salivary glands of insect vectors of these diseases. Observation in sieve tubes is facilitated by fluorescence microscopy with the DNA-specific fluorochrome 4′,6-diamidino-2-phenyl indole (DAPI), since healthy conducting sieve tubes do not contain nuclei. Plants affected by these diseases are susceptible to tetracycline therapy as are many other mycoplasmas. MLOs often have extensive branching resembling that of *Mycoplasma mycoides*. These putative plant yellows agents are also found, and appear to multiply in, their insect hosts, where, after an incubation period of one to several weeks, they reach a high titre in the salivary glands and the insect becomes capable of infecting the phloem of healthy plants.

It is worth stressing the difficulty in defining a 'unit taxon' among MLOs, since morphological criteria are limited, and both the criteria used in bacteriology and the serological methods used in virology are difficult to apply while MLOs have not been cultured. There has been some recent progress in using ELISA based on crude sap preparations (Clark *et al.* 1983; Clark & Davies 1984), and also in the use of monoclonal antibodies (Lin & Chen 1985). However, most characterization of MLOs still depends on symptoms, graft or dodder transmissibility to various hosts, and vector host range. *Vinca (Catharanthus) rosea* is a test plant to which many MLOs seem transmissible and in which they readily multiply.

These taxonomic difficulties complicate the presentation of MLO diseases in this handbook. In practice, the diseases of fruit trees and other woody plants (which are of greatest economic importance) have mostly been treated individually. The remainder, found mainly on herbaceous plants, are grouped under the heading 'aster yellows complex' (except clover phyllody MLO, which is well characterized and seems distinct). The elucidation of true relationships between the organisms concerned awaits the results of further research. Ultimately, there is every reason to expect that a bacterial type of nomenclature (even with pathovars ?) will be applicable. For further details of Mollicutes and MLOs consult Cousin 1972; Caudwell 1978; McCoy 1979; Freundt & Razin 1984; Razin & Freundt 1984; Tully 1984; Whitcomb & Tully 1984.

The Rickettsiales are an order of small, mainly rod-shaped, coccoid and often pleomorphic Gram-negative prokaryotes, with typical bacterial walls, that multiply only inside host cells. All are regarded as parasitic or mutualistic. The former are associated with vertebrates and arthropods; these may act as vectors or primary hosts, and cause disease in man or other vertebrate and invertebrate hosts. The mutualistic forms occur in insects. Rickettsia-like organisms (RLOs) have been reported as associated with a number of plant diseases, but the evidence that they are causal agents is usually scant. Indeed, so-called RLOs in plants may not be particularly closely related to rickettsias. Their main common characteristic is that they are phytopathogenic bacteria which are cultured with difficulty. On this basis, both MLOs and RLOs are coming to be referred to as 'fastidious prokaryotes'. Davis *et al.* (1978) found plant RLOs to be related antigenically to each other, but not to other phytopathogenic bacteria.

Finally, it should be stressed that the aetiology of some of the diseases listed in this chapter is uncertain, so that they are only included on the basis of similarity of symptoms. The same organisms could well be involved in many cases (cf. aster yellows). In addition, it is not necessarily possible at the present time to draw a sharp line between probable MLO-diseases and probable viral diseases (Chapter 4). As in Chapter 4, the organisms are considered in alphabetical order.

MYCOPLASMA-LIKE ORGANISMS AND SPIROPLASMAS

Apple chat fruit disease

The reported association of this disease with a mycoplasma-like organism (Beakbane *et al.* 1971) has not been confirmed by more recent investigations involving electron and light microscopy or transmission studies (Clark & Davies 1984). Chat fruit is known only in apple and has not been transmitted to other species. It has been recorded in most European countries, in S. Africa and possibly in New Zealand. Symptoms have been seen only in cvs Lord Lambourne and Tydeman's Early Worcester. On severely affected trees fruit development is retarded so that at harvest the apples are small and green. A small proportion of Lord Lambourne fruit develop dark green spots *c.* 5 mm in diameter, often bordered with red pigment. Affected fruit of Tydeman's Early Worcester are small, slightly flattened between calyx and peduncle and sometimes lop-sided. Infected Lord Lambourne trees tend to be upright and vigorous, while the leaves of diseased Tydeman's Early Worcester are small, giving the trees a sparse appearance.

The disease spreads slowly (Posnette & Cropley 1965) but no vector is known. Although the disease agent does not cause visible symptoms in most commercial cultivars, it may severely affect cropping (Sparks *et al.* 1983). Chat fruit may be more widespread than is suspected, as large-scale testing has taken place in only a few countries; moreover, a minimum period of three years is needed to test for its presence and the interpretation of tests is often difficult. Symptoms in the indicator cv. Lord Lambourne are sometimes indistinct and the pathogen is not fully systemic. It is one of the few virus-like diseases to have re-infected nursery material in the British EMLA scheme (Posnette *et al.* 1976) and until a quicker and more reliable method of detection is available it will be a threat to the production of healthy plants. For further information, see Posnette & Cropley (1969).

ADAMS

Apple proliferation MLO

(Syn. apple witches' broom MLO)

Basic description

Trees with symptoms of apple proliferation contain MLOs in the sieve tubes. Since these MLOs have been found in infected trees only and their occurrence has corresponded with the results of experiments to transmit apple proliferation, they are supposed to be the causal agents of the disease.

Host plants

Apple. Ornamental *Malus* species and *Vinca rosea* can be infected experimentally.

Geographical distribution

The disease was first reported from Italy in 1950 and later from the S. Netherlands, from Switzerland and S. FRG. Today it is known to occur in most countries of C. and S. Europe including France, Spain, and the

Balkan states. The northern border of distribution runs via Bonn, S. GDR, N. Czechoslovakia and S. Poland.

Disease

The first symptoms of apple proliferation are witches' brooms (Fig. 16), small fruit and late growth of some terminal buds at the end of the season. Beginning in July, secondary shoots grow from axillary buds on the upper part of some shoots. These secondary shoots are steeply erect and form a broom, through suppression of apical dominance. Leaves in the brooms often have enlarged stipules (Fig. 17). Infections with *Podosphaera leucotricha* are favoured by the development of young leaves in the brooms and the late growth of terminal buds. The economically most important symptom is reduced fruit size (mean fruit weight often diminished 30–60%). These small fruits have longer peduncles than healthy ones and often have a poor taste. Premature leaf reddening can occur in autumn. In spring the stipules on the first leaves of infected trees are very enlarged, dentate or notched; in healthy trees only small and narrow stipules are formed. Witches' brooms and enlarged stipules on leaves developed in spring are typical signs of apple proliferation, while small fruit, early leaf reddening and late shoot growth can occasionally be caused by other disorders.

Fig. 16. Witches' broom on apple (cv. Golden Delicious), due to apple proliferation MLO.

Fig. 17. Enlarged stipules of apple leaf (cv. Golden Delicious), due to apple proliferation MLO (healthy leaf on right).

Witches' brooms usually appear only during the first years of the disease, while enlarged stipules can be observed for a longer period. However, the typical symptoms are often absent in trees infected for a number of years. If trees in this chronic stage of the disease are cut back strongly, symptoms will develop again. The root system of diseased trees is also affected. The small roots are reduced in length, gnarled and crooked. If young trees are infected they will grow only poorly. Along with the disappearance of typical symptoms with age, there can be a recovery of damaged trees, but mostly the fruit remain smaller than on healthy trees and the weight of yield is lower. The chance of recovery is influenced by the virulence of the pathogen and the age of the tree at the time of infection.

Within infected trees, colonization of sieve tubes by MLOs undergoes seasonal fluctuations in the stem and shoots, while the organisms are always present in the roots. In this respect apple proliferation is similar to pear decline. During winter the MLOs are eliminated from the aerial parts of the tree, following degeneration of the sieve tubes. Recolonization of stem and shoots starts in April or May after the development of a new phloem circle, but it does not occur every year. In this case no witches' brooms appear (Schaper & Seemüller 1982). Staining microtome sections of bark samples with DAPI is a quick and reliable method for checking the distribution of MLO in apple trees. The correspondence between DAPI fluorescence in the sieve tubes and the presence of MLO has been proved by electron microscopy.

In infected trees without symptoms MLOs are most frequent in the roots, so such latent infections can be detected by testing roots either by the DAPI technique or by transmission experiments (tested root grafted to the root of an apple seedling top-grafted with the indicator cv. Golden Delicious, then observed for two seasons).

Epidemiology

Transmission by grafting buds or scions is very variable because of the irregular distribution of the MLOs in the aerial parts of the tree. Scions cut from the branches in February or March are not infectious unless they originate directly from a witches' broom. The disease is not seed-transmitted. In nurseries apple proliferation is very seldom observed. In young apple orchards, the first symptoms usually appear in the fifth year or later. Between the eighth and fourteenth year the number of diseased trees often rises considerably. Witches' brooms are mainly seen on trees with vigorous growth. Since latent infections reduce growth, such trees are presumably not infected via the propagating material, but by spread from other trees. Many infections do occur in orchards established with trees originating from healthy mother plants or in plantations of ungrafted apple seedlings. The causal agent is probably transmitted by sucking insects because in apple plantations with only few insecticide applications the incidence of infections is higher than in neighbouring orchards with a normal insecticide programme (Kunze 1976). Leafhoppers are probably the vector but neither the vector species nor the period during which transmission occurs are known.

Economic importance

Apple proliferation is one of the most serious plant diseases caused by MLOs in Europe. During the acute stage of the disease more than 80% of fruit is unmarketable because of its small size and poor taste. After a few years the trees can recover to some extent but a considerable percentage of fruit does remain undersized. Also in recovered trees with normal fruit size, crop losses of 20–40% can be caused by reduced weight of yield. Some cultivars like Golden Delicious, Cox's Orange Pippin and Jonathan are more sensitive to the disease than others, but damage is possible in all cultivars.

In S. FRG, Switzerland and Czechoslovakia the percentage of infections in orchard trees on strong-growing rootstocks often amounts to 20 or 30%, and occasionally up to 70%. The most intensive spread of the disease is observed under climatic conditions suitable for wine growing.

Control

As long as neither the vector nor the period of transmission is known, specific control of the spread of apple proliferation is not possible. However, fundamentally changing the insecticide programme in an orchard could lead to proliferation. In general, the occurrence of the disease is most frequent in orchards with strong-growing trees so the use of less vigorous rootstocks and moderate nitrogen fertilizer may reduce the rate of infection. For the same reason, vigorous pruning or lopping should be avoided. Young trees with disease symptoms should be removed because they do not recover, but this alone is not sufficient to stop further spread of the disease. Symptoms of proliferation can be suppressed by injection of tetracycline HCl into the stem. The size of fruit can be increased in this way for one or two seasons, although the MLOs remain alive in the roots. However, in many European countries the commercial application of antibiotics for the control of plant pathogens is not allowed for medical reasons. For further information see also EPPO 87 and Bovey (1963).

KUNZE

Apple rubbery wood disease

Apple rubbery wood disease was thought to be associated with an MLO (Beakbane *et al.* 1971) but, as with apple chat fruit disease, attempts to confirm this aetiology have not succeeded. The pathogen occurs in apples and also in pears where it does not cause visible symptoms. It has been associated with chlorotic blotch symptoms on quince C 7/1 and seedlings of quince E (Desvignes & Savio 1975). It has been recorded in most European countries and in Australia, Canada, India, New Zealand and S. Africa. Affected apple trees are stunted, often with strong shoots developing from the base of the trunk. Shoots and branches up to 3 years old are very flexible, due to incomplete lignification of the xylem, and are often bent to the ground, giving the tree a weeping appearance. Apple cv. Lord Lambourne appears to be sensitive to all isolates of the pathogen and is the standard indicator for the disease. Other sensitive cultivars include Golden Delicious and Gala, but many commercial cultivars, including Cox's

Orange Pippin, do not show any symptoms although vigour and cropping can be significantly reduced.

No vector is known and the disease spreads slowly in orchards (Luckwill & Crowdy 1950) or not at all. The disease was noticed in the 1930s in England as cv. Lord Lambourne became popular and was grafted onto trees and rootstocks that appeared to be healthy but were in fact frequently infected. During the last 30 years most countries in Europe have developed schemes for producing and distributing healthy planting material which does not readily become re-infected (Cropley 1963; Posnette *et al.* 1976).

ADAMS

Apricot chlorotic leaf roll MLO

This disease (Morvan 1977) occurs only in Europe, on apricot in France, and most often on *Prunus salicina* in other countries (Greece, Italy, Spain). Symptoms on apricot include early leaf-bud break (during winter dormancy or before the flower buds open); browning of the middle layer of the bark (especially after a severe winter); conical leaf roll, with irregular interveinal chlorosis (towards the end of summer) and proliferation of buds at the ends of short shoots. On *P. salicina*, decline symptoms are less characteristic (small, reddish leaves, with cylindrical rolling). Diagnosis can be confirmed by indexing apricot on a plum rootstock, or from the direct use of DAPI reagent to detect MLOs. Insect transmission certainly occurs but the vector has not yet been identified (*Fieberiella florii* is suspected).

In France, this pathogen is probably responsible for 60–70% of cases of apricot decline. When the trees come into fruit, 5% may be killed every year. Death normally follows 1–2 years after the appearance of the first symptoms, but much more rapidly if the rootstock is peach. Although the exact status of this disease in different parts of S. Europe is not completely clear, it certainly presents a risk to apricot orchards (EPPO 146) and requires careful use of budwood from healthy mother plants (preferably grafted on peach, since such trees are rapidly eliminated if infected).

SMITH

The aster yellows complex

Basic description

The MLO causing yellows in China asters (*Callistephus chinensis*) in N. America is transmissible to numerous other hosts, most of which are herbaceous. Other strain(s) (European aster yellows) occur in Europe. Many herbaceous hosts are infected by similar MLOs, causing somewhat different symptoms, or transmitted by different vectors, or with a different host range.

In the absence of clear criteria to distinguish these MLOs, it is not yet possible to characterize taxa, especially as they tend not to be host-specific. Accordingly, they are here considered as the aster yellows complex. A number of poorly characterized diseases probably caused by MLOs are also mentioned under this heading. The organisms involved are present as ovoid forms within the cytoplasm of young plant cells, and in the intestinal cells and salivary glands of insects. The typical highly polymorphic MLO structure appears when they invade adult sieve tubes and insect haemolymph. Degenerated MLOs are seen in senescent sieve tubes.

Host plants

The range of host plants affected is very wide. The forms which infect cultivated plants often occur in weeds or other wild plants, so that the crop plant may well be 'marginal' with respect to the natural host range of the MLO. Many of these MLOs are transmissible to *Vinca rosea* by insect vectors or dodder. Host plants of special interest are presented individually at the end of the section.

Geographical distribution

MLOs of the aster yellows complex occur throughout Europe. However, they are more characteristic of areas with warm summers (S. & E. Europe), probably because the insect vectors multiply more readily under these conditions.

Diseases

The symptoms caused in hosts of special interest are mentioned in the specific sections below. However, the MLOs of the aster yellows complex tend to cause very similar symptoms in different hosts, no doubt because of a common physiological (hormonal) action. Yellowing of the vegetative organs is the most characteristic symptom of infection by the typical aster yellows MLO. This discoloration varies with environmental conditions (outdoor or under glass) and cultivar. For example, potato cv. Bintje affected by 'purple top' shows a dark violet colour, cv. Ker Pondy shows a pink violet and cv. Ackersegen yellow.

MLO-yellows can be confused with virus-induced yellows, but tends to be visible mainly on the terminal leaves. Affected leaves show no mottling. Because of poor phloem function, an abnormal accumulation of starch occurs in the foliage. The diseased leaves tend to stand upright, and the terminal leaves may show epinasty. Inhibiting of apical dominance and shortening of internodes result in stunting, which can be more or less severe depending on the age of the plant and when it was infected. The axillary buds may sprout and grow upright, giving the bushy growth known as a witches' broom, or proliferation. In some cases, deficient lignification and wilting may result in death of the plant. Floral symptoms are very characteristic: sterility, virescence (greening of normally coloured floral parts), phyllody (transformation of floral parts, especially carpels, into proliferated leafy structures). In stolbur disease, a hypertrophied calyx surrounds aborted reproductive organs. The greening of floral organs tends to develop as the normal green colour of the vegetative organs disappears. The root system is also affected by MLO infection, becoming harder and more brittle. In perennial plants, the MLOs overwinter in the root system. In some cases, MLOs can be detected reliably in the root system during all seasons, but only sporadically in the aerial parts. A plant infected with an MLO of the aster yellows complex may be killed, remain alive but diseased (*Vinca rosea* reacts in this way to most MLOs) or may recover (as in the case of many crop plants).

Epidemiology

MLOs are transmitted by leafhoppers (Homoptera: Auchenorryncha, especially *Macrosteles* spp.), and sometimes Psyllids (Leclant *et al.* 1974). In contrast to the broad range of host plants the number of disease-carrying leafhopper families is limited. The insect inserts its stylet far enough to reach the sieve tubes and sucks MLOs into the digestive tract. MLOs attain the cells of the intestine and proliferate within the cytoplasm prior to being released into the haemolymph. At this point, they invade the entire body, reach the salivary gland cells and penetrate into the saliva. The insect can then inoculate a healthy plant by releasing MLOs into the sieve tubes from the salivary tract. The incubation period varies from two to several weeks in the insect. Another few weeks elapse between inoculation of the plant and manifestation of disease symptoms.

Two patterns can be distinguished in the epidemiology of aster yellows diseases. In the first case, leafhoppers start from the southern areas where they acquire MLOs and reach cultivated areas after a long migration period, and after incubation is completed. Such migrating insects capable of early crop inoculation are responsible for the epidemic character of some diseases. In the second case, vector populations remain local and tend to cause a moderate level of 'endemic' disease. In general, the incidence of 'aster yellows' on crop plants tends to be a side-effect of the behaviour of leafhoppers on their natural wild hosts. The aster yellows MLOs are not mechanically transmissible, but are transmitted in vegetatively propagated plants, in grafts and in cuttings. They are not seed-borne (infected plants generally set little, or sterile seed, in any case).

Economic importance

Infected plants may recover, but are very often killed. However, the level of incidence of aster yellows type diseases in Europe tends to be rather low and sporadic. They may appear unexpectedly in certain areas or years when conditions happen to favour spread of the vectors. Transmission in vegetative propagating material may present serious risks, and appropriate precautions have to be taken.

Control

Though the tetracycline antibiotics are effective and have been used against MLOs in N. America, they are prohibited for agricultural use in most European countries. In any case, remission of symptoms is often followed by a relapse. Heat treatment can be used in some cases for planting material. For example, gladiolus corms dipped in water at 50°C for 1 h will be freed of MLOs. Insecticide control of the vector is possible if the insects remain in the affected field. Control is most effective if the number of cycles of the vector is limited, the number of host plants available is restricted and the crop plant is able to recover (Cousin 1972).

Specific diseases

Vegetables and herbaceous ornamentals

Classic 'aster yellows' affects vegetables like lettuce (phyllody), carrot (proliferation and yellows) (Leclant *et al.* 1974), beet and spinach (yellows), onion (proliferation) (Cousin *et al.* 1970), celery (yellows), courgette (yellows) (Ragozzino 1978), cabbage and cauliflower (virescence) (Müller *et al.* 1973) and ornamentals, especially Compositae

(*Callistephus, Tagetes, Calendula, Chrysanthemum* (Verhoyen *et al.* 1979), *Gaillardia, Gerbera, Rudbeckia*) but also *Matthiola, Cheiranthus, Dianthus barbatus, Anemone, Delphinium* (Marwitz & Petzold 1976), *Primula* (Stevens & Spurden 1972), *Phlox, Gypsophila* (Ulrychová *et al.* 1983). Various herbaceous field crops are also affected by such diseases (virescence of flax), 'chlorantie' of oilseed rape, sunflower phyllody (Signoret *et al.* 1976), lupin witches' broom, necrotic crinkle mosaic of hops (Chod *et al.* 1971).

Solanaceae

Two distinct groups of MLOs affect potato. **Potato witches' broom MLO** causes a minor disease, with round leaves, proliferation, and sometimes flower symptoms. Known vectors (*Scleroracus* spp.) are different from those for stolbur (see below). The disease is commonly tuber-borne. The same MLO is apparently involved in lucerne witches' broom, and lucerne and other legumes may act as a reservoir for the infection of potato (APS Compendium). **Potato stolbur MLO** is more typically of the aster yellows type. Upper leaves are discoloured violet or yellow (purple top), sepals enlarge and flowers and fruits fail to develop. Infected tomatoes show proliferation (bushy growth) (Fig. 18). This MLO is reported to be transmitted by *Hyalesthes obsoletus*, and also by *Macrosteles* and *Lygus* spp., but there is no confirmation of this. It is rarely tuber-transmitted. It occurs widely in E. Europe where it can be damaging. However, it is not apparently transmitted within a potato or tomato crop, but only from adjoining weeds (especially the abundant *Convolvus arvensis*). Stolbur is only locally present in W. European countries, who seek to exclude it (EPPO 100); it occurs for example in France in tomato (Cousin & Moreau 1977); 'mal azul' in Portugal is probably the same disease (Rhodeia & Borges 1973/74). **Eggplant little leaf MLO** affects aubergine in India, but not apparently in Europe.

Fig. 18. Potato stolbur MLO in tomato (courtesy of Min. of Agriculture, Turkey).

Gramineae

Aster yellows occasionally affects wheat, barley and turf grasses (APS Compendium). In E. Europe, several MLO-diseases have been described, especially on oats, but also other cereals (pale green dwarf, Onishchenko 1984; pseudorosette). However, these seem of more interest as peculiarities for research work than as pathogens of any economic importance. In Italy (Amici *et al.* 1973), rice is affected by a yellows MLO (giallume), presumably distinct from the **rice yellow dwarf MLO** of the Far East. **Maize bushy stunt MLO** and **sorghum yellow stunt MLO** occurring in N. America (Bradfute *et al.* 1979) are not yet known in Europe.

Bulbs

Gladiolus may be affected by MLOs causing yellows, grassy-top or grassiness (proliferation) and virescence. Attacks are especially severe in Italy (Bellardi *et al.* 1985). Hyacinths are very occasionally affected by 'Lisser' disease, with stunting and aborted inflorescences, presumably a form of aster yellows (Van Slogteren & Muller 1972). Heat treatment of the corms or bulbs provides a solution in the Netherlands (Van Slogteren *et al.* 1976).

Strawberry

Besides 'green petal' caused by clover phyllody MLO (q.v.), strawberries are infected by more typical aster yellows strains, mainly in N. America. According to the time of infection in relation to flowering, petals, stamens and pistils may be virescent, or achenes may show phyllody (APS Compendium). In N.W. America there is also a **strawberry witches' broom MLO** of fairly minor importance, but which is considered a quarantine organism for Europe (EPPO 130).

Shrubby plants

The MLOs of trees (especially fruit trees) are mostly well characterized and seen as distinct from aster yellows. However, several MLO diseases are known on shrubby plants, which may be of the aster yellows type. **Hydrangea virescence MLO**, which turns the flowers of hydrangeas green is a serious disease of this important ornamental in France (Cousins *et al.* 1986), Belgium (Welvaert *et al.* 1975) and elsewhere,

spread by propagating material. 'Lavandin', a hybrid lavender widely grown in S. France for industrial perfume production, suffers from yellow decline due to an MLO (Moreau *et al.* 1974). Redcurrant (*Ribes houghtonianum*) is affected by MLO yellows in Czechoslovakia (Rakús *et al.* 1974). *Vaccinium* species, in Europe and N. America, are affected by witches' broom (Kegler *et al.* 1973; Blattný & Vána 1974) as are ornamental shrubs such as honeysuckle (Yakutkina 1979) and jasmine. Hazel commonly suffers in Campania (Italy) from 'maculatura lineare' disease which is probably an MLO disease (Ragozzino *et al.* 1973).
COUSIN & SMITH

Blackcurrant reversion disease

Blackcurrant reversion is of uncertain aetiology having been associated variously with bacteria (Silvere & Romeikis 1973), MLOs (Zirka *et al.* 1977) and with potato virus Y (Jacob 1976). Confirmatory evidence that any of these is the causal organism is lacking. *Ribes* species are the only known hosts for the disease which has been recorded in most countries where blackcurrants (*R. nigrum*) are grown. Affected bushes develop flat leaves with a smaller basal sinus and fewer main veins and marginal serrations than normal. Flower buds are hairless and appear redder than healthy ones. Severely malformed flowers are a feature of the disease in some countries, particularly those in Scandinavia and E. Europe. The disease takes several years to become completely systemic during which time yields are progressively reduced.

Reversion is transmitted by the eriophyid mite *Cecidophyopsis ribis* which feeds on meristematic tissue in the buds and disperses in late spring and early summer. Reverted bushes are much more susceptible to colonization by mites than healthy ones. The disease is effectively controlled by raising propagation material in isolation from commercial plantations and other sources of infection, by the routine inspection and removal of diseased bushes, and by spraying with acaricides. Cultivars may become available that are resistant both to the disease and to the vector (Keep *et al.* 1982). For further information, see Thresh (1970) and Jacob (1976).
ADAMS

Cherry little cherry disease

The disease is of uncertain aetiology although in Canada it has been associated with abnormal structures in the phloem, including elongate, flexuous virus-like particles (Raine *et al.* 1979; Ragetli *et al.* 1982). Susceptible plants include sweet and sour cherry, many ornamental cherries and some wild *Prunus* species. Little cherry has been recorded in most countries in Europe and in Australia, Canada, Japan, New Zealand and USA (Welsh & Cheney 1976). Symptoms in severely affected plants are premature reddening of the foliage in autumn and a reduction in the size, flavour and sweetness of fruit which may be pointed in shape. The vector is thought to be the apple mealybug *Phenacoccus aceris* and in British Columbia (Canada) the distribution of this insect parallels that of the disease (Slykhuis *et al.* 1980).

Many cherry cultivars are symptomless carriers of the disease but where sensitive cultivars are widely grown, as in the Kootenay region of British Columbia, severe crop losses occur. The disease was prevalent in English cherry orchards (Posnette *et al.* 1968) and probably in many other countries in Europe, but schemes to produce and distribute healthy planting material have reduced the numbers of infected trees to negligible levels in recently planted orchards. However, successful schemes could be jeopardized by natural spread and the slow and uncertain methods of diagnosis currently available using sensitive woody indicator plants such as cherry cv. Sam.
ADAMS

Cherry Molière disease

Cherry Molière disease is only known in S.W. France (Fos 1976). Diseased trees develop rosette foliage and wilted twigs, and the small fruits fall before maturity. There is now some evidence that this disease is associated with an MLO. Injections of the antibiotic oxytetracycline give very encouraging results. Treated trees recover well and return to full production in 2 years in most cases. Transmission by dodder from *Prunus mahaleb* rootstocks of diseased trees to *Vinca rosea* gives yellows symptoms. The use of DAPI shows fluorescent particles in the sieve tubes of infected *V. rosea* of diseased trees (*P. mahaleb* and sweet cherry). There are differences in susceptibilty (after inoculation by chip-budding) in cultivars and rootstocks. In particular, *P. mahaleb* rootstocks appear to be very susceptible. It seems that soil conditions and the state of health of the trees affect severity. Somewhat similar symptoms of cherry decline were reported earlier from the GDR and Moldavian SSR (Kegler *et al.* 1973).
BERNHARD

Clover phyllody MLO

Clover phyllody MLO has been distinguished serologically from MLOs of the aster yellows complex (Clark *et al.* 1983). It occurs widely throughout Europe, and in N. America, on *Trifolium repens* in particular, but also on *T. pratense*. In addition, it causes green petal disease on strawberry and is transmissible to many other hosts. On clover, characteristic symptoms are first stunting and vein chlorosis, followed after flowering by a striking hypertrophy of the inflorescences. Carpels develop into small green leaf-like structures (O'Rourke 1976). Symptoms develop best at higher temperatures (Nakamura *et al.* 1978). In strawberry, the MLO causes virescence of the petals, which later turn red. The fruits form 'button berries' (base of receptacle develops normally, but apex, with sterile ovules, remains undeveloped (APS Compendium)). The disease is common in old clover stands, which lose vigour in consequence. Clover seed crops may also be affected (Carr & Large 1963), and suffer serious losses since infected plants produce no seed. Root nodulation is also depressed on phyllody-affected plants, with a possible effect on the rhizobia even after re-isolation (Joshi & Carr 1967). Bacteroids are lysed and show rod-shaped inclusions (Smets *et al.* 1977). The disease on strawberry is only sporadic in occurrence, and less damaging than aster yellows (APS Compendium).

There is no very satisfactory control method. Heat, irradiation or antibiotics could be used to treat valuable breeding material. There is some variation in resistance between clover cultivars, but this is expressed differently to different isolates of the MLO, so there has been little progress in breeding for resistance (O'Rourke 1976). Other MLO-type diseases have been noted in clover (clover dwarf, clover degeneration) and are important particularly in E. Europe (e.g. in Lithuanian SSR, Genite & Stanyulis 1975). How they are related to the clover phyllody MLO or to members of the aster yellows complex remains to be demonstrated. Clover witches' broom disease (in the UK), in which flowers develop normally, is caused by the same agent as strawberry bronze leaf wilt (similar to the early stages of green petal but without flower symptoms) (APS Compendium).

SMITH

Elm phloem necrosis MLO infects *Ulmus* species in an area in N. America south of the Great Lakes where it has killed large numbers of trees. It is transmitted by the leafhopper *Scaphoideus luteolus*. Trees show premature yellowing and senescence of foliage and die the following year. The inner phloem of the lower trunk and roots (in which the MLO is found) is discoloured butterscotch yellow. European elms show some resistance, but this MLO does present a quarantine risk for Europe (EPPO 26).

Grapevine flavescence dorée MLO

Basic description

The disease is associated with an MLO, which is easily found in ultra-thin sections in phloem cells of the laboratory host plant broad bean, and in cells of infectious leafhopper vectors. It is rarely found in infected grapevines. It can be visualized by ISEM either in extracts of the stem apex of broad bean or in leafhopper extracts (Caudwell *et al.* 1982).

Host plants

In the field, only *Vitis vinifera* is known to be affected by flavescence dorée; no herbaceous plant is known to express symptoms of the disease or to harbour the pathogen. However, it seems likely that *V. labrusca* may be the original, symptomless host of flavescence dorée in N. America (Caudwell 1983). The sensitivity of American *Vitis* species is, in general, lower than that of *V. vinifera*. Within *V. vinifera*, some cultivars are resistant, others (e.g. Ugni Blanc, Baco 22A) show strong symptoms but recover the next spring. Others (e.g. Nieluccio) give strong symptoms and decline and die within 2–3 years. Baco 22A is the best indicator. Using the experimental vector *Euscelidius variegatus*, a large range of sensitive herbaceous plants, including *Vicia faba*, *Chrysanthemum carinatum*, *Lupinus* species, *Pisum sativum* and *Vinca rosea*, can be infected.

Geographical distribution

Since the first appearance of flavescence dorée in Europe, in two areas in Gascony (S.W. France) around 1954, the disease has spread to other districts of S. France (Corsica and Corbières) and to Pavia in N. Italy. The disease referred to as 'leaf curl and berry shrivel' (Uyemoto 1976) in N. New York State (USA) where the vector *Scaphoideus littoralis* is thought to have originated, may be flavescence dorée (Caudwell 1983). Symptoms similar to the disease have been reported in Romania and Sicily, but since *S. littoralis* has not been found there, the identity of the disease is in doubt. Other yellows diseases of grapevine, with similar symptoms, but not transmitted by *S. littoralis* have been described in Burgundy and Switzerland as bois noir (Caudwell 1964; Bovey 1972), in FRG as Vergilbungskrankheit (Gärtel 1959) and in many subtropical regions (e.g. Rhine Riesling problem in Australia, amarilliamento de Elqui in Chile).

Disease

During spring, newly affected grapevines exhibit growth inhibition at first (non-opening of buds or progressive shortening of internodes). The characteristic symptoms appear in summer. The most susceptible cultivars often adopt a weeping posture, the shoots bending over as though made of rubber. This is caused by the total lack of lignification throughout the whole length of the shoot. The leaves harden, roll towards their lower surfaces and tend to overlie one another like scales, thus giving the shoot a characteristic snake-like appearance. The hard, brittle leaves are at first a golden-yellow colour in white cultivars and red in black cultivars on the part most exposed to the sun. Creamy-yellow spots then appear along the veins, on black and white cultivars, most often along their whole length. These spots sometimes spread over areas precisely limited by two or three veins; they rapidly become necrotic, giving a characteristic appearance to the infected vein (Fig. 19). If symptoms appear early in the season, the inflorescence dries up, but if later, the berries will wrinkle and the pulp become fibrous and bitter. Phloem necrosis explains the clinical symptoms by blocking of assimilates in the leaves and a consequent lack of maturity of developing fruit and wood.

Fig. 19. Grapevine flavescence dorée MLO (courtesy of Biologische Bundesanstalt).

Recovery is a regular phenomenon in many cultivars, taking place during the winter after symptom expression. It is sudden, complete and final and the wood loses all vestige of infectivity. If a recovered vine is re-inoculated in the course of the next 3 years, it will develop only localized symptoms on a few shoots around the inoculation point. Several re-inoculations, each producing localized symptoms, may give an appearance of systemic infection. This defence reaction of the plant will disappear little by little over the next 4 or 5 years. After this delay re-inoculation will lead again to systemic symptoms. Localized symptoms are found only in cultivars showing recovery and are a useful characteristic for diagnosing flavescence dorée or other yellows diseases in many cultivars of grapevine.

Epidemiology

Only one species of vector, *Scaphoideus littoralis*, is known in vineyards (Schvester 1963). In N. America it feeds naturally on *V. labrusca*. Its introduction into Europe, before or together with the pathogen, was the cause of outbreak of flavescence dorée in France. For its development *S. littoralis* needs, in areas where its host occur, a summer long enough to allow egg laying by adults, and a winter cold enough for diapause. Its distribution in Europe does not yet cover all favourable climatic zones but already includes S. France and Corsica, Switzerland and N. Italy, much beyond the spread of flavescence dorée itself. Experimentally, it is possible to transmit flavescence dorée from *Vitis* species to broad bean via *S. littoralis* (Caudwell *et al.* 1970). Transmission from broad bean to broad bean can be achieved with another leafhopper, *Euscelidius variegatus*, which can be used for an infectivity test. In practice, the disease spreads very fast whether around foci in the vineyard or by means of flying adults helped by wind, to new areas up to 30 km away. The appearance of new foci in new regions often results from the transport of cuttings carrying either eggs of the vector in the bark and/or the disease.

Economic importance

Crop losses arise from the destruction of the flowers or of the bunches, and from the lack of ripening of the wood; this last feature is associated with an absence of lignification. In cultivars that show recovery, like Baco 22A, there is severe weakening of the plant initially, and, in the absence of re-inoculation, it can be a few years before the crop yield recovers. In very sensitive cultivars showing no recovery, like Nielluccio, the vine will give no harvest for 2 or 3 years before its sudden death. In the regions where the disease was not immediately identified, the disease often caused the ruin of many vine growers. Such was the case in Gascony (1955–1965) and Corsica (1968–75).

Control

Control of the disease is possible by reducing the vector population, and by the ability of the plant to recover. The problem of vector control is theoretically simple. *S. littoralis* lives on grapevines only; no other plant seems to harbour the vector, at least in Europe, and *S. littoralis* has only one generation a year. It is thus possible to cover the hatching period with insecticide treatment in the vineyard. In Gascony, the hatching period covers 5 weeks, and three treatments of insecticide are satisfactory. In Corsica, where the winter is mild, the breaking of diapause is partial and, consequently, the hatching period is as long as 3 months; thus six treatments or more would be necessary. An alternative solution is to apply an ovicide treatment in winter before the opening buds and two treatments in summer against the adult leafhoppers coming into the crop from the vicinity. For the cultivars that show recovery, such insecticide treatments enable the vineyard to recover little by little as the vector population is reduced. For cultivars where infection leads to death, the treatments should limit spread of infection and prolong the life of the vineyard.

CAUDWELL

Palm lethal yellowing MLO causes a very serious disease of coconut palms in tropical America and parts of Africa. Many palm species can be infected (e.g. in glasshouses or in S. Europe) and this MLO is a quarantine organism for Europe (EPPO 159).

In N. America, **peach yellows MLO** caused very serious losses last century. This is the classic fruit-tree yellows disease, transmitted by the leafhopper *Macropsis trimaculata*. It was one of the first on which heat therapy (10 min at 50°C) was used to produce pathogen-free material. It now occurs only sporadically. **Peach rosette MLO** and **peach X disease MLO** are two other N. American pathogens of the same group, both mostly present in wild *Prunus* species, from which orchards are liable to be infected. All three of these MLOs are considered quarantine organisms for Europe (EPPO 138, 139, 140).

Pear decline MLO

(Syn. pear leaf curl, pear moria)

Basic description

The MLO causal agent is spherical elongated, filamentous or irregular-shaped, non-cultivable and bounded by a trilaminar unit membrane, but lacking a rigid cell wall. It is distinctly different from the helical, motile, and cultivable spiroplasmas.

Host plants

The pear decline agent has been demonstrated in cultivars and rootstocks or pear, quince. *Pyrus ussuriensis* and in experimentally inoculated *Vinca rosea*.

Geographical distribution

Pear decline was first described from British Columbia (Canada) in 1948. It now occurs in all pear-growing areas along the Pacific coast of N. America, and in some areas of N.E. USA. There is evidence that the disease was present in Italy before it was found in N. America. More recently it has also been reported from other European countries including France, Spain, Switzerland, FRG, Austria, Czechoslovakia, USSR, Yugoslavia and Greece. It is likely to occur in other European countries (Seemüller 1981).

Disease

The most obvious symptoms of pear decline are either quick decline, slow decline, or leaf curl. The quick decline syndrome is characterized by a sudden wilting of the trees which die within a few days or weeks. It is favoured by hot, dry weather and is especially prevalent on trees on the oriental rootstocks *P. ussuriensis* and *P. pyrifolia*. Quick decline may also occur on trees on *P. communis* rootstocks, but the incidence is less and the trees often show red-leaf symptoms before they wilt. Slow decline, which is most often seen on the oriental stocks, is characterized by a progressive weakening of the trees and may fluctuate in severity. Terminal growth is reduced or nil, leaves are few, small, leathery and light green, the margins being slightly up-rolled. In autumn, leaves may become reddish rather than normal yellow and may drop prematurely; fruit size and fruit set is often reduced. The trees may live for many years or may die within a few years of infection. Leaf-curl symptoms have several features in common with slow decline symptoms. The major differences are that they occur on trees on the more tolerant rootstocks including *P. communis, P. calleryana*, or *P. betulifolia*. Also, the decline effect is usually less pronounced and the leaves are curled downward from tip to midrib, while the margins are rolled upward along the longitudinal axis (Raju *et al.* 1983).

The causal organisms are restricted to the phloem and seem to occur only within the sieve tubes. Colonized sieve tubes accumulate callose and often become necrotic. Necrosis may stimulate the cambium to produce replacement phloem. The pathological changes occur in the scion cultivar and in the rootstock, but on trees on sensitive rootstocks they are particularly obvious below the bud union. When there is severe blockage of phloem translocation, roots become starved and deteriorate, the feeder roots being the first affected. A distinct brown line may be observed on the outside of the cambium at the union of the scion and the rootstock tissue. However, this symptom is not common on *P. communis* or quince rootstocks.

The MLO colonization of the aerial part of the diseased trees is subject to seasonal fluctuation. The organisms are eliminated from the stem during winter due to the degeneration of the sieve tubes. In April or May the stem is recolonized from the roots where the organisms are present throughout the year. Recolonization of the stem is rather constant in the first few years of disease but later there may be periods of one or several years where it does not occur. The appearance of symptoms is closely related to stem colonization. Trees showing symptoms are, in most cases, colonized in the stem while symptomless trees are only slightly or not at all. However, diseased trees always remain infected in the roots, at least when grafted onto *P. communis* stocks. The macrosopic symptoms of the disease are usually not sufficiently specific for diagnosis or, as in case of the brown line, do not occur consistently enough. Therefore, the occurrence of anatomical alterations in the diseased phloem have been used for confirmation for many years. Although electron microscopy is now available, detection of organisms by this technique has often proved to be difficult, largely due to their low numbers and their uneven distribution in diseased trees. A quick, reliable, and widely used method is use of DAPI or the benzimidazole derivative Hoechst 33258. Another possibility for the detection of infection is transmission grafting to indicator plants. Good results have been obtained with the indicator cv. Precocious, which produces characteristic leaf symptoms. For further information on the disease see also Refatti (1967).

Epidemiology

Pear decline is readily transmitted by grafting infected tissue. In nature the causal agent is transmitted by *Psylla pyricola*, the pear psylla. The minimum feeding time of infectious psylla adults for transmission is a few hours. On young trees, symptoms may appear two months after feeding. On older trees, however, the incubation period seems to be considerably longer, perhaps more than one year. Whether the two other major psyllids occurring in Europe, *P. pyri* and *P. pyrisuga*, are vectors of the pear decline agent has not yet been proved.

Economic importance

The severity of the disease depends on the rootstock. In the pear-growing areas of the Pacific states of the USA where *P. ussuriensis* and *P. pyrifolia* were previously widely used, a great number of infected trees died, and pear production was reduced by half. Considerable differences in susceptibility were observed among *P. communis* rootstocks. Severe damage was reported from Italy, where about 50 000 trees were destroyed during the epidemic of 1945–1947. In Germany, tree losses from 20 to 40% and a high incidence of leaf curl have been recorded in some cases on seedling-grown trees. Damage seems to be less on trees worked on some *P. communis* stocks including own-rooted Williams, Anjou, Old Home and some Old Home × Farmingdale selections.

Control

The most effective method to reduce potential damage by the disease is the use of decline-resistant rootstocks (Westwood & Lombard 1983). A considerable level of resistance has been observed on Quince A, Quince C, *P. calleryana*, and especially on *P. betulifolia* rootstocks. Of the two major stocks used in Europe, quince is much less susceptible than the *P. communis* seedlings that are commercially available. Another important measure is the use of healthy planting material (EPPO 95). To prevent or minimize the introduction of the disease into a plantation, an effective spraying programme to control the psylla vector must be practiced. Diseased trees should be removed, at least in the nursery and in young orchards. In the USA, tetracycline antibiotics are successfully used on a large scale to control pear decline. Between the time of harvest and leaf fall the antibiotics are administered into the trunks of the trees by infusion or pressure injection. Due to the lack of registration, tetracyclines are not allowed to be used commercially in most of Europe.

SEEMÜLLER

Rubus stunt disease is associated with an MLO (Murant & Roberts 1971). It affects raspberry, blackberry and loganberry causing the production of numerous young, stunted shoots. Watering with aureomycin alleviates symptoms.

Spiroplasma citri Saglio *et al.*

Basic description

S. citri was the first mollicute to be cultured from plants (Saglio *et al.* 1971; Fudl-Allah *et al.* 1972) and shown to be a plant pathogen (Markham & Townsend 1974). *S. citri* was also the first spiroplasma to have been characterized (Saglio *et al.* 1973). Spiroplasmas are unique among Mollicutes in that they are motile and have a helical morphology. Since the discovery of *S. citri* more than 30 other spiroplasmas have been identified. At the present time they fall into 24 defined serogroups. *S. citri* is part of serogroup I. This group is further subdivided into 8 subgroups—*S. citri* represents subgroup I-1 (Bové 1984). Like other members of group I, *S. citri* has DNA with 26 mole % guanine plus cytosine and a genome size of 10^9 d. Plasmids are present in certain strains (Mouchès *et al.* 1984). It is serologically related only to spiroplasmas of group I and especially to those of subgroup I-2 (*S. melliferum*) and I-3 (*S. kunkelii*). By DNA-DNA hybridization it shows relatedness not only to these spiroplasmas but also to those of subgroup I-8 (*S. phoeniceum*). *S. citri* (subgroup I-1), *S. kunkelii* (subgroup I-3) and *S. phoeniceum* (subgroup I-8) are the only three sieve tube-restricted plant pathogenic spiroplasmas known today. *S. citri* can be identified unambiguously by a number of techniques including serology and analysis of its proteins by one and two dimensional polyacrylamide gel electrophoresis (Mouchès & Bové 1983). Spiralin (28 000 d) is the major protein of the *S. citri* membrane; it is specific to *S. citri* in that anti-spiralin monospecific IgGs do not recognize proteins of other spiroplasmas. The spiralin gene has been cloned in *E. coli* in which it is expressed as a 30 000 d protein (a preprotein?) (Mouchès *et al.* 1985). *S. citri* may be infected by three different viruses: SpV1, SpV2 and SpV3 (Cole 1979) but not by SpV4, a virus specific to spiroplasmas of subgroup I-2.

S. citri may be cultured in relatively simple mycoplasma media (Bové *et al.* 1983). The optimum temperature for growth is 32°C. The smallest viable *S. citri* cell is a 2-turn 'elementary' helix which grows essentially by one end into a longer parental helix. A 4-turn parental helix divides by constriction into two elementary helices. Longer parental helices undergo multiple divisions. Genome replication is not coupled to cell division (Garnier *et al.* 1984). *S. citri*, like eubacteria, contains three DNA polymerases but unlike eubacteria its RNA polymerase is insensitive to rifampicin. Like all other mollicutes, *S. citri* is insensitive to penicillin and other antibiotics interfering with bacterial cell-wall synthesis.

Host plants

In the Rutaceae (Calavan 1980), susceptible speices are calamondin, clementine, citrange, grapefruit, kumquat, lemon, mandarin limes, pummolo, rough lemon, satsuma, sour orange, sweet lime, sweet orange, and tangelo. Troyer and Cunningham citranges, and *Poncirus trifoliata*, remain symptomless. Tolerant rootstocks do not make susceptible scions tolerant. Susceptible species are sometimes symptomless carriers, especially under cool temperature conditions. Outside the Rutaceae, many cultivated and/or wild plant species were found in the S.W. USA to be naturally infected with *S. citri* (Oldfield & Calavan 1980). In E. USA, *S. citri* infection of horseradish causes brittle-root disease. In Morocco, Syria and Turkey, *Vinca rosea* was found to be naturally infected with *S. citri* (Bové 1984). In addition to natural infection, experimental transmission of *S. citri* to many plants by leafhoppers has increased the host range of the spiroplasma (Oldfield & Calavan 1980).

Geographical distribution

S. citri is widespread throughout the Mediterranean area, and the S.W. USA. It is present in the following countries or regions: Algeria, Arizona, California, Cyprus, Egypt, France (Corsica), Iraq, Iran, Israel,

Italy, Lebanon, Libya, Mexico, Morocco, Pakistan, Peru, Saudi Arabia, Spain, Syria, Tunisia, Turkey. It is probably more widespread than this.

Disease

The name 'stubborn' was applied as early as 1921 to non-productive navel orange trees in California. Less commonly used names in California include 'acorne' disease of oranges, stylar end greening, and blue albedo of grapefruits. In Palestine, it was known as 'little leaf' disease of citrus. Citrus stubborn disease as known in California and little leaf disease as described in Palestine are identical. The name 'stubborn' describes the disease that occurs in citrus (Calavan 1980; Gumpf & Calavan 1981); there is no special name for the disease in non-rutaceous hosts. Affected citrus trees appear slightly to severely stunted, frequently with abnormally dense bunches and upright foliation. Excessive dieback of shoots and branches occurs in severe cases. Development of multiple buds results in witches' broom. Affected trees have low yields. The proportion of fruits showing symptoms on a tree is variable. Fruits can be small and/or lopsided (curved columella); they sometimes show colour inversion, the peduncular end becoming coloured while the stylar end remains green. They may be acorn-shaped when, at the peduncular end, the albedo is thick and the flavedo coarse, while at the stylar end they are respectively thin and smooth. The vascular network in the albedo is sometimes prominent and the vascular bundles brownish-red. The type of fruit symptoms observed may depend on the environmental conditions in which the trees grow.

The presence of small, cupped leaves is typical of stubborn disease ('little leaf' disease in Israel). Leaves can also have abnormally upright positions, and the internodes are often short. Various types of mottle can affect the leaves. The distal portion of certain expanding leaves can be pinched-in and yellowish-green. This is of diagnostic value, especially on indicator plants. Under hot, dry conditions (e.g. in Iraq), leaves of certain shoots have misshapen, blunted or heart-shaped yellow tips. Such shoots are also highly diagnostic (Bové 1984).

With severely affected mature trees grafted on sour orange, there is sometimes pinholing (honeycombing) on the cambial side of the bark immediately below the bud union line. This symptom is common to both stubborn and tristeza diseases, and can be misleading. In the field, fruit symptoms are more useful than leaf symptoms for positive diagnosis. Culture of *S. citri* and ELISA (see *Indexing*) are most helpful in confirming a diagnosis based on symptoms.

Indexing

a) *Use of indicator plants*. The ideal indicator plants are vigorously growing Madame Vinous sweet orange seedlings kept at temperatures close to 32°C (optimum growth temperature of *S. citri*) in the day and near 27°C at night. Inoculum consists of young leaf patches including midrib (Calavan *et al.* 1972); side grafts have also been used as inoculum. Inoculation grafts are made directly below the indicator bud to be forced. Use 7 or more indicator seedlings per tree to be indexed, as well as healthy controls for comparison. Observable symptoms are slow growth, small cupped leaves, short internodes between leaves; mottled leaves with the distal portion pinched in and yellowish-green.
b) *Detection of* S. citri *by culture and ELISA*. There is a very good correlation between symptom expression of stubborn in Mediterranean countries and the detection of the causal agent (*S. citri*) by culture and ELISA (Bové *et al.* 1984). The plant material to use for isolating *S. citri* can be seeds with various degrees of abortion, the peduncular end of the fruit axis, and mature, mottled leaves from the summer flush of growth collected in October. *S. citri* is cultured on M1A or SP4 medium (Whitcomb 1983). Isolation of the stubborn spiroplasma is done by the filter method (Bové *et al.* 1983). The ELISA assay is carried out with anti-*S. citri* immunoglobulins as described by Saillard & Bové (1983).

Transmission

Natural transmission is through budwood, in highly varying percentages, and through leafhopper vectors. The following leafhoppers transmit *S. citri* in California: *Neoaliturus tenellus* (sugar beet leafhopper), *Scaphytopius nitridus* and *Scaphytopius delongi*. In the Mediterranean area, *S. citri* is spread by the leafhopper *Neoaliturus haematoceps* (*N. opacipennis*) from which *S. citri* can be consistently cultured (Bové 1984) and which has recently been shown to be a vector of *S. citri* (Fos *et al.* 1986).

Economic importance and control

Stubborn disease is destructive in most countries that grow citrus under hot, dry, desert or semi-desert conditions. Fruit quality is inferior; many fruits are malformed. The production of a diseased tree can be reduced 50–100%. In California 5–10% of sweet orange and grapefruit trees are estimated to be affected. In certain Mediterranean areas the damage is even more severe. The only practical control is

through production of healthy budwood and suitable siting of newly planted orchards to minimize the risk of vector-borne re-infection.
BOVÉ

RICKETTSIA-LIKE ORGANISMS

Beet latent rosette RLO was first observed in the FRG, causing the production of numerous small leaves on sugar beet. It is transmissible by *Piesma quadratum* to beet and spinach, or via micro-injection. Penicillin treatment caused degeneration of the rickettsia-like structures (Green & Nienhaus 1980/1). The disease has recently also been found in the GDR (Proeseler 1982).

Citrus greening bacterium may affect all or part of citrus trees, and cause stunting, fruit abnormalities and chlorotic leaf symptoms, particularly a blotchy mottle. Susceptibility of cultivars varies. The disease exists in two forms: a relatively mild form, sensitive to temperature over 32°C and transmitted by the psyllid *Trioza erytreae* occurring above certain altitudes in Africa (the African form), and a more severe form, tolerant of high temperatures, transmitted by the psyllid *Diaphorina citri*, and occurring down to sea-level in Asia (the Asian form). The organism responsible appears, by electron microscopy, to be a Gram-negative bacterium (especially in *Vinca* after dodder transmission from citrus), but it has yet to be cultured. Injection of 500–700 ppm tetracycline into trunks reduces symptoms by 50%. The African form has recently spread to N. Yemen and the Asian form to Saudi Arabia, where they have virtually wiped out mandarin and sweet orange. The Asian form was probably introduced into Saudi Arabia by pilgrims to Mecca coming from Asian countries, and there is now a similar risk of spread throughout the Near East and N. Africa and hence to Europe. Although the disease has been suspected in Europe (Sparapano *et al.* 1970), its presence has not been confirmed and citrus greening and its vectors remain serious quarantine organisms for Europe (EPPO 37, 46, 151).
SMITH

Clover club leaf RLO affects white clover producing leaves with long petioles and dwarfed leaflets with chlorotic margins. Severely affected leaves are twisted and clubbed. The RLO aetiology has been confirmed by electron microscopy and penicillin treatment (Markham *et al.* 1975). It occurs in the UK and USA, but is of no practical significance. An RLO-incited clover decline in France could be the same disease (Benhamou *et al.* 1978).

Grapevine infectious necrosis RLO has been reported from Czechoslovakia and Hungary (Ulrychová *et al.* 1975). Treatment with tetracycline causes almost total, and penicillin total remission of symptoms in newly developed shoots of mature, severly infected plants.

Grapevine Pierce's RLO causes a serious leaf-scald of grapevine in the USA, followed by dieback over 1–5 years (Hewitt 1970). It is transmitted by leafhoppers, and has caused four major epidemics this century, continuing to limit grapevine cultivation in the Gulf coastal plains area. The same organism also causes alfalfa dwarf and almond leaf scorch. A phytotoxin is produced (Lee *et al.* 1982). The RLO is not known in Europe, for which it presents a definite risk.

Grapevine yellows RLO is reported from the FRG (Küppers *et al.* 1975). RLOs have been recovered from the nematode *Xiphinema index* feeding on the roots, although transmission has not yet been demonstrated (Rumbos *et al.* 1977). The exact status of this disease relative to grapevine Pierce's disease on the one hand and grapevine flavescence dorée on the other, requires further study. For example, Rumbos (1979) found such RLOs in plants showing flavescence dorée symptoms in Greece.

Larch witches' broom disease is associated with an RLO. Infected seedlings develop smaller root systems than healthy ones and show early side-bud growth. Transmission is soil-borne on root debris (Nienhaus *et al.* 1976). RLOs have also been found in roots of spruce showing symptoms of decline in the FRG, but without any proof that they were in any way causative (Ebrahim-Nesbat & Heitefuss 1985). The possibility that decline or 'Waldsterben' involves an infectious agent continues to attract some support, but atmospheric pollution (acid rain) is more usually considered the primary cause.

Peach phoney RLO causes abnormal growth habit and poor yield in peach in S.E. USA. It is leaf-hopper-transmitted. It is a quarantine organism for Europe (EPPO 137).

REFERENCES

Amici A., Belli G., Corbetta G. & Osler R. (1973) Ricerche sull' epidemiologia del giallume del riso II. Indagini al microscopo electronico su pianti infestanti della risaia. *Riso* **22**(2), 111–118.

Beakbane A.B., Mishra M.D., Posnette A.F. & Slater C.H.W. (1971) Mycoplasma-like organisms associated with chat fruit and rubbery wood disease of apple, *Malus domestica*, compared with those in strawberry with green petal disease. *Journal of General Microbiology* **66**, 55–62.

Bellardi M.G., Vicchi V. & Bertaccini A. (1985) Micoplasmosi del gladiolo. *Informatore Fitopatologico* **35**(1), 35–39.

Benhamou N., Giannotti J. & Louis C. (1978) Transmission de germes de type rickettsoïde par la plante parasite *Cuscuta subinclusa. Acta Phytopathologica Academiae Scientiarum Hungaricae* **13**, 107–119.

Blattný C. & Vána V. (1974) Mycoplasma-like organisms in *Vaccinium myrtillus* infected with blueberry witches' broom. *Biologia Plantarum* **16**, 476–478.

Bové J.M. (1984) Wall-less prokaryotes of plants. *Annual Review of Phytopathology* **22**, 361–396.

Bové J.M., Whitcomb R.F. & McCoy R.E. (1983) Culture techniques for spiroplasmas from plants. In *Methods in Mycoplasmology. Diagnostic Mycoplasmology* (Eds S. Razin & J.G. Tully) Vol. 2, pp. 225–234. Academic Press, New York.

Bové J.M., Saillard C., Vignault J.C. & Fos A. (1984) Citrus stubborn disease in Irak and Syria: correlation between symptom expression and detection of *Spiroplasma citri* by culture and ELISA. In *Proceedings of the 9th Conference of the International Organization of Citrus Virologists Argentina.* University of California, Riverside.

Bovey R. (1963) Apple proliferation. In *Virus Diseases of Apples and Pears. Technical Communications no.* 30, pp. 63–67. Commonwealth Bureau of Horticulture, East Malling.

Bovey R. (1972) Présence de la flavescence dorée en Suisse et relations possibles de cette maladie avec le "Corky bark". In *Quatrième Conférence du Groupe International d'Etude des Virus et des Maladies à Virus de la Vigne, Colmar 1970*, pp. 167–170. INRA, Paris.

Bradfute O.E., Robertson D.C. & Poethig R.S. (1979) Detection and characterization of mollicutes in maize and sorghum by light and electron microscopy. In *National Science Council Symposium Series no.* 1, pp. 1–14. Taipei.

Calavan E.C. (1980) Stubborn. In *Description and Illustration of Virus and Virus-like Diseases of Citrus* (Eds J. Bové & R. Vogel) Vol. 2. SETCO-IRFA, Paris.

Calavan E.C., Olson E.O. & Christiansen D.W. (1972) Transmission of the stubborn pathogen in citrus by leaf-piece grafts. In *Proceedings of the 5th Conference of the International Organization of Citrus Virologists* (Ed. W.C. Price), pp. 11–14. University of Florida Press, Gainesville.

Carr A.J.H. & Large E.C. (1963) Surveys of phyllody in white clover seed crops, 1959–62. *Plant Pathology* **12**, 121–127.

Caudwell A. (1964) Identification d'une nouvelle maladie à virus de la vigne, la flavescence dorée. Etude des phénomènes de localisation des symptômes et de rétablissement. *Annales des Epiphyties* **15**, no. hors série.

Caudwell A. (1978) Etiologie des jaunisses des plantes. *Phytoma* **294**, 5–9.

Caudwell A. (1983) L'origine des jaunisses à mycoplasmes (MLO) des plantes et l'exemple des jaunisses de la vigne. *Agronomie* **3**, 103–111.

Caudwell A., Kuszala C., Bachelier J.C. & Larrue J. (1970) Transmission de la flavescence dorée de la vigne aux plantes herbacées par l'allongement du temps d'utilisation de la cicadelle *S. littoralis* et l'étude de sa survie sur un grand nombre d'espècies végétales. *Annales de Phytopathologie* **2**, 415–428.

Caudwell A., Meignoz R., Kuszala C,. Larrue J., Fleury A. & Boudon E. (1982) Serological purification and visualization in the electron microscope of the grapevine flavescence dorée pathogen (MLO) in infectious vectors extracts and in diseased plants extracts. *Comptes Rendus des Séances de al Société de Biologie* **176**, 723-729.

Chod J., Polák J., Novák M. & Křiž J. (1974) Vorkommen von mykoplasmaähnlichen Organismen in Hopfenblättern mit nekrotischen Kräuselmosaik. *Phytopathologische Zeitschrift* **80**, 54–59.

Clark M.F. & Davies D.L. (1984) Mycoplasma detection and characterisation. In *Report East Malling Research Station for 1983*, pp. 108–109.

Clark M.F., Barbara D.J. & Davies D. (1983) Production and characteristics of antisera to *Spiroplasma citri* and clover phyllody-associated antigens derived from plants. *Annals of Applied Biology* **103**, 251–259.

Cole R.M. (1979) Mycoplasma and spiroplasma viruses: ultrastructure. In *The Mycoplasmas: Cell Biology* (Eds M.F. Barile & S. Razin), pp. 385–410. Academic Press. New York.

Cousin M.T. (1972) Les mycoplasmes végétaux. *Le Sélectionneur français* **15**, 4–27.

Cousin M.T. & Moreau J.P. (1977) Les stolburs des Solanées. *Phytoma* **29**, 15–19.

Cousin M.T., Sharma A.K., Rousseau J., Poitevin J.P. & Savoure A. (1986) Hydrangea virescence: I. Description of the disease and its transmission to the differential host plant *Catharanthus roseus* by *Cuscuta subinclusa. Agronomie* **6**, 249–254.

Cropley R. (1963) Apple rubbery wood. In *Virus Diseases of Apples and Pears. Technical Communication no.* 30, pp. 69–72. Commonwealth Bureau of Horticulture, East Malling.

Davis M.J., Stassi D.L., French W.J. & Thomson S.V. (1978) Antigenic relationship of several rickettsia-like bacteria involved in plant diseases. In *Proceedings of the IVth International Conference on Plant Pathogenic Bacteria,* Vol. 1, pp. 311–315. INRA, Angers.

Desvignes J.C. & Savio A. (1975) Cydonia C7–1 and *Pyronia veitchii* two complementary indicators. *Acta Horticulturae* **44**, 139–146.

Ebrahim-Nesbat F. & Heitefuss R. (1985) Richettsienähnliche Bakterien (RLO) in Feinwurzeln erkrankter

Fichten unterschiedlichen Alters. *European Journal of Forest Pathology* **15,** 182–187.

Fos A. (1976) Observations entomologiques préliminaires sur le dépérissement du cerisier en Tarn-et-Garonne. *Revue de Zoologie Agricole et de Pathologie Végétale* **75,** 134–140.

Fos A., Bové J.M., Lallemand J., Saillard C., Vignault J.C., Ali Y., Brun P. & Vogel R. (1986) La cicadelle *Neoaliturus haematoceps* (Mulsant & Rey) est vecteur de *Spiroplasma citri* en Méditerranée. *Annales de Microbiologie (Institut Pasteur)* 137A, 97–101.

Freundt E.A. & Razin S. (1984) Genus I. Mycoplasma. In *Bergey's Manual of Systematic Bacteriology* (Ed. N.R. Krieg) Vol. 1, pp. 112–170. Williams & Wilkins, Baltimore.

Fudl-Allah A.E.-S.A., Calavan, E.C. & Igwegbe E.C.K. (1972) Culture of a mycoplasma-like organism associated with stubborn disease of citrus. *Phytopathology* **62,** 729–731.

Garnier M., Clerc M. & Bové J.M. (1984) Growth and division of *Spiroplasma citri:* elongation of elementary helices. *Journal of Bacteriology* **158,** 23–28.

Gärtel W. (1959) Die Flavescence dorée oder "Maladie du Bacco 22A". *Weinberg und Keller* **6,** 295–311.

Genite L.P. & Stanyulis Yu. P. (1975) (Clover diseases of the yellows type associated with mycoplasma-like organisms and their transmission by the leaf-hopper *Aphrodes bicinctus.*) *Trudy Biologo-Pochvennogo Instituta* **28,** 165–170.

Green S.K. & Nienhaus F. (1980/1) Further investigations on the latent rosette disease of sugar beet. I. Symptomatology, cytological and histological observations. II. Transmission, isolation and cultivation, serological tests. III. Treatment of plants with antibiotics and discussion of RLO pathogenicity. *Zeitschrift für Pflanzenkrankheiten und Pflanzenschutz* **87,** 747–755; **88,** 27–37, 142–147.

Gumpf D.J. & Calavan E.C. (1981) Stubborn disease of citrus. In *Mycoplasma Diseases of Trees and Shrubs* (Eds K. Maramorosch & S.P. Ráychaudhuri), pp. 97–134. Academic Press. New York.

Hewitt W.B. (1970) Pierce's disease of *Vitis* species. In *Virus Diseases of Small Fruits and Grapevines* (Ed N.W. Frazier), pp. 196–200. University of California, Berkeley.

Jacob H. (1976) Investigations on symptomatology, transmission, etiology and host specificity of blackcurrant reversion virus. *Acta Horticulturae* **66,** 99–104.

Joshi H.V. & Carr A.J.H. (1967) Effect of clover phyllody virus on nodulation of white clover (*Trifolium repens*) by *Rhizobium trifolii* in soil. *Journal of General Microbiology* **49,** 385–392.

Keep E., Knight V.H. & Parker J.H. (1982) Progress in the integration of characters in gall mite-resistant blackcurrants. *Journal of Horticultural Science* **57,** 189–196.

Kegler H., Müller H.M., Kleinhempel H. & Verderevskaya T.D. (1973) Untersuchungen über den Kirschverfall und die Hexenbesenkrankheit der Heidelbeere. *Nachrichtenblatt für den Pflanzenschutzdienst in der DDR* **27,** 5–8.

Kunze L. (1976) Spread of apple proliferation in a newly established apple plantation. *Acta Horticulturae* **67,** 121–127.

Küppers P., Nienhaus F. & Schinzer, U. (1975) Rickettsia-like organisms and virus-like structures in a yellows disease of grapevines. *Zeitschrift für Pflanzenkrankheiten und Pflanzenschutz* **82,** 183–187.

Leclant F., Marchoux G. & Giannotte J. (1974) Mise en évidence du rôle vecteur du psylle *Trioza nigricornis* dans la transmission d'une maladie à prolifération de *Daucus carota. Comptes Rendus Hebdomadaires des Séances de l'Académie des Sciences D* **278,** 57–59.

Lee R.F., Raju B.C., Nyland G. & Goheen A.C. (1982) Phytotoxin(s) produced in culture by the Pierce's disease bacterium. *Phytopathology* **72,** 886–888.

Lin C.P. & Chen T.A. (1985) Monoclonal antibodies against the aster yellows agent. *Science USA* **227,** 1233–1235.

Luckwill L.C. & Crowdy S.H. (1950) Virus diseases of fruit trees. 2. Observations on rubbery wood, chat fruit and mosaic in apples. In *Report of Long Ashton Research Station for 1949,* pp. 68–79.

Markham P.G. & Townsend R. (1974) Transmission of *Spiroplasma citri* to plants. In *Les Mycoplasmes* (Eds J.M. Bové & J.F. Duplan) Vol. 33, pp. 201–206. INSERM, Paris.

Markham P.G., Townsend R. & Plaskitt K.A. (1975) A rickettsia-like organism associated with diseased white clover. *Annals of Applied Biology* **81,** 91–93.

Marwitz R. & Petzold H. (1976) Elektronmikroskopischer Nachweis mycoplasma-ähnlicher Organismen in *Delphinium*-hybriden mit Blütenvergrünung und Verlaubung. *Phytopathologische Zeitschrift* **87,** 1–11.

McCoy R.E. (1979) Mycoplasms and yellows diseases. In *The Mycoplasmas. Vol. III. Plant and Insect Mycoplasmas* (Eds R.F. Whitcomb & J.G. Tully), pp. 229–264. Academic Press, New York.

McCoy R.E. (1984) Mycoplasma-like organisms of plants and invertebrates. In *Bergey's Manual of Systematic Bacteriology* (Ed. N.R. Krieg) Vol. 1, pp. 192–193. Williams & Wilkins, Baltimore.

Moreau J.P., Cousin M.T. & Leclant F. (1974) Le dépérissement jaune du lavandin (*Lavandula hybrida*). Diagnose et perspectives de lutte. In *Comptes Rendus des 4èmes Journées de Phytiatrie et de Phytopharmacie Circum-méditerranéennes,* pp. 108–112. Ruillière-Libercio, Avignon.

Morvan G. (1977) Apricot chlorotic leafroll; apricot decline today and tomorrow. *Bulletin OEPP/EPPO Bulletin* **7,** 37–55; 137–147.

Mouchès C. & Bové J.M. (1983) Electrophoretic characterization of mycoplasma proteins. In *Methods in Mycoplasmology* (Eds. S. Razin & J.G. Tully). pp. 241–255. Academic Press. New York.

Mouchès C., Candresse T., Barroso G., Saillard C., Wroblewski H. & Bové J.M. (1985) Gene for spiralin, the major membrane protein of the helical mollicute *Spiroplasma citri:* cloning and expression in *Escherichia coli. Journal of bacteriology* **164,** 1094–1099.

Müller H.M., Schmelzer K. & Kleinhempel H. (1973) Elektron-mikroskopischer Nachweis mykoplasma-ähnlicher Organismen in Siebzellen des Blumenkohls. *Archiv für Phytopathologie und Pflanzenschutz* **9,** 335–336.

Murant A.F. & Roberts I.M. (1971) Mycoplasma-like bodies associated with Rubus stunt disease. *Annals of Applied Biology* **67,** 389–393.

Nakamura S., Saito T. & Yamamoto M. (1978) (Studies on clover phyllody (part 2). Effects of temperature upon the appearance of symptoms). *Journal of Agricultural Science Tokyo Nogyo Diagaku* **23,** 97–101.

Nienhaus F., Brüsser H. & Schinzer U. (1976) Soil-borne transmission of rickettsia-like organisms found in stunted and witches' broom diseased larch trees *(Larix decidua). Zeitschrift für Pflanzenkrankheiten und Pflanzenschutz* **83,** 309–316.

Oldfield G.N. & Calavan E.C. (1980) *Spiroplasma citri:* non-rutaceous hosts. In *Description and Illustration of Virus and Virus-like Diseases of Citrus* (Eds J.M. Bové & R. Vogel) Vol. 4. SETCO-IRFA, Paris.

Onishchenko A.W. (1984) (The effect of different factors on the incubation period of phytopathogenic mycoplasmas in plants and leafhopper vectors). *Mikrobiologicheskii Zhurnal* **46,** 52–56.

O'Rourke C.J. (1976) *Diseases of Grasses and Fodder Legumes in Ireland.* An Foras Talúntais. Carlow.

Posnette A.F. & Cropley R. (1965) Field experiments with chat fruit diseases of apple. *Annals of Applied Biology* **55,** 439–445.

Posnette A.F. & Cropley R. (1969) Chat fruit disease of Tydeman's Early Worcester apple. *Report East Malling Research Station for 1968*, pp. 137–141.

Posnette A.F., Campbell A.I. & Cropley R. (1976) The re-indexing of EMLA scionwood trees and rootstocks growing on special stock nurseries. *Acta Horticulturae* **67,** 275–278.

Posnette A.F., Cropley R. & Swait A.A.J. (1968) The incidence of virus diseases in English sweet cherry orchards and their effect on yield. *Annals of Applied Biology* **61,** 351–360.

Proeseler G. (1982) Eine bisher unbekannte Krankheit der Beta-Rüben in der DDR. *Nachrichtenblatt für den Pflanzenschutz in der DDR* **36,** 20.

Ragetli H.W.J., Elder M. & Schroeder B.K. (1982) Isolation and properties of filamentous virus-like particles associated with little cherry disease in *Prunus avium. Canadian Journal of Botany* **60,** 1235–1248.

Ragozzino A. (1978) Il 'giallume' dello zucchino (*Cucurbita pepo*) in Campania. *Informatore Fitopatologico* **28** (9), 13–15.

Ragozzino A., Iaccarino F.M. & Viggiani G. (1973) Maculatura lineare of hazel. In *Comptes Rendus des 3èmes Journées de Phytiatrie et de Phytopharmacie Circum-méditerranéennes.* Sassari.

Raine J., Weintraub M. & Schroeder B.K. (1979) Hexagonal tubules in phloem cells of little cherry-infected trees. *Journal of Ultrastructural Research* **67,** 109–116.

Raju B.C., Nyland G. & Purcell A.H. (1983) Current status of the etiology of pear decline. *Phytopathology* **73,** 350–353.

Rakús D., Králík O. & Brčák J. (1974) (Mycoplasma in *Ribes houghtonianum* infected with a yellows disease). *Ochrona Rostlin* **10,** 307–309.

Razin S. & Freundt E.A. (1984) Class 1. Mollicutes. In *Bergey's Manual of Systematic Bacteriology* (Ed. N.R. Krieg) Vol. 1, pp. 140–141. Williams & Wilkins, Baltimore.

Refatti E. (1967) Pear decline and moria. In *Virus Diseases of Apples and Pears. Technical Communication* no. 30. Suppl. 1, pp. 108a–h. Commonwealth Bureau of Horticulture, East Malling.

Rodeia N. & Borges M. de L.V. (1973/4) Alterations in phosphorus and anthocyanins related to the presence of mycoplasmas in tomato plants. *Portugaliae Acta Biologica* **13,** 72–78.

Rumbos I. (1979) Studies on the etiology of a yellows disease of grapevine in Greece. *Zeitschrift für Pflanzenkrankheiten und Pflanzenschutz* **86,** 266–273.

Rumbos I., Sikora R.A. & Nienhaus F. (1977) Rickettsia-like organisms in *Xiphinema index* found associated with yellows disease of grapevine. *Zeitschrift für Pflanzenkrankheiten und Pflanzenschutz* **84,** 240–243.

Saglio P., Laflèche D., Bonissol C. & Bové J.M. (1971) Culture *in vitro* des mycoplasmes associées au stubborn des agrumes et leur observation au microscope électronique. *Comptes Rendus Hebdomadaires des Séances de l'Académie des Sciences D* **272,** 1387–1390.

Saglio P., L'Hospital M., Laflèche D., Dupont G., Bové J.M., Tully J.G. & Freundt E.A. (1973) *Spiroplasma citri* gen. and sp.n.: a mycoplasma-like organism associated with stubborn disease of citrus. *International Journal of Systematic Bacteriology* **23,** 191–204.

Saillard C. & Bové J.M. (1983) Application of ELISA to spiroplasma detection and classification. In *Methods in Mycoplasmology* (Eds S. Razin & J.G. Tully) Vol.1, pp. pp. 471–476. Academic Press, New York.

Schaper U. & Seemüller E. (1982) Condition of the phloem and the persistence of mycoplasma-like organisms associated with apple proliferation and pear decline. *Phytopathology* **72,** 736–742.

Schvester D. (1963) Genèse des symptômes et caractères de propagation de la flavescence dorée de la vigne. *Annales des Epiphyties* **14,** 167–174.

Seemüller E. (1981) Zum verstärkerten Auftreten des Birnenverfalls. *Erwerbsobstbau* **23,** 4–6.

Signoret P.A., Louis C. & Alliot B. (1976) Mycoplasma-like organisms associated with sunflower phyllody in France. *Phytopathologische Zeitschrift* **86,** 186–189.

Silvere A.P. & Romiekis M.A. (1973) (Formation and release of mesosome-like microbodies in the sporogenesis of the endophytic current revision bacillus.) *Eesti NSV Teadust Akadeemia Toimetised, Biologia* **22,** 274–277.

Slykhuis J.T., Yorston J., Raine J., McMullen R.D. & Li T.S.C. (1980) Current status of little cherry disease in British Columbia. *Canadian Plant Disease Survey* **60,** 37–42.

Smets G., Dekegel D. & Vanderveken J. (1977) Présence de structures anormales dans les bactéroïdes de trèfles blancs atteints de phyllodie. *Parasitica* **33,** 111–118.

Sparapano L., Majorana G. & Feldman A. (1970) Detection of a greening-like disease and of exocortis disease of Citrus in Italy by chromatographic methods. *Phytopathologia Mediterranea* **9,** 197–200.

Sparks T.R., Campbell A.I. & Parkinson S. (1983) Virus effects on growth and cropping of Cox and Golden Delicious. In *Report of the Long Ashton Research Station for 1982*, pp. 29–31.

Stevens W.A. & Spurden C. (1972) Green petal disease of primula. *Plant Pathology* **21**, 195–196.

Thresh J.M. (1970) Reversion of blackcurrant. In *Virus Diseases of Small Fruits and Grapevines* (Ed. N.W. Frazier), pp. 82–84. University of California, Berkley.

Tully J.G. (1984) Genus I. Acholeplasma. In *Bergey's Manual of Systematic Bacteriology* (Ed. N.R. Krieg) Vol. 1, pp. 115–181. Williams & Wilkins, Baltimore.

Ulrychová M., Petrů E., Jokeš M. & Jošková B. (1983) Mycoplasma-like organisms associated with stunting of *Gypsophila paniculata. biologia Plantarum* **25**, 385–388.

Ulrychová M., Vanek G., Jokeš M., Klobáska Z. & Králík O. (1975) Association of rickettsia-like organisms with infectious necrosis of grapevines and remission of symptoms after penicillin treatment. *Phytopathologische Zeitschrift* **82**, 254–265.

Uyemoto J.K. (1976) A new disease affecting the grapevine variety de Chaunac. *Proceedings of the American Phytopathological Society* **1**, 146.

Van Slogteren D.H.M. & Muller P.J. (1972) 'Lissers', a yellows disease in hyacinths, apparently caused by a mycoplasma. *Mededelingen van de Faculteit Landbouwwetenschappen Rijksuniversiteit Gent* **37**, 450–457.

Van Slogteren D.H.M., Bunt M.H. & Groen N.P.A. (1976) In *Jaarverslag 1976 Laboratorium voor Bloembollenonderzoek*. Lisse.

Verhoyen M., Genot M., Colin J. & Horvat F. (1979) Chrysanthemum virescence, a disorder of unknown etiology. *Phytopathologische Zeitschrift* **96**, 59–64.

Welsh M.F. & Cheney P.W. (1976) Little cherry. In *Virus Diseases and Non-infectious Disorders of Stone Fruits in North America. USDA Handbook no.* 437, pp. 231–237.

Welvaert W., Samyn G. & Lagasse A. (1975) Recherches sur les symptômes de la virulence chez *l'Hydrangea macrophylla. Phytopathologische Zeitschrift* **83**, 152–158.

Westwood M.N. & Lombard P.B. (1983) Pear rootstocks: present and future. *Fruit Varieties Journal* **37**, 24–28.

Whitcomb R.F. (1983) Culture media for spiroplasmas. In *Methods in Mycoplasmology* (Eds S. Razin & J.G. Tully) Vol. 1, pp. 147–158. Academic Press, New York.

Whitcomb R.F. & Tully J.G. (1984) Genus I. Spiroplasma In *Bergey's Manual of Systematic Bacteriology* (Ed. N.R. Krieg) Vol. 1, pp. 781–787. Williams & Wilkins, Baltimore.

Yakutkina T.A. (1979) (The complex use of basic methods of diagnosis in establishing the mycoplasma etiology of honeysuckle witches' broom.) In *Trudy Vsesoyuznogo Nauchno-Issledovatel'skogo Instituta Zashchity Rastenii*, pp. 52–56. Leningrad.

Zirka T.I., Skrypal I.G. & Alekseeko I.P. (1977) (Possible etiological factors of blackcurrant reversion.) *Tagungsberichte Akademie der Landwirtschaftwissenschaften zu Berlin, DDR* **152**, 101–105.

6: Bacteria I
Pseudomonas and *Xanthomonas*

The two genera *Pseudomonas* and *Xanthomonas* belong to the Pseudomonadaceae, a family of straight or slightly curved Gram-negative rods. They are motile by polar flagella, and are strictly aerobic and chemo-organotrophic with a respiratory metabolism. All are catalase-positive and nearly all are oxidase-positive.

PSEUDOMONAS

The type genus, *Pseudomonas*, has been extensively studied taxonomically in recent years and the work has been reviewed by Palleroni (1984). It contains animal and plant pathogens as well as saprophytes and has been divided into five sections. Sections I–III are differentiated from each other on the ability to accumulate poly-β-hydroxybutyrate (PHB) as a carbon reserve material, pathogenicity and cleavage of protocatechuate; they are plant pathogens, animal pathogens or saprophytes. Section IV comprises three non-phytopathogenic species with properties such as growth factor requirements and an inability to use nitrate as a nitrogen source that sets them apart from other pseudomonads. Section V comprises species of uncertain natural relationships with other well-characterized species in the genus; many are plant pathogens whose characters allow at least tentative inclusion in Sections I–II. Plant pathogens occur in Sections I, II and V. Diagnosis of plant diseases caused by pseudomonads (and the differentiation of the pathogens themselves) are treated in detail by Fahy & Persley (1983) and Lelliott & Stead (1987), and a general guide to the differentiation of most of the species included here is given in Table 6.1. In most cases this replaces the *Basic description* section of individual entries.

Most of the plant pathogens in the genus are in section I and most of these in the species *P. syringae*. There are still differing opinions on their taxonomy: some workers would include *P. viridiflava* within *P. syringae*, others distinguish groups within *P. syringae* based on phenotypic bacteriological criteria or on DNA/DNA hybridization. Here the following approach, which seems to command most acceptance among plant bacteriologists, has been adopted: i) *P. syringae* comprises a large number of pathogens most of which cannot yet be adequately differentiated on grounds other than differing host ranges. Those that can include the subspecies *savastanoi* and the pathovar *delphinii*. Even differentiation on host range is unsatisfactory since the full host range of many pathogens is not known (and in any practical sense probably unknowable) and may, in some cases, overlap. Almost all were differentiated nomenclaturally at species level until 1980 when their names became illegitimate under the rules of the international code. To meet this situation the International Society of Plant Pathology (ISPP) has recommended the adoption of the pathovar nomenclature put forward by Dye *et al.* (1980) as an expedient to provide means of naming the pathogens until a more rational determinative classification is made. This recommendation should be followed until the ISPP makes a new or modified recommendation. ii) *P. viridiflava*, which can be readily differentiated in the laboratory from *P. syringae* (Table 6.1) and which unlike the specialized pathogens of *P. syringae* is apparently a low-grade, opportunist pathogen, should be retained as a separate species. iii) There is adequate evidence from numerical analysis of phenotypic data and from DNA/DNA hybridization experiments that *P. cichorii* should be a separate species. iv) There are no known valid systematic grounds for separating *P. marginalis* from *P. fluorescens*, a species which has until recently been considered to consist solely of saprophytic strains. However within *P. fluorescens* there are strains that can rot plant storage tissues and the cultures of *P. marginalis* (and its synonym *P. pastinacae*) that have been studied form a reasonably homogeneous group, probably with somewhat more specialized pathogenic properties than the soft rotting strains of *P. fluorescens*. It is therefore empirically justified to keep the two groups separate, at least until further studies clarify their taxonomic relationship to each other and to the *P. fluorescens* complex.

Table 6.1. Differentiation of some European plant pathogenic species of *Pseudomonas*[a]

	PHB produced	Arginine-dihydrolase	Green fluorescent pigment	Oxidase	$NO_3 \rightarrow NO_2$	Denitrification	Pectinolysis (pH 7.0)	Acid from purple sucrose in 3 days	Levan colonies on sucrose nutrient agar
P. syringae	–	–	+[b]	–	–	–	–	+	+[c]
P. viridiflava	–	–	+[d]	–	–	–	+	–	–
P. cichorii	–	–	+	+	d	–	–	–	–
P. marginalis	–	+	+	+	+	+	+	+	+
P. fluorescens (soft rotting strains)	–	+	+	+	d	d	+	d	d
P. andropogonis	+	–	–	–	–	–	o	o	o
P. caryophylli	+	+	–	+	+	+	o	o	+
P. corrugata	+	–	–	+	+	+	o	o	+[e]
P. cepacia / *P. gladioli*	+	–	–	+	+	–	o	o	–
P. solanacearum	+	–	–	+	+	+	o	o	o

a) so far as is known, these determinants are invariable for freshly isolated strains within species except as indicated below and for pigment production and acid from sucrose where the symbols have the following meaning: '+', 95% of strains positive; '–', 95% of strains negative; 'd' = >5– < 95% strains positive; o = insufficient information.

b) some strains do not produce green fluorescent pigments on King's medium B and strains of pv. *mors-prunorum* and pv. *persicae* consistently do not produce it or produce it only on special media. The green fluorescent species generally can be differentiated by the LOPAT tests (Lelliott *et al.* 1966) alone.

c) pv. *delphinii* consistently fails to produce 'levan' colonies and different strains of subsp. *savastanoi* produce colonies varying between 'nearly full levan' to 'non-levan'.

d) also produces a characteristically green-centred colony on 5% sucrose nutrient agar.

e) mucoid colonies.

Pseudomonas amygdali Psallidas & Panagopoulos

The bacterium, a non-fluorescent member of the genus *Pseudomonas*, is distinguished from the other plant pathogenic species of the genus by its poor and slow growth on nutrient agar, forming colonies less than 0.1 mm in diameter in 3 days and 0.1–0.5 mm in 6 days at 26°C. On nutrient agar with 5% sucrose it forms distinctive colonies, pearly white in colour, round with a wrinkled surface, and convex with radial striations observable after 6 days. It also has different physiological and biochemical characteristics: it produces H_2S from cysteine or sodium thiosulphate and fails to utilize glycerol, xylose, melibiose, rhamnose, meso-inositol, malonic and maleic acids as carbon and energy sources (Psallidas & Panagopoulos 1975). The bacterium is highly host-specific, infecting only almond trees. Other stone fruit trees are not susceptible either naturally or experimentally.

In Greece, the disease has been found on the island of Crete and the Aegean islands of Kos, Rhodes and Chios where it is endemic. It has also been found in one orchard in the island of Euboea and, in the mainland, in the Atiki district. The disease has been reported in Turkey and also occurs in Afghanistan. *Pseudomonas amygdali* causes characteristic cankers on branches, twigs and trunks. Usually the cankers begin from leaf scars but any wound can serve as an entrance for the pathogen and for canker initiation. Cankers first appear as swellings of the bark. These crack open, and become surrounded by swollen cortex. The cankers are perennial being active not only throughout the year but also for many years. They may reach a length of 15–20 cm. The infected trees have dead branches, twigs and stems as a result of girdling. If the trunk is girdled by canker the whole tree dies. The bacterium does not live on the leaf surface as an epiphyte; it overwinters inside the cankers and is disseminated by the rain and wind. Over long distances the pathogen is disseminated with infected graftwood or seedlings. The pathogen has no specific environmental requirements; the disease develops under arid conditions and infection takes place during leaf fall, coinciding with the rainy season. Only preventive control measures can be recommended. These include pruning out diseased parts, uprooting badly affected trees and applying two copper compound sprays during the leaf fall period. To prevent the spread of the disease, measures should be taken to prohibit the movement of propagating material from infected areas to areas or countries free of the disease. Some cultivars, such as Ferragnes, Ferranduel and Marcona, are resistant and some such as Texas and Pagrati are tolerant.
PSALLIDAS

Pseudomonas andropogonis (Smith) Stapp has two pathovars: pv. *andropogonis* that causes a leaf striping and spotting of *Sorghum* species, Sudan grass, teosinte and maize, black spotting of clover, brown spotting of tulip bulbs, spotting of ornamental *Triplaris* and a rot of the terete vanda orchid; pv. *stizolobii* that causes a leaf spot of velvet bean, bougainvillea and broad bean. The disease can be of major economic importance to cereals in the Americas and Japan but in Europe has only been recorded incidentally in the USSR on sorghum and maize and from Hungary (CMI 372).

Pseudomonas caryophylli (Burkholder) Starr & Burkholder

P. caryophylli (EPPO 55) causes a wilt, foot and root rot of carnation that was at one time a limiting disease of that crop in the USA and caused damage to protected crops in Europe. The adoption of control measures during propagation (and particularly indexing and the abandoning of liquid rooting hormones) has reduced it to minor importance generally, although the potential for its resurgence exists if propagation methods are changed in a way that would favour it. Affected plants show a grey-green, later tan or brown wilting of the top or branches, and finally of the whole plant, accompanied by a yellowish to brown discoloration of the vascular strands and softening of the stem base cortex, the inner of which is characteristically sticky to the touch (Burkholder 1942). The disease is most severe at high soil temperatures (30–37°C). It is distinguished from the similar slow wilt of carnations (*E. chrysanthemi* pv. *dianthicola* q.v.) by its darker, sticky vascular discoloration, more rapid wilting and higher optimum temperature for symptom expression. Although the pathogen can survive in soil (probably in trash and rotted roots) for some months, it is mainly transmitted on the knife used to take cuttings from latently affected mother (stock) plants, or epidemically if the cuttings are stood in water or liquid rooting hormone before insertion in the propagation beds. Spread between plants in propagation or flowering beds is unimportant. Good control can be obtained by indexing cuttings used to generate stock plants (Nelson *et al.* 1959; see also *Xanthomonas campestris* pv. *pelargonii*), avoiding

standing cuttings in liquids and using a dust formulation of rooting hormone.
LELLIOTT

Pseudomonas cepacia Burkholder is an opportunist wound pathogen which also causes infection in humans and may be a rhizoplane organism. It commonly causes a yellow, slimy rot of the outer leaves of onion bulbs stored at high temperatures (*c.* 30°C) (Kawamoto & Lorbeer 1974). See also *Pseudomonas gladioli* pv. *alliicola*.

Pseudomonas cichorii (Swingle) Stapp

Host plants

P. cichorii (CMI 695) is an opportunist pathogen of many herbaceous plants on which it causes dark-brown to black necrotic leaf and stem lesions and probably exists as a migrant or resident epiphyte. The wide host range includes lettuce, endive, parsley, celery, mushroom, bean, *Brassica* species, pelargonium, chrysanthemum, gerbera, dahlia, coffee, tobacco, tomato, aubergine and sunflower. Tsuchiya *et al.* (1982) showed in tests with *c.* 100 species that 35 species of weed plant in 11 families were susceptible when 10^6–10^8 cell/ml were sprayed on intact leaves and 63 species in 18 families when sprayed on wounded leaves.

Geographical distribution

It is almost certainly of world-wide distribution both in field and protected crops.

Disease

Symptoms are variable but on lettuce they usually begin as small, dark-green water-soaked spots round the stomata, epidermal hairs or hydathodes that, under favourable environmental conditions of surface water and high humidity, rapidly spread and coalesce to form large, spreading, wet, dark-brown lesions. If the environment becomes dry, lesion extension usually ceases and the lesion can dry out and become lighter coloured. The symptoms on many hosts are similar to these but on some a narrow, yellow halo develops around the lesions. Variation in symptom expression includes: the development of small, colourless spots on cauliflower heads, rapidly turning brown and covering the head and, under humid conditions, leading to a wet rot associated with *Erwinia carotovora* and destruction of the whole head or, in dry conditions, establishment of *Alternaria brassicicola* (q.v.) (Coléno *et al.* 1971); zonate or concentric ring formation in leaf spots (on gerbera); severe leaf blight on coffee seedlings and defoliation of adult plants; varnish spot, a condition of lettuce in California in which dark-brown, shiny, firm, necrotic spots a few mm in diameter occur on blades and petioles of the leaves underneath the two or three outermost leaves of the head. On lettuce the later symptoms so resemble those caused by *P. marginalis* and *P. viridiflava* (q.v.) that Tsuchiya *et al. al.* (1979) have suggested that the three pseudomonads should be regarded as the joint cause of lettuce bacterial rot. It can invade vascular tissue.

Epidemiology

The pathogen can survive in dried, diseased lettuce leaves and in buried disease tissue over winter and in infested soil for 1 month in summer; it can be isolated from affected leaves of weeds in lettuce fields and from the rhizospheres of some weeds and lettuce (Ohata *et al.* 1982a). There is evidence for seed transmission in lettuce (Ohata *et al.* 1982a,b). In head lettuce the plant is more susceptible at later stages of growth with middle leaves, midribs and petioles more susceptible than inner or outer leaves, ribs and blades respectively (Shirata *et al.* 1982). Unwounded tissue can be readily infected and lesion development proceeds at 10–30°C (optimum 25°C). Ohata *et al.* (1979) suggested that, in Japan, the disease is closely related to climatic conditions.

Economic importance

Although no general quantitative assessments are available of the losses this disease causes, it is apparent that it is regarded of importance economically to lettuce production in Japan and to lettuce, celery and chrysanthemum production in parts of the USA. In Europe it is not generally regarded as a major disease but given the nature of the disease, its very wide host range and evidence that it can be the primary cause for later attacks by soft rot bacteria and *Alternaria brassicicola*, the losses are probably not inconsiderable; epidemic attacks on lettuce have been reported from Italy (Bazzi & Mazzucchi 1979).

Control

Although heat treatment can successfully free lettuce

seed of the pathogen, it would not of itself be expected to control the disease because of the general presence of the pathogen in the environment of the growing crop. Avoidance of conditions favourable for infection and wide rotations are recommended but are often impractical to implement for the culture of salad crops.
LELLIOTT

Pseudomonas corrugata Roberts & Scarlett

Basic description

P. corrugata almost certainly belongs to section II of the genus. It grows as characteristic yellowish to brown wrinkled colonies on many media, and produces a yellow to yellow-green, diffusible, non-fluorescent pigment (Scarlett *et al.* 1978).

Host plants

The only known host is tomato.

Geographical distribution

P. corrugata has now been isolated from crops affected by typical pith necrosis symptoms in FRG, GDR, California and Florida (USA) as well as many crops in England.

Disease

The pathogen is consistently associated in W. Europe with a commonly occurring pith necrosis of field and protected crops. There is little doubt that the disease there is caused by *P. corrugata* although a similar disease elsewhere has been sometimes attributed to other pseudomonads (*P. cichorii* by Wilkie and Dye (1974), *P. viridiflava* by Wilkie *et al.* (1973) in New Zealand and an unidentified member of LOPAT group Ia by Alivizatos (1984) in Greece). Koch's postulates have been fulfilled by at least one laboratory (Lukezic 1979) since the original work by Scarlett *et al.* (1978). Lukezic (1979) has also isolated *P. corrugata* from healthy roots of glasshouse-grown but not field-grown lucerne plants, thereby strengthening the suggestion that this pathogen may be a soil/water inhabitant. Doubts about the aetiology of tomato pith necrosis probably arise because *P. corrugata* can often not be isolated from plants with fully developed symptoms in which the disease appears to be no longer active.

The disease is typically one of mature plants particularly in crops that have been given high rates of nitrogen fertilizer and grown in high humidity and/or when there is free water on the plant surface. The first symptom is usually chlorosis of the youngest leaves when the fruit of the first truss is fully grown but still green. In severe cases the chlorosis intensifies affecting all of the top half of the plant which loses turgidity; the whole plant then collapses. A dark-brown to black, surface, stem lesion usually affects the point at which the first truss is attached and sometimes extends for up to 30 cm in length. The pith in the region of the external lesion is hollow and often has a laddered appearance towards the lesion's edge. Beyond the external lesion the pith is dark-brown and water-soaked but not soft; the leading edge of the pith discoloration is usually well defined. Sometimes, particularly in plants without external lesions, the pith is merely discoloured with or without cavities in the discoloured area and/or the vascular system is brown. Plants collapse and die or continue to grow very slowly. Less severely affected plants merely exhibit some chlorosis of the upper leaves, sometimes with moderate development of external stem lesions. In protected crops a dirty-white bacterial flux sometimes oozes from the stumps of the pedicels of lower leaves that have been trimmed off and dribbles down the stems. Following initial symptoms, surviving plants sometimes grow normally, show no further symptoms and produce a good crop. Necrosis and brown discoloration of the pith often extends from soil level to within a few cm of the growing points, into peduncles as far as the fruit and into leaf rachises. In older crops there is prolific development of adventitious roots on the stem in the area where the pith has been most markedly affected. The diameter of new stem growth in plants that have recovered from attack is markedly reduced.

Epidemiology

The epidemiology of the disease is little understood. The pathogen has been isolated from the mains water supply used to water crops and from a nursery reservoir but it may also exist in soil *per se* or as part of the rhizoplane flora. Studies in English glasshouses have not indicated spread from plant to plant but rather the near-simultaneous infection of plants from some unknown source when conditions are favourable (see above). However it is difficult to believe that some, probably significant, plant-to-plant spread does not occur during periods when bacterial flux is being produced.

Economic importance

Although the disease has been widespread in England since 1971 it has caused severe losses in comparatively few cases; it is described as a serious disease in most areas growing protected tomatoes in Greece.

Control

It can be controlled by avoiding excess use of nitrogen fertilizers and in protected crops by avoiding as far as possible the high humidities and free water on plant surfaces that follow indiscriminate watering and low night temperatures. Hygienic disposal of trash and debris from affected crops and cleaning of water tanks after attacks would seem sensible measures.
LELLIOTT

Pseudomonas gladioli Severini **pv. *alliicola*** (Burkholder) Young *et al.* causes a soft mushy rot of stored bulbs (Burkholder 1942) and necrotic lesions on seed stalks and leaves of onions (Hevesi & Viranyi 1975). Although common it is rarely severe except in onions stored at high temperatures. The species can be differentiated with difficulty from *P. cepacia* (q.v.) (Lelliott & Stead 1987).

Pseudomonas gladioli **pv. *gladioli*** Severini (syn. *P. marginata*) (Elliott 1951) causes superficial, smooth, 'lacquered' scabs with prominent raised margins on gladiolus corms. Similar symptoms have been reported on *Crocus* species and *Freesia refracta*. Other, probably less common, symptoms are brown pustules on the foliage promoted by heavy rainfall and injudicious watering, and a foot rot of the basal leaves especially on damp, heavy soils (Fischer 1942). Bacterial scab is common on gladiolus corms and although sometimes unsightly when the scale leaves are removed it is rarely serious. Control has been claimed for heat treatment and a variety of chemical corm dips and dusts but these are rarely practised specifically for this disease.

Pseudomonas marginalis (Brown) Stevens

Basic description

P. marginalis is a pectolytic fluorescent species closely related to the saprophytic *P. fluorescens*.

Hosts and host specificity

The pathogen is the usual cause (with *Erwinia carotovora* (q.v.)) of bacterial soft rots of plant storage tissue and of soft leaf and stem tissue, particularly in stored vegetables. It appears omnivorous to such tissues on any plant host if conditions favour its entry and multiplication. The crop plants most affected include lettuce, endive, chicory, cabbage, potato, carrot, celery and onion. A pathovar, pv. *pastinacae*, produces a pink diffusible pigment on 5% sucrose nutrient agar and causes a brown rot of parsnip roots (Burkholder 1960); however the validity of such separation is questionable (Hunter & Cigna 1981).

Geographical distribution

The pathogen appears to be ubiquitous although more common in temperate than tropical climates.

Disease

Symptoms on tubers, bulbs, storage roots and soft stems are a generalized, soft, water-soaked rot often discoloured dark-green, brown or black and indistinguishable from rots caused by other bacteria. On leaves, large, usually dark-coloured lesions spread from smaller, initial, marginal or interveinal spots. Depending on environmental conditions these lesions can continue to spread, causing complete rot in such vegetables as head lettuce or cabbage, or become inactive and dry out to leave brown necrotic patches. On some hosts narrow chlorotic margins frequently develop around lesions.

Epidemiology

P. marginalis appears to occur commonly on plant surfaces and in the rhizosphere and field soils (Sands & Hankin 1975; Cupels & Kelman 1980). Infection frequently occurs through wounds, although invasion of susceptible tissue through natural openings such as stomata, lenticels and hydathodes also occurs if free water is present on the tissue surface.

Economic importance

Because of the difficulty of visually differentiating soft rot caused by *P. marginalis* from that caused by other bacteria such as *Erwinia carotovora, E. chrysanthemi* and *Pseudomonas viridiflava* or leaf necrosis by *P. cichorii* (Lelliott & Stead 1987) it is not possible to do more than guess at its economic significance. However, bacterial soft rotting is generally considered the greatest single cause of loss from bacterial plant disease and there can be little doubt that *P. marginalis* is responsible for a considerable part of that loss. It can cause spoilage of vegetables stored at low temperature, conditions in which it is probably the most common cause of bacterial rot, and the increasing extension of washing, film wrapping and cold storage of vegetables will be expected to enhance its importance.

Control

Where practicable, measures to reduce inoculum potential by rotation and disposal of plant debris and to avoid water on tissue surfaces, particularly in store, are recommended.

LELLIOTT

Pseudomonas solanacearum (Smith) Smith

Basic description

This is a non-fluorescent, PHB-producing pseudomonad belonging to Section II.

Host plants

It attacks all the solanaceous crop plants as well as many non-solanaceous crops (the most important of which are banana and groundnut) and many weeds. Over 200 hosts are known.

Host specificity

Strains show varying degrees of host specificity. The most commonly accepted subdivisions of the species are the three pathovars of Buddenhagen & Kelman (1964) based on pathogenicity and the chemovars of Hayward (1964) based on biochemical characters. Given the extensive host range of the pathogen it is not implied that these are the only pathovars or that these three pathovars (and four chemovars) cannot be further subdivided; in fact there is evidence that this is so although much more work would be needed to establish such subdivision with sufficient precision to be generally useful. Work on serological differentiation within the species indicates that it is not correlated with host range or chemovar. The pathovars of Buddenhagen & Kelman are:
pv. 1—infects most members of the Solanaceae as well as many other plants;
pv. 2—infects triploid banana (causing moko disease) and *Heliconia* spp;
pv. 3—infects potato (naturally) and tomato (artificially) and may survive in some other *Solanum* spp.
Pathovars 1 and 2 both contain strains belonging to chemovars I, III or IV and pathovar 3 corresponds to chemovar II. For differentiation of pathovars and chemovars see Lelliott & Stead (1987).

Geographical distribution

The diseases caused by *P. solanacearum* are common on non-alkaline soils in those areas between latitudes 40°N/40°S that have a relatively high rainfall distributed fairly evenly throughout the year (Kelman 1953). Generally speaking, outside these latitudes and in mountainous regions within them, only pv. 3 (the pathovar adapted to lower temperatures) occurs significantly. In Europe (EPPO 58) the disease is locally distributed, mainly as pv. 3 in potato, in Greece, Portugal, Yugoslavia and USSR. Although unlikely to occur more than incidentally elsewhere in Europe, it is interesting that pv. 3 appears to have established itself in *Solanum dulcamara* growing aquatically or as a marginal in some streams in S. Sweden.

Disease

P. solanacearum causes wilt in its host, the symptoms of which vary to differing degrees between hosts. The literature up to the early 1950s is well reviewed by Kelman (1953). In this book, only the symptoms of the disease on potato, known as brown rot, are described. The first visible sign of infection is wilting of the leaves towards the top of the plant at high temperatures during the day with recovery at night; this eventually leads to permanent wilting and death. An external brown discoloration may develop as streaks on the stem starting at and extending from soil level. If stems are cut transversely, white, bacterial slime exudes from the vascular bundles, or can be expressed by squeezing the stem with pliers. In tubers (Fig. 20), an early symptom of the disease is necrosis of the vascular ring, visible when cut transversely, from which a pale yellow material can be extruded like toothpaste when pressure is applied to the side of the tuber. Later, a creamy fluid exudes from the ring of cut tubers. The necrotic area can extend into the parenchymatous tissue about 0.5 cm each side of the ring. Usually necrotic tissue is brown, but this symptom is not caused by all strains. Further

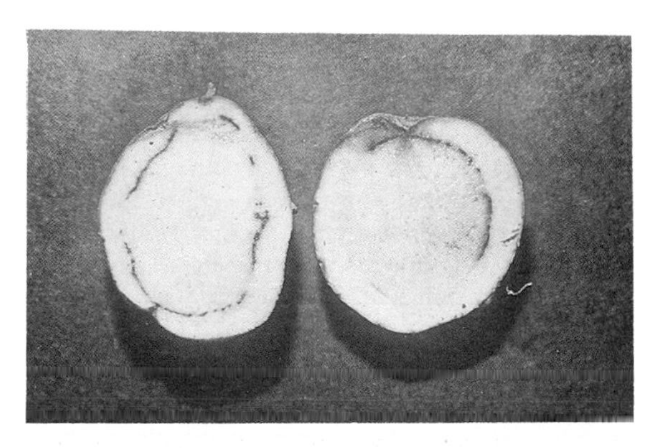

Fig. 20. Potato tubers infected by *Pseudomonas solanacearum* (courtesy of Min. of Agriculture, Turkey).

development of brown rot can result in a bacterial ooze to which soil adheres. This emerges from the eyes of tubers and from the point of attachment of the stolon. Infected but aerially symptomless plants can produce diseased tubers, and plants with affected tops can produce both diseased and non-diseased tubers; the latter are, however, often latently infected.

Epidemiology

In potatoes much transmission of disease is with mildly or latently infected tubers. The bacterium can however perennate in fields in weed hosts and survive between susceptible crops on weeds and probably in lesions on roots of non-host plants. In S. Sweden it overwinters in *Solanum dulcamara* growing in or on the edge of streams and can spread to potato crops irrigated with water from them (Olsson 1976).

Economic importance

The disease caused is limiting for the growth of many important tropical and subtropical host crops but in Europe it has caused substantial loss only in potato, and then only in Portugal, Greece and Yugoslavia.

Control

There is evidence for varying susceptibility to brown rot in potato cultivars and exploitation of this should mitigate losses in affected areas. Otherwise the use of 5–7 year rotations, cultivation to reduce susceptible weed populations and attention to the disease in seed certification schemes are the only control measures that can be recommended (EPPO 58).
LELLIOTT

Pseudomonas syringae van Hall **pv. *antirrhini*** (Takimoto) Young *et al.* causes a serious seedling blight (leaf and stem spots that coalesce) of antirrhinum grown commercially for the bedding trade in England (Lelliott & Stead 1987).

Pseudomonas syringae* pv. *aptata (Brown & Jamieson) Young *et al.* generally causes only slight to moderate loss to sugar beet crops as a leaf spot and foliar blight that also causes black streaking of seed stalks and internal root necrosis and vascular blackening (Mazzucchi 1975). Seed transmission (as contamination rather than by vascular invasion) can be controlled with mercury seed dressings. Heavy losses in warm humid parts of E. Georgia (USSR) have been reported. It has also been recorded on Swiss chard, lettuce and nasturtium. Its geographical distribution in Europe is uncertain but it has been recorded in Italy.

Pseudomonas syringae* pv. *atrofaciens (McCulloch) Young *et al.* causes brown spotting of leaves, stem, glumes and other aerial parts of wheat and barley and damages the grain. A similar disease of wheat is severe in parts of the USA and is attributed to *P. syringae* but it is not clear if the cause is pathogenically identical to pv. *atrofaciens*. In most of Europe the disease is insignificant but in the USSR it is severe on spring and winter wheat in the Ukraine (Koroleva & Novokhatka 1979; Min'ko & Koroleva 1980). It is seed-borne. There is resistance among wheat cultivars.

Pseudomonas syringae* pv. *avellanae Psallidas causes stem dieback of hazelnut in N. Greece (Psallidas 1987). Infection takes place through leaf scars and other wounds and spreads to the base of the bush and to other shoots. Unlike *X. campestris* pv. *corylina*, the bacterium does not infect leaves or husks. Rain and irrigation water facilitate spread, which also occurs with infected propagating material. Besides use of healthy material, control is based on preventive sprays with copper compounds during leaf-fall.
PSALLIDAS

Pseudomonas syringae* pv. *cannabina (Šutić & Dowson) Young *et al.* causes a leaf and stem spotting and bacteriosis of hemp which has been serious in Yugoslavia (Šutić & Dowson 1959) and has also been recorded in Bulgaria, Hungary, Italy and USSR. Many isolates are non-fluorescent and produce a brown, diffusible pigment in culture.

Pseudomonas syringae pv. *coronafaciens* (Elliott) Young *et al.*

P.s. pv. *coronafaciens* causes a common and sometimes major disease of oat crops (halo blight) wherever they are grown. Similar 'halo blight' diseases on other cereal and grass hosts have been described (e.g. *P. coronafaciens* pv. *zeae*, *P.c.* subsp. *atropurpurea*) which are best regarded as separate pathovars of *P. syringae*. *P.s.* pv. *striafaciens* is probably a synonym (Tessi 1953). The most common initial symptom is the appearance of small, water-soaked, dark-green, leaf spots that enlarge to 1–2 mm, sometimes becoming linear, usually accompanied by oval-shaped chlorotic haloes up to 1 cm diameter which result from the action of a toxin of the tabtoxin group. Haloes are not often produced when daily maximum temperatures exceed about 22°C. Glumes are sometimes spotted or the whole floret is blighted. The disease is seed-borne and plants can be attacked from the time the seed coat ruptures until the plant reaches maturity. In some infected seeds the plumule is necrosed and even destroyed and the radicle underdeveloped; occasionally the whole seed rots. The disease spreads in crops by rain splash from seed-borne primary foci; it infects mainly through the stomata and can be epidemic in crops when spread by heavy driving rain. Resistance in oats to the pathogen is uncommon and, in some crosses, governed by a single pair of dominant genes (Cheng & Roane 1968). Control through seed certification schemes is usually adequate.
LELLIOTT

Pseudomonas syringae pv. *glycinea* (Coerper) Young *et al.*

P.s. pv. *glycinea* is responsible for bacterial blight of soybean, a disease common to most if not all areas of major soybean production. At least 10 races are now known; they are determined by differential reaction on soybean cultivars (Fett & Sequeira 1981). In Europe the disease has been reported from France, Hungary, Romania, USSR and Yugoslavia. Its primary foci in crops derive from seed-borne infection that can inhibit germination and, on cotyledons, cause marginal lesions that enlarge and become dark-brown and necrotic. Often the lesions are covered, particularly on their underside, with a film of pale, greyish bacterial slime that can dry to a thin silvery crust; less commonly primary foci derive from over-wintered infected crop debris (Daft & Leben 1973). Secondary spread to infect young soybean leaves occurs by means of wind-driven rain (Daft & Leben 1972). There is evidence that the pathogen may be a resident epiphyte in buds; this could provide a continual source of inoculum (Leben *et al.* 1968). Infection occurs through natural openings on foliage and through wounds which occur commonly on sandy soils by abrasion with sand particles. Leaf lesions are at first small, angular, water-soaked, tan-coloured spots which enlarge to 1–2 mm diameter, become dark-brown to black with a dry centre and a water-soaked margin and surrounded with a narrow yellowish halo. Lesions can coalesce to produce large irregular areas of dead tissue. The centres of old lesions may drop out and the leaf appear ragged and torn. Older, lower leaves may drop, and young leaves and seedlings become stunted and chlorotic. If the growing point of seedlings is affected the plant usually dies. Large black lesions may develop on stems. Lesions on pods are initially small and water-soaked, later enlarging, coalescing and becoming dark brown to black; seeds in such pods may be covered with a bacterial slime. Harvested infected seeds may be shrivelled with sunken or raised lesions, merely discoloured or symptomless. The disease is essentially favoured by cool, wet weather.

In Europe (EPPO 56) the disease has not caused great loss but if the area of soybean production were to increase losses might be expected on the scale seen in the USA, where in the period 1975–77 it was by far the most damaging prokaryotic disease, causing an estimated annual average loss of $62 million (Kennedy & Alcorn 1980). By analogy with similar phaseolus bean diseases, control might be expected from raising seed in dry areas under furrow irrigation. Comparatively narrow rotation should be sufficient to eliminate trash-borne infection from fields since the pathogen does not apparently survive in it for two seasons. Seed certification and testing (Parashar & Leben 1972) should reduce primary foci. Durable resistance is difficult to achieve because of the many races of *P.s.* pv. *glycinea*.
LELLIOTT

Pseudomonas syringae* pv. *helianthi (Kawamura) Young *et al.* causes angular, brown leaf spots on sunflower (Piening 1976). Leaves with infected vascular tissue are frequently wrinkled due to the slower growth of the vascular than the laminar tissues.

Pseudomonas syringae* pv. *lachrymans* (Smith & Bryan) Young *et al.

P.s. pv. *lachrymans* (CMI 124) causes an angular leaf spot and fruit spotting of cucurbits, principally cucumber but also courgette, squash, melon and gherkin. It occurs generally wherever these crops are grown widely on a field scale; it is far less common in protected crops. It has also been reported on vegetable marrow and pumpkin. The disease first appears on the cotyledons as round to irregular, water-soaked spots and on true leaves as angular, water-soaked, brown spots which dry and may fall out to leave tattered leaf tissue. Water-soaked spots, sometimes covered with white exudate, occur on stems and petioles. Spots on fruit are at first minute, circular, water-soaked areas from which a translucent, whitish slime can exude under humid conditions, that dries to form a white crust. Fruits can rot if the pathogen or secondary rot-producing organisms penetrate deeply. Vascular tissue of stems and fruit may be invaded.

The pathogen is seed-borne and can overwinter in some climates in host plant debris. It is spread by rain splash, possibly by cucumber beetles and by workers' hands and implements, particularly when foliage is wet. Early infection causes significant yield reduction and fruit quality can be seriously reduced. In E. Europe and the USSR the disease can be of major importance in field crops. Seed treatment with mercury or heat treatment (Umekawa & Watanabe 1978) is effective and copper sprays give some control under field conditions. Where practicable, a three-year crop rotation, destruction of possibly affected host debris and avoidance of water films on plant surfaces are recommended.

LELLIOTT

Pseudomonas syringae* pv. *maculicola* (McCulloch) Young *et al.

P.s. pv. *maculicola* causes a bacterial spotting of brassicas (Shackleton 1966). It is most common on head cabbage but also occurs on cauliflower and Brussels sprout. It is seen not infrequently in most areas where these crops are grown. It has also been reported on radish, turnip and black mustard. Leaf spots are initially small, circular, water-soaked and centred on stomata; they become dark brown and necrotic in the centre and then turn brownish-grey with a darker border, are irregularly angular and 1–3 mm diameter. Spots can have a narrow light halo and can become coalescent. In wet weather there can be abundant blotching, and sometimes distortion of the leaves. Badly affected leaves turn yellow and drop. Discoloration of cauliflower heads occasionally occurs. The epidemiology of the disease has not been fully investigated but there is evidence that it is seed-borne and can survive in trash. In Europe the disease is only of significance during abnormally wet summers.

LELLIOTT

Pseudomonas syringae* pv. *mori* (Boyer & Lambert) Young *et al.

P.s. pv. *mori* causes a leaf spot and shoot blight of mulberry (Elliott 1951). It was first recorded in France but the only records in Europe since 1960 are in Hungary and the USSR, where the disease appears to be common and of sufficient importance for control by pesticide spraying to be recommended. Initially small, water-soaked leaf spots appear that enlarge, turn black or brown with a yellow halo and can coalesce. Leaves attacked when young become distorted. On midribs and veins spots are sunken. Translucent longitudinal stripes appear on young shoots becoming dark and sunken with translucent margins and a white to yellowish bacterial ooze may be extruded from lenticels as cirri. Both bark and wood are invaded and shoots die back or show one-sided growth. The girdling of young shoots can cause stunting. An aggressive form of the disease involving necrosis and destruction of bark and wood tissue of stems of seedlings has been described by Khurtsiya *et al.* (1979).

LELLIOTT

***Pseudomonas syringae* pv. *morsprunorum* (Wormald) Young et al.**

Basic description

P.s. pv. *morsprunorum* is distinguished from pv. *syringae*, with which it has been associated on stone-fruit trees, by its white growth in nutrient sucrose broth, inability to hydrolyse arbutin, utilization of tartrate but not lactate as a sole carbon source and sensitivity to certain specific phages (Garrett *et al.* 1966).

Host plants

P.s. pv. *morsprunorum* attacks *Prunus* species and is chiefly of importance on sweet cherry, sour cherry, plum and myrobalan. It can also be a problem on other stone-fruits such as almond, peach and apricot and may occur on ornamental *Prunus* species.

Host specialization

The bacterium occurs in specialized host-adapted strains. In England, those on cherry are distinguished from those on plum and each other by their sensitivity to certain specific (A7 group) phages and their capacity to invade through both leaf scars and wounds. In contrast, plum strains are insensitive to these phages and infect only through wounds. Two races differing in phage sensitivity and some biochemical characteristics show pathogenic specialization for individual cultivars of cherry. Host-adapted types have also been reported from apricot and, in Yugoslavia, from plum.

Geographical distribution

P.s. pv. *morsprunorum* occurs mainly in Europe. where it is widespread (CMI 132). Its presence has also been confirmed in recent years in Australia and S. Africa, and in the USA.

Disease

P.s. pv. *morsprunorum* causes a typical dieback and canker of its hosts, the symptoms being essentially indistinguishable from those caused by pv. *syringae*. The bacterium has a well defined life cycle with a winter canker stage alternating with a summer leaf-spot stage. In cherry and myrobalan, cankers originate in autumn or winter. They are usually located on branches, arising mainly from infection of leaf scars of fruiting spurs, or else are found in the crotch. Cankers first become visible in spring as shallow depressions in the bark and, on hosts such as cherry, globules of amber-coloured gum often appear on the surface of the canker. The production of gum increases later in the season and may be the first sign of infection on older, thickened branches. Dieback can occur at any time in the growing season when a canker girdles a branch or trunk; the time of its occurrence depends on the age of the branch affected and on how early it was girdled. Buds on young branches may fail to break or leaves wither very soon after bud break but, on older branches, or when a trunk is girdled, dieback may not be apparent until late July.

Plum trees are most susceptible for their first six years, particularly to stem infections. Leaf scar infection does not occur on this host and gum production is less common and less conspicuous than on cherry. Cankers may extend the entire length of the stem where, if confined to one side, they appear as strap-like depressions in the bark. If the stem is girdled, death of the tree ensues, usually by July at the latest.

Cankers are not normally perennial and trees are immune to canker infection in the summer when the host is meristematically most active. The first infections of the summer leaf-spot stage occur soon after petal fall on the young leaves of fruiting spurs of cherry and a little later on plum. Leaves developing on extension shoots may be infected during the summer. Once leaves mature they are resistant to further infection. Leaf spots are dark brown, circular or slightly angular and *c.* 2 mm diameter; adjacent spots sometimes coalesce to form larger patches of dead tissue, especially near margins or leaf apices. Spots are sometimes surrounded by a yellow 'halo', particularly on plum. In summer the spots drop out to give a 'shot-hole' effect.

Epidemiology

Like many other bacterial diseases, bacterial canker is favoured by wet weather. Rain, particularly when accompanied by strong winds, mobilizes and distributes the abundant epiphytic population of the pathogen. This may lead to stomatal infection giving rise to leaf spots; these occasionally reach epidemic proportions in early spring. In autumn, inoculum washed by rain onto fresh leaf scars is sucked into the broken ends of leaf trace vessels by negative tension in the vascular system, especially when strong winds have caused premature defoliation.

Leaf-surface populations tend to increase during cool, wet weather and to decrease during hot, dry conditions. In areas where such conditions prevail for long periods in the summer, e.g. in S. Europe and Mediterranean countries, numbers fall to undetectable levels until cooler conditions return. The bacteria overwinter in association with the host, either saprophytically in buds or in woody tissues e.g. following leaf-scar infection. Branches or stems may also be infected via wounds, for example from mechanical damage, tie chafing, animal grazing, cracks from winter injury or natural openings such as lenticels. The longest cankers result from infection during the period of maximum dormancy which apparently coincides with very small populations of the pathogen.

Economic importance

Losses are difficult to estimate but, in areas where conditions favour the disease, mortality among young trees, especially of plum, may be high. On other hosts infection in the early years may severely affect tree form. Much of the infection on older trees is slight, causing death of small branches and fruiting spurs, with some consequent crop loss that is difficult to evaluate. Where large branches of older trees are girdled the loss is more readily apparent.

Control

The dearth of bactericides and prohibition of the use of antibiotics in most European countries means that the only effective compounds known are copper-based fungicides, for example Bordeaux mixture in its various formulations. The spring leaf-spot phase of the disease may be controlled by weak (1:1½:100) Bordeaux mixture containing a vegetable oil (e.g. cotton seed oil or rape seed oil at 3 l per 400 l) to minimize phytotoxicity but rarely does the risk of damage by the leaf-spot stage of the disease warrant the risk of using a spray to which some hosts or cultivars are sensitive. Autumn leaf-scar infection of cherry can be controlled if the leaf surface population is severely reduced by three Bordeaux mixture sprays applied at approximately three-week intervals covering the period from the beginning to leaf fall using increasing copper concentrations as the time of natural defoliation approaches, viz. 4:6:100, 6:9:100 (both with vegetable oil to prevent premature defoliation and consequent deleterious effects on tree vigour), and finally, when leaf fall is imminent, a 10:15:100 Bordeaux mixture. There is no reliable recommended spray schedule for plum although winter sprays of Bordeaux mixture have been suggested.

No hosts are completely immune to the disease but trials in different countries using ranges of the most widely grown cultivars have revealed a considerable degree of resistance in some of them. Many countries have breeding programmes with bacterial canker resistance as one of the criteria on which selection is made. New semi-dwarfing rootstocks have greater resistance to canker than the older more vigorous rootstocks that they are replacing. Since, however, they have to be low-worked to be sufficiently productive, the losses from canker, especially on plum, will continue until new canker resistant cultivars are available. Vigorously growing shoots and branches on trees on good, deep, well-drained soils tend to be more susceptible than those on poorly growing trees or on trees deficient in nitrogen. There is some evidence that low potassium levels favour the disease. Prevention of stem injury by protective guards to prevent animal attacks, and attention to ties of staked trees to stop chafing will reduce possible entry sites for the bacterium.

Special research interests

The stable, host-adapted phage types within this organism have formed a useful system for cross- and mixed-infection studies on the epidemiology of the pathogen. They have also proved valuable in investigations of the genetics of the pathogen and on the role of cell wall components as determinants of virulence and host specificity factors in these organisms (Crosse & Garrett 1970; Garrett *et al.* 1974).

GARRETT

Pseudomonas syringae* pv. *persicae* (Prunier *et al.*) Young *et al.

Basic description

P.s. pv. *persicae* belongs to Group 1a of the fluorescent pseudomonads (Prunier *et al.* 1970). It does not produce fluorescent pigment in King B medium, but a proportion of strains (about 80%) produce a fluorescent pigment on an alternative medium, CSGA (Luisetti *et al.* 1972). It does not hydrolyse aesculin or arbutin, or, with the exception of two new strains, gelatin. In minimal medium it produces acids from glycerol, mannitol, sorbitol and sucrose but not from erythritol or inositol. Organic acids such as DL-lactate, D(−) tartrate and DL(+) tartrate, commonly used for species differentiation within Group 1a, are not metabolized. Malonate is used. *P.s.* pv. *persicae* has many antigens in common with other pathovars of *P. syringae*, namely pvs. *tomato, savastanoi, glycinea, garcae, papulans* and especially *syringae*. For this reason pv. *persicae* antisera have to be used carefully for diagnosis or detection.

Host plants

P.s. pv. *persicae* is pathogenic only to peaches and nectarines. Artificial inoculations of other *Prunus* spp. are unsuccessful.

Geographical distribution

The disease was first observed in S.E. France in 1966 as a limited focus in Ardèche. The disease spread steadily in Ardèche until 1973, then new foci occurred in 1975 on the other side of the Rhône and spread became continuous. Six departments are now concerned: Ardèche, Drôme, Gard, Isère, Loire and Rhône. The other French peach-growing areas have remained unaffected until now. There is no evidence of occurrence in other countries (EPPO 145).

Disease

P.s. pv. *persicae* causes bacterial dieback. The most typical symptom is the development during winter of a greenish to brownish discoloration around buds, mainly on young shoots. Then necrosis extends along the shoot, frequently progressing to older branches and sometimes the trunk. The girdling of the branches or trunk results in a more or less rapid dieback of the tree. This dieback can be observed in early spring or through the growing season, or even during the summer or autumn. On less susceptible cultivars, the extension of lesions is less and only results in cankers on the branches or trunk. Cankers can also occur after infection through pruning wounds. Buds contaminated by pv. *persicae* can be killed during winter allowing the bacteria to invade neighbouring tissues. When located on the trunk or in the crotch, diseased buds can be responsible for a rapid dieback of trees. Leaf spots can be observed especially during rainy springs: they are 1–2 mm wide necrotic spots surrounded by a chlorotic halo. During late summer they develop into leaf holes. On some cultivars, especially on nectarines, the pathogen can also initiate infection of fruits to produce small spots on the fruit surface which are frequently overlaid by gum. Severe attacks on fruit may significantly reduce the quality and quantity of yield.

Epidemiology

P.s. pv. *persicae* enters the tissues through leaf scars or through wounds. These can be pruning wounds or micro-lesions in buds during periods when air temperatures are less then 0°C. pv. *persicae* is able to induce ice nucleation at −3°C to −5°C. Peach tree tissues are susceptible to the pathogen during winter. The probability of infection is highest when contamination occurs at the end of autumn. During the growing season tissues are resistant to the bacterium. pv. *persicae* overwinters in buds and also in cankers. It also has an epiphytic phase; high population levels are recovered from flowers and leaves in spring (10^6–10^7 per leaf) and from leaves in autumn (10^5–10^6 per leaf). These bacteria on the leaf surface are responsible for leaf scar infections initiated during leaf fall. Temperatures below 0°C during winter and deficiency of the soil in some nutrients increase the severity of dieback. Rain in spring greatly favours the development of leaf spots and of spots on fruits. Rain and wind ensure an efficient spread of inoculum.

Economic importance

The disease is economically very serious, killing young trees (less than 5–6 years old), potentially within one winter; these are obviously the most susceptible. More than one million dead trees were recorded from 1966 to 1982.

Control

Prophylaxis is one major way of protecting peaches against infection, basically involving eradication and incineration of diseased trees or branches. There are no cultivars resistant to bacterial dieback, but some are tolerant. Peach growers have been advised against planting very susceptible cultivars. Since pruning wounds have been shown to be responsible for canker development, late pruning has been recommended. Tools have to be disinfected frequently to avoid contamination of wounds and large wounds have to be protected by a disinfectant mixture. Chemical control can be achieved by spraying copper oxychloride, sulphate or hydroxide during leaf fall. The most efficient protection is obtained with 4 or 5 sprays (copper concentration of 125 g/hl); the first spraying has to be made at just 5% leaf fall (Prunier *et al.* 1974).

Special research interest

A toxin is involved in pathogenicity. It is different from syringomycin or syringotoxin produced by strains of the syringae pathovar. Variation in pathogenicity between strains has been recorded. Strains are a mixture of avirulent and virulent cells and the proportion of virulent cells in a strain determines its virulence. Studies are being made on the effect of some factors (e.g. host, soil, climate) on this balance (Luisetti *et al.* 1981).

LUISETTI

Pseudomonas syringae pv. *phaseolicola* (Burkholder) Young *et al.*

Basic description

Pseudomonas syringae pv. *phaseolicola* is a typical member of LOPAT group Ia of Section I.

Host plants

Dwarf and scarlet runner beans are the important crop plants attacked but *Phaseolus lunatus var. macrocarpus, P. atropurpureus, P. radiatus, P. aureus, P. multiflorus, Pueraria thunbergiana* and *Pachyrrhizus erosus* are also reported to be natural hosts. However caution is needed in interpreting these reports since the host range of pathovars of *P. syringae* on leguminous hosts is uncertain and can overlap.

Host specialization

Two races (1 & 2) of the pathogen are generally recognized on the basis of pathogenicity to dwarf bean cv. Red Mexican U1-3; the races differ in reaction to phage 12S (Hale & Taylor 1973). Race 1 is much more common on scarlet runner and race 2 on dwarf bean in some countries. 'Halo-less' strains of the pathogen have been reported but failure to produce the characteristic halo symptom in plants can be due to high ambient temperatures rather than a non-toxin producing strain.

Geographical distribution

It occurs in all regions where dwarf and runner beans are grown commercially.

Disease

The most characteristic symptom of the disease (halo blight) is a small, angular, dark water-soaked spot on leaves that is often surrounded by a wide, pale-green or yellowish-green halo up to 2.5 cm diameter due to the production by the pathogen of phaseolotoxin. These spots can coalesce and later become brown and dry. Halo production is dependent on strain and on ambient temperature (optimum 16–20°C), is inhibited at more than 22°C and is apparently mediated by the inhibition of ornithine carbamyl transferase, possibly also the mechanism by which pv. *phaseolicola* severely inhibits nodulation of dwarf bean by *Rhizobium phaseoli* (Hale & Shanks 1983). On stems, spots are often sunken. On pods, round, water-soaked spots occur that are initially greasy (hence the common name grease spot) and up to about 1 cm diameter; they can coalesce and later become dry and irregular and brick-red to brown. These symptoms are often similar to those of common blight (*Xanthomonas campestris* pv. *phaseoli*, q.v.) which, however, never produces the haloes characteristic of halo blight and the exudate of which is yellow rather than silvery or creamy. Systemic symptoms are often limited to stunting, reversible wilting, chlorosis, foliar mosaic and leaf malformation particularly when they arise directly from seed-borne infection.

Another common systemic symptom in plants from infected seeds is girdling of the stem, or joint rot at early pod formation: small, water-soaked areas appear at the primary leaf node, enlarge to encircle the stem and later become amber coloured. Plants may break at this node. Seed infection can also produce seedlings with injured or dead growing tips or brown-black irregular lesions on the cotyledons. If systemic invasion is by way of small veins in the leaf, reddish water-soaked areas can develop adjacent to the veins; if it is by way of the petiole, the main leaf vein and its branches appear water-soaked at first and later become brick-red. Stem infection gives rise to reddish longitudinal streaks on the surface that often split. Pod spotting may involve the vascular elements of dorsal and ventral sutures causing elongated lesions of the surrounding tissue. Infected seeds may be symptomless, have discoloured areas on testas (which are only readily seen on white or light-coloured seeds), or by shrivelled and wrinkled. Symptoms are well described by Zaumeyer & Thomas (1957). Taylor (1970) has claimed that pv. *phaseolicola* can be readily and routinely identified by the use of three specific phages and a heat-stable antigen.

Epidemiology

The primary foci in crops are from infected seed or from infected weed hosts; overwintering in soil or in plant debris appears uncommon or absent in most climates. The disease is spread in the crop by rain splash from surface lesions on infected plants and this secondary spread can extend as much as 30 m downwind from a single primary focus in wind-driven rain. As few as one focus in 10000 plants can therefore cause epidemics in crops if conditions are favourable. However, not all infected seeds give rise to infected plants and under favourable weather some 1:2000 seed infection gave significant spread when 1:20000 gave no infection (Trigalet & Bidaud 1978). Although the disease is often systemic there is evidence (Taylor *et al.* 1979a) that seed infection is mainly by direct penetration of the pod from surface lesions. Seed contamination with infested debris also occurs and can be difficult to eliminate by seed disinfection (Grogan & Kimble 1967).

Economic importance

In N. America, where most cultivars carry some resistance to halo blight, the disease, although less important than common blight, is still a major disease of beans causing an estimated loss in 1976 of $2 million (Kennedy & Alcorn 1980). In Europe, where cultivars are more susceptible, it is probably the most important disease of dwarf and runner beans. Yield losses of 12.5% and 34% pod infection from 0.4% primary infection and 43% loss and 62.5% pod infection from 2.6% primary infectors are reported by Keyworth (1969).

Control

Treatment of seed with streptomycin or kasugamycin slurries (Taylor & Dudley 1977) or dry heat (Belletti & Tamietti 1982) gives good but not always adequate control of seed-borne inoculum. Testing seed by bacteriological methods or host assay is possible but the sample needed to detect say 1:10000 infected/infested seed is impracticably large. The size of samples of white-seeded cultivars can be reduced by initial screening with UV, which concentrates affected seed into samples small enough for testing (Wharton 1967; Parker & Dean 1968). The best method of control in high rainfall areas is still the production of seed stocks in arid climates using furrow irrigation and field inspection; but even when this is combined with seed treatment the elimination of inoculum in stocks cannot be achieved in a single growing season (Grogan & Kimble 1967). 2–4 sprays of copper compounds or streptomycin from flowering to pod set have given good control of foliar and pod infection in N. America, Europe and New Zealand but modelling by Taylor *et al.* (1979b) indicates that in some countries routine spraying may not be worthwhile economically if seed is treated. Breeding tolerant cultivars. is probably the best long-term control measure in Europe.

LELLIOTT

Pseuomonas syringae pv. *pisi* (Sackett) Young *et al.*

Host plants

The economically important hosts are field and garden pea. *P.s.* pv. *pisi* also attacks *Vicia benghalensis*. It has been reported on species of *Lablab*, *Lathyrus* and *Vigna*, but in view of the difficulty in differentiating the pathovars of *P. syringae* on Leguminosae these reports should be treated circumspectly.

Host specialization

Five races of pv. *pisi* are known based on differential reaction to pea cultivars.

Geographical distribution

The disease caused by pv. *pisi* occurs in most pea-growing areas and is particularly common in N. America. In Europe, it occurs locally in Bulgaria, Greece, Hungary, Italy, Romania, USSR and Yugoslavia. It has occurred sporadically in many countries of N.W. Europe (EPPO 57).

Disease and epidemiology

The disease (Young & Dye 1970) is seed-borne and primary foci arise in fields most commonly as plants growing from infected or infested seed. Less frequently, except in fields annually cropped with peas, the primary foci may derive from overwintered diseased plant debris, diseased volunteer pea plants or plants of the weed host, purple vetch, which can be severely attacked. Spread from primary foci can be rapid during driving rain, particularly when transient frost damage has increased susceptibility of plants to infection (Boelama 1972); spread also occurs in drainage and irrigation water, dust storms, and by contact, cultivation and harvesting machinery and possibly insect vectors. In rainy weather, lesions on leaflets and pods are initially small, round, oval or irregular, dark-green, water-soaked spots; they later enlarge and become angular being delimited by veins. Lesions may produce a creamy bacterial ooze that can dry to give a glossy sheen. On leaflets, lesions become yellowish, then brown and papery. On ripening pods, lesions are sunken and greenish-brown and may be limited to a narrow band along the sutures; invasion of the inside of the pod via the dorsal suture may cause the seed to become covered with bacterial slime or discoloured brown-yellow, and shrivelled with a water-soaked spot near the hilum. Infection of sepals often occurs and can spread to flowers; flower buds may be killed. On stems and petioles purple-chocolate streaks can develop, coalesce and cause shrivelling of stems. In dry weather with occasional frost, symptoms usually start on the stem near soil level as water-soaked, later olive-green to purple-brown spots that extend upwards to stipules and leaflets. The veins of these become brown to tan with a characcteristic fan-like pattern, and the interveinal tissues are water-soaked, then turn yellow to brown, finally drying out and becoming papery (Boelama 1972). Infection of foliage is by way of stomata and wounds; on stems the pathogen can invade not only the cortex but also the pith. Seed, which is invaded by way of the funicle and micropyle as well as externally contaminated, can also become infected or infested at harvesting by contact with diseased seed from contaminated viners and seed-dressing machines. The related *P. syringae* pv. *syringae* can cause symptoms under some autumn/winter conditions indistinguishable from some of those of pv. *pisi* (Taylor & Dye 1972).

Economic importance

In N. America, New Zealand and Argentina there have been periods when bacterial blight has caused considerable losses (in Wisconsin for example, it was second in importance only to root rots) and some severe attacks have been recorded in E. Europe. It is difficult to understand why the disease is not serious and occurs so rarely in W. Europe, unless it is that the races of pv. *pisi* common in regions where W. Europe multiplies and from which it imports much of its seed are of low virulence to the cultivars grown in W. Europe.

Control

Control is possible by seed treatment with streptomycin but this is prohibited by the health authorities of most countries. Failing this, seed certification and particularly growing seed in dry areas under furrow irrigation are the main recommended control measures.

LELLIOTT

Pseudomonas syringae pv. *porri* Samson & Benjama

P.s. pv. *porri* attacks only leek (*Allium porrum*). It has been reported in England, New Zealand and France (Lelliott 1952, Samson *et al.* 1981, Samson & Benjama 1986). The disease is a typical bacterial blight of the aerial parts of the plant. On young leaves, longitudinal water-soaked then yellow stripes are produced that may split and rot. As the healthy part of the blade goes on growing, it rolls up, giving the typical curly outline of young diseased plants. Old leaves show only water-soaked yellow spots, often associated with wounds. Flowering stems are very susceptible; they develop large, deep, green, water-soaked areas, often with a gummy exudate, giving yellow and brown, depressed lesions. In seed crops, stem lesions cause crank-shaped distortions and, if they girdle the stem, death of flower heads.

According to weather conditions, the disease can affect any stage of the plant. Inoculum sources would appear to be the seeds. In areas where leeks are produced all the year round, dissemination comes from the neighbouring diseased plots. However, whether infected seeds give rise to infected individual plants or whole beds depends on weather, soil conditions and fertilizer use. Losses are noticeable in large beds where extension of the first foci may lead to abandonment of parts of beds at the seedling stage. No estimates of yield loss are available. Disease incidence may be lowered by sprays of cupric products and cultural practices such as moderate fertilization.

SAMSON

Pseudomonas syringae* pv. *sesami (Malkoff) Young *et al.* causes a seed-borne leaf spot of sesame in E. Asia, the Americas, Africa, India, Bulgaria, Greece and Yugoslavia (Šutić & Dowson 1962). Two physiologic races have been reported.

Pseudomonas syringae* pv. *striafaciens (Elliott) Young *et al.* causes a disease very similar to that caused by *P.s.* pv. *coronafaciens* (q.v.) except that the lesions are more stripe-shaped than oval. Some workers (Tessi 1953; Harder & Harris 1973) consider the two to be synonymous and that the same organism causes a spectrum of syndromes.

Pseudomonas syringae pv. *syringae* van Hall

Basic description

The bacterium can be distinguished from saprophytic pseudomonads by the LOPAT tests (Lelliott *et al.* 1966) and from most other fluorescent plant pathogenic pseudomonads by utilization of L-lactate as the sole carbon source.

Host plants

P.s. pv. *syringae* attacks a very wide range of host plants. In Europe, it causes bacterial canker of stone fruits, the symptoms of which are indistinguishable from those caused by pv. *morsprunorum* (q.v.) (Cameron 1962); blossom blight of lilac, pear and citrus; leaf and flower blights of cereals including millets, sorghum, maize and wheat; blister spot of apple; leaf spots and blights of herbaceous hosts including clover, cowpea, lucerne, black mustard, coriander and pepper; shoot dieback of shrubs and trees including avocado, walnut, hawthorn and the ornamentals rhus, whitebeam, birch, cotoneaster, oak, laurel, pyracantha and magnolia.

Host specialization

Since isolates from different hosts vary in host range and no one isolate will infect all recorded hosts, it is likely that pv. *syringae* should be divided into several different pathovars. Host-related differences in phage sensitivity have been recorded. Isolates vary in their ability to produce syringomycin but *in vitro* toxin production is apparently not correlated with virulence in the field.

Geographical distribution

The pathogen is widely distributed on stone-fruits and citrus throughout Europe and other temperate fruit-growing areas of the world (CMI 336). Its distribution on some hosts, e.g. bean, is more restricted.

Disease and epidemiology

Symptoms on pear are sometimes confused with those caused by fireblight (*Erwinia amylovora* q.v.) especially when the infection spreads down the spur (Billing *et al.* 1960). Brown spot of bean caused by pv. *syringae* (Rudolph 1979) differs from halo-blight (pv. *phaseolicola*) in having no halo surrounding the reddish-brown spot. pv. *syringae* is readily spread in wind-driven rain and penetrates host tissues through stomata, nectaries or other openings such as wounds caused by pruning, insects or hail damage. It causes spotting of young leaves, stems, fruits, and cankers, shoot blight, and dieback of woody hosts. Brown spot of bean can be seed-borne. Low-temperature injury of blossom or fruitlets is a predisposing factor in citrus blast and blossom blight of pear and lilac. Sub-zero winter temperatures favour bacterial canker of stone-fruits, especially of apricot in E. Europe. The ice nucleating ability of pv. *syringae* increases frost damage on a number of hosts. The survival of the pathogen as an epiphyte is favoured by cool, moist conditions, and numbers decline when the weather is hot and dry.

Economic importance

Losses from this pathogen are not well documented and are difficult to ascertain but, when conditions are favourable for the bacterium, crop losses can be high. Cases of 40% death of 5- to 6-year-old apricots are reported from Hungary, and massive dieback in Yugoslavia; up to 25% loss of nursery cherry and plum are reported from Norway. In its totality, this pathogen causes more crop loss than is generally realized.

Control

The only effective control of the bacterium is by applying Bordeaux mixture in its various formulations. Timing and concentration will vary with crop but should be aimed at reducing the epiphytic population of the pathogen at those periods when the host is most susceptible to infection, and using concentrations that will not be phytotoxic. Pruning after bud burst may help to reduce incidence of canker and dieback on apricot and peach.

GARRETT

Pseudomonas syringae pv. *tabaci* (Wolf & Foster) Young *et al.*

P.s. pv. *tabaci* causes a major disease of tobacco in all the main tobacco-growing regions of the world. Physiologic races probably exist, since Ono (1979) described a strain pathogenic to resistant cultivars in Japan. Non-toxin producing strains occur naturally and were previously ascribed to '*P. angulata*'. Early accounts of pathogenicity to soybean should be discounted (Clayton 1950). The widely used name for the disease, wildfire, aptly describes the rate at which it can spread in seed beds in which it exists either as a rot of young leaves which spreads to the whole plant or, on older leaves, as a leaf spot surrounded by a chlorotic halo typical of diseases caused by the *P. syringae* pathovars.

The disease is seed-borne and can overwinter in plant debris. Spread and infection is promoted by cool, wet weather and strong winds. The most serious attacks usually occur in the seed and nursery beds where the disease can reach epidemic proportions. In the planted-out crop it reduces both quantity and quality of leaves. It is one of the most important diseases of tobacco in many parts of the world and in Europe has been rated a major tobacco disease in the USSR, Germany, Yugoslavia and Bulgaria. In those countries where its use is permitted, spraying with 200 ppm streptomycin can give almost complete control. Seed treatment, disinfection of seed bed frames and, where feasible, soil steaming of seed beds and protection of seed beds from wind, combined with three applications of Bordeaux mixture (4-2-50) at 40–50 l/100 m^2 of seed bed at 10- to 14-day intervals starting just before emergence, gives good control. Resistant cultivars have been bred (Raeber 1966).

LELLIOTT

Pseudomonas syringae pv. *tomato* (Okabe) Young *et al.*

Basic description

The bacterium is distinguishable from the other members of the LOPAT group Ia only by its host specialization.

Host plants

Tomato is the only host.

Geographical distribution

The bacterium has a world-wide distribution, presumably because of its seed-borne nature, and has been reported in almost every country where tomatoes are cultivated.

Disease

P.s. pv. *tomato* causes bacterial speck and infects almost every part of its host. On leaves, the symptoms are small (1–3 mm diameter) dark-brown to black spots which are usually surrounded by a distinct, yellow halo. This chlorotic halo can be very wide, when it is correlated with the production by the pathogen of the chlorosis-inducing toxin coronatin, or very narrow, when it is correlated with the production by the pathogen of another type of toxin chemically unrelated to coronatin. The spots, which occur all over the leaf surface, are more numerous at the periphery of the leaf blade. The lesions may coalesce to form irregular, dark-brown necrotic patches, mainly at the periphery, or necrotic streaks along the main or secondary veins of the lamina. On the stems, peduncles and leaf petioles, irregular, dark-brown, water-soaked lesions are formed. These lesions coalesce and form large blotches which are restricted to the epidermal tissues and do not penetrate into the cambial and vascular tissues. The blotches are not surrounded by a yellow halo and in the early stages resemble the symptoms caused by *Alternaria solani* or *Phytophthora infestans*, as do the blotches at the leaf edges. On fruits the symptoms are small, irregular, dark-brown to black slightly raised lesions which are superficial and restricted to the epidermal tissues. The bacterium can also invade the vascular tissues causing a severe stunting and a slight discoloration of the xylem vessels. It has also been reported that the pathogen infects the roots causing discoloration.

Epidemiology

P.s. pv. *tomato* is seed-borne and persists on infected seeds for as long as 20 years. It exists on seeds as cell aggregates inside holes and cavities and is capable of initiating disease if infected seeds germinate under favourable conditions (Devash *et al.* 1980). The pathogen is also said to be soil-borne although not necessarily soil-inhabiting; it has been found in the rhizosphere of many non-host plants (Schneider & Grogan 1977). From soil and infected seeds, the bacterium colonizes tomato leaves and other aerial surfaces where it can establish and maintain itself as a resident under adverse environmental conditions for extended periods of time without any visible disease symptoms. Leaf trichomes serve as a habitat during dry conditions for the survival of stable resident populations of the pathogen sufficient to cause an epidemic (Schneider & Grogan 1977). When the environmental conditions favour bacterial multiplication, the epiphytic population increases rapidly. The pathogen enters the plant tissue through the stomata, hydathodes or wounds and infects green, uninjured fruits through openings that remain after trichomes are shed and before the cuticle is fully developed. Prerequisite factors for disease initiation and symptom development are temperatures between 17°C and 25°C, and relative humidity above 80%, or free water on the leaf surface. The pathogen can invade the plant tissue when conditions cause water condensation on leaves at night.

Differences in the susceptibility of plant tissues have been reported. Plants at the cotyledonary or three-leaf stage are more susceptible than at older stages. The succulent tissues of greenhouse-grown tomatoes are more susceptible than the harder tissues of field-grown tomatoes. Fruits are infected only at the green stage, and red mature fruits are not susceptible. In the field, fruits are susceptible at the open corolla stage but green fruits (≤ 3 cm diameter) are only slightly susceptible. In the greenhouse, fruits are susceptible from the open corolla through all of the green fruit stage, green fruit (< 3 cm diameter) being the most susceptible stage under these conditions.

The spread of the disease in the crop is accomplished by rain or sprinkle-irrigation droplets which, on contact with infected plant surfaces or soil, become contaminated and transfer the pathogen to other plants for a distance that depends on the size of the droplet and its velocity.

Economic importance

The economic losses caused by bacterial speck arise from destruction of young plants, reduction in yield and reduction of market value both in processing and table tomatoes. Because of foliar damage, bacterial speck delays fruit maturity and reduces yield; disease-defoliated plants may produce small, sun-scalded fruits of poor quality. Yield losses depend on the developmental stage of the plant at the time of infection and the environmental conditions prevailing. Yield losses as high as 75% in plants infected at an early stage of growth have been reported from Israel. Because of the high sensitivity of the plants at the early stages of growth and the ability of the disease to spread rapidly from a point source of inoculum under favourable conditions, losses of seedlings in seed beds can be high: infection of tomato plants at the three-leaf stage resulted in 13% losses, while infections in younger seedlings caused much higher losses (Schneider *et al.* 1975).

Control

The first consideration in controlling bacterial speck of tomatoes is to obtain pathogen-free seed. In order to avoid or lower the probability of contamination, seeds should be produced in arid areas without overhead irrigation. Seeds should be disinfected by hot water treatment before sowing, classically 50°C for 25 min but other treatments have also been recommended, e.g. 48°C for 1 h (Devash *et al.* 1980) or 56°C for 30 min (Goode & Sasser 1980) for simultaneous control of *Corynebacterium michiganense*. Since some tomato cultivars are heat sensitive, the resistance of a cultivar to heat treatment should be tested before applying the method. Regular spray applications of bactericides at the early stages of plant growth or immediately after disease detection are recommended. Best results have been obtained with streptomycin sprays especially when the weather conditions are conducive for bacterial speck development. Copper compounds are recommended as a replacement for streptomycin, where its use is not allowed in agriculture. Alternate sprays with streptomycin and copper compounds should be considered because of the risk that streptomycin-resistant strains may develop if the former were used alone. Copper-maneb mixtures can give better control than copper alone. A three-year crop rotation with non-host plants and sanitation practices should be included in any control programme. Good aeration in greenhouses to reduce humidity, and restricting the incorporation of plant debris in the soil at the end of the growing season may help in reducing disease incidence. The use of resistant cultivars may also be considered (Gitaitis *et al.* 1982). Tomato cultivars exhibiting different degrees of resistance to bacterial speck have been found. Ont. 7710 and Rehovot-13 are resistant, Marmade and Cambell 28 exhibit a high degree of resistance while Hosen-Eiton and Ohio 7663 are moderately resistant cultivars.
PSALLIDAS

Pseudomonas syringae* pv. *tagetis (Hellmers) Young *et al.* causes large, necrotic leaf spots, with or without chlorotic haloes and an apical chlorosis on growing tissue of African marigold (Styer & Durbin 1981). The apical chlorosis appears to be caused by tagetitoxin, which is distinctly different from other *P. syringae* toxins (Mitchell & Durbin 1981).

Pseudomonas syringae* pv. *ulmi (Šutić & Tešić) Young *et al.* causes a leaf spot and shoot blight of elm in Yugoslavia that in wet seasons can cause complete withering of leaves and shoot tips (Šutić & Tešić 1958).

Pseudomonas syringae subsp. *savastanoi* (Smith) Janse

Basic description

The bacterium is a member of LOPAT group Ib and there is adequate evidence for considering it a subspecies rather than a pathovar of *P. syringae* (Janse 1982).

Host plants

It attacks olive (olive knot), ash (ash canker), oleander and Japanese privet, and although there is evidence from several workers (Šutić & Dowson 1963; Bottalico & Ercolani 1971; Urošević 1976) that isolates from these hosts are cross-infective to some or all of the hosts, Janse (1981, 1982) has proposed pvs *oleae, fraxini* & *nerii* on olive, ash and oleander respectively.

Geographical distribution

The diseases caused on olive, ash and oleander are widespread and probably occur whenever these hosts occur extensively; in Europe it is frequent on ash (Janse 1981), occurs generally and commonly on olive and is reported on oleander from Austria, Denmark, France, Greece, Italy, Yugoslavia and USSR and on privet in Italy (Bottalico & Ercolani 1971).

Disease

On all hosts, hyperplastic galls, associated with bacterial production of indoleacetic acid, develop on young twigs and shoots, almost exclusively at the point of emergence of new leaves or on leaf scars, pruning cuts, wounds and abrasions. Initially galls are small swellings which develop within a few months into more or less round, irregularly fissured, ochre knots, spongy at first and becoming hard and brown; on shoots and small branches galls are up to about 2 cm diameter. On olive petioles and leaf midribs of leaves, small knots or tubercles may develop and cause premature abscission. Roots and trunks can also be affected. Affected terminal shoots are stunted or killed and trees can be so debilitated by dieback as to be unproductive or die.

Epidemiology

Bacteria are abundant in and exude from developing galls and are spread from them by rain splash. On olive they have a resident epiphytic phase on leaves, reaching maximum populations in April and October in studies by Ercolani (1978). Infection of olive is directly related to the degree of wounding (from whatever cause); incidence is greatest in windy places, in small-fruited cultivars that are harvested by beating trees with sticks, after hailstorms and frost and where pruning is careless. There may be insect vectors. Moist winds in coastal areas favour infection.

Economic importance

Olive knot is a major, if not the most damaging, olive disease in most olive-growing regions. It can seriously reduce yield to a point at which plantations are no longer economically viable. There is evidence (Schroth *et al.* 1968) that the presence of knots on trees reduces the quality of olive fruits from the whole tree by imparting a detectable off-flavour. Ash trees can be badly disfigured.

Control

Without pruning, olive trees yield poorly and, since pruning wounds are common avenues of infection, control is difficult. In Sicily, and by inference much of the Mediterranean region, annual pruning, during which galled shoots are removed, should be completed by February (or before epiphytic populations increase). The main fertilizer application should be in January–February (Barratta & Marco 1981). Spraying with copper compounds in late autumn and in spring is advisable, although timing to reduce inoculum and to coincide with the main infection periods is important. Pruning tools should be disinfected on entering a relatively healthy orchard, but this practice is impractical between trees in a diseased orchard.

LELLIOTT

Pseudomonas viridiflava (Burkholder) Dowson

P. viridiflava is closely related to, but distinct from *P. syringae* (Billing 1970). Cultures in collections are frequently labelled *P. syringae* from which it is usually readily differentiated by its green-centred non-levan colonies on 5% sucrose nutrient agar. Pathogenicity and sometimes green pigment formation are rapidly lost in culture. It was originally isolated and described from dwarf bean (Burkholder 1930), causing stem dieback and galling above and below the point of inoculation. This symptom appears to be a laboratory curiosity never reported from the field. In fact, the pathogen is an opportunist pathogen of soft plant tissue, which is probably ubiquitous and causes diseases similar in many respects to the soft rot and necroses caused by *Pseudomonas marginalis* (q.v.) and necroses caused by *P. cichorii* (q.v.) from which they cannot be visibly distinguished. It is common in temperate climates as a cause of soft rots and leaf necroses of many plants and soft rots of storage tubers, and as an epiphyte and secondary invader of diseased tissue. See Wilkie *et al.* 1973; Lelliott & Stead 1987.

LELLIOTT

Pseudomonas woodsii (Smith) Stevens causes a leaf, stem and flower-bud spotting of carnation (Elliott 1951), occasionally reported from countries that grow carnations extensively commercially. It is of little economic significance. It is closely related to, if not a synonym of, *P. andropogonis* (q.v.).

XANTHOMONAS

The genus is closely related to *Pseudomonas* from which its members differ in that they only have one polar flagellum, in never denitrifying or reducing nitrate, in producing highly characteristic yellow pigments (xanthomonadins) which are brominated aryl polyenes, in not utilizing asparagine as a sole source of C and N, in requiring growth factors (usually including methionine, glutamic acid, nicotinic acid or a combination of these), in their growth being inhibited by 0.02 or 0.1% triphenyltetrazolium chloride and in being all plant pathogens. The molar percentage G+C of the DNA is 63–71 (Tm, Bd). Two of the species, particularly *X. campestris*, produce large amounts of extracellular polysaccharides known as xanthan gums, which are widely used in industry, particularly the food and oil-drilling industries. There are 6 species (if *X. populi* is included) which are readily differentiated from members of other genera with the exception of *Pseudomonas maltophilia** and from each other (Table 6.2) (Lelliott & Stead 1987). The data in Table 6.2 largely replaces the *Basic description* section of individual entries.

Table 6.2. Differentiation of European species of *Xanthomonas*

	X. campestris	*X. fragariae*	*X. ampelina*	*X. populi*
Mucoid growth on nutrient agar +1% glucose	+	+	−	+
Xanthomonadins produced	+	+	−	+
Hydrolysis of				
gelatin	d	+	−	−
aesculin	+	−	−	o
starch	d	+	−	o
H_2S from peptone	+	−	d	−
Urease activity	−	−	+	−
Growth on nutrient agar				
good	+	−	−	−
poor	−	+	+	−
no growth	−	−	−	+
Max. growth temperature (°C)	35–39	33	30	27.5

Symbols: + = 90% or more strains positive
− = 90% or more strains negative
d = 11–89% strains positive
o = no information

One species, *X. campestris*, exists as many pathovars most of which have high host specificity. Typically these are seed-borne pathogens that cause leaf spots but many can also invade the vascular system causing wilts and dieback. Some are known to exist epiphytically and this may be a common characteristic of the pathogens of this species. The name *X. populi* does not appear to have been validly published but there seems little doubt that this bacterium is a valid member of the genus. For a detailed treatment of the genus see Bradbury (1984). For diagnosis of the diseases see Fahy & Persley (1983) and Lelliott & Stead (1987).

**Pseudomonas maltophilia* is frequently isolated from human specimens and appears to be an opportunistic pathogen of man. It is also found in water, milk and frozen food and in the rhizosphere of plants, and seems identical with the type strain of *P. hibiscicola*, a reported plant pathogen; it is closely related to members of *Xanthomonas* (Bradbury 1984) and its inclusion in the genus has been proposed.

***Xanthomonas ampelina* Panagopoulos**

Basic description

X. ampelina is an atypical member of the genus *Xanthomonas*. It is most readily and quickly distinguished from the four other species of the genus by its production of a brown diffusible pigment on yeast-dextrose-chalk agar (also produced by a few strains of *X. campestris*), on which it grows less poorly than on other media, and by its positive urease activity. Other differential characters are given in Table 6.2. It grows very slowly; up to to 15 days incubation may be required before colonies appear which are 1 mm in diameter.

Host plants

The pathogen attacks only the European grapevine (*Vitis vinifera*).

Geographical distribution

The pathogen occurs in various regions of Greece, France, Spain, Italy (Sicily, Sardinia), Portugal, Turkey and S. Africa (EPPO 133). Moreover, a disease syndrome very similar to that due to *X. ampelina* has been reported from Austria, Switzerland, Yugoslavia, Bulgaria, Tunisia and the Canary Islands and attributed to *Erwinia vitivora*, a synonym of the ubiquitous saprophyte *E. herbicola*. The pathogen therefore very probably occurs in these countries too (CMI 378).

Disease

X. ampelina causes a vascular bacteriosis and leaf spotting of vine. It is known as bacterial blight or bacterial necrosis. It is also known as 'maladie d'Oléron', 'nécrose bacatérienne' and 'carbou' in France; 'tsilik marasi' in Greece; 'mal nero' in Italy; 'necrosis bacteriana' in Spain; 'mal negro' in Portugal; and 'vlamsiekte' in S. Africa. The disease was first reported in Italy in 1879 and in France in 1895. Its nature was confused (usually thought to be caused by *E. vitivora*, a saprophytic bacterium) until 1969 when the true cause was isolated and identified (Panagopoulos 1969). The most conspicuous symptoms of the disease are observed from early spring until mid-summer. In early spring, bud-break on affected spurs either does not occur because spur and/or branches are completely dead, or is delayed. Other spurs on the same plant grow normally. Bud breaking may occur in a few spurs only, and some of the developing shoots may be stunted, weak, chlorotic with one-sided dark-brown streaks, and eventually these will wilt and die. During this period, affected branches and spurs appear slightly swollen due to hyperplasia of the cambial tissue. The more vigorous shoots or canes show longitudinal cracks, starting from the lower internodes and extending slowly upwards. These internally borne cracks, which are surrounded by dark-brown to black necrotic tissue, become deeper, extend to the pith and develop into cankers. Such shoots or canes may show one-sided petiole cracking and sectorial or marginal necrosis of leaves. Cracks and cankers may also appear on main and secondary stalks of flowers and grapes. It is characteristic of the disease that the vascular tissue of shoots, canes, branches, petioles etc. exhibits an often one-sided, reddish or brown, linear discoloration.

Occasionally, on the young tender leaves, brown necrotic spots 1–2 mm in diameter develop, often surrounded by a halo. Symptoms vary according to cultivar and possibly the environmental conditions; in several cultivars shoot and cane cracking and cankers do not occur or occur very erratically. The symptoms of the disease can be confused with those caused by other pathogens such as *Phomopsis viticola*, *Eutypa lata* or the virus diseases fanleaf and black wood. Laboratory diagnosis, which can be accomplished by isolation and identification of the pathogen, is therefore necessary. It is important that small pieces of slightly discoloured vascular tissues are used for isolations and that the isolation plates are carefully examined for up to 10 days for the presence of the slow-growing, barely visible colonies of the pathogen which are often obscured by more vigorous saprophytic bacteria (e.g. *E. herbicola* and/or *Pseudomonas* spp.). Bacteriological tests for identification of the pathogen can be used (see Table 6.2) but the immunofluorescence indirect technique and sensitivity to phage ϕ 15 for rapid identification are also available (Ridé & Marcelin 1983). The pathogen can also be rapidly detected in vine sap using the immunofluorescence indirect technique. The latter method seems to be very useful in disease survey and pathological studies (Lopez *et al.* 1980; Ridé & Marcelin 1983). In addition, three other serological methods can be used for rapid identification of the pathogen, and of these the Ouchterlony gel diffusion test is the most specific (Erasmus *et al.* 1974).

Epidemiology

X. ampelina overwinters mainly in the vascular tissues of affected plants. Bacteria become mobilized in xylem vessels in late winter and spread in the vascular system into healthy branches and spurs and later into new, growing shoots which thus become infected. Cankers developing on new growth provide inoculum for direct leaf infections through stomata during rainy weather or overhead sprinkler irrigation in early spring. However, the most important source of inoculum is the vine sap exuding from pruning wounds in diseased plants and from infected branches and canes removed in pruning operations. The bacteria are disseminated locally by means of contaminated tools during pruning, and/or by rain especially in windy weather. They enter healthy plants through pruning or frost wounds. It has been found, at least under Cretan conditions, that the susceptibility of plant tissues to infection is higher from November to late January. In Greece the tissues are generally much less susceptible in February and March, although in France susceptibilty is lowest in January with peaks of susceptibility in December and February–March. Transmission of the pathogen by soil and roots during the control of phylloxera root aphid (*Daktulosphaira vitifoliae*) by flooding has been reported in France recently (Ridé & Marcelin 1983). Transmission of the pathogen over long distances and to unaffected regions is accomplished with planting or grafting material; up to 50% of healthy-looking and vigorous canes obtained from affected vineyards can be latently infected. The disease severity in systemically infected plants shows a considerable variation from season to season probably due to variable environmental conditions (Grasso *et al.* 1979). Prolonged wet weather, overhead irrigation and flooding are possibly some of the contributory factors for disease outbreaks.

Economic importance

The disease is of significant economic importance because no curative control measures are known and because it is endemic in several vine-growing regions, affecting commercially important cultivars. Losses arise from reduced productivity and longevity of the affected vineyards. In Greece (Crete and Peloponnese), the disease is most prevalent in areas where two of the most popular cultivars are grown (Sultanina and Corinthe noir). Several vineyards have been abandoned because of their progressive deterioration. In France thousands of hectares have been found to be contaminated (up to 80% of plants infected) and a 50% yield decrease estimated in the area of Charente-Maritime. The disease is widespread in Pyrénées Orientales, Aude, Ardèche, Vaucluse, Ile d'Oléron, and Vendée and serious damage has been reported on the cvs Alicante, Ugni Blanc, Grenache, Carignan, Macabeu and Valenci. In Spain, the disease is endemic and it causes serious damage in the regions of Zaragoza, Rioja and Ribeiro on cvs Garnacha and Macabeo. It has been suggested that the change from use of copper to organic fungicide sprays has largely contributed to the spread and severity of the disease.

Control

Control policy should be directed firstly to preventing the spread of the disease in newly established vineyards and to unaffected vine-growing regions, and secondly to minimizing disease incidence and severity within contaminated areas. It is very important that all planting and grafting material is obtained from non-contaminated areas and all nursery stock is under strict phytosanitary inspection. The following sanitary, cultural, and preventive measures are recommended in affected areas. All infected branches and canes should be removed in the course of pruning, collected and burned. Similarly, all dead or heavily infected plants should be uprooted and burned. In Greece, it is recommended that pruning should be carried out in dry weather during dormancy and as late as possible and in France that the first pruning should be in mid-January and the second as late as possible. All pruning tools should be repeatedly disinfected during use (see *Erwinia amylovora*). Spraying with Bordeaux mixture or fixed copper is recommended immediately after pruning and appears to be helpful in wet, rainy areas if applied at intervals up to the stage of half-expanded leaves. French workers suggest another copper application in autumn at 50% leaf-fall. Wounding the plants during cultivation and overhead sprinkler irrigation should be avoided.

PANAGOPOULOS

Xanthomonas campestris (Pammel) Dowson **pv. alfalfae** (Riker *et al.*) Dye (Moflett & Irwin 1975; CMI 698) causes leaf and stem spot, stem striping and seedling damping off of lucerne in Australia, India, Japan, Nicaragua, Romania, Sudan, USA and USSR. It overwinters in infected host debris and in soil in some climates; it is rarely economically important.

Xanthomonas campestris pv. *begoniae* (Takimoto) Dye

X. c. pv. *begoniae* causes a major disease of commercial cultivars of fibrous-rooted, tuberous and elatior begonias almost wherever they are grown (Strider 1975; Dye 1963; CMI 699). The Rex group appears resistant. The leaf spot phase begins as small water-soaked spots, most easily seen on the undersides of leaves near the margins. The spots enlarge and eventually coalesce if conditions are suitable, leading to soft rot and premature abscission. Stem spotting also occurs. Systemic infection can be latent in stock (mother) plants and spread by knives between cuttings and plants. Further spread to give the leaf and stem spot phase occurs by contact and by splash during watering, through wounds and natural openings. A further source of inoculum can be dried infested plant debris in which the pathogen can survive for at least 17 months. In the presence of the stem nematode *Aphelenchoides fragariae* the disease can be more severe and develop more rapidly (Riedel & Larsen 1974) as it also does in warm, moist conditions. Invasion of the vascular system occurs readily from leaf and stem lesions.

Although streptomycin sprays can give good control and copper sprays somewhat weaker control of the leaf and stem spot phase this is inadequate because even slight uncontrolled spotting can lead to unacceptable systemic infection (Strider 1975). Because of this the production of indexed cuttings using meristem-tip culture and culturing (Hakkaart & Versluijs 1982) and immunofluorescent microscopy (Digat 1978; Rattink & Vruggink 1979) has been developed which, with such other measures as irrigation from below instead of overhead and, where necessary, soil fumigation with methyl bromide (Strider 1975), should give good control.

LELLIOTT

Xanthomonas campestris pv. *campestris* (Pammel) Dowson

Hosts

The pathogen attacks all cultivated *Brassica* species, radish, ornamental stock (*Matthiola incana*) and many cruciferous weeds.

Host specialization

Two related pathovars, pv. *armoraciae* on horseradish, weakly pathogenic to cabbage and cauliflower and pv. *aberrans* of greater virulence to cauliflower, have been reported.

Geographical distribution

X.c. pv. *campestris* occurs generally wherever brassicas are grown commercially.

Disease

The pathogen primarily causes a vascular, seed-borne disease but Knösel (1961) and Moffett (1976) describe a leaf spot phase that the former ascribed to pv. *aberrans*, which is probably best regarded as part of the pathogenic spectrum of pv. *campestris*. Lesions begin on the under surface of the leaf as small, depressed, inconspicuous, water-soaked spots which soon become apparent on both surfaces as necrotic lesions with a distinct margin. The surrounding tissue is pale and slightly chlorotic. With the pathogen strains and host cultivars Moffett (1976) studied, there was little vascular infection; this could reflect host cultivar resistance or bacterial strain variation. The disease spreads rapidly in the seed bed from seed-borne infection or overwintered infested plant debris by wind-driven rain and rain splash to healthy seedlings, penetrating the cotyledons and lower leaves by way of hydathodes, wounded veins or the roots to produce marginal, chlorotic yellow to brown lesions with diffuse edges and blackening of the veins. Affected leaves often drop and the external symptoms of infection disappear for several weeks or are masked by downy mildew (*Peronospora parasitica* (q.v.)), during which time the pathogen develops systemically (Walker 1941). With increasing summer temperatures external lesions reappear, often at leaf margins, and further inoculum becomes available for secondary spread. At 20–28°C systemic marginal lesions predominate at first but later systemic lesions originate in the central areas of leaves; at 16°C most lesions are marginal (Cook *et al.* 1952). There is blackening of the vascular system. Cabbage plants infected during the vegetative period are frequently symptomless until the flowering stage when the pathogen is well established in the xylem of rachises, pedicels and pods, and often the funicles which frequently remain attached to the seed. Invasion of the fleshy petioles and head leaves is rapidly followed by soft-rotting bacteria, under suitable conditions.

Epidemiology

If temperatures for symptom expression are sub-optimal (as they are in temperate coastal climates where cool, mild winters are ideal for the vernalization required to induce flowering in such biennial brassicas as cabbage, Brussels sprouts, cauliflower and turnip), the disease can persist latently in the vascular system; infected but symptomless seed plants can therefore give rise to infected seed. The seed crop is most vulnerable to infection in late summer in the seed beds before transplanting and during late flowering and seed maturation. Black rot infection is difficult to detect during flowering and seed maturation not only because of latent infection but also because of the dense foliage of the crop, the presence of other diseases and senescence on lower leaves.

Economic importance

Black rot is considered the most important disease of crucifers world-wide (Williams 1980). It is most serious in tropical, subtropical and humid continental climates, but in the cool, maritime climates of N. Europe and N. America it seldom causes serious loss. In 1976 the estimated loss in the USA was $1 million (Kennedy & Alcorn 1980).

Control

Control is largely achieved by hot water treatment of fresh, well-developed seed at 50°C for 25–30 min for cabbage and for 15 min for most other crucifers. followed by organomercury dusting. This is combined with inspection and certification procedures for transplants, rotation of seed beds and avoidance where possible of the practices of clipping seed beds, soaking transplants in water and locating seed beds near production crops and fields. Where these precautions are partially or completely ignored epidemics can occur and losses can be very high. Resistance has been bred for cabbage; the site for resistance appears to operate in uninjured hydathodes (Staub & Williams 1972).
LELLIOTT

Xanthomonas campestris* pv. *cannabis Severin causes leaf spot of hemp in Romania (Severin 1978).

Xanthomonas campestris* pv. *cerealis (Hagborg) Dye See *X.c.* pv. *translucens*.

Xanthomonas campestris* pv. *citri (Hasse) Dye causes citrus canker in the Far East, and parts of Africa and S. America, and has been introduced several times into Florida (USA) where it is currently again under eradication. It is a major quarantine organism for Mediterranean citrus (EPPO 1).

Xanthomonas campestris* pv. *corylina (Miller *et al.*) Dye causes small (1–2 mm diameter) angular or irregular, yellow-green to dark-green water-soaked spots on leaves and husks, and dieback (Fig. 21) of branches and twigs with subepidermal lesions, of filbert and hazel nuts (*Corylus maxima* and *C. avellana*) (Prunier *et al.* 1976). It can cause severe damage in newly planted nurseries and orchards through death of young trees particularly those suffering drought stress (Moore *et al.* 1974). It probably occurs sporadically in European countries that grow the crop, including England, France, Yugoslavia and the Black Sea region of the USSR (EPPO 134).

Fig. 21. Dieback of a lateral shoot of hazel due to *Xanthomonas campestris* pv. *corylina* (courtesy of L. Gardan).

Xanthomonas campestris pv. *glycines* (Nakano) Dye

This bacterium, formerly '*X. phaseoli* var. *sojensis*', causes an economically significant seed-borne disease of soybean in most areas of the world in which the crop is grown. It is widespread in European USSR and Kazakhstan causing seed losses of up to 28% and incidence of up to 24% severe infection in some regions. It has been frequent in S. France and also reported from Romania and Bulgaria. Generally speaking it is far less economically important than

Pseudomonas syringae pv. *glycinea* (q.v.). The name bacterial pustule is commonly given to the disease, the first symptoms of which are small, inconspicuous, pale-green or reddish-brown spots slightly raised in the centre. These pustules may be on one or both surfaces of the leaf. They finally rupture the epidermis and, as the spot ages, usually collapse, shrivel or slough off. Lesions are never water-soaked (unlike those of bacterial blight (*P.s.* pv. *glycinea*)). They become angular and reddish-brown and may either remain small or fuse to form large, irregular, mottled, brown areas involving much of the leaf. Spots may be surrounded by a narrow, yellow border and parts of lesions may fall out leaving the leaf ragged. Defoliation may also occur.

Mixed infections by *X.c.* pv. *sojense* and *P.s.* pv. *glycinea* occur not infrequently on leaves and stems. The pathogen can overwinter in infested host debris on the surface of soil (but less well in buried host debris) or in volunteer plants from infected seed. The disease is favoured by warm wet weather, secondary spread occurring during rain storms. Control is best obtained by seed certification, growing seed in dry areas under furrow irrigation, ploughing in crop debris and where feasible, rotation with cereal crops. Resistance to the disease occurs in some cultivars (Nikitina & Korsakov 1978).

LELLIOTT

Xanthomonas campestris pv. *graminis* (Egli *et al.*) Dye

Including *X.c.* pv. *phlei* Egli & Schmidt, *X.c.* pv. *poae* Egli & Schmidt and *X.c.* pv. *arrhenatheri* Egli & Schmidt.

Host plants

A number of agronomically important grasses including *Lolium multiflorum* and *L. perenne, Festuca pratensis, Phleum pratense, Poa* spp., *Arrhenatherum elatius* and *Dactylis glomerata* are affected by the disease. *Alopecurus* sp., *Agrostis tenuis* & *Festuca ovina* are susceptible by artificial inoculation. None of six tested ornamental grasses was susceptible (Cleene *et al.* 1981).

Host specialization

There is evidence for a number of *X. campestris* pathovars on grasses. Isolates from *Phleum, Poa* and *Arrhenatherum* will only affect the genus from which they were isolated and the pathovars *graminis, phlei, poae* and *arrhenatheri* have therefore been proposed (Egli & Schmidt 1982). Future work may distinguish further host specialization in this group of grass pathogens.

Geographical distribution

Since 1975, when it was first described in Switzerland, the pathogen has also been reported from Belgium, France, Germany and UK. However the disease may be more widely distributed than indicated by the literature, given that it is almost certainly seed-borne (see below), the international nature of much of the world's seed trade and the possibility that the bacterium named '*X. translucens* var. *phlei-pratensis*' by Wallin & Reddy in the USA in 1945 (Wallin & Reddy 1945) is the same as *X.c.* pv. *phlei*.

Disease

The disease is a bacterial wilt (Egli *et al.* 1975). The flag leaves of flowering tillers characteristically show marginal, orangey stripes before wilting and become a light straw colour. Young leaves curl and wither and shoots become stunted and die at the heading stage. On less affected plants, inflorescences have difficulty in emerging and when they do are small and distorted. Sometimes inflorescences die without the leaves showing symptoms and such scattered white heads can be conspicuous in otherwise green crops. In vegetative tillers, affected leaves develop dark-green, water-soaked spots which often lengthen to embrace the distal part, or the whole of one leaf; the leaf eventually becomes wholly dark-green and flaccid before withering and dying. Isolation of the pathogen is facilitated if 0.25 ml of 1% actidione is added to 100 ml of the isolation medium to suppress fungal growth that can otherwise overgrow the rather slow-growing pathogen (Lelliott & Stead 1987).

Epidemiology

Little is known of the epidemiology of this disease and research into it is urgently required. Pathovars of *X. campestris* that have herbaceous hosts, and that have been thoroughly investigated in this respect, are seed-borne. In fact, the distribution and behaviour of bacterial wilt of grass in the UK is such that it would be difficult to explain if the pathogen were not seed-borne. By analogy with other pathovars of *X. campestris* it would be expected to be spread by rain splash, irrigation or drainage water, but observations of outbreaks in field variety trials strongly indicate that such spread is comparatively unimportant when compared with spread on the cutter bar of forage harvesters. Spread is also likely to occur on the mouthparts and feet of grazing animals.

Economic importance

A good correlation has been established in Switzerland between yield decrease of Italian ryegrass cultivars during hot, dry summers and their susceptibility to *X.c.* pv. *graminis* as determined by glasshouse inoculation and field observation (Schmidt & Nuesch 1980); a 40% yield loss (compared with the most resistant cultivars) was observed for the later cuts of very susceptible cultivars, corresponding to a 20% annual loss. Many of the newly bred, high-yielding hybrids of *Lolium multiflorum*×*perenne* are highly susceptible and have been quickly killed in field plots of comparative cultivar trials in the UK. The disease appears to be widely distributed in Switzerland and Belgium where disease symptoms were present in *c.* 80% of fields surveyed in 28 localities and in W. Scotland where they were present in 17 out of 24 fields examined with an incidence of 0.1–38.5% infected tillers/m^2. With the present trend to greater use of higher yielding-grass for production of beef, lamb and milk in W. Europe the potential for yield loss in both forage grass and grass for conservation is obvious.

Control

Control of bacterial wilt apears to lie in breeding resistant cultivars. In fact the disease has been so common in field cultivar trials in the UK and Switzerland that very susceptible cultivars are being eliminated during early stages of the trials. Some cultivars of Italian ryegrass are already highly resistant and in Switzerland inoculation in glasshouse screening is being used to accelerate breeding for resistance (Schmidt & Nuesch 1980). Given the potential of the disease for causing loss it would seem wise to screen for resistance in breeding even less susceptible forage grasses. In crops grown for conservation, disinfection of harvesting machines (see *Corynebacterium michiganense* pv. *insidiosum*) is advisable after use in affected crops or crops likely to be affected.

LELLIOTT

Xanthomonas campestris* pv. *gummisudans (McCulloch) Dye causes a bacterial blight of gladioli of minor importance in the USA (Elliott 1951). It has been reported from the Netherlands. It causes horizontal, water-soaked dark-green leaf spots that become brown and rectangular and can inhibit corm formation.

Xanthomonas campestris pv. *hederae* (Arnaud) Dye

X.c. pv. *hederae* causes a wide-spread and sometimes serious disease of ivies grown as pot plants under glass. It has been known as a minor disease in N. America for some 50 years (Elliott 1951), has been recorded in New Zealand and occurs in England. With the expansion of the houseplant trade in W. Europe and the widespread use of ivy in that trade, the disease became of some economic importance and is probably now widely distributed in the region. In recent years the general adoption of irrigation from below (usually a capillary matting) as cultural practice appears to have lessened its importance. Initial symptoms are small, circular, water-soaked, dark-green spots on leaves, more conspicuous on the lower surfaces. These enlarge to roughly circular or angular lesions, 2–10 mm diameter with greenish-brown, water-soaked margins and reddish-brown to black centres. Under humid conditions a sticky yellowish, bacterial exudate can appear onlesions, which dries to give them a grey appearance. Lesions may eventually dry out and crack. If infection is severe it can cause distortion, wilting and defoliation. Similar spots, but more elongated, can occur on petioles and stems (Dye 1967). Spread is presumably by water splash.

LELLIOTT

Xanthomonas campestris* pv. *holcicola (Elliott) Dye causes a common, sometimes serious leaf spot disease (Elliott 1951; Watson 1971) of sorghum, millet and sudan grass in most regions where these crops are grown. In Europe it has been reported from Romania and is common in the USSR but is less important than infection by *Pseudomonas syringae* (q.v.).

Xanthomonas campestris* pv. *hordei (Hagborg) Dye. See *X.c.* pv. *translucens*.

Xanthomonas campestris pv. *hyacinthi* (Wakker) Dye

X.c. pv. *hyacinthi* causes a widespread yellows disease of hyacinth. In the Netherlands it has been

described as the most important disease of that crop. In the field, it causes water-soaked, dark spots and long stripes on leaves from which inoculum is spread by rain splash and wind, cultivation machines and knives. Infection moves down the leaf as new infections result from bacteria being washed downwards by rain. Infection moves slowly when it reaches the subtending bulb scale, sometimes taking over 1 year to reach the basal plate of the bulb. From the basal plate the pathogen can invade and destroy the vascular and surrounding tissues, paving the way for bulb decay by secondary bacteria. This sequence of infection can, depending on the time of infection of the basal plate, result in total bulb loss, non-emergence after planting, emergence followed by leaf infection or growth followed by withering with leaf symptoms, or sudden death without leaf symptoms. If the vascular system of the leaf is invaded directly from the leaf spot phase, the bulb dies soon after lifting (Beijer 1972). Hot air treatment can cure infected bulbs (Kruijer & Vreeburg 1981).
LELLIOTT

Xanthomonas campestris **pv.** ***incanae*** (Kendrick & Baker) Dye causes bacterial blight of garden stocks (*Matthiola incana*) in N. America (Elliott 1951), characterized by soft, water-soaking of the stem and growing tip followed by general collapse. The disease is seed-borne and vascular. It has been reported in England.

Xanthomonas campestris pv. *juglandis* (Pierce) Dye

X.c. pv. *juglandis* (Miller & Bollen 1946, CMI 130) causes a bacterial blight of walnut and is common where the Persian walnut (*Juglans regia*) is grown commercially. Other *Juglans* species are also affected. It is characterized by black, necrotic spots on the leaves, green shoots and young nuts. Many nuts fall prematurely, but the husk, shell and kernel of others that reach full size are more or less blackened and destroyed. Cankers can form on twigs and branches but injury to the tree is much less serious economically than the loss of crop. It is generally regarded as a major disease of walnut. The pathogen overwinters in dormant and developing buds and catkins, from which leaves are infected in spring.

There is conflicting evidence as to whether blighted and cankered branches are also a source of overwintering inoculum. Spread by rain splash from infected foliage late in the season results in infection of nuts and infestation of developing buds and catkins. Pollen from diseased catkins is contaminated with the pathogen and is another possible source of inoculum for leaf infection. The disease is favoured by rainy weather particularly during and after flowering. Bacterial blight is the most destructive disease of walnuts in W. USA, causing an estimated mean annual loss of $2.2 million in 1976 (Kennedy & Alcorn 1980). It is similarly destructive in parts of the USSR. In the USA copper sprays (either as Bordeaux or the oxychloride) can give good control if regard is paid to weather conditions: few sprays are needed in years with little rain during the infection period, but up to three pre-blossom sprays (to which summer oil emulsion should be added to reduce injury to very young leaves) may be required in rainy seasons to minimize inoculum before the highly susceptible pistillate flowers are exposed, and to eliminate catkin infection and subsequent spread through infected pollen. A bloom spray and a post-bloom spray about three weeks after should follow. If rain is likely, additional sprays may be beneficial during the period of nut enlargement. Some cultivars are resistant.
LELLIOTT

Xanthomonas campestris pv. *malvacearum* (Smith) Dye

X.c. pv. *malvacearum* causes a bacterial blight of cotton (*Gossypium hirsutum* and *G. barbadense*). Some 30 races are known, differentiated on virulence to differential cultivars. None is related to phage susceptibility, which can be used to differentiate two phagovars. The disease occurs in all cotton-growing regions including those in S.E. Europe and the USSR. Symptoms depend on the age and part of the plant attacked so the disease caused has acquired different names. 'Seedling blight' involves stem and leaf spotting, yellowing and death. 'Angular leaf spot' begins as translucent, water-soaked spots between veins that enlarge, become angular and brown to black, and during intense infection coalesce causing leaves to yellow and fall. 'Bract spot' is similar. Discoloration of veins is uncommon. If bolls are small when attacked ('boll spot') they usually drop off; on larger bolls green, water-soaked spots become brown or black and shatter and the lint becomes wet, brown-stained and rotten ('boll rot'). Much boll rot is probably due to invasion by fungi through bacterial lesions. On stems the elongated water-soaked spots become long, sunken, black stripes; in susceptible cultivars the stem is girdled and the branch shrivels and dies ('blackarm') or breaks. Lesions can be covered with yellow bacterial exudate ('gummosis') that dries in the form of granules or a crust. Systemic infection is uncommon.

The pathogen is seed-borne as an external contaminant on the seed fuzz (with a small proportion of internal infection) and in adhering crop trash. It can also survive on volunteer cotton plants, and in dry crop trash for up to 17 years. Infectivity declines rapidly if the trash is weathered by rain or buried. Infection spreads mainly by rain splash and surface wash particularly during storms accompanied by spells of warm weather. Optimum conditions are an average daily maximum temperature of 35°C, 8 h daily sunshine and a weekly rain fall of 2.5–3.5 cm. Until the 1960s when resistant cultivars became available, bacterial blight was the most destructive disease of cotton in Africa. It is, however, only a moderate importance in European cotton-growing countries, although serious in Asian USSR. Control has traditionally been based on acid delinting, seed treatment with organomercury or copper formulations (or more recently bromopol), and measures to bury trash. Resistant cultivars are now available, with a known genetic basis (some 20 major genes, with minor genes playing a small role) and the emphasis must now be on durable resistance (Bird 1982). See also review by Ebbels (1976).
LELLIOTT

Xanthomonas campestris* pv. *oryzae (Ishiyama) Dye and ***X.c.* pv. *oryzicola*** (Fang *et al.*) Dye cause leaf blight and leaf streak of rice respectively. Neither occurs in the Mediterranean so they remain important quarantine organisms for Europe (EPPO 2, 3).

Xanthomonas campestris pv. *pelargonii* (Brown) Dye

Host plants

This bacterium attacks zonal, regal and ivy-leafed pelargonium (CMI 560). *P. graveolens* appears to be resistant. *Geranium* species are also hosts but attacks on them appear to be uncommon or at least unnoticed.

Geographical distribution

It occurs in most countries where pelargoniums are grown on any scale. In Europe it has been reported from Belgium, Czechoslovakia, Denmark, France, GDR, Greece, Hungary, Italy, the Netherlands, Portugal, Romania, Switzerland, Yugoslavia and UK; it also probably occurs in other W. European countries because of the extensive international trade in pelargonium cuttings.

Disease

The disease has two phases: leaf and stem spotting and a systemic wilt. Leaf and stem spots or blisters are initially small, dark-green and water-soaked. They enlarge, turn brown-red with a narrow greenish-yellow marginal zone and spread to create necrotic patches. If the bacterium invades the vascular system directly from the leaves or stems or by spreading to the cortex of the stem, affected leaves may wilt at high temperatures inwardly from their margins, to give an umbrella-like shape to the leaf. Wilting becomes irreversible and affected leaves shrivel and die, but characteristically remain attached. Eventually systemically affected plants develop a dry stem rot and die; often the stem is invaded before death by secondary fungi (including *Pythium* species) and bacteria to give a black wet rot from which *X.c.* pv. *pelargonii* can rarely be isolated. The wilting symptoms are not dissimilar from those of *Verticillium* wilt and the black stem rot from that caused by *Fusarium* species.

Epidemiology

The leaf and stem spotting phase is spread mainly by rain splash or by overhead watering, and infection usually takes place (through stomata and hydathodes) only when there is free water on the plant surface for some hours. Whitefly transmission is also possible (Bugbee & Anderson 1963) but this is probably insignificant. The systemic vascular phase is mainly spread by knives used to take cuttings since infection can be latent in mother plants underwatered in winter, and those with restricted roots from underpotting. Cuttings taken from latently infected plants, either from mother plants or as stoppings from plants raised for sale, are the chief source of inoculum in multiplication and growing-on beds or benches.

Economic importance

The disease is economically serious wherever pelargoniums are grown on any scale, although new propagation techniques have reduced losses.

Control

Control of the leaf and stem spotting phase is probably best achieved by avoiding surface water on plants, through such practices as capillary watering and regulation of temperature and ventilation to prevent water vapour condensing on plant surfaces, particularly at night. Attention to hygiene provides a further possible method of control, since the pathogen can survive for several years in dried plant debris. In outdoor cultivation spraying with streptomycin or with copper compounds gives incomplete control and rotation will mitigate or avoid reinfection from buried infested host debris in which the pathogen can survive for between 3 and 12 months under some soil conditions. There is evidence (Digat 1978) for the epiphytic survival of the pathogen for at least 2 months.

Unless cuttings can be raised with a very high possiblity of freedom from the pathogen, adequate control of the systemic phase is not obtainable. Where cutting production is on a large scale, particularly by specialist cutting raisers, the technique of indexing can be used. Cuttings destined to be used as nuclear stock plants are tested for the presence of the pathogen and those found free are grown in isolation to supply cuttings for further multiplication. These nuclear stock plants are replaced at intervals of 2–3 years by new indexed cuttings. Indexing is carried out aseptically, taking commonly five 1 mm transerse slices, including a node, from the middle of a surface-sterilized 2–3 cm stem piece. These are cultured in a tube of nutrient dextrose broth or agar and the cutting stored in a separate plastic sleeve at 0–2°C. If after 7–10 days there is no microbial growth in the tube the cutting is raised as a nuclear stock plant. Complete surface sterilization, followed by thorough washing, is crucial. No attempt is made to identify the pathogen and most cuttings fail because surface sterilization was not complete. An alternative method is to dispense with surface sterilization but to split the stem piece longitudinally, dissecting out the vascular tissues and culturing these.

The use of immunofluorescent microscopy to detect the pathogen itself in cuttings (Digat 1978) is not sufficiently sensitive and probably too laborious for commercial use but the possibility of using ELSIA could be further investigated. The indexing method will probably be outmoded by the use of tissue culture for propagation since with suitable growth media it should be possible to discard any infected culture.

LELLIOTT

Xanthomonas campestris pv. *phaseoli* (Smith) Dye

Host plants

The principal host is the French bean (*Phaseolus vulgaris*) in its various forms. There seems little doubt that lima bean and runner bean are natural hosts and that some other *Phaseolus* species are good artificial hosts but plants other than these should only be accepted as hosts if Koch's postulates have been clearly fulfilled. Many hosts were reported before reliable methods for testing pathogenicity had been established, and before it was realized that there are many pathovars of *X. campestris* attacking legumes, with overlapping or incompletely known host ranges. An example is the inclusion of cowpea (*Vigna sinensis*) in the list of hosts for *X.c.* pv. *phaseoli* despite the careful work of Burkholder (1930, 1944) confirmed by Vakili *et al.* (1975) that pv. *phaseoli* is pathogenic to French bean but not cowpea and that another pathovar pv. *vignicola* is pathogenic to both French bean (albeit with slightly different symptoms) and cowpea. Burkholder (1930) could not infect *P. angularis, Vigna sesquipedalis* or soybean with isolates from French bean. The suggestion that isolates of pvs *phaseoli* and *vignicola* are part of a pathogenic continuum (Schuster *et al.* 1978) needs further investigation before it can be accepted.

Host specialization

Some strains (formerly known as *X. phaseoli* var. *fuscans*) produce brown, melanin-like, diffusible pigments when cultured on complex media or media containing tyrosine. Such strains had at one time also been considered to have different pathogenic properties but they are now generally not considered to warrant separation from other strains.

Geographical distribution

The pathogen affects crops in most areas of the world where French beans are grown commercially as a field crop. Its geographical range overlaps that of the similar disease halo blight (see *Pseudomonas syringae* pv. *phaseolicola*) except that generally it is more common than that disease in warm temperate and subtropical climates and less common in cool temperate climates. It is widespread in N. America and the Moldavian SSR. In W. Europe it is sporadic and uncommon in the north but said to be widespread in Spain; it has been reported from Finland, France, the

Netherlands and Switzerland (EPPO 61). It is more common in E. Europe and locally established in Greece, Romania and other areas of the USSR. In Hungary, a survey in 1974 showed that, of isolates made from bean, 487 were *Pseudomonas syringae* pv. *phaseolicola* and only 22 *Xanthomonas phaseoli*.

Disease

The disease has two phases: a systemic phase resulting from invasion of the vascular system and a local phase of parenchyma infection.

Systemic infection gives rise to generalized symptoms, the severity of which varies with the proportion of the vascular system affected. If infection is light, plants will be slightly stunted and incipient reversible wilting may occur at the height of hot, dry days. If infection is heavier, a rapid wilt of seedlings often occurs, particularly at temperatures of 25–35°C, and in older plants a leaf, a branch or the entire plant may wilt and die, usually progressing in that order. The pathogen can invade tissues other than the vascular system. On the stem this can result in girdling lesions at the cotyledonary, or more frequently, higher leaf nodes, by invasion from leaf traces. Resultant lesions are reddish-brown and often weaken the stem causing breakage during storms. On leaves and pods lesions similar to those from stomatal invasions can occur (see below) particularly at temperatures of about 18°C, although on pods a long, irregular lesion along the suture usually indicates a vascular origin; on leaves the veins in lesions are often reddish-brown. Seeds can be invaded, via the pedicel and funiculus, causing a range of symptoms from a small yellow spot at the hilum to yellow discoloration of parts, or the whole, of the seed coat (Fig. 22). This is sometimes accompanied by wrinkling, due to invasion of the area beneath the coat. These symptoms are readily discernible on white seeds but not on coloured seeds. Seedlings from infected seed often have diseased growing tips which die leaving only the cotyledons—the 'snake's head' symptom.

On leaves local invasions through stomata and wounds just appear as small, water-soaked spots or light green, wilted areas which, as they enlarge, dry out and become brown and brittle. At the margins of such lesions is a yellow border, often surrounded by a narrow, pale-green zone. Unlike those caused by *Pseudomonas syringae* pv. *phaseolicola* (q.v.) a lesion often extends to cover much of the leaflet. After heavy rain or wind, affected leaves are often very ragged and torn. On stems, particularly of seedlings, water-soaked lesions may occur which later dry out and become reddish-brown; on older plants water-soaking is less evident and lesions are reddish, longitudinal streaks. Stem splitting, releasing of yellow ooze, may occur. On pods, small, water-soaked, dark-green spots gradually enlarge which, depending on the age of the pod and environmental factors, can become dry, sunken and brick-red, extending inwards from the outer edge. Yellow ooze that dries to a yellow encrustation often covers the lesion.

Fig. 22. Discoloration of dwarf bean seeds by *Xanthomonas campestris* pv. *phaseoli* (courtesy of V.R. Wallen, Agriculture Canada).

Epidemiology

Primary foci in crops usually originate from infected seeds in which the bacterium can survive for many years. In some regions, overwintering in plant debris occurs (and may occur in soil) but in others this is not always the case (Wallen & Galway 1979; Wimalajeewa & Nancarrow 1980). Spread from these foci is by wind-driven rain, overhead irrigation and possibly insects (Kaiser & Vakili 1978) although how significant insect vectors are generally and how great their role is in providing wounds through which bacteria subsequently enter is uncertain. Under experimental conditions, rifampicin-resistant pv. *phaseoli* can grow on leaves of non-host plants and can be recovered up to 12 days after inoculation (Cafati & Saettler 1980) but the epidemiological significance of this is undetermined. Apart from direct contamination or infection on the plant, seeds can be contaminated by infested trash at harvesting or dressing. Spread can also occur by soaking seeds in suspensions of root nodule bacteria. The disease is most severe in conditions of high rainfall and humidity and temperatures of about 28°C.

Economic importance

In the USA common blight was the most economically important bacterial disease of bean in 1976, causing an estimated loss of $4 million (Kennedy & Alcorn 1980). Recent estimates of losses in other countries are not available, and although it continues to cause important economic loss in Romania and Hungary and appears to be of importance in parts of the USSR, it is of little concern in other European countries.

Control

Seed certification and growing seed in dry areas under furrow irrigation are the most effective means of control. Seed testing by growing on samples (Alvarez *et al.* 1979) and isolation and serology (Trujillo & Saettler 1979) can only eliminate comparatively heavily affected seed lots, and seed treatment cannot reduce infection to levels below which epidemic attacks will be unlikely to occur. Spray programmes with copper compounds or streptomycin can give partial control. Where agronomically feasible, losses can be considerably reduced by timing planting to avoid growing the crop during periods of weather favourable to the disease. In some countries quite short rotations can markedly reduce the severity of attacks. In Ontario, Canada, a 10-year select seed programme using aerial photography to monitor crops was effective (Wallen & Galway 1979). Breeding for resistance and polygenic tolerance is hindered by differential genetic control of pod and leaf reactions such that high leaf tolerance is not associated with pod tolerance.

LELLIOTT

Xanthomonas campestris pv. *pruni* (Smith) Dye

X.c. pv. *pruni* (CMI 50; Elliott 1951) causes a major disease of peach in the USA (estimated loss in 1976 was $2 million) and has precluded the commercial production of Japanese plum in many regions. It is severe on plum and peach in Queensland, Australia, and losses can be high in some years on peach and apricot in New Zealand, and on peach in Taiwan. It also occurs in S. Africa, N. and S. America and many Asian countries. In Europe, it is locally established in Austria, Bulgaria, Italy (where it was severely epidemic in the north in 1980 on plum), the Netherlands, Romania (on plum) and USSR; it apparently rarely causes serious damage although the potential for considerable crop loss appears to exist in at least some of these countries (EPPO 62). Its principal hosts are peach, apricot and plum although almond, cherry and other *Prunus* species are also affected.

Leaf infection is first seen on the lower surface as small, pale-green to yellow circular or irregular areas usually concentrated towards the leaf tips, with a light-tan centre. As they enlarge, lesions are apparent on the upper surface, become angular and darken to purple, brown or black. There is a narrow yellow margin to the lesion. The necrosed tissue may drop out, particularly on plum, to give a shot-hole effect. A bacterial ooze may appear on lesions. Severely infected leaves turn yellow and fall. Defoliation can be severe and weaken trees. On peach, atypically, there can be a grey leaf spot on the upper surface. On peach fruit, small, circular, brown spots appear that become sunken, frequently with water-soaked margins and often with light-green haloes (Fig. 23). Pitting and cracking around spots occurs as fruit swell which may be insignificant or, when heavy infection of young fruit is involved, extensive and severely damaging. There may be flow of gum from lesions particularly after rain. On plum fruit, large, sunken, black lesions occur on some cultivars but only small, pit-like lesions on others, while on sweet and sour cherries early infection of fruit leads to invasion deep into the fruit tissue, which is distorted.

On peach twigs and watershoots, spring cankers occur on the tips of last year's wood as initially small, water-soaked, superficial blisters that extend 2–10 cm along the twig and may girdle it causing the 'black tip' symptom. Infection later in the season causes summer cankers: water-soaked, dark-purplish spots

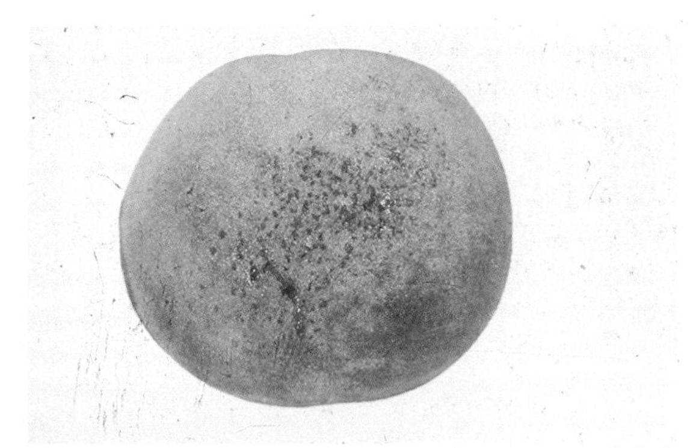

Fig. 23. Small circular sunken brown lesions on peach fruit due to *Xanthomonas campestris* pv. *pruni* (courtesy of H.L. Keil).

surrounding lenticels that dry out and become limited, and dark, sunken, circular to elliptical lesions with a water-soaked margin. On plum and apricot, cankers are perennial, unlike those on peach which are annual, and continue developing in 2–3 year old wood to produce deep-seated cankers that deform and kill twigs and branches and cause dieback that can render trees uneconomic to crop.

The disease can be spread in budwood and rootstocks. Inoculum in spring comes primarily from overwintering annual and perennial cankers (or possibly from buds in plum) and from fallen leaves. It is disseminated by wind-driven rain and infects leaves through stomata from which the bacteria can, after multiplication, egress and be further disseminated before symptoms are externally visible. Summer cankers following invasion through lenticels or wounds usually become sealed off and do not provide overwintering inoculum except in certain localities; this is provided by cankers forming late in the season during rains just before or during leaf fall. Conditions most favourable for severe infection are warm seasons (19–28°C) with high, frequent rains and heavy dews accompained by fairly high winds. The disease can be controlled by spring applications of copper or dodine; oxytetracycline sprays and trunk injection are also effective.
LELLIOTT

Xanthomonas campestris pv. *translucens* (Jones *et al.*) Dye

X.c. pv. *translucens* probably encompasses a group or mosaic of *X. campestris* pvs that attack Gramineae causing spotting of foliage and flower parts of wheat, barley, rye, oats, triticale and some grasses including *Bromus* species, *Elymus repens* and *Phleum pratense*. Early data on *formae speciales* is confused and conflicting: f.spp. *hordei, undulosa* and *secalis* of '*X. translucens*' were recognized on barley, wheat and rye respectively as principal natural hosts, but Hagborg (1942) in addition recognized f.sp. *hordei-avenae* on wheat and oats and f.sp. *cerealis* on wheat. Fang *et al.* (1950) considered grasses (*Bromus* spp., *Elymus repens*) as principal hosts of f.sp. *cerealis*, and added f.sp. *phlei-pratensis* on *Phleum pratense*. There was disagreement on the secondary (artificially inoculated) hosts of these forms, and Wallin (1946) further subdivided f.sp. *cerealis*. Cunfer & Scolari (1982) showed that strains from triticale are equally virulent to triticale, wheat and rye. Until the natural host ranges of these cereal pathogens and their relationship to *X.c.* pv. *graminis* (q.v.) has been further studied it is probably best to treat them, arbitrarily, as one pathovar, pv. *translucens*. The diseases caused are thought to be seed-borne and widely distributed in N. America, where they are usually of minor importance but occasionally severe when affecting barley and wheat. The position in the USSR and Romania appears similar. They have been noted in most other cereal-growing regions. On wheat, the disease commonly causes a blackening of the glumes known as black chaff, in which lesions are dark-brown to black, sunken stripes chiefly on the upper part of the outer glumes. In severe attacks grain may be shrivelled. The disease on other cereal hosts is similar. On wheat and other hosts water-soaked spots and streaks later becoming brown and translucent and then dried, occur on leaves and stems. Diseases are favoured by a wet, warm climate and are almost certainly mainly transmitted by wind-driven rain from seed-borne primary disease foci.
LELLIOTT

Xanthomonas campestris pv. *vesicatoria* (Doidge) Dye

X.c. pv. *vesicatoria* (CMI 20; Elliott 1951) Hayward & Waterston 1964) causes a leaf spot disease of tomato and pepper (including ornamental pepper). *Solanum nigrum* and the fruits of potato are also natural hosts. There is host specialization within the pathovar; it has been differentiated into four groups on pathogenicity to *Capsicum* cultivars (Cook & Stall 1969, 1982; Kimura *et al.* 1972). Some pepper strains are non-virulent, some are weakly virulent and some moderately virulent to tomato. A form that also affects radish and turnip (***X.c.* pv. *raphani*** (White) Dye) has been described in the USA and Brazil. It is a major disease of tomato and pepper in the USA (where it caused an estimated loss of $1.5 millions in 1976), India, Argentina, Sudan, Nigeria, Egypt and Australia. In Europe, it is apparently widespread and sometimes serious in Bulgaria, Italy and the USSR and has also been reported from Austria, Hungary, Romania and Yugoslavia. It is seed-borne and probably occurs wherever tomato and peppers are grown extensively as field crops. It is said to survive in the rhizosphere of some non-hosts. In certain climates it can overwinter in plant debris in the field. It is spread by wind-driven rain and the pathogen invades leaves through stomata, and fruit through wounds. Symptoms on leaves are at first small, water-soaked spots which become irregularly circular with yellow, translucent margins and brown to black, later

parchment-like centres. Spots may coalesce and form irregular streaks along veins or leaf margins. Edges and tips of leaves may become dead and dry and break away giving leaves a tattered appearance. Heavily infected leaves turn yellow or brown and young leaves become distorted and die. Small, brown or black raised dots or blisters form on fruits. Control is by seed treatment with mercury dressings (or hot water treatment for tomato only), 2–10 day sprays with copper compounds or streptomycin and 3–4 year rotations.
LELLIOTT

Xanthomonas campestris* pv. *vitians (Brown) Dye attacks lettuce and induces small, dark leaf spots with a well-defined edge that enlarge to 1–5 mm diameter and can coalesce to form large necrotic areas. A general wilting and rotting of leaves and stems with or without leaf spotting can occur. It has been reported from the USA, S. Africa, Japan, New Zealand and Italy (Elliott 1951) but its status in Europe is obscure.

Xanthomonas campestris* pv. *zinniae (Hopkins & Dowson) Dye, formerly *X. nigromaculans* f.sp. *zinniae*, causes a seed-borne leaf spot disease of zinnia. It was epiphytotic in N. America in the 1970s (Strider 1976) and has been reported from Australia, Bulgaria, India and Africa. Seed treatment with captan was said to be successful (Strider 1980).

Xanthomas fragariae Kennedy & King

X. fragariae (CMI 558) attacks strawberry (Kennedy & King 1962) in the USA, Brazil, Venezuela, Australasia, France, Greece, Italy (Sicily) and USSR (EPPO 135). Potentilla is an artificial host. The leaf spot phase appears as minute, water-soaked spots on the underside of leaves that enlarge, may coalesce, penetrate to the upper surface, darken, become angular (Fig. 24) and produce large, irregular necrotic areas and sometimes dark, water-soaked vein banding. Systemic invasion can occur (Hildebrand *et al.* 1967), probably under damp, nursery conditions, and crowns become infected causing decline or collapse of plants. Infection is usually intercellular in vascular tissue forming pockets in the xylem and cambium that often rupture the tissue of the crown; occasionally the bacterium occurs within the xylem vessels. Rapidly growing plants are much more susceptible than poorly growing ones. Transmission in crops is presumably by rain splash or overhead irrigation. The pathogen can overwinter in buried and possibly in surface leaf litter from which it could infect young leaves in spring. Control is through the production of disease-free planting material although this is difficult because crown symptoms may be hidden. Furrow is obviously preferable to overhead irrigation that can lead to severe losses where the disease is present.
LELLIOTT

Fig. 24. Strawberry leaf severely affected by *Xanthomonas fragariae* (courtesy of U. Mazzucchi).

'Xanthomonas populi' (Ridé) Ridé & Ridé*

Basic description

Previously known as *Aplanobacter populi* Ridé (Ridé 1958), this bacterium is now recognized as having the characteristics of a species of *Xanthomonas* (Ridé & Ridé 1978). The presence of a xanthomonadin has recently been shown. Mucous, cream to pale-yellow colonies appear in 2–10 days, according to season, on isolation from cankers on yeast extract-peptone-dextrose agar. Isolation by dilution is difficult. *'X. populi'* differs from other members of the genus by its sugar utilization spectrum (L-arabinose −, D-glucose +, mannose +, galactose +, fructose +, trehalose +, cellobiose −), its maximum temperature for growth (27.5°C) and its low NaCl tolerance (0.4–0.6%).

*This name is illegitimate under the International Code.

Host plants

'*X. populi*' is strictly specific to Salicaceae and principally *Populus*. It is endemic in a number of wild populations of *P. tremula* in N.W. Europe. However, it mainly damages industrial plantings of *P. euramericana* cultivars (cv. Blanc de Poitou, grandis, regenerata, I 45–51, more recently cv. Dorskamp and the hybrid *P. trichocarpa*×*deltoides* Barn). '*X. populi*' is not known on *Populus* in N. America, and many provenances of *P. trichocarpa, P. deltoides* and *P. tremuloides* from N. America, used in the last 30 years in European breeding programmes, have proved susceptible or highly susceptible to bacterial canker. An apparently separate pathovar has been described on *Salix dasyclada* by de Kam (1978, 1981) as a subspecies '*X.p.* subsp. *salicis*'; the name is illegitimate.

Geographical distribution

It occurs in Europe (Belgium, France, UK, the Netherlands, FRG, GDR, Czechoslovakia), probably also in Poland (where it may be confused with *Ceratocystis fimbriata* canker) and in USSR (EPPO 43). Its status in Asia is doubtful.

Disease

Susceptible clones near disease foci show symptoms within four years of planting. Slowly developing cankers (Plate 9) appear on small branches, main branches or trunks, which they may girdle and kill. In most cases, reaction-callus forms in which the bacterium persists in winter, and then renews mulitplication with cambial activity. The first signs of infection are seen in spring on the previous year's shoots: shortly before bud-burst, the tissues around infected buds swell, then burst, exuding a thick, whitish, bacterial slime which later turns brown. The exudate may also appear as droplets through the lenticels. When the branch is stripped, the cambium, phloem and cortical parenchyma appear glassy. Exudate may also appear later in the season (July–August) on highly susceptible clones, at the edge of spreading lesions. The final appearance of cankers depends on the species or clone of poplar; it may vary from closed cankers, to open cankers with more or less even callused margins, to a pseudo-tumorous appearance. As the cankers develop year by year, the structure of the wood is severely affected (Burdekin 1972; Gremmen & Koster 1972).

Epidemiology

The bacterial exudate produced in spring is dispersed between adjoining trees by wind and rain. The gradient of decreasing incidence develops in the direction of the prevailing wind in Atlantic regions. The viability and persistence of '*X. populi*' on the surfaces depends on temperature, the activity of the epiphytic microflora, and the release of any available inhibitory substances produced at the leaf and stipule surface. Bacterial penetration mainly occurs through fresh stipule scars, or through fresh petiole scars arising from storm damage or early leaf disease (*Marssonina brunnea*), or else through wounds inflicted by hail or insects. The larvae of the cambium miner *Phytobia cambii* are important in spreading the bacterium within the tree. It is carried along the mine formed in spring, and appears as cankers along the galleries in the following year.

Economic importance

Bacterial canker affects both timber production and its quality. Plantations of cv. Blanc de Poitou established in 1958–60 in N. France have suffered losses in commercial value of their timber of 5–60% according to situation. Where '*X. populi*' is endemic, it limits industrial poplar-growing by preventing the use of certain fast-growing hybrids.

Control

In regions suitable for growing poplars, but at risk because cankers are present on wild trees, only cultivars with good resistance can be grown. In France, the natural forest fund provides no subsidies for planting susceptible clones in Départements recognized to be at risk by ministerial order; all newly bred material is tested for susceptibility to bacterial canker before being released. Extensive resistance breeding work has been done in Belgium, France, the Netherlands and FRG to obtain intra- or inter-American (*P. trichocarpa*×*trichocarpa, P. deltoides*×*P. trichocarpa*) or Euramerican (*P. deltoides*×*P. nigra*) hybrids, showing good growth and also canker resistance, often also associated with adequate resistance to leaf diseases (*Drepanopeziza punctiformis, Melampsora larici-populina*). Little data is available on variation in pathogenicity in '*X. populi*'. The bacterium is known to show some variation in aggressiveness. In view of the importance of the risk to poplar-growing, possible changes in aggressiveness should be closely monitored. Until now, no real breakdown of resistance (development of new virulence) has been seen, presumably because resistance is polygenic. However, current work suggests some selection for greater aggressiveness and also for other pathogens (strains of *X. campestris*) on certain crosses with a common genetic background.

RIDÉ

REFERENCES

Alivizatos A.S. (1984) Aetiology of tomato pith necrosis in Greece. *Proceedings of the 2nd Working Group on Pseudomonas syringae Pathovars, 1984*, pp. 55–57. Hellenic Phytopathological Society, Athens.

Alvarez C.E., Vanegas G.G. & Victoria K.J.I. (1979) Transmission por semilla de bacterias fitopatogenas de frijol (*Phaseolus vulgaris* L.) en Colombia. *Acta Agronomica* **29**, 11–20.

Barratta B. & Marco L.Di (1981) Controllo degli attacchi di "rogna" nella cultivar Nocellara del Belice. *Informatore Fitopatologico* **31**, 115–116.

Bazzi C. & Mazzucchi U. (1979) Epidemie di *Pseudomonas cichorii* su lattuga. *Informatore Fitopatologico* **29**, 3–6.

Beijer J.J. (1972) Het verhop van de geelziekaantasting door *Xanthomonas hyacinthi* in blad en bol van de hyacint. *Mededelingen van de Landbouwhogeschol Wageningen* **72**(30).

Belletti P. & Tamietti G. (1982) L'impiego del calore secco nel risanamento del semi di fagiolo infetti da *Pseudomonas phaseolicola*. *Informatore Fitopatologico* **32**, 59–61.

Billing E. (1970) *Pseudomonas viridiflava* (Burkholder) Clara. *Journal of Applied Bacteriology* **33**, 492–500.

Billing E., Crosse J.E. & Garrett C.M.E. (1960) Laboratory diagnosis of fireblight and bacterial blossom blight of pear. *Plant Pathology* **9**, 19–25.

Bird L.S. (1982) The MAR (multi-adversity resistance) system for genetic improvement of cotton. *Plant Disease* **66**, 172–176.

Boelama B.H. (1972) Bacterial blight (*Pseudomonas pisi*) of peas in South Africa with special reference to frost as a predisposing factor. *Mededelingen van de Landbouwhogeschol Wageningen* **72**(13).

Bottalico A. & Ercolani G.L. (1971) *Pseudomonas savastanoi* su ligustro giapponense in Puglia. *Phytopathologia Mediterranea* **10**, 132–135.

Bradbury J.F. (1984) Genus II. *Xanthomonas* Dowson. In *Bergey's Manual of Systematic Bacteriology* Vol. 1 (Ed. N.R. Krieg), pp. 199–210. Williams & Wilkins, Baltimore.

Buddenhagen I. & Kelman A. (1964) Biological and physiological aspects of bacterial wilt caused by *Pseudomonas solanacearum*. *Annual Review of Phytopathology* **2**, 203–230.

Bugbee W.M. & Anderson N.A. (1963) Whitefly transmission of *Xanthomonas pelargonii* and histological examination of leaf spots of *Pelargonium hortorum*. *Phytopathology* **53**, 177–183.

Burdekin D.A. (1972) Bacterial canker of poplar. *Annals of Applied Biology* **72**, 295–299.

Burkholder W.H. (1930) The bacterial diseases of the bean, a comparative study. *Cornell Agricultural Experiment Station Memoir* no. 127.

Burkholder W.H. (1942) Three bacterial plant pathogens: *Phytomonas caryophylli* sp.n., *Phytomonas alliicola* sp.n. and *Phytomonas manihotis* (Arthaud-Berthet et Bondar) Viégas. *Phytopathology* **32**, 141–149.

Burkholder W.H. (1944) *Xanthomonas vignicola* sp. nov. pathogenic on cowpeas and beans. *Phytopathology* **34**, 430–432.

Burkholder W.H. (1960) A bacterial brown rot of parsnip roots. *Phytopathology* **50**, 280.

Cafati C.R. & Saettler A.W. (1980) Role of non-host species as alternate inoculum sources of *Xanthomonas phaseoli*. *Plant Disease* **64**, 194–196.

Cameron H.R. (1962) Diseases of deciduous fruit trees incited by *Pseudomonas syringae*. A review of the literature with additional data. *Oregon Agricultural Experiment Station Technical Bulletin* no. 66.

Cheng C.-P. & Roane C.W. (1968) Sources of resistance and inheritance of reaction to *Pseudomonas coronafaciens* in oats. *Phytopathology* **58**, 1402–1405.

Clayton E.E. (1950) Wildfire disease of tobacco and soybeans. *Plant Disease Reporter* **34**, 141–142.

Cleene M. de, Leyns F., Moster M. van den, Swings J. & Ley J. de (1981) Reaction of grass varieties grown in Belgium to *Xanthomonas campestris* pv. *graminis*. *Parasitica* **37**, 29–34.

Coléno A., Le Normand M. & Hingand L. (1971) Sur une affection bactérienne de la pomme de chou-fleur. *Comptes rendus des Séances de l'Académie d'Agriculture de France* **57**, 650–652.

Cook A.A. & Stall R.E. (1969) Differentiation of pathotypes among isolates of *Xanthomonas vesicatoria*. *Plant Disease Reporter* **53**, 617–619.

Cook A.A. & Stall R.E. (1982) Distribution of races of *Xanthomonas vesicatoria* pathogenic to pepper. *Plant Disease* **66**, 388–389.

Cook A.A., Walker J.C. & Larson R.H. (1952) Studies on the disease cycle of black rot of crucifers. *Phytopathology* **42**, 162–167.

Crosse J.E. & Garrett C.M.E. (1970) Pathogenicity of *Pseudomonas morsprunorum* in relation to host specificity. *Journal of General Microbiology* **62**, 315–327.

Cunfer B.M. & Scolari B.L. (1982) *Xanthomonas campestris* pv. *translucens* on triticale and other small grains. *Phytopathology* **72**, 683–686.

Cupels D.A. & Kelman A. (1980) Isolation of pectolytic fluorescent pseudomonads from soil and potatoes. *Phytopathology* **70**, 1110–1115.

Daft G.C. & Leben C. (1972) Bacterial blight of soybeans: seedling infection during and after emergence. *Phytopathology* **62**, 1167–1170.

Daft G.C. & Leben C. (1973) Bacterial blight of soybeans: field over-wintered *Pseudomonas glycinea* as possible primary inoculum. *Plant Disease Reporter* **57**, 156–157.

Devash Y., Okon Y. & Heins Y. (1980) Survival of *Pseudomonas tomato* in soil and seed. *Phytopathologische Zeitschrift* **99**, 175–185.

Digat B. (1978) Mise en évidence de la latence épiphylle du *Xanthomonas pelargonii* chez le pelargonium. Sélection sanitaire des boutures de pelargonium et de *Begonia×elatior* "Rieger" vis-à-vis des bactérioses par utilisation de l'immunofluorescence. *Annales de Phytopathologie* **10,** 61–66; 67–78.

Dye D.W. (1963) *Xanthomonas begoniae* in New Zealand. *New Zealand Journal of Science* **6,** 313–319.

Dye D.W. (1967) Bacterial spot of ivy caused by *Xanthomonas hederae* in New Zealand. *New Zealand Journal of Science* **10,** 481–485.

Dye D.W., Bradbury R.S., Goto M., Hayward A.C., Lelliott R.A. & Schroth M.N. (1980) International standards for naming pathovars of phytopathogenic bacteria and a list of pathovar names and pathotype strains. *Review of Plant Pathology* **59,** 153–168.

Ebbels D.L. (1976) Diseases of upland cotton in Africa. *Review of Plant Pathology* **55,** 747–763.

Egli T. & Schmidt D. (1982) Pathogenic variation among the causal agents of bacterial wilt of forage grasses. *Phytopathologische Zeitschrift* **104,** 138–150.

Egli T., Goto M. & Schmidt D. (1975) Bacterial wilt, a new forage grass disease. *Phytopathologische Zeitschrift* **82,** 111–121.

Elliott C. (1951) *Manual of Bacterial Plant Pathogens.* Chronica Botanica, Waltham.

Erasmus H.D., Matthee F.N. & Louw H.A. (1974) A comparison between plant pathogenic species of *Pseudomonas, Xanthomonas* and *Erwinia* with special reference to the bacterium responsible for bacterial blight of vines. *Phytophylactica* **6,** 1–18.

Ercolani G.L. (1978) *Pseudomonas savastanoi* and other bacteria colonizing the surface of olive leaves in the field. *Journal of General Microbiology* **109,** 245–257.

Fahy P.C. & Persley G.J. (1983) *Plant Bacterial Diseases: A Diagnostic Guide.* Academic Press, New York.

Fang C.T., Allen V.N., Riker A.J. & Dickson J.G. (1950) The pathogenic, physiological and serological reactions of the formae speciales of *Xanthomonas translucens. Phytopathology* **40,** 44–64.

Fett W.F. & Sequeira L. (1981) Further characterisation of the physiologic races of *Pseudomonas glycinea. Canadian Journal of Botany* **59,** 283–287.

Fischer R. (1942) Der Lackschorf der Gladiole. *Kranke Pflanze* **19,** 73–75.

Garrett C.M.E., Crosse J.E. & Sletten A. (1974) Relations between phage sensitivity and virulence in *Pseudomonas morsprunorum. Journal of General Microbiology* **80,** 475–483.

Garrett C.M.E., Panagopoulos C.G. & Crosse J.E. (1966) Comparison of plant pathogenic pseudomonads from fruit trees. *Journal of Applied Bacteriology* **29,** 342–356.

Gitaitis R.D., Phatak S.C., Jaworski S.A. & Smith M.W. (1982) Resistance in tomato transplants to bacterial speck. *Plant Disease* **66,** 210–211.

Goode M. & Sasser M. (1980) Prevention—the key to controlling bacterial spot and bacterial speck of tomato. *Plant Disease* **64,** 831–834.

Grasso S., Moller W.J., Refatti E., Magnano di San Lio G. & Granata G. (1979) The bacterium *Xanthomonas ampelina* as causal agent of a grape decline in Sicily. *Rivista di Patologia Vegetale,* S. IV, **15,** 91–106.

Gremmen J. & Koster R. (1972) Research on poplar canker in The Netherlands. *European Journal of Forest Pathology* **2,** 116–124.

Grogan R.G. & Kimble K.A. (1967) The role of seed contamination in the transmission of *Pseudomonas phaseolicola* in *Phaseolus vulgaris. Phytopathology* **57,** 28–31.

Hagborg W.A.F. (1942) Classification revision in *Xanthomonas translucens. Canadian Journal of Research Section C,* **20,** 312–326.

Hakkaart F.A. & Versluijs J.M.A. (1982) Control of leaf curl and *Xanthomonas begoniae* in 'Elatior' begonias by meristem tip culture and a cultural technique. *Abstracts, XXIst. International Horticultural Congress* Vol. II, no. 1675.

Hale C.N. & Shanks J.C. (1983) The halo blight bacterium inhibits nodulation of dwarf beans. *Plant Science Letters* **29,** 291–294.

Hale C.N. & Taylor J.D. (1973) Races of *Pseudomonas phaseolicola* causing halo blight of beans in New Zealand. *New Zealand Journal of Agricultural Research* **16,** 147–149.

Harder D.E. & Harris D.C. (1973) Halo blight of oats in Kenya. *East African Agricultural and Forestry Journal* **38,** 241–245.

Hayward A.C. (1964) Characteristics of *Pseudomonas solanacearum. Journal of Applied Bacteriology* **27,** 265–277.

Hevesi M. & Viranyi F. (1975) An unknown symptom on onion plants caused by *Pseudomonas alliicola. Acta Phytopathologica Academiae Scientiarum Hungaricae* **10,** 281–286.

Hildebrand D.C., Schroth M.N. & Wilhelm F. (1967) Systemic invasion of strawberry by *Xanthomonas fragariae* causing vascular collapse. *Phytopathology* **57,** 1260–1261.

Hunter J.E. & Cigna J.A. (1981) Bacterial blight incited in parsnip, *Pastinaca sativa,* by *Pseudomonas marginalis* and *Pseudomonas viridiflava. Phytopathology* **71,** 1238–1241.

Janse J.D. (1981) The bacterial disease of ash (*Fraxinus excelsior*), caused by *Pseudomonas syringae* subsp. *savastanoi* pv. *fraxini.* I. History, occurrence and symptoms. II. Etiology and taxonomic considerations. *European Journal of Forest Pathology* **11,** 306–315; 425–438.

Janse J.D. (1982) *Pseudomonas syringae* subsp. *savastanoi* (ex Smith) subsp. nov., nom. rev., the bacterium causing excrescenses on Oleaceae and *Nerium oleander* L. *International Journal of Systemic Bacteriology* **32,** 166–169.

Kaiser W.J. & Vakili N.G. (1978) Insect transmission of pathogenic xanthomonads to bean and cowpea in Puerto Rico. *Phytopathology* **68,** 1057–1063.

Kam M. de (1978) *Xanthomonas populi* subsp. *salicis,* cause of bacterial canker in *Salix dasyclada. European Journal of Forest Pathology* **8,** 334–337.

Kam M. de (1981) The identification of two subspecies of *Xanthomonas populi in vitro. European Journal of Forest Pathology* **11,** 25–29.

Kawamoto S.O. & Lorbeer J.W. (1974) Infection of onion by *Pseudomonas cepacia*. *Phytopathology* **64**, 1440–1445.

Kelman A. (1953) The bacterial wilt caused by *Pseudomonas solanacearum*. *North Carolina Agricultural Experimental Station Technical Bulletin* no. 99.

Kennedy B.W. & Alcorn S.M. (1980) Estimates of US crop losses to protocaryote plant pathogens. *Plant Disease* **64**, 674–676.

Kennedy B.W. & King T.H. (1962) Angular leaf spot of strawberry caused by *Xanthomonas fragariae* sp. nov. *Phytopathology* **52**, 873–875.

Keyworth W.G. (1969) Plant pathology. *Report of the National Vegetable Research Station, Wellesbourne 1968*, pp. 85–97.

Khurtsiya B.N., Gvinepadze M.Sh., Khutsishvili N.A. & Tukhareli A.R. (1979) (On a new form of bacteriosis of mulberry occurring in Georgia.) *Trudy Nauchno-Issledovatel'skogo Instituta Zashchity Rastenii Gruzinskoi SSR* **30**, 53–57.

Kimura O., Robbs C.F., Ribeiro R. de L.D., Akiba F. & Sudo S. (1972) Identifição de patotipos de *Xanthomonas vesicatoria*, occurrendo na região centro-sul do Brasil. *Arquivos de Instito Biológico* **39**, 43–49.

Knösel D. (1961) Eine an Kohl blattfleckenerzeugende Varietas von *Xanthomonas campestris*. *Zeitschrift für Pflanzenkrankheiten und Pflanzenschutz* **68**, 1–6.

Koroleva I.B. & Novokhatka V.G. (1979) (Resistance of winter wheat cultivars to basal bacteriosis.) *Selektsiya i Semenovodstvo* no. 2, 24–25.

Krieg N.R. (Ed) (1984) *Bergey's Manual of Systematic Bacteriology* Vol. 1. Williams & Wilkins, Baltimore.

Kruijer C.J. & Vreeburg P.J.M. (1981) Hoe kunnen hyacinten het best worden bewaard na de heetstook?. *Bloembollencultur* **92**, 224–225.

Leben C., Pusch V. & Schmitthenner A.F. (1968) The colonization of soybean buds by *Pseudomonas glycinea* and other bacteria. *Phytopathology* **58**, 1677–1681.

Lelliott R.A. (1952) A new bacterial disease of leeks. *Plant Pathology* **1**, 84–85.

Lelliott R.A. & Stead D. (1987) *Methods for Diagnosis of Bacterial Diseases of Plants*. Blackwell Scientific Publications, Oxford.

Lelliott R.A., Billing E. & Hayward A.C. (1966) A determinative scheme for fluorescent plant pathogenic bacteria. *Journal of Applied Bacteriology* **29**, 470–478.

Lopez M.M., Gracia M. & Sampayo M. (1980) Studies on *Xanthomonas ampelina* Panagopoulos in Spain. *Proceedings of the Fifth Congress of the Mediterranean Phytopathological Union* Patras, Greece, 6–57.

Luisetti J., Prunier J.P. & Gardan L. (1972) Un milieu pour la mise en évidence de la production d'un pigment fluorescent par *Ps. mors-prunorum* f. sp. *persicae*. *Annales de Phytopathologie* **4**, 295–296.

Luisetti J., Gaignard J.L., Lalande J.C., Drouhard A. & Lafreste J.P. (1981) Actualités sur le dépérissement bactérien du pêcher. *Arboriculture fruitière* nos 329/330, 23–27.

Lukezic F.L. (1979) *Pseudomonas corrugata*, a pathogen of tomato, isolated from symptomless alfalfa roots. *Phytopathology* **69**, 27–31.

Mazzucchi U. (1975) Vascular blackening in sugar beet taproots caused by *Pseudomonas syringae*. *Phytopathologische Zeitschrift* **84**, 289–299.

Miller P.W. & Bollen W.B. (1946) Walnut bacteriosis and its control. *Oregon Agricultural Experiment Station Technical Bulletin* no. 9.

Min'ko N.D. & Koroleva I.B. (1980) (*Pseudomonas atrofaciens*—the main agent of bacteriosis of spring wheat in the forest steppe of Ukraine.) *Mikrobiologischeskii Zhurnal* **42**, 415–419.

Mitchell R.E. & Durbin R.D. (1981) Tagetitoxin, a toxin produced by *Pseudomonas syringae* pv. *tagetis*: purification and partial characterization. *Physiological Plant Pathology* **18**, 157–168.

Moffett M.L. (1976) A bacterial leaf spot disease of several *Brassica* varieties. *Australian Plant Pathology Newsletter* **5**, 30–32.

Moffett M.L. & Irwin J.A.G. (1975) Bacterial leaf and stem spot (*Xanthomonas alfalfae*) of lucerne in Queensland. *Australian Journal of Experimental Agriculture and Animal Husbandry* **15**, 223–226.

Moore L.W., Lagerstedt H.B. & Hartmann N. (1974) Stress predisposes young filbert trees to bacterial blight. *Phytopathology* **64**, 1537–1540.

Nelson P.E., Tammen J. & Baker R.R. (1959) Culture indexing and control of vascular wilt diseases of carnation. *Phytopathology* **49**, 547.

Nikitina K.V. & Korsakov N.I. (1978) (Bacteriosis of soybean in the Far East and Southern regions of USSR and the search for resistance.) *Trudy po Prikladnoi Botanike, Genetike i Selektsii* **62**, 13–18.

Ohata K., Serizawa S. & Shirata A. (1982a) (Infection source of the bacterial rot of lettuce caused by *Pseudomonas cichorii*.) *Bulletin of the National Institute of Agricultural Sciences, C*, **36**, 75–80.

Ohata K., Tsuchiya Y. & Shirata A. (1979) (Differences in kinds of pathogenic bacteria causing head rot of lettuce of different cropping types.) *Annals of the Phytopathological Society of Japan* **45**, 333–338.

Ohata K., Serizawa S., Azegami K. & Shirata A. (1982b) (Possibility of seed transmission of *Xanthomonas campestris* pv. *vitians*, the pathogen of bacterial spot of lettuce.) *Bulletin of the National Institute of Agricultural Sciences, C*, **36**, 81–88.

Olsson K. (1976) Experience of brown rot caused by *Pseudomonas solanacearum* in Sweden. *Bulletin OEPP/EPPO Bulletin* **6**, 199–207.

Ono K. (1979) (Occurrence of a pathogenic strain of *Pseudomonas tabaci* to wildfire resistant tobacco in Japan.) *Bulletin of the Morioka Tobacco Experiment Station* **14**, 61–72.

Palleroni N.J. (1984) Genus I. Pseudomonas. In *Bergey's Manual of Systematic Bacteriology*, Vol. 1 (Ed. N.R. Krieg) pp. 141–199. Williams & Wilkins, Baltimore.

Panagopoulos C.G. (1969) The disease Tsilik marasi of grapevine: its description and identification of the causal agent (*Xanthomonas ampelina* sp. nov.). *Annales de l'Institut Phytopathologique Benaki* N.S. **9**, 59–81.

Parashar R.D. & Leben C. (1972) Detection of *Pseudomonas glycinea* in soybean seed lots. *Phytopathology* **62**, 1075–1077.

Parker M.C. & Dean L.L. (1968) Ultraviolet as a sampling

aid for detection of bean seed infected with *Pseudomonas phaseolicola. Plant Disease Reporter* **52,** 534–538.

Piening L.J. (1976) A new bacterial leaf spot of sunflowers in Canada. *Canadian Journal of Plant Science* **56,** 419–422.

Prunier J.P., Luisetti J. & Gardan L. (1970) Etudes sur les bactérioses des arbres fruitiers. II. Caractérisation d'un *Pseudomonas* non fluorescent agent d'une nouvelle bactériose de pêcher. *Annales de Phytopathologie* **2,** 181–197.

Prunier J.P., Luisetti J. & Gardan L. (1974) Etudes sur les bactérioses des arbres fruitiers. Essais de lutte chimique contre le dépérissement bactérien du pêcher en France. *Phytiatrie-Phytopharmacie* **23,** 71–88.

Prunier J.P., Luisetti J., Gardan L., Germain E. & Sarraquigne J. (1976) La bactériose du noisetier (*Xanthomonas corylina*). *Revue Horticole* no. 170, 31–40.

Psallidas P.G. (1987) The problem of bacterial canker of hazelnuts in Greece caused by *Pseudomonas syringae* pv. *avellanae. Bulletin OEPP/EPPO Bulletin* **17,** 257–262.

Psallidas P.G. & Panagopoulos C.G. (1975) A new bacteriosis of almond caused by *Pseudomonas amygdali* sp. nov. *Annales de l'Institut Phytopathologie Benaki* **11,** 94–108.

Raeber J.G. (1966) Susceptibility of tobacco varieties to wildfire (*Pseudomonas tabaci*) and observations on the resistance of Burley 21 and T.L. 106. *Proceedings of the 4th International Tobacco Scientific Congress, Athens, 1966,* pp. 581–588.

Rattink H. & Vruggink H. (1979) A method to obtain Xanthomonas-free begonia plants. *Mededelingen van de Faculteit Landbouwwetenschappen, Rijksuniversiteit Gent* **44,** 439–443.

Ridé M. (1958) Sur l'étiologie du chancre suintant du peuplier. *Comptes Rendus Hebdomadaires des Séances de l'Académie des Sciences* **246,** 2795–2798.

Ridé M. & Marcelin H. (Eds) (1983) La nécrose bactérienne de la vigne (*Xanthomonas ampelina*). *Bulletin Technique des Pyrénées-Orientales,* no. 106. INRA, Paris.

Ridé M. & Ridé S. (1978) *Xanthomonas populi* (Ridé) comb. nov. (syn *Aplanobacter populi* Ridé), spécificité, variabilité et absence de relations avec *Erwinia cancerogena* Urosevic. *European Journal of Forest Pathology* **8,** 310–333.

Riedel R.M. & Larsen P.O. (1974) Interrelationship of *Aphelenchoides fragariae* and *Xanthomonas begoniae* on Rieger begonia. *Journal of Nematology* **6,** 215–216.

Rudolph K. (1979) Die bakterielle Braunfleckenkrankheit an Buschbohnen (*Phaseolus vulgaris* L.) in Deutschland, hervorgerufen durch *Pseudomonas syringae* van Hall s.s. pathovar *phaseoli. Zeitschrift für Pflanzenkrankheiten und Pflanzenschutz* **86,** 75–85.

Samson R. & Benjama A. (1987) Description de la bactérie responsable de la graisse du poireau: *Pseudomonas syringae* pv. *porri* pv. nov. *Agronomie* (in press).

Samson R., Poutier F. & Rat B. (1981) Une nouvelle maladie du poireau: la graisse bactérienne à *Pseudomonas syringae. Revue Horticole* no. 219, 20–23.

Sands D.C. & Hankin L. (1975) Ecology and physiology of fluorescent pectolytic pseudomonads. *Phytopathology* **65,** 921–924.

Scarlett C.M., Fletcher J.T., Roberts P. & Lelliott R.A. (1978) Tomato pith necrosis caused by *Pseudomonas corrugata* n.sp. *Annals of Applied Biology* **88,** 105–114.

Schmidt D. & Nuesch B. (1980) Resistance to bacterial wilt (*Xanthomonas graminis*), increases yield and persistency of *Lolium multiflorum. Bulletin OEPP/EPPO Bulletin* **10,** 335–339.

Schneider R.W. & Grogan R.G. (1977) Bacterial speck of tomato: sources of inoculum and establishment of a resident population. Tomato leaf trichomes, a habitat for resident populations of *Pseudomonas tomato. Phytopathology* **67,** 388–394; 898–902.

Schneider R.W., Hall D.H. & Grogan R.G. (1975) Effect of bacterial speck on tomato yield and maturity. *Proceedings of the American Phytopathological Society* (Abs), **2,** 1118.

Schroth M.N., Hildebrand D.C. & O'Reilly H.J. (1968) Off-flavor of olives from trees with olive knot tumours. *Phytopathology* **58,** 524–525.

Schuster M.L., Coyne D.P., Hulloka M., Brezina L. & Kerr E.D. (1978) Characterization of bean bacterial diseases and implications in control by breeding for resistance. *Fitopatologia Brasileira* **3,** 149–161.

Severin V. (1978) Ein neues pathogenes Bakterium an Hanf—*Xanthomonas campestris* pv. *cannabis. Archiv für Phytopathologie und Pflanzenschutz* **14,** 7–15.

Shackleton D.A. (1966) A bacterial leaf spot of cauliflower in New Zealand caused by *Pseudomonas maculicola. New Zealand Journal of Science* **9,** 872–877.

Shirata A., Ohata K., Serizawa S. & Tsuchiya Y. (1982) (Relationship between the lesion development by *Pseudomonas cichorii* and growth stage and leaf position of lettuce and its infection mechanism.) *Bulletin of the National Institute of Agricultural Sciences, C* **36,** 61–73.

Staub T. & Williams P.H. (1972) Factors influencing blackrot lesion development in resistant and susceptible cabbages. *Phytopathology* **62,** 722–728.

Strider D.L. (1975) Chemical control of bacterial blight of Rieger Elatior begonias caused by *Xanthomonas begoniae*; Susceptibility of Rieger Elatior begonia cultivars to bacterial blight caused by *Xanthomonas begoniae. Plant Disease Reporter* **59,** 66–70; 70–73.

Strider D.L. (1976) An epiphytotic of bacterial leaf and flower spot (*Xanthomonas nigromaculans zinniae*) of Zinnia. *Plant Disease Reporter* **60,** 342–344.

Strider D.L. (1980) Control of bacterial leaf spot (caused by *Xanthomonas nigromaculans*) of zinnia with captan. *Plant Disease* **64,** 920–922.

Styer D.J. & Durbin R.D. (1981) Influence of growth stage and cultivar on symptom expression in marigold, *Tagetes* sp., infected by *Pseudomonas syringae* pv. *tagetis. Hort-Science* **16,** 768–769.

Šutić D. & Dowson W.J. (1959) An investigation of a serious disease of hemp (*Cannabis sativa* L.) in Yugoslavia. *Phytopathologische Zeitschrift* **34,** 307.

Šutić D. & Dowson W.J. (1962) Bacterial leaf spot of Sesamum in Yugoslavia. *Phytopathologische Zeitschrift* **45,** 57–65.

Šutić D. & Dowson W.J. (1963) The reactions of olive,

oleander and ash, cross-inoculated with some strains and forms of *Pseudomonas savastanoi. Phytopathologische Zeitschrift* **46,** 305–314.

Šutić D. & Tesić Z. (1958) Jedna nova bakterioza bresta izazivač *Pseudomonas ulmi* n.sp. *Zaštita Bilja* **45,** 13–25.

Taylor J.D. (1970) Bacteriophage and serological methods for the identification of *Pseudomonas phaseolicola. Annals of Applied Biology* **66,** 387–395.

Taylor J.D. & Dudley C.L. (1977) Effectiveness of late copper and streptomycin sprays for the control of halo-blight of beans (*Pseudomonas phaseolicola*). *Annals of Applied Biology* **85,** 217–221.

Taylor J.D. & Dye D.W. (1972) A survey of the organisms associated with bacterial blight of peas. *New Zealand Journal of Agricultural Research* **15,** 432–440.

Taylor J.D., Dudley C.L. & Presley L. (1979a) Studies of halo blight seed infection and disease transmission in dwarf beans. *Annals of Applied Biology* **93,** 267–277.

Taylor J.D., Phelps K. & Dudley C.L. (1979b) Epidemiology and strategy for the control of halo-blight of beans. *Annals of Applied Biology* **93,** 167–172.

Tessi J.L. (1953) Estudio comparativo de dos bacterios patógenos en avena y determinación de una toxina que origina sus diferencias. *Revista de Investigaciones Agricoles* **7,** 131–145.

Trigalet A. & Bidaud P. (1978) Some aspects of epidemiology of bean halo blight. *Proceedings of the IVth International Conference on Plant Pathogenic Bacteria, Angers,* Vol. II, pp.895–902.INRA, Angers.

Trujillo G.E. & Saettler A.W. (1979) A combined semi-selective medium and serology test for the detection of Xanthomonas blight bacteria in bean seed. *Journal of Seed Technology* **4,** 35–41.

Tsuchiya Y., Ohata K. & Azegami K. (1982) (Pathogenicity of the causal bacteria of rot of lettuce, *Pseudomonas cichorii, P. marginalis* pv. *marginalis* and *P. viridiflava* to various weeds.) *Bulletin of the National Institute of Agricultural Sciences, C* **36,** 41–59.

Tsuchiya Y., Ohata K., Iemura H., Sanematsu T., Shirata A. & Frijii H. (1979) (Identification of causal bacteria of head rot of lettuce.) *Bulletin of the National Institute of Agricultural Sciences C* **33,** 77–99.

Umekawa M. & Watanabe Y. (1978) (Dry heat and hot water treatments of cucumber seeds for control of angular leaf spot.) *Bulletin of the Vegetable and Ornamental Crops Research Station B* **2,** 55–61.

Urošević B. (1976) (Bacterial canker of ash, olive and oleander.) *Lesnictví* **22,** 133–144.

Vakili N.G., Kaiser W.J., Peréz J.E. & Cortés-Monllor A. (1975) A bacterial blight of beans caused by two *Xanthomonas* pathogenic types from Puerto Rico. *Phytopathology* **65,** 401–403.

Walker J.C. (1941) Origin of cabbage black rot epidemics. *Plant Disease Reporter* **25,** 91–94.

Wallen V.R. & Galway D.A. (1979) Effective management of bacterial blight of blight beans in Ontario—a 10-year program. *Canadian Journal of Plant Pathology* **1,** 42–46.

Wallin J.R. (1946) Parasitism of *Xanthomonas translucens* on grasses and cereals. *Iowa State College Journal of Science* **20,** 171–193.

Wallin J.R. & Reddy C.S. (1945) A bacterial disease of *Phleum pratense* L. *Phytopathology* **35,** 937.

Watson R.W. (1971) A bacterial pathogen of sorghum in New Zealand. *New Zealand Journal of Agricultural Research* **14,** 944–947.

Wharton A.L. (1967) Detection of infection by *Pseudomonas phaseolicola* in white-seeded dwarf bean seed stocks. *Annals of Applied Biology* **60,** 305–312.

Wilkie J.P. & Dye D.W. (1974) *Pseudomonas cichorii* causing tomato and celery diseases in New Zealand. *New Zealand Journal of Agricultural Research* **17,** 123–130.

Wilkie J.P., Dye D.W. & Watson D.R.W. (1973) Further hosts of *Pseudomonas viridiflava. New Zealand Journal of Agricultural Research* **16,** 315–323.

Williams P.H. (1980) Black rot (caused by *Xanthomonas campestris*): a continuing threat to world crucifers. *Plant Disease* **64,** 736–742.

Wimalajeewa D.L.S. & Nancarrow R.J. (1980) Survival in soil of bacteria causing common and halo blights on French bean in Victoria. *Australian Journal of Experimental Agriculture and Animal Husbandry* **20,** 102–104.

Young J.M. & Dye D.W. (1970) Bacterial blight of peas caused by *Pseudomonas pisi* in New Zealand. *New Zealand Journal of Agricultural Research* **13,** 315–324.

Zaumeyer W.J. & Thomas H.R. (1957) A monographic study of bean diseases and methods for their control. *United States Department of Agriculture Technical Bulletin* no. 868 (revised).

7: Bacteria II
Agrobacterium, Corynebacterium, Erwinia and Other Bacteria

Covered in this chapter are the important plant-pathogenic genera *Agrobacterium, Corynebacterium* and *Erwinia* and also the minor pathogens in *Bacillus* and *Cytophaga*. Finally included is the actinomycete *Streptomyces scabies*.

AGROBACTERIUM

With the exception of the species *A. radiobacter*, a non-pathogenic soil inhabitant, *Agrobacterium* species induce hypertrophies in their host plants. Speciation within the genus has been primarily based on pathogenicity characteristics: *A. radiobacter*, non-pathogenic; *A. tumefaciens*, causing galls; *A. rhizogenes*, rhizogenic causing hairy-root disease and *A. rubi* causing galls and originally described on *Rubus* species. However, it is now known that the pathogenicity of agrobacteria is determined by plasmids that can be readily transferred to other virulent and avirulent agrobacteria, and cannot therefore be used as a basis for taxonomy. At present, regardless of the phytopathogenic effect of the strains, three biovars of *Agrobacterium* are recognized on the basis of clear-cut phenotypic differentiation. Moreover, Holmes & Roberts (1981) conclude that these biovars should be given specific rank. To indicate the phytopathogenic effect of strains within each species the terms tumorigenic state, rhizogenic state and saprophytic state were proposed. Unfortunately, this solution cannot be effected at present because of nomenclatural difficulties stemming from the International Rules. Until or unless a previous ruling by the Judicial Commission is reversed, the existing nomenclature has to be followed, although it can be adapted, albeit clumsily, to accommodate the three biovars. The problem and possible solutions are exhaustively discussed by Kersters & De Ley (1984). Their proposal, used here, is:

A. tumefaciens (Smith & Townsend) Conn comprising tumorigenic strains belonging to biovars 1, 2 and 3;

A. radiobacter (Beijerinck & Van Delden) Conn comprising non-tumorigenic strains belonging to biovars 1 and 2;

A. rhizogenes (Riker *et al.*) Conn comprising rhizogenic strains belonging to biovars 1 and 2;

A rubi (Hildebrand) Starr & Weiss comprising 3 strains, two of which are Hildebrand's original cultures, with high per cent DNA homology among themselves but only 15% with other agrobacteria tested.

The members of the genus are Gram-negative, non-sporing rods, 0.6–1.0×1.5–3.0 μm occurring singly or in pairs, motile by 1–6 peritrichous flagella. They are aerobic with a respiratory type of metabolism. Growth on carbohydrate-containing media is usually accompanied by copious, extracellular polysaccharide slime. They are chemo-organotrophs utilizing a wide range of carbon sources. Differentiation of biovars and species is given in Table 7.1. For diagnosis of the diseases caused see Fahy & Persley (1983) and Lelliott & Stead (1987).

Agrobacterium tumefaciens (Smith & Townsend) Conn

(Syn. *A. radiobacter* var. *tumefaciens, A. radiobacter* subsp. *tumefaciens* (Smith & Townsend) Kean *et al.*)

Basic description

See generic description above, Table 7.1 and CMI 42.

Host plants

A. tumefaciens attacks many if not most dicotyledonous plants; 643 species from 331 genera are recorded as hosts of the pathogen. Stone and pome fruit trees, grapevines, roses and berries (*Rubus* spp.) are the most commonly and seriously affected crop plants.

Table 7.1. Differential characteristics of the species and biovars (Bv.) of the genus *Agrobacterium*

Characters	*A. tumefaciens*			*A. radiobacter*		*A. rhizogenes*		*A. rubi*
	Bv 1	Bv 2	Bv 3	Bv 1	Bv 2	Bv[2] 1	Bv 2	
Growth:								
at 35°C	+[1]	−[1]	d[1]	+	−		−	d
on medium of Schroth *et al.* (1965)	+	−	−	+	−		−	
on medium of New & Kerr (1971)	−	+	−	−	+		+	
in presence of 2% NaCl	+	−	+	+	−		−	
Production of 3-ketolactose	+	−	−	+	−	+	−	−
acid from:								
meso-Erythritol	−	+	−	−	+		+	+
Melezitose	+	−	−					
Ethanol	+	−	−	+	−		−	−
Alkali from:								
Na malonate	−	+	+	−	+		+	+
Na L-tartrate	d	+	+					
Na propionate	d	−	−					
Simmon's citrate +0.005% yeast extract	−	+		−	+		+	−
Reaction in litmus milk:								
alkaline	+	−	+	+	−		−	+
acid	−	+	−	−	+		+	−
Tumours produced on stems of young plants of *Helianthus annuus*, tomato, *Nicotiana tabacum* and/or discs of roots of *Daucus carota*	+	+	d[3]	−	−		+	+
Roots produced on discs of roots of *Daucus carota*	−	−	−	−	−	+	+	−

1: + = 90% or more of strains positive
− = 90% or more of strains negative
d = 11–89% of strains positive
2: rhizogenic biovar 1 strains have been little studied and are apparently rare.
3: most strains from grapevine have a very limited host range and isolates from this host should be tested on young shoots of grapevine.

Host specialization

Some strains of the pathogen appear to have a considerable host specificity which is primarily determined by the Ti plasmid (Bevan & Chilton 1982). Thus, the biovar 3 strains (Panagopoulos *et al.* 1978) have a narrow host range—they infect only grapevine, *Rubus* species and chrysanthemum, whereas the biovar 1 and 2 strains have a wide host range.

Geographical distribution

The bacterium has a world-wide distribution.

Disease

Crown gall or plant cancer is a typical neoplastic disease incited by a Ti plasmid harboured by the virulent strains of *A. tumefaciens*. Upon wounding and infection a piece of plasmid DNA (T-DNA) is transferred to the plant cell and is permanently integrated within the host nuclear DNA where it codes for tumorigenesis and also for the synthesis of tumour-specific compounds termed opines (Bevan & Chilton 1982). The disease first appears as small overgrowths or swellings, initiated from the transformed cells on any part of the plant, but particularly near the soil surface (crown), graft union, and roots. Aerial galling is quite common on certain plants (including grapevine and *Rubus* spp.). New galls appear during the growing season and are globular with smooth surfaces, whitish in colour, tender and soft. Later they become much larger (up to 30 cm in diameter), harder or woody, dark brown in colour and more or less fissured. Gall tissue is usually a confused mass of parenchyma and vascular elements. During dormancy tumours rot partially or completely, and in the case of perennial hosts often develop again in the same sites the following growing season. Since crown gall may be confused with tumour-like overgrowths due to a variety of causes such as insects, mites, fungi, viruses, genetic inbalance, wound healing and phytohormonic substances, a laboratory diagnosis is essential. It can be accomplished by: a) isolation and identification of the pathogen, b) tissue analysis for the presence of tumour-specific opines (octopine type, nopaline type, or agropine type) using high voltage paper electrophoresis, c) specific detection of T-DNA in the gall tissue using ^{32}P-labelled T-DNA as a probe (Fahy & Persley 1983). The diseased plants may be stunted with chlorotic leaves, of low productivity and may eventually die.

Epidemiology

A. tumefaciens mainly overwinters in tumours and in soil. The bacterium is generally considered to be a true soil inhabitant that can persist indefinitely. However, reports on survival in soil without plant debris are conflicting and the existence of biovars, notably biovar 3, has only been recognized recently. It is therefore not possible to assess and evaluate reports on the capacity of *Agrobacterium* to survive in soil and to determine which strains were investigated by earlier workers. Moreover, in recent ecological studies pathogenic strains were found only in close association with galled trees while almost all strains of agrobacteria from the soil and rhizosphere of healthy plants were non-pathogenic. There is also evidence that the pathogen survives saprophytically in the vascular tissues of grapevine and possibly of other plants. Transmission of the pathogen over long distances and into uninfested areas is usually accomplished with contaminated seed and planting or grafting material. In transmission over short distances, rain and irrigation or soil water are very important. The bacterium is also spread by means of soil, soil insects, animals and man, as well as by cultural practices. Fairly recent wounds, made by cultural practices, grafting, insects, hail or frost, appear to be a prerequisite for successful infection. Tumour induction is completed within about 20 h after wound inoculation at 25°C and thereafter bacterial cells are no longer necessary for the tumour development and its maintenance. Moreover, crown gall formation only occurs readily if the temperature does not exceed 30°C for a week after inoculation. The incubation period (time for gall appearance) is about 8–15 days during the growing season, but there is a delay of up to a few months when inoculation takes place during dormancy or at very low temperatures.

Economic importance

The disease constitutes a serious and economically important problem in nursery stock. Affected plants are stunted or killed and generally unmarketable because galled. Losses can also be severe in young orchards and vineyards. The effects of crown gall on plant vigour, yield and survival are dependent upon the location, size and number of the galls and the age of the plant when infection occurred. Crown gall was considered the most serious prokaryotic disease of California crops in 1975–77, with estimated annual loss of $23 million in 1976 for the whole USA (Kennedy & Alcorn 1980).

Control

A remarkably successful biological control of crown gall on fruit trees and roses has been developed by Kerr (1980). The method is cheap, simple and very effective and is now widely practised by commercial growers in several countries. It is based on the use of an avirulent strain of an agrobacterium (strain 84) that produces a nucleotide bacteriocin (known as agrocin 84) that inhibits most pathogenic agrobacteria (i.e. agrobacteria harbouring the nopaline-type Ti plasmid). Susceptible planting material (seeds, cuttings, roots of plants) is dipped in a cell suspension of the above strain and planted immediately. A few nopaline-producing, pathogenic strains however are resistant to agrocin 84 and consequently not subject to biological control (Kerr & Panagopoulos 1977). Also, all agrobacterium strains harbouring octopine and agropine Ti plasmids are resistant to agrocin 84 and therefore not controlled by it; these include the biovar 3 strains (octopine Ti plasmids) which are of considerable importance in causing crown gall on grapevine. Soil fumigation, root protection with chemicals and the use of eradicants such as Bacticin (a mixture of liquid hydrocarbons) have met with limited and inconsistent success. Other methods of control are based on cultural and sanitary practices. Thus, wounding of the trunk and roots during cultivation should be avoided. Planting and grafting material should only be obtained from disease-free stock plants. All tools, containers, and equipment should be sterilized prior to use, especially in propagation areas. The use of formalin or sodium hypochlorite solution has been recommended but these substances are corrosive and objectionable to operators; boiling water is an alternative but would frequently be impracticable. The disinfectants used in fireblight control (q.v.) would probably be most effective.

Special research interests

The new developments in crown gall research (Kahl & Schell 1982; Merlo 1982) have been the focus of considerable attention by plant pathologists, molecular biologists and plant genetic engineers. An understanding of the mechanism of control and the biology of the controlling organism could serve as a model for the development of biocontrol procedures against other bacterial diseases. Moreover, crown gall research provides model systems for studies of host–pathogen interactions (at organismal, genetic and molecular levels), plant cell growth and regulation, mammalian cancer, microbial evolution, and prokaryotic gene expression in a eukaryotic cell. Finally, one of the most exciting aspects of the crown gall system, which is a natural form of genetic engineering, concerns the potential use of the Ti plasmid as a vector for the transfer of foreign DNA into plant nuclear DNA. This has been achieved and offers the prospect of stable incorporation and expression of genes encoding beneficial traits (e.g. disease resistance, yield improvement, nitrogen fixation) into cultivated plants, all of which would be of great agricultural importance.

PANAGOPOULOS

Agrobacterium rhizogenes (Riker *et al.*) Conn (CMI 41) causes hairy root disease (so called because it induces root proliferation) and attacks a wide range of dicotyledonous plants (Cleene & De Ley 1981), but generally causes little economic damage. However considerable reduction in cucumber yields can be caused by a root disorder associated with this bacterium (Yarham & Perkins 1978).

BACILLUS

Bacillus polymyxa (Prazmowski) Mace is a widespread, pectolytic, spore-forming species that inhabits decomposing vegetation and can at high temperatures cause rots of fruits of pepper, aubergine, tomato, potato tubers, celery and cabbage, and bacterial pit of mushroom caps.

CORYNEBACTERIUM

The plant pathogenic members of the genus are a heterogeneous collection of species that recent taxonomic research has suggested should be distributed into other genera. Thus *C. fascians* has been transferred to *Rhodococcus*, *C. flaccumfaciens* and its pathovars to *Curtobacterium* and *C. michiganense* and its pathovars to *Clavibacter* as subspecies of *Clavibacter michiganense*. These taxonomic changes have not been adopted here since they are not yet generally accepted by plant pathologists. However the proposals that some nomen species are best regarded as pathovars has more general acceptance and has been adopted here. All the species are Gram-positive, pleomorphic. club-shaped (coryneform) bacteria that, because of their method of division, form characteristic 'V' 'Y' and 'W' arrangements of cells. Only *C. flaccumfaciens* is motile and all except this species grow slowly on artificial media. For more details of the genus refer to Lelliott (1974) and for diagnosis of the diseases they cause see Fahy & Persley (1983) and Lelliott & Stead (1987).

Corynebacterium fascians (Tilford) Dowson (syn. *Rhodococcus fascians* (Tilford) Goodfellow) (CMI 121) causes leafy gall disease characterized by proliferation, usually at or near ground level, of short stems which frequently become distorted or thickened. It is common and affects a wide range of herbaceous hosts including many ornamentals but only rarely causes serious economic damage. It is soil-borne, seed-borne and contagious, being mainly spread on hands and implements or by water splash during propagation.

Corynebacterium flaccumfaciens (Hedges) Dowson pv. ***betae*** (Keyworth *et al.*) Dye & Kemp (syn. *C. betae, Curtobacterium flaccumfaciens* pv. *betae* Collins & Jones) causes silvering disease of seed crops of red beet and fodder beet in the UK and Ireland which can be serious but is controlled by streptomycin seed treatment (Keyworth & Howell 1961).

Corynebacterium flaccumfaciens pv. *flaccumfaciens* (Hedges) Dowson

(Syn. *C. flaccumfaciens, Curtobacterium flaccumfacians* pv. *flaccumfaciens* (Hedges) Collins & Jones)

C.f. pv. *flaccumfaciens* is a slow-growing bacterium, producing cream- to butter-yellow growth on dextrose media, that causes an economically serious, seed-borne wilt of cultivars of *Phaseolus vulgaris* (CMI 43). Other species of *Phaseolus (P. coccineus, P. aureus, P. angularis* and *P. mungo)* are susceptible. It has also been recorded on soybean. Orange and violet variants of the bacterium occur. It is widely distributed in N. America and E. Europe, including the far-eastern USSR, but has not been confirmed in W. Europe (EPPO 48).

Bacterial wilt of bean is primarily a vascular disease. Affected plants are stunted and the leaves wilt. Wilting can occur in plants of all ages from seedling to maturity and is not confined to a single part of the plant. Wilted leaves become dry and brown and, after heavy rain or wind, ragged. Red lesions, which appear to originate from infected xylem, can develop at nodes and girdle the stem, weakening it and causing stems to break during windstorms. Dark-green, water-soaked lesions develop along the suture of pods, sometimes for the entire length, due to invasion from the underlying vascular tissue. Infected seed can be contaminated externally or systemically invaded; it is often symptomless, but severely affected seed can vary from a yellow spot on the hilum to patchy discoloration of the seed coat (which in white seeds may be bright yellow) and severe wrinkling with a varnished appearance to the coat. Most crop infection originates from infected seed and although there is some evidence for overwintering in soil (probably in trash), this is probably not important. It is not readily spread within the crop by rain splash and secondary spread occurs through wounds, particularly below soil level, caused by cultivation or by the nematode *Meloidogyne incognita*, and is assisted over short distances by irrigation water. Incidence is said to be higher in sandy loam than clay loam soils. Control is by use of disease-free seed and crop rotation; some cultivars are partially resistant.

LELLIOTT

Corynebacterium flaccumfaciens (Hedges) Dowson pv. ***oortii*** (Saaltink & Maas Geesteranus) Dye & Kemp (CMI 375) (syn. *C. oortii, Curtobacterium flaccumfaciens* pv. *oortii* Collins & Jones) causes geelpok (yellow pustule) and hel-vuur (hell-fire) of tulips, and is reported from the Netherlands, UK, GDR and Denmark, although probably it is of much wider distribution. It produces a yellow discoloration of the vascular tissue and small white spots on the outermost scale leaf of bulbs; affected leaves develop silver-grey spots up to 5 mm in diameter, and become brittle, rough and cracked. It is controlled by rogueing.

Corynebacterium michiganense **pv.** ***rathayi*** (Smith) Dye & Kemp (CMI 376) (syn. *C. rathayi, Clavibacter rathayi* (Smith) Davis *et al.*) causes yellow slime disease of *Dactylis glomerata*; it is common but rarely economically damaging on seed crops in N.W. Europe. It is very closely related, if not identical, to pathovars *tritici*, and *iranicum* of *C. michiganense* which cause similar diseases of wheat and some other grasses.

Corynebacterium michiganense pv. *insidiosum* (McCulloch) Dye & Kemp

(Syn. *C. insidiosum. Clavibacter michiganensis* ssp. *insidiosus* Davis *et al.)*

Basic description

C.m. pv. *insidiosum* (CMI 13) can be distinguished from other members of the genus by its production of dark-bluish granules of indigoidine on dextrose-containing media. Indigoidine production by some isolates is slow but can often be enhanced by incubation of streaked, plate cultures at 25–27°C until growth is clearly visible, followed by transfer to 15–20°C and incubation for up to 2 weeks.

Host plants

The disease this pathogen causes is economically important only in lucerne (*Medicago sativa*), although *Medicago falcata, Melilotus alba* and *Onobrychis viciifolia* are also said to be susceptible.

Geographical distribution

The bacterium is widespread in lucerne-growing areas of Britain and Italy and also apparently in Czechoslovakia, Poland and the USSR (EPPO 49). It is probably more widely distributed in Europe than is indicated by the literature. It was probably introduced into Europe with seed from N. America in the late 1940s.

Disease

The disease, known as bacterial wilt of lucerne, is often overlooked in the first year of a ley and by the second it can be distributed so uniformly throughout the ley that it is not realized that the general poor growth of the crop is due to disease—this insidious nature of the disease has given the causal organism of the disease its pathovar name *'insidiosum'*. In less severe outbreaks the disease is initially patchy in distribution. Symptoms are accompanied by chlorosis, reduction in size and often cupping of leaflets. Marginal, papery, whitish grey necrosis of leaflets is common and wilting may occur during hot weather. However the disease is not well named for N. European conditions as the characteristic symptoms there are not those of a wilt but of stunting and proliferation of stems; in this respect it differs from wilt caused by *Verticillium albo-atrum* in which wilting and rapid death of plants without stunting is characteristic. Death of plants infected by pv. *insidiosum* is slow and they often survive in a debilitated state until the next season. This appears to allow invasion of affected plants by *Sclerotinia* species which can mask the bacterial wilt symptoms effectively. A yellowish-brown discoloration of the vascular system of the tap root usually accompanies the aerial symptoms in some cultivars; in older plants this discoloration can often be seen as a moist or gummy ring in the outer wood in transversely or tangentially cut roots. With less advanced symptoms only pockets of discoloration are present in the wood and sometimes only slight flecking. Severe aerial symptoms are not always associated with marked root symptoms and vice versa and this appears to be related to host cultivar: in cvs. Du Puits and Eynsford, root symptoms can be most marked even when aerial symptoms are absent; in cvs. Europe and Vertus root symptoms can be far less evident and are often absent even in severely affected plants. Where root symptoms are present they can usually be distinguished from those of wilt caused by *Verticillium albo-atrum*, typically a grey discoloration that includes the centre of the root. However, both diseases can occur in the same plant.

The detection of large numbers of Gram-positive or immunofluorescent coryneforms in association with the characteristic symptoms is a sufficiently accurate diagnostic method for most purposes (Lelliott & Stead 1987). For greater accuracy, isolation (Close & Mulcock 1972) and slide agglutination with specific antiserum can be used.

Epidemiology

The disease is transmitted most rapidly within the crop by the cutter bars of harvesters and can spread by the movement of surface water. Harvesters, drainage water, wind-blown soil and trash may spread it to adjacent crops. Transport over longer distances is by seed or its accompanying debris, or in lucerne hay. There is good evidence that bacterial wilt can be transmitted by *Ditylenchus dipsaci* and that infection is much more extensive in the presence of *Meloidogyne hapla*.

The bacterium is said, largely on observational evidence, to survive in soil, either directly or in lucerne root debris, in the absence of its host, but there is no reliable evidence for the length of time it can do so; it would probably be unwise to assume that survival after ploughing in of an affected ley is less than 5 years.

Economic importance

In N. America, where it has been endemic for many years, pv. *insidiosum* was generally considered to be the most important pathogen of lucerne, having caused greater losses in some parts of the USA than all other lucerne pathogens together, until the breeding and adoption of resistant cultivars in the 1950s. Even so in 1976 losses were estimated at $17 million (Kennedy & Alcorn 1980). It is certainly a major pathogen of lucerne in Italy but its status in N. Europe is generally uncertain. What evidence there is indicates that it could be or become, a most important European pathogen ranking at least with *V. albo-atrum* as a cause of crop loss. The economic consequences of the pathogen arise from its reduction of vigour and growth of the crop so that yield is often considerably reduced in the second year and not economically worthwhile in the third. Generally speaking, if the disease is present, leys of susceptible cultivars are limited to a maximum economic life of 3 years.

Control

The only effective method of control has been by the use of resistant cultivars. Those bred in the USA are culturally unsuitable for growing in N.W. Europe but the resistant European cv. Maris Phoenix is suitable. Trials in the UK of this cultivar indicate that its dry matter yield is somewhat less than the standard cv. Europe, although its protein yield is said to be as high. However, Maris Phoenix is not resistant to *V. albo-atrum* and until a high-yielding cultivar with resistance to both *V. albo-atrum* and pv. *insidiosum* has been bred, and a programme to do this is well advanced, the use of Maris Phoenix as an insurance against bacterial wilt is probably only marginally justifiable, except on farms with a history of bacterial wilt.

C.m. pv. *insidiosum* is so widespread that any seed source must be suspect, although efficient seed cleaning and the growing of specialist seed crops in wide rotations can reduce the risk. If the disease is known to be present on a farm it is advisable, where possible, to cut the infected field or fields last and disinfect machinery after use in order to prevent further spread. It is also a sensible precaution to ensure that harvesting machinery brought onto a farm is disinfected before use. This is best achieved by high-pressure hosing of those parts likely to be in contact with the crop, particularly the cutter bar and wheels, to remove soil and plant debris and then spraying with a phenolic disinfectant.

Special research interest

Interactions between pv. *insidiosum* and other micro-organisms are of interest. Infection of plants with *Fusarium oxysporum* f.sp. *medicaginis* appears to inhibit symptoms of, and colonization of vascular tissue by pv. *insidiosum* (Johnson *et al.* 1982). Also the mechanism of resistance to bacterial wilt may be associated with certain characteristics of nitrogen fixation of *Rhizobium meliloti* (Bordeleau & Michaud 1981).

LELLIOTT

Corynebacterium michiganense (Smith) Jensen pv. *michiganense* Dye & Kemp

(Syn. *C. michiganense, Clavibacter michiganensis* ssp. *michiganensis* (Smith) Davis *et al.*)

Basic description

This (CMI 19) is a slow-growing, non-motile member of the genus. On nutrient glucose agar, on which growth is optimal, colonies are 1 mm in diameter after 5 days and 2–3 mm in 7–8 days; they are smooth, entire, convex, semi-fluidal when freshly isolated but become butyrous with prolonged sub-culturing, pale yellow becoming deeper yellow, opaque and glistening.

Host plants

Tomato and other *Lycopersicon* species, sweet pepper as well as *Solanum mammosum, S. douglassi, S. nigrum* and *Nicotiana glutinosa* are natural hosts. A number of solanaceous hosts are also susceptible on artificial inoculation (Thyr *et al.* 1975).

Host specialization

No specialized forms are generally accepted, although there are a few reports of strains with differential virulence to different hosts.

Geographical distribution

The bacterium is widespread in the main tomato-growing areas of the world, including Europe (EPPO 50).

Disease

The disease, known as bacterial canker or bird's eye spot of tomato, is a vascular and parenchymatal disease with a wide range of symptoms (well described in detail for glasshouse crops by Penna (1982) and for field crops by Strider (1969)), which vary depending on differences in cultural techniques used to grow the crop—particularly nutrition, whether they are field or protected crops, age of plants and the age at which they are affected—and probably the cultivar affected (Penna 1982). The vascular phase of the disease is probably far less common than the foliar blight phase.

Under protected cropping, vascular symptoms are usually not seen until plants reach or approach maturity. This is usually not before plants are 1.5–2 m high, symptoms appearing first on leaves in the region of the second or third truss above that being harvested. Small areas of interveinal tissue turn dull green and water-soaked then become desiccated and light brown; areas of necrosis extend and affected plants appear as if scorched by a chemical or to have some physiological disorder—conditions with which the symptoms can be easily mistaken. This desiccation is probably caused by a toxin produced by the pathogen that disrupts and alters the permeability of the cell membranes of host parenchyma. Eventually the plant wilts irreversibly, leaves first. Usually symptoms of wilt and necrosis move slowly up the stem so that the growing shoot is the last part to be affected; frequently, new vigorous shoots that often remain symptomless are produced from the base of the plants. Less commonly there is a rapid collapse of plants within a few days. Fruit may fail to develop and fall off, or ripen unevenly; it also often shows external marbling and internal bleaching of vascular and surrounding tissue. Plants with advanced symptoms almost always have some internal symptoms too. The cortex of young stem tissue may be easily peeled away from the stele, which is often straw-coloured and has a roughened texture. In older stem tissue discoloration of vascular tissue varies from straw-yellow to tan but the lower part of stems or roots may still appear healthy. Occasionally cavities develop in the pith associated with yellow bacterial ooze. Yellowing of vascular tissues and ooze production around them often occurs in petioles of affected leaves and in marbled fruit. Brown streaking and splitting of stems to produce cankers is not common. Systemic symptoms in field crops are not dissimilar to those under protected cropping although stem cankers rarely develop, and then only exceptionally during periods of high temperature and humidity; wilting is not always present.

Parenchymatal infection is generally more common under glass than in the field, particularly when syringing with water to aid pollination or high volume pesticide spraying is practised. Roughly circular, slightly raised, white spots about 1 mm in diameter develop on leaves, tending to be more numerous near the midrib and main veins; these may expand, coalesce and become surrounded by necrotic tissue that can drop out, particularly after spraying, giving the leaf a tattered appearance. Similar spots occur on stems, petioles, peduncles and calyces which, when packed close together form areas with a mealy appearance; such areas are rough and often turn light to dark brown and scabby. On fruits 'bird's eye' spotting begins as small white spots that develop to 3–6 mm diameter with a raised dark-brown centre; these can lose their white halo as the fruit matures, become wholly necrotic and crusty and can coalesce. Sometimes necrotic lesions develop on leaves, predominantly at their margins, consisting of a necrotic area up to 4×4 cm surrounded by a yellow-orange border 1–1.5 cm wide. Sometimes roughly circular yellow spots up to 1 cm in diameter, usually with a necrotic centre, develop over the leaf surface. These can coalesce and may be accompanied by necrotic streaking of petioles and leaf veins.

The detection of large numbers of Gram-positive, coryneform cells in association with characteristic symptoms is a sufficiently accurate diagnostic method for most purposes. Otherwise isolation of bacteria with the cell and colony morphology of *C. michiganense,* that produce characteristic symptoms on cotyledons of tomato seedlings is a more certain diagnosis (Lelliott & Stead 1987).

Epidemiology

The disease can be seed-borne, usually as a contaminant and rarely internally, and the pathogen can survive for at least 8 months in the seed although much seed contamination dies out before sowing. Where studied, not more than 1% seed transmission occurred (Grogan & Kendrick 1953). However when plants are grown for transplanting 1% seed transmission can result in 100% infection in the planted-out crop.

The bacteria can survive well in infected plant debris in or on the soil surface and on stakes and posts, and this can also provide primary foci of infection in crops, particularly in protected crops where the bacterium can also survive for months on the glasshouse structure or equipment. It probably does not survive long in soil *per se.* In some geographical areas, systemic infection is uncommon, probably because primary foci there derive from crop debris rather than seed infection and superficial infection proceeds for some months before invasion of the vascular system occurs, if it does at all before the crop senesces. In other areas systemic infection is the norm, probably reflecting comparatively high rates of seed contamination or infection, or rapid spread of infection in seedling beds before transplanting. Secondary spread of superficial infection occurs by rain splash, particularly when high winds drive the rain, or by spray droplets that have become contaminated with the bacterium by initially striking infected plants or other contaminated surfaces. High humidity during and after spread allows establishment of infection through wounds, stomata, hydathodes, lenticels or trichomes. Spread can also take place on contaminated clothing and hair of workers especially when the crop is wet and humidity high. In general such spread is less serious in field crops which are usually grown in climates with dry summers than in protected crops. Direct spread of systemic infection usually occurs during cultural operations such as trimming, defoliation and harvesting. Root infection occurring from soil-borne inoculum is not important unless there has been considerable root damage during planting.

Economic importance

Bacterial canker is a major disease of the outdoor tomato crop and can be very difficult to control under protected cropping (Shoemaker & Echandi 1976). Before the adoption of hygiene measures and specific seed extraction methods for control, the disease could commonly cause losses in yield of up to 70%. It is likely that in protected crops, where superficial symptoms can be difficult to diagnose without experience, losses are higher than realized.

Control

Bacterial canker is one of the most difficult tomato diseases to control once it is established in a crop as it can be extremely contagious. Detection of infected plants can be difficult and there are no really effective means of chemical treatment. Furthermore the pathogen can survive for long periods in the environment of the crop. Only seed that has been acid or fermentatively extracted (which markedly reduces contamination) should be used (Thyr *et al.* 1975). The general adoption of this technique by seed houses reduced the importance of the disease for many years but the recent trend, particularly in the USA, towards centrifugal extraction could hardly be better calculated to lead to high levels of seed contamination since comparatively small centrifugal forces (more than about 200 g) will precipitate most of the pathogen from the pulp with the seed.

Once the disease occurs in a field crop overhead irrigation and high volume spraying should be stopped and ultra low volume spraying with copper-based sprays should be substituted. Three sprays at 10-day intervals should be applied followed by sprays at 3–5 week intervals. Similar measures should be taken for protected crops but additional consideration should be given to rogueing, isolation of the infection and adoption of stricter hygiene measures (for details see Penna 1982). Whether hygiene measures or wide rotations are appropriate for field crops is a matter of judgement in individual cases. Removal of affected crop debris in protected crops is vital to prevent carry over to the next season. Sources of resistance are available but have not yet been incorporated to any significant degree in commercial cultivars.

LELLIOTT

Corynebacterium sepedonicum (Spieckermann & Kotthoff) Skaptason & Burkholder

(Syn. *C. michiganense* pv. *sepedonicum* Dye & Kemp, *Clavibacter michiganensis* ssp. *sepedonicus* Davis *et al.*)

Basic description

This (CMI 14) is a slow-growing, non-motile member of the genus that does not grow at 37°C. It is difficult to isolate on artificial media on which it is non-pigmented or pale yellow. It is best identified serologically or by host tests (see below).

Host plants

The only economically important host is potato but many species of *Solanum* and tomato can be infected experimentally. Some weed species of *Solanum* including *S. dulcamara* are natural hosts.

Geographical distribution

Ring rot was, until the last decade, comparatively widespread in Scandinavia, Finland and N. USSR, although rigorous attempts at control have reduced it to low levels in Denmark and to little economic importance in Sweden (EPPO 51). It has been reported, but not confirmed, from some mid- and E. European countries and it undoubtedly occurs in Poland and the GDR, though probably at low levels. From what is known of its world distribution it is unlikely to occur in S. Europe except in mountainous areas.

Disease

The bacterium causes bacterial ring rot, a wilt and tuber rot (Fig. 25). The aerial symptoms can be rather variable and, since they usually occur late in the growing season, can be obscured by symptoms of potato blight (*Phytophthora infestans*) or drought. The first indication of infection is a partial or complete wilting of one or more lower leaves or leaflets, the margins of which roll upwards and inwards. The leaves lose their colour and first become a dull light-green, then greyish-green with occasional mottling, and feel thin and smooth to the touch. They turn yellow and finally become brown and necrotic. Although the lower leaves are usually the first affected, sometimes some or all stems wilt from the top downwards. The daughter tubers may become infected without the

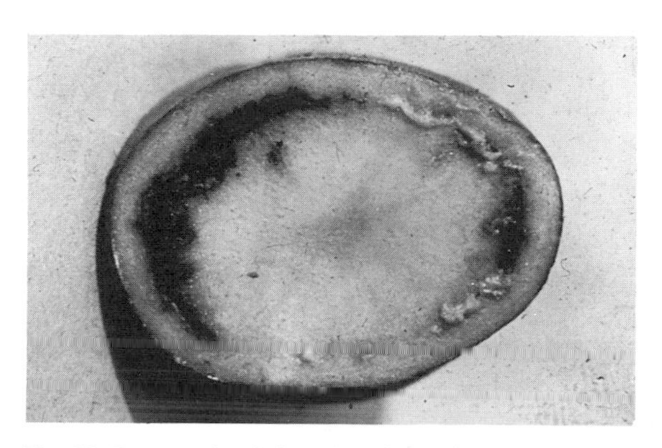

Fig. 25. Potato tuber infected by *Corynebacterium sepedonicum* (British Crown Copyright).

aerial part of the plant showing any sign of the disease. If affected stems are cut across at their base, but above any discoloration, a creamy or milky fluid which contains the ring rot bacteria can usually be squeezed with pliers from the area round the wood strands. This fluid is thicker than the normal sap, is not frothy and is confined to the woody tissues.

In plants showing aerial symptoms the bacteria normally penetrate the tubers as well as the stems, producing rather varied tuber symptoms. In lightly infected tubers the vascular ring and tissue immediately surrounding it is pale yellow or glassy and difficult to distinguish from that in healthy tubers. With more advanced infection, part or all of this vascular ring and some of the surrounding tissue rot, appearing white, cream, yellow or light brown. The rot is odourless and cheesy or crumbly in texture. Less commonly, a similar rot affects the centre of the tuber. The area of rotting may extend into the pith in the later stages and it is characteristic that the tissue outside the vascular ring can easily be separated from the inner tissues. Bacteria may break outwards from the vascular region, discolour the skin to reddish-brown and cause deep irregular fissures. These cracked tubers are very liable to invasion by secondary, soft rot bacteria which obscure the original ring rot symptoms and can reduce the tuber to a mere shell. The tuber symptoms as they appear in the field equally describe the symptoms in stored tubers. In the USSR subcortical, circular, yellowish areas have been described on tubers after harvesting, apparently derived from lenticular infection at lifting.

Typical symptoms are sufficiently characteristic for visual diagnosis to be adequate in most cases. However latent infection is an important factor in control of the disease and methods have been developed to detect it and to diagnose the disease in tubers when secondary bacterial rotting is also present. The most sensitive and reliable is the aubergine inoculation test (Lelliott & Sellar 1976) but this takes 6–40 days to complete and is demanding of glasshouse facilities. An immunofluorescence test (De Boer & Copeman 1980) is widely used and, if adequate positive and negative controls and optimum serum dilutions are employed, it is sufficiently specific for most tasks and can be completed in a few hours; however it can give false positives and would not be expected to detect very low levels of bacteria. Successful attempts to develop the ELISA technique have not yet been reported but it should be ideal for large surveys and in the testing of seed stocks.

Epidemiology

The spread of ring rot from plant to plant in the crop is negligible although there is experimental evidence that some insects, including the Colorado beetle (*Leptinotarsa decemlineata*) can transmit the disease. Most spread occurs when seed comes into contact with contaminated machinery, cutting knives or containers. These, when contaminated with diseased tissue from affected seed or ware tubers can cause extensive infection of seed tubers with which they come into contact. In machinery and storage sheds contaminated surfaces, including graders, planters, bin walls, sacks, boxes, chitting trays and vehicles remain contagious for at least six months and probably much longer (Nelson 1980). Experimentally, bacteria have been recovered by bioassay from tested surfaces for 14 months at 95% RH and 24 months at 12% RH at 5° or 20°C, and from dried infected potato stems for 26 months in an unheated machinery shed (Nelson 1980). If seed tubers are cut before planting, the knife, after cutting an affected tuber, can infect the next 20–30 healthy tubers cut. Similarly planting machines, particularly those in which the seed is drawn into the coulters by hooks or spikes, can spread infection through a seed lot. Tubers from infected volunteer (self-sown) plants, overwintering from a previous crop, could contaminate an otherwise healthy seed or ware crop at lifting; this could be of importance in maritime climates where volunteer tubers are not killed by freezing in winter.

In the growing crop, the bacterium passes from infected mother seed tuber to daughter tubers through the vascular system of the stolons and invades the vascular system of the developing daughter tuber. It can break out from the vascular system into surrounding parenchyma tissue and cause a progressive rot; the rate at which this happens presumably depends on environmental factors, particularly temperature and humidity, as well as the physiological condition of the tuber and the cultivar. Generally speaking increasing amounts of infection are expressed as rot with increased storage time, but latent infection frequently remains unexpressed even internally and can, in the case of seed, give rise to further infected crops if undetected. The degree of latency in stocks can be high—95% of infected tubers carrying latent infection have been detected in crops—and in tolerant cultivars. Latent infection can remain unexpressed through two seed generations before symptoms are produced. There is field evidence that the bacterium does not survive over winter in soil although it is assumed to do so in volunteer tubers in mild winters.

Economic importance

Losses arise mainly from rotting of tubers in the field and in storage; loss of yield from wilting of plants is generally small in comparison with that from tuber rot. For many decades ring rot was the major potato disease in the USA and Canada with estimates of annual losses as high as 10–15% of total crop a year in some USA states. Strenuous and costly efforts in recent years have reduced its importance there although it is still considered a major disease (Kennedy & Alcorn 1980). It is also a major disease in the USSR. For reasons that are not fully understood, it has caused little direct economic damage in mid- and W. Europe although it has been troublesome in Scandinavia. Undoubtedly the general absence of seed cutting and of pricker-type seed planters, which are potent means of spreading disease in N. America account in part for its comparative unimportance in many European countries where climatic conditions would otherwise be thought to favour its development. However, high and increasing mechanization in handling crops in these countries indicates that the potential for serious damage exists if the disease is allowed to become firmly established. If it did become established in such countries it would restrict the cultural options available to them for growing the crop.

Control

Control depends almost entirely on the production and planting of healthy seed; tolerant cultivars have been bred but these are not widely grown on a commercial basis. Production of healthy seed requires seed certification schemes with nil tolerances for ring rot in samples of tubers cut during inspections made for ring rot during storage. Because of the problem of latency, testing of at least high-grade stocks by the methods discussed under *Disease* (above) are necessary. The use of new, disposable bags, or disinfection of re-usable bags, and disinfection of bulk seed containers is probably also essential for moving and marketing seed. On affected farms all equipment and surfaces that could be contaminated should be rigorously disinfected after an outbreak of ring rot and all potatoes sold off the farm to the ware market before fresh healthy seed is brought on to the farm again. Fields that have had infected crops should not be planted with potatoes again while there is a risk of volunteer plants from the affected crop being present. Disinfection of seed merchant's premises and machinery and particularly those who also handle ware stocks should be routine. Quaternary ammonium compounds are probably the most practicable disinfectants for most uses.

LELLIOTT

CYTOPHAGA

Cytophaga johnsonae Stanier (syn. *Flavobacterium pectinovorum*) is an aerobic, yellow, pectinolytic, Gram-negative bacterium with gliding mobility. It occurs commonly on plant surfaces and appears under suitable conditions of high temperature and humidity to be capable of rotting and necrosis of soft stems and root tissue (Lelliott & Stead 1987).

ERWINIA

This is a somewhat heterogeneous genus of Gram-negative, rod-shaped bacteria ($0.5–1.0 \times 1.0–3.0$ μm) occurring singly, in pairs or sometimes in short chains. *Erwinia* species are facultative anaerobes, although some are only weakly so, utilizing a moderately wide range of carbon and energy-yielding sources; they are oxidase negative and catalase positive. Decarboxylases for arginine, lysine or ornithine, urease and lipases are only rarely detectable and glutamic acid is not decarboxylated. The mol % G+C of the DNA is 50–58%. Members are associated with plants as pathogens, saprophytes or as constituents of the epiphytic flora; at least one has also been isolated from human and animal hosts. Full descriptions are given by Lelliott & Dickey (1984). The genus can be somewhat arbitrarily and conveniently divided into three groups: the amylovora, herbicola and carotovora groups but such groupings are not wholly supported by cluster analysis and the presence of intermediate species between the groups indicates a continuum (Lelliott & Dickey 1984). The European members of the amylovora group, *Erwinia amylovora, E. tracheiphila* and *E. salicis,* all grow weakly anaerobically, require growth factors (except *E. salicis*), are non-pigmented on standard growth media, will not grow at 36C and are generally only moderately active biochemically. They are specialized pathogens. The European members of the herbicola group, the ubiquitous epiphyte and doubtful pathogen *E. herbicola* and the insect-transmitted *E. stewartii* are both usually yellow in culture but are not closely inter-related taxonomically. They are nutritionally undemanding. The European species of the soft rotting carotovora group, *E. carotovora* and *E. chrysanthemi,* are more active biochemically. They can occur in the same rotted tissue and are differentiated as in Table 7.2. For more details of diagnosis of the members of the genus consult Fahy & Persley (1983) and Lelliott & Stead (1987).

Erwinia amylovora (Burrill) Winslow *et al.*

Basic description

E. amylovora (CMI 44) is serologically distinct from most related bacteria found in association with tissue of its host. Colonies on 5% sucrose nutrient agar are levantype and characteristic of the pathogen. Slide agglutination of such colonies with appropriately diluted serum or the production by them of ooze on immature pear fruits is diagnostic.

Host plants

Within the family Rosaceae, the only important hosts are in the pome-fruit sub-family (the Pomoideae). The most commonly affected hosts in Europe have been pear, apple, hawthorn (*Crataegus* spp.), *Cotoneaster* spp. (particularly the larger species), *Sorbus aria* and *Pyracantha* ssp.; other less commonly affected hosts have been *Sorbus aucuparia, Stranvaesia davidiana*, ornamental apple and pear (*Malus* and *Pyrus* ssp.) and quince. Overwintering of the bacterium in hosts outside the Pomoideae is unknown. It has been isolated from many other hosts, including non-rosaceous ones, but this is of little economic or epidemiological significance.

Geographical distribution

It is widely distributed in N. America, New Zealand and in parts of the Far East (CMI 2). The first known infection of *E. amylovora* in Europe was in 1957 in England. It has since been found and is established in Belgium, Denmark, France, FRG, GDR, the Netherlands and Poland. It has reached the orchard areas of Aquitaine in S. France but is not yet known in the extensive apple- and pear-growing areas of Rhône, the Po valley of Italy or the Ebro valley of Spain. It is probably the most important intra-European quarantine organism (EPPO 52).

Disease

In pear and other highly susceptible hosts such as *Sorbus aria* and the large cotoneasters, *E. amylovora* typically causes the following symptoms. Blossoms and sometimes shoots of vigorously growing trees wilt and turn dark-brown or black (Plate 10). During summer, the bacterium spreads into spurs and rapidly (often up to 25–50 mm/day) along branches, forming cankers, and can affect and rot fruit. Branches are quickly killed. Usually the tree will die within one or a few seasons unless treated. During humid weather in spring and summer, a bacterial slime may ooze from affected shoots, blossom trusses, branches and fruits. Leaves on affected branches turn dark-brown and, like affected fruits, usually remain attached after leaf-fall in autumn. Typically the cortex of the affected branch is at first dark-green to brown and water-soaked but later becomes drier; the cambial region is discoloured reddish-brown. In autumn, or in periods of prolonged drought, the disease usually becomes inactive, cankers cease extending and cracks appear at their margins (in mild winters cankers can continue extending after leaf-fall). In less vigorous pear trees and in apples, hawthorn and less susceptible ornamental hosts the spread of bacteria within the tissue is less rapid and the damage proportionally less. Sometimes infection is restricted to blossoms or fruiting spurs and does not spread into the branches. The reddish discoloration of the cambial regions may be less apparent.

Epidemiology

Active, overwintering cankers produce inoculum in the following spring which is carried by rain splash or insects to blossoms or young shoots. In suitable weather, spread of infection between blossoms by pollinating insects can be very rapid. Rain splash, wind-borne dried slime or contaminated pollen are also vectors. Contaminated pruning implements can cause devastating spread of the disease. Risk of infection is greatest when temperatures exceed 18°C and there is rain. Blossom infection is particularly likely on warm (21–30°C), sunny days when insect activity is high. Damaging storms greatly increase the chances of shoot infection. Branches can be directly infected through wounds but bacteria usually invade shoots during wet, windy weather via the vascular system of leaves, particularly if damaged, or through leaf scars.

In N.W. Europe spread in pear orchards in spring is rarely important, almost certainly because inoculum levels high enough to produce widespread infection during the comparatively cold springs rarely coincide with blossom production. But more spread occurs in apple in which spring blossom production is later, i.e. when temperatures and inoculum levels are higher. However pear and, to a much lesser extent, apple can produce secondary, summer blossom which, in N.W. Europe, is a main avenue for futher infection of trees. Infected wild hawthorn or hawthorn hedges within 400 m of an orchard are potent sources of inoculum. Judged by experiences in N. America and from analysis of weather data, even in S. Europe epidemics in spring blossom should not be expected every year.

Long-distance spread occurs but its causes other than the movement of affected plants are largely speculative. Wind-borne pollen, dried slime or insects could be responsible for this spread and there is some evidence that after roosting on infected branches birds could carry infection to previously unaffected areas; however there is no concrete evidence that spread occurs by these means. There is contradictory evidence as to whether beehives can harbour *E. amylovora* for long enough to carry infection when moved to distant orchards or new areas but there is no field evidence in England that infection has ever been so introduced. Movement of infected planting material, particularly shrubs, is the main cause of spread to other areas and there is good circumstantial evidence that the bacterium can be introduced into orchards on contaminated wooden fruit boxes.

Unlike some other bacterial diseases of fruit trees, the pathogen does not exist on plant surfaces as an epiphyte, only as a migrant; it has a very short life on the leaves of pear or apple. There are considerable differences in tissue-susceptibility between cultivars of the same host, but there are often differences within the same cutivar between young and old wood. Often cultivars with high tissue susceptibility escape infection for many years in some areas because the time when this can occur most easily (blossoming and shoot extension) fails to coincide with weather and/or inoculum levels favourable for infection. Thus a cultivar may be susceptible in one area but not in another where it escapes infection. Generally speaking summer-blossoming pear cultivars and late-flowering apple and pear cultivars are most often affected and tissue susceptibility is much greater in pear than apple; there are also large differences in tissue susceptibility between susceptible shrubs.

Economic importance

In N.W. Europe direct orchard losses on a national scale have been minor, although some individual growers have suffered severely, but losses to nurserymen of ornamental hosts have been more serious and the effects on wild hawthorn and on hawthorn hedges have been not inconsiderable. There is however the potential for serious loss in orchards if unusually favourable spring weather coincides with unusually high amounts of overwintering inoculum. Very much greater orchard losses in S. Europe can be expected if or when the bacterium becomes widely established there. These are likely to be on the scale experienced in N. America, where the disease has limited pear production to areas less favourable to the disease and restricted the apple cultivars that can be grown in many areas. The outbreaks in S. France have already indicated that pear cv. Passe Crassane, which is also widely grown in Italy, is rarely likely to be economically worth growing in affected areas in S. Europe.

Control

In N.W. Europe, where spring blossom infection is infrequent and spasmodic, routine spraying of apple and pear as practised in N. America is not economically justified. Spraying to protect summer blossom is not feasible because of phytotoxicity to fruit and would also be of doubtful economic value for control. Spraying nursery stock with copper is often worthwhile as phytotoxicity is comparatively unimportant. However the disease can usually be adequately controlled in orchards if growers inspect orchard trees and adjacent hawthorns when symptoms of infection are likely to be seen, and then cut out infected branches or shoots quickly and with suitable hygienic precautions. Such periods can be predicted using warnings based on weather, and with knowledge of local disease behaviour (Billing 1984).

The amount of cutting-out needed depends on knowledge of many local factors and requires considerable experience to be successful without causing unnecessary damage. Gradual replacement of hawthorn hedges round orchards and nurseries is also advised. Control in the UK, which is also probably applicable to most of N.W. Europe, has been discussed in detail by Lelliott & Billing (1984). In S. Europe there is, as yet, insufficient experience to make firm recommendations for control. However the adoption of less susceptible cultivars and avoidance of practices that encourage soft, sappy growth will probably be necessary where these are practicable. Spraying with streptomycin during blossoming would probably be effective but is not at present permitted in most W. European countries for medical reasons. Cooper sprays, although less effective, may reduce losses particularly if application is based on risk assessment and warnings (Billing 1984).

Special research interest

Because of the comparative ease with which *E. amylovora* can be identified on isolation plates of 5% sucrose nutrient agar using colony characters and slide agglutination, it offers considerable scope for quantitative epidemiological studies.
LELLIOTT

Erwinia cancerogena Urošević* is said to cause necrosis and hypertrophy of the subcortical tissues, accompanied by marked fissures, of poplar (similar to bacterial canker (*Xanthomonas populi* pv. *populi*)) and a bark necrosis of *Picea abies* in Czechoslovakia. The neotype strain is closer to the non-phytopathogenic genus *Enterobacter* than to *Erwinia*.

Erwinia carotovora (Jones) Bergey *et al.* subsp. *atroseptica* (van Hall) Dye

(Syn. *E.c.* pv. *atroseptica*)

Basic description

This is a pectolytic member of the genus which is a strong facultative anaerobe and which utilizes a wide range of carbon sources. It can be distinguished from the closely related subspecies *carotovora* by the characteristics given in Table 7.2. For a fuller description see Lelliott & Dickey (1984) and CMI 551. Diagnosis of the disease it causes, potato blackleg, depends

*This name is not in the Approved Lists of Bacterial Names and is therefore illegitimate under the International Code.

Table 7.2. Differentiation of *Erwinia carotovora* and *E. chrysanthemi*

	E. carotovora subsp. *carotovora*	*E. carotovora* subsp. *atroseptica*	*E. chrysanthemi*
Growth at 36°C	+	–	+
Reducing substances from sucrose	–	+	+
Indole production	–	–	+
Phosphatase	–	–	+
Lecithinase	–	–	+
Acid from:			
lactose	+	+	(+)
trehalose	+	+	–
maltose	–	+	–
α-methyl glucoside	–	+	–

For methods see Lelliott & Dickey (1984).

primarily on the isolation of the Gram-negative, non-sporing bacteria from plants with typical disease symptoms. In warm temperate parts of Europe the possibility of blackleg caused by *E.c.* subsp. *carotovora* must be allowed for. For further details of diagnosis see Lelliott & Stead (1987).

Host plants

The only important host is potato. It has also been recorded as attacking *Delphinium* species, fodder beet, turnip, *Saintpaulia ionantha, Ficus elastica* and bean (*Phaseolus vulgaris*), although confirmation of these reports is needed because of the difficulty in differentiating this pathogen from *E.c.* subsp. *carotovora* whose host range is very wide.

Geographical distribution

The bacterium probably occurs wherever potatoes are grown extensively, but see also under *Disease* and *Economic importance* below.

Disease

The bacterium causes blackleg of potatoes of which it is the chief if not only cause in cool temperate climates such as those of N. and N.W. Europe. However, in warmer climates *E.c.* subsp. *carotovora* and *Erwinia chrysanthemi* can cause similar or identical symptoms. Under cool temperate conditions plants may sometimes die before, or soon after, they appear above the ground so that 'blanks' occur in the crop. More often though, the first signs of the disease are seen in plants early in the season, before the haulm meets across the drills; affected plants are then often stunted and 'hard', with pale green or yellowish foliage. The leaves, particularly those at the top of the plant, are stiff and erect, and their margins roll inwards. The base of the stem, for up to about 10 cm above and below soil level, is black or dark-brown and rotted, but still firm; during dry periods this basal rot is lighter-brown, dry and often shredded. Not all the stems of a plant may be affected; sometimes only one or two show symptoms. If affected stems are cut across about 5 cm above the rot, a brown discoloration of the three main strands of wood tissue can usually be seen. The old set or seed tuber is almost invariably completely rotted. As the season advances the leaves of affected plants turn brown and the haulam dies. In plants in which the disease appears later in the season, or after the haulm has met between the drills, or during spells of wet weather, the basal rot is blacker, wet and soft, and blackish streaks may sometimes be found externally, higher up the stems. Such plants often wilt and collapse rapidly.

Epidemiology

The subject of epidemiology is extensively reviewed by Pérombelon & Kelman (1980). In temperate regions where *E.c.* subsp. *atroseptica* is the main agent of blackleg early in the season most seed tubers are contaminated with the pathogen, usually in the lenticels, at planting. If it is present in the vascular system it can invade the stems and, through the stolons, the daughter tubers. Stem invasion may also take place through cortical tissue. Whether or not the contamination of seed tubers results in blackleg lesions on stems appears to depend on many factors including the inoculum level on the seed, soil temperatures, humidity and damage to seed. The relation of these factors to incidence is complex which makes the prediction of incidence difficult. In bad attacks practically all the tubers decay in the ground; however the disease does not usually advance so rapidly, and the tubers may remain sound for a time and then decay in storage. Tubers which have been affected only slightly survive the winter, and if planted may produce diseased plants. Extensive contamination of tubers occurs in the growing crop so that most tubers are carrying the bacteria externally when lifted; these bacteria can invade and rot tubers during storage if conditions favour them.

Considerable losses during storage may result from the infection of healthy tubers through contact with diseased ones. The soft, wet masses formed by the rotting, diseased tubers contain millions of bacteria which, under conditions of high humidity, can cause decay of sound tubers. The bacteria enter by way of the lenticels and through wounds. In dry storage conditions washed tubers may develop a hard rot. This takes the form of slightly sunken, dark brown, dry lesions that usually encircle lenticels. In severe cases the whole of one side of a tuber may be affected. The affected area is shallow and clearly differentiated from the underlying healthy tissue.

Studies on the survival of the pathogen(s) in soil have given conflicting results. But there appears little doubt that populations decline rapidly in fallow soil and that the rate of this decline is related to soil temperature, but not in a simple way. The bacteria probably survive in potato plant debris and certainly on volunteer tubers surviving the winter. They can also survive in the rhizosphere of weed hosts. However the significance of these as sources of overwintering inoculum is probably small in relation to survival as contaminants on/in seed tubers.

Attempts to produce seed stocks free of contamination has revealed other means by which crops can be contaminated during the growing season. Apart from movement of the bacteria in soil water from mother to daughter tubers and from stems to daughter tubers by rain, movement on insects (mainly dipterous flies) and from potato cull piles to crops is well documented. Dispersal over longer distances, in aerosols developed when the haulm is removed from affected crops by mechanical flailing before harvest, has been shown to occur in Scotland.

Economic importance

Blackleg is a major cause of loss not in the growing crop (where incidence is not usually above 2%) but in the stored crop. In N.W. Europe national overall losses as high as 5% can be expected in store mainly from rotting by *E.c.* subsp. *atroseptica*. In warmer parts of Europe higher losses, also involving *E.c.* subsp. *carotovora* are to be expected.

Control

Apart from the use of less susceptible cultivars, there has been little success in controlling blackleg in the field. However control of the major source of loss—in store—is possible. Crops should be lifted during dry weather, or if this is not possible, allowed to dry before storage. Adequate ventilation and temperature control of stores using forced ventilation to prevent or reduce surface water films on tubers reduces losses considerably. If parts of stores are affected they should be dressed and marketed first.
LELLIOTT

Erwinia carotovora (Jones) Bergey *et al.* subsp. *carotovora*

Basic description

See *E. carotovora* subsp. *atroseptica* above, Table 7.2., Lelliott & Dickey (1984), and CMI 552.

Host plants

The pathogen will attack most soft tissue of virtually any species of plant if the conditions are favourable.

Host specialization

No host specialization is known although strains are often more virulent to the host from which they were isolated than strains from other hosts when compared on a single host.

Geographical distribution

The bacterium is of world-wide distribution in both temperate and tropical climates.

Disease

E.c. subsp. *carotovora* causes a soft rot of plant tissue symptomatically indistinguishable from that caused by *E. chrysanthemi* and usually from that caused by *E.c.* subsp. *atroseptica*. Pathogenicity appears to be largely dependent on the production of the pectic enzyme endopolygalacturonate transeliminase (PGTE) in large quantities that breaks down the middle lamella between cells since strains of this pathogen and of *E. chrysanthemi* that have lost the ability to produce large quantities of PGTE are no longer virulent, but virulence can be restored by transferring this characteristic from wild, virulent strains. However this is unlikely to be the only determinant of pathogenicity since other bacteria that can produce copious PGTE are rarely associated with soft rots. Disorganization and death of potato cells occurs very early in the infection process, perhaps because of a reduction in the oxygen potential of the tissue, before PGTE can be detected in tissue or is likely to have caused significant cell wall damage (Lelliott, unpublished).

The pathogen is often present as a latent infection or more normally occurs extracellularly, as a contaminant or a commensal in many crop host plants. When conditions favour the pathogen its multiplication may be triggered and rotting ensue. The factors that disturb the balance between the pathogen and its host are poorly understood. One major factor known to affect initiation of rotting in potato tubers, and probably the tissues of many, if not all, other hosts, is the presence of free water on the tissue surface. This can increase tissue turgidity, consequently increasing its susceptibility to rotting and potentially depleting oxygen in the tuber, and causing cell leakage by affecting membrane integrity. Once leakage of cell solutes have flooded the intracellular spaces the pathogen can multiply rapidly in a nutrient solution in intimate contact with the host tissue.

Epidemiology

Generally speaking, *E. carotovora* subsp. *caratovora* does not appear to survive in soil *per se* although there appears little doubt that it can survive over winter in many climates in plant debris, on or in volunteer host plants, and in the rhizospheres of cultivated and weed plants, probably as part of the natural rhizosphere flora. Methods of spread have been investigated in more detail for the subsp. *atroseptica* but both subspecies are spread by insects, in aerosols, by grading, planting and harvesting machinery and during washing of vegetables before marketing.

Economic importance

This is one of the most economically important plant pathogens, particularly in lifted crops. World losses are estimated at $50–100 $\times 10^6$ a year for this species and *E. chrysanthemi*, although the latter itself is economically less important.

Control

Control of the field stage once it has begun is impractical. Rotation to avoid build-up of infected plant residues would be expected to reduce the inoculum available but is not usually agronomically feasible. Cull piles of diseased material in the field should be avoided where possible. Undoubtedly the single most important control measure is the avoidance of a film of moisture on the surface of packed or stored material; anything that can be done to reduce this in the design of plastic packs for vegetables and the design and operation of bulk stores will materially reduce losses.

Special research interests

This pathogen has been extensively studied as a model for simple host–parasite systems. Ecological studies of survival and dissemination and of the saprophytic phase are now also yielding interesting results.
LELLIOTT

Erwinia chrysanthemi Burkholder et al.

E. chrysanthemi (CMI 553) is closely related to *E. carotovora* from which it differs in the delayed utilization of lactose, utilization of trehalose, production of *a*-methyl indole, phophatase and lecithinase (Lelliott & Dickey 1984) (see Table 7.2). It causes various rotting, necrotic and systemic diseases of a wide range of tropical and subtropical crops (Dickey 1979), including blackleg of potato in both the tropics and subtropics. It is also common on ornamental protected crops such as saintpaulia, carnation, chrysanthemum, philodendron and dieffenbachia grown in temperature climates. It also attacks maize and pineapple and causes a foot rot in rice. There is evidence of the host specificity, and pathovars attacking chrysanthemum, carnation, dieffenbachia, and maize (among others) have been proposed. However although strains from chrysanthemum will not affect carnation, and most corn isolates will not attack chrysanthemum or other ornamental hosts, the patterns of host specificity are not simple (Dickey 1981) and probably form a mosaic with overlapping host ranges (Samson *et al.* 1987). Since some pathovars cause distinct diseases those likely to be found in Europe (EPPO 53) are treated individually (see below).
LELLIOTT

Erwinia chrysanthemi* pv. *chrysanthemi Burkholder *et al.* (syn. *E. chrysanthemi* Burkholder *et al.*) causes a blight of chrysanthemums which involves wilting, leaf and stem necrosis and vascular browning. A pith rot of rooting cuttings is not uncommon. Since the adoption of control measures at the propagation stage, including the use of indexed stock, the disease has become much less important economically. Almost certainly this pathovar attacks other crops naturally.
LELLIOTT

Erwinia chrysanthemi pv. *dianthicola* (Hellmers) Dickey

E.c. pv. *dianthicola* causes a widely distributed disease of protected and field-grown carnations known as slow wilt or bacterial stunt. It has been recorded in the UK, Netherlands, USA, New Zealand, France, Italy, FRG, Poland, Denmark, Sweden, and Greece and probably occurs wherever carnations are grown. Just after the beds are planted, infected plants can often be picked out by their unthrifty appearance, lighter colour, rather, staring habit, and flaccidity and loss of bloom of the leaves. A slow progressive wilt develops and plants become grey-green. These symptoms usually develop over many months before the plant is finally killed. Although plants often show the first external symptoms when about three months old, symptom development can be delayed for many months. Progressive wilting does not always occur and plants may be stunted and make little growth. A root rot accompanies the aerial symptoms but basal stem rotting is absent. The stems of infected plants are often extensively cracked, but this is an unreliable diagnostic characteristic because cracking is not uncommon, though usually less severe, on healthy plants. The inner wood of the infected stem is usually brown at the base. The discoloration tails off into streaks which become lighter coloured, dry and frayed towards the top of the stem, Sometimes, and particulary in young plants, the wood is discoloured only in streaks at the stem base. Where the stem is cracked, the underlying wood is often disorganized, leaving a dry necrotic pocket. Sections and teased preparations from affected plants show the bacteria mainly in the vessels, with little or no necrosis of surrounding tissue or dissolution of the vessels themselves. Heavily infected groups of vessels often extend to the tops of the stems and laterals. Bacteria are densely packed in the vessels, devoid of Brownian movement, and embedded in gelatinous matter, which is often extruded from the broken ends of the vessels. Teased preparations also show very light-brown rods of gelatinous matter, about 50 μm wide and up to 500 μm long, containing masses of bacteria from broken infected vessels.

The *dianthicola* pathovar is a vascular pathogen that rarely invades other tissue. The degree of wilting and stunting expressed would seem to be a consequence of the proportion of vascular tissue invaded and the physiological state of the plant. In plants with only a small proportion of xylem invaded grown in conditions that restrict growth, the disease often remains latent. When plants are growing rapidly, and much of the vascular system is invaded, wilting can be more rapid. The disease persists, often undetected, in stock mother plants and can be spread on the knife used to take cuttings. Spread is very rapid if cuttings are held in water or liquid rooting hormone before being placed in rooting beds, but does not usually occur between plants in growing beds; however the bacterium can survive in root debris in soil long enough to infect replants. In the 1950s and 1960s slow wilt was a major disease of carnations but is now far less common probably as a result of the adoption of cutting indexing schemes (see *Pseudomonas caryophylli* and *Xanthomonas campestris* pv. *pelargonii*) and of dry formulations of rooting hormones.

LELLIOTT

Erwinia chrysanthemi* pv. *dieffenbachiae (McFadden) Dye causes bacterial stem and leaf rot of *Dieffenbachia picta*. It has been recorded in FRG, France and Italy but probably occurs wherever the host is grown. More virulent strains spread in the vascular system and less virulent strains produce only local soft rot lesions. Losses can be heavy but the disease can be controlled in propagation material by treatment with antibiotics followed by hot water; hygiene precautions are needed to prevent spread in the growing crop (McFadden 1961)

Erwinia chrysanthemi* pv. *zeae Victoria *et al.* causes a bacterial top and stalk rot of maize involving wilting of whorls, stalk rot and final collapse of the plant. It has been serious in a few crops in Italy (Mazzucchi 1972). There appears to be resistance in some lines of maize.

Erwinia herbicola (Lohnis) Dye (CMI 232) is common on plant surfaces and as a secondary organism in lesions caused by many plant pathogens. It has been implicated in a number of diseases later shown to be caused by other pathogens but, interestingly, there is some reasonable evidence that some strains can cause galls. It is weakly pectolytic at pH 7.0–8.5 and this may account for its association with onion rots at high temperature.

Erwinia rhapontici (Millard) Burkholder (CMI 555) causes crown rot of rhubarb, pink grain of wheat and internal browning of hyacinth. It occurs epiphytically, and saprophytically in lesions caused by other bacteria. For details see Sellwood & Lelliott (1978).

Erwinia salicis (Day) Chester

E. salicis (CMI 122) causes watermark disease of the cricket-bat willow (*Salix alba* var. *caerulea*) a serious disease of local importance in the UK, and also in cultivars of white willow (*S. alba*) in the Netherlands where it has caused extensive and serious damage. For a fuller account, see Preece (1977). Typical symptoms have also been seen on *S. babylonica, S. fragilis, S. cinerea, S. purpurea* and *S. viridis*. Consistent disease production by inoculation with pure cultures is often difficult to achieve although easily achieved by parallel inoculations with infected sap. Some isolates may be non-virulent.

Typically, leaves wither and become reddish in late spring and during the summer; some affected branches lose their leaves and die. This process is repeated on other branches in succeeding years and trees become stag-headed but rarely die. During summer seemingly healthy shoots are produced along affected branches but these often later become affected. The wood of all affected shoots and branches is stained, as may be that of many larger limbs of the trunk and of the roots of diseased trees. In recently affected trees only shoots with withered leaves are stained, but extensive staining can occur in trees without obvious external symptoms. The wood of recently attacked trees is sodden with colourless bacterial slime which will exude from cut surfaces in late spring. With age or exposure to air such water-marked areas darken so that in trees that have been affected for two or three years the inner wood is dark brown to black; with further ageing stained wood becomes greyish.

E. salicis is a vascular pathogen that spreads in the xylem. It has been supposed that there are insect vectors for spread of the disease between trees but there is no good evidence for this. Spread does occur in propagating material. The pathogen can be detected in the phyllosphere of symptomless trees but there is no evidence as to whether it is a resident or a migrant part of the epiphytic flora.

Control through rogueing of visually affected trees has only been partially successful because the disease is often latent and undetectable visually. The use of simple, slide-agglutination tests to detect the bacterium in trees used for propagation (Wang & Preece 1973) and the disinfection of tools used for propagation could do much to limit the disease. Otherwise the use of less susceptible cultivars or clones where this is possible and propagating only from young stools are the only other measures that can be advised (Gremmen & de Kam 1981).

LELLIOTT

Erwinia stewartii (Smith) Dye

E. stewartii (CMI 123, EPPO 54) causes Stewart's disease or bacterial wilt of maize, a serious N. American disease that can cause total crop loss of sweet corn. It also affects dent, flint, flour and popcorn cultivars to a lesser extent. Some weeds can be symptomless carriers of the bacterium. It is locally established in Italy where at one time there were losses in the Venetian region. It has also occurred in USSR, Greece, Poland and Romania. Damage has been more extensive in Italy, where the entire plant is attacked than in the USA where the disease has been established and economically serious for many years and only the leaves are attacked in mature plants. Although the bacterium may be transmitted in seed and occasionally overwinters in soil, manure or plant debris, these are of little importance compared with transmission by insect vectors in which the bacteria survive during hibernation and are then transferred from plant to plant in the growing season. In the USA the chief vector is *Chaetocnema pulicaria* which migrates over considerable distances and which harbours and transmits the pathogen throughout its life. Seedlings may be directly affected from seed and die, but more usually they are infected from overwintering vectors that are then principally responsible for secondary spread throughout the growing season.

In Italy, *C. pulicaria* is not the vector but other vectors are known including adults and larvae of *Diabrotica undecimpunctata howarti*, *Chaetocnema denticulata*, larvae of *Delia platura*, *Agriotes mancus*, *Phyllophaga* sp. and larvae of *Diabrotica longicornis*. However, it is not certain that *E. stewartii* readily overwinters in Italy. Recent outbreaks, after a long gap, have been attributed to the import of infected seed (Mazzucchi 1984). In sweet corn the bacteria primarily colonize the vascular tissue throughout the plant and on leaves cause pale-green to yellow, longitudinal streaks with irregular or wavy margins that dry out and turn brown. Tassels are bleached, appear withered and die prematurely. Susceptible hybrids wilt rapidly. Cavities can form in the pith of the stalk near soil level in severe infections. Bacterial ooze can appear on the inner face of the husk and small water-soaked spots, later becoming dry and dark, can appear on both surfaces of the husk. The seed, but not the embryo, is invaded.

In dent corn the wilt phase is rare but the leaf blight condition appears, though less severely than in sweet corn, and streaks originate from the feeding marks of the beetle vector. Control is by the use of resistant cultivars and disease-free seed and by early spraying, based on prediction of the survival of the vector(s) over winter, with insecticides.

LELLIOTT

Erwinia tracheiphila (E.F. Smith) Bergey *et al.* (CMI 233) causes an economically serious bacterial wilt of Curcurbitaceae particularly in N. America, but appears to be uncommon in Europe. The striped cucumber beetle (*Acalymma vittatum*) is a vector.

STREPTOMYCES

Streptomyces scabies (Thaxter) Wakeman & Henrici

Basic description

Traditionally the actinomycete causing common scab of potatoes had been called *S. scabies,* but there is now doubt as to what this species is. The type strain no longer fits Waksman's description nor does it accurately describe the organism commonly isolated from diseased tissues. There appear to be many taxonomically different reference strains and Bergey classifies *S. scabies* as a 'species insertae sedis' within the Streptomycetaceae. The classification of the family is being revised and at present the position of the scab-forming isolates has not been determined. Isolates from potato generally have branched, hyaline mycelium, ash-grey aerial mycelium and a woolly colony; the branching being monopodial with sporophores wound into numerous regular spirals of smooth spores. Melanoid pigments are formed. However, isolates fitting the description are not necessarily pathogenic.

Host plants

Scab is most important on potato but can also occur on red beet, sugar beet, fodder beet, radish, carrot and perhaps other root crops. Whether these are caused by different *Streptomyces* species or '*S. scabies*' remains in doubt, although field soils producing scabby potatoes can also produce scabby root crops. Similarly it is not clear whether one or several *Streptomyces* species cause common scab of potato.

Geographical distribution

S. scabies has a world-wide distribution.

Disease

Lesions (scabs) of common scab (Lapwood 1973) are first seen as small (*c.* 1 mm diam.) discrete brown spots associated with lenticels which rapidly enlarge as the potato tuber swells. Later they can vary considerably in appearance, and many types have been described and named. Only 'superficial' scab, where lesions are merely superficial corky blemishes, merits distinction from 'normal' scab, in which lesions may be roundish or star-shaped with cracked or torn edges, or a series of irregular but concentric layers of wrinkled cork surrounding a central depression. Lesions may remain discrete but if the attack is severe normal lesions may coalesce and cracks or furrows develop cutting into the tuber tissue. Disease severity is generally assessed as the proportion of tuber surface affected, not by lesion type.

S. scabies probably penetrates young lenticels but the early stages of infection have not been convincingly recorded. However, penetration stimulates the tuber tissues to form a wound barrier a few cells below the surface and, if this stops further invasion, a 'superficial' scab, with little or no disruption of the tuber skin, is formed. If the barrier is breached, a second and sometimes third barrier is formed before penetration is stopped. More barriers and deeper penetration, coupled with the continued swelling of the tuber, result in increasingly severe forms of 'normal' scab. If the tuber stops growing so does the lesion. The main pathway of the mycelium seems to be intercellular and through the middle lamella, which makes it difficult to detect. Only in the older peripheral regions of lesions can the mycelium be found inside cells where it may sporulate abundantly.

S. scabies can be readily isolated by mixing 1:40 phenol with infested soil or macerated lesions before plating but the identification and pathogenicity tests can be slow and difficult. The actinomycete is strongly aerobic and grows successfully on a wide range of agar or shaken liquid media between 8 and 38°C (best at 27°C) and between pH 4.8 and 8.5. Cultures can be maintained under oil, in a freeze-dried state, or in sterilized soil. They tolerate desiccation.

Epidemiology

Although *S. scabies* is readily isolated from scabby tubers, the soil, where the organism is free-living, is considered to be the main source of inoculum. The disease is increased on quickly draining, gravelly or alkaline soils and by liming, while in some seasons rain or watering decreases it. Common scab appears unimportant when soil water potentials are −0.4 bar and above (pF 2.6 and below) but it is increasingly important as soils dry in the range −0.4 to −1.0 bar (pF 2.6–3.0). The disease develops at soil temperatures between 13 and 25°C and most lesions are produced at 20°C. Previous crops and management such as liming for barley, adding green manures, ploughing pasture (*S. scabies* may normally inhabit grass roots), frequency of growing susceptible crops and susceptibility of cultivars may affect soil populations and therefore scab incidence. Infection occurs only at young lenticels which form from stomata near the growing apex of the tuber. The young lenticel appears to become susceptible only after rupture of the stoma guard cells, and remains so for about a week, during which time, provided the soil is dry, suberin is deposited on the outermost cell walls (in wet soil, the cell division which has caused the rupture continues, and cells proliferate from the lenticel opening). Lenticels continue to be formed as long as nodes (eyes) separate from the apex. However, most of the mature tuber surface is composed of internodes formed during the first month. Therefore infections occurring during this period cover most of the tuber surface at harvest. This happens only if the weather remains dry. Infection does not occur in wet soil, even with large soil populations; the reason for this is not fully understood. Scanning electron microscopy has shown that the internodes that are nearest the apex and bear stomata have a sparse microflora. Older internodes with lenticels, when from dry soil, are increasingly colonized by actinomycete hyphae and perhaps discrete but often large bacterial colonies. In wet soils, by contrast, actinomycetes are rarely seen over such internodes and bacteria are generally scattered over them. The failure to infect such stomata may represent disease escape and in the case of young lenticels in wet soil, some form of microbial antagonism (Adams & Lapwood 1978).

Economic importance

Yield losses have been reported but, in its superficial form, scab decreases tuber quality rather than yield. During long-term storage there is a risk of water loss which is most severe in dry seasons when potato yields are small. In the UK it has been estimated that a dry June or July could result in 4% or more of the national yield being substantially affected by scab. Even a small amount of scab can render seed crops unsuitable for export, and pre-packaging table tubers has also tended to increase the economic importance of the disease.

Control

Methods of controlling common scab aim to decrease the amount of seed-tuber or soil-borne inoculum or to prevent infection. Potato cultivars differ in their resistance and selections have been made. Although the actinomycete sporulates profusely within the dead cells of scab lesions, disinfection of seed tubers with mercuric chloride, formalin or organomercury compounds is of doubtful value as the organism is present in most agricultural soils. It might be useful where virgin soils such as new polders or irrigated desert soils could become infested by planting infected seed. Most efforts to control scab aim to decrease the soil population of *S. scabies* and despite some success, no method is effective on all soils. Chemicals incorporated into the soil before planting include formaldehyde, urea formaldehyde, manganese sulphate and pentachloronitrobenzene (PCNB or Quintozene). PCNB is the best of these compounds but has certain disadvantages including possible carcinogenic breakdown products. Sulphur has been used to acidify soil to *c.* pH 5.2; this has decreased scab but can also decrease the yield of potatoes and of acid-sensitive rotation crops such as barley, and may select acid-tolerant strains of *S. scabies*. Organic manures such as green rye, rape seed, soybean meal or soybean crops have been ploughed into soils to decrease soil populations by increased acidity, moisture or multiplication of antagonistic organisms. Usually such treatments are uneconomic and are largely ineffective when populations are already large. Where irrigation is available it can be used to control the disease whatever the soil population. Irrigation regimes which prevent soil moisture from falling to critical levels aim to prevent infection during the 4 to 6 weeks following tuber initiation (Lapwood *et al.* 1973; Lapwood & Adams 1975).

LAPWOOD

REFERENCES

Adams M.J. & Lapwood D.H. (1978) Studies on the lenticel development, surface microflora and infection by common scab (*Streptomyces scabies*) of potato tubers growing in wet and dry soils. *Annals of Applied Biology* **90,** 335–343.

Bevan M.W. & Chilton M.D. (1982) T-DNA of the *Agrobacterium* Ti and Ri plasmids. *Annual Review of Genetics* **16,** 357–384.

Billing E.B. (1984) Principles and applications of fireblight risk assessment systems. *Acta Horticulturae* **151,** 15–22.

Bordeleau L.M. & Michaud R. (1981) Association between resistance to bacterial wilt and symbiotic nitrogen fixation in alfalfa. *Phytoprotection* **62,** 39–43.

Cleene M. & De Ley J. (1981) The host range of infectious hairy-root. *Botanical Review* **47,** 147–194.

Close R. & Mulcock A.P. (1972) Bacterial wilt, *Corynebacterium insidiosum*, of lucerne in New Zealand. *New Zealand Journal of Agricultural Research* **15,** 141–148.

De Boer S.H. & Copeman R.J. (1980) Bacterial ring rot testing with the indirect fluorescent antibody staining technique. *American Potato Journal* **57,** 457–465.

Dickey R.S. (1979) *Erwinia chrysanthemi:* a comparative study of phenotypic properties of strains from several hosts and other *Erwinia* species. *Phytopathology* **69,** 324–329.

Dickey R.S. (1981) *Erwinia chrysanthemi:* reaction of eight plant species to strains from several hosts and to strains of other *Erwinia* species. *Phytopathology* **71,** 23–29.

Fahy P.C. & Persley G.J. (1983) *Plant Bacterial Diseases: A Diagnostic Guide.* Academic Press, New York.

Gremmen J. & de Kam M. (1981) New development in research into the watermark disease of white willow (*Salix alba*) in Netherlands. *European Journal of Forest Pathology* **11,** 334–339.

Grogan R.G. & Kendrick J.B. (1953) Seed transmission mode of overwintering and spread of bacterial canker of tomato caused by *Corynebacterium michiganense. Phytopathology Abstracts* **43,** 473.

Holmes B. & Roberts P. (1981) The classification, identification and nomenclature of Agrobacteria, incorporating revised descriptions for each of *Agrobacterium tumefaciens. A. rhizogenes* and *A.rubi. Journal of Applied Bacteriology* **50,** 443–467.

Johnson L.E.B., Frosheiser F.I. & Wilcoxson R.D. (1982) Interaction between *Fusarium oxysporum* f.sp. *medicaginis* and *Corynebacterium insidiosum* in alfalfa. *Phytopathology* **72,** 517–522.

Kahl G. & Schell J.S. (1982) (Eds) *Molecular Biology of Plant Tumors.* Academic Press, New York.

Kennedy B.W. & Alcorn S.M. (1980) Estimates of US crop losses to procaryote plant pathogens. *Plant Disease* **64,** 674–676.

Kerr A. (1980) Biological control of crown gall through production of agrocin 84. *Plant Disease* **64,** 25–30.

Kerr A. & Panagopoulos C.G. (1977) Biotypes of *Agrobacterium radiobacter* var. *tumefaciens* and their biological control. *Phytopathologische Zeitschrift* **90,** 172–179.

Kersters K. & De Ley J. (1984) Genus III. *Agrobacterium* Conn 1942, 359. In *Bergey's Manual of Systematic Bacteriology,* Vol. 1 (Ed. N.R. Krieg), pp. 244–254. Williams & Wilkins, Baltimore.

Keyworth W.G. & Howell S.J. (1961) Studies on silvering disease of red beet. *Annals of Applied Biology* **49,** 173–195.

Labruyère R.E. (1971) Common scab and its control in seed-potato crops. *Verslagen van Lanbouwkundige Onderzoekingen,* no. 767, pp. 1–71.

Lapwood D.H. (1973) *Streptomyces scabies* and potato scab disease. *Actinomycetales: Characteristics and Practical Importance* (Eds G. Sykes & F.A. Skinner), pp. 253–260. Academic Press, London.

Lapwood D.H. & Adams M.J. (1975) Mechanisms of control of common scab by irrigation. *Biology and Control of Soil-borne Plant Pathogens* (Ed. G.W. Burrell), pp. 123–129.

Lapwood D.H., Wellings L.W. & Hawkins J.H. (1973) Irrigation as a practical means to control potato common scab (*Streptomyces scabies*): final experiment and conclusions. *Plant Pathology* **22,** 35–41.

Lelliott R.A. (1974) Section II. Plant Pathogenic Corynebacteria. In *Bergey's Manual of Determinative Bacteriology,* 8th Ed. (Eds R.E. Buchanan & N.E. Gibbons), pp. 611–617. Williams & Wilkins, Baltimore.

Lelliott R.A. & Billing E.B. (1984) Fireblight of apple and pear. *Ministry of Agriculture, Fisheries and Food Leaflet* no. 571 (Revised 1984). HMSO, London.

Lelliott R.A. & Dickey R.S. (1984) Genus VII. *Erwinia* Winslow, Broadhurst, Kuemwiede, Rogers & Smith 1920, 209. In *Bergey's Manual of Systematic Bacteriology,* Vol. 1 (Ed N.R. Krieg). pp. 469–476. Williams & Wilkins, Baltimore.

Lelliott R.A. & Sellar P.W. (1976) The detection of latent ring rot *Corynebacterium sepedonicum* in potato stocks. *Bulletin OEPP/EPPO Bulletin* **6,** 101–106.

Lelliott R.A. & Stead D. (1987) *Methods for the Diagnosis of Bacterial Diseases of Plants.* Blackwell Scientific Publications, Oxford.

McFadden L.A. (1961) Bacterial stem and leaf rot of *Dieffenbachia* in Florida. *Phytopathology* **51,** 663–668.

Mazzucchi U. (1972) A bacterial stalk rot of maize (*Zea mays*) in Emilia. *Phytopathologia Mediterranea* **11,** 1–5.

Mazzucchi U. (1984) L'avvizzimento batterico del mais. *Informatore Fitopatologico* **34** (1), 18–23.

Merlo D.J. (1982) Crown gall: a multipotential disease. *Advances in Plant Pathology,* Vol. 1 (Eds D.S. Ingram & P.H. Williams), pp. 139–177. Academic Press, London.

Nelson G.A. (1980) Long-term survival of *Corynebacterium sepedonicum* on contaminated surfaces and infected potato stems. *American Potato Journal* **57,** 595–600.

New P.B. & Kerr A. (1971) A selective medium for *Agrobacterium radiobacter* biotype 2. *Journal of Applied Bacteriology* **34,** 233–236.

Panagopoulos C.G., Psallidas P.G. & Alivizatos A.S. (1978) Studies in biotype 3 of *Agrobacterium radiobacter* var. *tumefaciens. Proceedings of the 4th International Conference on Plant Pathogenic Bacteria, Angers,* pp. 221–228. INRA, Angers.

Penna R.J. (1982) Bacterial canker of tomato. *Ministry of Agriculture, Fisheries and Food Leaflet* no. 793. HMSO, London.

Pérombelon M.C.M. & Kelman A. (1980) Ecology of soft rot Erwinias. *Annual Review of Phytopathology,* **18,** 361–387.

Preece T.F. (1977) Watermark disease of the cricket bat willow. *Forestry Commission Leaflet* no. 20. HMSO, London.

Samson R., Pautier F., Sailly M. & Jouan B. (1987) Nécessité de caractériser les *Erwinia chrysanthemi* isolées de *Solanum tuberosum* selon les biovars et serogroups connus. *Bulletin OEPP/EPPO Bulletin* **17,** 11–16.

Schroth M.N., Thompson J.P. & Hildebrand D.C. (1965) Isolation of *Agrobacterium tumefaciens-A. radiobacter* group from soil. *Phytopathology* **55,** 645–647.

Sellwood J.E. & Lelliott R.A. (1978) Internal browning of hyacinth caused by *Erwinia rhapontici. Plant Pathology* **27,** 120–124.

Shoemaker P.B. & Echandi E. (1976) Seed and plant bed treatments for bacterial canker of tomato. *Plant Disease Reporter* **60,** 163–166.

Strider D.L. (1969) Bacterial canker of tomato caused by

Corynebacterium michiganense: a literature review and bibliography. *Technical Bulletin North Carolina Agricultural Experiment Station* no. 193, pp. 1–110.
Thyr B.D., Samuel M.J. & Brown P.G. (1975) Tomato bacterial canker: control by seed treatment. *Plant Disease Reporter* **59,** 595–598.
Wang W.C. & Preece T.F. (1973) Infection of cricket bat willow (*Salix alba* var. *caerulea* Sm.) by *Erwinia salicis* detected in the field by use of a specific antiserum. *Plant Pathology* **22,** 95–97.
Yarham D.J. & Perkins S.W. (1978) Cucumber's mystery root mat disorder still spreading. *Grower* **90,** 18–22.

8: Oomycetes

The Oomycetes include some of the most widespread and destructive plant pathogens. The Saprolegniales are mainly saprophytes (water moulds), including only *Aphanomyces* as a plant pathogen. They produce zoospores in terminal filamentous sporangia, which remain *in situ* on the mycelium, and the zoospores form cysts which then release secondary zoospores. In the Peronosporales, a gradation in morphological specialization and parasitism can be seen, involving: i) the replacement of zoospore release from sporangia by germination by germ-tube, either according to environmental conditions, or exclusively; ii) the formation of discrete determinate sporangia, which separate from the parent hypha to be dispersed like conidia; iii) the formation of sporangia on aerial sporangiophores, rather than submerged, and their release by hygroscopic twisting; iv) the penetration of plant cells by increasingly specialized and determinate haustoria; v) biotrophic infection, either initially or throughout infection; vi) infection of leaves and stems rather than roots: vii) the formation of sporangia in chains within pustules.

These progressive specializations of the asexual reproductive state and of the mycelium infecting the plant stand in contrast to the fairly constant features of the sexual state, throughout the group: oogonia fertilized by antheridia and developing into oospores (many in Saprolegniales or one in Peronosporales), which provide the main means of persistence in soil or plant debris.

Biologically, this corresponds to the following broad sequence: i) *Pythium* (relatively unspecialized soil fungi attacking seedlings and roots, and causing some storage rots); ii) most *Phytophthora* species, resembling *Pythium* in biology but tending to be more host-specialized and to attack the root and collar of woody plants; iii) a few *Phytophthora* species which have 'escaped' from the soil to become leaf pathogens (notably *P. infestans*), and much resemble the next group in their biology—except that they are not obligate parasites and that their biotrophic phase is short; in many languages, e.g. French mildiou, they are given the same disease name as the next group; iv) the downy mildew fungi, obligate parasites of leaves and stems with a prolonged biotrophic phase and a high degree of host specialization; v) Albuginaceae, causing the 'white rust' diseases, similar to downy mildews but with sporangia formed in chains in pustules.

Finally it may be noticed that the Lagenidiales (with one plant pathogen *Lagena radicicola*) are now classed as Oomycetes rather than as Chytridiomycetes (q.v.), which they resemble in producing very little mycelium.

SAPROLEGNIALES

Aphanomyces cochlioides Drechsler

As a member of the Saprolegniales, *A. cochlioides* produces zoospores, in filamentous sporangia, which give rise to a second cycle of zoospores after encystment (dimorphism). It is associated with *Pythium* species in causing damping-off and black-root of sugar beet, and is very similar in biology to *Pythium* with which it has been confused. Seedlings may be killed, or the basal portion may be shrivelled and blackened with the plant surviving in a weakened state. *A. cochlioides* is relatively specific to beet. The use of soil fungicide treatments effective against *Pythium* but not against *A. cochlioides* has favoured a recent rise in importance of the latter, especially in monoculture (e.g. in France; Bouhot &Loridat 1985; in the FRG; Steudel 1979). Hymexazol is effective against it. *A. cochlioides* prefers rather high temperatures (20–25°C) and is more common in the southern beet-growing areas. While damping-off of beet was at one time attributed to *A. laevis* de Bary, this is now considered to be an error (Papavizas & Ayers 1974).
BOUHOT & SMITH

Aphanomyces euteiches Drechsler (CMI 600; APS Compendium; Papavizas & Ayers 1974) is an important root-rotting pathogen of pea in N. America. Its **f.sp. *pisi*** on pea has recently been distinguished from **f.sp. *phaseoli*** on bean (Pfender & Hagedorn 1982). Although there are reports of damage in Europe, in recent years it seems to be only in the nothern countries (USSR, Scandinavia) that the disease has some importance (Kotova &

Tsvetkova 1981). Fungicides are mostly ineffective and currently available resistance is only partial, so that the only practical means of control remains to check fields for their root rot potential before planting and avoid any more than only lightly infested (APS Compendium). ***A raphani*** Kendr. causes blacking disease of radish; significant damage has been reported in the FRG (Wendland 1976).

BOUHOT & SMITH

LAGENIDIALES

Lagena radicicola Vanterpool & Ledingham (syn. *Lagenocystis radicicola* (Vanterpool & Ledingham) Copeland) infects the roots of grasses (including cereals), producing a multinucleate sac-like thallus, very like an *Olpidium* (q.v.) (APS Compendia on Wheat and Turfgrass Diseases). However, its biflagellate zoospores, and oospores, clearly relate it to the Oomycetes. In the UK and probably elsewhere in Europe, it causes 'rootlet disease' of barley and wheat.

PERONOSPORALES

PYTHIACEAE

Pythium spp.

Basic description

The genus *Pythium* differs from the related genus *Phytophthora* by forming relatively undifferentiated sporangia, which are mostly not separated from their parent mycelium or dispersed, and which release their contents into a spherical vesicle in which zoospores develop. As in many other Peronosporales, germination by zoospores is in some species replaced entirely or at times by germination of the sporangium by germ-tube. At present, some 104 species have been described in the genus (Domsch *et al.* 1980). Of these ***Pythium ultimum*** Trow is the typical plant pathogen. It is now considered to be synonymous with ***P. debaryanum*** Hesse emend. Middleton and is also very close to ***P. irregulare*** Buisman. These polyphagous mostly temperate species can be said to form the *ultimum* group. ***P. aphanidermatum*** (Edson) Fitzp. (CMI 136) and ***P. butleri*** Subram. (CMI 37) represent another group of polyphagous but tropical or semi-tropical species. Thirdly, ***P. arrhenomanes*** Drechsler (CMI 39) and ***P. graminicola*** Subram. (CMI 38) tend to be associated with Gramineae in temperate zones (*arrhenomanes* group).

Other species occasionally found attacking plants are ***P. intermedium*** de Bary (CMI 40) and ***P. sylvaticum*** Campbell & Hendrix (CMI 120), ***P. mamillatum*** Meurs (CMI 117) and ***P. splendens*** Braun, the last two mainly in warm countries. In fact, numerous species (Waterhouse 1968) are principally saprophytic in soil, and may very occasionally be found attacking plants. A special case is ***P. oligandrum*** Drechsler, which is a hyperparasite on other *Pythium* species (Vesely 1978) and not pathogenic to the various hosts from which it has been isolated (CMI 119).

Host plants and specialization

The *ultimum* and *aphanidermatum* groups are essentially polyphagous and almost any plant may be attacked. However, some crop plants are very susceptible to attack, especially at the seedling stage, e.g. sugar beet, cucumber, *Pinus sylvestris*. Cultivars of the same species are known to vary in susceptibility. Although Gramineae may be attacked by the polyphagous species, they also tend to carry a distinctive *Pythium* flora, with predominance of the *arrhenomanes* group, which is not found on other plants.

Geographical distribution

Practically all species can be observed on all continents. Nevertheless, the *aphanidermatum* group mainly damages crops under high temperature conditions (tropical and equatorial areas), and *P. aphanidermatum* is principally found in Europe in Mediterranean areas (CMI map 309). The other species, exemplified by *P. ultimum*, are more damaging in cold climates and tend not to be present in the southern areas of Europe.

Disease

The first main type of disease caused by *Pythium* species is damping-off, due to sudden and fast developing attacks on young seedlings in the field or in nurseries, especially under damp conditions. The *ultimum* and *aphanidermatum* groups are those principally involved. If seeds or seedlings are attacked before emergence (pre-emergence damping-off), this is recognized by a complete failure to emerge or by a proportion of missing seedlings. If the attack occurs after emergence (post-emergence damping-off), the newly emerged seedlings collapse following attacks at ground level. Damping-off due to *Pythium* species is not necessarily easy to distinguish from similar infections by other fungi (e.g. *Thanatephorus cucumeris*, *Aphanomyces* species, various seed-borne pathogens). In general, soft-tissued seedlings of herbaceous plants show watery wet lesions, though this will not be seen so clearly in tree seedlings. *P. ultimum* also causes a watery wound rot of potato tubers.

Pythium species cause a second type of disease, whose importance has only been recognized fairly recently. This is a root necrosis, developing rather slowly from the root-tip backwards, with brown or black discoloration. The effect on the plant is a general slowing of growth, with possibly some yellowing and loss of yield. Plants may be able to compensate under favourable conditions by developing new roots. This type of attack has been recognized especially in cereals (wheat, maize), where the *arrhenomanes* group is especially involved. It is possible that many other plants can be affected in a similar way, not necessarily showing very pronounced symptoms. For example, specific apple replant disease (Sewell 1979), in which apples planted on old apple orchard land grow poorly, seems to be associated with soil pathogens of which *P. sylvaticum* is now considered the most important in the UK (Sewell *et al.* 1983). Specific replant disease is really a special case of so-called soil sickness, where continuous cropping with one host leads to a steady deterioration in crop growth, well reflected in the French name 'fatigue des sols'. Poorly growing plants show root lesions and recent work shows that *Pythium* species can very commonly be isolated from them (Bouhot 1982), especially in the case of beet, cereals and vegetable crops. However, many other fungi may be implicated in such effects (*Chalara elegans* (q.v.), *Fusarium* species), and indeed quite other factors.

Epidemiology

Pythium species mainly survive in the soil as oospores in decayed substrates, potentially for many years. These are present at rather low densities in the soil, compared, for example, with the chlamydospores of *Fusarium* species. Oospore germination is stimulated, under conditions of high soil moisture, by dissolved organic substances. A complex of proteins, lipids, carbohydrates and sterols seems to be needed and simple organic substances are ineffective. The source of these substances may be roots, germinating seeds or organic matter added to the soil (e.g. green manures) (Watson 1973). This effect can readily be exploited in baiting techniques for *Pythium* in the soil (e.g. with cucumber; Bouhot 1975). Oospores germinate to form sporangia, which may release zoospores (most species) or germinate by germ-tube (*ultimum* group). Liquid water is needed for zoospore dispersal, while *P. ultimum* will infect roots in moist soils even in the absence of liquid water. The fact that zoospores are water-dispersed creates a special risk in flooded soils. In addition, irrigation systems, in the field or in glasshouses, may easily spread *Pythium* attacks. The *ultimum* group is favoured by cool conditions (10–15°C) and high soil moisture, so in practice attacks are mainly seen in spring (and to a certain extent in autumn). The *aphanidermatum* group prefers temperatures over 25°C, and attacks may occur throughout the growing season in warmer countries.

Normal field soils always carry a low population of several pathogenic *Pythium* species. In a few cases, forest soils have been found to be suppressive to pathogenic *Pythium* species due to competition from Mucorales in particular (Bouhot & Perrin 1980), so that *Pythium* is apparently absent. The inoculum potential of a soil will only rise to a damaging level in the presence of a combination of favourable circumstances: intensive cropping, monoculture, high organic inputs (e.g. legumes). In fact, *Pythium*, as a good opportunist, colonizes fresh organic matter to build up high levels of inoculum within 24–48 h. It is a poor competitor with other fungi and bacteria however, and is rapidly replaced in such substrates, remaining only as a few scattered oospores. If a non-susceptible crop is grown (e.g. cereals), inoculum levels will tend to fall off sharply. These effects have principally been investigated in relation to damping-off. Little is yet known concerning the epidemiology of the root necrosis type of attack.

Economic importance

The main losses caused by *Pythium* in Europe are to field crops such as sugar beet. In principle, losses tend to be all-or-nothing. In some years, sowings of beet over large areas are completely destroyed (e.g. 15 000 ha in N. France in 1982; Richard-Molard 1982). These have to be resown, and it may even happen that successive resowings fail. The same may happen with various vegetable crops (cucurbits, Solanaceae, legumes). Similarly, nursery sowings of ornamentals and forest trees may be completely destroyed. It is difficult to estimate the overall economic impact of such incidents since total losses of this kind are not necessarily reported or analysed. The problem will certainly be greatest for species where the cost of the individual seed of a highly selected cultivar may represent a significant proportion of production costs.

The importance of *Pythium* in causing root necrosis of older plants is only beginning to become apparent, and is still difficult to estimate. Finally, *Pythium* may cause considerable damage in glasshouse crops if it is accidentally introduced into supposedly sterilized growing media, or if it contaminates irrigation water. In such circumstances, for example, beds for rooting chrysanthemum cuttings may be completely destroyed within a few days. Growing media which have been steam-sterilized are particularly liable to be heavily colonized by opportunistic soil fungi such as *Pythium* species and *Thanatephorus cucumeris*.

Control

Chemical control of damping-off can be assured very easily. Of the many modern compounds now available for seed treatment, or for soil treatment with granules, powders or drenches, can be cited: fenaminosulf, etridiazole, propamocarb, and hymexazol and possibly also metalaxyl and furalaxyl (though these systemic compounds should possibly be reserved for more specific uses). Such treatments are highly effective and, to date, no problems of resistance have arisen. Cultural practices are also important: soil drainage, delaying sowing after green manuring, sowing at temperatures above the *Pythium* optimum. In nurseries and glasshouses, where sterilized growing media are used, all precautions are needed to prevent contamination of tools, water, containers etc., in addition to the use of treated seeds. Good lighting will prevent etiolation, which predisposes to damping-off. The phenomenon of suppressiveness opens possibilities for biological control of *Pythium* (Bouhot & Perrin 1980). A proportion of suppressive soil can be introduced into a growing medium to 'transmit' the suppressiveness. In principle, once the specific biological antagonists are identified (e.g. Mucorales), it may be possible to introduce them directly. Certain organic soil amendments may have similar effects (Bouhot 1981). Another approach along the same lines is bacterization, where bacterial antagonists such as *Pseudomonas fluorescens* are applied to the seed (Suslow & Schroth 1982). Finally, deliberate mycorrhization of forest seedlings at their young, most susceptible stage, could protect against *Pythium* infections as well as ensuring better growth.

Special research interest

Pythium species have been an excellent material for the development of epidemiological models of potential infectivity of soils, in relation in particular to the relationship between *Pythium* and the soil microflora competing for the same soil substrate (Bouhot & Joannes 1982).

BOUHOT & SMITH

***Phytophthora infestans* (Mont.) de Bary**

Basic description

The genus *Phytophthora* is distinguished from the true downy mildews by its indeterminate branched sporangiophores and from *Pythium* by the complete differentiation of motile zoospores within sporangia before expulsion. *P. infestans* is atypical, being one of only two species that can shed sporangia easily into dry air. These may produce germ tubes directly but more often germinate indirectly releasing zoospores. Sporangiophores are swollen at the point of attachment of sporangia.

Host plants

P. infestans attacks various members of Solanaceae especially potato, tomato and some ornamentals.

Host specialization

Host-specific phenotypes virulent on potatoes carrying monogenically inherited resistance derived from *Solanum demissum* are well known (Shattock *et al.* 1977). Specificity towards general resistance (Caten 1974), which is polygenically inherited, remains a contentious issue. Monogenic resistance in tomatoes is also matched by compatible phenotypes of the fungus. The relationship between potato and tomato isolates of *P. infestans* is unclear but cross infection frequently occurs.

Geographical distribution

The centre of origin of *P. infestans* is C. America. Here both *A1* and *A2* compatibility types are found and thus sexual reproduction is common. Elsewhere only the *A1* compatibility type was thought to exist and plant health regulations in Europe were aimed at excluding *A2* isolates. Recently, however, *A2* strains of *P. infestans* have been identified in infected crops of potatoes in Switzerland (Hohl & Iselin 1984) and the UK (Tantius *et al.* 1986), and in blighted tubers imported from Egypt for public consumption in the UK (Shaw *et al.* 1985). The fungus was introduced into Belgium in the mid 1840s, probably in infected tubers from E. USA and subsequently infected growing potato crops throughout N.W. Europe and S. Scandinavia in a single season. Today, it is found in every potato-producing country of the world (CMI map 109).

Disease

P. infestans causes late blight of potatoes and tomatoes affecting all parts of the plant except the roots. Dark brown-black necrotic lesions develop on leaves especially at the tip and margin of leaflets. The fungus produces haustoria and exhibits limited biotrophic growth but rapid colonization of leaf tissue by intercellular hyphae and subsequent death of invaded cells produces extensive lesions in 5–7 days after infection. In humid conditions lesions are surrounded by a zone of dense, white sporangiophores. Dark necrotic lesions also occur on stem tissue. Tubers become infected by zoospores washed down by rain into the soil or by direct contact with blighted foliage at harvest. *P. infestans* produces a firm reddish-brown granular rot beneath a discoloured tuber skin. Blighted tubers invariably become secondarily infected by soft-rotting bacteria. Symptoms of late blight on tomato leaves are identical to those on potatoes. Stems may show dark streaks, and green fruits exhibit hard brown spreading patches. Fruit fails to ripen and becomes completely brown and shrivelled.

Epidemiology

In Europe the late-blight fungus overwinters as mycelium in infected tubers. These may be in seed stocks, in discard heaps or remaining in the ground from previous cropping. The relative importance of these sources varies between countries. In the UK piles of discarded ware potatoes, on which growth can be exceptionally early and prolific, are probably a more important source of infection than seed tubers (Boyd 1974). Stems and leaves initially become infected by contact with diseased tubers or from soil-borne inocula originating from blighted tubers. Occasionally systemically infected shoots from diseased tubers may develop but these usually deteriorate and fail to emerge above soil level.

Initial foci of disease in potato crops early in the season are often characterized by stem lesions. This is due to the greater retention of rain and dew drops in leaf axils rather than on leaf surfaces in young emerging crops. Stem lesions later serve to maintain the focus of disease in less favourable dry weather. *P. infestans* forms sporangia in abundance only at 100% RH but once formed they are readily detached even in dry conditions and can spread for many km and successfully infect other crops. Such sporangia dispersed from primary or secondary foci are often non-turgid and only germinate in free water. Germination is usually indirect over a wide temperature range (optimum 12°C), producing zoospores. These encyst and germlings penetrate epidermal cells directly, or occasionally via stomatal pores.

Under controlled conditions at 18°C and 100% RH, a lesion with an area of 6 cm^2 can develop in 7–8 days from infection, producing approximately 8.5×10^4 sporangia on susceptible cultivars. On a well grown potato plant with only 1% affected leaf area this represents 4.5×10^6 sporangia/plant and ensures rapid spread of late blight between crops under ideal conditions when free water is present on the upper leaf surfaces. Even if RH drops to 95%, approximately 1% of the sporangia may continue to germinate if they are fully turgid. However, relatively short exposure to drier conditions (30 min at 50–60% RH) will inactivate sporangia completely. A change to sunny, dry weather is sufficient seriously to affect the capacity of sporangia to germinate and persistent dry weather will restrict the spread of disease to other susceptible crops. Late blight may continue to spread in potato crops that have developed a canopy of foliage since the micro-climate within the crop may remain conducive to sporulation and germination.

The relationship between weather and disease has been established in many European countries (Germany, Netherlands, Scandinavia etc.) which use meteorological data to forecast the likely outbreak of late blight in potato crops. They all describe conditions in which free water remains on potato leaves for at least 15 h, thereby allowing non-turgid sporangia arriving from outside the crop time to imbibe water and germinate. These forecasting systems are now increasingly computerized, for example the UK has recently adopted a computer-based system which estimates sporulation in relation to synoptic meteorological data and predicts the number of lesions developing from these spores using a 4–8 day latent period depending on temperature.

Economic importance

P. infestans left an indelible mark on world history when it ravaged European potato fields in the middle of the last century. Ever since it has remained an important limiting factor in potato production world-wide. In Europe the first outbreaks of late blight typically occur on early maturing cultivars which are generally susceptible to infection. In the UK outbreaks occur in May–June on these cultivars in mild coastal areas of S.W. England and Wales. The disease is economically unimportant on these cultivars which are lifted early in the season for immediate consumption. Sporangia, however, spread eastwards infecting later maturing cultivars during July in wet seasons. On these cultivars severe defoliation indirectly affects yields of tuber but the relationship between yield potential of the crop, damage and crop loss by *P. infestans* and environmental factors is still unclear (Rotem *et al.* 1983). Infection early in the life of the crop can severely reduce yield but equally attacks late in the season can be very damaging because of the increased risk of tuber blight.

Because wet seasons encourage both crop growth and *P. infestans,* late blight usually affects the market price to individual growers since the incidence of tuber blight is high when potatoes are in abundant supply. In such years blighted tubers in store are at risk to secondary infection especially soft-rotting bacteria. Both indoor and outdoor tomato crops can be severely affected in August/September when blight is abundant on potatoes.

Control

The use of race-specific resistance to control *P. infestans* has been unsuccessful in both potato and tomato while general rate-reducing resistance is not popular in Europe (Umaerus *et al.* 1983). Instead, fungicides are regularly applied to late-maturing crops from mid-summer onwards. Dithiocarbamates, e.g. maneb, zineb and mancozeb, have largely replaced copper-based compounds as protective fungicides. Most growers apply fungicides every 10–14 days from the time plants meet across the rows or sooner if planting has been seriously delayed or outbreaks of late blight are forecast. Protective fungicides may also be applied to tomatoes or alternatively temperature can be increased and/or humidity reduced in protected crops. In 1978 systemic fungicides, e.g. metalaxyl and cymoxanil, became available and their curative action especially when combined with a dithiocarbamate, e.g. mancozeb, has given good control at longer spray intervals (Schwinn 1983). Resistant strains of *P. infestans* have been reported to some of these new compounds when used alone and in combination with mancozeb (Carter *et al.* 1982). To reduce tuber infection at harvest infected foliage is desiccated at least 14 days before lifting late-maturing cultivars.

Special research interest

The development of late blight in potato crops has been extensively studied and much of the data has been used in analytical epidemiology (Van der Plank 1963). From this pioneering study applied epidemiology and disease management has evolved with ongoing studies into integrated control (Fry 1977) and to disease forecasting systems, e.g. BLITECAST in the USA (Mackenzie *et al.* 1983). Molecular (Doke & Tomiyama 1980), biochemical (Preisig & Kuć 1985) and cytological (Coffey & Wilson 1983) aspects of resistance to potato late blight are areas of intensive research while genetical studies (Shattock *et al.* 1986) are beginning to investigate the inheritance of host-specific pathogenicity and fungicide resistance.
SHATTOCK

Phytophthora erythroseptica **Pethybr.**

P. erythroseptica produces ellipsoid/obpyriform, non-papillate sporangia and oogonia approximately 30×35 μm. Antheridia are amphigynous (CMI 593). The fungus occurs in most of Europe (except Spain and Portugal), N. and S. America, Australasia, Egypt, Iran, India, Indonesia (CMI map 83). Although occasionally recorded on a number of hosts (tomato, clover, asparagus, cineraria, raspberry, *Zantedeschia* etc.) behaving rather like *P. cryptogea* (q.v.), *P. erythroseptica* ony causes economically important diseases in Europe on potato (pink rot). Older literature mentions it in association with *P. cryptogea,* causing a dieback of forced tulips in the UK known as shanking, but this now seems of little importance. Infected portions of potato tubers have dark lenticels and skin discoloration. When cut, infected tuber tissue changes from white to pink in

20–30 min, and darkens on prolonged exposure. Rot usually develops from the stolon end. Above ground, plants may turn yellow and wilt in warm weather and lesions may form at the stem base.

The pathogen survives in soil for many years, probably as oospores; all underground parts of plants can be infected by mycelium or zoospores (Lonsdale *et al.* 1980). Zoospores are probably the main infective agent in soil. Sporulation is stimulated by aqueous soil extract. After infection of a potato plant, mycelium grows via stolons into daughter tubers which rot. Oospores formed in infected underground parts are released into soil by decay of debris and infested soil can be carried on healthy harvested tubers. Zoospores infecting the eyes of immature daughter tubers may not cause rotting but superficial infection of tubers, with mycelium and oospores in skin and eyes. These tubers may remain sound in storage and carry infection to new sites. Transmission of fungus may occur in storage, from rotting tubers to others which remain apparently healthy, but carry oospores and mycelium in skin and eyes from which infection can develop after planting (Cunliffe *et al.* 1977).

Losses can be severe in Ireland, but elsewhere are usually moderate and sporadic. They may be underestimated because secondary invaders, especially bacteria, may mask pink rot symptoms.

The fungus is difficult or impossible to isolate after secondary invasion of lesions. Susceptible crops should not be planted in infested soil. No potato cultivars are resistant. Transmission between plants on infested sites is reduced if drainage is improved.
EPTON

Phytophthora fragariae Hickman

Basic description

P. fragariae is a slow-growing species that grows best on complex media such as French bean agar, although it will not grow on malt agar. Cardinal temperatures are 3°, 22° and 30°C. When irrigated with mineral salt solutions it produces non-deciduous, large non-papillate sporangia. It forms few oospores in complex media but they can be abundant in the steles of infected strawberry roots.

Host plants and host specialization

Only the cultivated strawberry, by far the most important host, and loganberry have been found naturally infected by *P. fragariae*. Species of closely related genera in the family Rosaceae have been infected under experimental conditions. Pathotypes are recognized in the UK, Canada and the USA but different systems of nomenclature are used. One British race (B-66.11) has attacked all cultivars and clones on which it has been tested (Montgomerie 1967).

Geographical distribution

The disease was first recorded in Scotland in 1921 and soon after in N. America. Thereafter it spread rapidly, probably by transmission in infected runners, so that by the 1940s it was widely distributed in the UK. It has since been found in most European countries. Serious outbreaks have occurred in Switzerland and Austria and, since the late 1970s, there have been first records in Bulgaria, FRG, and Sweden in 1976, 1978 and 1979 respectively, but none yet in Norway or Denmark. The disease has also been recorded in France, Italy and the Lebanon, and probably in Japan, New Zealand and Australia.

Disease

P. fragariae causes a severe root rot of strawberries known as red core in Europe and red stele in N. America. Affected areas are commonly seen in late spring and early summer as patches of stunted, wilted or dead plants often in depressions or poorly drained parts of a field. The younger foliage may sometimes have a bluish coloration. Affected plants often have poorly developed and rotted root systems, the lack of laterals and rotting of the main adventitious roots from their tips upwards giving the roots a characteristic rat's tail appearance (Fig. 26). When the

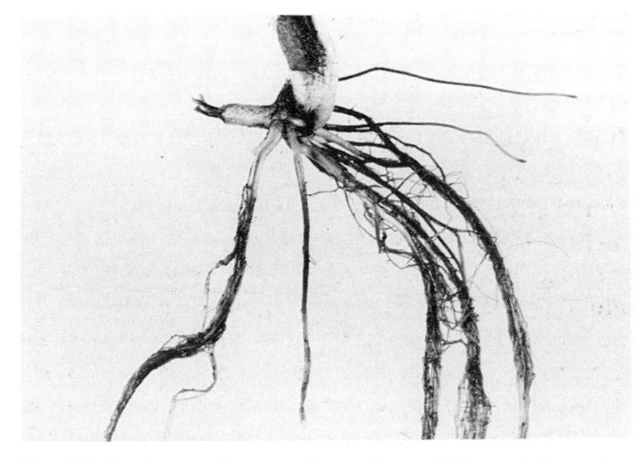

Fig. 26. Red core disease of strawberry (*Phytophthora fragariae*).

apparently healthy white parts of the roots are cut longitudinally, their centres are often a deep blood- or brick-red colour—hence the name of the disease. Stelar coloration appears to be a host response to infection and can extend well above the rotted part of the root, sometimes as far as the rootstock. The fungus can be isolated most readily from these non-rotted parts of affected roots using a selective medium (Montgomerie & Kennedy 1983). Abundant oospores are formed in the root steles and are most easily seen in early spring in the rotted parts of adventitious roots. Red steles and oospores together are taken as proof of infection by *P. fragariae* as no other fungus has been found to give similar symptoms. Affected plants may recover in the autumn if conditions are suitable for the production of new roots but these in turn are usually infected during the following winter.

Epidemiology

Initial infection of roots occurs in early autumn and is favoured by high soil moisture levels and low temperatures. The zoospores which infect roots can remain motile for several h at temperatures from just above freezing to *c.* 15°C, but above 15°C the motile period is greatly shortened. They are attracted to the root tips of white feeder roots where they encyst and penetrate; the fungus grows up the root producing sporangia on the outside and oospores in the stele. The sporangia release more zoospores to spread the infection for as long as soil temperatures and moisture levels are favourable. Spread is markedly affected by movement of surface water and is often very rapid and extensive in sloping fields. As temperatures rise in spring and early summer it becomes more difficult to isolate the fungus from roots, probably because it is mainly present as oospores. These are difficult to germinate, durable and can survive in soil for many years (Duncan & Cowan 1980). The disease can be spread by infested soil on farm implements or on the roots of plants. The pathogen is soil- and water-borne but its most important method of spread has undoubtedly been by the distribution of infected strawberry runner plants. Low levels of disease can readily escape detection and the movement of runner plants can lead to new outbreaks far distant from the original runner bed.

Economic importance

Large areas of plants can be destroyed and surviving plants often crop poorly producing numerous small fruit. In Canada, Gourley & Delbridge (1972) found that 78% of 9 ha of infected strawberries were destroyed with an attendant economic loss of $14000. Certification schemes for strawberry planting stocks usually preclude the sale of infected plants so that specialist producers of nursery stocks can face complete loss of a crop if part of it is infected. Fruit producers can control the disease by chemical and cultural methods, but these are costly. Because the disease is more severe in cool, wet conditions it should be worst in the more temperate regions of Europe, as evidenced by the serious decline in the strawberry acreage in Scotland between 1920 and 1936 (MAFF 1973).

Control

Where the disease is not widely established the most effective control is to prevent the dissemination of infected runner plants. Legislation aimed at preventing spread has existed in the UK for many years. In England and Wales the disease is notifiable and runner production is prohibited on scheduled land. However, certification is not mandatory and it is possible to sell uncertified runners. In Scotland, in contrast, land is not scheduled, but it is illegal to sell or even offer for sale runners which have not been inspected and received an official certificate; moreover, certification is not based on tolerance levels—if any infection is detected in a stock it is rejected. The scheme in Scotland used to depend on visual inspection of plants in the field, but recently a very sensitive Root Tip Bait Test (Duncan 1980) has been introduced, which has detected disease in stocks that had previously passed field inspection. Samples of root tips are removed in November from all stocks entered for certification; they are tested by growing the very susceptible alpine strawberry Baron Solemacher as a bait plant in a mixture of the root tips and compost. The test has also been used in other European countries.

Various control procedures can be used if the disease becomes established and these have been reviewed by Montgomerie (1977). Soil eradicants have generally given disappointing results, but economic control has been achieved with some fungicides, systemic compounds such as fosetyl-Al and metalaxyl being particularly promising. They are generally best applied in autumn when infection begins. Fungicides should not be used on nursery stocks as they may suppress symptoms without eradicating the disease. Improved soil structure and drainage can lessen disease severity; sub-soiling, drainage improvement and growing plants on ridges have all been used with some benefit. Poorly grown plants are usually more severely affected by disease, so attention should be paid to soil fertility. Cultivars vary in resistance; some have immunity to specific races, others have field resistance, others have both. Most resistant cultivars have been bred in the UK and N. America where the disease has been a problem for much longer than in continental Europe. Race-specific immunity has often failed and in breeding programmes emphasis is now placed on field resistance.

DUNCAN

Phytophthora nicotianae Breda de Haan

Basic description

P. nicotianae belongs to taxonomic group II of the genus (Newhook *et al.* 1978). It is a heterothallic species forming amphigynous antheridia. Papillate citriform sporangia and spherical chlamydospores are produced even on solid culture media. Waterhouse (1963) has differentiated two varieties (***P. nicotianae* var. *parasitica*** and **var. *nicotianae***) on purely morphological criteria. Other authors still use *P. parasitica* for isolates without strict host specificity and *P. parasitica* var. *nicotianae* (*sensu* Tucker) for the tobacco-specific isolates.

Host plants and specialization

The species, which occurs world-wide, attacks, as a whole, at least 72 genera of higher plants belonging to 42 families; among them are several Solanaceae (tobacco, tomato, aubergine), citrus, various tropical crops and ornamental flowers (anthurium, carnation, gloxinia, saintpaulia, peperomia etc.). The tobacco isolates seem to host-specific. Among the others some tendency towards host specialization can be noted, but to a far lesser extent (Bonnet *et al.* 1978).

Diseases

A thermophilic fungus (optimal growth around 30°C), *P. nicotianae* under temperate climates causes most of its damage in summer, or in crops grown under glass. Infection generally starts on the roots; root rot may provoke growth delay and leaf discoloration. On young herbaceous plants the fungus then progresses upwards; when it reaches the crown and the lower part of the stem the plant collapses and dies (tobacco black shank, tomato foot rot). Direct attack of the collar is also observed as a consequence of watering, since zoospore infection occurs if free water stands at the foot of the plant. Lower stalks, leaves and fruit can become infected from the soil by water-splash (e.g. brown rot of citrus fruit, wet buckeye rot of tomatoes). Infection by *P. nicotianae* on citrus is identical to that caused by the more frequent *P. citrophthora* (q.v.) (Boccas & Laville 1978). On tomato, *P. cryptogea* or *P. capsici* occasionally give the same symptoms as *P. nicotianae*, while late blight (leaf destruction and dry fruit rot) due to *P. infestans* is quite distinct (Weststeijn 1973).

Epidemiology

In the case of tobacco and tomato, epidemics mostly appear to originate from propagation beds. Soil temperature is a critical factor for infection: the fungus has maximum activity between 17 and 30°C; above 27°C, plant resistance has been observed to decrease with some tobacco cultivars (McCarter 1967) or to increase in the case of tomato (Weststeijn 1973). High soil moisture may reduce plant death if it allows better growth of roots to overcome root rot but excessive watering favours a massive release of zoospores from the sporangia present in the soil and their dispersal in the field. In the absence of the host plant, the fungus survives for three or four years, probably by means of chlamydospores which are produced abundantly in the soil.

Economic importance

Tobacco black shank is prevalent in the USA and other parts of the world. In Europe, it is known in Bulgaria, Greece, Spain and Italy, but losses are far less than those due to *Peronospora tabacina*. Black shank has not been reported in France, though the cultivars used are highly susceptible. On tomato, damage has been seen in several European countries, notably in Greece, the Netherlands, Romania and Spain. Many flower crops grown under glass are affected.

Control

Continuously cropping tobacco on the same land is a frequent practice which favours the build-up of a high soil inoculum. Crop rotations over 3 or 4 years have proved effective in controlling the disease (Flowers & Hendrix 1974), together with the use of resistant cultivars. The first tobacco cultivar with general resistance was constructed as early as 1930 and various levels of resistance of polygenic origin have been obtained from crosses with such cultivars of *Nicotiana tabacum* as Florida 301 or Beinhart 1000–1. Furthermore, resistance factors have been introduced by interspecific hybridization with *N. plumbaginifolia* and *N. longiflora*. In contrast only a few resistant lines of tomato are known at present (Boukema 1983), but good cultural practices and soil disinfection will generally minimize the risks. Efficient control on tobacco has also been obtained by soil drenches with fungicides such as prothiocarb and propamocarb (McIntyre & Lukens 1977) and, more recently, metalaxyl and furalaxyl which inhibit all stages of fungal development after penetration into the host (Staub & Young 1980). The same fungicides and also the systemic fosetyl-Al have been employed successfully against *P. nicotianae* on many of its other host plants.

RICCI

Phytophthora capsici Leonian

P. capsici belongs to taxonomic group II of the genus (Newhook *et al.* 1978). It is a heterothallic species forming amphigynous antheridia and elongated papillate sporangia, with a high optimum temperature for growth (26–32°C). It has a narrow host range including tomato, aubergine, courgette and watermelon, but it is mainly known as the causative agent of pepper blight (Polach & Webster 1972). In addition, a 'morphological form' of *P. palmivora* pathogenic on cocoa has been reassigned to *P. capsici*. On pepper, attacks occur on the roots or at the collar, and invasion proceeds upwards into the base of the stem. The external symptom is a rapid wilting of the plant which finally dies. In tropical conditions, all plant parts can be infected by rain splash. *P. capsici* is a limiting factor of pepper cultivation where irrigation is used, notably in Mexico and, for Europe, in Greece, Spain and Turkey; watering greatly enhances the dispersal of zoospores which spread from the first disease foci. Chemical treatments (for example soil drench with fosetyl-Al) may contribute to the control of pepper blight (Clerjeau & Beyries 1977), together with the use of available partially resistant cultivars. Different components of resistance (low receptivity, inducibility and stability) have been identified in separate lines of *Capsicum* and could be combined in the future into commercial cultivars (Pochard *et al.* 1983).

RICCI

Phytophthora cryptogea Pethybr. & Lafferty

Basic description

P. cryptogea belongs to taxonomic group VI of the genus (Newhook *et al.* 1978). It is characterized by ellipsoid non-papillate sporangia and hyphal swellings in clusters, nearly absent from solid media, but readily forming after immersion of the culture into mineral solutions or soil leachates. Upon confrontation with a strain of the opposite mating-type, it forms oogonia and amphigynous antheridia giving rise to thick-walled oospores. Cardinal temperatures (8–33°C) hardly distinguish it from *P. drechsleri* with which some authors have proposed to merge it. *P. cryptogea* is the only species causing expanding necrosis when inoculated on wounded surviving leaf discs of *Gerbera* (Bonnet *et al.* 1980).

Host plants

P. cryptogea attacks a wide range of trees and shrubs (*Prunus, Malus, Populus, Eucalyptus, Castanea, Vitis,* several Conifers and Ericaceae) and of herbaceous plants such as tomato, potato, chicory, safflower, bean, lucerne, melon, courgette, chick-pea, spinach, watercress. A number of flower crops are frequent hosts: *Gerbera* and other Compositae (*Aster, Callistephus, Chrysanthemum, Dahlia, Osteospermum, Senecio, Tagetes. Zinnia*) but also *Tulipa, Sinningia, Papaver, Cheiranthus, Matthiola, Antirrhinum, Gypsophila, Lupinus, Begonia,* for example.

Host specialization

Individual isolates may not have as large a host range as the species as a whole (Garibaldi 1974). A forma specialis *begoniae* has even been proposed (Kröber 1981).

Geographical distribution

P. cryptogea occurs world-wide.

Disease

Basically a root pathogen, *P. cryptogea* causes dieback of woody plants and wilt of herbaceous hosts by secondarily invading the collar and stem. In Europe, the most important damage is observed on ornamental trees and shrubs (Conifers and Ericaceae), on which *P. cryptogea* occurs together with other species, mainly *P. cinnamomi* (q.v.) and, more importantly, on *Gerbera jamesonii*. On this species, the first symptom is a sudden wilt, several leaves losing their turgor. This is followed by a black rot of the base of the petioles and floral scape. At this stage, the collar is largely invaded and breaks easily; part of the roots show browning or even complete destruction of the cortex. The symptoms may be confused with those due to other pathogens, notably *Rhizoctonia solani*, and isolation from the diseased tissues on a selective medium is often necessary (Bonnet *et al.* 1980).

Epidemiology

P. cryptogea may be introduced with infected plants especially in the case of cuttings from clones vegetatively propagated. It can survive for several years in soils or other substrates used in *Gerbera* cultivation (Orlikowski 1980). Using baits, such as pine or cedar needles, it has been detected in diseased plantations at high inoculum levels down to at least 30 cm in depth. At temperatures between 10 and 25°C, watering the substrate causes an inoculum increase due to sporangium formation and zoospore release; however, the sporangia formed retain their viability even at low soil matrix potential (Duniway 1983). High water content also favours zoospore dispersal and thus the spread of the disease; in optimal conditions, 15–20 *Gerbera* plants can be destroyed in 8 months starting from a single seat (Garibaldi 1974). Penetration probably occurs at wounds or in young roots. *Gerbera* is susceptible at any stage, but may be more so during flowering. The length of the incubation period depends upon inoculum density and soil temperature (optimum: 20–25°C); thus it is shorter between May and September (10–30 days) than in winter (20–50 days).

Economic importance

P. cryptogea is an important pathogen in ornamentals, capable of destroying a high proportion of plants by rapidly spreading in the crop; it is one of the limiting factors in the cultivation of *Gerbera*. The disease is favoured by moderate soil temperature, low light levels and high soil humidity. Plant susceptibility to penetration and tissue invasion seems to be increased by stress, such as temporary root asphyxia, excess salinity, unbalanced nutrition, etc. Synergism with attacks by other pathogens (*Erwinia carotovora* or *Meloidogyne arenaria*) has been observed. Because control methods are not fully satisfactory, losses may be high.

Control

The sanitary conditions of propagation beds and culture substrates are of prime importance. Fumigation of soils and substrates is used to eradicate the fungus; satisfactory results have been obtained with methyl bromide. Steam disinfection is also used especially in benches isolated from the soil. During cultivation, one can avoid creating favourable conditions for the disease by carefully monitoring the use of water and fertilizers. Maintaining the substrate temperature above 25°C decreases the risks of mortality in *Gerbera* (Scholten 1970). Numerous fungicides have been tried, generally as soil drenches. Even the most effective ones (etridiazole, metalaxyl and fungarid) can only slow down the rate of infection for a few months. A search for resistant cultivars of *Gerbera* has been undertaken in the Netherlands (Sparnaaij *et al.* 1975) and in France (Meynet 1980); only reduced susceptibility with quantitative inheritance has been found so far.

RICCI

Phytophthora drechsleri Tucker has a higher temperature optimum (30°C) than *P. cryptogea* (q.v.) but otherwise resembles it and may be synonymous. It is reported as damaging principally in N. America (on safflower) and in Asia (on cucurbits). In Europe, there are only scattered records (CMI map 281). Skidmore *et al.* (1984) who isolated it from potato in the UK stimulating oospore formation in *P. infestans* considered it a weak non-specific pathogen possibly causing no symptoms on infected plants.

Phytophthora cactorum (Lebert & Cohn) Schröter

Basic description

P. cactorum is homothallic with predominantly paragynous antheridia and markedly papillate, caducous sporangia. Some strains produce chlamydospores (CMI 111). Cardinal temperatures are approximately 4°, 25° and 30°C.

Host plants

Nienhaus (1960) listed more than 200 susceptible plants (including weeds) in over 60 families, notably the Rosaceae. However, these include many plants susceptible to inoculation but unknown as natural hosts, and reports in which *P. cactorum* has been confused with other *Phytophthora* species. In fact, relatively few crop plants suffer consistently from *P. cactorum* but, under appropriate conditions, a wide range of species is liable to attack.

Host specialization

There is evidence for the existence of pathotypes within *P. cactorum* of which the best documented is the distinct strain causing strawberry crown rot (Seemüller & Schmidle 1979). Apple isolates are also heterogeneous for virulence to different apple cultivars.

Geographical distribution

P. cactorum was first described in Germany in 1870 and has since been reported from most European countries. It is widespread through the temperate and warm-temperate regions of the world but has a very limited distribution in the tropics (CMI map 280).

Diseases

Diseases occurring in Europe are: seedling blights (conifers, beech and other trees); leaf and stem blights (lilac, rhododendron, cacti and other herbaceous ornamentals); fruit rots (apple, apricot, citrus, pear, plum and strawberry); crown rots (apple and strawberry); collar rots (apple, gooseberry, rhubarb, stone-fruits, sweet chestnut and other broad-leaved trees); and cankers (horse chestnut, *Acer* spp. and other hardwoods). *P. cactorum* is primarily an intercellular colonist of parenchyma. Small, sub-spherical or digitate 'haustoria' and toxins are produced, and a brief biotrophic phase is followed by a more prolonged necrotrophy. There is vascular colonization in some diseases, notably strawberry crown rot, where the main symptom is a rapid wilt. Attacks on seedlings, leaves, inflorescences and young shoots cause orange-brown discoloration, withering and collapse. Apple, pear and strawberry fruits develop a firm rot with a diffuse margin: green-brown, marbled to pale brown in apple; dark brown in pear; and, in strawberry, from brown to dark red depending on the ripeness of the berry when attacked. Vascular tissues are dark stained and stand out from the rotted flesh. Lesions on tree bark are characteristically mottled or zoned (Plate 11), becoming uniformly brown: drops of rusty-brown, watery liquid often exude from active lesions. Bark lesions may be checked or become perennial (especially in some apple scion cultivars) and ultimately cause girdling of large limbs or the whole tree. Girdling induces characteristic symptoms in the crown such as premature autumn leaf colour and defoliation.

Epidemiology

P. cactorum is soil-borne and widespread but not generally abundant; its ability to regenerate perennating inoculum is apparently low in some situations. Unusually severe attacks on several hosts have occurred when crops were planted among or after apples. The pathogen survives outside the living host as oospores formed in diseased tissue which enter the soil and remain viable for several, possibly many, years. In free water, light and at moderate to high temperatures (10–20°C) oospores germinate and zoospores are produced via sporangia. Zoospores are probably the main infective agents though deciduous sporangia may also initiate infection. Dissemination is by mass flow of soil water, irrigation, or by rain splash, and is thus associated with wet conditions or with poorly drained soils. The fungus may be introduced to apple orchards at planting with soil or in cryptic infections of tree roots; and latently infected strawberry runners are an effective vehicle for spreading crown rot between fields and between countries. Infection is mostly through wounds, unprotected surfaces, and natural openings such as stomata and lenticels. Sporangia may develop massively from diseased tissues into moist soil and give rise to further cycles of infection.

Host susceptibility is influenced by endogenous factors and probably also by external stress factors. In the two diseases most studied, apple collar rot and strawberry crown rot, the host undergoes marked seasonal changes of susceptibility. Apple scion cv. Cox's Orange Pippin only develops aggressive lesions from inoculation between bud burst and the onset of extension growth (Sewell & Wilson 1973), while strawberries are most susceptible during flowering and fruiting, summer and autumn infections frequently remaining latent until the following spring. In apple scions, susceptibility increases with tree age particularly after about 10–20 years and is positively correlated with the vigour of the rootstock to which the scion is grafted; resistant apple bark differs from susceptible bark in its ability to lay down impenetrable barriers by meristematic activity in advance of the invading pathogen. Apple fruits also become increasingly susceptible to infection with age. Cold-stored strawberry runners are particularly prone to crown rot. *P. cactorum* is only infectious in apple orchard soil when mean air temperatures exceed 10°C and its activity increases with temperatures up to 20°C (Sewell *et al.* 1974). A similar pattern of winter quiescence occurs in strawberry soils. The requirement for a coincidence of relatively warm, continuously wet conditions and susceptible tissues explains the sporadic nature of disease outbreaks.

Diseases similar to those caused by *P. cactorum* may be due to other *Phytophthora* species such as *P. syringae*, and in some situations one or more species may be involved with *P. cactorum*.

Economic importance

Apple collar rot was very destructive, mainly on Cox's Orange Pippin in Belgium, UK, the Netherlands and in Germany in the 1950s and 1960s but is less important in modern intensive orchards. The rootstock MM104 has been abandoned in Britain because of its susceptibility to crown rot, whilst the widely used MM106 suffers sporadic and sometimes serious losses, mainly in young orchards where trees are unstaked. *P. cactorum* is generally a more serious pathogen of clonal rootstocks in N. America than in Europe. Strawberry crown rot has spread from Germany to several other European countries since the 1950s and has become a major hazard to field crops in S. France (Molot & Nourrisseau 1966) and Italy, and to protected crops in more northerly areas. Post-harvest fruit rot of apple and pear due to *P. cactorum* is often serious in S. France but is less important in the north where *P. syringae* is more destructive (Bompeix & Mourichon 1977). Leather rot causes sporadic losses in strawberries, usually in unmulched crops, and may be aggravated by chemicals used for controlling *Botryotinia fuckeliana*.

Control

The unpredictability of diseases due to *P. cactorum* makes economic control difficult. Preventive methods employed are: disease avoidance by using resistant cultivars, judicious irrigation, avoiding poorly drained sites, high working and the use of resistant interstocks for collar rot of apple, mulching for leather rot of strawberries; and phytosanitary measures such as soil sterilization, decontamination of irrigation water and, in the case of strawberry crown rot in France, a system of healthy runner production. Apple trees partially girdled by collar rot or whose rootstocks are seriously damaged by crown rot can be restored by inarch grafting with resistant stocks.

Until recently, copper-based fungicides were used almost exclusively for controlling seedling diseases and blights, for preventing bark diseases and for protecting wounds left by surgical treatment of collar rots and cankers. Several synthetic fungicides, particularly captafol, captan, dithiocarbamates, etridiazole and propamocarb have proved effective against various *P. cactorum* diseases in trials, but it is the powerful systemics, metalaxyl and fosetyl-Al which offer the best prospects of efficient chemical control. Resistance is the ultimate answer to most diseases due to *P. cactorum*, and breeding improved resistant apple rootstocks and scion cultivars continues in several European countries and N. America.

Special research interest

P. cactorum is amenable in culture and has been used as a model species in studying aspects of the biology of homothallic *Phytophthora* species such as nutrition; spore production; genetics; zoospore release and behaviour; oospore germination and survival in soil.

HARRIS

Phytophthora syringae **Kleb.**

Basic description

P. syringae is homothallic with predominantly paragynous antheridia and semi-papillate, non-caducous sporangia (CMI 32). Monilioid hyphal swellings are produced by most isolates. Cardinal temperatures are approximately 1, 21 and 25°C.

Host plants

P. syringae is a pathogen of pome and stone fruits, citrus, lilac and fennel. Under certain conditions this fungus, like *P. cactorum*, may attack a wide range of species.

Geographical distribution

P. syringae occurs widely in Europe but has an otherwise limited distribution in exclusively temperate areas (CMI map 174).

Diseases

P. syringae causes fruit rot of apple (Fig 27), pear, peach and citrus, collar rot of apple and pear and stone fruits, leaf and shoot blight of lilac, and leaf blight of fennel. Soft tissues turn brown and collapse; the apple and pear fruit and bark rots are indistinguishable from those caused by *P. cactorum*, but collar rot lesions, on apple at least, never become perennial. Affected citrus fruits turn brown.

Fig. 27. Apple fruits (cv. Cox's Orange Pippin) rotted by *Phytophthora syringae*.

Epidemiology

P. syringae is soil-borne and generally widespread and abundant in apple orchard soils. It perennates as oospores which form in diseased tissue or, in the case of apple, in senescing fallen leaves, and may survive in soil for 15 years or more. In the presence of free water and light and at low to moderate temperatures oospores germinate and zoospores are produced via sporangia. Zoospores are disseminated by mass flow of soil water, irrigation or rain splash, and infection is mostly through wounds and natural openings such as stomata and lenticels. Sporangia develop from colonized tissue and can give rise to further cycles of disease. The activity of *P. syringae* in apple orchard soil is detectable when mean air temperatures are between 0 and 16°C and is highest at 12–14°C (Sewell *et al.* 1974). The bark of tree hosts is only susceptible to *P. syringae* during dormancy and in early spring. Epidemics are thus associated with excessive rainfall in cool or cold weather. Apple and citrus fruits infected at or just before harvest become the foci of massive secondary rotting in refrigerated storage. The emergence of *P. syringae* as a major fruit rot pathogen in apple and pear has followed the adoption of intensive orchard systems in which fruit is borne near the soil and herbicide use exposes fruit to soil splash (Harris 1981). Diseases due to *P. syringae* resemble those caused by other *Phytophthora* species, especially *P. cactorum*; both these pathogens may attack the same host simultaneously.

Economic importance

P. syringae was considered relatively unimportant until the 1970s when it emerged in several European countries as a major cause of post-harvest rot in apple and, to a lesser extent, in pear, particularly in long-term storage (Bolay 1977; Upstone 1978). Collar rot can be a very destructive disease of young apple trees exposed to wet soil between lifting and planting-out (Sewell & Wilson 1964), and is a common cause of the serious 'apoplexy' syndrome on apricots and other stone-fruits in Greece and possibly elsewhere in S. Europe and the Mediterranean (Kouyeas 1977). The lilac disease is widespread but of little significance now that forcing potted plants is no longer practised. Because of the difficulties of diagnosis, this pathogen may be much more generally important as a cause of winter diseases on perennial crops than is currently apparent.

Control

Copper compounds are used for protecting trees in autumn and winter against attack by *P. syringae* although there are no data to indicate how effective they are. However, Bordeaux mixture was superior to three synthetic fungicides for controlling fennel blight, and is still the standard for brown rot control of citrus (caused by *P. syringae* and other *Phytophthora* species) in the USA. Metalaxyl as a post-harvest dip has proved very effective in eradicating early primary infections on apple fruits and in controlling secondary rotting during storage (Edney & Chambers 1981); it is now used commercially for this purpose in Britain. Metalaxyl has a similar eradicative action on early infections of citrus fruits.

Reports on the effectiveness of fosetyl-Al for controlling fruit rot of apple are conflicting. In Switzerland, where post-harvest treatments are illegal, a late-season schedule of captan, dichlofluanid or folpet is recommended for simultaneous control of *Botryotinia fuckeliana*, *Perzicula* spp. and *P. syringae*. Other measures for controlling fruit rot are to modify pruning where possible to keep fruit away from the soil surface and, in wet harvests, to pick and market low-hanging fruit separately. General cultural practices aimed at avoiding excess water, injury to plants and contact between bark and soil during the period of susceptibility should also be employed to prevent diseases due to *P. syringae*.
HARRIS

Phytophthora cinnamomi Rands

Basic description

P. cinnamomi characteristically produces coralloid hyphae and botryose swellings on malt or V8 agar, and also forms thin-walled, mainly terminal chlamydospores (CMI 113, Newhook *et al.* 1978). The usual temperature range for growth is 5–34°C and mycelium in soil grows optimally at 24–28°C with a matric potential of *c.* −10 bar. *P. cinnamomi* produces large, non-papillate, non-deciduous sporangia (which usually proliferate internally) in certain liquid media, and non-sterile soil leachate is usually highly stimulatory to sporangium production. Sporangia are produced at soil matric potentials between −10 mb and −2500 mb at 15–35°C (with an optimum at 24°C, pH 5.5). The temperature, matric potential and pH of the soil also influence whether sporangia germinate directly or release 10–30 motile zoospores. *P. cinnamomi* is a heterothallic species forming oospores either by pairing *A1* and *A2* types or by interspecific crosses with the opposite mating type of, e.g., *P. cryptogea*. The *A2* type can also produce oospores 'homothallically' when provided with a stimulatory substance, e.g. volatiles from *Trichoderma viride*.

Host plants

P. cinnamomi attacks over 950 species and cultivars, mainly those of woody plants (Zentmyer 1980). Many of these are cultivated in Europe as ornamental trees and shrubs or as timber or fruiting trees. Among them are the following genera: *Acacia, Acer, Arbutus, Azalea, Betula, Calluna, Camellia, Castanea, Chamaecyparis, Cupressocyparis, Cupressus, Erica, Eucalyptus, Fagus, Hebe, Hibiscus, Juglans, Juniperus, Larix, Lavandula, Laurus, Magnolia, Malus, Morus, Myrtus, Nothofagus, Olea, Persea, Picea, Pinus, Platanus, Prunus, Pyrus, Quercus, Rhododendron, Robinia, Syringa, Taxus, Ulex, Vaccinium, Vitis*.

Geographical distribution

First described on *Cinnamomum burmannii* in Sumatra, the pathogen now has a world-wide distribution, including Europe (CMI map 302). The fungus is found in tropical and subtropical countries and in the Mediterranean and some mild, temperate regions where it has probably been introduced.

Disease

P. cinnamomi is primarily a root pathogen of woody species. It attacks young 'feeder' roots causing a rot which may extend into the base of the stem, forming a brown lesion on the white wood which can be seen if the bark is removed. If propagules are splashed onto aerial parts of the host, stems, branches, leaves and fruit can become infected. The many diseases caused are described as root or foot rots (e.g. rhododendron, walnut), collar or crown rots (e.g. banksia), or canker (e.g. plane). Root damage reduces uptake of water and nutrients and, therefore, *P. cinnamomi* diseases may also be described by foliar symptoms as a wilt or dieback (e.g. heather, rhododendron, and many other hosts) (Fig. 28) or by a reduction in leaf size (e.g. little leaf of pine). *P. cinnamomi* infection can also occur together with

Fig. 28. *Chamaecyparis lawsoniana* cv. Ellwoodii: wilting (centre) and death (right) due to *Phytophthora cinnamomi*.

other *Phytophthora* species, e.g. with *P. cambivora* (q.v.) causing a black root rot of chestnut (*Castanea* species) known as 'ink' disease and with *P. cryptogea*, *P. citricola* and *P. cactorum* in root or stem infections in a number of species.

Epidemiology

The rapid build-up and spread of diseases caused by *P. cinnamomi* are associated with high soil matric potentials and temperatures which favour the production of sporangia and spread by zoospores. These conditions exist with bad drainage and/or poor control of irrigation systems. Water flow spreads zoospores into contact with other hosts, and also into streams and lakes which may be used for irrigation. Zoospores are attracted to roots by chemotaxis and electrotaxis and after encystment and germination the fungus penetrates susceptible roots inter- and intra-cellularly. Young actively growing plants with a high proportion of 'feeder' roots are particularly prone to *P. cinnamomi* attack as are roots which have been damaged by severe water stress or wounding. Foliar symptoms reflect the imbalance between water uptake and transpiration, and loss of leaf colour and development in wilt and dieback may be rapid (within weeks of initial infection) if extensive root damage is followed by dry weather. However, symptoms may take several months or years to develop in a cool, damp climate and infected but symptomless plants can be a means of spreading the pathogen. Species and cultivars differ in susceptibility to *P. cinnamomi* attack (Végh & Le Berre 1982) and, therefore, infected but 'tolerant' plants may be a source of disease spread.

P. cinnamomi can invade soil organic matter and grow to a limited extent through moist non-sterile soil but the mycelium is readily lysed by *Trichoderma* species, bacteria and other soil micro-organisms. Sporangia and encysted zoospores can survive in moist soil for several weeks but, in the absence of a host and, when soil water potentials and temperatures are sub-optimal for sporangium production, chlamydospores and oospores serve as survival propagules. Germination of chlamydospores is stimulated by organic nitrogen and some root exudates, but the conditions influencing oospore germination have not been determined. A significant factor in the spread of *P. cinnamomi* diseases has been the distribution of infected plants by the world-wide trade in nursery stock and the intensification of production methods involving dense cropping and applied irrigation.

Economic importance

In Europe, *P. cinnamomi* is frequently reported to cause serious damage to high value ornamental trees and shrubs produced by the nursery stock industry and in amenity plantings, as well as species grown for timber or fruit, but there is little or no quantitative record of crop loss.

Control

Control of this pathogen is complicated by the very wide host range, the variable and sometimes lengthy period between infection and disease expression and the longevity (sometimes many years) of propagules in soil and root debris often at considerable soil depth. Disease control depends on the manipulation of cultural, biological and chemical measures as routine procedures to prevent infection and restrict spread from infested sites (MAFF 1979; Strouts 1981; Smith *et al.* 1983). Nursery stock should be grown with strict attention to hygiene using a clean water supply and propagating only from healthy seed or mother plants. Imported stock should be segregated on the nursery and preferably grown in containers for several months to check the health status of the plants. The composition of the growing medium for container-grown crops should ensure good aeration and drainage and the use of composted hardwood bark or pine bark has been reported to suppress *P. cinnamomi* attack in several ornamental shrub species. It is important to stand container-grown plants on a well-drained substrate (e.g. gravel, sand, bark) to prevent cross-contamination by zoospores, and to regulate the irrigation system to avoid splash and overwatering. Similarly, field crops should be planted in land which is well drained. Infected plants should be destroyed and the movement of infested soil on footwear, tools and machinery should be avoided. In infested field soils the incidence of *P. cinnamomi* can be reduced, if not eliminated, by cropping with a non-susceptible host, leaving the land fallow for at least four years or incorporating organic nitrogen into the soil. Steam sterilization or formaldehyde can be used to treat potting composts or to clean up propagation benches, beds, containers and tools.

The specific anti-Pythiaceae fungicides etridiazole, metalaxyl, furalaxyl, fosetyl-Al and propamocarb hydrochloride can be fungicidal to zoospores but act primarily as fungistats in suppressing hyphal growth and sporangium formation. They are particularly effective when applied as prophylactic treatments to the roots of healthy container-grown plants either as a drench or by incorporation into the rooting medium (Steekelenburg & Slavekoorde 1979; Smith 1980; Bruin & Edgington 1983). Applied as eradicants, they arrest but generally do not kill the fungus in infected roots and soil. The effectiveness of soil fumigants such as methyl bromide, metham-sodium, dazomet and dichloropropene is limited to the depth of fumigant penetration but they can give good control of the pathogen in the top soil, e.g. in the preparation of nursery seed beds. *P. cinnamomi* occurring in water from untreated sources can be controlled by filtering and chlorinating systems. Cankers and lesions on woody stems, if not too extensive, can be controlled by cutting out all the dead and dying tissue and applying a Bordeaux slurry to the wound (Strouts 1981).

Some biological systems show promise for the control of *P. cinnamomi*, and Zentmyer (1980) has reviewed the work on soil amendments, antagonistic organisms, mycorrhizae and 'suppressive' soils; but current data are generally insufficient to indicate, with precision, the appropriate biological system for a particular host/*P. cinnamomi* complex.

P. SMITH

Phytophthora cambivora (Petri) Buisman

P. cambivora forms long unbranched sporangiophores terminating in a single limoniform sporangium which germinates by releasing zoospores (CMI 112). Together with *P. cinnamomi*, it attacks chestnuts causing the 'ink disease'. *P. cambivora*, virtually restricted to Europe, has been recorded in Britain, France, Greece, Italy, Poland, Portugal, Spain, Switzerland and Yugoslavia (CMI map 70). The fungus enters the plant through fine roots. The surface of infected roots has a darkened appearance and this darkening often extends in tongues up the main stem above the dead main roots. Later, when the roots are dead, the tannin in them reacts and produces the characteristic inky discoloration.

P. cambivora spreads mainly by mycelium and by sporangia washed through the soil. It is difficult to estimate the losses caused. It has damaged many chestnut plantations in Spain, but, in other places, the reduction of chestnut areas may be more due to a decline in chestnut cultivation and to *Cryphonectria parasitica*. Root exposure, combined with chemical treatments of the surface of diseased roots and of the soil around them with copper salts, is the method used in Spain and Portugal to control the disease. The replacement of susceptible *C. sativa* by the resistant *C. crenata* has been suggested, but has not been carried out.

TURCHETTI

Phytophthora megasperma Drechsler

P. megasperma is homothallic and produces oogonia with paragynous (occasionally amphigynous) antheridia on solid media, and non-papillate internally proliferating sporangia in liquid cultures (CMI 115). It has been subdivided into **var. *megasperma*** and **var. *sojae*** Hildebrand on oogonial size, though not all isolates fall readily into one or other of these categories. It has a very wide host range, from stone fruits, ornamental and forest trees and shrubs to grapevine, sugarcane, raspberries, vegetables and pulse crops. The species is widespread in warm temperate and subtropical regions (CMI map 157) and appears to include various host-associated subgroups. Large-spored forms predominate in Europe. Distinct pathotypes of the small-spored forms occur on cultivars of lucerne, soybean and clover, especially in N. America. These have been referred to formae speciales by Kuan & Erwin (1980) but this approach has been questioned in recent experimental work (Hansen *et al.* 1987). The form specific to soybean is resticted to N. America and is considered a quarantine organism for Europe.

Attack is mainly via zoospores from sporangia on diseased host tissue or from germinating oospores. On woody hosts, infection may spread upwards from fine roots along the phloem and cambium, but more often begins at the root collar, sometimes spreading some distance up the stem beneath the bark. Active lesion edges are pale, mottled and water-soaked. Older lesions often appear ginger-brown. Diseased plants show small leaves, wilting, yellowing foliage, and dieback. Severe attacks may follow heavy rain, especially on poorly drained soils, and may progress rapidly in late autumn and early spring when host dormancy begins and ends. On herbaceous hosts attack is often at the stem base but also on roots, tubers and young shoots, producing a water-soaked lesion leading to wilting and foliar discoloration. Rapid local spread is largely via zoospores and long-term survival via oospores. The fungus is usually readily isolated from fresh lesion edges or infested soil using apple baiting methods. In Europe outbreaks are generally sporadic and of limited economic importance. On trees, attacks often associated with waterlogging and soil compaction range

from individual specimens to whole avenues (e.g. of *Aesculus*). Damage to brassicas is occasionally serious on poorly drained soils. In N. America serious losses frequently occur on some leguminous crops (APS Compendia on Soybean and Alfalfa Diseases). Control measures are similar to those for *P. cinnamomi* (q.v.).
BRASIER

Phytophthora citrophthora (R. & E. Sm.) Leonian

Basic description

P. citrophthora is a highly heterothallic species. Reproduction is mainly by sporangia and survival by chlamydospores (Tuset 1977). Cardinal temperatures are 5°C, 24–28°C and 32°C.

Host plants

Plants of over twelve families are attacked (CMI 33) in an unspecialized manner but citrus is the important host, most species being attacked, as are also *Poncirus* and various interspecific (mandarins, clementines) and intergeneric hybrids (citranges, citrumelos).

Geographical distribution

P. citrophthora has a world-wide distribution. It was introduced into Europe, possibly in the middle of the last century from the Azores, as a consequence of the general development of citriculture in the Mediterranean countries. It has specifically been recorded from Portugal, Spain, France, Italy, Greece, Cyprus and Malta, but probably occurs wherever citrus is cultivated.

Diseases

The following diseases are caused in citrus: gummosis, foot and collar rot, and brown rot of fruits. They are frequent in all citrus-growing areas and the most important fungal disease of the crop (Klotz 1973). *P. nicotianae* var. *parasitica* (q.v.) frequently causes very similar diseases in citrus and is the more important pathogen in some countries (Spain), especially causing gummosis and foot-rot. *P. citricola* and *P. hibernalis* (q.v.) generally cause only brown rot of fruits. Gummosis mainly affects the base of the trunk, but can affect any part of it. The main symptom is abundant formation of gum on the surface of the bark, which darkens in colour, while the internal tissues are killed through to the wood (which sometimes appears dark brown due to a thin layer of infiltrated gum). The lesions are variable in shape and extent and spread more rapidly in the vertical direction than laterally. The dead cortical tissues gradually dry out and the bark appears sunken and fissured, while the adjoining tissues appear healthy, thus giving a typical canker. The extent of gummosis depends on the species, on soil type and on climatic conditions (especially RH). Since atmospheric conditions are normally rather dry in Mediterranean countries, gummy exudations are rarely seen. It is more common for the fungus to cause foot or collar rot, not visible from the surface, either at a few cm depth in the soil, girdling roots at the neck, or at the soil surface itself. In the latter case, fungal development is usually restricted, although the hyperplastic growth of the edges of the canker may be conspicuous. Plants affected in either of these ways have pale-green leaves with yellowed veins. New shoots are few in number, short, have rather small leaves and a chlorotic appearance. Fruit production is reduced, with smaller than normal fruits. In extreme cases, the tree is killed. On fruits, *P. citrophthora* causes brown rot. Affected fruits lose their green or orange colour (according to their degree of ripeness at infection) and show various-sized dark green spots which rapidly change to clear brown and remain firm to the touch. At high RH, the fungus sporulates on the surface of the spots, giving a whitish bloom. Infected fruits drop prematurely. The spots are often secondarily infected by fungi such as *Penicillium italicum*, which causes softening after a week or so, turning them white then bluish.

Epidemiology

P. citrophthora has quite distinct saprophytic and parasitic phases (a typical soil-inhabiting fungus). The former occurs on plant debris in the soil and can last from several months to two or more years (according to conditions). The fungus persists in an essentially vegetative form (mycelium or chlamydospores). In temperate regions, sporangia are less important but contribute to persistence in the soil. *P. citrophthora* requires water to produce large numbers of sporangia and mature them fully, and also to ensure dispersal of sporangia and zoospores. Irrigation or rain water soaking into the soil in orchards is the main factor required to allow rapid (4–6 h at 16–18°C) production of sporangia, which immediately release zoospores (above 10–12°C). These remain mobile for 1–2 h at 18–20°C and move in soil water to plant surfaces, where they encyst and grow into the cortical tissues (through breaks in the epidermis). Inter- and intra-cellular mycelial growth leads to the dehydration and necrosis of these tissues. The incubation time is 3–12 days, according to the type, age and water content of the tissue, ambient temperature and RH. Fruits are infected by rain splash.

Economic importance

Losses arise from poor yields, death of trees (especially young ones) and fruit rot. The fungus is fairly common in all citrus orchard soils and few plantations will not have one affected tree. Attacks depend on flooding of the soil. Humid weather, with intermittent rain, at the time when fruits change colour, is highly favourable to brown fruit rot. In Mediterranean countries, with generally low atmospheric RH, only torrential rain, especially in autumn, provides the conditions for disease. Stocks considered to be resistant (sour orange, *Poncirus trifoliata*, citranges) can still be attacked under special cultural conditions (frequent heavy irrigation, shallow soil, abundant nitrogen fertilizer). The grafted cultivar is still liable to be attacked if the graft is too near the soil, any resistance in the rootstock being by-passed.

Control

Cultural methods must first be applied—use of tolerant or resistant rootstocks, grafting susceptible cultivars at least 15–20 cm above the soil, avoidance of soil flooding, moderate nitrogen fertilization, avoidance of heavy irrigation after long periods of drought etc. Fungicides may be used prophylactically (copper derivatives, dithiocarbamates, phthalimides such as captan, folpet or captafol) to block germination of sproangia and zoospores, by application to the base of shoots and trunks when heavy rain is expected or the first gummy exudates are seen. The systemic anti-oomycete products, applied to the leaves (fosetyl-Al) or roots (fosetyl-Al, metalaxyl, cyprofuram) have given good results in trials. Field treatments (three applications 21 days apart) have prevented canker development (Tuset 1983). No resistance to these fungicides has been observed up to 1983.

Control of brown rot involves both cultural measures (grass cover of plots, removal of lower branches to avoid water and soil splash) and fungicides. This is standard practice in all citrus-growing areas. Applications may be made to the soil in the droplet zone of the trees, with copper-based products, or mixtures of these with dithiocarbamates or phthalimides. Ideally, leaves are sprayed before or during the autumn rain period (October–November). These are the most effective control measures. Pre-harvest treatments with systemic (fosetyl-Al, metalaxyl) or penetrating fungicides (cymoxanil) do not check brown rot in the laboratory or in the field. Post-harvest treatments tried out in the USA, Israel and Greece, by immersion in metalaxyl at temperatures below 10°C, have given promising results. In Spain, identical trials at 15–18°C have not stopped the rot.

TUSET

***Phytophthora citricola* Saw.**

(Syn. *P. cactorum* var. *applanata* Chester)

P. citricola is highly homothallic, producing abundant oospores in agar culture. Sporangia have a broad papilla with a slightly swollen apex. The antheridium is always paragynous. Chlamydospores are rare or absent. It is an unspecialized parasite which has been isolated from over 16 species of crop plants, especially ornamentals (CMI 114). It behaves much like *C. cactorum* (q.v.) and also persists saprophytically on debris. In citrus, it has been isolated from fruits of sweet orange and lemon, and from citrus orchard soils (Favaloro & Sanmarco 1973). It occurs in all continents, but only in temperate climates. In Europe, it has been found in Italy (Sicily), Greece, UK, Germany and the Netherlands.

P. citricola causes typical brown rot of citrus fruits (cf. *P. citrophthora*) (Broadbent 1977), with conspicuous superficial mycelium and sporangia at high RH (over 80%). Infected fruits fall (particularly oranges) and are found in large numbers on the orchard floor, showing the typical brown spotting and having a characteristic smell. In cold stores in which humidity is generally high, fruits are commonly covered with masses of superficial white mycelium (at temperatures over 8°C, otherwise only the brown spotting is seen). *P. citricola* can also cause typical trunk cankers, with gummosis under certain circumstances. Infection occurs much as for *P. citrophthora* (q.v.) except that oospores play a role in survival. The incubation period in citrus fruit is 5–7 days at 12–14°C and 3–5 days at 18–20°C. Cardinal temperatures for growth are 3–4°C, 25–28°C and 30–32°C which means that the fungus is most often

isolated in autumn. Under normal rainfall conditions, *P. citricola* is a rather minor member of the brown rotting complex, and even less important as an occasional cause of foot rot and gummosis. It is only liable to become important as a consequence of heavy rain, or fog, in October. Control is as for *P. citrophthora*; spraying the lower foliage with copper oxychloride and a dithiocarbamate (zineb or mancozeb) has proved very effective. There is no information as yet on the effectiveness of systemic compounds.
TUSET

Phytophthora hibernalis Carne

P. hibernalis is a clearly homothallic species producing oospores with antheridia mainly of the paragynous type. The sporangia have a long pedicel, and hyphal swellings or chlamydospores are rare. The maximum temperature for development is around 25–26°C. Some isolates are difficult to distinguish from *P. syringae* (Tuset 1977; Ho 1981). *P. hibernalis* attacks practically all commercial citrus species (CMI 31), and is clearly preferentially adapted to citrus, especially orange and lemon. It is present in most temperate zones, e.g. in Europe in Italy, France, Portugal, Greece (Kouyeas & Chitzanidis 1968), Spain and UK. It mainly causes brown rot of fruit, but can cause a leaf blight under high RH conditions. It infects fruit, alone or in combination with other *Phytophthora* species under very wet conditions and preferentially at low tempertures (15–20°C). The fruit rot can continue in cold stores. Fruit infection occurs under the same conditions of soil flooding as for *P. citrophthora*, except that oospores also contribute to sproangium formation. Infection of fruit from zoospores splashed onto the skin is favoured by temperatures below 20°C, and the incubation period is 4–7 days.

Normally, *P. hibernalis* is a less important pathogen than the other *Phytophthora* species in Mediterranean countries. It requires daytime temperatures of 18–20°C and night-time temperatures not below 8°C, conditions which only occur in the winter months. It is then only liable to be a problem in orchards on heavy, poorly drained soils. Fruit infection is mainly limited to the lower branches, within 1.5 m of the soil (especially at 8–60 cm), with a high probability of soil and water splash by rain and wind. At high temperatures, *P. hibernalis* is replaced by *P. citrophthora*, which is the major cause of brown rot in Mediterranean countries. At most, 20% of brown rotting of fruit can be ascribed to *P. hibernalis*, except under special conditions in certain areas. Control is as for *P. citrophthora*, with special emphasis on cultural measures, and maintenance of prophylactic fungicide treatments into the winter months in regions where the rainfall justifies it. Systemic fungicides have not yet been tested against *P. hibernalis*.
TUSET

Phytophthora phaseoli Thaxter causes a leaf-blight of *Phaseolus* species, principally in N. America, but recorded from Italy and the USSR (CMI map 201). ***P. porri*** Foister (CMI 595) causes white tip disease of leek in S. France, the Netherlands and UK (Messiaen & Lafon 1970). It can also cause quite severe damage to salad onions in the UK (Griffin & Jones 1977). Semb (1971) found a form of this species rotting stored cabbage; it did not affect leek.

PERONOSPORACEAE

Peronospora parasitica (Pers.) Fr.

Basic description

P. parasitica forms dichotomously branched determinate sporangiophores which emerge from host stomata. The sporangia lack discharge pores and germination is always by means of a germ tube.

Host plants

P. parasitica occurs on most wild and cultivated members of the Cruciferae, including the agricultural and horticultural *Brassica* species, radish, *Sinapis* species, horseradish, and ornamentals such as *Cheiranthus*, *Matthiola*, *Draba*, and *Aubrieta* species.

Host specialization

Isolates from different host species differ in their virulence phenotype when tested on a range of crucifers. These differences in host range were originally used to delimit distinct species, but in the absence of supporting morphological criteria modern authors have regarded *P. parasitica* as a single aggregate species (Dickinson & Greenhalgh 1977). Pathotypes have been recorded, but the extent of host specificity is variable. Isolates frequently show virulence towards species other than their original host (Kluczewski & Lucas 1983).

Geographical distribution

P. parasitica occurs world-wide (Channon 1981).

Disease

The fungus infects from sporangia by forming a germ tube with appressorium and penetrating directly through the anticlinal walls of the epidermal cells, in which the first haustoria are produced. Subsequently the intercellular hyphae spread into the mesophyll tissues, where large lobed haustoria are formed within cells. In susceptible seedlings systemic invasion of the cotyledons, hypocotyls, and even roots can occur, leading to shrivelling, desiccation, and subsequent death of the plant (Fig. 29). On older leaves symptoms first appear as discrete chlorotic areas on the upper surface, frequently limited by veins. Below these lesions on the under surface a profuse downy growth of greenish-white sproangiophores develops (downy mildew). Such lesions eventually become necrotic and may dry out. In cauliflower broccoli and sprouts, infection of the curd, spears, or buttons can occur, discolouring tissues and predisposing them to spoilage by secondary soft-rot bacteria during storage. Much of the damage caused to the host during the later necrotic phase of the disease is due to the activity of secondary bacterial populations. On certain hosts e.g. *Capsella bursa-pastoris*, lesions give rise to hypertrophy and distortion of infected stems, but simultaneous infection by *Albugo candida* is frequently a complicating factor.

Fig. 29. *Peronospora parasitica* on *Brassica* seedlings.

Epidemiology

The fungus survives as oospores in host debris in the soil and primary infections probably originate from this source; conclusive experimental verification of host infection from oospores is not yet available. In localities where continuous cropping of brassicas ensures carry-over of inoculum, the oospore stage, while not necessary to initiate epidemics, may enhance variation in the pathogen population. Seed transmission of infection has also been suggested but is undoubtedly rare if it occurs at all. The disease is most prevalent under cool, moist conditions. The asexual cycle is extremely rapid and completed in 3–4 days at 20°C and high RH. Lower temperatures around 15°C seem to be most favourable for epidemic development as this favours sporulation, germination, and the infection process. Sporangia are mostly formed at night. Epidemic spread is through the aerial dispersal of spores released by hygroscopic twisting of the sporangiophore in response to changes in RH. Dispersal over short distances in water droplets can also occur. Sporangia survive for only a few days on leaves under typical field conditions, although at low temperatures in the absence of moisture they have been shown to remain viable for more than 100 days. The fungus may also survive in quiescent vegetative form within systemically infected plants; such latent infections may be a source of lesions on flowering parts like cauliflowers. Both homothallism and, more usually, heterothallism have been reported for *P. parasitica* (McMeekin 1960); oospores have been found in senescing leaves under field and glasshouse conditions and are probably more common than previously supposed.

Economic importance

The fungus is most damaging as a seed bed pathogen where early infection of seedlings can result in a substantial proportion of plants being killed outright. On adult plants the main effect is on quality, especially where the crop is intended for freezing or canning. Infection also predisposes produce to post-harvest bacterial spoilage while in store. The fungus is of major importance on the horticultural *Brassica* species; it is increasingly common on rape but while recent epidemics on this crop have been widespread enough to cause concern, the evidence for yield losses due to *P. parasitica* is inconclusive.

Control

The main requirement for control is in seedling brassicas, and at later stages in crops such as cauliflower where quality is the major concern. Traditionally the dithiocarbamates have given satisfactory control but these have been largely superseded by other fungicides such as chlorothalonil and dichlofluanid. Regular spraying is necessary to ensure good control. More effective systemic activity is shown by propamocarb hydrochloride which can be incorporated in the blocking compost or applied as a drench at sowing, pricking out, or transplanting. The acylalanine fungicides such as metalaxyl also give excellent systemic control as seed treatments, granular applications or sprays. Though these have not been cleared for use on brassica crops, during 1983 metalaxyl-resistant strains of *P. parasitica* have been isolated from eastern regions of the UK. Formulations of metalaxyl with dithiocarbamates have been recommended on rape at the establishment stage when disease is severe.

Cultural control measures are usually aimed at reducing humidity and persistent moisture films by good ventilation in glasshouses and cold frames, and reductions in planting density. Removal of crop debris may restrict sources of inoculum but it seems unlikely to have any impact in areas where continuous intensive cropping of brassicas is practised. Similarly the widespread cultivation of one or only a few rape cultivars has favoured epidemic development of the disease. Studies on a number of *Brassica* species have identified sources of resistance to *P. parasitica*, but in most cases only one or a few genes seem to be involved and it is unlikely that such resistance will prove durable in the field. More generalized types of resistance have been identified in *B. oleracea* and *B. napus* but require further evaluation.

LUCAS

Peronospora tabacina Adam

(Syn. *P. hyoscyami* Farlow, *P. nicotianae* Speg.)

Basic description

The sporangia of *P. tabacina* vary greatly in size (15–25 μm×10–17 μm), like the oospores (24–40 μm diameter) which are characterized by their brown colour and do not germinate *in vitro*.

Host plants

P. tabacina attacks *Nicotiana* species and, by artificial inoculation, various other Solanaceae (Schiltz 1981).

Host specialization

In Australia, Wark *et al.* (1960) described three ecotypes of *P. tabacina*: APT_1, APT_2 and APT_3, that differed in pathogenicity, host reaction, sporulation capacity on tobacco and temperature optima for development. Shepherd (1970) differentiated the three strains as follows: APT_2 attacks the hybrids of *N. tabacum* obtained with *N. debneyi* and *N. goodspeedii* and is called *P. hyoscyami* f.sp. *hybrida*. *P. tabacina* would be retained for APT_1 and *P. hyoscyami* f.sp. *velutina* for APT_3. Variations in pathogenicity have been observed in Europe, but not different races. A highly aggressive strain of the pathogen was observed by Schiltz & Coussirat (1969) which provoked some necrosis on resistant tobacco.

Geographical distribution

First reported on tobacco in Australia, *P. tabacina* was introduced to Europe from England in 1958, and spread very rapidly. It was found in the Netherlands and FRG in 1959. In 1960 it caused 80% destruction of tobacco in N.E. France, and in 1962 blue mould was present in all Europe, N. Africa and the Near East. Blue mould has also been reported in Canada, USA, S. America (Brazil, Argentina, Chile), Cuba and C. America, but does not occur in Japan, nor, until recently, in sub-Saharan Africa, where its progress was probably stopped by the Sahara (CMI map 23).

Disease

The symptoms of the downy mildew disease, also called blue mould, vary somewhat with the age and the genetical constitution of the plants, the severity of the attacks and the environmental conditions. In the seed bed, the fungus can infect very young plants and does not necessarily cause very distinct symptoms. Seedlings show clear round yellow spots on the upper surface with corresponding grey or bluish downy mould on the lower surface. This tissue disorder later turns into necrosis. Infection may become systemic at high temperatures or if fungicides are used. Blue mould, in the early stages, may easily be confused with cold injury or damping-off. In the field, various symptoms are found on adult plants (Plate 12). Most often, yellow spots appear on the leaves, accompanied by a typical sporulation on the lower part. Often the spots coalesce to form light brown, necrotic areas. The leaves partially disintegrate and become unusable. Systemic infection is also common on resistant cultivars. Plants commonly exhibit deformations of the leaves which become stunted or mottled. Discoloration occurs in the vascular system which becomes brown. No sporulation is visible. Lesions may occur not only on leaves but also on stem, buds and flowers.

Epidemiology

The development of *P. tabacina* and its dissemination (usually caused by air-borne sporangia) depends upon the inoculum (density, pathogenicity), and temperature, RH, light, host physiology, soil and wind. *P. tabacina* can survive as mycelium and overwinter in the host tissues (Schiltz 1967). Overwintering is usually effected by means of sporangia and oospores, but infections are not often attributed to the latter, except occasionally in the USA. Sporangia can be transported over a distance of 1600 km and still infect tobacco; they play the most important part in epidemiology. They are produced in great numbers on the surface of the leaves, and can survive for 4–5 months in dry conditions (Kröber 1965). Germination takes place in a few hours, by germ tube, and requires a moisture-saturated atmosphere (≥ 95% RH). The optimal temperature for germination is 15°C, but the conidia can germinate from 3.5°C to 30°C. The germ tube usually penetrates directly into the epidermal cells on the upper surface of the leaf. After this penetration, two definite phases can be distinguished: the 'expansion phase' during the first 48 h after inoculation, and the 'growth phase' between 3–7 days after infection. During the latter phase, the parasite considerably enlarges its surface in contact with the host; it produces conidiophores and sexual processes are initiated. When the conditions for infection are unfavourable (environment, resistant host, etc), the expansion of the hyphae is limited and confined close to the veins of the leaves, or the vascular systems of the stem. In this case, fructification does not occur and systemic infection takes place. In summary, the most favourable conditions for blue mould are an average temperature of 20°C (16°C at night and 24°C during the day), high RH, overcast weather and moving air that is favourable to dissemination of the parasite.

Economic importance

Losses can be very heavy; for example, in 1979, more than 95% of the tobacco crop in Cuba was destroyed by blue mould. In the USA, blue mould was principally a minor plant-bed disease, but in 1979 a loss of over $250 million to US and Canadian farmers was attributed to this disease. In Europe, it is one of the most serious diseases affecting tobacco. Every year, the parasite is reported in at least one of the countries of the Mediterranean basin. The disease has caused damage every year in Europe since 1959. In 1961, the evaluation was 75% of loss in Algeria and 65% in Italy. In 1969, the damage reached 25% in Morocco, 5% in Spain, 5% in France, and 15% in Belgium. The most favourable conditions in Europe are those of the Atlantic coast of S.W. France. However, with the appearance of resistant cultivars, and effective fungicides, the damage caused by blue mould is becoming less important.

Control

Traditional prophylactic measures (preventive treatments of seed beds) and resistant cultivars are generally recommended. In Australia and Europe several resistant cultivars have been developed and seem to give, in most cases, satisfactory results. The source of resistance is *N. debneyi* (trigenic heredity) and involves a hypersensitive reaction localized in the mesophyll. The host's reaction depends on its food supply, stage of development, temperature and light conditions (Schiltz 1981). However, resistance is not sufficient to control blue mould completely; the use of fungicides is necessary to prevent the disease, particularly at the seedling stage. At one time, the most practical method was the use of dusts or sprays containing dithiocarbamates (maneb, mancozeb, zineb, propineb) applied before the disease appeared, usually when the first two leaves were developed, and then twice weekly on the seed bed. This protection must be continued in the field, but the number of treatments depends on weather conditions and on the cultivars grown (Corbaz 1970). More recently, systemic fungicides of the acylalanine family have proved highly effective. In France, preventive spray applications of a mixture of metalaxyl and dithiocarbamate are recommended (Schiltz & Delon 1981). The use of a mixture with a protective compound ensures a wide action spectrum against other diseases and reduces the occurrence of tolerant strains of *P. tabacina.* This technique also allows a reduction in the number of treatments. In European and Mediterranean countries, the CORESTA (Cooperation Centre for Scientific Research Relative to Tobacco) has organized a warning system. Since 1961, any occurrence of blue mould is reported immediately to the Headquarters of CORESTA in Paris, which in turn informs by telex all members, instructing them to take appropriate measures to protect tobacco against the disease.

DELON & SCHILTZ

Peronospora destructor (Berk.) Caspary

P. destructor forms monopodially branching sporangiophores, ending in two sterigmata and bearing pyriform to fusiform sporangia. The mycelium is intercellular with simple or forked haustoria, and the oospores are spherical with a smooth thick wall. The fungus is confined to the genus *Allium*, mostly affecting *A. cepa*, *A. schoenoprasum* (chives) and *A. fistulosum*, but also other cultivated and wild species (CMI 456).

The fungus has a practically world-wide distribution causing severe epidemics in Europe (CMI map 76). The disease is a typical downy mildew appearing either in systemic or in local infection. Systemically infected plants are dwarfed with pale green leaves, while local lesions on leaves and seed-stalks appear as pale spots (Fig 30). Under moist conditions the fungus sporulates freely on the affected tissues giving a greyish-violet colour to the spots. Seed transmission has generally not been demonstrated, and attempts to induce infection with oospores have not been successful. The fungus, however, is capable of surviving in onion bulbs and these, although occurring scarcely, are considered the most important source of primary infection. Imported onion sets may introduce the disease. Survival on other *Allium* species is of minor importance except on *A. fistulosum* which overwinters outdoors and may serve as early inoculum in spring.

Fig. 30. A deformed seed-stalk of onion infected by *Peronospora destructor*.

Epidemiological factors include source of inoculum, duration of moisture, air temperature and size of onion field (Virányi 1981). Sporulation, germination of sporangia and host penetration are equally favoured at about 10–13°C with high RH, as well as prolonged wetness on the plant surface. It may be possible to determine critical periods for infection as has been done in some European countries (Virányi 1981). In addition, elaboration of a bioclimatological model by de Weille (1975) or the accurate analysis of weather variables in relation to disease spread by Hildebrand & Sutton (1982) provide further evidence of the environment-dependent disease pattern of *P. destructor*.

Yield may be reduced significantly by the pathogen, and 'bottle-necked' onion bulbs develop due to leaf damage. Seeds from affected plants germinate poorly. Practical measures to exclude sources of infection and to prevent disease spread are of great importance. Copper and dithiocarbamate fungicides are effective, if applied as protectants. Metalaxyl and mancozeb have also been used. Heat treatment has been used to eliminate inoculum from sets of onion and chives.

VIRÁNYI

Peronospora farinosa Fr.

P. farinosa forms an intercellular mycelium with branched haustoria in the host cells. Sporangiophores emerge through the stomata in small groups bearing elliptical sporangia. Oospores are globose and formation is numerous (CMI 765). Two formae speciales have been described: **f.sp. *beta*** Byford (syn. *P. schachtii* Fuckel) and **f.sp. *spinaciae*** Byford. The latter is sometimes considered as a separate species *P. effusa* (Grev.) Rabenh. or *P. spinaciae* Laubert.

f.sp. *betae*, confined to sugar beet, also attacks other cultivated forms of *Beta vulgaris*, and is widespread in the N. hemisphere (CMI map 28). During the 1970s it became more prevalent in several European countries reducing yield up to 50% in some areas (Byford 1981).

Young rosette leaves of systemically infected sugar beet plants become light green or yellowish, thickened and distorted. When moist conditions prevail, sporulation occurs abundantly on the lower leaf surfaces. The infected leaf rosette may become necrotic and die, forcing the plant to develop new leaves. Occasionally, local spots of downy mildew appear on the leaves of young plants. The seed crop may be attacked at any stage of growth. The fungus overwinters as oospores in the soil, or as mycelium in seeds and living beet plants. Affected seed crops are the most important source of inoculum. Sporulation and germination of sporangia are favoured at temperatures between 9 and 12°C combined with high humidity. Infection certainly takes place on seedlings at the cotyledon stage or on the growing point of older plants exposed to inoculum and covered with free water for at least 6 h (Byford 1981).

A number of observations suggest that downy-mildewed sugar beet is able to recover and that this recovery is influenced by plant age rather than weather conditions. There should be a balance between host and pathogen in this case (Byford 1981). Cultural practices such as separation of root crops from seed crops, or using resistant cultivars are suggested, and new systemic fungicides are considered to control the disease.

f.sp. *spinaciae* occurs widely on spinach (e.g. France, Germany, the Netherlands), causing yellow lesions on the older leaves (Messiaen & Lafon 1970; Byford 1981). It is most severe on protected crops or in wet seasons. Fungicide treatments are widely used, and a number of resistant cultivars are available (with corresponding pathotypes). The oospores are relatively short-lived (2–3 years) so crop rotation is possible.

VIRÁNYI

Peronospora manshurica (Naumov) H. Sydow

P. manshurica has sporangiophores with nearly straight sterigmata bearing hyaline, elliptical sporangia. Oospores are globose with a smooth to reticulated wall (Dunleavy 1981). The fungus is confined to soybean and has a wide range of variability in physiological specialization; at least 32 pathogenic races are known to exist (CMI 689). *P. manshurica*, not observed in Europe until 1942, is probably coextensive with its crop and was introduced to the continent by oospore-encrusted seeds. It has spread to a number of European countries so far, occurring throughout the soybean-growing areas.

Downy mildew of soybean is characterized by small, indefinite, pale green spots on the upper, and by a greyish purple down on the lower leaf surface. Pods can be infected with no external symptoms and seeds from such pods are often covered with a milky, white crust consisting of mycelial debris and of oospores of *P. manshurica*. Such seeds may give rise to systemically infected plants which are stunted and develop small rugose leaves (Dunleavy 1981).

The fungus overwinters as oospores in affected plant tissues or, more often, on the seed coat. Oospores remain viable for as long as 8 years (Pathak *et al.* 1978). The frequency of primary systemic infection depends on the rapidity of seed germination and temperature (Dunleavy 1981). Sporulation occurs on the leaves of systemically infected plants 1–2 weeks after emergence and is favoured by prolonged moisture and low temperatures at night. Sporangia are disseminated by wind to surrounding healthy leaves where they may cause secondary infection. Depending on environmental conditions the disease can spread very rapidly in areas of intensive soybean production. The leaf spot phase of the disease may reduce seed yield by 8% (Dunleavy 1981). When infection is more severe the whole leaf area becomes affected and if so, the plants die prematurely. Seeds from affected plants are oospore-encrusted to differing degrees.

Control measures include breeding for resistance, seed treatment with fungicides, seed inspection, deep ploughing of plant residues, and crop rotation (CMI 689).

VIRÁNYI

Peronospora trifoliorum de Bary

P. trifoliorum has pale violet sporangia borne on dichotomously branched sporangiophores; the oospores are pale yellow with a smooth or wrinkled epispore (CMI 768). *P. trifoliorum* causes downy mildew disease in a large number of genera in the Leguminosae, including *Melilotus*, *Trifolium*, *Lotus* and notably lucerne. There are many host-specific forms, and isolates from lucerne (**f.sp.** ***medicaginis-sativae*** (Sydow) Boerema & Verhoeven (syn. *P. aestivalis* Sydow) are non-pathogenic on other hosts. The fungus is widely distributed in temperate zones of the world and at higher altitudes in S. America (CMI map 343). In lucerne, penetration is direct or via stomata, and the intercellular mycelium may not produce haustoria. Infected leaves curl downwards and have twisted margins, with a pale violet downy growth of sporangiophores on the abaxial surface. Systemically infected plants are stunted and chlorotic.

The fungus overwinters in systemically infected crown buds and shoots; the importance of oospores in perennation is not clear. Sporangia are produced during periods of darkness and high RH at an optimum of 18°C (Fried & Stuteville 1977), and are dispersed by wind or rain splash. Evidence to date suggests that the fungus is homothallic and regularly produces oospores. Damage from the disease is mainly restricted to younger plants which may be weakened or killed in the first year.

Breeding for resistance is considered the most feasible means of controlling the disease. In lucerne, Flemish germplasm has proved a useful source of resistance, and cycles of recurrent phenotypic selection greatly increase the proportion of resistant plants (Stuteville 1981; APS Compendium of Alfalfa Diseases).

LUCAS

Peronospora viciae (Berk.) Caspary

P. viciae has sporangiophores which are unbranched for two-thirds or more of their length and with raised reticulations on the oospore surface. Host range covers *Lathyrus odoratus, L. sativus, Pisum arvense, P. sativum, Vicia faba* and *V. sativa,* and also wild species of *Lathyrus* and *Vicia*. Races of *P. viciae* have been recognized in the Netherlands (Hubbeling 1975) and Germany (von Heydendorff & Hoffman 1978) but not in the UK (Dixon 1981). *P. viciae* is a major pathogen of vining peas and broad beans. The form on *Pisum* has been separated as **f.sp. *pisi*** (Sydow) Boerema & Verhoeven (syn. *P. pisi* Sydow).

Symptoms caused by *P. viciae* on peas are of three types: systemic; localized sporulation on leaves, tendrils and flowers; and invasion of pods and seeds. Systemic symptoms first appear on emerging seedlings which are stunted and distorted. A dull mealy growth of sporulation is found on all infected organs, usually causing the young plant to wither and die. Infected tissues are heavily crowded with oospores. Localized foliar lesions form following secondary spread of *P. viciae* between and within crops. Infection is first noticed as yellow to brown blotches on the upper leaf surfaces with angular areas of fluffy white to bluish cottony mycelium on the under surfaces (Hagedorn 1974). Lesion size is 2–20 mm in diameter, larger lesions being limited by the main leaflet veins. Symptoms usually start on leaves 3–4 and progress up the plant, although on older plants they may be restricted to small necrotic flecks on the laminae which sporulate profusely under humid conditions. Symptoms on *Vicia* have not been investigated in detail but infection causes sporulating light grey lesions on the leaf under surface and chlorotic spots on the upper surface.

Infected haulm provides a means of perennation for *P. viciae.* Long-range transmission is likely to be by seed infection. Secondary spread is via rain and wind-transported sporangia for which optimal germination occurs at 4–8°C (Pegg & Mence 1970). Invasion from sporangia is by direct cuticular penetration. Sporangial formation requires 90% RH for more than 12 h and temperatures below 15°C. Oospore production is favoured by temperatures in excess of 20°C. Infections by *P. viciae* are encouraged by cool moist climatic conditions and disease becomes especially prevalent where coastal mists develop. The range of pea cultivars resistant to *P. viciae* has been described in detail by Dixon (1981). Chemical control can be achieved by using systemic seed dressings or by foliar sprays with dithiocarbamate compounds. Deep tillage and extended crop rotations are essential husbandry techniques for controlling *P. viciae.*

DIXON

Peronospora valerianellae Fuckel causes a serious seed-borne disease of lamb's lettuce (*Valerianella* spp.), widely grown as a salad plant in France and Germany. Fungicide treatments are used (Weinmann & Claussen 1979). ***P. jaapiana*** Magnus (CMI 687) causes downy mildew of rhubarb, but is rarely severe. ***P. arborescens*** (Berk.) Caspary attacks poppies, especially *Papaver somniferum* grown as an oilseed crop in C. and S. Europe. Use of clean seed, sanitation crop rotation and fungicide treatment are the main methods of control (CMI 686). ***P. lamii*** A. Braun (CMI 688) attacks many Labiatae, including cultivated herbs such as sage (*Salvia officinalis*) and basil (*Ocimum basilicum*). ***P. sparsa*** Berk. has for a long time been known as *Pseudoperonospora sparsa* (Berk.) Jacz., but it germinates by germ-tube. It is fairly common on roses throughout Europe, especially under glass, infection being favoured by RH of 90–100% and relatively low temperatures (CMI 690). Infected leaves frequently develop purplish to black spots, hence the name 'black mildew'. The leaves are liable to drop, so plants are defoliated at flowering time. Control is principally by avoiding high humidity (APS Compendium) but fungicide treatments are also used.

Many other *Peronospora* spp. occur on ornamentals of many different plant families (Wheeler 1981). Special mention may be made of ***P. anemones*** Tramier on the commercially important *Anemone coronaria* in France and Italy (CMI 684) and of ***P. antirrhini*** Schröter which causes very damaging systemic infections of *Antirrhinum* seedlings under glass (CMI 685). However, most of the other species occur on garden-grown plants of little commercial importance, and the major commercial ornamentals (chrysanthemums, pot plants) do not encounter downy mildew problems. ***P. dianthicola*** Barthelet does occur on carnation (CMI 764) and ***P. dianthi*** de Bary on annual *Dianthus* species (CMI 763).

Bremia lactucae Regel

Basic description

The genus *Bremia* is distinguished from other members of the Peronosporaceae by the swollen tips of its dichotomously branched sporangiophores which in preparations for microscopy collapse and appear 'disc-shaped' or 'cup-like' (CMI 682). The directly germinating sporangia ($12–26 \times 10–23\ \mu m$) are borne on 3–6 sterigmata and although collections of *Bremia* from different genera of Compositae have sometimes been assigned to separate species, there is little morphological justification for this (Crute & Dixon 1981).

Host plants

B. lactucae parasitizes species belonging to 36 different genera of Compositae. The most economically important host is lettuce, but commercially damaging attacks can also occur on chicory, endive, globe artichoke and ornamentals such as cineraria (Crute & Dixon 1981).

Host specialization

The fungus exhibits considerable host specialization; collections are only parasitic on their species of origin or closely related species belonging to the same genus. Isolates from lettuce vary for specific virulence as determined by their ability or inability to parasitize cultivars carrying different combinations of numerous specific resistance factors (Crute & Dixon 1981). Whilst 'physiologic races' have been classified taxonomically, pathogenic variation is now described in terms of the frequency of occurrence of specific virulence factors.

Geographical distribution

B. lactucae is distributed throughout all temperate and sub-tropical regions where lettuce is cultivated (CMI map 86).

Disease

The fungus causes localized symptoms of downy mildew on lettuce foliage. It is obligately biotrophic and white sporulation primarily on the lower leaf surface is the first symptom of infection under optimum conditions. More often, sporulation is preceded by the affected area turning pale green or yellow. On young leaves, sporulation may be diffuse over the whole lower surface whilst on mature leaves, lesions tend to be bounded by the larger veins. As lesions age, the tissues turn brown and may become brittle or rot. Both young seedlings and more rarely mature plants may become systemically colonized (Crute & Dixon 1981).

Epidemiology

Downy mildew first occurs as small, easily overlooked foci. Under suitable conditions, the fungus spreads by means of wind or splash-dispersed asexual spores to cause what sometimes appear as extensive 'overnight' outbreaks. The primary focus may result from incoming air-borne inoculum or soil inoculum as oospores from a previous crop. Oospores that form in lettuce tissue are infected either with strains of opposite mating type (B1 and B2) or with a homothallic strain (Michelmore 1981). Little is known about the conditions favouring infection by oospores and their precise role in disease epidemiology. The presence of lettuce is known to stimulate oospore germination (Morgan 1983) and root infection has been observed (Norwood & Crute 1983). In free water, the sporangia germinate to form a germ-tube at temperatures between 0 and 21°C with an optimum at 10°C. Penetration occurs as quickly as within 3 h and takes place directly into the epidermal cells or through the stomata. Depending on temperature and tissue susceptibility, a new crop of spores arises in 5–14 days. Sporulation is a nocturnal event which occurs at 4–20°C and requires 6 h darkness and $> 95\%$ RH. Spore release occurs as RH falls during the day, peaking at 12.00 h. Severe outbreaks of downy mildew occur when there are prolonged periods of cool, wet weather. In temperate regions, the disease causes few problems during mid-winter and mid-summer and is most troublesome in spring, late summer and autumn.

Economic importance

Direct assessment of economic losses due to *B. lactucae* is difficult although some attempts to do so have been made (Crute & Dixon 1981). It is reported to be the most important fungal disease of lettuce in several countries and to any directly attributed losses must be added the cost of prophylactic fungicide treatments and the effort expended on breeding for resistance. Infection by *B. lactucae* rarely results in plant death unless young seedlings are affected. Losses can be attributed to reduction in yield and quality brought about by the need to remove diseased leaves prior to marketing. In times of plentiful supply, even minor levels of disease can be sufficient to render a crop unmarketable. Infection by *B. lactucae* also allows the access of destructive secondary pathogens which can be of particular importance in causing post-harvest deterioration.

Control

Control measures aimed at avoiding soil-borne infection include removal or deep-ploughing of crop debris, crop rotation and, in protected crops, soil sterilization. Cultural procedures, aimed at minimizing free water retention on foliage can also be important in ensuring that air-borne infection is minimized, particularly in protected crops (Crute & Dixon 1981; Morgan 1984). Much effort has been directed towards breeding for resistance and there are many cultivars carrying genes for resistance. All the resistance employed has proved to be race-specific and control has been ephemeral. Nevertheless, correctly chosen resistant cultivars can provide a valuable component of a control strategy (Crute 1984) and are much relied upon when fungicide usage is restricted. Quantitatively expressed resistance which is apparently race non-specific is receiving attention and may be particularly valuable when integrated with fungicide usage (Crute 1984). Until the advent of the acylalanine fungicides chemical control of *B. lactucae* relied upon frequent dithiocarbamate applications, notably zineb, either as a dust or spray. In many countries, metalaxyl is now cleared for use on lettuce and provides a high degree of control. To try to avoid problems with fungicide resistance, only mixed formulations of this chemical are usually available (e.g. with mancozeb). In the UK, a fungicide-insensitive form appeared in 1983 and has since become widely distributed. Some control, inferior to that experienced with metalaxyl, is also exerted by other systemic chemicals specifically active against Oomycetes but unrelated to the acylanines (e.g. fosetyl-Al and propaminocarb).

Special research interest

The *B. lactucae*/*L. sativa* association is among the best genetically defined of any host/parasite relationship and provides a system for fundamental investigation of host/parasite genetics. Genetic definition has been aided by investigations into the control of sexual reproduction. Cytological and ultrastructural studies of events associated with both compatible and incompatible associations have provided a basis for biochemical investigations of fungal biotrophy and the molecular basis for host specificity.

CRUTE

***Plasmopara viticola* (Berk. & Curtis) Berl. & de Toni**

Basic description

The genus *Plasmopara*, of which *P. viticola* is a typical member, is distinguished from other Peronosporaceae by its monopodially branched sporophores with diverging tips, these bearing sporangia which germinate by releasing zoospores.

Host plants

P. viticola attacks *Vitis* species in general and also the related genera *Ampelopsis* and *Parthenocissus*. The European grapevine (*Vitis vinifera*) is very susceptible, but other *Vitis* species are less so (*V. labrusca*) or even very resistant (*V. riparia, V. rupestris, V. rotundifolia*).

Geographical distribution

P. viticola was introduced to France from the USA in 1878, probably with rootstocks of American *Vitis* species imported for their resistance to the phylloxera root aphid (*Daktulosphaira vitifoliae*). It spread extremely rapidly and disastrously to all grapevine-growing areas of Europe, reaching the Caucasus by 1887. The fungus now occurs worldwide (CMI map 221).

Disease

P. viticola attacks all parts of grapevine plants, particularly the leaves. According to incubation period and leaf age, lesions may show little damage, become yellowish and oily in appearance, or appear as restricted angular discoloured spots limited by the veins. Sporulation characteristically occurs on the lower leaf surface, giving the downy mildew symptom. Leaf infection as such reduces the plant's productivity to some extent but is most important as a source of inoculum for grape infection. Vine shoots are especially infested at the tips, becoming browned and curled; similar symptoms are seen on petioles, tendrils and young inflorescences, which, if attacked early enough, will dry up and drop. The young grapes are highly susceptible, appearing greyish when infected (grey rot) and covered with a downy felt of sporangiophores. Although the grapes become less susceptible as they ripen, infection of the bunch stalk can spread into older grapes (brown rot, without sporulation) and can also cause dropping of the bunches.

Epidemiology

P. viticola overwinters as oospores in fallen leaves in the soil, but, unlike most other economically important downy mildew fungi in Europe, it can survive on its perennial host as mycelium in buds and persistent leaves, especially in regions with mild winters. Sporangia from such regions in S. Europe can be wind-borne to vineyards in C. Europe, to act as initial inoculum there in the spring. Oospores survive best in the surface layers of moist soil, survival being little affected by temperature. They germinate in spring as soon as temperatures reach 11°C to give a sporangium from which primary dispersal of zoospores occurs by rain splash to adjacent plants. Production of sporangiophores and sporangia, through the stomata of infected organs, requires RH 95–100% and at least 4 h darkness, and has an optimum temperature of 18–22°C. The sporangia are detached from the sporangiophores by the dissolution of a cross-wall of callus, moisture therefore again being required. The sporangia are wind-dispersed to new leaves, where they germinate in free water to release zoospores (optimum temperature 22–25°C). The zoospores encyst near a stoma, which is penetrated by a germination tube from the cyst. Under optimal conditions, the phase from germination to penetration takes less than 90 min. Since sporangia are mostly formed at night and are inactivated by prolonged exposure to sunlight, infection generally occurs in the morning. The incubation period is 5–18 days, according to leaf age, temperature and humidity, and can be predicted reasonably accurately.

The disease is favoured by high humidity and rainfall, which facilitate spread of the pathogen and stimulate growth of new, highly susceptible, shoots. The ideal conditions are a wet, fairly warm spring, followed by a hot summer with fairly frequent rain storms. Thus, in more northern countries with a cool spring, the disease is unimportant. In the Mediterranean countries with a generally dry summer, the disease only presents a risk in particular years when spring and summer rainfall is heavier than average. The most favourable conditions in Europe are those of the Atlantic coasts of S.W. France, Spain and Portugal.

Economic importance

Losses arise from the destruction or dropping of bunches during the current year, through a fall in the quality of harvested grapes (if the proportion of diseased grapes is high), and through leaf and shoot infection which slows the ripening of the next year's wood and reduces productivity of the whole plant in the long term. Since control is mostly effective, actual losses in Europe due to the disease are low. However, control costs are substantial, and grapevine downy mildew is accordingly one of the most economically important plant diseases in Europe.

Control

Grapevine downy mildew in Europe has always been one of the world's major targets for fungicides, and is said to be one of the few diseases which in itself merits the development of a new compound. Copper-based fungicides are still widely used (the famous empirical discovery of Bordeaux mixture by Millardet in 1885 was made in relation to this disease), but these have more recently been replaced or combined with dithiocarbamates, folpet or other organic protectants. Some combinations (e.g. copper oxychloride/zineb) have a synergistic action. In France, 5–6 treatments are usually applied, but the number can vary considerably depending on weather conditions and region. The most important are just before and after flowering to protect the bunches, and in early August to prolong leaf-life and ensure good ripening of the wood. Forecasting models (based on meteorological and phenological data) have been developed and used to ensure optimum timing of chemical sprays (Maurin 1983). Warning systems operate in France and other countries to advise growers when to spray. The anti-oomycete products (especially benalaxyl, cymoxanil, metalaxyl, ofurace, oxadixyl and fosetyl-Al) have a curative and eradicant action, and a systemic protective action (limited for cymoxanil), which allows the number of treatments to be significantly reduced. They are mostly used in combination with purely protectant compounds to ensure greater persistence and a wide action spectrum against other diseases (Lafon & Bulit 1981). Resistance to acylalanines has been reported in *P. viticola* (Clerjeau & Simone 1982).

Other methods of control are limited in practical importance. Good soil drainage in winter or spring will tend not to favour oospore germination; pruning may be used to diminish a plant's susceptibility. Although resistant *Vitis* species exist (see host plants), they are of low viticultural quality. There has only been very limited success until now in Europe in breeding high-quality cultivars with downy mildew resistance.

Special research interest

Grapevine downy mildrew has been of considerable interest in modelling studies (see above), since it lends itself ideally to the application of meteorologically based forecasting systems. Studies on resistance to *P. viticola* in *V. riparia* have shown that this involves a delayed hypersensitive reaction, accompanied by accumulation of stress metabolites (ϵ- and α-viniferins) which are highly antifungal to sporangia and zoospores of *P. viticola* (Langcake 1981). These are not accumulated in the susceptible *V. vinifera*. Research on breeding resistant cultivars continues in several countries (e.g. France, Hungary). New sources of resistance against the main diseases and pests are being investigated (Bouquet 1980), and the use of partially resistant hybrids combined with a reduced spray programme is proposed (Doazan 1980).

LAFON & SMITH

Plasmopara halstedii **(Farlow) Berl. & de Toni**

(Syn. *P. helianthi* Novotelnova)

Basic description

P. halstedii is a typical member of the genus *Plasmopara* with its monopodially branched, slender sporangiophores, usually ending in 3 sterigmata. The sporangia are oval or elliptical in shape, often vary in size, and germinate by means of zoospores.

Host plants

P. halstedii, reported on over 80 species of the Compositae, mainly attacks the cultivated sunflower. Based on its wide host range, *P. halstedii* is considered a complex species and was renamed *P. helianthi* with three forms, f.sp. *helianthi*, f.sp. *perennis* and f.sp. *patens* (Novotelnova 1966). Because of some contradiction in the experimental data and recent findings on host specificity of the fungus however, the latter name is not generally accepted even in Europe (Sackston 1981).

Host specialization

At present three pathotypes of *P. halstedii* are evident, one of these occurring in Europe and the other two, which are more aggressive, in N. America. However, additional pathogenic forms of the fungus, similar to the Red River race in the USA were also reported from Italy, USSR and Romania (Sackston 1981).

Geographical distribution

Both the fungus and sunflower are native to N. America and were introduced through various routes to Europe where the disease was first found in Yugoslavia in 1941. Then it spread rapidly to all sunflower-growing areas of the continent (Novotelnova 1966), and became world-wide by 1977 (Sackston 1981; CMI map 286).

Disease

Affected sunflower shows a wide range of symptoms depending on the time and location of infection. The fungus mainly attacks young tissues and grows systemically within the host. Such plants may be stunted to various degrees and the leaves show a characteristic chlorotic pattern which increases in areas and intensity from lower to upper leaves (Fig 31). Under moist conditions sporulation occurs on the lower surface of chlorotic areas. The heads of stunted plants are stiff and face upwards, and only partially bear viable achenes. The roots of affected plants are usually smaller and darker than those of healthy plants. Severely mildewed sunflowers are damped-off or short-lived with poorly developed roots, completely chlorotic leaves and a tiny capitulum without seeds. All the symptoms described above result from systemic infection. Secondary localized infection of leaves may also occur resulting

Fig. 31. Typical chlorosis (left) and 'downy' cover (right) of sunflower leaves infected by *Plasmopara halstedii.*

in small, angular, pale green spots with a white cover of sporangia on the lower surface if conditions are favourable. Plants with no apparent external symptoms may latently harbour the pathogen in or on underground tissues or in the stem (Sackston 1981). *P. halstedii* may also cause less familiar symptoms such as pith discoloration, basal galls or unusual inflorescences.

Epidemiology

P. halstedii is a soil-borne pathogen, its oospores serving as primary inoculum to young underground tissues of sunflower. It may also be wind-borne, causing secondary infection of the leaves, or even seed-borne when seeds produced by affected plants carry mycelium and/or oospores of the pathogen. Oospores are long-lived and can survive for a number of years. Under favourable conditions they germinate, producing a sporangium from which zoospores are released. To germinate, oospores require a low temperature shock followed by moisture. The presence of host tissue may also promote this process (Delanoë 1972). The zoospores encyst on root or hypocotyl surfaces of sunflower seedlings and form a germination tube prior to penetration through the cuticle or, rarely, via stomata. The extent of fungal invasion mostly depends on infection site, age of tissues and expression of genetic resistance. Plants reaching the 4-leaf stage become resistant to systemic colonization if penetration occurs through underground tissues, while plants exposed to infection on their leaves retain susceptibility much longer. Similarly, different sunflower genotypes may be infected by *P. halstedii* to a variable extent according to the resistance gene(s) they possess.

Intercellular mycelium of the fungus may be present from the roots to the capitulum of susceptible plants, only the densely packed active meristem remaining free. Sporangiophores and sporangia arise from the stomata of infected leaves or develop on the surface of invaded roots. The significance of wind-borne sporangia in outbreaks of downy mildew is debatable, though successful secondary infections of above ground tissues may occur. Secondary infections result in either diseased, or, more probably, latently infected symptomless plants which then give rise to apparently healthy seedlings or produce infected seeds (Sackston 1981).

Economic importance

Severity of downy mildew attack is usually estimated by the percentage of plants with typical systemic symptoms, a figure which varies greatly from place to place and from year to year. Infection up to 90% has been reported from several European countries such as Yugoslavia, Bulgaria and the USSR (Novotelnova 1966). By 1977 it was rated a major disease in almost all sunflower-growing countries of Europe (Sackston 1981). Severely affected plants which reached maturity yielded little or no viable seed, and the achenes harvested from such plants were significantly smaller, lighter, and lower in oil content than those from healthy plants.

Control

With the introduction of high-oil cultivars to Europe a subsequent expansion of the downy mildew pathogen could be observed which has very soon become a limiting factor of sunflower production. No effective control was available at that time and the only possibility of decreasing yield losses was the use of cultural practices like crop rotation, removal of mildewed plants or choice of seeding date. Due to intensive selection work in many parts of the world, sunflowers resistant to *P. halstedii* have been developed and by the end of the 1970s these hybrids were mostly grown all over Europe. Although genes conferring resistance to the European race of the pathogen are still available, the probability that more virulent forms should arise or, more simply, be introduced from another continent remains a major practical risk.

Since the systemic fungicide metalaxyl has been developed and successfully applied as a seed dressing against downy mildews of various crops, it provides a good protection for sunflower growers as well. Resistance to this compound may arise however, as was the case with other downy mildews. For this reason and because genetic resistance is apparently incomplete, metaxyl combined with one of the protectant fungicides (Schwinn 1981) should be applied for seed treatment even if a downy mildew-resistant cultivar is grown.

Special research interest

The cause of severe stunting of downy mildewed sunflowers has been a subject of special interest stimulating plant pathologists to physiological studies. A correlation was found between disease severity and the ability of plants to reduce IAA activity in test solution (Cohen & Sackston 1974). Similarly, quantitative changes in phenolic compounds was observed in relation to reduced IAA content of mildewed plants (Cohen & Ibrahim 1975). However, recent investigations on the photosynthetic activity of infected sunflowers and on the distribution of assimilates within such plants may allow another explanation for this phenomenon (Virányi & Oros 1981).
VIRÁNYI

Plasmopara crustosa (Fr.) Jørstad (syn. *P. nivea* (Unger) Schröter) causes a frequent but minor disease on carrots, celery and other Umbelliferae (Viennot-Bourgin 1949). ***P. pygmaea*** (Unger) Schröter is widespread on *Anemone* but less important in practice than *Peronospora anemones*. ***P. ribicola*** Schröter is of minor importance on Ribes (Viennot-Bourgin 1949).

Pseudoperonospora cubensis (Berk. & Curtis) Rostovtsev

Basic description

The genus *Pseudoperonospora* is characterized by its dichotomously branched sporangiophores in their upper third bearing greyish to olivaceous purple papillated sporangia which germinate by releasing zoospores (CMI 457).

Host plants and host specialization

P. cubensis causes the downy mildew of cucurbits. The main cultivated hosts affected are species of *Cucumis* (cucumber, melon), *Cucurbita* (courgette, marrow, pumpkin) and *Citrullus* (watermelon). Cross-inoculation experiments and field observations showed the existence of various races of the fungus in different countries, specialized to individual hosts of the family Cucurbitaceae. Fungal races from *Cucumis* species are usually less pathogenic or avirulent to *Cucurbita* and *Citrullus* species.

Geographical distribution

P. cubensis has a world-wide distribution (CMI map 285), mainly on *Cucumis* species. With the exception of Yugoslavia and the USSR, where the pathogen was also found on *Cucurbita* species, in the rest of Europe it was reported only on cucumber and melon. Distribution on crops of the genus *Cucurbita* was recorded in about 40 countries but mainly in the USA, C. America and S. Africa. On *Citrullus*, the distribution is limited to about 25 countries of the southern hemisphere, mostly in America.

Diseases

P. cubensis grows biotrophically in the intercellular space of the host tissues and gives rise to small ovate haustoria in the cells. It mainly attacks the leaves but in some hosts other parts (e.g. cotyledons, stems, leaf petioles) may also be affected. Symptoms on leaves vary among different plant species or cultivars within the same species. On cucumber the symptoms on the upper leaf surface are bright, yellowish, angular lesions restricted by the veins (Fig. 32). On melons and watermelons the lesions are not so distinct, rather roundish with indefinite edges. Characteristic purplish sporulation appears on the lower leaf surface of the infected areas under humid conditions. The affected leaf blades may become dry

Fig. 32. Downy mildew symptoms on upper surface of cucumber leaf (*Pseudoperonospora cubensis*).

and curl but their petioles remain green and keep them attached to the stem. Heavy leaf infections cause stunting and death of the plant. Cucumber is particularly susceptible.

Epidemiology

P. cubensis mainly overwinters as an active mycelium either on wild cucurbits or on cultivated host plants growing in southern regions or in greenhouses. Thus the pathogen survives throughout the year by passing from winter to summer crops. The sporangia may be transferred by the wind over long distances to act as the initial inoculum for primary infections. Oospores have been observed in some countries (East Asia, Italy) but their importance in the pathogen's life cycle is obscure.

The duration of leaf wetness is the limiting factor for infection to occur. Sporangia germinate only in free water, by releasing 2–15 zoospores and infection takes place via the stomata of the lower leaf surface. At temperatures of 20–25°C, 2 h wetness is enough for infection to establish, provided the concentration of inoculum is > 100 sporangia/cm^2. High RH is also required for sporulation but prolonged moisture inhibits spore formation and reduces spore viability. Sporulation mainly occurs at night and maximal spore dispersal takes place in the morning. Disease is favoured by damp weather with temperatures of 15–25°C and 18 h of daylight. Epidemics can be expected under conditions when leaf wetness persists for at least 5–6 h and much inoculum is available for the primary infection. The incubation period varies from 3 to 12 days depending on temperature, humidity and inoculum concentration. Temperatures exceeding 30°C check the spread of the disease (Palti & Cohen 1980; Cohen 1981).

Economic importance

Heavy infections of the foliage badly affect fruit quality and yield. Diseased cucumber plants produce short malformed unmarketable fruit. Early infections may also result in a total loss of the crop. Downy mildew epidemics can be destructive to cucumber and melon fields in the Mediterranean countries when high humidities and unusually low temperatures prevail during late spring and summer months. The disease is particularly important in glasshouse crops during autumn/winter.

Control

The disease is mainly controlled by prophylactic sprays applied at regular time intervals to cover the new growth. The frequency and number of applications depends on the weather conditions and the rate of foliage development. In the open, sprays usually start with transplanting and are repeated at 10–15 day intervals. Fungicide application immediately after overhead irrigation or rainfall is also recommended. In glasshouse crops, treatments must be repeated at shorter intervals (7–10 days) particularly in the early stages of plant growth. Copper-based fungicides (e.g. copper oxychloride, copper carbonate) are still widely used. They should, however, be avoided on young plants because they cause retardation of plant growth and leaf scorching. Dithiocarbamates (e.g. mancozeb, maneb, propineb) and chlorothalonil give very good control of the disease. The systemic fungicides fosetyl-Al and metalaxyl provide no better control in comparison with dithiocarbamates when both are applied at weekly intervals. The mixtures of these specific fungicides with mancozeb may give better control due to combined protective and curative action. In experiments on potted plants propamocarb and prothiocarb gave promising results in controlling *P. cubensis* via soil application. Resistance to metalaxyl in cucumber grown in greenhouses was developed in Greece and Israel in 1979 (Pappas 1982). Other measures to eliminate the disease include: sowing in lines parallel to wind direction and at proper plant density; avoiding the growth of new plants next to older infected crops and overhead irrigation; providing good air circulation in glasshouses.

PAPPAS

Pseudoperonospora humuli **(Miyabe & Takahashi) G. Wilson**

Basic description

Like other members of the genus, and of the related *Plasmopara*, *P. humuli* is characterized by monopodial branching of determinate sporangiophores bearing sporangia singly at the tips (CMI 769). Sporangia germinate in water only indirectly, by zoospores, and in this respect *P. humuli* is primitive among downy mildew fungi.

Host plants

P. humuli attacks cultivated and wild hedgerow hops (*Humulus lupulus*). It has been found on nettles (*Urtica* spp.) and on *Celtis*, but as different specialized forms since there has been little success with cross-inoculation. No involvement of alternate hosts in the disease cycle has ever seriously been proposed.

Geographical distribution

P. humuli was reported in 1905 in Japan, then on wild hops in Wisconsin (USA) in 1909, and in 1920 it first appeared in Europe, on experimental hops at Wye College, Kent (England). Spread was then pandemic across Europe and N. America and nowadays the pathogen is encountered wherever hops are grown in the northern hemisphere (CMI map 14). However, in southern regions it is restricted to Argentina, never having spread to Australasia or to S. Africa.

Disease

On annual, above-ground hop growth, infection is localized on leaves ('angular leaf spot'), flowers and cones, and is systemic within shoots ('spikes'). The pathogen perpetuates as mycelium with haustoria in perennial root-stocks which it can kill. Each spring some shoots arise stunted and sickly ('primary spikes') and bear on their abaxial leaf surfaces dark, purplish-brown masses of sporangia; these provide the initial inoculum for the summer chain of disease. During subsequent growth, other basal shoots and lateral and main climbing shoots can all become diseased ('secondary spikes'), distorting the plant and allowing inoculum to increase. Leaf infection occurs throughout the season, but flowers can also become diseased in July, and cones up to harvest in September, giving direct losses in yield and quality. Although oospores often arise abundantly in diseased leaves and cones, they have rarely been germinated artificially and are thought to play no part in the disease cycle.

Epidemiology

Infection of shoots requires a longer minimum wet period (3 h) and occurs within a more limited temperature range (8–23°C) than leaf infection (1.5 h, 5–29°C). Significant infection of leaves and shoots requires 4–8 h wetness with moderate temperatures. The pathogen can enter only through stomata whose regulated movement by light determines zoospore (inoculum) efficiency. In light, zoospores in water settle directly on open stomata but in darkness, when stomata are mostly closed, they settle randomly over the leaf (Royle & Thomas 1973). Outbreaks of downy mildew are thus dependent on daytime rain and never occur in rain or dew at night. Whilst leaf symptoms develop in 3–10 days according to temperature over a wide range (7–28°C), spikes require 7–22 days over 9–20°C. Sporulation is diurnally rhythmic as is the release of sporangia into air. High RH (> 90%) conditions spore production whilst falling RH triggers air-borne release. Release is also brought about by the shock action of raindrops on upper, non-sporing leaf surfaces and by the entraining of sporangia by rain. Much dispersal occurs by air and although the efficiency of rain-dispersed inoculum is likely to be high, the importance of rain dispersal is not clearly established (Royle 1973).

Economic importance

Downy mildew has always been a serious factor limiting hop production. Direct loss in weight of the crop and in its content of alpha acid (which gives beer its bitterness) are caused by attacks on the cones which can both devalue and destroy crops. In the 1920s in Europe there were devastating losses, one, for example, in Germany in 1926 was valued at over 30 million marks. In the USA the disease forced a major shift in cultivation from the high rainfall areas of the east to the arid regions of Washington and Idaho. Nowadays, serious attacks of downy mildew are less common due mainly to resistant cultivars and the use of efficient, systemic fungicides. However, the pathogen is still widely prevalent and remains a threat wherever control is relaxed (Royle & Kremheller 1981).

Control

In susceptible cultivars control has always relied heavily on the routine use of protectant fungicides. Copper-based materials are still widely used but have largely been superseded by dithiocarbamates in the UK. In the 1960s streptomycin, and then more recently oomycete-specific fungicides (notably metalaxyl in mixture with copper), have been adopted in the early season, and by effectively reducing inoculum levels have greatly contributed to lower seasonal intensities of disease. Concern about copper and dithiocarbamate residues and the demand for high quality hops have encouraged the use of cosmetic late treatments and metalaxyl is now also used after flowering (July). Recent introduction of disease warning schemes in the FRG and the UK give new opportunities for more discriminate use of fungicides than before (Kremheller & Diercks 1983).

Resistant cultivars, derived mainly from wild German lines, have only been introduced since about

1970 and are now widely planted in a few countries. These require fewer sprays than susceptible cultivars. Of other means of control, cultural and sanitation practices can be valuable even though modern farm practices sometimes prevent their adoption. In particular, healthy planting material, mechanical removal of diseased tissues and suitable treatment of unharvested cones are important.

Special research interest

Like grapevine downy mildew, the hop disease has been the subject of recent attempts at forecasting, using weather and inoculum factors. Models of various kinds have been proposed and epidemiological studies have been chosen as an example in debates on modelling philosophy (Waggoner 1974). The curious zoospore behaviour to stomata differs from that in *Plasmopara viticola* and, in having a combined physiological and physical explanation, appears to be a unique phenomenon in mycology.
ROYLE

Pseudoperonospora cannabina (Otth) Curzi occurs on hemp throughout Europe, and may be damaging on hemp crops in E. Europe (CMI map 478).

Sclerophthora macrospora (Sacc.) Thirum., Shaw & Narasimhan

(Syn. *Sclerospora macrospora* Sacc.)

S. macrospora is the type species of the genus *Sclerophthora*, the perfect stage of which is close to that of a *Sclerospora*, while the sporangial stage resembles that of a *Phytophthora* (Kenneth 1981). It has now been recorded on more than 140 Gramineae, notably maize, but also wheat, rice, barley, oats, sorghum, *Eleusine, Pennisetum* and sugar cane (Safeeulla 1976) in most areas with temperate or warm temperate climate in Europe, America, Africa, Asia and Australia. In Europe it has been recorded in Austria, Bulgaria, Greece, Italy, Poland, Switzerland and Yugoslavia (CMI map 287). Individual isolates with limited range have been observed; for instance maize, the main host in most regions, may not be a host in other regions.

S. macrospora causes systemic infection. Symptoms vary greatly with the host, the time of infection and the degree of colonization. On maize severely affected plants show excessive tillering, and narrow chlorotic leaves, but the most characteristic symptom, usually known as crazy top, consists in partial or complete proliferation of the tassel which is finally transformed into a mass of leafy structures. Phyllody may also occur in the ears. On wheat the infected seedlings develop numerous tillers which are chlorotic and liable to wither rapidly. Many plants wither before heading and develop culms most of which are twisted and show characteristic deformations in leafy structures, especially in the last leaf enveloping the ear and in the ear itself; complete sterility may occur. On rice (on which the pathogen was named *Sclerospora oryzae* by Brizi) similar symptoms, but more pronounced at flowering time, may be observed. The panicles appear contorted, being unable to emerge freely from the leaf sheath, and are often reduced in size and remain green for longer than the normal time. The floral parts may be abortive and partially replaced by leaf-like structures.

In Europe *S. macrospora* usually occurs only in restricted, low-lying, inadequately drained areas. Infections on a large scale may, however, be observed following severe floods during the first 4–6 weeks after sowing and, on wheat, at the beginning of the renewed spring growth. The infections start from the resting oospores of *S. macrospora* which are frequently formed on perennial grasses. Seed-borne mycelium can also, in some cases, be a source of infection. *S. macrospora* is not considered to be of economic importance in Europe, except in isolated cases of water-logged crops. Owing to its sporadic occurrence, differences in varietal susceptibility of the affected crops have not been assessed. Usually the disease does not warrant control measures. Soil drainage and any other means of avoiding water-logging are suggested.
CASTELLANI

Of the other downy mildew pathogens of cereals, *Peronosclerospora* species do not occur in Europe, while ***Sclerospora graminicola*** (Sacc.) Schröter (CMI 770) is widespread in C. and S. Europe (CMI map 431) on minor crops such as *Setaria* and *Panicum* and also on wild grasses, but without any economic significance.

ALBUGINACEAE

Albugo candida (Pers.) Kuntze

A. candida forms club-shaped sporangiophores bearing basipetal chains of sporangia in white to

Fig. 33. White rust due to *Albugo candida*.

cream sori. Oospores have a thick, ornamented epispore wall (CMI 460). The fungus occurs on numerous members of the Cruciferae, including the cultivated *Brassica* species, *Raphanus sativus*, and many ornamentals. The specialization of isolates has been reported (Pound & Williams 1963). Causing white blister or white rust disease (Fig. 33), the fungus grows biotrophically as inter-cellular mycelium forming small spherical haustoria. Infected plants become covered in white, chalky, blister-like pustules, frequently with hypertrophy, distortion, and abnormal pigmentation of the affected tissues. Systemic infection of meristems and inflorescences gives rise to galls known as stagheads. Primary infections are by zoospores arising from oospores in soil or plant debris. Thereafter spread is by airborne sporangia that release zoospores on the surface of the host which encyst and penetrate directly. *A. candida* commonly occurs in close association with the downy mildew *Peronospora parasitica*, and the latter is often found parasitizing galls formed through infection by the white blister fungus. Damage is occasionally severe enough to cause yield reductions. The pathogen can be controlled by preventive sprays of copper-based and dithiocarbamate fungicides, while the acylalanines have good eradicant activity against both foliar and systemic infections. Race-specific resistance conditioned by single dominant genes is known in several *Brassica* species, and *Raphanus*. Quantitative inheritance of disease reaction type has been demonstrated in *B. campestris* (Fan *et al.* 1983).

LUCAS

Albugo tragopogonis (DC.) Gray (syn. *A. tragopogi* (Pers.) DC.) (CMI 458) causes white rust of Compositae, and is the most important disease of salsify (*Tragopogon porrifolius*) and black salsify (*Scorzonera hispanica*) (Messiaen & Lafon 1970). The pathogen also occurs on ornamentals (*Gerbera* spp.) and sunflower, and could have potential for the biological control of the composite weed *Ambrosia artemisiifolia* (Vyalykh & Zheryagin 1977). Fungicides are used for control.

REFERENCES

Boccas B. & Laville E. (1978) *Les Maladies à Phytophthora des Agrumes.* GERDAT-IRFA, Montpellier.

Bolay A. (1977) *Phytophthora syringae*, agent d'une grave pourriture des pommes en conservation. *Revue Suisse de Viticulture, d'Arboriculture et d'Horticulture* **9**, 161–169.

Bompeix G. & Mourichon X. (1977) *Phytophthora cactorum, P. syringae, Phytophthora* sp. parasites des pommes et des poires. *Compte Rendu Hebdomadaire des Séances de l'Académie d'Agriculture de France* **63**, 493–501.

Bonnet P., Ricci P. & Mercier S. (1980) Le dépérissement du gerbéra causé par *Phytophthora cryptogea*: diagnostic à partir du matériel végétal par isolement et détermination rapide. *Annales de Phytopathologie* **12**, 109–118.

Bonnet P., Maïa N., Tello-Marquina J. & Vernard P. (1978) Pouvoir pathogène du *Phytophthora parasitica* (Dastur): facteurs de variabilité et notion de spécialisation parasitaire. *Annales de Phytopathologie* **10**, 15–29.

Bouhot D. (1975) Recherches sur l'ecologie des champignons parasites dans le sol. V. Une technique sélective d'estimation du potentiel infectieux des sols, terreaux et substrats infestés par *Pythium* spp. Etudes qualitatives. *Annales de Phytopathologie* **7**, 9–18.

Bouhot D. (1981) Induction d'une résistance biologique aux *Pythium* dans les sols par l'apport d'une matière organique. *Soil Biology and Biochemistry* **13**, 269–274.

Bouhot D. (1982) La fatigue des sols. Position du problème et principes du diagnostic. *Colloques de l'INRA* no. 17, 9–12. INRA, Paris.

Bouhot D. & Joannes H. (1982) Potentiel infectieux des sols—concept et modèles. *Bulletin OEPP/EPPO Bulletin* **13**, 291–296.

Bouhot D. & Loridat P. (1985) Nouvelles donnèes sur les agents de fonte de semis, de pied noir et de nécroses des racines de la betterave. *Premières Journées d'Etudes sur les Maladies des Plantes*, pp. 83–92. ANPP, Paris.

Bouhot D. & Perrin R. (1980) Mise en évidence de résistances biologiques aux *Pythium* en sols forestiers. *European Journal of Forest Pathology* **10**, 77–89.

Boukema I.W. (1983) Inheritance of resistance to foot and root rot caused by *Phytophthora nicotianae* var. *nicotianae* in tomato. *Euphytica* **32**, 103–109.

Bouquet A. (1980) *Vitis×muscadinia* hybridization: a new way in grape breeding for disease resistance in France. *Proceedings of the Third International Symposium of Grape Breeding, June 15–18, Davis, California*, p.42.

Boyd A.E.W. (1974) Sources of potato blight (*Phytophthora infestans*) in the East of Scotland. *Plant Pathology* **23,** 30–36.

Broadbent P. (1977) *Phytophthora* diseases of citrus: a review. *Proceedings of the International Society for Citriculture* Vol. 3, 986–992.

Bruin G.C.A. & Edgington L.V. (1983) The chemical control of diseases caused by zoosporic fungi. In *Zoosporic Plant Pathogens. A Modern Perspective* (Ed. S.T. Buczacki), pp. 193–230. Academic Press, London.

Byford W.J. (1981) Downy mildews of beet and spinach. In *The Downy Mildews* (Ed. D.M. Spencer), pp. 531–543. Academic Press, London.

Carter G.A., Smith R.M. & Brent K.M. (1982) Sensitivity to metalaxyl of *Phytophthora infestans* populations in potato crops in south-west England in 1980 and 1981. *Annals of Applied Biology* **100,** 433–441.

Caten C.E. (1974) Intra-racial variation in *Phytophthora infestans* and adaptation to field resistance for potato blight. *Annals of Applied Biology* **77,** 259–270.

Channon A.G. (1981) Downy mildew of brassicas. In *The Downy Mildews* (Ed. D.M. Spencer), pp. 321–339. Academic Press, London.

Clerjeau M. & Beyries A. (1977) Etude comparée de l'action préventive et du pouvoir systémique de quelques fongicides nouveaux (phosphite-prothiocarbe-pyroxychlore) sur poivron vis-à-vis de *Phytophthora capsici*. *Phytiatrie-Phytopharmacie* **26,** 73–84.

Clerjeau M. & Simone Y. (1982) Apparition en France de souches de mildiou (*Plasmopara viticola*) résistantes aux fongicides de la famille des anilides (metalaxyl, milfurame). *Le Progrès Agricole et Viticole* **99,** 59–61.

Coffey M.D. & Wilson V.E. (1983) Histology and cytology of infection and disease caused by *Phytophthora*. In *Phytophthora: its Biology, Taxonomy, Ecology and Pathology* (Eds D.C. Erwin, S. Bartnicki-Garcia & P.H. Tsao), pp. 289–301. American Phytopathological Society, St Paul.

Cohen Y. (1981) Downy mildew of cucurbits. In *The Downy Mildews* (Ed. D.M. Spencer), pp. 341–354. Academic Press, London.

Cohen Y. & Ibrahim R.K. (1975) Changes in phenolic compounds of sunflowers infected by *Plasmopara halstedii*. *Canadian Journal of Botany* **53,** 2625–2630.

Cohen Y. & Sackston W.E. (1974) Disappearance of IAA in the presence of tissue of sunflowers infected by *Plasmopara halstedii*. *Canadian Journal of Botany* **52,** 861–865.

Corbaz R. (1970) Dix ans de lutte contre le mildiou. *Revue Suisse d'Agriculture* **2,** 90–92.

Crute I.R. (1984) The integrated use of genetic and chemical methods for the control of lettuce downy mildew (*Bremia lactucae*). *Crop Protection* **3,** 223–242.

Crute I.R. & Dixon G.R. (1981) Downy mildew diseases caused by the genus *Bremia*. In *The Downy Mildews* (Ed. D.M. Spencer), pp. 421–460. Academic Press, London.

Cunliffe C., Lonsdale D. & Epton H.A.S. (1977) Transmission of *Phytophthora erythroseptica* on stored potatoes. *Transactions of the British Mycological Society* **69,** 27–30.

Delanoë D. (1972) Biologie et épidémiologie du mildiou du tournesol (*Plasmopara helianthi*). *CETIOM Informations Techniques* no. 29, 1–49.

Dickinson C.H. & Greenhalgh J.R. (1977) Host range and taxonomy of *Peronospora* on crucifers. *Transactions of the British Mycological Society* **69,** 111–116.

Dixon G.R. (1981) Downy mildews of peas and beans. In *The Downy Mildews* (Ed. D.M. Spencer), pp. 487–514, Academic Press, London.

Doazan J.P. (1980) The selection of grapevine genotypes resistant to fungus diseases and their use under field conditions. *Proceedings of the Third International Symposium of Grape Breeding, June 15–18, Davis, California*, pp. 324–331.

Doke N. & Tomiyama K. (1980) Suppression of the hypersensitive response of potato tuber protoplasts to hyphal wall components by water soluble glucans isolated from *Phytophthora infestans*. *Physiological Plant Pathology* **16,** 177–186.

Domsch K.H., Gams W. & Anderson T.H. (1980) *Compendium of Soil Fungi*. Academic Press, London.

Duncan J.M. (1980) A technique for detecting red stele (*Phytophthora fragariae*) infection of strawberry stocks before planting. *Plant Disease* **64,** 1023–1025.

Duncan J.M. & Cowan J.B. (1980) Effect of temperature and soil moisture content on persistence of infectivity of *Phytophthora fragariae* in naturally infested field soil. *Transactions of the British Mycological Society* **75,** 133–139.

Duniway J.M. (1983) Role of physical factors in the development of *Phytophthora* diseases. In *Phytophthora, its Biology, Taxonomy, Ecology and Pathology* (Eds D.C. Erwin, S. Bartnicki-Garcia & P.H. Tsao), pp. 175–187. American Phytopathological Society, St Paul.

Dunleavy J.M. (1981) Downy mildew of soybean. In *The Downy Mildews* (Ed. D.M. Spencer), pp. 515–529. Academic Press, London.

Edney K.L. & Chambers D.A. (1981) The use of metalaxyl to control *Phytophthora syringae* rot of apple fruits. *Plant Pathology* **30,** 167–170.

Fan Z., Rimmer S.R. & Stefansson B.R. (1983) Inheritance of resistance to *Albugo candida* in rape (*Brassica napus*). *Canadian Journal of Genetics and Cytology* **25,** 420–424.

Favaloro M. & Sanmarco G. (1973) Ricerche sul marciume del colletto e radicale degli agrumi. Specie di *Phytophthora* presenti negli agrumeti della Sicilia Orientale. *Phytopathologia Mediterranea* **12,** 105–107.

Flowers R.A. & Hendrix J.W. (1974) Host and non-host effects on soil populations of *Phytophthora parasitica* var. *nicotianae*. *Phytopathology* **64,** 718–720.

Fried P.M. & Stuteville D.L. (1977) *Peronospora trifoliorum* sporangium development and effects of humidity and light on discharge. *Phytopathology* **67,** 890–894.

Fry W.E. (1977) Integrated control of potato late blight—effects of polygenic resistance and techniques of timing fungicide applications. *Phytopathology* **67**, 415–420.

Garibaldi A. (1974) Osservazioni sul marciume pedale della Gerbera causato da *Phytophthora cryptogea. Rivista della Ortoflorofrutticoltura Italiana* **58**, 108–120.

Gourley C.O. & Delbridge R.W. (1972) Economic loss from strawberry red stele disease in Nova Scotia. *Report Research Station Kentville, Nova Scotia, Canada for 1971*, pp. 63–64.

Griffin M.J. & Jones O.W. (1977) *Phytophthora porri* on autumn-sown salad onions. *Plant Pathology* **26**, 149–150.

Hagedorn D.J. (1974) Recent pea anthracnose and downy mildew epiphytotics in Wisconsin. *Plant Disease Reporter* **58**, 226–229.

Hansen E.M., Brasier C.M., Shaw D.S. & Hamm P.B. (1987) The taxonomic structure of *Phytophthora megasperma*: evidence for emerging biological species groups. *Transactions of the British Mycological Society* (in press).

Harris D.C. (1981) Herbicide management in apple orchards and the fruit rot caused by *Phytophthora syringae*. In *Pests, Pathogens and Vegetation* (Ed. J.M. Thresh), pp. 429–436. Pitman, London.

Heydendorff R.C. von & Hoffman G.M. (1978). Zur physiologischen Spezialisierung von *Peronospora pisi* Syd. *Zeitschrift für Pflanzenkrankheiten und Pflanzenschutz* **85**, 561–569.

Hildebrand P.D. & Sutton J.C. (1982) Weather variables in relation to an epidemic of onion downy mildew. *Phytopathology* **72**, 219–224.

Ho H.H. (1981) Synoptic keys to the species of *Phytophthora. Mycologia* **73**, 705–711.

Hohl H.R. & Iselin K. (1984) Strains of *Phytophthora infestans* from Switzerland with A2 mating type behaviour. *Transactions of the British Mycological Society* **83**, 529–530.

Hubbeling N. (1975) Resistance of peas to downy mildew and distinction of races of *Peronospora pisi* Syd. *Mededelingen van de Faculteit Landbouwwetenschappen Rijksuniversiteit Gent* **40**, 539–543.

Kenneth R.G. (1981) Downy mildew of graminaceous crops. In *The Downy Mildews* (Ed. D.M. Spencer), pp. 367–394. Academic Press, London.

Klotz L.J. (1973) *Color Handbook of Citrus Diseases*, pp. 8–15. Citrus Research Centre, Riverside, California.

Kluczewski S.M. & Lucas J.A. (1983) Host infection and oospore formation by *Peronospora parasitica* in agricultural and horticultural *Brassica* species. *Transactions of the British Mycological Society* **81**, 591–596.

Kotova V.V. & Tsvetkova N.A. (1981) (Agrotechnical measures for control of *Aphanomyces* root rot of pea). In *Trudy Vsesoyuznogo Nauchno-Issledovatel'skogo Instituta Zashchity Rastenii 1981*, pp. 61–65. All-Union Plant Protection Institute, Leningrad.

Kouyeas H. (1977) Stone fruit tree apoplexy caused by *Phytophthora* collar rot. *Bulletin OEPP/EPPO Bulletin* **7**, 117–124.

Kouyeas H. & Chitzanidis A. (1968) Notes on Greek species of *Phytophthora. Annales de l'Institut Phytopathologique Benaki* **8**, 175–192.

Kremheller H. Th. & Diercks R. (1983) Epidemiologie und Prognose des Falschen Mehltaues (*Pseudoperonospora humuli*) an Hopfen. *Zeitschrift für Pflanzenkrankheiten und Pflanzenschutz* **90**, 599–616.

Kröber H. (1965) Über die Lebensdauer des Konidien von *Peronospora tabacina. Phytopathologische Zeitschrift* **54**, 328–334.

Kröber H. (1981) Vergleichende Untersuchungen vom Grundtyp *Phytophthora cryptogea* und *P. drechsleri* abweichender Isolate. *Phytopathologische Zeitschrift* **102**, 219–231.

Kuan T.L. & Erwin D.C. (1980) Formae speciales differentiation of *Phytophthora megasperma* isolates from soybean and alfalfa. *Phytopathology* **70**, 333–338.

Lafon R. & Bulit J. (1981) Downy mildew of the vine. In *The Downy Mildews* (Ed. D.M. Spencer), pp. 601–604. Academic Press, London.

Langcake P. (1981) Disease resistance of *Vitis* species and the production of the stress metabolites resveratrol, ε-viniferin, α-viniferin and pterostilbene. *Physiological Plant Pathology* **18**, 213–226.

Lonsdale D., Cunliffe C. & Epton H.A.S. (1980) Possible routes of entry of *Phytophthora erythroseptica* and its growth within potato plants. *Phytopathologische Zeitscrift* **97**, 109–117.

MAFF (1973) Red core of strawberry. *MAFF Advisory Leaflet* no. 410, MAFF, London.

MAFF (1979) *Phytophthora* foot rot of hardy nursery stock. *MAFF Leaflet* no. 635. MAFF, London.

Mackenzie D.R., Elliott V.J., Kidney B.A., King E.D., Royer M.H. & Theberge R.L. (1983) Application of modern approaches to the study of epidemiology of diseases caused by *Phytophthora*. In *Phytophthora: its Biology, Taxonomy, Ecology and Pathology* (Eds D.C. Erwin, S. Bartnicki-Garcia & P.H. Tsao), pp. 303–313. American Phytopathological Society, St Paul.

Maurin G. (1983) Application d'un modèle d'état potentiel d'infection à *Plasmopara viticola. Bulletin OEPP/EPPO Bulletin* **13**, 263–269.

McCarter S.M. (1967) Effect of soil moisture and soil temperature on black shank disease development in tobacco. *Phytopathology* **57**, 691–695.

McIntyre J.L. & Lukens R.J. (1977) Screening chemicals to control *Phytophthora parasitica* var. *nicotianae. Plant Disease Reporter* **61**, 366–370.

McMeekin D. (1960) The role of oospores of *Peronospora parasitica* in downy mildew of crucifers. *Phytopathology* **50**, 93–97.

Messiaen C.M. & Lafon R. (1970) *Les Maladies des Plantes Maraîchères*. INRA, Paris.

Meynet J. (1980) *Rapport d'activité 1977–1980 de la Station d'Amélioration des Plantes Florales, INRA, Fréjus (France)*, 39–45.

Michelmore R.W. (1981) Sexual and asexual sporulation in the downy mildews. In *The Downy Mildews* (Ed. D.M. Spencer), pp. 165–181. Academic Press, London.

Molot P.M. & Nourrisseau J.G. (1966) Dessèchement printanier du fraisier provoqué par *Phytophthora cactorum. Compte Rendu Hebdomadaire des Séances de l'Académie d'Agriculture de France* **52**, 1001–1005.

Montgomerie I.G. (1967) Pathogenicity of British isolates of *Phytophthora fragariae* and their relationship with American and Canadian races. *Transactions of the British Mycological Society* **50**, 57–67.

Montgomerie I.G. (1977) Red core disease in strawberry. *CAB Horticultural Review* no. 5. CAB, Slough.

Montgomerie I.G. & Kennedy D.L. (1983) An improved

method of isolating *Phytophthora fragariae. Transactions of the British Mycological Society* **80,** 178–183.

Morgan W. (1983) Viability of *Bremia lactucae* oospores and stimulation of their germination by lettuce seedlings. *Transactions of the British Mycological Society* **80,** 403–408.

Morgan W. (1984) Integration of environmental and fungicidal control of *Bremia lactucae* in a glasshouse lettuce crop. *Crop Protection* **3,** 349–361.

Newhook F.J., Waterhouse G.M. & Stamps D.J. (1978) Tabular key of the species of *Phytophthora. CMI Mycological Papers* no. 143. CMI, Kew.

Nienhaus F. (1960) Das Wirtspektrum von *Phytophthora cactorum. Phytopathologische Zeitschrift* **38,** 33–68.

Norwood J.M. & Crute I.R. (1983) Infection of lettuce by oospores of *Bremia lactucae. Transactions of the British Mycological Society* **81,** 144–147.

Novotelnova N.S. (1966) *Downy Mildew of Sunflower* (in Russian). Nauka, Moscow.

Orlikowski L.B. (1980) Persistence of *Phytophthora cryptogea* in greenhouse substrate used for Gerbera growing. *Prace Instytutu Sadownictwa i Kwiacierstwa w Skierniewicach B* **5,** 131–140.

Palti J. & Cohen Y. (1980) Downy mildew of cucurbits (*Pseudoperonospora cubensis*). The fungus and its hosts, distribution, epidemiology and control. *Phytoparasitica* **8,** 109–147.

Papavizas G.C. & Ayers W.A. (1974) *Aphanomyces* species and their root diseases in pea and sugarbeet. *USDA Technical Bulletin* no. 1485. USDA, Washington.

Pappas A.C. (1982) Metalaxyl resistance and control of cucumber downy mildew with oomycete-fungicides. *Annales de l'Institut Phytopathologique Benaki* **13,** 194–212.

Pathak V.K., Mathur S.B. & Neergaard P. (1978) Detection of *Peronospora manshurica* in seeds of soybean *Glycine max. Bulletin OEPP/EPPO Bulletin* **8,** 21–28.

Pegg G.F. & Mence M.J. (1970) The biology of *Peronospora viciae* on pea; laboratory experiments on the effect of temperature, relative humidity and light on the production, germination and infectivity of sporangia. *Annals of Applied Biology* **66,** 417–428.

Pfender W.F. & Hagedorn D.J. (1982) *Aphanomyces euteiches* f.sp. *phaseoli* a causal agent of bean root and hypocotyl rot. *Phytopathology* **72,** 306–310.

Pochard E., Molot P.M. & Domingez G. (1983) Etude de deux nouvelles sources de résistance à *Phytophthora capsici* Leonian chez le piment: confirmation de l'existence de trois composantes distinctes dans la résistance. *Agronomie* **3,** 333–342.

Polach F.J. & Webster R.K. (1972) Identification of strains and inheritance of pathogenicity in *Phytophthora capsici. Phytopathology* **62,** 20–26.

Pound G.S. & Williams P.H. (1963) Biological races of *Albugo candida. Phytopathology* **53,** 1146–1149.

Preisig C.L. & Kuć J.A. (1985) Arachidonic acid-related elicitors of the hypersensitive response in potato and enhancement of their activities by glucan from *Phytophthora infestans. Archives of Biochemistry and Biophysics* **236,** 379–389.

Richard-Molard M. (1982) La fatigue des terres à betterave. *Les Colloques de l'INRA* no. 17, pp. 23–36. INRA, Paris.

Rotem J., Bashi E. & Kranz J. (1983) Studies of crop loss in potato blight caused by *Phytophthora infestans. Plant Pathology* **32,** 117–122.

Royle D.J. (1973) Quantitative relationships between infection by the hop downy mildew pathogen, *Pseudoperonospora humuli*, and weather and inoculum factors. *Annals of Applied Biology* **73,** 19–30.

Royle D.J. & Kremheller H. Th. (1981) Downy mildew of the hop. In *The Downy Mildews* (Ed. D.M. Spencer) pp. 395–419. Academic Press, London.

Royle D.J. & Thomas G.G. (1973). Factors affecting zoospore responses towards stomata in hop downy mildew (*Pseudoperonospora humuli*) including some comparisons with grapevine downy mildew (*Plasmopara viticola*). *Physiological Plant Pathology* **3,** 405–417.

Sackston W.E. (1981) Downy mildew of sunflower. In *The Downy Mildews* (Ed. D.M. Spencer), pp. 545–575. Academic Press, London.

Safeeulla K.M. (1976) *Biology and Control of the Downy Mildew of Pearl-millet, Sorghum and Finger-millet.* Westley Press, Mysore.

Schiltz P. (1967) Création de *Nicotiana tabacum* résistants à *P. tabacina* Adam. Analyse histologique et biologique de la résistance. *Annales SEITA-DEE,* sect. 2, no. 4, pp. 5–146.

Schiltz P. (1981) Downy mildew of tobacco. In *The Downy Mildews* (Ed. D.M. Spencer), pp. 577–599. Academic Press, London.

Schiltz P. & Coussirat J.C. (1969) Mise en évidence de la résistance des *Nicotiana* aux lignées virulentes de *Peronospora tabacina* et détermination du pouvoir pathogène du parasite. *Annales SEITA-DEE*, section 2, no. 6, p. 145.

Schiltz P. & Delon R. (1981) Le mildiou du tabac. Historique et moyens actuels de lutte. *Phytoma* no. 327, 37–41.

Scholten G. (1970) Wilt diseases in Gerbera. *Netherlands Journal of Plant Pathology* **76,** 212–218.

Schwinn F.J. (1981) Chemical control. In *The Downy Mildews* (Ed. D.M. Spencer), pp. 305–320. Academic Press, London.

Schwinn F.J. (1983) New developments in chemical control of *Phytophthora* In *Phytophthora: its Biology, Taxonomy, Ecology and Pathology* (Eds D.C. Erwin, S. Bartnicki-Garcia & P.H. Tsao), pp. 327–334. American Phytopathological Society, St Paul.

Seemüller E. & Schmidle A. (1979) Einfluss der Herkunft von *Phytophthora cactorum*-Isolates auf ihre Virulenz an Apfelrinde, Erdbeerrhizomen und Erdbeerfrüchten. *Phytopathologische Zeitschrift* **94,** 218–225.

Semb L. (1971) A rot of stored cabbage caused by a *Phytophthora* sp. *Acta Horticulturae* no. 20, 32–35.

Sewell G.W.F. (1979) Reappraisal of the nature of the specific replant disease of apple. *Review of Plant Pathology* **58,** 209–211.

Sewell G.W.F. & Wilson J.F. (1964) Death of maiden apple trees caused by *Phytophthora syringae* and a comparison of the pathogen with *P. cactorum. Annals of*

Applied Biology **53**, 275–280.

Sewell G.W.F. & Wilson J.F. (1973) Phytophthora collar rot of apple: seasonal effects on infection and disease development. *Annals of Applied Biology* **74**, 149–158.

Sewell G.W.F., Sivakadadcham B. & Pullinger J. (1983) Replant disease and poor growth of apple. *Report East Malling Research Station for 1982*, pp. 84–85.

Sewell G.W.F., Wilson J.F. & Dakwa J.T. (1974) Seasonal variations in the activity in soil of *Phytophthora cactorum*, *P. syringae* and *P. citricola* in relation to collar rot disease of apple. *Annals of Applied Biology* **76**, 179–186.

Shattock R.C., Janssen B.D., Whitbread R. & Shaw D.S. (1977) An interpretation of the frequencies of host-specific phenotypes of *Phytophthora infestans* in North Wales. *Annals of Applied Biology* **86**, 249–260.

Shattock R.C., Tooley P.W., Sweigard J. & Fry W.E. (1986) Genetic studies on *Phytophthora infestans*. In *Genetics and Plant Pathogenesis* (Eds P.R. Day & G.J. Jellis), pp. 175–185. Blackwell Scientific Publications, Oxford.

Shaw D.S., Fyfe A.M., Hibberd P.G. & Abdel-Sattar M.A. (1985) Occurrence of the rare A2 mating type of *Phytophthora infestans* on imported Egyptian potatoes and the production of sexual progeny with A1 mating types from the UK. *Plant Pathology* **34**, 552–556.

Shepherd C.J. (1970) Nomenclature of the tobacco blue mould fungus. *Transactions of the British Mycological Society* **55**, 253–256.

Skidmore D.I., Shattock R.C. & Shaw D.S. (1984) Oospores in cultures of *Phytophthora infestans* resulting from selfing induced by the presence of *Phytophthora drechsleri* isolated from blighted potato foliage. *Plant Pathology* **33**, 173–183.

Smith P.M. (1980) An assessment of fungicides for the control of wilt and die-back caused by *Phytophthora cinnamoni* in container-grown *Chamaecyparis lawsoniana* cv. Ellwoodii. *Annals of Applied Biology* **94**, 225–234.

Smith P.M., Brooks A.V., Evans E.J. & Halstead A.J. (1983) Pests and diseases of hardy nursery stock, bedding plants and turf. In *Pest and Disease Handbook*, 2nd ed. (Eds N. Scopes & M. Ledieu), pp. 473–556. British Crop Protection Council, Croydon.

Sparnaaij L.D., Garretsen F. & Bekker W. (1975) Additive inheritance of resistance to *Phytophthora cryptogea* in *Gerbera jamesonii*. *Euphytica* **24**, 551–556.

Staub T.H. & Young T.R. (1980) Fungitoxicity of metalaxyl against *Phytophthora parasitica* var. *nicotianae*. *Phytopathology* **70**, 797–801.

Steekelenburg N.A.M. van & Slavekoorde S.M. (1979) Chemical control of *Phytophthora* root rot in hardy nursery stock. *Mededelingen van de Facultiet Landbouwwetenschappen Rijksuniversiteit Gent* **44**, 545–551.

Steudel W. (1979) Der Einfluss von dreijähriger Fruchtfolge und Monokultur auf die durch wurzelparasitische Pilze bei Zuckerrüben verursachten Schäden. *Zuckerindustrie* **104**, 840–844.

Strouts R.G. (1981) *Phytophthora* diseases of trees and shrubs. *Arboricultural Leaflet* no. 8. Forestry Commission Research Station, Farnham.

Stuteville D.L. (1981) Downy mildew of forage legumes. In *The Downy Mildews* (Ed. D.M. Spencer), pp. 355–366. Academic Press, London.

Suslow T.V. & Schroth M.N. (1982) Rhizobacteria of sugar beets: effects of seed application and root colonization on yield. *Phytopathology* **72**, 199–206.

Tantius P.H., Fyfe A.M., Shaw D.S. & Shattock R.C. (1986) Occurrence of the A2 mating type and self-fertile isolates of *Phytophthora infestans* in England and Wales. *Plant Pathology* **35**, 578–581.

Tuest J.J. (1977) Contribución al conocimiento del género *Phytophthora* en España. *Anales del Instituto Nacional de Investigaciones Agrarias Serie Proteccion Vegetal* **7**, 11–106.

Tuset J.J. (1983) La 'gomosis' y 'podredumbre del cuello de la raiz' de nuestros agrios. II. Posibilidades actuales de lucha. *Levante Agricola* no. 247–248, 130–135.

Umaerus V., Umaerus M., Erjefält L. & Nilsson B.A. (1983) Control of *Phytophthora* by host resistance: problems and progress. In *Phytophthora: its Biology, Taxonomy, Ecology and Pathology* (Eds D.C. Erwin, S. Bartnicki-Garcia & P.H. Tsao), pp. 327–334. American Phytopathological Society, St Paul.

Upstone M.E. (1978) *Phytophthora syringae* fruit rot of apples. *Plant Pathology* **27**, 24–30.

Van der Plank J.E. (1963) *Plant Diseases: Epidemics and Control*. Academic Press, New York.

Végh I. & Le Berre A. (1982) Etude expérimentale de la sensibilité de quelques cultivars de bruyerès et de conifères d'ornement vis-a-vis du *Phytophthora cinnamomi*. *Phytopathologische Zeitschrift* **103**, 301–305.

Vesely D. (1978) Studies on mycoparasitism in the rhizosphere of emerging sugar beet. Parasitic relationship between *Pythium oligandrum* and some other species of the oomycetes class. *Zentralblatt für Bakteriologie, Parasitenkunde, Infektionskrankheiten und Hygiene* **133**, 195–200; 341–349.

Viennot-Bourgin G. (1949) *Les Champignons Parasites des Plantes Cultivées*. Masson, Paris.

Virányi F. (1981) Downy mildew of onion. In *The Downy Mildews* (Ed. D.M. Spencer), pp. 461–472. Academic Press, London.

Virányi F. & Oros G. (1981) Changes in the development and metabolism of sunflowers infected by *Plasmopara halstedii*. *Acta Phytopathologica Academiae Scientiarum Hungaricae* **16**, 273–279.

Vyalykh A.K. & Zheryagin V.G. (1977) (Conditions for infection of *Ambrosia artemisiifolia* by the white rust pathogen *Albugo tragopogonis*). *Mikologiya i Fitopatologiya* **11**, 135–140.

Waggoner P.E. (1974) Simulation of epidemics. In *Epidemics of Plant Diseases: Mathematical Analysis and Modelling* (Ed. J. Kranz), pp. 137–160. Springer-Verlag, Berlin.

Wark D.C., Hill A.V., Mandryk M. & Cruickshank I.A.M. (1960) Differentiation in *Peronospora tabacina*. *Nature* **187**, 710–711.

Waterhouse G.M. (1963) Key to the species of *Phytophthora* de Bary. *CMI Mycological Papers* no. 92. CMI, Kew.

Waterhouse G.M. (1968) The Genus *Pythium* Pringsheim. *CMI Mycological Papers* no. 110. CMI, Kew.

Watson A.G. (1973) Lutte biologique contre la fonte des semis de la laitue causée par *Pythium ultimum. Revue Suisse de Viticulture, d'Arboriculture et d'Horticulture* **5,** 93–96.

Weille G.A. de (1975) An approach to the possibilities of forecasting downy mildew infection in onion crops. *Mededelingen en Verhandelingen Koninklijk Nederlands Meteorologisch Instituut,* no. 97, 1–83.

Weinmann W. & Claussen K. (1979) Die Rückstandssituation in Feldsalatanbau nach Bekämpfung von falschem Mehltau und *Botrytis* mit Dichofluanid oder Folpet. *Nachrichtenblatt des Deutschen Pflanzenschutzdienstes* **33,** 33–38.

Wendland E.J. (1976) Zur Rettichschwärze-Bekämpfung unter besondere Versuchsergebnisse. *Gemüse* **12,** 78–80.

Weststeijn G. (1973) *Phytophthora nicotianae* var. *nicotianae* on tomatoes. *Netherlands Journal of Plant Pathology* **79,** (suppl. 1).

Wheeler B.E.J. (1981) Downy mildew of ornamentals. In *The Downy Mildews* (Ed. D.M. Spencer), pp. 473–486. Academic Press, London.

Zentmyer G.A. (1980) *Phytophthora cinnamomi* and the diseases it causes. *APS Monograph* no. 10. American Phytopathological Society, St. Paul.

9: Chytridiomycetes; Plasmodiophoromycetes; Zygomycetes

Like the Oomycetes (Chapter 8), the Chytridiomycetes and Plasmodiophoromycetes have coenocytic thalli and produce flagellate zoospores (posteriorly uniflagellate in the former, biflagellate in the latter). There is much variation within and between these groups in basic thallus structure and life cycle; some have no true mycelium, the whole thallus maturing into a sporangium, while others have some form of mycelium; the Plasmodiophoromycetes form naked plasmodia and are more like slime-moulds than fungi. However, the plant pathogens share many common features. They are soil-borne parasites with long-lived resting sporangia, producing zoospores to be dispersed in soil water, and infecting individual root, tuber or occasionally leaf or stem cells biotrophically and obligately. They may cause either only slight symptoms (e.g. *Olpidium, Polymyxa;* many species probably pass unnoticed) or cause some kind of gall (hyperplasia and/or hypertrophy), as for *Plasmodiophora, Synchytrium, Physoderma, Spongospora.* With a few notable exceptions, they are of rather minor economic importance. Some species are more important as virus vectors—these being the only fungi known to transmit plant viruses.

The Zygomycetes have only coenocytic mycelium and lack of a complex fructification in common with the other 'phycomycetes', or 'lower fungi'. The Mucorales are mostly soil saprophytes, dispersed by non-flagellate sporangiospores, and persisting as zygospores or chlamydospores. A few species cause non-specific fruit or tuber rots. The Endogonales (*Endogone, Glomus*) infect roots, forming endotrophic 'vesicular-arbuscular' mycorrhiza on numerous hosts (Mosse 1973); they are very widespread and are receiving increasing research attention. There is no suggestion that their presence is ever harmful to the host plant.

CHYTRIDIOMYCETES

Olpidium brassicae (Voronin) Dang.

(Syn. *Asterocystis radicis* de Wild.)

Basic description

The genus *Olpidium* contains about 50 species which are parasites of fresh water algae, fungi, moss protonema, pollen grains, flowering plants and small aquatic animals. They possess an endobiotic, holocarpic thallus which may completely fill a host cell. Thalli become entirely converted into either a zoosporangium or a resting sporangium (spore). These liberate zoospores (planospores) from the cell through exit tubes which either end at the host periphery or extend beyond it. Zoospores of *Olpidium* have a spherical head and trailing flagellum and encyst when attached to a host surface. An infection tube projects into the host cell through which cytoplasm passes from the zoospore. Penetration takes less than 1 h; 2 days after invasion small spherical thalli develop within the host root hairs and epidermal cells. The thalli enlarge to become multinucleate, form discharge tubes and within 4–5 days zoospores may be released. *Olpidium brassicae* produces thalli which are spherical and 12–20 μm in diameter or cylindrical and 20–45×25–220 μm; there may be several per host cell. Resting sporangia are 8–30 μm in diameter and characteristically verrucose or wrinkled. Zoospores are 3 μm in diameter, hyaline, posteriorly uniflagellate and 17 μm long (Sparrow 1960).

Host plants

O. brassicae is commonly found on cabbage roots and less frequently on those of lettuce; other hosts may be celery, cucumber, flax and melon (Ogilvie 1969). An isolate of *O. brassicae* from lettuce was shown to be able to infect vegetables such as leek, spinach beet, lettuce, pea, rhubarb and spinach, and also various wild plants, weeds and ornamentals (Tomlinson & Garrett 1964).

Host specialization

The strains of *O. brassicae* on lettuce and cabbage may not be identical since attempts at cross-inoculation between these hosts have failed.

Disease

Infection by *O. brassicae* apparently has little adverse effect on lettuce or cabbage. An *Olpidium* species is reported by Ogilvie (1969) to cause a blotchy yellow mottle of celery, pre- and post-emergence damping-off of cucumber and melon and also infection of the roots of more mature plants after potting-on, resulting in a dark green discoloration and wilting. On flax, under very wet conditions, *O. brassicae* causes a damping-off or 'burning', similar to that which can be caused by *Pythium* species (Viennot-Bourgin 1949). Plants dry out precociously (at 10–20 cm) and appear 'burnt', especially if a very dry period follows a rainy period favouring infection; this drying out prevents the plant from producing new roots as it normally would.

Epidemiology

The resting sporangia can survive seven or more years in the soil. The fungus spreads by zoospores in soil water, notably in glasshouse watering systems (cf. below).

Economic importance

O. brassicae is generally of minor economic importance except as a vector of tobacco necrosis virus (q.v.) and lettuce big vein organism (q.v.). The latter has become a major problem where lettuce is grown by the nutrient film technique. This provides the means for *O. brassicae* zoospores to be rapidly circulated around an entire crop, carrying and spreading the big vein organism (Campbell & Fry 1966). Particularly in winter, the crop may be quickly devastated. Control is only required in areas where *O. brassicae* is liable to transmit viruses.

Control

Soil sterilization with steam or chloropicrin may provide a measure of control; methyl bromide has been shown to produce limited control (White 1982). At the propagation stage seedlings are protected by incorporation of carbendazim into the peat blocks, the active material of which is a surfactant added by the manufacturers to carbendazim. Addition of freshly prepared surfactants such as Agral (90% v/v alkyl phenol ethylene oxide condensate) at 20 ppm into nutrient film solutions inhibits the development of lettuce big vein. The concrete channels in which lettuces are grown using the nutrient film technique are disinfected to kill resting sporangia of *O. brassicae* by treating with 1% v/v solution of an iodophor for 1 h (Tomlinson *et al.* 1981). Partial resistance in lettuce to *O. brassicae* has been noted by Tomlinson *et al.* (1982).
DIXON

Other *Olpidium* species have been described infecting the roots of many plants in a similar way. According to Lange & Insunza (1977), many of those with smooth resting spores (in contrast to the verrucose spores of *O. brassicae*) can probably be grouped in ***O. radicale*** Schwartz & Cook. Although this species is of little economic importance as such, isolates are suspected to transmit viruses (red clover necrotic mosaic virus; cucumber necrosis virus, if *O. cucurbitacearum* Barr is considered synonymous). ***O. trifolii*** (Pass.) Schröter shows the peculiarity of attacking the above-ground parts of *Trifolium repens* (Viennot-Bourgin 1949). ***Cladochytrium caespitis*** Griffon & Maubl. forms a fine mycelium bearing zoosporangia and resting spores. It infects the roots and sheaths of turfgrasses, causing yellow patches in turf (Viennot-Bourgin 1949).

Physoderma alfalfae (Pat. & Lagerh.) Karling

(Syn. *Urophlyctis alfalfae* (Pat. & Lagerh.) Magnus)

P. alfalfae causes crown wart of lucerne. Spores penetrating the host surface at soil level give rise to a narrow intracellular rhizomycelium bearing terminal septate swellings at the apex of which digitate haustoria form. Hemispherical resting sporangia, the most frequently observed stage, are eventually formed at their tips. These finally germinate to release zoospores (CMI 751). The infected crown buds just below the soil surface develop into large white, and then marbled brown galls which finally disintegrate to release the resting sporangia which can survive for several decades. Affected plants are killed. *P. alfalfae* is widespread in lucerne-growing countries throughout the world (CMI map 130), but it only causes losses in poorly drained or heavily irrigated soils. Providing good drainage and avoiding repeated cropping at sites known to be infested are the main control measures (APS Compendium). Some cultivars may show a degree of resistance (Raynal 1982).

CARR & SMITH

Other *Physoderma* species also cause galls on roots and stems, but are of minor importance: ***P. leproides*** (Trabut) Karling on sugar beet in some Mediterranean countries (CMI 752); ***P. graminis*** (Büsgen) de Wild. on turfgrasses (APS Compendium). ***P. maydis*** (Miyabe) Miyabe, causing brown spot of maize (CMI 753), does not occur in Europe.

Synchytrium endobioticum (Schilbersky) Percival

Basic description

S. endobioticum is a holocarpic obligately biotrophic parasite. The fungus colonizes single host cells, the sori being perceptible as golden brown bodies in the tissue. Summer sori, containing 2–7 sporangia, develop from haploid zoospores. Resting sori with one sporangium only emerge from diploid zygotes after copulation of two zoospores (isogametes); a single sporangium contains about 100–300 zoospores.

Host plants

S. endobioticum lives parasitically only on Solanaceae, in particular many tuber-bearing species and subspecies of the genus *Solanum*. Important host plants are tomato and common weeds like *Solanum nigrum* and *S. dulcamara*, but the principal host is potato.

Host specialization

After the extensive introduction of potato resistant only to pathotype 1 (D_1; common race) between 1920 and 1940, at least 10 new pathotypes with some 1000 foci emerged in Europe, India, S. America and Canada (Newfoundland). A low number of potato cultivars is resistant to all known European pathotypes; other cultivars show differential reactions. Most of the tuber-bearing species and subspecies of the genus *Solanum* include strains resistant or susceptible to new pathotypes.

Geographical distribution

Most, if not all, potato-growing countries have recorded at least one case of *S. endobioticum* (CMI map 1; EPPO 1977; Langerfeld 1984). The basic requirements for permanent endemic establishment are however climatic conditions with relatively cool, wet summers and long, cold winters, such as those prevailing in the central and northern areas of Europe (Bojnansky 1960). The prevailing opinion is that *S. endobioticum* originates from the S. American Andes, the gene centre of the potato. Hampson & Proudfoot (1974) believe that the introduction to Europe results from the importation of S. American breeding material after the *Phytophthora infestans* catastrophe of 1840–1850. There is an alternative opinion however that the fungus originally lived in Europe on wild Solanaceae and changed to its new host in the course of the mass propagation of potatoes.

Disease

All superficial meristematic zones of the plant are susceptible to wart disease but roots cannot be attacked. As a rule attack occurs on underground plant organs like the sprout axis, stolons, young tubers, and on eyes of older tubers. Typical symptoms are ivory white to dark brown tumours; these are pea-sized to walnut-sized or larger, and cauliflower shaped (Fig. 34). Zoospores and zygotes invade the meristem cells and develop summer and resting sori respectively (see above). Around the invaded cell, probably stimulated by the production of chemical substances, neighbouring cells divide and multiply. In the case of high infection density, the attacked plant parts are converted into proliferating, deformed leaf or sprout organs. Successive infections by zoospores from maturing summer sori, and consequently successive cell divisions, give rise to the formation of further tumours. Single infections or very low infection density result in the small galls typical of the Synchytriaceae. After decay of the tumours by rotting, the resting sori remain as long-time infestation in the soil.

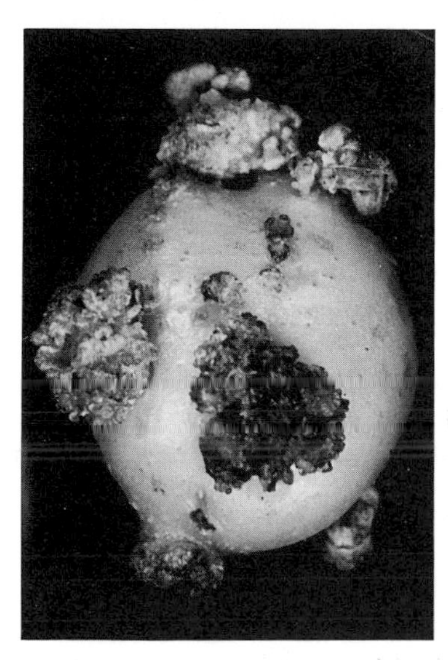

Fig. 34. Potato tuber attacked by *Synchytrium endobioticum*.

Epidemiology

The initial infection of potato plants or tubers usually starts from resting sori in the soil. Equally important are infections from resting sori on the surface of infested seed tubers or from infected, dormant eyes. A basic requirement for infection is the coincidence of susceptible host tissue, liquid soil water and released zoospores as inoculum. The number of generations per season (re-infections from summer sori) depends on the duration of this coincidence. In areas with moderate climatic conditions (see above) viable resting sori can exist for more than 30 years in the soil. The principal cause of spread is attacked or contaminated seed tubers. Over shorter distances, resting sori are frequently spread by soil on shoes, tools, and agricultural machinery. Another cause is the spread by manure—resting sori survive the digestion of domestic animals. Moreover, resting sori survive most of the procedures of potato-processing factories, and may be spread by irrigation with processing water.

Economic importance

Depending on the degree of attack the amount of damage ranges between total yield loss and a slight reduction of quality. Most national regulations, however, decree that potatoes from wart foci have to be consumed directly by the grower on his farm or household. An absolute estimation of total damage is difficult, as most wart foci occur in gardens or small fields. According to several reports, the average loss in commercially managed agriculture is very low. Of greater importance are the administrative restrictions following the discovery of wart foci. Apart from the prohibition of potato growing in wart foci, most national orders for the control of potato wart make additional restrictions on the surrounding areas (e.g. no seed growing, or no permission to grow cultivars susceptible to the local pathotype). Moreover, all quarantine orders require 'zero tolerance' for *S. endobioticum*. Occurrence of the wart fungus (above all new pathotypes) generally causes export limitations for extensive potato-growing areas.

Control

The thick-walled resting sori of *S. endobioticum* are extremely resistant to chemical treatment, the dosage of which often exceeds the phytotoxic threshold before most resting sori are killed. There are some positive records on the effect of chemicals in soil treatments, above all dinitro-cresols, carbamide, cyanamide, and methyl bromide. The safest means of controlling potato wart remains the breeding and growing of resistant cultivars. The re-formation of new pathotypes seems to proceed rather slowly. According to the literature, new 'multi-resistant' cultivars are resistant to most or all known European pathotypes of *S. endobioticum*. An important control measure is the prevention of spread by legislation—above all the prohibition of potato growing in wart foci and 'zero-tolerance' in import regulations (EPPO 82).

LANGERFELD

Other *Synchytrium* spp. cause leaf or stem galls of many wild plants (Viennot-Bourgin 1949), or crop plants in the tropics (CMI 756–760), but *S. endobioticum* is the only important species in Europe.

PLASMODIOPHOROMYCETES

Plasmodiophora brassicae Voronin

Basic description

Plasmodiophora is a genus of about seven species of vascular plant parasites, differing from other members of the Plasmodiophoromycetes in their resting spores which are not aggregated into spore balls. *P. brassicae* is the only economically important member of the genus.

Host plants

P. brassicae induces disease symptoms on all members of the Cruciferae although some are more susceptible than others; for example, among cultivated forms, *Raphanus* species are less severely affected. These statements must be qualified however, as host specialization by pathotypes on wild species has been little studied. Moreover, primary infection of root hairs can occur on a wide range of non-cruciferous plants but with no apparent adverse effects.

Host specialization

Many physiologic races, pathotypes or populations are known to occur. Some progress towards rationalization of the nomenclature and classification of these was made by Buczacki *et al.* (1975) and the subject has been reviewed more recently by Crute *et al.* (1979).

Geographical distribution

P. brassicae has a widespread distribution, especially in temperate regions; it probably occurs wherever crucifers are cultivated. It is generally believed that *P. brassicae* has been spread by the movement of plants and produce from an original centre in the Mediterranean area (Watson & Baker 1969).

Disease

P. brassicae causes the disease generally known as clubroot. It is manifest typically by swelling of the roots, either as a single large gall (the 'club' symptom) or as clusters of swollen lateral roots (the 'finger and toe' effect), somewhat reminiscent of the appearance of dahlia tubers. On plants, such as turnips and rutabagas, with naturally bulbous roots, the fibrous lateral roots swell. The severity of the effects of the disease on the host plant depends on a number of factors; infection of the tap root to produce the single large gall is almost invariably more serious than lateral root attack, whereas the time over which the gall remains firm and undecayed will significantly affect the stage at which the plant ultimately succumbs. Occasionally galls may develop on the aerial parts of plants and, on watercress (*Nasturtium* species), swellings are normally seen on the shoots.

Epidemiology

P. brassicae survives in the soil as uninucleate resting spores produced in the root galls and liberated when the galls decay. They can probably persist for at least 20 years in the absence of host plants although the possibility of the pathogen undergoing an abbreviated life cycle in the roots of non-crucifers should be borne in mind when such data are cited. Resting spores germinate to produce single, uninucleate biflagellate primary zoospores which encyst on the surface of root hairs or epidermal cells, penetrate the cell wall and form a plasmodium within. After nuclear division and growth, the plasmodium cleaves to form a group of zoosporangia within each of which several secondary zoospores develop. These are morphologically indistinguishable from primary zoospores and their precise function is uncertain. It is believed that they penetrate from the epidermal to the cortical cells for, within the latter, secondary plasmodia form, grow, undergo karyogamy and meiosis and ultimately cleave to produce the resting spores (Ingram & Tommerup 1972). Apart from its survival as resting spores *P. brassicae* may persist in wild crucifers, although there is considerable variability in susceptibility within and between species (Buczacki & Ockendon 1979). Quite commonly, weeds such as *Capsella bursa-pastoris* will be found free from disease on land where cultivated brassicas are severely affected. The disease is always most severe on acid, poorly drained soils, in seasons of high rainfall and moderate warmth.

As it is a pathogen with no air-borne phase, the dispersal of *P. brassicae* is by movement of infected plants and contaminated soil. Transplants infected in the seed bed often show no gall development by the time they are lifted, and thus provide an insidious method of disease transfer. Soil containing resting spores may be transported by wind, flood water or, most commonly by adhering to farm implements or boots. It is probable, although unproven, that clubroot may also spread when contaminated soil adheres to the feet of birds and mammals, whilst the passage of viable spores through the alimentary tract of cattle has been demonstrated, indicating that dispersal in dung is another possibility. The rare reports of seed transmission may be explained as arising when particles of contaminated soil adhere to seed, an event so unlikely in normal commercial practice as to be discounted.

Economic importance

When the level of soil inoculum is high and environmental conditions are favourable for infection of young plants, death of the hosts may occur at an early age. More commonly, death occurs when the galls on mature plants begin to decay but, more commonly still, plants are not killed outright and simply develop in a stunted and relatively worthless fashion. The precise economic losses due to clubroot are exceedingly difficult to determine and, further, the proportion of land contaminated is known only imperfectly because the disease is not one reported routinely to advisory officers; the possibility of its being present is accepted as a fact of the brassica grower's life. Knowing the effects on a crop that a severe attack can have, the commonest consequence will be to completely avoid this land for brassica cropping, thus the problem will result in a reduced flexibility of cropping options.

Control

Clubroot has always been one of the most difficult of plant diseases to control; firstly because *P. brassicae* is wholly confined to the soil, secondly because it is unaffected by most fungicides, and, thirdly, because there is very little host plant resistance to the pathogen. The numerous procedures that have been tried but failed are listed by Karling (1968). Methods that are presently fairly widely practised may be summarized thus: crop rotation is of limited value because of the longevity of the pathogen in the soil, although the general advice is that any rotation is better than none; liming to increase the soil pH to at least 7.5 is often very effective although the possibility of adverse effects upon other crops in a rotation should be borne in mind; the only effective fungicides are PCNB, mercurous chloride and benzimidazoles such as thiophanate-methyl and benomyl, whereas several soil partial sterilants including methyl bromide, chloropicrin, dazomet and D-D have given good, if costly, control. Apart from PCNB none of the orthodox fungicides are of value in direct-sown crops and their use is thus limited to the treatment of bare-rooted transplants or of modules of soil-less growing medium. The problems of breeding crops resistant to clubroot are confounded by the paucity of resistance sources and compounded by the complexity of pathogenic variation in *P. brassicae*. While slight success has been achieved in breeding, especially in some cultivars of *Brassica campestris* and *B. napus*, the resistance is limited in effectiveness to sites where complex and virulent mixtures of pathotypes do not occur.

BUCZACKI

Spongospora subterranea (Wallr.) Lagerh.

S. subterranea is the only economically important member of a small genus of plant-parasitic Plasmodiophoromycetes (Karling 1968). It occurs as two formae speciales. **f.sp. *subterranea*** Tomlinson and **f.sp. *nasturtii*** Tomlinson which are virtually indistinguishable morphologically but have different host ranges. f.sp. *subterranea* is the more important of the two and infects potatoes, tomatoes, other members of the Solanaceae, and possibly a few species in other families, while f.sp. *nasturtii* is confined to *Nasturtium* species. The possibility of pathotype specialization existing in *S. subterranea* has been debated for many years; conclusive proof is lacking although collections of f.sp. *subterranea* with differing size spore balls have been reported. f.sp. *subterranea* is of world-wide distribution and although f.sp. *nasturtii* has only been reported from W. Europe and the USA, it would be surprising if it did not occur more widely.

f.sp. *subterranea* causes powdery scab of potatoes, a disease that is more frequently disfiguring than damaging. The most typical symptoms occur in association with lenticels, wounds or, sometimes eyes on potato tubers as irregular, brown, crater-like depressions with raised and ragged margins containing a loose, powdery mass of spore balls. The latter serve to distinguish the disease from the lesions of common scab caused by *Streptomyces scabies* although experience and a microscope are needed before the identifications can really be made with certainty. Under conditions of high moisture or of alternating wet and dry periods, the lesions may extend to become large, cankerous swellings which render the tubers worthless. On roots and stolons, small, less obvious tissue proliferations may arise. The fungus can also damage tomato roots on which quite large, irregular galls are formed; the spore balls are very difficult to find, however.

f.sp. *nasturtii* causes malformation of watercress roots: the primary roots are swollen, stunted and often curved in a crook-like manner, while the secondary roots are swollen at the tips. Eventually, the roots decay, the plants become stunted and yellowed and may lose their anchorage.

The life cycle of *S. subterranea* is broadly similar to that of *Plasmodiophora brassicae*. Infection occurs

pore balls in soil or water germinate to produce primary zoospores which invade root hairs or epidermal cells and develop to produce plasmodia. Zoosporangia formed from these plasmodia produce secondary zoospores which either invade cortical tissues and form secondary plasmodia and more spore balls, concurrent with the appearance of disease symptoms, or reinvade root hairs to repeat the primary stages of the life cycle. Powdery scab is a disease of wet seasons and waterlogged soils, although it does not appear to be greatly influenced by soil pH. Many fewer fungicides have been evaluated against scab than against clubroot; chemical control seems similarly elusive and impracticable. Improvement of drainage, maintenance of long rotations and the destruction of affected tubers are probably all that are feasible. Crook root is controlled very successfully in commercial watercress beds by the addition to the water of zinc-containing frits. While no variations in varietal resistance to crook root have been reported, some potato cultivars are certainly highly susceptible to powdery scab, Pentland Crown being affected especially severely.
BUCZACKI

Polymyxa graminis Ledingham forms biflagellate zoospores which infect the roots of Gramineae to form plasmodia in root cells (APS Compendium of Wheat Diseases). The plasmodia finally develop into grape-like clusters of resting sporangia (cystosori). Since cell proliferation is not stimulated, damage is insignificant. However, as in the case of many related organisms (Giunchedi & Pollini 1984), this fungus acquires importance as a vector of viruses: wheat soil-borne mosaic virus (q.v.) (Lapierre 1980), and barley yellow mosaic virus (q.v.), both recently appeared in Europe. An even more important case of virus transmission is that of beet necrotic yellow vein virus (q.v.) by ***P. betae*** Keskin, to cause the disease known as 'rhizomania'. The fungus, which behaves like other *Polymyxa* species, is of no importance in itself, and its distribution is thus probably not known exactly. It certainly occurs more widely (CMI map 522) than the virus, for example in the UK where it is clearly a potential vector of a major disease that has not yet been introduced. The resting sporangia survive for many years in the soil and the virus apparently survives for as long.

ZYGOMYCETES

Mucor

The genus *Mucor* is typified within the Mucoraceae by producing dark columellate sporangia on single or branched (monopodially or sympodially) sporangiophores. The most important member of this genus responsible for causing plant disease, especially post-harvest disease, is ***M. piriformis*** Fischer which causes fruit rots of strawberry, raspberry, blackberry, loganberry, gooseberry, pears, apples, peaches and nectarines while ***M. hiemalis*** Wehmer (raspberries, guava), ***M. strictus*** Hagem (pear), ***M. racemosus*** Fres. (various fruits and vegetables) and ***M. circinelloides*** v. Tieghem (tomato) have also been reported to cause post-harvest rots (Smith *et al.* 1979; Dennis 1983). *M. piriformis* and the other species have a world-wide distribution but there is no evidence of strain specificity with respect to host. Infection of ripe soft fruit by *M. piriformis* can occur prior to harvest and in certain seasons this fungus has caused substantial losses in strawberry crops in the UK. However, in common with the other *Mucor* species the most severe infection occurs after harvest and especially when mechanical damage has occurred so that entry is gained via the wounds. On apple, pear, peach and nectarine, infection is accompanied by browning of the infected tissue whereas with the soft fruits no such discoloration occurs. In the latter cases infected areas have a water-soaked appearance and juice exudes from the rotted tissues forming a similar 'leak disease' (Plate 13) to that described for *Rhizopus* species. In all cases sporangiophores and black sporangia are formed on the surface of infected fruits, especially under conditions of high humidity.

M. piriformis is a wet-spored fungus that relies primarily on rain splash for the dispersal of spores; rain storms can substantially increase the contamination of fruit by this fungus, especially on soft fruit where soil splash occurs. From these initial infections further spread in the field can occur with dispersal of the sporangiospores (Harris & Dennis 1980). In crops such as strawberries which are harvested over a period of a few weeks in Europe, rotting by *M. piriformis* usually increases as the season progresses due to the increased inoculum levels. However heavy rain early in the season can result in high levels of infection throughout the season.

In common with *Rhizopus* species, *Mucor* species are most important as the cause of rotting during storage and distribution of fruits and vegetables, and, in the case of sulphited soft fruit, disintegration can be brought about by the acid-tolerant polygalacturonases similar to that described for *Rhizopus* species (Dennis 1983). Initial post-harvest infections usually occur via wounds, while subsequent spread is by infected fruit contacting healthy fruit or by the dissemination of sporangiospores in exuded juice. Infection by *M. piriformis* occurs at temperatures as low as 0°C. Thus although rapid cooling and maintenance of a low temperature reduces rotting, it does not prevent infection (cf. *Rhizopus* spp.), and the viability of sporangiospores is not reduced by exposure to chill temperatures (Dennis & Blijham 1980). Current commercial fungicides are ineffective against *M. piriformis* and Zygomycetes generally. The fungus is highly resistant to fungicides (benzimidazoles, dicarboximides) commonly used on fruit crops against other diseases such as grey mould (*Botryotinia fuckeliana*) (Dennis 1983). Thus the control of *Mucor* species can only be achieved by careful regular harvesting, storage at low temperatures and the avoidance of mechanical damage.

DENNIS

Rhizopus

The genus *Rhizopus* is distinguished from other Mucoraceae by the fact that the sporangiophores usually develop in clusters from the ends of the stolons at the point of origin of the rhizoids. ***R. stolonifer*** (Ehrenb.) Lind (CMI 524), ***R. sexualis*** (Sm.) Callen (CMI 526), ***R. oryzae*** Went & Prinsen Geerligs (CMI 525) and ***R. arrhizus*** Fischer are the main species responsible for causing plant diseases. *R. stolonifer* attacks a variety of fruits and vegetables (stone fruits, pome fruits, soft fruits, carrot, tomato, pepper, aubergine, avocado and sweet potato) while *R. sexualis* (soft fruit), *R. oryzae* and *R. arrhizus* (stone fruits) are more restricted in their hosts. Apart from *R. sexualis* which has only been reported on soft fruit from the UK and Japan, the other species have a world-wide distribution.

Rhizopus species are most important as causing rotting of harvested fruit and vegetables during storage and distribution through marketing channels. Such rotting can be devastating for crops exposed to high ambient temperatures (15–25°C) during marketing and where mechanical damage has occurred. Thus *Rhizopus* species are generally considered to be wound-invading fungi, although in the case of soft fruits infection of ripe undamaged fruit can occur both pre- and post-harvest (Harris & Dennis 1980). Infection by *Rhizopus* species initially results in a water-soaked appearance, but rapid maceration of the plant tissue follows. In fruits such as strawberries, tomatoes, peppers and sweet potato no change of colour of the infected tissue occurs whereas in other fruits and vegetables (plums, cherries, apricots, peaches, carrot and aubergine) the infected tissue is typified by a brown discoloration. In all cases the rot is very soft and watery and may result in total collapse of the fruit or vegetable. In addition juice exudes from the rotted areas, especially in soft fruit, hence the name 'leak disease'. As the rot progresses a fluffy white mycelium develops on the surface of the infected tissue, but the extent of this varies between species and is dependent on environmental conditions, especially RH (Harris & Dennis 1980). Masses of black sporangia are rapidly produced at RH below 80% for all the species mentioned while sporangial production is much reduced at higher RH, at which, in the case of the homothallic species *R. sexualis*, masses of zygospores are produced on the fruit surface. Infection of ripe fruit in the field arises from air-borne sporangiospores and from propagules in adhering dust particles or those transmitted by fruit flies. Subsequent sporulation on infected fruit then forms in inoculum source for further contamination of fruit, especially in periods of warm, dry weather. Post-harvest infections can also result from inocula in packhouse and storage rooms if debris and decaying fruit or vegetables are present. Under conditions of high ambient temperatures rapid spread within a container of fruit or vegetables occurs either by contact, dissemination of sporangiospores in exuded juice or air currents or by the rapid stoloniferous growth of these fungi.

Apart from causing rotting of fresh fruit and vegetables *Rhizopus* species can cause serious loss of quality of certain sulphited and canned fruits since the polygalacturonases produced during rotting are highly acid- and thermo-tolerant (Dennis 1983). In the absence of effective fungicides, control of pre-harvest infection of fruit has to be achieved by frequent harvesting so that overripe fruit do not remain in the field. Control of post-harvest infection and rotting is possible by rapid cooling and by maintaining a storage and distribution temperature of below 5°C. Such temperatures not only inhibit germination and growth of the fungi but also result in reduced viability of sporangiospores (Dennis & Blijham 1980; Dennis & Höcker 1981).

DENNIS

REFERENCES

Bojnansky V. (1960) Potato wart disease, *Synchytrium endobioticum*. *Nature* **185,** 367–368.

Buczacki S.T. & Ockendon J.G. (1979) Preliminary observations on variation in susceptibility to clubroot among collections of some wild crucifers. *Annals of Applied Biology* **92,** 113–118.

Buczacki S.T., Toxopeus H., Mattusch P., Johnston T.D., Dixon G.R. & Hobolth L.A. (1975) Study of physiologic specialization in *Plasmodiophora brassicae:* proposals for attempted rationalization through an international approach. *Transactions of the British Mycological Society* **65,** 295–303.

Campbell R.N. & Fry P.R. (1966) The nature of the association between *Olpidium brassicae* and lettuce big-vein and tobacco necrosis viruses. *Virology* **29,** 222–233.

Crute I.R., Gray A.R., Crisp P. & Buczacki S.T. (1979) Variation in *Plasmodiophora brassicae* and resistance to clubroot disease in brassicas and allied crops—a critical review. *Plant Breeding Abstracts* **50,** 91–104.

Dennis C. (1983) Soft fruits. *Post-Harvest Pathology of Fruits and Vegetables* (Ed. C. Dennis), pp. 23–42. Academic Press, London.

Dennis C. & Blijham J.M. (1980) Effect of temperature on viability of sporangiospores of *Rhizopus* and *Mucor* species. *Transactions of the British Mycological Society* **74,** 89–94.

Dennis C. & Höcker J. (1981) Effect of relative humidity on chilling sensitivity of sporangiospores of *Rhizopus* species. *Transactions of the British Mycological Society* **77,** 179–222.

EPPO (1977) First report of the working party on potato wart disease. *EPPO Publications Series C* no. 50.

Giunchedi L. & Pollini C.P. (1984) Funghi quali vettori di virus dei vegetali. *Informatore Fitopatologico* **34** (6), 41–54.

Hampson M.C. & Proudfoot K.G. (1974) Potato wart disease, its introduction to North America, distribution and control problems in Newfoundland. *FAO Plant Protection Bulletin* **22,** 53–64.

Harris J.E. & Dennis C. (1980) Distribution of *Mucor piriformis, Rhizopus sexualis* and *R. stolonifer* in relation to their spoilage of strawberries. *Transactions of the British Mycological Society* **75,** 445–450.

Ingram D.S. & Tommerup I.C. (1972) The life history of *Plasmodiophora brassicae. Proceedings of the Royal Society* (B) **180,** 103–112.

Karling J.S. (1968) *The Plasmodiophorales* 2nd Edn. Hafner, New York.

Lange L. & Insunza V. (1977) Root-inhabiting *Olpidium* species: the *O. radicale* complex. *Transactions of the British Mycological Society* **69,** 377–384.

Langerfeld E. (1984) *Synchytrium endobioticum.* Zusammenfassende Darstellung des Erregers des Kartoffelkrebses anhand von Literaturberichten. *Mitteilungen aus der Biologische Bundesanstalt für Land und Forstwirtschaft. Berlin-Dahlem* no. 219.

Lapierre H. (1980) Les virus des céréales à paille. *Phytoma* no. 321, 34–38.

Mosse B. (1973) Advances in the study of vesicular-arbuscular mycorrhiza. *Annual Review of Phytopathology* **11,** 171–196.

Ogilvie L. (1969) *Diseases of Vegetables. MAFF Bulletin* no. 123. HMSO, London.

Raynal G. (1982) Comportement au champ de quelques cultivars de luzerne vis-à-vis d'*Urophlyctis alfalfae. Agronomie* **2,** 197–200.

Smith W.L., Moline H.E. & Johnson K.S. (1979) Studies on *Mucor* species causing post-harvest decay of fresh produce. *Phytopathology* **69,** 865–869.

Sparrow F.K. (1960) *Aquatic Phycomycetes.* University of Michigan Press, Ann Arbor.

Tomlinson J.A. & Garrett R.G. (1964) Studies on the lettuce big-vein virus and its vector *Olpidium brassicae. Annals of Applied Biology* **54,** 45–61.

Tomlinson J.A., Faithfull E.M. & Clay C.M. (1981) Big-vein disease of lettuce. *Report of National Vegetable Research Station, Wellesbourne 1980,* pp. 82–84.

Tomlinson J.A., Faithfull E.M., Clay C.M. & Taylor J.D. (1982) Big-vein disease of lettuce. *Report of National Vegetable Research Station Wellesbourne 1981,* pp. 82–83.

Viennot-Bourgin G. (1949) *Les Champignons Parasites des Plantes Cultivées.* Masson, Paris.

Watson A.G. & Baker K.F. (1969) Possible gene centers for resistance in the genus *Brassica* to *Plasmodiophora brassicae. Economic Botany* **23,** 245–252.

White J.G. (1982) Control by soil partial sterilisation. *Report of National Vegetable Research Station Wellesbourne 1981,* p.83.

10: Ascomycetes I
General; Endomycetales, Taphrinales, Eurotiales, Erysiphales

ASCOMYCETES (GENERAL)

A vast number of species of Ascomycetes infect plants, causing important diseases or more often than not minor local infections of little significance. As 'higher fungi', they form septate hyphae and sexual spores in complex fructifications, following a nuclear cycle in which the main vegetative mycelium is haploid and monokaryotic. Almost all plant pathogenic species produce asexual conidia; this is the spore form most frequently seen on infected tissue or in culture. The type of conidial sporulation (wind-dispersed dry spores, splash-dispersed slime-spores from acervuli or pycnidia) is of great significance in the epidemiology of the diseases caused. The sexual fructification (ascocarp, asci, ascospores), though very diverse in structure and serving as the principal basis for distinguishing taxonomic groups, is in many cases rarely seen and tends to form only as an overwintering structure on dead plant tissue. The structure and function of the teleomorph state is often of little significance in pathological or epidemiological terms. In many cases it is not known, the perennation function being taken over by chlamydospores or sclerotia, or by survival on or in the living host plant. The sexual function may be replaced by parasexual recombination. It is this situation which has led to the use of a special nomenclature for the imperfect states of Ascomycetes (anamorphs). This handbook recognizes the necessity of using anamorph names, but avoids any use of the classification of the so-called Fungi Imperfecti (see *Preface*). In view of the great diversity of the group, the Ascomycetes are treated in several chapters and further aspects of the pathogenicity of the different orders are developed in each case.

A few Ascomycetes form no ascocarp. These include the Endomycetales—the yeasts and yeast-like fungi, a few of which are associated with rotting of fruits. The Taphrinales are specialized plant pathogens, existing in an epiphytic yeast-like form or as a mycelial form infecting leaves and shoots and forming a continuous 'palisade' of naked asci on the host surface. They characteristically cause either severe leaf distortion and discoloration (leaf curl) or witches' brooms of perennially infected shoots. Of the Ascomycetes which do form ascocarps, the first main group is those with more or less globose non-ostiolate fruiting bodies (cleistothecia). The Eurotiales include the fungi with anamorphs in *Penicillium* and *Aspergillus*, many of which have no known teleomorph. They produce huge quantities of dry air-dispersed conidia, and are among the main saprophytic colonizers of plant substrates, as they are able to develop under rather dry conditions. A few will infect the living storage tissues of plants, i.e. fruits, bulbs and tubers, usually through wounds.

The Erysiphales, one of the most important groups of plant pathogens, are obligate parasites, most forming an epiphytic mycelium penetrating only the epidermal cells of the host by lobed haustoria. A few (*Leveillula*) develop a mycelium within the leaf tissues. The superficial mycelium bears upright conidiophores with characteristic large rectangular conidia (*Oidium*) in chains (arthrospores). The combination of superficial white or greyish mycelium, with masses of powdery conidia, is the basis for the common name 'powdery mildew'. Most European species form cleistothecia on the mycelium towards the end of the season. These mature over the winter and dehisce, in spring, by an apical split through which the ascospores are discharged. The asci are characteristically few in number, globose and few-spored. The surface of the cleistothecia bears appendages of various shapes which distinguish the genera. The species which infect perennial hosts may also survive as mycelium in dormant buds.

The nomenclature of the Erysiphales is complicated by the existence of species with wide and overlapping host ranges, often difficult to distinguish by conidial and mycelial characteristics alone (*Erysiphe cichoracearum*/*Sphaerotheca fuliginea; Podosphaera tridactyla*/*Sphaerotheca pannosa*). In some cases, clear host specialization by *forma specialis* is the rule (*Erysiphe graminis; E. pisi*), while in many others the situation remains poorly defined. The powdery mildews have recently been treated in a detailed monograph (Spencer 1978).

ENDOMYCETALES

Typical yeasts (Saccharomycetaceae) are not normally associated with plant disease, though they often develop secondarily on rotting tissues. On grapevine, *Hanseniaspora* sp. (anamorph: ***Kloeckera apiculata*** (Reess emend. Klöcker) Janke) and ***Saccharomycopsis vini*** Kreiger-van Rij are among the species most active, with acetic acid bacteria, in causing sour rot of grapes, entering through wounds and transmitted by *Drosophila* species (Bisiach *et al.* 1982). ***Nematospora coryli*** Peglion (CMI 184) causes yeast spot or stigmatomycosis of legumes, cotton and other seeds, principally in the tropics. It can be introduced into fruits by the feeding punctures of pentatomid bugs. It was originally described on hazel nuts in Italy and has been reported locally causing up to 80% loss of pistachio nuts in Greece (Kouyeas 1979).

Geotrichum candidum Link

(Syn. *Oospora citri-aurantii* (Ferraris) Sacc. & Sydow)

G. candidum is easily identified by the chains of arthrospores and the dichotomous branching of the mycelium at the periphery of the colony (Butler *et al.* 1965). Some *Geotrichum* species are anamorphs of *Dipodascus*. The fungus causes decay of citrus fruits but it also attacks tomatoes and other vegetables. On citrus (**var.** ***citri-aurantii*** (Ferraris) Sacc. & Sydow) the disease is known as sour rot. It is characterized by a watery, readily smeared rot of the fruit with a very sharp and clearly defined margin. As decay develops a thin white fungal growth appears. The infected fruits omit a sour odour and attract ferment flies. Spores of *G. candidum* are present in most soils and are splashed onto the fruit with soil water. In packing houses the spores are washed from infected fruit and contaminate the treatment solutions, the brushes and conveyor belts. The pathogen enters fruits only through wounds which have at least reached the albedo. High temperatures favour the disease, the optimum for the fungus being 30°C. The disease is more prevalent on mature fruit or fruit stored for long periods (Eckert 1978). Control is achieved by reduction in fruit injuries, improved sanitation, elimination of over-mature fruit and avoidance of long storage. Sour rot is effectively controlled with guazatine as a post-harvest fungicide (Rippon & Morris 1981). Among the other approved fungicides for citrus fruit, only sodium orthophenylphenate gives some control.
CHITZANIDIS

TAPHRINALES

Taphrina deformans (Berk.) Tul.

(Syn. *Exoascus deformans* (Berk.) Fuckel)

Basic description

T. deformans has a mycelial parasitic phase and a yeast-like epiphytic phase. Growth of an intercellular mycelium (from budded conidia) leads to the characteristic hypertrophy and hyperplasia of infected tissues. Asci form naked at the leaf surface without any fruiting body, and germinate before and after discharge by budding (CMI 711; Mix 1949).

Host plants

The hosts are most commonly peach and nectarine, but the disease occurs less severely on almond and only rarely on apricot. Schneider (1971) has shown the existence of two forms: **var.** ***persicae*** on peach and nectarine, and **var.** ***amygdali*** on almond.

Geographical distribution

T. deformans occurs wherever the host plants are grown.

Disease

T. deformans most frequently causes leaf curl, the leaves bearing large blister-like deformations. Part or all of the lamina becomes hypertrophic. The diseased portions are thickened, convoluted and brittle, usually whitish-green in colour, but may also be bright pink or red. The blisters are finally covered by a whitish bloom when the asci form and liberate ascospores. Finally, the leaves brown when colonized by saprophytic fungi and fall. Young shoots may also be attacked, and similarly show hypertrophy with reddish discoloration. A diseased shoot generally arises from an infected bud, and all its parts (stem and leaves) are hypertrophied. Fruits are sometimes infected if leaves are heavily attacked during or after flowering. The irregular reddish blistered lesions can cover up to half the fruit surface.

Epidemiology

The problem of how *T. deformans* overwinters is still not fully resolved. One must reject any idea that mycelium persists, since leaves attacked in spring fall and decay, while leaves formed from June onwards remain uninfected. Mycelium on infected shoots disappears before winter, the infected tissues being exfoliated by the plant. It would therefore seem that *T. deformans* survives as the epiphytic yeast-like phase, arising from ascospores. These may readily persist between the bud scales of the host (making it necessary to spray in autumn and winter to ensure control). As soon as the buds burst, the fungus starts to develop, and this may explain the little blisters which can be seen on young leaves as soon as they spread. The overwintering and primary infection phase thus seems similar to that of the Erysiphales on woody hosts. However, secondary infection ceases in early June.

Economic importance

T. deformans causes potentially the most serious disease of peach and nectarine and requires regular chemical control.

Control

In the light of epidemiological data, two main periods can be recognized for treatment. At leaf-fall, trees are sprayed with a copper fungicide, possibly combined with another of the compounds below. At the beginning of bud swelling (late January to late March according to region), one spray is applied (and possibly another after a further 20 days) of a fungicide such as captan, captafol, ferbam, thiram or ziram. Burchill *et al.* (1976) have successfully used a mixture of fatty acid alcohols, as have been employed to treat *Podosphaera leucotricha* on apple. There are differences in susceptibility between peach and nectarine cultivars, and in particular the semi-wild cultivars known in France as 'pêchers de vigne' are rather resistant. However, no real resistance is known.
BONDOUX

Other fruit trees are attacked by *Taphrina* species, for example ***T. wiesneri*** (Rathay) Mix (syn. *T. cerasi* (Fuckel) Sadebeck) causes a leaf curl and witches' broom of cherry and apricot (CMI 712), but is of much less importance than *T. deformans*. It can be controlled by pruning away the brooms and by chemical treatment. ***T. pruni*** Tul. infects the fruits of plum (CMI 713), which become elongated with pocket-like invaginations (bladder or pocket plums). The fruit shrivels, becomes covered with a whitish bloom of ascospores and finally falls, without a developed stone. The fungus persists as mycelium in the twigs and can also cause witches' brooms; the form responsible for this is sometimes distinguished as ***T. insititiae*** (Sadebeck) Johansson. The fungus is widespread in Europe, occurring also on wild *Prunus* species, but is important mainly in northern countries, where up to 90% of fruits may be attacked. Affected fruits should be removed and destroyed; spraying with fungicides (as for *T. deformans*) 3 weeks before flowering is advisable in affected orchards. ***T. bullata*** (Berk.) Tul. causes blisters on the leaves of pear and quince and is common, for example, in S.W. France (Viennot-Bourgin 1949). It is controlled by treatments against *Venturia pirina*.

Taphrina populina **Fr.**

(Syn. *T. aurea* Pers.)

T. populina has asci varying in size on different poplar species and varieties (Mix 1949) but which are generally about 70–90×18–22 μm. Their ascospores form numerous bud spores (blastospores) measuring about 2×1 μm. It attacks poplars, especially clones of *P.*×*euramericana* (*P.*×*canadensis*), and less often, *P. deltoides* and *P. nigra*. Other poplars, including *P. alba* and *P. balsamifera* (*P. tacamahaca*) are also occasionally affected. The common cultivated poplar hybrids vary in susceptibility. Thus some clones of *P.*×*euramericana* are very susceptible, others very resistant (Taris 1970).

T. populina is widespread in Europe, and occurs locally in eastern N. America and parts of India, China and Japan (Mix 1949). It causes large blisters on poplar leaves. The blisters form rounded convex projections above the leaf surface, with a concave underside which is soon lined with a bright golden-yellow layer of asci. This concave side is usually on the underside of the leaf blade. The fungus overwinters as blastospores on the buds and as mycelium within them, and in damp weather in spring the young leaves are infected as the buds burst (Schneider & Sutra 1969). The leaf blisters are usually found from May or June until September. Leaf blister is common and striking, but the effect on the tree is generally slight. Control measures are not usually required, but if necessary young trees (especially in the nursery) may be sprayed with captan or a copper fungicide as the buds begin to swell.
PHILLIPS

A number of *Taphrina* species occur on forest trees in Europe sometimes causing conspicuous but relatively unimportant diseases. ***T. betulina*** Rostrup (syn. *T. turgida* (Rostrup) Sadebeck) is a common cause of witches' brooms on *Betula* species, while ***T. betulae*** (Fuckel) Johansson causes a leaf spot. ***T. caerulescens*** (Desm. & Mont.) Tul. causes yellowish, pale green or pinkish, slightly swollen leaf spots on many *Quercus* species. ***T. ulmi*** (Fuckel) Johansson causes diffuse yellowish or brown blotches on leaves of the suckers of various *Ulmi* species. Alder (*Alnus* species) has as many as four species: ***T. sadebeckii*** Johansson causing yellow, unthickened leaf spots about 1 cm across, ***T. tosquinetii*** (Westend.) Magnus (syn. *T. alnitorqua* Tul.) distorting leaves and twigs, ***T. alni*** (Berk. & Broome) Gjaerum (syn. *T. amentorum* (Sadebeck) Rostrup) causing catkin blight, and ***T. epiphylla*** (Sadebeck) Sadebeck causing witches' brooms (the last is probably a late form of the first and has identical asci).

EUROTIALES

Penicillium digitatum Sacc.

Basic description

Colonies on PDA amended with 0.2% yeast extract and 0.2% peptone, on which the fungus is usually isolated, grow rapidly, have a smooth velvety appearance and a pale olive-green colour; the reverse is colourless. Conidiophores are short (30–100μm), conidia variable, smooth, sub-globose to cylindrical, usually elliptical 3.4–12×3–8 μm (CMI 96).

Host plants

P. digitatum attacks citrus fruits.

Geographical distribution

P. digitatum is common in all citrus-growing areas all over the world.

Disease

The disease is known as green mould of citrus fruit (Plate 14). The first symptom is a round, slightly sunken, watery spot, 2–3 mm in diameter on the surface of the fruit. If the temperature is favourable the spot enlarges rapidly and is covered with a white mould, which soon becomes olive green by the production of spores. On the infected fruit the sporulating part of the fungus is surrounded by a large (1–2 cm) white margin of mycelium, ahead of which is a zone of soft watery rind. Later the whole fruit decays, shrinks and finally becomes mummified (Eckert 1978).

Epidemiology

The spores of *P. digitatum* can survive for long periods in the soil and in contaminated packing houses and storage rooms. They are transported by air currents onto sound fruits. The spores do not germinate on the surface of the fruit until the peel is injured; free water and nutrients are required for spore germination. Poor handling practices during harvesting and subsequent processing are the main causes of wounding. Frost damage, oleocellosis, water spot and insect injuries are also infection sites for *P. digitatum*. An important factor influencing the development of green mould is temperature. The optimum temperature for the growth of the fungus is 24°C. At this temperature irreversible infections take place within 48 h and the first symptoms appear after three days. At 15°C and at 10°C the first symptoms appear after 5–6 and 10–12 days respectively. At 5°C the development of symptoms is very slow. So if the period of handling and storage is short, fruit rot is negligible. If, however, the storage period is long, the fruit will eventually rot. High humidity is also a factor that favours the disease. Thus fruit picked in wet weather is more prone to decay than fruit picked under dry conditions.

Economic importance

Losses arise from the decay of harvested fruit. Further losses are caused by 'soilage' of sound fruits with spores from adjacent infected fruits.

Control

The most effective method of controlling green mould consists of careful handling of the fruit during harvesting, packing and marketing in order to avoid skin injuries. To help prevent injuries during harvesting, pickers should be equipped with gloves and special clippers, picking bags should be emptied in clean boxes free of twigs, soil and gravel, and loading and unloading should be carried out with care. Fruit should ideally be harvested during dry weather. However, due to the rising cost of labour, the application of such procedures is not always feasible. For this reason the use of fungicides after harvest is unavoidable. Fungicides are effective before the infection process is complete. Most infections take place during harvesting, so fruit should be treated promptly after harvest. The fungicides used for post-harvest treatment of citrus fruit are: the benzimaidazoles (mainly benomyl and thiabendazole), sodium orthophenylphenate, diphenyl, 2-aminobutane and imazalil. In Australia guazatine is widely used, since it is also effective against sour rot of citrus (*Geotrichum candidum*). Fungicides reduce the development of decay, and some of them like imazalil, thiabendazole and diphenyl suppress the production of spores on the infected fruit and prevent soilage. When fruit is to be degreened before packing, or if a delay occurs, the containers with harvested fruit should be drenched in a fungicide solution on their arrival in the packing house. During the processing of fruits fungicides can be applied in the bath or as a foam wash. They can also be incorporated into the water wax emulsion with which the fruit is coated at the end of processing. Diphenyl is impregnated in the wrapping papers or the pads that line the continers. A combination of these methods gives the best results. After extensive use of benomyl, thiabendazole, sodium orthophenylphenate, diphenyl and guazatine, resistant strains to one or more of these fungicides may be observed (Eckert 1982; Wild 1983). Benomyl-resistant strains are cross-resistant to thiabendazole, and sodium orthophenylphenate-resistant strains are cross-resistant to diphenyl. Resistant strains exist in low frequencies in natural populations and, through selection pressure by the use of fungicides, they become predominant in the populations of packing houses and storage rooms causing serious decay problems. Resistant strains from the packing houses can spread to orchards on contaminated equipment or treated culled fruits discarded in areas around orchards.

Strategies to avoid the development and spread of resistant strains consist of: i) application in the orchard of fungicides different from those that are applied in the packing houses; ii) alternation of chemically unrelated post-harvest fungicides; iii) monitoring of the *Penicillium* population in the packing house, and alteration of the fungicide programme if resistant strains are detected; iv) avoiding storage of treated fruit in the packing house; and v) good sanitation conditions in the packing house. Sanitation measures are essential to reducc the inoculum of *P. digitatum* in packing houses. Where possible, fruit arriving at the packing house should be unloaded and cleaned in a separate area to reduce contamination of the processing area and equipment. The processing area should be kept clean and all decayed fruit frequently removed and destroyed. Packing areas should be regularly disinfected with a fog of 1% formaldehyde solution or driol. Field boxes should be disinfected with formaldehyde or another disinfectant. Low temperatures during transit and storage maintain the quality of fresh fruit and retard the development and spread of green mould and other decays. Varietal differences must be considered when selecting temperatures for storing because of differential susceptibility to chilling injury.
CHITZANIDIS

Penicillium italicum Wehmer

P. italicum forms fasciculate, irregularly sporing blue-green colonies on PDA amended with yeast extract and peptone, with an orange-tan reverse. Its conidiophores are variable in length and occur singly or in bundles. Conidia are elliptical to sub-globose, 4–5×5–3.5 μm (CMI 99). *P. italicum* is pathogenic to citrus fruit. It is common in all citrus-growing areas, causing post-harvest rot. The disease is known as blue mould and like green mould it causes a soft rot. The spore-producing area on decayed fruit has a bright blue colour and is surrounded by a narrow (1–2 mm) margin of white mycelium. In an advanced stage of decay the fruit becomes a slimy shapeless mass (Eckert 1978). The epidemiology of *P. italicum* is very similar to that of *P. digitatum*. The fungus enters the fruit through wounds, but it also spreads from infected to healthy fruit by contact, thus forming 'pockets' of decayed fruit in the containers. At temperatures above 10°C it grows slower than *P. digitatum* but below 10°C faster. As a post-harvest rot blue mould is less important than green mould and becomes serious only under conditions that do not favour green mould. Control is achieved through careful handling and packaging of fruit, treatment with fungicides (see *P. digitatum*), sanitation in the packing house and low temperatures during transport and storage. The occurrence of benzimidazole--resistant strains is more frequent among populations of *P. italicum* than among *P. digitatum*.
CHITZANIDIS

Penicillium expansum Link

Basic description

Colonies on Czapek agar grow rapidly forming a heavily sporing velvety carpet of surface mycelium up to 2 mm deep. These colonies often show radial zonation, and are initially white becoming dull yellow-green to grey-green, with a reverse colourless to pale brown. The conidial heads are asymmetric and once or twice branched. The conidiophores are smooth, moderately long, 400–700 μm, with smooth conidia, usually elliptical and generally 4–5×2.5–3.5 μm (CMI 97).

Host plants

P. expansum attacks apples and other pomaceous fruit, as well as occasionally grapes, olives, cherries, pineapple, avocado and a range of vegetable storage organs. It also grows saprophytically on damp cereal grains and cereal products.

Geographical distribution

P. expansum has a world-wide distribution.

Disease

P. expansum causes the storage rot commonly known as blue mould of apples and pears. The disease is a classic soft rot whose typical external manifestation is slightly sunken lesions which spread rapidly; complete rotting of fruit can take as little as 5–7 days at orchard temperatures. Chilled storage retards but does not arrest spread. Rotted tissue is watery in texture and light in colour. During storage, conidia are produced in coremia, often in concentric rings on the fruit surface. Conidial tufts are initially white, becoming blue-green as the spores mature thus giving rise to the common name of the disease.

Epidemiology

Fruit infection is mainly via lenticels, especially if the fruit is bruised, or via wounds caused by fruit splitting, insects and birds, or poor handling. *P. expansum* spores are readily wind dispersed, and initial contamination of fruit occurs during the growing season from spores originating from saprophytic cultures in soil or orchard debris. Quiescent infections in lenticels are a major means of transport between field and the packing shed. Much secondary spread of infection can result from poor handling practices during harvesting, grading and marketing of fruit. In particular the use of washing water contaminated with spores of *P. expansum* can lead to marked losses during storage. The environment of the packing shed is often a major source of inoculum. Spores formed on discarded fruit and juice-contaminated packaging are readily transported to healthy fruit via air currents, insects and packing shed staff. Unlike some other post-harvest pathogens of pome fruits there is no evidence that *P. expansum* infects parts of the plant other than the fruit.

Economic importance

On a world scale, *P. expansum* is one of the most important causes of post-harvest decay of pome fruit. In N. Europe the scale of loss varies widely between seasons and cultivars, typically being in the range 5–20%, of which a substantial proportion would be directly due to *P. expansum*. In N. America *P. expansum* is regarded as the most important single cause of post-harvest loss.

Control

Current control measures rely on a combination of hygiene, avoidance of damage to the fruit surface, and fungicides. Strict sanitation should be enforced within and around packing sheds and all grading and packaging machinery kept scrupulously clean. This is especially important for fruit destined for long-term controlled atmosphere or low temperature storage. Even for fruit destined for immediate marketing care should be exercised due to the rapidity with which rots can develop. Careful handling of fruit is important both to avoid rupturing the skin, which provides a direct avenue for infection, and bruising the flesh. Latent infections in lenticels are much more likely to develop into spreading lesions following bruising. Applications of fungicides during the growing season aimed to control other diseases, notably scab and canker, provide some measure of fruit protection. Much more important has been the development of post-harvest dip applications of fungicides during the grading process. Benomyl, thiabendazole, iprodione, vinclozolin, mancozeb and imazalil have all given good protection. Unfortunately resistance to some of these fungicides has developed very rapidly. Resistance to benomyl is widespread in many growing regions, isolates insensitive to 500 ppm being common. Resistance to the dicarboximides is readily induced experimentally, the mutation rate being similar to that of *Botrytis cinerea*.

Special research interest

The very soft, light-coloured lesions of *P. expansum* on apple contrast strongly with the dark brown firmer lesions caused by *Monilinia* species. Both pathogens produce polygalacturonase and pectin lyase (Cole & Wood 1961; Spalding *et al.* 1973) but in lesions caused by *Monilinia* species phenolic oxidation and protein

precipitation proceed rapidly and little enzyme is recoverable from the relatively firm lesion. According to Walker (1970) *P. expansum* produces inhibitors of apple *o*-diphenol oxidase, so preventing both browning and inactivation of its pectolytic enzymes. *P. expansum* is one of several fungi known to produce the antibiotic patulin.
ARCHER

Eupenicillium crustaceum Ludwig

(Anamorph ***Penicillium gladioli*** McCulloch & Thom)

E. crustaceum (Pitt 1979) forms blue-green colonies on Czapek or malt agar which are predominantly conidial at 15°C and below. At 20°C and above colonies produce abundant white to pink and finally brown sclerotia up to 300 μm in diameter. Conidiophores are simple to terverticilliate with only a few elements at each stage; conidia are elliptical and are often found adhering in long chains.

The pathogen, which has a world-wide distribution, causes a storage rot of a range of bulb and corm-forming ornamentals, notably *Gladiolus, Tigridia, Freesia, Crocus, Cyclamen, Scilla, Iris, Montbretia* and *Tritonia.* Affected corms exhibit lesions up to 3 cm in diameter, deeply sunken, reddish-brown in colour and lacking a definite margin. Small buff or pinkish sclerotia may be present, often covered by surface scale leaves. The rot is only serious under moist conditions when blue-green tufts of spores may develop. Infection probably originates from the soil and spreads from corm to corm in storage especially under conditions of high humidity and low temperature. Control is possible by curing wounded corms for 10 days at 29°C. The disease is rarely serious provided that corms are handled with care to avoid wounding and are stored under dry, well ventilated conditions.
ARCHER

Penicillium corymbiferum Westling

P. corymbiferum rots bulbous irises, and will infect undamaged bulbs via the natural ruptures made by emerging roots if high humidity prevails during the cooling (9°C) and post-cooling (17°C) periods used to prepare the bulbs for forcing. Benomyl dipping is the normal method of control, but resistant strains now occur (Hill 1977). *P. corymbiferum*, together with ***P. cyclopium*** Westling, are the main causes of bulb rot of garlic in store (Messiaen & Lafon 1970), infected cloves becoming completely filled with *Penicillium* spores. Rotting is favoured by wounding, and tends to occur under the same conditions as favour the breaking of dormancy (5–10°C). Bulbs planted in winter may be infected in the soil. Attacks are also sometimes seen before harvest, especially on wounded bulbs or those already attacked by *Sclerotium cepivorum*. It is noteworthy that garlic bulbs contain substances (allicin and its derivatives) which are highly antifungal, but that these *Penicillium* species are insensitive to them.

Various other, generally saprophytic, *Penicillium* species are occasionally reported rotting fruits, bulbs or grain (e.g. ***P. oxalicum*** Currie & Thom causing blue-eye rot of maize cobs—APS Compendium). *Penicillium* species are also found to be antagonistic to various plant pathogens, and may be exploited in biological control, e.g. *P. funiculosum* Thom against *Fusarium oxysporum* f.sp. *radicis-lycopersici* (Marois *et al.* 1981); *P. lilacinum* Thom against *Phomopsis sclerotioides* (Ebben & Spencer 1979); *P. oxalicum* against seed-borne pathogens of pea (Windels & Kommedahl 1982); also *Talaromyces flavus* (Klöcker) Stolk & Samson, anamorph *Penicillium vermiculatum* Dang., against *Verticillum dahliae* (Marois *et al.* 1984).

Aspergillus species are almost all saprophytes, but some will infect stored fruits and tubers, especially in warm climates. In Europe, the only case of any significance is ***A. niger*** v. Tieghem rotting onion bulbs in store: masses of black spores form on the surface of infected bulb scales, penetration occurring through wounds (Messiaen & Lafon 1970) (Fig. 35). ***A. flavus*** Link is of great economic importance in tropical and subtropical countries in forming aflatoxins on stored

Fig. 35. Onion sets infected by *Aspergillus niger*.

plant produce, particularly groundnuts and maize (APS Compendia), infection being possible in the field before harvest. On crops grown in Europe, the problem is a minor one, though still of potential concern, for example in Yugoslavia (Brodnik 1976).

ERYSIPHALES

Erysiphe cichoracearum DC.

Basic description

E. cichoracearum has flexuous hypha-like appendages on black cleistothecia and many asci each containing 2 (perhaps 4) ascospores, but is generally seen as its anamorph (***Oidium erysiphoides*** Fr.). It is the typical member of the section *Euerysiphe* of the genus *Erysiphe*. Its chain-forming conidia are distinguished from those of *Sphaerotheca fuliginea* by their cylindrical form, lack of fibrosine bodies and by germ tubes formed from a corner of the conidium (Nagy 1970). It can be considered to include fungi attributed to *E. polyphaga* Hammarlund.

Host plants

E. cichoracearum is a collective species; it occurs in Europe on about 230 plant species, in Compositae, Cucurbitaceae and seven other families. Both cultivated plants (cucurbits, sunflower, ornamentals) and non-cultivated plants are found among its hosts, the majority of which are common hosts of *E. cichoracearum* and *S. fuliginea* (Blumer 1967). Tobacco is another important host (Cole 1978) and potato is occasionally infected in hot dry seasons.

Host specialization

E. cichoracearum is neither morphologically nor biologically uniform. There is great variability in size of both cleistothecia and conidia on different hosts. A number of forms differ in ability to attack particular host species, and these different strains may vary greatly in their host range. A strong biological specialization was established on composites (Blumer 1967), but not on cucurbits (Nagy 1972). The tobacco strain behaves differently on different *Nicotiana* species and seems to be composed of morphologically indistinguishable strains differing in pathogenicity (Delon 1983). The potato strain will infect tobacco, but is less virulent than the tobacco strain (Hopkins 1956). In actual crop production one plant may or may not be considered a source of inoculum for other species.

Geographical distribution

E. cichoracearum occurs throughout the world, but the form occurring on tobacco is not known in the USA (Lucas 1975).

Disease

E. cichoracearum causes a typical powdery mildew disease, with a white powdery cover on all aerial parts of the plant, including both leaf surfaces. Severely affected leaves become brown and shrivelled, resulting in premature defoliation. On tobacco, the disease always appears on leaves which have almost stopped expanding, and so is not visible for at least six weeks after transplanting. Cleistothecia in the form of black pinpoints occur infrequently but conspicuously. Symptoms due to *E. cichoracearum* are identical with those due to *S. fuliginea*, which may be distinguished on common hosts by the cleistothecial or conidial characteristics.

Epidemiology

Much of the data on the life cycle and epidemiology of *E. cichoracearum* is based on studies of cucurbit powdery mildew. Because two pathogens are involved (cf. *S. fuliginea*), the data require to be critically revised. Parallel studies have proved that the life cycle of the two pathogens is similar, but depending on weather and cultural conditions, one or the other species becomes dominant, according to their ecological requirements (Nagy 1976). Initial field infections are usually caused by airborne conidia that appear in profusion during the growing season, or by ascospores released from cleistothecia. *E. cichoracearum* is favoured by dry atmospheric and soil conditions, moderate temperature (20–25°C), reduced light intensity, fertile soil and succulent plant growth (Yarwood 1957). Germination of conidia occurs at temperatures of 15–30°C, with an optimum at 25°C, in the absence of free water. The conidiophores develop about 5–6 days after infection occurs. Sporulation is favoured by sunny weather and in cool, shady conditions. Wide variations in diurnal and nocturnal temperatures favour rapid mycelial growth. No colonies develop under saturated conditions (Nagy 1976), and rain prevents sporulation. Powdery mildew is also dependent on mineral nutrition, leaf age, susceptibility of the host and interaction with other pathogens. Production of cleistothecia is determined by the nutritional conditions of the host and the sexuality of the fungus. Some strains producing cleistothecia may overwinter as ascospores, while others without cleistothecia survive in the form of mycelia or conidia.

Economic importance

E. cichoracearum is of primary importance on cucurbits (cf. *S. fuliginea* for details), but only in the field, under dry conditions. The susceptibility of the different cucurbits is much as for S. *fuliginea,* except that *Cucumis melo* is tolerant (Nagy 1977). On tobacco, *E. cichoracearum* causes the major disease in some countries (Japan, S. Africa, Zimbabwe). Annual losses of 20–30% are estimated in S. Africa (Cole 1978). In Europe, it tends to be of rather secondary importance, occurring mainly in the Mediterranean countries (Renaud 1959). Attacks are sporadic, but the development of new cultivars more susceptible to the disease, and the use of new fungicides against *Peronospora tabacina*, have tended to favour the occurrence of powdery mildew. Ornamental plants infected with powdery mildew have a reduced decorative value because of mildew cover and premature defoliation.

Control

Cucurbit powdery mildew can be effectively and safely controlled by chemicals if attention is given to correct dosages, proper time of application and adequate leaf coverage. Chemical control with sulphur is very effective, but cantaloupes and some cucumber cultivars are sensitive to sulphur. Some organic compounds (dinocap, chinomethionate) have been developed. These act as eradicants with some curative effects. The first trials with systemic fungicides (benomyl) for the control of cucumber powdery mildew were conducted in 1966, and already in 1968 the resistance of the organism to benomyl was reported. Similarly dimethirimol rapidly became ineffective (Brent 1982). Evolution of resistance has stimulated investigation of other systemic chemicals (bupirimate, fenarimol, pyrazophos, tridemorph) as potential curative fungicides. Essentially the same chemicals are effective on other crops, e.g. on tobacco (dinocap at 12–15 day intervals, benzimidazoles at 10 day intervals). The other way of controlling cucurbit powdery mildew, through the development of resistant cultivars, has been successful for cucumbers and cantaloupes (Sitterly 1978). Resistant tobacco cultivars are widely grown in countries where the disease is most serious.

DELON & SCHILTZ; NAGY

Erysiphe cruciferarum Opix ex Junell

E. cruciferarum attacks members of the Cruciferae and some Papaveraceae. The collective name *E. polygoni* DC. was used by Salmon (1900) for the pathogen of rutabaga, turnip and other crop hosts. This species was reclassified by Blumer (1933) who used the name *E. communis* for those forms whose morphology was indistinctly known and where cleistothecia seldom developed. Further revision was proposed by Junell (1967), who regarded *E. communis* as a *nomen ambiguum*, and used *E. cruciferarum* for the powdery mildew pathogen of a wide range of wild and cultivated crucifers. This species is distinguished from *E. betae* by the length of the cleistothecial appendages and from *E. trifolii* by conidial shape (CMI 251). ***E. polygoni*** in the modern sense is restricted to the Polygonaceae and so is of no economic importance to crop plants (CMI 509). The mycelium of *E. cruciferarum* is white, dense, persistent and spreading; barrel-shaped conidia form from the mycelium either singly or in short chains; cleistothecia are scattered, globose, initially yellow coloured but become brown-black at maturity. The pathogen has a world-wide distribution.

Symptoms begin as star-shaped white lesions on the upper surfaces of foliage. These gradually coalesce until the entire plant is covered by an off-white mealy stroma (Dixon 1978; 1984). Infection of Brussels sprout by *E. cruciferarum* causes loss of the visual appeal of the buttons. Cabbage and cauliflower exhibit two forms of syndrome, first, a necrosis of the outer leaves with obvious powdery mildew lesions accompanied by much reduced curd or head size and second, necrosis of the inner leaves starting at the tips and progressing towards the centre which begins to rot. Rutabaga and turnip are regularly defoliated by *E. cruciferarum.* Infected foliage rapidly turns chlorotic and abscisses. The infection process and possible relationships to rainfall have been studied by Purnell & Preece (1971). Levels of resistance to *E. cruciferarum* in a wide range of crop hosts have been reported by Dixon (1978). Breeding work in the USA resulted in the production of resistant cabbage cultivars (Walker & Williams 1973). Resistance in rutabaga to *E. cruciferarum* may be of two types. One form is expressed early in the invasion process which suppresses fungal development beyond the appressorium stage, the second form is expressed as reduced

or delayed spore production (Munro & Lennard 1982). A range of systemic and eradicant fungicides are available for control of *E. cruciferarum*. The sowing date of various crops may be manipulated in order to avoid the more devastating effects of infection but such measures may be accompanied by a significant loss in yield (Doling & Willey 1969).
DIXON

Erysiphe betae (Vañha) Weltzien

E. betae was also formerly considered to be part of the collective species *E. communis* or *E. polygoni*. It forms an arachnoid to dense persistent mycelium with hyaline conidia in short chains. Cleistothecia form in groups and are globose, yellow at first, becoming dark brown later, with numerous irregular appendages. The ellipsoid to obovate asci contain 3–5 elliptic ascospores (CMI 151). The fungus, which is strictly specific to *Beta* species, is distributed world-wide but is especially important in Europe. The sexual form, known from C. Europe, has only recently been seen in France. The disease is rapidly spread by wind-borne conidia. The cleistothecia appear later, and carry the disease from year to year by persisting in debris, or on the seeds of a seed crop. Conidia germinate best at 70–100% RH and about 20°C. Once infection has occurred, further development on the leaf is favoured by dry air. Accordingly, alternating humidity and dryness favours the spread of *E. betae* (Drandarevski 1969, 1978). Losses of up to 16–20% of sugar yield can be caused in some years. Control can be achieved by spraying immediately after the first infection foci appear (generally by June or July) with a mixture of benomyl and micronized sulphur, or else of carbendazim, maneb and sulphur. Fenarimol is also effective. Two or three further applications may be needed, and these treatments also help control *Ramularia beticola* and *Cercospora beticola*. At present, such fungicide treatments give satisfactory protection of both root and seed crops of sugar beet.
LEBRUN

Erysiphe pisi DC.

E. pisi develops a thin mycelium on the host surface with mainly single barrel-shaped conidia. Globular cleistothecia develop only infrequently, but when produced, may be either gregarious or scattered. Each cleistothecium carries 10–30 appendages which are basally inserted, frequently of a knotty shape but not branched. They are brown coloured and at least as long as the ascocarp diameter and often much longer. Within the cleistothecium are 3–10 asci of varying shapes from ovate to sub-globose and within these are 3–6 ascospores (CMI 155). *E. pisi* can be distinguished from *E. trifolii* (also commonly found on members of the Leguminosae) by the length and shape of the cleistothecial appendages. A wide range of legume crops is attacked by *E. pisi* including: pea (*Pisum arvense* and *P. sativum*), lucerne, vetches, chick pea, pigeon pea, and lentil, as well as various tropical species. Different strains of *E. pisi* attack these crops, and Hammarlund's (1925) biological forms can now be considered as formae speciales of *E. pisi:* **f.sp. *medicaginis-sativae*** on lucerne, **f.sp. *pisi*** on pea and **f.sp. *viciae-sativae*** on vetch. *E. pisi* is widely reported on crops in Australia, Canada, France, India, Peru, S. Africa, UK, USA and Zimbabwe (Dixon 1978).

Symptoms caused by *E. pisi* are similar to those of other powdery mildews; a greyish-white mycelium develops as discrete lesions on the upper leaf surface, and gradually coalesces until the whole leaf is completely colonized and turns chlorotic and necrotic. As infection progresses the pathogen spreads to stems and pods. This powdery mildew can be transmitted by seed. Epidemics are encouraged by prolonged warm dry days and nights cool enough for dew formation (Hagedorn 1973). Several sources of resistance to *E. pisi* have been identified (Heringa *et al.* 1969). Breeding work is in progress in the USA (Gritton & Ebert 1975) and the UK (Dow 1978). In India *E. pisi* causes losses of 20–30% in pod number and 25% reduction in pod weight. Application of sulphur fungicides to seed can reduce losses and seed may also be treated with captan or thiram. Seed production in pathogen-free areas is also recommended. From the research point of view, *E. pisi* is of special interest because its haustorial complexes (haustoria with surrounding host-derived sheath) can be isolated and used for physiological studies (Gil & Gay 1977).
DIXON

Erysiphe trifolii Grev.
(Syn. *E. martii* Lév.)

E. trifolii resembles *E. pisi* but has longer, occasionally branched, cleistothecial appendages (CMI 156). It causes powdery mildew on a distinct group of leguminous hosts (*Trifolium* especially *T. pratense, Lathyrus* including *L. odoratus, Onobrychis*, fodder species of *Lupinus* etc.). Formae speciales have been recognized on *Trifolium, Lathyrus, Melilotus* and

Lotus, and pathotypes on *T. pratense*. Infected leaves carry a conspicuous heavy load of greyish fungal growth bearing conidial chains, but remain green until almost completely covered. They later turn yellow and shrivel. Plants are stunted and may wilt in hot weather (O'Rourke 1976). Cleistothecia appear late in the season in large numbers, but may be more important for genetic reassortment than for overwintering, since *E. trifolii* can readily survive on its perennial hosts (Carr 1971). Infection by conidia is possible over a wide humidity range and sporulation is favoured by moderately high temperatures and low humidity, and inhibited by rain (Carr 1971). The disease is widespread in Europe, on clover, causing damage mainly in warm dry weather in the summer months (O'Rourke 1976; Dixon 1978). Possibilities for control lie mainly in the use of resistant clover cultivars (Dixon 1978). *E. trifolii* is also common on the ornamental *Lathyrus odoratus*, grown under glass or outdoors, on which it can be controlled with fungicides; it is potentially damaging to the increasingly widely cultivated fodder lupins (Wheeler 1978).
CARR & SMITH

Erysiphe graminis DC.

Basic description

E. graminis differs from other *Erysiphe* species in its large cleistothecia (over 150 μm) with rudimentary appendages, containing up to 25 large, mostly 8-spored, asci. The haustorial appendages project in a digitate manner from either end of the haustoria, which lie along the length of the epidermal cells, with the haustorial neck asymmetrically placed towards one end. The secondary mycelium bears numerous thick-walled, rather rigid bristles (CMI 153).

Host plants

Probably almost all of the festucoid grasses and cereals of Europe are hosts of *E. graminis*. Other groups of Gramineae, exemplified by maize, millet, rice, sorghum, are not.

Host specialization

Formae speciales can be clearly distinguished on cultivated wheat, barley, oats and rye (**f.sp.** ***tritici, hordei, avenae*** and ***secalis*** Marchal respectively) and do not cross infect on these hosts. Many other formae speciales have been described on wild and pasture grasses. However, extensive cross-inoculation experiments in Israel (Wahl *et al.* 1978) have shown that specialization is not so strict in *E. graminis* from natural stands of wild grasses, where mildew cultures can frequently be found capable of infecting a wide range of species and genera. Since formae speciales can be intercrossed (Hiura 1978), there seems to be no reason to assume that the genetic mechanisms involved in specificity of formae speciales are different in nature from those controlling specific host-parasite interactions within formae speciales. European cereal mildew populations appear to have a simpler genetic structure and behave more as formae speciales. In general, the forms on wild grasses do not infect cereals, except in the case of a close genetic relationship between a cereal and its wild relative, as for example between *Hordeum spontaneum* and cultivated barley.

Geographical distribution

E. graminis occurs throughout Europe and worldwide.

Disease

Powdery mildew colonies occur most often on the leaves, especially on the upper surface of lower leaves (APS Compendia). They can occur on any of the above-ground parts (sheaths, glumes, even awns). Depending on the host and parasite genes actually interacting, a wide range of host responses to infection is possible, ranging from complete absence of macroscopically visible symptoms, differences in number, size, colour and shape of lesions, different levels of susceptibility at different stages and of different organs etc., to complete cover of the leaf surface by mildew. However, *E. graminis* does not generally kill its host.

Epidemiology

Cereal mildew in Europe overwinters principally on living host plants (Jenkyn & Bainbridge 1978). Winter crops provide a 'green bridge' from one season to the next. The primary infection of spring crops can originate from nearby winter crops, but also from distant sources. The resistance genes, present in the crop, select from the aerial pool spores with matching virulence. The race structure of a forma specialis in a given region thus reflects to a large extent the resistance genes present in the local cultivars. However, winds also modify the composition of *E. graminis* populations, due to a steady drift-in of races from other regions, so that unnecessary virulences may accumulate to a considerable extent. Refined and efficient techniques are now available to trace such effects (Limpert *et al.* 1983). On mature diseased plants, cleistothecia can be found in abundance. The ascospores within the cleistothecia can survive in a dry state for a long time and be released under humid conditions. This mechanism ensures the survival of mildew on wild grasses during the hot dry summers in the Mediterranean region. It is, however, not known to what extent ascospores contribute to the survival of cereal mildew in temperate climates between the harvest of spring crops and emergence of winter crops, since volunteer plants and long-distance spore transport by wind also provide a bridge between seasons.

Infection by *E. graminis* is optimal at 15–20°C, but can occur over a wide range from 5 to 30°C. The infection cycle from conidial germination to sporulation is normally 7–10 days. Infection by conidia requires 98–100% humidity but is inhibited by free water. Sporulation and spore dispersal occur most readily under relatively dry conditions. Powdery mildew is thus favoured by an alternation of wet and dry conditions such as occur very typically in N.W. Europe. In more continental or southern climates, conditions tend to be too hot and dry in the summer months for *E. graminis* to spread or be seriously damaging. On grasses, as on cereals, powdery mildew frequently develops in spring and summer and is favoured by high soil nitrogen, poor air circulation, and shading. Mildewed plants become more sensitive to stress (drought, low temperature, etc.) (O'Rourke 1976; APS Compendium of Turfgrass Diseases). In the winter crop mildew attack decreases frost hardiness due to the shortage of assimilates that are necessary for hardening. In general, early attacks, if serious, can reduce tillering and thus finally ear number while later attacks will reduce the green leaf area available at the time of grain-filling and so grain yield. These effects have been extensively studied in barley (and to some extent in wheat and oats) particularly in relation to the timing of fungicide applications (Griffiths 1981). They are extensively reviewed by Jenkyn & Bainbridge (1978), who also analysed the research on the effects of mildew on yield. The Large and Doling formula gives percentage yield loss as about 2.5 (barley or oats) or 2.0 (wheat) times the square root of the percentage leaf area infected at a late growth stage. This may overestimate losses at high mildew levels, and Jenkins & Storey (1975) have developed a quadratic curve based on a cumulative mildew index through the season, which can account for 76% of the total variation.

Economic importance

In Europe, barley is most seriously affected. Effective control with fungicides (cf. below) can increase ear number by 20% or more, and grain size by 5–10%. Losses in wheat, oats and rye can be as severe but the incidence of powdery mildew is not generally so high on these crops. *E. graminis* is also common on pasture and turf grasses, but the extent of losses is difficult to evaluate. Seed yields can be seriously reduced, and yield and quality of ryegrass has been shown to be affected (Davies *et al.* 1970).

Control

For many years, the use of resistant cultivars was the only satisfactory method for the control of *E. graminis*. Since the Second World War, a number of resistance genes have been used in barley, and have been successively overcome by the appearance of pathotypes with matching virulence. This is one of the classic examples of this type of relationship between host and parasite (Wolfe & Schwarzbach 1978). In general, new resistant cultivars have been regularly available but farmers' preference for a limited number of successful cultivars has tended to 'over-expose' them, with corresponding large shifts in the *E. graminis* population and a conspicuous loss of resistance in the field. Barley powdery mildew is thus one of the main diseases for which diversification strategies are proposed based on the principle of deploying many cultivars with different resistance genes in space (or time). Only a proportion of the *E. graminis* population can then attack each. In practice, selection appears to operate against the forms with multiple virulence, on cultivars with single resistance. Priestley (1981) recommends field-by-field diversification, with cultivars chosen from different diversification groups, while Wolfe *et al.* (1981) have shown that use of mixtures of cultivars from these groups will reduce powdery mildew incidence, and increase and stabilize yields, by comparison with the mean of the cultivars separately. Cultivar mixtures are now being marketed and used to a limited extent in the UK, Denmark and GDR. Another strategy is directed towards the 'green bridge' between the growing seasons, based on different resistance genes in spring and winter crops (Slootmaker *et al.* 1984).

Since 1969, fungicide treatment against *E. graminis* has become routine in N.W. Europe. The availability of first tridemorph and ethirimol, and subsequently a great variety of compounds and formulations aimed at overall disease control on cereals, has resulted in a transformation of the whole pattern of fungicide use in Europe. The problems of timing and number of applications on individual fields, in relation to the inherent risk, to observations in relation to thresholds, and to meteorological data have led to a variety of different types of warning system in European countries (Zadoks 1981; Cook & Yarham 1985; Curé 1985) in which *E. graminis* takes its place alongside the other cereal diseases. *E. graminis*, shows variation in sensitivity to fungicides and forms with moderate resistance to the sterol-biosynthesis inhibitors are now becoming common (Wolfe 1985a), so that problems of loss of control are now beginning to arise. Wolfe (1985b) proposes that the diversification strategies advocated for cultivars with powdery mildew resistance should be integrated with diversification of fungicide use.

Any long-term control strategy, including diversification schemes, breeding activities and fungicide use, should be based on adequate knowledge of the regional and large-scale genetic composition of the mildew population and its changes in time. Methods and devices have been developed that make description and monitoring efficient. The concept of 'virulence analysis' (Wolfe & Schwarzbach 1975) describes the populations in terms of relative frequencies of particular virulences making the definition and determination of pathotypes largely unnecessary. The Jet spore trap collects mildew spores from large air volumes onto living leaves. When mounted on a vehicle, representative samples of mildew from different regions can be obtained in a short time (Limpert & Schwarzbach 1982). These techniques are now being used in a European Community project in several countries.

SCHWARZBACH & SMITH

Erysiphe heraclei DC. (syn. *E. umbelliferarum* de Bary) infects umbellifers, in particular carrot, celery, fennel, parsnip, parsley and various herbs. It causes little damage in N. Europe, but can be serious on winter crops in Mediterranean countries (Dixon 1978; 1984). It may exist in host-specialized forms, and there is potential for resistance breeding in carrot (Bonnet 1983). ***E. ranunculi*** Grev. occurs most frequently on wild Ranunculaceae but has recently attracted attention on anemones (Price & Linfield 1982).

Leveillula taurica (Lev.) Arnaud

L. taurica is a powdery mildew with an endoparasitic habit, entering the host plant through stomata and not via the cuticle (CMI 182). The pathogen is widely distributed, but is more frequently found in dry areas of Europe, C. and W. Asia, and notably around the countries of the Mediterranean basin. The pathogen is not physiologically specialized since several Solanaceae (green pepper, aubergine, potato and tomato), Cucurbitaceae (cucumber, courgette), Malvaceae (cotton okra), Compositae (artichoke) and some woody perennials such as olive are affected. The most comspicuous symptom of the disease appears on the upper leaf surface of the mature leaves as scattered chlorotic spots among the veins, while the white powdery mass of the pathogen occupies the lower leaf surface. Chlorotic spots, followed by necrotic ones, sometimes lead to leaf fall (pepper and olives in nurseries can be severly defoliated). The pathogen is wind-borne and many outbreaks of the disease are related to dry conditions. However humid weather does not necessarily prevent the spread of the pathogen. Foliar sprays at 20-day intervals with fenarimol, triadimefon, triforine or nuarimol give excellent control.

TJAMOS

Microsphaera alphitoides Griffon & Maubl.

M. alphitoides forms elliptical or barrel-shaped conidia, and cleistothecia about 180–200 μm across, each containing up to 15 asci with up to 8 colourless, ellipsoid ascospores. It causes powdery mildew mainly on *Quercus* species, but is also found occasionally on *Fagus sylvatica* and *Castanea sativa*. Species of *Quercus* vary in susceptibility to *M. alphitoides*; among the European oaks, *Q. robur* is very susceptible, and *Q. petraea* less so. Within species, only slight differences in susceptibility occur between provenances (Leibundgut 1969). This mildew occurs throughout Europe and into Asia as far as China. It forms a powdery white covering of mycelium and conidia on young twigs and on the leaves, which, in severe attacks, shrivel, become brown, and fall prematurely. Cleistothecia usually appear only at the end of hot dry summers. The fungus overwinters as mycelium in the buds. In spring this mycelium forms conidia which infect the young oak leaves. Further crops of conidia spread the disease throughout the summer.

M. alphitoides may cause serious damage to nursery plants, reducing growth, increasing the number of unusable plants, and even killing seedling trees. It may also be serious on young trees newly planted in forests, affect coppice regrowth, and sometimes inhibit natural regeneration. It is usually unimportant on older trees, though periodically it has played a part in a complex of attack in which weather conditions, various insects and fungi have caused the dieback and death of oaks in many areas. Control, apart from the possible replacement of *Q. robur* by the less susceptible *Q. petraea*, is generally feasible only in the nursery, where the plants may be sprayed with colloidal or wettable sulphur, dinocap or benomyl (Marchetti & D'Aulerio 1980).
PHILLIPS

Microsphaera hypophylla Nevodovskii (syn. *M. sylvatica* Vlasov) is another powdery mildew of oak, attacking the leaves of *Quercus robur* in Scandinavia, Austria, Switzerland and USSR. It originally seemed quite distinct from *M. alphitoides* but this difference is no longer aboslutely clear.

Microsphaera grossulariae (Wallr.) Lév

M. grossulariae attacks gooseberry and is known as European gooseberry mildew. It is less common than *Sphaerotheca mors-uvae*. It is rarely found on currant bushes, but has also been recorded on some *Sambucus* species. The pathogen occurs mainly on the upper surfaces of leaves but may occasionally be found on the lower leaf surface or on berries. Symptoms may be seen from early May. The mycelium persists throughout the season but always remains thin and scanty. It is usually found on densely planted, shaded bushes. Black cleistothecia, scattered or in groups, may be found fully exposed on the mycelial patches. In autumn, they fall off leaves or drop to the ground with leaves and remain dormant during winter. *M. grossulariae* is almost innocuous causing little damage to bushes, but inducing some premature leaf-fall. It can be controlled following the cultural and chemical treatments recommended for control of *Sphaerotheca mors-uvae*.
JORDAN

Microsphaera begoniae Sivan is better known as its anamorph ***Oidium begoniae*** Puttemans, and in particular its large single-spored form **var. *macrosporum*** Mendonca & Sequeira. Infection of *Begonia* species can be severe, and requires fungicide treatment. The status of the small-spored form of *O. begoniae,* and its association with an *Erysiphe* or *Microsphaera* teleomorph, remains confused (Wheeler 1978). A number of other *Microsphaera* species occur on woody plants, and the forms on each host are differently attributed to the various species by different authors. ***M. penicillata*** (Wallr.) Lév. (syn. *M. alni* of various authors) can be taken in a broad sense (CMI 183) to cover the forms on alder, lilac, birch etc. It is widespread on alder in N. America and Europe. A serious outbreak occurred in 1948–50 on lilac in Europe, probably due to importation of a specialized form from N. America (Wheeler 1978). It is interesting to note a similar outbreak, similarly explained, on oak in Europe in 1907 (CMI 183), and now the appearance in Europe of ***M. platani*** Howe (probably the same as the fungus attacking plane in N. America and there called *M. alni*), which has become extremely widespread on plane in S. Europe (France, Greece, Italy) in the last few years (Gullino & Rapetti 1978). Individual forms may be specialized to plane species (Ialongo 1981a). This species has become a serious problem in Mediterranean countries, causing unsightly damage and defoliation. Triadimefon sprays have been recommended for control. *Euonymus japonica*, a widely grown ornamental shrub is frequently and conspicuously attacked by ***M. euonymi-japonici*** Viennot-Bourgin, and hydrangeas by ***M. polonica*** Siemaczko (anamorph ***Oidium hortensiae*** Jørstad) (Wheeler 1978). The former is said also to attack citrus (Boesewinkel 1981). ***M. viburni*** (Duby) Blumer is common on *Viburnum* and ***M. lonicerae*** (DC.) Winter on ornamental *Lonicera* species, especially *L. tatarica*.

Phyllactinia guttata (Wallr.) Lév

(Syn. *P. corylea* (Pers.) P. Karsten, *P. suffulta* (Rebent.) Sacc.)

P. guttata, which can be divided into numerous formae speciales, forms sparse mycelium on the leaves of various trees and shrubs. The cleistothecia, which are black at maturity, contain many asci, which usually each contain two ascospores (CMI 157). *P.*

guttata has a wide host range on many broad-leaved trees. In Europe, it is most common on *Corylus,* but may also occur on *Acer, Alnus, Betula, Castanea, Diospyros, Fagus, Morus, Pistacia, Populus, Prunus, Pyrus, Salix, Ulmus* and other broad-leaved trees. It occurs throughout Europe and across N. America.

The fungus may cause some defoliation, especially on hazel and beech hedges. The disease usually first becomes visible in late August as white mycelial patches, most often on the undersides of the leaves. A powder of white conidia is formed (and further spreads the disease) and the patches spread until often the whole leaf is covered. Later, towards the end of September, numerous cleistothecia arise all over the underside of the leaf. The cleistothecia carry the fungus over the winter. They are levered off the leaf surface by movements of the equatorial appendages, and become attached to adjacent substrates, including the buds of host plants, by the mucilage formed by their apical hyphal branches. When mature, they split open at the equator to reveal the asci, which then shed their spores (Cullum & Webster 1977). *P. guttata* rarely causes severe damage to its host plants. Control measures are therefore seldom required, but if necessary infected plants may be sprayed with colloidal or wettable sulphur or with dinocap.

PHILLIPS

Podosphaera leucotricha (Ell. & Ev.) Salmon

Basic description

The cleistothecium of *P. leucotricha* has long, bristly apical appendages and a single ascus. The ellipsoid-globose conidia contan fibrosin bodies and develop as chains on conidiophores with simple, non-swollen bases (CMI 158).

Host plants

P. leucotricha attacks *Malus* species, including cultivated apples, pear, quince, *Mespilus germanica* and *Photinia* species. There are reports of attack on almond and peach.

Host specialization

Despite regional differences in the relative susceptibilities of some apple cultivars there is no evidence of specialization on species or cultivars.

Geographical distribution

P. leucotricha occurs world-wide and is common in most apple-growing areas (CMI map 118).

Disease

On apple, *P. leucotricha* grows ectophytically on leaves, shoots, blossoms and young fruits. In spring, floral and foliar organs emerge totally or partially coated with powdery mildew from late-opening buds in which conidia and hyphae have overwintered; this is *primary mildew.* Infected fruit buds produce dwarf leaves and small tight flower buds that remain closed, and these organs wither in late spring. Infected axillary buds produce mildewed leaf clusters which soon wither. Infected vegetative terminal buds produce totally mildewed shoots; those on twigs that were not totally colonized the previous year manage to grow as 'white' shoots for several weeks until, much weakened, they shed leaves. During spring and summer, new conidial infections give rise to colonies of *secondary mildew;* these are discrete, variously shaped, powdery in dry weather, coloured white--buff-grey and occur mainly on the lower surface of leaves, although individual colonies may extend onto the upper surface (causing chlorotic spotting on some cultivars). Leaf infections first become visible on young unrolled leaves. Attacked leaves may be distorted, have a wavy margin and sometimes a yellow spot on the upper surface opposite a colony. Severely mildewed leaves turn brown and brittle, and fall. Secondary mildew sometimes colonizes the shoot tip, causing all subsequent growth to be totally mildewed; this is a serious problem on nursery plants. Newly-formed buds may be attacked by hyphae or conidia throughout the growing season, but only before bud scales harden and seal. Fruitlets of some cultivars (e.g. Jonathan) are susceptible, colonies leaving a web-like russet on the skin in this case. In winter, shoots that were totally colonized by primary or secondary mildew the previous spring and summer show as 'silvered' twigs and spurs with thin, pointed and loosely scaled buds. Infected buds on otherwise healthy twigs may be similarly distorted. 'Silvered' twigs and spurs are dead or have many dead buds; their viable buds in spring produce only feeble growth. Dark cleistothecia, partially embedded in mycelium, may form during the growing season, mostly on any severely mildewed wood from the current season but also in terminal buds. *P. leucotricha* is heterothallic.

Epidemiology

Many aspects of apple powdery mildew are discussed by Roosje & Besemer (1965) and Butt (1978). Conidia from primary mildew initiate the annual epidemic of secondary mildew on apple, and infected terminal buds are usually the source of most of this primary inoculum. The residence of primary mildew on the trees ensures early epidemics. Cleistothecia have no role. The quantity of primary mildew depends upon the intensity of bud infection—and hence the intensity of secondary mildew—in the previous year, and upon winter temperatures: a few hours at −20 to −25°C (or even −12 to −15°C in late winter) can almost eradicate the overwintering fungus. Infection periods cannot easily be characterized because liquid water is not needed; in England, conidia infect daily in summer if susceptible leaves are present, even in exceptionally cool, wet periods. This independence from wetness makes disease forecasting difficult. Leaf resistance increases rapidly with leaf age, most new infections occurring at the shoot tip on rolled leaves. Infection is favoured by high humidity, with an optimum near saturation. Although liquid water is harmful to conidia and hyphae, leaf infection is favoured by rain and dew, presumably because hairs shield deposited conidia from the droplets. The colonies grow as the young leaves expand and become visible after about 17 days.

Sporulation begins during incubation, with latent periods as short as 5 days. The dynamics of the shoot and the pathogen result in a chronological succession of colonies down the shoot, most sporulating colonies being on unrolled leaves near the shoot tip. Conidia mature faster in light than in darkness, so that they are mostly available for dispersal at times of turbulent air movement and low humidity. Accordingly, the concentration of airborne conidia changes diurnally. The onset of rain can cause a temporary increase of airborne conidia at any time, but the main effect of heavy rain is to damage the superficial mycelium, and consequently to reduce sporulation for the next 2–3 days.

Surveys in England have revealed large differences between neighbouring orchards in the intensity of mildew due to differences in tree age and size, cultivar and control measures. The course of a secondary mildew epidemic is determined by the quantity of primary mildew, times of shoot growth, weather in spring and summer and the resistance of the cultivar. In England, apple mildew spreads rapidly when anticyclonic conditions give warm, sunny days and dew; high night humidity particularly favours the disease whereas frequent rain is harmful. Epidemics are slow to start when April and May are cool, wet months and end suddenly when the daily maximum screen temperature is *c.* 30°C and the atmosphere becomes unusually dry. Descriptions of the type of weather associated with severe mildew range from hot and dry, to warm and wet and to cool and dry; these differences probably reflect the sensitivity of atmospheric humidity to crop environment and regional topography. Cultivars of apple differ in partial resistance, with Discovery being highly resistant and Crispin and Jonathan very susceptible.

Economic importance

Although there are regional differences in apple mildew epidemics in Europe, particularly due to climatic differences in winter temperature, summer temperature and summer humidity, fungicides are generally used intensively to control this disease. On a spur-bearing cultivar an infected fruit bud usually results in invasions of the bourse shoot in spring and subsequent colonization and death of the spur. This direct loss of fruit-bearing sites is additional to the reduction in vegetative growth) and hence potential cropping sites, due to leaf damage and the killing of twigs and buds. Mildew-induced russet of fruit skins reduces crop quality. Damage to cv. Cox on M.9 rootstock has been studied over eight years (Butt *et al.* 1983): an increase in the mean annual midsummer incidence of disease from 2% to 19% mildewed leaves resulted in reductions of 32% shoots/tree, 19% leaves/shoot, 27% fruit size, 15% crop weight and 33% crop value. The damage thresholds for these variates on Cox are below 10% mildewed leaves. The study indicated higher damage thresholds for Golden Delicious and Jonathan. Clearly, while damage soon becomes obvious when trees are left unprotected from mildew, losses can be suffered at relatively low intensities of mildew. Reduced fruit size is especially serious in the market.

Control

Sulphur was formerly used in apple orchards but modern chemicals have proved 'kinder' to cultivars like Cox, although Golden Delicious tolerates sulphur. Some newer fungicides are binapacryl, bupirimate, fenarimol, nitrothal-isopropyl, triadimefon, myclobutanil and penconazole. There is no evidence of *P. leucotricha* developing tolerance to fungicides. Sprays are generally applied every 7, 10 or 14 days from before flowering until late summer when most extension growth has ceased. This intensive schedule is necessary on a susceptible cultivar to cope with the resident primary mildew, the long period of production of susceptible leaves and buds and with the ability of conidia to infect almost daily. Although routine, many spray programmes are rational: in England, antisporulant fungicides applied before flowering reduce primary inoculum, thiophanate-methyl during flowering minimizes risks to pollen, and fungicides from flowering until late June are chosen to avoid the russeting of fruit skins; after June the skins are less sensitive to chemical damage. Also, spray intervals are usually reduced when shoots are growing fast and when buds (as well as leaves) need protection (Butt 1972). It is possible to both reduce the cost and improve the standard of control by using fungicides in accordance with weather, shoot growth and the incidence of infectious mildew (Butt & Barlow 1979). Side-effects of fungicides influence their choice. A chemical with acaricidal properties (e.g. binapacryl) is, for example, not compatible with the biological control of fruit tree red spider mite (*Panonychus ulmi*) using predacious mites. Also, some ergosterol-biosynthesis inhibitors (e.g. triadimefon) can have undesirable growth regulatory effects.

A winter spray with a bud-penetrating surfactant, alone or mixed with a proprietary fungicide, markedly reduces overwintering mildew. In former times much overwintering mildew was removed during drastic winter pruning. Nowadays, pruning is much lighter. Tree size is much smaller, however, making feasible the efficient removal by hand of primary mildew in spring; in view of the pathogen's high multiplication rate, sanitation has to be extremely thorough to significantly delay an epidemic of secondary mildew. It is important to site new orchards away from neglected plantings. Improved mist-blowers, nozzles producing fine drops and trees on dwarfing and semi-dwarfing rootstocks have made possible a shift from high-volume spraying (*c.* 2000–3000 l/ha) to medium (*c.* 500 l/ha) and very low volumes (*c.* 50 l/ha). Control is possible at 50 l/ha in intensive orchards when environmental and host factors are mitigating. There is growing interest in electrostatic orchard sprayers. A revolution in chemical control is promised by a new fungicide capable of controlling apple mildew for a complete season following a single soil drench in spring.

Interspecific crosses have successfully transferred major resistance genes from *Malus zumi, M. robusta* and other species into apple, but derived selections have not yet entered commercial production.

BUTT

Podosphaera tridactyla (Wallr.) de Bary differs from *P. leucotricha* in having very long, branched appendages (CMI 187). It is of little importance, occasionally causing damage to plum in nurseries (Burchill 1978) and occurring quite commonly on apricot, where it may be found with *Sphaerotheca pannosa*, from which it can be distinguished by conidial characteristics (Boesewinkel 1979) and by abundant cleistothecium formation (Ialongo & Simeone 1984). ***P. clandestina*** (Wallr.) Lév. (syn. *P. oxyacanthae* (DC.) de Bary), of which the previous species is sometimes considered a variety, can cause severe defoliation of hawthorn and occasionally attacks cherry, pear and quince (CMI 478, Burchill 1978). Its epidemiology has recently been extensively studied (Khairi & Preece 1978).

Sphaerotheca fuliginea (Schlecht.) Pollacci

Basic description

S. fuliginea is a typical member of the genus *Sphaerotheca*, with hypha-like appendages on the cleistothecia and one 8-spored ascus in each. It occurs generally as the anamorph (***Oidium erysiphoides*** Fr.) very similar to that of *Erysiphe cichoracearum*. The conidia are distinguished by their elliptical form, by having fibrosine bodies and extra, often forked germ tubes growing out from the side of conidia (Nagy 1970).

Host plants and specialization

The hosts cover about 150 plant species, mostly in the same family and in common with *E. cichoracearum* (q.v.). Like this fungus, *S. fuliginea* shows great variability in the size of cleistothecia and conidia on different hosts. It also has a number of forms differing in the ability to attack particular hosts. *S. fuliginea* was divided by Junell (1966) into twelve species based on morphological characteristics correlated with host specialization, *S. fuliginea* being reserved only for forms on *Veronica* species. Cucurbits, the most important host plants, were not, however, included in Junell's list.

Geographical distribution

S. fuliginea occurs throughout the world.

Disease

cf. *E. cichoracearum* (p. 256).

Epidemiology

The life cycle of *S. fuliginea* is similar to that of *E. cichoracearum* but the ecological requirements of the two pathogens are quite different. *S. fuliginea* requires a temperature of 20–30°C with an optimum at 22°C and 100% RH for the germination of conidia; during infection and sporulation, it tolerates high RH (Nagy 1976). Conditions for producing cleistothecia and overwintering are identical with those described for *E. cichoracearum*.

Economic importance

The most striking aspect of the loss caused to cucurbits by *S. fuliginea* (and *E. cichoracearum*) is a reduction of quality rather than yield, consisting of reduction in the assimilation surface and premature defoliation of plants. When the foliage is heavily infested, fruits may ripen prematurely and lack flavour. This reduced quality constitutes the chief damage for cantaloupe. Reduction in yield is proportional to time and severity of disease development. *S. fuliginea* infects cucurbits both in the field and in glasshouses, since it tolerates high RH. In humid weather conditions and on irrigated fields, *S. fuliginea* becomes the dominant species, and in glasshouses it is the only powdery mildew pathogen on cucurbitaceous plants. This species is of world-wide importance in glasshouses (Nagy 1976). Susceptibilities on the cucurbits are different: melon is very susceptible, cucumber and courgette are susceptible, watermelon is very resistant, *Cucurbita ficifolia* and *C. maxima* are immune to *S. fuliginea*. Cucumber and courgette are susceptible to both pathogens to the same extent, melon is more susceptible to *S. fuliginea* than to *E. cichoracearum* and watermelon is more susceptible to *E. cichoracearum* than to *S. fuliginea* (Nagy 1977). On ornamental plants, *S. fuliginea* causes reduction in aesthetic value. Sunflower is attacked by *E. cichoracearum* and *S. fuliginea* only at the end of the vegetation period.

Control

cf. *E. cichoracearum*.

Special research interest

Although cucurbit powdery mildew has been very intensively investigated, there is still confusion about its correct identification. Very few investigations of the two pathogens have been carried out in parallel. The genetic background of resistance to cucurbit powdery mildew has been widely investigated but it is still not known whether or not there is a correlation between the inheritance of resistance to the two pathogens.

NAGY

Sphaerotheca mors-uvae (Schwein.) Berk. & Curtis

S. mors-uvae attacks most *Ribes* species being most severe on gooseberry and frequently damaging on blackcurrant (Fig. 36). The pathogen reached Europe (S.W. USSR) in 1890, from American gooseberry nursery stock (hence its common name of 'American gooseberry mildew'). It was recorded on gooseberry in N. Ireland in 1900 and on black- and redcurrent in England in 1908. On gooseberry, symptoms may be seen from April as a white weft of fungal threads, occurring in patches on young leaves, shoots and berries. Similar symptoms occur on blackcurrant, but the disease is seldom damaging on berries. With age, the mycelium becomes thickly felted and brown. During the summer and autumn, small black cleistothecia may be seen embedded in this felted mycelium. If infection is severe, shoot growth is restricted, shoot tips die, leaves become distorted and small, and fall prematurely. On gooseberry, berries may become completely enveloped by mycelium.

Fig. 36. *Sphaerotheca mors-uvae* on blackcurrant cv. Baldwin.

On gooseberry, *S. mors-uvae* overwinters mainly as mycelium within buds. It can sometimes persist as mycelial strands on previously infected plant parts. Infections from ascospores have also been reported. On blackcurrant, primary infections are initiated in spring by ascospores discharged from cleistothecia formed on previously diseased stems or fallen leaves. The cleistothecia require moisture for differentiation and discharge (Merriman & Wheeler 1968) and an atmospheric RH of at least 97% for germination of ascospores. Once first infections are established, subsequent spread is rapid due to readily dispersed conidia. Conidia of *S. mors-uvae* need more than 10 h for germination to be completed, and germinate best at 18°C in a saturated atmosphere. Disease is favoured by relatively dry atmospheric and soil conditions, moderate temperatures and good illumination. Severity of attack is positively correlated with increased plant vigour. Losses on gooseberry are mainly due to decreased berry size and disease-induced skin blemishes that reduce their market value. On blackcurrant, attacks during flowering may cause the young berries to fall. However, losses mainly arise from decreased productivity of fruiting shoots in the following year. Severe leaf infection decreases flower initiation and the number of fruiting buds. Thus, few flowers or fruit develop on shoots severely diseased the previous summer. Selective pruning of diseased gooseberry shoots decreases disease. Control by the use of chemicals on both hosts can be achieved in two ways: by dormant season sprays of tar/petroleum or phenolic materials to decrease the surviving inoculum, or by repeated sprays during spring and summer to protect leaves and stems. A range of fungicides is currently used for protection: there is an increasing use of triadimefon and triforine balanced by a decreasing use of thiophanate-methyl, dinocap and quinomethionate. Resistant cultivars are also available.
JORDAN

Sphaerotheca humuli (DC.) Burrill

Basic description

Originally attributed to the aggregate species *S. castagnei* Lév., the fungus causing hop powdery mildew was firmly ascribed to *S. humuli* by Salmon (1900). It is often combined with the fungus causing strawberry powdery mildew (*S. alchemillae* q.v.) into *S. macularis* (Wallr.) Jacz. but it can clearly be distinguished by slender, rather than flexuous cleistocarp appendages and by its high degree of specialization to the hop. A typical powdery mildew, the conidia of *S. humuli* are barrel-shaped each germinating by up to four germ tubes.

Host plants and specialization

S. humuli is specific to wild and cultivated forms of the hop. Apparently without selection pressure, marked variation in virulence occurred within natural populations before the introduction of resistant cultivars to English hop growing in 1971. Five races are now recognized from their differential responses to major resistance genes in hop cultivars and breeding lines.

Geographical distribution

S. humuli is known to have been prevalent in England at least as long ago as 1700 and was destructive to hops throughout W. Europe and Russia in the late 1800s. In 1912 it caused serious crop losses in New York State (USA) but has apparently never occurred in the Pacific North West where hops are now grown. Throughout Europe it has since frequently been troublesome, especially in the last 20 years, due to changes in hop management practices and the cultivars grown.

Disease

S. humuli causes hop powdery mildew or 'mould' (Royle 1978). The pathogen overwinters both as cleistothecia in leaf and cone debris scattered over the ground and, increasingly with lack of tillage and the use of herbicides, as a mycelium within dormant rootstock buds at the soil surface (Liyanage & Royle 1976). It attacks all parts of the annual climbing growth producing typical white, powdery sporulating pustules on leaves, smothering shoots with mycelium, and distorting the flowers and young cones (the 'white mould' form of disease). Pustules on young leaves are often associated with raised blisters which, when coalescent, also cause distortion. Leaves become less susceptible as they age. Cleistothecia (causing the 'red mould' form) appear mainly on cones from July, interspersed among conidiophores. In late attacks, from mid-August, maturing cones can become discoloured and prematurely ripen, and abundant cleistocarps are found associated with a thin mycelium.

Epidemiology

The summer disease cycle originates either from

ascospores discharged in wet conditions from cleistocarps on the soil surface, or from conidia produced on diseased basal shoots which have arisen from infected rootstock buds. If not controlled, secondary infections in spring allow a gradual build-up of inoculum until July when infection frequency increases markedly due to higher temperatures. Like other powdery mildews, conidia of *S. humuli* can infect on most days, and are not dependent upon specific weather factors. Occasions when there is no infection are uncommon and they coincide with conditions (e.g. in rain) which inhibit inoculum production. Sporulation in *S. humuli* fluctuates diurnally and on dry days conidial release peaks at 13.00 h, with usually none at night.

Economic importance

Signs of powdery mildew in a hop garden have always alarmed growers because of its ability to render a crop unsaleable. As recently as 1971, before selective systemic fungicides were introduced, the disease caused many growers to abandon large proportions of the crops. Even though it continues to be a problem throughout W. Europe, the true scale of losses has never been measured satisfactorily and there is little information on damage relations. Direct attacks on flowers and young cones probably cause largest reductions in crop weight, in their content of alpha acid (the most important hop ingredient, which gives beer its bitterness) and in their market valuation. However, later attacks also impair yield and quality and in addition cause cones to disintegrate during machine harvesting.

Control

Fungicidal control against *S. humuli* has been practised since the mid-19th century when up to three applications of sulphur dusts were made in a season. Sulphur remained the basis of control until the 1960s, and, though still popular, was eventually superseded by dinocap which, in response to a worsening disease situation, increased steadily in usage until the early 1970s. Introduction of pyrazophos in 1972 helped to stem serious annual attacks, and in recent years further new materials have been introduced to control programmes which are still based on frequent, regular applications starting early in spring. Considerations of yield depressions and fungicide residues now encourage fungicides to be chosen according to the times of season when they are both effective and otherwise acceptable. Thus, a typical programme may involve early applications of pyrazophos, then of triadimefon up to flowering and triforine to protect the cones until harvest. Although not used directly for control of powdery mildew, copper fungicides used against hop downy mildew (*Pseudoperonospora humuli*), are known to have some useful side-effects against *S. humuli*. Conversely, dithiocarbamates favour the disease and their adoption in place of copper in the 1960s is held as one factor which favoured less effective powdery mildew control at that time. In recent years control has been greatly aided by resistant cultivars which were introduced in 1971 and now occupy a large proportion of the English hop acreage. Resistance in these cultivars is based mainly on major genes one of which, in some areas, has now succumbed to virulent pathogen races.

ROYLE

***Sphaerotheca alchemillae* (Grev.) Junell**

S. alchemillae infects strawberry, raspberry and other Rosaceae. It has frequently been combined with *S. humuli* (q.v.) into *S. macularis* (Wallr.) Jacz., in which case the form on strawberry is its f.sp. *fragariae* Peries. Leaf infection of strawberry is very common, the mildew being found mainly on the lower surface. The leaves then roll up, characteristically exposing the white mildewed area. Infected flowers are killed and infected fruit fails to ripen or store normally. Though cleistothecia are occasionally found, perennation is mainly as mycelium on overwintering strawberry leaves. The disease is most serious on early cultivars, which are most susceptible. Under hot dry conditions (15–27°C), which favour sporulation and spore dispersal, plants in production fields can be nearly white with mildew, and more than half the yield spoiled.

Fungicides (dinocap, sulphur) may be applied as protectants from just before first flowering and at 10–14 day intervals thereafter. Systemic compounds are also effective, but should ideally also control *Botryotinia fuckeliana*. Control of leaf mildew after harvest does not seem to affect yields in the following season. However, removal or herbicide destruction of overwintering leaves could provide a means of eliminating the initial inoculum (Cooke & Jordan 1978: APS Compendium). Raspberry is also affected, possibly by another form, and may need to be treated (e.g. in Scotland; Brokenshire 1984). Resistant cultivars are known in both crops.

SMITH

Sphaerotheca pannosa (Wallr.) Lév.

S. pannosa, occurring on various Rosaceae (CMI 189), is mainly important on rose, as **var. *pannosa*** (syn. var. *rosae* Voronichin), and on peach, as **var. *persicae*** Voronichin. Var. *pannosa* forms ellipsoid hyaline conidia in chains about 48–72 h after infection, initially on the lower surface of young leaves of rose, causing typical powdery mildew symptoms. A secondary mycelium frequently forms, consisting of straight hyphae about 6 μm wide with thickened walls; this persists in felty patches, the so-called 'pannose mycelium', at first white, then becoming grey to buff (Wheeler 1978). Cleistothecia are only rarely found, either in the mycelial felt or on stems around thorns. They are globose to pyriform with few mycelioid appendages, hyaline to light brown and with a single ascus. The disease is known world-wide and some biological specialization has been described (Mence & Hildebrandt 1966), and races identified (Bender & Coyier 1984). New rose cultivars with some resistance are frequently produced by commercial growers but few retain this resistance for long (Wheeler 1978). Infection causes leaf distortion, curling and premature leaf-fall. Attacks on stems just below flower buds prevent their opening. Fungal growth may develop on stems, and external mycelium may overwinter in mild seasons, and can perennate within dormant buds (Price 1970). There seems little evidence that cleistothecia play a significant role in spread and perennation.

Environmental conditions favouring the disease appear to be diurnal fluctuations; *c.* 15°C and 90–99% RH at night, allowing optimum conidial formation, germination and infection; and around 26°C and 40–70% RH during the day, favouring the release of conidia (Tammen 1973; Wheeler 1973). Repeated sprays of fungicides are needed to maintain control on developing young shoots. Fungicides used include dodemorph, bupirimate, triforine, imazalil, carbendazim, benomyl and drazoxolon; those with systemic activity are absorbed into leaves and although not extensively translocated are less affected by weathering than protectant fungicides.

Var. *persicae* occurs on various other *Prunus* species as well as peach (apricot, almond, nectarine). Forms specialized to peach occur (Ialongo 1981b). It infects leaves, young shoots and fruits (Burchill 1978). The disease is important in Italy and especially in E. Europe (Bulgaria, USSR), where fungicide treatment is needed and cultivars are screened for resistance. *S. pannosa* also attacks *Prunus laurocerasus*, widely grown as a hedging plant in W. Europe, on which it is important especially in France (Boudier 1978); older reports of powdery mildew on this species mainly concerned *Podosphaera tridactyla* (Wheeler 1978).

EBBEN; SMITH

Uncinula necator (Schwein.) Burrill

Basic description

U. necator has a typical external powdery mildew mycelium which at times bears cleistothecia with setae formed like a cross at the tip, usually with 6 asci each containing 4–7 ovoid ascospores (CMI 160).

Host plants

All Vitaceae are attacked (Boubals 1961) but only some *Vitis* species are very susceptible, especially the Asiatic species *V. armata* and *V. davidii*, and of course European grapevine. Some cultivars are extremely susceptible (Carignan, Cabernet-Sauvignon, Muscat à petits grains; Galet 1977). Non host-specialized forms have been reported.

Geographical distribution

U. necator originated in N. America, and first appeared in Europe in 1845 (under glass in S. England). The first French record was in 1847, and the whole country was invaded by 1850. The fungus is now found everywhere that grapevine is grown.

Disease

On susceptible grapevines, the fungus first appears on a few young shoots, which are stunted and whitened. Powdery mildew symptoms are then seen on leaves and vegetative shoots, but it is the berries which are most severely damaged. Infected epidermal cells of the berries show brown necrosis; if they are very numerous, the grapes burst during ripening, and then dry up or rot, being lost at harvest. Leaf and shoot symptoms are of minor importance, but losses of berries can reach 80–90%. Normally, fungal development stops on the berries in summer as soon as ripening begins (change of colour), but may continue on untreated leaves.

Epidemiology

U. necator overwinters as mycelium in the bud-scales of susceptible grapevine cultivars. When these buds develop in spring, they are heavily mildewed. Conidia are carried by wind and germinate rapidly under moist conditions (but not in water). The mildew mycelium spreads rapidly from the time the first haustoria are formed, and sporulates within 5 days (27°C), 11 days (15°C), or 25 days (10°C). Conidial germination is optimal at 25–28°C and occurs from 6–33°C (at 34°C for 10 h, conidia are killed). In late summer and autumn, the mycelium on leaves or heavily infected berries may develop the cleistothecia, whose role in epidemiology is uncertain.

Economic importance

U. necator caused massive damage when it first arrived in Europe. For example, wine production in France fell from 54 million hectolitres to only 10 million in 1854. However, sulphur was rapidly (1853–4) found to provide control (Galet 1977) and powdery mildew ceased to be a problem, with growers obtaining excellent control with up to 3–5 treatments during vegetative growth. In the last five years, however, attacks have suddenly become more severe on cultivars reported to have limited susceptibility (Aramon). The critical period for infection seems to be 15–20 days after flowering (Heritier 1983) and losses are quite significant.

Control

For over a century grapevine powdery mildew has been controlled with sulphur, which can be formulated as dustable powders, by trituration or sublimation. Adjuvants are added to ensure free flow. Doses of 15–60 kg/ha are applied according to type of sulphur and time of treatment. Sulphur can also be formulated as a wettable powder (micronized particles of 1–6 μm) to be used in spring mixed with fungicides against *Plasmopara viticola*, especially when the weather is not dry or warm enough for dusting. Dosages are then 10.0–12.5 kg/ha. These sprays can be used throughout the season, but at least seven must be applied. Dustable powders are mainly applied at three periods: i) at the 2–5 leaf stage of young buds (GS 09–12); ii) at flowering (GS 21–25); iii) several times between fruit set (GS 27) and beginning of berry ripening (GS 35). Very susceptible cultivars may need to be treated every 10 days. In recent years, various new compounds have been found to give good control (e.g. in France, triadimefon, fenarimol, triforine, diclobutrazol, penconazole, nuarimol). These are sprayed once on the leaves and then every 2 weeks on the berries.
BOUBALS

Uncinula adunca (Wallr.) Lév. (syn. *U. salicis* (DC.) Winter) forms white powdery patches on the leaves of poplars and willows in Europe and N. America. ***U. bicornis*** (Wallr.) Lév. (syn. *U. aceris* (DC.) Sacc.) is frequently seen on *Acer* species but is of no practical importance. ***U. tulasnei*** Fuckel on the same host has hardly forked appendages and smaller conidia (Viennot-Bourgin 1956).

A few powdery mildews are caused by conidial fungi with no known teleomorphs, a situation which is frequent in tropical countries. In Europe, these are all *Oidium* species. ***O. chrysanthemi*** Rabenh. is common on glasshouse chrysanthemums and requires fungicide applications (Fletcher 1984). It is doubtfully distinct from *E. cichoracearum* (Wheeler 1978). *Oidium* species, possibly in the same situation, also occur on saintpaulias, gloxinias, Kalanchoe etc. (Fletcher 1984). Others are: ***O. arachidis*** Babajan on groundnut in Bulgaria, ***O. carthami*** Jacz. on safflower in the USSR, ***O. cyclaminis*** Wenzl. on cyclamen, ***O. lini*** Skoric on linseed flax, ***O. ricini*** Jacz. on *Ricinus communis*, ***O. sesami*** Schembel on sesame and ***O. valerianellae*** Fuckel on lamb's lettuce.

REFERENCES

Bender C.L. & Coyier D.L. (1984) Isolation and identification of races of *Sphaerotheca pannosa* var. *rosae*. *Phytopathology* **74**, 100–103.

Bisiach M., Minervini G. & Salomone M.C. (1982) Recherches expérimentales sur la pourriture acide de la grappe et sur ses rapports avec la pourriture grise. *Bulletin OEPP/EPPO Bulletin* **12**, 15–27.

Blumer S. (1933) Die Erysiphaceen Mitteleuropas mit besonderer Berücksichtigung der Schweiz. *Beiträge Kryptogamenflora Schweiz* **7**, 1–483.

Blumer S. (1967) *Echte Mehltaupilze (Erysiphaceae)*. Gustav Fischer Verlag, Jena.

Boesewinkel H.J. (1979) Differences between the conidial states of *Podosphaera tridactyla* and *Sphaerotheca pannosa*. *Annales de Phytopathologie* **11**, 525–527.

Boesewinkel H.J. (1981) The identity of powdery mildew of citrus. *Nova Hedwigia* **34**, 731–741.

Bonnet A. (1983) *Daucus carota* subsp. *dentatus* géniteur de résistance à l'oïdium pour l'amélioration de la carotte cultivée. *Agronomie* **3**, 33–37.

Boubals D. (1961) Etude des causes de la résistance des Vitacées à l'oïdium de la vigne et de leur mode de transmission héréditaire. *Annales de l'Amélioration des Plantes*, **11**, 401–500.

Boudier B. (1978) Essais de lutte contre l'oïdium du laurier-cerise. *Phytiatrie-Phytopharmacie* **27**, 215–220.

Brent K.J. (1982) Case study 4: powdery mildew of barley and cucumber. In *Fungicide Resistance in Crop Protection* (Eds J. Dekker & S.G. Georgopoulos), pp. 219–230. PUDOC, Washington.

Brodnik T. (1976) The susceptibility of maize seed to invasion by *Aspergillus* spp. and their effects on germinability. *Zbornik Biotehniške Fakultete Univerze v Ljubljani Kmetijstvo* **26,** 91–96.

Brokenshire T. (1984) The control of powdery mildew on strawberries and raspberries. In *Crop Protection in Northern Britain 1984*. SCRI, Dundee.

Burchill R.T. (1978) Powdery mildews of tree crops. In *The Powdery Mildews* (Ed. D.M. Spencer), pp. 473–493. Academic Press, London.

Burchill R.T., Frick E.L. & Swait A.A.J. (1976) The control of peach leaf curl (*Taphrina deformans*) with off-shoot T. *Annals of Applied Biology* **82,** 379–380.

Butler E.E., Webster R.K. & Eckert J.W. (1965) Taxonomy, pathogenicity and physiological properties of the fungus causing sour rot of citrus. *Phytopathology* **55,** 1262–1268.

Butt D.J. (1972) The timing of sprays for the protection of terminal buds on apple shoots from powdery mildew. *Annals of Applied Biology* **72,** 239–248.

Butt D.J. (1978) Epidemiology of powdery mildew. In *The Powdery Mildews* (Ed. D.M. Spencer), pp. 51–81. Academic Press, London.

Butt D.J. & Barlow G.P. (1979) The management of apple powdery mildew: a disease assessment method for growers. *Proceedings of the 1979 British Crop Protection Conference, Pests and Diseases,* pp. 77–86.

Butt D.J., Martin K.J. & Swait A.A.J. (1983) Apple powdery mildew: damage, loss and economic injury level. *Proceedings 10th International Congress of Plant Conference, Pests and Diseases,* pp. 77–86. BCPC, Croydon.

Carr A.J.H. (1971) Herbage legume diseases. In *Diseases of Crop Plants* (Ed. J.H. Western). Macmillan, London.

Cole J.S. (1978) Powdery mildew of tobacco. In *The Powdery Mildews* (Ed. D.M. Spencer), pp. 447–472. Academic Press, London.

Cole M. & Wood R.K.S. (1961) Pectic enzymes and phenolic substances in apples rotted by fungi. *Annals of Botany* **25,** 435–452.

Cook R.J. & Yarham D.J. (1985) Fungicide use in cereal disease control in England and Wales. In *Fungicides for Crop Protection: 100 Years of Progress* (Ed. I.M. Smith), pp. 151–160. BCPC, Croydon.

Cooke A.T.K. & Jordan V.W.L. (1978) Powdery mildews bush and soft fruits. In *The Powdery Mildews* (Ed. D.M. Spencer), pp. 347–358. Academic Press, London.

Cullum F.J. & Webster J. (1977) Cleistocarp dehiscence in *Phyllactinia*. *Transactions of the British Mycological Society* **68,** 316–320.

Curé B. (1985) Utilisation des fongicides en France et stratégies de traitement sur blé d'hiver. In *Fungicides for Crop Protection: 100 Years of Progress* (Ed. I.M. Smith), pp. 161–170. BCPC, Croydon.

Davies H., Williams A.E. & Morgan W.A. (1970) The effect of mildew and leaf blotch on yield and quality of Lior Italian ryegrass. *Plant Pathology* **19,** 135–138.

Delon R. (1983) Comportement d'un assortiment d'espèces du genre *Nicotiana* vis à vis d'un isolat d'*Erysiphe cichoracearum* DC. provenant d'un tabac cultivé dans la région de Bergerac (France). *Communications de la Réunion des Groupes Phytopathologie et Agronomie du CORESTA, 18–22 September 1983, Bergerac.*

Dixon G.R. (1978) Powdery mildews on vegetable and allied crops. In *The Powdery Mildews* (Ed. D.M. Spencer), pp. 495–524. Academic Press, London.

Dixon G.R. (1984) *Vegetable Crop Diseases.* Macmillan, London.

Doling D.A. & Willey L.A. (1969) Date of sowing and the yield of swedes. *Experimental Husbandry* **18,** 87–90.

Dow P. (1978) Disease resistance in peas. *Annual Report John Innes Institute for 1977,* pp. 31–32.

Drandarevski C.A. (1969) Untersuchungen über den echten Rübenmehltau *Erysiphe betae. Phytopathologische Zeitschrift* **65,** 4–68, 124–154, 201–210.

Drandarevski C.A. (1978) Powdery mildew of beet crops. In *The Powdery Mildews* (Ed. D.M. Spencer), pp. 323–346. Academic Press, London.

Ebben M.H. & Spencer D.M. (1979) The use of antagonistic organisms for the control of black root rot of cucumber *Phomopsis sclerotioides. Annals of Applied Biology* **89,** 103–106.

Eckert J.W. (1978) Post-harvest diseases of citrus fruits. *Outlook on Agriculture* **9** (5), 225–232.

Eckert J.W. (1982) Case study 5: Penicillium decay in citrus fruit. In *Fungicide Resistance in Crop Protection* (eds J. Dekker & S.G. Georgopoulos), pp. 231–256. PUDOC, Wageningen.

Fletcher J.T. (1984) *Diseases of Greenhouse Plants.* Longman, London.

Galet P. (1977) *Les Maladies et Parasites de la Vigne,* pp. 15–87. Paysan du Midi, Montpellier.

Gil F. & Gay J.L. (1977) Ultrastructural and physiological properties of the host interfacial components of haustoria of *Erysiphe pisi in vitro* and *in vivo. Physiological Plant Pathology* **10,** 1–12.

Griffiths E. (1981) Role of fungicides in maximizing grain yield in barley. *Bulletin OEPP/EPPO Bulletin* **11,** 347–354.

Gritton E.T. & Ebert R.D. (1975) Interaction of planting date and powdery mildew on pea plant performance. *Journal of the American Society for Horticultural Science* **100,** 137–142.

Gullino G. & Rapetti S. (1978) Un mal bianco del platano. *Informatore Fitopatologico* **28** (11/12), 65–66.

Hagedorn D.J. (1973) Peas. In *Breeding Plants for Disease Resistance: Concepts and Applications* (ed. R.R. Nelson), pp. 326–343. State University Press, Pennsylvania.

Hammarlund C. (1925) Zur Genetik, Biologie und Physiologie einiger Erysiphalen. *Hereditas* **6,** 1–126.

Heringa R.J., van Norel A. & Tazelaar M.F. (1969) Resistance to powdery mildew (*Erysiphe polygoni* DC.) in peas (*Pisum sativum*). *Euphytica* **18,** 163–169.

Héritier J. (1983) Oïdium 1983 dans les vignobles de l'Aude. *Progrès Agricole et Viticole* **21,** 554–557.

Hill S.A. (1977) *Penicillium corymbiferum* on prepared Dutch iris bulbs: tolerance of benomyl and control. *Plant Pathology* **26,** 94–95.

Hiura U. (1978) Genetic basis of formae speciales in *Erysiphe graminis.* In *The Powdery Mildews* (Ed. D.M. Spencer), pp. 101–128. Academic Press, London.

Hopkins J.C.F. (1956) Mildew, white mould or white rust. In *Tobacco Diseases,* pp. 93–98. CMI, Kew.

Ialongo M.T. (1981a) Indizi di specializzazione in un mal

bianco del platano commune. *Annali dell'Istituto Sperimentale per la Patologia Vegetale Roma* **7,** 103–114.

Ialongo M.T. (1981b) Indagini sulla specializzazione biologica di una popolazione di *Sphaerotheca pannosa* var. *persicae. Annali dell'Istituto Sperimentale per la Patologia Vegetale Roma* **7,** 59–63.

Ialongo M.T. & Simeone A.M. (1984) Attacco di mal bianco su albicocco dovuto a due diverse e successive Erysiphaceae. *Informatore Fitopatologico* **34** (12), 31–33.

Jenkins J.E.E. & Storey I.F. (1975) Influence of spray timing for the control of powdery mildew on the yield of spring barley. *Plant Pathology* **24,** 125–134.

Jenkyn J.F. & Bainbridge A. (1978) Biology and pathology of cereal powdery mildew. In *The Powdery Mildews* (Ed. D.M. Spencer), pp. 283–321. Academic Press, London.

Junell L. (1966) A revision of *Sphaerotheca fuliginea* (Schlecht. Fr.) Poll. sensu lato. *Svensk Botanisk Tidskrift* **60,** 365–392.

Junell L. (1967) A revision of *Erysiphe communis* (Wallr.) Fr. *sensu* Blumer. *Svensk Botanisk Tidskrift* **61,** 209–230.

Khairi S.M. & Preece T.F. (1978) Hawthorn powdery mildew: overwintering mycelium in buds and the effect of clipping hedges on disease epidemiology. *Transactions of the British Mycological Society* **71,** 399–404.

Kouyeas H. (1979) Stigmatomycosis of the nuts of the pistachio tree. *Annales de l'Institut Phytopathologique Benaki* **12,** 147–148.

Leibundgut H. (1969) Untersuchungen über die Anfälligkeit verschiedener Eichenherkünfte für die Erkrankung an Mehltau. *Schweizerische Zeitschrift für Forstwesen,* **120,** 486–493.

Limpert E. & Schwarzbach E. (1982) Virulence analysis of powdery mildew of barley in different European regions in 1979 and 1980. In *Barley Genetics IV,* pp. 458–465. Edinburgh.

Limpert E., Schwarzbach E. & Fischbeck G. (1983) Influence of weather and climate on epidemics of barley mildew, *E. graminis* f.sp. *hordei.* In *Progress in Biometeorology 3* (Eds H. Lieth, R. Fantecki & H. Schnitzler). Swets & Zeitlinger, Lisse.

Liyanage A. de S. & Royle D.J. (1976) Overwintering of *Sphaerotheca humuli,* the cause of hop powdery mildew. *Annals of Applied Biology* **83,** 381–394.

Lucas G.B. (1975) *Diseases of Tobacco,* 3rd edition, pp. 297–303. Biological Consulting Associates, Raleigh, North Carolina.

Marchetti L. & D'Aulerio A.Z. (1980) L'oidio della quercia. *Informatore Fitopatologico* **30,** 23–24.

Marois J.J., Fravel D.R. & Papavizas G.C. (1984) Ability of *Talaromyces flavus* to occupy rhizosphere and its interaction with *Verticillium dahliae. Soil Biology and Biochemistry* **16,** 387–390.

Marois J.J,. Mitchell D.J. & Sonoda R.M. (1981) Biological control of *Fusarium* crown rot of tomato under field conditions. *Phytopathology* **71,** 1257–1260.

Mence M.J. & Hildebrandt A.C. (1966) Resistance to powdery mildew in rose. *Annals of Applied Biology* **58,** 309–320.

Merriman P.R. & Wheeler B.E.J. (1968) Overwintering of *Sphaerotheca mors-uvae* on blackcurrant and gooseberry. *Annals of Applied Biology* **61,** 387–397.

Messiaen C.M. & Lafon R. (1970) *Les Maladies des Plantes Maraîchères.* INRA, Paris.

Mix A.J. (1949) Monograph of the genus *Taphrina. University of Kansas Science Bulletin* **33,** 3–167.

Munro J.M. & Lennard J.H. (1982) Variation in the development of *Erysiphe cruciferarum* on two cultivars of *Brassica napus. Cruciferae Newsletter Eucarpia* **7,** 68–69.

Nagy G.S. (1970) Die Identifizierung des Mehltaus der Kürbisgewächse auf Grund der Konidienmerkmale. *Acta Phytopathologica Academiae Scientiarum Hungaricae* **5,** 231–248.

Nagy G.S. (1972) Studies on powdery mildews of cucurbits. I. Host range and maintenance of *Sphaerotheca fuliginea* and *Erysiphe* sp. under laboratory and glasshouse conditions. *Acta Phytopathologica Academiae Scientiarum Hungaricae* **7,** 415–420.

Nagy G.S. (1976) Studies on powdery mildews of cucurbits. II. Life cycle and epidemiology of *Erysiphe cichoracearum* and *Sphaerotheca fuliginea. Acta Phytopathologica Academiae Scientiarum Hungaricae* **11,** 205–210.

Nagy G.S. (1977) The susceptibility of cucurbitaceous plants against powdery mildews. *Növényvédelem* **13,** 433–440.

O'Rourke C.J. (1976) *Diseases of Grasses and Forage Legumes in Ireland.* An Foras Talúntais, Carlow.

Pitt J.I. (1979) *The Genus* Penicillium *and its Teleomorphic states* Eupenicillium *and* Talaromyces. Academic Press, London.

Price D. & Linfield C.A. (1982) Powdery mildew of anemones. *Transactions of the British Mycological Society* **78,** 378–379.

Price T.V. (1970) Epidemiology and control of powdery mildew (*Sphaerotheca pannosa*) on roses. *Annals of Applied Biology* **65,** 231–248.

Priestley R.H. (1981) Choice and deployment of resistant cultivars for cereal disease control. In *Strategies for the Control of Cereal Disease* (Eds J.F. Jenkyn & R.T. Plumb), pp. 65–72. Blackwell Scientific Publications, Oxford.

Purnell T.J. & Preece T.F. (1971) Effects of foliar washing on subsequent infection of leaves of swede (*Brassica napus*) by *Erysiphe cruciferarum. Physiological Plant Pathology* **1,** 123–132.

Renaud R. (1959) L'oïdium du tabac. *Revue Internationale des Tabacs* **34,** 151–160.

Rippon L.E. & Morris S.C. (1981) Guazatine control of sour rot in lemons, oranges and tangors under various storage conditions. *Scientia Horticulturae* **14,** 245–251.

Roosje G.S. & Besemer A.F.H. (1965) Waarnemingen en onderzoek over appelmeeldauw in Nederland van 1953 tot 1963. *Mededelingen van de Institut voor Plantenziektenkundig Onderzoek Wageningen* no. 369.

Royle D.J. (1978) Powdery mildew of the hop. In *The Powdery Mildews* (Ed. D.M. Spencer), pp. 381–409. Academic Press, London.

Salmon E.S. (1900) A monograph of the Erysiphaceae. *Memoirs of the Torrey Botanical Club* **9,** 1–292.

Schneider A. (1971) Mise en évidence de 2 variétés de *Taphrina deformans,* parasites l'une du pêcher, l'autre de l'amandier. *Compte Rendu Hebdomadaire des Séances de l'Académie des Sciences de Paris, Série D* **273,** 685–688.

Schneider A. & Sutra G. (1969) Les modalités de l'infection de *Populus nigra* L. par *Taphrina populina. Compte Rendu Hebdomadaire des Séances de l'Academie des Sciences de Paris Série D* **269,** 1056–1059.

Sitterly W.R. (1978) Powdery mildews of cucurbits. In *The Powdery Mildews* (Ed. D.M. Spencer), pp. 359–379. Academic Press, London.

Slootmaker L.A.J., Fischbeck G., Schwarzbach E. & Wolfe M.S. (1984) Gene deployment for mildew resistance of barley in Europe. In *Vorträg Pflanzenzüchtung* 6, pp. 72–84. Freising.

Spalding D.H. & Abdul-Baki A.A. (1973) *In vitro* and *in vivo* production of pectin lyase by *Penicillium expansum. Phytopathology* **63,** 231–235.

Spencer D.M. (Ed.) (1978) *The Powdery Mildews.* Academic Press, London.

Tammen J.F. (1973) Rose powdery mildew studied for epidemics. *Science in Agriculture* **20,** 10.

Taris B. (1970) Contribution a l'étude du *Taphrina aurea* (Pers.) Fr. agent de la cloque dorée du Peuplier. *Bulletin du Service de Cultures et d'Etudes du Peuplier et du Saule,* pp. 55–155.

Viennot-Bourgin G. (1956) *Les Champignons Parasites des Plantes Cultivées.* Masson, Paris.

Viennot-Bourgin G. (1956) *Mildious, Oïdiums, Caries, Charbons, Rouilles des Plantes de France. Encyclopédie Mycologique Vol. XXVI.* Lechevalier, Paris.

Wahl I., Eshed N., Segal A. & Sobel Z. (1978) Significance of wild relatives of small grains and other wild grasses in cereal powdery mildews. In *The Powdery Mildews* (Ed. D.M. Spencer), pp. 84–100. Academic Press, London.

Walker J.C. & Williams P.H. (1973) Crucifers. In *Breeding Plants for Disease Resistance: Concepts and Applications* (Ed. R.R. Nelson), pp. 307–325. Pennsylvania State University Press.

Walker J.R.L. (1970) Phenolase inhibitor from cultures of *Penicillium expansum* which may play a part in fruit rotting. *Nature* **227,** 298–299.

Wheeler B.E.J. (1973) Research on rose powdery mildew at Imperial College. *Journal of the Royal Horticultural Society* **98,** 225–230.

Wheeler B.E.J. (1978) Powdery mildews of ornamentals. In *The Powdery Mildews* (Ed. D.M. Spencer), pp. 411–445. Academic Press, London.

Wild B.L. (1983) Double resistance by citrus green mould *Penicillium digitatum* to the fungicides guazatine and benomyl. *Annals of Applied Biology* **103,** 237–241.

Windels C.E. & Kommedahl T. (1982) Pea cultivar effect on seed treatment with *Penicillium oxalicum* in the field. *Phytopathology* **72,** 541–543.

Wolfe M.S. (1985a) Dynamics of the response of barley mildew to the use of sterol biosynthesis inhibitors. *Bulletin OEPP/EPPO Bulletin* **15,** 451–457.

Wolfe M.S. (1985b) Integration of host resistance and fungicide use. *Bulletin OEPP/EPPO Bulletin* **15**, 563–570.

Wolfe M.S. & Schwarzbach E. (1975) The use of virulence analysis in cereal mildew. *Phytopathologische Zeitschrift* **82,** 297–307.

Wolfe M.S. & Schwarzbach E. (1978) The recent history of the evolution of barley powdery mildew in Europe. In *The Powdery Mildews* (Ed. D.M. Spencer), pp. 129–157. Academic Press, London.

Wolfe M.S., Barrett J.A. & Jenkins J.E.E. (1981) The use of cultivar mixtures for disease control. In *Strategies for the Control of Cereal Diseases* (Eds J.F. Jenkyn & R.T. Plumb), pp. 73–80. Blackwell Scientific Publications, Oxford.

Yarwood C.E. (1957) Powdery mildews. *Botanical Reviews* **13,** 235–500.

Zadoks J.C. (1981) EPIPRE: a disease and pest management system for winter wheat developed in the Netherlands. *Bulletin OEPP/EPPO Bulletin* **11,** 365–370.

11: Ascomycetes II
Clavicipitales, Hypocreales

The fungi covered by this chapter have in common the production of generally brightly coloured fleshy perithecia, with unitunicate asci. As pathogens, they fall into five broad categories: i) *Claviceps* and *Epichlöe*: specialized parasites of Gramineae; ii) fungi causing cankers on woody plants, the perithecial state (e.g. *Nectria*) mostly being a characteristic and distinctive stage in the disease cycle; iii) fungi causing root, collar and sometimes storage rots, generally soil-borne but sometimes also seed-borne, with anamorphs in *Fusarium* or similar form-genera, and the teleomorph known but often having little importance in the disease cycle; iv) as iii) but the teleomorph is not known, and survival is often by chlamydospores; v) as iv) but characteristically infecting, via the root system, the xylem of their hosts, causing the vascular wilt diseases. The very important form genus *Fusarium* has been specially monographed by Booth (1971), Nelson *et al.* (1981), Gerlach & Nirenberg (1982) and Nelson *et al.* (1983). Some familiar *Fusarium* species are now assigned to *Gerlachia*, with teleomorph *Monographella* in Sphaeriales (see Chapter 12).

CLAVICIPITALES

Claviceps purpurea (Fr.) Tul.

Basic description

Of the two *Claviceps* species occurring in Europe, ***C. paspali*** Stev. & Hall is restricted to *Paspalum* species (e.g. *P. distichum*, which is a common pasture grass in parts of Italy), appearing as roughly spherical buff-coloured bodies (≃ 3 mm in diameter) which force the floral parts to gape wide apart. *C. purpurea* is typified by dark spur-like sclerotia on a wide range of other cereals and grasses and is assumed to include various so-called species trivially deviating in character. It is recognized macroscopically by: i) a conidia-bearing sugary exudate from florets (the anamorph ***Sphacelia segetum*** Lév.)—indirect evidence may be insects feeding on the sugars, or the characteristic smell of fructose sweetness; ii) a hard black/purple, spur or banana-shaped outgrowth from within florets usually protruding from and larger than the seed it replaces (the sclerotial stage); iii) free sclerotia either mixed with harvested seed (and contrasting in colour and shape) or lying on the soil; iv) sclerotia in soil bearing one or more ascogenous stromata (Plate 15).

Host plants

Rye, wheat, triticale and barley are readily parasitized. Infection of oats is rare. Maize is not a host of ergot in Europe although it has its own pathogen (***C. gigantea*** Fuentes *et al.*) in America (APS Compendium). Many temperate pasture grasses may become infected (e.g. *Lolium, Dactylis, Festuca, Alopecurus*) and many weed or non-cultivated grasses.

Host specialization

Most isolates of *C. purpurea* parasitize a wide range of hosts. Host restriction operates under field conditions on a local scale but not necessarily in a different location. Probably most arable weed grasses are potentially hosts for *C. purpurea* from cereals.

Geographical distribution

C. purpurea is widespread in Europe across latitudes from Finland to Spain.

Disease

Under normal circumstances *C. purpurea* is restricted to the ovary. Since the host is essentially fully grown at the time of infection there is no obvious debilitating effect. When the ovary has been fully colonized, about 7–10 days after initial infection, plant sap rich in sugars exudes into the floral cavity and spills out as honeydew. Honeydew is an ideal growth medium for the sphacelial stage and also for saprophytic moulds (*Trichothecium roseum, Gibberella zeae, Cladosporium* species). This may cause biodeterioration of adjacent healthy grain or affect its marketability by spoiling its appearance. Large ergots, having a dry weight considerably greater than a seed, may act as a metabolic sink drawing nutrients away from adjacent developing seeds.

Epidemiology

In Europe, *C. purpurea* generally overwinters as sclerotia, either in grain about to be sown or already incorporated into surface layers of the soil. Inoculum may persist as sphacelial fructifications in such grasses as *Lolium perenne* and *Poa annua* but this is probably of little epidemiological significance. Sclerotia require a dormant period at low temperatures before germinating in late spring to yield ascosporogenous stromata. This sexual stage often coincides, by mechanisms that are poorly understood, with the flowering of an appropriate host which may in the case of wheat become infected directly. However, in S. England, for example, wheat may receive an inoculum of greater potential in the form of conidia in honeydew borne on precociously flowering arable weeds such as *Alopecurus myosuroides*. Ascospore release requires adequate soil moisture to maintain turgid stromata. Open-flowering cereals and grasses are more readily infected, since pathogenic hyphae follow the same route as a pollen tube through the stigma to colonize the ovule. If pollination precedes the arrival of the pathogen by 2–3 days, infection of the ovary is rarely achieved. The duration of open flowering or the extent of protogyny are the most influential factors predisposing the host to ergot infection. Consequently ergot is one of the principal factors potentially threatening the production of F1 hybrid cereals, since male-sterile lines in crossing programmes are at particular risk. Also F1 hybrids in which male sterility has been incompletely restored may behave, at least in part, as male-steriles. Dry weather for 2–3 weeks after flowering will greatly attenuate foci of primary infection, but high humidity will favour profuse exudation of conidia-bearing honeydew and facilitate an epidemic. Conidia are transferred to uninfected florets by rain splash, head-to-head contact and by insects. Ergot epidemics do not appear to spread laterally for more than a few metres, unlike wind-borne pathogens. Of course, the ascospores are wind-borne but they do not initiate epidemics, only foci for epidemics, and they do not move much laterally in the crop since the foliage acts as an efficient spore trap. Hence, cereal crops may often have their ergot infection confined mainly to the margins, initially infected from headland grasses.

Economic importance

Ovary infection invariably reduces grain yield by eliminating ovaries. Sclerotia also contain toxic metabolites (the ergot alkaloids) which reduce crop value (some European countries have very low statutory limits on the ergot content of milling wheat), or necessitate costly extra cleaning. Ergot may also present a hazard to cattle grazing infected grasses; the sphacelial stage is harmless but sclerotia may impair peripheral circulation and/or cause reproductive failure on account of the alkaloids. There is little experimental evidence concerning tolerance in practice, and statutory limits are probably unnecessarily low. In some European countries rye is deliberately infected to produce therapeutic alkaloids and *C. purpurea* must be the only plant pathogen which may have an economic value which exceeds the intrinsic value of the parasitized host.

Control

The use of clean seed, ploughing of land contaminated with ergots in order to bury them to a depth ($\simeq$ 7 cm) which precludes stromatal development, control of arable weed grasses and prevention of headland grasses from flowering generally protects against other than spasmodic infections. No immune cultivars of wheat, barley, rye or triticale have been found. Control in F1 hybrid production may be optimized by providing the maximum amount of a pollinator over the critical first days of flower opening of the male-sterile line. Pasture grasses should not be allowed to run to seed, particularly in the autumn when low night temperatures in temperate latitudes may predispose grazing cattle to impaired peripheral circulation induced by the ergot alkaloids.

Mercurial seed dressings were relatively ineffective in suppressing ascocarp formation although recent preliminary tests with a new fungicide show promise. The problem with seed dressings is that their action must persist against leaching for approximately 6 months after sowing winter cereals. Significant control at the time of flowering has been obtained using benomyl, but such measures may only be practicable in breeding material.

Special research interest

The following under-explored topics all impinge on the disease—mechanisms regulating germination of

sclerotia, the extent of host restriction of the pathogen, design of systemic fungicides, the mechanism of honeydew release, control of differentiation of the sclerotial phase and the tolerance of ergot by animals.
MANTLE

Epichloë typhina (Pers.) Tul. & C. Tul.

(Anamorph ***Sphacelia typhina*** Sacc.)

E. typhina causes choke disease of more than 50 species of grasses but seldom occurs on cereals. It is well-known for forming particular strains. It occurs predominantly in the northern hemisphere (Europe, USSR, USA, Canada, China) (CMI 639). The parasite can affect the development of the inflorescence, increase tillering and cause stem elongation. On the surface of the stems a cylindrical stroma is formed around the sheath of the leaf which covers the inflorescence. The fungus can entirely inhibit the emergence of the inflorescence so that no seed is formed; hence the disease has been named choke disease. The extent of damage caused by *E. typhina* is not the same on all grasses; for instance *Dactylis glomerata* and *Phleum pratense* are particularly susceptible. The infection is systemic. The fungus can be present in all parts of the host plant except the roots. Even the seed can be infected. The mycelium grows mainly intercellularly in all tissues. *E. typhina* perennates in the plant where it may remain latent for several years without visible symptoms of the disease. The stroma usually first produces conidia abundantly at the time of flowering of the grasses, followed by the appearance of ascospores. However, in spite of the presence of spores the inflorescences may not be infected. In some years and localities the fungus can cause considerable damage to grasses particularly to those grasses grown for seed. Its economic importance lies chiefly in causing losses in seed production in some species and in others by infecting the seed. In addition to this it produces in *Festuca arundinacea* a mycotoxin which is toxic to cattle (Porter 1981).

Little is known about the control of *E. typhina*. The production and use of healthy seed is the main recommendation. Since infected plants and seeds often do not exhibit external symptoms, detection of the fungus is important. Hitherto detection has required time-consuming microscopic investigations. By means of the ELISA test, however, *E. typhina* can be detected more rapidly on stems and seeds of *Festuca arundinacea* (Johnson *et al.* 1982). A host-specific fly *(Phorbia phrenione)* often occurs wherever the fungus is found. The larvae of the hyperparasitic fly graze upon the stromata of *E. typhina* (Kohlmeyer & Kohlmeyer 1974).
BARTELS

HYPOCREALES

Gibberella zeae (Schwein.) Petch

(Anamorph ***Fusarium graminearum*** Schwäbe)

Basic description

No microconidia are formed, and generally no chlamydospores. While the macroconidia are commonly found in nature, the perithecia are also common and play an important part in the disease cycle.

Host plants

G. zeae is important as a pathogen on cereals (maize, wheat, rice) but reported from various other hosts (CMI 384).

Geographical distribution

G. zeae is found world-wide, and generally in warmer climates than *Fusarium culmorum* (q.v.).

Disease

In Europe, *G. zeae* principally attacks maize and wheat. On maize, the foot is attacked at the mesocotyl level (between primary root and the crown roots), and infection spreads up the stalk, which shows browning of the lower internodes, internal pink to reddish discoloration and 'shredding' of the pith (APS Compendium). Infected stems are liable to lodge. *G. zeae* behaves essentially as a weak parasite of senescent plants, accelerating their senescence further. Host-plant susceptibility depends greatly on prior growth conditions; a plant which has been stressed earlier, by low night temperature or drought, is more liable to be heavily attacked. The timing of attack relative to grain-filling is critical (Barrière *et al.* 1981). An early attack will interrupt the water supply during grain-filling and cause serious losses, while a late attack is of minor importance. Cobs may be infected at a late stage by splashed conidia, and may be seriously rotted, but cultivars vary in their susceptibility to cob infection. On wheat, *G. zeae* causes foot rot and ear scab very like *Fusarium culmorum* (q.v.), but tends to be more important in the milder Atlantic climate of W. Europe than in more continental European areas.

Epidemiology

G. zeae can be seed-borne, especially on wheat, where infection of the ear via the stamens is characteristic. The fungus is especially important on seed-crops of wheat. In general, however, survival is mainly on crop debris, as perithecia. This is in clear contrast to *F. culmorum*, which survives saprophytically or as chlamydospores in soil or on debris. Indeed, perithecia form on the stem bases and ears, and ascospores may play an important role in secondary spread as well as primary infection. *G. zeae* is generally favoured by rather warmer conditions than *F. culmorum* (causing infection at temperatures over 12°C, instead of 7°C for the latter).

Economic importance

G. zeae causes serious losses in much the same way as *F. culmorum* (q.v.), though it is relatively less important on the small-grain cereals and more important on maize. It is the most serious economic pathogen of maize in W. Europe, causing losses of about 10% per year in France (Cassini 1983). Infected plants produce less grain; harvesting is made more difficult and less efficient by lodging, and the grain itself may be directly infected. As in the case of *F. culmorum,* mycotoxins may be formed in infected grain.

Control

While *F. culmorum* survives readily on debris, and cannot be controlled readily by rotation or cultural measures, *G. zeae* normally has to form perithecia to survive, and can to a substantial extent be controlled by ploughing in debris. Seed treatments against fungi causing seedling blight will be effective against *G. zeae* at the seedling stage. In other respects, the problems of control are essentially the same as for *F. culmorum.* Attempts are being made (e.g. in France) to include stalk-rot resistance in breeding programmes. Barrière *et al.* (1981) have shown that this depends very much on physiological characteristics of the plant, as well as intrinsic resistance to infection. Thus, plants will be more resistant to stalk rot if they are late in senescing, if they are tolerant of physiological stress, and if they have a source/sink relationship giving the right ear/whole plant ratio. In fact, modern hybrids are in general more resistant to stalk-rot than the cultivars first grown in Europe, against a background of steadily increasing inoculum of *G. zeae*. In recent trials, older hybrids were completely destroyed by the fungus.
CASSINI & SMITH

Gibberella avenacea R.J. Cooke

G. avenacea (CMI 25) has dark-red or blackish perithecia with hyaline oval ascospores, formed on the surface of cereal haulm-bases. The anamorph ***Fusarium avenaceum*** (Corda) Sacc. predominates, with longish, 5–7 septate, curved conidia. Microconidia are very rare, and no chlamydospores are formed. ***F. arthrosporioides*** Sherbakoff is very similar (Gerlach & Nirenberg 1982). *G. avenacea* is a weak parasite of hundreds of hosts from different families (e.g. maize, carnations, tulips) but it is best known as a cereal pathogen. Isolates from any one host are pathogenic to any other. The fungus is cosmopolitan in distribution. In cereals, it causes seedling blight, foot rot of the haulm-base, and it attacks ears and/or kernels, but is much less aggressive than *Fusarium culmorum* or *G. zeae,* though more adaptable to cooler temperatures. In maize, it is a minor cause of seedling blight and stalk rot, which are usually caused by other fusaria. *G. avenacea* is a very competitive soil saprophyte, but mainly causes seedling blight from heavily infected seed (Colhoun 1970). Haulm-bases are probably mainly colonized via roots from the soil, and ears from conidia on the haulm-base or decaying lower leaves, or by ascospores from perithecia on lower nodes. The fungus is easily spread by air movements. Barley yellow dwarf virus predisposes plants to infection. In cereals, reduced emergence and seedling blight are only sporadically and locally important, and foot rot usually unimportant. Serious ear/kernel infections (humid conditions, rainy weather) may cause minor crop losses, but are liable to be more important if heavily infected grains are used for seed.

Seed contamination is (was) perfectly covered by mercury treatment, but fungicide mixtures aimed at other fusaria should also be sufficiently active. Chemical control of *Pseudocercosporella herpotrichoides* does not cover *G. avenacea* at the haulm-base. There is no convincing way to control *G. avenacea* after ear infection. Although a relationship between crop sequence and population level of the fungus has been established, crop rotation procedures or intercropping cannot reduce soil inoculum (Kollmorgen 1974);

there are no resistant cereal cultivars, and differences in susceptibility are usually only minor.
FEHRMANN

Gibberella cyanogena (Desm.) Sacc.

(Anamorph ***Fusarium sulphureum*** Schlecht., syn. *F. sambucinum* Fuckel f.6 Wollenw.)

G. cyanogena is economically only important as a causal agent of potato rots, though isolated from at least 20 plant species. It appears on agar medium as a pink, salmon- to orange-coloured culture with masses of typical, *Fusarium*-like macroconidia. Microconidia are not formed. The external disease symptom on potato is a dark brown, slightly sunken rot zone with irregular edging. On cut tubers, the outer zone of the rotted tissue is of black-brown colour (Fig. 37). Inwards, by enzyme maceration, a whitish 'powdery dry rot' (the classical name of the symptom) with cavities develops. The edge of the rotted zone shows irregular protrusions. Confusion with symptoms caused by *Phoma exigua* var. *foveata* is possible. Most reports come from the northern parts of N. America and Europe, where the fungus seems to occur in all potato-growing countries. Epidemiological details and control measures are similar to those cited under *F. coeruleum*. Losses may be considerable. Seed tubers are more affected than ware potatoes. Several authors have observed synergistic effects between *G. cyanogena* and *Erwinia carotovora* ssp. *atroseptica*.
LANGERFELD

Gibberella fujikuroi (Saw.) Wollenw.

G. fujikuroi is the teleomorph of fusaria in section Liseola, generally known as ***Fusarium moniliforme*** Sheldon (CMI 22). According to Gerlach & Nirenberg (1982), its **var. *fujikuroi*** (anamorph ***F. fujikuroi*** Nirenberg) is the fungus which produces gibberellic acid in high quantities and thus causes the classic 'bakanae' disease of rice in the Far East; it does not occur in Europe. These authors distinguish it from ***G. moniliformis*** Wineland (anamorph ***F. verticillioides*** (Sacc.) Nirenberg), attacking mainly maize but also sugarcane and rice, which does not cause the seedling elongation symptoms and occurs widely in Europe. Nelson *et al.* (1983) consider these fungi as one species—*G. fujikuroi* with anamorph *F. moniliforme*. Whichever view is taken, the distinctive feature is the production of characteristic moniliform chains of microconidia on monophialides. Forms in which the chains are borne on polyphialides are referred to ***F. proliferatum*** (Matsushima) Nirenberg. ***Gibberella fujikuroi* var. *subglutinans*** Edwards (anamorph ***F. sacchari*** (Butler) Gams **var. *subglutinans*** (Wollenw. & Reinking) Nirenberg) is another maize pathogen, distinguished by forming microconidia only in 'false heads', and not in chains (CMI 23). Nelson *et al.* (1983) make a new species of it—*G. subglutinans* with anamorph *F. subglutinans*. ***F. sacchari* var. *sacchari*** is a non-European species causing curly top disease of sugarcane. In summary, from the viewpoint of a pathologist, it would seem that maize is attacked, in Europe, by two forms (*G. moniliformis* and *G. fujikuroi* var. *subglutinans*), and rice by the first of these. The status of *F. proliferatum* as a pathogen in Europe is not clear. In any case, it will be difficult to make sense of information on these various species in older literature, and research is now needed to clarify their status as pathogens. In the further treatment below, the name *F. moniliforme* is used loosely. The teleomorph is in any case not known to play a significant role in the disease cycle.

F. moniliforme is widespread in soils and is frequently found attacking the foot and roots of maize, though it is not altogether clear to what extent it is a primary parasite (artificial inoculation gives poor results). The 'mesocotyl' region (the stem section above the primary root but below the surface feeder roots) appears to be especially susceptible, and plants infected at this point are liable to wilt and die if put under stress. It is possible that special strains are involved in this type of attack; certainly, some cultivars are resistant to it. Infection may spread to the base of the stalk, and also through wounds on the sheaths (APS Compendium). *F. moniliforme* is also found causing a cob rot, although it is not entirely clear how the infection arises. Though the fungus

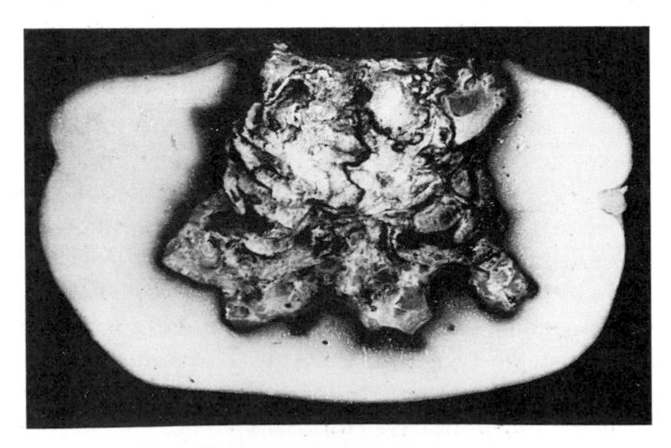

Fig. 37. *Gibberella cyanogena* in potato cv. Maritta, 6 weeks after artificial infection.

may be seed-borne, it is so common in soil that this is quite unimportant. Seedling blight is of minor significance. In Europe, *F. moniliforme* is a very minor pathogen on maize, especially compared with *Gibberella zeae*. It causes practically no losses on normal crops, but does occasionally damage seed-crops, since some parent lines are more susceptible to the wilting syndrome than their hybrid progenies. No special control is needed. Finally, it should be stressed that *F. moniliforme* is one of the group of widespread rather unspecialized fusaria in soils which may attack roots and seedlings of a great many host plants, and also cause fruit rots, especially in the tropics (CMI 22, 23). Attacks on rice and sorghum are only of minor importance in Europe.
CASSINI & SMITH

Gibberella baccata (Wallr.) Sacc.

(Anamorph ***Fusarium lateritium*** Nees)

G. baccata produces floccose, felted or slimy mycelium in culture. Yellow, orange or reddish-brown colours predominate, with deep orange spore masses arising from sporodochia. Cultures often form blue-black stromata in which perithecia may be initiated; they rarely develop further on agar, but on sterile wheat straw mature perithecia can be formed by homothallic strains and by appropriate combinations of mating types in heterothallic strains. Microconidia are usually absent; macroconidia are straight or slightly curved, 3–7 septate, beaked at the apex and with a pedicellate foot-cell (Booth 1971; CMI 310).

The pathogen has a wide host range and has been associated with dieback, cankers and bud rots (bud blight, bud blast) of woody trees and shrubs. Within *F. lateritium*, **f.sp. *cerealis*** Matuo & Sato is associated particularly (but not exclusively) with cereals and **f.sp. *mori*** Matuo & Sato (***G. baccata*** var. ***moricola*** (de Not.) Wollenw.) with mulberry. **f.sp. *pini*** Hepting is apparently confined to pines, causing pitch canker in N. America (but is a form of *G. fujikuroi* var. *subglutinans* according to Dunnell (1970)). The species has a world-wide distribution. A bud rot of apples has long been attributed to infection by *F. lateritium* (Wormald 1955) and the disease has been successfully reproduced at very high humidity. However, the main evidence has been that of a constant association with the disease, the pinkish pustules being frequently reported on buds which had died shortly before opening. Swinburne (1973) showed that *F. lateritium* was one of several fungi that could frequently be isolated from apparently healthy leaf-scars and buds. It seems likely, therefore, that this organism only becomes pathogenic when other factors increase the susceptibility of the host tissues to infection, e.g. through physical damage or physiological dysfunction. The disease is sporadic, and control measures are usually not required other than removal of badly affected shoots.
TALBOYS

Gibberella pulicaris (Fr.) Sacc.

G. pulicaris (CMI 385) is more commonly encountered as its anamorph, ***Fusarium sambucinum*** Fuckel. Cultures in slightly acid media are initially rose-coloured, becoming blood red, with conidiophores (mostly in sporodochia) producing septate fusoid macroconidia. Perithecia develop in sterile culture on straw if appropriate mating types are co-inoculated. In the field the perithecia are usually found only on cankers on woody plants. The pathogen has been reported as the causal agent of a storage rot of potatoes; root rots and seedling rots of various cereals, legumes, Solanaceae, strawberries and forest trees; and stem cankers on trees and shrubs (e.g. hops, *Sambucus*). It is not clear how frequently strains isolated from one type of lesion or from one host will infect other hosts or cause other kinds of lesions. The species is common in northern temperate and Mediterranean regions.

G. pulicaris is widely known as the cause of basal canker of hops. In established plants the lesions develop just below the soil surface, at or near the point where the herbaceous stem emerges from the perennial rootstock, and some form of damage seems to be a prerequisite for infection. Girdling of the stem is often accompanied by hyperplasia and periderm formation immediately above the lesion, and occurrence of pinkish-white conidial pustules under moist conditions. When girdling is complete the stem may remain attached by residual vascular tissue, but can easily be pulled away. There is rapid wilting of the leaves, with chlorosis and necrosis. Vascular discoloration is limited to a few cm above the lesion. The fungus also rots rootstock tissues and planting material overwintered under unfavourable conditions (Royle & Liyanage 1976). The disease is sporadic and unpredictable and causes only minor economic losses. Spring treatments with benomyl, thiabendazole, fentin, certain mercurial preparations etc., have given promising reductions in disease (Darby 1983), but control mainly depends on field hygiene.
TALBOYS

Gibberella heterochroma Wollenw. (anamorph ***Fusarium flocciferum*** Corda) is a similar polyphagous species of very minor importance (Gerlach & Nirenberg 1982). ***G. gordonii*** C. Booth (anamorph ***F. heterosporum*** Nees) (CMI 572) may be synonymous with ***G. cyanea*** (Sollm.) Wollenw. (anamorph ***F. reticulatum*** Mont.) (Gerlach & Nirenberg 1982). Soil- and possibly seed-borne, it causes a head-blight of cereals in the tropics but is a minor pathogen in Europe. Like the related ***F. graminum*** Corda, it is often associated with cereal inflorescences infected with ergot (Hornok & Walcz 1983). ***G. intricans*** Wollenw. (anamorph ***F. equiseti*** (Corda) Sacc.) (CMI 571) occurs world-wide in soil and is a minor cause of storage rotting of fruits and tubers, and also of root and stem rots in cereals (e.g. maize). It is more important in the tropics and subtropics. ***G. acuminata*** Booth (anamorph ***F. acuminatum*** Ell. & Kellermann) is a similar species, while ***Fusarium semitectum*** Berk. & Rav. is a major cause of storage rots on tropical crops but minor in Europe. ***G. tricincta*** El-Gholl *et al.* (anamorph ***F. tricinctum*** (Corda) Sacc.) is reported attacking seedlings of Gramineae and other hosts (El-Gholl *et al.* 1978). Other fusaria in the *Sporotrichiella* group are: i) ***F. poae*** (Peck) Wollenw., widespread in soils in temperate regions, causing minor head blights of cereals and grasses and a bud rot of carnation and chrysanthemum (CMI 308). Associated with the mite *Siteroptes graminum* which feeds on it, it causes silver top or whitehead of cereals; ii) ***F. sporotrichioides*** Sherbakoff is another widespread soil fungus causing seedling or storage rots. It is most significant as a producer of mycotoxins in stored grain (Seemüller 1968).

Nectria galligena Bresad.

(Anamorph ***Cylindrocarpon heteronemum*** (Berk. & Broome) Wollenw., syn. *C. mali* (Allescher) Wollenw.)

Basic description

N. galligena can be separated from the closely related species *N. ditissima* by perithecial characters and by the morphology of macro- and microconidia. It produces bright red globose perithecia which become darker and warted with age. Hyaline macroconidia are formed in sporodochia (CMI 147).

Host plants

The fungus attacks apple (Fig. 38), pear, and other *Malus* and *Pyrus* species. It has also been recorded

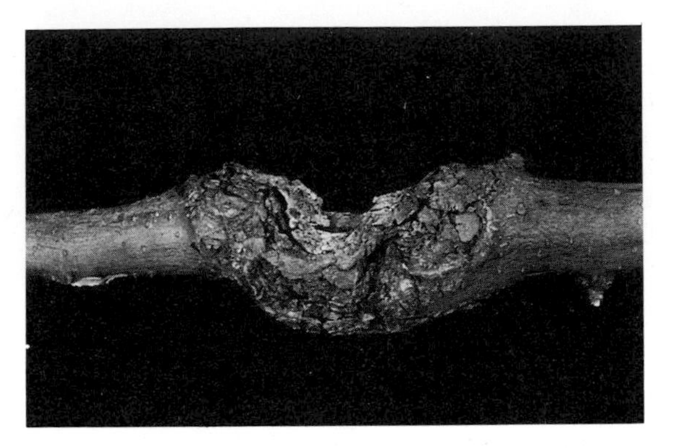

Fig. 38. *Nectria galligena* canker on apple.

on a large number of other woody hosts including *Acer, Betula, Carpinus, Carya, Crataegus, Cydonia, Fagus, Fraxinus, Juglans, Magnolia, Populus, Prunus, Quercus, Salix,* and *Sorbus.*

Host specialization

Differences between the size of ascospores of *N. galligena* from different host plants have been detected but cross-inoculation experiments with isolates from different hosts yielded conflicting results. Further investigations are required before the various isolates can be described as specialized pathotypes.

Geographical distribution

The origin of the disease is not clear. It was certainly known in Europe as an apple canker before 1866, and is referred to as European canker. The first report from N. America dates from 1899. The disease now occurs in virtually all the apple-growing areas of the world.

Disease (Swinburne 1975)

The young canker starts as a sunken area which increases in size and becomes elliptical. The edges of the canker rise above the surrounding healthy bark. The cankers are usually centred around leaf scars, wounds, twig stubs, or in crotches of the limbs. As the causal organism spreads, it encircles and kills the twig or branch. In cankers on older wood, it may spread more slowly and the annual peripheral expansion of the diseased area produces an elongated canker with a series of roughly concentric ridges. Removal of the adhering flakes of bark will reveal callus folds surrounding a central cavity extending into the wood. In addition to this open type of canker, a closed canker may be formed which may be considerably swollen. The bark above the invaded area is roughened and cracked but often remains intact for several years. All gradations may be found between these two symptom types. White to cream-coloured sporodochia may be observed on the surface of the folds and bright red perithecia may occur scattered or in clusters along the margin of the canker. Mainly in moist regions, *N. galligena* may also cause a fruit rot on apple and pear. It is called 'eye rot', because it typically occurs around the calyx. Infection takes place through the open calyx, lenticels, scab lesions, or wounds. The disease may occur on unripe fruit, before harvest, or only after several months of storage.

Epidemiology

The first spores to be produced by new cankers are conidia. Perithecia are not usually formed in the first year after infection. On older cankers, conidia and ascospores can be found throughout the year but their production and release depend on rainfall and hence vary with the climate. In temperate European countries, the period of maximum discharge in the different areas can be either in spring, early summer, late summer or in the autumn. In summer-dry climates, however, conidia are produced largely during the rainy autumn and winter months and ascospores are usually discharged at the same time. The conidia are disseminated by spattering rain whilst ascospores may be either forcibly ejected from the perithecia and disseminated by wind or extruded in a sticky mass and splash-dispersed. Infections most often occur through leaf scars which are highly susceptible during the first few days after leaf fall. After this period, the leaf scars become increasingly resistant, mainly due to the formation of a wound periderm. Frost, or the onset of bud burst and stem growth, may make leaf scars susceptible again during periods in winter or spring. Additional points of entry are the scars of bud scales, scars left after the removal of the fruits, tree-tie wounds, pruning cuts, frost damage, bark fissures at branch crotches, wounds of boring insects, and lesions caused by other fungi. Experiments under controlled conditions showed that at temperatures of 14–15.5°C the trees had to be wet for at least six hours before appreciable infection of leaf scars occurred. The percentage of infection increased rapidly with an increase in the length of the moisture period. The incubation period of the fungus in the twig may range from a few days to several weeks or even months, depending on temperature and moisture. Inoculum sources for the infection of young orchards are often infested neighbouring orchards, in some cases also windbreaks or hedges of infested host plants like *Populus* or *Crataegus*.

Economic importance

N. galligena canker is one of the most destructive diseases of apple in Europe (Kennel 1976). Pears are usually not as badly affected as apple but in some areas the disease can be serious. Severe damage may also occur on broad-leaved forest trees. The disease is reported from stone-fruit trees but it is of minor importance on these. Losses arise by the reduction of vigour due to the presence of the cankers and by girdling of twigs, branches, scaffold limbs and trunks. The disease is most serious in regions which are characterized by relatively mild autumn and winter temperature and by heavy autumn and spring precipitation or high humidity. Such climatic conditions favour infection and the growth of the invading fungus. Trees grown on heavy or poorly drained soil are usually more susceptible than those on lighter soils. All apple cultivars are susceptible to infection but there are marked differences in the degree of susceptibility.

Control

Control is still an unsolved problem particularly in areas where the disease is severe. In these cases satisfactory success can only be obtained with an integrated control programme which includes appropriate orchard management, sanitary measures, excision of cankers, removal of diseased branches and the application of fungicides. Because vigorously growing trees are more susceptible to canker than less vigorous ones, the use of excessive nitrogenous fertilizers should be avoided. Pruning of at least the most susceptible cultivars should be carried out during dry weather, at frost temperatures, or in mid-winter since the likelihood of infection is less under these conditions. Protection of pruning cuts by paints or by fungicide application immediately after pruning has also been suggested. More recently a device was described which applies a small quantity of fungicide or a suspension of a microbial antagonist to the wound surface during the cutting action.

Cankers should be removed from the trees whenever possible, to reduce fungal inoculum. In some cases the loss of larger branches, scaffold limbs, or the entire tree can be avoided by the application of fungicide-containing paints as an eradicant. The efficiency of paints is increased by scarifying the surface of the canker with a knife or a wire brush or when the canker is thoroughly cleaned out and cut well back to healthy tissue. Since the fungus lives and fruits for some time in dead wood, all prunings in the orchard should be destroyed. In the majority of apple-growing areas the most important measure to control *N. galligena* canker is the application of fungicides in autumn during leaf-fall. For maximum protection of the newly exposed leaf scars, the fungicides should be applied at the onset of leaf-fall and at 50% leaf-fall. Additional sprays might be useful if leaf-fall is delayed, or under rainy conditions. A further application before bud burst is recommended to protect cracks which develop in spring. It has been known for many years that control can be obtained with organomercury and copper compounds. Of the newer fungicides, thiophanate methyl, benomyl, thiabendazole, captafol, and pyridinitril proved also to give satisfactory results. Fungicides applied for scab control can give considerable reduction in the number of cankers.

In recent years some attention has been paid to biological control. Isolates of *Bacillus subtilis* and *Trichoderma viride* Fr. were found to be antagonistic to *N. galligena*. Suspensions of *B. subtilis* applied to scars immediately after removal of the leaves protected them against infections until the following spring. However, the apparent inability of *B. subtilis* to persist on bark or redistribute between scars must be overcome before biological control becomes commercially acceptable. Fruit infections are controlled by eradicating wood cankers, by reducing fungus inoculum with autumn sprays, and by the use of fungicides against *Venturia inaequalis* that happen also to be effective against *N. galligena*. Also, the application of fungicides before harvest or dipping harvested fruits has been proved to be effective.
SEEMÜLLER

Nectria haematococca Berk. & Broome

N. haematococca is the teleomorph of *Fusarium solani* and other *Fusarium* species. Individual concepts reveal various approaches to splitting into species, varieties, forms or races. In order to avoid confusion, serious plant pathogens on potatoes and legumes, cited as *Fusarium solani* or a related form and considered as anamorphs of *N. haematococca* by Booth (1971) or Gerlach & Nirenberg (1982) are treated as follows:

1 ***Fusarium solani*** (Martius) Sacc.
(syn. *F. solani* var. *martii* (Appel & Wollenw.) Wollenw.; var. *striatum* (Sherbakoff) Wollenw.). Associated with ***N. haematococca* var. *brevicona*** (Wollenw.) Gerlach.
2 ***Fusarium eumartii*** C.W. Carp.
(syn. *F. solani* var. *eumartii* (C.W. Carp.) Wollenw.; f. *eumartii* (C.W. Carp.) Snyder & Hansen). Associated with ***N. haematococca* var. *haematococca***.
3 ***Fusarium javanicum*** Koord.
(syn. *F. radicicola* Wollenw.; *F. javanicum* var. *radicicola* Wollenw.). According to Gerlach & Nirenberg (1982), associated with ***Nectria ipomoeae*** Halsted.
4 ***Fusarium coeruleum*** (Lib.) ex Sacc.
(syn. *F. solani* var. *coeruleum* (Sacc.) C. Booth). Association with a teleomorph uncertain.

The teleomorph stage is not known to play any significant role in the disease cycle.

Basic description

F. solani, F. eumartii and *F. javanicum* form typical, slightly curved, 1–8 celled macroconidia, and 1–2 celled microconidia. Agar cultures are white to creamy, sometimes with a green to blue-green coloration of the stroma. *F. coeruleum* forms similar 1–6 celled macroconidia but not microconidia. Agar cultures appear beige to pink, later blue or blue-violet. Thick-walled, 1–3 celled chlamydospores are formed in ageing macroconidia and mycelium by all species.

Host plants

All the anamorphs of *N. haematococca* can be found on many plants and, because of this wide host range, diseases on different major hosts are treated separately below. *F. solani sensu lato* occurs everywhere in soil and can be responsible for damping-off, foot rot and stem cankers of almost any plant. Like *F. oxysporum* and other *Fusarium* species, it will attack wounded or weakened tissues, or accompany other pathogens as a secondary invader. Host plants include: *Allium, Citrus,* Cucurbitaceae, *Ficus, Gladiolus, Glycine, Gossypium, Lupinus, Lycopersicon, Morus, Phaseolus, Pisum, Populus, Robinia, Solanum, Trifolium, Vicia.* The host range of the four species distinguished here remains somewhat confused by the difficulty of attributing older records. It seems clear that *F. solani* and *F. javanicum* (under warmer conditions) behave essentially as above (although some host-specific forms exist, cf. below), while *F. coeruleum* and *F. eumartii* are more especially potato pathogens (at least in Europe). *F. eumartii* seems to be associated with stem canker on its various tropical and subtropical hosts (cacao, cassava, coffee) but there seems to be some doubt whether this is the same fungus as the potato pathogen (Gerlach & Nirenberg 1982).

Host specialization

Some of the variation in host range of *F. solani* in the widest sense is covered by its present treatment as four *Fusarium* species. *F. eumartii* and *F. coeruleum* do not appear to produce any host-specific forms. Significant host specialization is found especially in *F. solani* itself, including **f.sp. *pisi*** (Jones) Snyder & Hansen affecting peas, **f.sp. *fabae*** Yu & Fang affecting *Vicia*, **f.sp. *phaseoli*** (Burkholder) Snyder & Hansen affecting *Phaseolus*, and **f.sp. *cucurbitae*** Snyder & Hansen affecting cucurbits. The forms on pea and *Phaseolus* are capable of cross-infection, while isolates from *Vicia* will occasionally attack pea. Isolates from pea will attack forage legumes such as *Trifolium pratense* and *Medicago sativa*, while those from *Trifolium* will infect pea (Clarkson 1978). Some of the other formae speciales described within *F. solani* (Booth 1971) may actually concern *F. javanicum* (Gerlach & Nirenberg 1982).

Geographical distribution

On account of the various taxonomic concepts, reports on the occurrence of the above-mentioned species in literature must be interpreted with some caution. *F. solani* seems to occur world-wide, but preferably in temperate zones. *F. javanicum* and *F. eumartii* are recorded from areas with long hot summers in all continents and from glasshouses elsewhere. *F. coeruleum* occurs in the traditional potato-growing areas, i.e. C. and N. Europe, N. USA and Canada.

A. *Diseases on potatoes*

Diseases

F. solani seems to be a secondary potato tuber invader, following primary pathogens like *Phytophthora infestans* and *Erwinia carotovora*. In older reports the varieties *'martii'* and *'striatum'* are described as 'true' causes of potato rots. *F. coeruleum* (Boyd 1972; Langerfeld 1978) is a typical 'dry rot' pathogen of potato tubers (Fig. 39). Wounds and broken or damaged sprouts are entrances for the fungal hyphae. The external rot symptom is a greyish brown zone, frequently marked by concentric shrinkings of the skin. White to bluish cushions of mycelium emerge sometimes from lenticels and wounds. On cut tubers, the rotted tissue changes from a pale brown within a few minutes to a darker brown. Zones with progressed rots show cavities, sometimes lined by a bluish stroma. The front of the rot zone grades into healthy tissue. Above-ground parts are not directly affected. *F. javanicum* and *F. eumartii* are causes of wilts of potato plants. Both fungi infect roots, above-ground parts, and tubers. Plant invasion by *F. eumartii* proceeds via roots and infected mother tubers, the symptom on tubers being a glassy to brown vascular ring and a dry, sunken rot of the stolon end. *F. javanicum* invades plants only from infected mother tubers. Tuber attack is known as 'jelly end rot'.

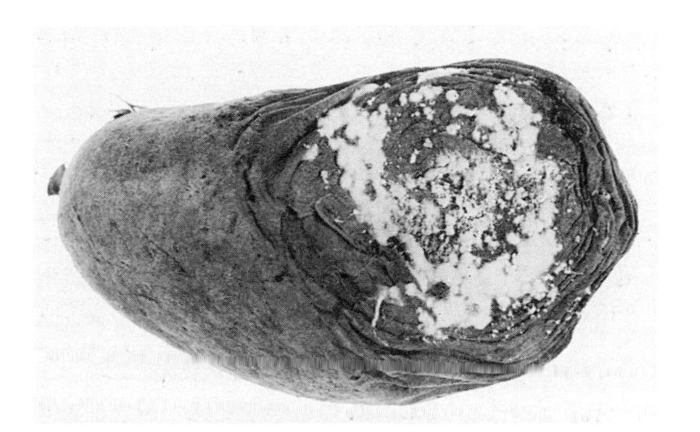

Fig. 39. External symptoms of dry rot of potato by *Fusarium coeruleum*.

Epidemiology

Permanent infestation of soil by thick-walled chlamydospores is characteristic but the main sources of transmission are infected seed tubers, which carry the pathogens to the next season and to new, previously uninfested locations. The increase of losses by *F. coeruleum* in the last half century correlates with the increase of tuber injury by machinery, in the course of mechanization of potato production. Another cause of this increase is the modern storage house as such: frost-free grading, preparation, and shipping of potatoes has become possible over the whole winter period. *F. coeruleum*, like the wound parasites *Gibberella cyanogena* and *Phoma exigua* var. *foveata*, infects and grows well at temperatures only a few degrees above zero. Furthermore, the rate of tuber injury (entrances for the fungi) is inversely correlated with the extent of defence reactions in wounds (suberization, periderm formation, etc.) in the lower temperature range.

Economic importance

Losses by *F. coeruleum* exceed in many cases the margin of economic significance. Seed is more affected than ware potatoes, above all as a result of longer storage time and higher risk of wounding by grading. Plants emerging from mother tubers with dry rot show high incidence of black-leg (*Erwinia carotovora* ssp. *atroseptica*), obviously by activation of latent bacteria in the tubers. There are no reports of severe damage due to *F. javanicum* or *F. eumartii* in the European or Mediterranean regions; cases of economically significant losses are mainly reported from America (APS Compendium on Potato Diseases). Indirect losses in seed production, by all mentioned species, occur by exceeding the legislative limits for rotted tubers, non-emerged or stunted plants, and black-leg.

Control

Recommendations for the reduction of tuber attack by *F. coeruleum* aim at technical and agricultural improvements during growth, harvesting and storage. The following measures are recommended: i) a curing period after harvest (1–2 weeks, tuber temperature above 12°C); ii) careful tuber handling (without wounding); iii) no tuber treatment below 8–10°C; iv) periodic ventilation to avoid water surplus and oxygen deficiency; v) use of cultivars with physiological and mechanical resistance. The most frequently used chemical compound for tuber treatment is thiabendazole, applied during harvesting or at storage house entrances. Thiabendazole persists for several months in its active form on tuber surfaces. Seed treatment interrupts the 'infection chain' to the next tuber generation.

B. Diseases on legumes

Diseases

Infection of pea plants by *F. solani* f. sp. *pisi* causes the root and hypocotyl cortex to become blackened and rotten, resulting in foliar chlorosis and general stunting of the plant. There is often a brilliant red discoloration of the root vascular system. In contrast to the effects of invasion by *F. oxysporum* f. sp. *pisi* this discoloration does not extend to the stem. However, isolates of the two fungi in fact closely resemble each other both culturally and in their ability to cause root rotting and wilting syndromes (Bolton & Donaldson 1972) and it has been suggested that they may constitute a single entity. Symptoms caused by f. sp. *fabae* and f. sp. *phaseoli* to their respective hosts consist of dry rotting of the upper tap-root region and stem collar at or just above soil level. The stem tissue takes on a reddish colour which may gradually darken and eventually become necrotic, particularly on *Vicia*. In sub-lethal attacks, secondary roots may be formed by *Phaseolus* plants. The foliage usually shows slow chlorosis and desiccation due to a reduction in water uptake by the root system.

Epidemiology

Optimal soil temperatures for disease development caused by *F. solani* are 26-28°C. Disease in peas is encouraged by high soil moisture deficits. Perennation takes place by saprophytic growth on decaying debris. Chlamydospores will germinate in the presence of root exudates from many non-hosts. In the rhizosphere of host plants there can be a saprophytic growth phase prior to penetration of the root cortex. Chlamydospore formation is promoted by soil aeration whereas host invasion and symptom development are encouraged by soil compaction (Burke *et al.* 1980). Dissemination occurs by movement of infected debris and in drainage water.

Economic importance

Foot-rot syndromes caused by *F. solani* have been reported for many years in Europe and N. America as major factors resulting in substantial yield losses to pea and bean crops.

Control

Pea lines with some resistance to f. sp. *pisi* have been identified (Muehlbauer & Kraft 1978). Chemical soil fumigation will control *F. solani* but is not economic. The main means of control is by use of 6-8 year rotations which avoid all legume species.

C. Diseases on cucurbits

F. solani f. sp. *cucurbitae* causes wilt symptoms of cucurbits (especially courgette), with rotting and maceration at the stem base (unlike the wilt caused by *F. oxysporum*). It is important especially in the Mediterranean region. Infection occurs from residues in soil or via infected seed, so control is through rotation and the use of clean seed (Dixon 1984).

Special research interest

For potato, recent research work points above all at early biochemical processes in wounded and infected tubers. Substances with antifungal activities against other pathogens (e.g. chlorogenic and coffeic acid) do not appear to have a decisive influence on defence reactions in potato tubers after infection by *F. coeruleum*. Another point of research interest is the production of resistant clones by conventional breeding and testing methods as well as by testing meristem cultures of strains with different ploidy states directly in culture exudates of the fungi. Studies of the pathogenicity of *F. solani* to legumes reveal that phytoalexins such as kievitone and pisatin are either detoxified or metabolized (Kuhn & Smith 1979). Soils suppressive to the development of *F. solani* f. sp. *phaseoli* populations have been identified in Japan. In these soils macroconidia fail to germinate successfully and chlamydospore formation is inhibited (Furuya *et al.* 1979). The incorporation of barley straw into infected soil has been associated with lysis of *F. solani* propagules. The addition of either linseed, cotton seed, soybean meals or chitin was associated with a reduction in populations of *F. solani* (Zakaria & Lockwood 1980). Cross-inoculation of peas with avirulent strains of *Fusarium* reduced symptom development caused by *F. solani* f. sp. *pisi* (Kováčiková 1980). It may, therefore, be possible to develop forms of biological control.
LANGERFELD; DIXON

Nectria cinnabarina (Tode) Fr.

(Anamorph ***Tubercularia vulgaris*** Tode)

N. cinnabarina produces coral-pink conidial pustules with elliptical conidia and red perithecia. It has been recorded on over 100 broad-leaved trees and shrubs, and also on *Larix, Picea* and *Pinus*. As a parasite it occurs mainly on *Acer* and *Ulmus,* and on beech hedges. *A. pseudoplatanus* is very susceptible but *A. negundo* appears to be immune. Elm cv. Christine Buisman, bred in the Netherlands for resistance to Dutch elm disease, was withdrawn because of its susceptibility to coral spot (Heybroek 1983). The fungal form commonly occurring on *Ribes* has been named ***N. ribis*** (Tode) Oudem. The fungus is worldwide in its distribution.

N. cinnabarina usually occurs only as a saprophyte, but sometimes spreads to living tissues and causes dieback and bark necrosis. The characteristic conidial pustules (coral spot) and later, perithecia, arise at all times of the year on the dead tissues. The wood of affected plants may stain brown or green in *Acer* and violet in *Fraxinus. N. cinnabarina* is mainly a wound parasite, spreading from wounds caused by pruning and hedge clipping, and by storms, hail and frost. Unsterilized pruning tools may carry its spores. In most areas *N. cinnabarina* causes serious damage only sporadically, but in some places, for example, the Netherlands, it can be an important pathogen against which control measures need to be taken. In nurseries, parks, gardens and arboreta, coral spot may be controlled by pruning and hedge clipping only in dry weather, or treating wounds with wound protectants, removing and burning woody plant debris, and spraying with a copper fungicide.
PHILLIPS

Nectria coccinea (Pers.) Fr.

N. coccinea (CMI 532) occurs on many broad-leaved trees, including *Alnus, Carpinus, Fagus, Populus, Quercus,* and *Ulmus*, as well as *Taxus*. It is widespread in Europe (also in Asia, Australia, and as **var. *faginata*** Lohman *et al.* (CMI 533) in N. America). It causes cankers on *Acer* and *Populus*, but is much more important in connection with beech bark disease (Houston & Wainhouse 1983) of *Fagus sylvatica* (and in N. America, of *F. grandifolia*). The leaves of parts of the crowns of affected trees become small, sparse and chlorotic, and black, sap-exuding, tarry spots appear on the trunks. The tissue under these spots is orange-brown, with a characteristic smell. Later, fruit bodies of *N. coccinea* may appear on and around the lesions. Rough cankers form on the trunk, or the bark may disintegrate and fall off, and the tree dies.

N. coccinea is a weak parasite attacking beech bark only in the presence of other adverse factors. The most important of these is infestation by the beech scale, *Cryptococcus fagisuga*; drought and inadequate nutrition may be others (Lonsdale 1980). The disease may sometimes destroy more than half the trees in a beech crop, and much reduce the value of the remainder. The diseased trees are soon also attacked by decay fungi, which quickly cause extensive rot. On a small scale, beech bark disease can be prevented by removing *C. fagisuga* infestations

before infection by *N. coccinea* occurs, by scrubbing the tree trunks or spraying with tar-oil winter wash in winter or with diazinon in spring or autumn. At present little can be done on a forest scale, except to fell and use affected trees as soon as possible, before decay fungi begin to affect the wood.
PHILLIPS

Nectria fuckeliana C. Booth

(Anamorph ***Cylindrocarpon cylindroides*** Wollenw. var. *tenue* Wollenw.)

N. fuckeliana forms red perithecia, *Cephalosporium*-type colourless microconidia and *Fusarium*-like macroconidia (CMI 624). It is almost entirely confined to spruces, especially *Picea sitchensis* and *P. abies*. It has occasionally been found on *Abies, Pinus* and other conifers. The form on pine has been distinguished as ***Scoleconectria cucurbitula*** (Tode) C. Booth. It occurs throughout Europe and N. America. The related N. American ***N. macrospora*** (Wollenw.) Ouellette (anamorph ***C. cylindroides* var. *cylindroides***) has been found on *Abies* in Norway (CMI 623). *N. fuckeliana* causes stem cankers with varying symptoms. Sometimes it gives rise to dead, sunken patches round the bases of branches. Sometimes it kills vertical strips of cambium, often at the base of the trunk (which then becomes fluted), or the strips may stretch several metres up even into the base of the crown. The dead bark usually remains attached to the fluted strips. Sometimes small branches are girdled, and parts of the crown die back, or the whole crown or even the entire tree may die. In moist conditions, pale yellowish conidial pustules and the red perithecia may form on the diseased tissues. Resin may exude from the damaged stems. *N. fuckeliana* appears to be a weak wound parasite, which can enter the tree only following damage by some other agent, such as winter cold, and wounding by brashing or extraction, or by insects, deer or other animals. It usually kills or deforms the stems of only a few scattered trees, and so overall it has not caused heavy losses. Diseased stems should be removed when the crop is thinned.
PHILLIPS

Nectria mammoidea Phill. & Plowr. causes minor cankers on various trees and shrubs. Its **var. *rubi*** (Osterwalder) Weese (syn. *N. rubi* Osterwalder) (anamorph ***Cylindrocarpon ianthothele*** Wollenw.) causes crown rot of raspberry in the UK but is only weakly pathogenic.

Nectria radicicola Gerlach & Nilsson

(Anamorph ***Cylindrocarpon destructans*** (Zinssmeister) Scholten, syn. *Cylindrocarpon radicicola* Wollenw., *Ramularia destructans* Zinssmeister)

N. radicicola occurs on a very large number of woody and herbaceous plants (Gymnosperms, Monocotyledons and Dicotyledons). Results obtained from trials with ornamental plants indicate that parasitic strains seem to be host-specific. It has even been found in cysts of some nematodes and in an earthworm (*Allolobophora caliginosa*) (Domsch *et al.* 1980). The fungus occurs particularly in temperate regions. It is widespread throughout Europe and has been reported from N. America, E. and S. Africa, Australia and New Zealand (CMI 14). The fungus has frequently been isolated from the root surface, rhizosphere and mycorrhiza of different plants (Domsch *et al.* 1980). Gerlach (1961) reviewed the disease symptoms caused by *N. radicicola* which ranged from lesions on roots or bulbs, root rot, storage rot and dry rot on various plants, to black rot on grapevines. *N. radicicola* is soil-borne and found mainly in cool, moist soils with low acidity; it can occur at considerable soil depths (to 100 cm). The optimum temperature for its growth is 20–21°C or even lower. Crop rotation affects the occurrence of the pathogen, e.g. the fungus occurs more frequently in wheat fields after cultivating wheat than after pea or rape. It can exist as a saprophyte in the soil and will form chlamydospores as a survival organ.

N. radicicola is considered to be a weak pathogen but damage caused by it can be of economic importance particularly in ornamental plants such as *Cyclamen, Begonia, Hydrangea* and *Saintpaulia*. The control of soil-inhabiting fungi such as *N. radicicola* is still problematic. The application of chemicals (e.g. formalin, chloropicrin, methylisothiocyanate) to the soil, steam treatment of the soil and the amendment of soil conditions contribute to the reduction of the fungus in infected soil. Farmyard manure also reduces its frequency in the soil, whereas nitrogen fertilization increases it.
EHLE

Nectria ditissima Tul. & C. Tul. (anamorph ***Cylindrocarpon willkommii*** (Lindau) Wollenw.) is known in many parts of Europe. It is almost confined to beech, on which it sometimes causes a severe canker and dieback. ***Pseudonectria rousseliana*** (Mont.) Wollenw. (anamorph ***Volutella buxi*** (Corda) Berk.)

causes conspicuous dead brown patches on plants of *Buxus sempervirens*.

Calonectria kyotensis Terashita

(Syn. *Calonectria uniseptata* Gerlach, *Calonectria floridana* Sobers)
(Anamorph ***Cylindrocladium floridanum*** Sobers & Seym.)

C. kyotensis attacks peach and various ornamentals: *Azalea, Dianthus barbatus, Paphiopedilum callosum, Pinus, Rhododendron* and *Syringa* (CMI 421). In pathogenicity trials the range of host plants was even wider (Sobers 1972). The fungus was first described from Florida (USA) in 1967 and has been reported in England, Germany, Japan and Mauritius. *C. kyotensis* causes root rot and is also associated with wilting, decline symptoms and the death of plants. Seedlings and cuttings are particularly endangered. Leaves of infected plants may exhibit lesions and can be chlorotic and stunted. The fungus is soil-borne and hence attacks mainly roots and seldom aerial parts of plants. Disease development can be rapid. Peach seedlings died 20–26 days after inoculating the soil with ascospore suspensions. Factors affecting disease severity in ornamentals are: varietal differences, temperature and amount of rainfall (Horst & Hoitink 1968). The pathogen seems to be of only regional economic importance, although severe losses (mainly seedlings and cuttings) can occur. Reports of satisfactory control measures are scanty. Soil drenches with benomyl controlled the fungus occurring on ornamental plants (Horst & Hoitink 1968). A method of control on tea bushes in Mauritius consists of a combination of chemicals (e.g. benomyl applied as soil drenches), fertilizers and earthing-up (CMI 421).
EHLE

Cylindrocladium scoparium Morgan

(Syn. *Diplocladium cylindrosporum* Ell. & Ev., *Cylindrocladium pithecolobii* Petch)

C. scoparium forms penicillate conidiophores with long sterile appendages characteristic of the genus (Domsch *et al.* 1980). It is a polyphagous plant parasite occurring on a very wide range of host plants particularly on ornamental woody plants. It has a world-wide distribution and has been reported from N. and S. America, Europe, Africa, Asia, Australia and New Zealand (CMI 362). *C. scoparium* is one of the most pathogenic species of its genus and causes damping-off, root rot and blight, stem lesions, stem canker, stem dieback, stunting, and also leaf spots, wilted leaves, defoliation and fruit rot. The optimum temperature for mycelial growth ranges from 25 to 30°C. The production and germination of conidia demands high humidity. These spores are dispersed by air currents and water splash. They rapidly lose their viability under dry conditions. Chlamydospores, in chains or clumps (microsclerotia), are the main source for spread, infection and survival and these can survive low temperatures for several weeks and remain viable in natural soil for several years (Domsch *et al.* 1980). The population of microsclerotia is affected by the cover crop. The cultivation and incorporation of maize in the soil caused it to decrease, whereas soybean used as a cover crop had the opposite effect (Thies & Patton 1970). The fungus is also spread by infected plants and infected debris in the soil.

The fungus can cause considerable damage to ornamental plants, especially azaleas (*Rhododendron indicum*). Losses of plants are particularly high in nurseries, in young seedlings and in cuttings. The key to disease control is the use of healthy cuttings, seedlings and plants, the sterilization of nursery soil and fungicide applications. Where steam treatment of the soil is impractical, chemical disinfection (chloropicrin, dichloropropene) of potting soil and seed beds is useful. Sprays and drenches with benomyl or chlorothalonil may also be effective. Dipping of infected cuttings and plants in benomyl can also provide disease control.
EHLE

Fusarium culmorum (W.G.Sm.) Sacc.

Basic description

Only the anamorph occurs (CMI 26), with fairly stable morphological characters. It has slightly curved conidia, usually 4–5 (up to 9) septate, but smaller conidia with fewer septa are also frequent, developing from monophialides on very fast-growing yellowish/reddish (to dark-brown) mycelium (Nirenberg 1981). There are no microconidia. The mainly intercalary chlamydospores are oval or globose, and brown in colour.

Host plants

F. culmorum attacks not only cereals, maize and other grasses, but also a vast range of dicotyledons, e.g. carnations, asparagus, young sugar beet plants, and even mature sugar beet roots under water stress.

Host specialization

There is some variability among isolates, but isolates from different host species are pathogenic on others as well, with no indication of specialization at species or variety level.

Geographical distribution

Optimum temperature for growth of the fungus is around 25°C, with a maximum at about 32°C, but the fungus is predominantly found in areas of moderate or warmer climate, and less in tropical regions.

Disease

F. culmorum is part of the foot-rot complex, with *Gibberella zeae* on maize and wheat and *Monographella nivalis* on wheat (Duben & Fehrmann 1979). In maize, *F. culmorum* (together with other *Fusarium* spp., *Pythium* and *Drechslera*) enters very young plants via the hypocotyls or roots to cause damping-off symptoms (yellowing of primary leaves, stunted growth and finally death and decay), or stem-breaking with intensive browning of the lower nodes. Before flowering, this type of symptom is mostly unimportant. However, later attack of the stem base by *F. culmorum* may cause very serious losses. Colonization of the haulm-base and succeeding stalk rot lead to premature wilting, browning and drying up of the leaves, and the plants usually then break at one of the lower nodes and ripen prematurely.

In wheat and other cereals, foot-rot symptoms, i.e. brown discoloration of the haulm-base, cannot be seen before flowering or even milky-ripe stage, except in very dry and warm areas or spring seasons, when browning may develop quite soon after tillering (Papendick & Cook 1974). Usually colonization of the haulm-base can easily be traced back to an infection of the roots. Severe foot rot together with water stress may culminate in white-heads and more serious crop losses. In wheat, leaf-blades may show sporadic dark brown or chocolate brown, oval-shaped necrotic spots of up to 2 cm length, darker than for *M. nivalis* leaf spots.

F. culmorum can cause disastrous cob rots in maize. After infection, cobs and sheathing leaves may appear rose- or pink-coloured due to mycelium and sporodochia. In wheat, ear scab predominates in rainy seasons and humid areas. After infection single and scattered spikelets or even larger portions of the ears may bleach, sometimes leading to white-head symptoms again. Poor photosynthesis in the glumes leads to insufficient grain-filling and drastically reduced single-grain weight. Ears, or parts of them, may be covered by pink, rose or reddish, fluffy mycelium, but, in contrast to *G. zeae*, no perithecia can be seen.

Other important diseases caused by *F. culmorum* are brown-patch of turf-grasses, and storage rot of apples and sugar beet. The fungus occurs very frequently in grassland and pastures. A foot rot of asparagus after infection with *F. culmorum* from the soil causes dark-brown lesions at the stem-base; the tissue becomes softened, and stems may die (cf. also *F. oxysporum* f.sp. *asparagi*). Similar foot rots are of importance in carnations and asters.

Epidemiology

In soil, the fungus may persist for a long time without a host plant, mainly as a competitive soil saprophyte on plant debris (straw residues), or as chlamydospores. Previous host crops usually increase the soil inoculum potential and subsequent attack of cereals (or maize). Seedling disease is mainly due to seed infection, but in contrast to *Monographella nivalis,* this process requires soil temperatures over 15°C. The fungus is able to enter the roots, coleoptiles and hypocotyl of seedlings from the soil also. Foot rot and stalk rot, mostly caused by root infection, are followed by conidia formation on leaf spots or even haulm lesions. Such conidia lead to ear infection in cereals and cob infection in maize. In wheat, the use of short cultivars or shortening the plants with cycocel (CCC) promotes the infection of the upper parts of the plants. There is no proof of ear infection by systemic growth of the fungus within the plant. Timing for ear infection by conidia is optimal at or just after anthesis, the decaying tissue of the anthers being a good substrate for the fungus to start colonization of the glumes.

Economic importance

Seedling diseases of wheat and barley caused by *F. culmorum* are only of minor importance; the situation might become different after early sowing and subsequent exceptionally dry, warm weather. Crop losses due to haulm-base attack of cereals of 10% or even more predominate in warmer Mediterranean countries or in C. and W. Europe in warmer, dry areas or years. Losses from stalk rot of maize—again very much depending on weather and climate—can be enormous, with up to the possibility of total yield loss; on average, the figure will probably be in the range of 10% (cf. *G. zeae* also). Ear scab of wheat and barley is considered to be more important than foot rot due to *F. culmorum* attack. Kernel infection lowers the germination rate and baking quality. Artificial inoculation at, or after, flowering may reduce yield by up to 60%, mainly by decreasing the single-grain weight and the number of grains per ear. Similar figures on yield losses should be true for maize cob rot too. Heavy mycotoxin production after severe colonization of grains, and also after harvest in storage, may prevent their use for milling or animal feed (Lásztity & Woeller 1975).

Control

There is no way to suppress *F. culmorum* attack by crop rotation or intercropping (e.g. with rape). Healthy seed must be used especially for early sowing and in warmer areas or seasons, and on sandier soils. Seed treatment for wheat and other cereals is not as promising as for *M. nivalis* control; with maize, however, seed coating with protectants such as captan, ziram or thiram is recommended. Every measure to diminish or to avoid water stress is of advantage. In wheat there are no resistant cultivars, but some differences in susceptibility. With respect to ear-scab, short-straw cultivars are as inappropriate as applications of cycocel (CCC) made too late. In maize, early-maturing cultivars are usually more susceptible. Sprays against *Pseudocercosporella herpotrichoides* (carbendazim-generators, prochloraz) do not control *F. culmorum* attack at the haulm-base. In experiments with winter wheat, it was shown that spraying guazatine before an artificial inoculation of ears at or just after anthesis, and pre- or post-infectional application of prochloraz could lower yield loss in the order of 20% only. As this is no more than a good side-effect of the mentioned chemicals, one should admit that no potent fungicide is available for the control of ear-scab in cereals or cob rot in maize. For the use of carbendazim-type fungicides against *F. culmorum* ear attack, see *M. nivalis*.
FEHRMANN

Fusarium oxysporum Schlecht.

Basic description

This most important species of the genus *Fusarium* has been described recently by Booth (1971), Messiaen & Cassini (1981), Gerlach & Nirenberg (1982) and Nelson *et al.* (1983). On PDA, colonies have a variable appearance, depending on the strain: generally, the aerial mycelium is at first white, and the medium discolours to various shades from violet to deep purple; colonies may appear cream or orange if sporodochia are abundant. Microconidia, which are always present, are oval-ellipsoid, mono- or bicellular, and formed on short unbranched phialides, never in chains, but grouped in false heads. Macroconidia, usually 3–5 septate, are fusoid, slightly curved, and often with a foot-shaped basal cell. They form first from individual phialides, then from sporodochia. Chlamydospores are solitary or in short chains. No teleomorph is known.

Host specialization

F. oxysporum is first and foremost an abundant and active saprophyte in soil and on organic matter. In addition, some strains have specific pathogenic activity. These constitute only a very small proportion of the total soil population, but cause serious crop diseases. Some are poorly specialized and cause seedling blight, necrosis or rots. However, the forms responsible for vascular wilts are the most important. In accordance with the concepts of Snyder & Hansen (1940), these strains can be considered as pathotypes specific either to species (formae speciales) or to some cultivars within a species (races). About 80 formae speciales have now been identified, and several are subdivided into races (Armstrong & Armstrong 1981). They cause losses on plants in all major Angiosperm families (except Gramineae), whether in temperate or tropical regions. The most important are considered in separate sections below. Table 11.1 summarizes the other formae speciales which occur in or present a risk to Europe.

Disease symptoms

Symptoms of 'fusarium wilt' vary according to host plant, pathotype and infection conditions. In general, the older leaves first show light vein-clearing, chlorosis of the lamina and/or wilting, these symptoms progressing to younger leaves, and often starting unilaterally on some leaves, corresponding to a localized infection of part of the root and stem vascular system. Affected parts of the plant turn brown; longitudinal necrotic streaks appear on the stems and spread towards the stem apex. Internal symptoms can be seen with the naked eye in sections of roots, stems or petioles. Browning starts in the vascular tissue and spreads to the cortex in the later

Table 11.1. Formae speciales of *Fusarium oxysporum* of relatively minor importance in Europe

Forma specialis	Hosts	Races	Optimum temperature*	Remarks	References
albedinis (Killian & Maire) Gordon	*Phoenix*		HT (28)	Serious quarantine organism causing Bayoud disease of date palm in N. Africa	EPPO 70
apii (Nelson & Sherbakoff) Snyder & Hansen	Celery etc.	+	HT (28)	USA. Only recently recorded in Europe	Bouhot & Olivier 1977
betae (Stewart) Snyder & Hansen	Beet	–	–	USA, USSR	
callistephi (Beach) Snyder & Hansen	*Callistephus chinensis* etc.	+	HT (28)		Fantino & Mazzucchi 1975
chrysanthemi Armstrong, Armstrong & Littrell	chrysanthemum	+	HT (28)	USA. Only recently recorded in Europe (Germany, Italy)	Gerlach *et al.* 1980
ciceri (Padwick) Matuo & Sato	chickpea		HT	An important disease in the tropics	Trapero Casas & Jimenez Diaz 1984
cucumerinum Owen	cucumber	+	HT (30)		CMI 215
cyclaminis Gerlach	cyclamen	+	HT (30)	A major constraint on cyclamen production in Europe	Rouxel & Grouet 1975
fabae Yu & Fang	vicia beans		?		Salt 1982
fragariae Winks & Williams	strawberry		?	Not yet in Europe	APS Compendium
lilii Imle	*Lilium* spp.		?		Bollen 1977
lupini Snyder & Hansen	lupin	+	?	Especially E. Europe (Poland)	Salleh & Owen 1983
medicaginis (Weimer) Snyder & Hansen)	lucerne		?		APS Compendium Bakheit & Toth 1983
phaseoli Kendr. & Snyder	phaseolus beans etc.	+	?	Serious in N. Greece	Aloj *et al.* 1983
raphani Kendr. & Snyder	radish etc.		HT (26)		Couteaudier *et al.* 1981
radicis-lycopersici Jarvis & Shoem.	tomato		LT (18)	Recent records in Europe	Couteaudier *et al.* 1985a
trifolii (Jacz.) Bilai	clover			Host-specialization status doubtful	
tuberosi Snyder & Hansen	potato		HT	Mainly in tropics	Thanassoulopoulos & Kitsos 1984

*HT = high temperature
LT = low temperature

phases of infection. Fusarium wilt develops especially fast at flowering or fruiting. It appears in the crop as foci which spread gradually, causing premature death of affected plants. Because of the consequences for control and of the possibility of confusion with other diseases, it is important to check carefully the symptoms of fusarium wilt. Even if *F. oxysporum* can be isolated from internal tissues, it may be useful to re-inoculate to check that the characteristic symptoms are obtained again.

Disease cycle

Fusarium wilts are classic soil-borne diseases. Infected plant debris is the main source of inoculum in the soil. Chlamydospores can persist in an inactive form for several years and germinate when nutrients are available, as in the vicinity of the young parts of roots. The germinated chlamydospore gives rise to inoculum by formation of hyphae, conidia and new chlamydospores. In this respect, the pathogen strains behave just like certain other *Fusarium* species, including non-pathogenic strains of *F. oxysporum*. In the rhizosphere, there is intense competition with other micro-organisms and in particular with other fusaria. Although many *F. oxysporum* strains are able to penetrate into the cortical tissues of the root, the host-specific strains are the only ones able to penetrate to the vascular tissue of a host plant and cause fusarium wilt. Penetration mainly occurs in the root elongation zone and may be assisted by wounds or nematode attack (especially by *Meloidogyne* species). The fungus then spreads up the xylem vessels by mycelial growth and by formation of microconidia which are carried by the transpiration stream. At later stages, it may spread to adjacent tissue causing externally visible necrosis. Pathogenesis is linked to blocking of the vessels and to the formation of enzymes and toxins. Fusarium wilt pathogens can also invade and colonize non-host plants, causing slight disease symptoms. Such symptomless carriers contribute to the carry-over and multiplication of inoculum.

Epidemiology and control

The fusarium wilt pathogens, which are of major importance for many crops (except cereals), can be spread in soil, dust, irrigation water, and via infected plants (but rarely by seed). Soil and vegetative cuttings are the most important agents in practice, and must be subjected to phytosanitary checks to prevent spread. Although the chlamydospores are very resistant, it is possible to use heat therapy of bulbs and disinfestation treatments for soil. The latter tend, however, to be rather ineffective for soils *in situ*, since even small quantities of residual inoculum can cause serious disease and inoculum remaining deep in the soil will rapidly recolonize the disinfested zone, in which a kind of microbiological vacuum exists. In intensive cropping systems, roots are grown in a substrate of limited volume, which can be kept free from *F. oxysporum* initially; however, re-infestation by plants, dust and irrigation water must be prevented. Some soils are suppressive to fusarium wilts, because a high level of competition from other micro-organisms prevents the formae speciales from damaging plants grown on them. This suppressive effect can even be transferred to a soil and established there, which opens prospects for microbiological control (Alabouvette *et al.* 1979, 1985; Couteaudier *et al.* 1985b).

Since acid soil, potassium deficiency and fertilization with ammonium nitrogen tend to favour the disease, liming, potassium fertilization and use of nitrates can sometimes reduce losses. High temperature, rapid growth and heavy plant transpiration all favour fusarium wilt, but these factors cannot readily be altered without directly affecting yield. Early success in control with systemic fungicides (benzimidazoles) was short-lived, since resistant strains of *F. oxysporum* rapidly appeared. The main prospects for control in fact lie in breeding resistant cultivars, now available commercially for most important host plants. Such cultivars may be highly resistant to only some races, or may be broadly tolerant to different races. In some cases, grafting onto a rootstock with a high level of broadly based resistance can give good results.

Special research interest

Interesting research areas have been reviewed in the compilations of Mace *et al.* (1981) and Nelson *et al.* (1983). They include: i) new rapid methods for characterizing *F. oxysporum* strains by their activities in soil or on hosts. Further work on the genetics, variability and stability of strains may allow one to anticipate the appearance of new pathogenic or fungicide-resistant strains or to select non-pathogenic strains with biological control potential; ii) population dynamics of pathogenic and non-pathogenic strains in different soils; mechanism of soil suppressiveness or conduciveness, and search for microbial populations which will protect the crop and can be used in integrated control procedures;

iii) nature of specific pathogenicity and of host resistance/susceptibility. The biochemistry and physiology of pathogenicity and defence reactions continue to be of considerable interest, and open possibilities for control by pre-inoculation or by effects on host nutrition. Finally, the search continues for cultivars with high, broad and durable resistance.
LOUVET

The following sections refer specifically to the most important formae speciales of *F. oxysporum* in Europe.

F. oxysporum f.sp. *asparagi* Cohen

f.sp. *asparagi* is part of a complex causing dieback of asparagus (Cassini *et al.* 1983; Fantino & Pavesi 1984). The other elements in the complex are *F. moniliforme* and unspecialized *F. oxysporum,* which alone cause root rot and crown rot, infecting especially through cutting wounds; *F. culmorum* (q.v.) has also been implicated. Only f.sp. *asparagi* spreads in the vascular system and causes wilting, but it will only cause serious infection if accompanying the other fungi, which are in any case widespread in soil. Plants show yellowing, dieback, and wilting. They may be killed, but otherwise produce smaller and fewer shoots, deteriorating over several years. The disease occurs world-wide, but has only recently become important, especially in France, because of changes in cultural practices. Breeding programmes have now made high-yielding hybrid cultivars available, and planting material is bought in from specialized nurseries rather than produced locally. Infection has spread with the nursery stock. In addition, growers now start harvesting within 2 years, instead of 3–4 years, and continue later through the season. This is feasible with the new cultivars but the plants are weakened and predisposed to infection in consequence. Control has to be applied in the nurseries. Soil disinfestation (e.g. with methyl bromide) is possible but expensive. Since one source of infection of planting material is the presence of inoculum on the outside of seed, seed treatment is required. Finally, the planting material can be dipped in a fungicide mixture before sale (e.g. carbendazim+captafol). A combination of these measures ensures that healthy planting material can be made available, which basically solves the problem.
CASSINI & SMITH

F. oxysporum f.sp. *cepae* (Hanz.) Snyder & Hansen

f.sp. *cepae* causes a basal rot of onion world-wide and may also be pathogenic to garlic and *Allium fistulosum.* However, two other *Fusarium* species have recently been reported as onion pathogens, i.e. *Nectia haematococca* and *Gibberella intricans,* from Bulgaria, Israel and Italy. Infection is initiated in the field from soil-borne propagules of the fungus capable of surviving as a saprophyte for years. By the time the early symptoms of the disease, such as leaf yellowing and/or dieback, become visible, decay has already started at the stem plate. The roots commonly turn pink and then decay; the basal portion of the bulb scales becomes rotted and this continues acropetally. Such affected plants are easy to lift from the soil. The pathogen is able to infect onion plants both through the uninjured root surface and wounds in underground tissues. Fungal invasion and disease progress is favoured at high temperatures (25–35°C) and moderate moisture conditions. Continuous cropping of onions or poor crop rotation equally enhances the occurrence and severity of basal rot, particularly in regions of semi-arid or hot continental climate. Although successful attempts have been made to produce resistant onion cultivars (Fantino & Badino 1981), chemical control seems to be the most reliable way of reducing yield losses significantly. For this, seed dressing and preplanting dips of onion sets with fungicides are recommended. A future strategy might be biological control with *Pseudomonas cepacia.* The use of a suspension of this bacterium for seed dressing or soil treatment gave promising results in some countries (Fantino & Bazzi 1981).
VIRÁNYI

F. oxysporum f.sp. *conglutinans* (Wollenw.) Snyder & Hansen

f.sp. *conglutinans* attacks a wide range of crucifers throughout warm temperate and tropical regions. There is some differential pathogenicity towards species of crucifers. The fungus causes a yellows wilting syndrome. Symptom expression is temperature-dependent, consequently infected plants may be apparently healthy until the temperature rises in early summer. Initially the foliage of infected plants turns a dull, yellow-green colour. All or only part of individual leaves may be affected. In the latter case the lamina may curl upwards due to uneven growth effects. Where the lower leaves are affected first, the syndrome gradually progresses towards the plant apex. Infection usually causes leaves to die prematurely and abscission to take place. As the chlorotic leaf tissue ages it becomes brown, dead and brittle. Typically the vascular system takes on a dark brown or yellow discoloration. Symptoms are intensified by potassium deficiency, and disease expression increases with temperatures in excess of 17°C. Transmission is by soil particles on implements, wind-borne soil, surface drainage water, water-borne soil, infected transplants and infected soil adhering to non-host material such as seed potato tubers (CMI 213). Chlamydospores provide a means of perennation.

Resistance is the most effective method of control and has not eroded despite the variability of the organism *in vitro*. Two types of resistance are known: type A is monogenically inherited and excludes the fungus from host vessels; it is unmodified by nutritional factors and is only rendered inoperative at high temperatures; type B is qualitatively inherited, allows limited entry to the vascular system, can be affected by nutritional deficiencies and is rendered inoperative by relatively small increases in temperature (Walker 1959). Only very limited means of chemical control are available, for instance, root dipping in ziram. Application of potassium fertilizers is an obvious method for alleviating symptoms. Forms of biological control may be devised since cross-protection has been achieved by prior inoculation with other formae speciales of *F. oxysporum* (Davis 1967).
DIXON

F. oxysporum f.sp. *dianthi* (Prill. & Delacr.) Snyder & Hansen

Host plants

f.sp. *dianthi* attacks *Dianthus* in general, with the exception of some species (*arenarius, brachyanthus, chinensis, petraeus*).

Host specialization

Eight pathotypes were found by Garibaldi (1981) in European countries on inoculation of 10 carnation cultivars. The most frequent were numbers 1 and 2, which represent 23 and 56% respectively of all isolates. Number 2 is common on American carnation, but the discovery of new resistant cultivars will probably increase the frequency of the other pathotypes.

Geographical distribution

Since the first reports in Connecticut (USA) and France in 1898–9 the disease has spread quickly and disastrously to all growing areas in the world.

Disease

The first external symptom appears on the leaf bases as a discoloration, gradually followed by withering. These symptoms slowly spread upwards into the lateral branches and shoots, characteristically on one side of the plant, causing curling of the stem. Stunting of the shoot or whole plant occurs especially if infection is initiated early. Following the initial one-sided development in the main stem, the entire plant finally wilts if environmental conditions are suitable. Within the plant, the affected tissue stands out sharply from neighbouring healthy tissue and appears whitish when pulled out. As the disorganization continues, the stem feels hollow to the touch. In hot and humid environmental conditions, white mycelium appears including many macroconidia and sometimes pink- to orange-coloured sporodochia. The hollow stems are often covered with mycelium.

Epidemiology

There are two main sources of inoculum—glasshouse soil and infected cuttings (possibly from symptomless stock). In the rooting substrate, the fungus sporulates and surrounding cuttings are infected. A commercial lot of cuttings may carry the fungus only in the substrate, or possibly in the roots as well. Infection of cuttings at the stem level is less frequent. According to the inoculum level and temperature, the incubation period varies from one month to several months. The optimum temperature is 25–30°C and in natural conditions a latent period of several months is observed consecutive to a decreasing temperature in winter. In the Mediterranean area all cuttings taken from symptomless stock in January–May are practically disease-free, the low temperatures stopping the development of the fungus in the vascular bundles of the mother plant. After planting in mid-June on infested soils, the first symptoms appear in early August and the spread of the disease slows down in October, remaining unchanged until early June. The dust from lanes between infested benches has been found to contain more than 10 000 propagules/g (Ebben & Spencer 1976) indicating an enormous inoculum potential in a glasshouse where the disease is present. The fungus can survive in soil (probably as chlamydospores) for at least 14 years and to a depth of 80 cm. Most infections start in the roots, or occasionally through wounds in the stem (but not on mature plants).

Economic importance

Fusarium wilt is the most important carnation disease in the world. For instance in S. France, all carnation areas over 15 years old were found so infested that economic crops could no longer be produced and other ornamental plants had to be substituted.

Control

Basically, the disease must be prevented by the use of healthy cuttings under clean conditions (uninfested substrates). Cultivation in raised benches, in steamed or fumigated soil, or else in soil-free growing media has long proved satisfactory, but this has tended to be replaced, for economic reasons, by use of ground beds in which control is much more difficult (Baker 1980). Disinfestation of infested greenhouse soil with fumigants (methyl bromide or metam sodium) sometimes results in a reduction in fusarium wilt but not an effective one. This is due to several factors: adsorption of fumigants, poor penetration to mycelium in fragments or at depth. Another approach to control is the adoption of benches with soil suppressive to fusarium wilt or substrate containing antagonists. In France, perlite becomes suppressive after colonization by a strain of *Fusarium roseum* isolated from suppressive soil (Tramier *et al.* 1979). Tolerant cultivars are known mainly in the 'spray' group (Elsy, Teddy, Juanica, Exquisite etc.). Among the large-flowered Mediterranean ecotypes, Athena, Diano, Pallas, Orion, Castella, Monte Carlo and Amapola are the most resistant in severe conditions. But in bench culture with a low inoculum level, cultivars with medium tolerance like Salome, Sacha, Pamela, Angela and Valero have a good record against fusarium wilt.
TRAMIER

F. oxysporum f.sp. *gladioli* (Massey) Snyder & Hansen

f.sp. *gladioli* causes corm rot and leaf yellowing in *Gladiolus* and similar diseases in *Crocus, Freesia,* bulbous *Iris, Ixia,* and some other Iridaceae. Isolates from *Gladiolus* are in general highly pathogenic to *Iris*, but isolates from *Iris* usually show only slight pathogenicity to *Gladiolus*. Similarly, isolates from large-flowering *Gladiolus* cultivars readily infect plants of the early-flowering *G.* x *nanus* and *G.* x *colvillei*, whereas isolates from the latter hardly infect large-flowering gladioli. Such observations suggest a considerable variability in pathogenicity with respect to the various host plants, within which distinct differences in susceptibility between cultivars have also been observed. The pathogen is present wherever *Gladiolus* and *Iris* are cultivated. In the Mediterranean countries, as in Florida (USA) and other subtropical regions, the disease constitutes one of the major problems in the production of these crops. In all host genera the pathogen causes premature leaf yellowing, often starting with the outermost leaf tips. Leaf curvature is seen frequently, especially in *Iris*. In corms of *Crocus, Gladiolus, Freesia,* and *Ixia*, rot develops in the corm base and rapidly enlarging yellowish to dark brown lesions are often found elsewhere. The central core turns brown, and this discoloration may spread into the lateral vascular bundles. The affected tissues shrink and harden during corm storage, and often show more or less distinct concentric ridges. In *Iris* a greyish-brown basal soft rot appears, sharply demarcated from the healthy tissue. In all corms or bulbs, external white or pinkish mycelium with many conidia may be formed during storage. Root rot is also evident, in *Gladiolus* combined with decay of the contractile roots.

The pathogen survives in soil for several years and is easily spread by conidia during storage. Outgrowth from infected corms or bulbs, causing infection of neighbouring plants, has been demonstrated in several host crops, as has the existence of latent corm infections in *Gladiolus*. Relatively high soil temperatures strongly favour disease development and therefore the disease is particularly severe during warm summers and in glasshouse cultivation. Under these conditions the economic importance may be considerable. Fungicidal disinfection of corms or bulbs should be combined with such cultural methods as crop rotation, rapid drying and dry storage, early harvest (especially for *Iris*) and glasshouse soil disinfection (Scholten 1971).
BERGMAN

F. oxysporum f.sp. *lini* (Bolley) Snyder & Hansen

In vitro, f.sp. *lini* produces macro- and microconidia, chlamydospores, and a mauve to violet pigment, according to the strain and medium used. The fungus is specifically pathogenic to flax, but can persist in soil or in various weeds or crop plants without causing symptoms. It occurs world-wide, and throughout most of Europe (CMI map 32), being most serious in USA, Japan, USSR and India. In France, it is widespread and of variable importance in the different flax-growing areas. Symptoms are very variable as are the times of appearance and intensity; these depend upon the inoculum level in the soil, climatic and crop conditions, cultivar etc. Infected plants become chlorotic, then wilt and dry out while remaining upright. The stems brown and the roots blacken. Whitish mycelium and sporodochia may occasionally be seen at the stem base. In serious early attacks, the whole plot may be affected. In later more moderate attacks, brown areas of variable size are seen in the crop. The harvested flax gives short fibres, which tend to break and are unsuitable for industrial processing; in addition, retting of infected flax proceeds poorly. The disease is certainly one of the most serious for the crop because it is common and because it causes losses to both fibre and linseed crops. The effects on fibre quality are particularly important. The fungus can survive in soil up to 1 m depth, and also survive and be dispersed in flax debris (fragments from the harvester, tow). It appears to prefer acid soils, and to be selectively affected by certain conducive soils. Climatic conditions are very important for disease development, in particular high humidity. The cardinal temperatures are 15°C, 25–28°C and 38°C. While specific races have been described in the literature, only differences in aggressiveness are known in France.

The main method of control is to avoid soils known to be infested and/or to follow a sufficiently long rotation: 7 years is recommended, and clover or oats are sometimes recommended as preceding crops. Liming can reduce the spread of outbreaks, while micro-elements will reduce severity. All residues from infected crops should be destroyed, and accidental transport from field to field by harvesting machinery must be avoided. On infested soils, resistant cultivars provide the only solution.
JOUAN

F. oxysporum f.sp. *lycopersici* (Sacc.) Snyder & Hansen

Host specialization

The fungus infects only *Lycopersicon* (CMI 217). Two pathotypes of the fungus are known in Europe: races 0 and 1. A new third race was reported in 1982 in Queensland (Australia).

Geographical distribution

The disease was first described in Europe in 1895 in Guernsey and the Isle of Wight (UK) but is now of world-wide importance. It occurs in Europe on tomatoes primarily under glass.

Disease

In young plants, one of the earliest symptoms is drooping of the petioles (epinasty). Yellowing of the lower leaves appears first, usually affecting the leaflets unilaterally. The affected leaves die, and the symptoms continue to appear on successively younger leaves. One or more branches may be affected while others remain symptomless. After the disease has advanced for a few weeks, browning of the vascular system may be seen in cross-sections of the lower stem (Walker 1971; Jones & Woltz 1981). Since the symptoms of verticillium wilt have much in common with those of fusarium wilt, isolation and identification of the pathogen is thus necessary.

Epidemiology

Soil and air temperatures are extremely important environmental factors in the development and severity of the disease. The optimum soil temperature for disease development is about 28°C and the above-ground symptoms are most rapid and severe when air temperature is also at that level. The fungus tolerates a wide range of pH in soil; it becomes established readily in many soil types; such soils remain infested almost indefinitely. Repeated planting of tomato crops in a soil generally increases the risks of disease occurring. Dissemination occurs via soil, infected transplants and infected crop residues. In soil-less culture, the main source of contamination is infested soil introduced with transplants, substrate, or irrigation water. Race 1 seems not to be spreading as rapidly as race 0 did early this century.

Economic importance

Fusarium wilt has been reported in at least 32 countries and is potentially one of the most damaging diseases of tomato, since infected plants wilt and die. However, the current availability of resistant cultivars (cf. below) keeps actual losses fairly low.

Control

Disease resistance was early found to be a promising means of control. Numerous cultivars and hybrids are commercially available with resistance to both races. Nevertheless, in various localities the use of resistant cultivars is not well adapted to requirements and some attempts at control by other means should be mentioned. Successful experiments were reported with the use of a resistant rootstock upon which scions of a desired type are grafted. Soil treatments by steam or fumigation (methyl bromide) can alleviate disease losses. The original objective of 'sterilization' in the greenhouse is eradication of soil-borne pathogens, but fears have been expressed that recontamination of pasteurized soil will cause greater losses than if the soil had not been treated. Application of chemicals during plant growth is expensive and does not give good control. Current studies (Alabouvette *et al.* 1985; Couteaudier *et al.* 1985b) on the introduction into the soil of selected specific antagonists from suppressive soils offer possibilities of biological control. Tomato breeding programmes throughout the world are concerned with the development of cultivars resistant to both pathotypes.
COUTEAUDIER

F. oxsporum f.sp. *melonis* (Leach & Currence) Snyder & Hansen

Host plants and specialization

F. oxysporum f.sp. *melonis* (CMI 218) is strictly specialized to melon. It does not attack other Cucurbitaceae although some isolates do sometimes induce symptoms on watermelon. There presently exist, world-wide, four pathotypes of f.sp. *melonis*: R 0, R 1, R 2 and R 1–2 (Risser *et al.* 1976) determined according to their virulence on two independent resistance genes *Fom* 1 and 2 (Mas *et al.* 1981). Their existence was confirmed in France by Risser (1973). Strains can also differ in aggressiveness.

Geographical distribution

f.sp. *melonis* is a very common and destructive pathogen in most melon-growing areas, widely reported in Europe, America and Asia. In Europe, pathotype 0 is commonest and pathotype 2 rarest (found only once in the Netherlands). Areas that become heavily infested are unsuitable for growing many common melon varieties.

Disease

The fungus can attack melon at any stage. Two typical syndromes are distinguished: either a slow and progressive yellowing (in most cases) or a sudden wilting. In the first case, the veins of some leaves turn yellow on one side, and these leaves become more and more yellow, thick and brittle. During this stage, the leaves exhale a typical odour and the stems are streaked by longitudinal necrosis from which gum exudes. Subsequently, the fungus sporulates on the necrotic zones forming pinkish-coloured sporodochia. These symptoms occur with pathotypes 0, 1, 2 and certain isolates of 1–2 (causing necrotic yellowing before the plant dies). In the second case, other isolates of race 1–2 cause a sudden wilting of the plants without prior yellowing or odour. Especially in the first case, a transverse section of the stems shows that the attacked vessels are discoloured to a clear brown. The plants show external symptoms especially when the fruits are maturing.

Epidemiology

f.sp. *melonis* forms abundant chlamydospores in soil, and sporodochia on stems of diseased plants provide a large source of inoculum. The inoculum level rapidly increases in the rhizosphere of a first crop, so that the second crop is severely attacked. A new pathotype (e.g. race 1 in France) may be widespread at low inoculum levels before susceptible cultivars are used (Bouhot 1981), occurring on organic matter in the soil or on the roots of symptomless carriers (Banihashemi & de Zeeuw 1975). The most severe symptoms are generally observed at 18–22°C. Contrary to most of the fusarium wilts, melon wilt is a disease that occurs in cool soils, and early in the season. Pathotype 1 seems less susceptible to higher temperatures. In addition, insufficient illumination and short day lengths increase disease, and the disease intensity is higher under dry conditions: in artificial inoculations symptoms occur more quickly when the RH of the air is 50–65%. High nitrogen also influences wilt development, possibly explaining its severity on liberally fertilized greenhouse plants.

Alluvial soils of the Chateaurenard area in France are naturally suppressive to f.sp. *melonis* (Louvet *et al.* 1981) and this was subsequently shown to be due to particular fungistatic properties of these soils. Insofar as the fungus reportedly survives indefinitely in the soil, low winter temperatures do not much affect the population of the pathogen especially if humus particles or intact melon tissues provide substrates which shelter the chlamydospores against environmental stress. The pathogen may also be able to colonize roots of non-host plants.

Economic importance

In France fusarium wilt rapidly increased after intensive cropping to such a point that melon crops regressed or even disappeared in certain growing areas (Saône valley, Orléans, Rennes basin). Most damage is observed in Europe on the Atlantic coast and in northern countries. During years with cool temperatures, early crops grown under glass are affected even in southern countries.

Control

This fungus, well adapted to long-term survival in soil, is particularly difficult to eliminate by crop rotation or long fallow. Disinfection of soil is feasible only for early planting where financial return is important. Technical difficulties in application have impeded the use of steam, steam–air mixtures or fumigants. Besides, following the application of fumigants, the risk of recolonization of the soil by the pathogen is serious. The hope that products with a therapeutic action (e.g. benomyl or thiophanate) could be used to prevent infection in early stages of plant growth has not been realized and applications giving a curative action require quantities which are nearly phytotoxic. Melon can be grafted on *Benincasa cerifera* which is resistant to penetration by all pathotypes. In spite of susceptibility to *Pyrenochaeta lycopersici* and *Verticillium albo-atrum* (to a lesser degree than melon), grafted plants are used by some producers in France and Italy. Growing resistant cultivars is the cheapest and easiest method of protection from fusarium wilt. Genes *Fom* 1 and *Fom* 2, together, give a very good protection against pathotypes 0, 1, and 2. In France and in protected crops, almost every cultivar has the gene *Fom* 1 (e.g. Doublon, Vedrantais) and may have genes *Fom* 1 and *Fom* 2 (e.g. Printadou, Alpha). In Spain, Italy, and in the field, susceptible cultivars can still be grown. Some non-specific, polygenic tolerance has been found, especially in Asiatic cultivars and has been introduced into some European cultivars (e.g. Piboule, Jador). This tolerance protects from pathotype 1–2 when the environment is not too favourable to the disease and when strains of race 1–2 are not too aggressive. Moreover, the transmissibility of suppressiveness to fusarium wilt of melon (Louvet *et al.* 1981) from soil to soil could offer an improved biological method of control of this disease.

Special research interest

Studies on cross protection between pathotypes (Molot *et al.* 1979) have provided an interesting model for the study of host–parasite interactions and defence reactions of the plant.

MAS & MOLOT

F. oxysporum f. sp. *narcissi* Snyder & Hansen

f.sp. *narcissi* attacks only narcissus cultivars and some species, causing two distinct diseases: basal rot and neck rot. The disease was first considered to be a storage rot because that was when the disease was most easily seen, but infection takes place when bulbs are actively growing. With basal rot the root plate is colonized via the roots and a soft chocolate brown rot develops among the fleshy scales. Under dry storage conditions the colour darkens and the bulb mummifies, but when conditions are damp a circle of sporodochia develops on the root plate which appear pink *en masse*. The fungus may be found in soils, even where bulbs have not been grown before (Price 1975a) and this represents the primary reservoir of infection. However, apparently healthy bulbs also carry the spores (Price 1975b). Infection can take place at any time during the growing season, either through the wounds caused by adventitious roots bursting through the root plate or into the bulb via the vascular system of the roots. Although infection can take place at any time of the year most occurs when soils are warm and roots begin to senesce from early summer onwards. Symptoms of early infection in the field are not particularly distinctive but yellowing of leaf tips and premature senescence are typical. Neck rot, the other expression of the disease, is typically a progressive rot of the base of the flower stalk which occurs after the flower has died or been removed. The rot penetrates the bulb and may be seen in June and July in longitudinally cut bulbs. By August the rot usually reaches the root plate and cannot then be distinguished from basal rot. Most cultivars have a degree of field resistance to basal rot especially *N. tazetta* cultivars and the trumpet daffodil St Keverne but Golden Harvest and Carlton are particularly susceptible. Few cultivars have any resistance to neck rot.

Control is based on good horticultural practices.

Long crop rotations are important. Growing narcissi as an annual crop instead of a biennial one lessens the chances of infection within a stock, and bulbs should be lifted before soils become warm in summer. Hot water treatment to control *Ditylenchus dipsaci* is effective against basal rot if formaldehyde is added to the hot water. Bulbs should be treated carefully to avoid bruising and wounding.

PRICE

F. oxysporum f.sp. *niveum* (E.F. Sm.) Snyder & Hansen

f.sp. *niveum* (CMI 219) is a widespread wilt pathogen attacking watermelon and also some *Cucurbita* species such as courgette. It causes damping-off and root rot of young seedlings and partial or general wilting, chlorosis, and collapse of older plants. This is accompanied by a yellow-orange to pale brown discoloration of the vascular tissue. The disease is most severe during hot summer days in plants carrying high yield. The pathogen persists for several years in the soil and seed transmission has also been reported. Although two pathotypes have been so far identified (possibly a third one exists in Israel), there is not sufficient evidence to distinguish them clearly. Control with resistant cultivars or hybrids (e.g. Calhoun Gray, Smokylee and Summit) can be achieved in areas where no virulent pathotypes occur. Crop rotation with non-cucurbitaceous hosts could also be practiced.

TJAMOS

F. oxysporum f.sp. *pisi* (Linford) Snyder & Hansen

f.sp. *pisi* is widely distributed in the pea-growing areas of N. America, Europe and Australia. Both *Pisum sativum* and *P. arvense* are hosts. At least 11 pathotypes have been identified (Hubbeling 1974). Two disease syndromes have been observed, 'wilt' and 'near wilt'. Wilt is also known as 'St. John's Disease' because symptoms are most noticeable around St. John's Day, 24 June. Host foliage becomes yellow and the leaflets and stipules curl downwards and inwards—a typical symptom of vascular disease in hosts with compound leaves. The foliage withers from the base of the plant upwards, and death ensues before pod formation or before swelling. Wilt may also be unilateral, affecting only one side of the host. The vascular tissues show a brown to orange or deep red discoloration which extends throughout the invaded root system. After death of the host the fungus grows out from the vessels and a white stromatic mycelium heavy with spores is found on the stem surface, especially under conditions of high humidity. Near wilt symptoms are similar to those of wilt but develop more slowly, frequently reaching maximum expression after the pods have developed. Race 2 is considered to cause 'near wilt', while race 1 and some other races are the causes of wilt (Dixon 1984). See also *F. solani* f.sp. *pisi*.

Control has been achieved by a long succession of resistant cultivars bred in response to the appearance of pathotypes. Resistance to race 1 is governed by gene F_n, to race 2 by F_{nw} and to races 4 and 5 by several recessive genes. The importance of this pathogen, at least in the UK, has greatly diminished with the use of resistant cultivars and geographical changes in the pea industry. The susceptible pulling-pea crop grown in Essex has largely disappeared, being replaced by a vining pea industry located in Lincolnshire and Norfolk.

DIXON

F. oxysporum f.sp. *tulipae* Apt

Pathogenic only to *Tulipa,* f.sp. *tulipae* is found in all tulip-growing countries in Europe and elsewhere. The fungus causes a bulb rot which may start from the basal plate, but frequently originates from the side or the bulb top. Infection often becomes manifest only after the bulbs have been stored for some weeks after lifting. The first signs of infection are small, yellowish, slightly sunken specks on the outer scales, which rapidly enlarge and spread into the inner scales and are often covered by pale mauve or white mycelium with many microconidia. The tissue turns brown, shrinks considerably, and then hardens. Frequently, the formation of blisters containing a gum-like substance is induced by the ethylene which is produced by the fungus in large quantities. Ethylene may also cause bud necrosis and other physiological disorders in otherwise healthy tulips. In the field, symptoms are only seen when, after a warm period, the bulb base is destroyed. Plants die off suddenly and leaves turn purple without yellowing. This field symptom is uncommon under moderate temperature conditions. During flower forcing in a heated glasshouse, a basal rot develops in the planted bulbs and leads to retarded growth, premature leaf yellowing, and in many cases flower-bud blasting. The fungus can survive in soil for several years. Other sources of infection are

conidia dispersed during storage, unnoticed minor infections and latent infections in the outer scale. Infection is promoted by high soil temperatures during the last part of the growth period, and especially when the tunic is turning brown. In flower forcing, disease incidence is increased at soil temperatures above about 15°C. In countries with a warm late spring, the disease is one of the main factors limiting profitable tulip bulb cultivation. In the glasshouse, it can cause severe losses in cut-flower production. Control is concentrated on cultural methods (Bergman & Bakker 1980), supported by disinfection of bulbs before planting. Emphasis should be put on such measures as crop rotation, rapid drying of bulbs after lifting, well-ventilated storage at low RH (60–80%), removal of affected bulbs from the planting stock, and in particular, late planting and early harvesting. Late planting avoids early infection of the basal plate via the outgrowing roots, and early lifting, i.e. when the tunics are turning yellow, takes advantage of the natural resistance factor, viz. the precursor of the fungitoxic lactone tulipalin, a compound which is present in the white tunic but vanishes as soon as the tunic turns brown and dies off (Bergman & Beijersbergen 1968).
BERGMAN

F. oxysporum f.sp. *vasinfectum* (Atk.) Snyder & Hansen

f.sp. *vasinfectum* principally attacks cotton but also tobacco, soybean and lucerne. In Europe it has been recorded in C. Greece, but not in Spain (CMI map 362). Although verticillium wilt is generally more important on cotton, fusarium wilt is exceptionally dangerous in hot cotton-growing areas during summer. Since symptom development is almost identical, disease identification can only be achieved after isolation of the pathogen. Fusarium wilt may cause seedling blight or wilt of older plants. Transient or permanent leaf flaccidity initially at the tips and later in the whole leaf-blade restricted to the lower leaves or expanded to the canopy of the affected plant is the first symptom. This is followed by interveinal chlorosis and eventually leaf desiccation and shedding. Stem or root sections reveal the presence of brown to black discoloration of the xylem vessels. f.sp. *vasinfectum* survives in the soil or on plant remnants as mycelium or chlamydospores easily dispersed by cultural practices. The pathogen, which is also seed-borne, gains entrance through the roots. This process is facilitated in wounds created by nematodes (*Meloidogyne* spp.). Six pathotypes have been differentiated so far. Race 1 has been identified in the cottonbelt areas of the USA, E. Africa and possibly Italy and race 2 in several regions of the USA (Alabama, California, S. Carolina). Race 3 occurs in Egypt, race 4 in India and possibly the USSR, race 5 in Sudan and race 6 in Brazil and Paraguay. The establishment of quarantine measures to prevent the spread of the fungus via cotton seed is recommended. Rotation programmes with corn, sorghum and other gramineous hosts could diminish the incidence of the disease. Resistant cotton cultivars are efficient in controlling the pathogen in the absence of corresponding virulent pathotypes.
TJAMOS

Fusarium redolens Wollenw. (syn. *F. oxysporum* var. *redolens* (Wollenw.) Gordon) is maintained as a species separate from *F. oxysporum* by some authors (Gerlach & Nirenberg 1982). Two specialized forms causing wilt of carnation and spinach have been described. Its most distinctive feature is a characteristic odour (resembling lilac) in culture. Other authors doubt the importance of this characteristic, and prefer to consider all isolates of *F. redolens* as simply forms of *F. oxysporum*. On carnation, for example, it is not at all clear whether there is any significant difference in biology and pathogenicity between isolates called *F. redolens* and isolates called *F. oxysporum* f.sp. *dianthi*.

Verticillium dahliae Kleb. and *Verticillium albo-atrum* Reinke & Berthold

Basic description

Verticillium dahliae and *V. albo-atrum* are two rather distinct species of the genus *Verticillium*. They are characterized by the hyaline, verticillately branched conidiophores and the hyaline ellipsoidal to subcylindrical mostly one-celled conidia, and differentiated by the formation of hyaline, dark brown or black microsclerotia in the former and by dark brown to blackish resting mycelium in the latter. No teleomorph is known.

Host plants

V. dahliae attacks an extremely wide range of annual and perennial dicotyledonous plants (citrus and pome trees are an exception). Among the herbaceous hosts are various species of: Solanaceae (aubergine, pepper, potato and tomato); Cucurbitaceae (melon and watermelon); Malvaceae (cotton and okra); Rosaceae (rose and strawberry); Compositae (artichoke). Many ornamentals and weeds are also affected. Woody hosts such as avocado, olive, pistachio and Prunus are also included, as well as seedling oaks and ornamental trees, especially species of *Acer*, *Catalpa* and *Koelreuteria*. Monocotyledonous plants (e.g. wheat and barley) have been reported as symptomless carriers of *V. dahliae*. *V. albo-atrum* causes economically important diseases in lucerne, cucumber, hop, potato and tomato.

Host specialization

There are few reports of host specialization in *Verticillium* (pepper or peppermint for *V. dahliae*; lucerne and hops for *V. albo-atrum*), but it is well documented that isolates from one host can cause disease in some crops but infect others without developing apparent symptoms (symptomless carriers).

Geographical distribution

Both species are widely distributed all over the world. *V. albo-atrum* is widespread in the temperate zone, while *V. dahliae* is present in temperate as well as subtropical regions.

Disease

Infection hyphae from *Verticillium* propagules can gain entrance through intact rootlets or roots, or at sites of lateral root emergence or wounds created by nematodes or cultural practices (Mace *et al.* 1981). The fungus advances inter- or intracellularly through the epidermis, cortex and endodermis and reaches the xylem tissue without causing obvious root-rot damage. Once vascular tissue is invaded, fungal growth is limited to the lumen of vessels. Conidia are produced in localized colonies, become dislodged and are transported upwards with the transpiration stream to form new colonies. Only during the late stages of the disease, when the tissues surrounding the vessels become moribund (mainly in herbaceous plants) is there any substantial growth of the fungus out of the vascular elements. Tylosis formation (outgrowth of the parenchymatous cells into the adjacent xylem vessels) is a very common host response. Other host reactions are gel formation and deposition, currently attributed to the action of various hydrolytic enzymes, and oxidation and polymerization of phenolic compounds leading to the deposition of melanin and the appearance of vascular discoloration (Bishop & Cooper 1983).

The pattern of symptom development is largely dependent on the botanical species of the diseased host. Initial outbreaks of verticillium wilt in a field are rather mild. Diseased plants are scattered or in groups but very rarely in a row. *V. dahliae* causes seedling blight in cotton or tomato and can also infect olive trees of over 100 years old. Symptoms in herbaceous hosts include loss of turgidity seen as diurnal flaccidity followed by transient or permanent wilting. Wilting (Fig. 40) develops mainly acropetally and is accompanied by marginal or interveinal chlorosis. Desiccation of the leaves is a common phenomenon followed by premature defoliation in heavy infections (e.g. in cotton). Diseased plants may be stunted and develop easily detectable discoloration of the vascular bundles (yellow, orange or pale brown in Cucurbitaceae, brown to dark brown in Solanaceae, Malvaceae or other hosts; in *Acer* spp. it is often dark green) (Fig. 41). Severe infections lead to the premature death of the plants. Infected stone-fruit trees show wilting of individual branches, leaf yellowing, necrosis, occasional defoliation and eventual death. The vascular browning is intensely dark. Symptoms in diseased branches of olive trees (Plate 16) first appear in leaves, which lose their silver-green

Fig. 40. *Verticillium dahliae* wilt of aubergine.

Fig. 41. Tiger-striping of hop leaves due to the progressive wilt strain of *Verticillium albo-atrum* (courtesy of P.W. Talboys).

lustre, changing to dull grey, then pale brown to brown. The rate of symptom expression in woody plants is very variable, from a slow progression to sudden collapse. Partially affected trees tend to recover by forming uninfected new growth if root re-infection does not take place in the next growing season.

Epidemiology

Verticillium wilt is considered as a single-cycle disease, since inoculum rarely produces new inoculum that could be effective within the same growing period. *Verticillium* propagules (microsclerotia) are able to withstand adverse environmental conditions and survive for over 12–14 years. Beyond their capacity for long survival, factors such as inoculum density and inoculum potential are important for the build-up or decline of epidemics. It has also been shown that strains of *Verticillium* vary in their virulence, thus affecting the incidence of the disease accordingly. The occurrence of a progressive and a fluctuating strain of *V. albo-atrum* have been proved for hop, the mild SS_4 and the defoliating T-1 for *V. dahliae* of cotton (the latter recorded in Spain) and race 1 and 2 for *V. dahliae* of tomato. Population synthesis also affects symptom development in a way analogous to the aggressiveness of the individual strains. The migratory nematodes (*Pratylenchus* group) can strongly enhance disease expression, and other genera (*Heterodera, Meloidogyne* and *Tylenchorhynchus*) may also do so. Similar positive interactions between verticillium wilt epidemics and other soil fungi (*Thanatephorus cucumeris, Chalara elegans, Macrophomina phaseolina* or *Alternaria radicina*) have been implicated. Disease development is critically affected by soil and air temperatures. Temperatures not exceeding an average of 21–24°C favour *V. albo-atrum*, while somewhat higher (21–25°C) facilitate *V. dahliae* infections. Irrigation or rain water negatively affects the ability of a host to overcome the disease due to the decrease of soil temperature during warm seasons. Dispersal of *Verticillium* propagules occurs by irrigation water, diseased plant debris (including weeds) or soil particles removed by agricultural tools and machinery. An important role in pathogen dissemination is also attributed to diseased propagative material (cuttings, tubers), while seed transmission is so far restricted to lucerne, safflower and sunflower. Cotton seed also transmits *V. dahliae* by desiccated leaves or petioles attached to the seed lint.

Economic importance

Heavy losses are inflicted in those areas where the pathogens are endemic. Restricted plant growth, desiccation or fall of leaves and eventual death lead to a pronounced reduction of expected crop production. Factors favouring the impact of the disease include the locally prevailing environmental conditions, the susceptibility of the host and the control measures taken. Among major crops, hop and lucerne suffer in northern parts of Europe, while potato-growing areas all over Europe and cotton-growing areas in Bulgaria, Greece, Russia, Spain and Turkey are extremely vulnerable. Losses in vegetables and ornamentals growing indoors are restricted if soil fumigation is practised. The effect of the disease on stone fruit or olive orchards ranges from limited crop losses of the affected branches to the total destruction of the whole tree.

Control

Verticillium wilt constitutes a real menace for world agriculture, since there are no therapeutic measures to control the disease. Root dipping or soil drenching with benomyl, carbendazim or thiophanate methyl have proved quite ineffective in practice (except for strawberries), although some symptom amelioration has been noticed in pot applications. Furthermore, the existing preventive measures are applicable to annual rather than to perennial crops. Vegetables and ornamentals can be protected by prior soil fumigation with steam, methyl bromide or other fumigants, but cost limitations do not allow large-scale application of the technique. Metham-sodium however, was shown to be promising in controlling the pathogen outdoors (Ben-Yephet & Frank 1983), when applied with the irrigation water in potato fields (60 ml/m^2 with the first 10% of irrigation volume followed by the rest of the water). Soil solarization (by covering the soil with transparent polythene sheets during summer) is also recommended for several herbaceous plants (Katan 1981). The method, which could be applied alone or in combination with low doses of soil fumigants, is especially applicable to countries around the Mediterranean basin. There are few cases of commercially used *Verticillium*-resistant or tolerant cultivars. The VF tomato cultivars are widely used in Europe, since race 2 of *V. dahliae,* capable of overcoming their resistance, is rather rare in the continent. Todd's Mitcham and Murray Mitcham are two peppermint selections carrying a high level of resistance against the pathogen. Wilt-tolerant hop cultivars are also available. The American or Russian *Verticillium*-tolerant cotton cultivars (the Acala SJ and the Tashkent series respectively) could be useful in cotton-growing regions with acute wilt problems. Some American potato cultivars with significant verticillium-wilt resistance have been reported. Allegra and Oblonga olive rootstocks might be of future hope for olive growers if their resistance to *V. dahliae* remains durable and effective outside the USA. Among the recommended cultural measures for preventing the disease are: chemical control of nematodes and weeds, removal of diseased plants or plant debris and the careful use of nitrogen fertilizers. It should also be stressed that disease intensity may be higher when wilt-susceptible orchards are intercropped with susceptible annual hosts. Frequent discing and irrigation should also be avoided.

Special research interest

There is a voluminous literature on verticillium wilts, but the problem of controlling the disease remains unsolved. It has become evident that the target of the control measures must be the pathogen prior to its entrance and establishment in the host. Thus, factors affecting the survival of the fungus in the soil should be of special research interest (e.g. the antagonistic fungus *Talaromyces flavus,* which may have potential for biological control). The development of evenly distributed fungicides, moving basipetally rather than acropetally into the vessels could protect susceptible hosts from consecutive infections. Chemicals which can intervene at the level of microsclerotium formation in plant remnants should also be sought and tested (Christias *et al.* 1981). Finally breeding or selection for new resistant cultivars, hybrids or root-stocks, is considered to be extremely important.

TJAMOS

Verticillium theobromae (Turc.) Mason & Hughes (syn. *Stachylidium theobromae* Turc.) causes a very different kind of disease—cigar-end of banana, a storage rot that shrivels and blackens the end of the fruit. It has been found in Greece and Italy (CMI 259; CMI map 146). Several other *Verticillium* species reported as weak root pathogens (*V. nigrescens* Pethybr., *V. nubilum* Pethybr., *V. tricorpus* Isaac; CMI 257, 258, 260) have no practical importance.

Phialophora cinerescens (Wollenw.) v. Beyma

P. cinerescens (CMI 503) is distinguished from *Verticillium* by conidiophores of a *Penicillium*-like structure, bearing at the tip bottle-shaped phialides. Conidia are cylindrical–ellipsoidal, at first hyaline and later smoke-coloured. The pathogen is known only on *Dianthus*: *D. chinensis, D. barbatus* and *D. caryophyllus* (*D. deltoides* and *D. caesius* are resistant). Originally described in Germany, *P. cinerescens* is now reported in Europe and the USA. The fungus causes a typical vascular disease (Moreau 1957) with gradual wilting during which the leaves sometimes become bluish before dying. As a result, the usual name of the disease in France is blue disease. Transverse sections of the stems show a brown coloration immediately below the surface. *P. cinerescens* grows at low temperatures from 10°C, with an optimum at 17–20°C. Dissemination mainly occurs by taking cuttings from symptomless carriers. In ground beds the fungus was recovered from soil depths reaching 80 cm. As with fusarium wilt, losses may be reduced in ground beds by soil fumigation with metham-sodium. However the only completely effective control is to steam the soil in raised benches to eradicate the pathogen and to prevent recontamination by using pathogen-free propagative material (Baker 1980). No current cultivars have adequate tolerance.

TRAMIER

Phialophora asteris (Dowson) Burge & Isaac (syn. *Verticillium vilmorinii* (Guégen) Westerd. & v. Luijk) causes a similar wilt of *Aster* and *Callistephus* in Denmark, Netherlands and UK. It causes significant losses only very sporadically (CMI 505). ***P. (Cephalosporium) gregata*** (Allington & Chamberlain) W. Gams causes brown stem rot of soybean in N. America (APS Compendium) and presents a significant danger to Europe, where it has been recorded in Hungary.

Hymenella cerealis Ell. & Ev.

H. cerealis is a sporodochial fungus which also has a moniliaceous state *Cephalosporium gramineum* Nisikado & Ikata (Gams 1971; CMI 301). The fungus attacks winter wheat most severely, but infection has been recorded on barley, oats, rye and many grasses including *Lolium, Agropyron* and *Dactylis.* A wide range of species in the Gramineae has been infected by inoculation. No pathogenic variation has been reported in Europe. In N. America an extensive study of 25 isolates and 1000 wheat lines indicated that there was a small range of virulence, with most isolates being highly virulent, and some range of resistance, but with no cultivars showing immunity (Mathre *et al.* 1977). Although first described from N. America, as *Hymenula*, in 1894 the fungus was not associated with disease until the description of its *Cephalosporium* state from Japan, where it was first recorded in 1931. It was subsequently reported from Europe in 1952 and N. America in 1955. In Europe it is recorded from UK, Netherlands, Italy, FRG and GDR.

H. cerealis causes cephalosporium stripe disease of cereals and grasses, especially winter wheat (APS Compendium). The fungus gains entry to the plant either at germination or through sites of root damage. Infection is vascular and the physical presence of mycelium, spores and extracellular polysaccharides in the xylem vessels interferes with translocation. In addition toxins are produced by the fungus. Usually only one or two adjacent vascular bundles are infected. The bundle itself becomes discoloured reddish-purple-brown and the adjacent leaf-blade and leaf sheath tissues become yellow, then white and finally necrotic. All tissues associated with an infected vascular bundle are affected, producing stripes that extend up through successive nodes and along the length of the leaf blade to the tip. When broken across at the node the vascular bundles can readily be seen to be discoloured. Infected plants grow more slowly than healthy plants, become stunted, and 'ripen' early to produce whiteheads containing little useful seed. The fungus infects mainly via sites of root damage which can be caused by mechanical means (harrowing, rolling or tractor movements during spraying); by insects, e.g. wireworms (Slope & Bardner 1965); or by frost heaving, which explains the tendency for autumn-sown crops to have higher levels of infection. There is some evidence that inoculum can be seed-borne at very low levels, but this must be considered unimportant in disease development. The main source of inoculum is infected straw. Inoculum on straw survives better if left on the surface than if ploughed in, so there is a tendency for disease levels to be higher where reduced tillage is practised. Disease levels are also higher where fields are poorly drained, have high rainfall and a short rotation. Early-sown autumn crops, especially those with high fertilizer applications, are at risk since both factors encourage the development of large root systems more subject to damage. Losses caused by cephalosporium stripe are generally unimportant. The useful yield of infected tillers is only 10–15% that of uninfected tillers but levels of incidence are usually so low as to make the loss insignificant in absolute terms (Richardson & Rennie 1970). No data are available for mainland Europe, but surveys in the UK have shown that up to 70% of winter wheat crops, 20% of spring barley crops and 8% of spring oat crops are affected, but at very low mean levels of incidence (0.4%, 0.05% and 0.02% respectively).

RICHARDSON

Other *Cephalosporium* or *Acremonium* species, which resemble microconidial fusaria, are associated with vascular infections in warmer countries and occur widely in soil and as secondary invaders of rotted plant tissues. ***C. acremonium*** Corda (syn. *Acremonium strictum* W. Gams) causes black bundle disease of maize (APS Compendium) and is noted in the USSR. ***Acremonium apii*** (M.A. Sm. & Ramsey) W. Gams causes brown spot of celery.

REFERENCES

Alabouvette C., Couteaudier Y. & Louvet J. (1985) Fusarium wilt-suppressive soils: mechanisms of suppression and management of suppressiveness. In *Ecology and Management of Soil-borne Plant Pathogens* (Eds C.A. Parker, K.J. Moore, P.T.W. Wong, A.D. Rovina & J.F. Kollmorgen). APS, St Paul.

Alabouvette C., Rouxel F. & Louvet J. (1979) Characteristics of Fusarium wilt-suppressive soils and prospects for their utilization in biological control. In *Soil-Borne Plant Pathogens* (Eds B. Schippers & W. Gams), pp. 165–182. Academic Press, New York.

Aloj B., Marziano F., Zoina A. & Noviello C. (1983) La tracheofusariosi del fagiolo in Italia. *Informatore Fitopatologico* **33** (11), 63–66.

Armstrong G.M. & Armstrong J.K. (1981) Formae speciales and races of *Fusarium oxysporum* causing wilt diseases. In *Fusarium: Diseases, Biology and Taxonomy* (Eds P.R. Nelson, T.A. Toussoun & R.J. Cook), pp. 391–399. Pennsylvania State University Press.

Baker R. (1980) Measures to control *Fusarium* and *Phialophora* wilt pathogens of carnations. *Plant Disease* **64,** 743–749.

Bakheit B.R. & Toth E. (1983) Results of resistance breeding on alfalfa. I. Resistance to Fusarium wilt. *Acta Agronomica Academiae Scientiarum Hungaricae* **32,** 424–429.

Banihashemi Z. & de Zeeuw D.J. (1975) The behaviour of *Fusarium oxysporum* f. sp. *melonis* in the presence and absence of host plants. *Phytopathology* **65,** 1212–1217.

Barrière Y., Panouillé A. & Cassini R. (1981) Relations source-puits et sélection du maïs pour la résistance à la pourriture des tiges. *Agronomie* **1,** 707–711.

Ben-Yephet Y. & Frank Z. (1983) Metham sodium: factors affecting its penetration into soil and its toxicity to soil-borne pathogens. *Acta Horticulturae* **152,** 183.

Bergman B.H.H. & Bakker M.A.M, (1980) Consequences and control of latent *Fusarium oxysporum* infections in tulip bulbs. *Acta Horticulturae* **109,** 381–386.

Bergman B.H.H. & Beijersbergen J.C.M. (1968) A fungitoxic substance extracted from tulips and its possible role as a protectant against disease. *Netherlands Journal of Plant Pathology* **74,** 157–162.

Bishop C.D. & Cooper R.M. (1983) An ultrastructural study of root invasion in three vascular wilt diseases. *Physiological Plant Pathology* **22,** 15–27.

Bollen G.J. (1977) Pathogenicity of fungi isolated from stems and bulbs of lilies and their sensitivity to benomyl. *Netherlands Journal of Plant Pathology* **83,** 317–329.

Bolton A.T. & Donaldson A.G. (1972) Variability in *Fusarium solani* f. *pisi* and *F. oxysporum* f. *pisi*. *Canadian Journal of Plant Science* **52,** 189–196.

Booth C. (1971) *The Genus Fusarium.* CMI, Kew.

Bouhot D. (1981) Some aspects of the pathogenic potential in formae speciales and races of *Fusarium oxysporum* on Cucurbitaceae. In *Fusarium: Diseases, Biology and Taxonomy* (Eds P.E. Nelson, T.A. Toussoun & R.J. Cook), pp. 318–326. Pennsylvania State University Press.

Bouhot D. & Olivier J.M. (1977) Première observation en France de la fusariose vasculaire du céleri à côtes due à *Fusarium oxysporum* f. sp. *apii*. *Annales de Phytopathologie* **9,** 515–520.

Boyd A.E.W. (1972) Potato storage diseases. *Review of Plant Pathology* **51,** 305–308.

Burke D.W., Miller D.E. & Barker A.W. (1980) Effects of soil temperature on growth of beans in relation to soil compaction and *Fusarium* root rot. *Phytopathology* **70,** 1047–1049.

Cassini R. (1983) Fusarium diseases of cereals in Western Europe. In *Fusarium: Diseases, Biology and Taxonomy* (Eds P.E. Nelson, T.A. Toussoun & R.J. Cook), pp. 56–63. Pennsylvania State University Press.

Cassini R., Nourriseau J.G. & Cassini R. (1983) Le dépérissement fusarien des aspergeraies. *Comptes Rendus des Séances de l'Académie d'Agriculture de France* **69,** 1355–1361.

Christias C., Tjamos E.M., Ziourdou C. & Kornaros E. (1981) *In vitro* inhibition of microsclerotia formation by cysteine hydrochloride in *Verticillium dahliae. Proceedings of the 3rd International Verticillium Symposium, Bari, abstract 21.*

Clarkson J.D.S. (1978) Pathogenicity of *Fusarium* spp. associated with foot rots of peas and beans. *Plant Pathology* **27,** 110–117.

Colhoun J. (1970) Epidemiology of seed-borne *Fusarium* diseases of cereals. *Annales Academiae Scientiarum Fennicae A IV Biologia* **168,** 31–36.

Couteaudier Y., Alabouvette C. & Soulas M.L. (1985a) Nécrose du collet et pourriture des racines de tomate. *Revue Horticole* no. 254, 39–42.

Couteaudier Y., Alabouvette C., Vegh I. & Louvet J. (1981) Manifestation de la fusariose vasculaire du radis dans la région parisienne. *Revue Horticole* no. 217, 43–44.

Couteaudier Y., Letard M., Alabouvette C. & Louvet J. (1985b) Lutte biologique contre la fusariose vasculaire de la tomate. Résultats en serre de production. *Agronomie* **5,** 151–155.

Darby P. (1983) Fusarium canker. *Report of the Department of Hop Research, Wye College 1982,* p. 24.

Davis D. (1967) Cross-protection in *Fusarium* wilt diseases. *Phytopathology* **57,** 311–314.

Dixon G.R. (1984) *Vegetable Crop Diseases,* 2nd edn. Macmillan, London.

Domsch K.H., Gams W. & Anderson T.H. (1980) *Compendium of Soil Fungi.* Academic Press, New York.

Duben J. & Fehrmann H. (1979) Vorkommen und Pathogenität von *Fusarium*-Arten an Winterweizen in der Bundesrepublik Deutschland. I. Artenspektrum und jahreszeitliche Sukzession an der Halmbasis; II. Vergleich der Pathogenität als Erreger von Keimlings-, Halmbasis- und Ährenkrankheiten. *Zeitschrift für Pflanzenkrankheiten und Pflanzenschutz* **86,** 638–652, 705–728.

Dunnell L.D. (1970) Susceptibility of southern pines to infection by *Fusarium moniliforme* var. *subglutinans*. *Plant Disease Reporter* **62,** 108–111.

Ebben M.H. & Spencer D.M. (1976) *Fusarium* wilt of carnation, caused by *Fusarium oxysporum* f. sp. *dianthi*. Relation of soil inoculum level to disease severity and control. *Annual Report of the Glasshouse Research Institute,* pp. 113–115.

El-Gholl N.E., McRitchie J.J., Schoulties C.L. & Ridings W.H. (1978) The identification, induction of perithecia and pathogenicity of *Gibberella (Fusarium) tricincta* n.sp. *Canadian Journal of Botany* **56,** 2203–2206.

Fantino M.G. & Badino M. (1981) Valutazione della tolleranza a *Fusarium oxysporum* f. sp. *cepae* in linee e varietà di cipolla. *Informatore Fitopatologico* **31** (7/8), 3–8.

Fantino M.G. & Bazzi C. (1981) Azione antagonista di *Pseudomonas cepacia* verso *Fusarium oxysporum* f. sp. *cepae*. *Informatore Fitopatologico* **32** (4), 55–58.

Fantino M.G. & Mazzucchi U. (1975)La fusariosi del astro. *Informatore Fitopatologico* **25** (6), 21–24.

Fantino M.G. & Pavesi C. (1984) Infezioni di Fusaria su asparago. *Informatore Fitopatologico* **34** (5), 15–18.

Furuya H., Owada M. & Vi T. (1979) A suppressive soil of common bean root rot in Kitami District, Hokkaido. *Annals of the Phytopathological Society of Japan* **45,** 608–617.

Gams W. (1971) *Cephalosporium-artige Schimmelpilze (Hyphomycetes)*. Gustav Fischer Verlag, Stuttgart.

Garibaldi A. (1981) Ulteriori ricerche sulla specializzazione biologica di *Fusarium oxysporum* f. sp. *dianthi*. *Rivista Ortoflorofrutti Italia* **65,** 353–358.

Gerlach W. (1961) Beiträge zur Kenntnis der Gattung *Cylindrocarpon* IV. *Cylindrocarpon radicicola*, seine phytopathologische Bedeutung und sein Auftreten als Erreger einer Fäule des Usambaraveilchens. *Phytopathologische Zeitschrift* **41,** 361–369.

Gerlach W. & Nirenberg H. (1982) The genus *Fusarium*—a pictorial atlas. *Mitteilungen aus der Biologische Bundesanstalt für Land und Forstwirtschaft Berlin-Dahlem* no. 209.

Gerlach W., Schickedanz F. & Dalchow J. (1980) Erstes Auftreten einer Fusarium-welke der Chrysantheme in Deutschland. *Nachrichtenblatt des Deutschen Pflanzenschutzdienstes* **32,** 1–4.

Heybroek H.M. (1983) In *Research on Dutch Elm Disease in Europe* (Ed. D.A. Burdekin). *Forestry Commission Bulletin* no. 60, pp. 108–113. HMSO, London.

Hornok L. & Walcz I. (1983) *Fusarium heterosporum*, a highly specialized hyperparasite of *Claviceps purpurea*. *Transactions of the British Mycological Society* **80,** 377–580.

Horst R.K. & Hoitink A.J. (1968) Occurrence of *Cylindrocladium* blights on nursery crops and control with fungicide 1991 on Azalea. *Plant Disease Reporter* **52,** 615–617.

Houston D.R. & Wainhouse D. (Eds) (1983) *Proceedings of the IUFRO Beech Bark Disease Working Party Conference, USDA Forest Service General Technical Report.*

Hubbeling N. (1974) Testing for resistance to wilt and near-wilt of peas, caused by race 1 and race 2 of *Fusarium oxysporum* f. sp. *pisi*. *Mededelingen van de Faculteit Landbouwwetenschappen Rijksuniversiteit Gent* **39,** 991–1000.

Johnson M.C., Pirone T.P., Siegel M.R. & Varney D.R. (1982) Detection of *Epichloe typhina* in tall fescue by means of enzyme-linked immunosorbent assay. *Phytopathology* **72,** 647–650.

Jones J.P. & Woltz S.S. (1981) Fusarium-incited diseases of tomato and potato and their control. In *Fusarium: Diseases, Biology and Taxonomy* (Eds P.E. Nelson, T.A. Toussoun & R.J. Cook), pp. 157–162. Pennsylvania State University Press.

Katan J. (1981) Solar heating (solarization) of soil for control of soil borne pests. *Annual Review of Phytopathology* **19,** 211–236.

Kennel W. (1976) Zur Situation bei Obstbaumkrebs (*Nectria galligena*). *Erwerbsobstbau* **18,** 36–39.

Kohlmeyer J. & Kohlmeyer E. (1974) Distribution of *Epichloe typhina* and its parasitic fly. *Mycologia* **66,** 77–86.

Kollmorgen J.F. (1974) The pathogenicity of *Fusarium avenaceum* to wheat and legumes and its association with crop rotations. *Australian Journal of Experimental Agriculture and Animal Husbandry* **14,** 572–576.

Kováčiková E. (1980) (*In vitro* studies on the relationship among *Fusarium* species from pea). *Ochrana Rostlin* **16,** 25–31.

Kuhn P.J. & Smith D.A. (1979) Isolation from *Fusarium solani* f. sp. *phaseoli* of an enzymic system responsible for kievitone and phaseollin detoxification. *Physiological Plant Pathology* **14,** 179–190.

Langerfeld E. (1978) *Fusarium coeruleum* als Ursache von Lagerfäulen an Kartoffelknollen. *Mitteilungen aus der Biologische Bundesanstalt für Land- und Forstwirtschaft Berlin-Dahlem* no. 184.

Lásztity R. & Woeller L. (1975) Toxinbildung bei *Fusarium*-Arten und Vorkommen der Toxine in landwirtschaftlichen Produkten. *Die Nahrung* **19,** 537–546.

Lonsdale D. (1980) *Nectria coccinea* infection of beech: variations in disease in relation to predisposing factors. *Annales des Sciences Forestières* **37,** 307–317.

Louvet J., Alabouvette C. & Rouxel F. (1981) Microbiological suppressiveness of some soils to *Fusarium* wilts. In *Fusarium: Disease, Biology and Taxonomy* (Eds P.E. Nelson, T.A. Toussoun & R.J. Cooke), pp. 262–275. Pennsylvania State University Press.

Mace M.E., Bell A.A. & Beckman C.H. (Eds) (1981) *Fungal Wilt Diseases of Plants.* Academic Press, New York.

Mas P., Molot P.M. & Risser G. (1981) *Fusarium* wilt of muskmelon. In *Fusarium: Disease, Biology and Taxonomy* (Eds P.E. Nelson, T.A. Toussoun & R.J. Cook), pp. 169–177. Pennsylvania State University Press.

Mathre D.E., Johnston R.H. & McGuire C.F. (1977) Cephalosporium stripe of winter wheat: pathogen virulence, sources of resistance and effect on grain quality. *Phytopathology* **67,** 1142–1148.

Messiaen C.M. & Cassini R. (1981) Taxonomy of *Fusarium*. In *Fusarium: Diseases, Biology and Taxonomy* (Eds P.E. Nelson, T.A. Tousson & R.J. Cook), pp. 427–445. Pennsylvania State University Press.

Molot P.M., Mas P. & Ferrière H. (1979) Etude de la prémunition à l'aide d'une technique de dosage biologique du *Fusarium oxysporum* f. sp. *melonis* dans les tissus de melon. *Annales de Phytopathologie* **11,** 209–222.

Moreau M. (1957) Recherches sur les maladies de dépérissement causées par les champignons. Etude particulière du dépérissement des oeillets. *Encyclopédie Mycologique Vol. XXX*. Lechevalier, Paris.

Muehlbauer F.J. & Kraft J.M. (1978) Effect of pea seed

genotype on pre-emergence damping-off and resistance to *Fusarium* and *Pythium* root rot. *Crop Science* **18,** 321–323.

Nelson P.E., Toussoun T.A. & Cook R.J. (1981) *Fusarium: Diseases, Biology and Taxonomy.* Pennsylvania State University Press.

Nelson P.E., Toussoun T.A. & Marasas W.F.O. (1983) *Fusarium Species: an Illustrated Manual for Identification.* Pennsylvania State University Press.

Nirenberg H. (1981) A simplified method for identifying *Fusarium* spp. occurring on wheat. *Canadian Journal of Botany* **59,** 1599–1609.

Papendick R.I. & Cook R.J. (1974) Plant water and development of *Fusarium* foot rot in wheat subjected to different cultural practices. *Phytopathology* **64,** 358–363.

Porter J.K. (1981) Ergot alkaloid identification in Clavicipitaceae, systemic fungi of pasture grasses. *Journal of Agricultural and Food Chemistry* **29,** 653–657.

Price D. (1975a) The occurrence of *Fusarium oxysporum* in soils, and on narcissus and tulip. *Acta Horticulturae* **47,** 113–118.

Price D. (1975b) Pathogenicity of *Fusarium oxysporum* found on narcissus bulbs and in soil. *Transactions of the British Mycological Society* **64,** 137–142.

Richardson M.J. & Rennie W.J. (1970) An estimate of the loss of yield caused by *Cephalosporium gramineum* in wheat. *Plant Pathology* **19,** 138–140.

Risser G. (1973) Etude de l'hérédité de la résistance du melon (*Cucumis melo*) aux races 1 et 2 de *Fusarium oxysporum* f. *melonis. Annales de l'Amélioration des Plantes* **23,** 259–263.

Risser G., Banihashemi Z. & Davis D.W. (1976) A proposed nomenclature of *Fusarium oxysporum* f. sp *melonis* races and resistance genes in *Cucumis melo. Phytopathology* **66,** 1105–1106.

Rouxel F. & Grouet D. (1975) Présence en France de la fusariose vasculaire du cyclamen. *Revue Horticole* **155,** 21–24.

Royle D.J. & Liyanage G.W. (1976) Infection of planting material by the hop canker organism, *Fusarium sambucinum. Report of the Department of Hop Research, Wye College 1975,* pp. 39–46.

Salleh B. & Owen H. (1983) Resistance of white lupin (*Lupinus albus*) cultivars to wilt caused by *Fusarium oxysporum* f. sp. *lupini. Phytopathologische Zeitschrift* **107,** 70–80.

Salt G.A. (1982) Factors affecting resistance to root rot and wilt diseases. In *Faba Bean Improvement* (Eds G. Hawtin & C. Webb), pp. 259–270. Martinus Nijhoff, The Hague.

Scholten G. (1971) Control of *Fusarium* in some bulbous and tuberous plants grown for cutflower production under glass. *Acta Horticulturae* **23,** 187–193.

Seemüller E. (1968) Untersuchungen über die morphologische und biologische Differenzierung in der *Fusarium*-Sektion *Sporotrichiella. Mitteilungen aus der Biologischen Bundesanstalt für Land- und Forstwirtschaft Berlin-Dahlem* no. 127.

Slope D.B. & Bardner R. (1965) Cephalosporium stripe of wheat and root damage by insects. *Plant Pathology* **14,** 184–187.

Snyder W.C. & Hansen H.N. (1940) The species concept in *Fusarium. American Journal of Botany* **27,** 64–67.

Sobers E.K. (1972) Morphology and pathogenicity of *Calonectria floridana, C. kyotensis,* and *C. uniseptata. Phytopathology* **62,** 485–487.

Swinburne T.R. (1973) Microflora of apple leaf scars in relation to infection by *Nectria galligena. Transactions of the British Mycological Society* **60,** 389–403.

Swinburne T.R. (1975) European canker of apple (*Nectria galligena*). *Review of Plant Pathology* **54,** 787–799.

Thanassoulopoulos G.C. & Kitsos G.T. (1984) Some aspects of Fusarium wilt of potatoes. *Proceedings of the Sixth Congress of the Mediterranean Phytopathological Union, Cairo,* pp. 275–277.

Thies W.G. & Patton R.F. (1970) An evaluation of propagules of *Cylindrocladium scoparium* in soil by direct isolation. *Phytopathology* **60,** 599–601.

Tramier R., Pionnat J.C., Bettachini A. & Antonini C. (1979) Recherche sur la résistance des sols aux maladies. IV. Evolution de la fusariose vasculaire de l'oeillet en fonction des substrats de culture. *Annales de Phytopathologie* **11,** 477–482.

Trapero Casas A. & Jimenez Diaz R.M. (1984) Fusarium wilt and root rot of chickpea in Southern Spain. *Proceedings of the Sixth Congress of the Mediterranean Phytopathological Union, Cairo,* pp. 252–256.

Walker J.C. (1959) Progress and problems in controlling plant disease by host resistance. In *Plant Pathology, Problems and Progress, 1908–1958.* (Eds C.S. Holton, G.W. Fischer, R.W. Fulton, H. Hart & S.E.A. McCallan), pp. 31–45. Wisconsin University Press.

Walker J.C. (1971) *Fusarium Wilt of Tomato. APS Monograph* no. 6. American Phytopathological Society, St Paul.

Wormald H. (1955) *Diseases of Fruit and Hops.* Crosby Lockwood, London.

Zakaria M.A. & Lockwood J.L. (1980) Reduction in *Fusarium* populations in soil by oilseed meal amendments. *Phytopathology* **70,** 240-243.

Plate 17 Virulent *Cryphonectria parasitica* canker on a chestnut shoot.

Plate 18 Acervuli of the '*Gloeosporium olivarum*' anamorph of *Glomerella cingulata* on olive.

Plate 19 *Seiridium cardinale* canker on cypress, with small black acervuli on bark surface.

Plate 20 Apricot branch killed by *Eutypa lata*.

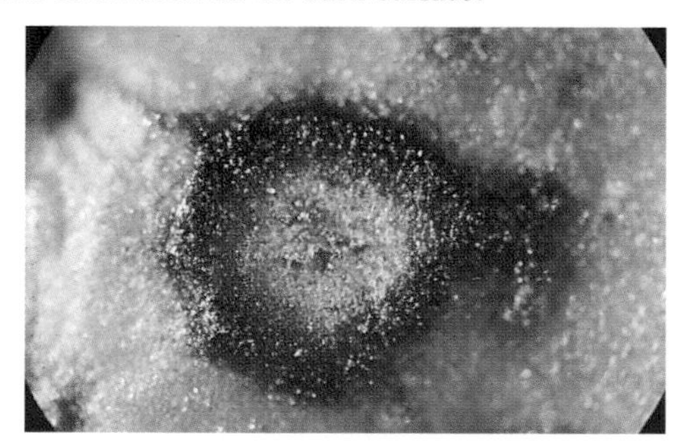

Plate 21 *Cercospora beticola* spot on sugar beet leaf.

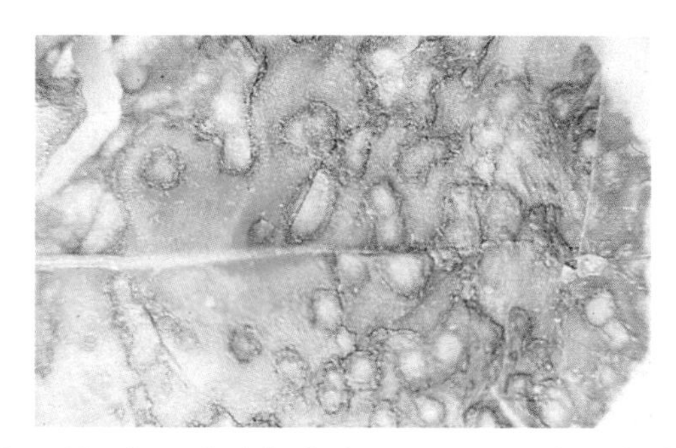

Plate 22 *Ramularia beticola* spots on sugar beet leaf.

Plate 23 *Pleospora bjoerlingii* spots on sugar beet leaf.

Plate 24 *Pyrenophora avenae* on oat leaf.

Plate 25 *Stigmina carpophila* on sweet cherry.

Plate 26 *Lophodermella sulcigena*.

Plate 27 Bulb onions showing rotted roots and white fungal mycelium typical of white rot (*Sclerotium cepivorum*).

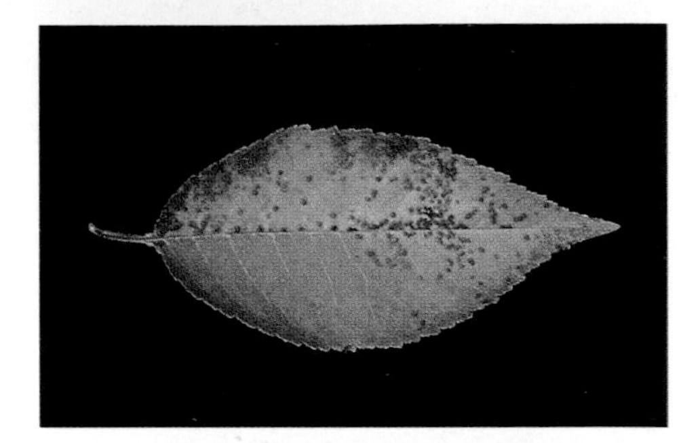

Plate 28 *Blumeriella jaapii* on sour cherry.

Plate 29 *Diplocarpon rosae*.

Plate 30 Aecia of *Pucciniastrum areolatum* on a leader of spruce killed by the fungus.

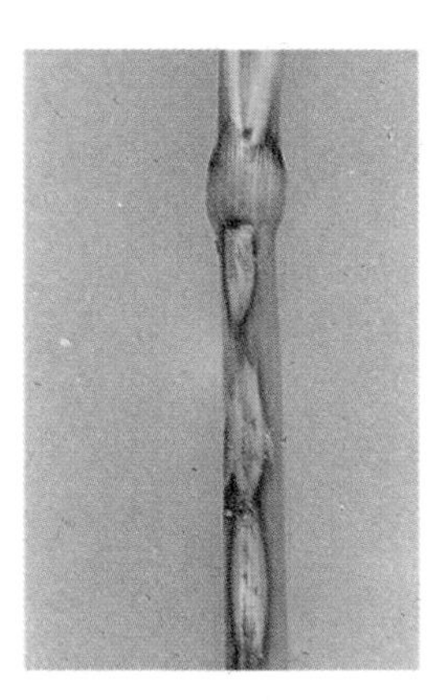

Plate 31 Sharp eyeshot lesions (*Ceratobasidium cereale*).

Plate 32 Silver-leaf disease (*Chondrostereum purpureum*) on peach.

12: Ascomycetes III
Ophiostomatales, Diaporthales, Polystigmatales, Sphaeriales, Diatrypales, Coryneliales

The Ascomycetes covered in this chapter are essentially the remaining 'pyrenomycetes', with unitunicate asci and generally black, hard-walled perithecia. Collectively, they are often associated with cankers of woody plants and the teleomorph stage is generally considered to be a conspicuous and regular element in the disease cycle. Many are somewhat marginal plant parasites and related species are purely saprophytic.

The Ophiostomatales produce long-beaked perithecia and have anamorphs of the *Graphium, Chalara, Cephalosporium* or *Verticillium* type (it is possible that the main wilt-producing *Verticillium* species in Chapter 11 should be included here). The main genera of pathogens are *Ceratocystis* and *Ophiostoma*, the latter now distinguished (de Hoog & Scheffer 1984) by having anamorphs other than in *Chalara*, rhamnose in the cell wall and resistance to cycloheximide. They characteristically cause wilt or canker diseases of trees, while *Chalara* species are important root-rot pathogens.

The Diaporthales, usually with beaked perithecia in a stroma, include a number of important plant pathogens. *Diaporthe*, with *Phomopsis* anamorphs (pycnidia with ovoid α-conidia and filiform β-conidia), cause stem cankers of many herbaceous hosts, and bark cankers of woody hosts. *Cryptodiaporthe, Cryphonectria* and *Valsa* are also associated with bark cankers. *Gnomonia* species cause leaf blotch or anthracnose mainly on woody hosts, while *Gaeumannomyces* (anamorph *Phialophora*) causes serious root diseases of Gramineae. It may be noted that some *Phomopsis*, now reclassified as *Phacidiopycnis*, are anamorphs of Rhytismatales (e.g. *Potebniamyces* q.v.).

The Polystigmatales is a recently created order grouping together in particular *Phyllachora* (causing tar spots of Gramineae) and *Glomerella*, causing cankers on woody hosts, and as its numerous *Colletotrichum* anamorphs, anthracnose on many herbaceous crops. The name *Gloeosporium*, formerly also used for many of these fungi (but also for quite unrelated ones) is now rejected. Conidial germination by appressorium is characteristic.

The Sphaeriales formerly included many perithecial Ascomycetes now classified in other orders. Its most characteristic members form perithecia in a carbonaceous stroma, and grow on bark as saprophytes (*Daldinia, Xylaria*) or parasites (some species of *Hypoxylon, Ustulina*). Other genera provisionally grouped here, following the *Dictionary of the Fungi*, very probably belong elsewhere. *Monographella* has anamorphs in *Gerlachia*, but previously in *Fusarium* or *Rhynchosporium*. *Discostroma* has *Seimatosporium* anamorphs and the related *Lepteutypa* has *Seiridium, Pestalotia* or *Pestalotiopsis* anamorphs (Von Arx 1984). Fungi in the latter imperfect genera are considered here for convenience. *Physalospora* is clearly separated from *Botryosphaeria* in the Dothideales, but both are considered to have *Botryodiplodia* anamorphs and further clarification is required.

The Diatrypales, with perithecia immersed in a carbonaceous stroma, are mostly saprophytes in the bark of dead wood. *Eutypa lata* is the only pathogenic species. The Coryneliales are mainly saprophytic or parasitic on tropical conifers, forming botuliform ascomata opening by clefts. In Europe, ***Caliciopsis pinea*** Peck causes a minor branch canker of *Pinus pinaster* in S.W. France (Lanier *et al.* 1976).

OPHIOSTOMATALES

Ceratocystis ulmi (Buisman) C. Moreau

(Syn. *Ceratostomella ulmi* (Schwartz) Buisman, *Ophiostoma ulmi* (Buisman) Nannf.)
(Anamorph ***Pesotum ulmi*** (Schwartz) Crane & Schoknecht, syn. *Graphium ulmi* Schwartz)

Basic description

C. ulmi is a typical *Ceratocystis* species, with a black, long-necked perithecium. The ascospores are colourless and slightly curved. They measure 4.5–6×1.5 μm, and ooze out to form a droplet around the top of the perithecial neck. The coremial conidial stage *P. ulmi* forms single-celled, colourless spores

(2–5×1–3 μm) at the tips of its black stalks which are 1–2 mm in height. A *Cephalosporium* or *Sporotrichum* stage also occurs, with colourless spores measuring 4–6×2–3 μm, formed in mucilaginous droplets. Finally, a yeast-like stage buds off cells of variable size in cultures and in the xylem of diseased trees.

Fig. 42. Elm shoots variously attacked by *Ceratocystis ulmi* (courtesy of Forestry Commission, GB).

Host plants

C. ulmi is almost entirely confined to species of *Ulmus* and the related genus *Zelkova*. Among the species of *Ulmus*, those of N. America are highly susceptible (APS Compendium), and those of Europe only slightly less so. Some Asian elms, including *U. pumila, U. japonica, U. parvifolia* and *U. wallichiana*, are resistant, and have been used in attempts to produce resistant hybrids for use in Europe, N. America and the USSR.

Strains

C. ulmi exists as two main strains. The first, a highly pathogenic 'aggressive' strain, unknown until the 1960s, itself has two races, a N. American form (NAN) and a Eurasian one (EAN). The second is a less pathogenic 'non-aggressive' strain (Brasier 1983). The strains can be differentiated by cultural characteristics and by pathogenicity tests on elms.

Geographical distribution

C. ulmi occurs in Europe, N. America, and parts of Asia (CMI map 36). The N. American form of the 'aggressive' strain apparently reached Europe from N. America in the 1960s and is now widespread in W. Europe. The European race of this strain seems to have spread more recently eastwards from the USSR and Iran, into much of E. Europe and Denmark, Norway and Ireland (Brasier 1983).

Disease

The disease caused by *C. ulmi* is usually called Dutch elm disease. Foliar symptoms are usually seen from early summer, when leaves in parts of the crown wilt and turn yellow and later die (Fig. 42). Tips of fast-growing twigs often curl to form 'shepherds crooks', which help to reveal affected trees in winter. Affected trees become stag-headed, and die during the summer or early the following growing season. In the European outbreak of the 1920s and 1930s, trees often recovered from attack (Peace 1960), but since the emergence of the 'aggressive' strain recovery has become uncommon. Internal symptoms are most easily seen in affected twigs, which if cut transversely show rings of dark-brown spots in the spring wood. If instead the bark and outer wood is peeled off these twigs, the spots show as discontinuous brown streaks. These indications occur only in annual rings of years in which active disease was present. The conducting vessels and some living cells become filled with brown gum and colourless, bladder-like tyloses. The vascular system of susceptible elms is soon blocked, and wilting in affected trees is at least partly due to this blockage and partly caused by toxins produced by the fungus.

Epidemiology

Primary infection takes place only through wounds, entirely or almost entirely through the activities of elm bark beetles. In much of Europe transmission is by *Scolytus scolytus* and *S. multistriatus* (also the main carrier in N. America) but *S. laevis* is an important carrier in Sweden. In N. America, *Hylurgopinus rufipes* also transmits the disease. The beetles emerge in spring and summer from galleries under the bark of infested wood. They fly to the tops of elms, and form feeding grooves in the crotches of the twigs, and in so doing transmit spores of *C. ulmi*, carried on and in their bodies, to many of the wounds they make. They then breed under the bark of dead and dying elms and elm material. The fungus spreads throughout the tree mainly in its yeast-like form, and later produces its fructifications, most often under dead elm bark separating it from the wood, and in the galleries of the bark beetles. The coremia form on newly dead tissues in all but the coldest weather. The perithecia are most common in winter, and form only when two compatible mating types occur together. Many elms form root grafts and reproduce clonally by root suckers. Hence whole elm populations are interconnected by their root systems, along which secondary spread of *C. ulmi* often takes place.

Some biological control of the disease is affected by organisms such as *Phomopsis oblonga* (Desm.) Trav. (which hinders the breeding of the bark beetles) and *Pseudomonas fluorescens* (Trevisan) Migula (which reduces the growth of *C. ulmi*) (Webber 1981; Yde-Andersen 1983). Apart from control by nuclear genes, pathogenicity of the fungus may be affected by other factors, such as the presence of dsRNA components known to occur in isolates of *C. ulmi* (Pusey & Wilson 1982; Brasier 1983).

Economic importance

Dutch elm disease is one of the most damaging of all tree diseases. In Europe in the 1920s and 1930s losses varied. Thus in Hilversum in the Netherlands, 70% of the elms were destroyed in 9 years, whereas in the UK, where many trees recovered, only about 10% of the trees were lost in some 30 years. European losses in the epidemic caused by the 'aggressive' strain of the fungus have been very heavy, and between 1970 and 1978 over 75% of elms in S. England were destroyed. Since then in the UK spread northwards has caused additional losses, and much destruction has since been experienced in other countries (France, Scandinavia). Damage has also been great in N. America.

Control

Dutch elm disease has proved very difficult to control. The most important control measures aim to remove the breeding grounds of the insect vectors by the felling and destruction of the wood and bark of diseased trees. Determined efforts to eradicate the disease in the USA soon after it was first found there were unsuccessful. Nevertheless these sanitation felling campaigns (often with the addition of insecticidal sprays) frequently slowed the progress of the disease. If sanitation felling is to have its full effect, it must be supported by legislative controls to prevent the movement of elm material from infected to non-infected areas and to stop the accumulation of heaps of infested material for use as firewood etc. These legislative controls are very difficult to operate. In the 1960s in S. England, the aggressive strain of *C. ulmi* gained a hold when its existence and characteristics were still unknown. A sanitation campaign combined with appropriate legislation slowed the spread to some degree, but had a marked controlling effect mainly in a few favoured places with elm populations geographically isolated from infested areas. Even there, losses have increased in recent years.

The use of insecticides, such as methoxyclor, which have had some controlling effect, has generally ceased because of environmental objections. Efforts at control by the injection of fungicides such as carbendazim and thiabendazole show promise, but are so expensive that they can be justified only on valuable specimen trees.

Current research interest

Much work on Dutch elm disease is in progress in N. America and Europe, especially on the causal fungus and its vectors, fungicides, control by the use of trap trees, the use of pheromones and arboricides, and on the selection and breeding of resistant elms. European work has been recently reviewed (Burdekin 1983).

PHILLIPS

Ceratocystis fimbriata Ell. & Halsted

(Syn. *Endoconidiophora fimbriata* (Ell. & Halsted) Davidson)

C. fimbriata forms bowler-hat shaped ascospores in long-necked perithecia, chlamydospores and endoconidia (CMI 141). It is widespread mainly in warmer parts of the world (CMI map 9), where it causes canker and wilt on various trees and tuber rot of sweet potato. In Europe, it has recently been reported as a cause of serious canker in poplars in Poland (Przybył 1984). Cross-inoculation experiments have shown a degree of host specialization,

Fig. 43. Characteristic brownish-black radial spindles in plane infected by *Ceratocystis fimbriata* f. sp. *platani* (courtesy of A. Vigouroux).

and in particular the existence of **f.sp. *platani*** Walker, which is confined to *Platanus* species, and is most destructive on *P. acerifolia* (London Plane) (Fig. 43). It occurs in E. USA, and in France (near Marseille), Italy, Spain, and Armenia (USSR). It causes canker stain of plane (Gibbs 1981). Diseased trees show slightly sunken, elongated or lens-shaped areas which later become irregular rough and blackened cankers. Affected wood turns red-brown or bluish black. Girdled branches die above the cankers, and the whole tree dies when the trunk is girdled. The first symptoms are usually those in the crown. The leaves are often dwarfed and yellow, and often wilt and fall.

C. fimbriata f.sp. *platani* is a wound parasite. It may be spread by wind, rain and insects, but almost all transmission is by contaminated pruning tools and wound paints. In Europe some spread also takes place through root grafts (Vigouroux 1979). Contaminated sawdust may also spread the disease. Stressed trees are especially severely affected. Canker stain has caused serious damage to *Platanus acerifolia* in the USA, and locally severe losses have occurred in France and Italy. The disease is readily controlled by keeping pruning to a minimum, doing it when growth is vigorous and healing rapid, avoiding other wounding where possible, and disinfecting tools between trees, destroying contaminated sawdust, and rapidly removing diseased trees.
PHILLIPS

Ceratocystis fagacearum (Bretz) Hunt (anamorph ***Chalara quercina*** Henry) causes wilt of oak in the USA and is a serious potential threat to European oaks. It is one of the most important quarantine organisms for Europe (EPPO 6; Gibbs 1981). Other *Ceratocystis* and *Ophiostoma* species, including some of those associated with blue-staining of wood, have *Verticicladiella* anamorphs. Some (e.g. ***O. piceaperdum*** (Rumbold) V. Arx, anamorph ***V. procera*** Kendr.) may be associated with dieback of pines in France (Morelet 1986).

Ophiostoma roboris Georgescu & Teodoru

(Anamorph ***Graphium roboris*** Georgescu *et al.*)

O. roboris bears a great resemblance to the saprophyte *Ceratocystis piceae* (Munch) Bakshi. It occurs mainly on oaks, but has also been recorded on other trees, including birch, elm and pine. It is known in Romania and the USSR, where it is associated with a disease known as vascular mycosis of oak, the symptoms of which much resemble the N. American oak wilt caused by *C. fagacearum*. Branches die, and the dried, dead leaves may fall, or remain hanging on the tree, while other leaves become twisted and light brown. Sometimes affected trees wilt suddenly in the middle of the summer. The growth of epicormic shoots is stimulated on the trunks and lower crown. If affected branches and twigs are cut, longitudinal brown streaks may be found in the outer sapwood. The disease has been briefly reviewed by Gibbs (1981) and Delatour (1986).

O. roboris is a weak parasite which causes vascular mycosis only in combination with other adverse factors, including poor soil conditions, damage by drought and various defoliating insects, and by other fungi, such as *Microsphaera alphitoides* and *Armillaria* species. It is probably transmitted by the yellow spotted longhorn beetle, *Mesosa myops*; if *O. roboris* spread to other parts of Europe, it might then be spread by the oak bark beetle *Scolytus intricatus*.

So far the losses caused by vascular mycosis have been relatively small and local. However, the disease might become more important if it spread to other areas (EPPO 76). To prevent the spread of *O. roboris*, when oak is moved from areas in which the fungus occurs the bark should be stripped off. The wood should also be dried to a maximum moisture content of 20% by weight, or fumigated with chloropicrin or methyl bromide. In outbreak areas, affected trees should be felled and the bark removed and burnt. Careful study of the pathogenicity, morphology and cultural characteristics of *O. roboris* and related fungi associated with vascular mycosis of oak is needed to facilitate their identification and evaluation as pathogens. Recently (Kryukova & Plotnikova 1982; EPPO 76), ***O. kubanicum*** Shcherbin-Parfenenko has been considered the most important in the USSR.
PHILLIPS

Chalara elegans Nag Raj & Kendr.

(Syn. *Thielaviopsis basicola* (Berk. & Broome) Ferraris, *Torula basicola* Berk. & Broome)

Basic description

The fungus produces two distinctive spore-types: i) phialoconidia, endogenously produced in long, subulate phialides (endoconidiophores) in fragile linear chains (corresponding to its name *C. elegans*); they are hyaline or light brown, thin-walled, cylindrical and measure 9–16.5×3–3.8 μm; ii) chlamydospores (aleuriospores), corresponding to the names *Thielaviopsis* or *Torula,* made up of 1–3 thin-walled basal cells surmounted by 1–8 brown or black cells (21–57 μm×13.5 μm). Some isolates produce giant chlamydospores with many basal cells (100–140 μm×8–9 μm) (Delon & Kiffer 1978; CMI 170).

Host plants

C. elegans attacks over 120 species from different families. It is most important on tobacco (and other *Nicotiana* species), other Solanaceae (tomato, aubergine), cotton, legumes (pea, dwarf bean, lupin, soybean, lucerne), cucurbits, and in particular on ornamentals, as a seedling pathogen of bedding plants and as a root pathogen of pot plants (*Chrysanthemum, Pelargonium* etc.).

Host specialization

C. elegans does not seem to have any forms or pathotypes specialized to some species or cultivars, but does exhibit considerable variability principally in artificial conditions. Stover (1950) found that two forms, brown and grey, exist in nature. The grey, wilt type is less pathogenic than the brown type. The latter occasionally mutates to the grey in soil.

Geographical distribution

As a soil-inhabiting fungus, *C. elegans* is very common in many countries throughout the world.

Disease

C. elegans causes a disease on tobacco named black-root rot (Fig. 44). The symptoms of the disease vary greatly according to the stage of development of tobacco. In the seed bed, very young seedlings may 'damp-off'. In this case the symptoms may easily be confused with those of other diseases. When the seedlings are older, they exhibit retarded growth and a discoloration of the leaves which turn pale green to yellow; roots are decayed and black. In the field, the disease generally occurs only in certain areas and causes varying degrees of stunting of the plants. Under severe infections, the plants become yellow and flower prematurely. Black lesions form on the roots and the fungus present in these localized lesions can easily be identified microscopically. The severity of the disease increases in poorly drained soil, especially if the weather remains cold and wet. On cotton (APS Compendium), *C. elegans* causes black-root rot especially of seedlings under cool moist conditions (principally in the USA, but reported in Europe). On vegetables (Dixon 1984), *C. elegans* causes root rot, penetrating via the ruptures caused by secondary root emergence, and producing purple to black lesions. The plants are stunted in consequence. On ornamentals (Fletcher 1984), *C. elegans* causes root rot of seedlings mainly after transplanting, but before planting out, under cool conditions. Black discoloration of the roots is characteristic. Violas, sweet peas and delphiniums are commonly affected. Pot plants (*Begonia, Senecio* x *hybridus, Cyclamen, Sinningia, Kalanchoë, Pelargonium, Euphorbia pulcherrima, Primula*) show rotting of the roots and stem base, and are stunted and unthrifty. Finally, while *C. elegans* is not normally of any great significance on woody plants, it is implicated in the specific replant disease of cherry and plum (Sewell & Wilson 1975). Certain rootstock clones (susceptible to *C. elegans*) are severely dwarfed when planted on old cherry land, while other resistant ones grow well.

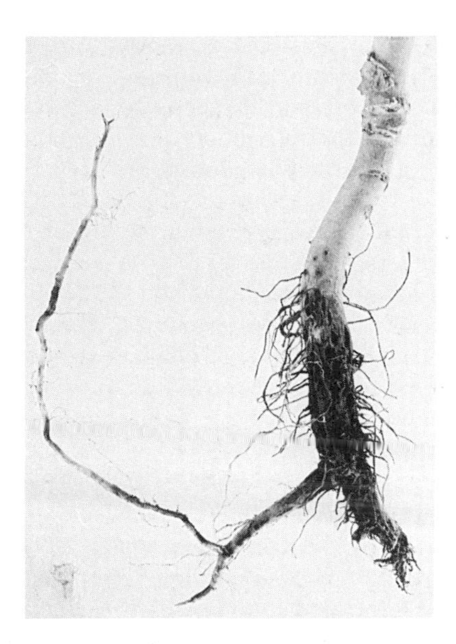

Fig. 44. Root system of adult tobacco affected by black-root rot (*Chalara elegans*).

Epidemiology

The development of black-root rot depends upon the inoculum potential of the soil and the pathogenicity of the isolate of *C. elegans*. The infective potential of soils can be determined using the 'carrot disk test'. This method has been extended to measure quantitatively the infestation of *C. elegans* in soil. Chlamydospores of the parasite can persist for 4 to 5 years in root tissue in the soil and three years in the soil itself (Lucas 1975), but the fungus may live indefinitely in the soil as a saprophyte. The severity of the disease depends largely on soil temperature; it is favoured by low temperatures. The soil pH is also very important in controlling the disease; with a pH of 5.6 or lower there is little or no disease at any temperature. Other conditions also contribute: high soil moisture, especially when near the saturation point, favours the disease, but the parasite is capable of developing in relatively dry soils. The severity of the disease is also conditioned by the behaviour of the host (Genève 1972), some cultivars of, for example, tobacco, being very susceptible.

Economic importance

The disease is probably one of the most serious with which tobacco growers have to contend. In Europe, according to a recent inquiry by CORESTA, *C. elegans* ranks second in importance only to *Peronospora tabacina* for the losses it causes and is considered to be an increasing problem. This situation is due partly to new blue-mould resistant cultivars more susceptible to *C. elegans*. In addition to the loss in yield when the disease occurs in the field, the quality of the tobacco leaves is reduced. Losses have been recorded in France, the FRG, Italy, Poland, Spain, Switzerland and Yugoslavia. *C. elegans* is of fairly minor importance on vegetables in Europe, but is among the commonest and most important causes of root rot in ornamentals, especially in glasshouses. Assessment of its importance is complicated by the difficulty of recognizing attacks and distinguishing them from those due to other root-rotting pathogens.

Control

On tobacco, the most effective control of black-root rot is obtained with the use of resistant cultivars. Sources of resistance are present in *Nicotiana* species such as *N. debneyi* where the resistance is controlled genetically by a single gene pair, or in *N. tabacum* where several factors are involved. Control is very different on seed beds or in the field. In the seed bed, the production of healthy transplants is one of the essential operations in tobacco cultivation. Soil disinfestation which eliminates weeds and soil-borne diseases is the principal method of control. Steam treatment is less used than in the past; chemicals like methyl bromide, chloropicrin, or sodium methyl dithiocarbamate are generally used. Benzimidazole fungicides used as a soil drench may also provide effective protection. In the field, besides planting resistant cultivars, it is essential to use disease-free transplants, to avoid transplanting on alkaline soils, and to follow recommended rotations for growing tobacco, avoiding susceptible hosts. Cold, wet, heavy soils are favourable for the disease and should be avoided if possible. Essentially the same principles apply to the control of the disease on vegetables and ornamentals.

DELON & SCHILTZ; SMITH

Chalara thielavioides (Peyr.) Nag Raj & Kend. (syn. *Chalaropsis thielavioides* Peyr.) is very similar to *C. elegans*, but differs in that it produces its chlamydospores singly rather than in chains. It is widespread in soil as a saprophyte behaving as a weaker pathogen than *C. elegans*, on similar hosts. It is associated with a minor storage rot of carrot roots (surface blackening) and also graft disease or black mould of rose and walnut. The black mycelium of the fungus overgrows the wound surfaces of stock and scion and prevents union. On walnut, the disease is easily controlled by the use of a formalin dip and is no longer a problem; on rose, control methods are not recorded. The disease has been more important in N. America (APS Compendium); it is only occasionally a problem in Europe (Brokenshire 1980).

DIAPORTHALES

Cryphonectria parasitica (Murrill) Barr

(Syn. *Endothia parasitica* (Murrill) Anderson & Anderson)

Basic description

The genus *Endothia* or *Cryphonectria* (CMI 704) is characterized by stromata, containing perithecia, immersed in the periderm. Each perithecium communicates with the exterior by means of a long neck. The ascospores, hyaline to pale yellowish, oblong, ovoid, fusoid or sub-ellipsoid are aseptate or uniseptate and not or very slightly constricted at the septum. Conidia (pycnospores) are also formed in pycnidium-like cavities in the stroma (cf. *Epidemiology* below).

Host plants

In Europe *C. parasitica* principally attacks *Castanea sativa. Quercus ilex, Q. pubescens* and *Q. petraea* are also susceptible (Biraghi 1951). Artificial inoculation tests showed the susceptibility of *Fagus sylvatica* to blight infection (Bazzigher 1953).

Geographical distribution

Chestnut blight disastrously damaged *C. dentata* in the USA and is found in many European countries (EPPO 69). France, Greece, Italy, Portugal, Spain and Yugoslavia were rapidly infested by the disease. In the Mediterranean area the blight is also present in Turkey. *E. parasitica* has also spread to Belgium, Hungary and Switzerland and has reached the Ukraine (USSR) (CMI map 66).

Disease

The general effect of the blight is to girdle and kill the part of sprouts or branches beyond the lesion, thus the early pronounced symptom of the disease in a chestnut tree or in a stump is dead branches or sprouts with yellow or brown wilted leaves. The young cankers are generally elliptical in outline at first and yellowish-brown on the surface. With increasing age cankers generally become rough, split and cracked with peeling bark (Plate 17) and the infected part is typically found to contain numerous flattened layers of branching mycelium known as 'fans'. Their presence is one of the best diagnostic symptoms of chestnut blight, in addition to the wilting foliage, cankers and fruiting pustules. The development of rapidly growing shoots from a point just below the girdled area is another evident symptom of blight which can be noted at any period of the year. Since the early 1950s, however, many blight-affected chestnut trees in Italy have remained alive and have regrown, with cankers that do not completely kill surrounding shoots or branches. In such cases, bark lesions are few, pycnidial production is low and perithecia are not observed. The buds below the cankers do not develop and so epicormic shoots do not develop. Under the bark, the tissues remain alive because the mycelium of these so-called hypovirulent strains grows externally and is very often not organized into compact fans.

Epidemiology

New infections arise from pycnospores and ascospores. At a temperature range of 15–25°C during damp weather, spore masses exude from the yellow-orange or reddish-brown pycnidia. A single spore horn of average size has been found to contain as many as 115 million minute one-celled pycnospores (Heald 1913). Each perithecial pustule contains 1–60 distinct flask-like cavities; the two-celled ascospores are shot into the air through the neck of each flask which opens at the top of a surface papilla. Maturing perithecia are more abundant in autumn and winter. The ascospores are wind-borne and may often be carried great distances. The pycnospores, produced abundantly in spring and autumn, are disseminated by birds, mites, crawling or flying insects and splashing rain (Russin & Shain 1983; Wendt *et al.* 1983). For pycnospores to germinate the temperature must range between 3° and 38°C while the germination of ascospores occurs between 18° and 38°C. The fungal spores lodge in wounds and after 8 h (at 22°C) the pycnospores begin to germinate, the ascospores more quickly. Upon germination the developed mycelium rapidly penetrates the inner bark and cambial layer. A large percentage of the new infections appear to be related to mechanical injury, but there is some evidence that natural cracks and fissures may also be an avenue of entrance. It is not clear how the hypovirulent strains spread in spite of their low pycnidial production.

Economic importance

Branches, sprouts or the whole tree may be killed when the infection has developed on the stem. The first appearance of blight in Europe was announced in Italy in 1938, and by 1951 many Italian chestnut areas were affected with the disease as was Switzerland, Yugoslavia and Spain. The rapid spread of *C. parasitica* in Italy can be related to the extensive planting of chestnut. The blight has caused serious losses of nut-crops and has killed many plants, but there is now a decrease in mortality in Italy, Switzerland and France. Even though this encourages a revival of chestnut growing, blight remains one of the most economically important plant diseases in Europe.

Control

Destruction of the infected branches, sprouts or whole tree is advisable to reduce inoculum potential. Chemical control of the fungus has proved difficult (Goidanich & Minghelli 1966). Utilization of resistant chestnuts is desirable, but the main new hope is based on hypovirulence. The hypovirulent strains produce healing cankers, and retard or stop canker growth when naturally or artificially inoculated. Hypovirulence appears to be transmitted from hypovirulent strains to virulent ones by hyphal anastomosis. This very efficient system of biological control is based on the compatibility between hypovirulent and healthy strains. In Italy the natural spread of hypovirulence is causing not only a general regression of the disease but also a decrease in mortality caused by *C. parasitica*, while in France Grente (1975) proposes that biological control of blight is possible by means of artificial inoculations.

Special research interest

Exploitation of the phenomenon of hypovirulence is a striking example of biological control. Studies on the hypovirulent strains have shown them to contain double-stranded RNA (Day *et al.* 1977) but have not explained the nature or spontaneous source of this possibly virus-like material which modifies the behaviour of the fungus. The mechanisms by which hypovirulent strains spread in nature are not clear but pycnospores produced by the blight fungus are one possibility and vectors such as insects or other animals that may visit *C. parasitica* cankers are another.

TURCHETTI

Diaporthe citri Wolf

Better known as its anamorph ***Phomopsis citri*** Fawcett, *D. citri* attacks all Rutaceae, but especially citrus (CMI 396) and is widespread in all citrus-growing areas. In Europe, it has been recorded in Spain, Italy, France, Greece, Portugal, Cyprus and USSR. *D. citri* causes melanose (very rarely seen in Mediterranean countries) on young leaves, shoots, stems and fruits. Small (0.1–1.2 mm in diameter) spots appear, reddish-brown to black in colour, surrounded by a whitish halo of suberized tissue. In other cases, if the RH is higher, the spots are smooth, slightly sunken and dark brown (Yamato 1977). More importantly, *D. citri* causes a fruit rot, which normally originates from the point of insertion of the peduncle (stem end). The pulp becomes tougher and dark brown, without altogether losing its flexibility.

D. citri is only a weak parasite, and rain is needed to disperse its conidia and permit infection and disease. Under Mediterranean conditions, the dry climate prevailing during fruit growth and maturation makes infection rare, so that *D. citri* behaves as a clear secondary pathogen. It can be readily isolated from desiccated branches, but is rather rarely found on fruit, and only then causing stem end rot. Specific control is hardly needed, although copper products (possibly mixed with dithiocarbamate, captan, folpet, captafol etc.) and also benzimidazoles have been found effective against melanose.

The teleomorph has not definitely been reported in Europe. It is separated with difficulty from ***D. medusaea*** Nitschke, which has led some authors (Fisher 1972) to consider them synonyms, although Yamato (1977) keeps them distinct. *D. medusaea* has been reported to cause a dieback of walnut in Italy (Vercesi 1982).

TUSET

Diaporthe eres Nitschke

(Syn. *D. perniciosa* Marchal)

D. eres is ubiquitous and represents a large species complex with a wide range of host-specific forms, the right taxonomic arrangement of which is not at all certain. More than 70 plant genera are listed as hosts, mostly deciduous trees, but some conifers and herbs are also apparently seriously damaged. The forms of the fungus occurring on fruit trees are in general called *D. perniciosa* (anamorph ***Phomopsis mali*** Roberts) by European authors. Woody parts and fruit are subject to attack. Infections on leaves have up to now only been observed in the USA. Forms causing a seedling blight and dieback of young conifers (*Picea, Pinus* etc.) are also known as *D. conorum* (Desm.) Niessl (anamorph *Phomopsis occulta* (Sacc.) Traverso). Attacks of the woody parts occur on apple, pear, cherry, plum and peach. On pome-fruit trees, the infections cause bark damage which is localized and more or less superficial. Occasionally a symptom arises which is called 'rough bark'. In Mediterranean countries (Greece), the pathogen has occasionally caused very severe damage in nurseries or newly established plantations of pear or quince, with necrosis of the bark and wood around the scion–stock union. On stone-fruit trees, *D. eres* is in the first place responsible for dieback, particularly on thinner branches or stems. Attacks on the fruit of stone-fruit trees are rarely observed. Only pome-fruit trees sometimes show substantial damage. On ripe or over-ripe fruit, the fungus causes a rot which normally appears at the stalk- end cavity and rapidly spreads to the whole fruit. In an advanced stage, the rots are soft and watery. *D. eres* survives in lesions on trees. Pycnidia with elliptical biguttulate 'α-conidia', and filamentous 'β-conidia' appear in these lesions.

The fungus commonly survives as a saprophyte on prunings, which after lying on the ground for a longer period, may bear perithecia with hyaline, fusoid ascospores which are one-celled when young and immature, and two-celled at maturity. Infection of the wood probably takes place only in winter, through wounds or possibly even leaf scars. In general, only weakened trees are attacked. The infection of fruit takes place before harvest. The fungus may gain entrance through lenticels or directly through the fruit skin. To control the fungus everything should be done to prevent any weakening of the trees and to remove the sources of inoculum (e.g. prunings). In orchards treated several times with benzimidazole fungicides a reduction of the infection pressure can be expected. Exceptionally long storage periods should be avoided.

KENNEL

Diaporthe helianthi Muntanola-Cvetkovic, Mihaljcevic & Petrov

(Anamorph ***Phomopsis helianthi*** Muntanola-Cvetkovic, Mihaljcevic & Petrov)

D. helianthi forms perithecia on overwintered plant residues, 350–450 μm in diameter, with a prominent beak 350–600 μm in length. Pycnidia are stromatic, 170–320 μm in diameter. During the vegetation period on the host plant as well as in artificial culture only filiform, β-conidia are formed. Pycnidia developed on overwintered sunflower stems produce both α- and β-conidia. The only known host plant is sunflower. Most of the cultivars, hybrids and breeding materials tested till now in Hungary, Yugoslavia and Romania, proved to be susceptible, but some of the recently developed NSH hybrids are highly tolerant or resistant. The pathogen was reported for the first time in 1979, from Yugoslavia (Acimović & Straser 1982). One year later the pathogen was found in the USA (Herr *et al.* 1983). *D. helianthi* occurred simultaneously in S.E. Hungary and N.W. Romania in 1981, and caused serious losses in France in 1985. The most striking symptoms are stem spots, originating from the attachment points of petioles, but the pathogen usually attacks leaves first. Part of a leaf bordered by main veins collapses, and the fungus invades the stem through the petiole. The initially brown stem spots spread rapidly and soon become dark grey to black. The upper part of the infected plant, above the girdling stem spot, turns dry. Pith along the infected part dissolves, and as a consequence of this, the hollow stem breaks easily. Yield loss is due partly to lodging and partly to small capitula with undeveloped seeds. *D. helianthi* also causes brown to black spots on the heads. Seeds from infected capitula carry the pathogen.

Sunflower is usually attacked after flowering. β-conidia, which develop *en masse* during the vegetation period do not germinate, at least under laboratory conditions, while α-conidia and ascospores germinate in 2–3 h. Cardinal temperatures are 10°C, 27°C, and 30°C (Vörös & Léranthné Szilágyi 1983). *D. helianthi* remains visible only on infected plant residues which overwinter above the soil surface. No pathogen was re-isolated from, and artificial inoculations were negative with, infected stem pieces buried 5–30 cm deep in the ground (Vörös *et al.* 1983). Moderately high temperatures (25–27°C), rain, dew, and high humidity increase the epidemic spread of the disease. Careful and proper ploughing of the infected plant residues into the soil prevents overwintering of the pathogen, and eliminates the primary inoculum from infected areas (Vörös *et al.* 1983). Seed treatment with benomyl and mancozeb combination, or with carboxin and oxine-copper eliminates the seed-borne pathogen. Less dense planting (about 50 000 plant/ha) decreases the humid microclimate and the epidemic spread of the disease, and at the same time promotes effective chemical control with repeated sprays with benomyl and contact fungicide combinations. The pathogen infects volunteer sunflower plants too, so their eradication is essential.

VÖRÖS

Phomopsis viticola (Sacc.) Sacc.

Basic description

The pycnidia contain ovoid to fusoid α-conidia and under certain conditions filiform slightly curved β-conidia, which can neither germinate nor infect. The teleomorph ***Cryptosporella viticola*** Shear was reported many years ago in N. America, but there is no recent confirmation that authentic *P. viticola* ever produces it (CMI 635). It may be noted that, formerly, *P. viticola* (syn. *Fusicoccum viticolum* Reddick) was considered the cause of dead-arm disease in N. America, while the disease in Europe was attributed to *Phoma flaccida* Viala & Ravaz and *Phoma reniformis* Prill. & Delacr. (cf. also *Guignardia baccae*) (e.g. Galet 1977). It is now accepted that no such distinction can be made, *P. viticola* being the primary pathogen everywhere, and the *Phoma* species merely secondary.

Host plants

P. viticola occurs only on *Vitis vinifera* (European grapevine), the cultivars of which vary in susceptibility. American and Asian *Vitis* species are not attacked, but their resistance has not been studied in any detail.

Geographical distribution

It is widespread in all grapevine-growing areas, except possibly S. America, but varying in importance according to the climate and cultivars planted.

Disease

Symptoms of dead-arm disease, or excoriosis, are seen only on the branches 3–4 internodes from the base of the plant. In early spring, the first discrete symptoms are small brown or black cortical spots, punctiform or linear, contrasting clearly in colour with the pale green of the young shoots at the 3–4 leaf stage. In late spring and early summer the symptoms become more striking and characteristic. Cortical lesions of three types are seen: i) violet-black or deep wine-coloured necrotic streaks, spindle-shaped, slightly sunken, isolated or confluent, about 1 cm long; ii) isolated blackish crusts; iii) superficial corky structures, pale to dark brown, striated (like a bar of chocolate), stretching over 1 or 2 internodes. During summer, when the branches begin to mature and lignify, longitudinal cracks split the bark following the line of necrosis, and the base of the branches may be girdled. In autumn, a final symptom is seen—whitening of the bark, often in patches, along the whole length of the current year's growth. *P. viticola* also attacks young leaves (petioles and main veins) and the stalk of the young bunches. The lesions resemble those on young shoots.

Epidemiology

P. viticola survives the winter as pycnidia on the basal branches and as mycelium in dormant buds (Bulit *et al.* 1973). The pycnidia form in autumn and mature over the winter. Spore release begins in February, but is most abundant in spring at the time of bud-burst in the crop. Mycelium is limited to the first buds at the base of the current year's growth. These are infected as they form during the preceding season, and they may be destroyed by *P. viticola* at bud-burst. Conidia, formed in cirri from the pycnidia, are splash-dispersed by rain. This explains the localization of necrosis on the first internodes of the branches, the only ones which lie within splashing distance. It also accounts for the slow spread of the disease in foci. Pycnospores germinate on the shoots in about 4 h at 25°C (optimal conditions), or 6 h at 20°C and 8 h at 15°C. Infection follows germination in about 5 h at 25°C (optimal conditions) but at spring temperatures of 15–18°C, a period of surface wetness of 7–10 consecutive hours is necessary (sufficient) (Bulit & Bugaret 1975). The young grapevine shoots are only liable to be infected for a short period, from leaf emergence to leaf unfolding. Infection occurs directly, without wounds.

Economic importance

Infection at the shoot-base causes serious damage by: i) killing infected buds or causing the formation of stunted shoots which fruit poorly; ii) girdling shoots which are then liable to be broken by wind or agricultural operations. Young bunches attacked at the stalk level dry out. Overall, infection may cause yield losses of up to 30% (for heavily infected plants), and also disturb the pruning of the plant, by making it necessary to leave longer shoots to compensate for the buds which do not grow. In addition, growers have difficulty in practice in coping with the disease, because the actual attack occurs so early (susceptible cultivars have to be treated at the beginning of leaf emergence) and so easily (a few mm of rain is sufficient to allow infection). Vineyards in areas with a wet spring are most at risk. Cultivars vary considerably in susceptibility. In France, Cabernet-Sauvignon, Muscat de Hambourg and Syrah, Aligoté, Chasselas, Chenin, Colombard and Muscadelle are susceptible, while Carignan, Cabernet Franc, Grenache, Merlot, Riesling, Semillon, Traminer and Ugni Blanc are rather resistant. Pinot Meunier is almost entirely resistant.

Control

In an infected vineyard, the first essential measure is to keep only healthy wood at pruning, after which chemical control can give further protection. Two approaches to chemical control are possible (Bulit *et al.* 1972). The first is to destroy, before bud-burst, the inoculum present on the shoots, especially the pycnidia on the surface and, to some extent, the mycelium in the buds. An eradicant treatment with sodium arsenite (625 g As per hl) is applied to dry or wiped shoots. This treatment is most effective if it wets the pruning wood abundantly (300–500 l per ha) and if it is carried out when most pycnidia have matured (i.e. 2–3 weeks before bud-burst, to avoid phytotoxicity). It may be noted that sodium arsenite is not authorized for use in many European countries, and that in France it has to be used in combination with a wildlife repellant. The second approach is to treat young shoots after bud-burst, to prevent infection at the time of maximum susceptibility. With dithiocarbamates, phthalimides, quinones or sulphamides, two preventive treatments are needed, one when 30% of the buds have reached the leaf emergence stage, and the next when 30% have reached the leaf-unfolding stage, so as to cover the susceptible stage entirely. For the systemic fosetyl-Al, combined with folpet or mancozeb, a single treatment at the beginning of leaf emergence is sufficient. This gives greater flexibility in timing, and is also very effective (Bugaret *et al.* 1980). Finally, a phytosanitary approach is possible, by producing budwood mother plants free from *P. viticola.* The general use of budwood from these plants would eliminate the first introduction of the disease into a vineyard.

BULIT & BUGARET

Phomopsis juniperivora Hahn

P. juniperivora occurs especially on a large number of species of *Juniperus*, and on *Cupressus* species (including *C. arizonica* and *C. macrocarpa*), and sometimes also on *Chamaecyparis, Cryptomeria, Cupressocyparis leylandii, Thuja, Thujopsis* and *Larix*. It is best known on *Juniperus virginiana* in N. America, but it is also present in Europe, and in S. Africa. It is associated with a shoot dieback, mainly of nursery plants, but sometimes of older trees. The shoot tips are first affected, and turn brown. The disease may then spread down the shoots and into the branches, on which perennial cankers may be formed. Further spread into the main stem may also take place, and in severe cases the whole plant dies. Long spells of damp weather, particularly in spring and autumn, encourage development of the disease (Caroselli 1957). So far in Europe, damage to nursery stock by *P. juniperivora* has been only occasional. Nevertheless considerable losses have sometimes been caused, especially in nurseries in France, where many species of Cupressaceae have been affected (Morelet 1982). *Phomopsis* dieback is best controlled by cultivating the nursery stock concerned only in dry weather, avoiding overhead irrigation, and spraying if necessary with benomyl (Morelet 1982).

PHILLIPS

Phomopsis sclerotioides v. Kesteren

P. sclerotioides is distinguished by the absence of β-conidia, the presence of sclerotia, and its apparent specialization on Cucurbitaceae (CMI 470). It was discovered in the Netherlands in 1962 (van Kesteren 1967). It has been observed in UK, Germany, France, Denmark, Norway and Switzerland. It is probably present in most areas where cucumbers are grown under glass. It has also been found in non-European countries like Canada and Malaysia.

P. sclerotioides causes black-root rot of glasshouse cucumbers. Symptoms are usually detected at the ripening of the first fruits. Plants wilt when air temperature is high, then recover in the evening. After a few days, wilting becomes permanent. In severe attacks plants die within a few days. In other cases, plants are stunted, with partial development or early ripening of fruits. These symptoms are associated with a root rot characterized by two types of damage: i) lesions consisting of dark 'zone lines' of thick-walled fungus cells enclosing necrotic root tissue and mostly oriented along the longitudinal axis of the root; and ii) small, dark pseudosclerotia distributed in the cortical cells, resulting in a speckled aspect of the roots. Pycnidia are rarely found on the host plant.

The fungus survives in the soil, probably as pseudosclerotia, pseudostromata, or dark hyphal cells in plant debris. Infection by cucumber mosaic virus sensitizes cucumber seedlings to root infection by *P. sclerotioides.* Powdery mildew infection is earlier and heavier on plants with black-root rot. *P. sclerotioides* is a major pathogen in the production of glasshouse cucumbers in several countries. Yield losses up to 50% have been recorded. An annual soil disinfestation by steam to 45 cm deep is the best means by which to grow healthy cucumbers in contaminated soil. Drenches with benzimidazole fungicides may prevent or stop disease development. Grafting on *Cucurbita ficifolia*, a slightly susceptible host, gives some protection in poorly infested soils. Soil fumigation with methyl bromide has been recommended, but rapid and disastrous recolonization of treated soil may occur. Growing plants in soil-less culture might be an appropriate control method. *P. sclerotioides* has been a promising tool for biological control studies with antagonistic micro-organisms (Gindrat *et al.* 1977).

GINDRAT

Several other species of *Diaporthe* and *Phomopsis* are of minor importance in Europe. ***D. cinerescens*** Sacc. (anamorph ***P. cinerescens*** (Sacc.) Traverso) is widespread causing canker of fig, penetrating through pruning cuts (Viennot-Bourgin 1949). ***D. phaseolorum*** (Cooke & Ell.) Sacc. occurs principally in N. America, where specialized forms attack *Phaseolus lunatus* and sweet potato. More important are two varieties on soybean **var. *sojae*** (Lehman) Wehmeyer (anamorph ***Phomopsis sojae*** Lehman), causing pod and stem blight and **var. *caulivora*** Athow & Caldwell, with no pycnidia, causing a much more damaging stem canker (APS Compendium). Var. *sojae* is readily seed-transmitted, but of minor practical importance; it has now been found in most soybean-growing countries in Europe, probably imported with seed. Var. *caulivora* is considered a quarantine organism for Europe, and is more liable to be carried over on residues. It has now been recorded in France, Spain and Yugoslavia (Signoret 1984). ***D. taleola*** (Fr.) Sacc. (syn. *Caudospora taleola* (Fr.) Starb.) and ***D. leiphaemia*** (Fr.) Sacc. (syn. *Amphiporthe leiphaemia* (Fr.) Butin) (anamorph ***Phomopsis quercella*** (Sacc. & Roum.) Died.) are common saprophytes on *Quercus* species, on which they may sometimes act as secondary parasites causing dieback of the shoots. ***D. woodii*** Punith. (anamorph ***P. leptostromiformis*** (Kühn) Bubák) causes stem blight of lupins (CMI 476). Inoculum is carried over as stromata in crop residues, and on seeds. Seedlings may be killed, or, on older plants, yellow patches form on the stem, with black stromata containing pycnidia and perithecia. Pods and seeds may finally be infected. Control is essentially by sanitation. ***D. vexans*** (Sacc. & Sydow) Gratz (anamorph ***P. vexans*** (Sacc. & Sydow) Harter) causes a stem and leaf blight, and fruit rot of aubergine (CMI 338), in warmer countries. In Europe, it has apparently been recorded only from Romania.

Phomopsis obscurans (Ell. & Ev.) B. Sutton (syn. *Dendrophoma obscurans* (Ell. & Ev.) Anderson) has globose or flask-like unilocular dark-brown or black pycnidia with a central opening. Conidiophores are filiform and branched. The fungus causes leaf spot disease on different *Fragaria* species and is common on cultivated *Fragaria* cultivars in many countries of Europe, Asia, N. America, Africa and Australia. Necrotic spots on leaves are round or elongated, and brown with a darker brown margin. Under wet conditions the fungus may totally destroy a high percentage of the leaves of a strawberry crop. The pathogen overwinters in infected leaf tissue in the form of

Fig. 45. Spots on strawberry leaves caused by *Phomopsis obscurans*.

mycelia or pycnidia. Copper fungicides and benzimidazoles are effective.
VAJNA

Phomopsis foeniculi has recently been described causing decline of fennel in France (Manoir & Vegh 1981). ***P. cucurbitae*** McKeen (CMI 469) causes black rot of the above-ground parts of cucurbits, but is of very minor importance. It is distinguished from *P. sclerotioides* (q.v.) by the absence of sclerotia and the presence of β-conidia. ***P. theae*** Petch (CMI 330), mainly known on tea in the tropics, also causes problems in the UK on *Rhododendron* during propagation and is being found more frequently on *Camellia*.

Fusicoccum quercus Oudem. causes a bark canker of oak. Pycnidia formed in spring and summer are smaller (0.5–1.5 mm) than those of *Phomopsis* type, while those appearing in the autumn are larger (up to 3 mm) stromatic and multilocular. α- and β-conidia are formed in both cases. Reddish-yellow discoloured lesions appear in spring, often around a lateral shoot. Callus forms through the season, in most cases sealing off the lesion. In some cases, the lesion can spread further the following year, and seedlings may be killed. Lesions also occur on the shoots of mature oaks, causing dieback, often after sub-lethal frost damage. The fungus is also a saprophyte on the bark of dead branches, which thus serve as a source of inoculum. Young oaks are particularly liable to attack on sandy soils with poor water-retaining capacity and these conditions should accordingly be avoided (Butin 1981, 1983).

Fusicoccum amygdali Delacr. causes a twig canker of peach and almond, especially in Greece, Italy, S.W. France and Spain. Infection mainly occurs in the autumn, via leaf-scars (Jailloux & Froidefond 1978) and brown lesions spread around the buds in spring. The distal portions of the shoot may wither, due to a translocatable toxin (fusicoccin), which causes excessive transpiration by stimulating the opening of stomata (Ballio 1978). Losses from dieback or defoliation may be quite severe and budwood production may be particularly affected. Infected parts should be pruned away, followed by treatment with dithiocarbamates or captafol (Marchetti 1977).

Cryptodiaporthe populea (Sacc.) Butin

(Anamorph ***Discosporium populeum*** (Sacc.) B. Sutton, syn. *Dothichiza populea* Sacc. & Briard, *Chondroplea populea* (Sacc.) Kleb.)

C. populea (CMI 364) affects many species and varieties of *Populus*, which however vary greatly in susceptibility. Thus the Lombardy poplar, *P. nigra* 'italica', is very susceptible. Among the numerous hybrid black poplars, *P.* x *canadensis* 'robusta' is highly susceptible, while at the other extreme, *P.* x *canadensis* 'gelrica' is resistant. Poplars of the subgenus *Leuce* tend to be resistant. Strains of *C. populea* vary in pathogenicity (Hubbes 1959). The fungus is found throughout Europe and in eastern N. America (CMI map 344).

C. populea causes a bark necrosis or trunk scab, especially on shoots in nursery stool beds, but sometimes on mature trees as well. It produces sunken, oval, grey, brown or black patches around the bases of twigs and branches and around buds. Pycnidia may be found on lesions from May to October, but the fungus is often rapidly displaced by secondary organisms. Perithecia may sometimes be found in winter. Infection takes place through wounds, including bud scale scars (especially in spring), leaf scars (particularly in autumn) (Gremmen 1978), and lenticels. Little infection occurs when temperatures are above about 15°C, and the fungus usually only attacks plants weakened by other factors, such as drought, leaf disease, excess nitrogen, or the shock of transplantation.

C. populea occasionally causes severe damage in nurseries, and in Italy and Germany very large numbers of young trees have been killed by the fungus even after the nursery stage. The disease can usually be avoided by good spacing in the nursery, and pruning before mid-August to promote the healing of wounds before the end of the growing season. If the disease is known to be troublesome in a given area, only resistant clones should be grown, and the plants may be sprayed with a copper fungicide.

PHILLIPS

Cryptodiaporthe castanea (Tul.) Wehmeyer (anamorph ***Discella castanea*** (Sacc.) v. Arx, syn. *Fusicoccum castaneum* Sacc.) sometimes causes a dieback and canker on *Castanea sativa*. ***C. salicella*** (Fr.) Petrak (anamorph ***Discella salicis*** (Westend.) Boerema) causes a minor dieback of *Salix* species. It is probably the same as ***C. salicina*** (Pers.) Wehmeyer (anamorph ***Discella carbonacea*** (Fr.) Berk. & Broome) (Butin 1958). Other *Cryptodiaporthe* species have *Diplodina* anamorphs. ***D. castanea*** Prill. & Delacr. is the cause of a common but unimportant canker and dieback disease of *castanea* in Europe. It is often found in association with *C. castanea* (above) (Lanier *et al.* 1976); there seems a possibility that the relationship is a closer one. ***D. passerinii*** Allescher is quoted in older books (Brooks 1953) as causing wilt of *Antirrhinum* and other ornamentals by infecting the stem base. However, there seems to be no recent information on this pathogen.

Valsa cincta Fr.

(Syn. *Leucostoma cinctum* (Fr.) Höhnel)
(Anamorph ***Cytospora rubescens*** Fr., syn. *C. cincta* Sacc.)

Basic description

V. cincta is characterized by valsoid ascostromata immersed in the periderm of dead trunks, stems and branches. Conidiomata with allantoid, hyaline conidia develop separately or sometimes in the ascostroma.

Host plants

V. cincta has a wide range of hosts belonging to the genera *Cotoneaster, Prunus, Sorbus, Swida* etc. However, the fungus occurs most frequently on peach and apricot.

Host specialization

The existence of strains or isolates of different virulence has been described (Helton & Konicek 1961; Rozsnyay 1977).

Geographical distribution

The fungus occurs in Europe, Asia (excluding tropics and humid subtropics) and N. America.

Disease

V. cincta is considered as a necrotrophic parasite which attacks damaged (stressed) plants causing infection and necrosis of phloem, cambium and xylem followed by apoplexy, dieback or perennial canker, and gum exudation.

Epidemiology

Ascospores, and especially conidia, are the main sources of infection. They develop mainly during spring and autumn and sometimes in winter. Spore release, distribution, penetration into the host and infection occur only under wet conditions. Furthermore mycelium can survive for a long time in outwardly healthy plants in a latent state. Thus, in some cases the appearance of necrosis must be considered as a result of awakening latent inoculum when the whole plants or their separate parts are stressed by adverse factors. Taking into account the available information it must be concluded that decline is caused by different biotic (fungi, bacteria, mycoplasma, viruses) and abiotic factors, effecting together or separately in various combinations, i.e. the disease has a complex aetiology.

Economic importance

Losses from decline reach 10–15% or sometimes more, notably in E. Europe, but not all such losses are caused only by *V. cincta*.

Control

The cultural practices that make the plants resistant are the following: growing of cultivars adapted for given locations and sites; using rootstocks with corresponding compatibility; correct soil management; balanced use of mineral fertilizers; late summer, autumn and winter irrigation; protection from injury, etc. (Babajan 1980; Paunović 1980). Because the plants, especially apricots and peaches, are most susceptible to infection in the dormant period, pruning must be done in spring. Spraying after pruning, using 2% Bordeaux mixture and protective paints with fungicides and treatment of cankers are all recommended (Morvan 1977; Rozsnyay 1977). Application of captafol and benomyl during leaf-fall and late winter gives significant reduction in *Cytospora* infection on twigs (Royse & Ries 1978).
GVRTISHVILI

In addition to *V. cincta*, several other *Valsa* species cause similar canker diseases of fruit trees in Europe. ***V. leucostoma*** (Pers.) Fr. (syn. *Leucostoma persoonii* (Nitschke) Höhnel; anamorph ***Cytospora leucostoma*** (Pers.) Sacc.) can be differentiated by conidial dimensions, and by host range (attacking cherry as well as peach and plum) (Schmidle *et al.* 1979). In practice, the two species are very often considered together. Schulz & Schmidle (1983) have shown that the size of necrosis is inversely related to the number of days with an average temperature over 10°C, while sporulation is correlated with the number of hours at 10–15°C and RH 90–100%. These indications can be used to time applications mostly made when the tree is in leaf. Schmidle & Schulz (1978) found that best protection was obtained with carbendazim+pyracarbolid, benomyl or thiophanate-methyl. The first two also showed curative effects. Schulz (1981) found prophylactic treatments with various antagonistic fungi (e.g. *Trichoderma* spp.) to give good inhibition of *V. leucostoma* and *V. cincta*. The two species are also recorded attacking apple, especially through pruning wounds (Kastirr & Ficke 1983), together with other species: ***Cytospora personata*** Fr. and ***C. schulzeri*** Sacc. & Sydow.

Valsa sordida Nitschke

(Anamorph ***Cytospora chrysosperma*** (Pers.) Fr.)

V. sordida occurs on various broad-leaved trees, but is important as a cause of disease almost solely on *Populus* species. Information on resistance of various poplars to its attack is conflicting, though Bloomberg (1962) correlated resistance to the fungus with resistance to water loss. *V. sordida* is common throughout much of Europe into the USSR, in the Americas, Africa, and Australasia (CMI map 416). The fungus, commonly saprophytic, sometimes acts as a weak parasite, causing a dieback by girdling twigs and small branches. It also causes a bark necrosis, producing sunken lesions on trunks and larger branches. It fruits on the dead tissues, on which its golden conidial tendrils are often conspicuous. It invades through wounds and lenticels, and may spread to living tissues from dead parts on which it is growing saprophytically. It mainly attacks cuttings, young trees on poor sites, and those affected by moisture stress, sunscorch, or prior damage by leaf fungi.

Serious attacks are uncommon, but the fungus sometimes destroys many cuttings, and spoils the shape of older trees by inducing excessive branching and sucker production. The disease is best controlled by siting poplar nurseries and plantations only on suitable fertile and water-retentive soils, and avoiding overcrowding and damage to plants. Spraying with copper fungicides and with dinitro-*o*-cresol, and dipping plants in thiram before winter storage (Walla & Stack 1980) have been recommended.
PHILLIPS

Various other *Valsa* species occur on trees as saprophytes or very weak parasites, e.g. ***V. nivea*** Fr. (syn. *Leucostoma niveum* (Pers.) Höhnel) on *Populus* and *Salix*, ***V. kunzei*** Fr., ***V. curreyi*** Nitschke (syn. *L. curreyi* (Nitschke) Défago) and ***V. abietis*** Fr. on various conifers. *V. abietis* has even been suggested as an antagonistic agent for biological control of *Lachnellula willkommii* on larch (Dorozhkin & Fedorov 1982).

Apiognomonia errabunda (Roberge) Höhnel

(Syn. *Gnomonia errabunda* (Roberge) Auersw.)
(Anamorph ***Discula umbrinella*** (Berk. & Broome) B. Sutton, syn. *Gloeosporium umbrinellum* Berk. & Broome)

A. errabunda causes leaf spot and shoot dieback on many broad-leaved trees and is widespread in Europe and N. America. It has been regarded as a complex of individual *Gnomonia* species on different hosts, transferred to *Apiognomonia* by Barr (1978). In particular, *A. errabunda sensu strictu* causes leaf spot of beech. *A. quercina* (Kleb.) Höhnel (syn. *G. quercina* Kleb., anamorph *D. quercina* (Westend.) v. Arx) causes leaf spot and dieback of oak, while *A. tiliae* (Rehm) Höhnel (syn. *G. tiliae* Rehm, anamorph *Gloeosporidium tiliae* (Oudem.) Petrak) produces leaf spots on young lime trees, possibly causing severe defoliation. The best known form of the fungus, on plane, has been known as ***A. veneta*** (Sacc. & Speg.) Höhnel (syn. *G. veneta* Kleb.; anamorph ***Discula platani*** (Peck) Sacc., syn. *Gloeosporium nervisequum* (Fuckel) Sacc.). In the latest revision of Monod (1983), *A. veneta* on plane is finally distinguished as a species from *A. errabunda* on beech and oak (thus including *A. quercina*). ***Discula betulina*** (Westend.) v. Arx (syn. *Gloeosporium betulinum* Westend.), widespread as a cause of leaf spot of birch in Europe and sometimes causing severe defoliation, no doubt belongs to the same complex, and similar species are found on other trees (*Acer, Carpinus, Corylus* etc.) (Cannon *et al.* 1985). The teleomorph is found on fallen leaves, while the fungus produces the anamorph stage on living leaves and petioles. The form of the fructification is variable, so that the fungus has been ascribed anamorphs in *Gloeosporidium, Discula* and *Sporonema* (the first two corresponding to the rejected *Gloeosporium*). However, these are just morphological variants (by position on leaf or petiole) of the same conidial fructification.

The most significant disease is that caused on plane. In Europe, the London plane, *Platanus* × *acerifolia*, is most often attacked, largely because it is so extensively grown. Its clones vary in susceptibility. The oriental plane, *P. orientalis*, is very resistant. Plane anthracnose has four forms (Neely 1976): i) bud blight, in which buds are killed and do not open in spring; ii) twig blight, in which whole twigs die back; iii) shoot blight, in which the buds begin to open, but the shoots die back before flushing is complete—many leaves soon fall, so that in severe cases the tree becomes almost leafless (but often produces a second flush); iv) leaf blight, in which throughout the summer infected leaves show dark brown dead zones bordering the main veins. Bud, twig and shoot blight all seem to arise in spring following infection in the previous year, and damp cool springs seem to favour the disease. Attacks of anthracnose are sometimes spectacular, but severe outbreaks are sporadic, and affected trees usually recover. Hence the impact of the disease is relatively small. Selection of resistant clones of *P.* × *acerifolia* would seem to offer the best means for the future control of anthracnose in Europe. Meanwhile, diseased shoots and fallen leaves should be collected and burnt. Nursery plants and small trees may be sprayed with benomyl or captafol (Marchetti *et al.* 1984; Spies *et al.* 1985).
PHILLIPS

Apiognomonia erythrostoma (Pers.) Höhnel (syn. *Gnomonia erythrostoma* (Pers.) Auersw.; anamorph ***Libertina effusa*** (Lib.) Höhnel, syn. *Phomopsis stipata* (Lib.) B. Sutton) attacks cherry and apricot, on which it causes large irregular leaf spots, which may be several cm in diameter, and which start yellow to reddish, and become brown. Serious attacks can cause premature leaf-fall. Perithecia developing in fallen leaves on the soil surface are the source of infection in spring. Infected leaves may also remain on the trees. The disease is mostly serious

on apricot in E. Europe. Chemical control is required there to eliminate ascospore discharge or prevent perithecium formation in winter. Fungicides applied against *Blumeriella jaapii* on cherry will readily control *A. erythrostoma*.
BONDOUX

Pleuroceras pseudoplatani (Tubeuf) Monod (syn. *Ophiognomonia pseudoplatani* (Tubeuf) Barrett & Pearce) is the cause of the giant leaf blotch disease of sycamore which is so far known in Denmark, Germany and UK.

Gnomonia leptostyla (Fr.) Ces. & de Not.

(Anamorph ***Marssoniella juglandis*** (Lib.) Höhnel, syn. *Marssonina juglandis* (Lib.) Magnus)

G. leptostyla attacks *Juglans* species generally, and in particular walnut (*J. regia*). It is widespread wherever the host is grown, but varies in importance according to region and year. Lesions (blotch or anthracnose) are formed on all green parts: leaves, petioles, young shoots and fruits. They appear as round brownish black spots several mm in diameter, fading in the centre. If attacks are heavy, leaves age and fall prematurely (August–September) and fruits dry up. *G. leptostyla* overwinters on dead leaves, on which perithecia form. Ascospores released in spring cause primary infection, and conidia formed on the lesions spread the disease throughout the season. Infection is favoured by temperatures around 21°C and high RH (96–100%), but conidia can survive the absence of free water for at least two weeks (Black & Neely 1978). If control is necessary, two sprays in May–June, at the time of ascospore discharge, are recommended. Benomyl has given best results (Berry 1977), while chlorothalonil, maneb, dodine and copper fungicides all give equally good control. In France, the disease is rather less important than bacterial blight caused by *Xanthomonas campestris* pv. *juglandis*. The copper treatments applied against the bacterium also control the fungus. Some cultivars are less susceptible but none fully resistant.
BONDOUX

Gnomonia comari P. Karsten (syn. *G. fragariae* Kleb.; anamorph ***Zythia fragariae*** Laibach) causes large, brownish leaf spots (Fig. 46) and also fruit rot of strawberry in N. America and many countries of Europe (CMI 737; APS Compendium). Under humid and relatively cold conditions, or in the glasshouse, the fungus can destroy 30–50% of leaf surface of strawberry and may cause severe fruit losses (stem-end rot). The pathogen overwinters in infected leaf tissue in the form of mycelia or pycnidia. The teleomorph is rarely found in nature but is readily formed in culture containing strawberry-leaf extract. Effective fungicides for control have been noted by Mappes & Locher (1981).
VAJNA

Fig. 46. Spots on strawberry leaves caused by *Gnomonia comari*.

Gnomonia rubi (Rehm) Winter occurs on *Rubus* spp. but also causes a dieback of cold-stored roses (Schneider *et al.* 1969). ***Cryptosporella umbrina*** (Jenkins) Jenkins & Wehmeyer (syn. *Diaporthe umbrina* Jenkins) causes brown canker of rose producing off-white necrotic lesions with a purplish margin, darkening to tan-coloured cankers dotted with pycnidia. Shoots may be girdled. The disease is controlled by pruning (APS Compendium). ***Melanconis juglandis*** (Ell. & Ev.) Groves (anamorph ***Melanconium juglandinum*** Kunze) causes a dieback of walnut and ***Melanconis modonia*** Tul. & C. Tul. (anamorph ***Coryneum modonium*** (Sacc.) Griffon & Maubl., syn. *C. kunzei* Corda) a dieback of chestnut, especially after coppicing (Viennot-Bourgin 1949). ***Greeneria uvicola*** (Berk. & Curtis) Punith. (syn. *Melanconium* *fuligineum* (Scribner & Viala) Cavara) causes bitter rot of grapes in N. America and dieback of grapevine shoots in Greece (CMI 538). It does not appear to be recorded elsewhere in Europe.

Gaeumannomyces graminis (Sacc.) v. Arx & H. Olivier

(Syn. *Ophiobolus graminis* Sacc.)
(Anamorph ***Phialophora radicicola*** Cain)

Basic description

Formerly considered as an *Ophiobolus* (Pleosporales), the fungus has now been transferred to *Gaeumannomyces* (Diaporthales). Its most characteristic feature is to form a network of brown runner hyphae along the surface of grass roots, sometimes in strands, with lateral hyphae forming hyphopodia (lobed in var. *graminis* or simple in others) from which penetration occurs (CMI 381–3). Curved to semicircular phialospores are also formed and this anamorph can probably be considered conspecific with *Phialophora radicicola*, which also exists as non-pathogenic strains in the soil.

Host plants and specialization

G. graminis causes take-all disease of many Gramineae, notably cereals. There are three varieties: **var. *tritici*** Walker, the most important, attacks wheat, barley, rye and triticale; **var. *avenae*** (Turner) Dennis causes severe disease in oats as well as in wheat (its ascospores are longer than those of var. *tritici*); **var. *graminis*** particularly attacks rice, its temperature optimum (25–30°C) being higher than that of the two others (20–25°C) (Walker 1975). All also attack cultivated and wild grasses.

Geographical distribution

G. graminis occurs world-wide, wherever cereals are grown intensively. The range of var. *graminis* in Europe remains to be confirmed. It is recorded on rice in Italy but could occur more widely on other cereals.

Disease

G. graminis attacks roots at all growth stages. Seedlings may be killed, but often attack occurs later, causing stunting and reduced tillering. As maturity approaches, infected plants may show dead, bleached, inflorescences, known as whiteheads. The base of infected plants sometimes shows blackening, with dark mycelium and perithecia under the leaf sheaths. When crown roots become infected and if weather is hot and dry, damage is considerable, and there is no yield, or only very shrivelled grain (take-all). Disease occurs in roughly circular patches, of a few plants or extending to nearly all the field. On rice, the disease is known as crown, black or brown sheath rot (CMI 381).

Epidemiology

Inoculum survives as saprophytic mycelium in rotted roots and the base of infected stems and will last as long as the crop residues. So in low pH soils where microbiological activity is weak, decomposition is slow and the survival of mycelium is long (2–3 years). In alkaline soils, survival is correspondingly short. Survival is also possible on the roots of volunteer cereal plants or on wild grasses (e.g. *Elymus repens*). Ascospore infection is possible but probably unimportant in the spread of take-all because of competition from other micro-organisms in the rhizosphere. Similarly, the phialospores do not seem to have a role in infection in the field. The first attacks come from saprophytic mycelium which is probably stimulated by exudates from young sensitive roots. Later progression of the fungus from root to root varies with the nature of soil. Root infection is favoured by soil temperatures of 10–23°C, alkaline pH and lighter soils. Wet soil conditions in spring increase disease. Dissemination of inoculum from one field to another is possible with soil and residues spread by agriculture practices. A field damaged by take-all may be harvested first because of its precocity, and may thus be a source of contamination for other fields.

Of the many factors which affect the development of take-all, the most important is probably soil micro-organisms. In sterilized soil, take-all is always heavy. When soil micro-organisms are destroyed in part by too high a calcium ratio, or after weed-clearing, damage is severe. Deficiency of nitrogen, phosphorus, or potassium increase the severity of take-all. However, too much nitrogen is also harmful and NH_4^+-N is better than NO_3^--N. Copper deficiency increases disease especially in barley. In fact when plants have a well-balanced nutrition, take-all is less severe. In the same way, over-use of herbicides against grass weeds increases take-all, as does deterioration of soil structure. Severity of take-all is influenced by the preceding crop and the disease very often occurs on the second or third crop of a susceptible cereal in succession. Generally, a break crop of legumes, potatoes or crucifers (in the absence of susceptible grass weeds or volunteer cereals) rapidly decreases inoculum density; maize has only a weak effect, while inoculum may survive for 1 year under beet.

Economic importance

Take-all disease is never important when the rotation only infrequently includes susceptible crops (e.g. wheat once every 3 or 4 years). On the other hand, on intensively grown cereals in all parts of the world (Europe, Australia, N. America especially), take-all is probably the most severe disease on wheat and barley because control is difficult and its effects not directly seen. Attacks of *G. graminis* are often confused with nutritional deficiencies of various kinds. It is difficult to estimate losses, but they may be 10–20% of harvest on second, third or fourth successive wheat crops.

Control

At present, the best method of control is rotation with break crops, particularly now that herbicides can eliminate the weed grasses which carry inoculum through the non-cereal crops. Straw burning has little or no effect on the level of inoculum. Direct drilling keeps inoculum at the surface and so is worse than deep ploughing. Early sowing may make crops more vulnerable to autumn infection. All cultural practices which destroy the microbiological balance of the soil increase the risk of take-all. At the present time, there are no resistant cultivars, and it is impossible to use chemical treatments in the field. In addition to rotational practices, it may be possible to develop specific biological control techniques. It has been known for many years that numerous elements of the soil microflora (fungi, bacteria) are antagonistic to *G. graminis*, and it is possible to encourage their development by green manuring or addition of organic materials to the soil. However, the best chance of successful biological control comes from the explanation and reproduction of 'take-all decline'. With monocultures of wheat or barley, severe attacks of take-all are followed by a spontaneous reduction in disease. The mechanisms involved in take-all decline are either modification of the soil microflora, reducing inoculum activity or loss of aggressiveness of the inoculum itself. It is possible to isolate avirulent strains from wheat in monoculture and to use these weakly pathogenic fungi for cross-protection against take-all. Cross-protection has been tried all over the world, either with bacteria (*Pseudomonas fluorescens* especially) or with fungi (hypovirulent strains of *G. graminis* or *Phialophora radicicola*). The antagonists could be pelleted onto the wheat seed. This method of biological control has raised new problems because it works more or less well on different cultivars. Breeding for this criterion is possible (Lemaire *et al.* 1982). The installation of a non-pathogen on the roots of wheat also tends to modify the growth and development of the plants, as in mycorrhiza (Lemaire *et al.* 1979).

Var. *avenae* also attacks turf grasses, causing a disease formerly known as Ophiobolus patch. *Agrostis* species are most affected, the patches (e.g. on putting greens) spreading most actively in wet years on poorly drained turf at high pH. In contrast to snow mould (*Monographella nivalis*), the patches continue to spread in summer and symptoms of bronzing and browning are most striking under hot dry weather conditions. Control is by proper drainage and irrigation, and the use of acidifying fertilizers (APS Compendium).

LEMAIRE

POLYSTIGMATALES

Polystigma rubrum (Pers.) DC. (anamorph ***Polystigmina rubra*** (Desm.) Sacc.) causes red leaf spot or leaf blotch of plum and other *Prunus* species. It occurs all over Europe (e.g. on *P. spinosa*) but is mainly important on plum in E. Europe (Bulgaria, Romania, USSR, Yugoslavia). Bright red cushion-like spots (stromata) 1–3 cm in diameter form on both sides of the leaf, with pycnidial ostioles occurring on the lower surface towards the end of the summer. Perithecia form in the stromata on fallen leaves over the winter and new infections are established by ascospores at the time of flowering. Damage is mainly due to premature leaf-fall and consequent weakening of the tree. The disease is readily controlled by fungicides. Resistant cultivars are reported in Bulgaria. A related species ***P. ochraceum*** (Wahlenb.) Sacc. forms similar yellow to brown spots on almond leaves, and occasionally causes extensive damage in Mediterranean countries. In Greece, winter sprays with oil, then maneb or zineb applications are recommended, at petal fall and 2 and 4 weeks later.

Phyllachora species cause tar spot of turf grasses; over 90 species have been described world-wide (APS Compendium). They are obligate parasites forming characteristic raised shiny black 'shields' (clypei) on both sides of grass leaves, surrounded by a chlorotic halo. The perithecia form under the clypeus, the ostioles opening through it. Infection is exclusively by ascospores. The disease is widespread in Europe, especially in wet, shaded, sites, but of no practical importance. Of two species considered by O'Rourke (1976) to be widespread in Ireland, ***P. graminis*** (Pers.) Fuckel is found on many grass genera, while ***P. dactylidis*** Delacr. is specific to *Dactylis glomerata*.

Glomerella cingulata (Stoneman) Spaulding & v. Schrenk

(Anamorph ***Colletotrichum gloeosporioides*** Penz.; syn. *Gloeosporium fructigenum* Berk. etc., see below)

Basic description

The dark perithecia are rarely seen in nature, *G. cingulata* mainly being encountered as its acervular anamorph (CMI 315). Probably several hundred described species of *Colletotrichum* and *Gloeosporium* can be considered to belong to this complex; some have been given names in *Glomerella*. Colonies produced *in vitro* are fast-growing and variously coloured according to strain. Some strains produce perithecia fairly regularly in culture, but greyish acervuli are mostly seen, bearing orange mucilaginous droplets of ovoid conidia. Cardinal temperatures are 0–5°C, 25–27°C, and 35–37°C.

Host plants and specialization

Plants in numerous families are attacked. The fungus also occurs saprophytically on many cultivated and wild plants which can act as reservoirs for infection. The numerous straight-spored *Colletotrichum* species described on specific hosts can virtually be considered to be formae speciales of *G. cingulata*, since they hardly differ morphologically from *C. gloeosporioides*. They are, however, clearly distinct biological entities, with a specific host range. In some cases numerous 'vertical' pathotypes have been described within these forms (cf. below). On coffee, special forms have even been described for different organs of the same host, but no equivalent situation is known in Europe. However, as with *C. fragariae* (see below), it may be possible for strains readily forming perithecia (and therefore known as *G. cingulata*) to occur and cause very similar disease to a host-specific conidial form. Such situations need to be clarified. The forms found on fruit crops (apple, cherry, almond, pear) seem less specialized, and will cross infect readily from one host to the other. *G. cingulata* also occurs on various ornamentals.

Geographical distribution

G. cingulata is ubiquitous, but it is generally more abundant in subtropical or tropical than in temperate zones.

Diseases

The main diseases caused by *G. cingulata* are considered below under two headings (fruit rots, diseases on woody hosts), except that the anthracnose diseases caused on various herbaceous hosts by host-specific *Colletotrichum* species are treated separately (*C. lindemuthianum* on phaseolus beans, *C. lagenarium* on cucumber etc.). Unspecialized forms of *G. cingulata* also cause minor leaf and stem spots on various herbaceous hosts. Fungi with names like *Gloeosporium cyclaminis* Sibilia, destroying the flowers of cyclamen in the bud, especially in Italy, may belong here (Viennot-Bourgin 1949; Garibaldi 1981).

Fruit rots

G. cingulata causes bitter rot of pip-fruit, similar in many respects to the rots caused by other 'Gloeosporiums' (cf. *Pezicula alba* and *P. malicorticis*). The lesions are centred on lenticels but are usually uniform in colour. The fungus apparently produces toxins which give the fruit a bitter taste. Acervuli are common and conspicuous. Due to its required temperature range for growth, *G. cingulata* will not rot fruit in low-temperature stores and so has little practical importance in Europe. It is of greater importance in the warmer wetter summers of N. America, and especially in subtropical conditions (in Brazil, it attacks fruit even in the orchard). The chemical treatments used against the *Pezicula* species (q.v.) are also effective against *G. cingulata*.

Diseases on woody hosts

G. cingulata attacks various woody hosts, including fruit and timber trees, in particular olive and willow (cf. below). It causes minor canker and dieback on rosaceous fruit trees and citrus. Fungi with names such as *Gloeosporium amygdalinum* Brizi and *G. limetticolor* Clausen possibly belong here. The former requires fungicidal control on almond in

Spain (Palazon-Español & Palazon-Español 1979), while the latter causes wither-tip of lime in the tropics, but in Europe has only been recorded from Yugoslavia (CMI map 119).

On olive, *G. cingulata* has been known as *Gloeosporium olivarum* Almeida. The disease occurs throughout the Mediterranean area and in Japan, Uruguay, Argentina, S. Africa and USA (California). The fungus can attack fruits, leaves and shoots. The main attack is on olives which initially show rounded depressions. With the development of the fungus the olives produce an orange exudate which includes masses of spores (Plate 20). Later the fruit becomes wrinkled and dry and falls prematurely. Attacked leaves show brown-yellow chlorotic spots with poorly defined margins. Acervuli may be observed on those spots (Zachos & Markis 1963). Leaf-fall may be heavy on diseased trees, and 2–3 year old shoots may be attacked and killed (Martelli 1960). Leaf and shoot infection has not been confirmed in Spain (Aspeitia 1975) but, in Portugal, Azevedo (1976) relates leaf-fall and death of shoots to intense attacks of the disease.

Primary infections result from the germination of spores from acervuli on olives or leaves (Zachos & Markis 1963) or 1-year shoots (Martelli 1960). Germination temperature ranges from 0 to 30°C with an optimum of 25°C (at the start of germination). According to the temperature, the incubation period varies between 6 and 27 days. For example, at 15–20°C and 25°C incubation lasts 10–7 days or 6 days respectively. Fungus development at 5°C is very slow and inhibition is observed at 0°C (Zachos & Markis 1959). A minimum 92% RH is needed for spore germination. Below that value, germination may occur whenever water is present. The most critical period for fungus attack occurs when fruit colour is changing, providing that climatic conditions are favourable (e.g. temperature and persistent rain, fog or mist). Considering such conditions, exposure of the olive tree field, together with soil drainage, may be important factors for disease occurrence. The attack of *Dacus oleae* may also increase incidence. Thus *D. oleae* control should be dealt with especially in areas wherever the disease is an important one.

The most striking effect of the disease is fruit damage or destruction; eventually leaf-fall or the death of shoots are other effects of the disease. Following fruit destruction olive oil production is reduced; poor quality, bad taste and high acidity will be characteristic of the oil produced from diseased olives which were not completely destroyed. Economic importance differs from one country to another and may also vary from one year to another. In Portugal, 1965 was important for the disease and losses higher than $3 million were registered (Coutinho 1968). More recently serious losses were observed too (Azevedo 1976) and, for example, in areas of high relative humidity more than 50% of fruit production may be affected (Valente 1979). In Greece and Italy important losses are also recorded. In the latter country, in Puglia, it was the most dangerous disease of olive trees (Martelli 1960); nevertheless, more recently there has been decreased attention from growers (Graniti & Laviola 1981).

At present disease control relies mainly on chemicals. Nevertheless other measures have to be taken in order to reduce inoculum and, in this way, reduce or improve chemical control. The first measures should be envisaged even before planting, i.e. exposure and soil conditions, proper aeration of the plants and drainage. Resistant cultivars or, at least, less susceptible cultivars. should be used. The use of late-ripening cultivars, in order to prevent most damage to fruit, when conditions for the development of the disease are not so favourable, should also be considered at a local or regional level. Proper annual pruning, whenever conditions favour high intensity of the disease, may play an important role in control by reducing inoculum and improving tree aeration. Copper fungicides are still the best products to use. The oldest product used is Bordeaux mixture. Oil formulations of copper can improve persistence of the product which may be of economic interest especially in areas where rain is more frequent. Organic fungicides such as zineb, thiram and captan are less effective than copper fungicides, lacking adequate persistence (Martelli & Piglionica 1961). Timing of fungicide treatment is important for control of the disease. In general, in Portugal and in Mediterranean countries, the first treatment is made between mid-September and the end of October, before autumn rains. Whenever conditions are favourable to the development of the disease, a second treatment is made 3 to 4 weeks after the first one. Eventually, in some special conditions a third treatment may be envisaged following a similar interval.

G. cingulata is also known as *G. miyabeana* (Fukushi) v. Arx & Miller (syn. *Physalospora miyabeana* Fukushi) on willow. It attacks *Salix* species, particularly cultivars and hybrids of *S. alba*, and the hybrid often called *S. americana*. It is known on willows throughout much of Europe, in N. America, and in Japan. The disease is known as black canker. The first symptoms show from May onwards in the form of red-brown, angular leaf spots. The spots spread over the whole leaf, which dies and becomes black and shrivelled. The dead leaves usually remain on the shoots, and the fungus then often spreads down into the stem, producing long oval flattened cankers usually cut off by a cork layer. The bark on the cankers becomes dry and splits, and fruit bodies of the fungus are produced on the dead tissues, chiefly on the cankers.

The fungus overwinters as mycelium in the cankers (on which it produces acervuli in the spring), and as perithecia, which form in the autumn, embedded within the host tissues. Black canker spreads most rapidly at high RH and at a temperature of about 25°C. Black canker is most important on willows used for basket making, as it makes the rods brittle and useless. It sometimes destroys cuttings in the nursery. Control measures can be taken only in nurseries and in beds of basket willows. Here, the plants may be sprayed with copper fungicides, or with zineb or a captan/brestan mixture, though frequent applications may be necessary, at least when weather conditions favour the disease (Diercks 1959).

BOMPEIX; JULIO; PHILLIPS

Colletotrichum lindemuthianum (Sacc. & Magn.) Scribner

Basic description

The genus *Colletotrichum* is characterized by the formation of acervuli, with hyaline conidia. Dark setae may be produced in the acervuli and this was formerly considered to distinguish *Colletotrichum* from *Gloeosporium*. This distinction proved untenable, however, the latter genus being a complex of many different genera, including *Colletotrichum*. Germinating spores produce appressoria before penetrating the host. *C. lindemuthianum* has aseptate spores, 11–20×2.5–5.5 μm, with obtuse ends. It is distinguished from *C. gloeosporioides* by host range (von Arx 1957). *G. cingulata* has been found as the teleomorph in Brazil (Chaves 1980), so *C. lindemuthianum* should be considered just as a host-specific form of this species (cf. *G. cingulata*). The fungus is heterothallic.

Host plants

C. lindemuthianum attacks French and runner beans (*Phaseolus vulgaris* and *P. coccineus* respectively), several other *Phaseolus* species such as *P. aureus, P. lunatus, P. mungo* and *P. radiatus*, cowpea (*Vigna unguiculata*), broad bean (*Vicia faba*) and other minor leguminous crops (CMI 316). Nevertheless *C. lindemuthianum* can only be considered important on *P. vulgaris*.

Host specialization

Early in this century races of *C. lindemuthianum* had already been distinguished on *Phaseolus* beans. Pathogenicity of the fungus is very unstable. Races, indicated by the letters alpha to epsilon, iota, kappa and lambda of the Greek alphabet, are in fact race groups, which might be further subdivided through an increase of the number of differential cultivars. Breeding for complete resistance to the different races has been complicated by the number of genes involved (Hubbeling 1957), and has always been followed sooner or later by the appearance of new races with matching virulence. Since its description by Mastenbroek (1960), the Are gene has become the most important source of resistance in Europe and in other regions of the world. It originates from a Venezuelan line, Cornell 49–242, is easy to incorporate in commercial cultivars, is dominant, and gives complete protection to the races alpha to epsilon and lambda. Notwithstanding the appearance of the virulent races kappa and iota, which match the Are gene, no breakdown of resistance of the Are gene in European agriculture has been recorded. In S. and C. America, however, races with virulence to the Are gene do occur in the field. Many more genes for resistance are available in germplasm for future use (Zaumeyer & Meiners 1965; Schwartz *et al.* 1982).

Geographical distribution

C. lindemuthianum is present in almost all bean-growing areas of the world (CMI map 177), except where beans are grown under furrow irrigation in hot, dry climates.

Disease

C. lindemuthianum causes anthracnose of leaves, stems and pods (Fig. 47). Symptoms on leaves consist of reddish spots and vein discoloration on the lower surface. The spots tend to be angular or radiate along the small veins. Later they turn dark brown or black. On stems, larger petioles and on pods sunken, long elliptical to round lesions are found. The lesion

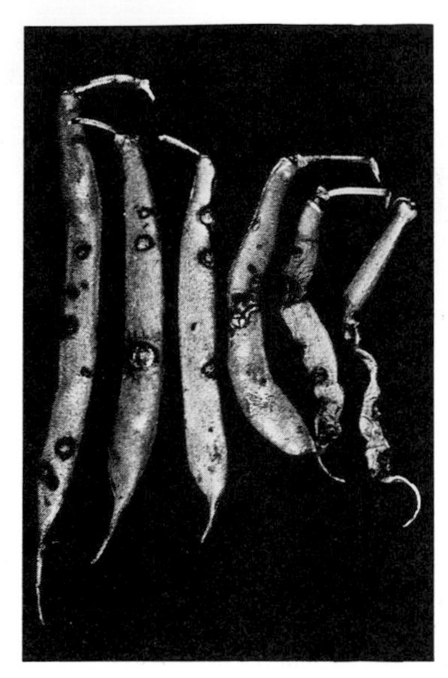

Fig. 47. Anthracnose (*Colletotrichum lindemuthianum*) on dwarf bean pods.

centre is brown, turning to black at the margin. The lesions on pods are normally surrounded by a reddish-brown border. In humid conditions a gelatinous pink mass of conidia may cover the centre of the lesion. In severe cases the disease causes leaf necrosis, defoliation and malformation of the pods. Lesions on pods may become pitted and extend to the seeds, which then show dark spots (Chaves 1980).

Epidemiology

Infected seeds from diseased pods can pass the disease from one year to the other. In humid weather sporulation occurs on infected seedlings, and the disease spreads all over the crop. Moderate temperatures are favourable for anthracnose. Apart from seed, infected stubble can also perpetuate the fungus, but only if kept dry. Under natural conditions with fluctuating humidity and temperature *C. lindemuthianum* survives poorly on dead plant material. Therefore, for all practical purposes only seed transfer is important (Tu 1983).

Economic importance

Before the introduction of resistant cultivars anthracnose was one of the most important bean diseases, often ranking first among fungal diseases. This is still the case in countries where resistant cultivars are not yet extensively grown. Theoretically complete crop failure might occur under temperate humid conditions. With snap and 'filet' beans minor attacks already represent big losses. With this crop a single lesion on a fraction of the pods is sufficient for economic 'total loss'. This is clearly not the case for dry beans. Nowadays nearly all snap beans grown in Europe are resistant to anthracnose, reducing the disease to minor importance.

Control

Three different options for disease control are available: i) chemical control during the vegetation period of the crop; ii) use of clean seed (including seed dressing with fungicides to protect the plant); iii) use of resistant cultivars. Chemical protection of the crop is possible with standard fungicides. However, when conditions favour anthracnose, fungicides need to be applied regularly, which makes the treatment costly and gives a risk of residues. In Europe field spraying is no longer practised on an appreciable scale, since more efficient means of controlling the disease are available. Clean seed is of utmost importance. In the USA seeds can be grown in the west, where anthracnose does not occur, but in Europe the climate generally favours the disease. Better quality seed is therefore often produced in other continents, e.g. E. Africa, but even then freedom from disease is not guaranteed. Seed transfer of anthracnose (and some other diseases) can be prevented through seed-dressing with systemic fungicides. Since the description and subsequent use of the dominant Are gene, resistance to anthracnose has proved to be very efficient in controlling the disease.

Special research interest

The Leguminosae have played a major role in research on phytoalexins. Several phytoalexins, e.g. phaseollin, are produced in bean plant tissue as a response to infection. In resistant cultivars the required level of phytoalexins to contain the pathogen is rapidly attained, and progress of *C. lindemuthianum* is limited to a few cells. Even in susceptible cultivars, after a successful start of the fungus, which may produce lesions of collapsed host tissue of some mm diameter, the stimulus emanating from the lesion suffices to elicit neighbouring healthy cells to produce phytoalexins in such quantities that the extension of the lesion is stopped (Theodorou *et al.* 1982). The fungus in the tissue of the lesion is

checked, not killed. The *P. vulgaris/C. lindemuthianum* system is an excellent tool for research on host/race specificity and on the triggering of production of non-specific phytoalexins.
GERLAGH

Colletotrichum lagenarium (Pass.) Ell. & Halsted

C. lagenarium is the causal agent of anthracnose disease of cucurbits. The fungus is widespread all over Europe, Asia, Africa, Australia and N. America, inflicting considerable losses on cucumber, melon, courgette and watermelon. The pathogen is seed-borne, although diseased plant debris in the soil may harbour the pathogen for at least a year. Wild cucurbits are also susceptible. Pale yellow to yellow watersoaked leaf spots, later becoming brown or black (watermelon) with irregular margins, are the main symptom of the disease. Petioles and stems are affected with the development of elongated, sunken streaks or cracks resulting in leaf desiccation. Young fruits are also attacked, failing to grow normally and frequently die. Infection of older fruits leads to the appearance of dark spots and severe breakdown particularly in watermelons. The production of orange pink masses from the acervuli of the fungus is obvious on the affected tissues under damp conditions. Consequently rainy or wet weather facilitates the dispersal of the pathogen and spread of the disease. Several pathogenic races of the fungus have been reported, while resistant cucumber or watermelon cultivars are currently available. Furthermore, control measures involving crop rotation with non-cucurbitaceous hosts and sprays with dithiocarbamates, chlorothalonil, benomyl, thiophanate methyl or carbendazim give sufficient control against anthracnose. Resistant strains may develop, however, after successive use of the systemic fungicides.
TJAMOS

Colletotrichum trifolii Bain & Essary is a straight-spored species which is the main cause of lucerne anthracnose in France (Raynal 1977), Italy, C. Europe and USSR. Stem lesions may girdle and kill stems, and spread to the crown to kill the whole plant (APS Compendium). Straw-coloured or white dead shoots are a conspicuous symptom. The fungus survives on the living crown, and may be spread by harvesting equipment. It is favoured by warm moist conditions, and so tends to be unimportant in N. Europe (O'Rourke 1976). It also attacks clover, notably *Trifolium pratense*. Resistant cultivars are available. Another species, ***C. destructivum*** O'Gara, with longer, slightly curved conidia, is less pathogenic (Raynal 1977) but can cause some defoliation; it is mainly isolated from leaves of otherwise diseased plants (*Verticillium albo-atrum, Phoma medicaginis*). A third species, variously regarded as *C. graminicola* (q.v.) or *C. truncatum* (q.v.) also occurs on these legumes, but is hardly pathogenic.

Colletotrichum lini (Westerd.) Tochinai causes anthracnose of flax world-wide (CMI map 159), is mainly seed-borne but can also spread from plant to plant. It is favoured by high RH and produces small spots on the cotyledons, yellow turning to brown. Spots in the collar region vary from orange to brownish-black, depending on the stage of the crop. The fungus can also spread to the capsules, which it deforms, and hence into the seeds.
JOUAN

Colletotrichum acutatum Simmonds

C. acutatum is distinguished from other *Colletotrichum* species mainly by having a proportion of conidia characteristically acute at both ends and relatively narrow (2.5 μm wide). However, it is very variable and cannot necessarily be easily distinguished from the unspecialized *C. gloeosporioides* (anamorph of *Glomerella cingulata* q.v.). No perfect state of *C. acutatum* has been reported (CMI 630). *C. acutatum*, recorded on over 20 genera, shows some host specialization and isolates from pine have been designated ***C. acutatum* f.sp. *pinea*** Dingley & Gilmour. The situation of *C. acutatum* in Europe remains confused. Recent attacks of leaf curl on cultivated anemone (*Anemone coronaria*) in France, Netherlands, Italy and UK have been ascribed to *C. acutatum*, although the fungus involved is considered by some to be identical with *C. gloeosporioides* (Linfield & Price 1983). The anemone disease was previously reported in Australia (Woodcock & Washington 1979). Symptoms were stunting, leaf distortion and the collapse of flower stalks. In Australia *C. acutatum* was originally described on ripening strawberry and tropical fruits (Simmonds 1965); the strawberry disease became known as black spot. In Florida (USA), a similar disease of strawberry is clearly caused by a different species, *C. fragariae* (q.v.) but *C. acutatum* has been intercepted in the UK on strawberry plants imported from California (USA). *C. acutatum* f. sp. *pinea* was found to cause

terminal crook disease on *Pinus radiata* and other *Pinus* species on nurseries in New Zealand (Dingley & Gilmour 1972) and also in Australia, Chile and Kenya. The fungus can survive at least two years as appressoria or chlamydospores on pine debris in soil. Control of anemone leaf curl involves disinfecting corms and spraying plants with either dithiocarbamate fungicides or a mixture of captan or captafol plus benomyl or carbendazim. These chemicals also control the pathogen on strawberry and a captan/carbendazim mixture or dichlofluanid plus oil have been used on pines.
COOK & DICKENS

Colletotrichum fragariae A.N. Brooks causes crown rot and fruit anthracnose in S.E. USA, while *C. dematium* and various *C. gloeosporioides* strains cause only fruit anthracnose. *C. fragariae* is probably to be regarded as part of the *G. cingulata* complex, although perithecium-producing strains of distinct morphology and appearance have also been isolated from crown-rotted strawberry (APS Compendium). Strawberry anthracnose has only recently been seen in Europe (France) and may present a threat, although there is still clarification needed on the fungi involved (cf. *C. acutatum*). ***C. tabacum*** Böning, causing tobacco anthracnose, though recorded in Germany (CMI map 307), is principally important in S. Africa.

Glomerella gossypii Edgerton (anamorph ***Colletotrichum gossypii*** Southworth) is close to the *G. cingulata* complex. It is a seed-borne pathogen of cotton, causing seedling anthracnose and boll rot. Badly diseased bolls become mummified and never open, and lint from diseased bolls is of inferior quality. Seed treatment virtually eliminates the seedling phase, but boll rot can be a problem in the higher rainfall areas of cotton production. Established in Europe only in Bulgaria and Romania, it is a threat to other cotton-growing countries (EPPO 71). ***G. glycines*** (Hori) Lehmann & Wolf, with short straight conidia, is distinct from the more important *Colletotrichum truncatum* (q.v.), and also causes soybean anthracnose. It mainly attacks mature plants (APS Compendium) and has been recorded in Spain (Signoret 1984).

Colletotrichum coccodes (Wallr.) Hughes

(Syn. *C. atramentarium* (Berk. & Broome) Taubenhaus)

C. coccodes is distinguished from the *C. gloeosporioides* type by its long straight acervular conidia with attenuated ends and by commonly forming sclerotia which can bear setae (CMI 131). *C. phomoides* (Sacc.) Chester is sometimes used as the name for the form on fruit but is probably synonymous (Messiaen & Lafon 1970). The host range includes species of 13 families, chiefly Solanaceae, Cucurbitaceae and Leguminosae. The pathogen is relatively unspecialized although isolates show variation in morphology and aggressiveness (Chesters & Hornby 1965). Its geographical distribution is worldwide (CMI map 190).

C. coccodes causes black dot of potato and tomato, anthracnose of fruit and brown root rot of tomato and other hosts (Fig. 48). Attacks can produce wilting. The fungus is soil-borne, surviving as sclerotia in decaying roots and other debris. It can be transmitted through infected potato tubers and occurs on tomato and aubergine seeds (Kendrick & Walker 1948; Porta-Puglia & Montorsi 1982). Water-transported conidia can spread the infection (Davet 1971). *C. coccodes* is a minor pathogen, but can cause severe losses under inappropriate growing conditions. Root rot was reported on hydroponically grown tomatoes (Schneider *et al.* 1978). Control measures include rotation, soil sterilization, clearing refuse, use of

Fig. 48. Black dots (sclerotia) and root rot of aubergine due to *Colletotrichum coccodes*.

healthy tubers and seeds. Trellising, mulching and spraying (with captafol, chlorothalonil, maneb) can prevent infection on tomato fruits. Sources of resistance, probably polygenic, are known for tomato (Dixon 1984).
PORTA-PUGLIA

Colletotrichum dematium (Pers.) Grove is a principally saprophytic species, with short curved conidia. It is itself recorded causing anthracnose of tomato (cf. *C. coccodes*) and a number of very similar species are plant pathogens. ***C. truncatum*** (Schwein.) Andrus & Moore (syn. *C. dematium* var. *truncatum* (Schwein.) v. Arx) causes soybean anthracnose, which resembles bean anthracnose in its biology (APS Compendium; Signoret 1981). It is recorded in France, Spain and Yugoslavia and is liable to be spread with seed (Signoret 1984). It may well occur more widely and not be specific to soybean (e.g. on lucerne). ***C. dematium* f.sp. *spinaciae*** (Ell. & Halsted) v. Arx (syn. *C. spinaciae* Ell. & Halsted) causes a seed-borne leaf spot of spinach, and occurs in France, Germany and Italy (Messiaen & Lafon 1970). ***C. capsici*** (Sydow) Butler & Bisby causes anthracnose of sweet pepper, but mainly in warm countries (CMI 317). ***C. circinans*** (Berk.) Vogl. (syn. *C. dematium* f.sp. *circinans* (Berk.) v. Arx) occasionally causes smudge or anthracnose of white-skinned onions; stromata formed beneath the cuticle of outer scales enlarge into smudgy patches under moist conditions in stores. Storage conditions have to be adjusted accordingly (Messiaen & Lafon 1970; Dixon 1984). Cultivars with coloured skins are in any case resistant, through the accumulation of catechol and protocatechuic acid in the outer scales (the classic instance of preformed antifungal compounds; Link & Walker 1933).

Glomerella graminicola Politis (anamorph ***Colletotrichum graminicola*** (Ces.) Wilson) occurs on Gramineae generally, and is widespread in Europe. It causes oval lesions on the sheaths and lower leaves of barley and wheat, resembling eyespot until the acervuli appear, with crescent-shaped conidia (CMI 131; APS Compendium), and surviving in crop debris in soil or on wild grasses. It is, however, of practically no economic importance with modern cultural practices. World-wide, it is more important on maize and sorghum, which may suffer leaf blight, top dieback and stalk rot (APS Compendium). In general, these crops seem to be little affected in Europe, although severe attacks are noted in Yugoslavia (Milatović *et al.* 1983). *G. graminicola* is also very common on turfgrasses, colonizing senescent tissues, and under some conditions killing patches of turf (APS Compendium). *Microdochium bolleyi* (q.v.) causes similar symptoms, and forms similar acervuli with crescent-shaped conidia but no setae (APS Compendium). *G. graminicola* can also be isolated from various legumes (e.g. lucerne), but is regarded in this case as essentially saprophytic by Raynal (1977), who also considers it to be practically synonymous with *C. truncatum* (q.v.). ***G. tucumanensis*** (Speg.) v. Arx & Müller (anamorph ***C. falcatum*** Went) is also very similar (CMI 133) but principally attacks sugarcane in the tropics (and *Cortadeira* in Europe, CMI map 186).

Magnaporthe grisea (Hebert) Barr

(Anamorphs ***Pyricularia grisea*** (Cooke) Sacc. and ***Pyricularia oryzae*** Briosi & Cavara)

Basic description

The fungus is characterized by obpyriform conidia, rounded below, attenuate above, and hyaline or pale-green in colour. These are two- to many-septate, and are produced singly at the successive growing points of the conidiophore. The conidia measure about 20–26×8–10 μm, depending on the isolate and the growth conditions. The anamorph name *P. oryzae* has been used for the fungus that occurs on rice, as distinct from *P. grisea*, described earlier (on *Digitaria sanguinalis*) and occurring on other cereals and grasses. However, many authors now consider these forms indistinguishable, though sometimes retaining the epithet *oryzae* as a matter of convenience. The identity of the two species is supported by the fact that the teleomorph obtained in culture by mating isolates of *P. oryzae* from rice and *Pyricularia* species from finger-millet and several grasses fits well with *Magnaporthe grisea,* earlier obtained by Hebert (1971) in a culture of two isolates of *P. grisea* from *D. sanguinalis* (Yaegashi & Nishihara, 1976). The name *Ceratosphaeria grisea* is used by the Japanese authors.

Host plants

P. oryzae sensu lato has been reported on a wide range of Gramineae, including cereals (rice, barley, oats, Italian, finger and pearl millet) and grasses (*Agropyron, Agrostis, Alopecurus, Dactylis, Digitaria, Echinochloa, Holcus, Phalaris, Poa, Setaria,* etc.).

Host specialization

Asuyama (1963) tentatively suggested the division of *P. oryzae* into 11 formae speciales on the basis of its hosts. More than 200 pathogenic races of *P. oryzae* have been demonstrated on an international set of differential cultivars of rice (Atkins *et al.* 1967; Ling & Ou 1969). A rigid race concept however, cannot be applied to this organism in which, owing to persistent heterokaryosis, new races may be produced by single conidia (Ou 1985).

Geographical distribution

P. oryzae is widespread in all rice-growing areas. In Europe it has been reported not only in the Mediterranean regions (Italy, S. France, Greece, Yugoslavia, Spain and Portugal), but also as far north as Bulgaria, Hungary and USSR (CMI map 51).

Disease

P. oryzae causes the widespread blast disease of rice (Fig. 49), first extensively studied in Italy (Farneti 1921), and more recently in Japan and other Asiatic countries (IRRI 1963). The pathogen produces spots or lesions on leaf-blades, nodes, parts of the panicle and grain, and, less frequently, on the leaf-sheath. On the leaves the spots begin as small, water-soaked, yellowish dots which, under moist conditions, enlarge quickly until 1–1.5 cm in length and 0.3–0.5 cm in width, becoming elliptical and somewhat pointed at both ends. The margins are usually brown to reddish-brown and the centre is often grey or whitish. Both the shape and colour of the spots vary, however, depending upon varietal reactions and environmental conditions. When the nodes are infected they become black and often break apart. Brown lesions may also occur on any part of the panicle. If the base of the panicle (neck) is infected (neck blast), the culm below usually breaks and the upper parts of the plant turn greyish-brown, looking as though they had been burned (hence the 'brusone' name given to the disease in Italy and other European countries). Plants attacked between the seedling and the maximum tillering stages often die (Baldacci 1971; Ou 1985).

Fig. 49. *Pyricularia oryzae* on rice leaves.

Epidemiology

Infection is caused by conidia which may be present in the air throughout the year in tropical regions, where several rice-growing cycles are possible in one year. In temperate regions conidia are formed by the pathogen which overwinters in rice stubble, straw-piles and grains, and may be found through the rice-growing season (usually from April to October). On susceptible plants they germinate and form appressoria whose germ tubes penetrate through the stomata or directly through the cuticle. Free water is required for germination and near humidity saturation for penetration. The minimum dew period for infection is 5–8 h at 25–26°C and 9–10 h at 21°C. The optimal temperatures for germination and penetration are a little below 28°C and 21–24°C respectively. The incubation period ranges from 4–5 days at 26–28°C to 7–8 days at 21°C. Sporulation is absent at RH less than 93%. Large spots produce 2000–6000 conidia each night for about two weeks. Tremendous numbers of conidia are therefore present in severely infected fields and especially in nurseries. All factors that weaken the plants, such as sudden variations in temperature or irrigation with water at excessively low temperatures, make them more susceptible to the attacks of the pathogen. A high level of nitrogen also favours the disease and this is a critical factor in growing high-yielding rice cultivars; farmers are urged to apply large doses of nitrogen fertilizers to obtain maximum yields.

Economic importance

In many Asiatic countries *P. oryzae* is regarded as the most serious menace to food production. In the 1930s and 1940s, it was responsible for famine in several districts of Japan. In Italy, where rice is grown on about 180 000 ha yielding more than 1 million tonnes of paddy, average reductions of 3–5% of yield are usually reported, but losses as high as 30% are not infrequent. Fortunately, in recent years, increased knowledge of the relationship between fertilization and disease severity and the availability of blast-resistant cultivars (see below) have worked together to reduce the losses caused by the pathogen.

Control

The following methods of control are suggested: burning stubble of diseased crops, avoiding fertilization with high levels of nitrogen and irrigation at too low a temperature, seed dressing with fungicides and employing resistant cultivars. For seed treatment, organomercury compounds and antibiotic have now been replaced by benomyl and other new systemic fungicides like edifenphos and tricyclazole (Froyd *et al.* 1976). The latter appears to be very interesting in that, by inducing host resistance, it provides good control until panicle emergence, when infection usually diminishes. Foliar sprays, although popular in Japan, are not applied in S. Europe for economical and social reasons. Among the rice cultivars extensively grown in Italy, Maratelli, Balilla, Vialone nano and Arborio are regarded as susceptible, and Roma, Razza 77, Cripto and Argo are regarded as sufficiently resistant even when grown on soils with a high level of nitrogen. Extensive breeding work is now being carried out at the Rice Research Center of the Italian Rice Agency (ENR), Mortara, in order to obtain rice cultivars suited to European conditions, largely employing Roncarolo and Gigante Vercelli as sources of resistance, and trying to overcome the difficulties of transferring blast resistance sources found in the subspecies *indica* into the combination *indica* x *japonica*. In breeding work, however, it is necessary to take into account the above-quoted high variability of *P. oryzae*.
CASTELLANI

Magnaporthe salvinii (Cattaneo) Krause & Webster (syn. *Leptosphaeria salvinii* Cattaneo; anamorph ***Nakataea sigmoidea*** Hara) causes stem rot of rice (CMI 344), and is most often seen as a sclerotial state (***Sclerotium oryzae*** Cattaneo), formed in infected tissues and surviving in the soil, or carried by irrigation water. It infects roots and lower leaf sheaths, both at the seedling stage and on older plants, causing lodging and killing a proportion of infected plants. Mainly serious in tropical countries (CMI map 448), it is of some importance in France (Camargue) (Bernaux 1977), Italy (where resistance testing is done; Moletti *et al.* 1983) and in Spain (Marin-Sanchez & Jiménez-Diaz 1981).

Gibellina cerealis Pass. causes white foot rot of wheat (CMI 534), spreading exclusively by ascospores. It occurs in Italy and E. Europe (Bulgaria, Hungary, Romania, USSR) but is of very minor importance.

SPHAERIALES

Rosellinia necatrix Prill.

(Anamorph ***Dematophora necatrix*** R. Hartig)

Basic description

R. necatrix is very similar to two other European *Rosellinia* species, *R. aquila* and *R. quercina* (q.v.). The mycelium of the three species is typified by the bulge seen in many cells near the septum. *R. necatrix* can be distinguished from the other two species by the different size of the asci and ascospores, its cultural characteristics, and its wide host range.

Host plants

In Europe, *R. necatrix* is found on many woody plants: grapevine, orchard trees (apple, pear, cherry, fig, olive, apricot, citrus), poplar, mulberry, soft fruits (currant, raspberry, strawberry), flowering shrubs (rose, jasmine, lavender, mimosa), and also on fleshy ornamentals (narcissus, peony, begonia, violet, cyclamen, carnation).

Host specialization

No host specialization exists. Isolates originating from apples, cherries, strawberries, peonies and jasmine are able to infect these five hosts with the same efficiency.

Geographical distribution

R. necatrix has been reported in Europe from FRG, Britain, France, Switzerland, Italy, Portugal, Spain, Greece, Hungary, Cyprus, Yugoslavia, Romania and Bulgaria. It seems to be more common in areas with a Mediterranean or oceanic climate than it is in continental areas. Outside Europe, it is often reported in Israel and Japan, and at higher altitudes in India and Iran.

Disease

R. necatrix causes a root and collar rot, known as white-root rot, indicating, like the French name 'pourridié laineux', the white, woolly mycelium around the affected roots. The fungus is a facultative parasite which is able to survive and grow in the soil in the absence of any host. It invades the entire underground parts of the fleshy hosts. In woody hosts (unlike *Armillaria mellea,* q.v.) it remains localized in the bark (parenchyma, phloem and cambium) and possibly the pith as well, but does not colonize the woody tissues. In the bark the fungus appears in the form of numerous little white finger-like fans. These organs, together with the external white or greyish mycelial sheets, allow one to distinguish *R. necatrix* from *Armillaria* spp., the fans of which are larger, more continuous, and localized in the cambium.

Trees attacked by *R. necatrix* show varying degrees of discoloration and dieback, together with a slowing down of growth. These symptoms are attributable partly to the destruction of the phloem and partly to the synthesis (by the fungus) of vivotoxins transmitted by the sap. In orchard trees, death may occur slowly, after several years of decline, or quickly, following a sudden shock such as that which occurs after a period of drought or the first onset of fruiting (Guillaumin *et al.* 1982).

Epidemiology

Three kinds of spores are produced: ascospores, conidia on coremia (the *Dematophora* stage), and chlamydospores. Although the conidia occur quite readily in culture, the three spore types are rarely seen in the field and are not reported to be involved in infection. The mycelium and aggregate organs can complete the whole infection cycle. The mycelium and mycelial strands ensure the dispersal of the parasite when growing in the soil. Upon meeting a root, the mycelium proliferates and creates a kind of muff around the root. Penetration takes place inside the muff. If the roots are young and do not have secondary tissues, the mycelium enters by infiltrating the outer cells. If the roots are older, entry is achieved by an 'infection cushion' identical to a sclerotium. The proliferation of the medulla of this 'cushion' causes breaking of the cork. Penetration seems to occur mainly by mechanical means (Mantell & Wheeler 1973; Tourvieille 1982). Inside plants the fungus colonizes the bark of the whole root by forming rhizomorphs which quickly branch and widen to form the typical finger-like fans.

The mycelium of *R. necatrix* grows well in soils with a high moisture content, and modern systems of irrigation have aggravated the disease in apple orchards. On the other hand, it requires a high level of oxygen, so that for most of the time the fungus is restricted to the topsoil. The fungus also requires a high level of organic matter and is not inhibited by a high pH (*in vitro*, the mycelium grows as well at pH 7.0 as at pH 5.0). In France, white-root rot is mainly found in two types of soil—in heavy calcareous soils (whatever the moisture content), and in moist organic soils in valleys (whatever the pH).

Economic importance

In France, *R. necatrix* was a major parasite on grapevine at the beginning of this century. For unknown reasons it is now rather mild on this host, but it has become a major problem on apple trees, in the same way as it has developed in Portugal. In Germany grapevine appears to remain the main host. In Italy the parasite is a threat mainly to poplars, while in Britain it has been reported mostly on bulbous flower plants.

Control

Prophylactic prevention includes: drainage, limiting irrigation and organic fertilization, and avoiding planting orchards on the sites of former vineyards. Eradication by soil disinfection before planting is easier with *R. necatrix* than with *Armillaria* spp. for two reasons: i) the mycelium is usually confined to the topsoil; and ii) the fungus survives in the bark of woody fragments but not in the xylem. Successful soil disinfection has been reported in Japan with chloropicrin, in India with vapam and formalin, and in France with methyl bromide. The most important consideration is to treat the whole of the surface where the fungus is present, otherwise, in the absence of competition from antagonists, the treated area will be rapidly recolonized from the margin. *R. necatrix* is known to be very sensitive to benzimidazoles and thiophanates. In contrast to the above-mentioned toxicants, these fungicides can be applied to living plants. In India, Gupta (1977) reported effective control of the fungus by applying carbendazim in the soil near the trunks of young apple trees. Although such a treatment might be expensive on a large scale, it could be suitable for nursery use. If a nursery becomes infested, it is possible to treat saplings by soaking their roots in a benzimidazole solution before planting them in an orchard. Physical treatment of the saplings with hot water has also been recommended for grapevine and mulberry.

Biological control has never been advocated, but *R. necatrix* is known to be very sensitive to various

kinds of antagonists, for example, bacteria, *Trichoderma*, basidiomycetes and nematodes. This low biological competitiveness may explain the absence of *R. necatrix* in forests although it can infect many kinds of forest trees if artificially inoculated. Tolerant rootstocks are known for jasmine (*Jasminum arborescens* and *J. dispersum*), citrus (*Poncirus trifoliata*), and stone fruits (*Prunus cerasifera*). The last is also tolerant to *A. mellea*. In apple, all available rootstocks are very susceptible. Certain wild species of *Malus* such as *M. toringoides* and *M. floribunda* are considered resistant. The Canadian stock Ottawa 11 (said to be tolerant) has not yet been tested in Europe.
GUILLAUMIN

Rosellinia aquila (Fr.) de Not, which is usually only a saprophyte on dead branches, has occasionally caused a root rot of spruce seedlings. ***R. quercina*** R. Hartig and ***R. thelena*** (Fr.) Rabenh. occasionally destroy patches of oak seedlings in forest nurseries in many parts of Europe.

Hypoxylon mammatum (Wahlenb.) J. Miller (syn. *H. pruinatum* (Klotzsch) Cooke) causes a serious canker of *Populus tremuloides* in the USA (CMI 356), tending to infect through insect wounds. Cankers on bark first appear sunken and yellowish, then lose the outer bark to expose the blackened cortex which stand out against the green trunks. Spread can be rapid and girdle the trees. A pathotoxin is involved in canker formation (Schipper 1978; Pinon 1984). Because of the severity of the disease in N. America, it was considered a quarantine hazard for Europe (EPPO 72). However, the fungus is quite widespread on *P. tremula* in mountain areas in France (Pinon 1979) and elsewhere in Europe, and most widely planted poplars in Europe are resistant. There is a danger that new hybrids with *P. tremuloides* ancestry could be susceptible and breeders should be aware of this.

Some other *Hypoxylon* species cause cankers on deciduous trees, and other species are bark saprophytes. ***H. rubiginosum*** (Pers.) Fr. is a widespread common species, e.g. on *Acer* and *Fraxinus* (CMI 357). ***H. serpens*** (Pers.) Kickx is common in Europe on beech (CMI 358). ***H. mediterraneum*** (de Not.) Ces. & de Not. (CMI 359) causes a bark canker of cork oak (*Quercus suber*), and other trees in Mediterranean countries known as charcoal disease. ***Ustulina deusta*** (Hoffm.) Lind (syn. *Hypoxylon deustum* (Hoffm.) Grev.) is a widespread decay fungus (CMI 360) which in Europe causes a serious root and butt rot of both living and dead trees and stumps, notably of beech, lime and elm.

Monographella nivalis (Schaffnit) E. Müller & v. Arx

(Syn. *Calonectria nivalis* Schaffnit, *Micronectriella nivalis* (Schaffnit) C. Booth)
(Anamorph ***Gerlachia nivalis*** (Ces.) W. Gams & E. Müller, syn. *Fusarium nivale* (Fr.) Ces.)

Basic description

Formerly considered to be related to the Hypocreales with *Fusarium* anamorphs (Chapter 11), *M. nivalis* is now classed in the Sphaeriales and has annellate, non-phialidic, conidiogenous cells (*Gerlachia*). Homothallic perithecia are produced *in vitro* on agar medium but more easily on sterilized straw; the ascospores are hyaline and 1–3 septate. Macroconidia are mostly (1-)3-septate. There are no chlamydospores. Colonies produced *in vitro* have white to pink or peach-coloured mycelium.

Host plants and specialization

The fungus attacks many Gramineae including cereals such as wheat, barley, rye and oats, but not maize. There is apparently no host specialization. Perithecium production from cereal isolates is frequent, but no perithecia develop in isolates taken from perennial grasses (J.D. Smith 1983).

Geographical distribution

Growth of the fungus is optimal at about 20°C (min: 0°; max: 28°C), but is highest in soils at 5–10°C or even lower (snow mould). Mainly for this reason, the fungus is limited to regions of cold or moderate climate. Areas with an unusually long cover of snow are most threatened, in C. Europe often at higher altitude; the fungus occurs in N. America as well as in Eurasia, and in the southern hemisphere also.

Disease

Losses occur during the emergence of young seedlings and surviving plantlets have brown lesions on the coleoptiles. The most important symptom is snow mould: young plants of wheat, rye and barley die under long-lasting snow cover with mild soil temperatures, or after snow melt. Leaves decay or become partially necrotic with distinct, dark brown edges around the necrotic spots. Diseased plants, at this stage, are usually covered by greyish or pink-coloured mycelium (Duben & Fehrmann 1979).

Foot-rot symptoms after attack of the haulm-base by *M. nivalis* are less distinct than for *Pseudocercosporella herpotrichoides* or *Ceratobasidium cereale. F. culmorum* infections are more frequent. During recent years, in C. Europe, attack of the (lower and) higher leaf-blades and sheaths became quite striking. At first sight such symptoms are similar to those caused by *Septoria* species. Longish, oval or egg-shaped, necrotic ('water-soaked') spots develop rapidly with typical chocolate-brown colour, often arising from an *Erysiphe graminis* pustule, mainly starting where the leaf-blade is bending (Forrer *et al.* 1982). On the necrotic tissue sporulation (conidia) is profuse. Very heavy attack of the ears and humid conditions may cause brownish necrotic spots on the glumes, encircled by distinct dark-brown tissue, but ear and kernel infection mostly stays latent. Ear scab, such as that caused by *Fusarium culmorum,* is rare, occurring in cool, humid and rainy seasons and areas, with kernel infections up to 100%.

Epidemiology

The fungus is a competitive soil saprophyte (Millar & Colhoun 1969). Infection of seedlings and young plants may originate from soil and from infected seed. Persistence in soil is mainly on straw debris from foregoing cereal crops; therefore, the crop rotation can influence the amount of soil inoculum to a certain extent. From the soil, young seedlings are entered through the roots or by direct infection of coleoptiles or leaf-sheaths. For development of snow mould, fairly constant temperatures around 0°C and CO_2 accumulation under crusted snow are especially promotive. After longer periods under such conditions, snow mould leads to serious losses. Mycelium growing on the soil surface is able to infect and colonize whole batches of plants very rapidly.

Perithecia develop on leaf-sheaths (and blades) in late spring and summer; conidia develop on decaying leaf residues as well as on the necrotic leaf spots described above. Conidia probably are of major importance for the successive infection of the upper plant parts. As with *F. culmorum,* ear infection is optimally timed in the period at or just after flowering, when decaying anthers can first be entered saprophytically before the living host tissue is infected by the fungus. Weather conditions during this time, preferably cool, rainy and wet, are highly important for the extent of kernel infection. Host reaction in the glumes is slow, not always leading to symptoms. Heavy attack may lead to 10 or even 20% loss in single-grain weight, but not in the number of kernels per ear. Infection of the haulm-base can take place via roots or from colonized leaf-sheaths. Development of symptoms is slow and late. In contrast to other foot-rot pathogens *M. nivalis* does not cause lodging.

Economic importance

Loss of seedlings on a minor scale may easily be compensated by tillering. Snow mould is by far the most important disease symptom. After long snow-cover, fields quite often have to be reploughed and re-sown in spring in some years or regions. Haulm-base attack is probably of no importance at all, as was also claimed for leaf spots. However, quite recently leaf symptoms even on the flag leaves, and sometimes also stem lesions below the ears, occur to such an extent that crop losses should be assumed. Very serious losses after ear attack are rare, even with a high percentage of infected kernels, but this becomes important in the next vegetation period.

Control

As far as possible, seed has to be free from *M. nivalis*, checked by laboratory seed testing or careful observation of seed production fields. Chemical control of *M. nivalis* has to aim at snow mould prevention in the first place (Haeni 1981). Conventional seed treatment with mercurial substances covered *M. nivalis* attack as well as *Tilletia caries, Pyrenophora teres* and *Ustilago avenae* with the same chemical. Replacement of mercury led to the introduction of different (and quite expensive) fungicide mixtures, in addition covering *Ustilago nuda,* and even (with triadimenol) *Erysiphe graminis*. Control of *M. nivalis* in seed with carbendazim-type products has recently been jeopardized by the selection of specifically resistant strains of the fungus. Iprodione-containing chemicals may be useful instead, but in Canadian golf greens the substance failed to control 'fusarium patch'—obviously due to iprodione resistance of the fungus. On the other hand, fungicides containing the protectant guazatine proved to be very effective and promising also with respect to fungicide resistance (multi-site inhibition of metabolism). Mixtures of carbendazim-type compounds such as thiabendazole with methfuroxam can also help to escape the carbendazim-resistance problem in a way. There is a promising tendency to replace dry powder treatment by coating the seed with liquid formulations. Seed

treatment mainly protects the plants from seedling diseases, but less from snow mould under conditions with high inoculum potential in the soil, and dangerous climatic conditions in winter. In certain areas therefore, winter crops may be replaced by spring-sown cereals. Sowing time in the autumn should not be too late. As yet, there is no way of protecting cereals chemically from infection of the haulm-base, leaves and ears. Ear treatment with carbendazim-type substances enhances the selection pressure for the establishment of carbendazim-resistant strains and populations of the fungus in the seed. There are no resistant cultivars available, although differences in susceptibility are observed. There is an urgent need for more breeding for resistance.
FEHRMANN

Monographella albescens (Thüm.) Parkinson *et al.* (anamorph ***Gerlachia oryzae*** (Hashioka & Yokogi) W. Gams, syn. *Rhynchosporium oryzae* Hashioka & Yokogi) causes leaf-scald of rice in the tropics (CMI 729). It has recently been reported from S. France (CMI map 492). ***M. cucumerina*** (Lindfors) v. Arx. (syn. *Plectosphaerella cucumerina* (Lindfors) W. Gams, *Micronectriella cucumeris* (Kleb.) C. Booth) is very common in soil in temperate regions, as a saprophyte on decaying plant material. Its anamorph, formerly *Fusarium tabacinum* (v. Beyma) W. Gams is now named ***Microdochium tabacinum*** (v. Beyma) v. Arx by Von Arx (1984). *M. cucumerina* is also a minor root pathogen of, for example, basil, cucumber, lucerne, tobacco, tomato and viola (Matta 1978; Gerlach & Nirenberg 1982). ***Microdochium bolleyi*** (Sprague) de Hoog & Hermanides-Nijhof (syn. *Aureobasidium bolleyi* (Sprague) v. Arx) is one of a complex of fungi infecting the roots of grasses and cereals. It very much resembles *Colletotrichum graminicola* (q.v.). Fitt & Hornby (1978) found *M. bolleyi* could penetrate to the root stele and cause disruption of transport processes and growth in wheat, while Reinecke (1978) found it to cause neither symptoms nor yield loss in winter wheat or rye in Germany. It is also found on maize (Krüger & Rogdaki-Papadaki 1980) and barley (Murray & Gadd 1981) in Europe. Reinecke & Fokkema (1981) found it to be an antagonist of *Pseudocercosporella herpotrichoides* (q.v.) and *M. bolleyi* can be seen as one element in the complex of interactions between well-known root pathogens, minor pathogens which may also antagonize them, and fungicides which act selectively on these different fungi, potentially making a disease worse by destroying antagonists or giving unexpected yield benefits by controlling minor pathogens (e.g. Bollen *et al.* 1983).

Seiridium cardinale **(Wagener) Sutton & Gibson**

(Syn. *Coryneum cardinale* Wagener)

Basic description

The subepidermal or subperidermal acervuli (Plate 18) appear as minute, black pustular bodies scattered on the surface of stem, branches, and fruits of affected trees (CMI 326; Sutton 1975). Hyaline, cylindrical, sterile hyphae intermingled with branched, septate conidiophores arise from the inner pseudoparenchyma. Conidia are formed at the apical end of hyaline, holoblastic, annellidic conidiogenous cells and subsequent proliferations. Most conidia are fusiform, smooth and straight, 4-euseptate, and the four thick-walled median cells are dark-brown. The two thin-walled end cells are hyaline. The apical cell is campanulate or conic and ends in a very short appendage or papilla. The basal cell is truncate with a marginal frill and usually bears a very short endogenous cellular appendage. A transient production of filiform, slightly curved, hyaline spermatia may occur within the acervuli (Motta 1979). A *Leptosphaeria* teleomorph has been observed in California on dead cypress branches and even in mixed cultures of heterothallic isolates (Hansen 1956), but it has never been described (and does not seem consistent with the current view of the relationships of either *Leptosphaeria* or *Seiridium*).

Host plants

Species of *Cupressus* and several species of *Chamaecyparis, Cryptomeria, Cupressocyparis, Juniperus, Thuja,* and other genera and hybrids are all host plants. Various degrees of susceptibility are shown by different species of *Cupressus*, ranging from the highly susceptible *C. macrocarpa* and other American species, through the moderately susceptible Mediterranean species, such as the varieties of *C. sempervirens*, to the relatively resistant N. Californian and Arizona cypresses, *C. bakeri* and *C. glabra*.

Geographical distribution

S. cardinale was described by Wagener (1939) in California (USA), where it had been observed since

1928 and was supposed to be an alien pathogen introduced from abroad. In Europe, it was first reported in France in 1944, then in Italy (1951), and later in other countries (UK, Greece, Portugal, Spain, USSR, Yugoslavia), where cypresses and other Cupressaceae are grown or form native woods. It has also spread to S. America, N. and S. Africa, Australia and New Zealand.

Disease

The first evidence of the disease is a browning or a reddening of the live bark around the point of entry of *S. cardinale.* The infected area sinks slightly, longitudinal cracks are formed and a resinous exudation is produced. Subsequently lentiform or elongated cankers develop around the infection sites and may girdle the small branches or the stem of young plants. The extension of cankers, however, is a long process on large branches or main stems of adult trees. Consistent flows of resin occur from these cankers. A cardinal-red discoloration of the layers of inner tissue (giving the specific epithet) is often shown by cuttings made through the resin-infiltrated bark. The foliage of the infected twigs and branches first shows a diffuse yellowing, and eventually turns brown or reddish-brown as the dieback progresses (Fig. 50). The extension of infection leads to the death of whole trees. The most conspicuous symptoms of the disease, i.e. the fading and death of twigs, branches and tree tops, which are noticeable at a distance, may help in disease surveys.

Fig. 50. Symptoms of *Seiridium cardinale* attack on branches and tops of trees of *Cupressus sempervirens* var. *stricta*.

Epidemiology

Conidia extruded from the acervuli are usually dispersed over short distances by rain water mostly in a downward direction, a limited lateral spread being assured by winds carrying conidia-laden droplets. Long-distance transport of inoculum, even to isolated areas, is assured by vectors like insects (the cypress bark beetle *Phloeosinus aubei* is a very efficient vector in the Mediterranean area) and birds, which can carry the inoculum up to the tree top. The world-wide diffusion of the disease has probably been favoured by intercontinental trade in infected nursery stock. Seeds may also carry the disease, even if they are collected from apparently healthy trees in infected areas. Infection usually occurs through wounds produced by various agents (wind, frost, insects), although penetration through natural openings seems to be possible. A relative humidity close to saturation (at 80% RH about half the conidia cannot germinate) is required for infection, whereas the temperature requirements for production and germination of conidia are wider: from about 5° to 32°C, with the optimum at 24–25°C.

Economic importance

Cypress canker is a destructive disease, fatal to the affected trees. At present, it is the major disease of cypresses in Europe. In the last decades, heavy losses have been caused to the nursery industry, particularly to ornamental Cupressaceae and to cypress plantations used both for forestation and as a windbreak. In fact, cypress is an almost irreplaceable replanting tree in the Mediterranean hilly areas with poor, calcareous or arid clay soils. Moreover, in areas where cypresses are a major element of the country, apart from gardens, parks and flanking roads, the death of a great number of trees has caused the devastation of the landscape. In central Tuscany (Italy), a survey made in 1979 showed that the disease had already killed about one million trees, i.e. one-quarter of the existing cypresses, and another quarter showed signs of infection. Historical places like Florence in Italy and Olympia in Greece have suffered changes in their scenery, decorated by cypresses for centuries, painted by famous artists and sung by poets. The consequences, which cannot be estimated in economic terms, are however appreciable for their implications for tourism.

Control

Localization and removal of infected individual trees and destruction of disease foci at their first appearance in a previously uncontaminated area is a fundamental measure for disease eradication. In areas where the disease is already established, early pruning of any limbs or tops showing symptoms and the felling of heavily infected or dead trees is recommended in order to reduce the sources of inoculum and to avoid spread of vectors to healthy trees. All the infected material, i.e. the cut plants if young, the pruned branches, and the bark removed from the severed stem of big trees, should be collected and burned. The surgical removal of incipient cankers from the branches or stems of trees, followed by protection of the exposed wood with healing preparations containing fungicides, may save individual trees. These control measures, which are usually applied to ornamental trees in gardens, parks and avenues, may be too expensive or difficult to apply when belts of windbreaks, large plantations or cypress woods are concerned. Due to the long infection period and to the height of the trees, chemical control by spraying trees with fungicides may also be uneconomic or impractical except for nurseries or valuable ornamental trees. Repeated (4–6 or more times per year) applications of protectant (e.g. captafol) or systemic (e.g. benomyl, carbendazim, thiophanate methyl) fungicides during the mild seasons are effective as preventive means of control. However, these sprays have little or no value if applied as a curative to already diseased trees. Control of beetle vectors is also advocated. Selection and cloning of naturally resistant cypresses, breeding for resistance to produce resistant hybrids or lines is thought to be the most promising, even if not immediate, way to achieve an efficient, lasting control of the disease (Grasso & Raddi 1979; Raddi & Panconesi 1981).
GRANITI

Lepteutypa cupressi (Nattrass *et al.*) Swart (anamorph ***Seiridium cupressi*** (Guba) Boesewinkel) has caused considerable losses on cypress plantations in Africa, Asia and Australasia (CMI 325). The fungus was then known as *Monochaetia unicornis* (Cooke & Ell.) Sacc., but Boesewinkel (1983) has shown the latter (now ***Seiridium unicorne*** (Cooke & Ellis) Sutton) to be distinct and have a wider host range (on Angiosperms as well as Cupressaceae). *L. cupressi* is the damaging pathogen of cypress, and has been recorded in Greece (Kos Island). This is of serious concern in Mediterranean countries, in view of the losses suffered from *S. cardinale* and the complications entailed for breeding programmes (Graniti 1986).

Pestalotia and *Pestalotiopsis* species are weak parasites or saprophytes, mainly in the tropics. ***Pestalotiopsis funerea*** (Desm.) Steyaert (CMI 514) forms five-celled conidia in acervuli, with characteristic appendages (3–6 at the apex, one at the base). It causes a needle blight and canker of conifers, especially in nurseries and is noted particularly on Cupressaceae in France (Morelet 1982). ***Pestalotia hartigii*** Tubeuf (syn. *Truncatella hartigii* (Speg.) Steyaert) similarly attacks conifer seedlings causing 'strangling disease' of *Picea* species. ***Pestalotiopsis guepinii*** (Desm.) Steyaert (CMI 320) causes a leaf spot and petal rot of Ericaceae, especially florists' *Rhododendron* species. Records of this species on *Camellia* can be attributed to the similar fungus ***Monochaetia karstenii*** (Sacc. & Sydow) B. Sutton, which has conspicuously branched appendages on the conidia, and frequently attacks young *Camellia* plants (P.M. Smith 1983). Finally, ***Pestalotia malorum*** Elenkin & Ohl. causes a sometimes serious bark necrosis of apple in the GDR (Senula & Ficke 1983) and the USSR (Khomyakov 1981).

Mastigosporium rubricosum (Dearn. & Barth.) Sprague

M. rubricosum causes leaf fleck of various pasture grasses (*Dactylis glomerata* and *Agrostis* species). Lesions appear first as elliptical water-soaked spots, which turn purplish brown in colour and are surrounded by an orange or reddish margin. Masses of three-septate conidia form whitish glistening spots at the centre of the lesions (O'Rourke 1976). The species mainly affected in N. Europe is *D. glomerata*, on which Gunnerbeck (1971) has distinguished ***M. muticum*** (Sacc.) Gunnerbeck, distinct from *M. rubricosum* on *Agrostis*. Spread occurs mainly in cool damp conditions in autumn or spring. In Britain and Ireland, losses in yield and quality can be quite severe, especially after heavy nitrogen applications. Early cutting or grazing may help to reduce infection, and some cultivars of *D. glomerata* are moderately resistant. ***M. album*** Riess causes a similar disease on *Alopecurus* species and ***M. kitzebergense*** Schlösser on *Phleum*.
CARR

Pleiochaeta setosa (E. Kirchner) Hughes (syn. *Ceratophorum setosum* E. Kirchner) forms 4 to 8-septate golden-brown conidia with terminal hyaline appendages, and occurs in most of Europe and throughout the world (CMI map 243) causing brown leaf spot of lupins and ornamental *Cytisus* (CMI 495). Lesions are circular and zonate. It can also attack stems, pods and roots and destroy whole plants. It forms microsclerotia in leaf tissues, which may be involved in overwintering, but is mainly transmitted by seed, which can be treated with fungicides or heat. It is mostly noted as important in E. Europe, and is one of the most destructive diseases of fodder lupins.

Spermospora lolii McGarvie & O'Rourke, ***S. ciliata*** (Sprague) Deighton and other *Spermospora* species cause leaf spot or blast of grasses (*Agrostis, Festuca, Lolium* etc.) in Germany, the UK, Ireland and the Nordic countries (O'Rourke 1976; APS Compendium). The fusiform conidia formed on a brown stroma bear a long sperm-like appendage. The disease is frequent but unimportant.

Discostroma corticola (Fuckel) Brackman (syn. *Clathridium corticola* (Fuckel) Shoem. & E. Müller, *Griphosphaeria corticola* (Fuckel) Höhnel) has the anamorph ***Seimatosporium lichenicola*** (Corda) Shoem. & E. Müller (syn. *Coryneopsis microsticta* (Berk. & Broome) Grove). It causes a minor canker of rose (APS Compendium), with smooth margins, occasionally girdling the stem and causing galls above. It is found on various other hosts (e.g. raspberry). It also occurs on grapevine and fruit trees in E. Europe.

Physalospora rhodina Berk. & Curtis (anamorph ***Botryodiplodia theobromae*** Pat., syn. *Diplodia natalensis* Pole Evans, *Lasiodiplodia theobromae* (Pat.) Griff. & Maubl.) (CMI 519) is a very widespread cause of unspecialized rots in the tropics, especially of stored tubers and fruit (e.g. citrus), including imports to Europe. It occurs in Mediterranean countries but is only of minor importance there.

Khuskia oryzae Huds. (anamorph ***Nigrospora oryzae*** (Berk. & Broome) Petch) is a common saprophyte on plant debris in warmer countries. It forms large black spherical to ovoid conidia singly on short conidiosphores (CMI 311). Occurring on many hosts throughout the world, in Europe it has mainly been associated with stalk and cob rot of maize (APS Compendium), e.g. in the USSR and Yugoslavia. It tends to attack plants already debilitated for other reasons.

DIATRYPALES

Eutypa lata (Pers.) Tul. & C. Tul.

(Anamorph ***Libertella blepharis*** A. Sm., syn. *Cytosporina lata* Höhnel)

Basic description

The genus *Eutypa* is composed of wood-inhabiting species which produce ascogenous stromata composed of a mixture of fungal and host tissues. The perithecia are immersed in the stroma with a polar ostiole erumpent more or less at the surface of the stroma. The asci are unitunicate, elongated, and with an apical pore. The ascospores are allantoid and more or less brown. Some species produce conidia of the form genus *Libertella* (Messner & Sutton 1982). The form on apricot was long considered a separate species *E. armeniacae* Hansf. & Carter, but this no longer seems justified (Carter *et al.* 1985).

Host plants

E. lata is responsible for canker and/or dieback of branches in apricot and in grapevine. It causes a similar disease in many other fruit trees, e.g. almond, apple, lemon, loquat, pear, plum, quince, walnut, also in *Ribes* species and in many forest and ornamental dicotyledonous tree species (Carter *et al.* 1983).

Host specialization

Recent studies of single-ascospore isolates indicate that host-specific pathotypes occur. Single-ascospore isolates collected from stromata in S. Australia, when inoculated to apricot, differ significantly in their virulence as indicated by the rate of canker development in 6 months. Furthermore, observations in certain regions of France, Greece, and Hungary have shown that while Eutypa dieback is abundant in grapevine it may be virtually absent from trees in nearby apricot orchards. In N.E. Greece, very severe cankers commonly occur in almond cv. Texas (= Mission) while other cultivars are less severely affected (Rumbos 1983).

Geographical distribution

The pathogen occurs in most regions of the world where the grapevine and the apricot are cultivated. It has been identified in most countries of Europe and in many regions bordering the Mediterranean (Carter *et al.* 1983).

Disease

The disease, Eutypa (or Cytosporina) canker and dieback, has also been referred to, in apricot, as apoplexy or gummosis and in grapevine as dead-arm disease. In apricot, the disease is indicated by a sudden wilting of the leaves on individual branches in summer. Fruit on these branches may ripen earlier than on the rest of the tree and the affected branches usually die before the end of that season (Plate 19). Gum is sometimes exuded from the affected portion of a branch. After the branch has died, the dry brown leaves usually remain attached for several months. In grapevine, the disease is indicated by cankers which extend from old pruning wounds and by weak and stunted shoots which, in mid-spring, contrast strikingly with adjacent healthy ones (Fig. 51). Early in the spring, leaves on diseased shoots are often small, chlorotic, and sometimes cupped, tattered or otherwise misshapen. They may be necrotic (i.e. brown) at the margins or spotted with small brown lesions. As they age they develop a tattered and scorched appearance. On these shoots the number and size of the bunches is smaller than on healthy plants. The berries very often fail to develop (dropping off) or remain small and seedless (shot berries, 'millerandage'). The shoots of affected arms will weaken in successive seasons until the arm dies.

Epidemiology

E. lata is dispersed entirely by air-borne ascospores produced in subcortical stromata on dead wood of the host (CMI 436). Perithecial stromata do not appear until 2 or more years after the death of the infected host branch, and their formation is rare in regions where the mean annual rainfall is less than 350 mm. Once formed, a stroma may continue to liberate ascospores for 5 years or longer, according to a regular cycle (Moller & Carter 1965). Spores are discharged from perithecia only during and after rainfall: the eight ascospores are ejected simultaneously from each mature ascus and dispersed as octads. Ramos *et al.* (1975) have now provided substantial evidence that, in certain geographical situations, viable ascospores may be carried for distances of more than 100 km to infect trees in drier regions where inoculum is not produced locally. It is occasionally possible to find pycnidia (*Libertella*) of the pathogen on infected host tissue, but the asexual spores are not infective and play no part in the disease cycle (Carter 1957; English *et al.* 1962). Ascospore infection takes place *via* wounds which expose the vascular tissue, the most common point of entry being fresh pruning wounds. Arrival of the inoculum is usually a two-stage process, initially involving aerial deposition of ascospores on branches and foliage, followed by a secondary dispersal by rain splash and rain washing (Carter 1965). Wounds usually lose their susceptibility within about two weeks after pruning, although Ramos *et al.* (1975) showed that, under Californian conditions, duration of pruning wound susceptibility depends on the time of the year. Susceptibility loss has been attributed to colonization of the surface layers of the pruned sapwood by other micro-organisms (Carter & Moller 1970; Price 1975).

Fig. 51. Stunted shoot symptoms due to *Eutypa lata* on grapevine cv. Chasselas.

Economic importance

In European apricot orchards, the percentage of trees attacked by *E. lata* varies between 5 and 30% depending on the climate, the age and the extent of pruning. The proportion of infected apricot trees may reach 80 and sometimes 100% in certain Australian or Californian orchards (Carter & Moller 1977). Important dieback symptoms are also observed on other fruit trees: almond cv. Texas in Greece (Rumbos 1983), apple cv. McIntosh in Austria (Messner & Jähnl 1981) and black currant in France and Switzerland (Baggiolini & Duperrex 1963). The damage to grapevines increases markedly with the age of the stocks. Infection is infrequent until the age of 10 years. It is often 5–20% on 16–20 year old stocks and may reach 30–60% in older grapevine plantations. The most important damage occurs in vineyards originally cultivated as 'gobelets' and later converted to the Guyot pruning system or to another shape better adapted to mechanical harvesting. In such reworked vineyards, infection rates of 40–80% are frequent and a constant renewal of the stocks or replanting of the vineyards becomes necessary (Bolay & Moller 1977). *E. lata* is also a wound parasite of numerous horticultural and ornamental tree and shrub species. The frequency of its occurrence is related to the extent to which the host species are subjected to pruning.

Control

Control measures for Eutypa dieback aim to reduce the likelihood of spores reaching unprotected wound surfaces: i) by reducing apricot tree pruning to the absolute minimum to maintain tree shape and size; ii) where large cuts are needed, and for all cuts on young apricot trees, by immediately applying a systemic benzimidazole fungicide (e.g. 2% benomyl). This can be done readily with a small hand-operated spray bottle and the dose should be sufficient to flood the cut surface thoroughly, ensuring penetration of the chemical into all the water-conducting cells. Pruning wounds less than 10 mm in diameter usually do not warrant treatment in mature trees. Note that whole-tree spraying with lower strength mixtures is not only wasteful, but ineffective. For the annual pruning of grapevines, maximum protection will be achieved if all cuts in 2 year and older wood are treated as on apricot trees; the wounds made in pruning 1-year old wood may be left untreated without serious risk. It is essential to treat large trunk wounds immediately when grapevines are being reworked: these constitute the greatest threat to the life of a vineyard.

BOLAY & CARTER

REFERENCES

Acimović M. & Straser N. (1982) Phomopsis sp.—a new parasite of sunflower. *Zaštita Bilja* **33,** 117–158.

Asher M.J.C. & Shipton P.J. (1981) *Biology and Control of Take-all.* Academic Press, London.

Apeitia E.M.S. (1975) Daños y enfermedades del olivo. In *Olivicultura Moderna,* pp. 213–234. Editorial Agrícola Española S.A., Madrid.

Asuyama H. (1963) Morphology, taxonomy, host range and life cycle of *Pyricularia oryzae.* In *IRRI (1963),* pp. 9–22.

Atkins J.G. *et al.* (1967) An international set of rice varieties for differentiating races of *Pyricularia oryzae. Phytopathology* **57,** 297–301.

Azevedo A.R. (1976) *A Defesa Sanitaria da Oliveira em Portugal.* Instituto Nacional de Investigação Agrária, Oeiras.

Babajan A.A. (1980) Factors promoting decline development of apricot trees in Armenia. *Acta Horticulturae* **85,** 251–255.

Baggiolini M. & Duperrex H. (1963) Observations sur la biologie et la nuisibilité de la sésie du groseillier et du cassis: *Synanthedon tipuliformis* Clerk. *Recherche Agronomique Suisse* **2,** 12–32.

Baldacci E. (1971) Le malattie del riso in Italia. *Compte rendu des 3es Journées de Phytiatrie et de Phytopharmacie circum-méditerranéennes.*

Ballio A. (1978) Fusicoccin the vivotoxin of *Fusicoccum amygdali.* Chemical properties and biological activity. *Annales de Phytopathologie* **10,** 145–156.

Barr M.E. (1978) *The Diaporthales in North America. Mycologia Memoir* no. 7. J. Cramer.

Bazzigher G. (1953) Beitrag zur Kenntnis der *Endothia parasitica* dem Erreger des Kastaniensterbens. *Phytopathologische Zeitschrift* **21,** 105–132.

Bernaux P. (1977) Localisation et pouvoir pathogène de *Sclerotium oryzae* Catt. et *S. hydrophilum* Sacc. sur plantules de riz. *Annales de Phytopathologie* **9,** 205–209.

Berry F.H. (1977) Control of walnut anthracnose with fungicides in a black walnut plantation. *Plant Disease Reporter* **61,** 378–379.

Biraghi A. (1951) *Endothia parasitica* e gen. *Quercus. L'Italia Forestale e Montana* **6,** 15–16.

Black W.M. & Neely D. (1978) Effects of temperature, free moisture and relative humidity on the occurrence of walnut anthracnose. *Phytopathology* **68,** 1054–1056.

Bloomberg W.J. (1962) *Cytospora* canker of poplars: the moisture relations and anatomy of the host. *Canadian Journal of Botany* **40,** 1281–1292.

Boesewinkel H.J. (1983) New records of three fungi causing cypress canker in New Zealand, *Seiridium cupressi* and *S. cardinale* on *Cupressocyparis* and *S. unicorne* on *Cryptomeria* and *Cupressus. Transactions of the British Mycological Society* **80,** 544–547.

Bolay A. & Moller W.J. (1977) *Eutypa armeniacae,* agent d'un grave dépérissement de vignes en production. *Revue Suisse de Viticulture, d'Arboriculture et d'Horticulture* **9,** 241–251.

Bollen C.J., Hoeven E.P. van der, Lamers J.G. & Schoonen M.P.M. (1983) Effect of benomyl on soil fungi associated with rye. 2. Effect of fungi of culm bases and roots. *Netherlands Journal of Plant Pathology* **89,** 55–66.

Brasier C.M. (1983) The future of Dutch elm disease in Europe. In *Research on Dutch Elm Disease in Europe. Forestry Commission Bulletin* no. 60 (Ed. D.A. Burdekin), pp. 96–104. HMSO, London.

Brokenshire T. (1980) Black mould of roses in Scotland caused by *Chalaropsis thielavioides. Plant Pathology* **29,** 56.

Brooks F.T. (1953) *Plant Diseases.* Oxford University Press, Oxford.

Bugaret Y., Bulit J. & Lafon R. (1980) Amélioration du traitement de l'excoriose (*Phomopsis viticola*) par l'utilisation en post-débourrement de la vigne de l'association de phoséthyl-Al et de folpel. *Phytiatrie-Phytopharmacie* **29,** 45–56.

Bulit J. & Bugaret Y. (1975) Action conjuguée de la température et de l'humectation sur la manifestation de l'excoriose de la vigne due à *Phomopsis viticola. Annales de Phytopathologie* **7,** 358.

Bulit J., Bugaret Y. & Lafon R. (1972) L'excoriose de la vigne et ses traitements. *Revue de Zoologie Agricole et de Pathologie Végetale* **1,** 44–54.

Bulit J., Bugaret Y. & Verdu D. (1973) Sur les possibilités de conservation hivernale du *Botrytis cinerea* et du *Phomopsis viticola* dans les bourgeons de la vigne. *Revue de Zoologie Agricole et de Pathologie Végétale* **1,** 1–12.

Butin H. (1958) Über die auf *Salix* und *Populus* vorkommende Arten der Gattung *Cryptodiaporthe. Phytopathologische Zeitschrift* **32,** 399–415.

Butin H. (1981) Über den Rindenbranderreger *Fusicoccum quercus* und andere Rindenpilze der Eiche. *European Journal of Forest Pathology* **11,** 33–44.

Butin H. (1983) *Krankheiten der Wald- und Parkbäume.* Georg Thieme Verlag, Stuttgart.

Cannon P.F., Hawksworth D.L. & Sherwood-Pike M.A. (1985) *The British Ascomycotina, an Annotated Checklist.* CMI, Kew.

Caroselli N.E. (1957) Juniper blight and progress on its control. *Plant Disease Reporter* **41,** 216–218.

Carter M.V. (1957) *Eutypa armeniacae,* an airborne vascular pathogen of *Prunus armeniaca* in Southern Australia. *Australian Journal of Botany* **5,** 21–35.

Carter M.V. (1965) Ascospore deposition in *Eutypa armeniacae. Australian Journal of Agricultural Research* **16,** 825–836.

Carter M.V. & Moller W.J. (1970) Duration of susceptibility of apricot pruning wounds to infection by *Eutypa armeniacae. Australian Journal of Agricultural Research* **11,** 915–920.

Carter M.V. & Moller W.J. (1977) *Eutypa* canker and dieback of apricot. *Bulletin OEPP/EPPO Bulletin* **7,** 85–94.

Carter M.V., Bolay A. & Rappaz F. (1983) An annotated host list and bibliography of *Eutypa armeniacae. Review of Plant Pathology* **62,** 251–258.

Carter M.V., Bolay A., English H. & Rumbos I. (1985) Variation in the pathogenicity of *Eutypa lata* (= *E. armeniacae*). *Australian Journal of Botany* **33,** 361–366.

Chaves G. (1980) Anthracnose. In *Bean Production Problems: Disease, Insect, Soil and Climatic constraints of Phaseolus vulgaris* (Eds H.F. Schwartz & G.E. Gálvez), pp. 39–54. CIAT, Cali, Colombia.

Chesters C.G.C. & Hornby D. (1965) Studies on *Colletotrichum coccodes. Transactions of the British Mycological Society* **48,** 573–581.

Coutinho M.P. (1968) Algumas notas sobre a gafa da azeitona. *Ao Serviço da Lavoura* **85,** 7.

Davet P. (1971) Recherches sur le *Colletotrichum coccodes. Phytopathologia Mediterranea* **10,** 159–163.

Day P.R., Dodds J.A., Elliston J.E., Jaynes R.A. & Anagnostakis S.L. (1977) Double-stranded RNA in *Endothia parasitica. Phytopathology* **67,** 1393–1396.

De Hoog G.S. & Scheffer R.J. (1984) *Ceratocystis* versus *Ophiostoma*: a reappraisal. *Mycologia* **76,** 292–299.

Delatour C. (1986) Le problème des *Ceratocystis* européens des chênes. *Bulletin OEPP/EPPO Bulletin* **16,** 521–525.

Delon R. & Kiffer E. (1978) *Chalara elegans* (= *Thielaviopsis basicola*) et les espèces voisines. I. Généralités et pathologie. *Annales du Tabac, Section 2* **15,** 159–186.

Diercks R. (1959) Zur chemischen Bekämpfung der Korbweiden-Parasiten *Fusicladium saliciperdum* und *Glomerella miyabeana. Pflanzenschutz* **11,** 125–130.

Dingley J.M. & Gilmour J.W. (1972) *Colletrotrichum acutatum* f. sp. *pinea* associated with terminal crook disease of *Pinus* species. *New Zealand Journal of Forestry Science* **2,** 192–201.

Dixon G.R. (1984) *Vegetable Crop Diseases.* Macmillan, London.

Dorozhkin N.A. & Fedorov U.N. (1982) (Mycoflora of canker tumours on Siberian larch and some biological features of *Lachnellula willkommii*). *Mikologiya i Fitopatologiya* **16,** 273–276.

Duben J. & Fehrmann H. (1979) Vorkommen und Pathogenität von *Fusarium*-Arten an Winterweizen in der Bundesrepublik Deutschland. I. Artenspektrum und jahreszeitliche Sukzession an der Halmbasis; II. Vergleich der Pathogenität als Erreger von Keimlings-, Halmbasis- und Ährenkrankheiten. *Zeitschrift für Pflankrankheiten und Pflanzenschutz* **86,** 638–652; 707–728.

English H., Davis J.R. & Devay J.E. (1962) *Cytosporina dieback,* a new disease of apricot in N. America. *Phytopathology* **52,** 361.

Farneti R. (1921) Sopra il 'brusone' del riso. *Atti dell' Istituto Botanico dell' Università Pavia ser. II* **18,** 109–122.

Fisher F.E. (1972) *Diaporthe citri* and *Phomopsis citri*, a correction. *Mycologia* **64,** 422.

Fitt B.D.L. & Hornby D. (1978) Effects of root-infecting fungi on wheat transport processes and growth. *Physiological Plant Pathology* **13,** 335–346.

Fletcher J.T. (1984) *Diseases of Greenhouse Plants.* Longman, London.

Forrer H.R., Rijsdijk F.H. & Zadoks J.C. (1982) Can mildew assist in the entry of *Fusarium* fungi into wheat leaves? *Netherlands Journal of Plant Pathology* **88,** 123–125.

Froyd J.D., Paget C.J., Guse L.R., Dreikorn B.A. & Pafford J.L. (1976) Tricyclazole. A new systemic fungicide for control of *Pyricularia oryzae* on rice. *Phytopathology* **66,** 1135–1139.

Galet P. (1977) *Les Maladies et Parasites de la Vigne.* Le Paysan du Midi, Montpellier.

Garibaldi A. (1981) Le principali malattie fungine del ciclamino. In *Atti della Giorneta Floricola su il Ciclamino da Vaso Fiorito, Genova, 25 aprile 1981,* pp. 107–115.

Gerlach W. & Nirenberg H. (1982) The genus *Fusarium* —a pictorial atlas. *Mitteilungen aus der Biologische Bundesanstalt für Land und Forstwirtschaft Berlin-Dahlem,* no. 20.

Gibbs J.N. (1981) European forestry and *Ceratocystis* species. *Bulletin OEPP/EPPO Bulletin* **11,** 193–197.

Gindrat D., van der Hoeven E. & Moody A.R. (1977)

Control of *Phomopsis sclerotioides* with *Gliocladium* or *Trichoderma*. *Netherlands Journal of Plant Pathology* **83,** Suppl. 1, 429–438.

Goidanich G. & Minghelli F. (1966) Prove di lotta contro *Endothia parasitica* in provincia di Modena. *Atti del Convegno Internazional del Castagno, Cuneo,* pp. 273–277.

Graniti A. (1986) *Seiridium cardinale* and other cypress cankers. *Bulletin OEPP/EPPO Bulletin* **16,** 479–486.

Graniti A. & Laviola C. (1981) Squardo generale alle malattie parassitarie dell'olivo. *Informatore Fitopatologico* **31** (1–2), 77–92.

Grasso V. & Raddi P. (Eds) (1979) *Il Cipresso: Malattie e Difesa. Firenze, 23/24 November 1983.* Comunità Economica Europea—AGRIMED.

Gremmen J. (1978) Research on Dothichiza-bark necrosis (*Cryptodiaporthe populea*) in poplar. *European Journal of Forest Pathology* **8,** 362–368.

Grente J. (1975) La lutte biologique contre le chancre du chataignier par 'hypovirulence contagieuse'. *Annales de Phytopathologie* **7,** 216–218.

Guillaumin J.J., Dubos B. & Mercier S. (1982) Les pourridiés à *Armillariella* et *Rosellinia* en France sur vigne, arbres fruitiers et cultures florales. I. Etiologie et symptomatologie. *Agronomie* **2,** 71–80.

Gunnerbeck E. (1971) Studies on foliicolous deuteromycetes. I. The genus *Mastigosporium* in Sweden. *Svensk Botanisk Tidskrift* **65,** 39–52.

Gupta V.K. (1977) Root rot of apple and its control by carbendazim. *Pesticides* **11,** 49–52.

Haeni F. (1981) Zur Biologie und Bekämpfung von Fusariosen bei Weizen und Roggen. *Phytopathologische Zeitschrift* **100,** 44–87.

Hansen H.N. (1956) The perfect stage of *Coryneum cardinale. Phytopathology* **46,** 636–637.

Heald F.D. (1913) Some notes on the dissemination of *Diaporthe parasitica. Phytopathology* **3,** 68.

Hebert T.T. (1971) The perfect stage of *Pyricularia grisea. Phytopathology* **61,** 83–87.

Helton A.W. & Konicek D.E. (1961) Effects of selected *Cytospora* isolates from stone fruits on certain stone fruit varieties. *Phytopathology* **51,** 152–157.

Herr L.J., Lipps P.E. & Watters B.L. (1983) Diaporthe stem canker of sunflower. *Plant Disease* **67,** 911–913.

Hubbeling N. (1957) New aspects of breeding for disease resistance in beans. *Euphytica* **6,** 111–141.

Hubbes M. (1959) Untersuchungen über *Dothichiza populea,* den Erreger des Rindenbrandes der Pappeln. *Phytopathologische Zeitschrift* **35,** 58–96.

IRRI (1963) *The Rice Blast Disease. A Symposium.* Johns Hopkins Press, Baltimore.

Jailloux F. & Froidefond G. (1978) Influence de la concentration de l'inoculum et du stade de défeuillaison sur la réceptivité des plaies pétiolaires du pècher à l'égard du *Fusicoccum amygdali,* agent du chancre. *Annales de Phytopathologie* **10,** 39–44.

Kastirr U. & Ficke W. (1983) Neue Ergebnisse zur Bedeutung der Krötenhautkrankheit am Apfel. *Nachrichtenblatt für den Pflanzenschutz in der DDR* **37,** 251–254.

Kendrick J.B. & Walker J.C. (1948) Anthracnose of tomato. *Phytopathology* **38,** 247–260.

Khomyakov M.T. (1981) (Biological features of *Pestalotia malorum*). *Mikologiya i Fitopatologiya* **15,** 321–325.

Krüger W. & Rogdaki-Papadaki C. (1980) Über die Wirkung von Temperatur, Bodenart, Bodenverdichtung und Düngung auf die Wurzelfäule und das Pilzspectrum des Maises. *Zeitschrift für Pflanzenkrankheiten und Pflanzenschutz* **87,** 298–316.

Kryukova E.A. & Plotnikova T.S. (1982) (The control of vascular mycosis of oak). *Zashchita Rastenii* no. 8, 22.

Lanier L., Joly P., Bondoux P. & Bellemère A. (1976) *Mycologie et Pathologie Forestières.* Masson, Paris.

Lemaire J.M., Carpentier F., Dalle J.F. & Doussinault G. (1979) Lutte biologique contre le piétin-échaudage des céréales. Modifications physiologiques chez le blé inoculé par une souche atténuée d'*Ophiobolus graminis.* II. Changement de la teneur en chlorophylle. *Annales de Phytopathologie* **11,** 193–197.

Lemaire J.M., Doussinault G., Lucas P., Perraton B. & Messager A. (1982) Possibilités de sélection pour l'aptitude à la prémunition dans le cas du piétin-échaudage des céréales, *Gaeumannomyces graminis. Cryptogamie, Mycologie* **3,** 347–359.

Linfield C. & Price D. (1983) Anemone leaf curl. *Grower* **99,** 27–29.

Ling K.C. & Ou S.H. (1969) Standardization of the international race numbers of *Pyricularia oryzae. Phytopathology* **59,** 339–342.

Link K.P. & Walker J.C. (1933) The isolation of catechol from pigmented onion scales and its significance in relation to disease resistance in onions. *Journal of Biological Chemistry* **100,** 379–383.

Lucas G.B. (1975) *Diseases of Tobacco,* 3rd edn, pp. 144–160. Biological Consulting Associates, Raleigh.

Manoir J. & Vegh I. (1981) *Phomopsis foeniculi* sp. nov. sur fenouil. *Phytopathologische Zeitschrift* **100,** 319–330.

Mantell S.H. & Wheeler B.E.J. (1973) *Rosellinia* and white root rot of Narcissus in the Scilly Isles. *Transactions of the British Mycological Society* **60,** 23–35.

Mappes D. & Locher F. (1981) Versuchserfahrungen bei der Bekämpfung von Erdbeerfruchtfäulen. *Mededelingen van de Faculteit Landbouwwetenschappen Rijksuniversiteit Gent* **46,** 969–978.

Marchetti L. (1977) II fusicoccum della drupacee. *Informatore Fitopatologico* **27** (10), 8.

Marchetti L., D'Aulerio A.Z. & Badiali G. (1984) Un biennio di lotta chimica contro l'antracnosi del platano. *Informatore Fitopatologico* **34**(10), 35–36.

Marin-Sanchez J.P. & Jiménez-Diaz R.M. (1981) *Pyricularia oryzae* and *Nakataea sigmoidea,* pathogens of rice in southern Spain. *Phytopathologia Mediterranea* **20,** 89–95.

Martelli G.P. (1960) Primo contributo alla conoscenza della biologia di *Gloeosporium olivarum. Phytopathologia Mediterranea* **1,** 31–43.

Martelli G.P. & Piglionica V. (1961) Tre anni di lotta contra la 'lebbra' delle olive in Puglia. *Phytopathologia Mediter-*

ranea **3,** 101–112.

Mastenbroek C. (1960) A breeding programme for resistance to anthracnose in dry shell haricot beans, based on a new gene. *Euphytica* **9,** 177–184.

Matta A. (1978) *Fusarium tabacinum* patogeno in natura su basicilo e pomodoro. *Rivista di Patologia Vegetale, IV.* **14,** 119–125.

Messiaen C.M. & Lafon R. (1970) *Les Maladies des Plantes Maraîchères.* INRA, Paris.

Messner K. & Jaehnl G. (1981) *Libertella* sp.—Ursache von Erkrankungs und Absterbeerscheinungen des Apfelsorte McIntosh. *Zeitschrift für Pflanzenkrankheiten und Pflanzenschutz* **88,** 18–26.

Messner K. & Sutton B.C. (1982) *Libertella blepharis,* pathogenic on apple trees of the variety McIntosh. *Mycotaxon* **14,** 325–333.

Milatović I., Palaversić B. & Vlahović V. (1983) (Investigations over several years of resistance in maize to *Colletotrichum graminicolum*). *Zaštita Bilja* **34,** 15–26.

Millar C.S. & Colhoun J. (1969) *Fusarium* diseases of cereals. IV. Observations on *Fusarium nivale* on wheat; VI. Epidemiology of *Fusarium nivale* on wheat. *Transactions of the British Mycological Society* **52,** 57–66; 195–204.

Moletti M., Mazzini F., Villa B. & Baldi G. (1983) Valutazione della resistenza di genotipi di riso allo *Sclerotium oryzae,* mediante infezione artificiele e naturale. *Informatore Fitopatologico* **33** (1), 59–62.

Moller W.J. & Carter M.V. (1965) Production and dispersal of ascospores in *Eutypa armeniacae. Australian Journal of Biological Sciences* **18,** 67–80.

Monod M. (1983) *Monographie Taxonomique des Gnomoniaceae. Sydowia,* beiheft 315.

Morelet M. (1982) La brunissure cryptogamique des cupressacées en France. *Revue Horticole* no. 227, 35–39.

Morelet M. (1986) Les *Verticicladiella* des pins en liaison avec les phénomènes de dépérissement. *Bulletin OEPP/EPPO Bulletin* **16,** 473–478.

Morvan G. (1977) Apricot decline today and tomorrow. *Bulletin OEPP/EPPO Bulletin* **7,** 134–147.

Motta E. (1979) Presenza di spermazi in *Seiridium cardinale. Annali dell'Istituto Sperimentale di Patologia Vegetale, Roma* **5,** 39–47.

Murray D.I.L. & Gadd G.M. (1981) Preliminary studies on *Microdochium bolleyi* with special reference to colonization of barley. *Transactions of the British Mycological Society* **76,** 397–403.

Neely D. (1976) Sycamore anthracnose. *Journal of Arboriculture* **2,** 153–157.

O'Rourke C.J. (1976) *Diseases of Grasses and Forage Legumes in Ireland.* An Foras Talúntais, Carlow.

Ou S.H. (1985) *Rice Diseases.* CMI, Kew.

Palazon Español I.J. & Palazon Español C.F. (1979) Estudios sobre *Gloeosporium amygdalinum* en los almendros espanoles. *Anales del Instituto Nacional de Investigaciones Agrarias, Proteccion Vegetal* no. 11, 29–43.

Paunović S.A. (1980) Cultivar rootstocks and environments as potential factors for successful apricot growing. *Acta Horticulturae,* 37–52.

Peace T.R. (1960) *The Status and Development of the Elm Disease in Britain. Forestry Commission Bulletin* no. 33. HMSO, London.

Pinon J. (1979) Origine et principaux caractères des souches françaises d'*Hypoxylon mammatum. European Journal of Forest Pathology* **9,** 129–142.

Pinon J. (1984) Propriétés biologiques de la toxine d'*Hypoxylon mammatum,* parasite des peupliers de la section Leuce. *Revue de Cytologie et de Biologie Végétales—Le Botaniste* **7,** 271–277.

Porta-Puglia A. & Montorsi F. (1982) Osservazione sulla micoflora di semi di melanzana. *Informatore Fitopatologico* **32** (6), 37–41.

Price T.V. (1975) Studies on the microbial colonization of sapwood of pruned apricot trees. *Australian Journal of Biological Sciences* **26,** 379–388.

Przbył K. (1984) Development of the fungus *Ceratocystis fimbriata* in shoots of poplar clones with differing resistance; Pathological changes and defence responses in poplar tissues caused by *Ceratocystis fimbriata. European Journal of Forest Pathology* **14,** 177–183; 183–191.

Pusey P.L. & Wilson C.L. (1982) Detection of double-stranded RNA in *Ceratocystis ulmi. Phytopathology* **72,** 423–428.

Raddi P. & Panconesi A. (1981) Cypress canker disease in Italy: biology, control possibilities and genetic improvement for resistance. *European Journal of Forest Pathology* **11,** 340–347.

Ramos D.E., Moller W.J. & English H. (1975) Susceptibility of apricot tree pruning wounds to infection by *Eutypa armeniacae:* production and dispersal of ascospores of *Eutypa armeniacae* in California. *Phytopathology* **65,** 1359–1364; 1364–1371.

Raynal G. (1977) Comparaison, en contaminations artificielles, des pouvoirs pathogènes des *Colletotrichum* isolés en France sur la luzerne. *Annales de Phytopathologie* **9,** 193–203.

Reinecke P. (1978) *Microdochium bolleyi* at the stem base of cereals. *Zeitschrift für Pflanzenkrankheiten und Pflanzenschutz* **85,** 675–685.

Reinecke P. & Fokkema N.J. (1981) An evaluation of methods of screening fungi from the haulm base of cereals for antagonism to *Pseudocercosporella herpotrichoides* in wheat. *Transactions of the British Mycological Society* **77,** 343–350.

Royse D.J. & Ries S.M. (1978) Detection of *Cytospora* species in twig elements of peach and its relation to the incidence of perennial canker. *Phytopathology* **68,** 663–667.

Rozsnyay Zs. D. (1977) *Cytospora* canker and dieback of apricots. *Bulletin OEPP/EPPO Bulletin* **7,** 69–80.

Rumbos I. (1983) *Eutypa* canker and dieback of almond in Greece. *Zeitschrift für Pflanzenkrankheiten und Pflanzenschutz* **90,** 99–101.

Russin J.S. & Shain L. (1983) Insects as carriers of virulent and hypovirulent isolates of *Endothia parasitica.*

Phytopathology **73,** 837.

Schipper A.L. (1978) A *Hypoxylon mammatum* pathotoxin responsible for canker formation in quaking aspen. *Phytopathology* **68,** 866–872.

Schmidle A. & Schulz U. (1978) Versuche zur chemischen Bekämpfung der Valsa-Krankheit an Süsskirsche und Pfirsich. *Nachrichtenblatt des Deutschen Pflanzenschutzdienstes* **30,** 153–155.

Schmidle A., Krähmer H. & Brenner H. (1979) Ein Beitrag zur taxonomischen Abgrenzung von *Leucostoma persoonii* und *L. cincta. Phytopathologische Zeitschrift* **96,** 294–301.

Schneider R.W., Grogan R.G. & Kimble K.A. (1978) *Colletotrichum* root rot of greenhouse tomatoes in California. *Plant Disease Reporter* **62,** 969–971.

Schneider R.W., Paetzholdt M. & Willer K.H. (1969) *Gnomonia rubi* als Krankheitserreger an Kühlhausrosen und Brombeeren. *Nachrichtenblatt des Deutschen Pflanzenschutzdienstes* **21,** 17–21.

Schulz U. (1981) Untersuchungen zur biologische Bekämpfung von *Cytospora*-Arten. *Zeitschrift für Pflanzenkrankheiten und Pflanzenschutz* **88,** 132–141.

Schulz U. & Schmidle A. (1983) Zur Epidemiologie der Valsa-Krankheit. *Angewandte Botanik* **57,** 99–107.

Schwartz H.F., Pastor Corrales M.A. & Singh S.P. (1982) New sources of resistance to anthracnose and angular leaf spot of beans (*Phaseolus vulgaris*). *Euphytica* **31,** 741–754.

Senula A. & Ficke W. (1983) Biotest zur Diagnose pilzlicher Rindbranderreger des Kernobsts. *Archiv für Phytopathologie und Pflanzenschutz* **19,** 299–308.

Sewell G.W.F. & Wilson J.F. (1975) The role of *Thielaviopsis basicola* in the specific replant disorders of cherry and plum. *Annals of Applied Biology* **79,** 149–169.

Signoret P.A. (1981) Les principales maladies du soja dans la région méditerranéenne. *6es Journées de Phytiatrie et de Phytopharmacie Circum-méditerranées, Perpignan,* pp. 202–208.

Signoret P.A. (1984) Les maladies du soja en Europe. Situation, recherches effectuées. *Eurosoya* no. 2, 18–23.

Simmonds J.H. (1965) A study of the species of *Colletotrichum* causing ripe fruit rots in Queensland. *Queensland Journal of Agricultural and Animal Science* **22,** 437–459.

Smith J.D. (1983) *Fusarium nivale (Gerlachia nivalis)* from cereals and grasses: is it the same fungus? *Canadian Plant Disease Survey* **63,** 25–26.

Smith P.M. (1983) Diseases of hardy nursery stock. *Annual Report Glasshouse Crops Research Institute 1981,* pp. 131–133.

Spies J.L., Knösel D. & Meier D. (1985) Anthracnose und Hitzeschäden an Platanen im Stadtgebeit von Hamburg. *Nachrichtenblatt des Deutschen Pflanzenschutzdienstes* **37,** 17–21.

Stover R.H. (1950) The black root rot disease of tobacco. I. Studies on the causal organism *Thielaviopsis basicola. Canadian Journal of Research Series C* **28,** 445–470.

Sutton B.C. (1975) *Coelomycetes.* V. *Coryneum.* CMI Mycological Papers, no. 138. CMI, Kew.

Theodorou M.K., Scanlon J.C.M. & Smith I.M. (1982) Infection and phytoalexin accumulation in French bean leaves injected with spores of *Colletotrichum lindemuthianum. Phytopathologische Zeitschrift* **103,** 189–197.

Tourvieille de Labrouhe D. (1982) Pénétration de *Rosellinia necatrix* dans les racines du pommier en conditions de contamination artificielle. *Agronomie* **2,** 553–560.

Tu J.C. (1983) Epidemiology of anthracnose caused by *Colletotrichum lindemuthianum* on white bean (*Phaseolus vulgaris*) in Southern Ontario: survival of the pathogen. *Plant Disease* **67,** 402–404.

Valente A.P. (1979) *Problèmes Phytosanitaires les plus Importants de l'Olivier au Portugal.* Direcção Geral de Protecção da Produção Agrícola, Oeiras.

Van Kesteren H.A. (1967) 'Black root rot' in Cucurbitaceae caused by *Phomopsis sclerotioides* nov. spec. *Netherlands Journal of Plant Pathology* **73,** 112–116.

Vercesi A. (1982) Disseccamenti da *Phomopsis* su noce. *Informatore Fitopatologico* **32** (12), 51–54.

Viennot-Bourgin G. (1949) *Les Champignons Parasites des Plantes Cultivées.* Masson, Paris.

Vigouroux A. (1979) Les 'dépérissements' des platanes. *Revue Forestière Française* **31,** 28–39.

Von Arx J.A. (1957) Die Arten der Gattung *Colletotrichum. Phytopathologische Zeitschrift* **29,** 413–468.

Von Arx J.A. (1984) Notes on *Monographella* and *Microdochium. Transactions of the British Mycological Society* **82,** 373–374.

Vörös J. & Léranthné Szilágyi J. (1983) Overwintering and thermal requirements of *Diaporthe helianthi. Növényvédelem* **19,** 355.

Vörös J., Léránth J. & Vajna L. (1983) Overwintering of *Diaporthe helianthi,* a new destructive pathogen on sunflowers in Hungary. *Acta Phytopathologica Academiae Scientiarum Hungaricae* **18,** 303–305.

Wagener W.W. (1939) The canker of Cupressus induced by *Coryneum cardinale* n. sp. *Journal of Agricultural Research* **58,** 1–46.

Walker J. (1975) Take-all diseases of Gramineae: a review of recent work. *Review of Plant Pathology* **54,** 113–144.

Walla J.A. & Stack R.W. (1980) Dip treatment for control of blackstem on *Populus* cuttings. *Plant Disease* **64,** 1092–1095.

Webber J.F. (1981) A natural biological control of Dutch elm disease. *Nature* **292,** 449–451.

Wendt R., Weidhaas J., Griffin G.J. & Elkins J.R. (1983) Association of *Endothia parasitica* with mites isolated from cankers on American Chestnut trees. *Plant Disease* **67,** 757–758.

Woodcock T. & Washington W.S. (1979) *Colletotrichum acutatum* on Anemone and Ranunculus. *Australasian Plant Pathology* **8,** 10.

Yaegashi H. & Nishihara N. (1976) Production of the perfect stage in *Pyricularia* from cereals and grasses. *Annals of the Phytopathological Society of Japan* **42,** 511–516.

Yamato H. (1977) Citrus melanose and its related disease in Japan. *Proceedings of the International Society for Citriculture* **3,** 997–998.

Yde-Andersen A. (1983) *Pseudomonas fluorescens* and *Ceratocystis ulmi* in wych elm. In *Research on Dutch Elm Disease in Europe. Forestry Commission Bulletin* no. 60. (Ed. D.A. Burdekin), pp. 72–74. HMSO, London.

Zachos D.G. & Markis S.A. (1959) Recherches sur le *Gloeosporium olivarum* en Grèce. I. Biologie du champignon. *Annales de l'Institut Phytopathologique Benaki,* N.S. **2,** 24–42.

Zachos D.G. & Markis S.A. (1963) Recherches sur le *Gloeosporium olivarum* en Grèce. II. Symptomatologie de la maladie. III. Epidémiologie de la maladie. *Annales de l'Institut Phytopathologique Benaki,* N.S. **5,** 128–130; 238–259.

Zaumeyer W.J. & Meiners J.P. (1975) Disease resistance in beans, anthracnose. *Annual Review of Phytopathology* **32,** 318–320.

13: Ascomycetes IV
Dothideales

This group, as understood in the wide sense of the seventh edition of the *CMI Dictionary of the Fungi,* includes fungi with bitunicate asci, in most cases formed in locules in a stroma immersed in plant tissues, known as pseudothecia rather than perithecia. These ascocarps are generally small and inconspicuous and often infrequently formed, so that there is a problem of their full characterization. Some genera are still seeking a certain home in either Sphaeriales or Dothideales. The current CMI Dictionary lists, but does not seek to characterize, families within the Dothideales, and no attempt is made here to base distinctions on ascocarp morphology. In some cases, however, families have a characteristic biology and in particular have certain specific anamorphs. Indeed, a great many anamorphic fungi not specifically associated with known teleomorphs belong in 'genera' whose association with the Dothideales, or with families within the Dothideales, seem quite clear.

The Dothideales thus include a large number of species principally causing lesions (spots, blotches, stripes) on leaves or stems; a few infect roots. Overwintering is often as pseudothecia on dead plant material, and multiplication through the growing season by dispersal of conidia. They are often seed-borne and sometimes survive in soil as sclerotia or chlamydospores. While many are saprophytes, it is certainly this order of fungi which includes the greatest number of plant parasitic species, often with specific epithets relating them to one host. Many such host-defined species are probably simply forms of polyphagous species.

The anamorphs of these fungi are most frequently seen, and largely characterize them as plant pathogens. There are very broadly two types: i) 1- to multiseptate hyaline or dark conidia borne on erumpent tufts of, or single, conidiophores *(Cladosporium, Cercospora, Ramularia, Drechslera, Alternaria)*; ii) pycnidial forms *(Phoma, Ascochyta, Septoria)*. Species within a single genus *(Mycosphaerella, Leptosphaeria)* may have anamorphs of different type or different 'genus' within one type—a fact which may have taxonomic and evolutionary significance still to be understood, or may simply reveal that the taxonomy of either or both teleomorphs and anamorphs could be improved.

Because the Dothideales is taken in a broad sense, this chapter has to subdivide to the family level. The features of the principal families can be summarized as follows (with examples of the anamorphs typically associated):

1) Dothideaceae—e.g. *Didymella, Mycosphaerella.* Leaf-spotting fungi with anamorphs of the *Cladosporium, Cercospora, Ramularia* and *Ascochyta* types, besides *Phoma* and *Septoria.* Within *Mycosphaerella,* species have been grouped by their anamorphs.

2) Pleosporaceae—e.g. *Pleospora, Leptosphaeria.* Stem- and leaf-infecting fungi with anamorphs of the *Alternaria, Stemphylium, Phoma, Septoria* types.

3) Pyrenophoraceae—a clearly distinct group with teleomorph/anamorph combinations *Pyrenophora/Drechslera, Cochliobolus/Bipolaris, Setosphaeria/Exserohilum, Pseudocochliobolus/Curvularia* (if *Drechslera* is subdivided). Mostly on Gramineae. The anamorphs used to be known as *Helminthosporium,* a name sometimes still used in relation to the diseases.

4) Venturiaceae—with a characteristic subcuticular infection habit (scab) and *Spilocaea* or *Fusicladium* anamorphs.

5) Elsinoaceae—also causing diseases known as scab, because of a tendency for cork-formation by the host. The pseudothecia have globose asci in monascous locules, and the anamorph *(Sphaceloma)* is acervular.

6) Botryosphaeriaceae, Didymosphaeriaceae, Dothioraceae—not very clearly characterized at the present time by biology or anamorphs. The fact that *Botryosphaeria* (in this group) and *Physalospora* (in Sphaeriales) have *Botryodiplodia* anamorphs is an anomaly.

7) Hysteriaceae, Schizothyriaceae—with flattened superficial ascocarps. The former are at most very weakly pathogenic (e.g. ***Hysterographium fraxini*** (Pers.) de Not. on ash), while the latter are mainly tropical leaf parasites (only one species of any significance in Europe).

There remain a number of cases of anamorph

genera which 'bridge' these families—especially *Phoma* and *Septoria*. The general policy in this Handbook of associating 'imperfect' fungi with the groups to which they are obviously related breaks down in such cases, which will presumably be resolved in due course by further taxonomic research. However, there seems no doubt that these are Dothideales-related, rather than connected with some other Ascomycete group. Finally, there are a number of anamorphic Ascomycetes which it has not, at the present time, been possible to place anywhere with even a small measure of confidence. In line with our aim not to use such concepts as 'Deuteromycetes', these are placed here but only as a matter of convenience. They may be associated with any Ascomycetes (Chapters 10–14) and one day no doubt will be.

DOTHIDEACEAE

Didymella applanata (Niessl) Sacc.

(Anamorph ***Phoma*** sp.)

D. applanata produces separate immersed pseudothecia with erumpent ostioles and 2-celled ascospores, the upper cell broader than the lower one. The fungus also produces pycnidia with erumpent ostioles releasing hyaline single-celled conidia (CMI 735). It attacks *Rubus* cane fruits, principally *R. idaeus,* and has been recorded world-wide wherever these fruits grow. Two pathotypes have been reported in Canada, but none in the UK.

The fungus causes spur blight, a disease physiologically suppressing the growth of lateral shoots on fruiting canes in the year following infection of leaves on young canes. Dark brown lesions spread in late summer from leaves to canes. In Europe they are usually confined to the lower half of canes, where buds do not produce many lateral shoots, so that their removal causes no yield loss. Lesions become silvered in winter, and pseudothecia and pycnidia develop in the primary cortex in separate groups of similar maturity, pycnidia tending to mature later in the year (Blake 1980). Pycnospores are released from late March to early November with a peak in July and August coincident with the onset of leaf senescence and the appearance of infections at the nodes (Burchill & Beever 1975). Ascospores are mature in April and discharged mainly from May to June when infections are rare. Spur blight causes serious yield loss in overgrown and weed-infested plantations or in those receiving high levels of nitrogen fertilizer. Fungicides have reduced infection in numerous trials in many countries, but yield increases could be attributed to control of other diseases (Williamson & Hargreaves 1981) or to the growth-promoting effects of benzimidazole fungicides (Mason 1981). Pre-harvest sprays of several fungicides used to control *Botryotinia fuckeliana* also control spur blight when directed at the young canes.

WILLIAMSON

Didymella bryoniae (Auersw.) Rehm

(Syn. *D. melonis* Pass., *Mycosphaerella citrullina* (C.O.Sm.) Grossenbacher, *M. melonis* (Pass) Chiu & Walker)

(Anamorph ***Ascochyta cucumis*** Fautrey & Roum.)

D. bryoniae forms black ostiolate erumpent pseudothecia and more commonly dark brown-black pycnidia on stems, leaves and fruit of Cucurbitaceae. The pathogen occurs world-wide (CMI map 450). The disease is known as black-stem rot, or gummy stem blight, due to the masses of black pycnidia and perithecia developing on lesions and the gummy exudate oozing from active stem and fruit lesions. Symptoms also occur on leaves as water-soaked irregular patches which turn brown. Fruit infection frequently starts from infected flowers. Lesions may appear sunken and rotted below, or infection may remain latent until after harvesting when a rapid rot may follow an increase in temperature (Steekelenburg 1983). Spores from fruiting bodies can be spread aerially, by water splash, or on hands and pruning knives. Pruning wounds are frequent sites of infection. Severe stem lesions lead to plant death. Crop residues are a primary source of infection, and spores can survive in the soil, on canes, strings, wires and glasshouse structures. Optimum temperature for growth *in vitro* and lesion development is 23–25°C. Mechanical injury, high RH and high nitrogen fertilization of the host, can increase disease levels, high light intensities decrease them (Svedelius & Unestam 1978). The disease is of economic importance in the protected cucumber crop in Europe and elsewhere and on field crops when humidity and temperature are favourable. Crop hygiene is necessary if early infection is to be reduced or avoided. Humidity within a protected crop may be less when growing in artificial substrates compared to soil-grown crops, thus decreasing disease severity. Chemical control is difficult because of the vigorous and continuous growth of the crop, and routine sprays are necessary. Carbendazim fungicides have been used with some success as sprays or in stem lesion paints, and

chlorothalonil and triforine have also given some control. Tolerance to carbendazim can develop, and for this reason alternating sprays of different chemicals have been recommended.
EBBEN

Didymella lycopersici Kleb.

(Anamorph ***Ascochyta lycopersici*** (Plowr.) Brunaud, syn. *Diplodina lycopersici* Hollós)

D. lycopersici causes sporadic and sometimes serious outbreaks of stem and fruit rot in both field-grown and protected tomato crops (CMI 272) throughout Europe (CMI map 324). Dark pycnidia, initially immersed but becoming erumpent, are formed profusely on infected tissue. Conidia are hyaline, ellipsoid, unicellular or one-septate. Pseudothecia have been found only rarely, and the epidemiological significance of the ascospores is not known. First symptoms are usually brown stem lesions at soil level. Complete girdling causes wilting of affected plants. Lower leaves turn yellow and root initials appear immediately above the lesion. Stem lesions bear pycnidia, almost colourless at first but eventually brown or black from which exude pale pink masses of pycnospores. Lesions higher up the stem follow conidial infection through recently formed wounds, particularly where leaves and side shoots have been removed during routine trimming. Fruit are typically infected at the calyx causing them to fall, and fruit lesions develop a black surface within which are embedded numerous pycnidia. Fruit symptoms are more commonly found in field crops. Hyphae and pycnidia of the fungus are also present both on and within seeds from affected fruit (Knight & Keyworth 1960). Seed transmission is particularly important for glasshouse crops growing in sterilized soils or inert substrates. Otherwise, primary infection is usually from soil-borne trash. Rapid secondary spread, through water splash, is a feature of the disease. In protected crops, contaminated pruning knives, and handling affected plants, also contributes to the spread of the disease. Infection and subsequent lesion development (3 weeks later) occur within an optimum temperature range of 15–20°C (Verhoeff 1963). Stem lesions are similar in appearance to *Botryotinia fuckeliana* symptoms but *D. lycopersici* lesions should *not* be cut out and affected plants should be handled last when trimming the crop. In the glasshouse soil sterilization with steam or chemical sterilants prevents carry-over of inoculum (formaldehyde fumigation has been used). Fumigation of haulms does not kill the fungus inside the tissue which can subsequently sporulate. In an affected crop diseased plants should be removed promptly and burnt. The disease can be controlled by high volume protectant sprays of maneb, captan or certain benzimidazole fungicides. Application of a benzimidazole as a root drench may be less effective for control of aerial stem lesions. Dicarboximide fungicides (iprodione, vinclozolin) applied as protectant sprays or as stem-base drenches have been effective. Where practical, benomyl or iprodione mixed with mineral oil and painted onto established stem lesions limits spread and sporulation.
BRADSHAW

Didymella ligulicola (Baker, Dimock & Davis) v. Arx

(Syn. *Mycosphaerella ligulicola* Baker, Dimock & Davis)
(Anamorph ***Ascochyta chrysanthemi*** F. Stev.)

D. ligulicola causes American rayblight of chrysanthemum (CMI 662). Though the name *D. chrysanthemi* (Tassi) Garibaldi & Gullino has been widely used, it is based on a supposed identity between the organism described in the USA and a little-known fungus described in Italy (*Mycosphaerella chrysanthemi* (Tassi) Walker & Baker). Walker & Baker (1983) have shown that the American species is distinct, and did not reach Europe until the 1960s, where it spread rapidly to all chrysanthemum-growing areas (CMI map 407). ***Phoma chrysanthemi*** Vogl. is a distinct fungus of uncertain status. Though all parts of the plant are attacked, the fungus is particularly important in causing rotting of cuttings and flower blight (reddish-brown spotting rapidly spreading to the whole head). The disease is most serious under conditions of intensive glasshouse cultivation. Though primary infection by ascospores from pseudothecia on plant debris can frequently occur, in practice the fungus mainly spreads by splash dispersal of pycnospores from infected planting material. Use of healthy planting material is thus essential; since many countries principally import cuttings, this makes *D. ligulicola* a quarantine organism (EPPO 66). Some cultivars are not heavily attacked. Fungicide control with the benzimidazoles has been very successful, but resistant strains have been found.
SMITH

Didymella exitialis (Morini) Müller (syn. *Sphaerella ceres* Morini) with an *Ascochyta* anamorph (CMI 633) causes leaf scorch of barley and wheat and is part of the complex set of similar *Ascochyta* species on various cereals and grasses (q.v.). Several others have *Didymella* teleomorphs (CMI 661).

Mycosphaerella pinodes (Berk. & Bloxam) Vestergren

(Syn. *Didymella pinodes* (Berk. & Bloxam) Petrak)
(Anamorph ***Ascochyta pinodes*** L.K. Jones)

M. pinodes forms globose, dark brown pseudothecia on host stems and pods. The ascospores are irregularly bi-septate, hyaline and ellipsoidal with a constriction at the central septum and with rounded ends. The pycnidia form on any infected organ, initially immersed in host tissue but at maturity emerging above the surface. They are dark brown to black, containing hyaline septate pycnospores with a slight constriction at the septum (CMI 340). Hosts of *M. pinodes* are principally *Pisum arvense* and *P. sativum,* but also *Lathyrus, Phaseolus* and *Vicia*. The fungus is widely distributed in temperate and subtropical areas (CMI map 316). Initially the lesions produced by *M. pinodes* on pea leaves are small and purplish; they may remain less than 5 mm in diameter lacking a distinct margin or may enlarge becoming black to brown with a definite outer ring. Infection usually spreads from leaves to petioles and thence to the stems, causing girdling lesions which may coalesce and thus give the entire stem a blue-black hue. Floral infection starts at pinpoint lesions on the petals which senesce very quickly. Infection causes uneven development of pods; the seeds may show no outward signs of infection or may have various degrees of shrinkage and be discoloured by brown spots. When infected seed germinates lesions form at the point of cotyledonary attachment. These cause a foot-rotting symptom along the length of the developing plumule and radicle. Diseased seedlings may die before emergence; if not then lesions become evident on the aerial host organs and eventually most infected plants are killed (Neergaard 1977). In addition to being seed-transmitted, *M. pinodes* survives saprophytically on trash. Ascospores arising from host debris are important sources of new infection for succeeding crops (Carter & Moller 1961). Ascospores are an efficient means of short-range spread of *M. pinodes*; peak release takes place in the afternoon. Splash dispersal of conidia is also a means of transmission. Control can be achieved by use of pathogen-free seed for which seed production in arid areas is often recommended, although this does not eradicate all infection (Leach 1960). Seed sanitation may be achieved by the thiram-soak technique (Maude 1966a), and benomyl may be added to the seed soak and also used as a foliar spray. It is essential to destroy pea haulm by burning and/or deep ploughing, to use at least 4-year rotational breaks between crops and to give sufficient spatial separation between pea crops to control *M. pinodes.*

DIXON

Mycosphaerella rabiei Kovachevskii (anamorph ***Ascochyta rabiei*** (Pass.) Labrousse) causes a severe blight of all above-ground parts of chickpea (CMI 337), throughout the Mediterranean region. It is both seed-borne and persists on debris, mainly as pycnidia. It can be controlled by seed treatments and sanitation, but may still require fungicide treatments. There is much current interest in screening chickpeas for resistance (Boorsma 1980), e.g. at ICRISAT in India and ICARDA in Syria.

Ascochyta pisi Lib.

A. pisi is the type species of the genus *Ascochyta*. Globose brown pycnidia, without setae, form on leaves and pods, the pycnidial wall being composed of 1–4 layers of elongated yellow-brown thin-walled cells (CMI 334). Within each pycnidium short, hyaline, conidiophores arise from the wall cells carrying pycnospores which are hyaline, straight or slightly curved, usually monoseptate with a slight constriction at the septum and with rounded ends. The mode of conidiogenesis may be an important character which distinguishes *Ascochyta* from related genera such as *Phoma* (Boerema & Bollen 1975). Hosts include *Pisum,* but also *Lathyrus* and *Vicia* with widespread distribution (CMI map 273). Four pathotypes (I–IV) were recognized by Wallen (1957), which in Canada exhibited geographical separation. *A. pisi* causes pea blight and pod spot with light brown foliar lesions which each have a prominent dark margin and pale centre. The pathogen is seed-borne and, when infected seed germinates, primary lesions form on the first leaves. Pre- and post-emergence damping-off and dwarfing may result from infection by *A. pisi* but essentially attacks occur on the aerial plant parts. Unlike *Mycosphaerella pinodes* and *Phoma medicaginis* var. *pinodella, A. pisi* does not cause foot rotting symptoms.

Transmission of *A. pisi* is via rain-splashed pycnospores, infected host debris and infected seed. This

pathogen has a low saprophytic ability compared with *M. pinodes*. The formation of soil-borne chlamydospores is rare (Dickinson & Sheridan 1968). Studies of the phytoalexins produced by *P. sativum* cvs Meteor and Kelvedon Wonder have indicated that conidia of *A. pisi* race 1 are less tolerant to phytoalexins than race 2 (Harrower 1973). Control is achieved by production of pathogen-free seed in arid areas, treating seed with benomyl, captan or thiram at 30°C. Some levels of resistance have been identified in certain cultivars. Resistance is governed by at least one pair of dominant genes. Work with leafless and semi-leafless peas has been aimed at identifying further resistance to *A. pisi*.
DIXON

Ascochyta fabae Speg.

A. fabae resembles *A. pisi* but is largely specialized to *Vicia faba,* causing leaf and pod spot (CMI 461). Carriage of seed-borne inoculum has spread *A. fabae* world-wide (CMI map 513). Symptoms of leaf and pod spot are seen first on the primary foliage leaves of seedlings developing from infected seed. Lesions are elongated, up to 10 mm long, with chestnut-brown margins and greyish centres. They develop first on the tips and margins of leaves, gradually spreading towards the main veins of the compound leaves and petioles. Elongated red-brown stem lesions develop which weaken the stem and may cause lodging. In severe attacks, the foliage is totally destroyed and developing pods become covered in lesions. Pod lesions are usually darker and more deeply sunken than those on leaves. Infected seeds are covered with circular dark brown spots. The main seat of infection is the seed testa rather than the cotyledons. This pathogen can remain viable on seed for up to 3 years and for 4–5 months on infected trash (Dixon 1984). Short-range spread occurs by rain splash transmitting the pycnospores which ooze from pycnidia in wet weather. For successful re-infection of seed there must be rain early in the development of a crop so that the pathogen is carried sufficiently high in the leaf canopy to re-infect the pods and seeds as they develop (Hewett 1973). The rate of epidemic development from infected seed and re-establishment of seed infection from diseased plants varies considerably from year to year and site to site depending on environmental conditions. Seed certification schemes have been used with considerable success to reduce the incidence of *A. fabae*. Treatment of seed with benomyl and thiram slurries can be used to reduce infection to low levels but tends to impair subsequent germination. Control by spraying standing crops (e.g. with maneb) has had only limited success.
DIXON

Ascochyta boltshauseri Sacc. (syn. *Stagonospora hortensis* Sacc. & Malbr., *A. hortensis* (Sacc. & Malbr.) Jørstad, *Stagonosporopsis hortensis* (Sacc. & Malbr.) Petrak) causes a fairly minor leaf and pod spot on *Phaseolus,* although serious losses have occasionally been reported. In nature, the pycnidia contain at once *Phoma*-like conidia, once-septate *Ascochyta*-like conidia, and two or multi-septate *Stagonospora*-like conidia. The conidia are much larger than those of *Phoma exigua* var. *exigua* (q.v.), which used to be known as *A. phaseolorum* Sacc. The pathogen is seed-borne. ***A. hortorum*** (Speg.) Sm., reported on artichoke, aubergine and many other hosts (Viennot-Bourgin 1949) is probably also synonymous with *P. exigua*. Other species noted on legumes, especially in E. Europe, are ***A. lentis*** Bondartsev & Vassilevskii on lentil, ***A. caulicola*** Laubert on lupin and *Melilotus*, ***A. trifolii*** Bondartsev & Trus on clover, ***A. punctata*** Naumov and ***A. viciae-villosa*** Ondrej on *Vicia* species. ***A. sojaecola*** Abramov is a seed-borne leaf-spotting pathogen of soybean recorded in the USSR.

There are numerous *Ascochyta* species on garden plants. **A. gerberae** Maffei is common and damaging on *Gerbera* and **A. cinerariae** Tassi on *Senecio* x *hybridus*. ***A. clematidina*** Thüm. causes a serious wilt and dieback of garden clematis.

A. gossypii Voronichin causes wet weather blight of cotton (CMI 271; CMI map 259; APS Compendium). Symptoms include leaf spotting and stem canker; the pathogen is seed-borne and carried by debris. Important in S.E. USA when conditions are very wet, it has only been reported from Greece and the USSR in Europe. The fungus could well be yet another form of *Phoma exigua* (q.v.). Numerous *Ascochyta* species occur on Gramineae and were covered in the broad sense by *A. graminicola* Sacc. Punithalingam (1979) has distinguished the species in this complex, some of which have *Didymella* teleomorphs, e.g. *D. exitialis* (q.v.). ***A. avenae*** (Petrak) Sprague & Johnson causes leaf blotch of oats in various parts of Europe (CMI 731) and ***A. sorghi*** Sacc. rough leaf spot of sorghum in Italy (CMI 632). The latter name was previously used in a wider sense, like *A. graminicola*. ***A. desmazieresii*** Cavara is damaging on rye grass (O'Rourke 1976; CMI 661) and many other species are reported on turf grasses (APS Compendium). The diseases are not serious enough to merit special control measures.

Mycosphaerella holci Tehon (anamorph ***Phoma sorghina*** (Sacc.) Boerema *et al.*) causes leaf-spotting on numerous tropical and subtropical Gramineae (including *Oryza, Panicum, Setaria, Sorghum*) but is not reported to be of any particular significance in S. Europe (CMI 584). ***M. pomi*** (Pass.) Lindau (anamorph ***Phoma pomi*** Pass.), described in Italy, causes Brooks fruit spot of apple, but is mainly reported from N. America (Yoder 1982a). ***M. larici-leptolepis*** Ito *et al.* (anamorph ***Phoma yano-kubotae*** Kitajima) is the most important defoliator of larch in Japan, the European *Larix decidua* being more susceptible than the Japanese *L. leptolepis*. It presents a major quarantine risk for European larch (EPPO 16). ***M. laricina*** R. Hartig may cause a locally severe needle cast of larch in Germany, Poland and Sweden. Other *Phoma*, considered under '*Anamorphs of Dothideales*' below, possibly belong with this group of *Mycosphaerella* spp., although *Phoma* seems more likely to be truly associated with *Leptosphaeria*.

Mycosphaerella maculiformis (Pers.) Schröter (anamorph ***Phyllosticta maculiformis*** Sacc.) is widespread as a cause of a usually unimportant leaf spot disease of broad-leaved trees, especially of oak and sweet chestnut. ***M. zeae-maydis*** Mukunya & Boothroyd (anamorph ***Phyllosticta maydis*** Arny & Nelson) was first noted as a pathogen in the early 1970s in the USA, owing to the greater susceptibility of maize plants with Texas male-sterile cytoplasm (as in the case of *Cochliobolus heterostrophus*, q.v., but less specifically). It causes pale lesions (yellow leaf blight) on the leaves (0.3–1.3 cm), with distinctive pycnidia, and survives on maize and grass debris on which the pseudothecia form (APS Compendium). The disease was seen in a limited area of S.W. France in 1971 on an imported hybrid with Texas cytoplasm (Cassini 1973), and seemed to be favoured by cooler conditions than *C. heterostrophus*. However, the abandonment of hybrids susceptible to this latter fungus has also incidentally led to the disappearance of *M. zeae-maydis* in Europe.

Mycosphaerella graminicola (Fuckel) Schröter

(Anamorph ***Septoria tritici*** Roberge)

Basic description

Pseudothecia of *M. graminicola* can easily be confused with both *M. tassiana* and *Didymella exitialis* but these fungi are readily distinguished in culture by their anamorphs (*Cladosporium* and *Ascochyta* species respectively). Pseudothecia of *D. exitialis* occur within lesions on green leaves while pseudothecia of *M. graminicola* have only been recorded from dead plant material after harvest. Ascospores are 2-celled, elliptical, hyaline, 9–16× 2.5–4 μm (Sanderson 1976). Pycnospores are hyaline, thread-like, slightly curved, with 3–7 septa, 43–70×1.5–2 μm (CMI 90).

Host plants

M. graminicola is widespread on wheat and other *Triticum* species, and occurs occasionally on triticale, rye and on some wild grasses.

Host specialization

Isolates adapted to either *Triticum aestivum* or *T. durum* were reported by Eyal *et al.* (1973). It is less certain whether adaptation to wheat cultivars occurs.

Geographical distribution

It is found in all wheat-growing areas of the world (CMI map 397).

Disease

The disease is known as septoria tritici blotch or speckled leaf blotch. First symptoms usually occur on seedlings at about the 4/5-leaf stage. Initially lesions are light-green to yellow, water-soaked areas on the lowest or first true leaf of the seedling. Lesions spread rapidly to form light-brown, linear to irregular patches, which appear speckled as pycnidia develop. During the growing season, most lesions occur on the lowest green leaf. Associated with the leaf lesions are extensive areas of yellowing and tip burn, which can readily be confused with senescence. On maturity the pycnidia within the lesions exude white to buff-coloured spore masses (cirrhi).

Epidemiology

The primary inoculum is usually wind-borne ascospores produced in pseudothecia formed on the stubble left standing after harvest. The ascospores are forcefully liberated and once airborne can travel long distances. They have been recorded associated with periods of moisture from 6 weeks after harvest through the winter to late spring (Sanderson & Hampton 1978). The reduction in ascospore numbers during the spring could account for the absence of this disease in most spring-sown wheat crops. Once lesions appear on leaves of the young crop, and pycnidia develop, then secondary plant-to-plant spread is by rain-splashed pycnospores. In regions where weather conditions do not stimulate the production of pseudothecia, and/or where crops are sown into infected stubble or trash, primary inoculum may also be in the form of rain-splashed pycnospores. Seed-borne transmission has not been recorded.

Economic importance

In many European countries (especially in the West), several regions of the USA, in C. and S. America, Australia, S. Africa and New Zealand, septoria tritici blotch can reach epidemic proportions and become a limiting factor in production. Yield losses of up to 40% have been recorded (Eyal & Ziv 1974; King 1977). In some areas, such as N. America and S. Brazil, crop losses have been reported as devastating.

Control

Effective levels of resistance are available in commercial cultivars. Septoria tritici blotch affects yield in two ways: i) early seedling infection before or at spikelet initiation reduces floret numbers and hence grain number, and also subsequent shoot and root development; and ii) infection on the adult plant affects grain-filling and hence grain weight. When fungicidal control is considered, the early phase of the disease is usually ignored, and recommendations are confined to the application of a single spray to protect the flag leaf from infection late in the season. However, in New Zealand at least, it is clear that ascospores are the primary source of infection, and that ascospore showers tail off rapidly in the spring (Sanderson & Hampton 1978). Highly effective control is achieved by the application of a fungicide on the seedlings at the 4/5-leaf stage during early tillering. As infection is eliminated at an early stage, there is no further source of primary inoculum for late infections when conditions are favourable for a build-up of disease on the adult plant. Fungicides applied at early stem extension for the control of *Pseudocercosporella herpotrichoides* give some degree of control of septoria tritici blotch; however, this is too late for optimum control, and some secondary spread within the crop can occur. In drier climates where wheat is sown into stubble and the primary source of inoculum is pycnospores produced on the old stubble during the growing season, as in Israel, then multiple applications of a fungicide are required.
SANDERSON & SCOTT

Mycosphaerella linicola Naumov

(Syn. *M. linorum* (Wollenw.) Garcia Rada)
(Anamorph ***Septoria linicola*** (Speg.) Garassini)

M. linicola causes pasmo disease of flax (CMI 709). It occurs in most European countries where flax is grown (but not Denmark, Ireland, Poland, Portugal or Scotland), and has spread within this century to N. and S. America and Australasia (CMI map 18). Because its distribution in Europe is limited, it is regarded as a quarantine organism (EPPO 75). It is most important on linseed cultivars. *M. linicola* is mainly seed-borne but can also survive as pseudothecia on debris. It causes a seedling blight, but the economically important phase is stem infection of maturing plants. Pycnospores are rapidly dispersed and cause stem lesions giving a characteristically banded appearance by alternation of diffuse bluish then brown lesions, with finely dotted pycnidia and healthy green zones. Leaf infections lead to leaf fall. Lesions also appear on the capsules. High humidity and temperatures of 22–25°C favour spread. Though resistant cultivars have much reduced the importance of the disease, and the benzimidazole fungicides provide good control, *M. linicola* still causes a problem in France and Italy, and in E. Europe (GDR, Hungary, USSR, Yugoslavia). Seeds remain the main source of infection, and since the fungus cannot reliably be detected by seed-testing methods, disease control in the seed crop is essential. Ideally, seed should be grown in areas where the disease does not occur.
JOUAN

Mycosphaerella ribis (Fuckel) Lindau (anamorph ***Septoria ribis*** (Lib.) Desm.) causes angular leaf spot of *Ribes* species. It is of minor importance in W. Europe, but requires fungicide control in Bulgaria and the USSR. The same applies to ***M. rubi*** Roarck (anamorph ***S. rubi*** Westend.) on *Rubus* species, which may cause premature defoliation. Whether *M. rubi* is distinct from ***Sphaerulina rubi*** Demaree & M. Wilcox said to have the same *Septoria* anamorph on raspberry in N. America, remains unclear. ***Sphaerulina rehmiana*** Jaap (anamorph ***Septoria rosae*** Desm.) also causes a minor leaf spot of rose (APS Compendium). ***M. sentina*** (Fr.) Schröter (syn. *M. pyri* (Auersw.) Boerema; anamorph ***S. pyricola*** Desm.) is widespread but unimportant causing leaf fleck of pear. Leaves are primarily infected by ascospores from fallen leaves, and the disease is then spread by pycnospores. Normal control of *Venturia pirina* will eliminate leaf fleck. ***M. pistacina*** Chitzanidis (anamorph ***S. pistacina*** Allescher), ***M. pistaciarum*** Chitzanidis (anamorph ***S. pistaciarum***

Caracciolo) and ***S. pistaciae*** Desm. all cause leaf spots of pistachio. Sprays of copper fungicides or zineb applied at the stage of foliage appearance give good protection against the pathogens. ***M. populi*** Schröter (anamorph ***S. populi*** Desm.) causes a very minor leaf spot of *Populus nigra* and *P.* x *euramericana* in many European countries (Lanier *et al.* 1976). In contrast, ***M. populorum*** G.E. Thompson (anamorph ***S. musiva*** Peck), which occurs only in N. America, causes severe cankers on hybrid poplars, in addition to leaf spotting. It is the principal reason for prohibiting the import of poplar planting material into Europe (EPPO 17). Other *Septoria* species are covered under '*Anamorphs of Dothideales*' below. The simultaneous association of *Septoria* with *Mycosphaerella* and *Leptosphaeria* is a point which requires elucidation.

Mycosphaerella mori (Fuckel) Wolf (anamorph ***Phloeospora maculans*** (Bérenger) Allescher) is a very common leaf-spotting pathogen on mulberry. By making the leaves inedible for silkworms, it used to be of considerable importance in the Mediterranean region (Viennot-Bourgin 1949). **M. ulmi** Kleb. (anamorph ***Phloeospora ulmi*** (Fr.) Wallr.) is the cause of a common but unimportant leaf spot of elm.

Mycosphaerella brassicicola (Duby) Johanson ex Oudem.

M. brassicicola produces globular, dark brown pseudothecia with apical papillate ostioles. The ascospores are hyaline, cylindrical and bicellular with rounded ends and no constriction. An anamorph stage is produced (***Asteromella brassicae*** (Chev.) Boerema & v. Kesteren), but its status in the life cycle is unclear (CMI 468). Most *Brassica* crops have been recorded as hosts. It is found in all the cool, moist regions of the world (CMI map 129). Symptoms of ringspot caused by *M. brassicicola* develop on all aerial plant organs from the cotyledons to the seed pods. Mature foliage is usually most heavily affected and defoliation is a frequent result. Lesions commence as small dark spots, visible on both leaf surfaces expanding up to 20–30 mm in diameter. Concentric zones of fungal growth are conspicuous, hence the name 'ringspot', with grey-yellow and brown tints. Each lesion usually has a definite edge bounded by a narrow water-soaked area and chlorotic zone. Where lesions are numerous the entire leaf becomes yellowish, with curled, cracked and ragged edges. The ring-like appearance of lesions is enhanced by a zonate development of pseudothecia. Stored cabbage infected with *M. brassicicola* may develop dark-brown to black lesions.

Transmission is largely the result of airborne ascospores, although seed-borne infection has been established. Invasion takes place through the stomata resulting in an inter- and intracellular mycelium. Cultural control is of major significance, and seed beds must be placed well away from existing crops and be free from infected debris. Autumn application of potassium fertilizers may help inhibit winter-time infection. Spring application of nitrogenous compounds will stimulate growth, replacing that destroyed by *M. brassicicola* in winter. Seed may be treated successfully with hot water at 45°C for 20 min or soaked in thiram to combat infection. Control may be achieved by sprays of maneb, mancozeb or benomyl which incorporate large quantities of wetter (Wilson 1971). Resistance to *M. brassicicola* has been reported in Roscoff cauliflower cultivars and some Brussels sprout lines.

DIXON

Cercospora beticola Sacc.

Basic description

C. beticola is characterized by its hyaline, pluriseptate, club-shaped conidia, formed from blackish-brown conidiophores. No teleomorph is known (CMI 721).

Host plants

C. beticola exclusively attacks *Beta*; some cultivars of *B. vulgaris* are more susceptible than others.

Geographical distribution

The fungus originates in C. Europe and the Mediterranean area, preferring warm wet conditions. It spread to France in 1887 and now to all sugar beet-producing areas world-wide (CMI map 96).

Disease

Primary infection courts appear as small reddish dots on marginal leaves and develop into round, depressed spots surrounded by a red-brown halo (Plate 21). At adequate RH, the spots become covered with an irregular grey coat of conidia from nearly black conidiophores (cf. *Ramularia beticola* whose chains of white spores are supported by conidiophores in white cushions). The invaded leaves may dehydrate, thus causing major decay of the beet. Seed crops may also be invaded from the base of the plant up to the bracts and seeds. In this case, seed germination can be partially impaired and there is a risk of seed transmission.

Epidemiology

Conidia develop rapidly (within 24 h) on leaf spots (Lebrun 1973). They are usually dispersed by rain splash over a relatively small area. They germinate at 14–35°C over 90% RH. During the 'infection period', which lasts for any length of time from 48 h (optimum at 27°C) to 3–5 days, depending on RH, germ-tubes penetrate the stomata and enter the sub-stomatal cavities. This period, during which the fungus is exposed to fungicide treatment, terminates with penetration into leaf parenchyma, within which the fungus develops for 12–21 days (incubation period) before becoming apparent (first secondary leaf spots). Inoculum is conserved and dispersed by: i) pseudo-stromata forming on seeds and surviving 20–32 months; ii) spread from seed crops to nearby ordinary crops, which is a major means of carry-over from one year to the next; iii) sclerotia-bearing debris in the soil, which, like seeds, can carry the disease for over 2 years.

Economic importance

In France, trials have occasionally shown losses as high as 44% in root weight and 2–3 degrees in dry matter. In countries such as Italy, crops may even be totally lost.

Control

Control measures are a direct result of biological studies—maximum elimination of debris, wider spacing of seed crops, seed treatment (usually carried out simultaneously against *Peronospora farinosa, Pleospora bjoerlingii* and *Pythium* species). The number of fungicide applications to the crop can be kept down by 'agricultural warning' methods (Darpoux & Lebrun 1964; Darpoux *et al.* 1966): i) recording the appearance of primary infection courts which are not dangerous in themselves; ii) noting temperatures over 17°C (optimum 27°C) combined with 'infection' rainfall (2–3 mm), which allows the small spots of no economic importance to sporulate, and the spores to be detached, carried over to other leaves or plants, and to germinate and penetrate into the tissues. Knowing, e.g. in France, that 3–5 days elapse between germination and actual penetration, treatment (copper, maneb, captafol etc.) should be carried out within this period. A fairly persistent chemical might protect the plant for approximately 15 days. Thiabendazole has a systemic action (Solel 1970), which protects the plant until the first spots appear. If a further treatment is applied immediately, sporulation will be inhibited, while an application within the next 15 days will kill any secondarily infecting fungus. When associated with the more classical fungicides, the systemic compounds exert a preventive action, combined with some curative action, and can protect crops invaded by the fungus for more than 5 weeks. However, these treatments should be used cautiously—mixed or used alternately with compounds such as maneb, mancozeb, captafol, to avoid any adaptation of parasitic strains, a phenomenon which has become widespread all over Europe. Tin salts associated with other molecules are also recommended but they display poor systemic effectiveness. Current methods for the control of *C. beticola* seem effective under normal infection conditions but may not succeed if irrigation is excessive, preventing use of the normal warning systems.
LEBRUN

Cercospora nicotianae Ell. & Ev.

C. nicotianae appears identical in morphology with *C. beticola* and *C. apii.* For Sobers (1968), *C. nicotianae* is a form of *C. apii* specialized to the genus *Nicotiana,* on which it produces the leaf spot disease known as frog-eye or barn-spot. Spots are irregular; many are paper-white with dark dots in the centre that denote the masses of fructification of the fungus. They are usually not more than 6 mm in diameter, and may be confused, under certain conditions, with brown spot (*Alternaria longipes*), wildfire (*Pseudomonas syringae* pv. *tabaci*) or any of several physiological leaf spots.

Frog-eye is dependent on the climate. The disease spreads rapidly during warm weather, particularly with intermittent rain. The fungus may overwinter on infected tobacco stumps or other host plants. In spring, conidia are formed and can contaminate young plants in seed beds. Infected transplants may carry the fungus from plant bed to the field. *C. nicotianae* also causes black spots on tobacco in the barn, after harvest. The disease probably causes most damage to flue-cured tobacco crops and cigar wrappers in C. Africa.

In Europe, the disease has caused losses in Italy,

Poland, FRG and Yugoslavia, but is decreasing in importance. It also occurs in the USA and Canada. In those tobacco-growing areas where frog-eye is a problem, suitable crop rotation and fertilization is recommended. Over-fertilization with nitrogen must be avoided, and harvest of tobacco leaves must be made at optimum maturity. Fungicides like benomyl can be applied in the field.
DELON & SCHILTZ

Mycosphaerella bolleana Higg. (anamorph ***Cercospora bolleana*** (Thüm.) Speg.) causes a leaf spot of fig. ***M. cruenta*** (Sacc.) Lan. (anamorph ***C. cruenta*** Sacc.) occurs mainly on *Vigna* and *Phaseolus* species in the tropics (CMI 463) but on soybean in the USSR. ***M. cerasella*** Aderhold (anamorph ***C. cerasella*** Sacc., syn. *C. circumscissa* Sacc.) and ***M. pruni-persicae*** Deighton (anamorph ***Miuraea persica*** (Sacc.) Hara, syn. *C. persica* Sacc.) are minor pathogens of cherry and peach respectively, with a limited distribution in Europe (CMI map 442). ***M. gossypina*** (Atk.) Earle (anamorph ***C. gossypina*** Cooke) causes a very minor late-season reddish spotting of cotton leaves wherever the crop is grown (APS Compendium). ***M. rosicola*** Davis (anamorph ***C. rosicola*** Pass.) on *Rosa* species (APS Compendium) may cause premature defoliation.

Although numerous *Cercospora* species attack crop plants in the tropics, few of them are important or even occur in Europe. ***C. kikuchii*** (Matsushima & Tomoyasu) Gardner (causing purple seed stain and leaf blight) and ***C. sojina*** Hara (causing frog-eye leaf spot) are important pathogens of soybean in warmer countries, liable to appear in Europe since they are seed-borne (APS Compendium). Only *C. kikuchii* is likely to persist in European conditions. ***C. sorghi*** Ell. & Ev. (CMI 419) is widespread in the tropics causing grey leaf spot of sorghum but in Europe occurs only in Italy (CMI map 338). Several species occur on legumes: ***C. zonata*** Winter is quite widespread causing a minor leaf spot of Vicia beans, while ***C. medicaginis*** Ell. & Ev. and ***C. zebrina*** Pass. occur on lucerne and clover respectively, and are indistinguishable morphologically (APS Compendium on Alfalfa Diseases). ***C. vexans*** Massal. causes a minor leaf spot of strawberry (APS Compendium). Of several species on vegetables, ***C. apii*** Fres. on celery (cf. also *C. nicotianae* above) and ***C. carotae*** (Pass.) Kaznovski & Siemaszko on carrot occur in Europe but are more important elsewhere (Dixon 1984). ***C. asparagi*** Pass. on asparagus is more serious in hot climates, while ***C. capsici*** Heald & Wolf (CMI 723) on pepper has been found in Italy and could be a new risk to crops under glass (Messiaen & Lafon 1970). Finally, ***Cercoseptoria pini-densiflorae*** (Hori & Nambu) Deighton (syn. *Cercospora pini-densiflorae* Hori & Nambu) causes a needle blight of many *Pinus* species in Africa and the Far East, which is serious, especially on *P. radiata* in Japan. It is considered to present a quarantine danger for Europe (CMI 329; EPPO 7).

Mycovelosiella concors (Caspary) Deighton (syn. *Cercospora concors* (Caspary) Sacc.) causes leaf blotch of potato in the cooler parts of Europe. Yellowish leaf spots, with grey conidial fructifications on the lower surface, may fall out, leaving only holes. The disease is only of minor importance (CMI 724; APS Compendium).

Pseudocercospora vitis (Lév.) Speg. (syn. *Cercospora vitis* (Lév.) Sacc.) is said by Viennot-Bourgin (1949) to be widespread but unimportant on grapevine leaves throughout the world. Marziano *et al.* (1978) record it seriously attacking one cultivar in S. Italy.

Corynespora cassiicola (Berk. & Curtis) Weir (syn. *Cercospora melonis* Cooke) attracted attention many years ago in Europe, causing target spot of glasshouse cucumbers. Although the disease may still have some importance on this crop (Dixon 1984), the problem was largely solved by the extremely durable resistance of cv. Butcher's Disease Resister. *C. cassiicola* is in fact a very widespread polyphagous organism in warmer countries (CMI 303) on numerous crops (e.g. aubergine, soybean, sesame, tomato). It persists on crop debris and spreads especially under warm moist conditions. Since forms to some extent specialized to soybean are reported (CMI 313), the recent appearance of *C. cassiicola* on soybean in Hungary (Ersek 1978) may be a cause for concern.
SMITH

Mycocentrospora acerina (R. Hartig) Deighton

M. acerina is a hyphomycete with evenly thickened, broad scars which remain on the conidiophore after the pluriseptate, usually obclavate-acicular, conidia have been shed (Deighton 1971). It infects a large range of herbaceous plants including many cultivated species. Widely distributed in temperate regions of the world, it is prevalent in most parts of N. Europe. Infection is often confined to young seedlings or senescent parts of mature plants with diminished resistance. Post-harvest infections of cultivated species of Umbelliferae, notably celery and carrot, are particularly damaging and have been described in detail (Day *et al.* 1972; Davies *et al.* 1981). The pathogen penetrates the intact cuticle of the celery petiole by means of appressoria but the periderm and pericycle of mature carrot roots are highly resistant and infection occurs mainly through wounds. After a lag phase of localized infection, the fungus grows necrotrophically in tissue causing a progressive dark brown or blackish rot which contains characteristic chains of chlamydospores, a useful diagnostic feature. In the field, these chlamydospores survive in crop debris for many years, eventually germinating and colonizing roots. Young seedlings, which are particularly susceptible, are sometimes completely killed but, on older resistant roots, the fungus grows superficially and forms more chlamydospores which may later be harvested with the roots. On the surface of flooded soil, chlamydospores also germinate to produce conidia which float readily from the conidiophores and are carried in water currents, the most likely mechanism of short-distance dispersal; splash dispersal also occurs (Klewitz 1972) but is relatively inefficient.

Where prevalent, *M. acerina* limits the storage potential of celery and carrots, and there are as yet no obvious genetical differences in resistance amongst present commercial lines. Minimizing the amount of damage during harvest reduces the incidence of the disease on carrots, but with large-scale harvesting and handling considerable damage appears inevitable. Good control of the disease on carrots and celery was achieved by a post-harvest benomyl dip (Derbyshire & Crisp 1978).

LEWIS

Mycocentrospora cladosporioides (Sacc.) P. Costa ex Deighton (syn. *Cercospora cladosporioides* Sacc.) cases a minor leaf spot of olive (Graniti & Laviola 1981). ***Mycosphaerella berkeleyi*** W.A. Jenkins (anamorph ***Cercosporidium personatum*** (Berk. & Curtis) Deighton), which with ***M. arachidis*** Deighton (anamorph ***Cercospora arachidicola*** Hori) is very important on groundnuts in the tropics, occurs in some Mediterranean countries (CMI 412, map 152). ***M. carinthiaca*** Jaap causes midvein spot of red clover. ***Scolicotrichum graminis*** Fuckel (syn. *Passalora graminis* (Fuckel) Höhnel, *Cercosporidium graminis* (Fuckel) Deighton) is considered to be the anamorph of ***Mycosphaerella recutita*** Johanson by Russian authors. It causes brown stripe or brown streak of many pasture grasses (especially *Dactylis*, *Phleum* nd *Poa*). Prominent black dots, consisting of dense masses of conidiophores, lie along the leaf veins. The disease is of rather minor practical importance (O'Rourke 1976). ***Cercosporidium punctum*** (Lacroix) Deighton attacks leaves and generally all above-ground parts of fennel and parsley in France (Bougeard & Vegh 1980) and Italy (Sisto 1983).

Ramularia beticola Fautrey & Lambotte

(Syn. *Ramularia betae* Rostrup)

R. beticola forms chains of white hyaline conidia on conidiophores gathered in typical white 'cushions'. Some conidia are cylindric and biseptate (15×1.5 μm) others are ovoid/elliptic (9×5 μm). No teleomorph is known but *Ramularia* species in general are anamorphs of *Mycosphaerella.*

The fungus is widespread in France, C. and N. Europe (UK, Denmark), and N. America. It attacks only sugar and fodder beet. Some cultivars are more susceptible than others. The disease occurs in spring (June) on root crops or in autumn on seed crops in S.W. France, and often in these conditions everywhere in Europe. The necrotic lesions are larger than those of *Cercospora beticola* (2–3 mm diameter) circular to ovoid, sometimes irregular (Plate 22). They may be encircled or not by reddish-brown pale tissue. The surface of the leaf spot carries silvery white tufted fungal cushions on which conidia are borne in chains when humidity is sufficient (over 70% RH) and at an optimal temperature of 16–17°C (in fact, from 5 to 25°C). Penetration through stomata is irregular (cf. *C. beticola*) but in less than 24 h, in optimum conditions, 90% of conidia have germinated and penetrated into the tissues (in nature 2–3 days). The fungus can persist in the leaf cortex with or without forming conidia; pseudosclerotia in debris in the soil can persist for over 2 years (as in *C. beticola*). Proximity between seed-crops and root-crops is a major factor in dispersal.

The fungus causes 5–10% loss in weight and dry matter in root crops, but seed crops may have 10–20% flowers sterilized. Seed crops transmit the disease from one year to the next. Control (Byford 1976) may be achieved by sprays applied immediately after the first infection (primary spots), in June for root crops, and in September and March for seed crops, usually with a benzimidazole or thiophanate systemic fungicide mixed with a dithiocarbamate (maneb or mancozeb). A second application will be required 2–3 weeks later. It should be noted that sulphur combined with the benzimidazoles may help control *Erysiphe betae* at the same time. A third application may be profitable according to climatic

conditions. Tin salts and fosetyl-Al show good efficacy against *R. beticola,* but all these treatments should be mixed or used alternately to avoid any adaptation of strains, such as is now widespread in *C. beticola* all over Europe.
LEBRUN

Ramularia vallisumbrosae Cavara

R. vallisumbrosae attacks many *Narcissus* species and cultivars but the disease is essentially associated with the commercial cultivation of daffodils. It occurs in warm moist weather and it is generally the later flowering cultivars that are attacked. Soon after leaf emergence small, sunken, grey-green or yellowish spots or streaks develop on leaves, especially towards the tips and on the flower stalks. These lesions increase in size turning a dark yellowish brown (though not the dark brown of leaf scorch caused by *Stagonospora curtisii*) and in wet weather a white powdery mass of spores is formed, hence the English vernacular name 'white mould'. In warm wet springs the disease becomes epidemic causing early senescence and a subsequent loss of bulb and flower yield the following season. Spores are dispersed by water splash and wind but lose their viability rapidly in dry weather. When infected leaves die, masses of minute but just visible black sclerotia form. These remain dormant during summer and autumn and germinate in winter to produce spores which infect the emerging leaves.

R. vallisumbrosae is unusual in that three types of spore are formed (Gregory 1939): 1–3 septate conidia measuring 14–44×4 μm in powdery white masses on the leaves; spherical pycnidia 55–75 μm with single-celled spores 3×1 μm; 3–7-septate spores 40–95×2.5–3.5 μm formed on sclerotium-like bodies 75–150 μm. It is the spores from these sclerotia which permit the seasonal carry-over of the disease. *R. vallisumbrosae* is unusual in annual crops and no control measures are needed. In biennial or perennial crops, the carry-over of disease is caused by overwintering sclerotia in old leaves. It is important that any leaf symptoms seen in the first year are treated with dithiocarbamates or captafol, which seem promising fungicides. Four separate applications at fortnightly intervals starting just before flowering are usually sufficient. At the end of the first season remaking the ridges will help to bury infected leaf material and so lessen the amount of infection of the young shoots.
PRICE

Mycosphaerella fragariae (Tul.) Lindau (anamorph ***Ramularia grevilleana*** (Tul.) Jørstad, syn. *R. tulasnei* Sacc.) causes white spot of strawberry leaves and 'black seed' on the berries (CMI 708, map 110). Lesions have a central whitish zone of conidia surrounded by a purplish border. Primary infection occurs via ascospores or conidia from leaf debris in the soil, or via conidia from lesions in overwintered leaves. Conidial spread, by rain splash, is favoured by high rainfall. Although the disease is common, its effects on plant growth are generally not very severe, especially now that resistant cultivars are widely available. It is readily controlled, if necessary, with protectant fungicides (APS Compendium). ***R. armoraciae*** Fuckel causes a white spot of horseradish throughout Europe, the lesions often falling out to leave a 'shot-hole'. The fungus has recently been noted also on male-sterile lines of rape (Brun *et al.* 1979). ***R. alba*** (Dowson) Nannf. (CMI 869) causes white mould of *Lathyrus odoratus,* which may be seriously damaging under wet conditions (Brooks 1953). It is now distinguished from ***R. deusta*** (Fuckel) Karakulin on other *Lathyrus* spp. (CMI 868). ***R. lactea*** (Desm.) Sacc. is common on *Viola* species (Deportes *et al.* 1979). ***R. primulae*** Thüm commonly causes a light leaf spot on *Primula* pot plants under glass (Fletcher 1984). ***R. rhei*** Allescher, causing a similar type of leaf spot on rhubarb, may require fungicidal control.

Mycosphaerella tassiana (de Not.) Johanson

(Anamorph ***Cladosporium herbarum*** (Pers.) Link)

M. tassiana is one of a group of fungi, not normally regarded as plant pathogens, which can be found on plant surfaces and within plant tissues. Other commonly isolated species are ***C. cladosporioides*** (Fres.) de Vries and *Alternaria alternata* (q.v.). The pathogenicity of these fungi however has yet to be fully elucidated. Various studies have regarded them as pathogens of a variety of crops although in early work there is doubt as to the precise identity of the fungi involved (Dickinson 1981a). Available evidence suggests that the fungi may make substantial growth in the phylloplane although their ability to parasitize the host plant is generally limited. Under favourable conditions isolates of these fungi are able to penetrate green, healthy leaves but are normally confined to sub-stomatal cavities and only a few host cells are killed (O'Donnell & Dickinson 1980). Attempted direct penetration of healthy epidermal cells is generally unsuccessful and coincides with the activation of host defence mechanisms. *A. alternata* may establish itself in healthy wheat tissue by penetrating senescent tissue at the leaf apex and growing internally along leaf veins (Dickinson 1981b). The fungi are also able to exploit plant tissues that have been wounded in some way, e.g. by insect attack. Macroscopic symptoms are not normally produced following tissue inoculation with these phylloplane fungi but effects on host physiology have been recorded, e.g. increased RNA and hydrolytic enzyme levels. Where recognizable symptoms occur, they tend to be manifested as a local tissue necrosis, e.g. leaf spots. Detached leaf studies (Skidmore & Dickinson 1973) have indicated that a major effect of leaf colonization by phylloplane fungi is an acceleration of leaf senescence.

Indications of possible harmful effects on crops have come from field observations. *Cladosporium* species and *A. alternata* are major components of the sooty mould complex of fungi on cereal heads. In a wet growing season, these fungi may become sufficiently numerous to reduce plant photosynthesis by shading and thus decrease both grain and seed quality (Dickinson 1981a). Many field experiments evaluating fungicides for foliar disease control in cereals also suggest that phylloplane fungi may have effects on host plants. In the UK (Cook 1981), France and FRG (Fehrmann *et al.* 1978), levels of foliar disease were too low to explain significant yield increases in response to fungicide sprays. A possible explanation is that the fungicide may restrict the harmful effects of phylloplane fungi. Experiments specifically designed to study fungicide impact on sooty moulds showed that fungicide application late in the growth of a cereal crop can have marked effects on phylloplane fungi and may occasionally delay leaf senescence and increase yield (Dickinson & Wallace 1976). Such results however, are the subject of much controversy (Fokkema *et al.* 1979b). Thus, while the potential of filamentous phylloplane fungi to invade host tissue has been demonstrated, its significance under normal field conditions is unclear. The importance of these fungi may be slight in terms of each individual infection but collectively they may cause significant damage to the host plant. As well as the three fungi discussed above many other species occur on plant surfaces (e.g. *Epicoccum purpurascens* Ehrenb.—CMI 680), although very little is known of their phylloplane activities. Large populations of yeasts e.g. *Sporobolomyces roseus, Cryptococcus* species and *Aureobasidium pullulans* (q.v.) occur on leaf surfaces and, primarily through nutrient competition, are antagonistic to some pathogens (Fokkema *et al.* 1979a).

BALDWIN

Mycosphaerella dianthi (Burt) Jørstad (anamorph ***Cladosporium echinulatum*** (Burt) de Vries, syn. *Heterosporium echinulatum* (Berk.) Cooke) causes ring spot of carnation and *Dianthus barbatus.* The leaves, and sometimes epicalyx, bear pale lesions, with a clear reddish border, on which rings of blackish conidia eventually appear. The fungus overwinters in infected leaves. Control is achieved by sanitation, fungicide treatments (dithiocarbamates) and reduced humidity (Viennot-Bourgin 1949; Wittman & Bedlan 1980). ***M. macrospora*** (Kleb.) Jørstad (anamorph ***C. iridis*** (Fautrey & Roum.) de Vries) causes blotch of *Iris* and other Iridaceae (*Gladiolus, Freesia, Hemerocallis*) (CMI 435). Oval brown spots appear along the leaves, with blackish conidial pustules at their margins. If numerous, they may fuse to give the leaf a rusty look, especially near the tip. Conidia are dispersed under wet rainy conditions and the fungus mainly survives on debris. Infected plants show reduced vigour. The disease can be controlled by sanitation and fungicide treatments.

Mycosphaerella allii-cepae Jordan *et al.* (anamorph ***Cladosporium allii-cepae*** (Ranojevic) M.B. Ellis, syn. *Heterosporium allii-cepae* Ranojevic) (CMI 679, 842) has been known for many years to cause a leaf spot of onion, but attracted attention recently by causing severe outbreaks in Ireland (Ryan 1978) and the UK. ***C. allii*** (Ell. & G.W. Martin) P.M. Kirk & Crompton (syn. *C. allii-porri* (Sacc. & Briard) Boerema) is known from leek in the Netherlands and the UK (CMI 841). ***C. variabile*** (Cooke) de Vries causes a frequent but unimportant leaf spot of spinach. ***C. phlei*** (C.T. Gregory) de Vries is a common pathogen of *Phleum pratense* in N. Europe (France, Ireland, Norway, UK) causing pale-brown leaf-spots with purplish margins. Although it may sometimes cause significant losses, control is difficult and cultivars show little difference in resistance (O'Rourke 1976).

Fulvia fulva (Cooke) Cif.

(Syn. *Cladosporium fulvum* Cooke)

Basic description

Fulvia fulva is the only typical member of the genus *Fulvia*, characterized by the presence of conidiophores with unilateral nodose swellings which may proliferate as short lateral branches which bear olivaceous to brown conidia (CMI 487).

Host plants

F. fulva, the causal agent of leaf mould of tomato (*Lycopersicon esculentum*), attacks only this crop in Europe. Tomato is very susceptible to the fungus but other species of the genus occurring in S. America are often resistant.

Host specialization

Many races (pathotypes) of the fungus exist carrying different genes for virulence. The nomenclature of the races in Europe is different from that in N. America. In the European nomenclature the race number corresponds with the dominant gene(s) for the resistance (R) it can overcome (Hubbeling 1971). Breeders have incorporated many resistance genes into commercial tomato cultivars originating from *L. chilense, L. hirsutum, L. peruvianum* or *L. pimpinellifolium.*

Geographical distribution

F. fulva most probably originally came from S. America, the home of the tomato but now occurs world-wide (CMI map 77).

Disease

F. fulva causes pale-yellow spots on the upper surface of tomato leaves. On the lower surface under these pale-yellow spots the fungal mycelium is visible first downy and light grey in colour. Later when sporulation starts the colour changes to dark-grey or brown. Leaf penetration occurs through stomata and the fungus grows intercellularly without producing haustoria. The fungus grows biotrophically for a period of 2–4 weeks depending on the environmental conditions. Under natural conditions the fungus sporulates on the lower leaf surface. Fully expanded leaves having functional stomata are particularly attacked. Leaves with severe symptoms turn brown and may drop. Organs, like fruits, having no stomata or other natural openings are not susceptible; the fungus cannot penetrate the host directly.

Epidemiology

The fungus can survive from the time the old plants are cleared away until the planting of the next crop, on pieces of dried infected foliage, on soil or on windows or benches in greenhouses, perhaps as a saprophyte, but more probably as conidia which can survive extreme conditions (low temperatures, desiccation, etc.) for months. In glasshouses the disease affects tomato plants under conditions of high RH (> 70%) and at 5–25°C (optimum temperature 22°C). In E. and S. Europe where tomatoes are still often grown in the field the disease is mainly a problem in spring and autumn. Limiting the periods of high RH could significantly decrease disease incidence. At 65% RH conidia hardly germinate. Once the fungus has entered the leaf, colonization is not influenced by RH. Conidiophores emerge from stomata after an incubation time of 10–14 days, conidia are then disseminated by wind and can cause secondary infections. Light seems to have a retarding effect on spore germination. The environment presented by the expanded leaf, at its lower surface, appears to be ideal for infection, for here stomata are far more numerous than on the upper surface, RH is high and conidia are protected from direct light.

Economic importance

Around 1930 losses in tomato crops caused by *F. fulva* were high. At that time no resistance genes were present in the commercial cultivars. At present, in areas where tomatoes are grown intensively in glasshouses (W. Europe), resistant cultivars are generally used, but in E. and S. Europe field-grown tomatoes are often still susceptible and high losses may occur. However, as many pathotypes of the fungus exist and appear, resistance may break down and epidemics may appear in some years, even when resistant cultivars are grown. During recent years ventilation and isolation methods in glasshouses have changed to lower heating costs. This has very often led to environmental conditions with high RH, favouring the disease. In general, the fungus can be controlled and as a result actual losses in Europe due to the disease are low.

Control

The disease can be kept under control by cultural practices in glasshouses. As far as possible the temperature should not be allowed to rise above 21°C nor the RH to exceed 70%. Adequate ventilation is therefore of great importance especially in localities where high humidities are prevalent in the early summer. Watering should be done in the morning and on bright days to ensure that plants are relatively dry during the night. About 50 years ago the first resistant cultivars were introduced in N. America and found their way to W. Europe. As in N. America, resistance genes Cf1, Cf2 and Cf3 were soon matched by virulent races of *F. fulva*. In 1954, cv. Vagabond (Cf2Cf4) appeared and its resistance was overcomc in 1967 by races 124 and 234. The newly introduced gene Cf5 was overcome by race 5 in 1976 and race 2345 in 1977 in the Netherlands. However, many resistant sources are still available (Kanwar *et al.* 1980) and most of the cultivars grown in the Netherlands are resistant to the most complex race. These resistant sources are used in breeding programmes in other countries too. In glasshouses where leaf mould, despite cultural practices or use of resistant cultivars, is still a problem, fungicides should be applied as a preventive measure. Zineb, maneb, benomyl or carbendazim in the form of a dust can be blown into the air around the plants when they are still young at intervals of 10–14 days. Spraying with benomyl, carbendazim, thiophanate-methyl, tolylfluanide, maneb or zineb also gives excellent results.

Special research interest

A gene-for-gene relationship has been proposed for the interaction tomato—*F. fulva* (Day 1956). As so many genes for resistance are identified and nearly as many races matching these genes exist, this interaction is being studied intensively in research concerning resistance mechanisms in gene-for-gene systems. De Wit & Spikman (1982) found evidence for the production of specific elicitors by races *in vivo* which induce chlorosis and necrosis in resistant cultivars. *In vitro* the production of only non-specific elicitors was reported (De Wit & Kodde 1981).
DE WIT

Cladosporium cucumerinum Ell. & Arthur

(Syn. *Scolicotrichum melophthorum* Prill. & Delacr.)

C. cucumerinum is one of the few plant parasites in the mainly saprophytic genus *Cladosporium*. The conidiophores are macronematous and micronematous bearing pale greyish, olivaceous, velvety or felted conidia (CMI 348). The fungus causes scab of cucurbits, especially cucumber, melon and courgette. It is widespread in N. America and Europe (CMI map 310). *C. cucumerinum* preferentially attacks young fruits and forms sunken lesions up to 10 mm in diameter with a gummy exudate. Plants in the seedling stage are also very susceptible, giving water-soaked symptoms whilst leaves and stems die back. Older leaves and fruits are less sensitive. The pathogen survives between crops on host debris and conidia are probably dispersed by air. The optimum temperature for disease development is around 17°C. The fungus penetrates the cuticle directly and initially grows intercellularly through epidermal and palisade cell layers. Later the mycelium grows intracellularly and colonizes the whole host tissue (Paus & Raa 1973). Losses arise mainly from the fact that infected fruits cannot be traded if they show the typical scab symptoms. Presently losses in Europe due to the disease are small mainly because resistant cultivars are grown. Resistance in a number of cucumber cultivars is controlled by a single dominant gene and has been very durable. Of the fungicides the dithiocarbamates have been successful, if applications are made before fruit formation. Biochemical aspects of resistance have been described, especially the role of lignification (Hammerschmidt & Kuć 1982).
DE WIT

Leptosphaerulina trifolii (Rostrup) Petrak

In many respects *L. trifolii* resembles *Pleospora herbarum* (q.v.), but has lighter ascospores with fewer septa and produces no anamorph. It has a similarly wide host range, throughout Europe, on Cruciferae, Gramineae, Leguminosae and Solanaceae and shows a degree of host specialization on these, although isolates from different hosts are hardly morphologically distinguishable in culture (CMI 146). The forms on different hosts have been considered as separate species, but the only ones of significance in Europe are those on *Trifolium* species, especially *T. repens* (*L. trifolii* in the narrow sense) and on *Medicago* species, especially lucerne. Graham & Luttrell (1961), the APS Compendium of Alfalfa Diseases and Barrière *et al.* (1974) consider the latter to be morphologically and biologically distinct as *L. briosiana* (Pollich) Graham & Luttr. On forage legumes, *L. trifolii* causes a leaf spot known as pepper spot or burn. Numerous discrete black spots up to 3 mm across occur scattered over both sides of the leaves and on the petioles (the latter is characteristic). If lesions are few, they enlarge to eyespots with a tan centre and dark brown border. Under favourable conditions, such spots may coalesce to burn the leaves.

L. trifolii survives as pseudothecia on dead leaves, discharging large numbers of ascospores, formed

only in the light. Fallen dead leaves provide a constant source of new ascospore inoculum under moist conditions. The disease is most severe in dense pure stands. High light intensities tend to favour the eyespot type of symptom, and lower light intensities pepper spotting (especially on the lower shaded leaves). Infection reduces both yield and quality of forage legumes (Willis *et al.* 1969), reducing protein content and increasing leaf oestrogen content (O'Rourke 1976). There are no practical control measures except the search for resistant cultivars. Raynal (1978) found highly significant cultivar/isolate interactions in lucerne, i.e. no consistent resistance. Hill & Leath (1979) have been able to obtain resistant forms by several cycles of selection, and found resistance generally to be correlated with that to *P. herbarum,* but the degree of resistance of selected cultivars may not be agromically very useful (Michaud *et al.* 1984).
CARR & SMITH

Cymadothea trifolii (Pers.) Wolf

(Syn. *Mycosphaerella killiani* Petrak)
(Anamorph ***Polythrincium trifolii*** Kunze)

C. trifolii causes black or sooty blotch of *Trifolium* species, especially *T. repens, T. incarnatum* and *T. hybridum* and is common in many European countries. It is mainly seen as its conidial state — obovate once-septate pale brown conidia borne terminally on characteristic dark wavy conidiophores formed on a dark stroma. A spermogonial state appears in the autumn before pseudothecia form in the spermogonial stromata (CMI 393). Infected leaves show dark olive-green spots (stromata) about 1 mm in diameter, mainly on the lower side. Affected leaflets curl up, dry out and finally fall (O'Rourke 1976). Infection is established by ascospores from pseudothecia on fallen leaves, and spread by conidia dispersed by wind or rain. It is favoured by warm wet conditions. Herbage yields are affected by stunting and defoliation, and infected material contains increased oestrogen levels which may be poisonous to livestock. No special measures for control are noted.
CARR & SMITH

Scirrhia pini Funk & Parker

(Anamorph ***Dothistroma septospora*** (Dorogin) Morelet)

S. pini forms pycnidia in black stromata, containing filiform, 1 to 5-septate conidia measuring 25–60×2–3 μm (CMI 368). It occurs mainly on pines, but has also been found on *Larix decidua* and *Pseudotsuga menziesii*, which however are little affected. Among the many pines attacked by the fungus, *Pinus radiata* has been most affected; *P. nigra* var. *maritima, P. ponderosa* and *P. bungeana* are very susceptible, but *P. contorta* is moderately resistant. *P. sylvestris* and *P. patula* appear to be immune to attack. New Zealand populations of *P. radiata* vary in susceptibility (Wilcox 1982). *S. pini* is known throughout much of Europe, in N. and S. America, S. and E. Africa, S. India, and New Zealand (CMI map 419). The teleomorph has so far been recorded only in N. America and France.

The fungus causes a needle blight, which first shows as small yellow spots on the needles. The spots extend to become yellow and later reddish bands, so that the disease is known as 'red band' in some areas. Black stromata containing pycnidia (and sometimes pseudothecia) form on the bands, and later the whole needle may be killed, and extensive needle cast of susceptible species takes place. Dispersal of the conidia requires light rain or thick mist. Trees of *Pinus radiata* and some other pines become more resistant with age (Gibson 1972).

In Europe, *S. pini* has occasionally caused considerable damage to pines in nurseries. Elsewhere, in E. Africa, Chile and New Zealand, it has devastated young plantations of *Pinus radiata* over large areas. This needle blight may be controlled by spraying with copper fungicides, and in some areas it may be necessary to spray susceptible species 3 or 4 times each year for the first 15 years after planting.
PHILLIPS

Scirrhia acicola (Dearn.) Siggers, which is an important pine needle cast fungus in N. America, has also been found in Yugoslavia, where it has caused similar damage (but has been successfully controlled), and in Austria, where it has so far proved insignificant. It is considered a quarantine organism in Europe (EPPO 22).

PLEOSPORACEAE

Pleospora bjoerlingii Byford

(Syn. *P. betae* Björling)
(Anamorph ***Phoma betae*** Frank)

Basic description

The pseudothecia of *P. bjoerlingii* form in the cortex of its host and have a short ostiolate papilla (200–500 μm diameter). The ascospores are brown, ovoid and slightly constricted at the three transverse septa. When mature, they develop a single longitudinal septum in each of the two central cells. Pycnidia are also found in the cortex, often over seed teguments (200–300 μm diameter) with cells bearing hyaline phialides. Pycnidiospores are also hyaline, oval to subcylindrical (3–7.5 μm). The teleomorph resembles *Pleospora herbarum* (q.v.), which is generally saprophytic but able to cause black necrosis on sugar beet leaves in spring, alone or with *Pseudomonas syringae* pv. *aptata*.

Host plants and geographical distribution

Very common and often damaging on beet in Europe, N. America and Africa.

Disease

P. bjoerlingii is normally a seed-borne fungus but is sometimes transmitted by soil debris; it occurs at four stages in the growth of a sugar beet crop. i) Black leg is caused by pycnospore infection, principally post-emergence, of young seedlings. The tips, root and rootlets become brown dark or black, and the seedlings may die; only a small proportion of them survive, to transmit the fungus to the adult plant, directly by spores, or perhaps also by systemic infection. A severe seed-borne attack may often cause pre-emergence disease and death, like that due to *Pythium ultimum*. ii) Crown rot—the crown of the growing plant rots and turns dark brown or black; this may develop into a deep, spreading rot later when affected beets are clamped. The surface of the root may be slightly depressed, sometimes with deep splits. Black dots (pycnidia) can be found over the lesion, and masses of pink spores exude from them in moist conditions. iii) Pleospora leaf spot (Plate 23)—ascospore infections produce circular necrotic patches marked with light and dark brown concentric rings (pycnidia). This leaf spot occurs in autumn and does not cause damage. Lesions are often centred on a rust pustule on leaves of both root and seed crop plants. iv) Stem rot—ascospore infections also often give rise to lesions on the seed-bearing stems of sugar beet and mangold seed crops. When attacks are severe, lesions coalesce, and the stem breaks off at the rotten crown of the root and falls over. The lens-shaped lesions extend vertically and have brown margins and grey centres, spotted with numerous pycnidia. Pseudothecia form later and ascospore discharge leads to infection of the following year's root and seed crops.

Epidemiology

The fungus may overwinter as pycnidia or pseudothecia in the soil in dead leaves or survive on seed-crop plants. Mostly, however, the fungus survives as pycnidia and mycelium in or on the seed coat, which infects the young seedlings during their germination, a high RH being required and an optimum temperature of 14–18°C. Wind and rain disperse conidia onto fresh leaves, healthy seedlings, or the base of the seed-bearing stems. Systemic infection is apparently possible, leading to infection of flowers and seeds.

Economic importance

Losses arise from the destruction of young seedlings (sometimes 30%), and of seed-bearing stems through a fall in quality of harvested plants and seeds.

Control

The black-leg stage can be controlled by treating the seed with a fungicide; all sugar beet seed supplied by British or French factories is so treated. The products used include organomercurials (in some countries) or iprodione-containing mixtures (e.g. with metalaxyl to control *Pythium ultimum* and *Peronospora farinosa* also; q.v.) (Lebrun & Viard 1979). On seed crops, two or three treatments are usually applied at the beginning of growth (autumn and spring). Iprodione with sulphur or sometimes with benzimidazoles is used to control *P. bjoerlingii* together with foliar disease (*Erysiphe betae, Ramularia beticola, Cercospora beticola*), depending on climate. Maneb and mancozeb are also good protectants. It is not necessary to control leaf or crown rot which are generally unimportant.
LEBRUN

Pleospora herbarum (Pers.) Rabenh. ex Ces. & de Not.

(Anamorph ***Stemphylium botryosum*** Wallr.)

In culture on various complex natural media, *P. herbarum* tends to form the anamorph first, followed by more or less distinct pseudothecia (1-1.5 mm in diameter) several weeks later. The fungus produces a phytotoxin (stemphylotoxin) (Barash *et al.* 1983). It is a weak parasite found on numerous weakened hosts, causing leaf spots especially of *Allium* species, beet (cf. also *P. bjoerlingii*), endive, lettuce and lucerne. It also rots ripening fruit of apple, and more rarely pear. On apple, *P. herbarum* causes a dry, or

slightly soft rot, dark brown in colour and irregular in outline. Conidia are rarely found on fruit. Lesions generally start at lenticels or at the calyx end. The source of infection is saprophytic growth on dried vegetative parts (branches, leaves etc.) (Bondoux *et al.* 1969). In cool storage conditions or in a controlled atmosphere fruits are rarely rotted by *P. herbarum,* which is favoured by higher temperatures (15–20°C). Specialized forms of *S. botryosum* have been reported (e.g. **f.sp. *lactucae*** and **f.sp. *lycopersici*** Rotem *et al.*). The latter forms part of a complex causing grey leaf spot of tomato, together with ***S. solani*** Weber reported from Italy and UK, and ***S. lycopersici*** (Enyoji) Yamam. (syn. *S. floridanum* Hannon & Weber). The complex is principally important in the tropics (CMI 471, 472; CMI map 333; Kranz *et al.* 1977). *S. lycopersici* has also been reported causing a ray fleck of chrysanthemum in Hungary (Folk 1976). Finally, ***S. vesicarium*** (Wallr.) Simmons has been found causing necrotic leaf spotting of tomato in Italy (Porta-Puglia 1981). A number of *Pleospora* species described on various hosts are in all probability simply *P. herbarum*. In general, losses are insignificant and control measures are hardly needed.
BOMPEIX

Like *P. herbarum* on lucerne, ***S. sarciniforme*** (Cavara) Wiltshire causes a very widespread 'target spot' on leaves of clover, especially *Trifolium pratense*, most notably in the autumn, as it is favoured by moist conditions and dense stands. In severe attacks, the leaves become shrivelled and dark brown. It is reported to be seed-borne (O'Rourke 1976; CMI 671).

Pleospora papaveracea (de Not.) Sacc. (anamorph ***Dendryphion penicillatum*** (Corda) Fr. causes a leaf blight of *Papaver* species, particularly of oilseed poppy in E. Europe (Poland, Ukraine). Infection occurs both through infected seed or by ascospores from pseudothecia on soil debris and losses can be quite severe (up to 50%). Fungicidal seed treatments offer the best control (CMI 730). ***P. infectoria*** Fuckel (probably to be considered a form of the polyphagous ***P. scrophulariae*** (Desm.) Höhnel) has an *Alternaria* anamorph associated with a leaf spot of strawberry, requiring fungicide control in Belgium (Bal *et al.* 1983).

Alternaria radicina Meier, Drechsler & Eddy

(Syn. *Stemphylium radicinum* (Meier, Drechsler & Eddy) Neergaard)

A. radicina produces usually solitary, non-beaked conidia with 3–7 transverse and 1 to several longitudinal septa (CMI 346). It principally attacks members of the Umbelliferae. No pathotypes are known. There is a need for a comparative study with ***A. petroselini*** (Neergaard) Simmons, a species described on parsley. *A. radicina* has been reported in temperate regions throughout the world, causing pre- and post-emergence damping-off, leaf blight, crown and root rot and storage rot (Gindrat 1979). Damping-off and leaf blight resemble those caused by other pathogens (e.g. *A. dauci* on carrot). Root rot is more characteristic on carrot ('black rot') in the field and during storage. Root rot of celeriac caused by *A. radicina* (Fig. 52) is often misdiagnosed as *Phoma apiicola* root rot (q.v.), because of symptom similarity. *A. radicina* attacks the crown of celeriac, including petioles, and the tap root, while secondary roots remain apparently intact. In the field, diseased plants are recognized by their chlorotic and stunted appearance.

Fig. 52. Root rot of celeriac due to *Alternaria radicina*.

A. radicina is seed- and soil-borne. It has a longevity of more than 8 years in soil. Root and crown rot may become locally important on carrot, celery, celeriac, and parsley. Storage rot of carrots, as well as root rot of celeriac grown for industrial processing may cause severe yield losses. Various seed treatments (e.g. 0.2% thiram at 30°C for 24 h) prevent

disease development from seed inoculum and subsequent contamination of soil (Maude 1966b). Rotations with non-umbelliferous crops are another preventive control measure. Particular care must be taken in seed production, because of high susceptibility of seed crops. Fungicide sprays on carrot crops with iprodione or chlorothalonil control leaf blight and diminish subsequent storage rot. The latter is also prevented by maintaining temperature at 0°C. Certain celeriac cultivars are somewhat tolerant to *A. radicina* root rot under field conditions.
GINDRAT

Alternaria citri Ell. & Pierce

A. citri occurs in all citrus-growing areas causing black rot. Before harvest, the disease appears as a black external rot or as a stylar end rot. When the disease develops after harvest it produces a brown stem end rot or an internal core rot (brown discoloration of the central axis of the fruit) (CMI 242). In the orchard, *A. citri* grows saprophytically on dead plant tissues and spores are transported by air currents onto the fruit. The fungus enters through wounds, through the stylar end (navel oranges) or through the stem end. On the stem end the spores remain quiescent until after harvest. Senescence of the button and underlying tissue favours the infection process. In general *A. citri* infects physiologically weakened fruit and has a very long incubation period. Thus the disease is found mainly on over-ripe fruits and fruits that have been stored for a long period (Eckert 1978; Schiffman-Nadel *et al.* 1983).

The growth regulator 2,4-D in combination with imazalil effectively controls stem end rot and internal core rot. The chemicals can be added in the bath or incorporated in the wax. Other approved fungicides for citrus fruit are not effective.
CHITZANIDIS

Alternaria brassicae (Berk.) Sacc. & *A. brassicicola* (Schwein.) Wiltshire

Basic description

A. brassicae is distinguished by its characteristic, very long (75–350 μm) pale olive conidium with a pronounced beak up to half its length. Conidia are usually solitary but may occur in chains of up to four. *A. brassicicola* has smaller (18–130 μm), darker conidia without a pronounced beak in chains of up to 20 or more (CMI 162, 163).

Host plants

Both fungi attack a wide range of cultivated and wild crucifers. Most, if not all commercially grown cultivars are susceptible including vegetable brassicas, forage brassicas and mustard. In Europe, the most serious damage has been caused on rape in recent years, with increased cultivation of this crop.

Host specialization

Though specialization to *B. campestris* and *Eruca sativa* has recently been shown in India, there is no evidence of host specialization in Europe.

Geographical distribution

A. brassicae is widespread through Europe and has been reported from almost everywhere cruciferous crops are grown (CMI map 353). *A. brassicicola* has been reported from every continent but it is not as widespread (CMI map 457).

Disease

Both species cause dark leaf spot (black spot, blight, pod spot, silique mould) of crucifers. The symptoms produced by the two species are similar and often indistinguishable. Both fungi are seed-borne and occur superficially in seed-lots as spores and mycelium and also as internal mycelium within the testa and occasionally in the embryo tissues. In transplant beds where seedlings are grown close together in conditions of high soil moisture and high humidity, seed infection by either fungus may cause severe damping-off but this is not usually a problem in direct-drilled field crops. On infected seedlings the fungi produce dark brown necrotic areas on the cotyledons and similar coloured streaks on the hypocotyl. On older plants all above-ground parts are attacked including leaves, stems, Brussel sprout buttons, cauliflower curds and the inflorescences of seeding plants. On leaves, symptoms first appear as small dark brown, almost black spots, each surrounded by a halo of chlorotic tissue. Older lesions are circular, often zonate with a papery, thin centre and may be covered with a mat of spores which are yellow in the case of *A. brassicae* and dark olive-brown in the case of *A. brassicicola.* The centres of the lesions may fall out to give a shot-hole effect. Leaf infection does not usually affect plant productivity although severe defoliation by *A. brassicae* has been reported in some cultivars of stubble turnip.

On cauliflower curds, spots grow rapidly to produce an extensive brown rot. On seeding plants dark necrotic lesions occur on the main axis, the inflorescence branches and on the pods. These lesions coalesce rapidly and cause premature ripening and splitting of the pods and may result in high levels of seed infection. Harvested seed is small, shrunken and has low viability.

Epidemiology

Most biological studies on *A. brassicae* have been made on rape in which infected seed is a primary source of the fungus. In England up to 19% internal seed infection may occur in commercial rapeseed and in India 50% seed infection is not uncommon in this crop and in mustard. In *B. oleracea* seed, *A. brassicae* is less frequent, most infection in this species being caused by *A. brassicicola* (Maude & Humpherson-Jones 1980). Both fungi survive in infected seed for many years in cool, dry conditions. Once established in the crop, they spread by means of airborne spores. These are produced abundantly in wet weather and are dispersed locally by rain splash and over considerable distances by wind. Spore release, which is passive, is stimulated by falling humidity and restricted by high humidity so that air spore concentrations in a diseased crop exhibit a distinct diurnal periodicity with maximum concentrations occurring in the early afternoon and minimum concentrations in the early morning. Free water is needed for spore germination and infection occurs most frequently through stomata. The optimum temperatures for spore germination, mycelial growth and infection of rape by *A. brassicae* are 17–24°C, when severe infection may occur within 6 h; however at lower temperatures infection is much restricted so that at 12°C only low levels of disease occur within 15 h (Louvet & Billotte 1964). For this reason, in Europe, dewfall is probably not an important factor in disease development in rape and protracted rainy weather is required for epidemics to occur.

A. brassicicola has a higher temperature requirement for spore germination (33–35°C), mycelial growth (25–27°C) and infection (28–31°C) than *A. brassicae* which may explain its more restricted occurrence. *A. brassicae* survives saprophytically on plant debris so that where crops are grown sequentially, overwintered diseased detritus may serve as an inoculum source for new crops. In Sweden in areas where only summer crops of rape are grown the disease is less severe than in areas where summer and winter crops form a continuous green bridge. On infected detritus microsclerotia and chlamydospores may play a part in the perennation of the pathogen (Tsunedo & Skoropad 1977). *A. brassicae* occurs commonly on a range of cruciferous weeds and the strains of the fungus present on these are pathogenic on cultivated crucifers so it is possible that wild hosts may have epidemiological significance.

Economic importance

In rape and other cruciferous seed crops, *A. brassicae* causes yield loss due to premature ripening and shedding of seed before harvest and by reducing the 1000 grain weight. In Europe epidemics occur on rape about two years in every five and in these years losses may be as high as 60% in individual crops. Weather conditions are critical during May–July when pods are developing and ripening and severe attacks may be expected if prolonged warm wet weather occurs at this time. In England the disease first caused serious losses in rape in 1980/81 and since then an estimated 3–4 million pounds has been spent annually on fungicidal control.

On vegetable brassicas such as Brussels sprouts, cauliflower and cabbage, dark leaf spot on the buttons, curds and outer leaves downgrades the produce, or in extreme cases results in complete rejection. Both species are involved but *A. brassicicola* only occasionally. *A. brassicicola* is most important on *B. oleracea* seed crops. Reductions in seed yield may be as high as 80%, and the fungus may severely depress germination to the extent that infected seed may be unsaleable.

Control

Sources of partial resistance to *A. brassicae* have been identified in rape (Tewari & Skoropad 1976) and to *A. brassicicola* in Brussels sprouts and cauliflower (Braverman 1977), but all currently available cultivars appear to be susceptible. The leaves of cultivars with a thick epicuticular wax layer retain fewer water droplets so that the number of potential infection sites for *A. brassicae* is reduced. Although *A. brassicae* has been a serious problem in continental Europe for many years fungicide treatments to control seed-borne and foliar infection have only recently been used extensively, mainly due to the absence of effective chemicals. In the past, soak treatments in hot water or fungicide suspensions have been used to eliminate seed-borne *Alternaria*. However, problems in drying the seed after treatment and also impaired germination in some seed stocks have restricted these treatments to small quantities of high-value vegetable brassica seed. Recently iprodione was identified as being highly toxic to both species (Maude & Humpherson-Jones 1980) and is now used extensively in England on rape and vegetable brassicas both as a dry or slurry seed treatment and as a foliar spray. On rape a first spray is applied towards the end of flowering and a second about 3 weeks later if the disease continues to increase. Fungicide application at these times is based on a need to protect the vulnerable inflorescence branches and developing pods. Good cultural practices such as rapid disposal of diseased debris soon after harvest, effective control of cruciferous weeds and isolation of successive crops help in reducing the danger of cross-infection.

Special research interest

Only limited epidemiological data are available to enable rational disease control measures to be used. Extensive studies are being made on the biology of the fungus in rape and vegetable brassicas and the search continues for sources of novel disease resistance and for alternative fungicides to supplement the limited number available. Recently fenpropimorph has been shown to be highly active against *Alternaria* and *Leptosphaeria maculans* (canker) on brassica seed (Maude 1983). Because *A. brassicicola* retains is pathogenicity, growth characteristics and capacity for sporulation in culture for many years it is a useful organism for screening fungicides for activity against dark-spored fungi.
HUMPHERSON-JONES

Alternaria solani Sorauer

(Syn. *Macrosporium solani* Ell. & G. Martin)

A. solani produces muriform conidia, pale to olivaceous brown in colour, usually borne singly, rarely in chains. Conidia are characterized by a distinct beak equal in length to or longer than the spore body (CMI 475). It is primarily a pathogen of the Solanaceae but also recorded on *Brassica* species. Rotem (1966) examined single spore isolates from potato, tomato and aubergine and concluded that there was no justification for division of the species into races or formae speciales. Distribution is worldwide (CMI map 89); in Europe the pathogen is most troublesome in southern areas with high summer temperatures.

A. solani can attack seedlings, mature plants, flowers and fruits of tomato. Young seedlings (< 3 weeks) develop dark areas on the stem at soil level. These extend upwards, girdling the stem and killing the plant (collar rot). Lesions are found on both leaves and stems of mature plants (early blight or target spot). Leaf lesions are small (< 3–6 mm), circular or angular with pronounced, ridged concentric rings. On older leaves, lesions (which are frequently surrounded by a chlorotic halo) may coalesce and kill the leaf. Leaves that fail to fall droop, giving the basal part of the plant a wilted appearance which extends upwards as the disease progresses. Stem lesions are elongate and frequently develop in petiole or shoot axils; they may girdle the stem. The fungus cannot penetrate the undamaged skin of tomato fruit but may produce lesions around the calyx scar. These lesions, dark brown in colour and slightly sunken, resemble a thumbprint; the lesions may extend and the fruit rot. On potatoes, discrete target spot lesions formed on leaves (early blight) may coalesce, leading to blight symptoms similar to those caused by *Phytophthora infestans*. On potatoes *A. solani* epidemics usually precede *P. infestans* outbreaks. Potato tubers may also be extensively rotted. Sources of infection include older diseased crops, alternative hosts including solanaceous weeds, e.g. *Solanum nigrum*, infected seed, and diseased crop debris where the fungus can survive for at least a year as conidia, mycelium or possibly chlamydospores (Basu 1971). Transmission is by wind- or splash-dispersed conidia. Conidial release is favoured by dry conditions; peak air spore concentrations occur at times of maximum windspeed and minimum RH. Conidial germination and penetration are favoured by lower temperatures (optimum 22°C) than mycelial growth (optimum 28°C). Mycelium and conidia are highly resistant to drought and high temperatures. Moisture is essential for sporulation.

Field resistance to *A. solani* is available in tomato and potato cultivars but adequate control is dependent upon the use of disease-free seed, thorough crop hygiene, rotation and regular fungicide applications. Seed can be dressed with fungicides or given hot water treatments (50°C for 30 min). Soil fumigation reduces the levels of trash-borne inoculum (McCarter *et al.* 1976). Sprays of chlorothalonil, captafol, dithiocarbamates and copper-based formulations have all been used successfully; iprodione has shown promise, while benzimidazoles are not generally effective. A feature of early blight is the crucial role played by phytotoxins in the disease. One non host-specific toxin produced by *A. solani*, alternaric

acid, was at one time considered to have potential as an antifungal antibiotic. Recent work using plant tissue cultures has suggested that *in vitro* the fungus produces two host-specific toxins; both of these lipid-like compounds must be present for symptoms to develop (Materu *et al.* 1978). The reaction of plant tissue cultures to *A. solani* toxins mirrors the response of mature plants to the pathogen. This has enabled crop cultivars to be assessed rapidly for resistance in the laboratory (Shepard *et al.* 1980).
JEVES

Alternaria longipes (Ell. & Ev.) Mason

(Syn. *Alternaria tabacina* (Ell. & Ev.) Hori)

A. longipes is a ubiquitous fungus of variable pathogenicity, commonly isolated from different plants, seeds or substrates. It is a wound parasite of numerous families and species, including tomato and pepper. Many morphologically distinct strains are pathogenic to tobacco (Simmons 1981), on which it causes 'brown spot'. The disease is important in countries with a warm climate (C. Africa, S. America and USA). In Europe, *A. longipes* is reported from Czechoslovakia, FRG, GDR, Hungary, Italy, Poland and Yugoslavia, and causes relatively minor losses.

The disease starts as water-soaked circular spots formed on the mature lower leaves. These spots are brown or red brown with concentric rings surrounded by a halo, and enlarge to 1–3 cm. Brown spot may be confused with frog-eye (*Cercospora nicotianae*) or wildfire (*Pseudomonas syringae* pv. *tabaci*). Under warm and wet periods favourable to the disease, the lesions become so numerous that they coalesce and render the entire leaf worthless and ragged. Generally, brown spot occurs late in the season, only on the mature leaves. The fungus overwinters as mycelium in plant debris. Conidia are spread by wind or splashed from the soil, and weakened or overmature leaves are generally the first to be infected. Generally, temperature is not a limiting factor; wet weather and humidity appear to be most important, while potassium deficiency is a predisposing factor. Because the disease is most important at harvest, the use of fungicides poses residue problems. Suitable crop rotation, fertilization and growing tolerant cultivars is generally recommended. Fravel & Spurr (1977) proved that biocontrol of brown spot seems possible with the application of non-pathogenic bacteria (*Bacillus cereus* subsp. *mycoides*) on tobacco leaves to reduce the severity of the disease.
DELON & SCHILTZ

Alternaria alternata (Fr.) Keissler (syn. *A. tenuis* Nees) and ***A. tenuissima*** (Kunze) Wiltshire are widespread saprophytes on plant material, which may occasionally cause leaf spots (e.g. on pelargonium and rose) or attack flowers, fruits or seeds. In fact, many such infections have been loosely associated with *A. tenuis* and may really be due to more specific pathogens (Neergaard 1977). This and the other species are probably to be associated with *Pleospora*. Production of phytotoxic metabolites is frequently reported among these fungi. In particular, ***A. alternata* f.sp. *lycopersici*** Grogan *et al.* causes a stem canker, with epinasty and angular necrosis of the leaves, in tomato cultivars sensitive to its specific toxin (Gilchrist & Grogan 1976). *A. alternata* has been recorded on tomato in Greece and elsewhere in Europe. *A. alternata* is also a member of the sooty mould complex on cereals, considered under *Mycosphaerella tassiana* (q.v.).

Many *Alternaria* species are characteristically seed-borne, and controlled by fungicidal seed treatment (Neergaard 1977). ***A. dauci*** (Kühn) Groves & Skolko causes a seedling blight of carrot and also spreads secondarily in the crop to cause a fairly severe leaf blight, especially on older leaves (Dixon 1984). Its epidemiology has been reported by Strandberg (1977). ***A. linicola*** Groves & Skolko mainly attacks oil-flax cultivars causing damping-off and leaf spot. It is seed-borne, then spreads from plant to plant during early crop growth. Damage is rarely seen later, although the fruits may be infected, according to climatic conditions, and the harvested seed is more or less affected. Control is essentially by seed treatment, which is rather ineffective, dependent upon the intensity of attack and the depth of penetration. ***A. raphani*** Groves & Skolko causes a moderately important leaf blight of radish and other crucifers. ***A. sesami*** (Kawamura) Mohanty & Behera (CMI 250) causes a serious leaf blight of sesame in S.E. Europe (CMI map 410). ***A. zinniae*** Pape is widespread, causing leaf spot and flower blight on numerous ornamental Compositae.

Other species may occasionally be carried by seed, but more characteristically survive on plant debris in the soil. They tend to be more important in the tropics and subtropics. ***A. carthami*** Chowdhury is reported on safflower in Italy, Romania and USSR (CMI 241). ***A. cinerariae*** Hori & Enyoji causes a leaf spot of florist's cineraria (*Senecio cruentus*). ***A. cucumerina*** (Ell. & Ev.) Elliott causes a leaf spot of cucurbits, especially melon and watermelon, in France and Italy, and also damages fruits (CMI 244; Viennot-Bourgin 1949). ***A. pluriseptata*** attacks cucumbers in the field in Czechoslovakia (Hervert *et al.* 1980). ***A. dianthi*** F. Stev. & Hall causes a leaf and stem blight of glasshouse carnations which is widespread in Europe (France, Greece, Italy), with characteristic purple-bordered blotches. Its control has been reviewed by Strider (1978); fungicide treatment is widely needed in Italy. ***A. porri*** (Ell.) Cif. causes purple blotch of onion and other *Allium* species (CMI 248). It is present in Europe (CMI map 350) but only of real significance in warmer countries. Much the same applies to ***A. helianthi*** (Hansf.) Tubaki & Nishihara, which affects leaves, stems and flowerheads of sunflower (CMI 582), reported to be important in Romania and Yugoslavia. Its biology has been described by Allen *et al.* (1983). ***A. macrospora*** Zimmerm. causes a widespread leaf spot of cotton (CMI 246; APS Compendium of Cotton Diseases), mainly important in warmer countries. ***A. vitis*** Cavara requires fungicide applications on grapevine in Romania (Bechet & Sapta-Forda 1981) but is not reported from W. Europe.

Leptosphaeria nodorum E. Müller

(Anamorph ***Septoria nodorum*** (Berk.) Berk., syn. *Stagonospora nodorum* (Berk.) Castell. & Germano)

Basic description

Pseudothecia of *L. nodorum* are similar to those of other *Leptosphaeria* species attacking wheat, and are best distinguished on conidial morphology and pathogenicity of the isolates (CMI 86). Ascospores are fusoid, subhyaline to pale brown, 19.5–22×4 μm. 3-septate, constricted at the septa with a swollen penultimate cell. Pycnospores, which are hyaline, straight or slightly curved, measure 13–28×2.8–4.6 μm, (av. 19×3.6 μm) (Bissett 1983). They are shorter than those of both *L. avenaria* and *S. passerinii* (q.v.).

Host plants

Widespread on wheat, but also on triticale, rye and barley, *L. nodorum* has also been recorded on a number of wild grasses.

Host specialization

Isolates adapted to either wheat or barley occur (Smedegaard-Petersen 1974). There are few reports of adaptation to particular wheat cultivars (Rufty *et al.* 1981).

Geographical distribution

L. nodorum has a world-wide distribution in all wheat-growing areas (CMI map 283).

Disease

Symptoms of septoria nodorum blotch first appear on seedlings as water-soaked dark green areas rapidly enlarging to become yellowish to tan-brown with a slightly darker border, often with a darker centre. The lesions are not as restricted by the leaf veins as those of septoria tritici blotch (*Mycosphaerella graminicola*) and become oval to lens-shaped. As the disease progresses the lesions enlarge and merge, producing necrotic light-grey to buff areas. Pycnidia form within the diseased tissue; they are light brown, and usually less conspicuous than those of *M. graminicola*, requiring a hand lens to be seen. Following periods of rain, salmon-pink spore masses (cirrhi) exude from the pycnidia. The disease becomes more aggressive as the crop matures, attacking the leaf, leaf sheath, culm and nodes. Infection of susceptible cultivars can lead to a weakening of the stems and lodging. On the glumes the disease (glume blotch) often starts at the tip as a greyish area with a brownish lower border; this develops downwards and the brown border is replaced by a greyish discoloration. Pseudothecia can develop within lesions on the flag leaf late in the growing season. After harvest the fungus becomes saprophytic on the dead plant tissue and pseudothecia can be found scattered throughout the plant.

Epidemiology

In first-year wheat crops, in the absence of any crop stubble or crop debris, the primary inoculum is usually wind-borne ascospores, which under ideal weather conditions will travel long distances. In the USA, ascospores of *L. nodorum* are produced over a wider range of climatic conditions and time span than those of *M. graminicola*, initiating infection in spring-sown crops as well as autumn-sown crops. In fields with a previous history of wheat and in the presence of wheat stubble or crop debris, the primary inoculum can be either pycnospores or ascospores. Once infection is established within the crop, plant-to-plant spread is by rain-splashed pycnospores, with infection becoming more widespread as the season progresses. Infection on the heads can lead to seed-borne infection. Seed-borne infection is not considered to be of great practical importance but it must be considered a factor in long-range dispersal or where other forms of primary inoculum are not available.

Economic importance

The disease is more prevalent late in the season; thus the effect on yield is primarily on kernel weight, though severe infection at flowering would also reduce kernel numbers. Because *L. nodorum* only requires a 3 h period of high humidity to initiate infection (Holmes & Colhoun 1974) it produces less erratic infection than *M. graminicola* which requires a minimum period of 20 h. It is therefore probably more widespread and hence more important in causing crop losses. Yield losses as high as 50% in the UK (Jenkins & Morgan 1969) and 65% for Switzerland (Brönnimann 1968) have been reported.

Control

Commercial cultivars are available with moderate levels of resistance. Fungicides, or fungicide mixtures are also available for use on susceptible cultivars. The fungicide should be applied to prevent the disease from becoming established on the flag leaf. If the disease is well established on the leaves directly below the flag leaf, and frequent periods of moisture due to rain, mist or dew have occurred just prior to flag leaf emergence, then economic returns can be expected from the application of a fungicide.
SANDERSON & SCOTT

Leptosphaeria avenaria Weber

(Anamorph ***Septoria avenae*** Frank, syn. *Stagonospora avenae* (Frank) Bissett)

It is difficult to distinguish *L. avenaria* from other *Leptosphaeria* species on gramineous hosts other than on conidial morphology and pathogenicity of isolates (CMI 312). Pycnospores are hyaline, straight or slightly curved 17–46×2.6–4.4 μm (av. 33×3.5 μm) and longer than in *L. nodorum* (Bissett 1983). Because of the similarity in conidial morphology between these two organisms it is very likely that many records of *L. nodorum* are in reality *L. avenaria* as suggested by Kruger & Hoffmann (1978) in the FRG.

L. avenaria, causing septoria avenae blotch, occurs world-wide (CMI map 323) mainly on oats, but also on wheat, barley, rye, and triticale and is recorded on many grass weed species. **f.sp. *avenaria*** Weber is specialized to oats and **f.sp. *triticea*** T. Johnson to wheat. First signs of infection (blotch) are small chocolate-brown spots on the leaves which enlarge to become lens-shaped and finally merge. Lesions turn light greyish-brown and are difficult to distinguish from those of septoria nodorum blotch. In first-year crops and in the absence of any crop debris, primary inoculum is usually wind-borne ascospores. Seed-borne infection may also play a part in initiating primary infection. Secondary plant-to-plant spread within the crop is by rain-splashed pycnospores. The disease is mostly not serious enough to have merited the development of resistant cultivars or other control strategies, although Kiesling & Grafius (1956) found 9 of the 5343 entries in the World Oat Collections resistant to *L. avenaria* when tested at Michigan.
SANDERSON & SCOTT

Leptosphaeria oryzaecola Hara (anamorph ***Septoria oryzae*** Cattaneo) causes speckled blotch of rice in Italy and the USSR.

Leptosphaeria coniothyrium (Fuckel) Sacc.

(Anamorph *Coniothyrium fuckelii* Sacc.)

Basic description

L. coniothyrium produces black immersed pseudothecia with cylindrical asci, pale olive brown 3-septate ascospores and hyaline pseudoparaphyses (CMI 663). The immersed pycnidia release single-celled pycnospores which appear pale brown in transmitted light.

Host plants

The fungus attacks wounds on vegetative stems of *Rubus* and *Rosa* species. It has been recorded on a wide range of other woody plants.

Geographical distribution

L. coniothyrium occurs world-wide (CMI map 185).

Disease

Cane blight, a disease of red raspberry (*Rubus idaeus*), black raspberry (*R. occidentalis*), loganberry (*R. ursinus*×*R. idaeus*) and cultivated blackberry (*R. fruticosus*), is caused by infection of young stems at wounds. It was recognized early this century on red raspberry and blackberry in N. America and New Zealand, but it has recently become an important disease in mechanically picked raspberries. Canes infected in spring usually die during their first season, but unless the majority are affected, early infection is unimportant because other canes can be selected to fruit during the following year. When wounds are infected during harvest the fungus is initially less aggressive, but dark brown cryptic vascular lesions spread from wounds during winter to extend throughout several nodes of the cane. There are no external symptoms until the pycnospores are discharged onto the cane surface to appear, *en masse*, as dark grey blotches (Williamson & Hargreaves 1978, 1979). Although later infections are slow to establish, the yield losses are potentially serious because growers cannot differentiate between infected healthy canes during the winter when they are selecting canes for tying to support wires for fruiting the following summer. The death of individual axillary buds, wilt of lateral shoots or death of entire canes occurs in spring, the extent of the losses depending on the position of the infected wound and the size of the lesion. Canes become brittle and may snap at the point of initial infection.

Epidemiology

Mycelium of *L. coniothyrium* survives in the dead tissues of overwintering canes and in the stubs of old fruiting canes at the base of the plant. Pycnidia are produced in late winter which release pycnospores from March to October in wet weather and pycnospores may continue to be released from old lesions for at least four years. Pseudothecia are formed on old cane stubs, often intermixed with pycnidia, and ascospores are released mainly in April and May. Both pycnospores and ascospores can establish infection, but the pycnospores are probably the most important source of inoculum for wounds made during late summer.

Economic importance

Although cane blight is common, serious yield losses are uncommon in hand-picked plantations. Losses occur in the year following a wet summer, either when young canes have been damaged by abrasion against old cane stubs or when farm equipment has wounded canes. However, cane blight has been a major factor preventing the introduction of mechanized raspberry harvesting in Europe, although machine-harvesting is successful in the Pacific N.W. of America. The disease has caused consistent yield losses of 30–50% in seasons following tests of American harvesters in Scotland (Williamson & Hargreaves 1978) and it has occurred following similar tests in France, USSR and New Zealand.

Control

Sprays of benomyl or thiophanate-methyl applied at high volume before, during or after harvest have given good control of cane blight when directed over the full height of the canes and especially at the cane bases (Williamson & Hargreaves 1981). Vigour control by removal of the first flush of young canes with a contact herbicide ensures that the canes which grow later are shorter, less prone to injury and have fewer lesions than the first flush canes (Williamson *et al.* 1979). The removal and burning of old fruiting canes by cutting them at soil level reduces the risk of abrasion wounds and removes an important source of inoculum.

Special note on midge blight

L. coniothyrium is also the most aggressive of a wide range of fungi that colonize tissues damaged by the larvae of the raspberry cane midge (*Resseliella theobaldi*) to cause midge blight. The pest is widely distributed in Europe and midge blight has caused serious losses in many countries (Woodford & Gordon 1978). The first generation larvae attack few canes, which recover from any fungal damage by cankering. Losses are minimal, since any broken canes can be removed during winter pruning. However, canes infected as a result of second or third generation attack cannot be so distinguished, and will give rise to serious losses as for cane blight (above). In fact, the larvae remove the protective suberin from cells in the periderm so that damage cannot be repaired. This allows fungi to penetrate to the vascular tissues below during autumn and winter, where lobate brown patches are formed (visible when the cortex and periderm are removed). *Didymella applanata* and *Gibberella avenacea* are among the fungi which sporulate on, and can be isolated from these patches (Williamson & Hargreaves 1979), but do not spread beyond the area damaged by larvae. *L. coniothyrium* is more aggressive, spreading from this area to form brown stripe lesions. Both types of lesion can cause cane losses. *R. theobaldi* is not known to be a vector for the fungi, which are splash-dispersed.

Midge blight causes serious losses in a season following one in which the peak of midge emergence coincides with the formation of the natural splits in young canes. Losses have been particularly serious in districts (i.e. Scotland) where the first generation is difficult to control effectively and the second generation is present on canes during harvest when insecticides cannot be used. The disease has recently become prevalent and important because cultivars which produce abundant splits have been introduced.

Fungicides alone cannot control the disease. Contact herbicide treatment of the first flush of young canes (in a sufficiently vigorous plantation) will control midge blight on replacement canes (Williamson *et al.* 1979). Otherwise, the disease must be controlled by insecticides sprayed when the splits appear on canes and midges are emerging. Care should be taken to avoid introducing the pest on planting material to areas where it has not been recorded (e.g. Ireland).

Other diseases caused by L. coniothyrium

Finally, *L. coniothyrium* also causes common canker of rose. It readily colonizes wounded rose stems and will cause dieback after pruning unless this is done carefully just above the nodes. Cankers are often also formed at the graft union during propagation in warm, moist conditions under glass (graft canker). Sprays with captan, dichlorofluanid or folpet are recommended shortly after pruning. Benomyl, carbendazim or thiophanate methyl could be interchangeably used with chemicals of the previous group. A somewhat similar disease (brand canker, also widespread in Europe) is caused by ***Coniothyrium wernsdorffiae*** Laubert, which has larger pycnospores released through longitudinal slits rather than spread in a sooty mass under the epidermis. Infection of stems tends to occur under moist conditions in winter (APS Compendium).
WILLIAMSON

Coniella diplodiella (Speg.) Petrak & Sydow

(Syn. *Coniothyrium diplodiella* (Speg.) Sacc.)

C. diplodiella attacks *Vitis* species (CMI 82). Originally a European disease described in Switzerland in the 18th century, the fungus now occurs worldwide. *C. diplodiella* attacks mainly grapes damaged by hail (white rot, ripe rot). Infected berries show white spots surrounded by dark lines. They then wither and are covered with white pycnidia. Mycelium can also invade the main stem and attack the shoot (1–5 cm); the bark tears and the wood is covered with white pycnidia. The fungus overwinters as pycnidia on fallen berries in the soil, where it can survive for 15 years. Spores are dispersed to the berries by the stormy summer rain which comes with the hail. The incubation period is 4–5 days at 20°C. Six days after infection, pycnidia appear on the berries and the stalk. The main period of susceptibility is between the setting and the ripening of grapes. *C. diplodiella* can considerably increase damage caused by hail storms during ripening. Preventive measures such as destruction of inoculum in the soil cannot be applied. Control is based on treatments immediately after the storm. The time limit for application is 24 h for thiram or folpet and 48 h for benomyl or dichlofluanid.
MARCELIN

Leptosphaeria maculans (Desm.) Ces. & de Not.

(Anamorph ***Phoma lingam*** (Tode ex Schwein.) Desm., syn. *Plenodomus lingam* (Tode) Höhnel)

Basic description

The globose, black pseudothecia of *L. maculans* contain asci with multiseptate ascospores and hyaline pseudoparaphyses (CMI 331). The pycnospores are produced in either sclerotioid or globose pycnidia and are small, hyaline and unicellular.

Host plants

The fungus is parasitic on wild and cultivated genera of Cruciferae, particularly *Brassica* species (rutabaga, rape, turnip, cabbage etc.), *Sinapis alba* (white mustard) and *Raphanus* species.

Host specialization

Variation in pathogenicity has been reported from several countries (Australia, Canada, England, USA), but the practical implications are not clear. In Australia, Thurling & Venn (1977) found significant interactions between three populations of *L. maculans* from rape stubble and cultivars of *B. napus* (rape) and *B. campestris* (turnip rape) and concluded that different strains of *L. maculans* occurred at various sites. Further work suggested that differences in resistance between host cultivars and in virulence between pathogen isolates were polygenically determined. In Canada, the most common strain of *L. maculans* on rape and turnip rape is avirulent and produces only superficial stem lesions (McGee &

Petrie 1978). However, a virulent strain which causes severe stem cankers also exists, but is less widespread. In the USA, virulent and non-virulent strains of *L. maculans* have been isolated from cabbage and seem to be similar to those isolated in Canada (Bonman *et al.* 1981). In England, virulent and non-virulent strains of *L. maculans* have been isolated from various *Brassica* seed crops (Humpherson-Jones 1983). The virulent isolates were more commonly found on rape but infected other species non-specifically.

Geographical distribution

L. maculans occurs world-wide but causes damage as a pathogen only in temperate regions or at high altitudes in the tropics. In Europe, it is a damaging pathogen almost everywhere.

Disease

L. maculans causes symptoms known as black-leg, canker, dry rot, and root, collar and stem rot. The pathogen is seed-borne and causes disease at all stages of plant growth. Symptoms are initially seen as pale lesions on the cotyledons and first true leaves of seedlings. The presence of dark pycnidia within these lesions is an important diagnostic character. Severe infections can result in damping-off in seedlings. As the plants grow older, lesions may develop on the stem, leaves and pods. Infection of the stem base and tap root is the most damaging phase of the disease and infected roots turn black and die. On rape, cankers usually develop at the point of attachment of lower leaves at the base of the stem. Cankers are concave, with a dark margin and pale centre, and can weaken and girdle the stem. Destruction of the root and stem system results in poor growth, premature ripening, lodging and even death. In root crops, e.g. rutabaga and turnip, infection of the swollen root leads to a dry rot.

Epidemiology

On rape initial infections are caused predominantly by ascospores dispersed from pseudothecia on infected debris. According to McGee & Petrie (1979), plants are more susceptible before the six-leaf stage than later and severe infections are likely to occur if this coincides with ascospore discharge. In Canada, rape is sown in late May and has six leaves by the end of June. Since ascospore discharge does not occur until July, severe infections are largely avoided. However, in Australia, France and England, the period of ascospore discharge coincides with the development of rape plants up to the six-leaf stage, so that crops frequently become severely infected. Infected seed can also act as a source of inoculum but is of less importance than infected debris, except where the seed transports the pathogen into uninfected areas, or new strains into existing infected areas. Once established in the crop, the pathogen spreads by rain-splashed conidia. These are produced in pycnidia which develop in lesions on any of the aerial parts of the plant. Epidemic development is thought to be favoured by mild wet conditions although the precise details are not well established. Stem cankers usually become visible from mid-flowering, but a proportion of these are initiated many months previously, before the six-leaf stage. After harvest, the fungus rapidly colonizes any uninfected stem and root tissue, and pseudothecia subsequently develop. The fungus is able to survive for at least three years on crop debris in the soil. The epidemiology on other *Brassica* species is similar to that on rape. However, in cabbage, initial infections arise largely from infected seed. In commerce, *Brassica* seedlings are often raised in densely sown seed beds, so that a few seedlings which have become infected from the seed can act as a source of inoculum for other healthy seedlings. Seedlings are subsequently transplanted into the field and any which are infected may develop severe symptoms later in the season.

Economic importance

Traditionally, *L. maculans* has been regarded as an important pathogen of vegetable *Brassica* species and epidemics have occurred at intervals on cabbage in the USA. In recent years, the importance of *L. maculans* on rape has markedly increased as the cultivated area of that crop has expanded to meet the greater world demand for oilseed. Major epidemics have occurred in France, Western Australia, Germany and England and canker has been widespread in most other areas where rape is grown intensively, including Canada. Following the epidemic in Western Australia, the local area of rape declined from 49 000 ha in 1972 to 2000 ha in 1974. On rape, *L. maculans* causes lodging and premature senescence, and yield losses of up to 60% have occurred in crops with severe cankers on the stem base. Reductions in 1000 seed weight by *L. maculans* have been reported from England and Germany.

Control

Differences in resistance between rape cultivars have been noted in trials in several European countries including Britain, France, FRG, Netherlands, Sweden and Switzerland. Resistant cultivars are characterized by fewer, smaller cankers on the stem bases. Differences in resistance of the foliage have not so far been noted. In the late 1970s, outbreaks of canker were threatening to limit the commercial production of rape in W. Europe, but the introduction of the more resistant cvs Jet Neuf and Rafal has resulted in a reduced incidence of the disease, thus allowing the continued production of intensively grown rape. Various field and glasshouse techniques are being developed which may enable inoculated plants to be rapidly screened for resistance. Infected debris of rape must be disposed of as soon as possible after harvest and certainly before the emergence of the following crop (chopping or burning straw after harvest, burying the remaining debris by ploughing). Because *L. maculans* can survive on debris in the soil for at least 3 years, a rotation of at least four years should be practised. These procedures do not eliminate the risk of infection but reduce the likelihood of severe attack. Trials have also shown that late sowing of rape results in lower infection levels, but this is only of limited use as a disease control method because late sowing often reduces potential yield in the absence of disease. Plants of other *Brassica* species and weeds can act as reservoirs of inoculum for commercial crops, so these should be removed wherever practicable. Traditionally, a hot-water treatment has been used to control *L. maculans* in the seed of vegetable *Brassica* species. Although useful, it has been somewhat unreliable in practice and has been superseded by more effective fungicide seed-treatments. Recent reports suggest that the control of *L. maculans* by foliar applications of fungicide can lead to large yield responses in rape. However, responses have been inconsistent from site to site and further work is needed before any general recommendations can be made.

PRIESTLEY & KNIGHT

Leptosphaeria libanotis Sacc. (anamorph ***Phoma rostrupii*** Sacc.) used to be important on carrot seed crops in Denmark (Neergaard 1977). It continues to be an important seed-borne disease of carrot in the USSR. ***L. lindquistii*** Frezzi (anamorph ***P. macdonaldi*** Boerema) causes black spot on sunflower in Yugoslavia (Marić *et al.* 1981), stems and heads being affected (El-Sayed & Marić 1981). ***L. pratensis*** Sacc. & Briard (anamorph ***Stagonospora medicaginis*** Höhnel, syn. *S. meliloti* (Lasch) Petrak) causes a leaf spot and root rot of lucerne and clover in the USSR (APS Compendium).

Herpotrichia juniperi (Duby) Petrak

(Syn. *H. nigra* R. Hartig)

The brown felt (or black snow mould) fungus is characterized by a dark-brown felt of coarse hyphae around brown killed conifer shoots, which had been covered by snow during the winter. Pseudothecia may be found on 2-year-old or older mycelium; they are spherical, 0.3–0.5 mm in diameter, blackish brown, and partly covered by coarse, dark hyphae; the ascospores are hyaline or light-coloured when within the asci, 4-celled (CMI 328). *H. juniperi* attacks conifers (*Abies, Juniperus, Picea,* and *Pinus*). Of greatest economic importance in Europe is the damage caused to plants of *Picea abies*. The fungus is widely distributed in northern and alpine regions with heavy and long-lasting snow cover both in Eurasia and in America.

Ascospores may be discharged at any time of the snowless period and may germinate at once. Under the snow, the mycelium grows from needle to needle and from shoot to shoot, killing the host tissue. As a snow mould fungus, it needs a very high RH and can grow fairly rapidly at 0°C and even below zero. In humid winter periods with temperatures near 0°C the fungus may also develop above snow cover (Hanso & Tõrva 1975). Within the snow the mycelium remains relatively light-coloured and thin-walled, but in the light, when the snow has melted, it will soon become dark, thick-walled, and resistant to all kinds of climatic influences during the summer, for example to drought.

Brown felt blight is found in afforestations and in nurseries. Heavy damage in spruce afforestations has been reported from Switzerland by Bazzigher (1976). Venn (1983) found that the fungus can cause considerable damage on overwintered cold-storage spruce plants. Control in the field will usually be too difficult and expensive except by changing to a more resistant tree species. In nurseries and under cold storage, dithiocarbamates such as mancozeb may be useful.

ROLL-HANSEN

Herpotrichia coulteri (Peck) Bose causes a similar but minor disease on pines in S. Europe (CMI 327). ***H. parasitica*** (R. Hartig) Rostrup (syn. *Acanthostigma* *parasiticum* (R. Hartig) Sacc.) (anamorph ***Pyrenochaeta parasitica*** Freyer & v.d. Aa) causes white felt blight of *Abies* in the Alps.

Pyrenochaeta lycopersici Schneider & Gerlach

In the natural state this fungus produces only a sterile grey mycelium. Pycnidia were induced to form in culture by Gerlach & Schneider (1966). The distinguishing feature of *Pyrenochaeta* species is the formation of simple, setose, brown pycnidia which contain unicellular, ellipsoid pycnospores. Structurally unspecialized microsclerotia are formed by *P. lycopersici,* which survive in the soil for at least 2 years (White & Scott 1973). Isolates of *P. lycopersici* from sweet pepper and tobacco are capable of infecting tomato. There is considerable variation in aggressiveness between isolates. Distribution of *P. lycopersici* appears to be restricted to areas of glasshouse tomato production; field-grown crops are seldom affected.

Prior to the establishment of a causal link between brown root rot or corky root rot on tomato and *P. lycopersici*, other organisms, particularly *Colletotrichum coccodes*, were implicated in this syndrome. Symptoms caused by *P. lycopersici* infection are usually first noted as a lack of plant vigour, foliar chlorosis, stunting and loss of yield. These are associated with cortical rotting of fine and medium size roots; the larger roots become corky, swollen, and cracked with a corrugated surface and eventually there is rotting at the stem base. Three lesion types have been distinguished: brown lesions which are visible within a month after planting on young feeder roots, accompanied by darker lesions with corky splitting on older and larger roots; lengths of corky root which often extend several cm; dark brown cortical basal stem lesions (Ebben 1974). Yield losses of 8–20% have been associated with 10–15% infection by *P. lycopersici* 8 weeks after planting. Growth of *P. lycopersici* occurs over the range 8–32°C, but growth rate and lesion spread is slow. Eventually host cells die and are invaded by hyphae with simultaneous thickening of the cell wall and an appearance of crystalline microbodies in neighbouring tissues (Delon *et al.* 1973).

Control initially rested on the use of wild resistant *Lycopersicon* species as rootstocks onto which susceptible commercial cultivars could be bench grafted. Subsequently resistance has been transferred from *Lycopersicon* species to *L. esculentum* resulting in the availability of a range of resistant cultivars. Soil sterilization using methyl bromide or chloropicrin has been used successfully (Cirulli 1968). In Greece, Israel and Italy it has been shown that *P. lycopersici* can be effectively controlled by solarization of the soil during summer. A combination of soil solarization with low dosages of methyl bromide could also be applied (Tjamos 1984). Indications of potential biological control come from observations that non-pathogenic strains of soil-borne *F. oxysporum* inhibited invasion of tomato roots by *P. lycopersici.* In natural soil the balance between these organisms was temperature-dependent (Ebben 1974; Davet 1976). DIXON

Pyrenochaeta terrestris (Hansen) Gorenz *et al.* causes pink root of onion, garlic, leek and other *Allium* species (CMI 397). It is a common soil-inhabiting fungus associated with the roots of many crops but causing no disease. Affected *Allium* plants wilt and show a much reduced root system when lifted, the remaining roots being discoloured pink by the pigment from the fungus (Messiaen & Lafon 1970). Because the optimum temperature is 26°C, the disease is only important in the very south of Europe (e.g. only at the very end of the season in S. France). In Israel and potentially elsewhere in Mediterranean countries, solar heating by transparent polythene mulching provides a means of control (Rabinowitch *et al.* 1981). Some onion cultivars are resistant (Levy & Gornik 1981).

Gemmamyces piceae (Borthwick) Casagrande (syn. *Cucurbitaria piceae* Borthwick; anamorph ***Camarosporium strobilinum*** Bomm. *et al.*) is an occasional cause of a bud blight of various species of *Picea* in many areas of N. Europe. ***Camarosporium pistaciae*** Zachos *et al.* forms pycnidia on the branches and fruits of pistachio in Greece, especially in rainy humid summers (Zachos & Roubos 1977). Spring application of copper fungicides, continued up to full fruit expansion, is recommended. ***C. dalmaticum*** (Thüm.) Zachos & Tzavella-Klonari causes a brown spot of olive fruit, which occasionally spreads to wither and rot the whole fruit. It is common in Greece (Zachos & Tzavella-Klonari 1979). The same authors (1980) describe ***C. flaccidum*** (syn. *Phoma flaccida* Viala & Ravaz) as the cause of excoriosis of grapevine in Greece (see *Guignardia baccae*).

Dilophospora alopecuri (Fr.) Fr. forms hyaline septate pycnospores with irregularly branched appendages (CMI 490) and probably has ***Lidophia graminis*** (Sacc.) Walker & B. Sutton as its teleomorph.

Infected leaves of grasses and cereals are spotted and distorted (twist disease) and finally bear black stromata containing pycnidia. The disease used to have some importance on cereals, but is now insignificant, occurring occasionally on grasses (APS Compendium of Wheat Diseases).

PYRENOPHORACEAE

Pyrenophora avenae Ito & Kuribayashi

(Anamorph ***Drechslera avenae*** (Eidam) Scharif)

Basic description

Pale yellowish, muriform ascospores are formed in globose or semi-globose pseudothecia, 300–600×450–800 μm, which at maturity develop a short beak. The dark pseudothecial walls bear setae and conidiophores. Conidia are light olive brown, cylindrical 2–6 septate and rounded at the ends, 10–18×35–150 μm. They are formed at the tips of the olive brown conidiophores which appear singly or in groups. The mycelium of *P. avenae* is grey and produces typical white tufts similar to those formed by *P. teres* (CMI 389).

Host plants and specialization

P. avenae is usually confined to the genus *Avena* but it has also been reported on other Gramineae. Although experimental evidence is limited, the available data suggest substantial differences in resistance between oat cultivars. Also differences in pathogenicity between isolates of *P. avenae* have been found (Rekola *et al.* 1970).

Geographical distribution

P. avenae was first described from Italy in 1889, but today it occurs world-wide (CMI map 105).

Disease

P. avenae is a seed-borne pathogen. It usually persists as mycelium present in the glumes and outer tissues of the caryopsis. At seed germination, the mycelium grows into the coleoptile and enters the first leaf. Sometimes it continues to grow into the second and third leaf and only in rare and severe cases does it advance to the fourth leaf. Two distinct phases of symptoms are recognized. The primary phase is associated with seedling disease and appears in plants arising from infected seed. Blighted seedlings may be discoloured and killed either before they emerge from the soil or shortly after emergence. Primary symptoms appear on the first leaf and others as several chlorotic, later brown spots which soon coalesce to form a stripe (Plate 24). Often the leaf becomes twisted or distorted. Seedlings with severe striping on the lower leaves are often killed or develop into retarded, weak plants which are later overgrown by surrounding, healthy plants. Otherwise the plant may recover and develop without any further signs of disease until secondary symptoms eventually develop. The primary phase of the disease usually ends at the tillering stage.

The secondary phase of the disease begins when the fungus starts to sporulate on the primary lesions. The conidia infect the upper leaves and sheaths and produce dark-reddish leaf spots of varying size, often surrounded by a yellow, translucent halo. These are usually most prominent later in the growing season when the crop density and the presence of dead leaf tissue create favourable nutritional and microclimatic conditions for fungal propagation. Conidia produced in the secondary phase of the disease may, under humid conditions favourable for spore production, result in further spread of the disease in the crop and ultimately lead to seed infection.

Epidemiology

The percentage of primary infected seedlings which develop from infested seed lots is highly influenced by environmental conditions, primarily soil temperature at sowing-time. Soil temperatures below 6–8°C during the period of germination favour disease frequency and severity, whereas higher soil temperatures reduce or possibly prevent disease, even when the seed is heavily infested. Accordingly early sowing in cool soils usually promotes primary infections whereas seedlings from later sowings tend to escape disease. High soil moisture at germination seems to reduce disease frequency. Infested plant debris such as straw and stubble left over from one season to the next may serve as a source of inoculum because the fungus is able to sporulate on such dead plant debris. The conidia are similar to those of *P. teres* and do not spread over long distances. It is believed that most conidia settle within 10 m from the inoculum source. The development of epidemics is influenced by the amount of inoculum, periods with high humidity, temperature and host susceptibility. Longer periods of warm dry weather will reduce or stop the spread, even if all other factors are favourable.

Economic importance

While *P. avenae* causes minor leaf spots on oats in many regions (e.g. Scandinavia) it often becomes severe in Scotland, N. England, Wales and Ireland where cool and rainy conditions prevail during the early seedling stage. It is reported to be of moderate importance in the FRG, the coastal region of the north-western states of the USA and the eastern part of Canada. Primary infected plants often have a high mortality and surviving plants are retarded in growth with up to 40% reduction in grain yield. However, such losses are usually compensated for by better tillering and compensatory growth of the remaining plants. Swedish field trials with seed dressing of seed lots with 40–70% infected kernels resulted in grain yield increases of 3–5% (Olofsson 1976), but much higher yield reductions have been reported under climatic conditions which favour disease development.

Control

P. avenae is seed-borne and the primary phase of the disease is effectively controlled by treatment of the sowing seed with appropriate non-mercurial fungicides. *P. avenae* is the first example of a cereal pathogen becoming insensitive to a fungicide. From the beginning of the 1960s the fungus developed a widespread, natural resistance to normal doses of organomercury seed treatment compounds (e.g. Greenaway & Cowan 1970). However, with the appearance of new, non-mercurial and highly effective fungicides for seed treatment (carboxin, imazalil) the resistance to organomercury compounds does not present a problem today. The secondary spread of the disease can be controlled by foliar sprays with systemic fungicides (triadimefon, propiconazole). Since infested plant debris may be a source of inoculum, destruction of plant residues and careful ploughing is recommended for control. Substantial differences in resistance between cultivars are reported, but the efforts to breed for resistance seem to be limited.

SMEDEGAARD-PETERSEN

Pyrenophora graminea Ito & Kuribayashi

(Anamorph ***Drechslera graminea*** (Rabenh. ex Schlecht.) Shoem.)

Basic description

Pseudothecia are rare and play no important role in the disease cycle but have been reported from Denmark and Russia. They are large, sclerotium-like, superficial or partly submerged bodies with dark rigid setae bearing conidia at the tips. At maturity the pseudothecia develop a short beak with or without a fringe of setae. Asci are similar to those of *P. teres* but slightly larger (CMI 388). The conidial stage arises under humid conditions from necrotic leaves or from straws of infected crops. Conidia develop laterally and terminally on light brown conidiophores appearing singly or in groups of 2–6. The conidia are straight, cylindrical, with 1–6 transverse septa and measure $10–25 \times 40–150$ μm. They are hyaline or yellowish-brown with dark scars which are the sites of the former attachment on the conidiophores. Often short, secondary conidiophores are formed from the apical conidial cell giving rise to short conidial chains. Like *P. teres, P. graminea* produces pycnidia whose role in the disease cycle is obscure.

Host plants and specialization

In nature *P. graminea* causes disease only on species of the genus *Hordeum*. Barley cultivars differ widely in resistance to *P. graminea* and fully resistant cultivars are available (Kline 1972). Nevertheless the information on physiological specialization and host resistance is inconclusive (APS Compendium). Some studies indicate that *P. graminea* is specialized into pathogenic races and that resistance is major-gene controlled (e.g. Nilsson 1975). Other results indicate that the fungus is not separated into pathogenic races and that resistance is quantitatively inherited (Knudsen 1980).

Geographical distribution

P. graminea occurs wherever barley is grown.

Disease

P. graminea is strictly seed-borne. In contrast to most other leaf pathogens it cannot successfully cause disease by direct leaf infection. Attempts by the fungus to penetrate directly through the leaf surface always result in tiny, hardly visible, necrotic lesions. In a crop the disease is therefore always confined to the plants arising from infected seeds. Seed infection occurs at the time of heading when the fungus sporulates on infected leaves. The conidia are blown from diseased plants to the young heads of surrounding plants. Seeds can be infected at all stages of development from before head emergence to the hard-dough stage, although the most severe infection occurs during the early stages of kernel development. In infected seeds the mycelium is located in the hull, pericarp and seed coat, and the embryo is thus not infected. When the seed germinates, the mycelium invades the young seedlings by penetrating the coleorhiza or the seminal roots from which it grows upwards and becomes established in the embryonic leaf primordia.

Infection of the developing seedling is highly influenced by soil temperature and humidity. Soil temperatures below 8–10°C during the germination period favour infection whereas infection is reduced at temperatures above 12–15°C. It is a common experience that seedlings may entirely escape infection at high soil temperatures even if the seed is heavily infested with *P. graminea* (Teviotdale & Hall 1976). Infected seedlings are more or less retarded in growth and may die even before emergence. Usually, however, symptoms first appear on the second and third leaf which develop one or more long chlorotic stripes parallel to the leaf rib, often extending from the base to the tip of the leaf. Later the yellow stripes turn brown or grey as the tissues become necrotic, and the leaf blade may split free and take on a frayed appearance. Usually all or most of the leaves of an infected plant exhibit stripe symptoms.

Infected plants are usually stunted. The ears may be arrested during the emergence from the sheath, usually resulting in barren or deformed ears with improperly developed, shrivelled kernels. In some spikes, however, grains may develop almost normally. High humidity during the time of heading is a prerequisite for fungal sporulation and highly increased seed infection (Metz & Scharen 1979). The conidia are relatively large and are dispersed by the wind over relatively short distances. Although most conidia settle within a short distance of the source, some may be carried over longer distances by strong winds and air currents.

Economic importance and control

Leaf stripe is a potential threat everywhere that barley is grown. In the first decade of this century the stripe disease was one of the most widespread and yield-reducing diseases of barley. But with the introduction of mercurial fungicides, in the 1920s, leaf stripe dropped to a level of little economic importance except in areas where seed treatment is insufficient (Africa, Asia). Today effective systemic fungicides (carboxin, imazalil) have mostly replaced mercurial seed treatment. Experience has shown that omission of seed treatment results in a rapid increase of leaf stripe in the course of a few years. Since infected plants produce no or only few and shrivelled grains the yield decreases by approximately 0.5–1% for each per cent of infected plants in the field. Even a reduced outbreak thus results in reduced yield. Resistant cultivars are available, but since pathogenic races exist cultivars may differ in resistance according to the locality where they are grown. A number of spring barley cultivars are resistant to at least some races and cv. Zita has proved to be fully resistant to all Danish races tested against it (Smedegaard-Petersen & Jørgensen 1982). Most winter barley cultivars seem to be susceptible.
SMEDEGAARD-PETERSEN

Pyrenophora teres Drechsler*

(Anamorph ***Drechslera teres*** (Sacc.) Shoem.)

Basic description

The pseudothecia of *P. teres* are formed after harvest on straw and stubble of infected barley plants. They are first subepidermal, then erumpent, semi-globose when young but at maturity flask-shaped with one or more short ostiolar beaks. The blackish brown walls bear dark, rigid setae on top of which are often formed conidiophores and conidia. The asci are hyaline, clavate or cylindrical, rounded at the apex and 30–60×200–300 μm. They are bitunicate and contain 2–8, usually 8, pale yellowish ellipsoidal or oval ascospores with 3–5 transverse septa and one longitudinal septum which always occurs in one of the middle cells. The ascospores are clearly constricted at the septa (CMI 390). Conidia develop on the tips of successive lateral proliferations of light brown conidiophores that occur singly or in groups of 2–3. The conidia, which measures 10–25×40–160 μm are cylindrical, rounded at the ends, and often with the basal cell slightly inflated. They are subhyaline to straw-coloured, smooth and usually with 2–7 transverse septa. On nutrient-rich substrates in culture the mycelium often produces characteristic vertical erect mycelial tufts. *P. teres* also produces pycnidia which occur abundantly on leaves and straw in nature as well as in culture. The pycnidia are globose with a short ostiole and measure 60–170 μm. Pycnospores are hyaline, non-septate, ellipsoidal or spherical. At maturity they are exuded in a slimy matrix appearing as a grey or white drop on top of the pycnidium. Apparently the pycnospores are unable to infect plants directly, but they germinate

*According to Cannon *et al.* (1985), this should be *P. trichostoma* (Fr.) Fuckel, which is then not a synonym of *P. tritici-repentis*.

and produce mycelium in culture. Their role in the disease cycle is not known.

Host plants

In nature *P. teres* is usually confined to the genus *Hordeum* although other host genera have been reported (Shipton *et al.* 1973). In artificial infection trials both wheat and especially rye can develop severe symptoms when inoculated with isolates from barley.

Host specialization

Barley cultivars differ greatly in resistance to *P. teres* but total resistance seems to be rare. A number of specific genes for resistance in barley have been identified. Pathogen isolates vary widely in virulence to different cultivars as well as in cultural characteristics and the existence of pathotypes has been reported (Clifford & Jones 1981).

Geographical distribution

Net blotch is widely distributed in virtually all barley-growing countries (CMI map 364). It is most severe in temperate regions with high humidity, but has also been recorded as severe in drier regions of Australia.

Disease

P. teres causes net blotch of barley. The fungus exists in two forms, a net form and a spot form which differ only in the symptoms produced on barley (Fig. 53) (Smedegaard-Petersen 1971). The net form of *P. teres* produces symptoms which initially appear as minute necrotic spots and streaks which soon increase to form narrow, dark brown, longitudinal and transverse streaks to produce a characteristic net-like pattern. These net symptoms have given rise to the name of the disease, net blotch. The affected part of the leaf turns brown and the adjoining tissues become chlorotic. Occasionally the initial lesions develop as greenish, water-soaked areas in which the dark netting later becomes evident. Sometimes lesions develop as long necrotic stripes which may be mistaken for stripe symptoms caused by *P. graminea*. The spot form of *P. teres* was first described by the author in 1971 from Denmark, and has now been found in many other barley-growing countries. The symptoms of the spot form also occur on leaves and leaf sheaths and appear as dark brown elliptical or fusiform lesions measuring 3×6 mm and surrounded by a chlorotic zone which eventually extends until the entire leaf withers. On the kernels, infection appears as a dark, diffuse coloration of parts of the kernel. These symptoms are difficult to distinguish from those caused by a number of other

Fig. 53. Symptoms of net form (left) and spot form (right) of *Pyrenophora teres* on barley.

pathogens. The capacity of *P. teres* to produce net and spot symptoms is controlled by gene pairs in two loci. The symptoms produced by the spot form of *P. teres* closely resemble the spot blotch symptoms caused by *Cochliobolus sativus*, but microscopic examination of the conidia easily allow for identification. The yield reduction of barley plants infected with *P. teres* is mainly a result of reduced grain weight, whereas the number of grains per plant is usually less affected.

Epidemiology

The disease is seed borne and can therefore be introduced to new areas with infected seed. Locally, however, infested stubble and straw left on top of the soil from one growing season to the next is the most important source of inoculum from which primary infections are established on the spring barley (Jordan 1981). On such plant residues the fungus overwinters as mycelium or pseudothecia (Fig. 54). Under Scandinavian conditions, sporulating mycelium persists in infested plant residues for at least one year. Volunteer plants of infected winter barley or overwintering spring barley may also be important primary sources of inoculum. Sporulation and infection require 10–30 h of high humidity. Light promotes the production of conidiophores and sporulation is diurnal. Sporulation occurs over a wide temperature range with an optimum of 20°C. Conidia are relatively large and are dispersed with the wind over relatively short distances. Most conidia settle within 3–8 m from the source but some conidia may be transported over longer distances with strong winds and air-currents. The development of epidemics is dependent on the amount of inoculum, periods with high humidity and host susceptibility.

Fig. 54. Overwintered barley straw with mature pseudothecia of *Pyrenophora teres*.

Longer periods with warm and dry weather will stop or delay epidemics, even if all other factors are favourable.

In Denmark the severe outbreak of net blotch in 1981 resulted in yield losses of up to 30% on the most heavily infected crops of the highly susceptible barley cv. Welam. The attack was associated with high humidity and rainfall in the spring and early summer. In 1982 the dry weather of the summer prevented heavy attacks in spite of the high amount of inoculum present from the previous year.

Economic importance

In Europe, disease severity has increased considerably during the last decade, probably due to changes in cultural practices such as more common practice of continual barley growing in the same field, increasing areas of winter barley and the use of direct sowing in the stubble without ploughing. Recently severe outbreaks have been reported from many European countries including FRG, UK and Scandinavia. Typical losses due to net blotch range from 10 to 25%, but much higher losses are recorded. In California (USA), losses approaching 100% have been reported from severely affected crops (APS Compendium).

Control

The use of fungicides for seed treatment has almost eliminated seed as a primary source of inoculum. Doubtless infested plant debris is the most important inoculum source where mycelium may persist for at least one year. Destruction of plant residues or careful ploughing is therefore recommended for control. Crop rotation with at least one season with a non-host crop highly reduces the amount of inoculum, and thus prevents or reduces disease. Barley cultivars with considerable resistance against *P. teres* are available, and barley from Ethiopia and Manchuria seems to be valuable sources for resistance (Caddel & Wilcoxson 1975). Recently new, systemic fungicides (triadimefon, prochloraz, propiconazole) have appeared which are highly effective in controlling net blotch by foliar spraying.

Special research interest

P. teres is in many morphological and biological traits similar to the leaf stripe pathogen *Pyrenophora graminea*. The two species are able to hybridize and

produce fertile progeny. This interspecific fertility has allowed for genetical analyses of pathogenicity and symptom-producing capacity (Smedegaard-Petersen 1977a). Two near-host-specific toxins have been isolated from *P. teres*. The toxins are chemically characterized and reproduce the most important symptoms of the disease (Smedegaard-Petersen 1977b). The toxins are not decisive for basic pathogenicity but toxin A seems to be a determining factor in virulence. The primary effect of toxin A is on host plasmalemma and respiration.
SMEDEGAARD-PETERSEN

Pyrenophora tritici-repentis (Died.) Drechsler (syn. *P. trichostoma* (Fr.) Fuckel; anamorph ***Drechslera tritici-repentis*** (Died.) Shoem.) causes yellow leaf spot of cereals (wheat) and grasses (*Elymus repens*) (CMI 494) and is principally damaging in other continents (Australia). It has, however, recently increased in incidence in Europe, for example in France (Eschenbrenner 1983). Although the fungus can be seed-borne, infection arises most frequently from ascospores released by pseudothecia on stubble remaining on the soil surface. The increasing tendency towards wheat monoculture, associated with minimum tillage techniques, has accordingly increased the risk of infection. Though the disease, which can be confused with septoria blotch, is becoming more frequent, there is as yet no indication that it is particularly damaging.

Pyrenophora and *Drechslera* spp. on grasses

A number of other *Pyrenophora* and *Drechslera* species, many seed-borne, cause leaf spots on the festucoid turf and pasture grasses of Europe (O'Rourke 1976; APS Compendium). The situation regarding principal host range is summarized below (Table 13.1). In N. Europe, ***P. dictyoides*** is of importance in causing leaf spot and net blotch of *Lolium* and *Festuca* species. Formae speciales have been distinguished on these respective hosts (**f.sp *perenne*** and **f.sp. *dictyoides***); the former is also known as *D. andersenii*. Lesions remain small during active leaf growth in summer but spread in autumn and winter, causing constriction and yellowing. Temperatures of 15–18°C are optimal. The net-blotch symptom is characteristic—i.e. longitudinal and transverse necrotic bars. ***P. lolii (D. siccans)*** causes brown blight of *Lolium* and other turf grasses, and is frequent in lawns and sporting turf. It causes scattered dark brown spots or streaks, without transverse bars, with yellowing around the lesions. It was recently found to be the commonest *Drechslera* on *Lolium* in Britain (especially on *L. multiflorum*) (Lam 1984). Control with fungicide (organomercurials, dithiocarbamates, chlorothalonil) is recommended. ***P. erythrospila*** and ***D. catenaria*** cause leaf-blight and crown rot on fine turf (e.g. golf or bowling greens), causing large reddish brown patches in the turf. ***P. bromi*** is frequently noted on the pasture-grass *Bromus inermis* in E. Europe. ***D. poae*** causes so-called melting-out of *Poa* species. In general, there is much variation within and between cultivars in resistance to these diseases, tetraploid types being more resistant than diploid. According to Labruyère (1977), differences in the resistance of ryegrass to 3 species affected disease incidence and yield much more than differences in seed contamination, and seed treatment had little effect. Grasses in other groups (eragrostoid, panicoid), more characteristic of warmer countries, suffer similar diseases due to *Setosphaeria* species, e.g. ***S. rostrata*** Leonard (CMI 587) and *Cochliobolus* species, e.g. ***C. cynodontis*** Nelson. Their anamorphs are either considered as *Drechslera* (as in this Handbook) or segregated as *Exserohilum* and *Bipolaris* (respectively). *Curvularia* species, e.g. *C. lunatus* (q.v.), are also found on grasses, and are associated with *Cochliobolus* teleomorphs now separated as *Pseudocochliobolus* (APS Compendium).
CARR & SMITH

Cochliobolus carbonum Nelson (anamorph ***Drechslera zeicola*** (Stout) Subram. & Jain, syn. *Bipolaris zeicola* (Stout) Shoem.) is widespread in maize-growing areas throughout the world (CMI map 388), causing a very minor leaf spot of maize (CMI 349) and also occurring in a variety of other hosts. Race 1 causes significant damage on a few specifically susceptible maize lines, which are not used in Europe. Races 2 and 3 cause rather different symptoms (APS Compendium); race 2 causes very little damage and race 3 again is only of any significance on certain highly susceptible inbred lines. *C. carbonum* is widely present in small amounts on maize seed, but is only of extremely minor importance in Europe (EPPO 64). The fungus produces a host-specific toxin, similar to that of *C. victoriae*, with which it can be crossed as a basis for studying the genetics of toxin production. It has been used quite widely as a model organism in physiological and genetic studies (e.g. on fungicide resistance).
CASSINI & SMITH

Table 13.1. Principal host range of *Pyrenophora, Cochliobolus* and *Drechslera* on grasses

Pyrenophora	*Drechslera*	*Agrostis*	*Alopecurus*	*Bromus*	*Dactylis*	*Festuca*	*Lolium*	*Phleum*	*Poa*
bromi (Died.) Drechsler	***bromi*** Ito			+					
dactylidis Ammon	***dactylidis*** Shoem.				+				
dictyoides Paul & Parbery	***dictyoides*** (Drechsler) Shoem.					+			
erythrospila (Drechsler) Shoem.	***erythrospila*** (Drechsler) Shoem.	+							
lolii Dovaston	***siccans*** (Drechsler) Shoem.				+	+	+		
	andersenii Scharif						+		
	catenaria (Drechsler) Ito	+	+		+	+	+		
	festucae Scharif					+			
	fugax (Wallr.) Shoem.	+							
	nobleae McKenzie & Matthews						+		
	phlei (Graham) Shoem.				+			+	
	poae (Baudyš) Shoem.								+
Cochliobolus									
sativus (Ito & Kuribayashi) Drechsler ex Dastur	***sorokiniana*** (Sacc.) Subram. & Jain						+		

Cochliobolus heterostrophus (Drechsler) Drechsler

(Anamorph ***Drechslera maydis*** (Nisikado) Subram. & Jain, syn. *Bipolaris maydis* (Nisikado) Shoem.)

C. heterostrophus was the cause of the disastrous epidemics of southern corn leaf blight in the USA in the late 1960s, on maize cultivars carrying the Texas-type cytoplasm for male sterility (CMI 301; APS Compendium). Race O is widespread causing a minor interveinal leaf spot in warm moist climates, but has never been of great importance, and does not occur in Europe. Race T is specifically pathogenic to the cultivars with Texas cytoplasm, on which it is much more damaging, causing lesions surrounded by a pale halo (it produces a pathotoxin acting on the mitochondria), and also attacking sheaths, husks and cobs. Race T presumably through its greater infection potential occurs in more northerly zones than race O. The fungus persists as spores and mycelium in debris, but can also be seed-borne—which was a cause for concern when long-distance spread of race T was seen as a risk. In fact, race T was introduced to most maize-growing areas of the world quite rapidly (presumably on seed), and in particular appeared in France, Italy, Switzerland, Yugoslavia and USSR, causing quite serious losses in small areas planted with imported seed (Cassini 1973). It was at first rated by EPPO as a serious quarantine organism, but, in fact, the change in the genetic base of hybrid maize forced by the disease and still presenting major inconvenience in maize breeding, has since altogether reduced its importance everywhere. It may be noted that the abandonment of susceptible breeding lines, and the disposal of useless seed, was estimated to have cost some 50 million francs to French maize-seed producers in the early 1970s.

CASSINI & SMITH

Cochliobolus miyabeanus (Ito & Kuribayashi) Drechsler ex Dastur

(Anamorph ***Drechslera oryzae*** (Breda de Haan) Subram. & Jain, syn. *Bipolaris oryzae* (Breda de Haan) Shoem.)

C. miyabeanus commonly occurs only as the anamorph, distinguished by usually curved conidia, 6–14 pseudoseptate, with a minute, slightly protruding hilum (CMI 302). *C. miyabeanus* infects rice, wild rice (*Zizania aquatica*), *Leersia hexandra* and few other species, under natural conditions. Some degree of physiological specialization is reported. The fungus is widespread in rice-growing areas (CMI map 92). The pathogen causes spots on coleoptile and leaves, tip burn symptoms (Nanda & Gangopadhyay 1982) and seedling blight. It can produce abnormal seedlings in laboratory tests for germination (Imolehin 1983). Brown spots, oval to irregular, appear on adult leaves. Glumes can be

discoloured and seeds shrivelled. *C. miyabeanus* survives in seeds, alternate hosts and stubble. Its airborne conidia can spread infection. It is common in Europe where it occasionally causes severe losses. Continuous cloudiness, humid weather, poor growth of plants and temperature range of 25–30°C are associated with dramatic disease outbreaks recorded in Asia (Padmanabhan 1973). Control measures include the use of healthy or treated seeds, removal of stubble, herbicidal sprays against collateral hosts, correction of soil abnormalities and proper fertilization. Resistant cultivars are known (Singh & Sharma 1975).
PORTA-PUGLIA

Cochliobolus sativus (Ito & Kuribayashi) Drechsler ex Dastur

(Anamorph ***Drechslera sorokiniana*** (Sacc.) Subram. & Jain, syn. *Bipolaris sorokiniana* (Sacc.) Shoem.)

C. sativus is known mainly as the conidial form, which is widespread on grasses and cereals (wheat and barley) throughout the world (CMI 701, CMI map 322). Although the fungus can be seed-borne, it is frequent in soil, where it persists as thick-walled conidia on debris. It also occurs on numerous weed grasses and can be considered as ubiquitous as *Fusarium culmorum* and *Gibberella zeae*, with which it is associated in causing root rot, foot rot and seedling blight (APS Compendia of Barley and Wheat Diseases). Attack by *C. sativus* is characterized by a darkening of the lesions, and brownish black lesions (spot blotch) on leaves. In Europe, *C. sativus* is mostly a rather unimportant component of the root-rotting complex; it is only in warmer areas of cereal cultivation elsewhere in the world (N. America, Australia) that it is of real significance. However, up to 15% losses due to seed-borne infection have been reported on barley in Scotland (Whittle & Richardson 1978). Seedling resistance is known in barley (APS Compendium), and fungicidal seed treatments will reduce seedling blight. On turf grasses, *C. sativus* stands out as the only member of its genus attacking the festucoid cool season grasses, and causing disease mainly at higher temperatures (20–35°C) in summer, in contrast to the *Pyrenophora* spp. (q.v.) which mainly attack these grasses (APS Compendium).
SMITH

Cochliobolus victoriae Nelson (anamorph ***Drechslera victoriae*** (Meehan & Murphy) Subram. & Jain, syn. *Bipolaris victoriae* (Meehan & Murphy) Shoem.) caused 'Victoria blight' in the USA on oat cultivars susceptible to its much studied host-specific toxin (CMI 703). It occurred in several European countries, but has been of no economic significance since susceptible cultivars ceased to be grown. ***Drechslera iridis*** (Oudem.) M.B. Ellis (syn. *Bipolaris iridis* (Oudem.) Dickinson) causes ink disease of *Iris*, especially bulbous species (CMI 434) and is one of the most important diseases of this host. Chlorotic streaks on the leaves develop into black necrotic lesions, which may spread to the bulb. It can be controlled by fungicides and sanitation. ***Pseudocochliobolus lunatus*** (Nelson & Haasis) Tsuda (syn. *Cochliobolus lunatus* Nelson & Haasis; anamorph ***Curvularia lunata*** (Wakker) Boedijn) is widespread in the tropics as a weak parasite of many hosts (CMI 475). Under the name ***Curvularia trifolii*** (Kauffm.) Boedijn **f.sp.** ***gladioli*** Parmelee & Luttr. (a distinct fungus according to CMI 307), it is associated with a corm rot of *Gladiolus* in Europe.

Setosphaeria turcica (Luttr.) Leonard & Suggs

(Syn. *Trichometasphaeria turcica* Luttr.)
(Anamorph ***Drechslera turcica*** (Pass.) Subram. & Jain, syn. *Exserohilum turcicum* (Pass.) Leonard & Suggs)

S. turcica causes northern leaf blight of maize in the USA and occurs widely in Europe (e.g. Bulgaria, France, Greece, Italy, Romania, Spain, Yugoslavia) and world-wide (CMI 304, CMI map 257). Only the conidial stage is of any importance in the field, the teleomorph occurring only very rarely (APS Compendium). The fungus causes large (up to 4×10 cm) straw-coloured lesions on the leaves, and persists on leaf debris, in which it forms chlamydospores over the winter. Infection is favoured by warm moist conditions, about 10 h continuous leaf wetness being needed (Cassini 1973). In France, these conditions occur especially in the S.W., where high daytime temperatures are combined with lower night temperatures, and late morning mists tend to occur during the flowering period in July. Overhead irrigation also tends to create favourable conditions. In practice, it is only if such conditions occur sufficiently early in the season that losses are of any significance. However, occasionally (every 10 years or so), yield losses may locally reach 20–25%. In any case, most cultivars grown in Europe are resistant (carrying gene Ht 1), making the disease practically insignificant. However, if susceptible cultivars are again used (as in France in 1981), *S. turcica* reappears.

Pathotypes able to overcome gene Ht 1 occur in N. America, but not yet in Europe; further resistance genes are also available (APS Compendium). Inoculum should also be destroyed by ploughing in debris, after thoroughly breaking it up. Such buried debris can remain infective, as was found in S.W. France when later mechanical weeding operations brought it back to the surface.
CASSINI & SMITH

VENTURIACEAE

Venturia inaequalis (Cooke) Winter

(Anamorph *Spilocaea pomi* Fr., syn. *Fusicladium dentriticum* (Wallr.) Fuckel)

Basic description

Pseudothecia, with or without dark setae, contain ascospores (13–16×6–8 μm) slightly unequally once-septate, at first hyaline to greenish, eventually olive-brown. Conidiophores are formed on a subcuticular stroma, and produce conidia successively giving characteristic annellations (*Spilocaea*). Conidia are obpyriform-obclavate, pale to mid olive brown, smooth, 0–1 septate.

Host plants

V. inaequalis attacks *Malus*, but has also been found on *Pyrus, Sorbus, Pyracantha, Cotoneaster integerrima, Crataegus oxyacantha, Viburnum* etc. (CMI 401). Formae speciales have been proposed for strains showing special pathogenicity to apple (**f.sp. *mali***) and *Sorbus* (**f.sp. *aucupariae***) in cross-inoculation tests. Apple cultivars vary considerably in resistance, and often become more susceptible with time, in connection with variability of the pathogen.

Geographical distribution

The pathogen is found world-wide wherever apple is cultivated (CMI 401).

Disease

The fungus causes scab (or black spot) of leaves and fruits, and also, under favourable conditions, of shoots, buds and flowers. On leaves, light spots appear on either surface, becoming dark olive green and showing a visible mat of fungus. The spots expand, run into each other, and extend mainly along the veins. The fungus tends to die out in the middle of older lesions, but remains living at the margin. On fruits, the scabs are smaller and darker. The infected tissue becomes suberized and tears if the apple is still growing. These tears can be points of entry for other fungi (e.g. *Monilinia fructigena*). Scab lesions on fruit can be confused with sooty blotch (*Gloeodes pomigena*). The fungus can infect young shoots causing clear twig scab on some cultivars without very clear symptoms on others (Kennel 1981). Conidia may be released from such lesions to act as a supplementary source for new infections (Hill 1975). The shoot lesions eventually appear as pustules on the bark, finally tearing. Very late infections on fruit often do not show symptoms until after a period of storage. The spots are deep-sunken, brownish-black, with no conidial production (storage scab).

Epidemiology

The fungus overwinters in fallen infected leaves on the ground or isolated on the wood of host plants. In early spring (April in the FRG), the pseudothecia are fully formed and will release their ascospores (to 2–3 cm) when wetted. Spore trapping shows that the first spores are released within 30 min of wetting, and a maximum is reached after 1–2 h. The spores are then carried by the wind to leaves on which they will germinate provided the surface remains wet for a sufficient period (25 h at 6°C or 9 h at 16–24°C—the classic 'Mills period'). Conidia have the same germination requirements as ascospores, but need longer periods of wetness (Moore 1964). The germ tube penetrates the cuticle from an appressorium, and the mycelium grows between the cuticle and the epidermis, forming its conidiophores in that position. The cuticle tears away and the dark mycelium is exposed to release the conidia. The incubation time is dependent on temperature and atmospheric humidity.

Economic importance

All apple-growing areas run a risk of scab, whenever climatic conditions are suitable. Scabby fruit cannot be sold in high-quality grades. On storage, scabby apples are liable to be infected by various rotting organisms. Scab control is accordingly essential in all commercial production.

Control

Although some resistant cultivars are being developed, practically all commercial cultivars are susceptible. Infection pressure can be reduced by pruning, soil cultivation (favouring earthworms) and by urea

applications in autumn (favouring decay of the fallen leaves), but these measures will never eliminate the fungus altogether. Chemical control during the growing season is therefore essential (though costly and ecologically undesirable). Preventive applications are made from budburst on, and repeated at 1–3 week intervals, with the shortest intervals in spring. Examples of purely preventive fungicides are the dithiocarbamates and tolylfluanid. Treatments made up to 36 h after the beginning of an infection period will stop germ-tube growth before infection is established: fungicides like dodine and dithianon have this so-called 'stop' action without being systemic. Captan, mainly used as a preventive, also has a certain 'stopping' activity. Finally, its subcuticular mycelial growth habit allows *V. inaequalis* to be treated with the new systemic fungicides up to 96 h (bitertanol, fenarimol, penconazole) after the beginning of the infection period. Infection periods can be determined by combined use of Mills tables with a thermohygrograph and leaf-wetness recorder. A spore-trap is also needed to monitor ascospore release in the spring. Electronic warning devices (Richter & Häusserman 1975) can facilitate the determination of infection periods, and computer warning systems have been developed for whole regions, as in Michigan (USA) (Jones 1978). A biological approach to scab control is conceivable, but this implies modification of existing production systems: scab-resistant cultivars, special cultural practices for trees and soil, biological treatments. Antagonists such as *Chaetomium globosum* Kunze seem promising in trials (Andrews *et al.* 1983).
RICHTER

Venturia pirina Aderhold

(Anamorph ***Fusicladium pyrorum*** (Lib.) Fuckel)

V. pirina is very variable in morphology and pathogenicity. Several races are known, differing in aggressiveness; at least 5 pathotypes are reported from Israel (Shabi *et al.* 1973). The fungus attacks only pear (CMI 404) and occurs in temperate and subtropical zones wherever pears are grown. It causes scab (or black spot) of shoots, leaves, fruits and buds, very similar to apple scab. Infections of the wood are striking, giving conspicuous swellings which later burst. However, damage does not generally penetrate very deeply. In dry years, cork formation by the tree can prevent any further penetration. Spotting on fruit can be even more severe than for apple scab. Infected fruits are deformed and deeply cleft. Pear scab develops under much the same conditions as apple scab, surface wetness and temperature having the most effect. Because of the importance of shoot infections (unlike on apple), epidemics are likely to start even earlier than if they came from ascospores alone. The criteria for control are much the same as for apple scab, but treatments must start earlier. Phytotoxicity of products is more of a problem than with apple. Although there is no formal demonstration that Mills periods can be applied to pear scab, treatments based on them have not led to any noteworthy failures. Dew formation is particularly important for infection.
RICHTER

Venturia carpophila E.E. Fisher (anamorph ***Fusicladium carpophilum*** (Thüm.) Oudem., syn. *Cladosporium carpophilum* Thüm.) causes scab of almond, apricot, peach and plum (CMI 402, CMI map 198). Lesions occur principally on twigs rather than leaves, and these are the source of conidia for the characteristic infection of the fruits. Though generally of rather minor importance, *V. carpophila* requires well-timed fungicide sprays on mirabelle plums in E. France (Raymondaud *et al.* 1985). ***V. cerasi*** Aderhold (anamorph ***Fusicladium cerasi*** (Rabenh.) Sacc.) is widespread on sweet cherry (CMI 706; CMI map 196) on which it causes leaf and fruit scab, overwintering on dead leaves and on shoots. It is of very minor importance in commercial orchards but can be damaging on sour cherry in gardens in wet weather.

Venturia populina (Vuill.) Fabricius (anamorph ***Pollaccia elegans*** Servazzi) is widespread in Europe, including the USSR and is also known in Canada. It attacks many species of *Populus* (especially *P. nigra*), producing relatively minor leaf spotting, death of leaves and dieback of the shoots throughout the growing season.

Venturia tremulae Aderhold

(Anamorph ***Pollacia radiosa*** (Lib.) Baldacci & Cif.)

V. tremulae is found on *Populus* species within the section *Leuce,* thus on *P. alba, P.×canescens, P. grandidentata, P. tremula* and *P. tremuloides.* The fungus is widely distributed in Eurasia and N. America, and has commonly been referred to as *V. macularis* (Fr.) E. Müller & v. Arx. However, this name can no longer be used since *Sphaeria macularis* Fr. is another fungus. Morelet (1985, 1986) distinguishes three varieties of *V. tremulae,* **var. *populi-albae*** exclusively on *P. alba,* **var. *grandidentatae*** especially on the American species *P. grandidentata* and *P. tremuloides* and **var. *tremulae*** mostly on *P. tremulae* (with 2 pathotypes of the last). The account below relates principally to var. *tremulae*.

Young leaves and shoots are infected in the spring by ascospores and/or conidia. The ascospores are thrown out from ascomata in old leaves on the ground, while conidia are produced on shoots attacked the year before. The infected leaves develop black, necrotic spots, and are often completely killed. The upper part of the shoot or the whole shoot is also often killed and turns black. Conidia are produced in moist weather on the necrotic tissue; they may infect other leaves and shoots. Young trees, seedlings and suckers of *P. tremula* may be badly damaged; there are, however, individual differences in resistance. Care should be taken in breeding to work with fungal material of known variety and pathotype (Morelet 1986). In nurseries it is possible to spray against the fungus, but the best control measure may be to breed resistant clones.
ROLL-HANSEN

Venturia chlorospora (Ces.) Karsten

(Anamorph ***Fusicladium saliciperdum*** (Allescher & Tubeuf) Tubeuf)

V. chlorospora, which is widespread in Europe, and in N. America, attacks many willows, especially some forms of *Salix alba* (including the var. *vitellina*), *S. americana* hort, and *S. decipiens*. Some willows, including *S. alba* var. *coerulea, S. fragilis* and *S. viminalis,* appear to be resistant. *V. chlorospora* causes the disease called willow scab. Irregular spots form on the leaves, which die and fall. Black, scab-like lesions arise on the shoots, and girdled twigs and branches die back. Pustules with *F. saliciperdum* conidia develop on the leaves and twigs.

The fungus overwinters on dead twigs infected the previous spring and summer. In some areas the teleomorph arises on fallen leaves in autumn and winter. Infection is mainly from conidia which form on infected twigs in spring and attack the young unfolding leaves. Cool, wet weather encourages the disease. In some areas, *V. chlorospora* causes severe damage in osier beds, and may reduce the growth of susceptible tree willows by causing much premature leaf-fall. Severe attacks render ornamental willows very unsightly. Nurserymen should take cuttings only from healthy stools, and apply winter washes to the stool beds. Nursery plants, osier beds, and young trees in parks and gardens may also be sprayed in the growing season with maneb, zineb, captan or brestan (Diercks 1959).

Nüesch (1960) has shown that several other *Venturia* species occur on willows and ***V. saliciperda*** Nüesch (anamorph ***Pollaccia saliciperda*** (Allescher & Tubeuf) v. Arx) is considered by some to be a more important pathogen than *V. chlorospora* (CMI 481). The relative importance of these fungi in different countries needs to be evaluated.
PHILLIPS

Spilocaea oleagina (Castagne) Hughes

(Syn. *Cycloconium oleaginum* Castagne)

Basic description

S. oleagina produces subcuticular colonies essentially like those of other scab fungi. On the basis of conidiophore morphology, it is clearly to be placed in *Spilocaea* rather than *Cycloconium*. The teleomorph is not known but could well be *Venturia.*

Host plants

Species of *Olea* are infected especially the olive tree, *Olea europaea*.

Geographical distribution

S. oleagina is widespread in the Mediterranean region, as well as in the major olive-growing areas of the world (CMI map 183).

Disease

S. oleagina causes a leaf-spot disease with symptoms mostly on the upper surface of the leaves (Fig. 55). The initially inconspicuous, then scarcely detectable lesions, later enlarge to form dark-brown, circular, mostly annular (with a green centre) or zonate spots that become lightly velvety and are often surrounded by concentric, faint yellow, violet or pale brown haloes (the so-called 'peacock' or 'bird's eye spots'). Chlorosis may extend to a great part of the lamina, and the leaf falls. The pathogen may infect the fruit pedicels, causing fruit drop, but direct attacks on drupes (scab) and on the tender shoot tips are rare. The infecting hypha enters the leaf by piercing and enzymatically degrading the thick cuticle, then grows parallel to the leaf surface as hyaline, septate, branched subcuticular mycelium. The colonies remain localized in a cutinized layer of the epidermal

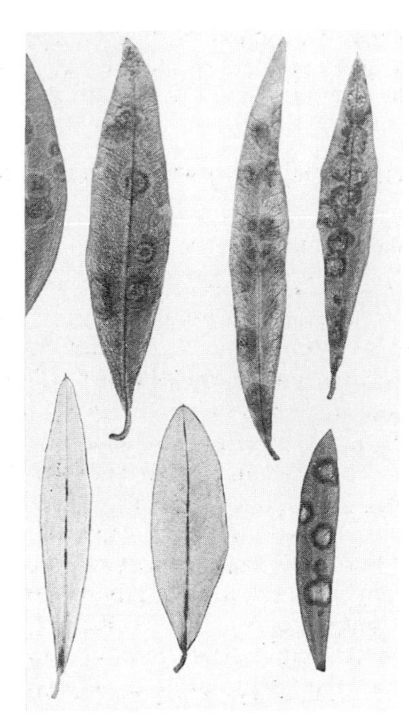

Fig. 55. Symptoms of leaf spot (*Spilocaea oleagina*) on the upper and lower surface of olive leaves.

cell wall until the leaf tissues decay. This habit has been associated with a defence reaction of the diseased leaf tissues (Graniti & De Leo 1966), that includes mobilization and hydrolysis of the phenolic glucoside oleuropin to give a phenolic aglycone, which in turn would inhibit the cell wall degrading enzymes produced by the fungus, thus hindering its further progression into the mesophyll. The affected leaves fall prematurely (a healthy leaf normally lives about one and a half years). Severe leaf-fall following heavy infection may defoliate trees. Recurrent infections cause poor growth and dieback of defoliated trees.

Epidemiology

If the environmental conditions are favourable (Laviola 1966; Saad & Marsi 1978), *S. oleagina* may grow throughout the year on its evergreen host. The disease is particularly severe in densely planted or poorly ventilated groves and in nurseries. The inoculum for primary infection usually comes from sporulating spots on the hanging leaves that have either overwintered or aestivated on trees. Conidia formed on these spots may be viable for several months. Once detached from their conidiophores, however, conidia lose their germinability in less than 1 week. Although 'dry spores', the conidia of *S. oleagina* are not effectively disseminated by air-currents. They are usually carried down by rain water, a limited lateral spread being probably assured by winds carrying conidia-laden droplets. Consequently, infection is more severe on the lower parts of the canopy, where humidity is also higher. Conidia on leaves that have fallen to the ground are practically insignificant for infection. Infection takes place within a wide range of temperatures with the optimum at 18–21°C (formation of appressoria seems to require 16–24°C), provided that wetness or a nearly saturated atmosphere is maintained on the leaves for 1–2 days, depending on the temperature. Hot summers, drought, and spells with no rains when other conditions are favourable, are in fact limiting factors for disease development. Usually there are one or two main infection periods: mostly during autumn and winter (in areas with dry summers and mild winters) or in spring and early summer (in areas with colder winters) or in both seasons, depending on local conditions and the seasonal growth of the trees. The incubation period is about 2 weeks under the most favourable conditions, but if infection is followed by a hot and dry (less frequently, cold) season, it may last several weeks and even months. For example, the appearance of leaf spots in autumn may be due to latent late spring infections*. Spots already formed in spring may stop growing in summer and resume their growth and sporulation, i.e. by extending their margins to a new ring, at the first autumn rains.

Economic importance

Apart from the dieback of twigs and branches defoliated by recurrent leaf-spot infection, losses arise mostly from a reduction of leaf coverage. Yield losses may also be a consequence of non-conversion of the axillary buds of leaves shed by the disease into apices developing flowering shoots, so that only macroblasts develop. Direct damage to fruit is rare and may only be important for table olives. Although data are scarce or not available for many countries, an average loss of 20–30% is thought to be a cautious estimate for areas where the disease is recurrent on

*Techniques have been devised (Loprieno & Tenerini 1959) for detecting latent infections, e.g. by dipping leaf samples in a hot (50–60°C) 5% KOH or NaOH solution for 2–3 minutes, until small, roundish, black spots appear where the colonies of *S. oleagina* are localized.

susceptible olive cultivars. However, higher losses have been reported.

Control

Pruning and other measures aimed at reducing humidity and shading are recommended for trees or groves subjected to recurrent attacks of leaf spot. Chemical control schedules include the application of fungicides before and during the main infection seasons. Bordeaux mixture, copper oxychlorides or some long-persistent organic fungicides (e.g. captafol) are generally used. In many areas with a dry Mediterranean climate, three sprays (at winter end, summer end and late autumn) are suggested. However, number (1–8) and time of applications vary considerably according to local and seasonal conditions. Aerial distribution of copper-based oil formulations by aircraft sprayers has been used in places where size and density of the olive groves, rocky or uneven ground, lack of water or other factors make treatments with the usual sprayers inapplicable or uneconomic. A phytotoxic effect shown by certain fungicides (especially copper) on infected leaves may be advantageous where infections occur in limited periods. After preventive sprays, most leaves with leaf-spot lesions fall to the ground, thus freeing trees from the existing sources of inoculum. Olive cultivars and clones often show a different susceptibility to the disease, and local selections are known to resist leaf spot, at least under certain environmental conditions. However, little has been done to improve the resistance of valuable but susceptible cultivars through selection or breeding procedures.
GRANITI

Spilocaea pyracanthae (Otth) v. Arx (syn. *S. eriobotryae* (Cavara) Hughes, *Fusicladium pyracanthae* (Otth) Viennot-Bourgin) causes scab on leaves and fruits of ornamental *Cotoneaster* and *Pyracantha*, and also on loquat (*Eriobotrya*) in the Mediterranean region and on medlar (*Mespilus germanica*) (CMI map 435). ***Apiosporina morbosa*** (Schwein.) v. Arx (syn. *Dibotryon morbosum* (Schwein.) Theiss. & H. Sydow) causes black knot of *Prunus* in N. America (CMI map 48). It is an ascospore-dispersed canker pathogen, especially important on plum, which is absent from Europe and should be excluded as a quarantine organism (CMI 224; EPPO 10). ***Platychora ulmi*** (Schleicher) Petrak (anamorph ***Piggotia ulmi*** (Grev.) Keissl.) is extremely common on the leaves of elm in the autumn, causing a conspicuous tar spot, which is however of no economic importance. The black stromata first develop pycnidia, then pseudothecia, over the autumn and winter (Viennot-Bourgin 1949).

Phaeocryptopus gaeumannii (Rohde) Petrak

(Syn. *Adelopus gaeumannii* Rohde)

P. gaeumannii is found only on species of *Pseudotsuga*. It affects all varieties of the Douglas fir, *Pseudotsuga menziesii*. Some differences in resistance to the fungus are shown by different provenances and by individual trees. It is known throughout Europe, in N. America, and in New Zealand and Tasmania (CMI map 42). The fungus causes the needle disease usually called Swiss needle cast. In late summer the affected needles become yellowish-green, and if badly diseased they later fall. The black pseudothecia of the fungus may be found in spring and early summer on the undersides of the leaves. Most diseased trees are between 10 and 40 years old, though older and younger trees may sometimes be affected.

The pseudothecia shed their spores in May, June and early July. The ascospores infect the young needles. The fungus grows within the needle tissues, and over the winter forms dark plug-like stromata in the cavities below the stomata on the undersides of the leaves. In spring the pseudothecia form in these stromata, projecting above the leaf surface outside the stomatal pores. Their asci begin to ripen in May, to shed their spores and so complete the life cycle. High humidity and high summer rainfall encourage the growth of the fungus (Merkle 1951). In some parts of Europe, *P. gaeumannii* sometimes causes severe defoliation. It then rarely kills trees, but may leave them open to attack by other fungi, especially *Armillaria* species. Nevertheless, although it is widespread and Douglas fir needles may be densely covered by its pseudothecia, in most cases it causes little damage. Control measures are not usually necessary, but in N. America Chastagner & Byther (1983) obtained excellent control on Christmas trees by spraying with chlorothalonil soon after bud break.
PHILLIPS

Rhizosphaera kalkhoffii Bubák

R. kalkhoffii forms black-stalked pycnidia (CMI 656), that occur mainly on spruces (especially *Picea pungens*, *P. abies* and *P. sitchensis*), but also on species of *Abies*, *Pinus* and *Pseudotsuga*. It occurs throughout most of Europe, and in N. America and Japan. Other *Rhizosphaera* spp. are anamorphs of *Phaeocryptopus*. *R. kalkhoffii* is usually only a saprophyte, but appears at times to cause a needle cast, especially of the blue spruce, *Picea pungens* cv. *glauca*. In May, the needles of affected trees become purple-brown. The small pycnidia of the fungus emerge through the stomatal openings, at first capped by a small white plug of wax. The diseased needles may then fall. In most areas, needle cast associated with *R. kalkhoffii* is of little or no importance. If necessary, affected trees may be sprayed with Bordeaux mixture, though this cannot be done on a forest scale (Waterman 1947).

PHILLIPS

ELSINOACEAE

Elsinoë ampelina Shear

(Anamorph ***Sphaceloma ampelinum*** de Bary)

E. ampelina has not been reported in its teleomorph state in Europe, though it originated there. All *Vitis* species are more or less susceptible. In Europe, the disease (anthracnose) is now of minor importance: early spring treatments and the disappearance of direct-producing hybrids needing only limited control of *Plasmopara viticola* have practically eliminated it. Outside Europe (S. America, S. Africa, S.E. Asia) it continues to be serious wherever the climate or cultivars favour it (CMI 439; CMI map 234).

E. ampelina attacks all green parts of the plant and can cause stunted growth and even reduce yield to zero. It causes perforating lesions on the leaves, cankers on the shoots, and grey, sunken circular spots on the berries, which finally dry out. Like *Guignardia bidwellii, E. ampelina* can develop early in the season even at low temperatures (9–10°C) and, like *P. viticola,* it is favoured by rainy springs, dew and fog. Sclerotia form on the shoots, overwinter and provide inoculum in spring. The disease spreads secondarily by conidial infection of the young shoots; adult leaves are hardly affected. The incubation period is short (5–6 days). More information is needed on survival and development of the fungus.

The classic treatment was to destroy the sclerotia in winter with a solution of 35 kg ferrous sulphate per litre of sulphuric acid. Sodium arsenite (1.25%) is now preferred, since it is also effective against *Phomopsis viticola* and *Stereum hirsutum*. Most fungicides applied in spring give some control, but probably more due to the date and number of treatments than to high inherent efficacy. Recent trials in Uruguay show captafol and folpet to be superior to ziram or copper products. Systemic fungicides (alone or combined with ziram) are not effective (Roussel 1984).

BULIT & ROUSSEL

Elsinoë fawcettii Bitancourt & Jenkins

(Anamorph ***Sphaceloma fawcettii*** Jenkins)

This and the related ***E. australis*** Bitancourt & Jenkins (anamorph ***S. australis*** Bitancourt & Jenkins) form epidermal or subepidermal pseudoparenchymatous dark-brown ascomata containing locules with elliptical asci. The oblong elliptical hyaline ascospores are 1–3 septate. Their anamorphs produce unicellular ellipsoid hyaline phialoconidia from hyaline or pale-brown enteroblastic integrated conidiogenous cells (CMI 438, 440). Both species cause wart-like lesions or corky eruptions (common scab; sour orange scab, Fig. 56) on fruit, leaves and stem of citrus in tropical and subtropical areas, where suitable rainfall conditions prevail. They are relatively unimportant in areas with a Mediterranean climate.

E. fawcettii occurs throughout the tropics and subtropics (CMI map 125). It has been found in Greece and Georgia and has been intercepted in the USA on fruits from Spain and Portugal. However, some European records are questionable. Besides fruits,

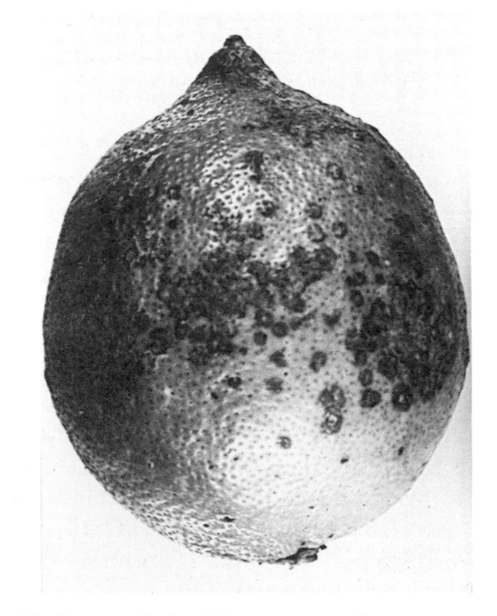

Fig. 56. Lemon fruit with symptoms of sour orange scab (*Elsinoë australis*).

leaves and stems of susceptible hosts (grapefruit, lemon, mandarin, sour orange and other citrus) show evident symptoms, whereas most cultivars of sweet orange are resistant.

E. australis causes sweet orange scab in S. America (CMI map 55). In Europe (Sicily) it has been reported on the fruit of lemon trees grown under humid and shaded conditions (Ciccarone 1956). Both scabs are usually controlled by sprays with fungicides (e.g. copper, captafol).

PERROTTA & GRANITI

Elsinoë veneta (Burkholder) Jenkins

(Anamorph ***Sphaceloma necator*** (Ell. & Ev.) Jenkins & Shear)

E. veneta forms subepidermal ascomata containing several monascous locules with ovoid thick-walled asci bearing 3-septate elliptical ascospores with gelatinous sheaths. Single-celled hyaline conidia are produced from subcuticular acervuli (CMI 484). The fungus attacks most *Rubus* cane fruits, but in Europe it is most prevalent on loganberry and raspberry. *E. veneta* occurs throughout Europe, N. and S. America and in Australasia; it causes cane spot, a potentially serious disease affecting all aerial parts of the plant. As the name suggests, the disease is best known for the deeply penetrating grey lesions with red margins on the canes, which in early and severe attacks may girdle them. However, in mild humid districts (e.g. Ireland) the fungus also severely attacks young leaves on the lateral shoots and destroys the interveinal tissues causing serious yield loss. Infection of developing fruits makes them unsightly and unmarketable.

The fungus survives on overwintering canes and old cane stubs at the base of the plant and releases ascospores and conidia, both of which may establish infections. Sprays of dichlofluanid or benomyl applied during the blossom period to control *Botryotinia fuckeliana* also control cane spot if the spray is directed to cover young canes. In areas of high risk, or with particularly susceptible cultivars sprays should commence when the axillary buds on fruiting canes open (O'Ríordáin 1969). Fewer cane spot lesions develop on the replacement canes after removing the first flush of young canes for cane vigour control but this method can be used only for tolerant cultivars growing vigorously in established plantations (Lawson & Wiseman 1983). This disease is most prevalent in overgrown and weed-infested plantations.

WILLIAMSON

Elsinoë pyri (Voronichin) Jenkins causes an unimportant anthracnose of apple and pear leaves and fruit throughout Europe and in USA (Viennot-Bourgin 1949). ***E. rosarum*** Jenkins & Bitancourt (anamorph ***Sphaceloma rosarum*** (Pass.) Jenkins) causes spot anthracnose of rose leaves. It is widespread in Europe and N. America and occasionally damaging under wet conditions (Viennot-Bourgin 1949; APS Compendium).

BOTRYOSPHAERIACEAE

Botryosphaeria obtusa (Schwein.) Shoem.

B. obtusa (syn. *Physalospora obtusa* (Schwein.) Cooke) with anamorph ***Sphaeropsis malorum*** (Schwein.) Cooke (syn. *Botryodiplodia malorum* (Berk.) Petrak & H. Sydow), attacking Rosaceae, is generally regarded as a separate species from ***B. quercuum*** (Schwein.) Sacc. (anamorph ***Sphaeropsis sumachi*** (Schwein.) Cooke & Ell.) on other woody hosts, though Boerema & Verhoeven include the former in the latter, with anamorph ***Botryodiplodia juglandicola*** (Schwein.) Sacc. It is found on numerous woody hosts, including fruit crops (apple, pear, peach, quince, grapevine, plum) and ornamentals, either as a weak parasite or as a saprophyte (CMI 394). For apple and pear, no resistant cultivars were found (Reis & Zanetti 1977). The fungus is very widely distributed on pip-fruit in temperate zones, but is mostly reported as damaging in Canada and the USA. In Europe, it is reported in Cyprus, France, Germany, Greece, Hungary, Italy, Spain, UK, USSR and Yugoslavia, attacking mainly fruits in northern areas, but also leaves and shoots further south. On apple, *B. obtusa* causes black rot of fruits, frog-eye leaf spot and shoot cankers (black-rot canker, New York apple canker, twig blight). Cardinal temperatures are 5°, 25–30° and 37°C (Cristinzio 1979). The teleomorph has mostly been considered of minor importance for disease, as it is mainly seen in saprophytic lesions on dead tissues. However, Sutton (1981) has shown that in N. Carolina (USA) both ascospores (air- and water-dispersed) and conidia (water-dispersed) are produced during the growing season, with a maximum in April–May. Pruning wounds are rapidly colonized and can produce ascospores and conidia to infect fruit during the same season. Losses are rather minor, except in the USA and the USSR. In W. Europe, the disease is seen rather sporadically on various fruit crops (especially pear and grapevine). Orchard treatments with ben-

zimidazole fungicides, or captafol, have proved effective (Matta & Mancini 1976) as have also benomyl+ziram (Reis & Zanetti 1977). Cankers can be painted with 5% $FeSo_4$ or 3% $CuSo_4$ (Parii 1978).
BOMPEIX

Botryosphaeria dothidea (Moug. ex Fr.) Ces. & de Not. causes a minor rose canker in Europe but is becoming increasingly serious on apple and peach in the S. USA (McGlohon 1982). ***B. ribis*** Grossenbacher & Duggar (anamorph ***Dothiorella gregaria*** Sacc.) (CMI 395) is a weak pathogen causing cankers on many woody plants. ***B. zeae*** (Stout) v. Arx & Müller (syn. *Physalospora zeae* Stout) causes grey ear rot of maize in the USA and has been recorded in France (CMI 774) and Hungary. Pseudothecia and pycnidia develop on leaves and sclerotia in rotted kernels and ears (APS Compendium). The anamorph described as *Macrophoma zeae* Tehon & Daniels is correctly a *Dothiorella*. With available resistant cultivars, this pathogen is no longer of any great importance.

Guignardia bidwellii (Ell.) Viala & Ravaz

(Anamorph **Phyllosticta ampelicida** (Engelmann) v. d. Aa, syn. *Phoma uvicola* Berk. & Curtis)

Basic description

Globose pseudothecia form in a stroma immersed in the plant tissue and contain hyaline one-celled ascospores. Pycnidia are abundant, with one-celled pycnospores bearing hyaline appendages. Spermatia are also formed (CMI 710).

Host plants

G. bidwellii is only found on Vitaceae (*Vitis, Ampelopsis, Parthenocissus*). Different *Vitis* species vary considerably in susceptibility. *V. vinifera* is among the most susceptible, but its cultivars show some resistance. Hybrids show intermediate resistance.

Host specialization

Races differing in aggressiveness have been reported.

Geographical distribution

In Europe, *G. bidwellii* is present only in France, especially along the Atlantic coast. Other areas are much less affected (Savoie, Ardèche, Allier). The disease has been reported in N. and S. America, but not in other continents (CMI map 81).

Disease

All green parts of the plant may be affected by black rot. The principal characteristic symptom is the presence of abundant black pycnidia in all lesions. On leaves, small discoloured lesions enlarge to form small irregular brick-red spots with a brown border. Pycnidia appear within these spots in 2–3 days. On petioles, tendrils, young shoots and inflorescences, elongated brown necrotic lesions appear, containing pycnidia, and may lead to desiccation of the organ attacked. Infected berries turn brown, are covered with pycnidia, and finally shrivel, remaining on the stalk. A sudden attack can destroy much of the grape harvest. Leaves are very susceptible when young, but become resistant as they mature. Berries remain fairly susceptible up to picking, but show a certain decrease after berry touch.

Epidemiology

Epidemics start early in spring, since the minimum temperature for development (9°C) is readily attained before the grapevines start to grow (e.g. in S.W. France). High temperatures do not destroy the conidia, protected by the pycnidia. Prolonged spring rain is the most important factor for infection ('pluies à black-rot'). The first spring rain moistens the last season's shrivelled berries and the ripe pseudothecia they contain. Repeated rain stimulates ascospore discharge and primary infection. According to the rainfall pattern in different years, ascospores may continue to be discharged for 1.5–3 months. Rainfall also favours secondary spread of pycnospores from the primary lesions, and the two processes may overlap (the incubation period of primary lesions being 16–20 days). The risks are especially high in this case, and close attention is needed. As for *Plasmopara viticola*, dew and fog do not seem sufficient to allow secondary spread. Finally, pseudothecia form in the affected berries, but their normal full development requires exposure to light (and may thus be affected by cultural practices).

Economic importance

Losses depend on climatic conditions and on the presence of the inoculum. Under favourable conditions for the fungus, attacks may lead to almost complete destruction of the grapes. In practice, the critical factor is the presence of the fungus in the plot or in adjoining abandoned plots.

Control

Special control measures are essential, and treatments against black rot have to be applied separately from the principal treatments against *P. viticola*. This entails extra costs, and it should be stressed that growers often make unnecessary 'insurance' treatments, through ignorance or excessive fear of the disease. In fact, an excellent warning system has been developed in France on the basis of pseudothecium maturation, ascospore release, and overlap of primary and secondary infection (Roussel 1970). Regular biological observations, coupled with observations on the crop and weather, can be used to define the times of major risk at which treatment is necessary. Chemical control has to be preventive, and timed separately from control of *P. viticola*. Even if the same products were effective, this would not remove the need for early treatment against black rot (and possibly also treatments after berry touch). In fact, chemical control presents many problems, due to the exceptionally long period of susceptibility, and to the need for really effective early treatment of the primary foci. These often pass unnoticed, but are extremely important in relation to later infection of the berries. In addition, the level of efficacy of contact fungicides against black rot is not very high (by comparison with other grapevine diseases), possibly because of very high inoculum levels at the time of primary infection. Organic fungicides are superior to copper compounds (which are quite inadequate), and the dithiocarbamates better than phthalimides such as folpet. The new systemic products used against *P. viticola* have little activity, and depend on the presence of an associated contact fungicide. In addition, it is essential to take various cultural measures to reduce the initial inoculum. In many French departments, official orders oblige growers to destroy the abandoned plots which often act as endemic sources of infection, or else to return them to normal production. If possible, infected berries should be collected and destroyed. Mechanical harvesting poses special problems in this connection, since the shrivelled berries are not detached from the stalks and remain on the plants.

Special research interest

Certain aspects of the biology of *G. bidwellii* remain obscure, e.g. factors controlling pseudothecium formation and involvement of pycnidia or spermatia in this. Better understanding might explain some very late primary attacks seen on berries in France in 1982–3. In addition, better fungicides are certainly needed. Certain sterol-synthesis inhibitors show promise, but, in view of their mode of action and persistence, would still need to be associated with contact products (Roussel *et al.* 1981).

ROUSSEL & BESSELAT

Guignardia aesculi (Peck) V.B. Stewart (anamorph ***Leptodothiorella aesculicola*** (Sacc.) Sivan., syn. *Phyllosticta sphaeropsoidea* Ell. & Ev., *Phyllosticta paviae* Desm.) is widespread in Europe and N. America, and sometimes causes a severe leaf blotch disease of horse chestnut trees. ***G. baccae*** (Cavara) Jacz. (anamorph ***Phoma flaccida*** Viala & Ravaz) was long held responsible for excoriosis or false black rot of grapevine in Europe, while *Phomopsis viticola* caused another form of the disease in America (Galet 1977). This distinction cannot be maintained (see section on *P. viticola*) and *G. baccae* is now considered merely saprophytic or secondary. ***G. citricarpa*** Kiely (anamorph ***Phyllostictina citricarpa*** (McAlp.) Petrak) causes black spot of citrus fruits, a serious disease in many parts of the world, but not apparently in the few European countries where it has been reported (Italy, Spain, USSR), where a weakly pathogenic form with a wider host range is found (CMI 85; CMI map 53). ***G. laricina*** (Saw.) Yamam. & Ito causes shoot blight and dieback of larch in China and Japan. In the last 20 years, with increasing larch acreages in Japan, it has been extremely damaging on nurseries and young plantations, requiring repeated cycloheximide treatment. It is accordingly an important quarantine organism for Europe (EPPO 12; CMI map 545).

Discosphaerina fulvida (Sanderson) Sivan. (syn. *Guignardia fulvida* Sanderson) is better known as its anamorph ***Aureobasidium lini*** (Lafferty) Hermanides-Nijhof (syn. *Polyspora lini* Lafferty, *Kabatiella lini* (Lafferty) Karakulin, *Aureobasidium pullulans* var. *lini* (Lafferty) W. Cooke). It causes browning and stem-break of flax in most of Europe (CMI map 88) and can attack the crop at any stage of development. Under conditions of high humidity, the flower and capsules may be affected. The cap-

sules take on a 'bronzed' appearance, and the seeds may finally be affected.
JOUAN

Aureobasidium pullulans (de Bary) Arnaud, of which the preceding fungus was once regarded as a variety, is an abundant plant surface epiphyte producing black yeast-like curved conidia. Since a number of fungi may grow in this way (see, for example, *Sydowia polyspora*) it is not altogether clear what are the true affinities of the species, and whether indeed it may not need further taxonomic attention. It is very frequently isolated from plant tissues and has been associated with various diseases, probably without justification. Its role in causing yellow discoloration and vein browning of grapevine leaves is discussed by Vercesi *et al.* (1982). More recently, it has been considered as a potential antagonist of plant pathogens in the phylloplane, with other yeast-like fungi (Fokkema & Van der Meulen 1976; Brame & Flood 1983). It is certainly very sensitive to fungicide treatments (Skajennikoff & Rapilly 1981; Fokkema & de Nooij 1981), which could thus disturb natural balances and favour certain pathogens, e.g. *Pyrenophora* species. However, the importance, in practice, of antagonism between plant pathogens and these yeast-like fungi still needs to be evaluated.

Kabatiella caulivora (E. Kirchner) Karakulin

(Syn. *Aureobasidium caulivorum* (Kirchner) W.B. Cooke)

K. caulivora produces curved brown conidia in acervuli, budding laterally or terminally to form further conidia. In culture, only this yeast-like conidial form is initially obtained, appearing white or pink then darkening to brown as mycelial growth begins. The fungus causes scorch (or anthracnose) of *Trifolium pratense* in temperate zones, e.g. in UK, Ireland, Netherlands, Scandinavia, USSR. In the FRG and Switzerland (Troxler *et al.* 1980), it has also been noted causing serious disease on *T. alexandrinum*. In more southern European countries (e.g. France, Italy), the pathogen has been recorded but does not appear to be of any great importance. Elsewhere in the world it is important on *T. subterraneum* (Australia) and *T. incarnatum* (N. America), and is also known in the USA as northern anthracnose to distinguish it from southern anthracnose due to *Colletotrichum trifolii* (q.v.).

The disease mainly affects the stems and petioles, on which small black spots appear, spreading to form dark-edged brown lesions up to 5×0.5 cm, with greyish acervuli finally appearing in the centre. Such lesions are common on the inflorescence stalks, causing the inflorescences to bend down in a 'shepherd's crook' form. The distal portions finally wilt. Leaflet stalks are very susceptible, and when infected, twist to turn the under surface of the leaflet upwards (Massenot & Raynal 1973). Heavily infected crops have a shrivelled and burnt appearance. Because a single lesion can kill a whole leaf or inflorescence, scorch is potentially a much more damaging disease than the leaf spots of clover. The fungus overwinters on infected plants and debris, from which rain-splashed conidia will start epidemics in mild wet weather in spring. Spread is most rapid in dense swards, at high humidity and low light intensity under the canopy (O'Rourke 1976). The fungus is also seed-borne, directly or on debris accompanying seed, and will then kill young plants. However, this is of relatively minor importance in practice.

General cultural measures for control include rotation, early cutting, removal of debris, and use of clean or treated seed. Fungicide treatments have been tried e.g. with benomyl (Helms & Andruska 1981) but use of resistant cultivars is the main control method, since there is considerable variation in resistance between available cultivars. Tetraploid *T. pratense* cultivars are generally more resistant than diploid. Breeding work has been carried out mainly in N. America (Smith & Maxwell 1973) and Australia.
CARR & SMITH

Kabatiella zeae Narita & Hirats.

K. zeae causes eyespot of maize and was reported from Japan and America before first appearing in Europe (Belgium, France, FRG) in the early 1970s. It causes spots on the leaves and sheaths, typically with a pale centre, a brown or purple ring and a yellowish halo (eyespots), fusing to form large necrotic areas. The long curved conidia are produced under very wet conditions and wind and splash-dispersed (APS Compendium). The fungus forms a tough stroma within the leaf tissues, which persists in leaf debris to sporulate and start new infections the following season (Cassini *et al.* 1972). No teleomorph is known. Serious outbreaks were observed in France in the early 1970s, apparently as a consequence of more frequent maize monoculture com-

bined with minimum tillage techniques (Cassini 1973). While some cultivars show a degree of resistance, and benzimidazole fungicides may have some potential, the only practical control lies in avoiding successive maize crops and in destroying the overwintering inoculum. Fungicide treatment of debris is ineffective, as the stromas are extremely resistant. Burying debris by ploughing, especially after sufficient crushing, will almost completely eliminate any risk of infection the following year. *K. zeae* has in fact caused no recent problems.
CASSINI & SMITH

DIDYMOSPHAERIACEAE

Didymosphaeria arachidicola (Khokhryakov) Alcorn *et al.* (anamorph ***Phoma arachidicola*** Marasas *et al.)* causes net blotch of groundnut and occurs in the USSR (CMI 736).

DOTHIORACEAE

Dothiora ribesia (Pers.) Barr (syn. *Plowrightia ribesia* (Pers) Sacc.) commonly produces black pustules (ascostromata) up to 3 mm wide on *Ribes*, especially redcurrant (CMI 707). It is essentially a wound parasite of little practical importance, unable to attack healthy tissue except from a saprophytic base on moribund tissue. ***Pleosphaerulina (Pringsheimia) sojaecola*** Miura (anamorph ***Phyllosticta sojaecola*** Massal.) causes a leaf spot of soybean in Bulgaria, Italy and USSR, as well as in N. America. It is seedborne and may also overwinter on crop debris (APS Compendium).

Sydowia polyspora (Bref. & Tav.) E. Müller

(Anamorph ***Sclerophoma pythiophila*** (Corda) Höhnel)

S. polyspora forms small black, globoid pseudothecia containing a few large asci. Each ascus contains from 20–26 colourless obovate ascospores each divided by up to three horizontal septa and sometimes also by one vertical septum, and measuring 6×17 μm (Miller 1974). The conidia measure 4–8×2–3 μm. In culture the fungus commonly produces an anamorph resembling *Aureobasidium pullulans*. *S. polyspora* occurs on the needles of many conifers, including those of *Abies, Larix, Picea, Pinus, Thuja* and *Tsuga* and is sometimes also found on conifer wood. It occurs widely in Europe into the USSR, and in N. America. *S. polyspora* is a common saprohyte on conifer needles, but sometimes causes blue stain in pine wood, and is associated with several conifer diseases (Batko *et al.* 1958), including shoot dieback, leaf cast, and (in the USSR) the production of witches' brooms. In some cases the fungus appears to have followed and increased the damage caused by the pine needle midge, *Contarinia baeri*. Outbreaks of disease involving *S. polyspora* have been mainly on pines (especially *P. sylvestris*), but in California (USA) Miller (1974) has described a tip dieback of *Abies concolor* (grown as Christmas trees) caused by this fungus. Damage by *S. polyspora* is generally local and sporadic. As outbreaks of the disease it can cause are relatively infrequent, control measures have not been evolved to deal with it.
PHILLIPS

SCHIZOTHYRIACEAE

Schizothyrium pomi (Mont.) v. Arx

S. pomi is the only member of the Microthyriales (forming flattened shield-like pseudothecia) to cause a significant plant disease in Europe. The pycnidial anamorph, long known as *Leptothyrium pomi* (Mont. & Fr.) Sacc., is now considered to be ***Zygophiala jamaicensis*** Mason, which causes greasy blotch of carnation (Baker *et al.* 1977). *S. pomi* is mainly known from apple, pear, and sometimes plum, but also attacks numerous woody plants which can act as reservoirs (e.g. in hedgerows around orchards). Orchards throughout Europe are attacked sporadically, according to season. The disease on carnation is important in the USA but not in Europe. *S. pomi* causes flyspeck on fruits, forming blackish stromata about 1 mm across firmly adhering to the surface of the epidermis. These specks often occur in clusters several cm in diameter. The disease was first described in France as long ago as 1834. It is favoured by warm wet summer weather (Geoffrion 1984). Although common in untreated orchards, normal spray programmes in commercial orchards readily eliminate it. Captan, dodine or copper in early September give good results. Captafol has been found better than zineb, when applied in mid-June and 15–20 days before harvest (Kondal & Dwedi 1979). Bitertanol has given better results than mancozeb and dinocap (Yoder 1982b).
BOMPEIX

ANAMORPHS OF DOTHIDEALES

Phoma tracheiphila (Petri) Kanchaveli & Gikashvili)

(Syn. *Deuterophoma tracheiphila* Petri)

Basic description

Immersed, flask-shaped or globose pycnidia appear as black points within lead- or ashen-grey areas of withered twigs, and occasionally on bark cracks, pruning cuts and leaf-scars. Mature ostiolate pycnidia bear a definite neck which separates and eventually tears the epidermis away from the underlying cortical tissues of the twig (Fig. 57). Within the pycnidial cavity, minute unicellular hyaline pycnospores are produced by enteroblastic phialidic conidiogenous cells. Larger phialoconidia are freely produced by hyphae grown on exposed wood surfaces (including woody debris on soil), wounded plant tissues, and—together with blastoconidia—within the xylem elements of the infected host (CMI 399; Ciccarone 1971).

Fig. 57. Ashy area of a lemon twig withered by mal secco, showing the ostiolate pycnidial necks of *Phoma tracheiphila* piercing and tearing the epidermis.

Host plants

Almost all citrus plants are susceptible to artificial infections of *P. tracheiphila*. Under field conditions, however, the disease caused by this fungus is highly destructive for lemon (*Citrus lemon*), although citron (*C. medica*), bergamot (*C. bergamia*), lime (*C. aurantifolia*), sour orange (*C. aurantium*), and rough lemon (*C. jambhiri*) are affected as well. Different degrees of resistance are shown by other species and hybrids of *Citrus* and related genera (*Eremocitrus, Fortunella, Poncirus, Severinia*). Most cultivars of sweet orange (*C. sinensis*), mandarin (*C. deliciosa*), clementine, tangerine (*C. reticulata*), and grapefruit (*C. paradisi*) are only occasionally affected by the disease. A number of rootstocks, such as Cleopatra mandarin (*C. reshni*), trifoliate orange (*Poncirus trifoliata*) and, to a lesser extent, Troyer citrange (*C. sinensis*×*P. trifoliata*) and *C. volkameriana*, have been reported to be resistant to the fungus.

Geographical distribution

P. tracheiphila is present throughout the Mediterranean region, including the Black Sea area, with the exception of a few countries (Spain, Portugal, Morocco) (CMI map 155). It first appeared on the island of Chios in Greece.

Disease

P. tracheiphila causes a vascular disease known as 'mal secco' (from the Italian 'male' = disease, and 'secco' = withered, dry). Infection may develop from twig tips and progress downwards on the affected branches. Symptoms appear in spring. Leaf and shoot chlorosis and defoliation are followed by a dieback of twigs and branches (Fig. 58). The growth of sprouts from the base of affected branches or stem, and the development of suckers from the rootstock, are a common response of the host to infection. These new growths are readily attacked

Fig. 58. Heavy defoliation and dieback of branches of lemon, due to *Phoma tracheiphila*.

by the pathogen. If not controlled, the disease progressively withers single branches and eventually kills the whole tree. An internal symptom of mal secco is a pink-salmon or orange-reddish discoloration of affected wood, seen on slanting cuts of recently infected twigs. It is associated with gum accumulation within the xylem elements. A rapid fatal form of the disease (the so-called 'mal fulminante') leads to a sudden wilting of branches or the whole tree. It is associated either with root infection or with the entry of the pathogen through wounds on the foot and trunk. Deep infections on trees may extend and cause a browning of the hard wood ('mal nero'), with no external symptoms. However, these abruptly appear when the invading fungus reaches the functioning xylem.

Epidemiology

Usually inoculum is provided by pycnospores extruded from pycnidia on withered twigs and suckers. Bits of torn epidermis, to which pycnidial necks and mycelial mats may firmly adhere, are easily removed by wind or rain splash, while the pycnidial bodies remain on the plant. Phialoconidia formed by mycelial hyphae on exposed woody surfaces of the tree or on debris may contribute to build up the inoculum. Both types of conidia are spread mainly by rain and wind, but accidental dissemination by vectors (birds and insects) has been reported. Direct leaf infection through natural openings (stomata) is questioned. The entry of the pathogen into host tissues is certainly facilitated by any kind of wounds on leaves, twigs, and other organs of the tree, as for example those produced at the insertion of lemon fruits to their peduncle. Consequently, wind, frost, hail, and cultivation practices that cause injuries to branches, trunk and roots favour infection. In the Mediterranean region, infection periods vary according to local climatic and seasonal conditions. In Sicily infections usually occur between November and February, but may occasionally occur later, e.g. following hail storms. In Israel, the infection period coincides with the rainy season (from mid-November to mid-April) with no correlation between amount of rain or number of rainy days and disease incidence. Conditioning factors for disease development are temperature (between 14 and 28°C) and RH (65–90%). The most favourable range of temperature both for the growth of the pathogen and for symptom expression is 20–25°C. Temperatures above 30°C stop mycelial growth but do not kill the fungus within the infected host tissues. Optimal temperatures for differentiation and development of pycnidia *in vitro* range between 10 and 15°C. On artificially infected sour orange seedlings, pycnidia were more abundant at a mean temperature of 10.5°C (range: 3–19°C) than at 20–22°C.

Economic importance

Mal secco is the most destructive fungal disease of lemon and other susceptible *Citrus* species in the Mediterranean region. Losses arise from a heavy reduction in yield due to: i) withering of infected twigs and branches; ii) the severe pruning that growers carry out in order to control the disease. Moreover, whole trees may be killed by the disease. If they are replaced by younger healthy trees, the orchard becomes uneven in age and size of trees. It is hard to estimate and to average the losses due to mal secco, because the disease incidence changes from year to year even in the same orchard. The disease is recurrent in orchards exposed to wind, but serious outbreaks of mal secco may follow frost spells and hail storms in spring. For all the above reasons, including the high cost of sanitary pruning, the growth of susceptible species and cultivars has been progressively reduced in the countries or areas where the pathogen is present. Moreover, the occurrence of *P. tracheiphila* discourages the introduction of new lemon cultivars or lines, including virus-free nucellar clones which are highly susceptible.

Control

Strict quarantine regulations operate in many countries to prevent the introduction of *P. tracheiphila* into mal secco-free areas. Where it occurs, mal secco is still kept under control by the costly practice of pruning infected twigs at the first appearance of symptoms (Salerno & Cutuli 1982). In order to save affected trees, growers may be compelled to cut whole branches, and sometimes they topwork pollarded trees by grafting with resistant cultivars or species. Careful pruning of withered shoots bearing pycnidia, timely removal of suckers, and burning pruned branches is recommended in order to reduce the amount of inoculum. Soil cultivation in late autumn or winter and prolonged non-tillage enhance the risk of root infection. The protection of orchards from wind and hail, e.g. with windbreaks and nets, effectively reduces infections. Chemical means of control are not widely used, except in nurseries. In fact, the repeated (e.g. 6–8) applications of protective fungicides (copper oxychlorides, captafol,

maneb+thiophanate-methyl, imazalil, etc.) which are recognized throughout the infection season (e.g. from October to February) are highly expensive, and may provide unsatisfactory results if wounds are produced on the trees. Spraying of trees is recommended immediately after hail or frost damage. Systemic fungicides (benomyl, carbendazim, thiophanate methyl), applied either as soil drenches or foliage treatments, have been tested with satisfactory results as preventives, but have at most transient effects when applied to already diseased trees, even if injected. Obviously, the best way to control mal secco would be to grow resistant cultivars or clones and to graft them onto resistant rootstocks, but unfortunately this is not easy to realize. For example, the susceptible lemon cv. Femminello has been replaced in several Sicilian areas either by cv. Monachello which is resistant but gives a qualitatively poor yield, or by other cultivars which do not adapt easily to different pedoclimatic conditions. Quite resistant clones of cv. Femminello, however, have been selected, but their diffusion is still limited.
PERROTTA & GRANITI

Phoma medicaginis Malbr. & Roum.

Basic description

Pycnidia form abundantly in culture releasing a light buff exudate of hyaline mostly aseptate conidia. Some of the conidia may be once-septate (which has led to classification in *Ascochyta*, cf. below), but this septation is secondary rather than primary 'distoseptation' (Boerema & Bollen 1975). Dark brown irregular chlamydospores are formed singly or in chains, in terminal or intercalary positions (CMI 518).

Host plants and specialization

Two varieties can be distinguished morphologically and by host range (Boerema *et al.* 1965). **Var. *medicaginis*** (syn. *P. herbarum* var. *medicaginis* Westend. ex Rabenh., *Ascochyta imperfecta* Peck) occurs on lucerne, and to a certain extent on other legumes but not on clover. **Var. *pinodella*** (L.K. Jones) Boerema (syn. *Ascochyta pinodella* Jones) occurs on pea and *Trifolium* species especially *T. pratense* and on various other legumes but not on lucerne.

Geographical distribution

The disease is widespread in Europe (and in temperate regions generally).

Disease

Var. *medicaginis* is responsible for 'black stem' of lucerne, causing long dark lesions on petioles and stems, and angular spots on leaves. The lesions may girdle whole shoots, and spread to cause a crown and root rot (O'Rourke 1976; APS Compendium). Var. *pinodella* causes similar symptoms on *Trifolium* species. On pea, it is part of the complex of seed-borne 'Ascochyta', and can be distinguished from *Mycosphaerella pinodes* and *A. pisi* (q.v.) by the fact that it principally causes foot rot rather than leaf symptoms (Dixon 1984) (and morphologically, by producing smaller mostly non-septate conidia and also chlamydospores).

Epidemiology

The fungus mainly survives on plant debris in the soil, where it can persist for at least 3 years, as pycnidia and/or chlamydospores. Plants are infected by rain-splashed conidia. Cool moist conditions favour infection, which is suppressed by warm dry weather. The fungus is also commonly seed-borne on all its hosts, following pod infection.

Economic importance

Black stem of lucerne is important in wetter, cooler regions e.g. in Ireland, UK, Sweden, Poland and USSR. Both yield and quality are reduced (Willis *et al.* 1969) and there is also an increase in the content of oestrogens potentially poisonous to livestock. The disease on clover and other forage legumes is of relatively minor importance. On pea, *P. medicaginis* is not always easy to separate in practice from the other 'Ascochyta' species, but is certainly less important than *M. pinodes* (APS Compendium). It is, however, more persistent in crop debris and soil.

Control

Clean seed can be produced in dry areas, and fungicidal seed treatment will protect from seed and soil-borne inoculum. Rotation, at more than 3-year intervals, will allow the organism to die out in the soil. Early cutting of forage legumes helps to avoid infection and removes infective debris. Cultivars of lucerne and pea show some variability in susceptibility, and screening programmes for resistance are conducted in several European countries (Poland—Szepieniec-Gajos 1983; France—Angevain 1983).
CARR & SMITH

Phoma exigua Desm.

Basic description

P. exigua produces dark brown to black globoid pycnidia (90–200 μm) from which hyaline, non-septate, cylindrical pycnospores (4–5×2–3 μm) are extruded in a gelatinous cirrhus. Its **var. *foveata*** (Foister) Boerema (syn. *P. foveata* Foister, *P. solanicola* Prill. & Delacr.), the most important pathogenic form, is distinguished from **var. *exigua***, a weak pathogen widespread in soil, on 2% malt agar by the production of anthraquinone pigments which form yellow crystals, whose production can be enhanced by the addition of thiophanate-methyl to the agar. Alternatively, exposing the colonies to ammonia vapour turns the anthraquinone pigments red. The colony growth of var. *foveata* is regular but that of var. *exigua* is scalloped. A number of fungi previously described as separate species are now considered to be synonymous with var. *exigua* (e.g. *Ascochyta phaseolorum* Sacc., *A. nicotianae* Pass.) or to belong to other varieties of *P. exigua* (cf. *Host plants*). With the exception of var. *linicola*, considered under a separate heading, var. *foveata* is the only significant pathogen, and this account relates almost exclusively to it.

Host plants

Var. *foveata* occurs mainly on potato but pycnidia have been found on the stems of weeds in potato fields. Var. *linicola* occurs on flax (cf. below). Var. *exigua* is found on numerous hosts, including in Europe potato (cf. below under var. *foveata*), chicory roots (Dennis & Davis 1978), aubergine (Logan & O'Neill 1981), on which var. *foveata* also occurs, and *Phaseolus vulgaris* on which it causes brown specks on the pods (Boerema *et al.* 1981). The same authors also describe **var. *diversispora*** causing severe damage to *P. vulgaris* in the FRG and the Netherlands, infecting nodes and petioles, which are discoloured and carry abundant pycnidia. Finally, **var. *lilacis*** has been distinguished causing damping-off of lilac seedlings, and also leaf necrosis and twig dieback (Boerema 1979).

Geographical distribution

Var. *exigua* occurs world-wide. Var. *foveata* was first isolated and described in Scotland in 1936 and has now been found in Belgium, Denmark, FRG, Finland, France, Netherlands, Norway, Ireland, Romania, Sweden and UK. It has also been reported in S. Australia, Tasmania and New Zealand (CMI map 210). In 1979, var. *foveata* was isolated from stems of *S. tuberosum* subsp. *andigena* and *Chenopodium quinoa* growing in the highlands of Peru.

Disease

Stems and leaves of potato become infected by var. *foveata* throughout the growing season but clusters of pycnidia develop only as the stems begin to senesce naturally or by chemical desiccation. Pycnidia are generally scattered on the lower part of the stem but elsewhere are largely confined to local spindle-shaped lesions at the junction of petiole and stem. Before and for about 1 week after harvest a degree of true latent infection of the tubers occurs where the fungus penetrates the periderm and forms a resting mycelial body in or just under the cork tissue. Such tubers appear healthy until the fungus is activated to develop by damage at lifting, grading or handling during transport, and by exposure to low temperatures (< 4°C). During storage var. *foveata* causes a tuber rot known as potato gangrene. Black circular 'thumb-mark' or larger, irregularly shaped, sharply black-edged lesions develop from small, round, dark depressions occurring at wounds, eyes or lenticels, to give extensive, dark brown or purplish cavities in which pycnidia later develop either singly or in clusters. Sometimes the rots are dark brown and soft with no or very small cavities but still with a definite margin. Var. *exigua* only rarely causes smaller and more restricted gangrene rots (Boyd 1972). A seed potato stock may be tested for the presence of var. *foveata* by damaging tubers using a standard method and incubating the tubers at 5°C for 8–10 weeks (Fig. 59). The identity of the fungus causing rots can be checked either by culturing from the margins of the rots on 2% malt agar or a semi-selective agar, or by detecting specific anthraquinone derivatives in rotted tissue using thin layer chromatography (EPPO 78).

Epidemiology

Var. *foveata* does not appear to survive in field soil for more than two years but can survive longer on potato groundkeepers and certain adjacent associated weeds. The fungus can survive for at least one year in dry soil in potato stores and can be dispersed into the air. However, the amount of infection occurring from such inoculum is very low and this

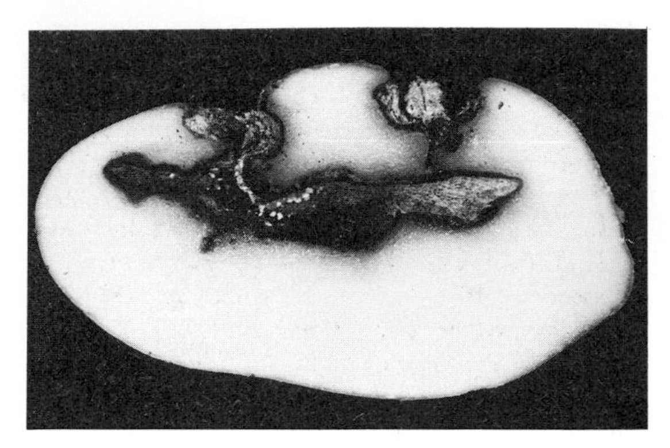

Fig. 59. *Phoma exigua* var. *foveata* on potato cv. Maritta, 6 weeks after artificial infection.

source will only be important in the production of healthy high-grade seed stocks. Var. *foveata* mainly persists through the planting of diseased or contaminated seed tubers. Infection of stems, leaves and developing tubers occurs throughout the growing season. After haulm destruction, both soil inoculum and potential tuber gangrene generally increase the longer the tubers remain in the ground although there may be some decrease if tubers are left for more than 6 weeks. Between haulm destruction and lifting, the principal source of infection for progeny tubers grown from diseased or contaminated seed tubers appears to be the underground plant parts such as stolons and stem base and not the mother tuber or pycnospores washed down from the above-ground stems (Adams 1980). Small numbers of air-borne propagules of var. *foveata* have been recovered during rainfall and when it was dry during the growing season before haulm destruction (Carnegie 1984). This inoculum is an important source of contamination for healthy seed stocks particularly as the extent of pycnidial lesions on dead stems is greater the earlier the green stems become infected. Pycnospores are exuded from stem pycnidia during rainfall and the impaction of raindrops on the stems disperses pycnospores in ballistic droplets to neighbouring plants or in aerosol droplets to plants distant from the source. Infection of healthy stem cutting plants has been detected over 1 km away from a diseased crop. Healthy tubers can become contaminated from infested soil on harvesting and grading machinery. Most gangrene develops after damage at lifting and at other handling operations. Wounding introduces inoculum from contaminated soil into the tuber flesh and activates the development of latent infection in the periderm. The incidence of gangrene is increased by lifting from cold wet soils and, after lifting, grading and handling, by low storage temperatures (2–6°C) which inhibit the formation of wound periderm. Tuber susceptibility to var. *foveata* also increases during storage (Boyd 1972).

Economic importance

Planting diseased seed tubers does not generally reduce yield significantly although a 20% reduction in yield has been reported when more than 60% of seed tubers were affected by gangrene. Plants from diseased seed are generally less vigorous, produce more stems and yield a higher proportion of small tubers. The occurrence of gangrene in seed stocks can also increase the incidence of *Erwinia carotovora* ssp. *atroseptica* in the growing crop. Gangrene is economically most important as a storage disease of seed tubers because after grading and transporting to the purchaser, a high incidence can subsequently develop during storage. In Europe the disease is presently of greatest importance in north temperate countries, particularly Ireland, Scotland and Sweden.

Control

The incidence of gangrene is reduced by burning down the haulms using chemical desiccants as early as possible and then harvesting the tubers as soon as practicable before soil conditions become cold. However, gangrene can be more prevalent after using certain haulm desiccants (Logan *et al.* 1976). The amount of tuber damage at lifting and grading should be minimized and tubers stored at *c.* 15°C for 10 days to encourage wound healing. Effective control of gangrene can be achieved by fumigating tubers with 2-aminobutane (sec-butylamine) 1–3 weeks after lifting. While in the past dipping tubers in organomercury solutions and more recently spraying tubers with thiabendazole at lifting have generally achieved good control, both fungicides have been less effective on stocks grown in the north-east of Scotland (Boyd 1972; Hamilton *et al.* 1981). In more recent trials there, spray applications on the harvester with a mixture of thiabendazole and a salt of 2-aminobutane have given a degree of control similar to that for fumigation by 2-aminobutane. Tubers propagated from stem cuttings are free of var. *foveata* but an annual fungicide treatment at harvest is necessary to control the contamination occurring during commercial multiplication.

CARNEGIE

Phoma exigua var. *linicola* (Naumov & Vassilevski) Maas

(Syn. *Ascochyta linicola* Naumov & Vassilevski)

P. exigua var. *linicola* causes foot rot of flax. The fungus grows irregularly in culture, fruiting poorly on acid medium but well on V8 medium. The pycnospores are mostly unicellular, but 1–3% of bicellular spores may be seen, according to strain. Some variation in aggressiveness and cultural characteristics has been observed. However, the fungus is very similar to *P. exigua* on other crops, being distinguished practically only by its specificity to flax. It may be carried on certain weeds (*Polygonum, Veronica*). It occurs in most flax-growing areas in Europe (France, Belgium, Netherlands, Czechoslovakia, USSR).

At seedling emergence, small brown spots are seen on the cotyledons and collar. Under favourable conditions, pycnidia form on these spores and conidia are dispersed by wind and rain to other plants. At a later stage, black patches bearing pycnidia form on the lower part of the stem, pushing up the epidermis in strips. Infected plants produce little seed. Infection occurs readily at 15–25°C at high RH (> 80%), but may develop even at 5°C. The pathogen is mainly transmitted on or in seed, where it can persist for several years. However, it can also survive on straw and on tow. Plants may be infected at any stage of development, but the seedling stage is by far the most susceptible (cotyledon to 4–8 leaf). The disease causes serious losses in some seasons. It affects linseed yield and fibre quality. The fungus is practically always present, so that suitable conditions in spring are all that are needed to ensure infection. Seed treatments provide satisfactory control, while fungicides applied to the growing crop have little effect.

JOUAN

Many other ***Phoma*** species occur on plants as parasites or saprophytes. Some are anamorphs of *Mycosphaerella* or *Leptosphaeria* (q.v.). Some of the previously described species can probably now be included in *P. exigua* (q.v.) to be distinguished from *P. herbarum* Westend., a widespread saprophyte. ***P. destructiva*** Plowr. causes stem and leaf symptoms on tomato and aubergine like those due to *Alternaria solani* (q.v.) but is more associated with wet conditions than the latter (Messiaen & Lafon 1970). It occurs for example in Greece, France, Italy and USSR. On fruits, it causes a stem-end black rot with conspicuous pycnidia. Bottalico *et al.* (1983) have shown that a toxin, phomenone, is responsible for symptoms of wilt and leaf necrosis. ***P. andina*** Turkensteen causes black potato blight in Andean countries (EPPO 141) and is an important quarantine organism for European countries. ***P. apiicola*** Kleb. causes a root and collar rot of celery and celeriac, the affected parts turning first golden-brown and then dark brown. The disease is transmitted on crop debris and on seed and can be controlled by the same measures as *Septoria apiicola* (Messiaen & Lafon 1970). ***P. chrysanthemicola*** Hollos has caused serious disease on glasshouse chrysanthemums in N. Italy (Garibaldi & Gullino 1981) (see also *Didymella ligulicola*). ***P. eupyrena*** Sacc., a common soil fungus, has also been isolated from potatoes showing gangrene-like symptoms in Germany (Janke & Zott 1983). ***P. glomerata*** (Corda) Wollenw. & Hochapfel, also found as a secondary invader of potato, is a widespread saprophyte, forming characteristic dictyochlamydospores as well as pycnidia. It has also been recorded blighting grapevine flowers and grapes and rotting various fruits (CMI 134). ***P. incompta*** Sacc. & Martius has been associated with a severe dieback of olive in Greece (Malathrakis 1980). ***P. pomorum*** Thüm. (syn. *P. prunicola* (Opiz) Wollenw. & Hochapfel) is similar to *P. glomerata* but with long chains of single chlamydospores as well as dictyochlamydospores (CMI 135). It is a polyphagous, weak or secondary parasite, causing, for example, leaf spot of apple, pear and *Prunus,* or root rot of strawberry (Mirkova & Kachamarzov 1978). ***P. macrostoma*** Mont. is another widespread saprophyte which is occasionally a weak parasite of woody plants. Finally, ***P. valerianellae*** Gindrat *et al.* is the main damping-off pathogen of lamb's lettuce (*Valerianella locusta*) in France and causes considerable damage. It is seed-borne and can readily be controlled by seed treatment (Végh *et al.* 1978). ***Plenodomus meliloti*** Dearn. & Sanford (syn. *Phoma sclerotioides* Boerema & v. Kesteren) causes brown root rot of lucerne and clover in N. America (APS Compendium) and has been recorded in Finland and the USSR.

Phyllosticta species, like *Phoma*, are anamorphs of fungi such as *Mycosphaerella* and *Leptosphaeria*. The name tends to be used for leaf-infecting forms, and innumerable species have been described by their host names. It seems likely that there is considerable synonymy, and in any case hardly any are of sufficient economic importance to be considered further. ***P. solitaria*** Ell. & Ev. causes apple blotch in N. America, on leaves, twigs and fruit. It is considered a quarantine organism for Europe (EPPO 20).

Septoria apiicola Speg.

(Syn. *S. apii* Rostrup, *S. apii-graveolentis* Dorogin, *S. petroseleni* Desm. var. *apii* Cavara)

S. apiicola forms pycnidia immersed in host tissue and containing filiform pycnospores with several transverse septa. Hosts include wild and cultivated celery and celeriac. It is likely that weed hosts form a reservoir of infection for cultivated crops. The pathogen is distributed world-wide (CMI 88) and causes a severe seed-borne leaf spot (also referred to as late blight). Mycelium penetrates directly into host leaves and petioles to establish an intracellular infection. Symptoms begin as chlorotic spots or flecks which later turn necrotic. Lesions may remain < 3 mm in diameter or expand up to 10 mm, with several lesions coalescing together. The necrotic areas have a definite margin and are surrounded by a chlorotic halo which merges gradually into uninfected tissue. Mycelium is not restricted to the lesion but ramifies into surrounding cells. Small black pycnidia form quickly once infection is established. Where the lesions remain small, pycnidia are numerous and scattered throughout the lesion but in larger lesions pycnidia are concentrated near to the lesion centre and are fewer in number. The necrotic tissue is reddish-brown towards the lesion centre and a darker reddish brown at the periphery.

The major means of transmission of *S. apiicola* is via infected seed but diseased debris is an important avenue of perennation and local dissemination. Disease is not normally seen until late in the growing season but may occasionally appear in the seed bed killing the young plants. Pycnospores have survived for up to 10 months on celery leaves under laboratory conditions and for at least 4 months on debris in soil. Pycnidia, mycelium and pycnospores are found on infected seed; the pathogen is absent from seed embryos and endosperms but mycelium penetrates pericarps and testas. Infected seed can carry viable pycnospores for at least 15 months (Sheridan 1966). Pycnospores germinate over the temperature range 9–28°C (optimum 20–25°C). Transmission from plant to plant is by water splash, consequently dissemination and epidemic development are favoured by rain and irrigation water.

Crop rotation is a significant means of control since Maude & Shuring (1970) showed that celery transplanted in June could become infected from crop debris carrying *S. apiicola* which had been placed in the soil nine months earlier. Viable pycnospores could be extracted from the buried debris for as long as visible plant material persisted. There should be at least a 2-year break between plantings of susceptible crops on the same land. Complete control is obtained by soaking seed for 24 h at 30°C in an aqueous suspension of 0.2% thiram (Maude 1970). Control in infected crops may be achieved with captafol, tin plus maneb and organo tin compounds. No commercial celery cultivars are resistant to *S. apiicola*.

DIXON

Septoria lycopersici Speg.

S. lycopersici forms simple, thin-walled pycnidia, developing subepidermally, becoming erumpent as the host tissue shrinks. Pycnospores are hyaline, 2- to 6-septate and filiform. The main hosts are tomato, potato and aubergine; weeds such as *Datura stramonium, Solanum carolinense* and *S. nigrum* are also infected. There is world-wide distribution, and races have been identified (CMI 89). The form on potato has been distinguished as ***var. malagutii*** Ciccarone & Boerema. As it is confined to S. America it is regarded as a quarantine organism for Europe (EPPO 142). **Var. *lycopersici*** causes leaf spot of tomato, mainly in E. Europe. Lesions form on the leaves and stems at any stage of host development. Older leaves usually exhibit the first symptoms. These are seen as circular to irregular, rarely confluent, often vein-limited and depressed water-soaked lesions. They are pale brown and later grey with dark margins up to 2 mm in diameter. Relatively few black pycnidia are formed within each lesion. Heavily infected leaves fall prematurely and this may expose ripening fruits to the effects of sunscald. Lesions will develop on stems but are rarely found on fruit.

Transmission of var. *lycopersici* is by rain-splash and wind-blown water. Localized movement within and between crops may take place on the hands and clothing of fruit pickers. There may be spread via vectors such as beetles. Seed transmission of var. *lycopersici* may take place but has not been fully proved. Perennation takes place on soil-borne debris from one season to the next (Marcinkowska 1977). Pathogen growth is encouraged by high RH for at least 48 h and temperatures of 20–26°C (Marcinkowska 1977; Coulombe 1979). Invasion occurs via the stomata forming an intercellular mycelium possessing intracellular haustoria. Mycelium is found in advance of the necrotic areas of each lesion. Resistance to var. *lycopersici* has been obtained from *Lycopersicon glandulosum, L. hirsutum* and *L. peruvianum*. Resistance functions either by restrict-

ing lesion size or in a hypersensitive manner. Both forms are controlled by single dominant genes (Barksdale & Stoner 1978). Fungicidal control is obtained with copper preparations, zineb or mancozeb (Dunenko & Belikova 1977). Methods of husbandry control include crop sanitation, rotation and the destruction of weed hosts.
DIXON

Septoria passerinii Sacc.

(Syn. *Septoria murina* Pass.)

S. passerinii is readily distinguished from other species on barley by the slightly curved subulate conidia which measure 21–52×1.5–2.2 μm (av. 35×1.8 μm) (Bissett 1983). *Mycosphaerella graminicola* is morphologically similar but has pycnospores up to 85 μm long. *Leptosphaeria nodorum* pycnospores are shorter, 13–20 μm long, and wider (2.5–4.5 μm). The fungus occurs on barley (*Hordeum vulgare*) and other wild barley grasses. Isolates are highly specialized, with those from *H. vulgare* infecting mainly cultivated *Hordeum* species, whereas isolates from *H. jubatum* infect *H. jubatum* and related wild *Hordeum,* but not cultivated *Hordeum*. Septoria passerinii blotch (speckled leaf blotch of barley) has been identified from Europe, N. America and Asia.

The first signs of infection are small chocolate-brown spots which enlarge to become lens-shaped as they expand and merge. Lesions turn light greyish-brown and are difficult to distinguish from those of both septoria nodorum blotch or septoria avenae blotch. The disease attacks all parts of the plant, becoming more prevalent as the season progresses. When the disease is severe, defoliation may occur and kernels may be light and deformed causing reduced yield. This disease has caused significant economic losses in Canada (Green & Bendelow 1961; Bissett 1983), but is of minor importance in Europe.

Although the teleomorph has not been identified, long-range dispersal and infection in first-year crops in the absence of any wheat crop debris should be considered as arising from wind-borne ascospores. Spread from old crop debris, and secondary plant-to-plant spread within the growing crop, is by rain-splashed pycnospores. Some attention has been given to the development of resistant cultivars with dominant genes for resistance being identified. Frequent applications of nabam and zineb, although giving some control, were uneconomic (Buchannon & Wallace 1962).
SANDERSON & SCOTT

Septoria ampelina Berk. & Curtis (probably synonymous with *S. melanosa* Elenkin) causes melanose (blackish-brown spotting) of certain grapevines used as rootstocks (*Vitis riparia*), but generally not *V. vinifera*, and has been noted in France, Spain (Boubals & Mur 1983) and the USSR. ***S. azaleae*** Vogl. causes leaf-scorch of azalea and can be severe on plants forced under glass. ***S. cannabis*** (Lasch) Sacc. causes a common leaf-blight of hemp in Italy through C. Europe to the USSR (CMI 668). ***S. chrysanthemella*** Sacc. causes black leaf spot of florists' chrysanthemums (CMI 137) throughout Europe. It occurs on outdoor crops under wet conditions, and most importantly in propagation beds, where it has to be controlled with high-volume fungicide sprays (Fletcher 1984). ***S. citri*** Pass., probably the anamorph of an associated *Mycosphaerella* species, causes greasy spot disease of citrus in New South Wales (Australia) (Wellings 1981). Rather similar symptoms (gummy intumescences) were seen by Grasso & Catara (1982) in Italy—and in both cases *Glomerella cingulata* seems associated as a secondary invader. *S. citri* also causes a serious brown spotting of lemon fruits in Sicily (Grasso & Rosa 1983) and has occasionally been found on oranges in Greece. ***S. cucurbitacearum*** Sacc. causes a leaf spot of most cultivated Cucurbitaceae, but is infrequent and unimportant (CMI 740). ***S. dianthi*** Desm. mainly attacks the older leaves of carnations on which it causes spots with a less pronounced purple border than *Alternaria dianthi* and lacking the black sooty growth characteristic of the latter. The disease is uncommon (Fletcher 1984). ***S. gladioli*** Pass. is widespread on *Gladiolus* and also occurs on *Freesia, Crocus* etc., causing purple-brown spotting of the leaves, and brown or black spotting of the corms—spreading to form large sunken lesions (hard rot). The disease is most serious in wet seasons (Brooks 1953). ***S. glycines*** Hemmi causes brown spot of soybean and can be severe in N. America. It has been recorded from a few European countries (Hungary, Yugoslavia) but reports may arise simply from use of infected imported seed. Since the disease is mainly prevalent in warm moist conditions (APS Compendium), it seems unlikely to cause serious problems in Europe. ***S. helianthi*** Ell. & Kellerman causes a leaf spot of sunflower which is widespread and damaging in E. Europe (CMI 276). The pathogen persists on debris in the soil or with seeds. It is not clear whether it is truly seed-borne, but seed treatment is in any case recommended. ***S. humuli*** Westend. causes a leaf spot of hops, on which little information seems available in Europe (Viennot-

Bourgin 1949). ***S. lactucae*** Pass. causes a minor, seed-borne leaf spot of lettuce in Europe (CMI 335) and is much more important in the tropics (Messiaen & Lafon 1970). ***S. petroselini*** Desm. is very common on parsley in Europe, causing pale leaf spots with a brown border, and with clearly visible pycnidia. It is very similar to *S. apiicola* but does not attack celery. Whereas other *Septoria* diseases tend to have minor and indirect effects on yield, the quality of marketed parsley is seriously and directly affected by *S. petroselini*. It can be controlled by use of clean or treated seed, and by fungicide sprays of the foliage, although there is then a problem of safety intervals (Messiaen & Lafon 1970). Attention to the disease on the seed crop could probably eliminate it.

OTHER ASCOMYCETE ANAMORPHS

The fungi included under this heading are probably many of them anamorphs of Dothideales (especially the pycnidial forms). However, their correct position could well be in Chapter 12, or conceivably in any of the Ascomycete chapters. Pycnidial and hyphomycete forms are presented separately.

PYCNIDIAL FORMS

Gloeodes pomigena (Schwein.) Colby

G. pomigena is a pycnidial fungus causing sooty blotch especially of apple and to some extent of pear. First found in N. America, it has been found throughout Europe since the 1930s, more or less sporadically according to season. Fruits bear sooty blackish spots 0.5–1 cm in diameter, rounded or irregular in shape. The spots may spread to cover the whole fruit surface. The mycelium is entirely superficial, confined to the cuticle which shows cracking. It can be removed by rubbing, leaving pale brownish marks. These symptoms should not be confused with sooty moulds, which stand out more clearly and are blacker, and which are entirely removed by rubbing. As for flyspeck (*Schizothyrium pomi*), warm wet summers favour spread. *G. pomigena* causes no real damage, but decreases the market quality of fruit. Physical methods were once used to remove the blemishes (e.g. shaking with moist sawdust). Dilute hypochlorite has also been used to bleach them. Chemical treatments against *S. pomi* (q.v.) effectively control *G. pomigena*.

BOMPEIX

Hapalosphaeria deformans Sydow causes hypertrophy of the flowers of raspberry and *Rubus fruticosus* in Scotland, forming pycnidia in the anthers which break down to fill the flowers with white conidia. It is mainly of interest as a curiosity (Brooks 1953). ***Hendersonia acicola*** Münch & Tubeuf is a common secondary pathogen following *Lophodermella sulcigena* on pine needles (q.v.) (Watson & Millar 1971). ***Kabatina juniperi*** R. Schneider & v. Arx is a wound parasite which sometimes causes a severe shoot dieback of wild and ornamental *Juniperus* species and varieties in Europe and N. America, while ***K. thujae*** R. Schneider & v. Arx causes a sometimes severe leaf browning, canker and dieback of various members of the Cupressaceae, including *Chamaecyparis, Cupressus* and *Thuja*.

Macrophomina phaseolina (Tassi) Goidanich

M. phaseolina forms sclerotia known as ***Rhizoctonia bataticola*** (Taubenhaus) Briton-Jones and dark brown pycnidia, containing large (14–30×5–10 μm) hyaline aseptate pycnospores (CMI 275). O'Brien & Thirumalachar (1978) note that another pycnidial fungus, with dark 2-celled pycnospores, is often associated and has been considered a mature form; they consider this fungus distinct, as ***Botryodiplodia solani-tuberosi***. ***M. phaseolina*** attacks innumerable hosts, especially in hot countries (favoured by soil temperatures over 28°C) and is widely distributed in S. and E. Europe, though it is not in general as important a pathogen in Europe as in tropical countries. Economically important disease is caused to sunflower (France, Spain, USSR, Yugoslavia), maize (Austria, France, Yugoslavia), soybean (France, Hungary, Spain) and cotton (Greece). Other hosts in Europe include fruit trees (citrus, quince, pear, olive, *Prunus*), legumes (groundnut, *Phaseolus, Vicia,* lentil, chickpea, lucerne), Solanaceae (aubergine, potato, tomato), conifers (*Pinus, Cupressus*) and others (safflower, saffron, sesame, strawberry, watermelon). *M. phaseolina* initially causes dark brown to black lesions of the roots. Infected plants may show a characteristic silvery discoloration of the epidermal and subepidermal layer of the stem base and tap root (ashy blight). The fungus spreads up the vascular and pith tissues of the stem, finally forming numerous small sclerotia, like finely powdered charcoal (charcoal rot) giving the infected tissues a greyish-black colour. Sclerotia are found along the vascular elements and bordering the pith cavity. Affected plants are stunted and ripen prematurely. Sunflower, for example, shows small distorted heads with a central zone of aborted flowers (Jimenez-Diaz *et al.* 1983), reduced yield, 1000-seed weight and seed and oil quality (Pustovoit & Borodin 1983).

The fungus persists as sclerotia in the soil, or may be seed-borne. It can grow saprophytically but has rather poor competitive saprophytic ability. Infection by *M. phaseolina* is greatly affected by predisposing factors. Thus, attacks are seen most characteristically when hot dry conditions follow a period of normal growth, for example when irrigation of soybeans is stopped to accelerate ripening (APS Compendium). In Spain, attacks on sunflower vary considerably from year to year, mainly according to temperature and rainfall in June and July (Jimenez-Diaz *et al.* 1983), while in France they tend to occur late and after drought (Saumon *et al.* 1984). On maize, *M. phaseolina* tends to be associated with *Fusarium moniliforme* in the general stalk-rot complex.

Control is mainly by cultural methods, especially maintenance of good irrigation under high temperature conditions (APS Compendium), though Montes Agusti *et al.* (1975) found little difference in incidence on soybean in Spain at different irrigation levels. Short *et al.* (1980) in America found that the number of sclerotia in soil builds up under continuous cropping of maize or soybean, and severity is related to this. Approaches to control include the use of various soil amendments, N-enriched amendments seeming most effective in causing population decline (Filho & Dhingra 1980), and also seed bacterization or similar biological methods. The disease has been extensively reviewed by Dhingra & Sinclair (1978).
SMITH

Phaeocytostroma ambiguum (Mont.) Petrak (syn. *Phaeocytosporella zeae* Stout) causes a stalk blotch of maize, with one-celled brown conidia in multilocular pycnidia (APS Compendium). Apart from the USA, it has been found in France, USSR and Yugoslavia, in the last of which lines and hybrids are tested for resistance, the pathogen being potentially more destructive than *Gibberella zeae* (Draganić 1983). The related ***P. sacchari*** (Ell. & Ev.) B. Sutton causes rind disease or sour rot of sugarcane (CMI 87) and has been recorded in Portugal.

Plectophomella concentrica, described by Redfern & Sutton (1981), causes cankers on small shoots of elm (and occasionally of lime) in Scotland and N.E. England. It is distinct from ***P. ulmi*** (Verrall & May) Redfern & Sutton (syn. *Dothiorella ulmi* Verrall & May) causing 'native elm wilt' in N. America (APS Compendium).

Pseudoseptoria stomaticola (Bäuml.) B. Sutton

(Syn. *Selenophoma donacis* Pass. var. *stomaticola* (Bäuml.) Sprague & Johnson)

P. stomaticola produces one-celled sickle-shaped pycnospores (CMI 400) which are smaller than those of *P. donacis* (Pass.) B. Sutton (syn. *S. donacis* var. *donacis*). It causes halo spot of several grasses and cereals. Irregular grey spots surrounded by dark brown margins are formed on leaves, stems and awns. Mostly black pycnidia form in rows in the centre of the lesions. Barley, *Phleum pratense* and *Dactylis glomerata* can be severely attacked. Further work is needed to elucidate host specialization. *P. donacis* attacks many other grasses (including *Arundo donax* (Frisullo 1984)) and is of less importance on crop plants. *P. stomaticola* has been reported from Australia, New Zealand, N. America, N. Europe, Spain and Algeria (Sprague 1950). It is spread by rain splash and a very humid climate is necessary for the disease to prevail in a region. Infected seed, stubble and volunteer plants are means of overwintering for the fungus in barley. In grasses the disease is perpetuated on the living plant tissue. In barley heavy attacks can lead to poor filling of the grains due to the destruction of the upper leaves and the awns, but economic importance is mostly very restricted. The relationship between per cent attacked plant tissue and yield reduction is considered to be somewhat similar to that of *Rhynchosporium secalis* on barley (James 1969). Benomyl and thiophanate methyl have shown good control of halo spot in barley (Brokenshire & Cooke 1977). Zineb likewise has shown good effect against *P. stomaticola* (Jenkins & Melville 1972).
MAGNUS

Septocyta ruborum (Lib.) Petrak (syn. *Rhabdospora ramealis* (Rob.) Sacc., *R. ruborum* (Lib.) Jørstad) causes purple blotch of the canes of *Rubus fruticosus* in several European countries (especially Switzerland) but is of little practical importance (CMI 667). ***Sirococcus strobilinus*** Preuss (syn. *Ascochyta piniperda* Lindau), which occurs in Europe and N. America, causes a dieback of young shoots of spruce plants (and sometimes of pines), mainly in forest nurseries. ***Sphaeropsis sapinea*** (Fr.) Dyko & Sutton (syn. *Diplodia pinea* (Desm.) Kickx) (CMI 273), a fungus of world-wide distribution, may sometimes cause a locally severe dieback of young pines (and occasionally of other conifers) in plantations and (less often) in nurseries. ***Stagonospora curtisii*** (Berk.) Sacc. causes leaf-scorch of *Narcissus, Galanthus* and *Hippeastrum* in much of Europe (and the USA). It mainly affects the leaf tips, and is rarely important except in wetter western districts. Control with fungicides is currently studied in the Netherlands and the UK. ***S. fragariae*** Briard & Har. (syn. *Ascochyta fragariae* Lib., *Septoria fragariae* (Lib.) Desm., *Septogloeum fragariae* (Lib.) Höhnel) is widespread in Europe causing brown spots with purplish margins on strawberry leaves rather similar to those due to *Mycosphaerella fragariae* (q.v.) of which it is not the anamorph. Abundant pycnidia form in the centre of the lesions, and infect the fruits, on which lesions appear as hard brown sunken areas with seeds highly clustered (hard rot). Control with fungicides may be needed (APS Compendium).

Stenocarpella maydis (Berk.) B. Sutton

(Syn. *Diplodia maydis* (Berk.) Sacc.)

S. maydis and ***S. macrospora*** (Earle) B. Sutton (syn. *D. macrospora* Earle) form pycnidia with once-septate pale brown pycnospores much larger (7.5–11.5×44–82 μm) in *S. macrospora* than in *S. maydis* (5–8×15–34 μm) (CMI 83, 84). Both fungi are widely distributed on maize in the warmer areas of the world, but are hardly established in Europe and so are considered to be quarantine organisms; *S. maydis* is established in Italy, but neither species in other countries (France, Portugal, Romania, USSR) where they have been reported (EPPO 67). The two species cause very similar diseases and may occur together. They are seed-borne, but also survive on debris. They cause seedling blight, crown and root rot, then stalk and ear rot. The tissue of lesions on stalks becomes brown to straw-coloured and spongy. These lesions may spread into the ears, which may be completely rotted and bleached (early infection), showing extensive white mycelium and scattered pycnidia. Losses arise from poor grain filling and from lodging, and can commonly reach 10–20% (APS Compendium). *S. macrospora* is favoured by warm, humid conditions mostly not prevailing in Europe but there is much more substantial risk from *S. maydis,* which can occur in cooler regions. Some resistant cultivars exist, but control is mostly through seed treatment. Good control of the quality of imported maize seed is accordingly needed in countries at risk in Europe.

SMITH

Wojnowicia hirta Sacc. (syn. *W. graminis* (McAlp.) Sacc.) produces large setose pycnidia with 7-septate conidia on the roots of Gramineae (CMI 773) but though frequently isolated seems to be a weak secondary pathogen. According to Fitt & Hornby (1978), it does not penetrate to the stele of wheat roots and does not affect the growth of the wheat plants.

HYPHOMYCETE FORMS

Cheilaria agrostis Lib. (syn. *Septogloeum oxysporum* Bommer, Rousseau & Sacc.) is widespread but unimportant causing blotch or char spot of many grasses (*Lolium perenne, Phleum pratense, Agrostis* species etc.). Reddish brown blotches bear an off-white mass of fusiform 1- to 4-septate conidia on a stroma which finally becomes raised and black (char spot) (CMI 488; O'Rourke 1976).

Cryptostroma corticale (Ell. & Ev.) Gregory & Waller

C. corticale forms thin black stromata in the cambium and sometimes in the bark. The stromata split parallel to the surface, forming a roof, supported by vertical columns, and a sporing floor. The intervening space, about 1 mm deep, becomes filled with a brown mass of oval conidia which measure 4–6.5×3.5–4 μm (Gregory & Waller 1951; CMI 539). In N. America, *C. corticale* is known mainly as a saprophyte on maple logs and dying trees of *Carya* and *Tilia*. In Europe it occurs entirely on *Acer* species, on which it is often saprophytic, but in some years it is parasitic (mainly in France and S. England), almost entirely on *Acer pseudoplatanus,* but occasionally also on *A. campestre, A. platanoides* and *A. negundo*. It causes the wilt called sooty bark disease. Symptoms first appear between May and September, when patches of leaves wilt, the brown leaves remaining attached to the affected branches. Later, in the following spring, stromata form large blisters in and under the bark, which breaks off to reveal the sooty spores. The blisters spread to cover much of the trunk surface, by which time the tree is usually already dead. Inside affected trees, the heartwood is stained yellow or brown, and this stain may extend into the root system, and through the sapwood to the cambium, where the stromatal blisters then form. This staining fades after the death of the tree. Epidemics of sooty bark disease occur only periodically, developing in years closely following those in which in any of the summer months from June to

August the mean daily maximum temperature is 23°C or more (Dickenson & Wheeler 1981). So far, sooty bark disease has been sporadic and limited in distribution, but in severe epidemics it has killed up to 20% of the sycamore trees in the affected areas. No control measures have been developed to deal with it.
PHILLIPS

Helminthosporium solani Durieu & Mont.

(Syn. *Spondylocladium atrovirens* Harz)

H. solani has slow-growing hyphae that turn brown to dark grey with age. Dark conidiophores arise singly or in groups either from mycelium or from stromata consisting of a few thick-walled cells, and dark 2–8 pseudoseptate conidia, rounded and broad at the base and pointed at the apex, are borne in verticils. The fungus is known only as a pathogen of potato tubers; although infection may be more prevalent on some cultivars than others, resistance to infection has not been reported. Silver scurf occurs widely in Europe and N. America and is a superficial blemishing disease that discolours the tuber skin; the light brown patches enlarge over the tuber surface, darken with age and appear silvery when wet. Water loss is increased and in dry conditions tubers shrivel. The disease originates from infections on seed tubers and after planting the fungus sporulates on the tuber surface. As no other parts of the potato plant are affected it is likely that conidia dispersed into the soil infect developing tubers, initially at the stolon end. The disease is often most severe on crops from slightly affected tubers because they have a greater ability to produce conidia as lesions enlarge than seed with severe disease. Old lesions produce fewer conidia. Silver scurf is often visible as discrete light brown spots at harvest but most disease development occurs in warm (15–20°C) and humid stores; the whole of the surface becomes affected and production of conidiophores gives a sooty appearance (Jellis & Taylor 1977).

Silver scurf is of minor importance and although eating quality is not affected it can have a detrimental effect on the appearance and saleability of washed pre-packed potatoes and especially of red-skinned cultivars. When shrivelled seed potatoes are used, plant emergence and growth are initially delayed but, as the seed soon becomes turgid in moist soil, plants recover and yields are not affected. Much severe scurf can be avoided by harvesting early and storing tubers in cool (2–4°C) and dry conditions which prevent enlargement of lesions. The disease is also decreased by irrigation. Seed tuber treatment with fungicides that persist on the skin and prevent sporulation control the disease in the stored progeny tubers and among the most effective are imazalil, thiabendazole and carbendazim (Cayley *et al.* 1983). These materials are also effective when applied to tubers before storage.
HIDE

Helminthosporium allii Campanile causes 'soot' disease of white garlic cultivars due to the black conidia forming on the outer scales of the bulbs. It is widespread and very common and fungicide treatment has been recommended e.g. in Austria and Bulgaria. However, Messiaen & Lafon (1970) consider it purely superficial and unimportant unless the bulbs are left too long in wet soil.

Meria laricis Vuill.

M. laricis has spores measuring 8–10×2–3 μm, becoming 2-celled just before they germinate. It is confined to larch. It may attack many *Larix* species, but causes serious damage only on *L. decidua* and occasionally on the hybrid *L.×eurolepis*. The fungus is widespread throughout Europe and into Asia, and also occurs in western N. America and in New Zealand (CMI map 379).

M. laricis causes needle cast of larch nursery plants. It may occur in older trees in the forest, but does negligible damage to trees more than eight years old. Diseased needles turn brown from their tips downwards to the points of infection, the first symptoms usually becoming visible in May. The needles at the shoot tips usually remain healthy. The fungus fructifies on the diseased needles, generally on the undersides, and small pustular bundles of conidiophores emerge through the stomata. On fallen needles the fungus usually dies out before the spring, but it persists on those left hanging on the plants over the winter. In spring, spores produced on these needles infect the young foliage of the new shoots. These spores spread infection for only a few hundred metres. They germinate only at RH above 90% (Biggs 1959), and at temperatures between 0 and 25°C, and are soon killed by drying. *Meria* needle cast causes serious damage only in relatively warm wet summers. Initial infections in previously unaffected nurseries almost always arise through the importation of diseased transplants.

M. laricis seldom kills plants, but may increase

the number of poor, stunted and useless plants, and hence decrease the final yield in the affected beds. Larch should be grown only in nurseries at some distance from larch plantations which may cause initial infections, and transplants should not be brought on from nurseries elsewhere. If the disease appears in spite of precautions, the plants should be sprayed with colloidal or wettable sulphur, captan or zineb.
PHILLIPS

Myrothecium roridum Tode (CMI 253), forming shiny black masses of rod-shaped conidia on white sporodochia, is a widespread soil saprophyte found on decaying plant material and occasionally causing spotting of leaves or fruit, especially in warmer countries. Recorded hosts include *Antirrhinum* and *Viola*, and many glasshouse ornamentals (Chase 1983) but it seems to have little practical importance in Europe.

Phaeoisariopsis griseola (Sacc.) Ferraris (syn. *Isariopsis griseola* Sacc.) is widespread and damaging in the tropics causing angular leaf spot of *Phaseolus* species. The greyish lesions bear conspicuous coremia (1 mm in length), and the fungus can persist as sclerotia in debris or on seed (EPPO 73). *P. griseola* has been reported from many European countries but possibly only because it is seed-borne and much *Phaseolus* seed is imported from the tropics. Attacks of some importance have been noted in S.W. France under moist conditions in autumn (Messiaen & Lafon 1970).

Polyscytalum pustulans (Owen & Wakef.) M.B. Ellis

(Syn. *Oospora pustulans* Owen & Wakef.)

Basic description

P. pustulans forms erect, generally branched conidiophores which are 2–4 μm thick and up to 140 μm long. The lower part is pale brown and sometimes swollen at the base. Conidia (6–8×2–3 μm) are dry, hyaline, cylindrical, mainly non-septate and develop in long, often branched chains. Colonies in culture are grey and powdery.

Host plants

P. pustulans develops on tubers, underground stems and roots of potato. The fungus has, however, been isolated from brown lesions on the roots of other Solanaceae (tomato, *Solanum, Nicotiana*) (APS Compendium).

Geographical distribution

P. pustulans is common in the cool temperate countries of N. Europe and the USSR. It has also been found in N. America, S. Africa, Australia and New Zealand.

Disease

P. pustulans causes a tuber skin disease known as skin spot which is commonly seen after several months storage as small (1–3 mm diameter), discrete, black or purplish pimples occurring either singly or in groups on the tuber surface. These may be aggregated around eyes, stolon scars or damaged skin. The fungus penetrates the tuber tissue through lenticels, eyes and skin abrasions to a depth of 1–2 mm before its spread is checked by the formation of cork periderm. After 2–3 months the cork layer around the infected tissue breaks down and typical 'skin spots' are produced. If the formation of cork cambium is prevented by low temperatures (< 4°C) or by using certain sprout inhibitors the fungus can penetrate further into the tissue resulting in deeper and larger pustules. In the USSR, symptoms of black superficial stains and large depressed patches have been recorded.

The fungus also affects the buds of tuber eyes and in severe cases all the buds in all the eyes of a tuber may be killed. The incidence of eye infection is not necessarily related to the occurrence of skin spot on the tuber. Infection of stem bases, stolons and roots by *P. pustulans* produces small, light brown spots which later coalesce to form large, brown, largely superficial patches often with deep longitudinal cracks. The incidence of eye infection by *P. pustulans* can be measured by microscopic examination of excised eye plugs incubated at 15°C and a high humidity for 5 days (Hide *et al.* 1968) or of small pieces of eye tissue cultured on FHC selective medium at 15°C for 2 weeks (Bannon 1975). These tests give a reliable prediction of disease.

Epidemiology

P. pustulans can readily be detected in field soil up to four years after a potato crop but only rarely afterwards. The fungus can survive in soil as microsclerotia in plant debris for at least 8 years. In potato stores, *P. pustulans* survives for at least 6 months in dry soil which can be dispersed into the air (Carnegie *et al.* 1978). Significant amounts of infection can occur from this airborne inoculum which is likely to be an important source of inoculum for the contamination of healthy tubers.

Infected seed tubers are the main source of inoculum in most seed and ware potato crops although the amount of skin spot developing on progeny tubers does not necessarily reflect the severity of skin spot on the seed tubers (Boyd 1972). *P. pustulans* spreads and sporulates first at the base of stems and then higher up the stem base and on stolons. Infection of tubers occurs initially in the buds and bud scales of eyes rather than in skin and increases throughout the growing season. Tubers nearest the mother tuber and stem base are generally more heavily infected than those further away. *P. pustulans* spreads and develops more rapidly in heavy than in light soils. Spread and development of skin spot is also enhanced by wet soils during the lifting period and by low temperatures during storage immediately after lifting. Infected tubers are usually symptomless at harvest. Further infection of tubers, particularly eyes, can occur in bulk stores under damp conditions from conidia produced on infected skin and sprouts and then dispersed into the air (Hide *et al.* 1969). Cultivars with thick-skinned tubers are generally resistant to the development of skin spot.

Economic importance

Skin spot is a problem in the seed potato industry because apparently healthy tubers can develop symptoms during storage on a purchaser's farm before planting. The subsequent complaints and publicity can affect the reputation and hence the sales of both grower and country. The principal importance of *P. pustulans* is in causing the death of eyes on seed tubers. This results in blanking and uneven, delayed emergence in the growing crop particularly if tubers are unsprouted and planted in cool, wet soil. The amount of blanking varies with cultivar but differences appear to be associated with the vigour of the sprouts rather than the susceptibility of cultivars to skin and eye infection. Seed infection tends to reduce stem numbers. This reduction or delay in emergence can alter the ratio of ware and seed although the total yield may only be slightly reduced. Severe skin spot makes washed ware tubers unattractive and sometimes unsaleable. Potato processors who store tubers at low temperatures and use sprout suppressants incur increased peeling losses when tubers are infected by *P. pustulans*.

Control

Infection can be reduced by lifting early, during warm conditions and by storing tubers dry in boxes. Before planting tubers should be sprouted in warm, dry conditions. The occurrence of skin spot can also be reduced by growing cultivars which are relatively resistant to eye and skin infection. Good control has been achieved by dipping tubers in an organomercury solution at time of lifting although its effectiveness decreases as the date of lifting is delayed. This treatment is now rarely used because of the toxicity of mercury compounds and because it is often difficult to dry the tubers afterwards. Fumigation with 2-aminobutane (sec-butylamine) of tubers soon after lifting has given excellent control (Graham *et al.* 1973). Although good control of skin spot has been reported using thiabendazole applied mainly as a dip and benomyl applied as a dust, less effective control has been recorded in N.E. Scotland when thiabendazole was applied as a spray. However, spray applications with a mixture of thiabendazole and a salt of 2-aminobutane have given much better control of skin spot in recent trials in N.E. Scotland. Thiabendazole and benomyl applied to tubers before planting reduces the subsequent infection of plants and progeny tubers. Tubers free of *P. pustulans* can be produced by growing potato stem cuttings or microplants but an annual fungicide treatment is necessary to maintain the health of the seed crop during subsequent commercial multiplication.

CARNEGIE

Pseudocercosporella herpotrichoides (Fron) Deighton

(Syn. *Cercosporella herpotrichoides* Fron)

Basic description

Conidia are formed in tufts on free mycelium, hyaline, mostly slightly curved, 3–7 septate, 26.5–47.0×1.0–2.0 μm. *In vitro*, mycelium may differ in colour and growth type (and also in pathogenicity; see below): 'W-type', usually with whitish or greyish colonies with sharp and even edge; the slower growing 'R-type' with feathery-edged, darker colonies. Using mainly spore characteristics and colony characters *in vitro*, Nirenberg (1981) divided the species into two varieties and added two new species to the genus:

i) ***P. herpotrichoides* var. *herpotrichoides*** (conidia 35–80 μm×1.5–2 μm, curved);

ii) ***P. herpotrichoides* var. *acuformis*** (conidia 43–120 μm×1.2–2.3 μm, straight);

iii) ***P. anguioides*** (conidia 80–260 μm);

iv) ***P. aestiva*** (conidia 15–32 μm).
As yet, the identity with the classical W-, R- (and C-) types is not clear. There are no resting spores, but sclerotia-like structures (stroma) are formed on the haulm-surface.

Host plants

Cereals are the main economic hosts, but many grass species may also be attacked. Wheat is attacked most severely, followed by barley and rye.

Host specialization

The well-known W- and R-types (most clearly distinguished in the UK—Scott & Hollins 1980) both attack wheat but similarly severe symptoms on rye are produced by the R-type only. A more recently described 'C-type' (Cunningham 1981) is able to parasitize wheat and *Elymus repens*, but not rye. *Aegilops squarrosa* is attacked by the R- and C-types, but not by the W-type; *Aegilops ventricosa* is resistant in any case. Significant cultivar/isolate interactions were found by Scott & Hollins (1977), but they were of small magnitude and environmentally labile, so it seems unlikely that isolates virulent on certain cultivar could arise or persist.

Geographical distribution

P. herpotrichoides occurs in cooler regions only: C., N. and W. Europe, and most severe crop losses are observed in coastal areas due to high humidity and longer infection periods than under continental conditions. In Mediterranean countries, the fungus is virtually unknown. Due to different temperature requirements the W- and R-types may occur with different frequency in different regions.

Disease

P. herpotrichoides causes eyespot: longish-oval, pale-brownish lesions on the lowest leaf-sheaths and underlying haulm surface surrounded by a darker brown, but quite diffuse margin (in contrast to 'sharp eyespot' with a more black-brown, very sharply edged margin, caused by *Ceratobasidium cereale*). In summer, black resting structures (similar to those formed by *C. cereale*) are usually formed on the haulm surface. More diffuse symptoms on the haulm-base caused by *Fusarium* species and *Monographella nivalis* mostly appear later in summer; these fungi mostly depend on higher temperatures. Inside the haulm, greyish mycelium with conidia may be found. In coastal or other areas with fairly mild winter conditions the first symptoms can sometimes already be seen in spring, due to early infection and long periods with conditions suitable for infection and pathogenesis. On the other hand, in more continental climates with long-lasting frost and snow-cover, distinct symptoms mostly do not develop before the end of May or even later. At milky- or dough-ripe stage, attack can lead to serious 'foot-rot' symptoms: colonization all round the haulm-base causes softening of the host tissues, which may result in breaking the plants at their stem-base and even lodging, i.e. rotting of and breaking at the haulm base, of whole fields ('straw breaker'). This in turn may facilitate severe ear and leaf attack by different fungi, mainly *Septoria* species, and consequent additional losses. Roots, leaf blades and ears are not attacked by *P. herpotrichoides*, but after early lodging upper internodes may sometimes show eyespot symptoms.

Epidemiology

The fungus is a poorly competitive soil saprophyte, persisting on stubble residues for several years. Occasionally, persistence on weed grasses may also be important. Depending on climatic conditions and sowing date, sporulation and infection by conidia may take place during the whole autumn, winter and spring until about May/June. Sporulation is optimal below 10°C and at high air humidity, and conidial formation on straw residues on the soil surface is the most important inoculum source. After late sowing, continuous frost conditions and/or long snow cover, spring infections become more important (previous crops with late harvest such as sugar beet; continental climate).

In traditional crop rotations with 33–50% cereals only, serious eyespot damage used to be quite rare. Intensification of cereal production since the early 1950s has led to a considerable build-up of inoculum in the fields. In addition, the disease is promoted by topsoil quality (e.g. high pH value), early sowing date, sowing rate, number of tillers per plant, number of haulms per m^2, and other factors. Plants are usually infected through the leaf-sheaths surrounding the haulm-base. When the fungus has established in the outer layer, the leaf-sheaths are penetrated successively, finally leading to the infection of the haulm itself. The succeeding foot rot is most serious if this process is complete long before heading (GS 31/32). If a high percentage of plants in a field is parasitized in this way, the extent of crop loss mainly depends on the conditions for further pathogenesis (Schrödter & Fehrmann 1971). Reports from the literature and experimental results on temperature dependency of pathogenesis are somewhat contradictory at present, but according to experience fairly cool rainy summers and areas with moderately warm and fairly humid (maritime) climates are very much in favour of lodging. Only such progress of the disease leads to high or very high losses in yield; eyespots without foot rot at about milky/dough-ripe stage are without or of minor influence on yield. If the weather in May to July is warm and dry, losses are usually low.

Economic importance

Yield losses on winter cereals probably average between 5 and 10% in those areas of N./N.W. France, UK, Netherlands, S. Scandinavia and Germany which are not too far from the coast; this figure probably also applies to neighbouring eastern countries. Under conditions of continuous wheat cultivation and after severe lodging due to heavy eyespot, crop losses of up to 60% have exceptionally been observed. Wheat suffers from higher losses than barley and rye; losses in oats are less significant. Losses in spring-sown crops are mostly quite low, not justifying any chemical protection.

Control

There are very limited possibilities for preventive control, such as late sowing and less intensive crop rotations: in a 50% rotation a two-year-change between cereals and non-cereals is more advisable than a crop change each year. Intercropping has been shown to be of no advantage. There is no hope of breeding for completely resistant cultivars but neither is there any need since the plant tolerates a moderate level of attack. A number of cultivars show a certain degree of resistance, mainly due to genetic sources such as cvs Cappelle Desprez, Roazon, VPM 1 and *Aegilops ventricosa*. In cvs Cappelle Desprez and Roazon this is at least partially due to delayed establishment of the fungus on the outer leaf-sheath and delayed penetration of leaf-sheath layers. Chemical control with protective compounds did not prove useful, but systemics like the quite cheap benzimidazole-type fungicides (carbendazim, thiophanate methyl, benomyl) and recently the somewhat more expensive imidazole-derivative prochloraz, are quite effective (Fehrmann & Schrödter 1972). Application does not aim at eradication, but at avoiding more serious attacks. At average yields, increases after spraying in the order of 1.5 (−2.5)% cover the expenses. Spraying a haulm-stabilizing growth regulator like cycocel (CCC) prevents lodging and allows higher nitrogen fertilization, but does not help the plant to escape from physiological damage by the eyespot fungus. Effects of cycocel and benzimidazole-type fungicides on yield mostly turned out to be additive. The optimal time for spraying systemics is usually at GS 31/32. However, some delay up to GS 37 is possible. Earlier application—even in autumn—is exceptional, later application useless.

In low-risk situations with respect to eyespot control (crop rotation, sowing date, cultivar, plant density, weather etc.), spraying has become a routine procedure. In such cases financial output (yield increase) from fungicide application is 2–4 times as high as the input (on average). In many situations, however, decision making is a real problem. This is mainly due to the fact that at application time symptoms are mostly not yet visible: it takes about 6–8 weeks from infection to symptom development. After mild winters followed by warmer weather in early spring, e.g. in coastal areas, massive infections may be so early and symptom development so fast, that 20–25% of young plants showing symptoms at GS 31/32 can already be considered as a kind of threshold value for spraying; however, this rule of thumb does not apply to all regions. At the time of spraying, there may be 100% infected plants, but without any (macroscopically) visible symptoms present. It is possible to compute the probability of infection during a given period of time from temperature and air humidity records. This system helps in decision making, since the systemic fungicides act curatively as well as protectively and stop infections several weeks before; the computation of infection probability is used, in various ways, for warning purposes in different European countries. It takes into account sporulation and infection, but not the process of pathogenesis of the young plant after infection; the first time of haulm penetration is of major importance. Hence, in addition to the calculation of infection probability from meteorological data two other characteristics should be added to the computation model: i) influence of local parameters on field inoculum (crop rotation, soil quality, date and depth of sowing etc.); ii) speed of pathogenesis as influenced by temperature. This system has now been tested in the FRG. A quarter of the fungicide applications turned out to be superfluous, and about 90% of the decisions of the computation model were correct. In the near future, an inexpensive electronic warning device with a comprehensive model will be available in the FRG.

Benzimidazole-type fungicides were introduced into practice in the early 1970s, and for more than 6 years there was no serious indication of resistance in field populations of *P. herpotrichoides* (Fehrmann *et al.* 1982). In England, however, between 1981–83 benzimidazole-resistance problems gradually came up, locally at first, but finally leading to regional problems (Griffin & King 1985). Field populations in N. Germany, N. France, England and neighbouring countries are now largely benzimidazole-resistant. Official recommendations proposed replacement by prochloraz or a mixture of benzimidazoles and prochloraz. This makes spraying quite expensive, but there is no other choice. For the further future of eyespot control, mixing the two systemics should be more advisable than alternating use.
FEHRMANN

Pseudocercosporella capsellae (Ell. & Ev.) Deighton (syn. *Cercospora bloxamii* Berk. & Broome) causes small white spots with a faint purple border on *Brassica* species, especially turnip and rutabaga. It is common in wetter areas of N.W. Europe (CMI map 197) but is of minor importance. Small sclerotia ensure overwintering.

Pycnostysanus azaleae (Peck) Mason, originating from N. America on native azaleas, is widespread on evergreen *Rhododendron* species in the open in S. England (Brooks 1953). Buds infected in autumn become silvery grey and fail to open; in spring they are covered with the very distinctive coremia of the fungus. The leafhopper *Graphocephala coccinea* lays eggs in slits on the flower buds, providing a means of entry and possibly also acting as a vector. Viennot-Bourgin (1981) confirms the latter in the Paris region, where the disease is now common. Insecticides or fungicides can be used for control, but the disease seems to be of no importance on commercial glasshouse azaleas.

Ramichloridium pini de Hoog & Rahman

R. pini has recently been described causing a shoot dieback disease of *Pinus contorta* (de Hoog *et al.* 1983). Early symptoms are dead terminal buds with yellow needles on the previous year's shoot. These needles turn brown by autumn and are retained on the shoot until the following spring, when *Lophodermium conigenum* may fruit on them. In less severe attacks the bud flushes weakly and the needles turn yellow and fall prematurely. Internally the pith and leaf traces are discoloured rusty-brown. In the earlier stages of infection *R. pini* can be isolated onto malt agar. Later it is necessary to use a selective medium of modified Czapek Dox agar to suppress the growth of *Sydowia polyspora*, which always colonizes diseased shoots, and to which the dieback has previously been attributed by some authors. *R. pini* conidia develop on needle fascicle scars in spring, and infection probably occurs then on bud scales and new needle sheaths. Dieback due to *R. pini* has been found in Ireland, UK and possibly in Scandinavia.
MILLAR

Rhynchosporium secalis (Oudem.) J. Davis

Basic description

R. secalis has hyaline to light grey mycelium, developing sparsely as a compact stroma under the cuticle of the host. The sessile conidia, borne on cells of the fertile stroma, are hyaline, cylindrical to ovate, with a short apical beak on most spores, and measure 12–20×2–4 μm (CMI 387). ***R. orthosporum*** Caldwell, occurs on a number of grasses and is differentiated from *R. secalis* by the longer, more cylindrical conidia without the apical beak.

Host plants

In addition to barley and rye, *R. secalis* attacks many wild and cultivated grasses including species of *Agropyron, Bromus, Dactylis, Lolium, Phleum* and *Hordeum*.

Host specialization

There is evidence that forms of *R. secalis* are specialized to particular genera. Formae speciales *hordei, secalis, agropyri, phalaridis* have been distinguished but their distinctness has been questioned. Pathotypes occur on barley: four have been distinguished in Argentina, nine in the USA, ten in Japan, three in W. Australia, four in Bulgaria and two in the UK.

Geographical distribution

R. secalis is widely distributed in temperate regions. It has been recorded in nearly fifty countries, including all European countries except Italy, Spain and Poland (CMI map 383).

Disease

R. secalis causes leaf blotch or scald of barley, rye and various other Gramineae (Shipton *et al.* 1974). Leaves and leaf-sheaths are the principal organs affected but glumes and awns of barley are also attacked. Lesions on leaves are lenticular, usually 1–2 cm long and often confluent. Many infections start at the junction of the leaf and leaf-sheath forming a lesion at the base of the leaf: one such lesion can kill the rest of the leaf. The first sign of infection is water soaking, then the area becomes bluish-grey and later the central part becomes pale greyish-brown with a distinct dark-brown or violet-brown border, except on rye where no dark border is formed. Conidia occur abundantly in the centre of the lesions but are produced over practically the entire lesion area. Conidia germinate readily on the leaf surface and may produce one or more germ tubes from each cell. Appressoria may be produced at the end of germ tubes but penetration, which is normally direct and not through stomata, is not dependent upon appressorial formation. Following penetration of the cuticle a subcuticular mycelium develops, subsequently forming a stroma one to several cells in thickness. Later, hyphae penetrate the epidermal cell layer and at this time collapse of mesophyll cells occurs, an event which is evident externally as water soaking and scalding of tissues. Toxins are considered to be implicated in symptom development (Auriol *et al.* 1978) but there is still doubt about the precise identity of the compound(s) involved. Optimum temperature for growth in culture, conidial germination, infection and sporulation is 15–20°C. Infection of barley leaves occurs readily over the temperature range 6–20°C and is often well established in about 9–10 h under cool moist conditions. Lesion development proceeds normally when post-inoculation temperature is 12–24°C but few lesions develop outside this range.

Epidemiology

Seedlings developing from infected grain may develop typical scald lesions at the tip of the coleoptile 4–6 days after emergence, or remain symptomless. However, the practical importance of seed transmission and of alternative grass hosts is unknown and most primary infections are considered to arise from spores produced on infected debris of a preceding crop. Depending upon environmental conditions, *R. secalis* survives in infected debris for up to 12 months; alternate wetting and drying, and microbial antagonism are apparently the major factors restricting survival. Conidia are produced on stromata or sclerotia on infected debris where cool, moist conditions prevail. Free moisture or an RH of 95–98% is necessary and conidial production is most abundant in the temperature range 10–20°C.

R. secalis is heavily dependent on free water or high RH at all stages of its development except the incubation period; periods of dry weather have a marked inhibitory effect. Conidia are dispersed by rain splash and infection requires leaf surface wetness. Data for the duration of leaf surface wetness and temperature permit some assessment of the risk of infection. With a wetness period lasting 24 h there is a high risk of infection at any temperature from 6 to 24°C; with a period of 12 h high risk is confined to temperatures 16 to 24°C. Application of nitrogen fertilizer increases the severity of epidemics. This is apparently due to enhancement of sporulation rather than to an increase in susceptibility to infection *per se*. Race non-specific resistance in barley cultivars is correlated with a high ratio of water-soluble carbohydrate to total nitrogen in leaves (Jenkyn & Griffiths 1978).

Economic importance

Leaf blotch or scald is a common disease of barley especially in the cooler and semi-humid barley growing areas. It can cause severe losses, yield reductions of 30–40% having been reported from many parts of the world. The disease has caused concern in several European countries and has increased its geographical range in the UK since about 1960 although it is still most severe in western and south-western areas. Yield loss may arise from reduction in the numbers of tillers and numbers of grains per tiller but most commonly from reduction in grain size. A linear relationship has been reported (James *et al.* 1968) between yield loss and percentage of the laminar area affected by leaf blotch on the flag and second leaf at growth stage 75. The percentage loss in yield is equivalent to approximately two-thirds of the percentage infection on the flag leaf, or one-half of the percentage infection on the second leaf.

The importance of the disease on rye is difficult to assess but severe attacks have been reported from the USSR and Switzerland. Leaf blotch is not normally considered to be important on grasses although occasionally severe outbreaks have been reported on rye grasses (*Lolium perenne* and *L. multiflorum*). Both *R. secalis* and *R. orthosporum* occur on rye grasses and the latter may be the most common, at least in the UK. It should be noted however, that isolates of *R. secalis* may also infect barley: *R. orthosporum* is restricted to grass hosts.

Control

Destruction of stubble and infected straw, which can be achieved by burning and deep ploughing, significantly reduces the severity of leaf-blotch attack. Similarly, volunteer barley, which can act as a bridge between successive crops, should be destroyed. Balanced fertilizer applications and, in particular, avoidance of excessive use of nitrogen fertilizer are helpful in reducing disease severity. The use of resistant cultivars plays an important role in disease control. Eleven major resistance genes have been identified in barley, five existing as alleles or pseudoalleles at one locus on chromosome 3. Expression of these resistance genes is, however, modified by complementary genes and by environmental factors (Habgood & Hayes 1971). Major gene resistance has not been widely used in commercial cultivars and where it has been used experience indicates that it is not durable. More important is race non-specific resistance, which, in most situations, can provide adequate protection. Leaf blotch may also be controlled by fungicide application. Of the fungicides currently in use on cereals the following are highly effective: captafol, carbendazim, thiophanate methyl, prochloraz, propiconazole, triadimefon (MAFF 1983). In most locations, leaf-blotch epidemics do not occur with sufficient regularity to justify routine use of fungicides but applications are generally economic if the disease can be found on any of the top three leaves especially at or soon after flag leaf emergence. Fungicide application after ear emergence is not recommended.

GRIFFITHS

Stigmina carpophila (Lév.) M.B. Ellis

(Syn. *Coryneum beijerinckii* Oudem., *Clasterosporium carpophilum* (Lév.) Aderhold)

Basic description

The fungus does not form an acervulus in the true sense of the term. The conidia are elongate-ovoid, light greenish-yellow, 4- to 6-celled and offer a reliable means for identification. A teleomorph, ***Ascospora beijerinckii***, was described by Vuillemin in France but has not been confirmed by others.

Host plants

S. carpophila attacks many cultivated *Prunus* species including sweet cherry, sour cherry, plum, peach and nectarine, apricot and almond. It also occurs on wild or ornamental *Prunus* species like *P. laurocerasus, P. davidiana, P. serotina,* and *P. virginiana.*

Geographical distribution

The disease was first recorded from France in 1853, but was probably present in Europe before that time. Later it was reported from several other countries and now occurs in all stone-fruit growing areas.

Disease

The disease is known as shot-hole, fruit spot, coryneum blight or peach blight (Plate 25). Leaves, fruits, twigs, buds and blossoms are attacked by the fungus (Naef-Roth 1949). Leaf lesions are at first small circular, purplish or brown areas which expand to spots 3–10 mm in diameter. They are frequently surrounded by a narrow zone, light-green to yellow in colour. The diseased area drops out resulting in a shot-hole effect or a ragged appearance when the lesions are large. Diseased leaves often fall off early in the season. If petioles are infected, the entire leaf is soon killed. Fruit lesions are of the same colour and size as leaf lesions. On peach, nectarine and apricot, they later become rough and corky. On cherry, infection leads to depressed lesions and malformed fruits which tend to dry up. Twig infection results in the formation at first of purplish or brown spots 2–3 mm in diameter which later expand into elongated necrotic cankers which are often covered with gummy exudate. Severely attacked twigs are killed. Hypertrophic cankers up to 10 cm in diameter may be formed on twigs and larger branches of peach. Diseased buds are dark-brown to black and often show a glistening layer of gum. Blossom blight is the result of bud infection or of canker development at the base of the flower pedicel. The flowers wither before they are fully expanded. The symptoms vary on different stone fruits. On cherries and plums, leaf spotting is the principal symptom (also on cherry fruit in wet years). On peaches and nectarines, twig and bud blighting are predominant, while in apricot bud blighting, fruit and leaf spotting occur most often. On almonds, the principal symptoms are leaf spotting and blossom and fruit infections.

Epidemiology

In spring, the fungus produces the first inoculum mainly on lesions on twigs, buds and leaf-scars where it survives from one season to the next. The conidia remain viable for several months when kept dry. They are not easily detached from the conidiophores by moving air, but are readily detached by water and so are largely disseminated by rain. The conidia germinate over a relatively wide range of temperature and as low as 2–4°C (the optimum is 18–21°C). An important factor for infection is moisture. At least 24 h of continuous moisture is necessary for twig infection in peach. The incubation period is 3–8 days for twigs, depending on temperature. The activity of the fungus depends on climatic conditions. In temperate regions with regular summer precipitation, infections may start as the flower and leaf buds open, resulting in blossom blight and leaf infection. Under these conditions, infections may occur during a great portion of the growing season. As the fungus attacks mainly young tissue, leaves are infected predominantly from April to June. Twigs are infected mainly in June and July but also later in the summer and autumn. Under warmer conditions, the fungus is inactive during the dry, hot summer months. As soon as the autumn rains begin, activity is renewed. Infection of buds and twigs may occur at this time and continue throughout the winter when temperatures are above freezing. In spring, the new growth is prone to infection. Thus, under these conditions the critical infection periods are late autumn and early winter, and in the spring as the buds open. Direct penetration of leaves by germ tubes from the conidia has been observed. The conidia germinate equally well on both surfaces, but the underside of leaves is more prone to infection than the upper side. In all probability, the fungus is capable of penetrating the twig, fruit and blossoms in the same manner. Infections also occur via stomata and leaf-scars.

Economic importance

The disease is greatly enhanced by moist conditions. Therefore, it is most severe in areas where periods of high humidity and rainfall occur at a time when susceptible tissue is present. Also, sprinkler irrigation increases the incidence of the blight. Losses arise from fruit infection, killing the twigs, buds and blossoms and from the damage caused to the leaves. Defoliation, especially if it happens every year, weakens the tree considerably and reduces yield, vigour and hardiness. The disease is a major problem on peaches, nectarines, apricots and almonds in some areas while others are obviously less affected. Reductions in yield up to 50% have been reported for peaches and apricots. Cherries and plums are severely damaged only in humid areas. On sour cherry the disease generally seems to be a minor problem.

Control

The disease is relatively easy to control (English & Davis 1962) but timing of protective spray applications must be varied according to the principal damage caused to the crop and to the infection conditions in the different climatic areas. For peach and nectarine, one autumn application during or after leaf-fall but before winter rains begin is sufficient to protect against twig and/or bud infections in summer-dry areas. In areas with substantial precipitation in spring and summer, additional sprays are necessary at bud swell, petal fall and after bloom. For apricots the same schedule applies as for peaches with the exception that, even under dry summer conditions, one application at full bloom or petal-fall stage is recommended to prevent fruit and leaf infections. For sweet cherries, plums and almonds, where protection of leaves and/or fruits is of prime importance, sprays at bud swell, at the end of the blossoming period, and, depending on the climatic and weather conditions, one or two applications after bloom are necessary. An application during or after leaf-fall is also recommended for these crops to reduce the disease.

In the past, Bordeaux mixture and fixed copper material have been the standard fungicide to control the disease. Because these materials often cause leaf injury if applied after leaves emerge, they are now only recommended for autumn applications and as dormant or delayed dormant sprays. In nectarines and peaches, however, organic fungicides should be preferred for the dormant or delayed dormant sprays because of their greater effect against leaf curl (*Taphrina deformans*). Organic fungicides should be used for all applications when the trees are in leaf. Captan, ziram, ferbam, dithianon, folpet, metiram and propineb have proved to be effective.

SEEMÜLLER

REFERENCES

Adams M.J. (1980) The role of seed-tuber and stem inoculum in the development of gangrene in potatoes. *Annals of Applied Biology* **96,** 17–28.

Allen S.J., Brown J.F. & Kochman J.K. (1983) The infection process, sporulation and survival of *Alternaria helianthi* on sunflower. *Annals of Applied Biology* **102,** 413–419.

Andrews J.H., Berbee F.M. & Nordheim E.V. (1983) Microbial antagonism to the imperfect stage of the apple scab pathogen, *Venturia inaequalis. Phytopathology* **73,** 228–234.

Angevain M. (1983) Méthode d'infection artificielle pour la sélection de la luzerne contre *Phoma medicaginis. Agronomie* **3,** 911–916.

Auriol P., Strobel G., Pio Beltran J. & Gray G. (1978) Rhynchosporoside, a host-selective toxin produced by *Rhynchosporium secalis*, the causal agent of scald disease of barley. *Proceedings of the National Academy of Sciences USA* **75,** 4339–4343.

Baker K.F., Davis L.H., Durbin R.D. & Snyder W.C. (1977) Greasy blotch of carnation and flyspeck of apple: diseases caused by *Zygophiala jamaicensis. Phytopathology* **67,** 580–588.

Bal E., Gilles G., Greemers P. & Verheyden C. (1983) Problems and control of *Alternaria* leaf spot on strawberries. *Mededelingen van de Faculteit Landbouwwetenschappen Rijksuniversiteit Gent* **48,** 637–646.

Bannon E. (1975) Infection of potato tubers by *Oospora pustulans* during the growing season in relation to the distribution of tubers in the ridge. *Potato Research* **18,** 588–596.

Barash I., Manulis S., Kashman Y., Springer J.P., Chen M.H.M., Clardy J. & Strobel G.A. (1983) Crystallization and X-ray analysis of stemphyloxin I, a phytotoxin from *Stemphylium botryosum. Science* **220,** 1065–1066.

Barksdale T.H. & Stoner A.K. (1978) Resistance in tomato to *Septoria lycopersici. Plant Disease Reporter* **62,** 844–847.

Barrière Y., Massenot M. & Raynal G. (1974) Les maladies des légumineuses fourragères. II. Les maladies du feuillage. Taches foliaires à *Leptosphaerulina* (pepper spot). *Annales de Phytopathologie* **6,** 341–347.

Basu P.K (1971) Existence of chlamydospores of *Alternaria porri* f. sp. *solani* as overwintering propagules in soil. *Phytopathology* **61,** 1347–1350.

Batko S., Murray J.S. & Peace T.R. (1958) *Sclerophoma pithyophila* associated with needle-cast of pines and its connexion with *Pullularia pullulans. Transactions of the British Mycological Society* **41,** 126–128.

Bazzigher G. (1976) Der schwarze Schneeschimmel der Koniferen (*Herpotrichia juniperi* und *Herpotrichia coulteri*). *European Journal of Forest Pathology* **6,** 109–122.

Bechet M. & Sapta-Forda A. (1981) (Investigations on the growth, development and control of *Alternaria vitis*, the pathogen of alternariosis of grapevine). In *Contributii Botanice*, pp. 147–156. Universitatea Babes-Bolyai, Cluj-Napoca.

Biggs P. (1959) Studies on *Meria laricis*, needle cast disease of larch. *Report on Forest Research, 1958,* pp. 112–113. HMSO, London.

Bissett J. (1983) *Stagonospora avenae; S. nodorum; Septoria passerinii.* In *Fungi Canadenses* nos. 239, 240, 243. Agriculture Canada, Ottawa.

Blake C.M. (1980) Development of perithecia and pycnidia of *Didymella applanata* (causing spur blight) on raspberry canes. *Transactions of the British Mycological Society* **74,** 101–105.

Boerema G.H. (1979) Damping off of lilac seedlings caused by *Phoma exigua* var. *lilacis. Phytopathologia Mediterranea* **18,** 105–106.

Boerema G.H. & Bollen G.J. (1975) Conidiogenesis and conidial septation as differentiating criteria between *Phoma* and *Ascochyta. Persoonia* **8,** 111–114.

Boerema G.H., Dorenbosch M.M.J. & Leffring L. (1965) A comparative study of the black stem fungi on lucerne and red clover and the footrot fungus on pea. *Netherlands Journal of Plant Pathology* **71,** 79–89.

Boerema G.H., Crüger G., Gerlagh M. & Nirenberg H. (1981) *Phoma exigua* var. *diversispora* and related fungi on *Phaseolus beans. Zeitschrift für Pflanzenkrankheiten und Pflanzenschutz* **88,** 597–607.

Bondoux P., Bompeix G., Morgat F. & Viard P. (1969) *Les Principales Pourritures des Pommes et Poires en Conservation.* CTIFL, Paris.

Bonman J.M., Gabrielson R.L., Williams P.H. & Delwiche P.A. (1981) Virulence of *Phoma lingam* to cabbage. *Plant Disease* **65,** 865–867.

Boorsma P.A. (1980) Variability in chickpea for blight resistance. *FAO Plant Protection Bulletin* **28,** 110–113.

Bottalico A., Frisullo S., Iacobellis N.S., Capasso R., Corrado E. & Randazzo G. (1983) Sulla presenza di fomenone nelle piante di pomodoro con infezioni di *Phoma destructiva. Phytopathologia Mediterranea* **22,** 116–119.

Boubals D. & Mur G. (1983) Une autre maladie de la vigne sévit dans le Penedes (Espagne). *Progrès Agricole et Viticole* **100,** 453.

Bougeard M. & Vegh I. (1980) Etude préliminaire sur le *Cercosporidium punctum* agent de la cercosporidiose du fenouil. *Cryptogamie-Mycologie* **1,** 205–221.

Boyd A.E.W. (1972) Potato storage diseases. *Review of Plant Pathology* **51,** 297–321.

Brame C. & Flood F. (1983) Antagonism of *Aureobasidium pullulans* towards *Alternaria solani. Transactions of the British Mycological Society* **81,** 621–624.

Braverman S.W. (1977) Reaction of Brussels sprout introductions to artificial inoculation with *Alternaria brassicicola. Plant Disease Reporter* **61,** 360–362.

Brokenshire T. & Cooke B.M. (1977) The effect of six systemic fungicides on halo spot disease of barley caused by *Selenophoma donacis. Annals of Applied Biology* **87,** 41–45.

Brönnimann A. (1968) Zur Kenntnis von *Septoria nodorum*, dem Erreger der Spelzenbraune und einer Blattdurre des Weizens. *Phytopathologische Zeitschrift* **61,** 101–146.

Brooks F.T. (1953) *Plant Diseases.* Oxford University Press, Oxford.

Brun H., Renard M., Jouan B., Tanguy X. & Lamarque C. (1979) Observations préliminaires sur quelques maladies du colza en France: *Sclerotinia sclerotiorum, Cylindrosporium concentricum, Ramularia armoraciae. Sciences Agronomiques Rennes (1979),* 7–77.

Buchannon K.W. & Wallace H.A.H. (1962) Note on the effect of leaf disease on yield, bushel weight, and thousand-kernel weight of Parkland Barley. *Canadian Journal of Plant Science* **42,** 534–536.

Burchill R.T. & Beever D.J. (1975) Seasonal fluctuations in ascospore concentrations of *Didymella applanata* in relation to raspberry spur blight incidence. *Annals of Applied Biology* **81,** 299–304.

Byford W.J. (1976) Experiments with fungicide sprays to control *Ramularia beticola* in sugar-beet seed crops. *Annals of Applied Biology* **82,** 291–297.

Caddel J.L. & Wilcoxson R.D. (1975) Sources of resistance to net blotch of barley in Morocco. *Plant Disease Reporter* **59,** 491–494.

Cannon P.F., Hawksworth D.L. & Sherwood-Pike M.A. (1985) *The British Ascomycotina. An Annotated Checklist.* CMI, Kew.

Carnegie S.F. (1984) Seasonal occurrence of the potato gangrene pathogen, *Phoma exigua* var. *foveata,* in the open air. *Annals of Applied Biology* **104,** 443–449.

Carnegie S.F., Adam J.W. & Symonds C. (1978) Persistence of *Phoma exigua* var. *foveata* and *Polyscytalum pustulans* in dry soils from potato stores in relation to reinfection of stocks derived from stem cuttings. *Annals of Applied Biology* **90,** 179–186.

Carter M.V. & Moller W.J. (1961) Factors affecting the survival and dissemination of *Mycosphaerella pinodes* in South Australian irrigated pea fields. *Australian Journal of Agricultural Research* **12,** 879–888.

Cassini R. (1973) Etat actuel des principales maladies du maïs. *Phytiatrie-Phytopharmacie* **22,** 1–18.

Cassini R., Gay J.-P. & Cassini R. (1972) Observations sur le cycle de développement et sur les organes de conservation de *Kabatiella zeae. Annales de Phytopathologie* **4,** 367–371.

Cayley G.R., Hide G.A., Read P.J. & Dunne Y. (1983) Treatment of potato seed and ware tubers with imazalil and thiabendazole for control of silver scurf and other storage diseases. *Potato Research* **26,** 163–173.

Chase A.R. (1983) Influence of host plants and isolate source on *Myrothecium* leaf spot of foliage plants. *Plant Disease* **67,** 668 –671.

Chastagner G.A. & Byther R.S. (1983) Infection period of *Phaeocryptopus gaeumannii* on Douglas-fir needles in Western Washington. *Plant Disease* **67,** 811–813.

Ciccarone A. (1956) *Elsinoë australis* agente di una 'scabbia' degli agrumi, in Sicilia. *Rivista di Agrumicoltura* **2,** 31–35.

Ciccarone A. (1971) Il fungo del 'mal secco' degli agrumi. *Phytopathologia Mediterranea* **10,** 68–75.

Cirulli M. (1968) Chemical control trials against tomato corky root. *Annali Facoltà Agraria Università Bari* **22,** 453–460.

Clifford B.C. & Jones D. (1981) Net blotch of barley. *UK Cereal Pathogen Virulence Survey 1980. Annual Report,* pp. 71–77.

Cook R.J. (1981) Unexpected effects of fungicides on cereal yields. *Bulletin OEPP/EPPO Bulletin* **11,** 277–285.

Coulombe L.J. (1979) Mesure de paramètres concernant la sensibilité à deux maladies, ainsi que la production et la qualité des fruits chez deux cultivars de tomate. *Phytoprotection* **60,** 79–92.

Cristinzio G. (1979) Studi su di un ceppo di *Botryosphaeria obtusa* patogeno su vite. *Annali della Facoltà di Scienze Agrarie della Università degli Studi di Napoli Portici* **13,** 88–100.

Cunningham P.C. (1981) Occurrence, role and pathogenic traits of a distinct pathotype of *Pseudocercosporella herpotrichoides. Transactions of the British Mycological Society* **76,** 3–15.

Darpoux H. & Lebrun A. (1964) *Etudes réalisées sur le Cercospora de la Betterave de 1953 à 1964.* Publications ITB-INRA, Paris.

Darpoux H., Staron T. & Lebrun A. (1966) Action du thiabendazole sur le *Cercospora beticola. Revue IIRB* no. 150.

Davet P. (1976) Etude d'interactions entre le *Fusarium lycopersici* et le *Pyrenochaeta lycopersici* sur les racines de la tomate. *Annales de Phytopathologie* **8,** 191–202.

Davies W.P., Lewis B.G. & Day J.R. (1981) Observations on infection of stored carrot roots by *Mycocentrospora acerina. Transactions of the British Mycological Society* **77,** 139–151.

Day J.R., Lewis B.G. & Martin S. (1972) Infection of stored celery plants by *Centrospora acerina. Annals of Applied Biology* **71,** 201–210.

Day P.R. (1956) Race names of *Cladosporium fulvum. Report of the Tomato Genetics Cooperative* **6,** 13–14.

Deighton F.C. (1971). Studies on *Cercospora* and allied genera. III *Centrospora.* CMI Mycological Papers no. 124. CMI, Kew.

Delon R., Reisinger O. & Mangenot F. (1973) Etude aux microscopes photonique et électronique de racines de tomates var. Marmande atteinte de maladie liégeuse. *Annales de Phytopathologie* **5,** 141–162.

Dennis C. & Davis R.P. (1978) Storage rots of chicory roots caused by *Phoma* and *Botrytis. Plant Pathology* **27,** 49.

Deportes L., Gilly G., Cuany A., Mercier S., Poupet A. & Marais A. (1979) Contribution à l'étude de quelques problèmes techniques posés par la culture de la violette. *Revue Horticole* no. 200, 13–21.

Derbyshire D.M. & Crisp A.F. (1978) Studies on treatments to prolong the storage life of carrots. *Experimental Horticulture* **30,** 23–28.

Dhingra O.D. & Sinclair J.B. (1978) *Biology and Pathology of Macrophoma phaseolina.* Universidade Federal de Vicosa, Minas Gerais (Brazil).

Dickenson S. & Wheeler B.E.J. (1981) Effects of temperature, and water stress in sycamore, on growth of *Cryptostroma corticale. Transactions of the British Mycological Society* **76,** 181–185.

Dickinson C.H. (1981a) Leaf surface microorganisms as pathogen antagonists and minor pathogens. In *Strategies for the Control of Cereal Disease.* (Eds J.F. Jenkyn & R.T. Plumb), pp. 109–122. Blackwell Scientific Publications, Oxford.

Dickinson C.H. (1981b) Biology of *Alternaria alternata,*

Cladosporium cladosporioides and *C. herbarum* in respect of their activity on green plants. In *Microbial Ecology of the Phylloplane.* (Ed. J.P. Blakeman), pp. 169–184. Academic Press, London.

Dickinson C.H. & Sheridan J.J. (1968) Studies on the survival of *Mycosphaerella pinodes* and *Ascochyta pisi. Annals of Applied Biology* **62,** 473–483.

Dickinson C.H. & Wallace B. (1976) Effect of late applications of foliar fungicides on activity of microorganisms on winter wheat flag leaves. *Transactions of the British Mycological Society* **67,** 103–112.

Diercks R. (1959) Zur chemischen Bekämpfung der Korbweiden-parasiten *Fusicladium saliciperdum* and *Glomerella miyabeana. Pflanzenschutz* **11,** 123–130.

Dixon G.R. (1984) *Vegetable Crop Diseases.* Macmillan, London.

Draganić M. (1983) (Investigation of stem resistance of lines and hybrids of maize to rot *Phaeocytosporella zeae* and *Gibberella zeae* on artificial inoculation). *Zaštita Bilja* **34,** 405–410.

Dunenko M.A. & Belikova V.K. (1977) Septoriosis of tomato. *Zashchita Rastenii* no. 7, 35–36.

Ebben M.H. (1974) Brown root rot of tomato. *Annual Report Glasshouse Crops Research Institute Littlehampton UK 1973,* pp. 127–135.

Eckert J.W. (1978) Post-harvest diseases of citrus fruits. *Outlook on Agriculture* **9**(5), 225–232.

El-Sayed F. & Marić A. (1981) (Contribution to the study of the biology and epidemiology of *Phoma macdonaldi,* pathogen of black spot of sunflower). *Zaštita Bilja* **32,** 13–27.

English H. & Davis J.R. (1962) Efficacy of fall applications of copper and organic fungicides for the control of *Coryneum* blight of peach in California. *Plant Disease Reporter* **46,** 688–691.

Ersek T. (1978) Occurrence of *Corynespora cassiicola* on soyabean in Hungary. *Acta Phytopathologica Academiae Scientiarum Hungaricae* **13,** 175–178.

Eschenbrenner P. (1983) L'helminthosporiose du blé. *Phytoma* no. 345, 13–14.

Eyal Z. & Ziv O. (1974) The relationship between epidemics of septoria leaf blotch and yield losses in spring wheat. *Phytopathology* **64,** 1385–1389.

Eyal Z., Amiri Z. & Wahl I. (1973) Physiological specialization of *Septoria tritici. Phytopathology* **63,** 1087–1091.

Fehrmann H. & Schroedter H. (1972) Ökologische Untersuchungen zur Epidemiologie von *Cercosporella herpotrichoides.* IV. Erarbeitung eines praxisnahen Verfahrens zur Bekämpfung der Halmbruchkrankheit des Weizens mit systemischen Fungiziden. *Phytopathologische Zeitschrift* **74,** 161–174.

Fehrmann H., Horsten J. & Siebrasse G. (1982) Five years' results from a long-term experiment on carbendazim resistance of *Pseudocercosporella herpotrichoides. Crop Protection* **1,** 165–168.

Fehrmann H., Reinecke P. & Weihafen U. (1978) Yield increases in winter wheat by unknown effects of MBC-fungicides and captafol. *Phytopathologische Zeitschrift* **93,** 359–362.

Filho E.X. & Dhingra O.D. (1980) Population changes of *Macrophomina phaseolina* in amended soils. *Transactions of the British Mycological Society* **74,** 471–481.

Fitt B.D.L. & Hornby D. (1978) Effects of root-infecting fungi on wheat transport processes and growth. *Physiological Plant Pathology* **13,** 335–346.

Fletcher J.T. (1984) *Diseases of Greenhouse Plants.* Longman, London.

Fokkema N.J. & Meulen F. van der (1976) Antagonism of yeast-like phyllosphere fungi against *Septoria nodorum* on wheat leaves. *Netherlands Journal of Plant Pathology* **82,** 13–16.

Fokkema N.J. & Nooij M.P. de (1981) The effect of fungicides on the microbial balance in the phyllosphere. *Bulletin OEPP/EPPO Bulletin* **11,** 303–310.

Fokkema N.J., den Hauter J.G., Kosterman Y.J.C. & Nelis A.L. (1979a) Manipulation of yeasts on field-grown wheat leaves and their antagonistic effect on *Cochliobolus sativus* and *Septoria nodorum. Transactions of the British Mycological Society* **72,** 19–29.

Fokkema N.J., Kastelein P. & Post D.J. (1979b) No evidence for acceleration of leaf senescence by phylloplane saprophytes of wheat. *Transactions of the British Mycological Society* **72,** 312–315.

Folk G. (1976) Stemphylium ray speck of chrysanthemum in Hungary. *Kertészeti Egyetem Közleményei* **40,** 403–410.

Fravel D.R. & Spurr H.W. (1977) Biocontrol of tobacco brown-spot disease by *Bacillus cereus* subsp. *mycoides* in a controlled environment. *Phytopathology* **67,** 930–932.

Frisullo S. (1984) Parassiti funghini dell'Italia meridionale. IV. *Pseudoseptoria donacis* su *Arundo donax. Phytopathologia Mediterranea* **23,** 49–51.

Galet P. (1977) *Les Maladies et les Parasites de la Vigne.* Paysan du Midi, Montpellier.

Garibaldi A. & Gullino M.L. (1981) Gravi attacchi di *Phoma chrysanthemicola* f.sp. *chrysanthemicola* su crisantemo in Italia. *Informatore Fitopatologico* **31** (5), 27–30.

Geoffrion R. (1984) Les maladies des arbres fruitiers à pépins au cours de ces dernières années. *Arboriculture fruitière* **364,** 34–39.

Gerlach W. & Schneider R. (1966) Infectionsversuche mit *Pyrenochaeta terrestris* und der Kerkwurzel—*Pyrenochaeta* an Tomate und Küchenwiebel. *Phytopathologische Zeitschrift* **56,** 19–24.

Gibson I.A.S. (1972) *Dothistroma* blight of *Pinus radiata. Annual Review of Phytopathology* **10,** 51–72.

Gilchrist D.G. & Grogan R.G. (1976) Production and nature of a host-specific toxin from *Alternaria alternata* f.sp. *lycopersici. Phytopathology* **66,** 165–171.

Gindrat D. (1979) *Alternaria radicina,* un parasite important des ombellifères maraîchères. *Revue Suisse de Viticulture, d'Arboriculture et d'Horticulture* **11,** 257–267.

Graham D.C., Hamilton G.A., Quinn C.E. & Ruthven A.D. (1973) Use of 2-aminobutane as a fumigant for control of gangrene, skin spot and silver scurf diseases of potato tubers. *Potato Research* **16,** 109–125.

Graham J.H. & Luttrell E.S. (1961) Species of *Leptosphaerulina* on forage plants. *Phytopathology* **51,** 680–693.

Graniti A. & Laviola C. (1981) Sguardo generale alle malattie parassitarie dell'olivo. *Informatore Fitopatologico* **31** (1/2), 77–92.

Graniti A. & De Leo P. (1966) Osservazioni su *Spilocaea oleagina* IV. Resistenza alla macerazione enzimatica come probabile reazione di difesa delle foglie di olivo al parassita. *Phytopathologia Mediterranea* **5,** 65–79.

Grasso A. & Catara A. (1982) Osservazioni su intumescenze gommose delle foglie di agrumi. *Informatore Fitopatologico* **32** (9/10), 43–46.

Grasso S. & Rosa R. La (1983) Gravi attacchi di *Septoria citri* su frutti di limone. *Rivista di Patologia Vegetale* **19,** 15–20.

Green G.J. & Bendelow V.M. (1961) Effect of speckled leaf blotch, *Septoria passerinii* on the yield and malting quality of barley. *Canadian Journal of Plant Science* **41,** 431–435.

Greenaway W. & Cowan J.W. (1970) The stability of mercury resistance in *Pyrenophora avenae. Transactions of the British Mycological Society* **54,** 127–138.

Gregory P.H. (1939) The life history of *Ramularia vallisumbrosae* on narcissus. *Transactions of the British Mycological Society* **23,** 24–54.

Gregory P.H. & Waller S. (1951) *Cryptostroma corticale* and sooty bark disease of sycamore (*Acer pseudoplatanus*). *Transactions of the British Mycological Society* **34,** 579–597.

Griffin M.J. & King J.E. (1985) Benzimidazole resistance in *Pseudocercosporella herpotrichoides*: results of ADAS random surveys and fungicide trials in England and Wales, 1982–84. *Bulletin OEPP/EPPO Bulletin* **15,** 485–494.

Habgood R.M. & Hayes J.D. (1971) The inheritance of resistance to *Rhynchosporium secalis* in barley. *Heredity* **27,** 25–37.

Hamilton G.A., Lindsay D.A., Quinn C.E., Ruthven A.D., Shipton P.J. & Gray W.J. (1981) Comparison of the fungicides thiabendazole and 2-aminobutane used for the control of gangrene and skin spot on potatoes grown under Scottish conditions. *Proceedings of the 1981 British Crop Protection Conference–Pests and Diseases,* pp. 355–362. BCPC, Croydon.

Hammerschmidt R. & Kuć J. (1982) Lignification as a mechanism for induced systemic resistance in cucumber. *Physiological Plant Pathology* **20,** 61–71.

Hanso M. & Tõrva A. (1975) Black snow mould in Estonia I. On ecology and morphology of *Herpotrichia juniperi. Metsanduslikud Uurimused* **12,** 262–279.

Harrower K.M. (1973) Differential effects of phytoalexin from *Pisum sativum* on two races of *Ascochyta pisi. Transactions of the British Mycological Society* **61,** 383–386.

Helms K. & Andruska A. (1981) Benomyl and protection of *Trifolium subterraneum* against clover scorch. *Phytopathologische Zeitschrift* **101,** 110–122.

Hervert V., Marvanová L. & Kazda V. (1980) (*Alternaria pluriseptata* on cucumber and notes on its classification). *Česká Mykologie* **34,** 13–20.

Hewett P.D. (1973) The field behaviour of seed-borne *Ascochyta fabae* and disease control in field beans. *Annals of Applied Biology* **74,** 287–295.

Hide G.A., Hirst J.M. & Mundy E.J. (1969) The phenology of skin spot (*Oospora pustulans*) and other fungal diseases of potato tubers. *Annals of Applied Biology* **64,** 265–279.

Hide G.A., Hirst J.M. & Salt G.A. (1968) Methods of measuring the prevalence of pathogenic fungi on potato tubers. *Annals of Applied Biology* **62,** 309–318.

Hill R.R. & Leath K.T. (1979) Comparison of four methods of selection for resistance to *Leptosphaerulina briosiana* in alfalfa. *Canadian Journal of Genetics and Cytology* **21,** 179–186.

Hill S.A. (1975) The importance of wood scab caused by *Venturia inaequalis* as a source of infection for apple leaves in the spring. *Phytopathologische Zeitschrift* **82,** 216–223.

Holmes S.J.I. & Colhoun J. (1974) Infection of wheat by *Septoria nodorum* and *S. tritici* in relation to plant age, air temperature and relative humidity. *Transactions of the British Mycological Society* **63,** 329–338.

Hoog G.S. de, Rahman M.A. & Boekhout T. (1983) *Ramichloridium, Veronaea* and *Stenella*: generic delimitation, new combinations and two new species. *Transactions of the British Mycological Society* **81,** 485–490.

Hubbeling N. (1971) Determination trouble with new races of *Cladosporium fulvum. Mededelingen van de Faculteit Landbouwwetenschappen Rijksuniversiteit Gent* **36,** 300–305.

Humpherson-Jones F.M. (1983) Pathogenicity studies on isolates of *Leptosphaeria maculans* from brassica seed production crops in south east England. *Annals of Applied Biology* **103,** 37–44.

Imolehin E.D. (1983) Rice seedborne fungi and their effect on seed germination. *Plant Disease* **67,** 1334–1336.

James W.C. (1969) A survey of foliar diseases of spring barley in England and Wales in 1967. *Annals of Applied Biology* **63,** 253–263.

James W.C., Jenkins J.E.E. & Jemmett J.L. (1968) The relationship between leaf blotch caused by *Rhynchosporium secalis* and losses in grain yield in spring barley. *Annals of Applied Biology* **62,** 273–288.

Janke C. & Zott A. (1983) Pathogenität von *Phoma eupyrena* an Kartoffelknollen. *Archiv für Phytopathologie und Pflanzenschutz* **19,** 115–119.

Jellis G.J. & Taylor G.S. (1977) The development of silver scurf (*Helminthosporium solani*) disease of potato. *Annals of Applied Biology* **86,** 19–28.

Jenkins J.E.E. & Melville S.C. (1972) The effect of fungicides on leaf diseases and yield in spring barley in South-west England. *Plant Pathology* **21,** 49–58.

Jenkins J.E.E. & Morgan W. (1969) The effect of *Septoria* diseases on yield of winter wheat. *Plant Pathology* **18,** 152–156.

Jenkyn J.F. & Griffiths E. (1978) Relationships between the severity of leaf blotch (*Rhynchosporium secalis*) and the water-soluble carbohydrate and nitrogen contents of barley plants. *Annals of Applied Biology* **90,** 35–44.

Jimenez-Diaz R.M., Blanco-Lopez M.A. & Sackston W.E. (1983) Incidence and distribution of charcoal rot of sunflower caused by *Macrophomina phaseolina* in Spain. *Plant Disease* **67,** 1033–1036.

Jones A.L. (1978) Analysis of apple scab epidemics and attempts at improved disease prediction. In *Special Report New York State Agricultural Experiment Station Geneva* no. 28, pp. 19–22.

Jordan V.W.L. (1981) Aetiology of barley net blotch caused by *Pyrenophora teres* and some effects on yield. *Plant Pathology* **30,** 77–87.

Kanwar J.S., Kerr E.A. & Harney P.M. (1980). Linkage of Cf-1 to Cf-24 genes for resistance to tomato leaf mould, *Cladosporium fulvum. Report of the Tomato Genetics Cooperative* **30,** 20–23.

Kennel W. (1981) Zum Auftreten von Schorfkonidien auf äusserlich unversehrter Rinde von Apfelzweigen. *Mitteilungen aus der Biologischen Bundesanstalt für Land- und Fortstwirtschaft Berlin-Dahlem* no. 203, 117.

Kiesling R.L. & Grafius J.E. (1956) Several oat varieties resistant to natural infection by *Leptosphaeria avenaria. Phytopathology* **46,** 305–306.

King J.E. (1977) Surveys of diseases of winter wheat in England and Wales 1970–75. *Plant Pathology* **26,** 8–20.

Klewitz R. (1972) Schäden an Pastinak durch *Centrospora acerina. Nachrichtenblatt des Deutschen Pflanzenschutzdienstes* **24,** 166–168.

Kline D.M. (1972) *Helminthosporium* stripe resistance in spring barley cultivars. *Plant Disease Reporter* **56,** 891–893.

Knight D.E. & Keyworth W.G. (1960) Didymella stem-rot of outdoor tomatoes. I. Studies on sources of infection and their elimination. *Annals of Applied Biology* **48,** 245–258.

Knudsen J.C.N. (1980) Resistance to *Pyrenophora graminea* in 145 barley entries subjected to uniform natural inoculum. *Royal Veterinary and Agricultural University, Copenhagen, Yearbook 1980,* 81–95.

Kondal M.R. & Dwedi A.K. (1979) Comparative efficacy of fungicides for the control of sooty blotch and flyspeck diseases of apple. *Indian Journal of Mycology and Plant Pathology* **9,** 113–114.

Kranz J., Schmutterer H. & Koch W. (1977) *Diseases, Pests and Weeds in Tropical Crops.* Paul Parey, Berlin.

Kruger J. & Hoffmann G.M. (1978) Differentiation of *Septoria nodorum* and *Septoria avenae* f.sp. *triticea. Zeitschrift für Pflanzenkrankheiten und Pflanzenschutz* **85,** 413–418.

Labruyère R.E. (1977) Contamination of ryegrass seed with *Drechslera* spp. and its effect on disease incidence in the ensuing crop. *Netherlands Journal of Plant Pathology* **83,** 205–215.

Lam A. (1984) *Drechslera siccans* from ryegrass fields in England and Wales. *Transactions of the British Mycological Society* **83,** 305–311.

Lanier L., Joly P., Bondoux P. & Bellemère A. (1976) *Mycologie et Pathologie Forestières.* Masson, Paris.

Laviola C. (1966) Contributo alla conoscenza della biologia di *Spilocaea oleagina* in Puglia. In *Atti 1° Congresso Unione Fitopatologica Mediterranea,* pp. 327–339. Bari.

Lawson H.M. & Wiseman J.S. (1983) Techniques for the control of cane vigour in red raspberry in Scotland: effects of timing and frequency of cane removal treatments on growth and yield in cv. Glen Clova. *Journal of Horticultural Science* **58,** 247–260.

Leach C.M. (1960) Phytopathogenic and saprophytic fungi associated with forage legume seed. *Plant Disease Reporter* **44,** 364–369.

Lebrun A. & Viard G. (1979) Etude de l'efficacité de deux nouveaux fongicides utilisés en désinfection de semences. *Phytiatrie-Phytopharmacie* **28,** 29–40.

Levy D. & Gornik A. (1981) Tolerance of onions to the pink root disease caused by *Pyrenochaeta terrestris. Phytoparasitica* **9,** 51–57.

Logan C. & O'Neill R. (1981) A study of eggplant infection by the potato gangrene pathogen. *Record of Agricultural Research* **29,** 1–3.

Logan C., Copeland R.B. & Little G. (1976) The effects of various chemical and physical haulm treatments on the incidence of potato gangrene. *Annals of Applied Biology* **84,** 221–229.

Loprieno N. & Tenerini I. (1959) Metodo per la diagnosi precoce dell' occhio di pavone dell'olivo (*Cycloconium oleaginum*). *Phytopathologische Zeitschrift* **34,** 385–392.

Louvet J. & Billotte J.M. (1964) Influence des facteurs climatiques sur les infections du colza par *l'Alternaria brassicae* et consequences pour la lutte. *Annales des Epiphyties* **15,** 229–243.

MAFF (1983) *Use of Fungicides and Insecticides on Cereals 1983.* Booklet 2257, MAFF, London.

Malathrakis N.E. (1980) A disease of olive trees caused by the fungus *Phoma incompta. Proceedings of the 5th Congress of the Mediterranean Phytopathological Union,* pp. 48–50. Hellenic Phytopathological Society, Athens.

Marcinkowska J. (1977) Septoria leaf spot of tomato. I The development of the disease under glasshouse and field conditions; II Overwintering of *S. lycopersici* and viability of its pycnospores. *Acta Agrobotanica* **30,** 341–358; 384–393.

Marić A., Maširević S. & Fayzalla S. (1981) The occurrence of *Leptosphaeria lindquistii,* the perfect state of *Phoma macdonaldi,* pathogen of black spot of sunflower in Yugoslavia. *Zaštita Bilja* **32,** 329–334.

Marziano F., Aloj B. & Noviello C. (1978) Osservazioni su un violento attacco di *Pseudocercospora vitis. Annali della Facoltà di Scienze Agrarie della Università degli Studi di Napoli Portici* **12,** 281–291.

Mason D.T. (1981) Effects of benomyl on yield components of red raspberry (*Rubus idaeus*) in relation to the incidence of spur blight (*Didymella applanata*) and cane botrytis (*Botrytis cinerea*). *Journal of Horticultural Science* **56,** 193–198.

Massenot M. & Raynal G. (1973) Les maladies des légumineuses fourragères. I. Les anthracnoses provoquées par les Melanconiales. *Annales de Phytopathologie* **5,** 83–100.

Materu U., Strobel G. & Shepard J. (1978) Reaction to phytotoxins in a potato population derived from mesophyll protoplasts. *Proceedings of the National Academy of Sciences of the USA* **75,** 4935–4939.

Matta A. & Mancini G. (1976) Osservazioni sul cancro del pero da *Botryodiplodia malorum* in Piemonte. *Coltivatore e Giornale Vinicolo* **122,** 1–9.

Maude R.B. (1966a) Pea seed infection by *Mycosphaerella pinodes* and *Ascochyta pisi* and its control by seed soaks in thiram and captan suspensions. *Annals of Applied Biology* **57,** 193–200.

Maude R.B. (1966b) Studies on the etiology of black rot, *Stemphylium radicinum* and leaf blight, *Alternaria dauci* on carrot crops; and on fungicide control of their seed-borne infection phase. *Annals of Applied Biology* **57,** 83–93.

Maude R.B. (1970) The control of *Septoria* on celery seed. *Annals of Applied Biology* **65,** 249–254.

Maude R.B. (1983) Seed treatment control of *Alternaria* infections of brassica seed. *Proceedings of the 10th International Congress of Plant Protection, 1983,* p. 1202. BCPC, Croydon.

Maude R.B. & Humpherson-Jones F.M. (1980) Studies on the seed-borne phases of dark leaf spot (*Alternaria brassicicola*) and grey leaf spot (*Alternaria brassicae*) of brassicas. The effect of iprodione on the seed-borne phase of *A. brassicicola. Annals of Applied Biology* **95,** 311–319; 321–327.

Maude R.B. & Shuring C.G. (1970) The persistence of *Septoria apiicola* on diseased celery debris in soil. *Plant Pathology* **19,** 177–179.

McCarter S.M., Jaworski C.A. & Johnson A.W. (1976) Soil fumigation effects on early blight of tomato transplants. *Phytopathology* **66,** 1122–1124.

McGee D.C. & Petrie G.A. (1978) Variability of *Leptosphaeria maculans* in relation to blackleg of oilseed rape. *Phytopathology* **68,** 625–630.

McGee D.G. & Petrie G.A. (1979) Seasonal patterns of ascospore discharge by *Leptosphaeria maculans* in relation to blackleg of oilseed rape. *Phytopathology* **69,** 586–589.

McGlohon N.E. (1982) *Botryosphaeria dothidea*—where will it stop? *Plant Disease* **66,** 1202–1203.

Merkle R. (1951) Über die Douglasien-Vorkommen und die Ausbreitung der Adelopus-Nadelschutte in Wurttemberg-Hohenzollern. *Forst und Jagd* **122,** 161–191.

Messiaen C.M. & Lafon R. (1970) *Les Maladies des Plantes Maraîchères.* INRA, Paris.

Metz S.G. & Scharen A.L. (1979) Potential for the development of *Pyrenophora graminea* on barley in a semi-arid environment. *Plant Disease Reporter* **63,** 671–675.

Michaud R., Richard C. & Surprenant J. (1984) Selection for resistance to *Leptosphaerulina* leafspot in alfalfa. *Phytopathologische Zeitschrift* **110,** 69–77.

Miller D.R. (1974) *Sydowia polyspora* found on white fir twigs in California. *Plant Disease Reporter* **58,** 94–95.

Mirkova E. & Kachamarzov V. (1978) (Root rot of strawberry caused by *Phoma prunicola.*) *Gradinarska i Lozarska Nauka* **15,** 58–64.

Montes Agusti F., Calatrava J. & Romero Munoz F. (1975) Relacion entre dosis de riego y desarrollo de la podredumbre carbonosa en plantas de soja. *Anales del Instituto Nacional de Investigaciones Agrarias, Proteccion Vegetal* no. 5, 65–74.

Moore M.H. (1964) Glasshouse experiments in apple scab. I. Foliage infection in relation to wet and dry periods. *Annals of Applied Biology* **53,** 423–435.

Morelet M. (1985) Les *Venturia* des peupliers de la section *Leuce.* I. Taxinomie. *Cryptogamie-Mycologie* **6,** 101–117.

Morelet M. (1986) Les risques d'adaptation de *Venturia tremulae* aux trembles sélectionnés. *Bulletin OEPP/EPPO Bulletin* **16,** 589–592.

Naef-Roth S. (1949) Untersuchungen über den Erreger der Schrotschusskrankheit des Steinobstes, *Clasterosporium carpophilum. Phytopathologische Zeitschrift* **15,** 1–39.

Nanda H.P. & Gangopadhyay S. (1982) Tip burn by *Bipolaris leersiae* of rice and its sporulating ability in India. *Annals Phytopathological Society Japan* **48,** 261–266.

Neergaard P. (1977) *Seed Pathology.* Macmillan, London.

Nilsson B. (1975) Resistance to stripe (*Helminthosporium gramineum*) in barley. In *Barley Genetics III,* pp. 470–475. Garching (FRG).

Nirenberg H.I. (1981) Differenzierung der Erreger der Halmbruchkrankheit. I. Morphologie. *Zeitschrift für Pflanzenkrankheiten und Pflanzenschutz* **88,** 241–248.

Nüesch J. (1960) Beitrag zur Kenntnis der Weidenbewohnenden Venturiaceae. *Phytopathologische Zeitschrift* **39,** 329–360.

O'Brien M.J. & Thirumalachar M.J. (1978) Identity of fungi inciting charcoal rot disease. *Sydowia* **30,** 141–144.

O'Donnell J. & Dickinson C.H. (1980) Pathogenicity of *Alternaria* and *Cladosporium* isolates on *Phaseolus. Transactions of the British Mycological Society* **74,** 335–342.

Olofsson B. (1976) Undersökninger rörande *Dreschslera*-arter hos korn och havre. *Meddelanden Statens Växtskyddsanstalt* **16,** 323–425.

O'Ríordáin F. (1969) Control of cane spot and spur blight of raspberry with fungicides. *Proceedings 5th British Insecticide and Fungicide Conference, Brighton* pp. 155–158. BCPC, Croydon.

O'Rourke C.J. (1976) *Diseases of Grasses and Forage Legumes in Ireland.* An Foras Talúntais, Carlow.

Padmanabhan S.Y. (1973) The great Bengal famine. *Annual Review of Phytopathology* **11,** 11–26.

Parii I.F. (1978) The prevention of a bark disease of apple. *Zashchita Rastenii* no. 2, 56–57.

Paus F. & Raa J. (1973) An electron microscope study of infection and disease development in cucumber hypocotyls inoculated with *Cladosporium cucumerinum. Physiological Plant Pathology* **3,** 461–464.

Porta-Puglia A. (1981) *Stemphylium vesicarium* su pomodoro nelle Marche. *Annali dell'Istituto Sperimentale per la Patologia Vegetale Roma* **7,** 39–46.

Punithalingam E. (1979) Graminicolous *Ascochyta* species. *Mycological Papers* no. 142, CMI, Kew.

Pustovoit G.V. & Borodin S.G. (1983) (Harmfulness of grey rot of sunflower.) *Zashchita Rastenii* no. 9, 41.

Rabinowitch H.D., Katan J. & Rotem I. (1981) The response of onions to solar heating, agricultural practices and pink root disease. *Scientia Horticulturae* **15**, 331–348.

Raymondaud H., Pineau R. & Martin M. (1985) Contribution à la connaissance de la biologie et de l'épidémiologie de *Cladosporium carpophilum* agent de la tavelure du mirabellier en Lorraine. *Agronomie* **5**, 563.

Raynal G. (1978) Contamination artificielle de la luzerne avec l'agent du pepper-spot, *Leptosphaerulina briosiana.* Mise au point, sensibilité de 30 cultivars et variation du pouvoir pathogène. *Revue de Zoologie Agricole et de Pathologie Végétale* **77**, 1–13.

Redfern D.B. & Sutton B.C. (1981) Canker and dieback of *Ulmus glabra* caused by *Plectophomella concentrica* and its relationship to *P. ulmi. Transactions of the British Mycological Society* **77**, 381–390.

Reis E.M. & Zanetti R. (1977) O cancro da macieira. 2. Controle de doença. *Summa Phytopathologica* **3**, 221–225.

Rekola O., Ruokola A-L. & Kurtto J. (1970) Damage caused by *Helminthosporium avenae* Eidam on the crop yield of oats in Finland. *Acta Agriculturae Scandinavica* **20**, 225–229.

Richter J. & Häussermann R. (1975) Ein elektronisches Schorfwarngerät. *Anzeiger für Schädlingskunde, Pflanzenschutz, Umweltschutz* **48**, 107–109.

Rotem J. (1966) Variability in *Alternaria porri* f.sp. *solani. Israel Journal of Botany* **15**, 48–57.

Roussel C. (1970) Importance de la connaissance de la biologie dans les avertissements agricoles. Etude comparative de l'évolution du mildiou et du black-rot. *Phytoma-Défense des Végétaux* **228**, 19–24.

Roussel C. (1984) L'anthracnose de la vigne en Uruguay. *Progrès Agricole et Viticole* **9**, 251–253.

Roussel C., Sommier J.Y. & Mansencal C. (1981) Nouvelles possibilités dans la lutte contre le black-rot de la vigne. *Phytiatrie-Phytopharmacie* **30**, 165–176.

Rufty R.C., Herbert T.T. & Murphy C.F. (1981) Variation in virulence in isolates of *Septoria nodorum. Phytopathology* **71**, 593–596.

Ryan E.W. (1978) Leaf spot of onions caused by *Cladosporium allii-cepae. Plant Pathology* **27**, 200.

Saad A.T. & Marsi S. (1978) Epidemiological studies on olive leaf spot incited by *Spilocaea oleaginea. Phytopathologia Mediterranea* **17**, 170–173.

Salerno M. & Cutuli C. (1982) The management of fungal and bacterial diseases of Citrus in Italy. *Proceedings of the International Society of Citriculture 1981*, vol. 1, 360–362. Fruit Tree Research Station, Shimizu (Japan).

Sanderson F.R. (1976) *Mycosphaerella graminicola* (Fuckel) Sanderson comb. nov., the ascogenous state of *Septoria tritici* Rob. apud Desm. *New Zealand Journal of Botany* **14**, 359–360.

Sanderson F.R. & Hampton J.G. (1978) Role of the perfect states in the epidemiology of the common septoria diseases of wheat. *New Zealand Journal of Agricultural Research* **21**, 277–281.

Saumon E., Herbach M., Goore B.K. & Davet P. (1984) Le desséchement précoce des tournesols. Dynamique de la colonisation des plantes par les champignons du sol et envahissement tardif par *Macrophoma phaseolina. Agronomie* **4**, 805–812.

Schiffman-Nadel M., Waks J., Gutter Y. & Chalutz E. (1983) Alternaria rot of citrus fruit. In *Proceedings of the International Society of Citriculture 1981 Volume 2,* pp. 791–793. Fruit Tree Research Station, Shimizu (Japan).

Schroedter H. & Fehrmann H. (1971) Ökologische Untersuchungen zur Epidemiologie von *Cercosporella herpotrichoides.* III. Die relative Bedeutung der meteorologischen Parameter und die komplexe Wirkung ihrer Konstellationen auf den Infektionserfolg. *Phytopathologische Zeitschrift* **71**, 203–222.

Scott P.R. & Hollins T.W. (1977) Interactions between cultivars of wheat and isolates of *Cercosporella herpotrichoides. Transactions of the British Mycological Society* **69**, 397–403.

Scott P.R. & Hollins T.W. (1980) Pathogenic variation in *Pseudocercosporella herpotrichoides. Annals of Applied Biology* **94**, 297–300.

Shabi E., Rotem J. & Loebenstein G. (1973) Physiological races of *Venturia pirina* on pear. *Phytopathology* **63**, 41–43.

Shepard J., Bidney D. & Shahin S. (1980) Potato protoplasts in crop improvement. *Science* **208**, 17–24.

Sheridan J.E. (1966) Celery leaf spot: sources of inoculum. *Annals of Applied Biology* **57**, 75–81.

Shipton W.A., Boyd W.J.R. & Ali S.M. (1974) Scald of barley. *Review of Plant Pathology* **53**, 839–861.

Shipton W.A., Khan T.N. & Boyd W.J.R. (1973) Net blotch of barley. *Review of Plant Pathology* **52**, 269–290.

Short G.E., Wyllie T.D. & Brislow P.R. (1980) Survival of *Macrophomina phaseolina* in soil and in residue of soybean. *Phytopathology* **70**, 13–17.

Simmons E.G. (1981) *Alternaria* themes and variations. *Mycotaxon* **13**, 16–34.

Singh R.A. & Sharma V.V. (1975) Evaluation of rice germplasm and varieties for resistance to 'brown spot'. *Il Riso* **24**, 383–386.

Sisto D. (1983) *Cercosporidium punctum* su finocchio in Italia meridionale. *Informatore Fitopatologico* **33** (7/8), 55–58.

Skajennikoff M. & Rapilly F. (1981) Variations de la phylloflore fongique du blé liées à des traitements fongicides. *Bulletin OEPP/EPPO Bulletin* **11**, 285–302.

Skidmore A.M. & Dickinson C.H. (1973) Effect of phylloplane fungi on the senescence of excised barley leaves. *Transactions of the British Mycological Society* **60**, 107–116.

Smedegaard-Petersen V. (1971) *Pyrenophora teres* f. *maculata* f. nov. and *Pyrenophora teres* f. *teres* on barley in Denmark. *Royal Veterinary and Agricultural University, Copenhagen, Yearbook 1971,* 124–144.

Smedegaard-Petersen V. (1974) *Leptosphaeria nodorum (Septoria nodorum),* a new pathogen in barley in Denmark and its physiologic specialization on barley and wheat. *Friesia* **10,** 251–264.

Smedegaard-Petersen V. (1977a) Inheritance of genetic factors for symptoms and pathogenicity in hybrids of *Pyrenophora teres* and *Pyrenophora graminea. Phytopathologische Zeitschrift* **89,** 193–202.

Smedegaard-Petersen V. (1977b) Isolation of two toxins produced by *Pyrenophora teres* and their significance in disease development of net-spot blotch of barley. *Physiological Plant Pathology* **10,** 203–211.

Smedegaard-Petersen V. & Jørgensen J. (1982) Resistance to barley leaf stripe caused by *Pyrenophora graminea. Phytopathologische Zeitschrift* **105,** 183–191.

Smith R.R. & Maxwell D.P. (1973) Northern anthracnose resistance in red clover. *Crop Science* **13,** 271–273.

Sobers E.K. (1968) Morphology and pathogenicity of *Cercospora apii* f.sp. *nicotianae. Phytopathology* **58,** 1713–1714.

Solel Z. (1970) Systemic effect of benzimidazole against *Cercospora beticola. Phytopathology* **60,** 1196–1198.

Sprague R. (1950) *Diseases of Cereals and Grasses in North America.* Ronald Press, New York.

Steekelenburg N.A.M. van (1983) Epidemiological aspects of *Didymella bryoniae,* the cause of stem and fruit rot of cucumber. *Netherlands Journal of Plant Pathology* **89,** 75–86.

Strandberg J.O. (1977) Spore production and dispersal of *A. dauci. Phytopathology* **67,** 1262–1266.

Strider D.L. (1978) Alternaria blight of carnation in the greenhouse and its control. *Plant Disease Reporter* **62,** 24–28.

Sutton T.B. (1981) Production and dispersal of ascospores and conidia by *Physalospora obtusa* and *Botryosphaeria dothidea* in apple orchards. *Phytopathology* **71,** 584–589.

Svedelius G. & Unestam T. (1978) Experimental factors favouring infection of attached cucumber leaves by *Didymella bryoniae. Transactions of the British Mycological Society* **71,** 89–97.

Szepieniec-Gajos A. (1983) Investigations into the resistance of lucerne to leaf-spot diseases. *Biuletyn Instytutu Hodowli i Aklimatyzacji Roslin* **150,** 117–121.

Teviotdale B.L. & Hall D.H. (1976) Factors affecting inoculum development and seed transmission of *Helminthosporium gramineum. Phytopathology* **66,** 295–301.

Tewari J.P. & Skoropad W.P. (1976) Relationship between epicuticular wax and blackspot caused by *Alternaria brassicae* in three lines of rapeseed. *Canadian Journal of Plant Science* **56,** 781–785.

Thurling N. & Venn L.A. (1977) Variation in the responses of rapeseed (*Brassica napus* and *B. campestris*) cultivars to blackleg (*Leptosphaeria maculans*) infection. *Australian Journal of Experimental Agriculture and Animal Husbandry* **17,** 445–451.

Tjamos E.C. (1984) Control of *Pyrenochaeta lycopersici* by combined soil solarization and low dose of methyl bromide in Greece. *Acta Horticulturae* **152,** 253–258.

Troxler J., Lehman J. & Briner H.U. (1980) Résultats d'essais de variétés de trèfle d'Alexandrie (*Trifolium alexandrinum*) et de trèfle de Perse (*T. resupinatum*). *Revue Suisse d'Agriculture* **12,** 235–239.

Tsunedo A. & Skoropad W.P. (1977) Formation of microsclerotia and chlamydospores from conidia of *Alternaria brassicae. Canadian Journal of Botany* **55,** 1276–1281.

Végh I., Champion R., Bourgeois M. & Brunet D. (1978) Présence en France de *Phoma valerianellae* sur mâche. *Revue Horticole* **184,** 39–43.

Venn K. (1983) Winter vigour in *Picea abies* (L.) Karst. IX. Fungi isolated from mouldy nursery stock held in overwinter cold storage. *Meddelelser fra Norsk Institutt for Skogforskning* **38**(7), 1–32

Vercesi A., Minervi G. & Bisiach M. (1982) *Aureobasidum pullulans* su foglie di *Vitis vinifera. Rivista di Patologia Vegetale* **18,** 77–81.

Verhoeff K. (1963) Voetrot en 'kanker' bij tomat, veroorzaakt door *Didymella lycopersici. Netherlands Journal of Plant Pathology* **69,** 298–313.

Viennot-Bourgin G. (1949) *Les Champignons Parasites des Plantes Cultivées.* Masson, Paris.

Viennot-Bourgin G. (1981) Observation simultanée en France du bud blast du rhododendron et d'une cicadelle jouant le rôle de vecteur. *Agronomie* **1,** 87–92.

Walker J. & Baker K.F. (1983) The correct name for the chrysanthemum ray blight pathogen in relation to its geographical distribution. *Transaction of the British Mycological Society* **80,** 31–38.

Wallen V.R. (1957) The identification and distribution of physiologic races of *Ascochyta pisi* in Canada. *Canadian Journal of Plant Science* **37,** 337–341.

Waterman A.M. (1947) *Rhizosphaera kalkhoffii* associated with a needle cast of *Picea pungens. Phytopathology* **37,** 507–511.

Watson A.R. & Millar C.A. (1971) Hypodermataceous needle-inhabiting fungi in pines in Scotland. *Transactions of the Botanical Society of Edinburgh* **41,** 250.

Wellings C.R. (1981) Pathogenicity of fungi associated with citrus greasy spot in New South Wales. *Transactions of the British Mycological Society* **76,** 495–499.

White J.G. & Scott A.C. (1973) Formation and ultrastructure of microsclerotia of *Pyrenochaeta lycopersici. Annals of Applied Biology* **73,** 163–166.

Whittle A.M. & Richardson M.J. (1978) Yield loss caused by *Cochliobolus sativus* on Clermont barley. *Phytopathologische Zeitschrift* **91,** 238–256.

Wilcox M.D. (1982) Genetic variation and inheritance of resistance to *Dothistroma* needle blight in *Pinus radiata. New Zealand Journal of Forestry Science* **12,** 14–35.

Williamson B. & Hargreaves A.J. (1978) Cane blight (*Leptosphaeria coniothyrium*) in mechanically harvested red raspberry (*Rubus idaeus*). *Annals of Applied Biology* **88,** 37–43.

Williamson B. & Hargreaves A.J. (1979) Fungi on red raspberry from lesions associated with feeding wounds of cane midge (*Resseliella theobaldi*). *Annals of Applied Biology* **91,** 303–307.

Williamson B. & Hargreaves A.J. (1981) The effect of sprays of thiophanate-methyl on cane diseases and yield in red raspberry, with particular reference to cane blight (*Leptosphaeria coniothyrium*). *Annals of Applied Biology* **97,** 165–174.

Williamson B., Lawson H.M., Woodford J.A.T., Hargreaves A.J., Wiseman J.S. & Gordon S.C. (1979) Vigour control, an integrated approach to cane, pest and disease management in red raspberry (*Rubus idaeus*). *Annals of Applied Biology* **92**, 359–368.

Willis W.G., Stuteville D.L. & Sorensen E.L. (1969) Effects of leaf and stem diseases on yield and quality of alfalfa forage. *Crop Science* **9**, 637–640.

Wilson G.J. (1971) New system of ringspot control looks hopeful. *New Zealand Commercial Grower* **26**, 11.

Wit P.J.G.M. De & Kodde E. (1981) Further characterization and cultivar-specificity of glycoprotein elicitors from culture filtrates and cell walls of *Cladosporium fulvum*. *Physiological Plant Pathology* **18**, 297–314.

Wit P.J.G.M. De & Spikman G. (1982) Evidence for the occurrence of race and cultivar-specific elicitors of necrosis in intercellular fluids of compatible interactions of *Cladosporium fulvum* and tomato. *Physiological Plant Pathology* **21**, 1–11.

Wittman W. & Bedlan G. (1980) Die Nelkenschwärze. *Pflanzenarzt* **33**, 17.

Woodford J.A.T. & Gordon S.C. (1978) The history and distribution of raspberry cane midge (*Resseliella theobaldi*), a new pest in Scotland. *Horticultural Research* **17**, 87–97.

Yoder K.S. (1982a) Fungicide control of Brooks fruit spot of apple. *Plant Disease* **66**, 564–566.

Yoder K.S. (1982b) Broad spectrum apple disease control with bitertanol. *Plant Disease* **66** (7), 580–583.

Zachos D.G. & Roubos I. (1977) Recherches sur la biologie du champignon *Camarosporium pistaciae*. *Annales de l'Institut Phytopathologique Benaki* **11**, 346–353.

Zachos D.G. & Tzavella-Klonari K. (1979) Recherches sur l'identité et la position systématique du champignon qui provoque la maladie des olives attribuée au champignon *Macrophoma* on *Sphaeropsis dalmatica*. *Annales de l'Institut Phytopathologique Benaki* **12**, 59–71.

Zachos D.G. & Tzavella-Klonari K. (1980) Recherches sur les caractères et la position systématique du champignon qui cause la maladie de la vigne connus sous les noms 'faux black-rot' et 'excoriose'. *Annales de l'Institut Phytopathologique Benaki* **12**, 208–215.

14: Ascomycetes V
Rhytismatales, Pezizales, Caliciales, Helotiales

The apothecial ascomycetes (discomycetes) include a number of important plant pathogens. With the recent transfer of *Phacidium* to Helotiales, the name Rhytismatales now replaces the more familiar 'Phacidiales'. In the Rhytismatales the hymenium of the apothecium is embedded in the host tissues and is covered by a membrane which splits at maturity. The Pezizales differ from the Helotiales in having operculate asci and most often larger more complex apothecia. In the Helotiales the asci are inoperculate and are present as a hymenium in a generally rather insignificant apothecium that is frequently associated with, but not embedded in plant tissue. The Caliciales are principally lichenized forms but include some soil saprophytes or weak root parasites. Many Rhytismatales characteristically infect the needles of conifers, causing needle cast (*Lophodermium, Lophodermella, Didymascella, Rhabdocline*). *Rhytisma* causes very distinctive tar spots on leaves of deciduous trees. In the Helotiales, the family Sclerotiniaceae contains the important genera *Sclerotinia, Monilinia,* and *Botryotinia* (the last with the common anamorphic form *Botrytis*). All are aggressive necrotrophic parasites causing extensive damage to parenchymatous tissues of a wide range of crops. Many perennate as sclerotia, or stromata, from which the apothecia are formed. In other Helotiales, leaf spots (*Drepanopeziza, Diplocarpon, Pseudopeziza, Pyrenopeziza* and cankers (*Crumenulopsis, Lachnellula, Pezicula*) are the most frequent symptom types.

RHYTISMATALES

Lophodermium seditiosum Minter, Staley & Millar

Basic description

In *Lophodermium seditiosum* the ascocarps are totally subepidermal. Occasionally brown, but never black stromatic lines are formed in the pine needle. It is the only European *Lophodermium* with these features. The *Leptostroma* state is subepidermal on secondary needles (*Leptostroma rostrupii*) but subcuticular on primary needles (*Leptostroma austriacum*). The conidia have no known function. In the past the fungus has frequently been confused with *Lophodermium pinastri* (q.v.) (CMI 568).

Host plants

L. seditiosum has been recorded on secondary needles of a range of 2-, 3- and 5-needled pines, including *Pinus cembra, P. contorta* var. *contorta* and var. *latifolia, P. densiflora, P. durangensis, P. halepensis, P. montezumae, P. mugo, P. nigra* var. *maritima* and var. *nigra, P. palustris, P. pinaster, P. pinea, P. radiata, P. resinosa, P. sylvestris* and *P. virginiana,* and on primary needles of *P. sylvestris.* It is probably present on other pines.

Geographical distribution

Although reported in Europe only from the Netherlands, Poland, Scandinavia, UK and Yugoslavia, it is undoubtedly widespread in N. and C. Europe (including the USSR) and possibly spread from there to N. America.

Disease

L. seditiosum causes severe browning and needle cast of seedlings, transplants and young plantation trees, and premature cast of green needles of older trees. In the nursery the main symptoms are brown, often drooping, secondary needles in spring, and fallen needles with ascocarps in summer and autumn. In severe infections, attached primary needles may also have ascocarps of *L. seditiosum.* In young trees, infection levels vary from occasional brown attached needles to total browning. In older plantations, the disease shows up on stressed trees or branches affected by, for example, cold winds. Sometimes only the needles on one side of a shoot are browned. Where disease is chronic, a spring fall of slightly yellowed diseased needles occurs. As these turn brown through the summer they develop a few brown lines, brown conidiomata and ascocarps which mature in autumn and are a major source of inoculum for future infections. Ascocarps also form on attached needles on living branches, on brashing and thinning debris and on cone scales.

Epidemiology

Ascospores of *L. seditiosum* are ejected sequentially from the asci in high humidity in autumn. They adhere strongly to the needle surface where they normally germinate by producing an appressorium directly from the spore. Germination can occur within 12 h at 20°C and 100% RH. Establishment of infection in a needle takes several weeks. Penetration of the needle is through the cuticle by means of an infection peg from the underside of the melanized appressorium in the fourth week after spore deposition. Penetration through stomata is uncommon. Infection hyphae colonize epidermal cells causing minute yellow spots which soon turn brown. These have been used to forecast disease severity (Rack 1959). A hyphal complex of intertwining hyphae spreads into the nearby hypodermal and mesophyll cells, and the fungus can then rest as a latent infection or, if the infection is heavy, may colonize the stelar tissue progressively and kill the needle. Heavy infections seem to be associated with warm wet conditions at the time of sporulation of the fungus. When needles with latent infections are weakened by other causes such as wind or cold, with or without water stress, the uninfected parts of the needles may turn brown and be colonized rapidly by fast-growing hyphae of *L. seditiosum* from the existing hyphal complexes. There are some indications that infected needles are more susceptible to browning by climatic factors but this is not yet proven.

Economic importance

In nurseries in bad years, whole beds can be killed, often in a few weeks, with no obvious warning symptoms. Infected plants with no obvious symptoms have sometimes been sold, only to develop symptoms later, resulting in claims for compensation. In some years, young plantations are defoliated heavily by *L. seditiosum*. The American Christmas tree industry has suffered particularly from this disease. In older stands the disease seems to be of little consequence in terms of affecting growth. The severity of the disease varies from year to year and from country to country.

Control

Bordeaux mixture and lime sulphur were traditionally used for chemical control in nurseries. At present, in the USA, maneb and chlorothalonil are most used (4 applications in August–September), and benomyl and chlorothalonil+cycloheximide have been found to be effective (Nicholls & Skilling 1974). In Sweden, one cycloheximide spray at 10 ppm in late autumn, or one or 2 sprays with chlorothalonil+cycloheximide, give acceptable control without causing chlorosis (Klingström & Lundeberg 1978). In Yugoslavia, 5 sprays with zineb or captan are effective in most seasons (Uscuplić 1981). The trend has therefore been towards fewer but better timed sprays. In addition to normal hygiene, nurseries should be sited away from older pines (plantations or wind breaks). A two year break between pine crops is recommended. Because the fungus may be latent in green needles and disease levels vary from year to year, it is prudent to spray all nursery stock if there is any history of the disease in the area. Otherwise, the disease may be carried by apparently healthy plants. Any cultural malpractice, such as shallow planting, is liable to 'bring out' disease.

In general, northern provenances of *Pinus sylvestris* show greater resistance than southern provenances, but the relation is not always consistent. *P. contorta* appears to be relatively resistant, and in the USA *P. resinosa,* although very susceptible in nurseries, is resistant in plantations.

Special research interest

More work is needed to clarify the relationship between the strains and species of *Lophodermium* and their distribution. The conidia, but not ascocarps, form readily in culture. This needs solving, as does the role of conidia. Most required are studies of the factors which allow *L. seditiosum* to colonize needles from latent infections and thus cause severe disease.
MILLAR

Lophodermium pinastri **(Schrader) Chev.**

L. pinastri differs from *L. seditiosum,* with which it has generally been confused, in producing black stromatic lines in infected needles and having ascocarps which are only partly subepidermal (Minter 1981). Although it infects green needles, it spreads extensively in them only when they senesce or if they are damaged by other fungi or insects. It is thus a weak pathogen. It is very common and found fruiting in the litter of most temperate 2- and 3- and some 5-needled pines both on needles and cones. It is widespread in Europe, Asia, W. USA, Japan, Australia and New Zealand.

In the litter, black stromatic lines first delimit separate strains of the fungus in the needle, then the conidial state ***Leptostroma pinastri*** Desm. develops over winter, and finally the black ascocarps, measuring 700–1200 μm long, often with red labia, develop to maturity in spring (CMI 567; Minter & Millar 1980). The ascospores are ejected in humid conditions. They germinate on needles by a germ tube with a terminal appressorium, and infection is normally by cuticle penetration. As needles age the number of infections increases, so that when the needles senesce, their tissues are normally dominated by *L. pinastri*. It is not proven that *L. pinastri* causes premature needle cast, but it seems likely that heavily infected trees may cast their needles early, especially if growing in stress conditions.

MILLAR

Lophodermella sulcigena (Rostrup) Höhnel

(Syn. *Hypodermella sulcigena* (Rostrup) Tubeuf)

L. sulcigena causes browning and premature needle cast of *Pinus mugo, P. nigra* var. *nigra* and var. *maritima, P. sylvestris* and *P. contorta* (CMI 562) (Jalkanen 1981). It appears to be confined to N. and C. Europe, but to occur there wherever its hosts are found. The main symptom of this disease, known as pine needle blight, lophodermella needle cast or Swedish pine cast, is discoloration of the current year's needles, which turn a light pinkish-brown and develop violet striations which are the elongated ascocarps (Plate 26). All ages of trees can be infected, from ground level to the top of the crown, but young trees appear to be more susceptible. Susceptibility varies considerably between trees in any one stand, and is influenced by soil and needle nutrient status. Heavy infections reduce growth. Nursery infections are rare, perhaps because of the lack of inoculum.

The disease starts in early summer, when, in rain or humid conditions, ascospores, released from the previous year's needles, infect the base of a new growing needle just above the sheath. Because the needle grows from a basal meristem, this infection spot is carried away from the sheath, and visible symptoms appear about one month after infection as yellow or pink spots at various points on the needles. Browning extends from these spots distally and proximally but the base of the needle usually remains green. One or both needles of a pair may be infected. Elongated ascocarps up to 20 mm long develop over winter to mature the following spring and thus complete the life cycle (Terrier 1944). Needle cast occurs in autumn about 15 months after the initial infection.

Natural control is affected locally by other fungi and by dry summers, so that considerable annual variation in disease incidence occurs. Many needles infected by *L. sulcigena* are infected and colonized by *Hendersonia acicola* (q.v.), especially in the lower part of the crown (Mitchell *et al.* 1976) in the autumn of the first year, and *L. sulcigena* fails to fruit. The needles become brittle and are often broken, leaving a green needle stub on the tree. In some years, needles infected by *L. sulcigena* are colonized secondarily by *Lophodermium seditiosum,* in which case the primary pathogen fails to fruit. The needles are retained, and *L. seditiosum* fruits in the following summer and autumn. Chemical control using zinc or copper-based fungicides is effective but, because the needles are growing during the infection period, the frequency of application required is likely to make chemical control uneconomic.

MILLAR

Lophodermella conjuncta (Darker) Darker

(Syn. *Hypodermella conjuncta* Darker)

L. conjuncta infects two-needled pines, causing premature cast of 2-year and older needles. Infection of first-year and older needles shows first as bright yellow bands in which small brown ascocarps are embedded. Later, the whole infected area of the needle becomes brown, and when dry and closed the ascocarps are difficult to see. The period from infection to maturation of ascocarps varies from 12 months in Scotland, to 19 months in Switzerland, to 24 months in Finland. Needle cast is not as pronounced as in *L. sulcigena* and, because first-year needles remain green, it is unlikely that significant tree growth reduction occurs. Zineb application to new foliage reduces the incidence of *L. conjuncta* (CMI 658).

MILLAR

Lophodermium conigenum Hilitzer is common on two-needled pines throughout Europe, infecting green needles without causing any significant damage (CMI 565; Minter & Millar 1980). ***L. pini-excelsae***

S. Ahmad is similarly a weak pathogen on five-needled pines (CMI 785) and ***Meloderma desmazierii*** (Duby) Darker on *P. strobus* (CMI 569). ***L. juniperinum*** (Fr.) de Not. is common on juniper, but doubtfully parasitic (CMI 797). ***Lirula macrospora*** (R. Hartig) Darker (syn. *Lophodermium macrosporum* (R. Hartig) Rehm) and ***Lophodermium piceae*** (Fuckel) Höhnel infect needles of spruce but are only weakly pathogenic (CMI 794; Butin 1983). ***Lirula nervisequia*** (Fr.) Darker (syn. *Hypodermella nervisequia* (Fr.) Lagerb.) causes a locally important needle cast of *Abies alba* in Europe (CMI 783).

Didymascella thujina (E. Durand) Maire

(Syn. *Keithia thujina* E. Durand)

D. thujina forms red-brown apothecia which become dull brown when dry and contain sticky, colourless, but later brown, unequally two-celled ascospores. *D. thujina* causes serious damage only to *Thuja plicata,* though it has also been recorded on *Chamaecyparis lawsoniana. Thuja standishii* and its Fl hybrids with *T. plicata* are not affected. ***D. tsugae*** (Farlow) Maire causes a similar but unimportant disease on *Tsuga canadensis* in Scotland.

The fungus occurs in much of Europe, and in N. America (CMI map 149). It causes a needle blight which is very damaging to nursery plants. On older trees it is important only as a source of nursery infection. From April onwards, affected leaflets become brown and die, and on each scale leaf one to three apothecia form below the epidermis, which splits open to form a scale-like flap. The old apothecia dry, to leave small round depressions. Stems are sometimes girdled. Young seedlings are usually little affected, but large parts of the foliage of second-year seedlings and transplants may be destroyed. Ascospores are shed from May to October at RH above 90% and temperatures of 12°C or more. Overwintering is by ascospores shed in the autumn, and by apothecia which form in autumn but ripen and shed their spores in the following spring. These spring-shed and overwintered ascospores infect the new foliage. Plants affected by *D. thujina* needle cast become small and stunted, and many are killed. In some areas the disease makes it difficult to raise plants of *Thuja plicata.* Nursery infection may be hindered by growing *Thuja* stock on a rotation system in nurseries sited 2–3 km from other *Thuja* plants (Pawsey 1963). If infection still occurs, the crop should be sprayed up to three times with cycloheximide (Burdekin & Phillips 1970) or up to twelve times with triadimefon, benomyl or mancozeb (Mathon 1982).

PHILLIPS

Potebniamyces pyri (Berk. & Broome) Dennis

(Syn. *Potebniamyces discolor* (Mouton & Sacc.) Smerlis, *Phacidiella discolor* (Mouton & Sacc.) Potebnja, *P. pyri* (Fuckel) Weindlmayr)
(Anamorph ***Phacidiopycnis malorum*** Potebnja)

P. pyri forms immersed, fleshy-gelatinous, pseudoparenchymatous, black ascocarps with a violet covering layer. The hymenium is exposed by rupture of the upper stromatal layer. The ascospores are ovoid, hyaline or rarely brown, non-septate or with one or two transverse septa. The eustromatic pycnidia are immersed, uni- or multilocular, with ellipsoid conidia formed on enteroblastic phialides. *P. pyri* attacks pear, apple, quince and *Crataegus* species. It is a very common pathogen on quince and pear. First described from Belgium, it has been reported from the USSR (Ukraine, Moldavia), Germany, UK, Hungary, Norway and probably occurs in other N. and C. European countries.

Fig. 60. Canker of pear due to *Potebniamyces pyri.*

P. pyri is a facultative parasite infecting twigs, branches, limbs and trunks of fruit trees (Fig. 60). Penetration can occur through wounds or small dead twigs. The dead area of host tissue is irregular in

form, collapsed and reddish-brown. If a twig or branch is girdled, the distal part dies back, with the dried leaves persisting. Cankers forming on limbs and trunks are perennial. The fungus also causes brown rot of fruits. Eustromatic pycnidia develop immersed in the bark throughout the year and produce numerous pycnospores. Apothecia develop immersed in the dead bark tissue from September onwards. Infection begins in spring and continues through the summer months. The fungus overwinters in the form of mycelia in the host tissue, as well as in the form of pycnospores and ascospores. The incubation period is about 4–5 weeks. Losses (dieback, fruit rot) are mostly moderate, but in the 1960–70s in some countries (USSR, Moldavia, Hungary) intensive spread of the pathogen and considerable losses have been recorded (Popushoi 1971; Vajna 1971). In Hungary, growing conditions unfavourable for the host plants are major factors stimulating the spread of the pathogen and severity of the disease. Major preventive measures are the selection of an adequate planting and proper management (irrigation, fertilization, and so on). Dead branches should be removed and burned. Perennial cankers can be cured with fungicide preparations for wound treatment containing captafol or benomyl. More data are needed on the influence of ecological factors on disease development.
VAJNA

Rhabdocline pseudotsugae Sydow

R. pseudotsugae forms ascocarps measuring up to 3×0.3 mm. They open by a split to show the orange-brown (finally brown) hymenium. The long, sticky ascospores with slightly swollen ends become two-celled after discharge, one cell being dark brown with a thick wall (CMI 651). The fungus is restricted to species of *Pseudotsuga*. Of the varieties of Douglas fir (*P. menziesii*) (the only important host), the varieties *caesia* and *glauca* (the grey and blue Douglas firs) are much more susceptible than the green form, var. *viridis*. Resistant individuals occur even in populations of vars *caesia* and *glauca*. *R. pseudotsugae* is widespread throughout Europe and in N. America (CMI map 52), and has recently been reported in Australia. The fungus causes a needle cast. Infection first becomes evident in the autumn, when pale yellow blotches about 1–2 mm across appear mainly on the undersides of the needles. The blotches later become red- or purple-brown, and bear the elongated ripe ascocarps with their slit-like openings in May and June or early July. The disease is usually troublesome only on trees up to about 30 years of age, and is rare in the nursery. The ascospores are shed from May onwards, and the sticky spores adhere to the young needles. Infection occurs only when RH is near 100% and the temperature is between 1 and 15°C. Growth within the needles takes place throughout the summer, the resulting symptoms showing themselves in autumn. On badly affected trees of susceptible varieties, the needles are quickly shed, but on others they fall off more gradually throughout the year. Successive heavy annual needle losses may stop growth and kill trees. Hence in many areas it is possible to grow only the more resistant green Douglas fir on any scale. In N. America, Douglas fir grown for Christmas trees may be rendered unsaleable by this disease. In the forest this disease may be controlled only by growing the resistant green Douglas fir. On Christmas trees, benomyl gives good control (Morton & Miller 1982).
PHILLIPS

Rhytisma acerinum (Pers.) Fr.

(Anamorph ***Melasmia acerina*** Lév.)

R. acerinum attacks *Acer* species (CMI 791). In Europe the sycamore is most often affected, but the fungus also occurs on *Acer campestre, A. platanoides* and *A. negundo*. Many other *Acer* species are attacked in N. America. Several strains of *R. acerinum* are known, and one, sometimes separated as *R. pseudoplatani* Müller, occurs only on the sycamore.

The fungus causes the disease tar spot. Infection first becomes apparent in June or July as pale, water-soaked spots on the upper sides of the leaves. These spots become larger, and yellow in colour, and a large black, round stroma with a shiny outer surface quickly forms in and below the epidermal cells. Each stroma grows until it is about 15 mm across, and is slightly raised above the leaf surface. Small rounded spermogonia develop from June to August in the centre of the stroma, and shed their spermatia through round pores. Elongated, sometimes branched ascocarps then form, radiating around the old spermogonial area. Ascocarp development is incomplete at leaf-fall, but continues over the winter. Asci are formed in January or February, and the spores are shed in April or May through long slits in the ascocarp. Infection of the young leaves then takes place.

Successive defoliation by *R. acerinum* may weaken

young trees, especially in the nursery, but on larger trees the fungus causes negligible damage. On a small scale, infection can be prevented by raking up and burning infected leaves in autumn. If necessary, nursery stock may be sprayed with a copper fungicide or with one of the dithiocarbamates, starting at bud burst, when infection begins (Wittmann 1980).
PHILLIPS

Rhytisma punctatum (Pers.) Fr. is a rather uncommon, but similar species found occasionally on sycamore, on which it forms large black leaf spots, each made up of many smaller ones. ***R. salicinum*** (Pers.) Fr. causes tar spot on willows. It is common, for example, in Scandinavia. ***Ascodichaena rugosa*** Butin (anamorph ***Polymorphum quercinum*** (Pers.) Chev., syn. *P. rugosum* Hawksw. & Punith.) causes black bark scurf on beech. Though often considered as a bark saprophyte, the fungus has been shown to attack the phellogen layer (Butin 1983). A black crust of accumulated pycnidia builds up and the conidia are dispersed by browsing slugs. The disease is of no economic importance. ***Colpoma quercinum*** (Pers.) Wallr. (anamorph ***Conostroma didymum*** (Fautrey & Roum.) Moesz forms narrow leathery ascocarps below the bark of oak twigs, which open with a longitudinal slit to reveal a light yellow hymenium. It is common on dead twigs and occasionally causes a minor dieback. ***Cryptomyces maximus*** (Fr.) Rehm is a rather rare pathogen on willow branches and twigs, on which it forms black blister-like stromatic cushions.

Naemacyclus minor Butin

(Syn. *Cyclaneusma minus* Di Cosmo *et al.*)

N. minor causes a needle cast of 2-, 3- and 5-needled pines on all continents and appears to be more pathogenic than *N. niveus* (Pers.) Sacc. with which it was previously confused. The short concolorous apothecia which lack a clypeus are quite distinctive in the way that they open by two well-defined flaps. The pycnidia are inconspicuous (CMI 659, 668). In some areas it causes severe needle cast both in nurseries and in plantations, whilst in others it appears to be secondary to other needle disease fungi. Ascospores infect first-year and older needles of *Pinus sylvestris*, in spring but also in autumn, and symptoms normally appear about one year later. First symptoms are light green then yellow spots which develop into brown transverse bands on otherwise yellowed needles. Pycnidia develop below stomata in these bands followed by one or several ascocarps in each band. Infected needles are cast in the autumn of their second year. Ascocarps may form on attached needles but are more commonly found in the litter.

Evidence from plantations and from inoculation experiments suggests that some trees are resistant to *N. minor*. In New Zealand, in preliminary experiments, two-and-a-half-year old seedlings of *P. radiata* were not infected but two-year old rooted cuttings were. However, in Scotland (Karadzic 1981) and the USA (Kistler & Merrill 1978) seedlings and young plants of *P. sylvestris* were readily infected. In the USA mancozeb has been shown to control effectively the fungus in the field, and in New Zealand dodine, anilazine and benomyl are effective.
MILLAR

PEZIZALES

Rhizina undulata Fr.

(Syn. *R. inflata* (Schäffer) Quélet)

R. undulata forms chestnut to chocolate brown, hollow apothecia connected to the soil by white, hollow, root-like structures (CMI 324). It is confined to conifers, of which it attacks many species. It is especially damaging to *Picea sitchensis*, but also affects other spruces, species of *Pinus* and *Larix*, and occasionally also Douglas fir and species of *Abies* and *Tsuga*. It is found over much of N. and C. Europe, in N. America and in S. Africa.

R. undulata causes the disease known as group dying. In newly planted areas, patches of trees will die. On older, pole-stage trees, the foliage becomes thin, and may suddenly be shed, prolific coning takes place, and on the trunk the lenticels enlarge and exude resin. By this time many of the roots of the tree will already be dead, and if the bark is stripped off, they will show rounded lesions which often coalesce to produce large dead areas. Affected trees soon die as the fungus grows outwards, producing groups of dead trees over areas as large as 0.1 ha. These groups may then be much enlarged by windblow. The fungus grows only on the sites of fires, normally those made when thinning trees or when burning lop and top to prepare land for replanting. Colonization only takes place on acid soils and in the presence of fresh conifer roots (Jalaluddin 1967). In the past, *R. undulata* has sometimes caused severe losses, but these can now be readily prevented

by prohibiting the lighting of fires in conifer plantations.
PHILLIPS

CALICIALES

Coniocybe pallida (Pers.) Fr. (syn. *Roesleria hypogea* Thüm & Pass.) is one of the fungi causing foot rot of grapevine. Less common than *Armillaria mellea* and *Rosellinia necatrix*, it occurs mostly in the FRG, E. France and Austria. The fungus can be identified by the presence, on the dead wood, of little carpophores, which are 1 cm high, and resemble nails. This species is generally regarded as a weak parasite, occurring only in moist clayey soils.

HELOTIALES

Botrytis allii Munn.

Basic description

B. allii is probably identical with the earlier described *B. aclada* Fres. The conidiophores are small (less than 1 mm high), compact and occur in dense mats on leaf and bulb substrates. Individually, they are single-stemmed bearing a head of branched phialides with single-celled conidia. Conidia are narrow and ellipsoidal (7–11×5–6 μm). The small size of the spores distinguishes this fungus from *B. squamosa* which is of similar habit but has larger spores (10–14×6–9 μm). The short conidiophores of both species serve to distinguish them from *Botryotinia fuckeliana* which has a much branched conidiophore up to 2 mm high.

Host plants

B. allii attacks most onion cultivars causing neck rot in the stored bulbs. Shallot and garlic are also susceptible but leek and some lines of *A. fistulosum* (Japanese bunching onion) are moderately resistant at all stages of growth. Seedling resistance has been noted in *A. schoenoprasum* (chives) and in a number of wild species of *Allium*.

Geographical distribution

B. allii occurs on onions world-wide (Maude 1983) but it causes greatest disease problems in temperate climates (CMI map 169).

Disease

The disease, known as neck rot, only becomes evident when onion bulbs have been in store for eight or more weeks. Typically the upper parts of affected bulbs are soft when pressed and removal of the dry brown wrapper scales (leaves) reveals a black mass of sclerotia enclosing the neck tissues (Fig. 61). The individual sclerotia are up to 5 mm in diameter. A grey mould sometimes extends below the sclerotia and bears the conidiophores of the fungus. The fungus invades downwards in the food scales of the onion causing a brown discoloration of the affected tissues which become granular and have a 'cooked' appearance. Severely diseased bulbs decay and are covered with a sporulating grey mould. The sides and bases of bulbs are rarely attacked. However, attacks may occur where the protective wrapper scales are split before or during harvest. In these cases the grey mould phase develops. Infection of, and the development of sclerotia on and in the onion necks distinguishes neck rot from white rot (*Sclerotium cepivorum*) which produces a white mould on the roots and bases of bulbs in which there are the minute (0.3–0.5 mm) black sclerotia of the fungus.

Epidemiology

In Europe the seed-borne phase is most important in initiating outbreaks of the disease. The fungus invades the tip of the cotyledon from the seed coat tissues during germination and grows downwards in the leaf. It sporulates on the necrotic parts of the cotyledon leaf and conidia are released to infect the apices of the newly formed true leaves of that plant and neighbouring plants. This pattern is repeated as new leaves are formed and old ones die until the

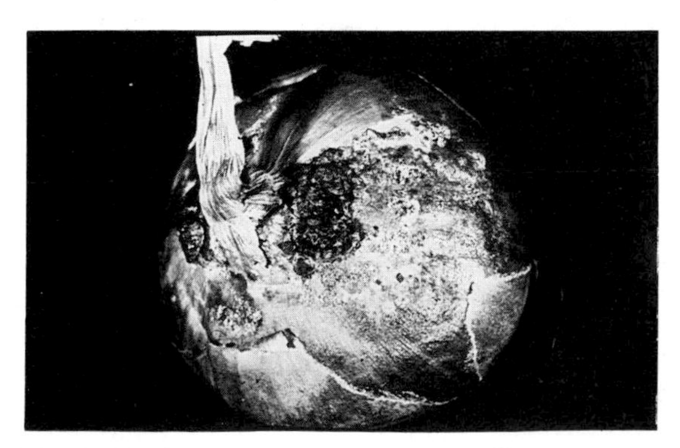

Fig. 61. Neck rot of onion due to *Botrytis allii.*

fungus infects the leaves continuous with the internal tissues of the onion neck and grows downwards in them to invade the bulb storage tissues, thus causing neck rot. Because the pathogen produced no symptoms on onion leaves the significance of infected seeds in the biology of the disease was not realized for many years. This relationship now has been established (Maude & Presly 1977).

A secondary source of *B. allii* is contaminated soil. Although sclerotia which persist for up to 2 years in unsterile sandy loam soil fail to produce conidiophores after 6 months they are probably capable of carrying the pathogen through from the end of one growing season to the beginning of the next. The soil-borne sources of neck rot may be extremely important where onion crops are grown in close rotation or continuously on the same patch of land. Other sources of the pathogen, such as onion dumps (cull piles) and the infection bridges produced by the overlapping of autumn with spring-sown crops may also perpetuate the disease in certain areas (Maude 1983).

Economic importance

Neck rot is the main disease of bulb onions in temperate parts of the world. In the early 1970s in Britain, the disease went unchecked and losses in excess of 50% were recorded in some stores. Disease losses on a similar scale have been reported from many countries (Maude 1983). The disease is favoured by cool wet summers which facilitate the production and dissemination of spores and infection of leaves.

Control

Virtual elimination of the disease can be obtained by dusting onion seeds with benomyl which will eradicate seed-borne infection. In the UK, neck rot is controlled to an acceptable level of less than 5% each year by growing main crop onions on wide rotations (3–4 years between crops) from onion seeds commercially treated with benomyl (1 g a.i./kg seed) or benomyl/thiram mixtures (1 g a.i. of each fungicide/kg seed) and by direct harvesting of maturing crops which are then post-harvest dried at 30°C in a high air flow (Maude 1983). No field control measures are used in the UK but in the FRG the effectiveness of seed treatment control is increased by a single crop spray with benomyl just before death of the onion foliage. To prevent incursion of the pathogen from secondary sources it is important to avoid infection bridges either from contaminated soil to new crops or from crop to crop by the overlap in time of bulb or seed production crops with newly emerged onions. Wherever possible rotations of more than two years should be used and onion crops should be sited at not less than 300 m from each other.
MAUDE

Botryotinia fuckeliana (de Bary) Whetzel

(Syn. *Sclerotinia fuckeliana* (de Bary) Fuckel)
(Anamorph ***Botrytis cinerea*** Pers.)

Basic description

Botryotinia is separated from *Sclerotinia* by having a more compact, gelatinized sclerotial medulla and a *Botrytis* anamorph. The teleomorph does occur in the field, but the anamorph *Botrytis cinerea* is predominant. Sclerotia vary in form and size, but are usually at least 3 mm diameter. A period of cold storage is needed for apothecia to be produced. The apothecial stalks are up to 3 cm long and 1–2 mm thick, the discs are concave, yellowish-brown, and up to 8 mm in diameter. Asci are cylindrical, ascospores ellipsoidal to fusiform 9–15×4–7 μm, and uninucleate. Conidiophores are more or less straight, branched and brown, but paler near the apex. The terminal branches bear smooth, single-celled conidia which are obovate or ellipsoidal, hyaline to pale brown, or in mass, grey-brown. Conidia measure on average 10×7.5 μm but show large variations (CMI 431). *B. cinerea* is a species aggregate which could correspond to several teleomorphs in *Botryotinia.* Many aspects of the biology of *Botrytis* species are described by Jarvis (1977) and Coley-Smith *et al.* (1980).

Host plants and specialization

B. fuckeliana is a non-specific parasite, attacking a vast number of plant species (vegetables, ornamentals, field crops, fruit trees), especially on annuals. It also attacks forest tree seedlings. It can behave both as a pathogen and as a saprophyte. Although there is a large variation in cultural morphology between isolates and also large variations in aggressiveness, host specialization has not been found.

Geographical distribution

B. fuckeliana occurs world-wide, but mainly in temperate and subtropical areas. Very high temperatures prevent its development.

Grey mould of grapevine

Disease

One of the most significant diseases caused by *B. fuckeliana* is grey mould of grapevine (Gartel 1970; Galet 1977), which is treated separately here as a characteristic example (other hosts are treated in the next section). The fungus attacks especially the bunches, but also any herbaceous part of the grapevine, which is susceptible at all stages of development. Before flowering, *B. fuckeliana* can destroy complete inflorescences, or attack the stalk or single sectors. After flowering, the withered corollas tend to be caught in the developing bunches and serve as a saprophytic base for infection of the fruit stalks or bunch stalk, leading to berry drop. After change in colour and during ripening, the grapes are infected directly or through wounds, and show the characteristic soft rotting with abundant conidial sporulation (grey mould).

Herbaceous parts of the grapevine may be attacked from the very beginning of the season. Infected buds and young shoots necrose and dry out. In late May–early June, a few leaves show large necrotic patches which are reddish brown in colour and irregular in outline, often at the edge of the lamina. However, this type of damage is of no practical importance. At the end of the season, shoots which have failed to mature and form wood satisfactorily through the summer are readily attacked and show whitening of the surface with characteristic black sclerotia or greyish patches of conidia.

Epidemiology

Infection is heavily influenced by weather conditions (Bulit & Dubos 1982). Rainfall and RH in particular affect all stages of development while temperature, though of lesser importance, is still significant. *B. fuckeliana* mainly overwinters as sclerotia, which form on the woody shoots from autumn onwards. It can also persist as mycelium in bark cavities or in dormant buds. In spring, in mild wet weather, the sclerotia and mycelium sporulate to produce conidia. The peak of sporulation from sclerotia is in mid-April to late May, and this inoculum is presumably responsible for early infections on young bunches and on leaves.

Rain and wind spread the conidia within the vineyards and beyond them. The presence of airborne inoculum corresponds well with the time of sporulation of sclerotia, and later (after a change in colour and during ripening) with the presence of abundant infection on the susceptible grapes. The optimum temperature for conidial germination is 18°C. Germination in water is much stimulated by the presence of various organic solutes (sugars, amino acids, organic acids). Germination may occur without free water, if RH exceeds 90%. In practice, however, infection will occur in strict relation to the effects of water and temperature. Free water must be present for at least 15 h at 15–20°C; the lowest temperature for infection is around 8°C, but the period of surface wetness must then be longer than 15 h. Conidia and mycelium can infect susceptible tissues directly, but natural or artificial wounds facilitate entry. To attack green tissues in active growth, *B. fuckeliana* must first colonize saprophytic substrates as a base. These may be various fragments of plant debris, such as withered corollas, and also sections of leaf or petiole caught in the bunches as they mature.

All cultivars of grapevine are susceptible to attack, but there are some differences in reaction which have been linked to the tightness of the bunches, the thickness or anatomy of the grape skin, and the chemical composition of the berries (presence of anthocyanins and other phenolic compounds; Nyerges *et al.* 1975). Grapevine is also known to produce phytoalexins (viniferins), in quantities which may be linked to the relative resistance of different cultivars (Langcake & McCarthy 1979). In addition to inherent genetic differences, grapes are more or less susceptible according to their physiological status. Before change in colour, infection is rare unless there is physical damage (Bulit & Lafon 1977); in contrast, the ripe berries are very readily infected both by conidia and by mycelium.

Economic importance

Attacks by *B. fuckeliana* cause damage of two types. The first is essentially quantitative: yields are reduced by a fall in the number of bunches (due to stalk rot) and in the amount of juice which can be obtained from the remaining bunches. The other is qualitative: grey mould alters the quality of wine, giving taints and poor vinification properties. The effect of the latter is to make the wine unstable and unsuitable for long-term maturation especially in the case of red wines. It is only in special circumstances on certain cultivars that *B. fuckeliana* acts as a 'noble rot', increasing the permeability of the skin of intact grapes so that the juice is concentrated, with accompanying chemical changes, to provide the basis for the renowned Sauternes wines in France and Tokays in Hungary.

Control

Grey mould control depends on prophylactic cultural measures and on chemical treatments (Agulhon 1973). Disease development is increased on vines with heavy dense foliage, and this can be avoided by choosing a less vigorous rootstock and limiting nitrogen fertilization. The grapes should be kept well ventilated and in sunlight, and this can be achieved by suitable training and pruning systems. Protection is needed against any agents liable to wound the grapes (especially the berry moths *Eupoecilia ambiguella* and *Lobesia botrana*), and any treatments applied against other pests should as far as possible also act against *B. fuckeliana* or at least not favour it.

Chemical treatment is frequently essential and can only be preventive. In France, it is normally based on a so-called standard programme:

treatment A—at end of flowering and beginning of fruit set
treatment B—just after berry-touch
treatment C—when the grapes change colour
treatment D—3–4 weeks before picking.

Specific fungicides are now available (dicarboximide group), especially vinclozolin and iprodione. The equipment must ensure good coverage of the grapes. Chemical control may be complicated by the appearance of strains resistant to fungicides such as the benzimidazoles and dicarboximides. Resistance to the former is now so persistent and widespread that this type of fungicide has become quite ineffective against *B. fuckeliana*. Dicarboximide resistance is less persistent and is still restricted in distribution (Pommer & Lorentz 1982). Resistant strains seem to be less competitive than sensitive ones, and there would seem to be good prospects for management of this group of fungicides to avoid selection pressure on resistant strains.

Recent research in France (Molot *et al.* 1983) has led to the development of a mathematical model of the 'infection potential' of *B. fuckeliana,* to simulate the course of epidemics on grapevine. This has given good results in forecasting the risk of grey mould and deciding whether treatments are necessary. Over several years, good control has been obtained with only 2 or 3 of the normal 4 treatments of the standard programme. Other research aims to use the antagonistic fungus *Trichoderma harzianum* Rifai as a biological control agent (Dubos *et al.* 1978). Encouraging results have been obtained, especially in a strategy combining the antagonist with normal chemical treatments. It should be stressed finally that, provided there is no local problem of dicarboximide resistance, the available control measures against grey mould can be applied with considerable success, and are fully compatible with the production of high-quality wines. Actual losses due to *B. fuckeliana* in vineyards are thus fairly low.

Attacks on other hosts

Disease

On other hosts, *B. fuckeliana* produces a soft rot (grey mould), blight or brown lesions. The fungus almost exclusively infects plants from colonized dead or senescent tissues or through wounds. It can cause pre- and post-emergence damping-off of seedlings (ornamentals, vegetables, forest trees). On stems and petioles, brown lesions occur, which spread slowly in the stem tissues (e.g. in vegetables such as tomato, sweet pepper, aubergine or cucumber, pea, French bean, chickpea, or in ornamental bedding plants, or in glasshouse ornamentals such as begonia, carnation, chrysanthemum, cineraria, cyclamen, fuchsia, pelargonium, poinsettia, rose, saintpaulia or tradescantia). On leaves, brown, sometimes water-soaked, lesions develop, spreading rapidly, especially under humid conditions (e.g. in lettuce, corn salad). On flowers, *B. fuckeliana* causes spotting (e.g. in freesia) and grey mould blight (e.g. on carnation, chrysanthemum and numerous garden plants). Flower buds are destroyed in various vegetable crops. Fruits are very commonly infected (strawberry, *Rubus,* pea, French bean, tomato, sweet pepper, apple, pear). *B. fuckeliana* typically produces a soft rot spreading rapidly, especially in soft fruits such as strawberry. An exception is the infection of tomato fruits from conidia, leading to the formation of small local lesions usually surrounded by a yellow halo ('ghost spot'). On all infected parts, except for the 'ghost spot', the fungus sporulates abundantly, especially on soft tissues. On a number of fruits, latent or quiescent infections can occur, e.g. on strawberry, raspberry and apple. The fungus penetrates young tissues at or shortly after blossom, but further colonization of the tissue takes place at fruit ripening. Finally, *B. fuckeliana* can also cause rot of stored vegetables (cabbage, artichoke, onion, potato, beet, carrot).

Epidemiology

The fungus survives mainly on infected plant debris. Although a number of isolates produce sclerotia *in vitro,* their role *in vivo* is not well understood. Seed infection does occur, but seems to be unimportant*. The main source of inoculum is conidia, which are wind-borne and are also spread by rain (splash dispersal). Alternating high and low RH seem to favour release of conidia. Recent research has shown that wind-borne conidia can survive on the petals of glasshouse-grown flowers for long periods. Under conditions of high humidity, leading to the formation of a water film on the petals, these conidia germinate and germ-tube penetration occurs. Plant debris colonized by *B. fuckeliana* is also spread by wind and rain and provides a large saprophytically based inoculum sticking to wet plant surfaces.

Economic importance

Losses arise from damping-off of seedlings, from the killing of plant parts above stem lesions and from the destruction of fruits. The latter is often considered the most important. The fungus can spread rapidly under a wide range of temperatures, even in cold storage. In glasshouses throughout Europe, *B. fuckeliana* is usually the single greatest cause of losses.

Control

Until recently, thiram and zineb were mainly used. These are now replaced by the fungicides described in the section on grey mould of grapevine and also dichlofluanid, tolyfluanid and chlorothalonid. The same problems of fungicide resistance arise on other crops as on grapevine. Indeed, because of the lack of host specificity, resistant strains can readily spread from one crop to another. In greenhouses, it is recommended to use thiram, dichlofluanid, chlorothalonil or captan when the infection pressure is low. Only at very high disease pressure may one or two dicarboximide treatments be used. In everbearing strawberries, a maximum of four dicarboximide treatments should be used.

*One exception is on flax, which is readily infected at the seedling stage by seed-borne *B. fuckeliana* which produces orange lesions at collar level, and may dry out and kill the seedlings. In this case, seed treatment is effective.

Special research interest

No single area of research stands out as being dominated by *B. fuckeliana* but because of its overall importance and the ease with which it can be cultured and conidia obtained the species has featured in a wide range of investigations covering almost every aspect of plant pathology.

VERHOEFF; BULIT & DUBOS

Botrytis fabae **Sardiña**

Basic description

B. fabae is distinguished from other *Botrytis* species by its limited host range, its pathogenicity on *Vicia* beans and by conidial size. Only one other species of *Botrytis, B. cinerea* (anamorph of *Botryotinia fuckeliana*), infects *V. faba. B. fabae* always produces larger conidia both *in vivo* and *in vitro* than does *B. cinerea* (Harrison 1983), and is usually the more pathogenic. No teleomorph is known.

Host plants

The host plants are *Vicia faba* and *V. sativa.*

Geographical distribution

B. fabae occurs throughout the world wherever field or broad beans are grown.

Disease

B. fabae is a virulent pathogen causing chocolate spot disease of *Vicia* beans (Gaunt 1983) and infecting all above-ground parts of the plant. In dry conditions the fungus remains dormant in the host or grows only slowly and the disease is then characterized by the presence of rust-coloured to dark-brown or grey, more or less circular spots up to several mm in diameter on the leaves, petals and pods, but elongated on the stems. A single leaflet may bear hundreds of lesions on each surface. Under wet or humid conditions the fungus spreads rapidly, both inter- and intracellularly, producing both pectolytic enzymes and non-enzymic phytotoxins. Lesions blacken, increase in size and coalesce, leading to a destructive blight (Fig. 62). This is referred to as the aggressive stage of the disease and most of the leaves drop off and the stems bend causing the plants to fall over. The fungus sporulates on the dead

Fig. 62. Non-aggressive (left) and aggressive (right) chocolate spot lesions on broad bean, due to *Botrytis fabae*.

tissues. Under alternately wet and dry conditions concentric rings may be formed as the growth of a lesion starts and stops. These lesions are distinguished from those caused by *Ascochyta fabae* by the absence of pycnidia. Black sclerotia 0.5–4.0×0.5–1.0 mm are formed in stem tissues when *B. fabae* has become well established.

Although *B. fabae* also infects *V. sativa* (common vetch) causing pale- to medium-brown leaf and stem lesions, it has not been extensively studied on this crop.

Epidemiology

B. fabae overwinters principally as sclerotia in dead bean stems. In mild humid weather sclerotia produce wind-dispersed conidia which can infect bean plants. Sclerotia do not survive for long in soil, often becoming parasitized by *Gliocladium roseum* Bain. (Harrison 1979). Alternatively the pathogen can overwinter in lesions on volunteer bean plants, in lesions on plants of an autumn-sown bean crop, or in lesions on *V. sativa,* which, besides being grown as a crop, also commonly occurs as a weed (Harrison 1979). In wet conditions the fungus sporulates on these lesions.

During the growing season the disease is spread within a crop and from one field to another by conidia, which may remain infective after being blown over large distances (Harrison 1983). One conidium can initiate a lesion (Wastie 1962). An RH of at least 85–90% is required for infection but free water is not necessary (Harrison 1984). At an RH of less than 70% lesions do not increase in size, but the rate of increase is directly proportional to RH at 70–100%. Conidia are produced if the RH is at least 80–90%, depending on other conditions. The fungus thrives and the disease can progress over a wide temperature range (3–28°C), 15–20°C being the optimum for most ontogenetic processes.

Although *B. fabae* commonly occurs in commercial seed stocks, chocolate spot has not been demonstrated conclusively to be seed-borne (Sode & Jørgensen 1974), the pathogen usually dying either during seed storage or on the epidermis of the young shoot and root soon after sowing, without causing any disease symptoms. Microconidia are sometimes produced but their role is not clear.

Economic importance

Damage to the crop depends largely on when the disease becomes aggressive, if indeed it does. If the aggressive stage appears before pods swell then all the crop may be destroyed, but if the disease remains non-aggressive then the loss of photosynthetic tissue may cause only a slight yield loss (Williams 1975).

If the disease does not appear until after the pods are fully swollen, yields are unaffected. Losses seem to be heavier in Mediterranean countries than in the cooler countries of N. Europe. In Scotland the disease rarely becomes aggressive (Harrison 1981). In northern countries autumn-sown crops are more seriously affected by chocolate spot than are spring-sown crops. Dense crops are more prone to the disease than those in which air can circulate freely.

Control

Benomyl or carbendazim are commonly used to control chocolate spot, although strains of *B. fabae* tolerant to these fungicides do occur. A spray application as the flowers start to open, followed by a second treatment three weeks later usually provides adequate protection. However the timing and frequency of spraying for optimum disease control can vary considerably depending on weather conditions and the symptoms present. Alternatively, two to five sprays of chlorothalonil at fortnightly intervals give good control. Attention to hygiene, especially the efficient removal of debris from a previous bean crop, either by burning or ploughing, reduces the inoculum and lessens the risk of serious disease. All commercial bean cultivars are at least moderately susceptible to chocolate spot. Attempts to breed for increased resistance are continuing.

During the planning of control measures, consideration should be given to the possibility of substantial economic losses resulting from the use of tractor-mounted spraying equipment, particularly in a tall crop, and aerial spraying should be considered.

Special research interest

There have been intensive studies on the role of phytoalexins in resistance to both *B. fabae* and *B. cinerea*. The difference in pathogenicity between these two species on broad beans can be explained largely in terms of a much more rapid rate of metabolism of phytoalexins by *B. fabae* than by *B. cinerea*. This work has been reviewed by Mansfield (1982).
HARRISON

Botryotinia convoluta (Drayton) Whetzel (anamorph ***Botrytis convoluta*** Whetzel & Drayton) causes a rot of iris rhizomes and spreads into leaves which become yellow, wilt and die. Parts of the plant above ground become covered with a mass of conidiophores and conidia. Characteristic convolute shiny black sclerotia occur on the surface of rhizomes and in the soil. Wounded rhizomes are particularly susceptible to infection. The disease has been reported from several European countries (Gerlach 1961).
HARRISON

Botryotinia draytoni (Buddin & Wakef.) Seaver

(Anamorph ***Botrytis gladiolorum*** Timmermans, syn. *B. gladioli* Kleb.)

B. draytoni causes the most destructive and widespread disease of gladiolus. Small, brown, sunken lesions occur on the upper part of the corm. Unless stored at high temperatures and with good ventilation such corms usually rot and the whole corm may be destroyed, although the outermost tissues sometimes remain sound. The characteristic sponginess of a rotting corm has led to the name 'spongy rot' for this stage of the disease. After planting the pathogen spreads to the leaf bases and the shoots become yellow and may die. Sporulation occurs on necrotic tissue and the disease is further spread by airborne conidia, 8–16×5–8 μm, which cause spotting on otherwise healthy leaves and flowers. Leaf spots are brown, but those on flowers are colourless with a water-soaked appearance. These small, roughly circular lesions expand at high humidities. When the flowers are infected the disease is known as petal spot and it often renders the blooms worthless. The pathogen grows from mother to daughter corm (Scangerla 1956) and overwinters as sclerotia 1–9 mm across or as mycelium on corms, dead leaves and stems and as sclerotia in the soil.

Careful attention to crop hygiene can reduce the risk of infection where gladioli are grown during successive years. The disease can be controlled by treating the corms with fungicides (Jacob 1969) or submerging them in hot water. Spraying the foliage and flowers with dithiocarbamate fungicides or with chlorothalonil or captan gives good control of leaf and petal spot. The disease also affects freesia, iris and *Ixia* species.
HARRISON

Botrytis elliptica (Berk.) Cooke is a widespread pathogen of lilies which can cause severe damage under humid conditions during the summer. Many species of lily are susceptible including *Lilium candidum, L. longiflorum* and *L. speciosum* although some, such as *L. regale*, are more resistant. Usually the disease appears first on the leaves as grey-brown to orange spots with a yellow-green halo, but buds, flowers and stems may also become infected. Sclerotia may appear on old plant parts but are not common on the bulbs. ***B. galanthina*** (Berk. & Broome) Sacc. causes grey mould on various species of snowdrop (*Galanthus*). The disease appears to be most common in the cooler, northern countries of Europe, where, in some seasons, it can completely kill clumps of snowdrops. Shoots become covered with conidiophores, often under a cover of snow, giving them a velvety appearance, and the fungus spreads to the bulbs which then rot. Sclerotia in diseased bulbs are probably the main source of infection. ***B. hyacinthi*** Westerd. & v. Beyma causes fire disease, and has been recorded in England and the Netherlands on outdoor hyacinths. Sclerotia, 1–2 mm in diameter, and air-dispersed conidia are produced on infected tissue. It also occurs on *Lilium regale* in Belgium and on *Muscari* species in the Netherlands.

Botryotinia narcissicola (Gregory) Buchw.

(Anamorph ***Botrytis narcissicola*** Kleb.)

B. narcissicola causes smoulder of *Narcissus* species. *B. narcissicola* and *B. fuckeliana* induce similar lesions from which either is readily isolated. The size and shape of their conidia are similar and the two species are most readily distinguished by the size and distribution of their sclerotia when grown on agar media. Sclerotia of *B. narcissicola* are smaller (1–2 mm diameter), of a more regular spherical shape and are more evenly distributed across the medium than those of *B. fuckeliana* (O'Neill & Mansfield 1982).

Smoulder is one of the most important diseases of daffodils and other species of *Narcissus,* especially in the second season of biennial crops, and occurs throughout Europe. Snowdrops are also reported to be susceptible. Narcissus bulbs bearing sclerotia typically produce leaves with a dark brown lesion at the tip. Lesions also frequently occur along the leaf edges and give rise to characteristic sickle-shaped leaves. Usually several leaves on a plant are infected. Lesions are less frequently found in flower buds and occasionally shoots fail to emerge. Symptoms of secondary infection are dark brown rots which develop after flower picking in the ends of flower stalks and in broken leaves, brown lesions near the bases of leaves and stalks and a rusty-brown leaf or flower flecking. Sclerotia may be found in dying leaves and flower stalks and on bulb scales.

Infected bulbs and sclerotia in the soil are probably the major sources of smoulder outbreaks. Conidia, which develop on necrotic tissue, usually fail to colonize healthy tissue (O'Neill & Mansfield 1982). However, senescent or wounded tissue such as broken leaves or flower stalk ends can readily become invaded and the fungus may then progress into the neck of the bulb. Extensively infected bulbs may rot. Disease control measures consist of rogueing plants with primary symptoms, avoiding the introduction of plant debris into the soil and using a rotation of at least 3 years between plantings of narcissus bulbs.

HARRISON

Botrytis paeoniae Oudem. has been recorded on paeony in Denmark, England, Germany, Italy, Poland and Scotland and can cause a blight in humid conditions. The disease often starts below the soil surface, originating from overwintered sclerotia or hyphae on rootstocks. The fungus spreads and invades stems and leaves producing grey-brown lesions on which infective conidia are produced in wet weather. Sclerotia develop on infected tissue, overwinter on dead stems and, unless below ground, produce conidia in the spring. ***Botryotinia polyblastis*** (Gregory) Buchw. (anamorph ***Botrytis polyblastis*** Dowson) causes fire disease of narcissus. Ascospores infect the flowers which become spotted and then shrivel. Conidia produced on the dead tissue cause leaf spotting and as the leaves die at the end of the season large sclerotia up to 1.5 cm in length are formed. These overwinter in crop debris and give rise to apothecia and ascospores during spring. The fungus can readily be distinguished from *B. narcissicola* and *B. fuckeliana* by its exceptionally large conidia, from 30 to 60 μm in diameter. Some cultivars, e.g. Grand Soleil d'Or are particularly susceptible to this disease while others such as Carlton, Golden Harvest and King Alfred are resistant (Moore 1949). ***Botryotinia porri*** (v. Beyma) Whetzel (anamorph ***Botrytis byssoidea*** J. Walker, syn. *Botrytis porri* Buchw.) causes neck and bulb rot of onions but is rarer than *B. allii* and *B. squamosa* in Europe (CMI map 165). It has also been reported to cause brown lesions on the blanched area of the leaves of leek in the UK and is a common and economically significant pathogen of cold-stored leeks in Finland (Tahvonen 1980). Spraying leeks with benomyl or thiophanate-methyl before harvest reduces damage during storage. ***Botryotinia squamosa*** Viennot-Bourgin (anamorph ***Botrytis squamosa*** J. Walker) has been

reported from several European countries where it causes yellow-white spots on leaves of onions and necrosis of leaf tips (Tichelaar 1966). If leaves remain wet, lesions expand causing leaf-rot disease and the whole plant may be killed. Conidia are produced on lesions and the fungus overwinters as sclerotia on bulbs and crop debris. Green salad onions may also be attacked, but damage to this crop is usually limited to white leaf lesions. It should be noted that *B. allii* (q.v.) is much more destructive than *B. squamosa* or *B. porri*.
HARRISON

Botrytis tulipae Lind

B. tulipae has more or less straight conidiophores, branched alternately and mostly in the apical region. They are brown, but paler near the apex. The conidia are hyaline, measuring 10×17 μm, with large variations, but larger than those of *B. fuckeliana* and often scantily produced. Sclerotia are irregular in shape, small, about 1 mm in diameter, and black to brown-black. *B. tulipae* occurs mainly on *Tulipa* species, rarely on *Lilium* species. It has been recorded from Europe, N. America, Argentina, Japan, Australia and New Zealand.

On aerial plant parts, two distinct types of lesions are formed (Price 1970). 'Spotting' refers to small yellowish-white, non-spreading lesions, usually formed after short periods of high humidity. 'Fire' refers to spreading lesions, formed under conditions of prolonged high humidity. These lesions are greyish, slightly sunken, bearing conidiophores with conidia in the centre. On bulbs, lesions of variable size are formed, sometimes bearing sclerotia. Sclerotia are also produced on decaying leaves and flower stalks (Doornik & Bergman 1974). Infections start from sclerotia, present either in the soil or on bulbs. Mycelium grows from lesions on bulb scales after planting. Infected leaves and infected flower stalks develop from infected bulbs, producing lesions bearing conidia. Conidia are spread by wind and rain. From leaves with 'fire' lesions mycelium spreads to bulbs. Mycelium from mother bulbs can grow directly into the scales of daughter bulbs. Temperatures of 10–15°C are optimal for disease development. Infected bulbs may produce malformed stalks, leading to a decrease in the number of flowers. Infected bulbs as well as infected flowers lose their market value. Treatment with mancozeb or captafol before planting will give some control. In the field zinc- and manganese-dithiocarbamates are used, and recently also vinclozolin, iprodione or procymidone. In case of spotting on flower stalks one treatment with benomyl is recommended.
VERHOEFF

Gloeotinia granigena (Quélet) Schumacher

(Syn. *G. temulenta* (Prill. & Delacr.) Wilson *et al.*, *Phialea temulenta* Prill. & Delacr.)
(Anamorph ***Endoconidium temulentum*** Prill. & Delacr.)

G. granigena has a life cycle somewhat similar to *Claviceps purpurea* (q.v.). Apothecia formed on infected seeds release ascospores which, at the time of flowering, infect the ovaries of grasses (especially *Lolium* spp.) via the stigmata. Conidia formed as a slime on the seed surface can spread infection to adjacent florets. Finally, infected seeds, when harvested, are 'blind' i.e. unable to germinate if early infection has destroyed the embryo. The blind seeds are shrunken and covered with a waxy-brown deposit of dried conidia, below the glumes. The disease was originally described on rye, the infected grain being toxic to man and livestock (Viennot-Bourgin 1949). The disease is widespread and occasionally serious in N. Europe (CMI map 348), and also France, Germany and E. Europe. It is favoured by cool wet conditions at the time of flowering, allowing apothecium formation and ascospore infection (O'Rourke 1976). An unusual epidemic in the Netherlands was caused by wet summer conditions (Tempe 1966). It is important especially in seed crops of ryegrass. Storing seed for over 2 years, or hot water treatment, will eliminate the fungus. Various fungicides can be used to prevent formation of apothecia, and nitrogen fertilization has also been found to reduce infection, more by effects on the plant than on apothecium formation. Resistance breeding (into S24 perennial ryegrass) has given good results.
CARR & SMITH

Monilinia fructigena (Aderhold & Ruhl.) Honey ex Whetzel

(Syn. *Sclerotinia fructigena* Aderhold & Ruhl.)
(Anamorph ***Monilia fructigena*** Pers.)

M. fructigena is closely related to *M. laxa* with which it shares many features and from which it can be differentiated by a combination of minor characteristics (see *M. laxa*). It is the commonest cause of fruit rot in top fruit orchards causing characteristic brown rot symptoms with erumpent conidial pustules often in concentric circles. The main hosts are apple, pear, plum, quince and cherry; peach, apricot and nectarine are less affected. In susceptible cultivars infection spreads into young twigs causing local cankers and wilt of extension shoots. Losses continue post-harvest but decline in importance with increasing periods of storage. *M. fructigena* is found throughout the fruit-growing regions of Europe, Asia, N. Africa and in parts of S. America.

M. fructigena overwinters mainly in small mummified fruit, either attached to the tree or on the ground. During spring and summer of the following year, given suitable moisture, sporodochia form on the surface of mummies from which conidia are transferred by wind and insects to young fruit. Initial infection is always via wounds, usually scab lesions or sites of insect damage, but subsequent spread by contact between adjacent fruit is possible. Secondary spread is by further crops of conidia from newly affected fruit produced as soon as seven days after infection. Twig infections are less common but can provide an additional means of overwintering. Harvested fruit are a dead-end in the disease cycle, and the teleomorph is invariably absent.

Losses from brown rot are highly visible to the grower, but are rarely worth specific control measures in their own right. On balance early-maturing cultivars are most affected, but the majority of diseased fruits are those which would in any case be rejected for other reasons such as bruising, or bird and insect damage. Apart from avoiding very susceptible cultivars in disease-prone districts, few control measures are specifically aimed at brown rot. By reducing the amount of inoculum produced, fungicides used for the routine control of foliar diseases provide an element of incidental control that in practice is often sufficient. Prospects for resistance breeding are limited at least in apple since most resistant cultivars are of the cider type with fruit characteristics unsuitable for dessert use.

M. fructigena has been much used as a model pathogen in studies on the role of cell wall-degrading enzymes in pathogenesis and on their inhibition by plant phenolic compounds as a mechanism of resistance (Byrde & Willette 1977).

ARCHER

Monilinia laxa **(Aderhold & Ruhl.) Honey ex Whetzel**

(Syn. *Sclerotinia laxa* Aderhold & Ruhl., *S. cinerea* (Bonorden) Schröter)
(Anamorph ***Monilia laxa*** (Ehrenb.) Sacc.)

Basic description

In culture the mycelium is initially hyaline but develops a dark irregular stromatal crust as the colonies age; this corresponds to the more massive stroma (mummified fruit) in nature. Aerial hyphae in culture and sporodochial tufts on fruit become zoned in response to diurnal illumination. Characteristic monilinioid macroconidia are ovoid to lemon-shaped and occur in branched chains clumped into slightly smoke-grey sporodochia (cf. cream-white to buff in *M. fructigena*). *M. laxa* may be distinguished from the closely related species *M. fructigena* and *M. fructicola* on the basis of host range, colour of fresh conidial pustules, mode of conidial germ-tube branching, colony characteristics and the patterns of soluble proteins (Byrde & Willetts 1977; Willetts *et al.* 1977). ***M. fructicola*** (Winter) Honey, affecting mainly *Prunus* in N. America and Australasia is absent from Europe and considered as a quarantine organism (EPPO 154).

Host plants

The host range is primarily restricted to the Rosaceae. The main commercial crops infected in Europe are apple, pear, peach and nectarine, apricot, plum, sweet cherry, sour cherry and almond.

Host specialization

Infection of apple blossom has been ascribed to **f. sp. *mali*** Wormald *sensu* Harrison. This form does not naturally attack the other hosts, and other strains of *M. laxa* do not normally attack apple (Wormald 1954). f.sp.*mali* also differs from typical *M. laxa* in cultural characteristics on some media, in its ability to secrete large amounts of polyphenol oxidase (Wormald 1954), and in its polygalacturonase isoenzyme pattern (Willetts *et al.* 1977).

Geographical distribution

M. laxa is widespread in fruit-growing areas of Europe and only in the far north has it not been recorded. It also occurs world-wide in temperate and subtropical regions (CMI map 44).

Diseases

M. laxa primarily causes a blossom wilt of pears and stone fruits. The infection often spreads to cause twig blight and cankering, especially on stone fruit. *M. laxa* f.sp. *mali* causes similar symptoms on apple. A 'wither tip' form of the disease results from direct infection of young shoot tissue by *M. laxa*, generally on plum. *M. laxa* also causes a fruit rot (brown rot) of stone fruits, which can be severe. The fungus kills the cells and the tissue becomes brown and, though still firm, a little softer. *M. fructigena* also causes a brown rot of stone fruit, and both species may cause fruit losses on a single tree. Neither *M. laxa* nor its f.sp. *mali* normally rot apple or pear fruit.

M. laxa is readily identified in Europe by the production of grey cushions of monilioid conidia on infected fruit, blossoms or spurs, since the rather similar species *M. fructicola* does not occur here. In general *M. laxa* is more a pathogen of blossom and twigs than of fruit.

Epidemiology

M. laxa overwinters as mycelium in infected cankers and spurs, or (except for f.sp. *mali*) in mummified fruits. Conidia are produced during humid weather in winter and spring: sporulation occurs readily at fairly low temperatures (e.g. 10°C). The conidia infect open blossoms via the stigma and style. Later in the season fruitlets and fruits of *Prunus* species can be infected by conidia produced either on overwintered inoculum or from more recent blossom infections. Conidia may be dispersed by air or water, or carried on insects attracted to infected or injured fruit. Infection by spores is almost always through an injury, but secondary mycelial spread can occur from an infected fruit to another in contact. Direct leaf infection can occur on plum. The teleomorph, although very occasionally reported in Europe, is of negligible importance in the normal life cycle of *M. laxa*.

Economic importance

This varies greatly from season to season and from site to site. In wet springs blossom wilt infection of stone fruits can be severe—losses of about 50% of trusses are occasionally reported (Byrde & Willetts 1977). Sarejanni *et al.* (1952) mentioned severe damage to almond blosson and, around Athens, the destruction of the entire apricot crop in 1951 as a result of blossom infection. *M. laxa* f.sp. *mali* can cause damage to apple in localized orchards, those in damp localities (e.g. on hills or near the sea) being most at risk. In a fruit control trial in 1968, Byrde & Melville (1971) recorded a mean of 16 infected spurs per tree on naturally infected unsprayed trees. Little information is available on fruit losses caused by *M. laxa*, perhaps partly because *M. fructigena* is also involved in brown rot of stone fruits.

Control

There is now a range of fungicides normally effective against *M. laxa*. Among carbendazim-generating compounds, carbendazim itself, benomyl and thiophanate-methyl have been very successful. In the dicarboximide group, vinclozolin, iprodione and procymidone are very effective. With both groups of compounds, however, the emergence of resistant variants of the pathogen is a problem, especially with the carbendazim-generating compounds. Resistance to them has already occurred in *M. laxa* and in *M. fructicola*, in which resistant isolates are known to be as fit as carbendazim-sensitive forms (Gilpatrick 1982). The older fungicides captan, captafol, chlorothalonil, dithianon and sulphur often give partial control, while triforine is a useful fungicide that may help to avoid resistance problems. The use of resistant cultivars is of some value. Byrde & Willetts (1977) listed cultivars of stone and pome fruits that showed some resistance. None is immune, however, and many partially resistant cultivars are unsuited for commercial culture because of limitations in yield or quality.

General hygiene is important in the control of *M. laxa*; this includes the removal and destruction of mummified fruits and infected spurs. Control of insect vectors to restrict fruit infection is also useful, when appropriate.

Special research interest

Unlike the related species *M. fructicola* and *M. fructigena, M. laxa* has been little used as a 'model pathogen' in research. This may partly reflect the greater world-wide economic importance of *M. fructicola*, and the ease of obtaining apple fruit (in contrast to stone fruit) for work with *M. fructigena*. Willetts *et al.* (1977) have however, compared the isoenzyme pattern of *M. laxa* cell-wall degrading enzymes with that in related species, and Byrde & Willetts (1977) speculated on its possible relationship to these species and a postulated ancestral form of 'monilinioid' *Sclerotinia*.

BYRDE

Several related ***Monilinia*** species occur as minor pathogens generally causing leaf blotch symptoms. ***M. johnsonii*** (Ell. & Ev.) Honey (syn. *Sclerotinia crataegi* Magnus) is found on *Crataegus*, ***M. linhartiana*** (Prill. & Delacr.) Buchw. on quince and ***M. mespili*** (Schellenberg) Whetzel on medlar. Several species including ***M. vaccinii-corymbosi*** (Reade) Honey, ***M. urnula*** (Weinman) Whetzel and ***M. baccarum*** (Schröter) Whetzel attack *Vaccinium* species producing mummified berries. ***M. ledi*** (Nawaschin) Whetzel has curiosity value as an obligately heteroecious ascomycete, forming its conidia on the leaves of *Vaccinium uliginosum* and its ascocarps on the fruits of *Ledum palustre*. ***Ovulinia azaleae*** Weiss, the causal agent of azalea petal blight, is locally damaging in wet seasons (Wormald 1954).

Sclerotinia borealis Bubak & Vleugel

(Syn. *Myriosclerotinia borealis* (Bubak & Vleugel) Kohn)

S. borealis attacks cereals and grasses during overwintering (Fig. 63). Its geographical distribution is restricted to northern parts of the Nordic countries, USSR, Japan and N. America, to regions having an extremely stable, cold winter climate with prolonged, dry snow cover. Severe attacks rarely occur at localities having less than about 5–6 months of snow cover. In regions with more than about 6 months of snow cover, this pathogen is regarded as one of the most important snow mould fungi, regularly causing severe damage. After snow melt in spring the disease appears as bleached patches of dead or dying plants, covered with sparse, greyish white mycelium. Greyish white, later black, sub-globose to oval, elongated, flattened, or irregularly shaped sclerotia, 2–8×1–4 mm diameter, are formed on or within infected tissue. Germination of sclerotia takes place during late autumn. Sporulation is dependent on moderately low temperatures (optimum at about 9°C) and excess moisture conditions. Ascospores spread to surrounding plants before permanent snowfall serves as the main source of inoculum. Infection may also occur through direct mycelial growth into predisposed host tissues, but this is probably of minor importance for disease development. Sclerotia buried in the soil may survive for several years.

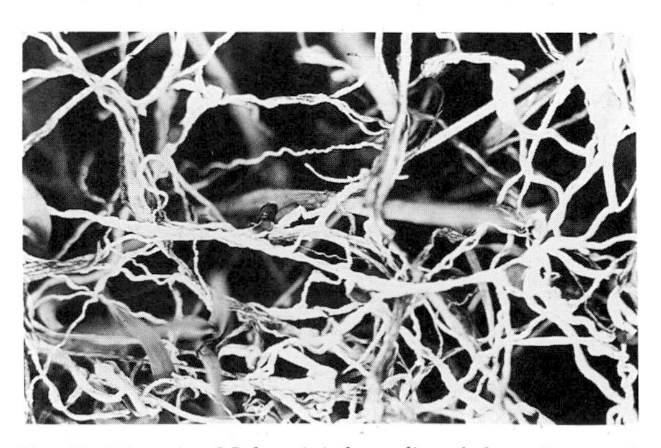

Fig. 63. Sclerotia of *Sclerotinia borealis* on infected leaves of *Phleum pratense*.

The fungus is well adapted to conditions prevailing under a snow cover. Mycelial growth occurs from below −6°C, with an optimum at 3–6°C. At −3 and 0°C the growth capacity is about 55 and 80% of the optimum, respectively. Predisposing factors are early snow in autumn or prolonged, dry snow cover on unfrozen or slightly frozen ground, and acid soils. Unbalanced or excessive nitrogenous fertilizer, particularly applied late in the season should be avoided. Fungicides, e.g. quintozene, applied late in the autumn before snowfall may control the disease.

Great variation in resistance between different cultivars is known. Northern cultivars of good winter hardiness are usually less susceptible than those of southern origin. Host resistance is assumed to be non-specific and quantitative, governed by several genes, each possessing a slight additive effect. Laboratory methods for large-scale screening for resistance in breeding material have been developed. ÅRSVOLL

Sclerotinia bulborum (Wakker) Rehm

S. bulborum causes black slime of hyacinths, although other bulb plants are also attacked, including anemone, chionodoxa, crocus, fritillary, iris, muscari, narcissus, scilla and tulip. The disease is recorded occasionally in England and in the Netherlands where it formerly caused severe losses. The bases of hyacinth bulbs rot so that the leaves are easily removed. Outer bulb scales often rot completely while the inner ones become thin and dark grey. White mycelium within the bulb gives rise to sclerotia 5–20 mm long which blacken as they age. Heavily infected bulbs disintegrate in the soil leaving the sclerotia intact. These produce infective mycelium in the soil, and apothecia and ascospores above ground the following season. The disease is

controlled by removing infected plants and the soil immediately surrounding them, by sterilizing infested land, or by spraying plants with procymidone.
HARRISON

Sclerotinia gladioli Drayton

(Syn. *Stromatinia gladioli* (Drayton) Whetzel)

S. gladioli is the commonest pathogen of gladioli, causing dry rot disease. It is widespread throughout Europe. Soon after emergence shoots begin to turn yellow and then brown as they die, by which time the underground parts of the plant are already heavily infected. Minute black sclerotia 0.1–0.3 mm across appear on necrotic tissue, often including the corms and roots. The fungus penetrates the corm at the base of the scale leaves, producing brown-black rings, and spots appear elsewhere on the corm. These lesions often develop into an extensive dry rot. The disease is introduced when lightly infected corms are planted. The fungus grows through the soil and infects nearby healthy corms. *S. gladioli* can survive in soil for many years. Acidanthera, crocus, freesia, narcissus, montbretia and snowdrops are alternative hosts. Control measures include hot water treatment of the corms, dusting them with thiram, destroying infected plants and soil sterilization with methyl bromide or dichloran.
HARRISON

Sclerotinia homoeocarpa F. Bennett

S. homoeocarpa causes dollar spot of turf (CMI 618). The fungus produces thin rind-like black stromatic sheets in culture, rather than true sclerotia, and apothecia have only rarely been seen. Some authors (APS Compendium) do not consider that it is a true *Sclerotinia*, but rather that it corresponds to one or even several different species of *Lanzia* and/or *Moellerodiscus*. The disease is most important in N. America, where many turf grass species are attacked. In Europe, it has been recorded in the UK and in the Netherlands, especially on *Festuca rubra*. Irregular bleached patches 2–15 cm across appear on normal turf; on golf greens, the spots are more typically dollar-sized. The fungus spreads by mycelial growth from leaf to leaf under wet conditions and persists as stromata. The disease is worst on turf which has been water- or nutrient-stressed. Control is essentially by good management; many fungicides may be effective (mercury or cadmium-based preparations, benzimidazoles etc.). Populations tolerant to particular fungicides are known in N. America (Warren *et al.* 1977) and may in part correspond to different species.
CARR & SMITH

Sclerotinia minor Jagger

S. minor is closely related to *S. sclerotiorum*, and although taxonomically distinct, there is much overlap between the diseases caused by the two organisms. *S. minor* has smaller sclerotia (0.5–2.0 mm diam.) and rarely forms apothecia. Infection depends upon the juxtaposition of sclerotia with susceptible host tissues which are infected via myceliogenic germination. 'Lettuce drop', characterized by basal stem infection, wilting of lower leaves and progressive collapse of the whole plant is the classic symptom type. Apart from a reduced emphasis on ascospore inoculum, the entry for *S. sclerotiorum* should be read as applying also to *S. minor*.
ARCHER

Sclerotinia sclerotiorum (Lib.) de Bary

(Syn. *S. libertiana* Fuckel, *Whetzelinia sclerotiorum* (Lib.) Korf & Dumont)

Basic description

In culture the fungus grows rapidly producing abundant white cottony mycelium on which black sclerotia develop, up to 1 cm diameter with a slightly pitted surface often exuding droplets of liquid in a saturated atmosphere. In hypocotyls of *Phaseolus vulgaris*, two morphologically distinct hyphal types have been reported (Lumsden & Dow 1973). Large intercellular infection hyphae (mean diameter 17 μm), oriented predominantly parallel to the long axis of the hypocotyl, form first, followed by a second morphological type, ramifying hyphae (mean diameter 8.5 μm). These develop sub-apically from the infection hyphae, branch profusely and penetrate the host both inter- and intracellularly. Unicellular, phialidic microconidia often form on old colonies. There have been various claims for their germination *in vitro* or for a function as spermatia, but their role if any in the life cycle remains obscure. There are no macroconidia. Sclerotia exhibit either myceliogenic or carpogenic germination, the former giving rise to vegetative hyphae, the latter to apothecia. Apothecia are cup-shaped, yellowish-brown, up to 10 mm in diameter on a smooth cylindrical stalk which emanates from the sclerotium. Asci are cylindric-clavate, up to 130×10 μm, 8-spored; the ascospores are uniseriate and elliptical, 9–13×4–6.5 μm.

S. sclerotiorum has often been confused with *S. minor* and with *S. trifoliorum*. Willetts & Wong (1980) after an extensive comparative review concluded that the three species were distinct and could be reliably distinguished on the basis of numerous morphological, physiological and biochemical characteristics.

Host plants

S. sclerotiorum has one of the widest host ranges of any pathogen. Purdy, in APS (1979), lists 225 genera from 64 families of higher plants as hosts, mainly angiosperms. Parasitism of Solanaceae (tomato), Cruciferae (rape, cabbage), Umbelliferae (carrot, celery), Compositae (lettuce, chicory, sunflower), Cucurbitaceae (cucumber, melon) and Leguminosae (dwarf bean) is especially common. Woody plants, grasses and cereals are rarely attacked. Price & Colhoun (1975) demonstrated that individual isolates could attack a wide range of hosts.

Geographical distribution

The pathogen is of world-wide occurrence but the disease is mainly one of cool moist conditions. It is a pathogen of the winter season in subtropical and Mediterranean climates and of high altitude in the tropics.

Disease

S. sclerotiorum causes a progressive soft rot on unlignified tissues of a wide range of hosts. The stems of herbaceous plants and vegetable storage organs are most frequently damaged. Lesions on leaves also occur but only spread if conditions are very favourable. In sunflower, the capitula are also affected. On different crops names such as 'watery soft rot', 'white mould' and 'cottony soft rot' are used. Characteristic features are the formation of extensive, soft, usually light coloured lesions and the growth of a white, fluffy carpet of mycelium on the surface and within cavities of the host plant. Later, prominent black sclerotia form within pith cavities of stems and petioles, the occurrence of which is a reliable diagnostic feature. Plants that are infected in the main axis near soil level frequently exhibit wilting as the first major symptom, rapidly followed by total collapse and lodging.

Epidemiology

Sclerotia are the most important means of perennation. Their survival time in soil is very variable, but 6–8 years is thought to be an upper limit. Survival of mycelium in seeds may also occur but epidemiologically it is of little consequence. Under most conditions myceliogenic germination is of limited importance because only limited saprophytic spread occurs in non-sterile soil. However where sclerotia and susceptible plants are in close proximity devastating stem base infections may result. A prerequisite for carpogenic germination is a period of chilling to break dormancy followed by rising temperatures and a high humidity. In temperate latitudes apothecia typically mature during spring and early summer, although there are many reports relating to other seasons of the year. The apothecial stipes elongate in response to light and the ascospores are wind-dispersed. Spores landing on potential hosts need water for germination, a requirement of 16–24 h being typical. Germination is possible throughout the range 0–25°C, with an optimum at 15–20°C. On sunflower in France, 42 h wetness is needed for ascospore infection of the capitulum, and symptoms appear about 5 weeks later. This threshold can be used to define regions at risk (Lamarque 1983). Also it seems that an exogenous nutrient base is required for infection. Wounded, dead, or senescent tissues are readily colonized and serve as a food base from which infection of healthy tissues can take place. For crops such as dwarf bean and rape, senescent petals provide a major avenue of infection, and the coincidence of flowering and ascospore release is of major epidemiological significance.

Germinated ascospores produce appressoria which can vary from simple lobed forms to complex multibranched cushion-like structures. Entry is usually by direct penetration through the cuticle, assisted by extensive pectolytic and cellulolytic dissolution of the host cell structure. Sclerotia are returned to the soil as the host decomposes, or they may be distributed by cultural operations, harvesting, grazing etc. In most regions the absence of a conidial stage and the environmental requirements for apothecium formation restrict *S. sclerotiorum* to a single annual infection cycle.

Economic importance

The damage caused by *S. sclerotiorum* tends to be sporadic and for most of the crops that it affects *S. sclerotiorum* is usually regarded as a lesser pathogen. However, due to the nature of the disease, where outbreaks do occur losses can be serious. Most reports (in declining order of frequency) refer to sunflower (serious losses in C. France), rape, dwarf bean and lettuce (especially when grown under glass). Ornamental bedding plants are also commonly affected. In N. America it is considered one of the most important pathogens of *Phaseolus* species, and also attacks soybean. In Europe the disease has become more common in recent years due to the increased cultivation of rape.

Control

Chemical control has been directed both at inhibiting sclerotial development and at protecting plants from ascospore infection. The former has been attempted with a variety of soil sterilants and broad-spectrum fungicides. Protective spraying has generally been with benomyl, a dicarboximide compound or dichloran. Somewhat surprisingly in view of the nature of the pathogen, resistant germplasm has been identified in *Phaseolus* species and breeding programmes are in progress in the USA. Biological control with microbial antagonists which colonize the sclerotia holds some promise for the future. Cultural methods of control must be practised with care: deep ploughing 2 years in succession can re-expose viable buried sclerotia, and crop rotation is of limited value in view of the wide host range.

Special research interest

S. sclerotiorum has been used as a model pathogen to study host cell wall degradation during infection. Studies have encompassed changes in tissue permeability, tissue softening and cell death, and the roles of pathogen-produced enzymes and of oxalic acid in achieving these effects. Many aspects of sclerotinia disease in N. America were discussed in an American Phytopathological Society Symposium in 1978 (APS 1979).

ARCHER

Sclerotinia trifoliorum Eriksson

S. trifoliorum resembles *S. sclerotiorum* but, besides its different host preferences, has smaller sclerotia, (which develop more slowly), and germinates to form apothecia in the autumn rather than in the spring or summer. Willetts & Wong (1980) concluded, on the basis of the overall biology of these two species and *S. minor*, that they are distinct and bounded by intersterility. This has been confirmed by studies on immunoelectrophoresis (Scott 1981) and pectic enzymes (Cruikshank 1983). Uhm & Fujii (1983) have shown that *S. trifoliorum* is heterothallic, that microconidia from one strain spermatize sclerotia from another, and that mating type segregates with spore size in the ascus (four large and four small ascospores are formed). *Trifolium pratense* is the main host but other *Trifolium* species are also attacked, including *T. repens,* previously considered resistant, on which new virulent biotypes may be developing. **Var. *fabae*** Keay was described many years ago on *Vicia faba* but its present status is uncertain. Other hosts of **var. *trifoliorum*** include lucerne and other *Medicago* species, *Onobrychis viciifolia, Vicia sativa* and to a lesser extent *V. villosa,* and *Coronilla varia*. In general, Leguminosae are attacked by *S. trifoliorum* rather than *S. sclerotiorum,* but there are exceptions (*Phaseolus* spp.). The pathogen is widespread in temperate zones throughout the world (CMI map 274).

The disease on clover is variously referred to as rot, crown rot, canker etc. It has been recently reviewed in detail by Scott (1984). Ascospores from apothecia arising in the autumn from germinating sclerotia cause minute (less than 1 mm) spots on the leaves. These lesions remain latent during the cold winter months but spread as soon as conditions become milder to destroy the entire leaves and spread by mycelial growth to the crown and roots which are rotted. White, then black sclerotia form in the rotted tissues in late winter and spring and persist in the soil. Direct infection from sclerotia is not generally reported to occur in the field, although it can readily be obtained in the laboratory or glasshouse, and may be used in resistance testing. Sclerotia may produce apothecia several years in succession, or remain dormant for several years before first producing apothecia (O'Rourke 1976). Freezing sclerotia at −18°C for 24–48 h stimulates germination 4–6 weeks later, especially if late autumn temperatures and light regimes are simulated. Accumulation of sclerotia in the soil under continuous clover cropping will cause progressively more serious disease (clover sickness). Initial ascospore infection is favoured by mild humid conditions in autumn, and later spread by mild spells in winter. Sclerotia may contaminate seed, and this provides a means of introducing the disease into new areas. *T. pratense* suffers serious losses in N. and E. Europe. Attacks on other forage legumes tend to be relatively minor.

Control is mainly by cultural means. A 4–5 year rotation is recommended on infested land. Early cut-

ting or grazing reduces the risk of losses during the following winter. Heavy fertilization favours disease, and should accordingly be avoided on infested land. *T. pratense* cultivars vary considerably in susceptibility, tetraploids being generally considered more resistant. Testing for resistance has been carried out in many European countries (Czechoslovakia, France, Poland, Switzerland, UK, USSR), and is complicated by the choice of type of inoculum and conditions for the testing (e.g. Raynal 1981).
CARR & SMITH

Sclerotinia candolleana (Lév.) Fuckel (syn. *Ciborinia candolleana* (Lev.) Whetzel) causes circular brown spots on the leaves of oak from July onwards. ***S. pseudotuberosa*** Rehm (syn. *Ciboria botschiana* (Zopf) Buchw., *Stromatinia pseudotuberosa* (Rehm) Boud.; anamorph ***Rhacodiella castaneae*** Peyr.) rots acorns and the nuts of *Castanea sativa*. ***S. spermophila*** Noble is widespread on white clover seed, but rarely causes any extensive damage. ***S. tuberosa*** (Hedw.) Fuckel (syn. *Dumontinia tuberosa* (Hedw.) Kohn) causes black rot of *Anemone* species and has been recorded from England, France and FRG. Infection of the rhizomes, which become covered with hyphae, occurs through the mycelium in soil that arise from ascospores (Pepin 1980). ***Cristulariella depraedans*** (Cooke) Höhnel in some years causes a spectacular leaf spot disease of sycamores and maples (*Acer* spp.) in N. Europe and N. America, but it occurs too late in the season to cause serious damage. The related ***C. moricola*** (Hino) Redhead (syn. *C. pyramidalis* Waterman & Marshall) has a *Sclerotinia*-like teleomorph ***Grovesinia pyramidalis*** (Cline *et al.* 1983).

Sclerotium cepivorum Berk.

Basic description

S. cepivorum produces black, nearly spherical sclerotia about 0.2–0.5 mm in diameter or, less frequently, much larger irregularly shaped stromatic masses. It also has a phialidic spermatial state but there is no sexual mechanism nor conidial state.

Host plants

S. cepivorum is a pathogen of the genus *Allium* and the few reports of its occurrence on plants other than this genus are probably erroneous. There is some evidence that leeks are less susceptible than onions, garlic and *A. fistulosum*.

Geographical distribution

S. cepivorum was first described in England in 1841 and subsequently reported from many other countries. It now has a world-wide distribution.

Disease

In the disease known as white rot, plants are typically invaded through their root systems. The fungus grows necrotrophically in the roots and may invade the bulbs or shoot bases causing a soft rot. Plants that are attacked early often die in the seedling stage. Since the root systems are often completely rotted infected plants can easily be pulled out of the ground. Infected bulbs are at first covered with a fluffy white mycelium (Plate 27) but this subsequently dies down to form a mat in which large numbers of sclerotia are embedded. Above-ground symptoms usually show first in onions as a yellowing of the older leaves which progressively wilt and fall over. The symptoms are usually less striking in leek and garlic and wilting may not always occur. If conditions are not ideal, rotting of bulbs may continue in storage.

Epidemiology

S. cepivorum perennates as sclerotia in soil. These are known to be capable of surviving for at least 15 years but may not do so under all conditions. They can survive passage through the digestive tracts of grazing animals and may be spread in this way or by transfer of soil on farm implements or boots. Sclerotia are held in check by soil fungistasis and only germinate in response to specific factors released by the roots of host plants (Esler & Coley-Smith 1983). They normally only germinate once, as their nutrient reserves are totally depleted during the process. Primary infections arise from mycelium produced by germinating sclerotia. Amongst closely spaced plants such as salad onions or bulb onions raised in multiseeded peat blocks, secondary infections resulting from plant to plant spread may be significant. Mycelial growth in soil is negligible unless sustained by a food base such as an infected plant or sclerotium. Sclerotial germination, mycelial growth and root infection are all markedly influenced by temperature, the optimum lying between 10 and

20°C. Mycelial growth and germination of sclerotia are restricted by higher temperatures and in consequence in countries with a hot climate white rot is confined to the winter months. In certain countries crops can be grown successfully on infested land in summer where, when grown in winter, severe losses would occur.

Economic importance

Losses from white rot vary from small percentages to the complete destruction of a crop. They are of greatest importance in cool wet summers and of least importance under hot dry conditions. On the whole the disease is more severe in summer crops in N. Europe than in Mediterranean countries but in the latter it may be severe in winter crops. In certain countries the severe losses sustained have caused the abandonment of *Allium* crops.

Control

Protectant fungicides have been used for controlling white rot in salad onions. Seed treatments with calomel, dicloran, benzimidazoles and dicarboximides have given varying degrees of control. At present the most effective of these chemicals appear to be the dicarboximides. Good control of white rot in salad onions has been achieved with a seed treatment with iprodione (50 g a.i. per kg seed) followed by at least one stem basal spray with the same chemical at 0.05–0.15 g a.i. per metre of row (Entwistle & Munasinghe 1981). The use of dicarboximides in bulb onions and garlic has been less successful so far. Recently, use of these fungicides has encountered problems through stimulation of a soil microflora which rapidly degrades them (Walker *et al.* 1984). Soil partial-sterilants have shown promise in controlling white rot but their cost is too high to justify their use on a large scale. Recent results from the USA indicate that metham sodium applied at a relatively low concentration in irrigation water gives a worthwhile degree of control in salad onions. Other possible control procedures such as the use of sclerotium germination stimulants and biological methods are still at an experimental stage but show some promise.

Special research interest

S. cepivorum is of particular interest because of the unique nature of the specific response of its sclerotia to stimulants of host origin. This has led to the identification of the stimulants as flavour and odour compounds of *Allium* species and to an exploration of their use in the reduction of sclerotial populations.
COLEY-SMITH

Ustilaginoidea virens (Cooke) Takahashi causes false smut of rice and maize (CMI 299). The black conidia, formed on hard sclerotia replacing the grain, superficially resemble smut spores. The sclerotia are said to germinate to produce apothecia, giving a life cycle like *Claviceps purpurea*. Widespread, but of minor importance in most rice-growing areas (CMI map 347), the fungus has been reported in Italy.

Blumeriella jaapii (Rehm) v. Arx

(Syn. *Coccomyces hiemalis* Higg.)
(Anamorph ***Phloeosporella padi*** (Lib.) v. Arx, syn. *Cylindrosporium padi* (Lib.) P. Karsten ex Sacc.)

Basic description

There are different opinions about the taxonomy of the pathogen(s). In Europe, *B. jaapii* is regarded as the causal fungus. In America, however, three fungi are considered to infect the same host range as *B. jaapii*, namely *Coccomyces hiemalis* Higg., *C. prunophorae* Higg., and *C. lutescens* Higg (Anderson 1956). In this entry the European concept will be followed, because all three *Coccomyces* species seem to be synonyms of *B. jaapii*. The teleomorph was first described in N. America in 1887. In Europe, it was not observed on cultivated stone fruits until 1964, but is now very common (Burkowics 1964).

Host plants

Until 1942, *B. jaapii* was observed in Europe only on *Prunus padus*. Since that time the fungus has extended its host range to several cultivated, ornamental, and wild *Prunus* species. Sour cherry, sweet cherry and *P. mahaleb* have proved to be very susceptible. The fungus also attacks plum, apricot, *P. cerasifera*, *P. serrulata*, and *P. pennsylvanica*. Peach is resistant.

Geographical distribution

The fungus was first described at about the same time in Europe (1885) and the USA (1886). The disease is now reported from all continents and its distribution appears to coincide with the major stone-fruit growing areas where humid conditions prevail.

Disease

The disease is known as cherry leaf spot (Plate 28). Other names are yellow leaf, shot-hole disease, and leaf drop. Its characteristic symptoms on the leaves are minute purple spots on the upper surfaces of the leaves first appearing in late May to early June. The outline of the lesion is not usually well defined in its early stage. Later the area in the centre of the spot becomes necrotic and may fall out, especially on plums and apricots. The spots on the under surfaces of the leaves are brown. Following heavy dews or rains a white mildew-like growth may be observed. Also, acervuli are formed near the centre of the lesions which release conidia as a creamy mass seen as a whitish speck. The infected leaves may turn yellow, but frequently there is an area around the spots which remains green, giving the leaves a mottled appearance. The leaves may fall early in the season and complete defoliation frequently occurs in nursery stocks or young trees in the orchard. Petioles, fruits and twigs are sometimes infected.

Epidemiology

The fungus forms dark, leathery or fleshy apothecia on fallen leaves which produce ascospores that are mature by the time the first young leaves have opened. They are forcibly discharged during any rainy period from that time until 6 or 7 weeks after petal fall. The ascospores are responsible for the majority of primary infections but some of those may also be caused by conidia produced in the old acervuli on fallen leaves. The spores become attached to the surface of the leaves and germinate in a few hours if moisture is present. The fungus penetrates through stomata. Young leaves are susceptible as soon as they emerge. In contrast, fully expanded and aged leaves are less susceptible. Temperatures of 16–19°C are most favourable for symptom development. Under these conditions the first evidence of infection is as early as five days after inoculation. At lower temperatures and in the absence of rainfall or dew, symptom expression may be delayed for 10 or 15 days. Soon after the appearance of the lesions, acervuli are formed which release conidia and are responsible for secondary infections. The rate of build-up of leaf spotting is dependent on the rainfall, since conidia are mainly transported by splash and by wind-blown mist. Conidia remain viable after a long period of drying.

Economic importance

Leaf spot is a very serious disease of sweet and sour cherries especially in wet seasons and in areas where humid conditions prevail. Plums and apricots are usually not as badly affected as cherries. The disease has become more severe in Europe because ascospores are readily transported by wind and are a much more efficient inoculum for primary infections than conidia. An early epidemic leads to premature defoliation of the trees causing dwarfed, unevenly ripened fruit with insipid taste. The most serious effect of the disease occurs in the year after an extensive defoliation. Freezing and death of trees due to the lack of stored sugars leads to a reduction in fruit set, fruit size and vigour. These effects are more pronounced when the disease is prevalent for 2 or 3 years consecutively.

Control

The aim of control measures is to prevent primary and early secondary infections. Sprays should start at petal-fall and should be repeated at least twice at 10–14 day intervals. Under more severe infection conditions, four or five sprays might be necessary, the last applied immediately following harvest. As a promising alternative to this fixed time-interval spray schedule, a predictive system for timing sprays according to the infection risk has recently been developed. By using this system in combination with eradicant fungicides, fewer sprays might be necessary to control the disease. Captan, dodine, dithianon, propineb, and triforine give good leaf spot control. Promising results were also obtained with compounds of the triazole group e.g. bitertanol. Resistant cultivars are not available for either sweet or sour cherries.
SEEMÜLLER

Cenangium ferruginosum Fr. (syn. *C. abietis* (Pers.) Duby) is a weak parasite occasionally causing dieback of the lower branches of *Pinus* species, especially in poor physiological condition or after insect attacks (Lanier *et al.* 1976). In N. America, the related ***Atropellis piniphila*** (Weir) Lohman & Cash and ***A. pinicola*** Zeller & Goodding cause serious trunk and branch cankers of *Pinus contorta* and are considered as quarantine organisms for Europe (EPPO 5).

Crumenulopsis sororia **(P. Karsten) Groves**

(Syn. *Crumenula sororia* P. Karsten)
(Anamorph ***Digitisporium piniphilum*** Gremmen)

C. sororia has black apothecia with a grey hymenium, about 1–1.5 mm diameter. The asci contain colourless, (0–)3(–5)-septate ascospores usually measuring 17–23×5–6 μm (Batko & Pawsey 1964). The black, sub-globoid pycnidia are about 0.5 mm diameter. Their multicellular conidia have finger-like branches, and usually measure 30–50×3.5–5.5 μm. Isolates of *C. sororia* differ somewhat in pathogenicity. *C. sororia* has been most troublesome on *Pinus nigra* var. *maritima* and *P. contorta*, but it also affects other varieties of *P. nigra*, and *P. sylvestris, P. halepensis, P. cembra* and *P. griffithii.* Provenances of *P. contorta* differ in susceptibility, those from coastal Washington (USA) and British Columbia (Canada) are less affected than those from the Californian mountains (Hayes 1975; Stephan & Butin 1980). The fungus occurs in Finland, France, FRG, UK, Netherlands and USSR. It causes rather vaguely defined, resin-soaked cankers, chiefly on main stems. On some pine species, especially *P. nigra* var. *maritima* most cankers form around dead or dying side branches, though this is not so in *P. contorta*. The inconspicuous black apothecia and pycnidia of the fungus occur on the diseased bark. If cankers are cross-sectioned, the infected wood and bark of most pines (but not of *P. contorta*) will show a characteristic black stain.

The fungus sheds its spores mainly in August and September, canker formation probably beginning in autumn, chiefly at the nodes, and proceeding more rapidly in the following spring. Small drops of resin then appear on the bark, and as the fungus grows and spreads, more resin runs down over the trunk below as the bark splits over the rather indeterminate canker. Resin production is especially copious on *Pinus contorta.* In later stages of infection of this species, internodal cankers may also form. When cankers are well developed, death of cambial cells leads to a flattening of the stem. Small trees may be girdled and killed. In other cases, cankers may be occluded. The disease develops most rapidly after wounding, though undamaged trees may be affected. Attack is mainly confined to young, poorly grown trees suffering from nutrient deficiency, poor drainage, inadequate soil depth and structure, and exposure (Batko & Pawsey 1964). *C. sororia* is potentially damaging because it may reduce the increment of individual trees by up to 40%. So far, however, its overall effect on whole stands does not seem to have been very great, partly because it occurs mainly in crops already debilitated by other adverse factors. To avoid the disease, when growing susceptible pines, steps should be taken to ameliorate unsuitable soil conditions by the use of fertilizers, drainage and other cultural techniques. No other measures are yet available, although in future it may be possible to use resistant provenances of some species.

PHILLIPS

Diplocarpon rosae **Wolf**

(Anamorph ***Marssonina rosae*** (Lib.) Died., syn. *Actinonema rosase* (Lib.) Fr.)

D. rosae may be distinguished from related species by its occurrence on *Rosa,* to which it is confined. Species and cultivars of *Rosa* differ in susceptibility, and races of *D. rosae* differ in pathogenicity (Bolton & Svejda 1979). The origins of *D. rosae* are obscure. Its distribution is world-wide in temperate and tropical zones (CMI map 226). *D. rosae* causes blackspot disease of roses (Plate 29). Circular lesions with ragged margins, dark brown or black, up to 15 mm diameter, usually develop on the upper surface of leaflets. Close inspection reveals radiating strands of subcuticular mycelium and small, wart-like acervuli. Minute lesions may form on the stem, leaf rachis and bud scales. Mycelial strands of contiguous parallel hyphae form a network between the cuticle and epidermis. Club-shaped haustoria are formed in epidermal cells. In summer, hyaline conidia are produced in subcuticular, erumpent acervuli. They are oval to elliptical in shape (20–25×5–6 μm) and are divided by a transverse septum marked by a constriction. In autumn, oblong, single-celled, hyaline spermatia (2–3 μm long) may be formed in the same acervuli as the conidia, or separately in similar stromata.

The fungus persists saprophytically on fallen leaves, where conidia are formed in subepidermal acervuli. Apothecia sometimes form on dead leaves in spring. These consist of a subepidermal ascogenous layer and a subcuticular cover which ruptures to expose the asci. Asci are subclavate (70–80×12–18 μm) and interspersed by capitate paraphyses. They contain eight hyaline, two-celled ascospores which resemble conidia in size and shape. Ascospores are forcibly ejected. Conidia are dispersed by rain splash and possibly by insects. Under favourable conditions (15–25°C with moisture), conidia may germinate

within 9 hours. The germ tube forms an appressorium with a central infection peg that penetrates the cuticle. Acervuli may appear within nine days. The fungus overwinters on dormant buds and stems, and in dead leaves. Conidia from these sources initiate fresh infections in spring (Cook 1981). A paucity of apothecia records indicates that ascospores are an infrequent source of infection. Sprays of benomyl, propiconazole, thiophanate-methyl and triforine are effective against blackspot, if applied frequently. Loss of potential sales of roses through the hardening of attitudes to blackspot has renewed attempts to breed resistant cultivars.
ROBERTS

Diplocarpon earliana (Ell. & Ev.) Wolf (anamorph ***Marssonina fragariae*** (Lib.) Kleb.) (CMI 486) causes leaf-scorch of strawberry. Leaves are covered with dark-centred purplish blotches, on which dark shining acervuli form before the leaf shrivels. Leaf and flower stalks may be girdled, and calyces may be disfigured by infection. The disease is favoured by warm conditions (20–25°C). Apothecia and acervuli form on overwintered leaves, and both ascospores and conidia may be involved in primary infection. In view of this, perennial crops are much more liable to be damaged than strawberries grown as an annual crop (APS Compendium). The disease is of relatively minor importance in Europe, since currently used cultivars are more resistant.
SMITH

Drepanopeziza punctiformis Gremmen

(Anamorph ***Marssonina brunnea*** (Ell. & Ev.) Magnus)

D. punctiformis forms dark brown apothecia containing club-shaped asci, with ascospores 6–17×4–7 μm. The conidia, formed in pustules, measure 15–18×4–6 μm (Byrom & Burdekin 1970). *D. punctiformis* attacks poplars in the subgenera *Aegeiros* (black poplars) and *Tacamahaca* (balsam poplars). Most affected are clones of *Populus×canadensis* (*P.×euramericana*). The fungus occurs in the greater part of Europe, and in Japan and New Zealand, causing a leaf spot. Its apothecia form on both surfaces of fallen leaves, and shed their ascospores in May. These infect young poplar leaves (including their petioles), and sometimes young green twigs. They give rise to many small dark brown spots which often coalesce over the entire leaf surface. Conidial pustules form on these spots over the summer. Badly diseased leaves fall prematurely. *D. punctiformis* has caused severe defoliation of poplars in many years, especially in the Netherlands, Italy and Japan.

On a small scale, in parks and gardens, leaf spot can be controlled by raking up and destroying or composting the fallen leaves. Some control may be obtained by spraying with copper fungicides, materials of the dithane group, or with difolotan or dodine, beginning early in the season, when infection takes place (Gojković 1970). Ultimately the most effective means of control is by growing resistant clones such as *Populus trichocarpa* cv. Fritzi Pauley.
PHILLIPS

Drepanopeziza ribis (Kleb.) Höhnel

(Syn. *Pseudopeziza ribis* Kleb.)
(Anamorph ***Gloeosporidiella ribis*** (Lib.) Petrak)

D. ribis causes leaf spot or anthracnose of *Ribes* species, being common especially on blackcurrant, but occurring also on redcurrant, gooseberry and ornamental species (CMI 638). Formae speciales ***rubri, nigri*** and ***grossulariae*** have been respectively distinguished. It occurs throughout most of Europe, and also in N. America, Australasia and Japan (CMI map 187). Dark brown spots appear on the leaves from May onwards, and as they multiply may coalesce and kill large areas of leaf surface, causing premature leaf fall. Infected leaves may turn yellow, with a narrow green zone remaining around each brown spot. Spots also appear less frequently on petioles, young shoots, fruits and peduncles. Serious attacks may defoliate plants entirely. The fungus overwinters in dead leaves on which small apothecia (300–400 μm) form in spring to establish primary ascospore infection. Most spread is by splash-dispersal of conidia formed in acervuli at the centre of the leaf spots. Sporulation and dispersal are optimal at 16–24°C under moist conditions.

The disease is potentially very damaging wherever blackcurrants are grown, especially in N. Europe (from Ireland across to the USSR), but is fairly readily controlled by fungicide treatments (copper, dithiocarbamates, benzimidazoles, sterol-inhibitors) soon after flowering, and at least once more at 14-day intervals, with possibly a final treatment after harvest. Resistance to benzimidazoles has been observed, for example in Poland, where mixtures or alternation with dithiocarbamates maintain control (Nowacka *et al.* 1980). In any case, dead leaves, which are the potential source of inoculum, should

be removed and destroyed. Cultivars range from very susceptible to moderately resistant.
SMITH

Drepanopeziza sphaeroides (Pers.) Höhnel

(Anamorph ***Marssonina salicicola*** (Bresad.) Magnus)

D. sphaeroides forms small black apothecia with colourless, 1-celled ascospores 13–15×6–7 µm. The colourless, clavate or pyriform, unequally 2-celled conidia of the anamorph measure 11–19×3–7 µm. The fungus is confined to *Salix* species, mainly *Salix×sepulcralis*, but sometimes other species including *S. alba* var. *coerulea* and *S. fragilis*. It occurs throughout Europe, and in Argentina, Egypt and New Zealand. *D. sphaeroides* causes the disease described as anthracnose. Elongated, sunken, irregular cankers with a dark rim and paler centre form on the young shoots. Often purplish or black spots also appear on the leaves (which may become distorted and fall from the tree), and sometimes on the catkins. Whitish acervuli develop on the cankers and leaf spots, and apothecia form on the fallen leaves. The fungus persists over the winter mainly on the cankers, but also on fallen leaves and perhaps on bud scales. Spores are formed in spring, and infect the new growth. Wet summer weather encourages development of the disease.

In severe attacks *Salix×sepulcralis* becomes covered with unsightly cankers and leaf spots, and may lose its weeping habit, so that it no longer has any value as an ornamental tree. Only healthy shoots should be used in propagation. On small trees anthracnose may be checked by spraying with captafol or quinomethionate (Gaggini 1970), or with benomyl, maneb or mancozeb, used from bud swelling (Anselmi 1979). Ultimately, the selection and breeding of resistant trees is likely to provide the best means of control.
PHILLIPS

Drepanopeziza populi-albae (Kleb.) Nannf. (anamorph ***Marssonina castagnei*** (Desm. & Mont.) Magnus) causes a leaf spot disease of *Populus* (mainly of *P. alba*) in Europe, and in Iran and Canada. ***D. populorum*** (Desm.) Höhnel (anamorph ***Marssonina populi*** (Lib.) Magnus, syn. *M. populi-nigrae* Kleb.), which is known over much of Europe, causes leaf spots on poplars of the sections *Aigeiros* and *Tacamahaca*. It causes irregular brown spots on the leaves (cf. *D. punctiformis*) and is especially damaging to Lombardy poplar. ***D. salicis*** (Tul. & C. Tul.) Höhnel forms spots on the leaves of *Salix* species. Acervuli of its conidial state ***Monostichella salicis*** (Westend.) v. Arx form on the lesions. Another species ***D. triandrae*** Rimpau (anamorph ***Marssonina kriegeriana*** (Bresad.) Magnus) also occurs on *Salix* species (Rimpau 1962)

Gremmeniella abietina (Lagerb.) Morelet

(Syn. *Scleroderris lagerbergii* Gremmen, *Crumenula abietina* Lagerb.)
(Anamorph ***Brunchorstia pinea*** (P. Karsten) Höhnel)

Basic description

The 3-septate ascospores measure 15–22×3–5 µm, and the sickle-shaped, several-celled conidia 25–40×3–3.5 µm (CMI 369). The fungus is markedly variable. Thus in Europe, a form with long, 5–7 septate conidia up to 73 µm long, and known as ***B. pinea* var. *cembrae*** (Morelet 1980) appears to be confined to *Pinus cembra* in the Alps, while the form on other hosts, ***B. pinea* var. *pini***, has shorter, 3-septate conidia not more than 55 µm long. Growth in culture of var. *cembrae* is faster than that of var. *pini*. In parts of Europe, the fungus rarely produces apothecia. In recent years in N. America, two strains of the fungus have been identified. One, called the 'European strain', is serologically similar to the fungus in Europe, and produces no apothecia. The other, known as the 'N. American strain', forms many apothecia, and is less damaging than the 'European strain' (Skilling 1981). A third, 'Asian strain', now known from Japan (Yokota 1983), was found on *Abies* species; in some respects this resembles *B. pinea* var. *cembrae*.

Host plants

G. abietina is important mainly on pines. The varieties of *Pinus nigra* have been most affected, but in recent years *P. sylvestris* has sometimes been badly damaged. Many other pines have been attacked, though so far damage to *P. contorta* has been slight. *Picea abies* has sometimes been severely affected, and from time to time species of *Abies, Larix, Pseudotsuga* and *Tsuga* have suffered minor attack.

Geographical distribution

G. abietina occurs in Europe into the USSR, N. America and Japan (CMI map 423).

Disease

G. abietina causes a disease variously called scleroderris canker and brunchorstia dieback; these names give some indication of the range of symptoms and of the fact that in some areas the fungus is found only or mainly as the anamorph. It chiefly affects young trees (8–25 years old), but has sometimes caused serious damage to nursery plants, especially *P. sylvestris*. In many areas it occurs mainly as a dieback. Infection usually takes place in May or June, and the fungus then develops chiefly as a saprophyte in dead cells on bud scales and in the outer tissues of long shoots. It also persists as pycnidia hidden in the bark, and the so called cryptopycnidia of Cauchon & Lachange (1980). Spread to living tissues takes place in the following autumn and winter, first affecting the bases of the current year's buds, which soon show resin bleeding. More conspicuous symptoms then arise. The fungus grows upwards further into the buds, which either fail to flush or produce only thin, distorted shoots. It also grows downwards into the shoots and into the adjacent needles; the bases of these turn brown, and they finally die and fall. Many affected shoots soon show dead, leafless tips with brown needles at the base, succeeded by still healthy needles below. If the diseased shoots are cut, a yellow or greenish stain may be found around the cambium. In severe attacks, some trees may die, but in many cases new shoots grow out from below the dead shoot ends, though in later years forking or bending at these points remains to indicate earlier damage. The European strain of the fungus causes dieback throughout the crown, but damage by the N. American strain is usually confined to the lower crown only.

In some areas, especially in Scandinavia and the USA, the dieback may be accompanied by or replaced by another form of the disease, in which cankers form on the branches and stems. In Norway, where the disease is found mainly on *P. sylvestris*, the cankers may girdle and kill branches of young trees, while in the USA, long cankers on main stems may deform older trees, and younger trees may be girdled and killed. These stem cankers appear to arise when the fungus invades the trunk from infected side branches.

Epidemiology

Many factors affect the incidence of the disease, and not all of them are as yet well understood. The fungus is nearly always present either in a latent form or growing as a saprophyte on susceptible trees. Hence inoculum is usually available and the fungus can enter the parasitic phase when other factors are favourable. It then sometimes enters the tree through wounds, though it can also readily invade undamaged plants. Most infection takes place in spring and early summer, when inoculum is abundant, and the trees appear to be most susceptible. The varieties of *P. nigra* are especially affected when grown outside their natural range. Thus *P. nigra* var. *maritima* is affected in the UK (especially at high elevations), the Netherlands, N. Germany, and the Scandinavian countries.

Weather, microclimate and topographical conditions are important. Thus high humidity, high summer rainfall and low summer temperatures encourage the development of the disease, which also often develops on north-facing slopes and in valleys and depressions, or in crops shaded by adjacent older plantations. Frost damage sometimes also seems to play a part, and Yokota (1983) in Japan (where the disease occurs only in cold areas subject to deep winter snow) found that he could most easily induce infection by wounding with dry ice.

Economic importance

G. abietina seems to be an increasing threat to susceptible pines (and to a lesser extent to other conifers) in the more northerly, damper and cooler parts of Europe, as well as in eastern N. America. It is in particular inhibiting the use of the varieties of *P. nigra* in N. Europe, as in recent years damaging attacks on these have sometimes occurred unexpectedly in areas in which the crops have previously been unaffected.

Control

In nurseries the disease can be controlled by spraying with chlorothalonil, though up to seven applications may be needed (Skilling & Waddell 1974). In areas where the disease is known to occur, plantations, especially of the susceptible varieties of *P. nigra*, should be carefully sited to avoid high elevations, north-facing slopes, damp valleys and depressions, and the shading of the young crops by already estab-

lished stands. In the case of *P. sylvestris*, it may be possible to select resistant origins in some places.
PHILLIPS

Heteropatella antirrhini Buddin & Wakef. is a minor pathogen causing shot-hole disease of *Antirrhinum*. Of rather greater importance is ***H. valtellinensis*** (Traverso) Wollenw. which causes sporadic outbreaks of leaf rot in carnation and other *Dianthus* species. Other *Heteropatella* species have teleomorphs in *Heterosphaeria*.

Lachnellula willkommii (R. Hartig) Dennis

(Syn. *Trichoscyphella willkommii* (R. Hartig) Nannf., *Dasyscypha willkommii* (R. Hartig) Rehm)

Basic description

The white, hairy, stalked apothecia have an orange hymenial disc, and measure up to about 4 mm in diameter. The colourless, elliptical ascospores measure 15–28×6–9 μm. Small white waxy stromata produce hyaline spermatia measuring 2–8×1–2 μm (CMI 450).

Host plants and specialization

L. willkommii is confined to species of *Larix*. The most affected is *L. decidua,* although provenances within this species vary in their degree of susceptibility. At one extreme, most high alpine provenances are highly susceptible, while on the other hand the Carpathian larches are among the most resistant; some Sudeten provenances show both high resistance and good growth. *L. kaempferi* is generally resistant, but comparatively recently it has been badly attacked by the fungus in Japan (which suggests that the fungus may exist in more than one strain). *L.×eurolepis* (the hybrid between *L. decidua* and *L. kaempferi*) is usually resistant, and so is *L. russica* (*L. sibirica*).

Geographical distribution

L. willkommii occurs across Europe from the western seaboard eastwards into the USSR, and in Japan; it has recently been rediscovered in the eastern parts of N. America.

Disease

L. willkommii causes a severe canker of stems and branches, a disease usually called larch canker or European larch canker. Cankers first appear (usually in winter) as rounded or oval depressions. In time they become oval, stepped, blackened areas, depressed at the centre but swollen around their edges. The affected tissues usually exude resin, which runs down over the stem or branch below. The apothecia of *L. willkommii* form around the edges of the cankers. Sometimes cankers are occluded by the vigorous growth of callus tissues.

Epidemiology

Infection begins, mainly in winter, around the short and long shoots, around dormant buds, and perhaps also through leaf-scars. Having penetrated the bark, the fungus kills the cortex and cambium, so that no further cambial growth occurs in the affected area. The tissues around the diseased zone are stimulated to abnormally vigorous growth so that growth of the diseased stem or branch becomes excentric. As the fungus grows outwards from the infection centre, the host reacts by the production of a cork barrier. This prevents further extension of the canker in the active growing season of the tree. The fungus, however, is capable of growth at low temperatures, and in winter it usually proceeds through or round the cork wall, and destroys a further ring of cambium. The canker therefore exhibits annual growth rings, and is surrounded by swollen tissue, on which the fungus produces its conidial pustules, which generally arise in the first half of the year. They are succeeded by the apothecia, which may subsequently be found at any time of the year.

It has proved difficult to define the factors affecting the incidence of larch canker. In certain years, canker initiation reaches epidemic proportions, but the factors common to these years remain obscure. Ascospore discharge is associated with rainfall, and occurs at any time of the year if the temperature is above freezing (Sylvestre-Guinot 1981). Germination takes place only when the RH exceeds 92% (Ito *et al.* 1963), and there is some evidence that canker is especially common on damp sites. High March temperatures may encourage the disease (Leibundgut *et al.* 1964), and years with a high incidence of canker seem in some areas to be ones with high adelgid infestations. Alternating cold and warm temperatures in winter appear to increase the extension of cankers.

Economic importance

Larch canker causes severe damage chiefly on *Larix decidua*, on which it is second in importance only to the root and butt rot caused by *Heterobasidion annosum*. On other species it is usually of little importance, though it sometimes affects *L.* × *eurolepis* to a lesser degree. On *L. decidua,* the degree of damage varies from site to site, and with the provenance. The amount of loss varies also with the position of the cankers, for if these are mostly confined to the branches, loss will be small, whereas if many are found on the trunks of the trees, the value of the stand will be greatly reduced.

Control

In Europe, if *L. kaempferi* or *L.* × *eurolepis* are grown, larch canker is unlikely to cause significant loss. On some sites it may be impossible to grow many provenances of *L. decidua* without serious risk of canker. The disease is then best avoided by growing a resistant provenance.
PHILLIPS

Marssonina panattoniana (Berl.) Magnus

M. panattoniana produces club-shaped, uniseptate, slightly curved conidia (12–18×3–5 μm) borne on a short sporophore, but there is some doubt about its position in the genus *Marssonina* (Moline & Pollack 1976). The fungus causes ringspot, anthracnose or rust on lettuce, fire on chicory and endive and can infect related Compositae including weeds. Host specialization is suspected since isolates from lettuce are apparently not pathogenic on chicory and vice versa but detailed information is lacking (Moline & Pollack 1976). The disease is found in most temperate regions where the host crops are extensively cultivated (CMI map 82). The first signs of disease on lettuce are small water-soaked spots on the underside of leaves or petioles which turn yellow and finally brown. These spots range from pin-pricks to 5 mm in diameter and may be round or angular. On the surfaces of petioles and larger veins, lesions are more irregular and sunken, turning a reddish-brown and comparable to slug-damage. Under conditions of high humidity, whitish-pink sporulation may be visible at the margins of lesions. Later, the centres of leaf spots may dry and be lost so that the leaves become perforated. In severe cases, heart leaves may also become affected. The disease causes problems only infrequently; losses result from death or stunting if plants are infected when young, from a reduction in crop quality as a consequence of the need to trim off infected outer leaves and from deterioration of heads during shipment. Overwintered field crops and those raised in unheated glasshouses are most at risk. Cool, wet, poorly drained conditions favour infection. The primary source of inoculum is leaf debris but composite weeds may also play a role and seed infection has been demonstrated. Plant to plant spread occurs by means of conidia which are dispersed by large-water droplets and in fine mists. Routine crop rotations and soil sterilization play an important part in preventing disease outbreaks but for crops at risk, regular foliar applications of maneb, thiram, captafol or captan provide an adequate prophylactic treatment.
CRUTE

Pezicula alba Guthrie and *P. malicorticis* (H. Jackson) Nannf.

(Syn. *Neofabraea perennans* Kienholz)

Basic description

Both species are mostly known as their anamorphs. For *P. alba*, this has often been wrongly known as *Gloeosporium album* Osterwalder. In fact, typical pycnidia are formed (Bompeix 1973), correctly named as ***Phlyctaena vagabunda*** Desm. (syn. *Trichoseptoria fructigena* Maubl.). For *P. malicorticis*, the anamorph is ***Cryptosporiopsis curvispora*** (Peck) Gremmen (syn. *Gloeosporium perennans* Zeller & Childs).

Host plants

Both species principally attack apple and are rarely found on pear or quince (Wormald 1955). They also occur on many wild species (von Arx 1958).

Geographical distribution

In the USA and Canada, *P. alba* is much less frequent than *P. malicorticis*, but in Europe *P. alba* is most often reported (Belgium, Finland, FRG, Italy, Netherlands, Poland, Sweden, Switzerland). It is particularly important in the UK (on cv. Cox's Orange Pippin) and in France on cv. Golden Delicious (both are very susceptible). *P. malicorticis* also occurs in Australia, New Zealand and Zimbabwe (CMI map 128).

Disease

On the vegetative parts of the apple tree, *P. alba* is purely saprophytic, while *P. malicorticis* is able to attack and destroy young shoots and cause minor cankers. These attacks are often difficult to detect in the orchard, and cause only negligible damage in Europe. Penetration and establishment of the two species occurs much as for *Nectria galligena,* mainly involving wounds or leaf and peduncle scars (leaf-fall, picking). The main importance of the *Pezicula* species is their attack on apples, whether they are on the tree or fallen, or on cold-stored fruit. The point of entry is almost always a lenticel. Typical symptoms consist of perfectly concentric brown soft lesions, centred on a lenticel, with a more or less discoloured central part (especially for *P. malicorticis*). Tissues are rotted essentially by cell-wall degrading enzymes; no toxins have been reported. The fungi are not capable of rotting immature fruit tissues. Fruit may therefore carry latent infections (Edney 1958) which only develop in store several months later (mostly in February, or a little earlier for *P. malicorticis*). The growth rate in fruit (cv. Golden Delicious) is very low (0.7–1.4 mm per day at 0°C) but it rises as the fruits ripen. Controlled atmospheres (e.g. 3% O_2 – 5% CO_2) hardly affects the growth of the *Pezicula* species *in vitro* or *in vivo*, but do reduce the number of lesions appearing, especially for *P. alba* (Bompeix 1978). This effect is only transitory, however, since more lesions appear when the fruit is removed from the controlled atmosphere.

Epidemiology

Cankers on the branches provide a permanent source of inoculum in the orchard. Pycnidia (*P. alba*) or acervuli (*P. malicorticis*) release conidia by water splash over relatively short distances. Apothecia are rare in Europe and presumably have only a minor role; conidia but no ascospores were found in spore trapping over several years. Conidia are principally released in autumn and winter, with a minimum in July. Fruits are mainly infected in the weeks immediately before harvest. Fruits are only infected on the tree. Even though pycnidia or acervuli may be formed on rotted tissue in store, there is no spread from fruit to fruit. This contrasts with *Penicillium* species, *Botryotinia fuckeliana* or *Rhizopus stolonifer*. However, any processing which involves dipping in water could spread the conidia.

Economic importance

Until 1968, storage losses were very high and the 'Gloeosporiums' were the main problem in fruit stores. Rotting increased from February to April, and could reach 88%, losses of 30–60% being quite common. Adequate fungicides have reduced these losses, but damage is still frequently recorded throughout Europe, usually because of poorly applied treatments. The appearance of benzimidazole-insensitive strains (cf. below) is liable to cause further problems. The Mediterranean area is generally less affected. One of the most widely planted cultivars, Golden Delicious, is very susceptible; on the whole, red cultivars are more resistant.

Control

Pre-harvest treatments with benzimidazole fungicides now provide good control: thiabendazole, benomyl, carbendazim, thiophanates (the first two are also registered for post-harvest use in France). Orchard treatments are effective in controlling fruit rot, but do not eradicate the pathogens from the trees (Bompeix & Morgat 1969). They create a risk of fungicide resistance in the natural population. Since, in any case, losses in the orchard are negligible, post-harvest treatment under controlled conditions is preferable (Bompeix & Morgat 1969). Simple dipping, or drenching, are the techniques most often used. Thiabendazole and benomyl are mostly used at 500–1000 mg/l, but new formulations may be as effective at lower concentrations. The obvious advantages of post-harvest treatment are better control over treatment conditions, and the elimination of any risk of appearance of resistant strains.

Special research interest

The two *Pezicula* species have served as models for the study of latent infection in fruits. Like *N. galligena,* they stimulate immature apple fruits to form the extremely simple phytoalexin benzoic acid.
BOMPEIX

Pezicula corticola (C. Jørgensen) Nannf. causes a minor surface canker of apple, and infrequently, pear.

Phacidium coniferarum (Hahn) Di Cosmo

(Syn. *Potebniamyces coniferarum* (Hahn) Smerlis, *Phacidiella coniferarum* Hahn)
(Anamorph ***Phacidiopycnis pseudotsugae*** (Wilson) Hahn, syn. *Phomopsis pseudotsugae* Wilson)

P. coniferarum produces ascocarps in erumpent black stromata (Hahn 1957). The pycnidia of the much commoner anamorph contain colourless spores measuring 4.5–8.5×2–4 μm (Wilson 1921). *P. coniferarum* occurs mainly on Douglas fir (*Pseudotsuga menziesii*, especially var. *viridis*), but may also affect species of *Cedrus, Larix, Pinus* and *Tsuga*. It is present throughout W. and C. Europe into the USSR, and in the Pacific Northwest of the USA, and New Zealand (CMI map 320). On young Douglas fir it causes dieback of shoots and girdles young stems. Downwards extension of the fungus on affected shoots is eventually arrested by a layer of cork. The healthy stem below continues growth, and a clear junction develops between the wider living stem and the narrower dead shoot above. Spread from lateral shoots may lead to girdling of the main stem, and death of the shoot above the girdled zone. Badly affected trees then die, and survivors become bush-like in growth. On older trees (up to about 25 years old), oval cankers may develop on the trunks when the fungus grows in from lateral shoots or colonizes wounds. On Japanese larch, the fungus causes resinous cankers. Pycnidia of the fungus, and often the fructifications of other fungi, usually form in the dead bark.

P. coniferarum is a weak wound parasite which enters plants mainly in the dormant season through wounds made in nursery cultivation and in pruning and brashing, and perhaps through leaf-scars. The overall damage caused by *P. coniferarum* is relatively small, though occasional severe losses arise in nurseries and young plantations, in which up to half the trees may be destroyed. Cankers on older trees lead only to a small loss in the timber value. To guard against infection, nurseries should be sited away from affected plantations. Any diseased trees in nurseries and young plantations should be removed and burnt.
PHILLIPS

Phacidium infestans P. Karsten

P. infestans forms both epiphyllous and hypophyllous apothecia which are intrahypodermal and lens-shaped. The hymenium is uncovered by radial splitting of the overlying epidermis. The ascospores are hyaline, one-celled, oval-elongate, and slightly flattened on one side (CMI 652). *P. abietis* (Dearness) Reid & Cain, from N. America and Japan, and *P. pini-cembrae* (Rehm) Terrier from the Alps (CMI 653) are here included in *P. infestans*.

The main host is *Pinus sylvestris*, but attacks are found on many pine species, both the native *P. cembra, P. sibirica*, and *P. mugo* and to some extent the introduced *P. contorta*. Attacks on seedlings of *Picea abies* have been recorded from the USSR (Kossinskaja 1974), while in Norwegian nurseries, damaging attacks have been found on *Abies concolor, A. grandis, A. procera, A. nordmanniana*, and *Picea pungens*. Mature apothecia mostly do not form on these species. The form known as *P. abietis* causes snow mould on *Abies* species and on *Pseudotsuga* in N. America (Reid & Cain 1962), and on *Abies sachalinensis* in Japan, but it has apparently not been found on *Pinus* species, thus clearly differing from the European *P. infestans* as to host spectrum.

The disease caused by *P. infestans* is a typical snow mould. The spores are discharged from the asci in late autumn after the first frost, in cool moist periods, often when snow falls and melts again before there is a lasting snow cover. Dead infected needles, often broken and spread by wind, are also infection sources. The mycelium can grow relatively fast at 0°C and even develop at −5°C; but it is extremely sensitive to drought, and therefore develops just under snow cover. There is usually some growth even in the middle of winter, but fastest growth takes place later in winter, when the snow starts to melt and the temperature in the snow is around 0°C or even a little higher in cavities in the snow and especially in needles absorbing sunlight that penetrates from above. In the snow the mycelium is white or light grey. It grows from needle to needle, from branch to branch, and even short distances through the snow from plant to plant. The hyphae infect and kill the needles. The killed needles will keep a dirty green or olive-greenish colour as long as they are inside the snow. But after the snow has melted, the needles will soon be brown or reddish-brown. On the brown needles tiny brownish, later darker, spots will usually be found: the young developing apothecia (Fig. 64). During the summer the needles

Fig. 64. Apothecia of *Phacidium infestans* on last season's needles of *Pinus sylvestris*, killed last winter.

are bleached and become a grey colour, and the unripe apothecia are more conspicuous. In late autumn the asci and ascospores are ripe, and the hymenium is uncovered by radial splitting of the epidermis. Within great areas of northern and alpine Europe *P. infestans* causes more damage to pine species, especially *P. sylvestris,* but also *P. cembra* and *P. sibirica,* than any other fungus. The greater part of the regeneration may be killed before the young trees reach above the snow cover. The fungus may also cause great economic losses in nurseries where control measures have not been carried out; numerous, often merging, plots with dead plants may develop during the winter.

The fungus is fairly easy to control in nurseries by spraying in late autumn before there is a lasting snow cover, with, for example, lime sulphur or a dithiocarbamate such as maneb. Out in the forest the best control measure is the use of resistant trees, for example spruce on northern slopes with long-lasting snow cover, or planting of the more resistant N. American *P. contorta,* instead of *P. sylvestris.* Small regeneration plots, where the snow cover is higher and lasts longer than elsewhere, should be avoided. Northern provenances of *P. sylvestris* have been found to be more resistant to *P. infestans* than more southerly ones.

ROLL-HANSEN

Pseudopeziza jonesii (syn. *Leptotrochila medicaginis* (Fuckel) Schüepp; anamorph ***Sporonema phacidioides*** Desm.) causes yellow leaf blotch of lucerne (APS Compendium) causing losses much like *P. medicaginis* (q.v.). Fan-shaped chlorotic lesions form between the veins, bearing first pycnidia (spermatia?) then black apothecia on a central blackish stroma. The disease is more important in N. America than in Europe, where its distribution is apparently rather restricted (CMI map 141).

Pseudopeziza medicaginis (Lib.) Sacc. causes common leaf spot or brown spot of lucerne. Leaflets show small circular brown to black spots with a characteristic fringed margin, at the centre of which, on the lower surface, one or several small 1-mm apothecia form. Ascospores discharged from these re-infect leaves; there is no conidial stage (CMI 637; APS Compendium; O'Rourke 1976). Formae speciales have been distinguished on lucerne (**f.sp. *medicaginis-sativae*** Schmiedeknecht) and *Medicago lupulina* (**f.sp. *medicaginis-lupulinae*** Schmiedeknecht). A similar species ***P. trifolii*** (Bivona-Bernardi) Fuckel (with formae speciales ***trifolii-pratensis*** Schüepp on *Trifolium pratense* and ***trifolii-repentis*** Schüepp on *T. repens* and other *Trifolium* species) is of rather less importance on clovers (CMI 636). Schüepp (1959) regarded all of these fungi, together with ***P. meliloti*** Sydow, as forms of *P. trifolii.*

P. medicaginis can cause serious losses on lucerne in N. (UK, Ireland) and E. Europe. The fungus persists on fallen leaves and debris, or else in lesions on the host, and spreads especially in cool wet weather favouring infection. Ascospores are naturally discharged as pairs within a common membrane, and infect much more readily in pairs. Infected leaves eventually shrivel and dry. Losses arise from defoliation, reduced quality and increased oestrogen content (Morgan & Parbery 1980). Early harvesting avoids losses. Fungicides can be used but are doubtfully economic. Cultivars vary in resistance (Barrière 1977), those of the N. European-Flamande type being attacked less than the Grimm type (Carr 1971).

CARR & SMITH

Pseudopeziza tracheiphila Müller-Thurgau

P. tracheiphila attacks *Vitis* species (*V. labrusca, V. riparia, V. berlandieri, V. vinifera*) as well as *Parthenocissus* and related genera. Some cultivars of *V. vinifera* are very susceptible notably Chasselas, Carignan, Pinot, Chardonnay, Müller-Thurgau, Gamay, Riesling, Aramon and Sylvaner. The disease is mainly important in N. European vineyards, mainly in Germany and France (Sondey & Moncomble 1983).

P. tracheiphila particularly attacks the foliage giving rise to symptoms referred to as 'rot brenner' or 'rougeot parasitaire'. The bunches are very rarely damaged directly. On leaves, the disease first appears in the shape of livid spots, like downy mildew oil spots. These enlarge and colour very quickly, red for red cultivars and yellow for the white ones. Very often spots appear only between the principal and secondary veins. The fungus overwinters in dead fallen leaves on the soil as mycelium which continues to grow saprophytically during spring, when apothecia develop on the leaf surface. They contain asci with eight ascospores. The first ascospores are generally mature at the time of bud burst. Production of spores can continue until fruit set and sometimes longer. The spores are wind-dispersed and germination takes place in free water on the young leaves. Infection needs 16 h (temperature 20°C) rising to 48 h (temperature 15°C) with a period of rain or high humidity of 2–4 days. The incubation period is long (3–5 weeks).

Severe infections lead to premature defoliation which weakens the vines. Indirect effects on the fruit, bunch withering, and poor ripening lead to serious losses in some regions. Control is by protective spraying, but no product has a sufficient curative action. The risk of infection exists from bud burst, hence there is a need for early application. The method chosen in France and particularly in Champagne consists of treatment after the expansion of every third leaf (3rd, 6th, 9th, etc.), with an extra treatment for heavy rainfall (about 20 mm). The most efficient purely protectant fungicides are copper or dithiocarbamate products (mancozeb is mainly used in France). Carbendazim-generating compounds are becoming increasingly popular due to their greater rain fastness. Anti-oomycete products (cymoxanil, fosetyl-Al, metalaxyl and ofurace) have little effect on the pathogen.

MONCOMBLE

Pyrenopeziza brassicae B. Sutton & Rawlinson

(Anamorph ***Cylindrosporium concentricum*** Grev.)

P. brassicae is characterized by white subcuticular acervuli with aseptate, hyaline, cylindrical conidia formed only on *Brassica* species. Mature apothecia are rare on host debris, although readily obtained in culture. The fungus is heterothallic, with two mating types. All major forms of cultivated brassicas are potential hosts (Rawlinson *et al.* 1978). Considerable heterogeneity of resistance is found between *Brassica* varieties, subspecies, cultivars and individuals. The disease is distributed through much of N. Europe with infrequent records of attack on vegetable brassicas elsewhere (CMI map 193). However, since 1974 it has occurred regularly, with occasional severe attacks, on winter rape in Britain and since 1983 in France. It grows biotrophically beneath the cuticle and later intercellularly, until mature acervuli rupture the cuticle exposing white conidia, sometimes grouped in concentric circles on leaves. On rape these spread to give a mealy appearance. Leaves then yellow and lesions become bleached; those formed during leaf expansion can cause gross distortion and eventual withering. Acervuli can occur in winter on apical tissues deep within the crown of rosette-stage plants. In spring young unexpanded leaves, bracts, pedicels and buds on primordial branches can be infected long before symptoms become readily visible during stem extension. Infection spreads to stems causing superficial fawn lesions with black speckling at the edges and later splitting of the epidermis. Severe infection can kill or stunt susceptible plants in winter and damage buds, flowers and pods in spring.

The main source of inoculum is conidia splashed and dispersed in wind-borne droplets from infected plants and crop debris; conidia can also be carried on seed and rarely in acervuli beneath the seed coat. Survival of conidia on dried debris is possible for at least 10 months (Maddock & Ingram 1981). Free water is necessary for efficient germination and penetration. Spread in rape crops is favoured by mild, wet weather in autumn and spring; herbicides which remove leaf wax increase disease; dry conditions and high temperatures in spring usually check disease development. Economic losses in rape arise from loss of photosynthetic area over winter, and fewer pods because of damaged buds or premature splitting of infected pods. Control is by early application of fungicides (e.g. benomyl, prochloraz); late applications may be effective but are often uneconomic. With its prevalence on the rapidly expanding area of winter rape in Europe, *P. brassicae* is no longer an infrequent minor pathogen. Its apparent state of balanced co-evolution with many brassica hosts may change with this large-scale agricultural exploitation of one host. Increased selection pressure on the pathogen population could provide a unique opportunity to study host–parasite evolution.

RAWLINSON

REFERENCES

Agulhon R. (1973) Quelques aspects de la lutte contre la pourriture grise en 1972. *Vignes et Vins* **220,** 5–11.

Anderson H.W. (1956) *Diseases of Fruit Crops.* McGraw-Hill, London.

Anselmi N. (1979) La 'bronzatura' del salice causata da *Marssonina salicicola* (Bres.) Magn. *Cellulosa e Carta* **30,** 3–19.

APS (1979) Symposium on *Sclerotinia* (= *Whetzelinia*): taxonomy, biology and pathology. *Phytopathology* **69,** 873–910.

Arx J.A. von (1958) The 'Gloeosporiums' of stone fruit. *Phytopathologische Zeitschrift* **33,** 108–114.

Barrière Y. (1977) Contaminations artificielles de la luzerne avec des cultures pures de *Pseudopeziza medicaginis.* Classement de 24 cultivars suivant leur résistance au parasite. *Annales de Phytopathologie* **9,** 33–43.

Batko S. & Pawsey R.G. (1964) Stem canker of pine caused by *Crumenula sororia. Transactions of the British Mycological Society* **47,** 257–261.

Bolton A.T. & Svejda F.J. (1979) A new race of *Diplocarpon rosae* capable of causing severe black spot on *Rosa rugosa* hybrids. *Canadian Plant Disease Survey* **59,** 38–40.

Bompeix G. (1973) Ecologie du *Pezicula alba* sur *Pyrus malus* en France. *Fruits* **28,** 757–773; 863–886.

Bompeix G. (1978) The comparative development of *Pezicula alba* and *P. malicorticis* on apples and *in vitro* (air and controlled atmosphere). *Phytopathologische Zeitschrift* **91,** 97–109.

Bompeix G. & Morgat F. (1969) Lutte chimique contre les pourritures des pommes en conservation: efficacité du bénomyl et du thiabendazole. *Comptes Rendus des Séances de l'Académie d'Agriculture de France* **55,** 776–783.

Bulit J. & Dubos B. (1982) Epidémiologie de la pourriture grise. *Bulletin OEPP/EPPO Bulletin* **12,** 37–48.

Bulit J. & Lafon R. (1977) Observations sur la contamination des raisins par le *Botrytis cinerea* Pers. In *Travaux dédiés à Georges Viennot-Bourgin,* pp. 61–69. Société française de phytopathologie, Paris.

Burdekin D.A. & Phillips D.H. (1970) Chemical control of *Didymascella thujina* on western red cedar in forest nurseries. *Annals of Applied Biology* **67,** 131–136.

Burkowics A. (1964) *Blumeriella jaapii* (Rehm) von Arx on cultivated stone fruits in Poland. *Phytopathologische Zeitschrift* **51,** 419–424.

Butin H. (1983) *Krankheiten der Wald- und Parkbäume.* Georg Thieme Verlag, Stuttgart.

Byrde R.J.W. & Melville S.C. (1971) Chemical control of blossom wilt of apple. *Plant Pathology* **32,** 48–50.

Byrde R.J.W. & Willetts H.J. (1977) *The Brown Rot Fungi of Fruit: their Biology and Control.* Pergamon Press, Oxford.

Byrom N.A. & Burdekin D.A. (1970) *Drepanopeziza punctiformis.* British records no. 101. *Transactions of the British Mycological Society* **54,** 139–141.

Carr A.J.H. (1971) Herbage legumes. In *Diseases of Crop Plants* (Ed. J.H. Western). Macmillan, London.

Cauchon R. & Lachange D. (1980) Recherche de cryptopycnides, pour un diagnostic précoce de *Gremmeniella abietina. Canadian Journal of Plant Pathology* **2,** 232–234.

Cline M.N., Crane J.L. & Cline S.D. (1983) The teleomorph of *Cristulariella moricola. Mycologia* **75,** 988–994.

Coley-Smith J.R., Verhoeff K. & Jarvis W.R. (1980) *The Biology of Botrytis.* Academic Press, London.

Cook R.T.A. (1981) Overwintering of *Diplocarpon rosae* at Wisley. *Transactions of the British Mycological Society* **77,** 549–556.

Cruikshank R.H. (1983) Distinction between *Sclerotinia* species by their pectic enzymes. *Transactions of the British Mycological Society* **80,** 117–119.

Doornik A.W. & Bergman B.H.H. (1974) Infection of tulip bulbs by *Botrytis tulipae* originating from spores or contaminated soil. *Journal of Horticultural Science* **49,** 203–207.

Dubos B., Bulit J., Bugaret Y. & Verdu D. (1978) Possibilités d'utilisation du *Trichoderma viride* Pers. comme moyen biologique de lutte contre la pourriture grise (*Botrytis cinerea* Pers.) et l'excoriose (*Phomopsis viticola* Sacc.) de la vigne. *Comptes Rendus des Séances de l'Académie d'Agriculture de France* **64,** 1159–1168.

Edney K.L. (1958) Observations on the infection of Cox's Orange Pippin apples by *Gloeosporium perennans. Annals of Applied Biology* **46,** 622–629.

Entwistle A.R. & Munasinghe H.L. (1981) The effect of seed and stem base spray treatment with iprodione on white rot disease (*Sclerotium cepivorum*) in autumn-sown salad onions. *Annals of Applied Biology* **97,** 269–276.

Esler G. & Coley-Smith J.R. (1983) Flavour and odour characteristics of species of *Allium* in relation to their capacity to stimulate germination of sclerotia of *Sclerotium cepivorum. Plant Pathology* **32,** 13–22.

Gaggini J.B. (1970) Anthracnose of weeping willows. *Gardeners' Chronicle* **168,** 13.

Galet P. (1977) *Les Maladies et les Parasites de la Vigne,* Tome I. Paysan du Midi, Montpellier.

Cartel W. (1970) Potentialités de *Botrytis cinerea* en tant que parasite de la vigne, notamment en ce qui concerne l'infection et l'incubation. *Weinberg und Keller* **17,** 15–52.

Gaunt R.E. (1983) Shoot diseases caused by fungal pathogens. In *The Faba Bean* (Vicia faba *L.*). *A Basis for Improvement* (Ed. P.D. Hebblethwaite), pp. 463–492. Butterworth, London.

Gerlach W. (1961) Über die *Botrytis*-Wurzelstockfäule der *Iris* und ihr Vorkommen in Deutschland. *Nachrichtenblatt des Deutschen Pflanzenschutzdienstes* **13,** 7–9.

Gilpatrick J.D. (1982) Case study 2: Venturia of pome fruits and Monilinia of stone fruits. In *Fungicide Resistance in Crop Protection* (Eds J. Dekker & S.G. Georgopoulos), pp. 195–206. PUDOC, Wageningen.

Gojković G. (1970) Chemical protection of poplars against *M. brunnea* in Jugoslavia. *Topola* **14,** 3–68.

Hahn G.G. (1957) *Phacidiopycnis (Phomopsis)* canker and dieback of conifers. *Plant Disease Reporter* **41,** 623–633.

Harrison J.G. (1979) Overwintering of *Botrytis fabae. Transactions of the British Mycological Society* **72,** 389–394.

Harrison J.G. (1981) Chocolate spot of field beans in Scotland. *Plant Pathology* **30,** 111–115.

Harrison J.G. (1983) Survival of *Botrytis fabae* in air. Distinguishing between lesions caused by *Botrytis fabae* and *B. cinerea* on field bean leaves. *Transactions of the British Mycological Society* **80,** 263–269; **81,** 663–664.

Harrison J.G. (1984) Effect of humidity on infection of field bean leaves by *Botrytis fabae* and on germination of conidia. *Transactions of the British Mycological Society* **82,** 245–248.

Hayes A.J. (1975) The mode of infection of Lodgepole pine by *Crumenula sororia* Karst. and the susceptibility of different provenances to attack. *Forestry* **48,** 99–113.

Ito K., Zinno Y. & Kobayashi T. (1963) Larch canker in Japan. *Bulletin of the Government Forest Experiment Station Tokyo* **155,** 23–47.

Jacob M. (1969) Der Einfluss unterschiedlicher Trocknungs- und Lagerverhältnisse auf die Wirkung einer Feuchtbeizung von Gladiolenpflanzgut. *Nachrichtenblatt des Deutschen Pflanzenschutzdienstes* **23,** 94–98.

Jalaluddin M. (1967) Studies on *Rhizina undulata.* I. & II. *Transactions of the British Mycological Society* **50,** 449–459; 461–472.

Jalkanen R. (1981) *Lophodermella sulcigena* on pines. A literature review *Folia Forestalia* no. 476, 1–15.

Jarvis W.R. (1977) *Botryotinia* and *Botrytis* species:

taxonomy, physiology and pathogenicity: a guide to the literature. *Canada Department of Agriculture Monograph* no. 15.

Karadzic D. (1981) Infection of *Pinus sylvestris* by *Naemacyclus minor.* In *Current Research on Conifer Needle Diseases, Proceedings of the IUFRO Working Party on Needle Diseases, Sarajevo 1980* (Ed. C.S. Millar). Aberdeen.

Kistler B.R. & Merrill W. (1978) Etiology, symptomology, epidemiology and control of *Naemacyclus* needle cast of Scotch pine. *Phytopathology* **68,** 267–271.

Klingström A. & Lundeberg G. (1978) Control of *Lophodermium* and *Phacidium* needle cast and *Scleroderris* canker in *Pinus sylvestris. European Journal of Forest Pathology* **8,** 20–25.

Kossinskaja I.S. (1974) *Facidioz Sosny (Phacidiosis in Pinus sylvestris).* Novosibirsk.

Lamarque C. (1983) Conditions climatiques nécessaires à la contamination du tournesol par *Sclerotinia sclerotiorum;* prévision des épidémies locales. *Bulletin OEPP/EPPO Bulletin* **13,** 75–78.

Langcake P. & McCarthy W.V. (1979) The relationship of resveratrol production to infection of grapevine leaves by *Botrytis cinerea. Vitis* **18,** 244–253.

Lanier L., Joly P., Bondoux P. & Bellemère A. (1976) *Mycologie et Pathologie Forestières.* Masson, Paris.

Leibundgut H., Dafis S. & Bezançon M. (1964) Etudes sur diverses provenances de melèze européen (*Larix decidua* L.) et la variabilité de leur infection par le chancre du melèze (*Dasyscypha willkommii* Hart.). *Schweizerische Zeitschrift für Forstwesen* **115,** 255–260.

Lumsden R.D. & Dow R.L. (1973) Histopathology of *Sclerotinia sclerotiorum* infection of bean. *Phytopathology* **63,** 708–715.

Maddock S.E. & Ingram D.S. (1981) Studies of survival and longevity of the light leaf spot pathogen of brassicas, *Pyrenopeziza brassicae. Transactions of the British Mycological Society* **77,** 153–159.

Mansfield J.W. (1982) The role of phytoalexins in disease resistance. In *Phytoalexins* (Eds J.A. Bailey & J.W. Mansfield), pp. 253–288. Blackie, Glasgow.

Mathon B. (1982) Les dépérissements des cyprès et des thujas dus à deux champignons. *Revue Horticole* no. 228, 41–44.

Maude R.B. (1983) Onions. In *Post-harvest Pathology of Fruits and Vegetables* (Ed. C. Dennis), pp. 73–101. Academic Press, New York.

Maude R.B. & Presly A.H. (1977) Neck rot (*Botrytis allii*) of bulb onions. I. Seed-borne infection and its relationship to the disease in the onion crop. II. Seed-borne infection in relationship to the disease in store and the effect of seed treatment. *Annals of Applied Biology* **86,** 163–180; 181–188.

Minter D.W. (1981) *Lophodermium on pines. CMI Mycological Papers* no. 147. CMI, Kew.

Minter D.W. & Millar C.S. (1980) Ecology and biology of three *Lophodermium* species on secondary needles of *Pinus sylvestris. European Journal of Forest Pathology* **10,** 169–181.

Mitchell C.P., Williamson B. & Millar C.S. (1976) *Hendersonia acicola* on pine needles infected by *Lophodermella sulcigena. European Journal of Forest Pathology* **6,** 92–102.

Moline H.E. & Pollack F.G. (1976) Conidiogenesis of *Marssonina panattoniana* and its potential as a serious post-harvest pathogen of lettuce. *Phytopathology* **66,** 669–674.

Molot B., Agulhon R. & Boniface J.C. (1983) Application d'un modèle à la lutte contre *Botrytis cinerea* sur vigne. *Bulletin OEPP/EPPO Bulletin* **13,** 271–279.

Moore W.C. (1949) *Diseases of Bulbs.* MAFF Bulletin no. 117. HMSO, London.

Morelet M. (1980) La maladie à Brunchorstia. 1. Position systématique et nomenclature du pathogène. *European Journal of Forest Pathology* **10,** 268–277.

Morgan W.C. & Parbery D.G. (1980) Depressed fodder quality and increased oestrogenic activity of lucerne infected with *Pseudopeziza medicaginis. Australian Journal of Agricultural Research* **31,** 1103–1110.

Morton H.L. & Miller R.E. (1982) Chemical control of *Rhabdocline* needle cast of Douglas-fir. *Plant Disease* **66,** 999–1000.

Nicholls T.H. & Skilling D.D. (1974) Control of *Lophodermium* needle-cast disease in nurseries and Christmas tree plantations. *USDA Forest Service Research Paper* NC-110, 11 pp.

Nowacka H., Cimanowski J., Karolozak W. & Puchaia Z. (1980) (The use of thiophanate-methyl derived fungicides for the control of anthracnose of currant.) *Prace Instytutu Sadownistwe i Kwiaciarstwa w Skierniewicach A* **22,** 129–135.

Nyerges P., Szabo E. & Donko E. (1975) The role of anthocyan and phenol compounds in the resistance of grapes against *Botrytis* infection. *Acta Phytopathologica Academiae Scientiarum Hungaricae* **10,** 65–78.

O'Neill T.M. & Mansfield J.W. (1982) The cause of smoulder and infection of narcissus by species of *Botrytis. Plant Pathology* **31,** 65–78.

O'Rourke C.J. (1976) *Diseases of Grasses and Forage Legumes in Ireland.* An Foras Talúntais, Carlow.

Ouwehand J. (1981) Wortelrot in hyacinten: geen probleem meer. *Bloembollencultuur* **91** (51), 1354–1355.

Pawsey R.G. (1963) Rotation sowing of *Thuja* in selected nurseries to avoid infection by *Keithia thujina. Quarterly Journal of Forestry* **56,** 206–209.

Pepin R. (1980) Le comportement parasitaire de *Sclerotinia tuberosa* (Hedw.) Fuckel sur *Anemone nemorosa* L. Étude en microscopie photonique et électronique à balayage. *Mycopathologia* **72,** 89–99.

Pommer E.H. & Lorentz G. (1982) Resistance of *Botrytis cinerea* Pers. to dicarboximide fungicides—a literature review. *Crop Protection* **1,** 221–230.

Popushoi I.S. (1971) *Mikoflora Plodovykh Derev'ev SSSR (The Mycoflora of Fruit Trees in USSR).* Nauka, Moscow.

Price D. (1970) Tulip fire caused by *Botrytis tulipae* (Lib.) Lind; the leaf spotting phase. *Journal of Horticultural Science* **45,** 233–238.

Price K. & Colhoun J. (1975) A study of variability of isolates of *Sclerotinia sclerotiorum* (Lib.) de Bary from different hosts. *Phytopathologische Zeitschrift* **83,** 159–166.

Rack X. (1959) Beziehungen zwischen Infektionsdichte und Nadelverlust bei der Kiefernschütte. *Nachrichtenblatt des Deutschen Pflanzenschutzdienstes* **12,** 177–181.

Rawlinson C.J., Sutton B.C. & Muthyalu G. (1978) Taxonomy and biology of *Pyrenopeziza brassicae* sp.

nov. (*Cylindrosporium concentricum*), a pathogen of winter oilseed rape (*Brassica napus* ssp. *oleifera*). *Transactions of the British Mycological Society* **71,** 425–439.

Raynal G. (1981) La sclérotiniose des trèfles et luzernes à *Sclerotinia trifoliorum*. I. Choix d'une méthode de contamination artificielle. II. Variabilité du parasite, résistance des plantes en conditions contrôlées. *Agronomie* **1,** 565–572; 573–578.

Reid J. & Cain R.F. (1962) Studies on the organisms associated with 'snow-blight' of conifers in North America. II. Some species of the genera *Phacidium, Lophophacidium, Sarcotrichila,* and *Hemiphacidium*. *Mycologia* **54,** 481–497.

Rimpau R.H. (1962) Untersuchungen über die Gattung *Drepanopeziza* (Kleb.) v. Höhnel. *Phytopathologische Zeitschrift* **43,** 257–306.

Sarejanni J.A., Demetriades S.D. & Zachos D.G. (1952) Rapport sommaire sur les principales maladies des plantes observées en Grèce au cours de l'année 1951. *Annales de l'Institut Phytopathologique Benaki* **6,** 5–9.

Scangerla M. (1956) Ricerche sulla patogenicita e sull' epidemia di *Botrytis gladiolorum* nei gladioli. *Phytopathologische Zeitschrift* **27,** 41–54.

Schüepp H. (1959) Untersuchungen über Pseudopezizoideae sensu Nannfeldt. *Phytopathologische Zeitschrift* **36,** 213–269.

Scott S.W. (1981) Serological relationships of three *Sclerotinia* species. *Transactions of the British Mycological Society* **77,** 674–676.

Scott S.W. (1984) Clover rot. *Botanical Review* **50,** 491–504.

Skilling D.D. (1981) *Scleroderris* canker—the situation in 1980. *Journal of Forestry* **79,** 95–97.

Skilling D.D. & Waddell C.D. (1974) Fungicides for the control of *Scleroderris* canker. *Plant Disease Reporter* **58,** 1097–1100.

Sondey J. & Moncomble D. (1983) Le brenner ou rougeot parasitaire. *Vititechnique* **66,** 15–16.

Stephan B.R. & Butin H. (1980) Krebsartige Erkrankungen an *Pinus contorta*-Herkünften. *European Journal of Forest Pathology* **10,** 410–419.

Sylvestre-Guinot G. (1981) Etude de l'émission des ascospores du *Lachnellula willkommii* (Hartig) Dennis dans l'Est de la France. *European Journal of Forest Pathology* **11,** 275–283.

Tahvonen R. (1980) *Botrytis porri* Buchw. on leeks as an important storage fungus in Finland. *Journal of the Scientific Agricultural Society of Finland* **52,** 331–338.

Tempe J. de (1966) Blind seed disease of ryegrass in the Netherlands. *Netherlands Journal of Plant Pathology* **72,** 299–310.

Terrier C.A. (1944) Über zwei in der Schweiz bisher wenig bekannte Schüttepilze der Kiefern *Hypodermella sulcigena* und *Hypodermella conjuncta*. *Phytopathologische Zeitschrift* **14,** 442–449.

Tichelaar G.M. (1966) Een bladvlekkenziekte op *Allium* soorten, veroorzaakt door *Botrytis squamosa*. *Netherlands Journal of Plant Pathology* **72,** 31–32.

Uhm J.Y. & Fujii H. (1983) Heterothallism and mating type mutation in *Sclerotinia trifoliorum*. *Phytopathology* **73,** 569–572.

Uscuplić M. (1981) The infection period of *Lophodermium seditiosum* and the possibility of its control in nurseries. *Zaštita Bilja* **32,** 375–382.

Vajna L. (1971) (Occurrence of canker disease of apple and pear trees in Hungary, caused by *Phacidiella discolor*). *Növényvédelem* **7,** 345–350.

Viennot-Bourgin G. (1949) *Les Champignons Parasites des Plantes Cultivées*. Masson, Paris.

Walker A., Entwhistle A.R. & Dearnaley N.J. (1984) Evidence for enhanced degradation of iprodione in soils treated previously with this fungicide. In *Soils and Crop Protection Chemicals* (Ed. R.J. Hance), pp. 117–123. BCPC, Croydon.

Warren C.G., Sanders P.L., Cole H. & Duich J.M. (1977) Relative fitness of benzimidazole and cadmium-tolerant populations of *Sclerotinia homoeocarpa* in the absence and presence of fungicides. *Phytopathology* **67,** 704–708.

Wastie R.L. (1962) Mechanism of action of an infective dose of *Botrytis* spores on bean leaves. *Transactions of the British Mycological Society* **45,** 465–473.

Willetts H.J. & Wong J.A.-L. (1980) The biology of *Sclerotinia sclerotiorum, S. trifoliorum,* and *S. minor* with emphasis on specific nomenclature. *Botanical Review* **46,** 101–165.

Willetts H.J., Byrde R.J.W., Fielding A.H. & Wong A.-L. (1977) The taxonomy of the brown rot fungi related to their extracellular cell wall degrading enzymes. *Journal of General Microbiology* **103,** 77–83.

Williams P.F. (1975) Growth of broad beans infected by *Botrytis fabae*. *Journal of Horticultural Science* **50,** 415–424.

Wilson M. (1921) *The Phomopsis Disease of Conifers*. *Forestry Commission Bulletin* no. 6. HMSO, London.

Wittmann W. (1980) *Rhytisma acerinum*. *Pflanzenarzt* **33,** 89.

Wormald H. (1954) *The Brown Rot Diseases of Fruit Trees*. *Technical Bulletin of the Ministry of Agriculture no. 3*. HMSO, London.

Wormald H. (1955) *Diseases of Fruits and Hops*. Crosby Lockwood, London.

Yokota S. (1983) Etiological and pathological studies on *Scleroderris* canker in Hokkaido, Japan. *Bulletin of the Forestry and Forest Products Research Institute* no. 321, 89–116.

15: Basidiomycetes I
Ustilaginales

The Ustilaginales cause the smut diseases, characterized by the production of masses of usually black teliospores (brand spores, ustilospores) in infected tissues. These teliospores, like those of Uredinales, germinate to produce a short septate basidium on which basidiospores are formed. These either: i) germinate to produce primary mycelium which infects the host and is dikaryotized by mycelial fusion in the host tissue (e.g. in *Ustilago*); or ii) fuse in pairs on the basidium and then give rise to sporidia, dispersed as ballistospores (e.g. in *Tilletia*). There may be several cycles of sporidial formation. In fact, the 'shadow or mirror yeasts' (Sporobolomycetaceae), probably related to Ustilaginales, grow only in this way (*Itersonilia* spp. are pathogens). The primary mycelia can readily be cultured, which makes smut fungi very convenient material for genetic studies.

A few species form local lesions on their hosts, producing teliospores at the centre of the lesion. In this case, the life cycle is virtually that of a microcyclic rust. However, most smuts infect their hosts systemically, or, more specifically, infect meristematic tissues and grow with them as the plant grows. Mycelial growth is mostly intercellular (without haustoria) and often very sparse until sporulation. Teliospores then form in sori on stems, petioles or leaves (in herbaceous perennial dicotyledons, in which the smut perennates in the stock, and in some monocotyledons), in anthers or, most characteristically, in the ovaries of Gramineae. Woody plants are not infected by smuts. The teliospores are wind-dispersed but only over short distances. They readily fall to the soil, where they can persist for long periods, to infect germinating seedlings. In the cereal smuts, seed transmission is developed to a very specialized degree. In 'covered' smuts, teliospores develop late relative to grain maturation, and remain more or less enclosed by the seed coat until grain is shed and falls to the soil. Harvesting and threshing provides for even more effective dispersal of teliospores to the surface of healthy grain. In 'loose' smuts, teliospores form early in infected tillers at the same time as healthy tillers flower, and are readily released and wind dispersed to the open flowers. The developing grain then carries a latent internal infection of the embryo.

The smuts of annual grasses (the only ones of major economic importance) are the classic 'simple interest' diseases, reproducing once a season in phase with their hosts. Since a contaminated seed generally gives an infected plant, and an infected plant produces no grain, yield losses can be predicted relatively accurately from levels of seed contamination or from the observed percentage infection in the field. The fact that smuts are seed-transmitted makes them relatively easy to control, so that actual losses are low. However, they are likely to reappear rapidly if control standards are slackened, and they are among the most serious cereal diseases in third world agriculture.

Mordue & Ainsworth (1984), in a recent critical account of the British smuts, revised the nomenclature of cereal smuts. The new names have been indicated in this account but not adopted.

Angiosorus solani Thirum. & O'Brien (syn. *Thecaphora solani* Barrus) causes potato smut in C. and S. America (EPPO 4; CMI map 214). Tubers are malformed and become filled with a mass of smut spores. It is potentially a very serious disease in Europe, and is one of the reasons for the imposition of strict quarantine precautions on potato tubers imported as germplasm from S. America.

Entyloma calendulae (Oudem.) de Bary **f. *calendulae*** causes leaf spot of *Calendula,* and **f. *dahliae*** (Sydow) Viégas of *Dahlia* (CMI 801, 802). The life cycle differs from that of most smuts, resembling more that of a microcyclic rust. Sporidia (basidiospores) infect leaves to form pale yellow lesions, within which masses of nearly colourless teliospores develop. These germinate *in situ* to form further sporidia. The fungus persists in plant debris and soil. Control is by sanitation and fungicides. Similar species occur on various cultivated plants (e.g. ***E. fuscum*** Schröter on poppy, CMI 803; ***E. dactylidis*** Cif. and other spp. on turfgrasses, APS Compendium) but are of very minor importance. ***E. oryzae*** H. & P. Sydow (CMI 296) causing leaf smut of rice, is mainly found in the tropics and is of concern in India. It occurs in S. France (CMI map 451).

Graphiola phoenicis (Moug.) Poit. infects the leaves of palms (*Phoenix, Chamaerops* etc.) in Mediterranean countries (and in glasshouses) forming small blackish pustules containing a tuft of fertile hyphae bearing small yellowish smooth spores (Viennot-Bourgin 1949). Recent studies confirm that this smut-like fungus, which can be damaging to glasshouse palms, is a Basidiomycete but place it in a separate order Graphiolales (Cole 1983).

Itersonilia perplexans Derx. is a yeast-like fungus, reproducing by wind dispersal of ballistospores resembling the sporidia of the Ustilaginales. The formation of clamp connections is a further indication that its group (Sporobolomycetaceae) should be considered in the Basidiomycetes. *I. perplexans* is widespread as a leaf surface saprophyte, but also causes petal blight of chrysanthemums, especially outdoors and in the humid conditions of autumn and early winter. Lesions start as pinpoint brown spots on outer florets, which spread and are covered with a characteristic dull white bloom (Fletcher 1984). Control is mainly by avoidance of humid conditions, though fungicides have been used. The fungus has also been recorded on sunflower and globe artichoke (Ponchet *et al.* 1977). A similar species with larger ballistospores, also forming chlamydospores, is ***I. pastinacae*** Channon causing root canker of parsnip and dill (Ponchet *et al.* 1977).

Sphacelotheca cruenta (Kühn) Potter

S. cruenta is one of the most important species of the genus, which can be separated from *Ustilago* by having sori enveloped in a peridium formed around a central columella of host tissue. In *S. cruenta* the sori are covered by a very delicate peridium and the spores are globose to sub-globose, light yellowish or olivaceous brown, thin walled, 6.5–10 μm in diameter, and minutely echinulate. *S. cruenta* affects *Sorghum bicolor, S. halepense, S. sudanense, Sorgastrum nutans* and certain cultivars of sugar cane derived from crosses in which *S. halepense* was one parent. It is widespread in tropical and temperate areas but has not been reported from Australia, Oceania or the Malaysian-Indonesian archipelago; in Europe it is known in Cyprus, Germany, Hungary, Italy and USSR (CMI map 408). Three races have been described, whose reaction to five cultivars has been tabulated by Tarr (1962). *S. holci* Jackson is now regarded as a pathogenic race of *S. cruenta*. Hybridization between *S. cruenta* and *S. reiliana* and *S. sorghi* (q.v.) has been recorded.

In plants infected at the seedling stage, grains are replaced by more or less conical sori, varying considerably in size (3–4×3–18 mm), at first enclosed in a delicate grey peridium which usually ruptures before the infected head emerges, releasing the powdery, black mass of the spores (loose smut, Fig. 65), and showing a central columella, consisting of host tissue, extending the length of the sorus, which remains as a characteristic black, curved spike after the spores have been blown away. Sometimes sori can also be formed on the rachis and branches of the panicle and, occasionally, on the stalk. The affected plants head prematurely and produce thin stalks with numerous slender tillers which give them a growth habit resembling that of Sudan grass. When heads are directly infected by airborne spores, systemic infection does not occur and the growth habit of the plant is not affected (Viennot-Bourgin 1978). Infection by seed-borne spores occurs as in *S. sorghi* (q.v.). Spore germination is optimal at about 25°C; high infection occurs at 20–25°C. Survival of the spores in soil is low and probably too short to ensure viable inoculum for a succeeding crop, except in dry conditions.

Losses caused by *S. cruenta* in Europe are usually low but marked yield reductions have, however, been recorded in fodder and sugar sorghum in Italy (Goidanich 1939). Seed dressing (see *S. sorghi*) is effective against seedling infection. Early destruction of smutted heads and crop rotation are suggested in order to reduce head infection by airborne spores.
CASTELLANI

Fig. 65. Loose smut of sorghum (*Sphacelotheca cruenta*).

Sphacelotheca destruens (Schlecht.) Stevenson & Johnson causes head smut of *Panicum miliaceum*. It is a loose smut, the disease being transmitted by teliospores carried on seed (CMI 72). It is present in most of Europe (CMI map 219) but millet is a very minor crop except in the USSR, where *S. destruens* causes the most important disease of the crop and is controlled by fungicidal seed treatments or by selection of resistant cultivars.

Sphacelotheca reiliana (Kühn) Clinton

(Syn. *Sorosporium reiliana* McAlp., *Sphacelotheca holci-sorghi* (Riv.) Cif.)

S. reiliana replaces the inflorescences by large sori at first covered by a peridium, but which soon rupture to release a dusty mass of globose to sub-globose, brown, conspicuous echinulate (9–12 μm diameter) spores. It is widely distributed in sorghum- and maize-growing areas in Europe (CMI map 19). Two formae speciales infecting sorghum and maize are known; the former appears to be more widely distributed (Al Sohaily 1963). A hybrid of *S. reiliana* infects both maize and sorghum. Several races have been reported on sorghum. Hybridization has been recorded between *S. reiliana, S. sorghi* and *S. cruenta*.

Individual flowers of the tassel, and whole cobs in maize and panicles in sorghum, may be replaced by large galls (smut sori), consisting of the vascular tissues surrounded by the mass of the spores of the pathogen, protected by a whitish-grey peridium which soon flakes away disclosing the black-brown spore mass, intermingled with long, dark-coloured remnants of vascular bundles, remaining after spore dispersal. This symptom easily distinguishes head smut from the other two (*S. sorghi* and *S. cruenta*), in which each affected grain is replaced by a small sorus, the head as a whole retaining its inflorescence structure (Goidanich 1939; Tarr 1962). Although some infection may be induced by spores present in contaminated seed, most is caused by spores formed in the preceding crop, surviving in the soil where they can remain viable for more than eight months. They germinate (the optimum temperature is 21–31°C) as in *S. sorghi* (q.v.). Infections take place in young plants even after the seedling stage (Fenwick & Simpson 1969), resulting in a systemic mycelium which develops into the aerial parts of the plant and invades the undifferentiated floral tissues. Infection is favoured by relatively low soil temperature and moisture, delaying the growth of the seedlings.

In general *S. reiliana* does not cause crop losses as severe as those caused by *S. sorghi,* but up to 30–40% infected plants have been observed on susceptible hybrids of sorghum and maize. In Italy sugar sorghums appear particularly susceptible. Seed treatment with non-systemic fungicides may achieve only partial control because infection can occur in contaminated soil over a relatively long period after germination. Good results, however, have been obtained (Simpson & Fenwick 1971) by seed treatment with carboxin (150–200 g/hl). Early destruction of the heads before spore dispersal and crop rotation are suggested in order to reduce inoculum. Head-smut resistant cultivars of sorghum and maize are now available.

CASTELLANI

Sphacelotheca sorghi (Link) Clinton

S. sorghi has sori which are oblong to subconical, covered by a firm peridium and filled with a powdery mass of globose to sub-globose, olivaceous brown, smooth to minutely echinulate spores 5–8.5 μm in diameter. It has been reported on the different types of *Sorghum bicolor* and other species of the genus like *S. halepense* and *S. sudanense* and is co-extensive with the cultivation of *Sorghum*; in Europe it is long known in Mediterranean countries and as far north as Denmark (CMI map 220). Several races of *S. sorghi* have been identified on the basis of the reactions shown by different types of grain and hybrid sorghums. A key for their identification was proposed by Vaheeduddin (1951). Hybridization has been recorded between *S. sorghi* and *S. cruenta*.

In infected plants nearly all, but more frequently only a certain number of grains, localized at the top, bottom or side of the head, are replaced by cylindrical or conical, whitish to grey-brown smut sori, 4–13×2–4 mm diameter. They are covered by a rather thick peridium which may persist unbroken until threshing (covered smut, Fig. 66), and are filled with a dark-brown, powdery mass of spores around a slender central columella consisting of remnants of the vascular bundles. Infection is caused by seed-borne spores which can survive several years under the conditions in which grains are normally stored. When contaminated seeds are sown, the spores, as soon as soil conditions become favourable, germinate producing a dikaryotic germ tube or a basidium forming lateral and apical, small, hyaline sporidia. Before invading the host, sporidia with complementary sex factors fuse. The mycelium of the pathogen

Fig. 66. Covered smut of sorghum (*Sphacelotheca sorghi*).

grows thereafter within the plant and ultimately forms its sori in the inflorescences. As only very young seedlings can be infected, the environmental conditions prevailing between seed germination and emergence of the seedlings above ground markedly influence disease incidence (Tarr 1962). Infection is favoured by moderately low temperatures (20–25°C) which, besides favouring a rapid germination of the spores, slow down the growth of the seedling, keeping them susceptible for a longer time. *S. sorghi* is not known to be soil-transmitted; its spores in fact germinate readily over a fairly wide range of soil temperatures and moisture levels, and rarely survive between successive sorghum crops (Ciccarone & Malaguti 1950). *S. sorghi* is responsible for yield losses in grain sorghum that may average 5–10% annually with localized peaks of more than 60%. It also reduces the yield of fodder and sugar sorghum and depreciates that of broom corn by deforming the central axis and branches of the panicles and dusting them with its spores. Good control may be obtained by seed dressing with fungicides such as copper oxychloride, chloranil, thiram, carboxin and, where still permitted, organomercurials.

CASTELLANI

Tilletia caries (DC.) Tul.

(Syn. *T. tritici* (Bjerk.) Wolff)

and *Tilletia foetida* (Wallr.) Liro

(Syn. *T. laevis* Kühn)

Basic description

As the two species may cross, and there is overlapping of pathogenicity, they are treated jointly. There is considerable variation in teliospore morphology, the spores of *T. caries* being reticulate, 14–23 μm, those of *T. foetida* smooth, 17–20×18–22 μm; spore masses are brown to black.

Host plants

Cultivated wheat, including *T. aestivum, T. dicoccum, T. durum* and *T. spelta*, rye, and many other Gramineae, including species of *Agropyron, Bromus* and *Lolium* are all susceptible to attack by either pathogen.

Host specialization

Both species contain many pathogenic races, the numbers varying according to the different cultivars used, but making a total of at least 150 races (Neergaard 1979). Pathogenicity is often more different between pathogenic races within one of the species than between the two species.

Geographical distribution

Distribution is world-wide (CMI maps 294, 295) but pathogenic races have different distributions, some being world-wide, some confined to restricted regions. The risk of introduction of races to new areas should thus be considered in quarantine policy.

Disease

Tilletia belongs to the type of covered smuts in which the mature spore mass is kept for some time within the sorus, often until the sorus becomes free from the host. The disease is systemic, the teliospores replacing the ovuliferous tissue. Emerging infected ears look normal but symptoms become apparent when the ears become older. They have a loose, open appearance, the spike is elongated and the bunt balls fatter than healthy grains. The haulms of affected plants are shortened. Crushed bunt balls have a fishy smell caused by the presence of trimethylamine.

Epidemiology

The teliospores overwinter as contaminants on the grain and are capable of surviving in soil without the host for a few weeks only. The spores germinate simultaneously with the grain, and the seedling is susceptible until the coleoptile splits. As soon as the primary leaf protrudes the seedling becomes practically immune. Because of this very short period of seedling susceptibility, the seed transmission rates of the pathogens are subject to great variation depending on soil factors (compare loose smut).

Economic importance

The disease causes losses of grain in varying percentages and additional losses by lowering in quality of 'smutty' grain. In the 1920s there were often dramatic losses, in some fields more than 80% in N. America and Europe. Modern control measures (certification schemes and consistent seed treatment) have reduced the importance of the disease substantially in technically advanced agricultural countries, whereas in developing countries losses are still high. In Nepal and Afghanistan fields with 40–60% infection have been observed (Saari & Wilcoxson 1974). In Turkey only half of the wheat sown is treated and where untreated seed is sown losses generally amount to 10–15%, but may reach 70–95% in some fields (Parlak 1981). When seed is treated losses fall to 0.5% (Iren 1981).

Control

Control measures include seed certification, seed treatment and appropriate crop management, including the selection of adjusted sowing dates, breeding for resistance, and quarantine precautions directed against foreign pathogenic races. In seed certification the normal standard for field approval for infection of smuts (loose and covered) is one smutted ear in 10000 but standards vary according to the susceptibility of cultivars. Seed treatments used for decades all over Europe have effectively reduced bunt to usually negligible levels. Organomercurial compounds, now generally abandoned, are used in certification schemes for nuclear seed stocks. Standard commercial treatments use hexachlorobenzene, quintozene or fuberidazole. The systemic fungicides thiabendazole and carboxin are mainly aimed at the specific control of loose smut (*Ustilago tritici*) but also control *T. caries*. Resistance of wheat to bunt is based on a number of major genes and some minor genes. The dominant genes control resistance against a certain group of races. Crossing between resistant cultivars may result in transgressive segregation (Fisher & Holton 1957).
NEERGAARD

Tilletia controversa Kühn

T. controversa causes dwarf bunt of winter wheat and some grasses. The spores produced in spherical sori are typically surrounded by a hyaline gelatinous sheath which distinguishes them from those of *T. caries*, produced in elliptical sori. Their ornamentation distinguishes them from those of *T. indica* and *T. foetida* (CMI 746). A close relationship with *T. caries* and *T. foetida* is indicated by the fact that pathotypes of *T. controversa* carry the same virulence factors. Symptoms of infection are very similar to those of other bunt diseases, but with characteristic stunting. Although the fungus can be seed-borne (and this is the most likely means of long-distance dispersal), infection typically arises from spores in the soil, which can remain viable for up to 10 years. The best conditions for infection occur during winter under persistent snow cover. For this reason, the disease is of no significance on spring wheat. Grasses are only susceptible during their first winter, which is the only time that they normally have exposed stem buds susceptible to infection. The disease is currently important mainly in E. Europe (Hungary, Poland, USSR), and also in N. America. It occurs in C. Europe, but in the west, where winter conditions do not favour it, it is apparently absent (CMI map 297); though once reported from France, it has now disappeared from there (EPPO 83).
SMITH

Tilletia indica Mitra (syn. *Neovossia indica* (Mitra) Mundkur) causes Karnal bunt of wheat, and occurs in India, Iraq and Pakistan (EPPO 23). It is principally seed-borne and could cause serious problems in most warm temperate parts of the world (Europe, N. America, Australia) if it spread. It is accordingly universally considered as a serious quarantine organism. Its accidental introduction to Mexico has highlighted the dangers from the exchange of germplasm (seed lots) between research and breeding stations in different continents.

Urocystis cepulae Frost

The spore balls of *Urocystis* are surrounded by a layer of sterile cells. Those of *Urocystis cepulae* are dark-brown (24–22 μm), with individual teliospores reddish-brown with smooth walls (4–6 μm). *U. cepulae* causes smut disease of bulb and salad (green) onions (Fig. 67), leek and garlic. *Allium fistulosum* is reported to be resistant and potential smut resistance has been identified in a few accessions from

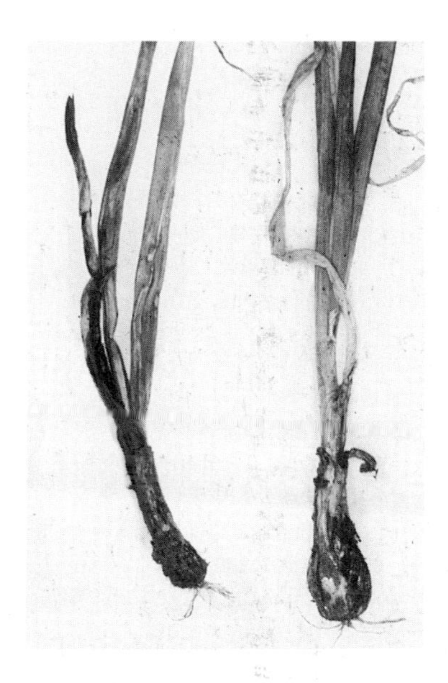

Fig. 67. Onion smut (*Urocystis cepulae*).

the world onion germplasm collection and in two commercial cultivars (Utkhede & Rahe 1980). *U. cepulae* occurs in onion-growing areas of Europe and in many other countries (CMI map 12).

Symptoms appear as silver to lead-coloured streaks along the length of the cotyledon leaves which may be bent. The streaks are the sori or pustules of the fungus which rupture to produce dark-brown lesions. Most young infected plants die within 3–4 weeks but some, although distorted and bearing lesions, may survive to harvest. The disease is probably not seed-borne but spore balls could occur as contaminants of onion seed samples. The main source of the disease is spore balls present in the soil. Infection occurs through the onion cotyledon at soil level. Infecting hyphae penetrate the leaf epidermis and enter the mesophyll where they keep pace with the growth of the cotyledon leaf. The sori are produced in these tissues. If the fungus penetrates the meristematic tissue at the base of the cotyledon leaf then the leaves may also become infected. The pustules (sori) of the pathogen mature 3–4 weeks after infection and the spore balls are released. These may survive for up to 20 years in the soil. Individual spores germinate at 13–22°C to produce oidial fragments. Most infection occurs at soil temperatures of 10–12°C but plants can be infected up to 25°C, but not at 29°C.

The disease occurs locally and sporadically in Europe and is of minor economic importance. It is controlled by seed treatment using large amounts of fungicide attached to onion seeds by means of a sticker. Fungicides including ferbam and thiram (Croxall & Hickman 1953), benomyl and carboxin/thiram mixtures (Crête & Tartier 1973) have been used. Hexachlorobenzene also reduces the disease (Duran & Fischer 1959).
MAUDE

Urocystis occulta (Wallr.) Rabenh.

U. occulta forms spore masses in different vegetative parts of the host, and few in the reproductive organs. The teliospores are firmly united in balls which are partially surrounded by a layer of sterile cells. *U. occulta* causes flag, stripe or stalk smut of rye. Seedling infection takes place from teliospores contaminating the grains, later young side shoots being often infected. The pathogen causes stunting, malformation and streaking of leaves and stems. It occurs in N. and S. America, Asia and Australia and is widespread in Europe where it is of economic importance in some regions. Seed transmission rates vary exceedingly according to conditions of soil moisture and temperature; most severe infections take place at low moisture levels. Previously flag smut caused heavy losses in Europe, up to 60% in Germany. Now the disease is sometimes severe in E. Europe. It is easily controlled by appropriate seed treatment. Organomercurial compounds, now generally abandoned, are used for nuclear seed stocks in certification programmes. Standard commercial treatments use thiabendazole, fuberidazole, triadimenol, benomyl and carboxin.
NEERGAARD

Urocystis agropyri (Preuss) Schröter causes flag smut of leaves of grasses (CMI 716), with linear black erumpent sori along the leaves. The teliospores can both contaminate seed and persist in the soil (cf. also *Ustilago striiformis*). In warmer countries, this species is also found on wheat, but is of no significance on this crop in Europe. ***U. anemones*** (Pers.) Winter (CMI 806) is common on wild *Anemone* and occasionally severe on cultivated species, causing dark blisters on leaves and petioles. The teliospores contaminate the soil. ***U. violae*** (Sow.) A. Fischer v. Waldh. causes similar symptoms on cultivated *Viola*. ***U. gladiolicola*** Ainsworth (CMI 807) causes black blisters on the leaves and corms of gladioli and is transmitted in soil and by infected corms. ***U. colchici*** (Schlecht.) Rabenh. behaves similarly on *Colchicum*. The fungus known as *U. gladioli* W.G.Sm. is not a smut at all, the 'spore-balls' being in reality bulbils of *Papulaspora dodgei* Conners (CMI 807).

Ustilago avenae (Pers.) Rostrup

(Syn. *U. segetum* (Bull.) Roussel var. *avenae* (Pers.) Brunaud)

Basic description

The teliospores are spherical to sub-spherical, pale greenish-brown to black, with fine echinulation, 4–8 μm diameter. The spore mass is dark greenish-brown. *U. avenae* is distinguishable from *U. hordei* by the surface echinulation of its spores, the spores of the latter species being smooth. The fungus is similar to *U. nigra* Tapke, semi-loose smut of barley, generally considered to be a synonym of *U. avenae*.

Host plants

The principal host is oat but the pathogen occurs on a wide range of *Avena* species including *A. fatua* and *A. nuda,* as well as on *Arrhenatherum elatius* on which it is common (Neergaard 1979) and sometimes known as *U. perennans* Rostrup.

Host specialization

Numerous pathogenic races have been reported: 29 races identified on 17 different cultivars in the USA (Reed 1940), plus additional races encountered later (Neergaard 1979). Crosses betweeen races may result in transgressive segregation with increased virulence.

Geographical distribution

Distribution of the species is world-wide (CMI map 238) whereas that of the pathogenic races is confined to regions, being dependent on the cultivars grown.

Disease

U. avenae causes loose smut. The dark greenish-brown powdery sori replace the kernels, the glumes being usually completely destroyed and the spore masses easily released. Seedlings may be killed; lesions and spore masses may develop on the flag leaf. Symptoms largely depend on the host cultivar and the pathogenic race.

Epidemiology

Two disease cycles may be distinguished. Wind-borne teliospores from smutted spikes lodge on open flower parts and produce infection of the pericarp and the inner wall of the glumes, or else later disseminated spores remain as contamination between the caryopsis and the glumes or lodge on the surface of the glumes during threshing. Systemic infection develops either from infected seed ('flower and seedling infection') or as seedling infection established from the spores carried as contamination. The seed infection type of disease cycle has a bearing on the method of seed treatment to be applied, and probably explains why in the USA and other places it has been more difficult to reduce infection of loose smut of oats than covered smut of oats (*U. hordei*).

Economic importance

Loose smut of oats is more prevalent in most parts of the world than covered smut. Severe infection occurs in developing countries but is infrequent in technically advanced agricultural countries, including those of Europe, due to consistent use of organomercurials since the 1920s.

Control

The disease has largely been eliminated or reduced by field inspection and certification of seed crops. Seed treatment with volatile mercurials is effective for killing mycelial infections (now mostly used for the first generations in certification programmes). Otherwise, various systemic fungicides may be used: benomyl, carboxin, pyracarbolid and thiabendazole.
NEERGAARD

Ustilago hordei (Pers.) Lagerh.

(Syn. *U. segetum* (Bull.) Roussel var. *hordei* (Pers.) Rabenh., *U. segetum* var. *segetum*)

Basic description

The teliospore mass replaces the ovary and is brownish to purple-black. Spores are spherical to sub-spherical, yellowish to brown, 5–11 μm diameter, the spore wall is thin and smooth (CMI 749).

Host plants

There are two formae speciales, one attacking various species of barley (**f.sp. *hordei***), another (**f.sp. *avenae*** Boerema & Verhoeven, syn. *U. kolleri* Wille, *U. levis* (Kellerman & Swingle) magnus) attacking species of oats. *Agropyron, Bromus, Elymus, Secale* and *Triticum* are also infected.

Host specialization

At least 15 ('vertical') pathogenic races occur on barley, and an equal number on oats, but numbers vary according to the different cultivars used (Neergaard 1979). Vertical pathogenic races may vary considerably in aggressiveness ('horizontal races').

Geographical distribution

The disease occurs world-wide (CMI map 460) but only a few of the pathogenic races have a wide geographical distribution themselves.

Disease

U. hordei causes covered smut of barley and oats. The membrane of the smut sorus that replaces grains of infected ears does not rupture until harvest. Smutted heads emerge later than healthy ones and sori may occur as elongated stripes on the flag leaf or as stem galls.

Epidemiology

The germ tubes of the sporidia, formed by the basidia developed from the teliospores on contaminated grain or in soil, penetrate into the coleoptile. This is susceptible only as long as it remains unsplit. Seed transmission rates vary considerably depending on soil temperature and humidity. Covered smut of barley requires higher soil temperatures than wheat bunt (*Tilletia caries*). The mycelium advances intercellularly into the seedling through the parenchyma and penetrates into the flower primordia. Diseased plants are smaller than healthy ones. Dissemination of the spores takes place during harvesting and culminates with the completion of threshing. Spores residing on the pericarp beneath the hull may germinate and start infection even during storage if the grain is kept under relatively humid conditions. This mode of infection is relevant to the technique of the seed treatment to be applied. However, using modern drying and storage techniques sub-hull germination of spores should rarely occur in present practice.

Economic importance

Covered smut of barley and oats is well under control in Europe wherever resistant cultivars, certified seed and seed treatment are used, severe attacks being encountered only where appropriate control is neglected. In Sweden 15% of 2180 seed samples were contaminated, 3–4% severely. The number of heavily contaminated lots increased with the number of years that untreated seed had been used (Fritz 1978).

Control

Certification programmes have contributed to control; seed treatment by volatile organomercurials is used in seed multiplication schemes. Commercial seed treatment uses carboxin, pyracarbolid or thiabendazole. However, resistance to benomyl has been noticed. Fuberidazole and imazalil are also used in seed treatments.

NEERGAARD

Ustilago maydis (DC.) Corda

(Syn. *U. zeae* (Beckm.) Unger)

U. maydis forms its spores in large irregular swellings 1–15 cm in size on the inflorescences, leaves and stems of maize (CMI 79). These appear whitish then greyish, and finally burst to release black spores. Originating in America, *U. maydis* now occurs in all parts of Europe where maize is grown (CMI map 93), and throughout most of the world. Infection of the maize plant can occur on all above-ground parts, particularly young, actively growing or meristematic tissues, but is not systemic (APS Compendium). Rain carries spores down the leaf sheaths to the meristems. Alternatively, callus tissue formed at points of wounding by cultural methods, hail or insects (e.g. *Oscinella frit*) is liable to be infected. The incubation period is 2–3 weeks, and spores are released a week later. The spores are carried by wind either to cause secondary infections in the crop, or to fall to the soil. Cobs are infected relatively late, and the infection can spread explosively in late summer under favourable conditions.

U. maydis mainly persists as spores in the soil where they can remain viable for many years. Spores can be carried on seed, but this is of minor practical importance. Seedlings can be attacked and killed in significant numbers, if the crop is sown in a heavily infested soil, and plants infected early may give few small cobs. The main damage comes from smut lesions on leaves, stems and cobs. However, in general, losses in Europe are fairly minor: if 10% of plants are infected, only 2% loss of yield can be expected (Legros & Geenen 1979). It is only in somewhat exceptional circumstances that serious attacks occur, as in the hot dry summer of 1976. High nitrogen fertilization and insect damage will also favour infection. There has been concern that smut balls harvested in forage maize could be toxic to livestock, but recent experience in Europe has not confirmed this. Any problems which arose in the past are more likely to have been due to accompanying fusaria.

There are no very satisfactory control methods. The fungus persists in soil too long for rotation to be useful, and fungicide treatments (of seed, soil or plants) are expensive and rather ineffective. Destruction and deep ploughing of debris would remove inoculum, but would have to be practised over wide areas, since crops can be infected by wind-borne spores. Many maize cultivars have vertical resistance, and correspondingly numerous pathotypes of the fungus exist, making this type of resistance useless in practice. Some cultivars may show useful degrees of field or horizontal resistance.

CASSINI & SMITH

Ustilago nuda (Jensen) Rostrup

(Syn. *U. segetum* (Bull.) Roussel var. *tritici* (Pers.) Brunaud)

Basic description

Since there are distinct differences between the loose smut fungi on wheat and barley in their morphology and nuclear behaviour, the name *U. tritici* (Pers.) Rostrup is often retained for the former. However, the two pathogens have disease cycle, epidemiology and control in common. The teliospores are dark, finely echinulate, and 5–8 μm diameter.

Host plants

U. nuda attacks all cultivated species of barley, while *U. tritici* occurs in all cultivated wheats. ***U. vavilovi*** Jacz. is closely related, and infects rye as well as wheat.

Host specialization

Both species contain a considerable number of pathogenic races, the number varying according to the test cultivars used. Halisky (1965) recorded 20 races of the barley smut fungus, and Tusa & Rădulescu (1973) in Romania 22 race groups of the wheat smut fungus (Neergaard 1979).

Geographical distribution

The distribution of both species is world-wide (CMI map 368) but pathogenic races may have restricted distributions.

Disease

As in most smuts, infection is systemic. Symptoms may be visible before heading, the leaves becoming slightly discoloured (yellow, brown), but this condition is generally only apparent on certain cultivars (chrome stripe). The smutted ears extend earlier than the healthy spikes and are carried higher out of the leaf sheath than the heads of healthy plants. In wheat some infected plants may become stunted, with poor tillering, a condition considered as hypersensitivity, the pathogen being inhibited so that it cannot reach the heads. These plants may recover from the disease.

Epidemiology

The disease is most prevalent (Neergaard 1979) in countries with a humid climate and in years when moisture and inoculum were adequate at flowering time the previous year. The dormant mycelium is carried in any part of the embryo except the radicle. The seed transmission rate in barley is practically 100%. The hyphae grow intercellularly behind the growing point and penetrate into the leaves and the primordia of the ears, as well as into the roots. Some weeks before the ears emerge, the teliospores begin to form, replacing the tissues of the ovary. From the smutted heads the short-lived teliospores are dispersed by wind to healthy flowers where they germinate producing infective hyphae. They penetrate through the young ovary wall and eventually establish themselves as dormant mycelium in the embryo before the maturation of the seed. The flowers are susceptible to infection only for a few days following fertilization.

Economic losses

In many developing countries both pathogens cause heavy losses in barley and wheat crops comparable to losses often encountered in the first decades of this century in N. America and Europe. In N.W. India losses up to 40% were encountered in 1967–71, and in some developing countries losses of about 5% due to one or the other of these smuts are regarded as unimportant. In Europe losses are usually less than 1%, partly because of extensive use of resistant cultivars. In Turkey losses amount to 10–15%, in some fields up to 90% where seeds are sown without treatment. In a survey made in Turkey in 1972–74, average crop losses in infected fields amounted to 16% or at least 35000 tons of wheat which is the equivalent of the wheat requirement in Turkey of 200000 people (Parlak 1981). According to Iren (1981), estimates of the average annual loss in Turkey over the last 10 years did not exceed 5%. Losses vary from 0.1 to 2% depending on weather conditions and cultivars.

Control

In technically advanced agricultural countries, successful control is obtained by use of resistant cultivars, by certification programmes including seed-crop inspection, and regular laboratory testing of seed by the embryo-count procedure (ISTA 1976; Rennie 1981). Usual standards for loose smut in certification programmes are 0.1–0.2% infected embryos for the earliest three multiplication categories. In contrast to the barley test which gives practically complete positive correlation to attacks in the field, only 33–50% of the incidence estimated for wheat in the embryo tests appears in the field; this may be due to sowing conditions or to insufficiently refined testing methods (Hewett 1979) but may be due to the occurrence of hypersensitive resistance (see *Disease*). Especially for wheat, late seed-crop inspection and control plots are important measures in seed certification. The embryo-count procedure is also used to test for the need for seed treatment as in regular schemes in Bulgaria and India where a tolerance of 0.5–1.0% is applied. These 'need for treatment testing schemes' reduce the costs of seed treatment. Systemic fungicides (for barley, carboxin, thiabendazole or triadimenol and for wheat carboxin or pyracarbolid) have replaced the cumbersome hot water treatment, thus avoiding soaking and drying the seed. Selection of areas suitable for seed production takes into consideration the epidemiological conditions for flower infection. There are different types of resistance: 'embryo resistance' and 'embryo susceptibility linked with adult plant resistance' (barley cv. Keystone), both preventing seed transmission, and 'closed flower resistance' providing escape of infection.

NEERGAARD

Ustilago crameri Körn. (CMI 78) causes a serious head smut of foxtail millet (*Setaria italica*) in S. Europe. Infected grains are swollen with teliospores (Viennot-Bourgin 1949). The teliospores are externally seed-borne, seedlings being infected at germination, so control measures against covered smuts (e.g. *Tilletia caries*) are effective. Smuts also commonly affect the heads of pasture grasses. ***U. bullata*** Berk. (CMI 718) is common on *Bromus* and *Agropyron* species. Severe effects on seedling survival, yield and inflorescence formation in *B. catharticus* have been found (Falloon 1979). ***U. cynodontis*** (Pass.) Henn. is common and conspicuous on the heads (and leaves) of *Cynodon dactylon* in S. European countries (CMI 279). ***Ustilago striiformis*** (Westend.) Niessl (CMI 717) causes a stripe smut on the leaves of many grasses (like *Urocystis agropyri*, but with unaggregated teliospores). ***Ustilago hypodytes*** (Schlecht.) Fr. (CMI 809) forms its telia on the stem internodes, below the leaf sheaths. The leaf smuts of turf grasses (*U. agropyri, Ustilago* spp., *Entyloma* spp.) are generally of rather minor importance in Europe, but this may be attributed to the use of clean or treated seed. Fungicides may be used to treat infected turf. The diseases spread most readily on watered well-fertilized turf, and infected plants will tend to die out selectively if these treatments are withheld (APS Compendium).

Ustilago violacea (Pers.) Roussel

(Syn. *Microbotryum violaceum* Deml & Oberwinkler)

U. violacea causes anther smut of Caryophyllaceae (CMI 750). Teliospores infect buds and young shoots to establish systemic infections. The fungus sporulates in the anthers, replacing the pollen with a mass of conspicuous purple spores. Pollinating insects may transmit the disease (Jennersten 1983), but it is not seed-borne. Familiar on wild Caryophyllaceae throughout Europe, e.g. **var. *major*** Clinton on *Silene dioica* and **var. *stellariae*** (Sow.) Savile on *Stellaria*, *U. violacea* only occasionally attacks cultivated species such as carnation. Affected plants are stunted with many side-shoots. Since modern cultivars rarely produce anthers, the possibilities of transmission are limited (Fletcher 1984). However, sori also form on carnation petals. *U. violacea* has been intensively studied genetically and physiologically; plant extracts stimulate it to change from saprophytic yeast-like growth to the parasitic mycelial form, and α-tocopherol has been identified as a compound re-

sponsible for most of this activity (Castle & Day 1984).
SMITH

U. vaillantii Tul. & C. Tul. is another anther smut, occuring on small spring-flowering Liliaceae such as *Muscari, Chionodoxa* and *Scilla* (CMI 810). Finally, ***U. tragopogonis-pratensis*** (Pers.) Roussel and ***U. scorzonerae*** (Alb. & Schwein.) Schröter cause head smuts of salsify and black salsify respectively. They persist in the stocks and fill the whole capitulum with teliospores. Though common, they are of negligible economic importance (Viennot-Bourgin 1949).

REFERENCES

Al-Sohaily J. A. (1963) Physiological specialization of *Sphacelotheca reiliana* on sorghum and maize. *Phytopathology* **53,** 723–726.

Castle A.J. & Day A.W. (1984) Isolation and identification of α-tocopherol as an inducer of the parasitic phase of *Ustilago violacea. Phytopathology* **74,** 1194–1200.

Ciccarone A. & Malaguti G. (1950) Prime prove di campo e di laboratorio per la lotta contro *Sphacelotheca sorghi* e per la conoscenza della sua biologia in Venezuela. *Rivista di Agricoltura Tropicale e Subtropicale* **44,** 145–177.

Cole G.T. (1983) *Graphiola phoenicis:* a taxonomic enigma. *Mycologia* **75,** 93–116.

Crête R. & Tartier L. (1973) Trois années d'essais de lutte chimique contre le charbon de l'oignon *Urocystis magica. Phytoprotection* **54,** 31–40.

Croxall H.E. & Hickman C.J. (1953) The control of onion smut (*Urocystis cepulae*) by seed treatment. *Annals of Applied Biology* **40,** 176–183.

Duran R. & Fischer G.W. (1959) The efficacy and limitations of hexachlorobenzene for the control of onion smut. *Plant Disease Reporter* **43,** 880–888.

Falloon R.E. (1979) Further studies on the effects of infection by *Ustilago bullata* on vegetative growth of *Bromus catharticus. New Zealand Journal of Agricultural Research* **22,** 621–626.

Fenwick M.S. & Simpson W.R. (1969) Evidence of post-seedling infection by *Sphacelotheca reiliana. Phytopathology* **59,** 1026.

Fisher G.W. & Holton C.S. (1957) *Biology and Control of the Smut Fungi.* Ronald Press Co, New York.

Fletcher J.T. (1984) *Diseases of Greenhouse Plants.* Longman, London.

Fritz T. (1978) Förekomst av hårdsot hos korn. *Svensk Frötidning* **47,** 85–87.

Goidanich G. (1939) Le più importanti malattie del sorgo, con speciale riferimento a quelle del sorgo zuccherino. *Indicatore Saccarifero Italiano* **32,** 3–29; 77–102; 166–168.

Halisky P.M. (1965) Physiologic specialization and genetics of the smut fungi. *Botanical Reviews* **31,** 114–150.

Hewett P.D. (1979) Regulating seed-borne disease by certification. In *Plant Health* (Eds P.L. Ebbels & J.E. King) pp. 163–173. Blackwell Scientific Publications, Oxford.

Iren S. (1981) Wheat diseases in Turkey. *Bulletin OEPP/EPPO Bulletin* **11,** 47–52.

ISTA (1976) International rules for seed testing. *Seed Science & Technology* **4,** 3–49.

Jennersten O. (1983) Butterfly visitors as vectors of *Ustilago violacea* spores between Caryophyllaceous plants. *Oikos* **40,** 125–130.

Legros P. & Geenen J. (1979) Problèmes actuels relatifs à la protection des grandes cultures. 5. Le chabon du maïs. *Revue de l'Agriculture* **32,** 445–451.

Mordue J.E.M. & Ainsworth G.C. (1984) *Ustilaginales of the British Isles. Mycological Paper no. 154.* CMI, Kew.

Neergaard P. (1979) *Seed Pathology.* Macmillan, London.

Parlak Y. (1981) Seed-borne pathogens on wheat (particularly smuts) in Turkey. *Bulletin OEPP/EPPO Bulletin* **11,** 83–86.

Ponchet J., Mercier S. & Augé G. (1977) Variabilité de la morphologie et du pouvoir pathogène de quelques isolats du genre *Itersonilia.* In *Travaux dédiés à Georges Viennot-Bourgin.* Société Française de Phytopathologie, Paris.

Reed J.M. (1940) Physiologic races of the oat smut. *American Journal of Botany* **27,** 136–143.

Rennie N.J. (1981) Barley loose smut. Working sheet No. 25 (2nd ed.) In *ISTA Handbook on Seed Health Testing, Section 2.* ISTA, Zürich.

Saari E.E. & Wilcoxson R.D. (1974) Plant disease situation of high-yielding dwarf wheats in Africa and Asia. *Annual Review of Phytopathology* **12,** 49–68.

Simpson W.R. & Fenwick M.S. (1971) Suppression of corn head smut with carboxin seed treatments. *Plant Disease Reporter* **55,** 501–503.

Tarr S.A.Y. (1962) *Diseases of Sorghum, Sudan Grass and Broom Corn.* CMI, Kew.

Tusa C. & Rădulescu E. (1973) (Contribution to the knowledge of physiologic specialization of *Ustilago tritici* in Romania) *Analele Institutului de Cercetări pentru Cereale şi Plante Technice Fundulea C* **39,** 59–66.

Utkhede R.S. & Rahe J.E. (1980) Screening world onion germplasm collection and commercial cultivars for resistance to smut. *Canadian Journal of Plant Science* **60,** 157–161.

Vaheeduddin S. (1951) Two new physiologic races of *Sphacelotheca sorghi. Indian Phytopathology* **3,** 162.

Viennot-Bourgin G. (1949) *Les Champignons Parasites des Plantes Cultivées.* Masson, Paris.

Viennot-Bourgin G. (1978) Origine et nature de la mycocécidie provoquée par le *Sphacelotheca cruenta. Revue de Mycologie* **42,** 253–261.

16: Basidiomycetes II
Uredinales

The rust fungi, through the complexity of their life cycles, have long been one of the plant pathogenic groups of greatest interest to mycologists. The basic life cycle is well known, together with the fundamental distinction between autoecious and heteroecious rusts (the former producing all spore types on one host, the latter alternating between two hosts to complete the full cycle). Macrocyclic rusts produce all the spore types (spermatia in spermogonia, aeciospores in aecia, urediniospores in uredinia, teliospores in telia, basidiospores on basidia). Other rusts have shorter cycles, in which some of the spore types are missing. Though a terminology exists for this, we have preferred simply to indicate the spore types formed. One special case is worth noting: the endocyclic rusts with aeciospores that function as teliospores. The names of the spore types, and of the fructifications in which they are formed, continue to vary widely between authors. We have chosen to follow the usage of Cummins & Hiratsuka (1983), which is ontogenetic rather than morphologic.

Basically, the rust fungi are obligate parasites, mostly of leaf tissues (although a few have been cultured). Aeciospores or urediniospores germinate on the leaf surface to form germ-tubes which grow towards stomata. Over a stomatal aperture, an appressorium is formed and a penetration hypha enters the substomatal space where it forms a substomatal vesicle. Haustoria, bounded by a haustorial sheath, form in the adjoining mesophyll cells and the fungus spreads by intercellular mycelium and intracellular haustoria to form a normally circular colony in the leaf tissues. The mycelium gives rise to spores which erupt as rusty pustules through the epidermis (urediniospores are often followed by teliospores in the same pustule). The urediniospores are dispersed to re-infect the same host. The teliospores which survive on debris germinate *in situ* to give rise to basidia and basidiospores, normally early in the growing season. These provide primary inoculum to re-infect the same host (autoecious) or the alternate host (heteroecious). Basidiospore germ-tubes appear to penetrate directly and there is interest now in studying parasitic behaviour in the alternate host, in which the mycelium is, at least initially, monokaryotic (cf., for example, *Puccinia poarum*).

From a plant pathological point of view, we may note that rust fungi can perennate: i) as teliospores; ii) especially on woody hosts, as perennial mycelium; iii) in warmer climates or glasshouses where leaves are available for infection all year round, by simple urediniospore cycling. Infection on the alternate host is non-repeating, cycling only once a year, while infection of the telial host is generally repeated over and over by the urediniospores (like the conidia of Ascomycetes). A few rusts of the 'microcyclic' type simply repeat within a season by successive teliospore/basidiospore formation.

Alternation between hosts is a very characteristic element in the epidemiology of rust fungi, and has had obvious implications for control in a few cases (notably *Puccinia graminis*). However, heteroecism can be regarded as a constraint on multiplication of the pathogen, so that many heteroecious rusts are of less economic importance than autoecious ones. Some are important because they can bypass the alternate host, perennating in another way or in more southerly regions.

In any case, the leaf-infecting biotrophic habit of rusts means that crop losses due to rust infections are mainly indirect. So, although there are numerous species of Uredinales attacking almost every cultivated plant, the diseases caused are often of rather minor importance. Only cereals, forest trees and some vegetable and flower crops really suffer from serious rust diseases. Some major families are hardly affected (Cruciferae, Cucurbitaceae, Solanaceae). Introduced rusts may be particularly damaging. Finally, it may be noted that, like other obligate biotrophs, the rust fungi show a strong tendency to host specialization at generic, specific and subspecific levels, in particular to form resistance-gene-specific pathotypes. This is expressed mainly at the urediniospore stage. The subject is discussed further in the section on *Puccinia graminis*.

Arthuriomyces peckianus (Howe) Cummins & Y. Hiratsuka causes orange rust on many *Rubus* spp. (CMI 201). It lacks the uredinial stage. An endocyclic form (***Gymnoconia nitens*** (Schwein.) Kern & Thurston, syn. *Kunkelia nitens* (Schwein.) Arthur) occurs in N. America, with aeciospores behaving as teliospores (Cummins & Hiratsuka 1983). Though widespread in N. Europe, the disease is relatively unimportant.

Cerotelium fici (Butler) Arthur produces uredinia and telia on fig (CMI 281). It is widespread in Mediterranean countries (CMI map 399) but of little economic importance (Viennot-Bourgin 1949).

Chrysomyxa abietis (Wallr.) Unger

C. abietis has elongated yellow or orange telia up to 10 mm long, forming chains of orange or red-brown teliospores (CMI 576). The fungus is most commonly found on *Picea abies,* but it also occurs on *P. sitchensis, P. engelmannii, P. pungens* and *P. rubens.* It occurs widely in Europe and east into Asia. *C. abietis* infects the current needles through the stomata (Grill *et al.* 1980). In mid-summer, the diseased needles show bright, lemon-yellow transverse bands which later darken to a deep yellow. In late summer, elongated telia form on the undersides of the needles, on the yellow bands. They then remain dormant over the winter, and continue their development the following spring. In early summer (usually May–June) they open by a slit through the epidermis as yellow or orange pustules. The teliospores produce basidiospores which infect the young current year's needles. After teliospore germination, the old needles die and fall. Hence by the late summer, current needles remain on the affected branches, but those of the previous year are largely or altogether absent.

Epidemics caused by *C. abietis* are scattered and sporadic. This is because spruce needles are susceptible only when very young. Epidemics develop only when basidiospore production coincides with the flushing of the needles, and so occur only in occasional years, especially in young, dense stands. In spite, therefore, of the heavy loss of needles on affected trees, the overall economic importance of *C. abietis* is localized and relatively small, and usually no special measures are taken to control it. If the disease does cause trouble in young, dense stands, thinning to increase the air flow may be worthwhile.
PHILLIPS

Chrysomyxa arctostaphyli Dietel, with aecia on *Picea* and telia on *Arctostaphylos uva-ursi,* causes broom rust of spruce in N. America (CMI map 441), with symptoms similar to those of *Melampsorella caryophyllacearum* on fir. It presents a major quarantine threat to Europe, where its alternate host is widespread in spruce forests (EPPO 8).

Chrysomyxa ledi (Alb. & Schwein.) de Bary, like *C. abietis,* causes needle cast of *Picea,* but with only spermogonia and aecia on this host. The uredinia and telia form on *Ledum* (**var. *ledi***), e.g. in Scandinavia and USSR, or *Rhododendron* (**var. *rhododendri*** (de Bary) Savile, syn. *C. rhododendri* de Bary) e.g. in N.and C. Europe. The disease is of little importance on spruce but can be damaging on cultivated *Rhododendron.* ***C. woroninii*** Tranzschel also alternates between *Ledum* and *Picea* but affects only the leaf buds and female cones (spruce bud rust). It occurs in the USSR and N. America (McBeath 1984).

Chrysomyxa pirolata Winter forms spermogonia and aecia on the cone scales of *Picea* and uredinia and telia on Pyrolaceae. It is systemic in its uredinial stage and is therefore also found outside the spruce area. *C. pirolata* sometimes reduces the spruce seed crop, but the fungus is not of great importance.
ROLL-HANSEN

Coleosporium tussilaginis (Pers.) Lév.

C. tussilaginis produces white aecial blisters and spermogonia on pines of the sub-genus *Pinus*; in Europe, *P. sylvestris, P. nigra* and *P. halepensis* are most often affected, but in N. America many others are also attacked. The uredinia and telia occur on various mostly wild Compositae (*Tussilago farfara, Senecio* spp.), Campanulaceae, Ranunculaceae and Scrophulariaceae. On these alternate hosts the forms show slight morphological differences, and are therefore sometimes separated into about ten species (Lanier *et al.* 1978), or formae speciales. One such form (**f.sp. *senecionis-sylvatici*** (Wagner ex Gäum.) Boerema & Verhoeven, syn. *C. senecionis* Fr.) is occasionally damaging on cineraria in the glasshouse, requiring fungicide control (Fletcher 1984).

C. tussilaginis, common in much of C. and N. Europe, is usually called the pine needle rust. Basidiospores formed on one of the alternate hosts infects pine needles of the current year in the autumn. In the following spring, small yellow spermogonia may be found from April onwards, to be succeeded by the aecia, which burst open to shed their spores in early and mid-summer. The aeciospores infect the alternate hosts, on which the uredinia arise

as orange spots on the undersides of the leaves and on the stems; on winter-hardy plants such as *Senecio vulgaris* they produce their spores through the year. The telia are formed in autumn as bright orange-red crusts on the undersides of the leaves. If the alternate hosts are common, *C. tussilaginis* occasionally causes significant damage to nursery plants and to young, closely planted trees but usually the fungus is of negligible importance. Control measures for this disease are rarely necessary. Good weed control in the forest nursery to eliminate the common weeds which act as alternate hosts should keep infection to a low level.
PHILLIPS

Fig. 68. Telia of *Cronartium flaccidum* on *Vincetoxicum hirundinaria.*

Cronartium flaccidum (Alb. & Schwein.) Winter

(Syn. *C. asclepiadeum* (Willd.) Fr.)

Basic description

The genus *Cronartium,* to which *C. flaccidum* belongs, is characterized by intracortical and subepidermal spermogonia (pycnia). Aecia, subepidermal and of peridermioid type, erupt wrapped by a peridium; aeciospores are catenulate, verrucose and obovoid-ellipsoid. Uredinia, wrapped by a peridium, are subepidermal and open with a pore; urediniospores, which originate individually, are echinulate. Telia, originally subepidermal, begin to erupt in the shape of columns; teliospores are unicellular, catenulate and firmly adherent laterally and terminally. Basidia are external.

Host plants

C. flaccidum produces its uredinia and telia on *Vincetoxicum hirundinaria* (Fig. 68) and various other wild or garden herbaceous plants (*Impatiens, Nemesia, Paeonia, Pedicularis, Schizanthus, Tropaeolum, Verbena*). In Europe, the main aecial hosts are *Pinus halepensis, P. nigra* (in its different forms, including sspp. *laricio*, Fig. 69, and *nigra*), *P. pinaster, P. pinea* and *P. sylvestris,* though the last is considered resistant in Italy.

Fig. 69. Aecia of *Cronartium flaccidum* on *Pinus nigra* ssp. *laricio.*

Host specialization

Gäumann (1959) differentiated three formae speciales by telial hosts:
1) **f.sp. *typica*** on *Vincetoxicum* and *Paeonia* but not *Gentiana*;
2) **f.sp. *gentianae*** on *Gentiana* and *Paeonia* but not *Vincetoxicum*;
3) **f.sp. *ruelliae*** on *Ruellia* but not *Gentiana, Paeonia* or *Vincetoxicum*.

Geographical distribution

f.sp. *typica* is found in the countries of C. and S. Europe, in the southern regions of Scandinavia and the UK. f.sp. *gentianae* occurs over the Austrian Alps, the Carpatico-Danubian region, Caucasus and China. f.sp. *ruelliae* is present in the Baltic regions of the USSR.

Disease

The alternate hosts show orange uredinial pustules, and then telia (projecting reddish bristles) on the lower leaf surface. Pine needles, infected by basidiospores, show yellow-red spots 1–2 months later. The fungus forms a stroma, between mesophyll and endoderm, from which hyphae develop along the vascular bundles of the needle, reaching the stem. Through the leaf traces, mycelium reaches the wood and the pith. During the summer and autumn of the year following infection, in the case of 1–2 year old seedlings, spermogonia form on the stem (they look like drops of viscous liquid). In the case of plants more than 5 years old, spermogonia appear after 2–3 years. Their appearance is preceded by a slight swelling of the infected part. Aecia ripen at the beginning of the 2nd or 3rd year. After about a month, they are exhausted, and this stage is followed by necrosis of the surrounding cortical tissue, reaching the wood and phloem, so that the top can die.

Mycelium of *C. flaccidum* remains active in surviving cortical tissue and extends, in the following seasons, forming new spermogonia and aecia. So the necrotic area increases with the progression of the infection to the oldest branches and to the stem, where large cankers are formed. If they girdle the stem, the tree is killed.

Epidemiology

C. flaccidum survives as systemic mycelium in infected pine branches and trunks, and, in this genus, the telial stage has no perennating function. Basidiospores, formed on the alternate host, are relatively short-lived and delicate; they are dispersed most readily at night, by light winds which can carry them large distances to isolated woods, especially in mountain areas. In Italy, basidiospore dispersal mainly occurs in June–September (coastal areas) and July–October (mountains). Heavy rain brings down the airborne inoculum and prevents spread. Infection of pine is favoured by the same conditions as basidiospore spread. Basidiospores on the needle surface germinate with a promycelium which reaches and penetrates a stoma in 1–20 h, according to the susceptibility of the pine species. Aecia appear in March–April (coastal areas) or May–June (mountains). The wind-dispersed aeciospores germinate best in the early morning (18–22°C; RH 60–80%). Orange uredinial pustules appear 8–14 days after aeciospore infection of the alternate host, which can be reinfected by urediniospores, with the same requirements for dispersal and germination as aeciospores.

C. flaccidum manifests itself cyclically with epidemic waves in connection with favourable circumstances: i) rainy years, with mild temperatures; ii) the presence of new pine plantations in which the alternate host is present (pine is particularly susceptible up to 20 years old).

Economic importance

Damage arises from the death of the top or of the whole tree when systemic infection has spread to the stem. Italy is the country in which the most serious epidemics have taken place; in fact, since 1937, whole plantations of *P. pinea, P. pinaster* and *P. nigra* have been destroyed (Moriondo 1975). This epidemic spread in Italy is probably connected with the new planting of pine in areas where *V. hirundinaria* is widespread, or where pine was never present, or else with the renovation of old plantations. In other countries *C. flaccidum* is present in a sporadic way, presenting no economic problems, because the climate is less favourable.

Control

New planting of pine, especially in pure stands, should be avoided in areas where the alternate host is present. Infected plants should be destroyed to reduce the quantity of inoculum and resistant species should be used. *P. sylvestris* is resistant in Italy but not in N. Europe. In Italy an almost complete lack of genetic resistance has been ascertained in *P. pinea*. This is not so in *P. pinaster,* where a great variability of behaviour between provenances has been found (Raddi & Fagnani 1978).

Special research interest

Observations on basidiospore germination on the needle surface, with a fluorescence microscope (Ragazzi & Dellavalle 1982) show variation in behaviour on susceptible (*P. nigra* ssp. *laricio*) or resistant (*P. sylvestris*) species. Resistance factors may thus operate at the level of the epidermis and such observations could be used for the selection of resistant individuals.

RAGAZZI

Cronartium ribicola J.C. Fischer

Basic description

The yellow spermogonia of *C. ribicola* measure 2–3 mm across and exude golden droplets containing spermatia. The aecial blisters (2–4×1–3 mm) form orange, globoid, ellipsoid or polyhedroid aeciospores (22–29×18–20 μm). Small uredinia are produced in groups 1–5 mm across, with orange urediniospores measuring 21–24×14–18 μm. The orange or brownish, bristle-like telia are up to 2 mm long, and are made up of oblong teliospores measuring up to 70×21 μm (Wilson & Henderson 1966; CMI 283).

Host plants

C. ribicola produces its spermogonia and aecia on pines of the sub-genus *Strobus* (the soft or 5-needled pines) and its uredinia and telia on *Ribes*. In Europe, *C. ribicola* is most important on the introduced *Pinus strobus,* and to a lesser extent on *P. monticola.* In N. America it also damages other 5-needled pines. Some soft pine species, including the European *P. peuce* and the introduced *P. aristata, P. koraiensis,* and *P. wallichiana* are markedly resistant. The resistance of pines also increases with the age of the tree, and there is evidence that infection of pines often takes place in the nursery. Different species of *Ribes* also vary in susceptibility, blackcurrant being more susceptible than gooseberry, which is more susceptible than red and white currants. Blackcurrant cv. Viking has been bred for resistance to *C. ribicola.*

Geographical distribution

C. ribicola is thought to have originated in Asia, but it is economically important mainly in N. Europe and N. America (CMI map 6).

Disease

On pines, the fungus, usually called white pine blister rust, produces perennial cankers (like *C. flaccidum*). The earliest symptoms show as small discoloured spots on the needles. From the spring following infection, small yellowish swellings may be found on the shoots, on the bark at the bases of affected needles. Later still (some 2–4 years after infection) spermogonia and white aecial blisters form in spring and summer on the swollen tissues, and the aecia burst open to release clouds of orange aeciospores. The fungus is perennial in the pine, and forms cankers on the branches or girdles them and then spreads to the main stem. The girdled branches die, and their red-brown, dead needles (and in spring and summer the aecial pustules) form the most conspicuous symptoms of the disease. On the main stems, large cankers bearing aecia are also formed, and eventual girdling of the trunk commonly results in the death of the tree. On leaves of *Ribes,* the fungus produces its uredinia and a close felt of bristle-like telia. Attack is annual. In severe cases, the affected leaves become brown or black, die and fall early.

Epidemiology

The teliospores on leaves of *Ribes* germinate in summer and autumn to produce the basidiospores that infect pine needles through the stomata and directly through the epidermis. Basidiospore production requires high humidities, and is inhibited at temperatures above 20°C (Van Arsdel *et al.* 1956). Hence infection tends to occur in relatively cool spells with mist or heavy dew. Infection of pines is therefore intermittent, and in most areas does not occur every year. The mycelium of the fungus grows down into the shoot, on which spermogonia arise from March to September. Their appearance is usually delayed, however, until, at the earliest, the autumn of the first year after that of infection, and more often until the summer of the year after this. Further canker development then takes place, and aecia arise, usually first of all in the spring following the production of the first spermogonia. The considerable delay between infection and the production of readily visible symptoms makes it difficult to achieve any effective inspection of imported tree stock.

Aeciospores may be released at various times from March until mid-summer, and infect young leaves of *Ribes* species. Leaves of *Ribes* species are very susceptible for about the first 16 days after they begin to unfold, but their susceptibility then quickly declines. Uredinia and telia appear on the undersides of the leaves from about July–October. The urediniospores cause further infections on the *Ribes* leaves. The teliospores germinate on leaves on the plant and on the ground, and the basidiospores they produce complete the cycle, re-infecting 5-needled pines.

The basidiospores soon lose their viability and so cannot spread the fungus far to adjacent pines. The maximum spread is usually considered to be 5–6 km but effective spread is generally much less.

Urediniospores, however, may spread up to 300 km throughout large *Ribes* populations, thus enlarging the reservoir of infected plants which may then threaten 5-needled pines over a widened area. Aeciospores from pines may spread infection to *Ribes* plants for up to 600 km, though it is unusual for spread to exceed about 250 km.

The fungus *Tuberculina maxima* Rostrup sometimes parasitizes the aecia of *C. ribicola,* but seems to exercise little control.

Economic importance

Damage to *Ribes* is usually slight, as infection develops relatively late in the season. If frequent annual attacks occur and cause much premature defoliation, the plants are weakened and fruit production is reduced. *C. ribicola* has caused severe losses on 5-needled pines, however. In many areas, both in N. America and Europe, it has killed large numbers of trees, and cankers in surviving trees have produced serious defects in the wood. In many European countries damage by *C. ribicola* has led to the abandonment of *Pinus strobus* as a forest tree.

Control

Many attempts have been made to protect pines from *C. ribicola* either by growing them only at a distance of several km from *Ribes* species or by eradicating the latter from pine areas. Some success attended extensive eradication programmes in parts of N. America. In Europe, success has been very limited, partly because blackcurrant is an important crop in many countries, both in gardens and on commercial holdings. Indeed, in Norway, blackcurrant growers can demand the destruction of infected *P. strobus*. On a small scale, the disease on pines has been mitigated by pruning out all the lower branches, or (to less effect) cutting out obviously diseased branches before the fungus has had time to spread into the trunks of the affected trees (Lehrer 1982). Frequent spraying of young pines with maneb gives effective control but is scarcely economic (Gremmen & de Kam 1970). If the disease is troublesome on blackcurrants, the plants can be sprayed with copper fungicides such as Bordeaux mixture. Much work has been done by plant breeders to produce pines combining resistance to *C. ribicola* with other characters of value to the forester. Some resistant blackcurrant cultivars have been successfully bred.
PHILLIPS

Endocronartium pini (Pers.) Y. Hirats. probably has its origin as an autoecious endocyclic form of the *Peridermium* aecial stage of *Cronartium flaccidum*. Hiratsuka (1968) observed how aeciospores of such forms, on germinating, give rise to a basidium-like organ, instead of dikaryotic mycelium. Thus, pine can be re-infected directly via these spores, which are effectively teliospores. *E. pini* occurs in C. and S. Europe, Ireland, southern UK and Scandinavia and the European part of the USSR. It causes swellings on the branches of various pines: *P. nigra* ssp. *laricio, P. mugo, P. pinaster, P. sylvestris, P. uncinata.* Infection takes place through the needles or shoots of the current year and aecia appear after 2–3 years. Branches may be completely surrounded and from them the infection passes to the main stem. At this stage the whole plant is threatened. The most susceptible plants are those between 40 and 50 years old, which manifest the infection in large flows of resin. The disease never reaches an epidemic level, however, and is of little economic concern. It may be prudent to destroy infected trees.
RAGAZZI

A similar situation exists in N. America with ***E. harknessii*** (J. Moore) Y. Hirats., related to *C. coleosporioides* (q.v.). This, like the other non-European *Cronartium* species in N. America, is considered a quarantine pest for Europe (EPPO 11). Several species of *Cronartium* (with *Peridermium* aecial states) occur only in N. America (CMI 557–9; maps 444, 475–6, 541). The aecial hosts are *Pinus* species, while the telial hosts are either herbs or shrubs (Scrophulariaceae, *Comandra, Myrica*—in the case of ***C. coleosporioides*** Arthur, ***C. comandrae*** Peck and ***C. comptoniae*** Arthur), or trees (*Quercus, Castanea* in the case of ***C. fusiforme*** Hedgc. & Hunt ex Cummins and ***C. quercuum*** (Berk.) Miyabe ex Shirai). They present a serious potential risk to pines in Europe (EPPO 9,11) although it is not really known in many cases which European species could act as alternate hosts. Some do occur in Europe (*Myrica gale, Comandra umbellata*). Similar species occur in N. India (***C. himalayense*** Bagchee) and the Far East (***C. kamtschaticum*** Jørstad) (EPPO 11, 18), the latter being regarded as a *forma specialis* of *C. ribicola* by some authors. Most European countries prohibit import of *Pinus* from outside Europe because of this risk.

Cumminsiella mirabilissima (Peck) Nannf. (CMI 261), autoecious and macrocyclic, forms conspicuous uredinial pustules on the under surface of leaves of the ornamental shrub *Mahonia aquifolium*, appearing as red spots. It can be damaging to nursery plants. It is a N. American fungus, introduced to Scotland in 1922 and now widespread throughout Europe.

Gymnosporangium sabinae (Dickson) ex Winter (syn. *G. fuscum* DC.) forms its aecia on pear and its teliospores on *Juniperus* species of the *sabina* group (CMI 545). On pear, the leaves bear reddish-orange spots, variable in shape, usually 1–3 cm in diameter. Spermogonia appear on the upper surface as minute rounded bodies about 1 mm in diameter. The aecia, formed on the lower surface, appear as fibrous creamy-white bundles 2–5 mm long (*Roestelia*). The pathogen can also cause perennial cankers on pear branches, with the formation of aecia. These may be destructive on young trees of certain pear cultivars, particularly for the shape of the trees. On juniper, *G. sabinae* causes canker-like swelling of the shoots, on which brown telial 'horns' bear the teliospores. The fungus is widespread in Europe but has rather limited economic importance. Where the alternate host is scarce, it may be sufficient to avoid planting pear within 1 km of where it occurs, but this is impractical in southern countries where *J. sabina* is widespread. No specific control is needed, most treatments against *Venturia pirina* in any case being affected.
BONDOUX

G. tremelloides (A. Braun) ex R. Hartig causes a similar disease of apple (CMI 549), the alternate hosts being *Juniperus* species of the *oxycedrus* group, including the very widespread *J. communis*. It is in practice of minimal economic importance, for much the same reasons as *G. sabinae*. Several other species similarly alternate between *Juniperus* and woody Rosaceae: ***G. clavariiforme*** (Wulfen) DC. (CMI 542) on the *oxycedrus* group, with aecia mainly on *Crataegus* but also on pear and *Amelanchier*; ***G. confusum*** Plowr. (CMI 544) on *oxycedrus* and *sabina* groups with aecia mainly on *Crataegus, Chaenomeles, Cydonia, Mespilus* and rarely on pear; ***G. cornutum*** Arthur ex Kern (syn. *G. juniperi* Link) on *J. communis* and *Sorbus aucuparia*. All are of minor importance on their main hosts and occur on economically more important hosts such as pear only rarely.

Several *Gymnosporangium* species are restricted either to N. America (***G. juniperi-virginianae*** Schwein., ***G. globosum*** Farlow and ***G. clavipes***(Cooke & Peck) Cooke & Peck) or to the Far East (***G. shiraianum*** Hara, ***G. yamadae*** Miyabe ex Yamada) or both (***G. asiaticum*** Miyabe ex Yamada) (CMI 541, 542, 547, 550; maps 561, 121, 528, 530). Their telial hosts (various *Juniperus* species) occur naturally or as planted ornamentals in Europe, and these fungi accordingly present a risk to their aecial hosts apple and pear in Europe. Introduction on the latter is unlikely, since only plants in dormancy are likely to be traded. However, an increasing air-freight trade in Bonsai trees from the Far East presents a real risk of introduction on junipers (EPPO 13).

Hamaspora longissima (Thüm.) Körn occurs on many *Rubus* species in Africa and Asia and presents a certain risk for Europe (EPPO 14).

Melampsora populnea (Pers.) P. Karsten

(Syn. *M. tremulae* Tul.)

Basic description

The urediniospores produced in minute orange uredinia, measure 15–25×11–18 μm. The almost equally small telia contain teliospores measuring 22–60×7–12 μm. The fungus exists in three main formae speciales: **f.sp. *pinitorqua*** R. Hartig, **f.sp. *laricis*** R. Hartig and **f.sp. *rostrupii*** Wagner ex Kleb., often separated as individual species under the names *M. pinitorqua* Rostrup, *M. larici-tremulae* Kleb. and *M. rostrupii* Wagner ex Kleb. (syn. *M. aecidioides* Plowr.). The aeciospores form in caeomata; those of f.sp. *pinitorqua* measure 14–20×13–17 μm, those of f.sp. *laricis* 14–17×12–16 μm, and those of f.sp. *rostrupii* 13–24×11–17 μm. These rusts are further described by Wilson & Henderson (1966) and Lanier *et al.* (1978).

Host plants

The uredinia and telia of the rusts of this group are produced on the undersides of the leaves of *Populus tremula* and *P. alba*; those of f.sp. *pinitorqua* also form on *P. canescens*, and those of f.sp. *rostrupii* on *P. tremuloides*. Different poplar clones vary in their resistance to *Melampsora* species but clones resistant in one area may not be so in another, because of variations within the forms of the rust. The spermogonia and caeomata of f.sp. *pinitorqua* are formed on the stems of pine shoots, especially those of *Pinus sylvestris*, but also on those of *P. pinea* and *P. pinaster* (both of which are very susceptible), and sometimes other pines. *P. nigra* ssp. *laricio* and *P. contorta* are little affected. The spermogonia and caeomata of f.sp. *laricis* are formed on the needles of European larch, while those of f.sp. *rostrupii* are produced on the leaves of species of the wild herbaceous plant *Mercurialis*.

Geographical distribution

M. populnea is common throughout much of Europe, the Middle East, Kenya and S. Africa, and Uruguay (CMI map 389).

Disease

On poplar leaves the uredinia are found from August into the autumn, on the undersides, as orange spots about 1/3 mm across. They are closely followed by the slightly larger, dark brown telia. The sori often cover almost the entire under surfaces of the leaves, which may then fall prematurely. On pines, f.sp. *pinitorqua* is known as the pine twisting rust, because of its effects on the shoots of the tree. Infection of young pine shoots takes place laterally, some way below the tip, so that growth ceases on one side, and the shoot bends. Sometimes the stem is girdled, and the tip then hangs down and dies. Multiple leaders may develop below the damaged tip. Otherwise the tip continues growth, and straightens above the bent zone. At the point of infection within the bend, flat, irregular, blister-like spermogonia are formed, and large, elongated aecia (caeomata) up to about 20×3 mm then develop in the cortex. In May and June they burst to reveal the mass of orange-yellow spores. In pine nurseries in South Karelia, Krutov (1981) found that wet, warm spring weather encouraged infection by f.sp. *pinitorqua*. He also formulated a longer term forecasting system based on June, July and August temperatures and the total rain falling when the temperature was over 5°C in the August of the preceding year.

On larch, the small, pale yellow spermogonia of f.sp. *laricis* form on both sides of infected needles, and are followed by small, pale yellow, pustular aecia (caeomata), usually on the underside only. The fungus on larch is very inconspicuous, and does not appear to be common or of any importance. On *Mercurialis*, the spermogonia and the bright orange caeomata of f.sp. *rostrupi* appear on the undersides of the leaves and on the petioles and stems, usually from April to June. They occur on yellow spots which often form confluent rounded or elongated groups.

Economic importance

Although the uredinia and telia are numerous and conspicuous on poplar leaves, attack develops late in the season after growth of the host has ceased so there is little damage in most areas. However, considerable losses have sometimes resulted locally (though less than those caused by *M. larici-populina*). On pines, twisting rust may cause severe damage to young trees by killing or deforming their tops and inducing the production of multiple leaders. Attacks are local and sporadic, however, occurring almost entirely where pines have been planted on the sites of broad-leaved woodland containing residual suckering aspen (*Populus tremula*).

Control

Control measures need to be considered only in the case of attacks on *Populus* and *Pinus*. Where significant damage does occur in poplar nurseries, a copper fungicide may be applied to the plants, while a search is made for poplar clones resistant to the races of the fungus present in the area. Susceptible species of pine should not be planted on sites containing aspen. If the disease appears, aspen may be weeded out, but this is expensive and rarely justifiable, because as the trees reach the thicket stage the aspen is suppressed and further infection ceases.
PHILLIPS

Melampsora larici-populina Kleb.

M. larici-populina forms spermogonia and aecia on needles of *Larix*, and uredinia and telia on the leaves of species of *Populus*, mainly of the sections *Aigeiros* and *Tacamahaca* and their hybrids. The fungus is native to Eurasia and N. Africa. It has been introduced, together with *Populus* species, into regions where *Populus* and *Larix* are not native, for example into the rest of Africa, and into S. America, Australia, and New Zealand. The fungus can overwinter in the uredinial state, at least in relatively warm climates. The small, orange uredinia are formed on the lower side of poplar leaves, while the reddish-brown telia occur on the upper side (CMI 479). Poplar species, cultivars and hybrids, and even closely related clones, differ greatly in their resistance to *M. larici-populina* and there are pathotypes within the rust fungus that attack different ranges of poplar clones. Attack on larch species by *M. larici-populina* is of very little importance. Poplars, however, are sometimes badly damaged; the leaves may

be heavily attacked and fall prematurely. The best control is probably by selection and breeding of resistant clones. The work is, however, complicated by the existence of different races of the fungus and the probable formation of new races by recombination of genes, when the alternate host larch is found near poplars.
ROLL-HANSEN

Melampsora medusae Thüm. is of considerable importance in N. America, and also occurs in S. America, Japan and Australia, where it caused much concern when recently introduced (EPPO 74). The aecial host is larch, and the uredinia and telia occur on various poplars (CMI 480). In Europe, where it has been recorded sporadically, it has recently caused some severe attacks on poplar in S.W. France (Taris & Estoup 1975). In such warm areas, it may be able to overwinter in the uredinial state (Pinon 1986). A final poplar rust is ***M. allii-populina*** Kleb., which may cause defoliation of various poplars in section *Aigeiros*. It occurs in many European countries, as well as in parts of Africa, S. America and S.W. Asia. Uredinia and telia form on poplar and spermogonia and aecia on *Arum* and *Allium* species. There is also a whole series of *Melampsora* species on willows, as summarized in Table 16.1. Of these, ***M. amygdalinae*** and ***M. salicis-albae*** can be seriously damaging to their hosts, used as basket willows. Attacks on the aecial hosts are of no significance.

Melampsora farlowii (Arthur) J.J. Davis, an autoecious rust forming only telia, causes rust on *Tsuga* in N. America. With the increasing importance of *T. heterophylla* as a planted species in Europe, introduction of this disease must be prevented (EPPO 15).

Melampsora lini (Ehrenb.) Desm.

M. lini attacks cultivated flax and various wild *Linum* species (CMI 51). It is widespread throughout the world, wherever *Linum* occurs wild or is cultivated (Europe, Asia, America, Australia) (CMI map 68). The form on cultivated flax is **var. *liniperda*** Körn, and on wild species **var. *lini*.** In spring, attacks are first recognized by the appearance on both leaf surfaces of yellowish-orange aecial pustules, which are first subepidermal then erumpent. Uredinia then appear on the leaves, flower buds and sometimes stems, at first subepidermal, then pulverulent. Finally, on leaves but especially stems, the telia appear as very characteristic brownish-red then black patches or crusts.

The fungus mainly persists as teliospores in crop debris left in the field or contaminating seed lots. Under favourable conditions, these give rise to the basidiospores which re-infect flax (i.e. an autoecious, macrocyclic rust). High humidity and fairly high temperature (optimum 26–27°C) favour all stages in the life cycle, especially the germination of aeciospores and urediniospores. Heavy soils are reported to favour the disease. Numerous highly host-specific races (pathotypes) have been described on both wild and cultivated flax (14 in the USA and Canada, 4 in Europe). Of these, only two are known in France. There are correspondingly striking differences in sus-

Table 16.1. *Melampsora* species on *Salix* in Europe

	Spermogonia and aecia	Uredinia and telia
M. allii-fragilis Kleb.	*Allium* spp.	*S. fragilis* *S. pentandra*
M. amygdalinae Kleb.	—	*S. triandra* (autoecious)
M. capraearum Thüm.	*Larix* spp.	*S. capraea* etc.
M. epitea Thüm.		
(syn. *M. euonymi-capraearum*	*Ribes* spp.	*S. purpurea*
Kleb., *M. repentis* Kleb.,	*Euonymus europaeus*	*Salix* spp.
M. larici-epitea Kleb., *M. ribesii-*	Orchidaceae	*Salix* spp.
purpureae Kleb.)	*Larix* spp.	*S. viminalis* etc.
M. larici-pentandrae Kleb.	*Larix* spp.	*S. pentandra* *S. fragilis*
M. ribesii-viminalis Kleb.	*Ribes* spp.	*S. viminalis*
M. salicis-albae kleb.	*Allium ursinum*	*S. alba*

ceptibility between species and cultivars of flax. Flax rust forms the basis of Flor's classic gene-for-gene theory on the genetics of race-cultivar specificity. The formation of numerous telia on the stems at the time of most active growth can cause much damage: yield and quality are much reduced, the stems are fragile, and the fibres are irregular in length and unsuitable for processing. However, though once considered a very serious flax disease, rust is now of little importance in Europe due to the ready availability of resistant cultivars, which seem fairly stable. If attacks do occur, crop residues should be burnt.
JOUAN

Melampsora ricini Noronha (probably synonymous with *M. euphorbiae* (Schubert) Castagne, produces uredinia and telia only on *Ricinus communis*. Mainly important in India, it has caused losses from defoliation in Italy and Portugal (CMI 171).

Melampsorella caryophyllacearum Schröter

M. caryophyllacearum forms spermogonia (minute, honey-coloured) and aecia (shortly cylindrical with a torn white margin) on the needles of almost all *Abies* species, the spermogonia mainly on the upper sides, the aecia on the under surfaces. The numerous uredinia (small yellow pustules) and telia (yellowish or pinkish spots) form on the undersides of the leaves of species of *Arenaria, Cerastium* and *Stellaria*. The fungus is systemic in both the *Abies* species and the alternate hosts. *M. caryophyllacearum* is widespread in Europe into the USSR, and in N. America, parts of S. America, and in China and Japan.

Basidiospores from the alternate hosts infect buds of *Abies* in May and June. The infected shoots develop elongated swellings, which in the following summer produce many erect shoots (witches' brooms) bearing spirals of soft, pale yellowish needles. From June to September, spermogonia and aecia develop on these needles, which are shed by the end of the summer. The tumerous swellings and the witches' brooms (which may persist for 20 years or more) enlarge each year, the surface of the tumours becoming cracked and cankered. Infection is encouraged by moist spring weather, and occurs if basidiospores are released when young *Abies* shoots are at a susceptible stage (Pupavkin 1982). In some areas *M. caryophyllacearum* may cause severe damage, especially on mature and over-mature trees, reducing growth and sometimes killing trees. Decay fungi may invade the cankers, and wind, snow and ice may break trees where canker production has made the wood brittle. Where possible, some control of the fungus may be exercised by cutting out the witches' brooms. On a forest scale, however, it is possible only to remove affected trees when thinning.
PHILLIPS

Melampsoridium betulinum (Pers.) Kleb. may occasionally defoliate young birches in the nursery and the forest. It forms its uredinia and telia on the affected leaves and its aecia on larch needles. It also infects alder and is to be distinguished from ***M. alni*** Kaneko & N. Hiratsuka, which only occurs in the Far East (Roll-Hansen & Roll-Hansen 1981).

Species of ***Milesina*** are of interest in alternating between ferns (uredinia and telia) and *Abies* species (spermogonia and aecia). ***M. kriegeriana*** (Magnus) Magnus is quite common as a needle rust on *Abies* in Europe but of little importance.

Phragmidium mucronatum (Pers.) Schlecht.

P. mucronatum species belongs to a genus of autoecious, macrocyclic rusts, confined to the Rosaceae (CMI 204). All species have thick-walled, multiseptate, pedicellate teliospores. In Europe, four species attack roses but only *P. mucronatum* and ***P. tuberculatum*** J. Müller (CMI 208) are important (Wilson & Henderson 1966). *P. mucronatum* only attacks the Albi group of roses and species of Bracteatae and Caninae including *Rosa laxa* hort, while *P. tuberculatum* is restricted to hybrid tea and floribunda cultivars. *P. tuberculatum* overwinters solely by teliospores while *P. mucronatum* can produce systemic mycelia which give rise to aecia in hypertrophied buds in spring.

Rose rusts occur world-wide. They spread between leaves by windblown aeciospores and urediniospores. Teliospores which develop in late summer/autumn have hygroscopic pedicels and many adhere to stems germinating the following spring producing basidiospores which infect emerging leaves. Losses arise from premature abscission of rusted leaves. This is particularly important on *R. laxa* which is used as a rootstock for bud-grafting hybrid tea and floribunda cultivars. Defoliation reduces root and shoot development affecting subsequent growth and therefore the quality of cultivars budded on to rusted

stocks (Shattock & Rahbar Bhatti 1983). Curative fungicides, e.g. oxycarboxin, triforine, propiconazole, provide effective economic control—the actual choice of compound depends upon the presence of other pathogens, e.g. *Diplocarpon rosae, Sphaerotheca pannosa* (Shattock & Rahbar Bhatti 1983).
SHATTOCK

Phragmidium rubi-idaei (DC.) P. Karsten

P. rubi-idaei (CMI 207) resembles other species of *Phragmidium* but is characterized by having teliospores with up to 10 cells (Wilson & Henderson 1966). The related species ***P. violaceum*** (C. Schultz) Winter (CMI 209), ***P. bulbosum*** (Strauss) Schlecht. (CMI 203) and ***Kuehneola uredinis*** (Link) Arthur (CMI 202) attack only brambles (*Rubus fruticosus* and *R. caesius*) and are economically unimportant in Europe, although *P. violaceum* is being used as a biological control agent against *R. fruticosus* in Chile (Oehrens 1977) and Australia.

P. rubi-idaei attacks wild and cultivated raspberry and is widespread throughout temperate regions of the world. It overwinters, like *P. tuberculatum,* by teliospores. Spermogonia and aecia occur on adaxial surfaces of leaves of primocanes and fruiting canes in May–June followed by uredinia in June–July and telia in July–August. Premature defoliation of severely affected leaves usually occurs in August–September. There are no data on the effects of rust on yield of raspberry in Europe. Cane vigour control, involving removal of the first flush of primocane with dinoseb-in-oil, significantly reduces the incidence of rust (Anthony *et al.* 1987). British cultivars vary in rate-reducing resistance to rust. Malling Jewel is highly susceptible at the aecial stage but comparatively resistant at the uredinial stage (Anthony *et al.* 1985). The incidence of rust has increased with the introduction in the late 1960s of the susceptible cultivar Glen Clova. Curative fungicides, e.g. oxycarboxin, benodanil delay the development of rust in plantations but so far have not been used commercially (Wale 1981).
SHATTOCK

Pileolaria terebinthi Castagne (syn. *Uromyces terebinthi* (DC.) Winter) forms urediniospores and characteristically discoid teliospores in large encrusted pustules on the leaves of pistachio in E. Mediterranean countries (Viennot-Bourgin 1949). Infection of fruit clusters is also common, resulting in malformation of the pistachio nuts. Large pustules of uredinia are formed on the misshapen fruits. Protective sprays with dithiocarbamates could be effective, while current systemic fungicides against rust diseases could be tested.

Puccinia allii (DC.) Rudolph

Autoecious and macrocyclic, *P. allii* forms uredinia containing spherical to elliptical urediniospores with hyaline to yellowish, spiny walls (CMI 52). Aecia are rare. Telia, usually scattered in appearance and covered by the epidermis, contain chestnut-brown, elliptical to cylindrical, often irregular teliospores. In *Puccinia* these are typically once-septate, though some may be non-septate as in *Uromyces* (q.v.). There is no sharp distinction between these very similar genera. Indeed, forms of *P. allii* with only one-celled teliospores have been called *Uromyces ambiguus* (DC.) Lév. *P. allii* has a wide host range within the genus *Allium*, leek, garlic and onion being the most frequently affected hosts. Many authors separate the rust on leek as ***P. porri*** (Sow.) Winter, with slightly larger urediniospores. Isolates from leek do not cross-infect onion or chives (Gjaerum & Langnes 1981). The fungus is distributed world-wide.

Rust on leaves and stems appears as bright orange or somewhat brownish, circular to elongated uredinial pustules along the veins, followed by stromatic, blackish telia. On leek where only uredinia develop, another symptom may occur: chlorotic spotting of the leaves, which is probably due to unsuccessful fungal invasion (Dixon 1984). When rust infection is severe the leaves may be killed. Seed transmission of *P. allii* has been reported but not investigated thoroughly (Dixon 1984). The fungus is favoured by high humidity, moderate to low temperatures, dense planting, excessive soil nitrogen and deficiencies of potassium. *P. allii* has become a particular problem on leek in some regions of Europe where production is concentrated in small areas and plants are grown almost all the year round. The rust may also be a factor limiting cultivation of garlic, common onion, Welsh onion (*A. fistulosum*) and chives (*A. schoenoprasum*). A long rotation time is suggested to prevent build-up of the fungus and promising results can be achieved with dithiocarbamates and new systemic fungicides (Dixon 1984). Another possibility would be the application of *Bacillus cereus*, a bacterium which proved to be highly effective in inhibiting germination of urediniospores

and rust development in the host (Doherty & Preece 1978).
VIRÁNYI

Puccinia asparagi DC.

Macrocyclic and autoecious, *P. asparagi* attacks cultivated asparagus wherever it is grown as well as wild asparagus species (CMI 254). It has been reported naturally infecting onion and can be successfully inoculated onto various *Allium* species (Beraha 1955). Ornamental *Asparagus* species (e.g. *A. sprengeri*) are resistant. The first symptoms occur in the spring on the edible stalks in the form of oval, light-green lesions on which spermogonia and aecia are subsequently formed. These are followed 2–3 weeks later by the more conspicuous uredinial stage, which covers stems, twigs and leaves with small, rusty-red pustules. Primary uredinia can be surrounded on the stem by concentric rings of secondary and tertiary uredinia. Later in the season the uredinia give place to black telia. Uredinia form from substomatal primordia only, and the first formed urediniospores are occasionally extruded through the stoma, which is unusual for the genus (Lubani & Linn 1962). As the parasite remains localized in the superficial tissues of the stem, infections do not directly interfere with the growth of the plant. However, infected plants are progressively debilitated.

P. asparagi overwinters as teliospores on old stems. It develops over a fairly wide temperature range being particularly aggressive during the warmest summers (sporulation optimal at 25–30°C). Urediniospores germinate best at 10–15°C, and a wetting period of 3 h is sufficient for their germination at 15°C. Low night temperatures favouring dew formation and urediniospore germination, followed by high daytime temperatures favouring the development of uredinia are ideal conditions for epidemics of *P. asparagi* (Beraha *et al.* 1960). Control is based on eradication of initial inoculum. Stubble, volunteers and wild plants, should be burnt or sprayed with eradicants during the winter. In order to prevent dew formation, aeration of the crops should be favoured and irrigation reduced. Application of fungicides should be attempted, especially in seed beds, if rust infections are heavy.
MATTA

Puccinia caricina DC. var. *pringsheimiana* (Kleb.) D.M. Henderson

P. caricina var. *pringsheimiana* is macrocyclic and heteroecious, producing urediniospores and teliospores on several species of *Carex* and spermogonia and aecia on leaves and fruits of gooseberry, and occasionally on red and blackcurrant (which may be heavily attacked by other forms of *P. caricina*) (Wilson & Henderson 1966). Orange-red raised areas develop on leaves, petioles and fruits in May–June. Soon after the spots become covered with a close aggregation of aecia with long wide cylindrical peridia, giving the disease its name–cluster cup rust. The disease is widely distributed in Europe particularly in the proximity of sedges. It is common in Wales especially after a dry spring which delays germination of teliospores (Moore 1959). However, rust is only important if berries are affected causing considerable loss of crop. Control can be achieved by a single spray of protectant fungicide, e.g. maneb, applied 14 days prior to flowering. Rust can be avoided by not positioning plantations near sedges or by not using peaty mulches which may contain teliospores from infected sedges. In some places it is possible to eradicate the fungus by burning dead sedge early in the spring.
SHATTOCK

Puccinia coronata Corda

(Syn. *P. coronifera* Kleb., *P. lolii* Nielsen)

Basic description

A macrocyclic, heteroecious rust with spermogonia amphigenous, with projecting paraphyses. Aecia are amphigenous, in groups or scattered, cylindrical or cup-shaped, whitish, laciniate. The aeciospores are angular-globoid, finely verruculose, orange 16–25×12–20 μm. Uredinia are amphigenous, scattered or confluent, oblong, pulverulent, orange; and the urediniospores round to ovoid, yellow 14–39×10–35 μm. Telia are scattered or associated with uredinia, oblong or linear, long-covered by epidermis, black; with teliospores clavate, apex flattened with crown of 5–8 digitate projections (hence 'crown' rusts (Simons 1970).

Host plants

Uredinia and telia occur on cultivated and wild oats (*Avena* spp.) and on a wide range of grasses including species of *Lolium* and *Festuca, Dactylis glomerata* and *Phleum pratense*. Spermogonia and aecia are on *Rhamnus cathartica* and *Frangula alnus*.

Host specialization

The fungus exhibits a high level of specialization, and classification, based on both the primary and alternate host, is in a state of flux. Intensive investigation in both Europe and N. America has resulted in the separation of the rust into formae speciales of which at least seven are generally recognized. Brown (1938) distinguished seven varieties of *P. coronata* in Britain: **var. *alopecuri*** on *Alopecurus pratensis,* **var. *arrhenatheri*** on *Arrhenatherum elatius,* **var. *avenae*** on *Avena* spp., **var. *calamagrostidis*** on *Calamagrostis canescens* and *Phalaris arundinacea,* **var. *festucae*** on *Festuca elatior,* **var. *lolii*** on *Lolium perenne and* **var. *holci*** on *Holcus lanatus.* Each of these had previously been identified in continental Europe. She also reunited the two species *P. coronata* Kleb. and *P. coronifera* Kleb. (*P. lolii* Niels.) as there were no clearcut morphological or pathological differences. The varieties appear to be specialized on the alternate host with aecia of *P. coronata* var. *calamagrostidis* being produced on *Frangula alnus* only and those of the other varieties on *R. cathartica.* The additional **var. *gibberosa*** Lagerh. identified in Germany and Norway, appears to differ morphologically in having distinctive teliospores. An alternative approach retains var. *avenae* but regards the other varieties as formae speciales of var. *coronata.*

Within var. *avenae* there is extensive specialization. Specific cultivar/pathogen strain interactions are governed by major genes (designated Pc) in the *Avena* host and corresponding virulence genes in the pathogen. A large number (*c.* 300) of pathotypes each carrying different specific combinations of virulence genes has been identified on a world-wide basis.

Geographical distribution

The crown-rust fungus is of world-wide distribution in the uredinial state on *Avena* species and grasses, although the aecial state is somewhat more restricted. Var. *avenae* is widespread in N. America, Australia and Europe but is uncommon in the more arid areas of oat cultivation. Distribution is just as wide on non-cultivated *Avena* species and on wild and cultivated grasses.

Disease

On the gramineous host, the uredinial stage appears as bright orange, ragged, powdery pustules 1–5 mm in length. Secondary pustules may appear as satellites around the primary one and pustules may coalesce on susceptible cultivars to form irregular lesions. The powdery orange urediniospores are produced profusely and may appear to stain the leaf in wet conditions. Pustules form mainly on leaf blades giving the disease its common name of leaf rust in N. America. Uredinia also occur on leaf sheaths and floral parts and occasionally on culms. Telia are similarly distributed but are more common on leaf sheaths of oats late in the season.

Epidemiology

Infection of the alternate host, *Rhamnus,* is by basidiospores produced in the spring from germinating teliospores which have overwintered. The resulting aecia are generally produced in late spring to early summer and infection of the main grass host by aeciospores initiates the uredinial stage. Several uredinial generations are produced in a season, dissemination of the pathogen being by airborne urediniospores. In mild climates, the fungus may perennate in the uredinial stage either on grasses or autumn-sown oats. The limiting factor in the uredinial survival of var. *avenae* is probably survival of the oat host, in that autumn-sown oats are the least winter-hardy of the cereals. Spring infections are initiated either by aeciospores from neighbouring *Rhamnus* or by wind-blown urediniospores travelling from regions where the uredinial stage survives the winter. In eastern parts of Europe, the alternate host is relatively important whereas in maritime regions of N.W. Europe survival is probably by dormant uredinial mycelium. In the Nordic countries, such survival only occurs in W. Norway and long-distance dispersal of urediniospores is regarded as the most common means of spread. Urediniospore germination and host penetration require free moisture, a condition normally met by night-time dew. On oats, this means rapid development of disease in early summer, while on grasses climatic conditions associated with the autumn equinox are conducive to infection.

Economic importance

Damaging outbreaks of oat crown rust have been reported for a number of European countries, including the USSR and the UK. Severe damage also occurs in S. Africa, Israel, Australia, Canada, S. America and USA. Loss in yield is largely through a reduction in numbers of grains produced and in grain weight and this can be of the order of 10–20% although higher losses have been reported. Where grown for forage, for example in S. Africa, Australia and S. USA, severe damage can also occur. On forage grasses, crown rust can be particularly damaging to the aftermath growth especially in ryegrass (*Lolium* spp.). There may be considerable defoliation and loss of palatibility to stock. In addition, pure stands of *Lolium* species grown for seed may be severely affected.

Control

Chemical control of oat crown rust, although available, is rarely economic except on high-priced seed crops; this is even more true of forage grasses. Where other diseases such as powdery mildew (*Erysiphe graminis*) are also threatening it may be economic to use the modern broad-spectrum fungicides which are translocatable, prophylactic and therapeutic. These include diclobutrazol, fenpropimorph, propiconazole, triadimefon and triadimenol. The preferred method of control is through the use of resistant cultivars and there is a long history of breeding for resistance to var. *avenae* in oats in Europe, Australia and N. America. These efforts have relied largely on the use of major (Pc) genes which govern a hypersensitive response to the pathogen. Apart from the cultivated oat, *Avena sativa*, sources of such resistance include the hexaploid *A. sterilis,* the tetraploid *A. abyssinica* and the diploid *A. strigosa.* Such hypersensitive resistance may be temperature-sensitive in expression and may be expressed only in mature tissues although it is commonly insensitive to temperature and is expressed throughout the life of the plant. It is this resistance that is most readily overcome by pathotypes of the pathogen. This has led to the deployment of these resistances in multi-line cultivars in N. America (Browning & Frey 1969) as a means of effecting more stable control. In Europe, such resistances are present in some commercial cultivars, notably from Sweden and Czechoslovakia. Other quantitative resistance of a non-hypersensitive nature and which results in 'slow rusting' in the field is also being exploited in the hope that it will be more durable. In addition, susceptible cultivars vary in their degree of susceptibility and adequate control can be achieved in most situations by avoiding the growing of highly susceptible cultivars.

Resistance to var. *lolii* exists in *Lolium perenne* and cultivars of perennial ryegrass currently recommended in the UK vary widely in their degree of susceptibility. As in oats, the avoidance of highly susceptible cultivars should do much to control damaging outbreaks.

Special research interest

Apart from its economic importance, a number of biological features of *P.c.* var. *avenae* have made it attractive for research purposes. Genetical, physiological, histological and ultrastructural studies have given insight into host/parasite relationships and the nature of biotrophy. Current strategies for the control of plant diseases using mixtures of cultivars and multiline cultivars are founded in such proposals for the control of oat crown rust (Browning & Frey 1969). Resistance breeding and associated research dates from the early 1900s and continues up to the present day.

CLIFFORD

***Puccinia graminis* Pers.**

Basic description

Uredinial and telial pustules most typically occur on stems and are relatively large, about 2×5 mm, emerging from the parenchyma through the epidermis. The dry, recurved and frayed epidermis remains conspicuous. The pustules, longitudinally arranged are cinnamon brown in the beginning and become a dull black when the host plant ripens (hence the common names 'black rust' in Europe or 'stem rust' in the USA). *P. graminis* is heteroecious and macrocyclic (Lehmann *et al.* 1937; Wilson & Henderson 1966; Cummins 1971).

Host plants

P. graminis attacks barley, oats, rye, Triticale, wheat and numerous cultivated and wild grasses, except those from the tribes *Andropogoneae* and *Paniceae*. *P. graminis* stands as the example for the phenomenon of host alternation, completing the asexual phase of its life cycle on gramineous plants and the sexual cycle on *Berberis* species (barberry) and some *Mahonia* species. *B. vulgaris* has been responsible for numerous epidemics on cereals.

Host specialization

Different schools of thought exist according to evolutionary condition, environment or local scientific tradition. Evidence from Israel, in or near the gene centre, conflicts with information from peripheral areas such as N.W. Europe. Primarily, **ssp. *graminicola*** Urban infects many grass species (*Phleum, Dactylis, Festuca*) but not cereals. It has smaller urediniospores than **ssp. *graminis***, the black rust of cereals and grasses, within which **var. *graminis***, with somewhat smaller urediniospores, inhabits *Triticum, Aegilops* and *Elymus*, and **var. *stakmanii*** Guyot, Massenot & Saccas, with somewhat larger spores, infects *Avena, Hordeum* and *Secale*. Cummins (1971) recognized the subspecies but not the varieties.

Formae speciales are subspecific taxa representing the result of 'physiologic' specialization at the generic level (**f.sp. *avenae, secalis, tritici*** of Eriksson & E. Henning, etc.). Hybridization between formae speciales is possible in principle. The formae speciales are conspicuous and agronomically useful entitities in areas such as N.W. Europe, where cereal cultivation is evolutionarily young but their distinctive characters vanish in or near the gene centre, as in Israel. In Europe, barley is attacked mainly by f.sp. *secalis* but in N. America epidemics on barley are due to rust which also attacks wheat. Finally, races, taxa representing 'physiologic' specialization at the cultivar level, are numerous; several hundred are on record. Their distinction and identification is agronomically important since many are epidemiological entities readily recognizable by farmers. Recent advances allow every isolate to be characterized by its virulence and avirulence toward all known resistance genes in the host. This high resolution technique obviates the need for detailed race descriptions. New races or, in modern terms, new virulence combinations (pathotypes) appear frequently in response to the selection pressure exerted by cultivars resistant at the time of their introduction.

Geographical distribution

P. graminis occurs world-wide wherever susceptible hosts grow. The graminicolous forms are found mainly near barberry bushes. Typical pockets are (sub)alpine river flood areas heavily infested by *B. vulgaris*. In the Mediterranean area, other barberries (e.g. *B. aetnensis, B. cretica*) take over. In recent times, the major economic host is wheat, which can be attacked anywhere, but is especially endangered in S.W. and S.E. Europe.

Disease

All parts of the host are attacked with the exception of the roots and seeds. Uredinia produce cinnamon brown to dark brown wind-borne urediniospores. They occur on both sides of the leaves, especially on the lower (abaxial) side, but they are more common and quite conspicuous on the stems (peduncle and leaf-sheaths). The fungus also infects the inflorescences, especially the ears of wheat and rye. Uredinia are often surrounded by a light green halo. When the host ripens, the single-celled dikaryotic urediniospores are replaced by the two-celled diploid teliospores and the dark, brown uredinia turn into black telia, mainly on the stems. A wheat crop can appear as if damaged by fire. Black rust in cereals can appear in typical foci or hot spots, measuring 0.5–5 m in diameter, where the crop has a reddish glow and eventually turns brown. A person entering an infected crop returns with his clothes reddish-brown from adhering urediniospores. On barberries, the rust fungus forms conspicuous bright orange aecia, arranged in groups, which may disfigure ornamental species.

Epidemiology

Of the three wheat rusts (black, brown and yellow) black rust is the most thermophilic, usually appearing latest. *P. graminis* can overwinter in both the sexual and the asexual phase. On maturing gramineous plants, the fungus forms telia with teliospores, which usually germinate only after having been subjected to alternate freezing and thawing and/or wetting and drying. In spring, at about the time when the new, young barberry leaves are present, the basidiospores float through the air when the weather is moist. Intercepted by barberry leaves, they infect them to form inconspicuous lesions bearing 1 mm sized yellowish spermogonia on the adaxial leaf surface, which ooze a slime loaded with spermatia (haploid sperm cells). Insects transmit the slime (nectar or honey dew) from spermogonium to spermogonium, thus fertilizing heterothallic strains. The fertilized mycelium develops bright orange aecia, arranged in clusters (hence cluster cups) on the abaxial leaf surface. The aecia forcibly eject orange dikaryotic aeciospores which infect the gramineous host. As basidiospores seem to have a short flight range (< 10 m) and aeciospores usually have only a medium flight range (< 100 m), the sexual cycle is normally a localized affair. Nevertheless, many barberry-originated epidemics of black rust on wheat, rye and oats are on record.

Asexual overwintering in the uredinial phase is possible, but *P. graminis* is relatively thermophilic and asexual overwintering is rarely recorded north of the Alps. In the Mediterranean area, oversummering is a problem. Where differences in altitude provide suitable niches for survival in different seasons, the uredinial phase can persist all year round. Good evidence comes from Israel and Morocco, circumstantial evidence from Greece (Crete) and possibly Italy. Most of this overseasoning rust is graminicolous, but cerealicolous strains can, under suitable meteorological conditions, move north and eventually cause devastating epidemics, even in S.W. England and S.E. Sweden. Long-distance dispersal of urediniospores (with a flight range of hundreds of km) does occur in Europe, and when the crops in the target area are late the resulting infection can be disastrous. The rust appears either in small foci, which extend gradually, or evenly dispersed over the whole field. In Europe, there is no evidence for an annual southward migration of the rust, as in N. America (Zadoks 1965; Hermansen 1968; Zadoks & Bouwman 1986).

Economic importance

In wheat plants the fungus disturbs the normal sink-source relationships and functions as a strong sink, thus depriving the growing seeds of assimilates. As the epidermis is disrupted by the fungus, the plant may dry out so that the sap stream can no longer reach the ear. As a result, the plant will ripen prematurely, and form fewer and smaller seeds. Light to severe damage, and even total loss may follow. Stems may break where lesions are crowded. Even light damage can cause serious financial loss because of decreased milling and baking quality. In the Upper Danube Valley and in some areas of European USSR long-term averages for crop losses in wheat used to be around 10%, in other parts of Europe 1–5%.

Control

Barberry eradication, imposed by law, has effectively controlled annual outbreaks of black rust on wheat, rye and oats in Sweden and Denmark. In Bavaria, barberry-free zones bordering wheat areas have been reasonably effective. In the Netherlands, susceptible ornamental barberries were replaced by resistant cultivars. Resistance breeding, which can be highly effective, has not been practised much in Europe except in central, southern and south-eastern areas, and, for spring wheat, in Switzerland. Breeding for earliness to escape from infection has proved highly successful, e.g. in Italy. In Europe, chemical control has not passed beyond the experimental phase. Nickel salts, sulphur and dithiocarbamates, though reasonably effective, have not been registered for use. Systemic fungicides against black stem rust are not (yet) used in Europe, but have been successfully tested, e.g. in Hungary (triadimenol). Timely sowing, more or less simultaneously over relatively large areas, reduces risks; late crops are to be avoided.

Special research interest

For a long period, black stem rust of wheat was a popular research subject common to plant breeders, geneticists, epidemiologists, physiologists, and biochemists, and knowledge advanced with great strides. Recently, the topic has lost much of its glamour as the disease is considered to be under control, or at least to be controllable. A new research line is the comparison of results from virulence analysis and isoenzyme analysis.

ZADOKS

Puccinia helianthi Schwein.

P. helianthi, autoecious and macrocyclic, is common on cultivated sunflower and also attacks a number of other *Helianthus* species (Parmelee 1977). Pathotypes of the pathogen are known to occur in Canada and in Argentina but have not yet been noted from Europe. *P. helianthi* is distributed world-wide and has been recorded in almost all European and some adjacent Mediterranean countries engaged in sunflower production (Sackston 1978).

Symptoms of sunflower rust may appear on all the above-ground tissues of plants but are more prevalent on leaves. Small, orange to yellowish spots appear in compact circular groups followed by brownish, circular to elongated and pulverulent uredinia scattered over the leaf surface. Usually late in the season, deep brown to blackish, pulvinate telia develop on the affected, senescent tissues. *P. helianthi* overwinters as teliospores which germinate early in the spring, giving rise to primary infection. Penetration and successive developmental stages of the fungus are favoured by relatively high temperatures and moisture during spore germination.

Although well-distributed through the continent, *P. helianthi* causes considerable yield loss only in a few countries of Europe, e.g. in Bulgaria, USSR and Yugoslavia (Sackston 1978). Control measures to reduce build-up of inoculum and to prevent disease spread are of great practical importance. Resistant cultivars should be used, if available. If necessary, chemical control is suggested with either dithiocarbamates or triadimefon.

VIRÁNYI

Puccinia hordei Otth

(Syn. *P. anomala* Rostrup, *P. simplex* Eriksson & Henning)

Basic description

Macrocyclic and heteroecious, *P. hordei* has spermogonia amphigenous, numerous, in groups or scattered, spherical to ellipsoid, 86–144×99–158 μm, honey-coloured then blackish. Aecia are amphigenous, scattered, cupulate, yellow, 200–300 μm diameter. Uredinia on *Hordeum* species are amphigenous, scattered, minute, cinnamon brown, and the telia are amphigenous or on culms, blackish-brown, oblong, confluent, up to 3 mm, 6.5 mm on culms, remaining covered by epidermis (Clifford 1985). One-celled teliospores are quite common in this species.

Host plants

The uredinial and telial stages occur widely on cultivated barley, and commonly in Israel on the wild species *H. spontaneum* and *H. bulbosum*. The uredinial stage has also been reported on *H. murinum* in Norfolk, England. Spermogonia and aecia occur on *Ornithogalum, Leopoldia* and *Dipcadi* species (Liliaceae). In Israel, *Ornithogalum* species co-exist with wild *Hordeum* species and the alternate host is essential for the survival of the pathogen, as it is in Greece. In C. and N.W. Europe the alternate host is unimportant.

Host specialization

The forms on *H. murinum* and *H. bulbosum* have been considered an autonomous species, ***P. hordei-murini*** Buchw. but the evidence is that these forms are conspecific with *P. hordei*. The fungus exists as pathotypes on *H. vulgare* and these are differentiated by host genetic factors which are designated $Pa_{1\text{-}n}$.

Geographical distribution

Brown rust is the most important rust disease of barley and is widely distributed wherever the crop is grown, particularly in the cool temperate regions of barley cultivation. It occurs in N. America, Argentina, New Zealand and N.W. Europe at damaging levels. Intensification of barley cultivation in N.W. Europe has resulted in a considerable increase in the disease in the last 10–15 years. In the UK it was the most important disease of spring barley in 1970 and was also at high levels in 1971.

Disease

On the main barley host, uredinial infections first appear on the leaves as small, round, light yellowish-brown pustules which darken with age. A chlorotic halo is normally associated with the pustules and a more general yellowing of the leaf is common. Infection is normally confined to the leaf and leaf-sheaths although culm and head infection can occur late in the season. At this time, slate-grey telia may be formed. These often occur in stripes on leaf-sheaths and they are long-covered by the epidermis; they also occur on stems, heads and leaf-blades.

Epidemiology

The sexual stage is rare in the main areas of barley cultivation and is epidemiologically insignificant. Where endemic, the fungus perennates in the uredinial stage either on volunteer plants or autumn-sown crops. There has been a very large swing to autumn sowing in N.W. Europe in the 1970s and this, combined with the increased number of barley crops in the rotation and high fertility, has favoured the pathogen. Survival and development over winter are limited by both temperature and survival of host tissue. Cultivar susceptibility and competition with other leaf pathogens also affects survival. Free moisture, usually satisfied by night-time dew or light rain, is essential for germination and penetration. Germination occurs over the temperature range 5–25°C and is high at 10–20°C. Colonization is limited by temperature and increases to a maximum in the range 5–25°C. Under optimum conditions, sporulation begins 6–8 days after infection but can take up to 60 days at 5°C. The sporulation period is similar in the range 10–20° but is reduced at 25°C. Spore dispersal is by wind, and subsequent survival of spores is reduced by high light intensity. The uredinial stage can thus survive and develop under winter conditions prevalent in cool temperate regions, but rapid disease development only occurs in warm summer weather when free moisture is available overnight. These conditions are best satisfied by anticyclonic weather patterns during crop growth.

Economic importance

Although widespread where barley is grown, the disease has generally been considered unimportant in economic terms, although with changing cropping practices there have been some severe local outbreaks in N.W. Europe in the last 10–15 years. In trials, yield reductions of up to 30% have been reported but data on field crops is scarce: losses of the order of 10–20% have been recorded in England and New Zealand. Effects on the host depend on the duration and severity of infection. Spring barley, especially if late-sown, is particularly at risk. Early severe infections can result in reduced root and shoot growth, with stunting and a reduction in numbers of fertile tillers and grains per ear. Epidemics tend to occur late and consequently the most common effect is on grain size and quality. Yield reductions have been estimated at 0.5–1% for each 1% increment of rust assessed at the milky ripe stage of growth.

Control

The principal controls are through the use of genetically resistant cultivars or fungicides (Jones & Clifford 1983). Although the older contact eradicant fungicides such as the dithiocarbamates and sulphur are active against *P. hordei*, repeated applications are necessary which renders them uneconomic. During the 1960s and 1970s a number of systemic fungicides were developed and this, together with favourable economics of cereal growing, has resulted in their widespread use to control cereal diseases. Carboxin and benodanil, which are transported acropetally in the xylem and which are active against rusts including *P. hordei*, were the first to become available. More recently, compounds with a broader spectrum of activity against cereal pathogens have been developed and are widely used in N.W. Europe. These include benomyl, diclobutrazol, triadimefon, propiconazole, prochloraz and fenpropimorph.

Two main types of host resistance are recognized. The first (Type 1) is governed by major genes and, to date, nine such have been described and designated Pa_1-Pa_9. Few European cultivars carry such resistance although some modern winter barley cultivars appear to carry some unidentified factors. Virulence to all but Pa_7 is common and widespread in Europe, and similar spectra of virulence are found in Israel, N. Africa and New Zealand. In N. and S. America the pathogen population appears to carry fewer virulences. The second (Type 2) resistance is expressed quantitatively through the slow development of relatively few, small pustules. There is little or no necrotic host response which is typical of Type 1 resistant reactions. Such resistance, which is typified by cv. Vada, is derived from *Hordeum laevigatum* which, through a cross with cv. Gull, is in the parentage of many European cultivars. Although the inheritance is relatively complex it has been widely and effectively used in the development of resistant European cultivars since *c.* 1970. The resistance has remained stable and effective in widespread use and is thus durable.

Special research interest

Studies of Type 2 resistance, centred on descriptive histology, symptomatology and genetics, should now form the basis of physiological and biochemical investigations into the nature of durable resistance and its relation to non-host resistance and pathogen specificity.

CLIFFORD

Puccinia horiana **Henn.**

Basic description

P. horiana forms only teliospores (CMI 176; EPPO 80).

Host plants

Many species of *Chrysanthemum* are host plants to *P. horiana*.

Geographical distribution

P. horiana originated in China and Japan. Its spread across the world started in 1963, and it is now present in Europe, Australasia and S. Africa. In many European countries (France, FRG, Italy, etc.) it is now firmly established; however, several countries in which outbreaks have occurred have since succeeded in eradicating the disease (Denmark, Norway, Sweden, UK).

Disease

Leaves are infected by wind-borne basidiospores, which give rise to small (5 mm) pale green lesions. Telia form on the lower leaf surface, appearing pale buff or pinkish. The lesions become sunken on the upper surface and prominent on the lower. The telia turn whitish as basidiospores are produced (Fig. 70).

The symptoms are reflected in the name 'white rust' (in no way corresponding with the other so-called white rusts—*Albugo* spp.).

Fig. 70. Telia of *Puccinia horiana*, white with basidiospores, on the lower surface of a chrysanthemum leaf (British Crown Copyright).

Epidemiology

White rust is essentially a glasshouse disease, in which the pathogen survives on its living host throughout its life cycle. Teliospores can only survive on detached leaves for a few weeks (8 at 50% RH, or less at higher humidity). Almost certainly the fungus cannot survive outdoors in N. Europe. Infection of leaves by basidiospores requires a high RH and a film of moisture, and is optimal at 17°C. Basidiospore discharge will start within 3 h in favourable conditions, and penetration can occur in 2 h, so 5 h wetness is sufficient for new infection. The incubation period is normally 7–10 days, but growth is retarded by periods at higher temperature. The basidiospores are very sensitive to desiccation, and are therefore unlikely to be wind-borne for very long distances.

Economic importance

P. horiana is largely spread with planting material (chrysanthemum cuttings) and its importance in Europe follows from an increased importation of such material from specialist growers in Mediterranean or tropical countries where the disease was introduced. In countries where *P. horiana* is established, it causes considerable direct problems to growers, as well as the indirect problems which arise if the material is to be exported (Water 1981). Special intensive inspection and control procedures may be needed (Veenenbos 1984). Countries in which it is absent have to maintain equivalent inspection procedures at import and provide for costly eradication measures if outbreaks are found. In fact, an analysis of the relative cost of excluding or 'living with' the disease can be made (Lelliott 1984), and has led to the conclusion for the UK that exclusion continues to be economically viable on a 5-year average, although eradication costs in individual years have been very high. These results constitute a case-study for the economic justification of plant quarantine.

Control

The use of healthy planting material is fundamental. Destruction of leaf debris in an infested glasshouse will reduce the risks of carry-over. Various fungicides sprayed on the foliage have preventive, and even curative action (triadimefon, triforine, oxycarboxin); strains insensitive to oxycarboxin have been reported, but disappear in the absence of the fungicide. Heat therapy has been proposed as a method for destroying *P. horiana* in planting material (cuttings in water at 45°C for 5 min), but this remains to be tested in practice (Coutin & Grouet 1983).
SMITH

Puccinia pelargonii-zonalis Doidge

P. pelargonii-zonalis is an autoecious species producing urediniospores and teliospores on zonal pelargonium (CMI 266). It was first reported in S. Africa (1950–52) and Australia (1959) and has since spread to many parts of the world (CMI map 412). In Europe, it appeared in France in 1962 and had spread to Belgium, Italy, Switzerland and UK within 3 years, and now to over 20 countries (but not Ireland). It has been eradicated in Finland and Norway. First symptoms are chlorotic spotting visible on both leaf surfaces. Reddish-brown uredinia appear mostly on the lower surface in concentric rings (Fig. 71). Only urediniospores are known to germinate and they are accordingly the only source of infection. Since infected leaves tend to be fairly rapidly shed, the fungus is carried over as much by urediniospores (which can survive up to 11 weeks in detached leaves) as by mycelium in the plant. However, inconspicuous lesions may persist on stipules. The optimum temperature for urediniospore germination is 11–23°C, and 8 h of surface wetness or high humidity is required for infection. The incubation period is 11–14 days at 17°C, urediniospores being released a few days later. They are readily dispersed within the glasshouse, but may also be carried long distances by wind out of doors (EPPO 81).

Fig. 71. Uredia of *Puccinia pelargonii-zonalis* on lower surface of pelargonium leaf (British Crown Copyright).

Infected plants show reduced vigour through premature leaf-fall. Serious attacks may destroy young rooted cuttings. In most countries where it occurs. *P. pelargonii-zonalis* causes a serious problem for multiplication of pelargoniums. Cuttings carrying possibly latent lesions provide a major means of introduction of disease into a glasshouse, so checks on the quality of planting material are essential. Heat treatment (38°C for 48 h or 34°C for 4 days, at high RH) will destroy mycelium and urediniospores in planting material. Otherwise, the disease can be controlled by fungicides (Spencer 1976).
SMITH

Puccinia recondita Roberge

(Syn. *P. rubigo-vera* Winter, *P. triticina* Eriksson)

Basic description

A macrocyclic heteroecious rust with spermogonia, chiefly epiphyllous, scattered in small groups, and aecia hypophyllous, grouped, cupulate, rarely cylindrical, yellow, with aeciospores globoid-ellipsoidal, 13–26×19–29 μm, and a colourless, finely verrucose wall. Uredinia are mainly epiphyllous, scattered, round to oblong, 1×1–2 mm, cinnamon brown, with urediniospores globoid to broadly ellipsoidal, 13–24×16–34 μm; the wall is pale cinnamon brown to yellowish, 1–2 μm thick, echinulate, pores 4–8, scattered, and distinct. Telia are hypophyllous and on sheaths, small, oblong, long-covered by epidermis, black; paraphyses in thin layers surrounding or dividing sori; teliospores mostly 2-, occasionally 1- or 3-celled, oblong to clavoid, 13–24×36–65 μm, rounded, truncate or obliquely pointed above, slightly constricted, with a brown wall.

Host plants

Uredinia and telia occur on a wide range of grasses and on wheat (*Triticum* spp.) and rye (*Secale cereale*). Spermogonia and aecia are on a number of host genera in the families Ranunculaceae, Boraginaceae and Crassulaceae.

Host specialization

Classification of the brown or leaf rusts of cereals and grasses which are attributed to the complex species *P. recondita* presents difficulties. Within the species, Wilson & Henderson (1966) include a number of formae speciales based on the haplont and diplont host but this classification is not universally recognized. These include: **f.sp. *recondita*** Roberge (syn. *P. dispersa* Eriksson & E. Henning, *P. secalina* Grove) with spermogonia and aecia on *Anchusa officinalis* and *A. arvensis*, and uredinia and telia on *Secale cereale*; **f.sp. *tritici*** (Eriksson) C.O. Johnston (syn. *P. dispersa* f.sp. *tritici* Eriksson, *P. triticina* Eriksson, *P. rubigo-vera* f.sp. *tritici* (Eriksson) Carleton), with spermogonia and aecia on *Thalictrum* species, and uredinia and telia on *Triticum vulgare*. Within f.sp. *tritici* there is extensive specialization on specific cultivars of *Triticum* species. Differentiation of such pathotypes is by reference to their interaction with wheat cultivars carrying specific genetic factors (Lr genes) conditioning the response. Surveys of virulence have been carried out in Europe for many years as well as in N. America and Australia. These have identified a wide range of races carrying different combinations of virulences. Surveys in C. and E. Europe and the USSR indicate similar common races but these differ from those in W. and S. Europe. The virulences tend to reflect the resistance factors present in agriculturally important cultivars but unnecessary virulences occur commonly in the pathogen population in Britain, for example. Novel variation arises in the pathogen through mutation, and reassortment of these factors is principally via the sexual cycle on the alternate host.

A further 11 formae speciales are classified by Wilson & Henderson (1966) according to their grass host preference, but none appears to be of significance on economically important grass species. The brown rust of ryegrass (*Lolium* spp.), described by Sampson & Western (1954) as *Puccinia dispersa* and occurring in Britain, is not included by Wilson & Henderson (1966) in their classification. Carr (1971) suggests the possibility that this is really the closely related ***P. schismi*** Bubák. However, Wilkins (1973) reported that the brown-rust fungus which attacked strains of *L. multiflorum* from the Po Valley, Italy as well as *L. perenne* and *Festuca pratensis* was indeed *P. recondita*.

Geographical distribution

Brown or leaf rust of wheat is of world-wide distribution where wheat is grown. It is particularly prevalent in C., S. and N.E. Europe but is also common in Spain, France, the Netherlands and Britain. It is common on rye in C. and E. Europe but is more sporadic in N.W. Europe.

Disease

On wheat and rye the uredinial pustules are scattered, round to oblong, and up to 2 mm in length. The orange-brown colour deepens with age. The pustules occur primarily on the upper surfaces of leaf-blades, also on leaf-sheaths and floral parts, but rarely on culms. Rings of uredinia may occur around primary infection sites and the black telia may also appear in this way. Telia commonly develop in rows on leaf sheaths as small black sori, long covered by the epidermis (cf. stem rust).

Epidemiology

The aecial stage is unimportant epidemiologically and the fungus persists mainly in the uredinial stage which can survive relatively low temperatures. Uredinial mycelium survives on autumn-sown wheat under quite extreme conditions and is responsible for perennation of the fungus in E., C. and N.W. Europe and N. America. There is also movement of spores from more southern regions of these continents in spring to initiate new foci of infection further north. In regions of mild winters new infections may occur as spore germination and penetration occur in a temperature range of 5–25°C, although the optimum is between 10–20°C. The generation time is approximately 7–10 days under conditions of warm, moist weather. Anticyclonic conditions are most favourable for disease development as dry windy days favour spore dispersal, and cool nights with dew favour spore germination and penetration.

Economic importance

The disease is relatively unimportant on rye but is arguably the most important single disease of wheat on a world wide basis. It is important on both spring and autumn-sown cultivars especially in Europe, USSR, Australia, S., C. and mid-W. USA and in Canada. It is damaging in Britain, the Netherlands and France, and in Romania, Bulgaria, Yugoslavia, Poland and USSR. Yield losses in the order of 5–10% are common and may be much higher (up to 40%) in epidemic years on susceptible cultivars. It has become more common in N.W. Europe in recent years largely due to the growing of highly susceptible cultivars, some of which had previously been resistant.

Control

This is principally through the use of genetically resistant cultivars and there is a long history of breeding for resistance in N. America, Europe and Australia (Jones & Clifford 1983). Breeding has relied largely on the use of major genes (Lr genes) which confer resistance to specific races of the pathogen but which have been overcome by others. Typically, such resistance, when effective, is expressed throughout the life of the plant although some genes confer adult plant resistance and others are temperature-sensitive in their expression. Some resistances have been transferred to *T. aestivum* from related species and genera such as *Aegilops* and *Agropyron* species and *Secale cereale*. The search for more durable forms of resistance has resulted in the identification of partially expressed resistances which show reduced infection frequency and longer latent period, resulting in 'slow rusting' in the field. The selection away from ultra-susceptibility is advocated as the preferred approach in regions where the pathogen has limited epidemic potential such as in N.W. Europe. Cultural practices aimed at balanced host nutrition can be beneficial and it is reported that the application of chlorides of lithium, sodium and potassium can give good control. A number of modern systemic fungicides are effective in the control of f.sp. *tritici* and in intensive cereal-growing areas where high input systems are employed, these measures are increasing. Some such chemicals are highly specific in their control of the pathogen whereas others have a broad spectrum of activity which makes them attractive to use where other diseases such as mildew (*Erysiphe graminis* f.sp. *tritici*) or stripe (yellow) rust (*P. striiformis*) are present. Some of the more specific chemicals reported to be effective are oxycarboxin, benodanil and RH-124 (4-N-butyl-1,2,4-triazole). Chemicals with a broader spectrum of activity include triadimefon, triadimenol, diclobutrazol, propiconazole and fenpropimorph and prochloraz. Disease-forecasting systems are being developed to ensure more effective and efficient fungicide use.

Special research interest

The importance of the disease has stimulated a great deal of research and development throughout the world. Much of the pioneer work on interspecific and intergeneric transfer of resistance factors was carried out with this disease. It has been the subject of numerous investigations into, for example, the genetics, physiology, biochemistry, histology and ultrastructure of host–parasite relations resulting in a fundamental insight into the nature of parasitism. A number of features of the pathogen make it a convenient 'laboratory species' and it is also amenable to field studies of epidemiology, population genetics etc. Much technology developed for handling and studying *P. r. tritici* has general application to other *Puccinia* spp.
CLIFFORD

Puccinia sorghi **Schwein.**

(Syn. *P. maydis* Bérenger)

P. sorghi (CMI 3) is a macrocyclic and heteroecious species which forms spermogonia and aecia on *Oxalis*, and uredinia and telia on maize (main host) and, in America and S. Africa, on teosinte (*Euchlaena mexicana*). The uredinial and telial stages are practically co-extensive with the distribution of the main host. The spermogonial and aecial stages are infrequent in Europe, but have been recorded in Spain and in England (CMI map 279; Mahindapala 1978a). Several pathotypes of *P. sorghi* have been separated by their reactions on lines of maize having specific genes for resistance (Hagan & Hooker 1965).

During periods of high humidity, especially in thickly planted maize, both surfaces of the leaves, and less frequently the husk and the floral bracts of the tassel, may become covered with innumerable, circular to elongate, golden to reddish-brown pustules (uredinia) which soon burst disclosing the powdery, yellowish-brown mass of the urediniospores. Later, telia appear as larger black patches which remain covered by the epidermis for a long time. The leaves lose their green colour, dry up and the plants may fail to form their cobs. The aecial infections are limited in nature and can be bypassed where several growing cycles of maize, for grain or as green fodder, are possible in the same year. Consequently the infections of *P. sorghi* are essentially caused by wind-blown urediniospores. Migrations of rust urediniospores from continental Europe to S. England have been reported.

Cool temperature (15–23°C) and high RH (98–100%) favour rust development and spread. Optimum urediniospore germination occurs with 2 h in the dark followed by 4 h in the light at 24°C. The generation time is 10 days at 15°C and 5 days at 25°C with 70–80% RH (Mahindapala 1978b). As the older tissues are generally resistant to *P. sorghi*, disease severity is particularly related to the amount of immature foliage developing on the plants at the time of infection. Severe epidemics of *P. sorghi*, seldom observed in Europe, are not uncommon in the USA and in the coolest, wet, regions of the tropics, especially where maize is grown year round. In these areas *P. sorghi* can affect all components of yield causing grain yield losses as high as 25–30% (Groth *et al.* 1983). In Europe the disease does not warrant chemical control; in India (Singh & Musymi 1979), good control has been obtained with three foliar applications of triadimefon (50 g/ha). Many maize cultivars resistant to *P. sorghi* are now available. Mexican maize appears to be one of the best sources of rust resistance.
CASTELLANI

Puccinia striiformis **Westend.**

Basic description

An autoecious rust, with lesions consisting of flecks and stripes with chains of small, more or less confluent interveinal pustules, each about 0.5–1 mm in size, bright yellow to orange, later sometimes interspersed with black dots, often accompanied by chlorosis and necrosis. The common name is yellow rust (Europe) or stripe rust (USA) (Cummins 1971; Wilson & Henderson 1966; CMI 291).

Host plants

P. striiformis attacks barley, rye, Triticale, wheat and many cultivated and wild grasses. No alternate host is known.

Host specialization

In different areas of the world, host specialization has proceeded to different degrees (cf. *P. graminis*). The form on *Dactylis glomerata* has received varietal rank, **var. *dactylidis*** Manners, having slightly smaller urediniospores with a somewhat higher temperature optimum than the normal form **var. *striiformis***. Host specialization at the generic level is strong in N.W. Europe where the **f.sp. *agropyri, hordei, secalis*** and ***tritici*** have been distinguished. The status of various forms on other Gramineae is not so clear; var. *dactylidis* is highly specialized, as are several isolates from *Elymus repens* and *Poa* species. f.sp. *tritici* may occasionally infect individuals of grass species, and some isolates from grasses may, at times, infect wheat. In other areas, such as N.W. USA, C. Asia and China, the sharp divisions suggested by the distinction of formae speciales do not hold at all, and the rust can migrate from grasses to wheat and vice versa. Scores of races have been distinguished (Johnson *et al.* 1972), differentiated not only by differential sets of seedling plants but also by differential sets of adult plants. The 'field races' so distinguished fit better to farmers' experience than the races as identified on seedlings in the greenhouse. New races often appear.

Geographical distribution

P. striiformis can occur wherever susceptible hosts grow. In 1979 it appeared in Australia and in 1980 in New Zealand presumably carried by air travellers. The major economic host is wheat, which is especially endangered in cool and humid areas such as the Netherlands, the Romanian Black Sea coast, and (sub)alpine areas. Wheat in semi-arid areas is at risk when the cultivation of highly susceptible cultivars coincides with unusually wet periods.

Disease

The fungus can attack all parts of gramineous plants excluding roots. Rarely, non-sporulating pustules can be seen on growing seeds, but yellow rust is not seed-transmitted. The fungus grows in a semi-systemic way. A lesion grows into a large fleck in a tender leaf or a long stripe contained between veins in a leaf of an adult plant. Stripes can attain a length of 10 cm and the growth rate of flecks in seedling leaves can be up to 1 cm per day. Each lesion carries hundreds of minute pustules. The primary pustules are uredinia, concatenated and sometimes confluent in long interveinal lines. In tender leaves the fungus can cross the veins so that more or less oval colonies are formed, while in coarse leaves the colonies are in the form of long stripes of about 1 mm width. The fungus appears mainly on leaves, where it sporulates profusely on both surfaces, but more on the adaxial surface. On leaf-sheaths, long narrow stripes are formed, which sporulate on the adaxial side. Occasionally, the peduncle is infected but sporulation on the peduncle occurs only in exceptional cases, when the host cultivar is unusually susceptible and the weather is very favourable for rust development. Similarly, infection of the ears including the awns occurs in some cultivars only. The fungus can sporulate profusely on the adaxial side of the glumes, but not on the abaxial side, which turns yellow.

Towards ripening of the host, small dark telia may be formed on the abaxial leaf surface mainly, arranged in linear series. On the leaf sheaths and awns, telia can be conspicuous when about 0.5 mm sized uredinia and telia alternate in a zebra pattern. Yellow rust attacks in commercial fields often begin as foci or hot spots, 0.5–2 m in diameter, from where the disease spreads over the fields. With severe attacks, the plants shrivel and the crops appear drought-stricken. The clothing of a person entering the crop will turn orange-yellow from adhering urediniospores. The ground becomes yellowish from spores.

Epidemiology

Of the three wheat rusts, yellow rust is best adapted to low temperature. In the high-risk areas, it appears earliest. Though telia can germinate and form basidiospores, the sexual cycle and the alternate host have never been found (hemi-form). The rust oversinters and oversummers in the asexual form, the uredinio-stage. In the latent phase the rust can withstand severe winter cold as long as the host leaf survives, but sporulating lesions are killed by moderate frost. The rust survives equally well in the host under snow cover. Able to germinate and infect at near-zero temperature the rust can reproduce during mild winters, though latency periods will then be long, up to 3 months. In areas with summer drought, oversummering is a major problem, solved by invading volunteer wheat at places where some moisture is left, or by migrating to susceptible grasses (not in N.W. Europe). In mountainous areas, vertical migration on orographic winds is usual. In Europe, long-distance migration over hundreds of kilometres is normal; it occurs in a haphazard way, but its dangerous effects are damped by a kind of gene deployment practised unintentionally by various countries. In the Mediterranean area, yellow rust can attack unexpectedly and severely, when the growing season and the preceding host-free season have been cool and moist (Hassebrauk & Röbbelen 1975; Zadoks 1961; Zadoks & Bouwman 1986). The epidemiology of yellow rust on barley, *Dactylis glomerata* and other grasses follows the pattern described above, but few details are known. f.sp. *secalis* has not been recorded during the last 50 years.

Economic importance

The yellow rust fungus behaves as a strong sink, competing with the usual sinks in the plant. Early infection of young plants thus weakens the roots, so that the crop will be drought-susceptible at a later stage. Vigour, tillering and seed set are reduced. Later infections destroy the foliage and enhance evapotranspiration, so that few or no assimilates reach the growing seeds. Infections of rachis, glumes and awns are equally destructive. Losses in wheat and barley can, when only the foliage is infected, reach 40%. When peduncles and/or glumes are severely infected, total loss may follow. Even light but regular infections may impair the milling and baking quality of wheat or the malting quality of barley, at great financial loss. In areas ecologically favourable to yellow rust, long-term means of losses in wheat used to be over 5%, as in parts of the UK and the Netherlands. In many less favourable areas estimates of losses in wheat varied from 2 to 6%. Of the grasses, *Dactylis glomerata* is most severely affected: heavy infections weaken the plants and reduce their nutritional value.

Control

Cultural control has been attained by avoiding overlap of crops yet to be harvested and new crops already sown, as was customary in some northern and alpine areas. Avoidance of early sowing, selection of early maturing cultivars, and good stubble ploughing have promoted cultural control. The recent tendencies to sow earlier using direct drilling, as in the UK, and to undersow wheat with grasses (*Lolium* spp. mainly) omitting summer ploughing, as in the Netherlands, may lead to loss of control. Resistance breeding is highly effective, though many initially resistant cultivars become susceptible to newly emerging races (Röbbelen & Sharp 1978). Varietal diversification is recommended, with at least 2 cultivars per farm to spread the risk of losses due to the appearance of a new race. Multilines and varietal mixtures provide good protection in most years. Chemical control is successful and widely applied. Ni-salts are effective but rarely registered for use. Protective chemicals such as dithiocarbamates are inadequate when the epidemic is really severe. Among the systemic fungicides oxycarboxin is moderately effective. Ergosterol biosynthesis inhibitors (EBIs) are moderately (tridemorph) to highly (diclobutrazol, fenpropimorph, triadimefon, triadimenol, propiconazole) effective. Alternation between the two EBI groups (C-14-demethylation; isomerase) may reduce selection pressure for fungicide-resistant strains. Triadimenol, used as a seed dressing, may damage young seedlings.

ZADOKS

Puccinia antirrhini Dietel & Holway (CMI 262), indigenous to California on other hosts, proved able to cause serious damage to European *Antirrhinum* species introduced there as garden plants. The fungus spread to the UK by 1933, and then throughout Europe (CMI map 40). In Europe, only the urediniospore infection cycle occurs. Though teliospores may form in hot summers, the basidiospores do not re-infect *Antirrhinum*, so an alternate host is presumably missing in Europe. The fungus can be very damaging to all parts of the plant, including the capsules, so that seed production is directly affected. Control is by fungicides or, mainly, resistant cultivars.

Puccinia apii Desm. is an autoecious macrocyclic rust on celery (CMI 284) of very minor practical importance. ***P. rubiginosa*** Schröter (syn. *P. petroselini* (DC.) Liro) is a very similar species on parsley. ***P. arenariae*** (Schumacher) Winter forms teliospores only on Caryophyllaceae, and especially on *Dianthus barbatus*. Carnation is occasionally infected, but *Uromyces dianthi* is much more important on this crop.

Puccinia brachypodii Otth. has varieties on several grass genera, the most important being var. *poae-nemoralis* (Otth) Cummins & H. Greene (syn. *P.*

poae-sudeticae (Westend.) Jørstad) on *Poa* species (and, though rarely seen, *Berberis* as an alternative host). This rust is distinguished from the generally commoner *P. poarum* (q.v.) by the presence of numerous capitate paraphyses in the uredinia, and by relatively infrequent teliospore formation (O'Rourke 1976). In some countries (e.g. Czechoslovakia, Cagaš 1983), it is the most important rust on *Poa*. Fungicide control may be needed on seed-crops and resistant cultivars are available.

Puccinia buxi DC. produces teliospores only on *Buxus sempervirens*, causing hypertrophy of leaves and terminal dieback (Viennot Bourgin 1949). ***P. carthami*** Corda is a macrocyclic autoecious rust on safflower (CMI 174) occurring mainly in the tropics (CMI map 424), but also in Mediterranean countries. Severe attacks have been recently recorded in Italy (Cappelli & Zazzarini 1983). ***P. chrysanthemi*** Roze causes brown or black rust of glasshouse chrysanthemums, on which urediniospores are formed (and rarely teliospores) (CMI 175). It is more important in Japan (where it originated) and N. America than in Europe, where another Japanese species *P. horiana* (q.v.) causes far more problems. Sanitation, avoiding high humidity and fungicides can be used for control (Fletcher 1984).

Puccinia hieracii (Röhl.) Martius is a collective autoecious species occurring on many Compositae, and divided into four varieties. Rust of chicory and endive is caused by **f.sp. *cichorii*** (Bellynck ex Kickx) Boerema & Verhoeven (syn. *P. cichorii* Bellynck ex Kickx, *P. endiviae* Pass.) of **var. *hieracii***. It occurs sporadically, usually towards the end of the season; infected plants should be destroyed and a 2-year rotation imposed (Messiaen & Lafon 1970). ***P. hysterium*** Röhl. (syn. *P. tragopogonis* (Pers.) Corda) and ***P. jackyana*** Gaüm. are autoecious rusts on salsify and black salsify respectively. The mycelium perennates in the stocks forming abundant aecia which deform and stunt the young leaves. Unlike *P. jackyana, P. hysterium* forms no urediniospores. A form of *P. jackyana* forming no aecia has been called *P. scorzonerae* (Schumacher) Jacky (Viennot-Bourgin 1949, 1956).

Puccinia iridis Rabenh. is macrocyclic and heteroecious, alternating between *Urtica* species (aecial host) and *Iris* species (telial host) (CMI 285), but readily persisting by urediniospore formation. It is of little practical importance. ***P. lagenophorae*** Cooke, an autoecious macrocyclic rust introduced to Europe from Australia in the 1960s, is now widespread on the weed *Senecio vulgaris* and can attack glasshouse cinerarias (Fletcher 1984). ***P. malvacearum*** Mont., originating in Chile but now distributed world-wide (CMI 265), is common on wild and ornamental *Malva* and on hollyhock (*Althaea*). It is the classic microcyclic rust, perhaps more important as research material than as an economic plant pathogen.

Puccinia menthae Pers., autoecious and macrocyclic (CMI 7), causes the most important disease of cultivated *Mentha* species. Overwintering rhizomes are infected by teliospores in the soil and give rise to pale, swollen, distorted shoots in the spring (systemic infection). Aecia form on the leaves of these shoots, and the rust then spreads from leaf to leaf by aeciospores and then urediniospores. Infected rhizomes can be heat-treated (44°C for 10 min) before planting. There are many resistant cultivars, and correspondingly as many pathotypes.

Puccinia opizii Bubák has the unusual feature for a *Puccinia* that the aecial host (lettuce) is the economic one, while the uredinia and telia form on *Carex*. It occurs quite commonly in most parts of Europe if the alternate host is present near lettuce beds, but has little practical importance. It also occurs on wild *Lactuca* species. ***P. pittieriana*** Henn. (CMI 286) produces teliospores only on tomato and potato in S. and C. America. It is a serious potential hazard to potato in Europe (EPPO 155).

Puccinia poarum Nielsen, macrocyclic and heteroecious, is widespread causing leaf rust of *Poa* species. The main alternate host is *Tussilago farfara*, on which two generations of aecia may be produced each year (O'Rourke 1976). *P. poarum* is generally the commonest rust on *Poa* (but cf. also *P. brachypodii* var. *poae-nemoralis*). It has also been used extensively as a model heteroecious rust in recent comparative studies of infection behaviour on the two hosts (Al-Khesraji & Lösel 1980, 1981; Whipps & Lewis 1984).

Puccinia purpurea Cooke (CMI 8) causes large purplish spots on the leaves and sheaths of sorghum. It is widespread in Mediterranean countries, but of less importance there than in tropical countries, where resistant cultivars are being selected. ***P. ribis*** DC. forms teliospores only on *Ribes*, especially red-currant. It is widespread in Europe, but of importance only in Scandinavia, and then only in poorly managed plantings. ***P. trabutii*** Roum. & Sacc. (syn.

P. isiacae (Thüm.) Winter) forms aecia on many diverse hosts including sugarbeet (Viennot-Bourgin 1956) and exceptionally cucumber (Bräutigam & Kühn 1983) (cucurbits are mostly unaffected by rusts). The telial host is *Phragmites australis*. ***P. violae*** (Schumacher) DC. is an autoecious macrocyclic rust on wild and cultivated *Viola* species. The form particularly affecting pansies (formerly known as *P. aegra* Grove) establishes a persistent mycelium in the rootstock. Formation of aecia heavily deforms the leaves. ***P. vincae*** Berk. on *Vinca* species has a similar habit.

Pucciniastrum areolatum (Fr.) Otth

P. areolatum is a host-alternating rust forming spermogonia and aecia on species of the genus *Picea,* for example *P. abies*, and uredinia and telia on species of the genus *Prunus*, for example *P. padus*. The fungus is found in the greater part of Europe and eastwards through Asia to Korea and Japan. The spermogonia are usually formed on the outer side of the cone scales and sometimes on the leader and other spruce shoots. They are rather inconspicuous, dark, elongated and up to 1 mm long. Aecia are formed on the inner side of the cone scales or on the shoots (Plate 30); they are roundish, about 1 mm in diameter; the wall is thick and hard and opens by a lid the next spring or during a subsequent year. The infected cones will hang on the tree for many years. The aeciospores remain viable for more than one year; they infect young *Prunus* leaves. The uredinia are small, bullate, light yellowish, appearing on reddish spots on the underside of the leaves. The telia appear as reddish-brown, later dark brown, polygonal spots on the upper side of the leaves. It may be noted that, in plum, aeciospore infection cause necrotic spots which fall out leaving holes in the leaves. The teliospores overwinter, and next spring young spruce cones, and sometimes also shoots, are infected. Thus the life cycle is completed in 2–3 years. The spruce seed crop is sometimes greatly reduced where the two hosts occur in the same locality. Damage to spruce shoots is more rare.
ROLL-HANSEN

Pucciniastrum epilobii Otth. is a needle rust of world-wide distribution which locally causes marked damage to stands of *Abies* in which the alternate hosts are also present; all these (on which the uredinia and telia are formed) are members of the Onagraceae (e.g. *Epilobium* species). The spermogonia and aecia are produced on the needles of the *Abies* species. Two formae speciales have been distinguished: **f.sp. *abieti-chamaenerii*** (Kleb.) Gäum. with alternate host *E. angustifolium* and **f.sp. *palustris*** Gäum. with alternate host *Fuchsia*. The latter can also be damaging to ornamental *Fuchsia* in the glasshouse and may need fungicidal control (Fletcher 1984).

Tranzschelia pruni-spinosae (Pers.) Dietel

T. pruni-spinosae is a heteroecious macrocyclic rust producing spermogonia and aecia on *Anemone* and uredinia and telia on *Prunus*. The hypophyllous aecia are prominent with a broad revolute 3- to 5-lobed margin. Uredinia and telia are hypophyllous, sometimes causing cankers on twigs and young branches and occasionally infecting fruit. The golden-brown paraphysate uredinia are formed throughout the growing season and the darker telia towards leaf-fall. Urediniospores are broadly fusiform or clavate, 22–43×13–19 μm with a golden brown wall markedly thickened at the apex. In **var. *discolor*** (Fuckel) Dunegan, the two cells of the teliospores are markedly dissimilar, the lighter-coloured smooth proximal cell contrasting with the darker coarsely verrucose terminal one (cf. **var. *pruni-spinosae*,** in which they are similar).

T. pruni-spinosae has a world-wide distribution and causes losses throughout Europe wherever cultivated *Prunus* are grown. It is common on plum, apricot, nectarine, peach and almond, but infrequent on cherry. Many records fail to distinguish between var. *discolor* and var. *pruni-spinosae* and many references to the latter are probably in error. In the UK var. *discolor* is the common form on wild *Prunus* species, causing sporadic economic damage to *Anemone* and cultivated *Prunus* (Linfield & Price 1983). There is general agreement that var. *pruni-spinosae* is largely restricted to wild *Prunus* species, mainly *P. spinosa*, at least in N. Europe. There is evidence from several parts of the world for 'physiologic' specialization. This may in part be responsible for the failure of different authors to agree about the precise host range of the two forms.

Anemone probably plays an insignificant part in propagating the fungus in most parts of Europe. Overwintering mycelium in small cankers on 1 and 2 year old shoots, together with urediniospores surviving on overwintering leaves are the main means of survival. Severe infections on *Prunus* result in premature defoliation, a reduction in growth of

extension shoots, local dieback and a reduction in flowering the following season. Intensity varies greatly between seasons, although the local occurrence of infected *Anemone* often leads to an earlier establishment of the disease on *Prunus*. On *Anemone* the disease causes little damage in the first year but in the second and subsequent years considerable losses may result due to the ability of the fungus to perennate as dormant mycelium in corms. In spring, shoots on infected plants are distorted and usually die after the aecia have matured, but the most direct cause of loss is the failure of infected plants to flower. Control measures are rarely considered worthwhile, although in trials dithiocarbamates, oxycarboxin and benodanil have given promising results. In several countries, notably in E. Europe, attempts are being made to select for resistant clones.
ARCHER

Uredo quercus Brond. is an anamorph name for a rust which sometimes forms spots bearing uredinia on the leaves of *Quercus* species. It is widely distributed but not generally common, and is most often found in the Mediterranean area. Telia have rarely been seen, and it is sometimes considered to be a stage of *Cronartium quercuum* (q.v.).

Uromyces appendiculatus (Pers.) Unger

(Syn. *Uromyces phaseoli* (Pers.) Winter)

U. appendiculatus (CMI 57) has a macrocyclic autoecious life cycle, generally producing only urediniospores and teliospores in nature. Urediniospores are brown, spiny, unicellular and thin-walled, while teliospores are somewhat darker in colour, single-celled, globoid to ellipsoid with a wall thickness of 3–4 μm. Most *Phaseolus* species are susceptible to **var. *appendiculatus*** (syn. var. *typica* Arthur) but the fungus, segregated into over 35 pathotypes so far, is mostly common on *P. vulgaris* world-wide. Another form, **var. *vignae*** parasitizes *Vigna* species only. Rust symptoms develop on leaves and pods but not stems and branches. Small, white and slightly raised individual spots appear at first on the leaf under surface (uredinia). Secondary sori develop outside the primary infection site as a ring. With age, uredinia may be replaced by telia, which are black-brown in colour. Urediniospores and teliospores can be found in the same sorus. When infection is severe enough, complete defoliation may occur resulting in total loss of the crop. Urediniospores have been found to germinate best at temperatures between 17 and 23°C with at least 6–8 h of wetness on the plant surface. Further rust development, i.e. the appearance of secondary pustules on bean plants, seems to be related to inoculum concentration, as well as host factors such as photosynthetic activity and mineral nutrition (Galbiati 1978). In semi-arid climates, bean rust is disseminated only in humid seasons probably because the pathogen has a relatively slow infection rate and low rate of survival between short wet periods (Bashi & Rotem 1974). Breeding for resistance is a possible means of control, although its effectiveness is limited by new virulent forms arising within the fungal population. As with other rusts, the use of multiline cultivars or race non-specific resistance may partly solve this problem. Induced resistance of bean to rust might be another way of control by using either a weakly virulent race or some incompatible microbiological agents as inducers (Johnson & Allen 1975; Schönbeck *et al.* 1980). Effective chemical control can be achieved with systemic fungicides as either seed dressings or foliar sprays. The host–parasite system is being widely used in a variety of fundamental studies of rust fungi, the investigations mainly focusing on structural and physiological aspects of (in)compatibility (Littlefield 1981).
VIRÁNYI

Uromyces dianthi (Pers.) Niessl

U. dianthi (CMI 180) forms uredinia and telia on *Dianthus, Arenaria, Butonia, Gypsophila, Lychnis, Saponaria, Tunica* and *Silene*. Spermogonia and aecia are formed on *Euphorbia*. Uredinia occur on both leaf surfaces and stems, scattered or in groups, oval to irregular, often forming concentric rings and often confluent, 0.5–1.5 mm diameter, and rust brown. Urediniospores are golden brown, broadly ellipsoid 20–24×24–30 μm, strongly echinulate with 3–4 equatorial pores. Telia are as uredinia but dark brown. Teliospores are chestnut brown, ellipsoid 20–23×25–29 μm, finely verrucose with a hyaline papilla over the terminal germ pore. Spermogonia are abundant on both surfaces of lower leaves 130–160 μm diameter. Aecia are scattered on lower leaf surfaces, orange yellow, and up to 0.5 mm diameter (CMI 180).

Four specialized forms on *U. dianthi* have been recognized (Gäumann 1959). The organism is of world-wide distribution causing carnation rust (Fig. 72). This important disease occurs all the year round on the leaves and stems of the cultivated carnation. It appears first as light green spots from which the masses of urediniospores later erupt. Individual

Fig. 72. Carnation rust caused by *Uromyces dianthi.*

leaves may be killed and entire plants may be injured to the extent that they are unprofitable. The disease is transmitted by urediniospores which can remain viable for 6 months or more, and rust may be spread from one country to another in infected plant material. Aecia are not known outside Europe. Disease control is facilitated by avoidance of conditions in which free water remains on the leaves. Therefore, adequate ventilation is required, surface watering should be used and fungicides should be applied as dusts rather than sprays at times when ambient atmospheric humidity is high. Zineb and ferbam are used as protectant fungicides and the systemic fungicide, oxycarboxin, provided as a single drench has given complete control for six months (Spencer 1979). Resistant cultivars are available, particularly of spray carnations. Mycoparasites have been tested as biological control agents of *U. dianthi,* and *Verticillium lecanii* has shown potential (Spencer 1980).
SPENCER

Uromyces fabae (Grev.) de Bary ex Fuckel

(Syn. *Uromyces viciae-fabae* (Pers.) Schröter)

U. fabae (CMI 60) is a macrocyclic autoecious rust with spherical hyaline aeciospores and elliptical to obovoid yellowish urediniospores. The teliospores are similar in shape but with chestnut-coloured walls and long yellow pedicels. The fungus commonly attacks broad bean, pea and lentil crops but has been found to have a considerably wider host range on various species of *Vicia, Lathyrus* and *Lens* (Conner & Bernier 1982a). Race specialization is also evident from recent studies on broad bean and pea lines and cultivars (Conner & Bernier 1982b). Although geographically widespread, the pathogen usually causes slight injury, and the affected plants show typical rust pustules only on their leaves and stems.

Aeciospores may overwinter in Mediterranean climates but are unable to survive severe winters in northern regions. Other possible perennating stages of the fungus include uredinio-mycelium in leaves and stems of autumn-sown crops, as well as urediniospores and/or teliospores remaining viable for 1–2 years. Dissemination by seeds and plant debris has also been reported. High atmospheric humidity favours infection. Irrigation applied late in the season may promote the severity of the disease. Effective chemical control has been achieved in field trials against natural infection with both contact and systemic fungicides (Mohamed *et al.* 1981).
VIRÁNYI

Uromyces betae Kickx is an autoecious macrocyclic rust on sugar beet (CMI 177) spreading mainly by urediniospores, and surviving on overwintered seed crops, or as teliospores contaminating seed lots. In W. Europe, though *U. betae* is common, it usually appears too late in the season to cause any appreciable damage. In C. and E. Europe (e.g. the USSR), it may cause losses and justify fungicidal treatment. Resistant cultivars are available. ***U. dactylidis*** Otth causing leaf rust of *Dactylis glomerata* has *Ranunculus* species as alternate hosts (O'Rourke 1976). Its **var. *dactylidis*** also occurs on *Festuca* (syn. *U. festucae* Plowr.) and *Cynosurus*, while its **var. *poae*** (Rabenh.) Cummins (syn. *U. poae* Rabenh.) occurs on *Poa*. The *Uromyces* rusts of grasses are of very minor importance compared with *Puccinia* species. ***U. pisi*** (DC.) Otth (syn. *U. pisi-sativi* (Pers.) Liro) (CMI 58) alternates between *Euphorbia cyparissias* (aecial host) and pea, on which it is probably less common than the autoecious *U. fabae* (q.v.). Infection generally occurs too late in the season to be of any practical significance. ***U. striatus*** Schröter, like *U. pisi* (q.v.) with which it is sometimes equated, forms its spermogonia and aecia on *Euphorbia cyparissias*. The uredinia and telia are formed on *Medicago* species (lucerne rust). It is widespread (CMI 59, CMI map 342) but only of some importance in Europe in warmer southern regions (e.g. S. USSR), where, as in the USA, there is interest in resistant cultivars. ***U. lupinicola*** Bubák is also of some importance in the USSR on fodder lupins.

Uromyces transversalis (Thüm.) Winter is autoecious, producing urediniospores and teliospores. It attacks *Gladiolus* species and a number of other genera in the Iridaceae. It originates in S. Africa and has recently spread to the Mediterranean region—Malta (1969), Portugal (1975), and Morocco (1978). It has been reported from France, Italy and Spain but is probably not established there. Infected leaves show characteristic orange pustules elongated across the leaf (sometimes followed by smaller black teliospore pustules) (Fig. 73). It is doubtful whether the fungus can survive in corms; urediniospores are known to be short-lived but teliospores may provide for longer survival. Although *U. transversalis* could probably not persist outdoors in N. Europe, it constitutes a substantial threat to gladiolus production under glass or in the open (EPPO 84). It is potentially a much more serious disease than the unimportant rust caused by the indigenous ***Puccinia gladioli*** Castagne, which forms scattered brown-black pustules and produces aecia on *Valerianella* species (Viennot-Bourgin 1956).

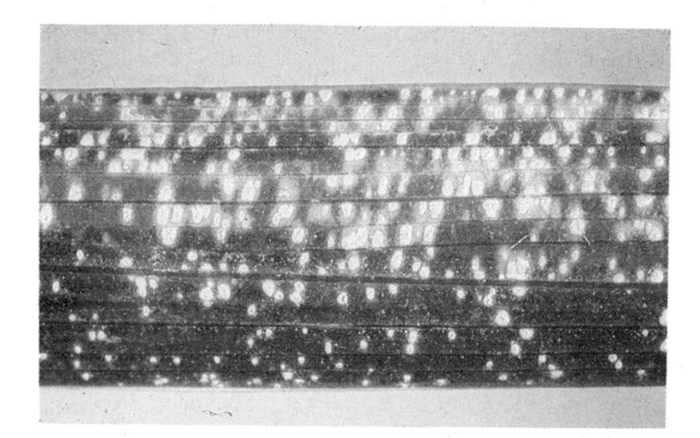

Fig. 73. Uredia of *Uromyces transversalis*, elongated across gladiolus leaf (British Crown Copyright).

Several *Uromyces* species occur on *Trifolium* and their nomenclature is somewhat confused. ***U. trifolii*** (R. Hedw. ex DC.) Fuckel is now considered to be the name for the microcyclic rust of *T. repens* (syn. *U. flectens* Lagerheim, *U. nerviphilus* (Grognot) Hotson). It forms telia aggregated along the veins, and on petioles and stems, and also overwinters in its host. It is probably the commonest rust on this species, on which a macrocyclic autoecious rust also occurs (***U. trifolii-repentis*** Liro **var.** ***trifolii-repentis***). **Var.** ***fallens*** (Arthur) Cummins of this species is common on *T. pratense*. ***U. anthyllidis*** (Grev.) Schröter (syn. *U. jaapianus* Kleb.) is widespread on yellow-flowered *Trifolium* and on many other genera of Leguminosae, while ***U. minor*** Schröter occasionally causes serious damage to shamrock (*T. dubium*) in Ireland (O'Rourke 1976). None of these rusts are of any great economic significance in Europe. ***U. ciceris-arietini*** (Grognot) Jacz. (CMI 178) is very similar to *U. anthyllidis*, but without a known aecial state. It is widespread on chickpea in Mediterranean Europe, and can seriously reduce yields.

REFERENCES

Al-Khesraji T.O. & Lösel D.M. (1980) Intracellular structures of *Puccinia poarum* on its alternate hosts. *Transactions of the British Mycological Society* **75,** 397–411.

Al-Khesraji T.O. & Lösel D.M. (1981) The fine structure of haustoria, intracellular hyphae and intercellular hyphae of *Puccinia poarum*. *Physiological Plant Pathology* **19,** 301–311.

Anthony V.M., Shattock R.C. & Williamson B. (1985) Interaction of red raspberry cultivars with isolates of *Phragmidium rubi-idaei*. *Plant Pathology* **34,** 521–527.

Anthony V.M., Williamson B. & Shattock R.C. (1987) The effect of cane management techniques on raspberry yellow rust (*Phragmidium rubi-idaei*). *Annals of Applied Biology* **110**, 263–273.

Bashi E. & Rotem J. (1974) Adaptation of four pathogens to semi-arid habitats as conditioned by penetration rate and germinating spore survival. *Phytopathology* **64,** 1035–1039.

Beraha L. (1955) Susceptibility of *Allium* species to *Puccinia asparagi*. *Plant Disease Reporter* **39,** 98–100.

Beraha L., Linn M.B. & Anderson H.W. (1960) Development of the asparagus rust pathogen in relation to temperature and moisture. *Plant Disease Reporter* **44,** 82–86.

Bräutigam S. & Kühn H.J. (1983) *Puccinia trabutii*—ein ungewöhnlicher Rostpilz an Gewächshausgurken (*Cucumis sativus*). *Nachrichtenblatt für den Pflanzenschutz in der DDR* **37,** 43–44.

Brown M.R. (1938) A study of crown rust, *Puccinia coronata* in Great Britain. II. The aecidial hosts of *Puccinia coronata*. *Annals of Applied Biology* **25,** 506–527.

Browning J.A. & Frey K.J. (1969) Multiline cultivars as a means of disease control. *Annual Review of Phytopathology* **7,** 355–382.

Cagaš B. (1983) (The resistance of *Poa* varieties to the rusts *Puccinia brachypodii* var. *poae-nemoralis* and *P. poarum*). *Ochrona Rostlin* **19,** 121–127.

Cappelli C. & Zazzerini A. (1983) Risultati di un biennio di ricerche su alcuni aspetti fitopatologici del cartamo. *Informatore Fitopatologico* **33** (1), 39–45.

Carr A.J.H. (1971) Grass diseases. In *Diseases of Crop Plants* (Ed. J.H. Western). Macmillan, London.

Clifford B.C. (1985) Barley leaf rust. In *The Cereal Rusts* Vol. 2 (Eds W.R. Bushnell & A.P. Roelfs), pp. 173–205. Academic Press, New York.

Conner R.L. & Bernier C.C. (1982a) Host range of *Uromyces viciae-fabae. Phytopathology* **72,** 687–689.

Conner R.L. & Bernier C.C. (1982b) Race identification in *Uromyces viciae-fabae. Plant Pathology* **4,** 157–160.

Coutin R. & Grouet D. (1983) La protection phytosanitaire du chrysanthème. *Phytoma* no. 351, 31–37.

Cummins G.B. (1971) *The Rust Fungi of Cereals, Grasses and Bamboos.* Springer, Berlin.

Cummins G.B. & Hiratsuka Y. (1983) *Illustrated Genera of Rust Fungi, Revised Edition.* APS, St Paul.

Dixon G.R. (1984) *Vegetable Crop Diseases.* Macmillan, London.

Doherty M.A. & Preece T.F. (1978) *Bacillus cereus* prevents germination of uredospores of *Puccinia allii* and the development of rust diseases of leek, *Allium porrum,* in controlled environments. *Physiological Plant Pathology* **12,** 123–132.

Fletcher J.T. (1984) *Diseases of Greenhouse Plants.* Longman, London.

Galbiati C. (1978) Sulla formazione delle corone uredosoriche di *Uromyces appendiculatus* in *Phaseolus vulgaris* e sulla relazione di queste con fotosintesi e metabolismo dell'amido. *Rivista di Patologia Vegetale* **4** (14), 47–57.

Gäumann E. (1959) *Die Rostpilze Mitteleuropas. Beiträge zur Kryptogamenflora der Schweiz 12.* Baxhdruckerei Buchler, Bern.

Gjaerum H.B. & Langnes R. (1981) Løk- og purrerust. *Gartneryrket* **71,** 482.

Gremmen J. & de Kam M. (1970) De bestrijding van blaasroest in kwekerijen van *Pinus strobus. Nederlandsch Boschbouwtijdschrift* **42,** 54–57.

Grill D., Hafellner J. & Waltinger H. (1980) Rasterelektronenmikroskopische Untersuchungen an *Chrysomyxa abietis*-befallenen Fichtennadeln. *Phyton, Austria,* **20,** 279–284.

Groth J.V., Zeyen R.J., Davis D.W. & Christ B.J. (1983) Yield and quality losses caused by common rust (*Puccinia sorghi*) in sweet corn (*Zea mays*) hybrids. *Crop Protection* **2,** 105–111.

Hagan W.L. & Hooker A.L. (1965) Genetics of reaction to *Puccinia sorghi* in eleven corn inbred lines from Central and South America. *Phytopathology* **55,** 193–197.

Hassebrauk K. & Röbbelen G. (1975) Der Gelbrost, Puccinia striifomis West. IV. Epidemiologie. Bekämpfungsmassnahmen. *Mitteilungen aus der Biologischen Bundesanstalt für Land- und Forstwirtschaft Berlin-Dahlem* no. 164.

Hermansen J.E. (1968) Studies on the spread and survival of cereal rust and mildew diseases in Denmark. *Friesia* **8,** 161–359.

Hiratsuka Y. (1968) Morphology and cytology of aeciospores and aeciospore germ tubes of host-alternating and pine-to-pine races of *Cronartium flaccidum* in northern Europe. *Canadian Journal of Botany* **46,** 1119–1122.

Johnson R. & Allen D.J. (1975) Induced resistance to rust diseases and its possible role in the resistance of multiline varieties. *Annals of Applied Biology* **80,** 359–363.

Johnson R., Stubbs R.W., Fuchs E. & Chamberlain N.H. (1972) Nomenclature for physiologic races of *Puccinia striiformis* infecting wheat. *Transactions of the British Mycological Society* **58,** 475–480.

Jones D.G. & Clifford B.C. (1983) *Cereal Diseases: their Pathology and Control* (2nd Ed.). John Wiley, London.

Krutov V.I. (1981) (Long-term forecast of pine rust, *Melampsora pinitorqua,* in felling areas of South Karelia). *Mikologiya i Fitopatologiya* **15,** 150–155.

Lanier L., Joly P., Bondoux P. & Bellemère A. (1978) *Mycologie et Pathologie Forestières.* Masson, Paris.

Lehmann E., Kummer H. & Dannenmann H. (1937) *Der Schwarzrost.* Lehmann, München.

Lehrer G.F. (1982) Pathological pruning: a useful tool in white pine blister rust control. *Plant Disease* **66,** 1138–1139.

Lelliott R.A. (1984) Cost-benefit analysis as used in the United Kingdom for eradication campaigns against alien pests and diseases. *Bulletin OEPP/EPPO Bulletin* **14,** 337–341.

Linfield C.A. & Price D. (1983) Host range of plum *Anemone* rust, *Tranzschelia discolor. Transactions of the British Mycological Society* **80,** 19–21.

Littlefield L.J. (1981) *Biology of the Plant Rusts. An Introduction.* Iowa State University Press, Ames.

Lubani K.R. & Linn M.B. (1962) Entrance and invasion of the asparagus plant by uredospore germ tubes and hyphae of *Puccinia asparagi. Phytopathology* **52,** 115–119.

Mahindapala R. (1978a) Occurrence of maize rust, *Puccinia sorghi,* in England. *Transactions of the British Mycological Society* **70,** 393–399.

Mahindapala R. (1978b) Host and environmental effects on the infection of maize by *Puccinia sorghi. Annals of Applied Biology* **89,** 411–421.

McBeath J.H. (1984) Symptomology on spruce trees and spore characteristics of a bud rust pathogen. *Phytopathology* **74,** 456–461.

Messiaen C.M. & Lafon R. (1970) *Les Maladies des Plantes Maraîchères.* INRA, Paris.

Mohamed H.A., Khalil S.A., Zeid N.A., El Sherbeeny M. & Ismail I.A. (1981) Effect of fungicides on rust reaction of faba beans. *FABIS Newsletter* **3,** 46–47.

Moriondo F. (1975) Caratteristiche epidemiche della ruggine vescicolosa del pino: *Cronartium flaccidum. Annali dell'Accademia Italiana di Scienze Forestali* **24,** 331–406.

Oehrens E. (1977) Biological control of the blackberry through the introduction of rust *Phragmidium violaceum* in Chile. *FAO Plant Protection Bulletin* **25,** 26–28.

O'Rourke C.J. (1976) *Diseases of Grasses and Forage Legumes in Ireland.* An Foras Talúntais, Carlow.

Parmelee J.A. (1977) Puccinia helianthi. *Fungi Canadenses* no. 95. Agriculture Canada, Ottawa.

Pinon J. (1986) Situation de *Melampsora medusae* en Europe. *Bulletin OEPP/EPPO Bulletin* **16,** 547–552.

Pupavkin D.M. (1982) (Rust canker of fir). *Zashchita Rastenii* no. 8, 24.

Raddi P. & Fagnani A. (1978) Miglioramento genetico del pino per la resistenza alla ruggine vescicolosa: controllo ed efficacia della densità di inoculo. *Phytopathologia Mediterranea* **17,** 8–13.

Ragazzi A. & Dellavalle Fedi I. (1982) Observation under fluorescence on progress of basidiospore germination in *Cronartium flaccidum* on the needle surface of certain pine species. *European Journal of Forest Pathology* **12,** 246–251.

Roebbelen G. & Sharp E.L. (1978) Mode of inheritance, interaction and application of genes conditioning resistance to yellow rust. *Advances in Plant Breeding* **9.** Paul Parey, Berlin.

Roll-Hansen F. & Roll-Hansen H. (1981) *Melampsoridium* on *Alnus* in Europe. *M. alni* conspecific with *M. betulinum. European Journal of Forest Pathology* **11,** 77–87.

Sackston W.E. (1978) Sunflower disease mapping in Europe and adjacent Mediterranean countries. In *Helia Information Bulletin of the Research Network on Sunflower* no. 1, pp. 21–31.

Sampson K. & Western J.H. (1954) *Diseases of British Grasses and Herbage Legumes* (2nd Ed.) Cambridge University Press.

Schönbeck F., Dehne H.-W. & Beicht W. (1980) Untersuchungen zur Aktivierung unspezifischer Resistenzmechanismen in Pflanzen. *Zeitschrift für Pflanzenkrankheiten and Pflanzenschutz* **87,** 654–666.

Shattock R.C. & Rahbar Bhatti M.H. (1983) The effect of *Phragmidium mucronatum* on rose understocks and maiden bush roses; fungicides for control of *Phragmidium mucronatum* on *Rosa laxa* hort. *Plant Pathology* **32,** 61–66; 67–72.

Simons M.D. (1970) Crown rust of oats and grasses. *APS Monograph* no. 5. APS, St Paul.

Singh J.P. & Musymi A.B.K. (1979) Control of rusts and powdery mildews by a new systemic fungicide, bayleton. *Pesticides* **13,** 51–53.

Spencer D.M. (1976) Pelargonium rust and its control by fungicides. *Plant Pathology* **25,** 156–161.

Spencer D.M. (1979) Carnation rust and its control by systemic fungicides. *Plant Pathology* **28,** 10–16.

Spencer D.M. (1980) Parasitism of carnation rust (*Uromyces dianthi*) by *Verticillium lecanii. Transactions of the British Mycological Society* **74,** 191–194.

Taris B. & Estoup G. (1975) Un nouveau danger pour la populiculture: le *Melampsora medusae. Revue Girondine Agricole* no. 156, 21.

Van Arsdel E.P., Riker A.J. & Patton R.F. (1956) The effects of temperature and moisture on the spread of white pine blister rust. *Phytopathology* **46,** 307–318.

Veenenbos J.A.J. (1984) The green corner: a pre-export inspection system for chrysanthemum cut flowers and pot plants in the Netherlands. *Bulletin OEPP/EPPO Bulletin* **14,** 369–372.

Viennot-Bourgin G. (1949) *Les Champignons Parasites des Plantes Cultivées,* Masson, Paris.

Viennot-Bourgin G. (1956) *Mildious, Oïdiums, Caries, Charbons, Rouilles des Plantes de France. Encyclopédie Mycologique* Vol. 16. Lechevalier, Paris.

Wale S.J. (1981) Raspberry rust–a potential threat? In *Crop Protection in Northern Britain.* West of Scotland Agricultural College, Auchincruive.

Water J.K. (1981) Chrysanthemum white rust. *Bulletin OEPP/EPPO Bulletin* **11,** 239–242.

Whipps J.M. & Lewis D.H. (1984) Infection and subsequent patterns of development of the pycnial-aecial stage of *Puccinia poarum* on *Tussilago farfara* and effect of the fungicide, oxycarboxin, on fungal growth. *Transactions of the British Mycological Society* **82,** 455–467.

Wilkins P.W. (1973) *Puccinia recondita* on ryegrass. *Plant Pathology* **22,** 198.

Wilson M. & Henderson D.M. (1966) *British Rust Fungi.* Cambridge University Press.

Zadoks J.C. (1961) Yellow rust on wheat, studies in epidemiology and physiologic specialization. *Netherlands Journal of Plant Pathology* **67,** 69–256.

Zadoks J.C. (1965) Epidemiology of wheat rusts in Europe. *FAO Plant Protection Bulletin* **13,** 97–108.

Zadoks J.C. & Bouwman J.J. (1986) Epidemiology in Europe. In *The Cereal Rusts* Vol. 2 (Eds W.R. Bushnell & A.P. Roelfs), pp. 330–369. Academic Press, New York.

17: Basidiomycetes III
Hymenomycetes

The majority of plant pathogens in the higher basidiomycetes are members of the Aphyllophorales, characterized by variously shaped complex basidiocarps, without gills. Most attack the woody tissues of plants, and are not necessarily very clearly separated in biology from related saprophytes causing rotting of timber, which attack living plants only rarely or when under stress. Some attack the roots and base of tree trunks, generally from a saprophytic base in the soil. The host is progressively killed, and the pathogen then continues life as a saprophyte. Others enter wounds in the trunk or branches, infection then arising from basidiospores. The Aphyllophorales attacking woody plants mostly fall into the families Ganodermataceae (dark-spored, with perennial poroid basidiocarps containing dark setae in the hymenium), Stereaceae (with non-poroid, mostly appressed or effused-reflexed basidiocarps) and Polyporaceae (poroid, hyaline-spored and without setae).

The rather few Agaricales (gill-fungi) which attack living plants are broadly similar in biology to the fungi considered above. The genus *Armillaria* is of particular importance, distinguished especially by its formation of rhizomorphs facilitating spread from saprophytic substrates to the roots of its hosts. It may be noted that numerous Agaricales (and Gasteromycetes) infect plant roots, but more or less mutualistically, forming ectotrophic mycorrhizae on the roots of forest trees.

Other Aphyllophorales attack herbaceous plants, mostly from the soil, causing root and collar rots and 'moulds' of turf (Corticiaceae, *Typhula* spp.). In this, they resemble the plant pathogens in the Auriculariales (*Helicobasidium*) and Tulasnellales (*Thanatephorus, Ceratobasidium*), which have rather simple short-lived basidiocarps and persist as sclerotia in the soil (*Sclerotium, Rhizoctonia*). These include some of the most important root pathogens of crops.

Finally, the rather aberrant order Exobasidiales are highly adapted parasites causing blister-like galls on leaves; their biology is superficially much like that of the Taphrinales (q.v.).

EXOBASIDIALES

Exobasidium vaccinii (Fuckel) Voronin causes a leaf and shoot gall of *Vaccinium* species known as red leaf disease. Leaves are infected by basidiospores, or by conidia budded from them, carried by wind or rain. Infected leaves are severely malformed, showing thickened reddish galls, with a white efflorescence below the point of development of the hymenium. The disease is more significant in N. America than in Europe (CMI 778). A very similar disease is widespread in Europe on horticultural azaleas (*Rhododendron* spp.), especially under glass, and is caused by **var. *japonicum*** (Shirai) McNabb (syn. *E. japonicum* Shirai) (CMI 780). However, control measures are scarcely needed.

Microstroma juglandis Sacc. causes a leaf spot of walnut, the lower leaf surface becoming covered with a whitish efflorescence of spores. The basidiomycete affinities (enteroblastic budding) of this genus are reviewed in a transmission electron microscope study by Arx *et al.* (1982).

AURICULARIALES

Helicobasidium brebissonii (Desm.) Donk

(Syn. *H. purpureum* Pat.)
(Anamorph ***Rhizoctonia crocorum*** (Pers.) DC., syn. *R. violacea* Tul.)

Basic description

Hyphae usually consist of binucleate cells, which when mature are approximately 4–8 μm diameter with septa 55–200 μm apart, and violet-brown in colour. Hyphae frequently aggregate into strands up to 1 mm in diameter. Infection cushions formed on root surfaces are 150–200 μm in diameter. Typical *Rhizoctonia* sclerotia are formed.

Host plants

Host plants include carrot, potato, sugar beet, asparagus, celery, lucerne and at least 163 other species in 43 families (Hull & Wilson 1946; Viennot-Bourgin 1949; Herring 1962).

Host specialization

The results of various workers are contradictory (Herring 1962; Molot & Simone 1965). It is probable that isolates have preference for certain hosts but do not constitute specific forms of the fungus.

Geographical distribution

H. brebissonii occurs in 47 countries world wide. This includes Europe, except Spain and Portugal, N. and S. America, Africa, Middle and Far East and Australia (CMI map 275).

Disease

The fungus causes violet root rot of numerous crops, trees and herbaceous plants, particularly root and tuber crops. It is especially important on carrots, asparagus and sugar beet. Roots and tubers become covered with strands of violet to brown mycelium. Infection cushions are formed, probably initiated in the host after penetration. The distribution of the cushions varies with the host. As the fungus penetrates roots more deeply, secondary invasion, especially by soft-rotting bacteria, occurs.

Epidemiology

The pathogen persists in soil for an unknown number of years, as sclerotia which are usually small (< 5 mm) but may be larger, up to several cm in some reports. Sclerotia may be formed on fine lateral roots (carrot), on tap root and laterals (sugar beet) or in a mycelial web in the soil. Sclerotia remain in the soil after harvest. The pathogen may also be harboured by a wide range of weeds, some perennial. With carrot, infection can occur anywhere on the root surface, usually on older roots. Conditions favouring infection and both local and long distance spread of the fungus are not fully understood. Spread by contaminated machinery is likely. Wind dispersal of mycelial strands and sclerotia may also occur on light soils subjected to high winds. The use of manure from animals fed on diseased roots and tubers could also be involved in spread of the pathogen. The teleomorph is rarely found in the field and the role of basidiospores in the epidemiology of the pathogen is unknown. They may form on mats of mycelium on the soil surface around heavily infected roots.

Control

A number of currently available chemicals did not control the disease (Dalton *et al.* 1980) and no resistance was evident in an extensive trial of carrot cultivars (Dalton *et al.* 1981). To reduce losses on infested sites: i) avoid growing susceptible crops in close rotation; ii) suppress growth of volunteer potatoes, root crops and susceptible weeds; and iii) do not plough infected crops back into the land or dump infected crops on land which may be used for future growth of susceptible crops.

EPTON

TULASNELLALES

Thanatephorus cucumeris (Frank) Donk

(Syn. *Corticium solani* (Prill. & Delacr.) Bourdot & Galzin, *Pellicularia filamentosa* (Pat.) Rogers)
(Anamorph ***Rhizoctonia solani*** Kühn)

Basic description

The rapidly growing multinucleate hyphae of relatively wide diameter are first hyaline and then become dark brown. Branches occur near a distal septum, often at right angles, and are constricted at the point of origin and septate shortly above. Some hyphae become swollen and monilioid and later darken in colour. Sclerotia are composed of dark brown cells and are almost uniform in texture. Ellipsoid and apiculate basidiospores develop on a hymenium and each basidium usually has four sterigmata (CMI 406).

Host plants

T. cucumeris attacks a wide range of crop plants including bean, beet, brassicas, potato and tomato as well as many herbaceous ornamentals. The stem canker disease of potato is considered below separately from the damping-off disease of numerous hosts.

Host specialization

The aggregate species *R. solani* contains fungi with a wide variation in morphology, nutritional requirements and pathology (Anderson 1982). Numerous stable variants exist and are probably adapted to different soils, climates and vegetation regions of the world, but there is continuous variation in character and much overlapping occurs between groups. Consequently, there is no single factor that distinguishes it from other similar fungi and insufficient information to consider division of the group into species or formae speciales. Designation of the teleomorph as *T. cucumeris* has aided precision and should avoid confusion between it and the mycelium of other basidiomycetes. One principle used in grouping is based on the ability to anastomose between members of the same group; four main anastomosis groups (AG) have been recognized. AG 1 contains isolates causing seed and hypocotyl rots, aerial blight and web blight. AG 2 isolates cause canker of root crops and root rots of conifers. AG 3 isolates are pathogens of potato and barley and can also cause seed rot. AG 4 isolates cause seed and hypocotyl rots of many angiosperm species. Other groups and subgroups have been proposed but as some isolates do not fall into any of the groups, the erection of varieties based on anastomosis remains questionable at present.

Geographical distribution

T. cucumeris is found world-wide and probably occurs in most arable soils.

Stem canker of potatoes

Disease

The infection of sprouts on seed tubers both during humid storage and in soil after planting causes the death of apices. Other shoots develop from tuber eyes or axillary buds and several generations of these may be pruned off but eventually some stems usually emerge. Most infection occurs before shoot emergence and lesions on the stem bases are dark brown, dry and corky with a well-defined margin. As the flow of photosynthates to tubers is interrupted, leaves become stiff and curl, although these symptoms are usually transient. Occasionally aerial tubers form in the axils of basal leaves. During growth *T. cucumeris* attacks stolons which may be pruned off and sever developing tubers (Hide *et al.* 1973). During humid weather in June and July, the teleomorph of *T. cucumeris* forms as a white felt over stems and leaves near the soil surface; this is usually an indication of stem canker although it may develop from mycelium growing epiphytically over the stem base. Tubers are infected with mycelium at any time during growth and sclerotia (black scurf) form towards the end of the season when haulms senesce.

Epidemiology

Most stem cankers originate from sclerotia or mycelium on seed tubers and in the UK about 25% of seed tubers are affected with black scurf. The disease is most common in dry, cool soils and its severity is probably related to the speed of shoot growth because stems become resistant to infection when they have emerged. Similarly, soil conditions may affect the time and severity of stolon infection. The disease can also originate from inoculum in soil, although compared with sclerotia on seed tubers, its quantity and position is likely to render it more important for the infection of stolons and tubers than of stem bases. The role of basidiospores in the disease is not known.

Economic importance

Losses in yield are related to the proportion of missing plants and also to the extent to which shoot emergence, plant growth and tuber bulking are delayed by infection. Although shoots may be severely affected, some usually do emerge but the plants are stunted and initially have decreased haulm weight and leaf area. However, plants compensate by increasing the size of lateral shoots arising from axillary buds on aerial stems. Similarly, although stolon pruning decreases the number of tubers, plants produce more on stolon branches so that the number of saleable tubers at harvest is seldom affected. There are often more large and small tubers, and fewer of intermediate size. Although the rate of tuber bulking is not affected, the start of bulking is delayed and therefore yields from early harvests can be seriously affected. By maturity, infected plants may yield up to 10% less than healthy ones, but in crops, losses are likely to be less because vigorous and healthy plants compensate in growth and yield for weak or absent neighbours.

Control

Much damage to shoots can be avoided by decreasing the time between planting and emergence, for example by sprouting seed and shallow planting. Seed tuber treatments are effective in controlling stem canker and depend on fungicide deposits over the whole tuber surface preventing mycelial growth from sclerotia (Hide & Cayley 1982). Organomercury compounds have been mostly superseded by less toxic materials including benzimidazoles (benomyl, carbendazim, thiabendazole, thiophanate methyl), iprodione and tolclofos methyl. Dithiocarbamates (mancozeb, maneb, thiram) and formaldehyde are also effective. Dipping tubers in fungicide suspension has been replaced by sprays and dusts, thus avoiding the risk of black-leg and soft rot resulting when tubers are wetted. Seed tubers may be treated at harvest or during loading into store but treatment after size grading has advantages because table potatoes can be removed and, less soil being present, more of the fungicide is deposited on the tuber skin.

Damping-off and other diseases

Disease

In crops such as sugar beet, *T. cucumeris* invades and kills seeds both before or soon after the start of germination. Infection during the juvenile seedling phase after emergence causes typical damping-off; cortical tissues of the hypocotyl are especially susceptible and production of polygalacturonase enzyme by the fungus results in tissue decay and the collapse of stems. After the juvenile phase, when stems have secondary thickening, brassica and tomato seedlings remain susceptible to infection and, although cortical tissues decay in well-defined areas encircling the stem, the central stele provides support and the plant remains upright, symptoms typical of wire-stem. The fungus infects tomato leaves in contact with the soil, spreads through the petiole and causes canker on the stem, and fruits infected through the epidermis or wounds develop brown spots. Similarly, crater spots develop on celery. Lower leaves of cabbage and lettuce plants decay (bottom rot) and infection spreads upwards to the heads which later become brown and slimy and eventually mummify (head rot). Infection of the hypocotyl and tap root of sugar beet can kill seedlings, and plants that survive are stunted and develop a fangy and proliferated root system. Leaf infection leads to decay of the crown and a dry rot of the fleshy root. Lesions also occur on carrot and turnip and root rots on clover, lucerne and cereals (but see also *Ceratobasidium cereale*).

Epidemiology

Soil is the major source of *T. cucumeris* inoculum (saprophytic mycelium, sclerotia) in the damping-off diseases; after the initial infection the disease spreads outwards in circular or irregular patches. Environmental conditions greatly influence all phases of the disease which is favoured by high air humidity, extreme soil temperatures and by over-moist or saline soil. Such conditions delay germination and permit invasion from soil and spread to surrounding plants. In field crops inoculum from soil is spread locally in rain splash, and seedling disease of sugar beet is prevalent in cold wet springs. Infection of sugar beet leaves occurs in hot weather; plants wilt but may apparently recover during cool dry weather. Clay soils and those with little organic matter favour the disease.

Economic importance

Damping-off and other diseases can cause important but sporadic loss of seedlings. Cabbage plants affected with wire-stem remain weak and yield losses may be considerable (up to 30%).

Control

Disease in nursery beds is prevented by good hygiene and by avoiding conditions that favour the disease. Disinfecting seed boxes and sterilizing compost with steam or nabam eradicates the fungus, and the disease can also be controlled by incorporating quintozene in compost. In protected crops, extremes of temperature and wet soil should be avoided and adequate aeration given. Also avoiding seed from diseased pods and shallow seed sowing helps to prevent serious outbreaks of the disease. In field crops, amounts of soil inoculum can be decreased by rotations with non-susceptible crops, disposal of diseased crop residues, incorporating cereal straw and improving drainage. Although some cultivars show tolerance to the disease, there are difficulties in breeding for resistance to such a variable pathogen.
HIDE

Rhizoctonia tuliparum (Kleb.) Whetzel & Arthur

(Syn. *Sclerotium tuliparum* Kleb.)

R. tuliparum can be distinguished from *Thanatephorus cucumeris* by its slower growth rate, its slightly narrower hyphae with binucleate cells and its more compact sclerotia. There is no known teleomorph. The fungus is principally a pathogen of *Tulipa* and bulbous *Iris* in which it causes grey bulb rot. Under appropriate conditions it will attack a wide range of plants but rarely does so in the field. *R. tuliparum* was first clearly distinguished in the Netherlands in 1884 but has since been reported from most countries of N. Europe and elsewhere (CMI 407). The disease causes the appearance of bare patches in fields of tulips. Soil adheres to bulbs that

have failed to emerge and the bulb scales are rotted from the top downwards. At first infected bulbs show grey or purplish-grey areas but these eventually turn dark brown. Large sclerotia, white at first but dark brown later on and up to about 7 mm in diameter, are found either on or near the nose of infected bulbs. The disease also occurs in forcing boxes of tulips and irises and causes gaps or produces plants with ragged and torn foliage and no flowers. The fungus persists in the soil as sclerotia which can survive for 5–10 years. They rarely germinate in natural soils at temperatures above 10°C and temperature rather than plant resistance appears to be a major determinant of host range (Gladders & Coley-Smith 1978). Mycelium can grow to a distance of about 15 cm from germinating sclerotia. Host penetration occurs predominantly by means of fine hyphae produced at the centre of infection cushions. Infection cushions only occur on cuticularized surfaces and in consequence *R. tuliparum* does not invade roots directly but does penetrate stems and bulbs. The disease can be controlled by long rotation or by the use of quintozene lightly forked into the soil at a rate of 50 g/m^2 (Moore 1979). Bulbs may be dipped for 1–8 h in 0.4% commercial formaldehyde.
COLEY-SMITH

Rhizoctonia carotae Rader is distinguished from *Thanatephorus cucumeris* by the presence of clamps on the mycelium. Due to its capacity to infect at 1°C, it causes crater rot of carrots in cold store, persisting on infested crates, which should therefore be disinfested (CMI 408).

Ceratobasidium cereale Murray & Burpee

(Anamorph ***Rhizoctonia cerealis*** v.d. Hoeven, syn. *R. solani* Kühn p.p.)

By virtue of its host range and morphology, *R. cerealis* is considered to be distinct from *R. solani*. This has been confirmed by the recent description in culture (but not yet in nature) of the teleomorph (Murray & Burpee 1984), which is quite distinct from *Thanatephorus cucumeris* (associated with *R. solani sensu stricto*). The early failure to distinguish *R. cerealis* from *R. solani* makes it difficult to establish the host range and distribution of *C. cereale* from the literature, but it is known to attack wheat, barley, oats, rye, maize and other Gramineae, and to occur in UK, Nordic countries, Netherlands, Italy, FRG, Romania, Yugoslavia and N. America. There is evidence of some variation in pathogenicity and host range of different isolates taken from both cereals and turf grasses. There are also differences in host susceptibility, generally reported as rye > oats > wheat > barley, and Hollins & Scott (1983) demonstrated that there is a range of resistance to *C. cereale* available in winter wheat cultivars grown in Britain.

C. cereale causes sharp eyespot of cereals. Lesions occur on the lower stem and leaf sheaths (Plate 31). The leaf sheath lesion is pale with an irregular but roughly elliptical purplish-brown margin. The lesion on the culm beneath the sheath is similar, but more regular in outline. The centre of the culm lesion has a brownish 'pupil' consisting of superficial brownish mycelium and sclerotia which can be readily removed with a finger nail. Lesions usually occur on the two lowest internodes, but can occur higher up the stem. In severe attacks adjacent lesions may coalesce and girdle the stem. Occasionally sclerotia and mycelium can be found inside the hollow stem (Clarkson & Griffin 1977). Plants affected early in their growth may develop 'whiteheads' or lodge, but the usual effect is a diminution of grain production by infected tillers, which produce fewer and smaller grains (Richardson *et al.* 1976). Infection is favoured by cool dry conditions (Pitt 1964) so autumn-sown cereals are more at risk than spring-sown ones. The source of inoculum is sclerotia and mycelium on straw and stubble.

Sharp eyespot does not usually cause severe losses, although it is widespread. Winter wheat is the crop most often affected in the UK, with various surveys over a 20-year period showing 34–100% of fields affected with a mean of 60%. Incidence within crops was, however, generally low with the mean infection in any one year ranging from 3 to 31% (overall mean incidence 10%). Only a small proportion of stems come into the category of severe infection, however, and losses in British wheat crops on average are estimated to be about 0.4% (Clarkson & Cook 1983). The disease is increasing in importance in France, where its effects on tillering and yield are currently being studied (Lucas & Cavelier 1985). No chemical control methods currently used on cereals appear to have any consistent effect on levels of sharp eyespot infection, and some, e.g. benomyl, tend to increase infection levels (Van der Hoeven & Bollen 1980). Levels in winter wheat are reduced by late sowing.
RICHARDSON

Rhizoctonia fragariae Husain & McKeen causes a root rot of strawberry and possibly many other species (APS Compendium). *Thanatephorus* *cucumeris* also attacks strawberry, but *R. fragariae* can be distinguished by its binucleate hyphae, a *Ceratobasidium* teleomorph and by forming masses of characteristic moniliform resting cells in culture, rather than sclerotia. Strawberry plants can be stunted by severe rootlet-pruning and the fungus also causes a soft rot of fruit and an anther and pistil blight. Mainly known from N. America and Japan, *R. fragariae* has been recorded in Italy (D'Ambra & Mutto 1969). Indistinguishable *Ceratobasidium* species are said to be widespread in soil and to resemble the *Rhizoctonia* species which form mycorrhizae on orchids.

APHYLLOPHORALES

Laetisaria fuciformis (McAlp.) Burds.

(Syn. *Corticium fuciforme* (Berk.) Wakef., *Phanerochaete fuciformis* (Berk.) Jülich)

L. fuciformis causes red thread disease of turf grasses (APS Compendium), which has only recently been distinguished from pink patch disease due to *Limonomyces roseipellis* (q.v.) and possibly other *Limonomyces* species (Stalpers & Loerakker 1982; Kaplan & Jackson 1983). *Agrostis, Festuca, Lolium* and *Poa* species are principally affected, and the disease is widespread in cool moist regions in Europe (and N. America). Leaves are overgrown by a reddish mycelium, which produces characteristic red thread-like structures at the apex of the leaves. The mycelium may also aggregate into pink cottony flocks, containing masses of arthroconidia. Minute basidiocarps are formed on the infected tissues or on the threads. Affected leaves may be bound together by the mycelium, and may themselves redden (anthocyanins) before drying out to tan colour. Infected turf shows irregular more or less circular patches, from a few centimetres to several metres in diameter, of characteristically reddened plants. The disease is mostly spread by mycelial growth from plant to plant, or by dispersal of the threads, arthroconidia or plant debris. The role of basidiospores is uncertain. Attacks occur mainly in the autumn, during periods of rather low temperature and high humidity (O'Rourke 1976). It is most serious in thick dense swards on soils with low nitrogen. However, serious attacks are fairly rare, only minor damage being caused to lawns, golf courses and permanent pastures. Traditionally, mercury- and cadmium-based preparations were used for control. Hims *et al.* (1984) found benodanil and anilazine to be the most successful fungicides of several systemic compounds tried, and also noted that nitrogen amendment had no effect on disease, at least within a few weeks after application.

CARR & SMITH

Limonomyces roseipellis Stalpers & Loerakker causes pink patch disease of turf grasses (APS Compendium) and was formerly confused with *Laetisaria fuciformis* (q.v.), from which it differs in having binucleate hyphae with clamp connections and in producing neither arthroconidia in cottony flocks nor red threads. It tends accordingly to spread more slowly and is rarely important on mown turf. It is also restricted to *Lolium perenne* and *Festuca rubra*. Though apparently widespread in Europe, the distinction from *L. fuciformis* is too recent for its individual importance to be fully assessed.

Corticium rolfsii Curzi

(Syn. *Athelia rolfsii* (Curzi) Tu & Kimbrough)
(Anamorph ***Sclerotium rolfsii*** Sacc.)

C. rolfsii forms distinctive rounded sclerotia with a sharply differentiated darker rind. These are the main means of survival in the soil (the basidial fructifications being of little significance in the disease cycle). It causes root and collar rots on numerous hosts in the warmer parts of the world (CMI 410). In C. and S. Europe (CMI map 311), it can under certain conditions be destructive to aubergine, pepper, potato, tobacco, tomato, groundnut, bean, lupin, sugar beet, strawberry, melon, watermelon, and woody hosts (for example olive, jasmine, myoporum and grapevine in Greece). However, it is hardly mentioned in the literature as a significant pathogen in Europe, even under glass. Concern has been expressed at the possibility of its introduction into glasshouses in N. Europe on ornamental bulbs from warmer countries. There are interesting reports from Israel on its biological control since sclerotia and mycelium are destroyed by *Trichoderma harzianum* (Elad *et al.* 1980). ***Sclerotium delphinii*** Welch has slightly larger sclerotia, but is otherwise practically indistinguishable from *C. rolfsii*. It has caused a serious crown rot of *Iris* in the Netherlands and attacks various ornamentals.

SMITH

Sclerotium hydrophilum Sacc. causes a minor root and stem rot of rice in France and Italy (Bernaux 1977), and attacks many aquatic macrophytes. Its basidiomycete affinities have been shown by TEM (Punter *et al.* 1984), and this clearly distinguishes it from *S. oryzae*, the sclerotial state of *Magnaporthe salvinii* (q.v.).

Trechispora brinkmanii (Bresad.) Rogers (anamorph ***Phymatotrichopsis omnivora*** (Duggar) Henneb., syn. *Phymatotrichum omnivorum* (Shear) Duggar) causes Texas root rot of cotton in the USA, persisting as sclerotia and infecting by mycelial strands. It is considered a serious quarantine threat to cotton-growing countries in Europe (EPPO 21). ***T. coharens*** (Schwein.) Jülich & Stalpers causes white patch disease of turf in England, while ***T. farinacea*** (Pers.) Lib. causes a similar disease in the Netherlands. The symptoms are due to the appearance of the mealy white fructifications. These are of minor importance, but may be favoured by the use of fungicides more active against ascomycetes than basidiomycetes (APS Compendium).

Typhula incarnata Lasch and *Typhula ishikariensis* Imai

T. incarnata and *T. ishikariensis* cause grey or speckled snow mould of winter cereals and grasses, the latter also forage legumes during overwintering. *T. incarnata* is widespread in the temperate and cooler regions, sometimes causing severe damage. The disease is most prevalent in areas having a relatively mild, wet winter climate. Winter cereals especially may be severely damaged, even in localities with a relatively short period of snow cover (2–3 months), and slight attacks may occur without any snow cover during the winter. *T. ishikariensis* is restricted to the northern parts of the Nordic countries, USSR, Japan, N. America, and to the Alps, to regions having a stable, cold winter climate with prolonged, deep snow cover. Severe attacks rarely occur at localities having less than about 4 months of snow cover. In regions having more than about 5 months of deep snow cover, this pathogen is regarded as one of the most important snow mould fungi, regularly causing severe damage.

The disease appears as bleached patches of dead or dying plants, sparsely covered with a greyish-white mycelium and speckled with sclerotia, reddish brown to dark brown (*T. incarnata*) or dark brown to almost black (*T. ishikariensis*), 0.5–2 mm in size (Fig. 74). Sclerotia remain dormant on the soil surface during the summer months and serve as a source of inoculum, mainly through direct mycelial growth into predisposed host tissues. Basidiospore infection is assumed to be of minor importance. Sclerotia buried in the soil may survive for several years. Mycelial growth occurs from below −6°C, with an optimum at 9–12°C. Predisposing factors (Årsvoll 1978) are early snow in autumn or prolonged deep snow cover on unfrozen ground. Unbalanced or excessive nitrogenous fertilizer, particularly if applied late in the season, and unhardy, susceptible cultivars, poorly adapted to the local environment should be avoided. Fungicides, e.g. quintozene, propiconazole, triadimefon, applied late in the autumn before snowfall, may control the disease. Great variation in resistance between cultivars is known (Jamalainen 1974). Host resistance is assumed to be non-specific and quantitative, governed by several genes, each possessing a slight additive effect. Laboratory methods for large-scale screening for resistance in breeding material have been developed.

ÅRSVOLL

Fig. 74. Sclerotia of *Typhula ishikariensis* on leaves of *Phleum pratense*.

Chondrostereum purpureum (Pers.) Pouzar

(Syn. *Stereum purpureum* (Pers.) Fr.)

Basic description

C. purpureum forms well-differentiated carpophores, at first flat on the host bark, and then with a free margin, up to several cm across. The upper surface is white-pilose and the lower bears a smooth hymenium without cystidia, pale lilac fading to pale brown.

Host plants

About 175 species in 26 families are recorded as hosts, including all pip-fruit and stone-fruit, plum, peach, cherry, apricot and pear being especially susceptible. Many forest trees are also affected, especially poplar, beech and birch. No host specialization is recorded.

Geographical distribution

The fungus is widespread and recorded in most European countries: Belgium, France, FRG, UK, Greece, Italy, Netherlands, Nordic countries, Poland, Portugal, Romania, Spain, Switzerland, USSR, Yugoslavia.

Disease

C. purpureum causes silver-leaf disease (Plate 32), characterized by the metallic sheen of the leaves of infected plants. At first, only leaves are affected, then necrosis and blisters appear on the shoots. In severely affected plants, the foliage is stunted and chlorotic. Sections of silver-leaved shoots often show browning of the woody tissue. Symptoms are often very irregularly distributed over the tree, one sector of foliage remains normal while others are diseased. Similarly, attacks in orchards tend to be irregular. This distinguishes silver-leaf from other causes of similar symptoms, especially those due to the mite *Vasates cornutus* on peach. Trees, or parts of trees, may die within months or may, after showing symptoms in spring, recover either temporarily or permanently.

Epidemiology

The disease is spread by basidiospores from the carpophores which form, rather irregularly, on dead or dying tissues, especially on some hosts (plum, cherry, poplar), under moist, shaded conditions (generally close to the soil at the end of summer). Carpophores can dry out, but recover and sporulate once humidity rises again. Wind-borne basidiospores are deposited on exposed woody tissue, especially pruning wounds. They penetrate to some depth in the wood and there germinate to give a parasitic mycelium in the host tissues. Toxins formed by the fungus are carried to the leaves to cause symptoms. The fungus itself is mainly limited to the larger branches, roots and trunk, and does not spread to the shoots or leaves. The need for high humidity for carpophore formation, sporulation and infection generally limits serious disease to oceanic regions with a humid climate and mild winters. However, local microclimatic conditions may favour *C. purpureum* in any region.

Economic importance

Silver leaf is a serious disease in many regions. Trees in orchards are often killed, and, even if the percentage of individuals affected is low, the gaps in the rows make good management difficult. Young vigorous trees are generally most affected. In nurseries, any affected tree has to be destroyed.

Control

Good protection is obtained by treating all pruning wounds immediately with fungicidal paints. Excellent results can also be obtained by applying a suspension of *Trichoderma harzianum* conidia (10^{0} per ml), this can be done at pruning with special secateurs carrying a spray attachment (Grosclaude *et al.* 1973; Grosclaude 1984). Pruning should be done only after spring, and the vicinity of poplars should be avoided. There is no certain way of curing already diseased trees, although numerous injection treatments have been tried. Some authors claim to have good results from injecting *Trichoderma* spores in nutrient medium (Dubos & Ricard 1974).

GROSCLAUDE

Stereum hirsutum Fr.

S. hirsutum forms thin, leathery brackets, which are sometimes resupinate, but more often imbricate, with a yellow-brown or greyish, often zoned upper surface. They are usually 2–10 cm across. The smooth lower surface is at first bright yellow, later dull yellow or greyish. The colourless spores are elliptical, flattened on one side, and are white in the mass. In young cultures, the hyphae produce large whorled clamp connections (Cartwright & Findlay 1958). The fruit bodies of the fungus may appear at all times of the year.

S. hirsutum grows on a wide range of broad-leaved trees, especially on oak and beech, and also attacks stone-fruit such as peach. It is common and widespread in Europe and N. America. It occurs mainly as a saprophyte on dead branches, fallen trees, and felled timber (especially sapwood) left lying on the forest floor. On such wood of oak (on which it is especially common), it produces a rather fibrous yellowish-white decay of the sapwood. Rarely it may cause a pipe rot of the heartwood of living trees. In addition, *S. hirsutum* is frequently implicated (with *Phellinus igniarius*) in a rot of the woody shoots of grapevine known as esca, black measles or apoplexy. Infection occurs through pruning wounds, and leads

to powdery rot of the trunk or branches, not normally spreading to the roots. Berry spotting and interveinal leaf chlorosis are the characteristic symptoms of the disease, resembling those caused by severe drought. Distal shoots may suddenly shrivel and die (apoplexy). Longitudinal sections of the diseased trunks show a yellow discoloration of the disintegrated wood. Fructifications may appear at the base of the plant, but basidiospore inoculum may well originate from sources outside the vineyard. The disease can be controlled by treatment of pruning wounds with sodium arsenite (Galet 1977).
PHILLIPS & SMITH

Stereum sanguinolentum Fr.

S. sanguinolentum forms thin often resupinate, fructifications but sometimes also small, often imbricate brackets with a greyish or fawn, rather zoned upper surface with radiating fibrils. The smooth hymenium is greyish to buff-coloured, and when fresh it bleeds if scratched or bruised. *S. sanguinolentum* is found on conifer stumps and on dead and living coniferous trees of many species. In Europe it is important as a cause of significant loss especially in *Picea abies*, but it may cause damage also in *P. sitchensis*, and sometimes in other conifers, especially the larches. It occurs widely in Europe, and in N. America, E. Africa and New Zealand. It may cause a trunk rot. This may show only after felling, though resin sometimes flows down the lower parts of the trunks of standing trees decayed by the fungus. Fruit bodies may also be found on wounds on the trees, and these wounds may also show a yellow or brownish stain. Internally, infection is first seen as a yellowish-brown stain which develops to the stage of incipient decay. The wood is then still firm, but with time becomes dry and fibrous, and orange-brown flecked with yellow streaks (Greig 1981). *S. sanguinolentum* is often found as a saprophyte on stump surfaces and the ends of trunks left on the ground. It may however enter living trees as a wound parasite through broken branches, wounds made when brashing and pruning and in produce extraction, and others made by cattle and game (Pawsey & Gladman 1965; Roll-Hansen & Roll-Hansen 1981). The incidence of infection and rate of decay increase with the size of the wound concerned. The fungus grows within the tree at rates of up to 40 cm/year. Nevertheless advanced decay develops slowly, and is usually found mainly in trees of from 50–80 or more years of age. Decay caused by *S. sanguinolentum* generally gives rise to major loss only where conifers are grown on very long rotations. Infection may be kept to a minimum by avoiding the wounding of trees when extracting produce, and by taking care to cut live branches cleanly and close to the stem when pruning and brashing. In parks, gardens and arboreta, damaged branches should be carefully trimmed out.
PHILLIPS

Stereum frustulatum Fr. (syn. *S. frustulosum* Fr.) is a widespread but relatively uncommon fungus that causes a white pocket rot of oaks, often high in the tree and in dead branches. ***S. gausapatum*** Fr. is a common cause of a yellowish-white, soft pipe rot of stems and branches of standing and felled broad-leaved trees, especially oaks and sweet chestnuts. ***S. rugosum*** Pers. is another fungus that may cause a white pipe rot of oaks, sometimes in standing trees, but mainly in dead wood and after felling; it occasionally also affects beech and other broad-leaved trees.

Ganoderma applanatum (Pers.) Pat.

G. applanatum produces rather flattened, sessile, often imbricate, perennial brackets up to 40 cm across. The upper surfaces of the sporophores are reddish-brown, zoned and lumpy, and finally form a hard varnished crust which is usually covered with a brown deposit of basidiospores fallen from above. The growing edge of the sporophore is white. The pores are white, but become brown with age (CMI 443). *G. applanatum* is especially common and damaging on old, over-mature beeches, but it also occurs on a great many other broad-leaved trees, including poplars and elms, willows, oaks, sycamores and horse chestnuts. It sometimes also attacks pines and other conifers. The fungus is common and occurs virtually world-wide (though forms in the tropics are sometimes regarded as a separate species). It enters through wounds, including those resulting from the fall of branches, and those on damaged roots. It causes a soft, white, spongy rot, and the decayed wood usually breaks up into rectangular blocks. In its earliest stages, however, the rot shows in wood sections only as a whitish mottle. In living trees the decayed area is surrounded by a thin, dark brown line containing tyloses and gum (Cartwright & Findlay 1958). The fruit bodies occur on tree trunks and buttress roots at all times of the year. In cultures of the fungus, numerous crystals are produced round the submerged hyphae and in the medium. On beech, *G. applanatum* is one of the common causes of decay, and through rot of the roots, butts and large branches

it is important because it renders the trees unstable and dangerous. Little can be done to control attacks on forest trees. Infection of parkland trees can be reduced by pruning damaged branches cleanly to the trunk and by protecting the trees from wounding by domestic stock.
PHILLIPS

Other species of *Ganoderma* are of less importance on various broad-leaved trees, especially those indicated: ***G. adspersum*** (Schultz) Donk (beech), ***G. lucidum*** (Leysser) Karsten with a usually stalked sporophore (oak, chestnut), ***G. pfeifferi*** Bresad. (oak, beech), ***G. resinaceum*** Bondartsev (oak, alder, beech, willow).

Phaeolus schweinitzii **(Fr.) Pat.**

(Syn. *Polyporus schweinitzii* Fr.)

Basic description

When growing on tree trunks and stumps, *P. schweinitzii* forms annual, bracket-like fructifications up to 30 cm across, but when growing on the ground attached by mycelial cords to buried wood or roots, it has a rounded cap and a stout velvety brown central stalk. The centre of the upper side of the fruit body is also brown and velvety, but the growing edge is bright yellow. The pore surface below is greenish yellow and water-soaked, and becomes brown if bruised. The colourless, elliptical or oval spores measure 7–8×3–4 μm (Wakefield & Dennis 1981). The sporophores are found in late summer and autumn but soon die and become dark brown and corky. These old fruit bodies persist for up to a year before decomposing. The velvety surface of the fresh fruit body gives the fungus its common English name of 'velvet top'. Cultures of *P. schweinitzii* are at first white, later yellowish or tawny, and they contain many rounded chlamydospores (Cartwright & Findlay 1958).

Host plants

P. schweinitzii is known almost exclusively as a cause of decay in conifers. Many conifer species may be affected, especially *Picea sitchensis* and other spruces, Douglas fir, the common larches and pines, as well as cedars and species of *Chamaecyparis*. On broad-leaved trees there have been records on oak and cherry, and on charred *Eucalyptus* stumps.

Host specialization and variation in the fungus

Cultural studies show that there are many strains of *P. schweinitzii,* which vary in growth rate and appearance. These strains are not known to vary markedly in pathogenicity.

Geographical distribution

P. schweinitzii occurs widely in Europe as well as in N. and C. America, parts of Asia, New Zealand, and some other areas (CMI map 182).

Disease

The fungus causes a brown cubical rot. In its earliest stages infection shows in sections of wood only as an inconspicuous stain, which is sometimes yellowish or pinkish brown. At a later stage the wood becomes brown, dry and soft, and breaks into cubical blocks with thin yellow or white mycelial sheets between. The blocks of decayed wood fall readily apart, and may easily be crumbled to a powder. The area of decay may be surrounded by a narrow, dark brown ring of resinous wood. Freshly infected wood has a strong smell of aniseed or turpentine. At least at the base, the rot may occupy most of the diameter of the trunk, and spread vertically several metres up the stem; sometimes it progresses upwards as isolated pipes of dark, brittle tissue. Living fruit bodies in summer and autumn (and later their dried dead remains) on and around a tree may indicate infection. Otherwise attack may be revealed only when a tree is felled, or its rotted trunk collapses in the wind.

Epidemiology

P. schweinitzii appears to enter through the roots (mainly after damage by some other agent such as drought or attack by *Armillaria* spp.), or sometimes through wounds (including fire scars); primary infection is probably by basidiospores.

Economic importance

P. schweinitzii may cause severe damage to individual trees. Even in its early stages it much reduces the strength of the wood, and eventually decay may hollow out most of the trunk for several metres from the ground. The overall damage is usually limited, however, because the fungus typically affects only scattered trees. Further, although it may be found in quite young, pole-stage trees, severe decay is most often confined to trees of fifty years of age, or more. Commercial conifer crops are now usually harvested before they reach this age. Locally, losses in old crops (especially of pines) may sometimes be considerable. Decay caused by *P. schweinitzii* is also important because of its great effect on the strength of the wood, and because the fungus is not killed by air-dry seasoning. Hence growth may recommence in infected timbers if these are not kept dry.

Control

Little can be done to control infection of living trees by *P. schweinitzii,* except where possible to protect the roots and trunks from wounding, for example by extraction or by fires. Because of the effects of the decay on converted timber, all obviously infected wood should be carefully cut out, and any wood which may possibly retain small traces of infection should be sterilized by kiln drying (Cartwright & Findlay 1958), or chemically impregnated.
PHILLIPS

Phellinus igniarius (L.) Quélet

(Syn. *Fomes igniarius* (L.) Kickx)

P. igniarius forms thick, sometimes hoof-shaped, woody brackets up to 20 cm across. The hard, grey to black crust covering the upper surface is deeply and coarsely furrowed and often cracked, and its obtuse growing margin is often paler. The pore surface is greyish or cinnamon (CMI 194). The fungus is common on a wide range of broad-leaved trees (willow, ash, birch). On poplars, it occurs more rarely and is replaced by *P. tremulae* (q.v.) and ***P. populicola*** Ryv. It is widespread in Europe, N. America and Asia. The fungus causes a soft, central white rot of the heartwood, from which it may sometimes spread to the living sapwood. In its earlier stages infected wood is yellowish-white, usually surrounded by a yellowish-green to deep brown or black zone. Later the decayed wood is soft, white and spongy, with irregular concentric black lines embedded in it (Cartwright & Findlay 1958). Invasion of the tree takes place through dead branch stubs and other wounds. *P. igniarius* causes considerable damage to willow and birch, and occasionally attacks other trees. Like *Stereum hirsutum* (q.v.), it can also cause esca or apoplexy of grapevine.

The most important control measure is common, rational silviculture. Young stands should be relatively dense without coarse branches. The stands should be kept in good growth by thinning to encourage over-growth of dead branches and wounds. Where possible, careful trimming of branch stubs and general avoidance of wounding is important.
PHILLIPS

Phellinus tremulae (Bondartsev) Bondartsev & Borisov

P. tremulae forms fruit bodies which are easy to distinguish from those of *P. igniarius* by the often smaller size, slower growth, sharper edge, finer radial cracks, and narrower concentric furrows; the fruit bodies are very often creeping from the stem outwards on the underside of dead branches. A detailed description of anatomy, cultural characters, hosts, distribution, and pathogenic characters was given by Niemelä (1974). The main hosts are *Populus* species of the section *Leuce: P. tremula* in Europe and N. Asia, *P. grandidentata* and *P. tremuloides* in N. America. It is seldom found on other *Populus* species. There are few reports from other genera of broad-leaved trees. The fungus is widely distributed in Europe except for the south-western parts and the Mediterranean. The main area in Eurasia is Fennoscandia, the northern parts of Poland and Russia, and northern parts of Asia. It is further widely distributed in Canada and N. USA.

The rot is central, and in a rather advanced stage it is light yellowish-brown or whitish, and fibrous. In a cross-section some very narrow, dark brown lines are often seen. Around the central rot is a 2–4 mm broad, dark brown ring, and outside the ring often a greenish zone up to 1 cm wide. *P. tremulae* is, over extensive areas, the most important rot fungus on the aspen species. This is the case in Fennoscandia for example. Aspen should only be grown at high sites. If the growth rate is too slow, the stems will be rotten before they reach log size.
ROLL-HANSEN

Phellinus pomaceus (Pers.) Maire

(Syn. *Fomes pomaceus* (Pers.) Lloyd)

P. pomaceus is sometimes regarded as a subspecies or variety of *P. igniarius*, to which it is undoubtedly closely related. It has rather small, hard and woody fructifications which are most often resupinate, but sometimes hoof- or cushion-shaped, and measure up to about 6 cm across. The hard upper crust, when present, is grey or brownish and concentrically furrowed. Its actively growing margin is velvety and greyish, later yellowish-brown. The pore surface is pale brown, at first with a greyish bloom (CMI 196).

The fungus appears to be confined to trees of the family Rosaceae, especially species of *Prunus* and *Crataegus*. It is most common on old neglected plum trees, on which it causes a white crumbly rot in the centre of branches or of the trunk. Around this area of advanced decay is a hard, firm zone of purple-brown incipient rot (Cartwright & Findlay 1958). The fungus appears to enter the heartwood through wounds made by the breakage of branches. Large trees may become subject to branch breakage and exposure of the heartwood. Some control is possible by careful pruning and the trimming back of broken branches close to the trunk to prevent infection. Infected branches may be removed and burnt.
PHILLIPS

Phellinus pini (Thore) Pilát

(Syn. *Fomes pini* (Thore) P. Karsten, *Trametes pini* (Thore) Fr.)

P. pini has perennial fruit bodies which are pileate, broadly attached, ungulate to triquetrous in section, 5–20 cm wide and up to 10 cm thick at the base, and with a rather sharp to obtuse margin, and are woody hard (Ryvarden 1978). The upper surface of the fruit body is rusty brown to dark brown, or on old specimens, dark grey to nearly black. The pores are irregular, sometimes daedaloid, and yellowish-brown to greyish-brown. *P. pini* occurs mainly on pine. It is widely distributed in N. Europe, Asia and N. America. It causes a trunk rot, mainly in trees 50 years or more of age. It enters through wounds and broken branches, and its fruit bodies (which may persist for many years) usually form on wound sites and branch stubs. Within the tree the fungus grows up and down the trunk, causing serious decay, most often in the upper part of the trunk around the site of infection. The decay takes the form of a red ring rot, which first shows as a pinkish or purplish stain in the heartwood (Cartwright & Findlay 1958). Later, small lens-shaped white pockets appear in the decayed wood. The decay extends in lines along the annual rings, and eventually forms concentric rings separated by more or less sound tissues. The wood therefore falls apart in the form of decay known as ring scale. The decayed wood is often highly resinous. Details of the decay have been described by Blanchette (1980). *P. pini* is one of the most important causes of decay in pines, especially in N. America. In Europe it mainly decays old trees 50–100 years of age, which is beyond the usual length of the forest rotation in many areas. In Nordic and alpine zones, however, where the rotation is often over 100 years, *P. pini* is an important cause of rot. Trees obviously decayed by *P. pini* should be removed when thinning, to salvage as much timber as possible before the fungus moves downwards into the valuable butt end of the trunk.
PHILLIPS

Phellinus chrysoloma (Fr.) Donk (syn. *Polyporus chrysoloma* Fr., *Fomes abietis* P. Karsten) is a species close to *P. pini*. The two are sometimes combined as *P. pini* (Thore) Pilát, in which case *P. chrysoloma* is simply a form. On standing living trees, the fruit bodies are usually perennial, resupinate to pileate, often very widely effused, or in dense imbricate clusters or rows. On lying trunks it may form long rows of partly fused pilei along the upper edge of an effused resupinate part (Ryvarden 1978). *P. chrysoloma* is an important cause of decay, mostly of *Picea*, rarely of *Pinus* and *Larix*, but also of other gymnosperms. The rot is not such a typical ring rot as that caused by *P. pini* in pine, and has a greater tendency to spread through the larger part of the stem and to attack the sapwood.

Phellinus robustus (P. Karsten) Bourdot & Galzin

(Syn. *Fomes robustus* P. Karsten)

P. robustus forms a thick, rather hoof-like, woody bracket with a dark grey to almost black crust which is at first velvety and later glabrous but cracked. The rounded edge is at first paler. The pore surface is cinnamon coloured. The fruit bodies may persist and enlarge for many years (CMI 197). The fungus mainly affects oaks, but is sometimes found on sweet chestnut and other broad-leaved trees. ***Phellinus hartigii*** (Sacc. & Schnabl.) Imazeki is a form of this fungus which attacks *Abies* species, and sometimes also species of *Larix*.

P. robustus has a world-wide distribution. In Europe it is known especially in France and Germany. It causes a rather slow-growing yellowish-white rot, that mainly destroys the sapwood. The pockets of decay, surrounded by a darker line, are usually some distance from the ground (Cartwright & Findlay 1958), and are eventually indicated by the presence of the long-lasting fruit bodies formed on excrescences on the bark. In oaks and firs the decay may be severe, and lead to stem breakage. Damage caused by this fungus is usually local, though it may be considerable. Trees in public places are potentially dangerous because of their liability to break

and fall. Hence affected trees should be felled and utilized as soon as possible.
PHILLIPS

Phellinus ribis (Schumacher) P. Karsten causes a collar rot of *Ribes* species, especially redcurrant, entering through wounds at the base of the plant and forming perennial fructifications around it, sometimes encircling the stem. Infected plants gradually die back and have to be destroyed. Attacks are generally found only on old plants in neglected plantations (Viennot-Bourgin 1949).

Inonotus dryadeus (Pers.) Murrill

(Syn. *Polyporus dryadeus* (Pers.) Fr.)

I. dryadeus forms a thick lumpy bracket up to about 30 cm across, with a yellowish (later brown) upper surface which exudes watery drops in moist weather. The pore surface has a whitish bloom. The fungus occurs only on species of *Quercus*. It is widespread in Europe, and is known also in N. America. It is a fairly common cause of decay of the heartwood in the roots and lower parts of the butts of oak trees, in which it causes a soft white rot. In its earliest stages the affected wood is dark brown streaked with yellow, but later the wood disintegrates until it resembles paper pulp. The rotten wood may be divided into sections separated by thick sheets of whitish mycelium (Cartwright & Findlay 1958). As only the heartwood is affected, the fungus has little effect on the vigour of the tree, which may remain alive for many years. Thus, detection often depends on the presence of fruit bodies of the fungus, which may be found in summer and autumn, but do not appear every year. The loss of timber resulting from the activities of *I. dryadeus* is relatively small. The fungus is important mainly because it makes the affected trees weak and liable to be blown over by gales, often with little or no prior warning.
PHILLIPS

Inonotus hispidus (Bull.) P. Karsten

(Syn. *Polyporus hispidus* Bull.)

I. hispidus has a thick, bracket-shaped sporophore measuring 10–30 cm across. The upper surface is densely hairy, at first orange-rust in colour, later a deeper brown, and finally black with age. The pore surface is yellowish, later rusty brown (CMI 193). *I. hispidus* may be found on many broad-leaved trees, notably ash, apple and walnut, and sometimes also plane, oak, willow and other trees. It is present in Europe, N. America, Asia and Australia. It causes an often spongy, whitish or yellowish rot, at first surrounded by a gummy brown zone. Later stages vary with the host tree. Thus in ash the decayed wood is soft and pale in colour, and quickly loses most of its strength, and as it dries, circular cracks appear along the annual rings. In walnut the rotten wood becomes yellow and spongy, while in plane it becomes filled with gum (Cartwright & Findlay 1958). *I. hispidus* enters the tree through wounds, especially those resulting from branch breakage. Hence it is very common in hedgerow ash, which is very susceptible to branch damage. Once within the tree it grows mainly in the heartwood, but may also spread into the living wood, and form long, narrow, inconspicuous cankers on the outside of the trunk (Toole 1955). Fruit bodies form in summer on these cankers, and as they die in the autumn they often fall to the ground. *I. hispidus* is most important on ash, the wood of which is used for sports goods, tool handles etc., which require strength and toughness. Ash wood with only slight infection must therefore be rejected. No reliable measures can be taken to prevent infection by this fungus.
PHILLIPS

Inonotus obliquus (Pers.) Pilát

I. obliquus forms annual, resupinate, widely effused fruit bodies, a few mm thick, at first fleshy and whitish, later hard, brittle and dark brown; the tubes are always oblique. The fruit bodies are formed under the bark of dying or recently killed trees. The spores are released after splitting of the bark. The fungus also forms peculiar sterile conks. These are black, deeply cracked and very hard. The conks penetrate the bark. They may develop in the course of many years and be up to 40 cm in diameter. *I. obliquus* attacks many species of deciduous trees, *Betula* species being the main hosts. It is widely distributed in Eurasia and N. America. The rot belongs to the white rot group. The colour is yellowish. In the advanced stages, the annual rings in the wood easily separate from each other, and the medullary rays are almost completely destroyed and filled with a weft of mycelium. The fungus probably infects through wounds and dead branches. In many areas *I. obliquus* is one of the most important fungi in birch, along with fungi such as *Fomes fomentarius, Phellinus igniarius*, and *Piptoporus betulinus*. The best control may be avoiding all kinds of wounds on the stem.

ROLL-HANSEN

Inonotus weirii (Murrill) Kotlaba & Pouzar (syn. *Phellinus weirii* (Murrill) Gilbertson) causes a serious butt rot of conifers, and especially Douglas fir in N. America and Japan (CMI 323; map 490). It is a quarantine organism for Europe (EPPO 19); all imported conifer logs should be debarked as a precaution against its introduction.

Heterobasidion annosum (Fr.) Bref.

(Syn. *Fomes annosus* (Fr.) Cooke)
(Anamorph ***Oedocephalum lineatum*** Bakshi)

Basic description

The fruit body of *H. annosum* is usually bracket-shaped, though when growing under roots or on the roofs of rodent burrows it is often irregular and resupinate. It is 5–10 (< 30) cm across. The upper surface is covered with a zoned reddish-brown crust which turns brown or black with age. The white or biscuit-coloured underside is pierced by fine pores. The growing margin is white or cream. The colourless, globose or sub-globose basidiospores measure 4–5×3–5 μm. The swollen heads of the anamorph *Oedocephalum lineatum*, bear sub-globose or ovoid, colourless conidia measuring 4.5–7.5 (< 10.5)×3–6 μm (CMI 192). James & Cobb (1982) found that isolates of the fungus differed in virulence, and Worrall *et al.* (1983) found evidence of some host specialization. Similarly, Korhonen (1978a) found two intersterility groups of the fungus, S and P. The S group typically causes a butt rot of *Picea abies,* but sometimes attacks *Pinus sylvestris* saplings. The P group typically causes damage in pine stands, killing pine of all ages. It also causes butt rot of spruce, and attacks, for example, *Juniperus communis,* and sometimes *Betula* species, *Alnus incana, Calluna vulgaris* and other woody plants. Korhonen also found slight, but statistically significant differences, in the morphology of the basidiocarps, size of conidia, growth rate, and in resistance against *Peniophora gigantea* on malt extract medium (see below).

Host plants

H. annosum attacks many broad-leaved trees, but is economically important only on conifers. Infested broad-leaved trees, themselves often little affected, may however transmit the fungus to subsequently planted conifers.

Geographical distribution

H. annosum is almost world-wide in temperate and some tropical areas (CMI map 271).

Disease

H. annosum kills pines of any age, especially on dry sites with a high soil pH. Affected trees first show a thinning of the foliage, with shortened needles. Fructifications of the fungus may develop near the bases and roots and on the stumps of the dead trees, and thin white mycelium may be found under the bark. Young trees of other conifers may also be killed, but older trees are more often affected by root and butt rot. Inside trees affected by root and butt rot, the fungus causes a pocket rot which becomes dark brown with time, with black flecks and small, narrow, rather lens-shaped pockets of white material. At an earlier stage, the rotting wood may show greyish, pink, lilac or bluish staining, while in the final stages the wood becomes light, dry and fibrous. The extent of root and butt rot varies with the species. On Douglas fir and pine, the rot is usually confined to the roots, which may be almost completely destroyed, and the trees are then often blown down by wind. In other conifers the rot progresses into the butt, and may eventually hollow out the trunks, which may snap in the wind. Among individual conifer species, *Abies alba* and *A. grandis* are generally resistant but may be severely attacked in some areas. *A. balsamea* is susceptible and *Tsuga heterophylla* is highly susceptible. Susceptibility has two main aspects, however; the first is indicated by the percentage of affected trees, the second by the extent of decay in the trunk. Thus pines do not usually suffer from butt rot, while among other conifers, crops of *Abies grandis* usually show few decayed trees and the rot moves only slowly up the trunk. On the other hand decay moves quickly up the stem in *Thuja plicata* and in spruces (*Picea* spp.). In old spruces, decay may reach a height of more than 8 m. Symptoms on broad-leaved trees are generally similar to but usually much less severe than those on conifers.

Epidemiology

Primary infection is almost always by the colonization of thinning stumps by basidiospores. The fungus grows down into the stump tissues and passes from infested stumps to surrounding healthy trees via root grafts and root contacts. Stump colonization may be reduced in the presence of competing saprophytes, especially that of *Peniophora gigantea* (Fr.) Massee, which also invades the stump surfaces; the growth of *H. annosum* in stumps may also be restricted by that of *Armillaria* species growing upwards from the roots or from the root collar. Rhizosphere fungi may also hinder the movement of the fungus in the soil from root to root. As a rule *H. annosum* appears soonest and develops most rapidly on sites of previous conifer crops, in which the fungus is already present in the residual stumps. For similar reasons it increases with each conifer rotation. On sites not previously under conifers, it causes most damage on sites that were formerly under agriculture. This is at least partly connected with former liming and associated high soil pH. On light soils with a pH above 6.0, losses in pine may be especially heavy; but in some areas pine may be killed at a lower soil pH, and butt rot of other conifers often occurs on relatively acid soils. Soil moisture conditions seem to have some effect. The losses on light soils appear at least in part to be connected with water stress, which may predispose trees to attack. For the same reason trees may be more often affected when the soil is low in organic matter.

Economic importance

Root and butt rot by *H. annosum* is the most serious cause of loss in conifer crops in Europe and N. America. Deaths in pine crops may cause a loss of up to 20% of the volume. Death of roots and gaps made in affected crops may lead to further loss through windblow. In other conifers, butt rot in susceptible species may affect 80% or more of the trees, and result in the loss of the valuable butt length of each affected tree. Economic loss may be reduced by careful conversion. Thus Pratt (1979) examined a *Picea sitchensis* crop in which *H. annosum* occupied about 9% of the volume of each tree. If whole decayed butt lengths were rejected, 33% of the volume and 43% of the value was lost. If, on the other hand, the sound wood of affected butts was converted to pallet boards, the loss in value could be reduced to 10%. Also, timber with only incipient decay may be used, for example, for fencing, provided that it is thoroughly impregnated with creosote to eliminate infection.

Control

In newly planted, uninfested conifer plantations, the aim of control measures must be to prevent the colonization of thinning stumps by *H. annosum* (Rishbeth 1952). Many chemicals can be applied to stump surfaces for this purpose; in the past, creosote, sodium nitrite, disodium octaborate and borax have been effectively used. In many areas, a 20% solution of urea, with a marker dye, is probably now most often used. In pine crops, an oidial suspension of the competing saprophyte *Peniophora gigantea* may be used instead of a chemical. If an already infested stand is to be replaced by another conifer crop, pines may be used, except on alkaline or ex-agricultural sites. In suitable areas, the resistant *Abies grandis* may be considered. In some areas in which *H. annosum* has built up to a high level, it has been possible to replant with conifers only if the infested stumps have first been removed by excavating machines. This is possible only in fairly light soils where the terrain is reasonably level.
PHILLIPS

Climacocystis borealis (Fr.) Kotlaba & Pouzar

(Syn. *Spongipellis borealis* (Wahlenb.) Pat.)

C. borealis forms annual, sessile, flat fruiting bodies up to 15 cm long, 8 cm wide, and 4 cm thick. They are white to cream-coloured when young, later straw-coloured; soft and watery when fresh, and hard and brittle when dry. The upper surface is shaggy. The species has been transferred from the genus *Spongipellis* to *Climacocystis* because of its characteristic hymenial cystidia. *C. borealis* attacks conifers, notably *Picea* species, seldom broad-leaved trees. It is widely distributed in Eurasia and N. America. It causes a kind of white mottled rot of a peculiar cubical appearance in the basal heartwood. The wood is split into minute cubes by furrows filled with white mycelium. The cubes are not easily separated from each other, even at a rather advanced stage. Even though *C. borealis* is widely distributed, it is nowhere of great importance and thus causes relatively small losses. Control lies in the avoidance of wounds on roots and the lower part of the stem and in a low rotation age.
ROLL-HANSEN

Fomes fomentarius (L.) Kickx

F. fomentarius has hoof-shaped fructifications which may be up to 40 cm across. The smooth, grey to black upper surface is concentrically zoned, with a blunt margin. and the pore surface is greyish. The fungus occurs on many broad-leaved trees (both living and dead), and on their stumps, but it is nearly always found on beech and birch. Over quite large areas it occurs exclusively on birch, and in others only on beech. It can continue its growth on dead fallen timber in damp conditions (Cartwright & Findlay 1958). It is widespread, occurring throughout Europe and N. America. *F. fomentarius* causes a white or yellowish-white rot in both the sapwood and heartwood, usually in the upper parts of the trunk. Dark zone lines occur in the decayed wood, which forms cracks often filled with yellowish or cream-coloured mycelium. The perennial fruit bodies may be found at any time of the year. The fungus enters the tree through wounds, especially those caused by the breakage of branches. Although nothing can protect trees in forests from infection by *F. fomentarius,* something can be done on a small scale in parks and gardens by cutting off broken limbs closely and cleanly at the trunk.

PHILLIPS

Fomitopsis pinicola (Sow.) P. Karsten

(Syn. *Fomes pinicola* (Schwartz) Cooke, *F. marginatus* (Pers.) Gill., *Ungulina marginata* (Pers.) Pat.)

F. pinicola is known as the red belt fungus and is the type of the genus *Fomitopsis* Karsten. Its conk is very characteristic: margin yellowish, then apricot to orange-red; upper surface blackish, concentrically sulcate, covered with a hard skin dissolving in KOH; the lower surface is straw-yellow, cream or apricot-coloured when handled or when old; the sourish smell is very typical. *F. pinicola* develops on a large number of host plants, both conifers and hardwoods, but it seems more frequent on conifers, such as *Abies* and *Picea* species. In spite of the results of Mounce (1929), from which he concluded that it showed no host specialization, the form on broad-leaved trees is considered by some authors (Lanier *et al.* 1976–78) to be distinct from that on conifers. *F. pinicola* is widely distributed in Europe (and in N. America and Asia), though it is very rare in the UK. It is responsible for a brown-reddish cubical rot of the wood, the cubical pieces being held together by thick wefts of whitish mycelium. Identification of the fungus from the attacked wood must be done after isolation, if necessary on a selective medium; specific identification of mycelial cultures is possible (Stalpers 1978). Little is known of its epidemiology. Basidiospores are produced in the air (in Poland) from April to December, with a maximum production in April, May and June. Its economic importance may be great in some cases because of the decay of the heartwood of living trees; the fungus is also very destructive to dead trees and to felled logs lying in the forest. So far no effective control measures are available. Because of its high capacity to destroy the wood, *F. pinicola* might be a good model fungus for studies on cellulose and lignin degradation.

DELATOUR

Fomitopsis cytisina (Berk.) Bondartsev & Singer (syn. *Fomes fraxineus* (Bull.) Fr.) is widespread in Europe and N. America, where it sometimes causes a severe white rot of the basal parts of the trunks of broad-leaved trees, especially of ash, elm, *Robinia*, poplar and beech.

Laetiporus sulphureus (Bull.) Bondartsev & Singer

(Syn. *Polyporus sulphureus* Bull.)

L. sulphureus is a common, imbricate, bracket-like polypore. Its brackets are at first bright orange above and pale sulphur-yellow below, though they become paler with age. The spores are white or yellowish in the mass (CMI 441). *L. sulphureus* causes decay in trees. It is particularly important on oak and sweet chestnut; it is also common on cherry, and less frequent on *Alnus, Fagus, Juglans, Populus, Robinia, Salix* and pear. Conifers affected include *Taxus*, and sometimes *Pinus*, *Picea* and *Larix*. The fungus is widespread in Europe and N. America. Its sporophores develop in summer and autumn on rotten stumps and on tree trunks, often on wounds and old branch stubs, which may show signs of decay. Otherwise no sign of attack may be seen until a tree is felled, or when hollow it collapses in the wind. Early stages of decay appear only as a yellow or reddish stain. Later the wood becomes a deep reddish-brown, and breaks up into cubical blocks, often separated by yellowish sheets of mycelium resembling chamois leather. In culture the fungus produces a sparse, white to pale buff, downy mycelium, covered by a powder of secondary spores. Chlamydospores form within cultures and in decayed wood. *L. sulphureus* enters trees through wounds, especially those made by the breaking of large branches; it may also sometimes enter through the roots.

L. sulphureus is important partly because of the severity of the decay it causes, particularly in oaks open-grown in parks and in hedgerows. It destroys a substantial amount of timber, and also renders the trees unstable and dangerous. It also continues to grow after felling, and so may cause progressive decay in structural timbers (Cartwright & Findlay 1958). To prevent infection, care should be taken to avoid wounding of standing trees, and, when possible, broken branches should be cut off smoothly close to the trunk. Before wood is seasoned, rotted parts should first be cut out, and so should adjacent, apparently healthy wood, to about 30 cm beyond the edges of visible decay.
PHILLIPS

Meripilus giganteus (Pers.) P. Karsten
(Syn. *Polyporus giganteus* (Pers.) Fr.)

M. giganteus has fan-shaped, annual fruit bodies which form imbricate tufts on branching stalks arising from a common basal mass, and which blacken when bruised. The individual pilei are usually up to about 25 cm across, but sometimes larger. The top of each pileus is yellow to dark-brown, with small hairy or scurfy scales. *M. giganteus* occurs on living and dead trees of many broad-leaved species, but is especially common on beech and oak. It is widespread in Europe, and is also present in N. America. It causes a white rot, mainly of the roots and the basal part of the butt. In the case of beech, the leaves of affected trees may be small and sparse. The deeper roots of such trees may be decayed by *M. giganteus*, and then often become victims of windblow. The large and conspicuous fruit bodies may be found from July to January, sometimes on roots around the bases of living trees, and often on tree stumps. *M. giganteus* is important chiefly because by rotting roots it renders beech trees dangerous, as they may become windblown with little or no warning. No effective measures can at present be taken to prevent infection by this fungus.
PHILLIPS

Piptoporus betulinus (Bull.) P. Karsten
(Syn. *Polyporus betulinus* Fr.)

P. betulinus forms annual brackets 8–20 cm across, rounded or kidney-shaped, often prolonged in the centre above into a boss resembling a small stalk. The upper side has a smooth, pale brownish, separable skin, which is incurved and projecting over the edge of the pore surface. The pores are small and white. The fungus occurs only on birch. It is found more or less throughout the areas occupied by *Betula* species in Eurasia and N. America. It causes a brown rot of birch, on which in many areas it is the commonest of all decay fungi. The fruit bodies are found on living and dead birch trees in summer and autumn. Wood in an advanced stage of decay breaks into regular, square-sided blocks, often with thin sheets of whitish mycelium between them. *P. betulinus* is a wound parasite (though quite small wounds allow it to gain entry). It often attacks trees with broken stems or branches, or which have been damaged by fire. The fungus continues to grow on infected trees for several years after their death, both on the standing trees and on fallen wood. It is the most important cause of death and decay of birch trees. Little can be done to control its activities, except where possible to fell and burn diseased trees.
PHILLIPS

A number of other genera of Aphyllophorales are of minor importance in attacking living trees. ***Bjerkandera adusta*** (Willd.) Karsten (syn. *Polyporus adustus* (Willd.) Fr., *Leptoporus adustus* (Willd.) Quélet) is an important cause of a rapid white fibrous rot of many living and dead broad-leaved trees, especially of beech trees affected by beech bark disease (see *Nectria coccinea*) and of oaks. ***Fistulina hepatica*** Fr., the 'beefsteak fungus', causes a rather dry, brown, cubical rot in many broad-leaved trees, but is found most often in oak and sweet chestnut. ***Grifola frondosa*** (Dickson) Gray sometimes causes a severe butt rot of oak, rendering the trees dangerous; in Europe it also occurs on hornbeam, and in N. America it affects various other broad-leaved trees. ***Polyporus squamosus*** Huds., the 'dryads' saddle', causes a white, fibrous rot of trunks and large branches of many broad-leaved trees, especially of elm, *Acer* species, walnut and oak. ***Rigidoporus ulmarius*** (Sow.) Imazeki (syn. *Polyporus ulmarius* Sow.) causes a brown cubical rot in the trunks of many broad-leaved trees (especially elm) in parts of Europe, Asia and N. America.

AGARICALES

The *Armillaria mellea* complex

General introduction

The *Armillaria* strains with ringed fruit bodies have long been regarded as belonging to a single species variously called *Armillaria mellea* (Vahl) Kummer, *Armillariella mellea* (Vahl) Karsten or *Clitocybe mellea* (Vahl) Ricken. However, this so-called species shows a high degree of variability in its biological behaviour as well as in the morphology of its fruit bodies in nature and its mycelium in culture. Many authors have tried to resolve this difficulty by establishing several species on the basis of fruit body morphology. In Europe the most recent attempt was made by Romagnesi (1973). But as the pattern of sexuality of the fungus was then unknown, it was not certain if any of these species were true biological entities. Hintikka (1973) explained the biological cycle of the fungus by showing that *Armillaria* behaves according to a heterothallic and tetrapolar pattern. It thus became possible to mate the different strains of the fungus, and to show the existence of intersterile groups, subsequently regarded as the foundation for the taxonomy of the '*Armillaria mellea* complex'. In Europe, Korhonen (1978b) showed that all existing strains could be divided into five intersterile groups, which he named A, B, C, D and E. These taxa coincide partly with Romagnesi's divisions (Romagnesi & Marxmüller 1983), and may be regarded as true Linnean species. They have been given Latin names as follows:

A) ***Armillaria borealis*** Marxmüller & Korhonen
B) ***A. cepaestipes*** Velenovsky
C) ***A. obscura*** (Pers.) Herink
D) ***A. mellea*** *sensu stricto*
E) ***A. bulbosa*** (Barla) Kile & Watling

Subsequent work has confirmed that these species can be distinguished in many European countries, including France (Guillaumin & Berthelay 1981). It has also been shown that many differences exist between the species as regards their geographical distribution, ecological specialization and pathogenic behaviour. It seems that species A, B and E are not very pathogenic, if not completely saprophytic. Conversely, species C and D appear to be strong parasites but differ in their host range. Species D is the sole species found on orchard trees and grapevine, and can also be a weak parasite of deciduous forest trees. By artificial inoculation it can be made to infect coniferous trees (Rishbeth 1982), but it is rarely found on conifers in nature. On the other hand, species C is in most cases the species responsible for the infection of conifers.

If this recent work is considered it can be seen that results obtained with '*Armillaria mellea*' before 1973 are difficult to interpret. Previous work done in nature can usually be related to *A. mellea sensu stricto* or *A. obscura* by simple examination of the hosts involved, but many laboratory results are difficult to identify precisely with a species. It must be emphasized that results obtained with one species must not be extrapolated to the entire European '*A. mellea* complex' since the five species show so many differences. Roll-Hansen (1985) has recently reviewed the situation in Europe. In other continents the situation is even more confused. Thus in N. America the *A. mellea* group is divided into ten intersterile groups (Anderson *et al.* 1980). Two of these correspond to the European *A. obscura* and *A. mellea sensu stricto* with which they are partly fertile. It is not yet known if they have the same function as regards their distribution and pathogenic behaviour. Some basic work on *Armillaria* carried out in the USA cannot be definitely attributed to any particular species.

GUILLAUMIN

Armillaria mellea (Vahl) Kummer *sensu stricto*

Basic description

This fungus corresponds well with Korhonen's *Armillaria* group D. In English-speaking countries it is known as the honey fungus (on account of the colour of its fruit body). It differs from the other ringed European *Armillaria* species as follows: i) carpophore morphology: a very clear cap (yellow to honey), almost without scales. Stipe fasciculate, very long, of constant diameter, without a bulb at the base, without scales. Ring well developed, ciliate, white to yellow; ii) hymenium: without clamp connections and dikaryons; iii) subterranean rhizomorphs: thin, contorted, very brittle; iv) rhizomorphs in pure culture, on 2% malt agar, ribbon-shaped, contorted, fast-growing, branching profusely; v) identifiable by immunology: the species has a specific antigen, 'T 1'.

Host plants

i) Grapevine; ii) all orchard trees: peach, apricot, almond, cherry, apple, pear (on quince as a rootstock), walnut, olive, fig, medlar and citrus; iii) soft fruit: blackcurrant, gooseberry, raspberry, and kiwi fruit (*Actinidia*); iv) ornamental trees and shrubs: rose, mimosa, privet, pyracantha, etc.; v) deciduous forest trees.

Beech, poplars, birches, willows, elms, maples and chestnuts are sometimes known to be attacked by

Armillaria, but up to now it is not known if *A. mellea (sensu stricto)* is the species involved. However, this has been established for beech, mulberry and various species of oak. *A. mellea* can sometimes attack conifers (for instance in parks where isolated conifers are surrounded by broad-leaved trees). But in most cases the species responsible for the infection of conifers is *A. obscura*.

In Mediterranean zones, the cork oak and *Eucalyptus* can also be attacked by a ringless species of *Armillaria*, ***A. tabescens*** (Scop.) Emel. Young *Eucalyptus* plantations were seriously affected in S.W. France in 1984 and the same species could be responsible for losses in Spain, where *Eucalyptus* is frequently attacked by an as yet unidentified *Armillaria*.

Geographical distribution

The species is widespread in W. and S. Europe. It is absent from Norway, Sweden and Finland (Korhonen 1978b). Its presence in Poland and N. Russia remains to be proved.

Disease

The behaviour of the fungus differs in deciduous forests and in agricultural woody plantations. In oak forests *A. mellea* is a weak parasite infecting trees enfeebled by unfavourable situations or climatic stresses. Conversely, on grapevine and orchard trees, it is a primary parasite able to attack healthy trees. Basically, *A. mellea* causes rot of the hard living tissues (bark and sapwood) in the roots of its host, by means of its cellulolytic and lignolytic enzymes. Although the fungus is known to produce phytotoxins in pure culture, chemical poisoning does not seem to play an important part in pathogenicity. The physical destruction of sapwood, cambium and phloem is sufficient explanation for the visible symptoms on the above-ground parts, without invoking a phytotoxin. On orchard trees, as in the case of *Rosellinia necatrix*, the fungus can cause either a decline (involving yellowing and loss of leaf turgidity, a slowing down of growth, and an early fall of leaves in autumn), or an apoplexy, a sudden wilting of the tree in a period of water stress.

Epidemiology

The basic infection cycle has been reviewed by Rykowski (1978). However, the classical ideas were established either in Europe on conifers or in N. America. In both cases the species concerned were probably not *A. mellea* as now understood. However, Guillaumin & Rykowski (1980) have infected the white walnut in artificial conditions with two isolates of *A. mellea sensu stricto* and obtained results very similar to the classical data of Thomas (1934). It seems that the basidiospores play no part in the infection cycle. The whole cycle takes place in the soil. The fungus is preserved as mycelial masses in the dead wood buried in the soil. This mycelium is often protected by 'black lines' (pseudosclerotia). It produces subterranean rhizomorphs (highly differentiated linear organs growing in the soil by a meristematic apical zone). These rhizomorphs come into contact with roots and attach themselves to the roots by means of mucilage, then they produce short branches which directly penetrate the root bark. In some cases, infection probably occurs by direct contact between colonized wood and healthy roots.

After penetration has occurred, the fungus extends in the root by means of flat rhizomorphs and fans (in the cambium) and in the form of undifferentiated mycelium (in the sapwood). Thus the disease spreads partly by the subterranean rhizomorphs in the soil and partly by subcortical rhizomorphs inside the roots. After the death of the tree, the rotten roots remaining in the soil can initiate rhizomorphs again, thus continuing the infection cycle.

The disease can be found in all soil types. Contrary to some presumptions, it is not restricted to heavy or damp soils, or to those with a particular pH, although the rhizomorphs of *A. mellea* are known to grow more actively in alkaline soils. In vineyards and orchards the appearance of the disease depends upon the presence of decomposing wood in the soil. This may be the case if the plantation succeeds a previous orchard, if it is sited on the banks of a river subject to flooding, or if it is planted after deforestation. The last situation is the most dangerous, especially if the natural vegetation consisted of hardwoods. The various oak species, although rather tolerant of *A. mellea* themselves, are a threat to subsequent plantings.

Economic importance

The disease caused by *A. mellea* is probably most severe in S. Europe; in France, Spain, Italy, Portugal and Greece. Unfortunately, citations concerning *Armillaria* in these countries are far less abundant than in northern countries where conifer forests and *A. obscura* are concerned. However, it has been established that the honey fungus is very serious on grapevine in S. France and Greece, on stone fruits in France (Rhône Valley) and Spain (Ebro Valley), on walnut in France and on citrus fruits in Spain. In France it has been proved that *A. mellea* is always the species involved. In other countries it is only a probability. In N. France, Britain, Benelux and Germany, the disease seems to be more sporadic. It mainly concerns apple, ornamental woody plants and, in some special cases, *Quercus robur*.

Control

Control of *Armillaria* root rot is very difficult because all the fungus organs are in the soil, sometimes at a considerable depth, and the mycelium is protected inside the dead wood. Prophylactic steps include: the avoidance of risky sites for plantations; as complete a removal of roots as possible, for instance by sub-soiling; rotation with annual crops (insufficient if the buried wood fragments are large); artificial depletion of the fungal root bases by poisoning stumps after felling the trees.

Soil disinfection can be carried out with fumigants such as methyl bromide, carbon disulphide and sodium methylisothiocyanate. The important point is to inject the fumigants at maximum depth, at least 60 cm. Successful treatment will result in a decrease in the number of foci rather than in the eradication of the fungus. Some chemicals, including mixtures of phenolic compounds, hinder the growth of rhizomorphs, but are effective only in light soils, and need repeated applications; they do not replace the removal of stumps and roots as the main method of control. Trenching, a method traditional in viticulture, aims to stop the growth of rhizomorphs and diseased roots. It can be improved by burying a polythene sheet in a vertical position in the trench. Tolerant rootstocks exist for walnuts (*Juglans nigra* and *J. hindsii*), pears (quince is susceptible, but rootstock 'pear' is tolerant), and for *Prunus*; rootstocks of *Prunus domestica, P. insititia* and *P. cerasifera* are more tolerant than the roots of peach, almond and apricot (Guillaumin & Pierson 1983). In certain cases, curative steps may save individual trees if the attack is not too far advanced. The collar and the upper parts of the main roots should be uncovered, the diseased parts cut off, and antiseptic applied to the wounds. It is also advisable to leave the main roots exposed to the air for a whole summer.
GUILLAUMIN

Armillaria obscura (Pers.) Herink

(Syn. *A. ostoyae* (Romagnesi) Herink, *Armillariella ostoyae* Romagnesi)

A. obscura corresponds to Korhonen's species C. The cap of its fruit body is covered with dark scales which also occur on the ring. The stipe is more or less cylindrical, without a bulbous base. In culture on gallic acid agar, the fungus produces dichotomously branched, curved rhizomorphs. These are yellowish-brown when young, and less brittle than those of *A. mellea* (Rishbeth 1982). Its fruit bodies form on stumps and at the foot of trees, and also on the ground between, arising from rhizomorphs. *A. obscura* is one of the highly pathogenic forms within the *A. mellea* complex. Its host range is narrower than that of *A. mellea sensu stricto:* though it sometimes affects broad-leaved trees (including *Sorbus, Betula* and *Fraxinus*), it is mainly restricted to conifers of family Pinaceae (Marxmüller 1982; Rishbeth 1982). It is especially common on stumps of spruce, but its range also includes *Albies, Pinus, Larix, Pseudotsuga* and *Tsuga*. The geographical distribution of *A. obscura* is still incompletely known, but the fungus appears to be widespread in Europe and in N. America. It may kill trees on acid soils, and is also an important cause of decay. The principles used in the control of *A. obscura* are the same as those already set out for *A. mellea*.
PHILLIPS

Armillaria bulbosa (Barla) Kile & Watling (syn. *Armillariella bulbosa* Romagnesi), Korhonen's species E, is the most common *Armillaria* species in deciduous forests, in which it is generally regarded as a saprophyte. However, some cases have been observed in which it has behaved as a weak parasite, on *Quercus robur* and on poplars. It does not occur in Nordic countries where the commonest form on deciduous trees is ***A. cepaestipes*** f. ***pseudobulbosa*** Marxmüller & Romagnesi.

Collybia fusipes Fr. is a very common species found in oak forests as a weak parasite, frequently occurring on the surface of the roots of healthy trees in the form of latent lesions. On trees (mainly *Quercus robur*) weakened by ecological causes, the lesions develop and the fungus can invade the whole root system, causing a wet, orange rot of bark and sapwood (Delatour & Guillaumin 1984). ***Pholiota squarrosa*** (Müller) Fr. sometimes causes a soft brown rot of various broad-leaved trees, including poplars, apples and cherries, and sometimes of conifers. ***Pleurotus ostreatus*** (Jacq.) Kummer and ***P. ulmarius***

(Bull.) Fr. can cause heart rot of standing trees (mainly beech and elm respectively), which they attack through wounds and branch stubs. ***Calyptella campanula*** (Nees) Cooke is an oddity as an agaric which can cause root rot and wilting of glasshouse tomatoes in the UK. Susceptibility is associated with high soil moisture, attacks being most severe along irrigation lines. The yellow fruiting bodies are often seen round the bases of the plants in July–August. The disease is only likely to occur if infested soil is used (Clark *et al.* 1983).

REFERENCES

Anderson J.B., Korhonen K. & Ullrich R.C. (1980) Relationships between European and North American biological species of *Armillaria mellea*. *Experimental Mycology* **4,** 87–95.

Anderson N.A. (1982) The genetics and pathology of *Rhizoctonia solani*. *Annual Review of Phytopathology* **20,** 329–347.

Årsvoll K. (1978) *Studies on factors causing winter damage in Norwegian grasslands, with special reference to snow mould fungi.* Dr. Agric. Thesis, Agricultural University of Norway.

Arx J.A. von, Walt J.P. van der & Liebenberg N.V.D.M. (1982) The classification of *Taphrina* and other fungi with yeast-like cultural states. *Mycologia* **74,** 285–286.

Bernaux P. (1977) Localisation et pouvoir pathogène de *Sclerotium oryzae* et *S. hydrophilum* sur plantules de riz. *Annales de Phytopathologie* **9,** 205–209.

Blanchette R.A. (1980) Wood decomposition by *Phellinus (Fomes) pini*: a scanning electron microscopy study. *Canadian Journal of Botany* **58,** 1496–1503.

Cartwright K. St. G. & Findlay W.P.K. (1958) *Decay of Timber and its Prevention.* HMSO, London.

Clark W.S., Richardson M.J. & Watling R. (1983) *Calyptella* root rot—a new fungal disease of tomatoes. *Plant Pathology* **32,** 95–99.

Clarkson J.D.S. & Cook R.J. (1983) Effect of sharp eyespot (*Rhizoctonia cerealis*) on yield loss in winter wheat, and of some agronomic factors on disease incidence. *Plant Pathology* **32,** 421–428.

Clarkson J.D.S. & Griffin M.J. (1977) Sclerotia of *Rhizoctonia solani* in wheat stems with sharp eyespot. *Plant Pathology* **26,** 98.

Dalton I.P., Epton H.A.S. & Bradshaw N.J. (1980) Field trials of fungicides for the control of violet root rot of carrots. *Annals of Applied Biology* **94** (Supplement, Tests of Agrochemicals and Cultivars, no. 1, 20–21).

Dalton I.P., Epton H.A.S. & Bradshaw N.J. (1981) The susceptibility of modern carrot cultivars to violet root rot caused by *Helicobasidium purpureum*. *Journal of Horticultural Science* **56,** 95–96.

D'Ambra V. & Mutto S. (1969) Ricerche su di una alterazione della fragola connessa alla presenza in Italia della *Rhizoctonia fragariae*. *Rivista di Patologia Vegetale, seria 4* **5,** 85–104.

Delatour C. & Guillaumin J.J. (1984) Un pourridié méconnu: le *Collybia fusipes*. *Comptes Rendus de l'Academie d'Agriculture de France* **70,** 123–126.

Dubos B. & Ricard J.L. (1974) Curative treatments of peach trees against silver leaf disease (*Stereum purpureum*) with *Trichoderma viride* preparations. *Plant Disease Reporter* **58,** 147–150.

Elad Y., Chet I. & Katan J. (1980) *Trichoderma harzianum*: a biocontrol agent effective against *Sclerotium rolfsii* and *Rhizoctonia solani*. *Phytopathology* **70,** 119–121.

Galet P. (1977) *Les Maladies et Parasites de la Vigne.* Paysan du Midi, Montpellier.

Gladders P. & Coley-Smith J.R. (1978) *Rhizoctonia tuliparum*: a winter-active pathogen. *Transactions of the British Mycological Society* **71,** 129–139.

Greig B.J.W. (1981) *Decay in Conifers. Forestry Commission Leaflet* no. 79. HMSO, London.

Grosclaude C. (1984) L'efficacité des produits de protection des blessures chez les végétaux ligneux. *Phytoma* no. 362, 38–42.

Grosclaude C., Richard J. & Dubos B. (1973) Inoculation of *Trichoderma viride* spores via pruning shears for biological control of *Stereum purpureum* on plum tree wounds. *Plant Disease Reporter* **57,** 25–28.

Guillaumin J.J. & Berthelay S. (1981) Détermination spécifique des armillaires par la méthode des groupes de compatibilité sexuelle. *Agronomie* **1,** 897–908.

Guillaumin J.J. & Pierson J. (1983) Le pourridié-agaric des arbres fruitiers à noyau; l'importance du choix des porte-greffes. *L'Arboriculture Fruitière* nos. 353–4, 38–42.

Guillaumin J.J. & Rykowski K. (1980) (Infection of walnut by *Armillaria mellea* under artificial conditions). *Folia Forestalia Polonica Series A* **24,** 191–213.

Herring T.F. (1962) Host range of the violet root rot fungus *Helicobasidium purpureum* Pat. *Transactions of the British Mycological Society* **45,** 488–494.

Hide G.A. & Cayley G.R. (1982) Chemical techniques for control of stem canker and black scurf (*Rhizoctonia solani*) disease of potatoes. *Annals of Applied Biology* **100,** 105–116.

Hide G.A., Hirst J.M. & Stedman O.J. (1973) Effects of black scurf (*Rhizoctonia solani*) on potatoes. *Annals of Applied Biology* **74,** 139–148.

Hims M.J., Dickinson C.H. & Fletcher J.T. (1984) Control of red thread, a disease of grasses caused by *Laetisaria fuciformis*. *Plant Pathology* **33,** 513–516.

Hintikka V. (1973) A note on the polarity of *Armillariella mellea*. *Karstenia* **12,** 32–39.

Hoeven E.P. Van der & Bollen G.J. (1980) Effect of benomyl on soil fungi associated with rye. 1. Effect on the incidence of sharp eyespot caused by *Rhizoctonia cerealis*. *Netherlands Journal of Plant Pathology* **86,** 163–180.

Hollins T.W. & Scott P.R. (1983) Resistance of wheat cultivars to sharp eyespot caused by *Rhizoctonia cerealis*. *Annals of Applied Biology* **102** (Supplement, Tests of Agrochemicals and Cultivars no. 4, 126–127).

Hull R. & Wilson A.R. (1946) Distribution of violet root rot (*Helicobasidium purpureum*) of sugar beet and preliminary experiments on factors affecting the disease. *Annals of Applied Biology* **33,** 420–432.

Jamalainen E.A. (1974) Resistance in winter cereals and

grasses to low-temperature parasitic fungi. *Annual Review of Phytopathology* **12,** 281–302.

James R.L. & Cobb F.W. (1982) Variability in virulence of *Heterobasidion annosum* isolates from ponderosa and Jeffrey pine in areas of high and low photochemical air pollution. *Plant Disease* **66,** 835–837.

Kaplan J.D. & Jackson N. (1983) Red thread and pink patch disease of turf grasses. *Plant Disease* **67,** 159–162.

Korhonen K. (1978a) Intersterility groups of *Heterobasidion annosum*. *Communicationes Institutionis Forestalis Fennici* **94**(6), 1–25.

Korhonen K. (1978b) Interfertility and clonal size in the *Armillaria mellea* complex. *Karstenia* **18,** 31–42.

Lanier L., Joly P., Bondoux P. & Bellemère A. (1976–1978) *Mycologie et Pathologie Forestières*. 2 vols. Masson, Paris.

Lucas P. & Cavelier N. (1985) *Rhizoctonia cerealis*, agent du rhizoctone des céréales. Premiers résultats sur les conditions de développement et la nusibilité. *Compte-rendu des Premières Journées sur les Maladies des Plantes*. pp. 107–115. ANPP, Paris.

Marxmüller H. (1982) Etude morphologique des *Armillaria* ss. str. à anneau. *Bulletin de la Société Mycologique de France* **98,** 87–124.

Molot P. & Simone J. (1965) Polyphagie ou specialisation parasitaire du *Rhizoctonia violacea* Tul. *Comptes Rendus Hebdomadaires des Seances de l'Académie des Sciences* **51,** 228–232.

Moore W.C. (1979) *Diseases of Bulbs*. HMSO, London.

Mounce I. (1929) Studies in forest pathology. II. The biology of *Fomes pinicola*. *Canadian Department of Agriculture Bulletin* **111,** NS.

Murray D.I.L. & Burpee L.L. (1984) *Ceratobasidium cereale* sp. nov. the teleomorph of *Rhizoctonia cerealis*. *Transactions of the British Mycological Society* **82,** 170–172.

Niemelä T. (1974) On Fennoscandian Polypores. III. *Phellinus tremulae* (Bond.) Bond. & Borisov. *Annales Botanici Fennici* **11,** 202–215.

O'Rourke C.J. (1976) *Diseases of Grasses and Forage Legumes in Ireland*. An Foras Talúntais, Carlow.

Pawsey R.G. & Gladman R.J. (1965) *Decay in Standing Conifers Developing from Extraction Damage. Forest Record* **54.** HMSO, London.

Pitt D. (1964) Studies on sharp eyespot of cereals. I. Disease symptoms and pathogenicity of isolates of *Rhizoctonia solani* Kühn and the influence of soil factors and temperature on disease development. *Annals of Applied Biology* **54,** 77–89.

Pratt J.E. (1979) *Fomes annosus* butt rot of Sitka spruce. III. Losses in yield and value of timber in diseased trees and stands. *Forestry* **52,** 113–127.

Punter D., Reid J. & Hopkin A.A. (1984) Notes on sclerotium-forming fungi from *Zizania aquatica* (wild rice) and other hosts. *Mycologia* **76,** 722–732.

Richardson M.J., Whittle A.M. & Jacks M. (1976) Yield-loss relationships in cereals. *Plant Pathology* **25,** 21–30.

Rishbeth J. (1952) Control of *Fomes annosus*. *Forestry* **25,** 41–50.

Rishbeth J. (1982) Species of *Armillaria* in southern England. *Plant Pathology* **31,** 9–17.

Roll-Hansen F. (1985) The *Armillaria* species in Europe—a literature review. *European Journal of Forest Pathology* **15,** 22–31.

Roll-Hansen F. & Roll-Hansen H. (1981) Root wound infection of *Picea abies* at three localities in southern Norway. *Meddelelser fra Norsk Institutt for Skogsforskning* no. 36.

Romagnesi H. (1973) Observations sur les *Armillariella* (II). *Bulletin de la Société Mycologique de France* **89,** 195–206.

Romagnesi H. & Marxmüller H. (1983) Etude complémentaire sur les armillaires annelées. *Bulletin de la Société Mycologique de France* **99,** 301–324.

Rykowski K. (1978) Infection biology of *Armillaria mellea*. In *5th International Conference on Problems of Root and Butt Rots in Conifers (Kassel), Proceedings,* pp. 215–218.

Ryvarden L. (1978) *The Polyporaceae of North Europe*, vol. 2, pp. 219–507. Oslo.

Stalpers J.A. (1978) *Identification of Wood-inhabiting Aphyllophorales in Pure Cultures. Studies in Mycology* no. 16. Baarn, Netherlands.

Stalpers J.A. & Loerakker W.M. (1982) *Laetisaria* and *Limonomyces* species (Corticiaceae) causing pink diseases in turf grasses. *Canadian Journal of Botany* **60,** 529–537.

Thomas H.E. (1934) Studies on *Armillaria mellea* (Vahl) Quél.; infection, parasitism and host resistance. *Journal of Agricultural Research* **48,** 187–218.

Toole E.R. (1955) *Polyporus hispidus* on southern bottomland oaks. *Phytopathology* **45,** 177–180.

Viennot-Bourgin G. (1949) *Les Champignons Parasites des Plantes Cultivées*. Masson, Paris.

Wakefield E.M. & Dennis R.W.G. (1981) *Common British Fungi*. Saiga Publishing Co. Ltd., Hindhead, Surrey.

Worrall J.J., Parmeter J.R. & Cobb F.W. (1983) Host specialization of *Heterobasidion annosum*. *Phytopathology* **73,** 304–307.

Pathogen Index

This index contains all the names and synonyms of pathogens mentioned in the Handbook. Disease names are not indexed, except in the case of some viruses, for which they are effectively synonyms, or in the case of diseases of unknown aetiology, for which they are the only names available.

Names in bold are the accepted names, while names in light type are synonyms. Page references in bold refer to the beginning of the main entry on the pathogen; page references in light type are cross-references.

Polyporus ulmarius 520
Polyscytalum pustulans 408
Polyspora lini 393
Polystigma ochraceum 324
Polystigma rubrum 324
Polystigmina rubra 324

Host Index

In this index, page references are given for pathogens for which a particular host is mentioned. Authors have frequently used broader categories, such as plant genus or family names. These are also included, with cross-references from the species as appropriate. Some very broad categories (conifers, deciduous trees, fruits, herbaceous hosts, ornamentals, plants (wide host range), seeds, woody hosts) also appear. It is stressed that, although this index approximates to a European host/pathogen list, it is necessarily very incomplete in this respect. In particular, it will often not indicate whether a pathogen with a wide host range has been recorded on a particular host. In any case, authors did not necessarily attempt to give comprehensive host ranges.

The hosts are given by their scientific names. The English names used in the text for most common crop plants are also included, with cross reference to the scientific name.